Fourth Interdisciplinary Scientific Forum with International Participation on New Materials and Promising Technologies (NMPT-4)

IOP Conference Series: Materials Science and Engineering
Volume 525

Moscow, Russia
27 – 30 November 2018

ISBN: 978-1-5108-8817-3
ISSN: 1757-8981

Printed from e-media with permission by:

Curran Associates, Inc.
57 Morehouse Lane
Red Hook, NY 12571

Some format issues inherent in the e-media version may also appear in this print version.

No changes have been made to the content of these proceedings. There may be changes to pagination and minor adjustments for aesthetics.

Printed by Curran Associates, Inc. (2019)

For permission requests, please contact the Institute of Physics at the address below.

Institute of Physics
Dirac House, Temple Back
Bristol BS1 6BE UK

Phone: 44 1 17 929 7481
Fax: 44 1 17 920 0979

techtracking@iop.org

Additional copies of this publication are available from:

Curran Associates, Inc.
57 Morehouse Lane
Red Hook, NY 12571 USA
Phone: 845-758-0400
Fax: 845-758-2633
Email: curran@proceedings.com
Web: www.proceedings.com

Table of contents

Volume 525

Fourth interdisciplinary scientific forum with international participation "New materials and promising technologies"

27–30 November 2018, Moscow, Russian Federation

Accepted papers received: 1 April 2019

Preface

Papers

IOP Conf. Series: Materials Science and Engineering **525** (2019) 011001 doi:10.1088/1757-899X/525/1/011001

Introduction

From November 27 to 30, 2018, the Fourth interdisciplinary scientific forum with international participation "New materials and promising technologies" was held in Moscow in the building of the Presidium of the Russian Academy of Sciences, where topical and applied problems of creating new materials for the future and promising technologies for their processing were discussed. The Forum was held for the fourth time. The Forum was organized by the Council of Young Scientists of the Russian Academy of Sciences and the Coordinating Council for Youth Affairs under the President of the Russian Federation. Co-organizers of the Forum were the Russian Academy of Sciences (RAS), The Russian Chemical Society D.I. Mendeleyev and Institute of Metallurgy and Materials Science. A.A. Baikova

The Ministry of Science and Higher Education of the Russian Federation and the Russian Foundation of Basic Research supported the Forum.

Thematic areas included in the program of the Forum, helped to gather in one place representatives of different interdisciplinary research groups in the area of interest which includes the creation of the fundamentals of materials and their application in various industries. The Forum program was expanded by discussing the problems of introducing scientific developments and interaction of scientists with consumers of scientific results, foundations and development institutions. Additionally, the role of international cooperation, public and educational organizations in the scientific process, and the strategy of popularizing scientific achievements were discussed at the Forum.

The scientific sections of the forum covered a wide range of topics on nanomaterials and nanotechnologies, metallurgy, experimental methods of researching materials and structures, ceramic materials, composite and functional materials, structural materials, biocompatible materials, and materials and technologies for green chemistry.

IOP Conf. Series: Materials Science and Engineering **525** (2019) 011001 doi:10.1088/1757-899X/525/1/011001

NM

Moscow 2018

IOP Conf. Series: Materials Science and Engineering **525** (2019) 011002 doi:10.1088/1757-899X/525/1/011002

Peer review statement

All papers published in this volume of *IOP Conference Series: Materials Science and Engineering* have been peer reviewed through processes administered by the proceedings Editors. Reviews were conducted by expert referees to the professional and scientific standards expected of a proceedings journal published by IOP Publishing.

IOP Conf. Series: Materials Science and Engineering **525** (2019) 012001 doi:10.1088/1757-899X/525/1/012001

Effect of type of ceramic particles on efficiency of gas dynamic spraying and hardness of hybrid coatings AlMg6/C60

A V Aborkin[1], A I Elkin[1], I A Evdokimov[2], N V Sachkova[3] and A E Sytschev[3]

[1] Vladimir State University named after Alexander and Nikolay Stoletovs, Vladimir, Russia
[2] Technological institute for superhard and novel carbon materials, Troitsk, Moscow, Russia
[3] Merzhanov Institute of Structural Macrokinetics and Materials Science, Russian Academy of Sciences, Chernogolovka, Russia

E-mail: aborkin@vlsu.ru

Abstract. The hybrid powders AlMg6/C60 + 10-70 wt.% Al_2O_3 and AlMg6/C60 + 10-70 wt.% AlN were obtained by the method of mechanical synthesis in a planetary ball mill. These powders were used to create coatings using cold gas-dynamic spraying. The influence of the type and concentration of ceramic reinforcement on the growth efficiency and microhardness of the formed hybrid coatings was studied. It has been found out that the use of ceramic particles of AlN (up to 50 wt.%) as a reinforcement contributes to a more efficient growth of the coating compared to Al_2O_3. At the same time, the microhardness of coatings reinforced with AlN is lower than that containing Al_2O_3.

1. Introduction

In the last decade, gas-thermal methods of forming protective coatings on the surfaces of machine structures and parts have been widely developed. These methods have played a decisive role in the development of fundamentally new high-tech methods for the synthesis of coatings that have significantly competed with traditional methods of hardening and coating.

One of the promising methods underlying the creation of such technologies and coatings should be considered the method of cold gas-dynamic spraying, developed at the Institute of Theoretical and Applied Mechanics of the Siberian Branch of Russian Academy of Sciences [1-3].

At present, both in the Russian Federation and abroad, intensive investigations are being conducted, which are aimed at the development of gas-dynamic spraying technologies, the study of the structure-phase composition and functional properties of coatings. The diversity of the research conducted and the results obtained can be judged by at least detailed reviews [4–7], a considerable part of which is the development of new powder materials for the synthesis of coatings with an increased complex of properties.

Nevertheless, an analysis of the current state of research shows that works on the preparation and study of heterogeneous coatings strengthened simultaneously by nano- and micro-sized particles are of a single character. At the same time, taking into account the experience of creating bulk hybrid composites based on aluminum and its alloys [8–12, etc.], the creation of heterogeneous coatings obtained by gas-dynamic spraying seems to be promising. However, the gas-dynamic spraying of powders with a particle size of less than 2 microns is difficult due to their inhibition in the compressed layer, which occurs when an ultrasonic gas stream flows onto an obstacle. In addition, when accelerating particles of a powder in a gas stream, there is a strong variation in velocity for particles of different sizes, while the nanoscale fraction can scatter without falling into the coating. Therefore, the use of polydisperse powder, which is a mechanical mixture of nano- and micro-sized particles, for gas-dynamic spraying seems to be unpromising. For the formation of heterogeneous coatings, in addition to micro-sized reinforcement containing nano-sized particles, it is necessary to use powder in the form of a mechanical mixture consisting of agglomerates of complex composition and ceramic particles. In this case, the agglomerates should be particles of a matrix material with nanosized reinforcement distributed in their volume, as well as micro-sized ceramic particles embedded in the agglomerates and located on the surface [13]. Use for gas-dynamic spraying of powders having such a structure, allows to obtain heterogeneous coatings with a uniform structure and a high level of mechanical properties [14]. Since the presence of nanoscale particles increases the mechanical properties of the matrix material [15], increasing the strength of fixing micro-size ceramic particles in the matrix. At the same time, the plasticity of the particles of the matrix material remains sufficient for the effective formation of a coating with a low porosity [16].

To obtain such powders, a method of mechanical processing in a high-energy ball mill can be successfully used, providing the synthesis of multicomponent powders with a particle size distribution necessary for gas-dynamic spraying (average particle size 15-20 μm) [13-16]. The obvious advantages of this method include the possibility of implementing the distribution of reinforcimg particles, nanostructuring and dispersing of a matrix material within one technological cycle [17].

The aim of this work is to study the effect of the type of ceramic particles on the efficiency of gas-dynamic spraying and the hardness of hybrid coatings AlMg6 + 0.3wt.% C60.

2. Materials and Methods

Powder for cold gas-dynamic spraying was obtained by mechanical synthesis in a planetary ball mill for two technological stages. At the first stage, nanocomposite powder AlMg6 + 0.3 wt.% C60 was obtained. At the second stage, 10, 25, 30, 35, 50, and 70 wt.% α-Al2O3 or AlN were added to the obtained nanocomposite powder (hereinafter the content of ceramic particles in the powder is indicated) and the treatment was continued. The average particle size of α-Al2O3 and AlN was 18.5 and 5.9 μm, respectively.

The granulometric composition of the obtained powders was determined on a Microsizer-201C instrument. The morphology of the obtained hybrid powders and the formed coatings, as well as their microstructures, was studied using an ultra-high resolution auto-emission scanning electron microscope Zeiss Ultra Plus based on Ultra 55.

The resulting powders were sprayed onto sheet substrates from the steel of 1008 grade. For spraying was used installation of cold gas-dynamic spraying DYMET-404. The sputtering mode is stationary (sprayed to a point) at an air flow temperature of 450 °C, the distance from the nozzle exit to the substrate surface is 10 mm, and the exposure time is 15 seconds. The effectiveness of the deposition of the coating was determined by the magnitude of the growth of the coating, which was calculated as the difference in the mass of the substrates before and after the cold gas-dynamic spraying.

The microhardness of the coatings was measured on transverse thin sections by the Vickers method using a Shimadzu HMV-2 microhardness tester with a load of 245 mN and a holding duration of 10 s.

3. Results

As a result of processing in a planetary ball mill, hybrid powders were obtained. Figure 1 a, b shows typical SEM images of synthesized powders characterizing their morphology. The synthesized hybrid powders were a mechanical mixture consisting of agglomerates of complex composition and micro-sized ceramic particles. Agglomerates of complex composition were particles of a nanocrystalline

IOP Conf. Series: Materials Science and Engineering **525** (2019) 012001 doi:10.1088/1757-899X/525/1/012001

matrix material containing fullerenes embedded in them, as well as micro-sized ceramic particles located on the surface. It is noted that with an increase in the content of ceramic particles in the mixture, their concentration on the surface of the particles of the matrix material increases markedly.

Figure 1. SEM images of AlMg6/C60 + Al_2O_3 and AlMg6/C60 + AlN powders: a, d – n = 10 wt.%; b, e – n = 30 wt.%; c, f – n = 50 wt.%;
Numbers indicate: 1 - agglomerates; 2 - ceramic particles

The study of the granulometric composition of the hybrid powder shows that it is characterized by the content of particles described by a bimodal distribution, with the first maximum with a particle size of about 5 microns corresponding to a ceramic reinforcement. It should be noted that with the adding of 10–70 wt.% of ceramic particles into the powder mixture, mechanical processing even for 15 minutes has a significant effect on the average size of the formed agglomerates, which decreases by more than 3 times and is ~17 μm. Depending on the content of ceramic particles, the average particle size of the synthesized powders is 12.3–17.8 μm and 8.2–16.4 μm for Al_2O_3 and AlN, respectively. It should be noted that 26.4% of AlMg6/C60 powder particles + 70 wt.% AlN have a size less than 5 microns. This circumstance allows us to predict that the deposition of this composition may be ineffective.

IOP Conf. Series: Materials Science and Engineering **525** (2019) 012001 doi:10.1088/1757-899X/525/1/012001

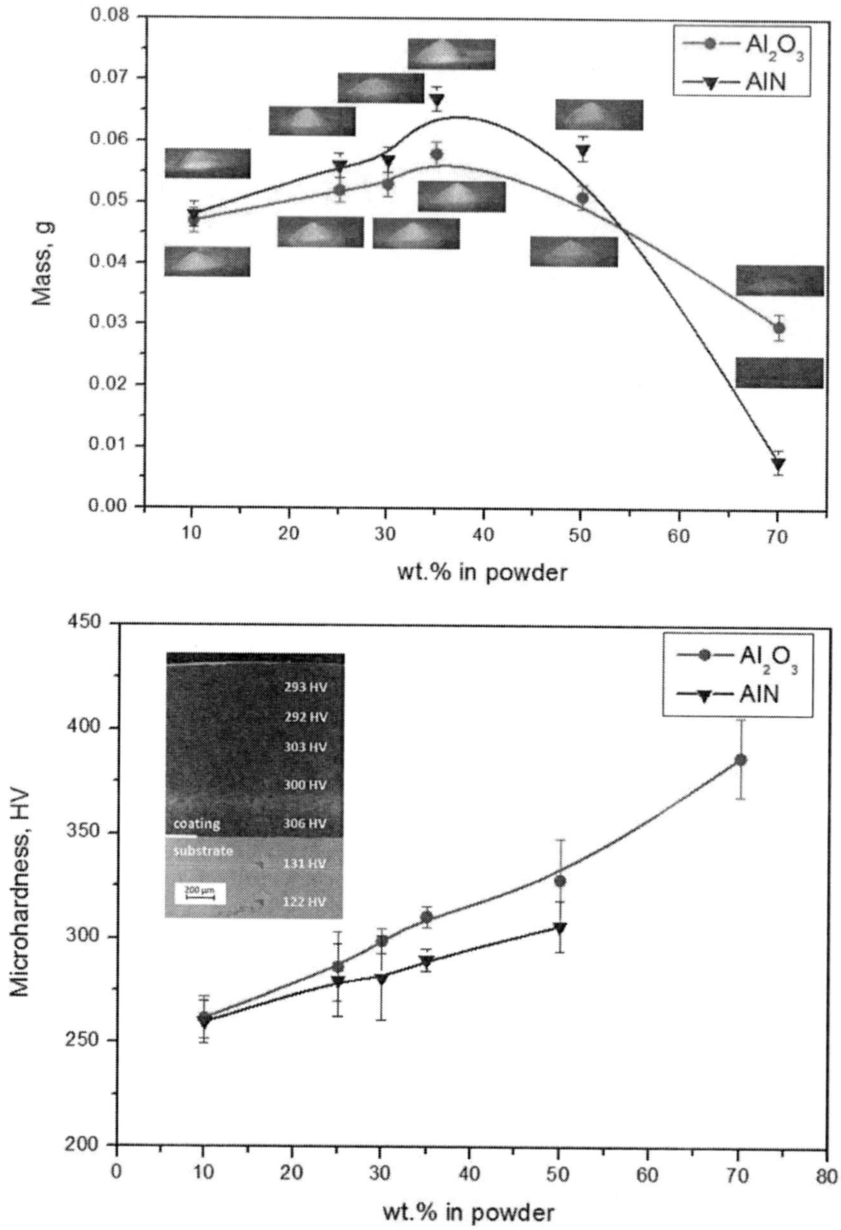

Figure 2. The effect of the content of ceramic particles on the efficiency of gas-dynamic spraying of powders (a) and the microhardness of the formed coatings (b)

Figure 2a presents the data characterizing the efficiency of gas-dynamic spraying of mechanically synthesized powder mixtures. The presence in the powder mixture of AlN particles contributes by 7–10% more effective growth of the coating compared to Al_2O_3. However, it should be noted that the use of Al_2O_3 particles as a ceramic additive allows to achieve good sputtering efficiency of powder mixtures with a high content of ceramic particles (70%). The highest sputtering efficiency, in the experiments performed, regardless of the type of ceramic particles, is achieved at 30-35 wt.% of ceramic particles in the powder mixture. A further increase in the content of ceramic particles leads to a decrease in the efficiency of cold gas-dynamic spraying.

Figure 2b shows the change in microhardness (obtained on transverse polished sections) of hybrid coatings depending on the content of ceramic particles in the powder. A comparative analysis of the data shows that addition of 10–70 wt.% Al_2O_3 into the powder leads to an increase in the hardness of the coatings from 262 to 387 HV. The hardness of coatings obtained by gas-dynamic spraying of powder containing AlN is slightly lower and varies in the range of 260-306 HV at 10-50 wt.% AlN.

References

[1] Patent USSR 1986 1246638.
[2] Alkhimov A P, Kosarev V F and Papyrin A N *Soviet Physics Doklady* 1990 35(12) 1047–1049.
[3] Papyrin A, Kosarev V, Klinkov S, Alkhimov A and Fomin V 2007 *Cold spray technology.* (Elsevier Science, Amsterdam) p 336.
[4] Raoelison R N, Verdy Ch and Liao H *Mater. Des.* 2017 **133** 266–287.
[5] Raoelison R N, Xiea Y, Sapanathan T, Planche M P, Kromer R, Costil S and Langlade C *Additive Manufacturing* 2018 **19** 134–159.
[6] Moridi A., Hassani-Gangaraj S. M., Guagliano M., Dao M. Cold spray coating: review of material systems and future perspectives // Surface Engineering 36 (2014) 369–395.
[7] Li W, Assadi H, Gaertner F and Yin S *Critical Reviews in Solid State and Material Sciences* 2018 **43** 1–48.
[8] Singh J and Chauhan A *J. Mater. Res.* 2016 **5** 159–169.
[9] Alidokht S A, Abdollah-Zadeh A, Soleymani S and Assadi H *Mater. Des.* 2011 **32** 2727–2733.
[10] Kanayo Alaneme K and Apata Olubambi P *J. Mater. Res.* 2013 **2** 188–194.
[11] Show B K, Mondal D K and Maity J *Metallogr. Microstruct. Anal.* 2014 **3** 11–29.
[12] Bodunrin M O, Alaneme K K and Chown L H *Journal of Materials Research and Technology,* 2015 **4**(4) 434–445.
[13] Aborkin A V, Alymov M I, Arkhipov V E and Khrenov D S *Doklady Physics* 2018 **63**(2) 50–54.
[14] Aborkin A V, Sobol'kov A V, Elkin A I and Arkhipov V E *J. Phys.: Conf. Ser.* 2018 **951** 012010.
[15] Aborkin A V, Alymov M I, Kireev A V, Sobol'kov A V and Arkhipov V E *Metallurgist* 2018 **62**(7-8) 809–814.
[16] Suryanarayana C, Al-Aqeeli N *Progress in Materials Science* 2013 **58**(4) 383–502.
[17] Aborkin A V, Alymov M I, Kireev A V, Elkin A I and Sobol'kov A V *Nanotechnologies in Russia* 2017 **12**(7-8) 395–399.

IOP Conf. Series: Materials Science and Engineering **525** (2019) 012002 doi:10.1088/1757-899X/525/1/012002

Analysis of the atomic structure of the amorphous metal alloy $Al_{84}Ni_6La_{10}$

K B Aleinikova, E N Zinchenko and A A Zmeikin

Voronezh State University, 394006, 1 Universitetskaya Ploschad, Voronezh, Russia

E-mail: xenale@mail.ru, zinchenko@vsu.ru, zmeykin@list.ru

Abstract. The paper presents an analysis of the atomic radial distribution function of the amorphous metal alloy $Al_{84}Ni_6La_{10}$ based on the fragmentary model. A comparative analysis of the most probable interatomic distances in the alloy and its compounds demonstrated the presence of crystalline nuclei of the following phases: Al_3La, $AlNi_3$, Al_4La and Al. The crystalline nuclei of the Al_4La compound are the smallest (~7-8 Å), while the minimal size of the most nuclei of the other phases is 1-1.5 nm. The following phases were observed in the crystallised alloy: $Al_{11}La_3$, Al_3Ni, and Al.

1. Introduction

Amorphous metal alloys (AMA) are becoming more and more popular in industry due to their atomic structure. AMAs have a number of advantages over their crystalline analogues. Aluminium-based amorphous alloys are widely used in the aircraft industry. AMA-based composite materials are lighter, and thus reduce the total weight of the plane when used for producing fuselage shells and load-bearing structures. The total portion of composite materials in the total weight of modern airplanes is up to 20% [1]. The world's leading producers of AMAs are the USA and China, as their production output is several times larger than in other countries. In Russia, the major manufacturer of AMAs is the Asha Metallurgical Plant. However, amorphous alloy ribbons of over 20-cm width are still being imported [2].

Properties of any substance are determined by its atomic structure. Although amorphous materials are industrially manufactured, there is still no unified understanding of their atomic structure. As they lack long-range order, it is impossible to study the structure of amorphous materials applying the methods used for analysing the structure of crystals. The only thing we can determine by a diffraction experiment with amorphous materials performed using the Fourier synthesis, is the atomic radial distribution function (ARDF). This function can be analysed using various models. The best way to assess the quality of any such model is to calculate the atomic radial distribution function (or most commonly, correlation function) using the Finback-Warren method based on atomic pair distribution functions [3,4]. For the calculated ARDF to be similar with the one determined experimentally, the initially selected model can be specified using the Monte Carlo method. However, this method is applied to a limited number of atoms or structural units. Any limitation in this case results in the interatomic distance of the atoms near the boundary being cut off. This, in turn, leads to errors while determining N_{ij} – the number of atoms on the *i*th coordination sphere of the radius r_{ij} of the *j*th atom. Using this method one should also imply that the amorphous substance is atomically homogeneous, which is often proved to be untrue by the high resolution electron microscopy [5,6]. The fragmentary model, on the other hand, allows for the analysis of the structure of heterogeneous amorphous materials [7-10]. This model assumes that an amorphous substance is a patchwork of fragments of compound structures or crystalline nuclei which may appear in substances with certain elemental composition. Model atomic radial distribution functions are calculated using complete structural data of the expected crystals-analogues and then compared by the position of their maximums to the atomic radial distribution functions determined experimentally. If he expected interatomic distances in the model correspond to the results of the experiment, we can assume that the studied alloy contains crystalline nuclei of the phase, whose structural data was used to calculate the model. Below we present the results of the analysis of the atomic structure of the amorphous metal alloy $Al_{84}Ni_6La_{10}$ carried out using the fragmentary model.

2. Experimental

The amorphous metal alloy was obtained from the A.A. Baikov Institute of Metallurgy and Material Science of the Russian Academy of Sciences following the methodology described in [11]. According to the results of X-ray fluorescence analysis, the composition of the alloy corresponded to the formula $Al_{84}Ni_6La_{10}$. The samples were ribbons of 3-mm width and 40 mkm thickness.

Experimental X-ray scattering intensity curves $I(2\theta)$ were determined at the Centre for Collective Use of Scientific Equipment of Voronezh State University using diffractometer ARL X'TRA (Cu Kα-radiation, λ = 1.5418 Å; θ-θ-scanning from 4 to 145° at 2θ) and DRON-3 (monochromatized Mo Kα-radiation, λ = 0.7107 Å, graphite monochromator in the secondary beam; θ-2θ-scanning from 7 to 135° at 2θ). Scanning was performed step by step (0.2° scanning steps 2θ each, pulse accumulation time - 30 sec.). For the scanning, 5 or 6 layers of 3×10 foils of the amorphous alloy were glued to a single-crystal Si-wafer, carved in such a way that it would not reflect anything at any rotation angle of the diffractometer. At each step the scanning was performed several times. The results of the repeated tests were used to determine the mean intensity curve.

In order to calculate the interference function $i(S)$ of the intensity curve, prior to the analysis of the experimental data $I(2\theta)$ a correction for polarisation of secondary X-rays was applied. The correction was different for the experiments with copper and molybdenum. Thus, $I(2\theta)$ was divided by the polarization factor

$$P(2\theta) = \frac{1+\cos^2(2\theta)}{2} \quad (1)$$

for the experiment with Cu radiation, and

$$P(2\theta) = \frac{1+\cos^2(2\theta)\cos^2(2\alpha)}{2} \quad (2)$$

for the experiment with Mo radiation, where 2θ is the diffraction angle, α = 6.08° is the reflection angle of the graphite monochromator. The experimental intensivities were then normalised by the scale factor m to the values similar to the sum of the squares of the atomic scattering factors $\sum_{i=1}^{f.u.} f_i^2(S)$, where $S = 4\pi\sin\theta/\lambda$ is the diffraction vector. The sums of the squares of the atomic scattering factors were calculated using the data from [12], assuming that the formula unit (f.u.) consisted of 100 atoms of the alloy.

Interference functions $i(S)$ were calculated using the method described in [13] according to the following expression

$$i(S) = \frac{mI(S)}{\sum_{i=1}^{f.u.} f_i^2} - \overline{\frac{mI(S)}{\sum_{i=1}^{f.u.} f_i^2}} \quad (3)$$

The sum of the squares of the atomic scattering factors at those points of S, where the values of the experimental intensity curve were set, was determined by means of a programme that applied the cubic spline interpolation technique [14] to the curve $\sum_{i=1}^{f.u.} f_i^2$. The average line $mI(S)\Big/\overline{\sum_{i=1}^{f.u.} f_i^2}$ was a polynomial of a 2-5 degree approximating the values of $mI(S)\Big/\sum_{i=1}^{f.u.} f_i^2$, whose coefficients were determined using the least square method [15]. The degree of the polynomial was selected so that the resulting interference function would oscillate near the S axis. The range of values of the normalizing constant j for the interference functions obtained for Cu and Mo radiations was determined as described in [13]. Normalized interference functions were then "sewn" together right after the largest first maximum of the interference function determined by Cu radiation. The function $i(S)$ is demonstrated in Figure1.

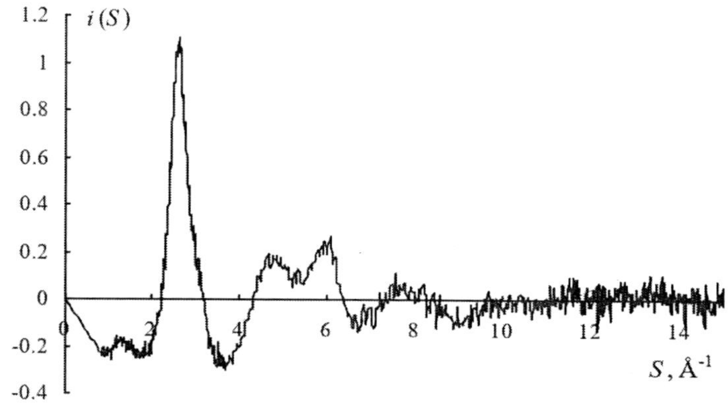

Figure1. Interference function $i(S)$ with the introduced attenuation coefficient α=0.07.

Experimental atomic radial distribution function $4\pi r^2\rho(r)$ was calculated using the formula

$$4\pi r^2\rho(r) = 4\pi r^2 \frac{D_x}{M1.65}\left(\sum_{i=1}^{f.u.} K_i\right)^2 + \frac{2r}{\pi}\sum_{i=1}^{f.u.} K_i^2 \int_0^{S_{max}} S\cdot i(S)\sin(Sr)dS. \quad (4)$$

$D_x = 2.7$ g/cm^3 is the X-ray density of aluminium, M is the mass fraction of all the formula unit of the alloy, K_i is the relative scattering ability of the atoms. The scattering unit was an Al atom, while the relative scattering abilities of other atoms were calculated as described in [16] using the formula:

$$K_i^2 = \frac{\int_0^{S_{max}} S^2 f_i^2(S)dS}{\int_0^{S_{max}} S^2 f_{Al}^2(S)dS} \quad (5).$$

Integration was performed at 0.01 Å^{-1} steps along S after each 0.01 Å on r. Figure 2 shows the experimental atomic radial distribution function normalized to a single atom by dividing it by the number of atoms in the formula unit. It is overlaid with a model ARDF of Al. The maximums of the experimental ARDF are located as follows: 2.58; 3.36; 4.54; ~5.05; 5.53; 6.18; 6.69; 7.2; 7.6; 8.08; 8.54; ~8.9; 9.41 Å.

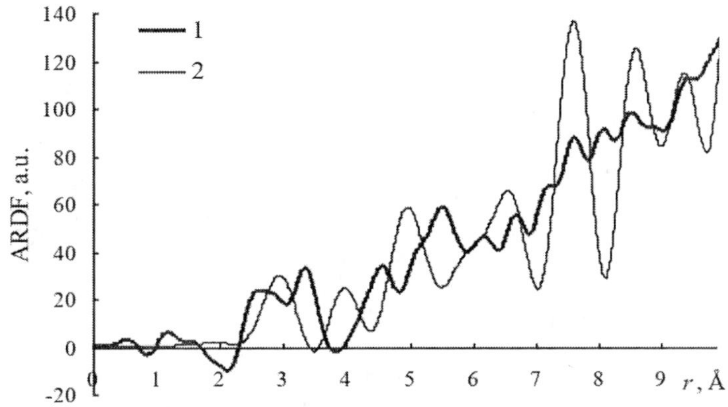

Figure 2. Experimental ARDF of the alloy (1) and model ARDF of Al (2).

3. Discussion

The experimental ARDF was analysed using the fragmentary model [7-10]. In metal alloys the fragments are crystalline nuclei. If 2 or 3 unit cells fit in the linear dimensions of a crystallite, it means that the diffraction pattern of this nanodispersed substance is similar to the diffraction pattern of glass, as demonstrated in the experiments with Al described in [10]. Although the diffraction pattern shows that the atoms are located randomly, the interatomic distances within 1nm of a crystalline nucleus conform to its atomic structure. We know, that maximums of the experimental ARDF correspond to the most probable interatomic distances in the observed sample, taking into account the scattering ability of the atoms at the ends of the interatomic vector. If the alloy consists of crystalline nuclei of the same phase, maximums of the experimental ARDF correspond to the most probable interatomic distances of this phase. If the alloy consists of crystalline nuclei of several phases, maximums of the experimental ARDF correspond to the superposition of the most probable interatomic distances of the crystalline nuclei of these phases. The analysis of the experimental ARDF using the fragmentary model includes comparison of the position of maximums of independently calculated curves. One of the curves is the experimental ARDF, the others - model ARDFs calculated using complete structural data of crystalline nuclei as described in [7, 8]. When calculating a model ARDF it is impossible to account for either the defects of the actual crystal structure, or the boundaries between the crystalline nuclei. Therefore, the experimental and model ARDFs do not coincide completely even when the amorphous sample consists of nanocrystallites of a single phase. This happens because model ARDFs oscillate about parabolas comprising their X-ray density, while the parabolas of experimental ARDFs comprise the actual density of the sample, which has a different value. Another explanation is that the maximums may change due to cut-offs in interatomic distances [17]. The position of the maximums, however, must remain constant. The values of the maximums are determined by the scattering abilities of the atoms and the formula of the compound. This means that the ARDFs in the calculated model coincide neither with the experimental ARDF, nor with each other. In order to compare and analyse these curves, they are normalised to single atoms (by dividing by the number of atoms in the formula), and to a single parabola (experimental). For some model ARDFs of compounds that contain elements with high scattering abilities, additional scaling can be used, however, that does not change the position of the maximums.

Model ARDFs calculated for the compounds of Al-La, Al-Ni, and normalised to a single atom are demonstrated in Figure 3 and 4. The figures show that they differ in the position of their maximums.

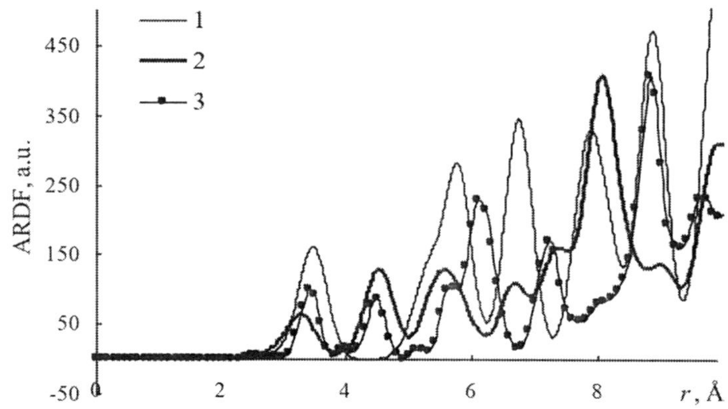

Figure 3. Model ARDFs of crystalline analogues Al_2La (1), Al_3La (2) and Al_4La (3).

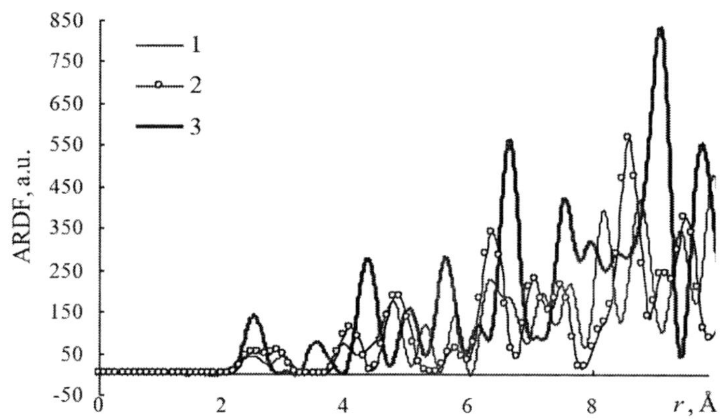

Figure 4. Model ARDFs of crystalline analogues Al_3Ni (1), AlNi (2) and $AlNi_3$ (3).

If the experimental ARDF has maximums typical for a certain crystalline phase or several phases, we can assume that crystalline nuclei of these phases are present in the analysed substance. This assumption underlies a kind of a phase analysis of amorphous substances [9, 10].

The easiest way to analyse the experimental ARDF is to overlay it with model ARDFs. If the largest maximum of a model ARDF coincides with the minimum of the experimental curve, we can conclude that nanocrystallites of the compound are not present in the studied sample. If the maximums coincide within the limits of experimental error (~0.05 Å), nanocrystallites of the substance are present in the studied alloy.

In our analysis, we first overlaid the experimental ARDF with a model ARDF calculated for pure aluminium (Figure 2). The amorphous alloy containing over 80% of aluminium is bound to contain its crystalline nuclei. However, Figure 2 demonstrates that the maximums of the two curves are different in the area of small r and coincide only after 7 Å. This can be accounted for by the fact that the scattering ability of Al atoms is lower compared to the scattering ability of heavier atoms (La and Ni). Therefore, the first coordination spheres are mainly formed by nanocrystallites of compounds of La and Ni with aluminium. We can assume that the amorphous alloy contains a large number of nanocrystallites of Al-La and Al-Ni phases with the size smaller than 1 nm. Indeed, the first maximums of the experimental ARDF are better described by the model ARDF calculated using the structural data of Al_3La (Figure 5).

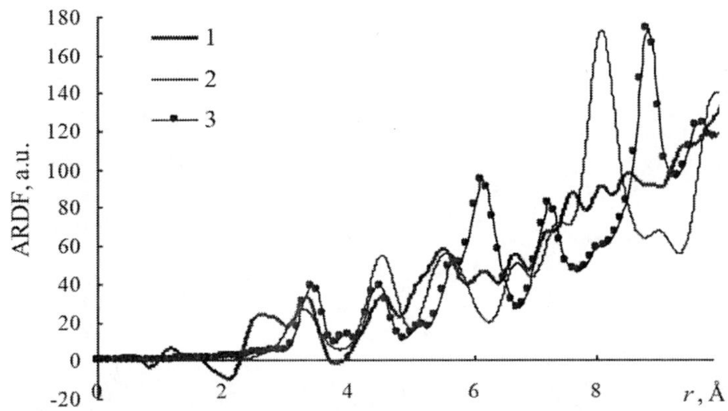

Figure 5. Experimental ARDF of the alloy (1) and model ARDF of crystalline analogues Al_3La (2) and Al_4La (3).

The first maximum of the experimental curve is 2.58 Å, which is smaller than the interaomic distance Al-Al in pure aluminium (2.92 Å). However, in Ni-Al compounds such interatomic distances correspond to Ni-Al bounds. With excess aluminium, it is natural that the alloy contains crystals of intermetallic compounds rich in Al. In earlier experiments we observed crystalline nuclei Al_3Ni in similar alloys [9, 10]. Here, however, the largest maximum of the model ARDF of the same compound positioned at 6.3 Å coincided exactly with the minimum of the experimental curve. Out of all the Al-Ni based compounds, the best model to describe the maximums of the experimental ARDF was the model calculated using the structural data of $AlNi_3$ (Figure 6). Earlier, we found similar crystalline nuclei ($Ni_{1.5}Al_{0.5}$) while analysing the experimental ARDF, based on a Cu radiation experiment with the alloy $Al_{83}Ni_{10}La_7$ [18].

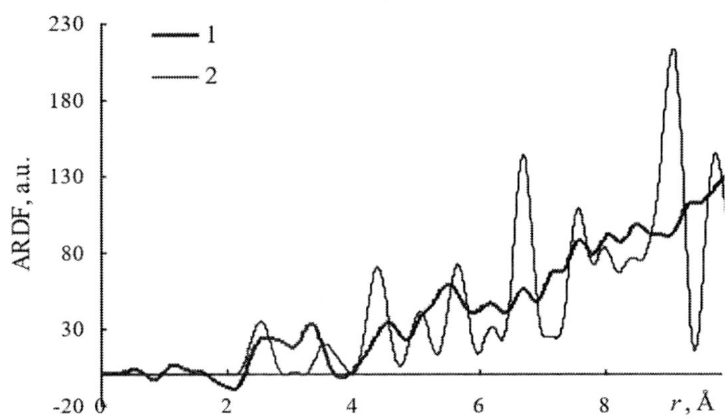

Figure 6. Experimental ARDF of the alloy (1) and model ARDF of the crystalline analogue $AlNi_3$ (2).

The three phases (Al, Al_3La and Ni_3Al) could account for practically every maximum of the experimental curve, except for clear maximums at 6.18Å and 7.2Å. These maximums coincided with the maximums of a model curve based on the structural data of Al_4La. Interatomic distances of this compound resulted in a shift towards larger *r* of the maximum of the experimental ARDF (3.36 Å) as compared to the interatomic distances of Al-La in Al_3La (3.29 Å). However, with the higher coordination spheres (over ~8Å) they did not coincide with the maximums of the experimental curve. This allowed us to conclude that the size of the crystalline nuclei of Al_4La did not exceed 7-7.5Å.

IOP Conf. Series: Materials Science and Engineering **525** (2019) 012002 doi:10.1088/1757-899X/525/1/012002

Figure 7 presents the experimental ARDF (based on the alloy's formula) and the sum of model ARDFs (Al, Al_3La, Ni_3Al и Al_4La) calculated for their formulas (not normalised and without additional scaling). The method allows using model ARDFs in various proportions depending on the formula of the alloy. The proportion of each model in the final ARDF was determined by the number of maximums coinciding with the maximums of the experimental curve. The best result was achieved using the ARDF of the following composition: $6Al_3La+4Al_4La+2Ni_3Al+48Al = Al_{84}Ni_6La_{10}$. There is an obvious correlation between the maximums of the experimental curve and the maximums of the final ARDF.

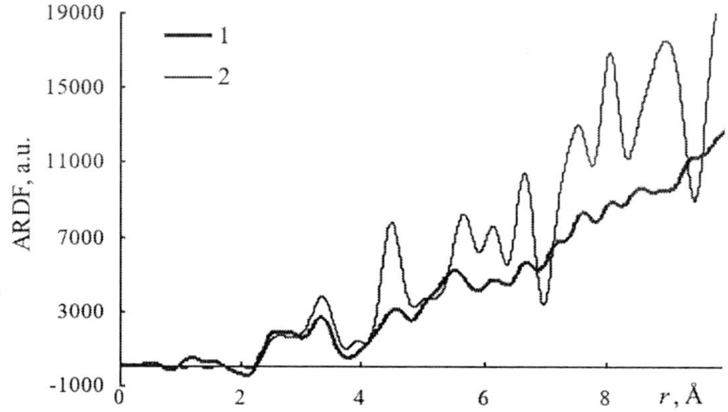

Figure 7. Experimental ARDF of the alloy (1) and the sum of model ARDFs (2).

The amorphous alloy was crystallised using pulse photon irradiation (PPI) in a high vacuum. The debyeogram produced by diffractometer ARL X'TRA (Figure 8) demonstrated the presence of three phases in the quenched sample: pure Al and compounds $Al_{11}La_3$ and Al_3Ni (Table). It was not the only time when we observed that the grown crystals were different from the crystalline nuclei found in the amorphous phase [9, 10].

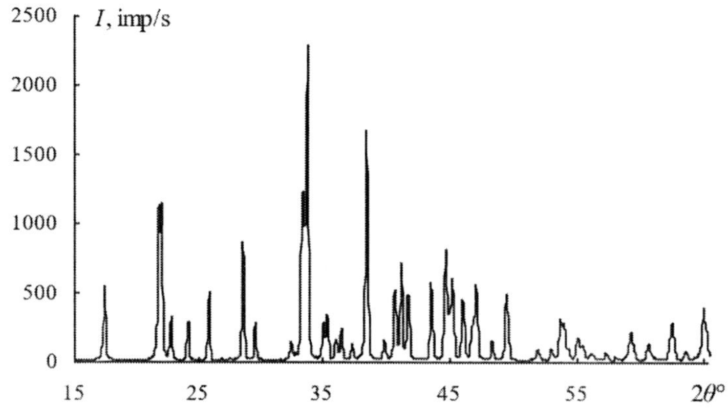

Figure 8. Diffractogram of the crystallised alloy $Al_{84}Ni_6La_{10}$.

IOP Conf. Series: Materials Science and Engineering **525** (2019) 012002 doi:10.1088/1757-899X/525/1/012002

Table 1. Interplanar space and intensivities (in brackets) of the crystallised alloy, $Al_{11}La_3$, Al_3Ni, and Al.

$Al_{84}Ni_6La_{10}$ d, Å(I/I_0, %)	$Al_{11}La_3$ [19] d, Å(I/I_0, %)	Al_3Ni [20] d, Å(I/I_0, %)	Al [21] d, Å(I/I_0, %)
5.067 (22)	5.07 (43)		
4.04 (42)	4.06 (60)		
	4.02 (57)	4.02 (40)	
3.895 (13)		3.89 (70)	
3.68 (11)		3.68 (50)	
3.44 (19)		3.44 (100)	
3.117 (37)	3.115 (4)		
3.017 (13)		3.01 (60)	
2.758 (6)	2.757 (7)		
2.684 (50)	2.686 (44)		
	2.675 (43)		
2.653 (100)	2.654 (100)		
2.556 (13)	2.554 (6)	2.55 (30)	
2.533 (16)	2.534 (6)		
2.487 (75)	2.486 (6)		
2.460 (12)		2.46 (40)	
2.406 (6)		2.40 (20)	
2.340 (80)			2.338 (100)
2.260 (7)	2.26 (4)	2.26 (5)	
2.217 (23)	2.215 (21)		
2.190 (34)	2.19 (15)	2.18 (70)	
2.160 (22)	2.169 (8)	2.16 (80)	
2.076 (27)	2.074 (2)	2.07 (100)	
2.029 (35)	2.03 (5)	2.01 (40)	2.024 (47)
	2.01 (5)		
2.002 (23)	1.9995 (2)	2.00 (90)	
1.97 (23)		1.97 (70)	
1.927 (22)	1.926 (7)	1.93 (100)	
1.880 (8)		1.88 (30)	
1.841 (23)	1.839 (13)	1.84 (30)	
1.760 (4)		1.76 (20)	
1.730 (4)	1.726 (3)		
1.707 (13)	1.698 (11)		
1.668 (8)	1.668 (16)		
1.655 (4)	1.657 (4)		
1.637 (2)	1.636 (2)		
1.558 (9)	1.558 (7)		
1.526 (5)	1.526 (2)	1.526 (30)	
1.484 (13)	1.484 (10)		
1.462 (4)	1.4615 (9)		
1.433 (16)	1.428 (3)		1.431 (22)

4. Conclusion.

Using the fragmentary model to analyse the experimental atomic radial distribution function allows to find certain interatomic distances in an amorphous metal alloy. These distances are typical for crystalline nuclei of the compounds whose formation allows for elemental composition of the alloy. Thus, when we compare experimental and model ARDFs, we can clearly observe crystalline nuclei of

IOP Conf. Series: Materials Science and Engineering **525** (2019) 012002 doi:10.1088/1757-899X/525/1/012002

Al_3La in the studied alloy. The experimental and model curves practically coincide in the first coordination spheres. The position of the maximum of the experimental curve at 3.36 Å as compared to 3.29Å for the model ARDF of the crystal-analogue Al_3La, is accounted for by the experimental error and the presence of crystalline nuclei of Al_4La (the position of the first maximum of the model ARDF is 3.43 Å). The presence of the nuclei also accounts for the maximums of the experimental curve at 6.18 Å and 7.2 Å. These most probable interatomic distances are not typical for any other compound of the systems Al-Ni and Al-La. The other maximums are best accounted for by the model ARDF calculated using the structural data of Ni_3Al. Compounds $Al_{11}La_3$, Al3Ni, and Al were observed in the crystallised alloy.

References

[1] The Boeing Company website: http://www.boeing.ru.
[2] Asha Metallurgical Plant website http://www.amet.ru/
[3] Warren B E 1971 *Kristallografiya* **16** 1264-70
[4] Wright A K 1998 *Fizika i Khimiya stekla* **24** 218-65
[5] Kaito C and Ahhl J 1987 *Phys.* **24** 117-19
[6] Glezer A M 2002 *Russian journal of chemistry* **XLVI** No 5 57-63
[7] Aleinikova K B and Zinchenko E 2009 *Journal of structural chemistry* **50** Supplement S93-99
[8] Aleinikova K B, Zinchenko E N and Likhach N I 2005 *Zavodskaya laboratoriya. Diagnostika materialov* **71** No 4 27-31
[9] Aleinikova K B, Zmeikin A A, Zinchenko E N and Ievlev V M 2012 *Glass Physics and Chemistry* **38** 71-76
[10] Aleinikova K B, Zinchenko E N and Zmeikin A A 2018 *Glass Physics and Chemistry* **44** 307-13
[11] Belousov O K, Kovneristyi Yu K and Popova E V 2005 *Deformatsiya i razruschenie materialov* No 8 29-37
[12] Cromer D T and Waber J T 1965 *Acta Cryst.* **18** 104-109
[13] Nabitovich I D, Stetsiv Ya I and Voloschuk Ya V 1967 *Kristallografiya* **12** 584-90
[14] MacCraken D and Dorn W 1977 *Numerical methods and FORTRAN programming* (Moscow: Mir) p 584
[15] Faddeev M A and Markov K A 2005 *Numerical methods* (Nizhny Novgorod: Nizhny Novgorod State University Publishing House) p 256
[16] Veinshtein B K 1957 *Kristallografiya* **2** 29-37
[17] Aleinikova K B, Zinchenko E N, Arakcheeva A.V and Grinevich V V 2002 *Crystallography Reports* **47** 555-58
[18] Aleinikova K B and Zmeikin A A 2009 Proceedings of Voronezh State University No 1 5-9
[19] *ICDD PDF-2 №* 00-024-0501.
[20] *ICDD PDF-2 №* 00-002-0416.
[21] *ICDD PDF-2 №* 00-004-0787.

IOP Conf. Series: Materials Science and Engineering **525** (2019) 012003 doi:10.1088/1757-899X/525/1/012003

NMR studies of 3D topological insulators over a large temperature range

A O Antonenko[1], E V Charnaya[1], V V Marchenkov[2,3] and S V Naumov[2]

[1]Physics Department, St. Petersburg State University, 1 Ulyanovskaya Street, St. Petersburg, 198504, Russia
[2]M.N. Mikheev Institute of Metal Physics of Ural Branch of Russian Academy of Sciences, 18 Sophya Kovalevskaya Street, Ekaterinburg, 620108, Russia
[3]Ural Federal University, 19 Mira Street, Ekaterinburg, 620002, Russia

E-mail: nastya5555@mail.ru

Abstract. Results of NMR measurements of the Bi2Te3 and Bi2Se3 3D topological insulators within the temperature range 11-293 K are presented. Two lines were demonstrated in the 125Te and 77Se spectra over the whole temperature range for single crystals in the c^B0 and c||B0 orientations as well as for Bi2Te3 and Bi2Se3 powders. The observed lines correlated to two non-equivalent positions of chalcogen atoms in the crystalline structure of investigated bismuth chalcogenides. The thermal activated character of charge carriers was found for the Bi2Te3 single crystal and powder. The activation energy was evaluated. It was shown for the Bi2Se3 powder that thermally activated charge carriers did not play the main role in conductivity. The nontrivial behaviour of the 125Te and 77Se NMR spectra for Bi2Te3 and Bi2Se3 single crystals in the orientation c||B0 was found at low temperatures which could not be explained by the orientation dependence of the line shapes and shifts in the topologically trivial semiconductors without structural phase transitions.

1. Introduction

Few years ago they started intensive studies of a new class of materials with nontrivial topology named "topological insulators" (TIs). The materials with unique properties, which appear due to topology of electron bands, were theoretically predicted by L. Fu and C. Kane in 2007 and later were experimentally found [1]. The nontrivial topology of electron bands in these materials emerges due to the strong spin-orbit coupling (SOC) and band inversion and is preserved by time reversal symmetry. It was shown that semiconductors like Bi_2Te_3 and Bi_2Se_3, which were earlier well known for their good thermoelectric properties, belonged to three-dimensional (3D) TIs. Such materials are considered as semiconductors or insulators, however, they show metallic surface conductivity. The electrons on the surface can move almost without any energy losses.

TIs bismuth chalcogenides with general formula $A_2^{V}B_3^{VI}$ are under enhanced attention of researchers due to their prospective applications in constructing elements and devices in different areas such as spintronics, magneto- and optoelectronics [2-4]. While the main properties of TIs are already confirmed, investigations of TIs are on the early stage of progress. Today there are only few studies of such TIs as Bi_2Te_3 and Bi_2Se_3 by nuclear magnetic resonance (NMR) [5-12].

This work reports the thermal evolution of the Bi_2Te_3 and Bi_2Se_3 properties over a large temperature range obtained by NMR.

2. Experiment

The Bi_2Te_3 and Bi_2Se_3 3D TIs were studied within a temperature range from 11.4 to 293 K using a Bruker Avance 400 NMR pulse spectrometer with the magnetic field 9.4 T. The high quality single crystals were grown by the Bridgman-Stockbarger method. The electrical resistance ratio at room and liquid helium temperatures, which characterizes the amount of lattice defects was equal to 25 for Bi_2Te_3 and 1.65 for Bi_2Se_3. Both crystals have the rhombohedral symmetry and the same layered crystal structure. The five individual atomic layers making up a quintuple layer lie in the sequence A1-Bi-A2-Bi-A1, where A1 and A2 denote the two different types of locations of chalcogen atoms (tellurium in case of Bi_2Te_3 or selenium in case of Bi_2Se_3). These atoms, A1 and A2, are crystallographically non-equivalent. The quintuple layers are coupled by the weak Van der Waals interaction. The layers are stacked along the *c*-axis. The unit cell contains three formula units. The X-ray diffraction (XRD) method was used to control the sample orientation.

The temperature evolution of NMR spectra was studied for the powder samples and for thin single crystalline plates in two different orientations, in which the crystallographic *c*-axis was parallel and perpendicular to the external magnetic field B_0. Powder samples were obtained from the single crystals using mortar and pestle. The thickness of plates was 0.6 mm for Bi_2Te_3 and 3 mm for Bi_2Se_3.

To detect the NMR signals from ^{125}Te in Bi_2Te_3 and ^{77}Se in Bi_2Se_3 a spin-echo pulse sequence $\pi/2$-τ-π was applied. Because of very wide NMR lines and quite low natural abundance of ^{125}Te and ^{77}Se (near 7 % for both isotopes) the signals were bad enough. To improve the signal to noise ratio we used the number of scans in each experiment no less than 2^{11}. Two types of NMR spectra were collected. First, we found the intensities of the spin echo signals at various frequencies and plotted their envelope. Second, we summed up the echo data at stepped offset to get varied offset cumulative spectra (VOCS).

3. Results and discussion

3.1 Spectra at room temperature

The ^{125}Te spectra showed two lines at room temperature for all single crystalline and powder samples (figure 1). The ^{77}Se NMR spectra demonstrated clearly two lines only for the powder sample while only single lines were seen for the crystalline plates in both orientations because of low signal-to-noise ratio (figure 1).

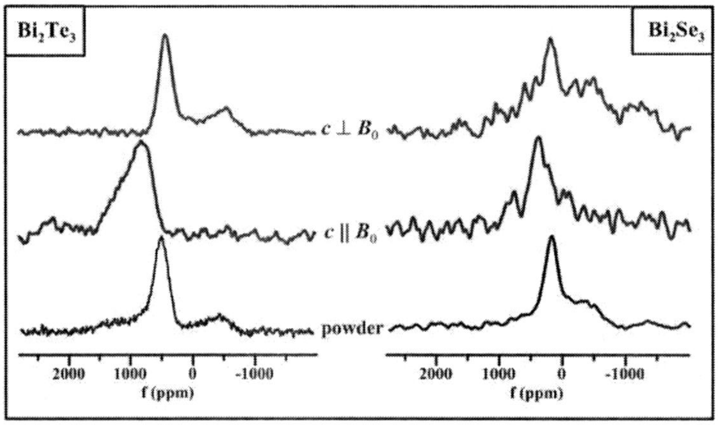

Figure 1. ^{125}Te (on the left) and ^{77}Se (on the right) VOCS for the Bi_2Te_3 and Bi_2Se_3 single crystals in two orientations and powders, respectively, at room temperature.

To explain the existence of these lines, simulations of the ^{125}Te and ^{77}Se powder NMR spectra were performed. The NMR data obtained for the Bi_2Te_3 and Bi_2Se_3 single crystals allowed us to calculate some parameters of the powder NMR spectra such as the isotropic shift δ_{iso}, anisotropy δ, and asymmetry η for the bismuth chalcogenides under study. We showed that the same values of these parameters could be found from the simulation of the powder NMR spectra. We demonstrated that the powder spectra could be deconvolved into two components (figure 2). The component intensity ratio was about 2:1 for the both compounds. In the crystallographic structure of the bismuth chalcogenides under study there are two non-equivalent crystallographic positions of chalcogen atoms. The amount of atoms in the A1 position is twice as much as in the A2 position. Using this and the results of the powder spectra simulation we could attributed the more intensive line to Te1 in Bi_2Te_3 and Se1 in Bi_2Se_3 and the less intensive one to Te2 in Bi_2Te_3 and Se2 in Bi_2Se_3. Earlier for nanoparticles of Bi_2Te_3 a little shoulder on the ^{125}Te NMR single line was shown [6]. This shoulder was treated as an influence of the sample surface.

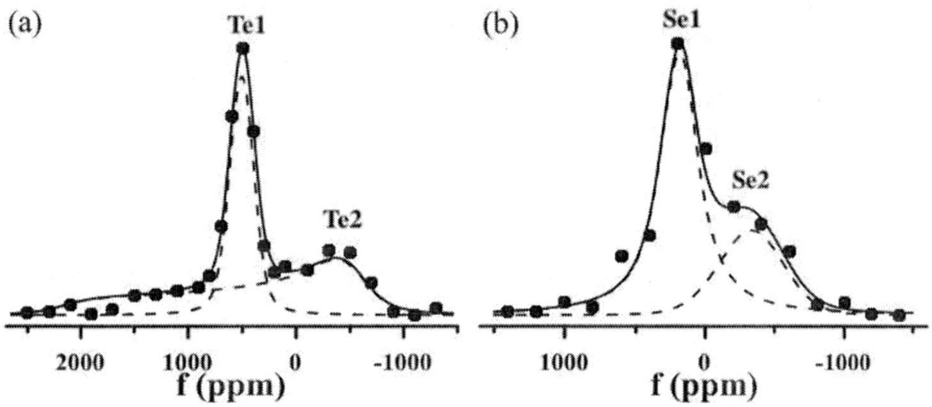

Figure 2. Deconvolution of the ^{125}Te (a) and ^{77}Se (b) NMR spectra at room temperature into two components (dashed lines) for the Bi_2Te_3 and Bi_2Se_3 powders, respectively. Solid lines – results of simulation of the powder NMR spectra, dots – experimental data from the envelope spectra.

3.2. Temperature evolution of the ^{125}Te NMR spectrum for Bi_2Te_3

We found almost the same behaviour of the ^{125}Te NMR spectra for the Bi_2Te_3 single crystal in the orientation $\boldsymbol{c} \perp \boldsymbol{B}_0$ and for powder with decreasing temperature. For both samples there was significant shift of the NMR lines to lower frequencies with decreasing temperature and the line width increased about two times (figure 3).

IOP Conf. Series: Materials Science and Engineering **525** (2019) 012003 doi:10.1088/1757-899X/525/1/012003

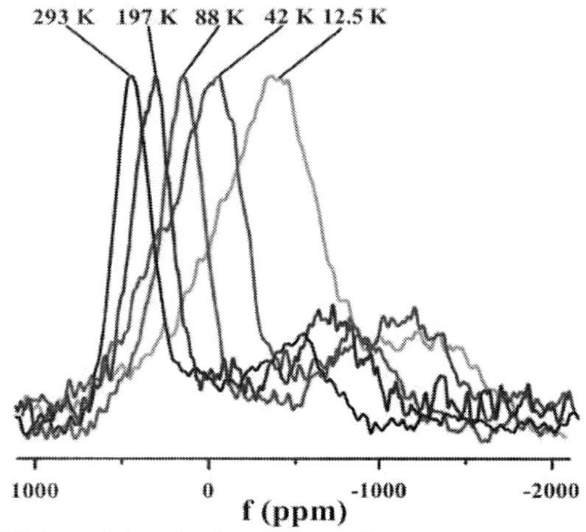

Figure 3. [125]Te VOCS for the Bi_2Te_3 single crystal in the orientation $\mathbf{c}\perp\mathbf{B}_0$ at different temperatures.

Using data obtained from the temperature evolution of the NMR spectra we can distinguish the contribution of the Knight shift to the total shift of NMR lines for both samples. It is known that for semiconductors the Knight shift has the following dependence on the charge carrier density n and temperature T [13]: $K_s \propto n/T$. For the intrinsic semiconductors this expression can be transformed into

$$K_s \propto \frac{n_0 e^{\frac{-E_a}{k_B T}}}{T} \tag{1}$$

where n_0 – charge carrier density in the high-temperature limit, E_a – activation energy and k_B – Bolzman constant.

The activation energies for charge carriers can be found using the temperature dependences of the Knight shift for Te1 and Te2 and the activation law (1). The values of E_a were the same for the single crystalline and powder samples and both lines within the framework of experimental error. The average value of the activation energy was 26 meV. It is one third of the energy gap for Bi_2Te_3 [14]. This means that the Fermi level in samples under study was shifted to the conduction band.

Extremely different situation was for single crystal Bi_2Te_3 in the orientation $\mathbf{c}||\mathbf{B}_0$. We found that the [125]Te NMR spectrum remained unchanged in general until 130 K. At lower temperatures the spectra changed drastically. The intensity of the line corresponded to Te2 increased significantly. At the lowest temperature this line became dominant (figure 4). Such inconsistency could not be caused by the orientation dependence for the topologically trivial semiconductor crystals without structural phase transition. Nowadays the nontrivial behaviour of the NMR spectra at low temperatures couldn't be explained within existing theoretical models because they do not take into account the strong SOC which plays the important role in TIs.

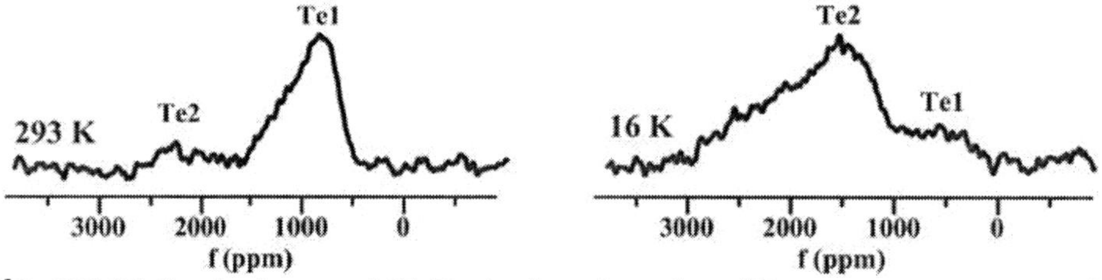

Figure 4. [125]Te VOCS for single crystal Bi_2Te_3 in the orientation $\mathbf{c}||\mathbf{B}_0$ at room temperature (on the left) and at the lowest measured temperature, 16 K, (on the right).

3.3. Temperature evolution of the ^{77}Se NMR spectrum for Bi_2Se_3

As Bi_2Se_3 studied had more defects which appeared during growing process than Bi_2Te_3 ($\rho_{293\,K}/\rho_{4.2\,K}$=1.65) we did not see such a pronounced shift of the ^{77}Se NMR spectra with decreasing temperature for the Bi_2Se_3 single crystal in the orientation $\boldsymbol{c}\perp\boldsymbol{B}_0$ and powder as it was for Bi_2Te_3 (figure 5). That means that charge carriers of defect nature play the main role in conductivity of Bi_2Se_3. However, for single crystalline Bi_2Se_3 in the orientation $\boldsymbol{c}\|\boldsymbol{B}_0$ we observed the nontrivial behaviour of ^{77}Se NMR spectra at low temperatures similar to that observed for bismuth telluride.

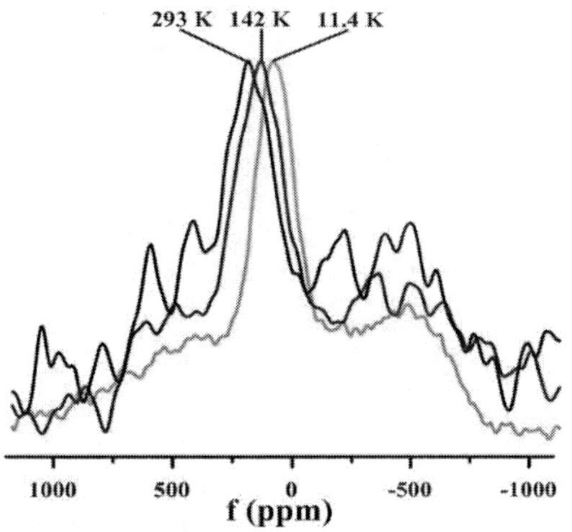

Figure 5. ^{77}Se VOCS for the Bi_2Se_3 single crystal in the orientation $\boldsymbol{c}\perp\boldsymbol{B}_0$ at different temperatures.

4. Conclusions

It was found that the ^{125}Te and ^{77}Se NMR spectra over the temperature range 11.4-293 K were generally consisted of two lines attributed to Te1 and Te2 in Bi_2Te_3 and Se1 and Se2 in Bi_2Se_3 for all samples studied. It was observed almost the same temperature evolution of ^{125}Te NMR spectra for the Bi_2Te_3 single crystal in the orientation $\boldsymbol{c}\perp\boldsymbol{B}_0$ and powder. The shift of NMR lines was determined mainly by the Knight shift. It was demonstrated the thermal activation character of charge carriers for the Bi_2Te_3 samples and the activation energy was found. It was shown that charge carriers of defect nature played the main role in conductivity for the Bi_2Se_3 single crystal in the orientation $\boldsymbol{c}\perp\boldsymbol{B}_0$ and powder. For bismuth chalcogenides single crystals in the orientation $\boldsymbol{c}\|\boldsymbol{B}_0$ the abnormal behaviour of the ^{125}Te and ^{77}Se NMR spectra was discovered. This anomaly can not be explained within the framework of existing theoretical models.

Acknowledgments

The studies were supported by RFBR, grant 19-57-52001. NMR and XRD measurements were carried.

References

[1] Fu L and Kane C L 2007 Topological insulators with inversion symmetry *Phys. Rev. B* **76** 045302

[2] Hasan M Z and Kane C L 2010 Colloquium: Topological insulators *Rev. Mod. Phys.* **82** 3045-67

[3] Scanlon D O, King P D C, Singh R P, De la Torre A, Walker S McK, Balakrishnan G, Baumberger F and Catlow C R A 2012 Controlling bulk conductivity in topological insulators: key role of anti-site defects *Adv. Mater.* **24** 2154-8

[4] Olbrich P, Golub L E, Herrmann T, Danilov S N, Plank H, Bel'kov V V, Mussler G, Weyrich Ch, Schneider C M, Kampmeier J, *et al.* 2014 Room-temperature high-frequency

transport of dirac fermions in epitaxially grown Sb_2Te_3-and Bi_2Te_3-based topological insulators *Phys. Rev. Lett.* **113** 096601
[5] Taylor R E, Leung B, Lake M P and Bouchard L-S 2012 Spin-lattice relaxation in bismuth chalcogenides *Phys. Chem. C* **116** 17300-5
[6] Koumoulis D, Chasapis T C, Taylor R E, Lake M P, King D, Jarenwattananon N N, Fiete G A, Kanatzidis M G and Bouchard L-S 2013 NMR probe of metallic states in nanoscale topological insulators *Phys. Rev. Lett.* **110** 026602
[7] Georgieva N M, Rybicki D, Guehne R, Williams G V M, Chong S V, Kadowaki K, Garate I and Haase J 2016 ^{77}Se nuclear magnetic resonance of topological insulator Bi_2Se_3 *Phys. Rev. B* **93** 195120
[8] Levin E M, Riedemann T M, Howard A, Jo N H, Bud'ko S L, Canfield P C and Lograsso T A 2016 ^{125}Te NMR and Seebeck effect in Bi_2Te_3 synthesized from stoichiometric and Te-rich melts *Phys. Chem. C* **120** 25196-202
[9] Podorozhkin D Yu, Charnaya E V, Antonenko A, Mukhamad'yarov R, Marchenkov V V, Naumov S V, Huang J C A, Weber H W and Bugaev A S 2015 Nuclear magnetic resonance study of a Bi_2Te_3 topological insulator *Phys. Solid State* **57** 1741-5
[10] Antonenko A O, Charnaya E V, Nefedov D Yu, Podorozhkin D Yu, Uskov A V, Bugaev A S, Lee M K, Chang L J, Naumov S V, Perevozchikova Yu A, *et al.* 2017 NMR studies of single crystals of the topological insulator Bi_2Te_3 at low temperatures *Phys. Solid State* **59** 855-9
[11] Antonenko A O, Charnaya E V, Nefedov D Yu, Podorozhkin D Yu, Uskov A V, Bugaev A S, Lee M K, Chang L J, Naumov S V, Perevozchikova Yu A, *et al.* 2017 NMR study of topological insulator Bi_2Te_3 in a wide temperature range *Phys. Solid State* **59** 2331-9
[12] Antonenko A O, Nefedov D Y, Charnaya E V, Naumov S V and Marchenkov V V 2018 ^{77}Se low-temperature NMR in the Bi_2Se_3 single crystalline topological insulator *Appl. Magn. Reson.* 49 599-605
[13] Selbach H, Kanert O and Wolf D 1979 NMR investigation of the diffusion and conduction properties of the semiconductor tellurium. I. Electronic properties *Phys. Rev. B* **19** 4435
[14] Yavorsky B Yu, Hinsche N F, Mertig I and Zahn P 2011 Electronic structure and transport anisotropy of Bi_2Te_3 and Sb_2Te *Phys. Rev. B* **84** 165208
out in the Research park of St. Petersburg State University.

IOP Conf. Series: Materials Science and Engineering **525** (2019) 012004 doi:10.1088/1757-899X/525/1/012004

The interaction of zirconium oxide nanoparticles with model binary and complex alloyed nickel alloys containing TiN

S N Anuchkin, V A Volchenkova, E K Kazenas, N A Andreeva and T N Penkina

A.A.Baikov Institute of Metallurgy and Materials Science, Russian Academy of Sciences (IMET RAS). Russia

E-mail: AnuchkinSN@yandex.ru; volch.v.a@mail.ru; ekazenas@ imet.ac.ru; andreeva150388@mail.ru; t-penka-01@yandex.ru

Abstract. Interaction of exogenous nanoparticles of refractory compounds ZrO_2 with model binary and multicomponent nickel alloys containing surfactants - tin were investigated. Shown that the removal of surfactants in the metal after ZrO_2 nanoparticles introduction depends from the time of isothermal holding in the induction furnace and alloying elements composition. The complete chemical composition of the alloys has been studied using the method of atomic emission spectrometry with inductively coupled plasma. The contents of various forms of zirconium have been determined. The effect of alloying elements on the change in the ratio of forms have been showed.

Development of nanotechnologies in metallurgy opens the ways to improve metal properties using exogenous nanoscale particles of refractory phases (NPRP) as reagents refining melts from harmful surfactants, and inoculation in metal crystallization processes. Currently, there are a large number of publications devoted to the interaction of ultrafine particles with melts based on iron, nickel and aluminum exclusively in the process of crystallization of industrial metal products. The technological process for treating of nickel melt by the modifying complex on the base of Ti(C,N) with 50-100 nm refractory particles is described in [1]. Nickel alloy GS3DK-VI was used for study. During the crystallization of the casting, the nanoparticles of titanium carbonitride served as additional centers of crystallization and contributed to the formation of a smaller grain structure. A homogeneous structure with inclusions of intermetallic phases along the boundaries and inside the grains was obtained after modification. A significant of grinding of the macrostructure (grinding grain 6-10 times) led to improvement in the tensile strength (8-10%); yield strength (10-13%); elongation (1.1-1.3 times); toughness (40-44%), and the durability of the alloy (6-30%) depending on the stress test. The influence of SiC nanoparticles on the structural and mechanical properties of nodular cast iron was studied in [2]. It was noted that the effect of SiC nanoparticles on the formation of ferrite, decreases of perlite, and the content of carbide contributes to the improvement in toughness of the cast structure. Increase in toughness and decrease in hardness after heat treatment were also observed. Modification of high-strength cast iron by nanoscale particles was investigated in [3]. Cerium oxide nanoparticles with a size of 10 nm were added to the melt before casting. Concentration of nanoparticles in the metal was ranged from 0.03 to 0.09 wt.%. The authors found that after the introduction of cerium oxide, globular graphite and carbides were crushed. It was shown that the modification led to an improvement in hardness and strength, mainly due to changes in the volume fraction and size of carbides. The abrasive wear resistance of the material increased with changes in hardness. It was noted that only small concentrations of nanoparticles were effective and with a high amount of cerium oxide in the metal, their advantages diminished. The alloying of 08Cr18Ni10T steel with nano- and microdispersed ZrO_2 particles during vacuum-arc remelting was described in [4]. Authors have shown that the nanosized ZrO_2 particles are more evenly distributed in the melt volume than micron sized ZrO_2 conglomerates. Values of strength and yield strength increase at concentrations of ZrO_2 particles up to 0.3 wt%, compared to the original steel, and decrease at concentrations above 0.4 wt%. Fatigue on alternating cantilever bending of steel in initial and modified states was investigated. In case of modification a significant (>30%) increase in fatigue strength of the low-cycle fatigue region, and slight (~10%) change in fatigue strength with a large number of loading cycles have been observed.

The presented review confirms relevance of application NPRP in melts during the crystallization process to improve service properties of products. But in these studies, the physicochemical regularities of the interaction nanoparticles with metal melt components are not fully considered. Results obtained differ significantly depending on the selected system NPRP-melt. The papers also do not consider the interaction between surface-active substances and NPRP in a liquid metal, and their possible effect on the redistribution of nanoparticles in the melt. Ultrafine particles of tungsten carbide, titanium nitride and carbonitride with microparticles of nickel and slag as protectors were used in [5]. Listed particles were introduced into powder, composite wires, agglomerated fluxes and coated electrodes for manual arc welding. Authors suggested that during the transition to the weld metal, on surface of the nanoparticles adsorb the surfactant – oxygen, sulfur, or other compounds. Under the influence of the circulation flow, the formed complexes of NPRP-surfactants are uniformly distributed in the peripheral area of the weld pool. On the cooled layer these complexes can form clusters that serve as additional crystallization centers, which contribute to the modified weld metal and its operational properties. Thus, the study of the physicochemical laws for the interaction of NPRP and surfactants in model binary and multicomponent nickel alloys could lead to more fully investigation of refractory nanoparticles on improving metal properties as a result of inoculation processes during metal crystallization and melt refining from harmful surfactants.

In [6] have been proposed and experimentally confirmed the hypothesis about the interaction of nanoparticles of refractory phases with metal melt surfactants. The mechanism was schematically presented as follows: introduction exogenous NPRP (disperse phase) into a liquid metal (dispersion system) leads to redistribution of surfactants on the boundary particle-melt (lyophilic or lyophobic) and their adsorption on the surface of the nanoparticles. Taking into account the quasichemical theory of the structure of melts, structural compounds are formed - ensembles Me + (NPRP-surfactants) as result of absorption. Due to the surfactant gradient, the processes of moving the ensembles in the melt, their possible merging or degradation, and further removal under the influence of adsorption nature forces on the Me-phase interface (ceramics, slag, gas) occur. As a result, one part of the ensemble is removed from the metal, i.e. the process of metal refining from harmful surfactants is implemented. Another part of the ensembles will further influence the crystallization processes and mechanical characteristics of the metal.

The purpose of this work is to study the interaction of ZrO_2 NPRP with binary and alloyed nickel alloys containing surfactants – impurity-tin, and to determine the forms of existence of zirconium in these alloys after entering ZrO_2. Surfactant impurity tin has been selected because of it is a typical non-ferrous metal impurity (NFMI) with surface-active properties [7]. The existing problem of NFMI removal [8] could also make possible to consider a new mechanism for removing NFMI by an adsorption mechanism from nickel alloys using NPRP. The choice of zirconium oxide is explained, firstly, to obtain new knowledge about the behavior of group IV metal oxides of the Periodic Table of Elements when interacting with liquid nickel and its impurities, taking into account the crystallography of materials. Secondly, in expanding the use of this oxide for practical purposes, for example, when producing ODS steels, cell material for measuring the EMF components of liquid alloys, etc.

Methods and materials of research.

ZrO_2 nanoparticles were obtained by a plasma-chemical method using a plasma generator of an electric arc plasma torch with a power of 25 kW, and the synthesis process took place during the interaction of the powder of zirconium chloride with an air plasma. The specific surface area 19.14 m^2/g was analyzed by the BET method on a Micrometrics TriStar 3000 analyzer. The average particle size $d_{m.}$ (BET) for ZrO_2 was 55 nm. X-ray phase analysis (Rigaku Ultima 4) showed the presence of 80% ZrO_2 (baddeleyite) and 20% - $ZrO_{1.88}$. The introduction of NPRP into a liquid melt was realized as a briquette of a composite material prepared from a mixture of NPRP powders ZrO_2 and Ni. The analysis of nickel powder showed the presence of 80% of particles with a size of 5–25 μm. Preparation of a mixture Ni powders (97.5 wt.%) + ZrO_2 (2.5 wt.%) [9] was carried out in a Fritsch Pulverisette 6 planetary mill in Ar atmosphere for 1 hour. Chemical analysis of the powder showed the presence 2.23 wt. % ZrO_2 in the resulting mixture. To refine the powder from oxygen, reductive annealing was carried out in an atmosphere of hydrogen at 800°C for 1 hour with a consumption of hydrogen of 25 l/h. After treatment, the mixture of powders was subjected to uniaxial pressing at a pressure of 35 MPa degassing in an Ar atmosphere at 300°C and in vacuum at 300°C, followed by vacuum storage at 25°C.

To carry out elemental analysis of the samples both on the main components (Cr, Co, Al, Mo, Nb, Ta, W, Re, Ti, etc.) and on microalloyed additives and impurities (La, Y, Si, Sn, Fe, Ca, Mg, B, P, Zr) and to determine the content of the forms of zirconium existence, the method of atomic emission spectrometry with inductively coupled plasma (AES with ICP) was used. When determining zirconium and these elements of an AES with ICP, it has a number of advantages over other methods: higher sensitivity than atomic absorption, atomic fluorescence spectrometry and photometric method of analysis and a wide range of calibration graphs [10]. The studies were carried out on a sequential atomic emission plasma spectrometer of ULTIMA 2 firms "HORIBA JOBIN YVON" (France-Japan). The high optical resolution of the spectrometer (5pm) provides a better signal to background ratio, which leads to low detection limits of elements. The following optimal analytical parameters for the simultaneous AES with ICP determination of the chemical elements were chosen: discharge power - 1.2 kW; flow rate of cooling argon is 14 l/min; transporting 0.80 l/min; plasma-forming 0.5 l/min; observation height of 15 mm above the upper coil of the induction coil; sample flow rate is 1.0 ml/min. The analytical lines for each element were found as free from spectral overlays of the nickel matrix element. The influence of nickel on the analytical signals of metals has been studied. Increasing the Ni concentration to 200 ppm did not affect the analytical signals of impurity elements with their content exceeding 10 ppb. The method of interactive matrix matching [10] was used to eliminate matrix noise caused by changing nickel concentration. It has been found that the content of determined elements at the levels of 0.1 -10 ppm mutual effects are not manifested. The concentration of the matrix element in the solution has been chosen, the stability of the burner - nebulizer system is not disturbed. For nickel favorable concentration is not more than 2000 ppm.

The method of AES with ICP requires conversion of analyzed samples into solution. Optimal schemes of sample dissolution have been found and the composition of acids has been selected to obtain stable solutions of easily hydrolysable determinable elements: Mo, Nb, Ta, Ti, W, Zr. To prevent the deposition of tungsten salts into the sediment, H_3PO_4 was added [11]. For acceleration of decomposition of samples modern systems were used: microwave mineralizer "Minotaur-2(Russia) and microwave system MARS 5(CEM Corporation, USA). Autoclave microwave dissolution using the MARS 5 module proved to be the most effective. The 100 mg samples were completely dissolved in 10 minutes in a mixture of HF + HNO_3 + $HClO_4$ at a temperature of 210°C. The method of "wet fusion" was used to determine silicon: heating of samples in the mixture H_2SO_4 and $(NH_4)_2SO_4$ [12]. All used reagents were of the high purity. We used twice distilled water. Standard solutions were prepared from metals of high purity [11] or applied standard multi-element solutions of Merk company. To determine the content of different forms of zirconium existence by AES with ICP, the difference in their solubility was used. Aqua regia: mixture of HCl conc. + HNO_3conc. (3:1) was applied for conversion into solution Zr dissolved in metal. Zirconium oxide (ZrO_2 - in the form of baddeleyite) is one of the most difficult compounds to be open. Dissolution of ZrO_2 was carried out in the microwave sample preparation system MARS 5 under different modes (P, T) by different acid mixtures according to the previously developed method [13].

Three series of experiments with model binary and multicomponent alloys were conducted. During the first series of experiments the interaction of ZrO_2 with model binary systems Ni-Sn (0.0840 wt.%) and Ni-Sn (0.0460 wt.%). These alloys were smelted in a vacuum resistance furnace with a graphite heater in a corundite crucible (98% Al_2O_3) with a capacity of 46 ml at $P_{He+10\%\ H2}$=0.1 MPa. Nickel was introduced into the melt without breaking the tightness of the Sn grade VHF-000 with a 20-minute isothermal exposure and crystallization of the metal in the crucible. The second series of experiments included investigation model multicomponent nickel alloys (MNA). As known the smelting of heat – resistant nickel alloys occurs in two periods-oxidative and reducing [14]. To study the efficiency of the input of exogenous refractory nanoparticles in different periods of melting, as a model MNA alloy was chosen, the composition of which corresponded to the oxidative period in the melting of heat-resistant alloys. These alloys were smelted in a resistance vacuum furnace with a graphite heater, where Sn of VHF-000 grade was also introduced into the melt to study the heterophase interaction NPRP of ZrO_2 with surfactant impurities. As two model systems, a multicomponent carbon alloy of composition (wt.%): Ni (basic) - Co (8.97) - Cr (4.93) - Mo (1.02) - W (8.54) - Re (3.96) - C (0.168) - Sn (0.0532) - S (0.0161) - O (0.0063) (hereinafter - MNA1) and a multicomponent carbon-free alloy of composition (wt.%): Ni (basic) - Co (9.04) - Cr (4.02) - Mo (1.18) - W (11.62) - Re (2.04) - C (0.005) - Sn (0.0487) - S (0,0146) - O (0,0029) (hereinafter - MNA2) were melted. During the third series of experiments, adsorption interactions in heat-resistant nickel industrial alloys were investigated. GS32 and GS36 were chosen as model alloys. The composition of these alloys (wt.%) was: GS32 - Ni (basic) - Cr (4.54) - Al (4.31) - Mo (1.10) - W (8.13) - Co (8.78) - Re (3.35) - Ta (4.21) - Nb (1.47) - C (0.222) - B (0.013) - Y (0.403) - Mg (0.0014) - Ca (0.0002) - S (0.0003) - O (0.0005) - N (0.0002) - Si (0.14) - P (0.0065); GS36 - Ni (basic) - Cr (3.81) - Al (5.47) - Mo (1.26) - W (11.42) - Co (8.27) - Re (2.01) - Nb (1.04) - Ti (1.13) - La (0.02) - Y (0.005) - Ca (0.005) - O (0.0016) - N (0.0004) - Si (0.061) - P (0.0017) - Fe (0.12). To study the heterophase interaction of ZrO_2 with tin, the tin content in the metal was increased to 0.0385 wt.% (GS 32) and up to 0.0428 wt.% (GS 36) when introduced into the melt of Sn brand VHP-000. Carbon content in all alloys was determined by oxidative melting using the analyzer of "LECO" model CS-400, and oxygen content – by reducing melting using the analyzer "LECO" model TS-436.

The study of the heterophase interaction of NPRP with surfactants was carried out in a vacuum induction furnace (VIF) with a special conical inductor, which provided a concentrated input of RF energy into the melt and intensive mixing of the metal with NPRP on the surface and inside the melt [15]. The experiments were carried out in an atmosphere He with 10% H_2 content at a pressure of 0.2 MPa. Heating and melting of the metal were controlled by an optical pyrometer EOP-66, calibrated at the melting points of pure metals, with data processing using the Ecochrome software. After attaining equilibrium with the gaseous phase of the liquid metal injected preform of the composite material (Ni+NPRP ZrO_2) with subsequent isothermal aging from 120 to 1200 second depending of the experience. The average temperature of the experiments ranged from 1519-1653°C and depended from the series of experiments. The content of NPRP ZrO_2 in the melt after the introduction of the briquette was 0.10 wt.%.

Results of experiments and their discussion.

The changes in the tin content in the metal were studied, analyzing the redistribution of nanoparticles and ensembles in the melt under the influence of adsorption forces [16]. More than 80 experiments with the following 12 types of alloys were conducted: Ni-Sn(0.0460 wt.%); Ni-Sn(0.0460 wt.%)-ZrO_2(55 nm); Ni-Sn(0.0840 wt.%); Ni-Sn(0.0840 wt.%)-ZrO_2(55 nm); MNA1; (MNA1)-ZrO_2(55 nm); MNA2; (MNA2)-ZrO_2(55 nm); GS32; (GS32)-ZrO_2(55 nm); GS36 and (GS36)-ZrO_2(55 nm). The results are presented in the form of dependences $[Sn]=f(\tau)$. The change in the content of Sn was presented as the degree of removal $\alpha = (([Sn]_{initial}-[Sn]_{ending})/[Sn]_{initial})\cdot 100$.

The results of the first series of experiments with Ni-Sn binary systems (Figure 1) allow to draw the following conclusions: 1) The introduction of ZrO_2 NPRP provided a decrease in the content tin and the α_{Sn} values were 14-22 rel.%. 2) Value comparison of the content tin and the α_{Sn} with and without the introduction of ZrO_2, makes it possible to consider the existence of heterophase interaction of ZrO_2 NPRP with surfactant in the Ni-Sn melt and the removal of Ni+(ZrO_2-Sn) ensembles. 3) Data processing of time dependence by the least squares method when describing by polynomial function ($[Sn]=0.0811-3.42\cdot10^{-5}\tau+1.86\cdot10^{-8}\tau^2$ for the Ni-Sn system (0.0840 wt.%) and $0.0451-1.87\cdot10^{-5}\tau+1.06\cdot10^{-8}\tau^2$ for Ni-Sn(0.0460 wt.%) showed the presence of an extremum at 720-860 seconds, which can serve as a proof of possible reversible processes of tin adsorption/desorption at the boundaries of ZrO_2-Sn.

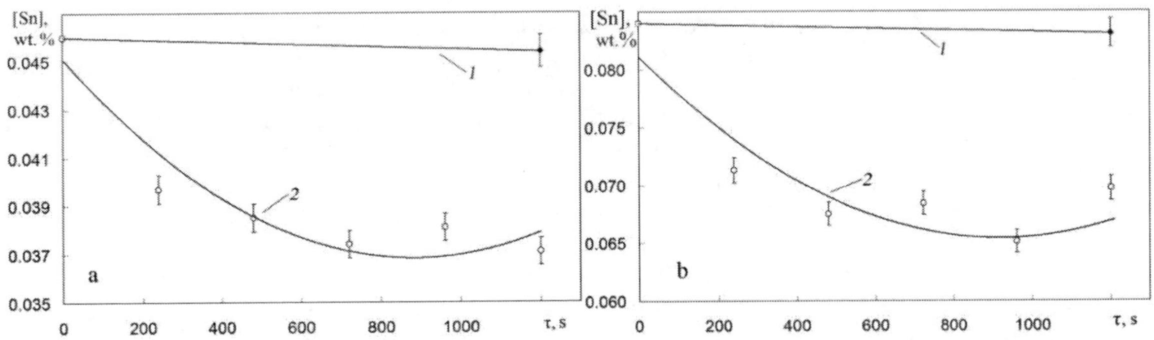

Figure 1. Dependence of tin content on the duration of exposure in melts Ni-Sn (0.0460 wt.%) (a) and Ni-Sn (0.0840 wt.%) (b): 1 - without introducing NPRP; 2 - ZrO_2 (55 nm).

Concentrations of different forms of zirconium in this alloy have been determined. It was found that under experimental conditions in the model binary systems the ratio of Zr, which was in the melt in the form of a solution and in the form of NPRP ZrO_2 is 1:80 or about 99% exists in the form of ZrO_2.

The results of the second series of experiments with model multicomponent nickel alloys (Figure 2) showed that, first of all, the introduction of ZrO_2 low-frequency nickel alloys at an isothermal exposure time 240-1200 s. The content of tin decreased, the values of α_{Sn} were 16-27 rel.% (MNA1) and 15-28 rel.% (MNA2). It may indicate heterophase interaction of NPRP ZrO_2 with surfactants in the melt and their removal in the form of ensembles (Ni-Me)+(ZrO_2-surfactants), where Me is a component of a complex alloy. Secondly, data processing by the least squares method describing by polynomial function ($[Sn]=0.0502-2.90\cdot10^{-5}\tau+2.14\cdot10^{-8}\tau^2$ for MNA1 and $[Sn]=0.0476-2.31\cdot10^{-5}\tau+1.09\cdot10^{-8}\tau^2$ for MNA2) showed the presence of an extremum, which can serve as evidence of possible processes of surfactant desorption. When comparing the obtained data with the model binary systems, it was found that in the MNA the processes of tin removal in experiments without inputting NPRP are more intensive. The values of α_{Sn} in MNA 1 and MNA2 after 1200 s. were 4.3 and 5.1 rel.%, respectively. After the introduction of ZrO_2, the degree of tin removal in MNA1 had the highest values at the initial stages of melting, and in MNA 2 - at longer exposures of about 1000 seconds (Figure 2), that can be connected with different influence of alloying elements of the melt on the change of the structure of ensembles and tin desorption from their surface.

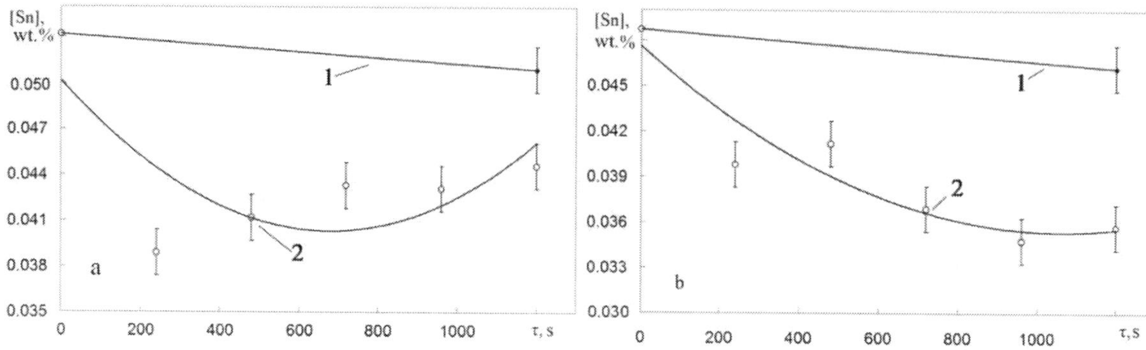

Figure 2. The dependence of the tin content on the duration of exposure in the MNA1 (a) and MNA2 (b): 1 - without introducing NPRP; 2 - ZrO_2 (55 nm).

The content of various forms of existence zirconium was determined by the method of AES with ICP: in the form of "easly opening" and "hardly opening" compounds. An assumption was made that the "easy-opening" compounds were zirconium dissolved in the metal or in the form of non-oxide inclusions, "hardly opening" as ZrO_2. Analysis of the results showed that, firstly, with an initial Zr content in the metal of 0.074 wt.% (in the form of ZrO_2), the amount of zirconium removed from the metal to the interface phases, presumably in the form of ensembles, is on average in MNA1 - 55 rel.% and in MNA2 - 64 rel.%, while 45-36 % remain in the metal and influence the crystallization process. The highest degree (71%) of zirconium removal from metal is observed at an exposure of 1200 seconds in MNA2. Secondly, there is a significant change in the composition of the particles that have not been removed during long exposures of the metal. In the MNA1 at 240 seconds exposure, the ratio of zirconium in the form of "easly opening" and "hardly opening" compounds were 1:9, i.e. 90% of the particles are presumably in the form of ZrO_2, At 1200 seconds exposure the ratio was 1:1, i.e. only half of the particles remain as oxide. In the MNA2, in contrast to the MNA1, there was no significant change in the composition of the particles that have not been removed during long metal exposures. The amount of zirconium in the form of "easly opening" compounds after 1200 seconds exposure is about 7%. These results indicate the significant role of the elemental composition of the MNA1 in the formation and change of non-metallic inclusions.

The change in carbon content was studied during experiments with MNA1 without introducing NPRP and after entering ZrO_2. It was found that after introducing ZrO_2 NPRP a significant decrease in carbon content was observed. After 1200 seconds exposure in VIF the value [C] in experiments without introducing NPRP and after entering ZrO_2, was 0.123 and 0.078 wt.% respectively. In [17] it was noted, that zirconium has a high affinity not only for oxygen, but also for other implantation elements (nitrogen, carbon, sulfur) and forms with them strong compounds ZrN, ZrS_2, ZrC. The possibility of reducing zirconium oxide with carbon can also be judged by the deoxidizing ability of carbon and zirconium in nickel. It was shown [18, 19], that with a content of 0.168 wt.% C and 0.074 wt.% Zr in a nickel melt, carbon has a greater deoxidizing ability than zirconium. From a thermodynamic point of view, the reaction of formation of carbide can be represented as a total reaction:

$$ZrO_{2(s)} + 3[C]_{Ni} = ZrC_{(s)} + 2CO_{(g)} \quad \Delta G_T^{O} = 609070 - 235{,}11 \cdot T \ [20,21]$$

Partial pressure P_{CO} (excluding the influence of alloying elements) was estimated. The model system Ni-C(0.168 wt.%)-O(0.0063 wt.%)-ZrO_2(0.1 wt.%) was considered. Dependence of the equilibrium constant from the reaction on temperature [6] was used to determine the value 31.56 Pa of P_{CO} at 1873 K what, taking into account the assumptions, may indicate the possible formation of zirconium carbide after the introduction of nanoparticles, which will lead to a change in the adsorption mechanism of the interaction of NPRP with tin.

The results of the third series of experiments with heat-resistant nickel alloys (Figure 3) showed that, firstly, the introduction of ZrO_2 NPRP at isothermal exposure of 120-600 s. resulted in a decrease in [Sn]. Values of α_{Sn} in the GS32 and GS36 alloys were 11–18 and 12–18 rel.%, respectively, which may indicate heterophase interaction of the NPRP ZrO_2 with tin in the melt and their removal as ensembles. Secondly, data processing using the least squares method when describing by a polynomial function (GS32 - $[Sn] = 0.0370\text{-}2.99\cdot10^{-5}\tau + 4.49\cdot10^{-8}\tau^2$ and GS36 - $[Sn] = 0.0417\ \text{-}3.91\cdot10^{-5}\tau + 5.63\cdot10^{-8}\tau^2$) showed the presence of an extremum, which can serve as evidence of possible processes for the desorption of surfactant tin. Comparing the obtained data, it can be seen that, first, a more intensive removal of tin was observed without introducing NPRP compared to binary alloys. Secondly, the degree of removal of tin has the highest values in the initial stages of smelting. This once again confirms the influence of alloying elements during long exposures on the change in the structure of ensembles and the desorption of tin from the surface of nanoparticles.

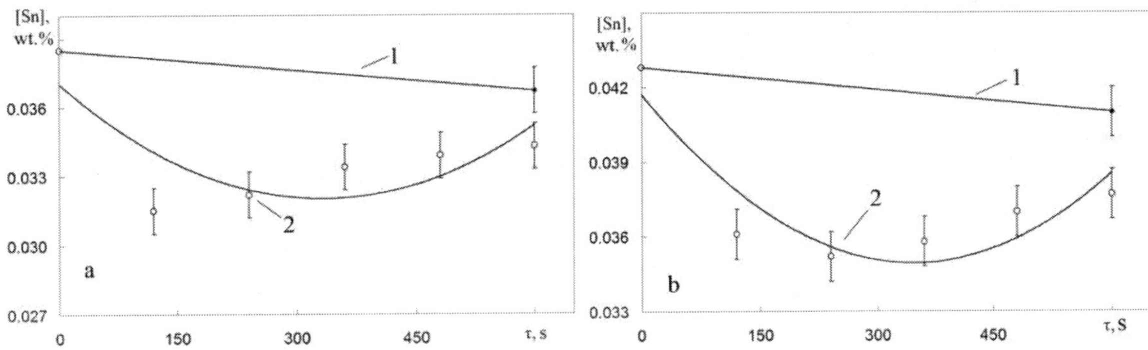

Figure 3. The dependence of the tin content on the duration of exposure in the heat-resistant alloys GS32 (a) and GS36 (b): 1 - without introducing NPRP; 2 - ZrO_2 (55 nm).

As before, the content of various forms of zirconium is determined: in the form of "easily opening" and "hardly opening" compounds. The analysis of the results showed that, firstly, at the initial 0.074 wt.% Zr content in the metal (in the form of ZrO_2), the amount of zirconium removed from the metal to the interface, presumably in the form of ensembles, in both alloys averages 50 rel.%. Secondly, the time of isothermal exposure in the VIF does not significantly affect the proportion of zirconium removed from the metal. Third, there is a significant change in the composition of the particles not removed. At the minimum exposure 120 seconds the ratio of zirconium in the form of "easily opening" and "hardly opening" compounds is 1:1 (GS32) and 1:1,5 (GS36). This indicates the effect of alloying elements of the alloy and changes in the composition of the oxide particles even at low exposures. During long exposures at 600 seconds the amount of the "easily opened" connections is 89 rel.% (GS32) and 84 rel.% (GS36), which confirms the possibility of reducing the zirconium oxide melt elements. All this once again indicates that nanoparticles with excess surface energy will interact with melt elements with changes in the composition of inclusions and desorption of tin from the surface of nanoparticles. Changes in the content of tin and zirconium show increasing concentration of tin with increasing exposure time but the proportion of removed zirconium remains unchanged. This suggests that the interaction of zirconium oxide with the melt elements may result in the formation of complex, spinel-like compounds which remain in the slag phase. The tin after desorption passes into the melt. Therefore, in order to achieve the greatest effect of nanoparticles, it is necessary to introduce them into the liquid metal at minimum holding times before crystallization.

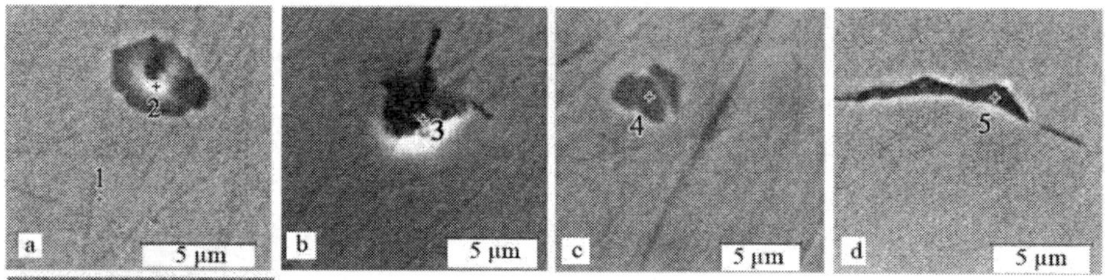

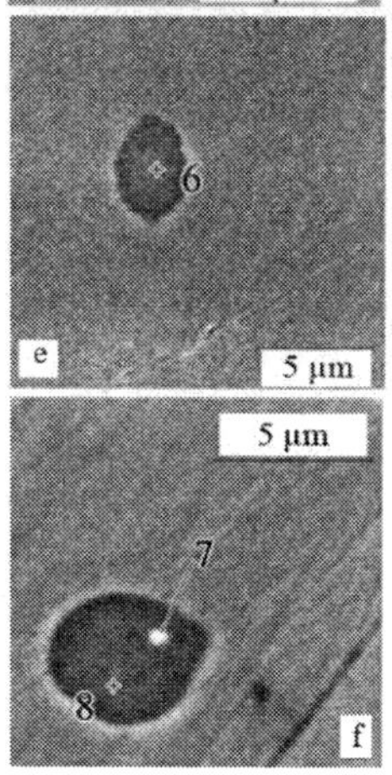

№	Ni	Cr	Co	Fe	W	Zr	Al	O	S	Sn
1	77.93	4.97	8.89		8.21					
2	2.82					74.64		22.54		
3	52.06	7.20	6.55	3.41	4.96	13.09	2.14	10.59		
4	18.69	3.62	1.76	0.91	1.30	53.95			19.77	
5	39.06	5.60	4.02	1.81	2.08	32.38			15.05	
6	26.58	2.64				46.97			23.81	
7	38.82	3.42	2.41	1.93		37.07			12.96	3.39
8	18.27	3.34	0.93	1.02	1.79	53.58			21.07	

Figure 4. REM micrographs of non-metallic inclusions MNA 1 after 240 s. (a, b, c, d) and 1200 seconds (e, f) exposure in VIF.

The analysis of the matrix, grain boundaries and non-metallic inclusions of MNA1 samples after experiments with different isothermal exposures on JEOL JSM-6700F and JSM-6480LV raster electron microscopes was carried out. Non-metallic 1-3 microns inclusions were fixed in the sample without input of nanoparticles. Elemental analysis was constant and characterized by the presence of Cr, O, Co and Ni. After the introduction of the ZrO_2 NPRP into samples, the following types of non-metallic inclusions were found - simple and complex inclusions containing oxygen and complex inclusions of spherical and trapezoidal shape containing sulfur. Simple inclusions containing oxygen are agglomerated zirconium oxide particles of micron-sized size (Figure 4a), complex ones have Zr Co, Cr, Ni and O in their composition (Figure 4b). Inclusions containing sulfur (Figure 4c, 4d, 4e and 4f), consist of Zr and S with the presence of Cr, Co, Ni, W, Fe and Sn. The presence of iron peaks in elemental analysis can be related to the steel cup/balls napkin during the preparation of composite material in a planetary mill for the introduction of ZrO_2 NPRP into the liquid metal. It is worth noting that elemental analysis always detects 20-30% of carbon inclusions, but as a rule, it does not considered when analyzing by this method. Figure 4f shows the inclusion containing a small fraction of tin, which is probably related to the adsorption of tin as a surfactant admixture at the interface between the phases of nanoparticle - melt with the formation of ensembles with an excess tin content. Thus, it can be assumed that these inclusions could be formed from the Ni-C-Me+(ZrO_2-Sn) ensembles, where Me is a component of a complex alloy formed as a result of interaction of exogenous nanoparticles with melt components at long exposures. Comparing the data related to different forms of zirconium, it is clear that the proportion of "hardly opening" compounds at 240 seconds is 90%, and at 1200 seconds - 50%. Assuming that only ZrO_2 is the "hardly opening" compound, the absence of a large number of oxide inclusions may indicate that a significant proportion of ZrO_2 contains in the form of nanoscale particles.

Conclusions:

1. Heterophase interaction of exogenous ZrO_2 nanoparticles (55 nm) with tin surfactants in model binary and multicomponent nickel alloys has been studied for the first time. An assumption was made about the redistribution of nanoparticles in the melt under the influence of adsorption forces.
2. It was found that tin removal was probably the result of heterophase interaction between NPRP ZrO_2 and surfactant, with forming ensembles and their removal to the the following phase interfaces: metal-gas, metal-slag or metal-lining. It was shown that the degree of tin removal depends on the time of isothermal maturation in VIF and composition of alloying elements. The values of α_{Sn} were 14-22 rel.% for Ni-Sn binary systems, 15-28 rel.% for multicomponent nickel alloys and 11-18 rel.% for heat-resistant alloys.
3. Using the method of atomic emission spectrometry with inductively coupled plasma, the contents of various forms of zirconium were determined. The influence of alloying elements on the change in the ratio of these forms was shown. It was found that average amount of zirconium (presumably in the form of ZrO_2) removed from the metal at the interface is 50 rel.%, and the remaining particles can influence the crystallization process.
4. The analysis of the matrix, grain boundaries and nonmetallic inclusions of MNA1 was carried out. Nonmetallic inclusions of the following several types were found – simple and complex inclusions containing oxygen and complex inclusions of spherical and trapezoidal shape containing sulfur. The influence of the holding time in the VIF on the composition and size of inclusions was shown.

Acknowledgments

The work was carried out according to the state assignment number 075-00746-19-00

References

[1] Kalinina N E,Yukhimenko A E and Kalinin V T 2014 Complex modification of the complex alloyed heat-resistant alloys *Vestnik engine-building* **2** pp 181-185.

[2] Ruixiang Zhu, Yong S and Xinyu Qin 2017 *Study on the Influence of Nano-SiC on the Structure and Properties of Nodular Cast Iron. IOP Conf. Series: Materials Science and Engineering* V.**250** pp 1-8 (012051).

[3] Sun Xiaoguang, Wang You, Li DYand Wang Guodong 2013 *Modification of carbidic austempered ductile iron with nano ceria for improved mechanical properties and abrasive wear resistance. Wear* V. **301** № 1–2 pp 116-121.

[4] Aksenova S I, Borts B V, Korotkov I M, Pakhomov A V and Sokolenko V I 2016 *Problems of atomic science and technology V I Fatigue of austenitic steels, modified with oxides of zirconium and yttrium* **№1 (101)** pp 160-166.

[5] Sokolov G N, Lysak V I, Zorin I V, Artemyev A A, Dubtsov Yu N, Antonov A A and Yelsukov S K 2016 *the Mechanism of the effect of ultrafine refractory components contained in the welding material, on the structure of the deposited metal. News of Volgograd state technical University* **№9 (188)** pp 116-120.

[6] Anuchkin S N, Burtsev V T and Samokhin A V 2010 *The interaction of nanoparticles of refractory compounds with a surfactant in a nickel melt: I. Heterophase interaction. Russian metallurgy (physical metallurgy)* **№ 11** pp 1014-1020.

[7] Nogi K, Chung W B, McLean A and Miller WA 1991 *Surface tension of liquid Fe-(Cu, Sn, Cr) and Ni-(Cu, Sn) binary alloys. Materials Transactions JIM* V.32 **№ 2** pp 164-168.

[8] Guzenkova A S, Ivanov S S, Isaev G A and Kudrin V A 2008 Production of steel, pure, non-ferrous metals. (Moscow: MGVMI) 118 p.

[9] Anuchkin S N, Gvozdkov I A, Samokhin A V, Serov G V and Burtsev V T 2011 *Properties of composite nanomaterial Al_2O_3/Ni, obtained by the method of Mechanochemistry. Physics and chemistry of materials processing* **№2** pp 71-78.

[10] Thompson M and Walsh D N 1988 Guide to inductively coupled plasma spectrometric analysis

(M.:Nedra) 288 p.

[11] Lazarev A I, Kharlamov I P and Yakovlev P Ya 1976 Handbook of chemist-analyst. M.: metallurgy 184 p.

[12] Bock R 1984 Decomposition methods in analytical chemistry (Moscow:Chemistry) 432 p.

[13] Andreeva N A, Anuchkin S N, Volchenkova V A, Kazenas E K, Penkina T N and Fomina A A 2018 *Development of phase analysis methods of impurity elements in alloys based on iron and nickel. IOP Conf. Series: Materials Science and Engineering* **V.347** pp 1-9 (012003).

[14] Sidorov V V, Kablov D E and Rigin V E 2016 *Metallurgy of foundry heat-resistant alloys: technology and equipment (Moscow: VIAM)* p 368.

[15] Glebovsky V G and Burtsev V T 1974 Melting of metals and alloys in the suspended state (Moscow: Metallurgy) 176 p.

[16] Minaev Yu A 1984 *Surface phenomena in metallurgical processes (Moscow: Metallurgy)* p 152

[17] Nasuta L Y and Saenko A 2017 *Peculiarities of application of zirconia as microlaserpeel element in the smelting high-quality steels: a literature review. Proc. abstracts of International scientific-technical conference "University science – 2017". Mariupol: state higher educational institution "Priazovskiy state technical University"* Vol.**1**. pp 122-126.

[18] Aleksandrov A A 2016 *Influence of zirconium on the solubility of oxygen in liquid nickel and melts of Ni-Fe. Metal.* No.**5** pp 61-68.

[19] Aleksandrov A A 2018 *Thermodynamics of solutions of oxygen in melts of the system Fe-Ni containing boron.* Aleksandrov A A, Dashevskiy V Ya and Leont'ev L I 2018 *Izv. higher educational. Ferrous metallurgy* No. **3** pp 201-210.

[20] Buzek Z 1979 Fundamental Thermodynamic Data on Metallurgical Reactions and Interactions of Elements in System Significant for Metallurgical Theory and Practice. Buzek Z (Ostrava: Vyzkumny ustav hutnictvi zeleza) 110 p.

[21] Turkdogan E T 1985 Physical chemistry of high-temperature processes (M.: Metallurgy) 344 p.

IOP Conf. Series: Materials Science and Engineering **525** (2019) 012005 doi:10.1088/1757-899X/525/1/012005

The hexahistidine containing organophosphorus hydrolase enzyme and bacterial cellulose based functional materials

A G Aslanli[1], N A Stepanov[1], O V Senko[1], O V Maslova[1], I V Lyagin[1] and E N Efremenko[1,2]

[1]Faculty of Chemistry, Lomonosov Moscow State University, Moscow 119991, Russia
[2]Correspondence: Leninskiye Gory, 1/3, Moscow 119991, Russia;

E-mail: elena_efremenko@list.ru

Abstract. Bacterial cellulose (BC) is a unique biopolymer for creation on its basis functional materials used as drugs, matrices for regenerative therapy, dressings, etc. In this study, the possibility of combining of BC with hexahistidine containing organophosphorus hydrolase (His_6-OPH), exhibiting lactonase activity against gram-negative pathogenic bacteria is considered to obtain composite materials with special properties. The dry and wet samples of fibrous BC, impregnated with solutions of the His_6-OPH enzyme and various antimicrobial agents were obtained; their catalytic characteristics were investigated. The preservation of the enzymatic activity of the His_6-OPH enzyme in the composition of the obtained samples with BC, including after the lyophilization process, has been established. The stabilization of the enzymatic activity of His_6-OPH in the composition of fibrous materials in the presence of poly(aspartic) acid and antimicrobial agents is shown.

1. Introduction

Bacterial cellulose (BC) is of particular interest as a biocompatible material for scientific research. BC is a fibrous biopolymer synthesized by *Acetobacter xylinum* cells and some other types of bacteria [1]. BC does not contain impurities of hemicellulose or lignin and does not require cleaning in contrast to plant cellulose. Due to this, the process of BC obtaining is significantly simplified, and BC molecules form a nanofibrillar structure. In addition, BC has high strength characteristics and crystallinity degree, which can reach 90%, good light transmittance factor, biocompatibility, cell adhesion, and is also characterized by high permeability to liquids and gases, which allows using this fibrous material to create various modern composite materials for different purposes [2]. Due to its unique properties, BC is widely used in such industries as water purification, cosmetic, the preparation of sensor technologies, etc. [3, 4, 5]. The BC is especially widely used in medicine as surface coatings for implants, dressings, etc. [6, 7]. It is known that BC is actively used as components for fibrous and other composite materials. For example, it was shown that when poly(vinyl) alcohol (PVA) cryogels are filled with BC nanofibers, the rigidity and heat resistance of the composites increase in comparison with unfilled samples. Nanomaterials containing particles of gold, silver and zinc oxide with bactericidal and conductive properties have already been developed on the basis of BC [8].

In this study an attempt was made to create a fibrous material using an enzyme such as hexahistidine containing organophosphorus hydrolase (His_6-OPH) as a partner for combining with BC. It is known that the His_6-OPH enzyme has a high catalytic activity in the processes of hydrolysis of a number of toxic organophosphorus compounds, and also exhibits lactonase activity against a number of signaling molecules of pathogenic gram-negative bacteria - N-acyl homoserine lactones (AHLs) [9, 10].

Since in recent years there has been a rapid increase in the number of antibiotic-resistant microorganisms and a significant decrease in the efficiency of many of the known antibiotics, it is necessary to find an alternative to these antibiotics [11]. For this purpose, antimicrobial preparations based on the His_6-OPH enzyme or its poly(electrolyte) complexes (PEC) and various antibiotics were obtained earlier, and their catalytic characteristics were studied [12, 13]. It has been established that antibiotics do not have a measurable negative impact on the His_6-OPH enzyme, and their enzyme inhibitory concentrations do not exceed the minimum inhibitory concentrations of antibiotics applied to bacterial cells. It also showed that the combination of the His_6-OPH enzyme and β-lactam antibiotics leads to the stabilization of the catalytic activity of His_6-OPH and an increase in the action efficiency of both the enzyme and antibiotics. These results were also confirmed by molecular docking method [13].

The objective of this paper was to assess the possibility of obtaining and studying the catalytic characteristics of biocompatible functional materials based on BC and the His_6-OPH enzyme containing various antibacterial agents for the manifestation of antimicrobial properties and the main catalytic activity of the enzyme itself.

2. Results and Discussion

The dry and wet samples of BC were exposed in distilled water for a certain time at T = 8 ° C. The mass of samples with BC nanofibers was measured periodically until the termination in weight change (Figure 1).

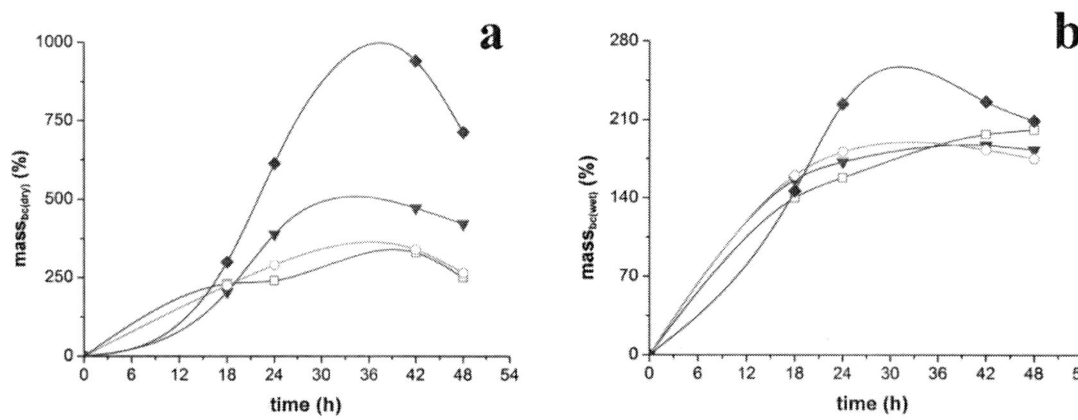

Figure 1. The swelling kinetic of dry (a) and wet (b) samples with BC, where ♦ - BC+His_6-OPH/meropenem/Ag; ▼ - BC+His_6-OPH/meropenem; ○ - BC+His_6-OPH/PLD/Ag and □ - meropenem.

It has been shown that the highest rate of swelling is observed in the case of dry sample with BC, as expected. At the same time, in both cases of dry and wet samples with BC, the swelling rate of the sample BC + His_6-OPH/meropenem/Ag was maximized. It should be noted that different kinetic of swelling was observed for the remaining samples, in the case of dry and wet samples of BC, which may be explained by the difference in the porosity of dry and wet BC nanofibers.

8 samples of different composition based on both dry and wet BC nanofibers were obtained to study their catalytic characteristics (Table 1).

Table 1. Composition of samples based on BC nanofibers.

#	Composition	
	Enzyme or EPC	Antimicrobial agent
1	His_6-OPH	none
2	His_6-OPH	meropenem
3	His_6-OPH	Ag
4	His_6-OPH	meropenem + Ag
5	His_6-OPH/PLD	none
6	His_6-OPH/PLD	meropenem
7	His_6-OPH/PLD	Ag
8	His_6-OPH/PLD	meropenem + Ag

The obtained samples were exposed for 16 hours at T = 8°C and then were lyophilized. The enzymatic activity of His_6-OPH in the composition of the obtained samples was determined before and after drying (Figure 2).

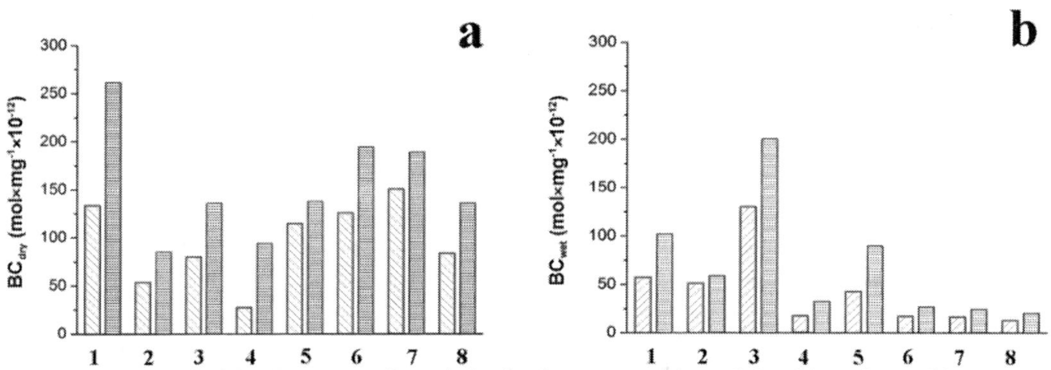

Figure 2. His_6-OPH enzymatic activity in the composition of dry (a) and wet (b) BC nanofibers, where orange (left) and green (right) columns are before and after drying, respectively. The numbering of the samples correspond to that indicated in table 1.

As a result, it has been found that there is a specific increase in the enzymatic activity of His_6-OPH in all BC samples, after drying. For samples with dry BC nanofibers, higher values of His_6-OPH activity were observed in comparison with wet samples of BC. As noted above, this may be due to the different porosity of dry and wet BC nanofibers, which in turn leads to differences in carrier capacity. As a result, a higher capacity of the dry BC fiber leads to better absorption of enzyme solutions and, therefore, greater enzymatic activity.

A higher enzymatic activity of samples containing His_6-OPH/Ag than with His_6-OPH/meropenem/Ag or His_6-OPH/meropenem was noted. At the same time, there was no difference in the effect of antimicrobial agents on the enzymatic activity for samples containing PEC of His_6-OPH with poly-(L-aspartic acid). However, in the case of dry BC, the enzymatic activity of samples with the pure His_6-OPH enzyme was lower than with its PEC in the presence of antimicrobial agents.

3. Conclusions

The results of the study showed that composite materials with stable enzymatic activity and potential antimicrobial properties can be obtained on the basis of bacterial cellulose and the His_6-OPH enzyme.

The use of the enzyme-poly(electrolyte) complex in the presence of antimicrobial agents contributes to a better stabilization of the His_6-OPH enzymatic activity.

4. Materials & Methods

4.1. Materials

The poly-(*L*-aspartic acid) sodium salt (PLD_{50}, MW = 6.8 kg·mol^{-1}) (Alamanda Polymers, Huntsville, AL, USA) was used for production of enzyme polyelectrolyte complexe. Antibiotic (meropenem) was acquired in the local pharmacy and was used as received. Paraoxon and other reagents were purchased from Sigma (Saint Louis, MO, USA). Recombinant *Escherichia coli* strain SG13009[pREP4] (Qiagen, Hilden, Germany) transformed by plasmid encoding His_6-OPH [14] was used for His_6-OPH production.

4.2. Preparation of enzyme

Recombinant cells of *Escherichia coli* SG13009[pREP4] were cultivated, and the enzyme was isolated and purified by methods published previously [15]. The purified preparation of His_6-OPH was characterized by enzymatic activity as described earlier [16]. The concentration of protein was determined by Bradford assay, and protein purity was analyzed by SDS-PAGE in 12% polyacrylamide gel using Mini-PROTEAN II cell (Bio-Rad, Hercules, CA, USA) followed by Coomassie Blue (R-250) staining. According to SDS-PAGE (data not shown), the purity of His_6-OPH preparation obtained (MW ≈ 37 kg·mol^{-1}) was ca. 98%. The specific organophosphorus hydrolase activity of the purified enzyme was 4300 U·mg^{-1}. One unit of enzymatic activity (U) was defined as the quantity of the enzyme necessary to hydrolyze 1 μmol of paraoxon per min at 25°C.

Enzyme-polyelectrolyte complexe (EPC) was produced similarly to the known procedure [17]. In brief, to produce an His_6-OPH/PLD_{50}, an aliquot of 20 g·L^{-1} PLD_{50} water solution was added to the purified His_6-OPH (1.6 ± 0.1 g·L^{-1}) in 50 mM phosphate buffer (pH 7.5) containing 150 mM NaCl. The aliquot volume was calculated so that "enzyme:polymer" molar ratio was 1:5. Next, the mixture was held for 30 min at 8°C. The effective hydrodynamic diameter of the particles of prepared complexe was determined at 25°C by dynamic light scattering using a Zetasizer Nano ZS (Malvern Instruments Ltd., Malvern, UK) as it was published previously [17] of necessity.

4.3. Measurement of swelling kinetic

The BC film was produced and purified according to the procedure described earlier [18]. After the BC-film was washed in distilled water, its properties were investigated. Initial solutions of meropenem at concentration of 2 g·L^{-1} (5.2 mM) in 50 mM phosphate buffer (pH 8.0) containing 150 mM NaCl, and 60 mg·L^{-1} colloidal silver nanoparticles were mixed with the 2 g·L^{-1} solution of enzyme (28 μM) 1:1 volumetric proportion and exposed for 30 min at room temperature. Then the mixture was added to BC pellicles and exposed at temperature 8°C for 48 h. The mass of the BC samples was periodically measured until the termination in its mass changes.

4.4. Measurement of enzyme activity

Initial solutions of meropenem, colloidal silver nanoparticles were mixed with the enzyme or EPC as described above and exposed for 30 min at room temperature. Then the mixture was added to BC pellicles and exposed at temperature 8°C for 16 h. Thereupon the samples were lyophilized. The organophosphorus hydrolase activity was determined before and after lyophilization as described earlier [19]. The required concentration of a Paraoxon was added to a glass container with a batch of the BC sample and 0.1 M carbonate buffer (pH 10.5). Aliquots were removed from the container at regular intervals and analyzed using an Agilent-UV spectrophotometer equipped with a thermostated analytical cell to study the accumulation of 4-nitrophenolate anion at 405 nm. Enzyme concentration hydrolyzing 1 μmol of the substrate per minute at 25°C was taken as the unit of enzymatic activity.

IOP Conf. Series: Materials Science and Engineering **525** (2019) 012005 doi:10.1088/1757-899X/525/1/012005

The rates of the enzymatic reaction were calculated by the initial linear sections of kinetic curves (v_0 = tgα).

The every data are presented as means of at least three independent experiments ± standard deviation (±SD).

Acknowledgement
This work was financially supported by RFRR project 18-29-17069.

References

[1] Nechyporchuk O, Belgacem M N and Bras J 2016 Production of cellulose nanofibrils: A review of recent advances. *Ind. Crop. Prod.* **93** 2-25

[2] Ul-Islam M, Khan S, Ullah M W, Park J K 2015 Bacterial cellulose composites: synthetic strategies and multiple applications in bio-medical and electro-conductive fields. *Biotechnol. J.* **10** 1847–61

[3] Jozala A F, de Lencastre-Novaes L C, Lopes A M, de Carvalho Santos-Ebinuma V, Mazzola P G, Pessoa-Jr A and Chaud M V 2016 Bacterial nanocellulose production and application: a 10-year overview. *Appl. Microbiol. Biotechnol.* **100(5)** 2063-72

[4] Ummartyotin S and Manuspiya H 2015 A critical review on cellulose: from fundamental to an approach on sensor technology. *Renew. Sust. Energ. Rev.* **41** 402-12

[5] Osong S H, Norgren S and Engstrand P 2016 Processing of wood-based microfibrillated cellulose and nanofibrillated cellulose, and applications relating to papermaking: a review. *Cellulose* **23(1)** 93-123

[6] Picheth G F, Pirich C L, Sierakowski M R, Woehl M A, Sakakibara C N, de Souza C F and de Freitas R A 2017 Bacterial cellulose in biomedical applications: A review. *Int. J. Biol. Macromol.* **104** 97-106

[7] Sulaeva I, Henniges U, Rosenau T and Potthast A 2015 Bacterial cellulose as a material for wound treatment: properties and modifications. A review. *Biotechnol. Adv.* **33(8)** 1547-71

[8] Shah N, Ul-Islam M, Khattak W A and Park J K 2013 Overview of bacterial cellulose composites: a multipurpose advanced material. *Carbohydr. Polym.* **98(2)** 1585-98

[9] Sirotkina M and Efremenko E N 2014 Rhodococcus lactonase with organophosphate hydrolase (OPH) activity and His_6-tagged OPH with lactonase activity: evolutionary proximity of the enzymes and new possibilities in their application. *Appl. Microbiol. Biotechnol.* **98** 2647–56

[10] Ventola C L 2015 The antibiotic resistance crisis: part 2: management strategies and new agents. *Pharm. Ther.* **40(5)** 344

[11] Maslova O, Aslanli A, Stepanov N, Lyagin I and Efremenko E 2017 Catalytic characteristics of new antibacterials based on hexahistidine-containing organophosphorus hydrolase. *Catalysts* **7** 271

[12] Aslanli A, Lyagin I and Efremenko E 2018 Novel approach to Quorum Quenching: rational design of antibacterials in combination with hexahistidine-tagged organophosphorus hydrolase. *J.Biol.Chem.* **399(8)** 869-79

[13] Efremenko E N, Votchitseva Y A, Aliev T K and Varfolomeev S D 2005 Recombinant plasmid DNA pTES-HIS-OPH and producer of oligohistidine-containing organophosphate hydrolase. RU Patent 2255975.

[14] Efremenko E, Votchitseva Y, Plieva F, Galaev I and Mattiasson B 2006 Purification of His_6-organophosphate hydrolase using monolithic supermacroporous polyacrylamide cryogels developed for immobilized metal affinity chromatography. *Appl. Microbiol. Biotechnol.* **70** 558–63

[15] Votchitseva Y A, Efremenko E N, Aliev T K and Varfolomeyev S D 2006 Properties of hexahistidine-tagged organophosphate hydrolase. *Biochemistry (Moscow)* **71** 167–72

[16] Efremenko E N, Lyagin I V, Klyachko N L , Bronich T, Zavyalova N V, Jiang Y and Kabanov A V 2017 A simple and highly effective catalytic nanozyme scavenger for organophosphorus neurotoxins. *J. Control. Release* **247** 175–81

IOP Conf. Series: Materials Science and Engineering **525** (2019) 012005 doi:10.1088/1757-899X/525/1/012005

[17] Stepanov N and Efremenko E 2018 "Deceived" Concentrated Immobilized Cells as Biocatalyst for Intensive Bacterial Cellulose Production from Various Sources. *Catalysts* **8(1)** 33
[18] Efremenko E, Lyagin I, Votchitseva Y, Sirotkina M, and Varfolomeyev S 2007 Polyhistidine-containing organophosphorus hydrolase with outstanding properties. *Biocatal Biotransfor* **25(1)** 103-8
[19] Sirotkina M, Lyagin I and Efremenko E 2012 Hydrolysis of organophosphorus pesticides in soil: New opportunities with ecocompatible immobilized His_6-OPH. *Int. Biodeterior. Biodegradation* **68** 18-23

IOP Conf. Series: Materials Science and Engineering **525** (2019) 012006 doi:10.1088/1757-899X/525/1/012006

Polypropylene composite material and its rheological and mechanical properties depending on the size of the filler $CaCO_3$

E A Bilalova[1,2], **E V Prut**[1], **O P Kuznetsova**[1]

[1]Department of Chemical Physics,ICP RAS, Moscow, Russia;
[2]MIPT, Dolgoprudny, Russia

E-mail: bilalova@phystech.edu

Abstract. The main idea of this paper is to investigate the influence of the filler sizes on the important composite material's properties, such as viscosity, MFR, shear stress, elongation at break and Youngs modulus. Ten compositions of PP/$CaCO_3$ composites were prepared in a two-rotor mixer (Brabender) with calcium carbonate content o 5 - 25 wt. (%). Laser diffraction method was used to analyze sizes of the powder filler, considering it to be approximately spherical. Found out, that with an increase in the concentration of powder filler increases the value of Youngs modulus (E, MPa) and also there is a significant decrease in the elongation at break (ϵ, %), and the strength also decreases, but rather weakly. Viscosity showed different trends for both sizes at the dependence of concentration with constant shear stress.

1. Introduction

The problem of the disperse filler in the composite material remains relevant to this day. The main task through the process of developing polymer composite materials (PCM) is to create an optimal structure that meets the requirements for the use of products from PCM. This is achieved by selecting the components of the PCM, their ratios, methods for obtaining and products from it. It is well-known that powder filler, such as $CaCO_3$, makes composite material more durable and reliable, decreases the end cost and has an impact on viscosity. Also, the choice to study PP and $CaCO_3$ is come from the article by Thriveni Thenepalli et al [1], where this combination enhances the mechanical properties of plastic parts used in automobiles.

Many factors of rheological and mechanical properties were studied. Typical yield stress behavior exhibited by suspensions is described in the reviews by YEH WANG [2]. Filling viscous thermoplastics with rigid fillers in an amount of more than 20 % refers to a transition from plastic flow to brittle fracture is observed [3]. In this case, there is a significant reduction in the impact strength, the work of destruction. The modulus of elasticity increases with the amount of filler, but it increases the size and number of cracks, like pores, arising in the loading process when the matrix is peeled off from the dispersed particles at the time of reaching the stresses corresponding to the adhesive strength of the system. Theoretical data shows that by reducing the size of the filler particles and the dispersion of their diameters can significantly reduce the likelihood of large defects.

From the study of AY Goldman, CJ Copsey [4] the toughness of the $CaCO_3$ filled materials

IOP Conf. Series: Materials Science and Engineering **525** (2019) 012006 doi:10.1088/1757-899X/525/1/012006

increasing at low temperatures is close to the toughness produced using rubber at 23°C, and even better at lower temperatures.

As known from T. Kaully's et al [5] research, the dominancy of particle size distribution was demonstrated by studying the relative suspensions viscosity as function of shear rate.

Having studied the recent discoveries in the field of filled composite materials, it is found that not much attention is paid to the effect of particle size on the properties, so this issue is relevant.

2. Experimental procedure

2.1. Materials preparation

The matrix consists an isotactic PP brand Moplen HP 5008 (Poly Murray, Basel, Corea), melt flow rate MFR=0.6 g/min (5 kg/1900 s), density ρ=0.9 g/cm^3, melting point T=190 °C. Powder $CaCO_3$ was used of two particle diameters. Particle sizes are ranged from 0.1 μm to 80 μm.

Composite with different compositions, viz. PP/ $CaCO_3$ 95/5, 90/10, 85/15, 80/20, 85/25, were prepared by mixing the components in a Plastograph EC (Brabender GmbH & Co. KG, Germany) at 190 °C, 100 rpm for 10 min. The stirring continued until the recorded torque reached equilibrium. To avoid the PP degradation, time of mixture was minimized. For further research, the sheets were prepared by forming the melt of mixed composites in a hot press at 190 °C for 10 min.

2.2. Dispersion analysis

To estimate the size Measurements were carried out on a laser diffraction analyzer of particle size distribution Analysette 22 MicroTec plus with a dispersion unit in the liquid (size range: 0.08 - 2000 μm). The results are shown in the Figure 1.

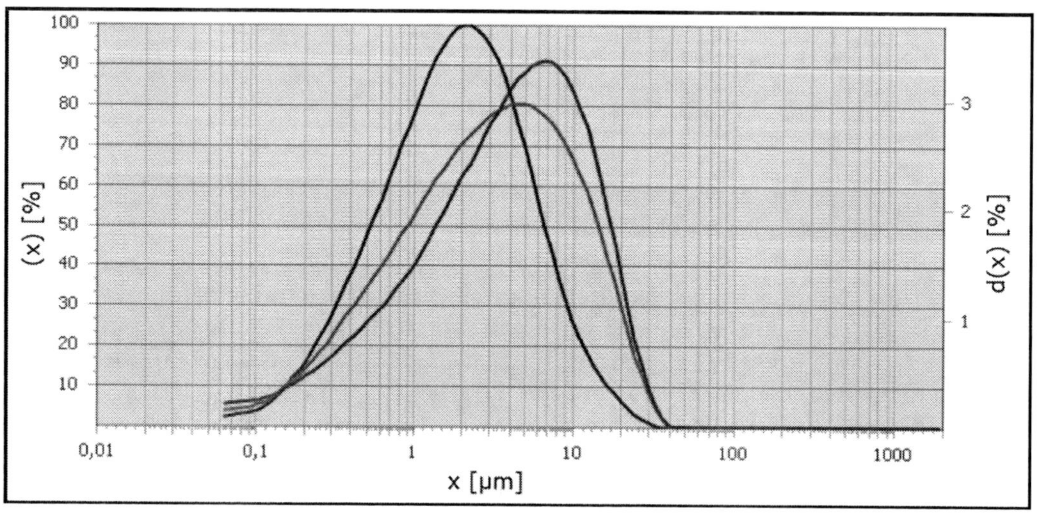

Figure 1. The dependence of the integral and differential curves on the size of the first $CaCO_3$ powder in the sample.

2.3. Rheological properties

Rheological tests were carried out on the viscometer IIRT-5 at a temperature of 190 °C and a load of 1.905 kg to 14.005 kg in increments of 0.5 kg, 1 kg and 1.5 kg to obtain average values, 3 samples were tested at each load. The test for MFR was performed with the load varied according to GOST by 2.16 kg, 5 kg and 10 kg at measurement times - 10 min, 2 min, 1 min and 30 sec (for 10 kg), respectively.

Results are shown in Figure 2 (a) and (b).

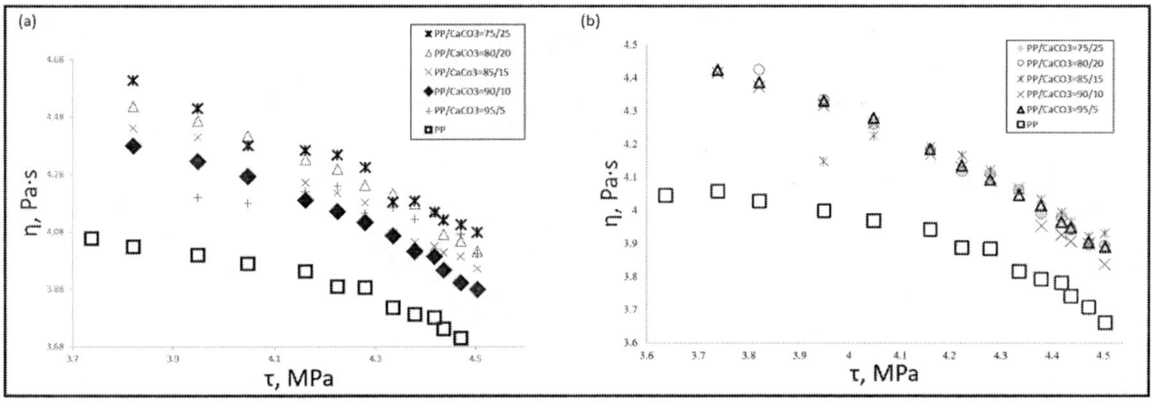

Figure 2. Plots of $\log|\eta|$ versus shear stress at 190 °C for (a) d_1 and (b) d_2 sizes of the filler.

2.4. Mechanical properties

The tensile test was carried out on a universal testing machine Instron, model 1122 of the company "INSTRON Ltd."(England.) For testing samples were prepared in the form of two-sided blades with a length of the working part l = 35 mm, a width a = 5 mm and a thickness b = 0.5 mm. The tests were carried out at a relative tensile speed of 50 mm/min at room temperature. The load-deformation diagrams obtained were used to calculate the main deformation-strength properties of materials: modulus of elasticity, deformation at yield point, strength and elongation at break. To obtain average values, 5 samples of each composition of different filler concentrations were tested.

3. Results and discussion

3.1. Dispersion analysis

Received results on Figure 3 (a) and (b) show us unimodal distribution and that most of the particles in the $CaCO_3$ filler have the following average sizes: d_1 = 4.1 μm and d_2 = 3.3 μm.

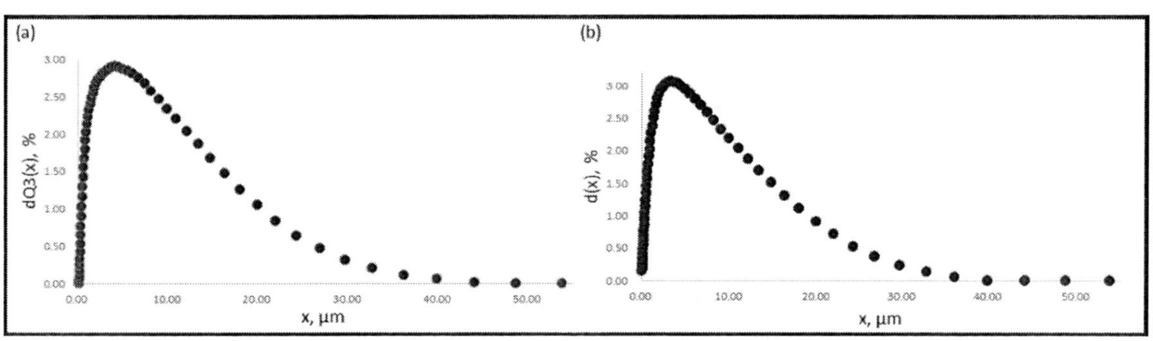

Figure 3. Unimodal volume distribution of the particles sizes for (a) d_1 = 4.1 μm and (b) d_2 = 3.3 μm, different colour corresponds to 3 measurements.

3.2. Rheological properties

In the studied shear range on Figures 2 (a) and (b), the viscosity of both PP and composites decreases monotonically with increasing shear stress, which is typical for non-Newtonian flow. At the same time, with the growth of the filler content, there is a slight deviation from the Newtonian flow. It can be assumed that the rheological behavior depends significantly on the agglomeration of the filler particles. However, having built the dependence of viscosity on the filler concentration at a constant shear stress, new results arise on Figures 4 (a) and (b).

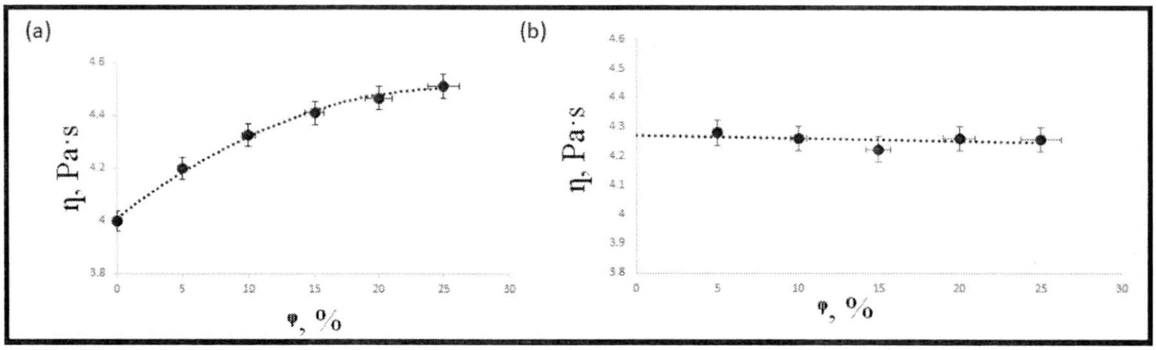

Figure 4. Plots of $\log|\eta|$ versus concentration of the filler $CaCO_3$ [wt.%] at constant shear stress $\tau = 4MPa$ for (a) $d_1 = 4.1\mu$m and (b) $d_2 = 3.3$ μm .

It is generally accepted [6] that the simplest dependence describing the viscosity of the filled composite is the Einstein equation (1):

$$\eta = \eta_0 \left(1 + \alpha\varphi\right)$$

where η_0 - viscosity of the matrix, φ - volume fraction of the filler, α - constant, and Guth equation (2)

$$\eta = \eta_0 \left(1 + \beta\varphi + \delta\varphi^2\right)$$

where β and δ are constants.

Equations (1) and (2) can describe the dependence on Figure 4 (a) on equal terms. For compositions on Figure 4 (b) with sizes of the filler d_2=3.3μm no dependencies are appeared. That can be explained by variations in the amount of $CaCO_3$, which could change little the morphological state of PP/$CaCO_3$ compositions [7].

3.3. Mechanical properties

Received mechanical characteristics show us typical deformation dependencies for filled polymers compositions. In the study of deformation stretching of samples, the formation of a 'neck' is observed. On the stress-elongation curves, yield stress is observed .

Found out, that with an increase in the concentration of powder filler goes up the value of Youngs modulus (E, MPa) and also there is a significant decrease in the elongation at break (ϵ, %), and the strength also decreases, but rather weakly.

Hence, amount of the filler and its size have a significant influence on rheological and mechanical properties, even if the diameters of particles are different on 20 %.

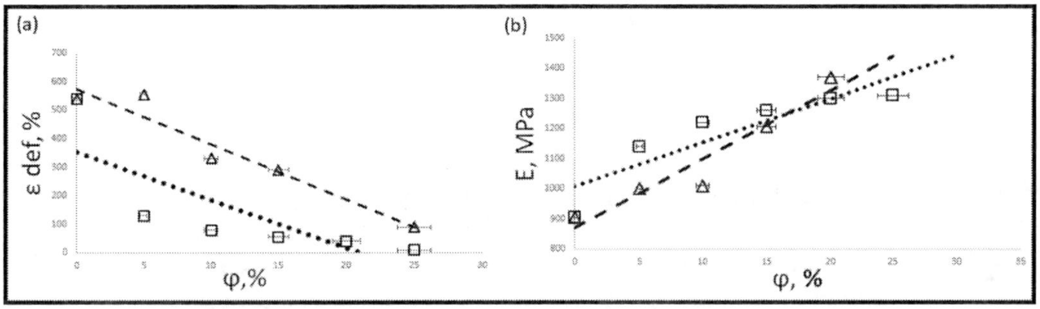

Figure 5. Graphics of (a) elongation at break (ϵ, %) and (b) Youngs modulus (E, MPa) versus concentration of the filler $CaCO_3$ [wt.%] for □ d_1=4.1μm and △ d_2=3.3μm with linear trend lines.

4. Conclusion

From the rheological tests it seems that viscosity depends on the concentration of the filler for d_1=4.1μm and can be explained by Einstein-Guth-Gold model.

Mechanical tests open a new interesting correlation between the size of the filler and elongation properties. With an increase in the concentration of powder filler increases the value of Young's modulus (E, MPa) and also there is a significant decrease in the elongation at break (ε, %), and the strength also decreases, but rather weakly. The cause may be the occurrence of agglomerations, or insufficient adhesion between PP and $CaCO_3$.

These results give an opportunity to understand better the reasons of unusual rheological properties, using SEM analysis and the other techniques.

Acknowledgments

Many thanks to Mezentseva T, Zhorina L and Zvereva U for assistance, advice and contribution to the development of science. The work was carried out within the framework of the state task of the FAO Russia (Topic 46.14, 0082-2014-0006, No. AAAA-A17-117-117032750202-6).

References

[1] Thenepalli T, Jun A Y, Han C, Ramakrishna C and Ahn J W 2015 *Korean Journal of Chemical Engineering* **32** 1009–1022
[2] Wang Y and Wang J J 1999 *Polymer Engineering & Science* **39** 190–198
[3] Prut E, Kuznetsova O, Karger-Kocsis J and Solomatin D 2012 *Journal of Reinforced Plastics and Composites* **31** 1758–1771
[4] Goldman A Y and Copsey C J 2000 *Materials Research Innovations* **3** 302–307
[5] Kaully T, Siegmann A and Shacham D 2007 *Polymer composites* **28** 524–533
[6] Nielsen L E 1967 *Journal of composite materials* **1** 100–119
[7] Han C D 2007 *Rheology and processing of polymeric materials: Volume 1: Polymer Rheology* vol 1 (Oxford University Press on Demand)

IOP Conf. Series: Materials Science and Engineering **525** (2019) 012007 doi:10.1088/1757-899X/525/1/012007

Improved tetrathienoacene synthesis

M S Skorotetcky[1], O V Borshchev[1] and S A Ponomarenko[1,2]

[1]Enikolopov Institute of Synthetic Polymeric Materials of Russian Academy of Sciences, Profsoyuznaya St. 70, Moscow 117393, Russia
[2]Chemistry Department, Moscow State University, Leninskie Gory 1-3, Moscow 119991, Russia
E-mail: skoroteckiy@ispm.ru

Annotation. This paper describes a method for preparation of 2,6-dibromo-thieno[3,2-*b*]thieno[2',3'':4,5]thieno[2,3-*d*]thiophene (tetrathienoacene), an important building block of conjugated organic materials. The synthesis route involves the preparation of 3-bromothieno[3,2-*b*]thiophene from 3,4-dibromothiophene, followed by the preparation of the corresponding disulfide and oxidative ring closure to produce tetrathienoacene. The combination and adaptation of various literature methods allowed to increase the total reaction yields in the case of 3-bromothieno[3,2-b]thiophene from 54% to 72% and tetrathienoacene from 27% to 45%.

1. Introduction

To date, there are many classes of conjugated molecules that are used in organic electronics devices. Organic field effect transistors (OFET), which a based on such materials, are a key element in organic microcircuits, radio frequency identification (RFID), optical and chemical sensors and so on. The main characteristic of organic semiconductor materials is charge carrier mobility, which is defined as the drift velocity of charge carriers at an electric field strength in a semiconductor of 1 V/m. The value of charge mobility determines the amount of current in field-effect transistors, as well as the maximum speed of operation of integrated circuits based on them. Research in this field shows that conjugated semiconductor materials based on tetrathienoacene derivatives shows their high practical potential due to strong intermolecular interactions, extensive intramolecular π-conjugation, and close intermolecular π–π stacking. [1-4]. However, they are still poorly studied, and synthetic methods for their preparation have a several significant drawbacks: the high cost of the initial reagents, the multi-stage and complexity of purification, low yields. This work is aimed at studying the methods of obtaining tetrathienoacene (**TTA**) - the main subject of semiconductor materials based on it.

2. Results and Discussion

For the first time, **TTA** synthesis consisting of 6 stages was described by Y Mazaki and K Kobayashi in 1989 [5]. The key precursor in this scheme is the preparation of 3-bromothieno[3,2-*b*]thiophene (**5**), which then, through the production of disulfide (**6**), was undergo of oxidative ring closure (Figure 1). The authors proposed a new method for its preparation, which turned out to be much easier and gave higher yields, as compared to previously known. However, in the article no methodologies and experimental data on the synthesis were given, which makes it difficult to reproduce the obtained results.

Later, a one-pot method for the preparation of **TTA** was proposed, which combined the preparation of disulfide and the oxidative ring closure in a single reaction vessel [6]. The main disadvantage of this method is a complex experimental technique, which requires constant evacuation of the reaction vessel, and a rather low **TTA** yield of 27%. For comparison, the total yield of **TTA** for the last 2 stages is given by K Kobayashi was 49%. On the other hand, later publications show that it is not so easy to achieve such high reaction yields at this stage [1]. There is also a longer method for the synthesis of **TTA**, which has 8 stages, but eliminates the stage of oxidative ring closure using an additional double-annulation reaction with ethyl 2-thioxoacetate [7]. In addition to the longer path, there is a problem with the double decarboxylation reaction, the low reaction yields of which adversely affect the total yield. Thus, it can be seen from the above that the method proposed by K Kobayashi with an optimized stage of oxidative ring closure at the last stage can be the optimal way to obtain **TTA**.

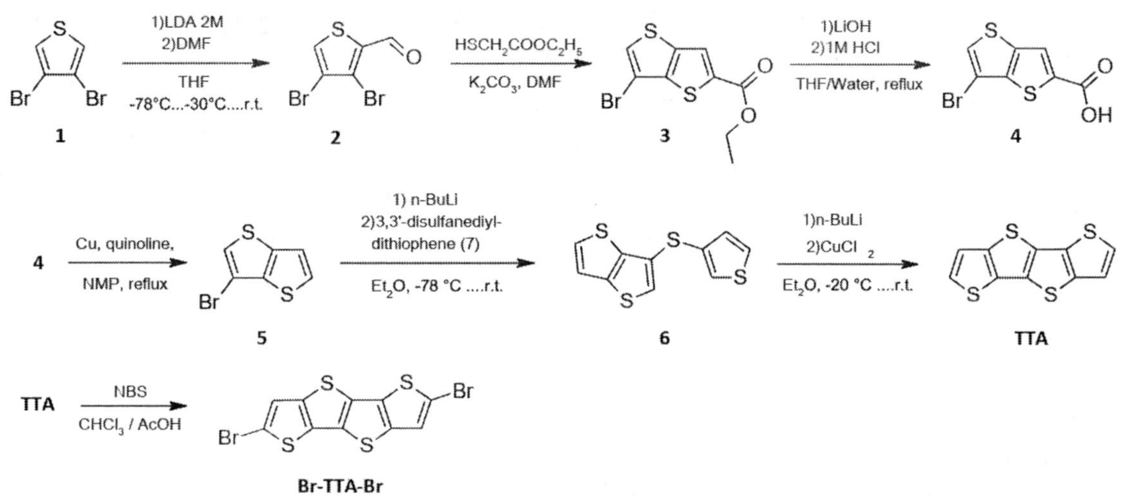

Figure 1. Adaptation of the literature scheme for the synthesis of **TTA** [8,5].

The synthesis of the key precursor **5** was prepared by adapting the techniques described earlier [8,5]. Initially, from commercially available 3,4-dibromothiophene (**1**) by lithiation followed by formylation with dimethylformamide, it was possible to obtain 3,4-dibromothiophene-2-carbaldehyde (**2**) which was introduced into the next step without further purification. Although this dibromide is more expensive than 3-bromothiophene, it allows to immediately get a bromine derivative of thieno[3,2-*b*]thiophene, avoiding two additional stages. Carbaldehyde **2** was introduced into the annulation reaction with ethyl 2-thioxoacetate. After separating the precipitate obtained, the yield for both stages of pure compound **3** was 84%. Cleavage of the ether protection group with lithium hydroxide and subsequent filtration of the precipitate formed allowed to obtain compound **4**. All stages, including the preparation of compound **4**, were performed without additional purification, and the results of GPC and NMR indicate their high purity and good reaction yields. Nevertheless, several attempts to carry out the last decarboxylation reaction according to the conditions given in the literature were not so successful and allowed to obtain pure compound **5** only after purification on a column with a yield of 47% (literature data 85.3%). It is also strange that the indicated temperature of the reaction is 260 °C, since the boiling point of the quinoline itself in which the reaction is carried out is only 237 °C. Apparently, at such a high temperature, a side reaction with the opening of the cycle occurs, as evidenced by the appearance of an impurity with a lower absorption maximum according to GPC. However, copper catalysts in decarboxylation reactions begin to work only at elevated temperatures, usually more than 170 °C. The variation of temperature and the selection of solvents made it possible to find a compromise between the reaction rate and the absence of side processes. It turned out that the use of *N*-methyl-2-pirrolidone with a boiling point of 202 °C as the main solvent and quinoline in catalytic amounts made it possible to reduce the reaction temperature and increase the reaction yield, which, considering the last stage, amounted to 86.5% in 30 minutes. Thus, the total yield of compound **5** for all 4 stages was 72%. At the next stage, 3,3'-disulfanediyl-dithiophene (**7**) was used as a source of sulfur. This approach is also not optimal, since at the next stage thiophene-2-thiol is formed, which has an unpleasant smell and requires its vacuum distillation. A promising analogue may be the replacement of compound **7** with *S*-thiophen-3-*yl* 4-methylbenzene-1-sulfonothioate. The preparation of **TTA** was carried out by oxidative ring closure in ether in the presence of copper chloride (II). During the reaction, oligomerization occurs due to non-selective lithiation. The amount of ether has a positive effect on the reaction yield: the more of it, the less oligomerization products and the easier extraction. The use of cyclohexane instead of *n*-hexane, as described in the literature, due to the much greater solubility of **TTA** in it simplifies its purification, especially when upscaling. At the last stage, we carried out the bromination of **TTA** with NBS in a mixture of chloroform/acetic acid. According by GPC, the reaction is almost quantitative, which for some applications allows the use of the resulting **Br-TTA-Br** without future purification. A pure compound can be obtained by recrystallization from toluene. The resulting functional dibromide can easily be used further to produce various conjugate materials using cross-coupling reactions.

3. Conclusion

Thus, by combining and optimizing various approaches for preparation of **TTA**, described in the literature, it was possible to significantly increase its total yield. One more advantage is that most of the stages do not require additional purification, and the procedures themselves are quite simple. Although some changes introduced are small, they turn out to be significant in terms of their impact, and also help to summarize the scattered literature materials.

Acknowledgement

This work was supported by Russian Science Foundation (grant 18-73-10182) and performed in the framework of leading science school NSh-5698.2018.3.

References

[1] Liu Y, Wang Y, Wu W, Liu Y, Xi H, Wang L, Qiu W, Lu K, Du C and Yu G Synthesis, Characterization, and Field-Effect Transistor Performance of Thieno[3,2-b]Thieno[2′,3′:4,5]Thieno[2,3-d]Thiophene Derivatives. *Adv. Funct. Mater.* **2009**, *19* (5), 772–778

[2] Liu Y, Di C, Du C, Liu Y, Lu K, Qiu W and Yu G Synthesis, Structures, and Properties of Fused Thiophenes for Organic Field-Effect Transistors. *Chem. - A Eur. J.* **2010**, *16* (7), 2231–2239

[3] Youn J, Huang P.-Y, Huang Y.-W, Chen M.-C, Lin Y.-J, Huang H, Ortiz R. P, Stern C, Chung M.-C, Feng C.-Y, et al. Versatile α,ω-Disubstituted Tetrathienoacene Semiconductors for High Performance Organic Thin-Film Transistors. *Adv. Funct. Mater.* **2012**, *22* (1), 48–60

[4] Vegiraju S, Lin C.-Y, Priyanka P, Huang D.-Y, Luo X.-L, Tsai H.-C, Hong S.-H, Yeh C.-J, Lien W.-C, Wang C.-L, et al. Solution-Processed High-Performance Tetrathienothiophene-Based Small Molecular Blends for Ambipolar Charge Transport. *Adv. Funct. Mater.* **2018**, *28* (28), 1801025

[5] Mazaki Y and Kobayashi K Synthesis of Tetrathieno-Acene and Pentathieno-Acene: UV-Spectral Trend in a Homologous Series of Thieno-Acenes. *Tetrahedron Lett.* **1989**, *30* (25), 3315–3318

[6] Chen M.-C, Chiang Y.-J, Kim C, Guo Y.-J, Chen S.-Y, Liang Y.-J, Huang Y.-W, Hu T.-S, Lee G.-H, Facchetti A, et al. One-Pot [1+1+1] Synthesis of Dithieno[2,3-b:3′,2′-d]Thiophene (DTT) and Their Functionalized Derivatives for Organic Thin-Film Transistors. *Chem. Commun.* **2009**, No. 14, 1846

[7] Frey J, Bond A D, and Holmes A B Improved Synthesis of Dithieno[3,2-b:2′,3′-d]Thiophene (DTT) and Derivatives for Cross Coupling. *Chem. Commun.* **2002**, No. 20, 2424–2425

IOP Conf. Series: Materials Science and Engineering **525** (2019) 012007 doi:10.1088/1757-899X/525/1/012007

[8] Han J, Qi J, Zheng X, Wang Y, Hu L, Guo C, Wang Y, Li Y, Ma D, Qiao W, et al. Low-Bandgap Donor–acceptor Polymers for Photodetectors with Photoresponsivity from 300 Nm to 1600 Nm. *J. Mater. Chem. C* **2017**, *5* (1), 159–165

IOP Conf. Series: Materials Science and Engineering **525** (2019) 012008 doi:10.1088/1757-899X/525/1/012008

Structural–crystallographic and morphological properties of silicon composition of phosphate ore

I A Pochitalkina[1] **and O V Vinokurova**[1]

[1] Department of Technology of Inorganic Substances and Electrochemical Processes, D I Mendeleev University of Chemical Technology of Russia, Moscow, 125480, Russia

E-mail: pochitalkina@list.ru

Abstract. In order to identify the prospects for complex processing of high-silicon phosphate by direct acid methods, the properties of the mineral raw material and its insoluble residue were investigated. Clarified literature data on the composition of the original sample. New data were obtained on the chemical composition, morphology, microstructure and optical properties of the siliceous component of the ore.

1. Introduction

Natural phosphate ores are represented by apatites and phosphorites. The minerals of the apatite group are expressed by the general formula $3Ca(PO_4)_2{\cdot}CaX_2$. The phosphate substance of phosphate ores is represented by fluorapatite $3Ca(PO_4)_2{\cdot}CaF_2$, fluorohydroxycarbonateapatite $Ca_{10}P_5CO_{23}(F,OH)_3$, francolite $Ca_{10}P_{5.2}C_{0.8}O_{23.2}F_{1.8}OH$, kurskit $Ca_{10}P_{4.8}C_{1.2}O_{22.8}F_2{\cdot}(OH)_{1.2}$. The composition of phosphate ores includes minerals impurities: quartz, glauconite, calcite, limonite, dolomite, magnesian silicates, kaolin, pyrite, feldspar, etc. [1].

In terms of phosphate reserves Russia ranks fourth in the world. The peculiarity of the domestic raw material base is the predominance of apatite ores in its structure, locally concentrated on the Kola Peninsula. The high content of the target component of 39.4% in terms of P_2O_5 and the low content of impurities, including up to 1.5% of the insoluble residue represented by alpha-quartz, ensured the active exploitation of deposits and the supply of quality raw materials for the phosphoric industry. For this reason, phosphorite ores, which are of low quality, widely distributed in the territory of the Russian Federation, belonged to reserve reserves, and their direct reagent processing was considered ineffective. Operation of the Egoryevskoye, Vyatsko-Kamskoye, Polpinskoye deposits consisted mainly in the production of phosphate rock from them – a product of primary enrichment, which was used with low efficiency to increase soil fertility [2].

Nevertheless, the long-term development of a rich Khibinsky massif leads to its depletion, deepening of the developed quarries, flooding of the developed horizons and, as a result, increasing the cost of raw materials production. The remoteness of deposits from the main consumers of apatite (enterprises for its processing by acid extraction methods) contributes an additional, very tangible share to the increase in the cost of raw materials and end products based on it.

With a view of rational nature management, already from the end of the last century, the issue of engaging in the processing of substandard phosphate raw materials has become topical. There is an industrial experience of replacing the share of apatite with phosphate rock from the Kingiseppskoye and Egoryevskoye fields in the production of phosphoric acid and products based on it [3].

In order to saturate the domestic market of the Russian Federation with affordable phosphorus-containing products, the prospect of direct processing of low-grade phosphate raw materials from the Bryansk region is estimated. The choice of the object of study is due to the presence of industrially significant reserves of phosphate raw materials, a developed network of highways and railways, the proximity of agricultural consumers, as well as the presence of an agricultural holding in the region [4]. It should be noted that the information on the chemical composition of phosphate ores of individual deposits, including the Polpinsky sample, available in literary sources was relevant in the eighties of the last century [1, 5], therefore, they require clarification, and the involvement in the processing of reserve types of raw materials requires further exploration.

Table 1. The chemical composition of deposits of phosphate raw materials.

Object	The content of the component, % mass									
	P_2O_5	CaO	Fe_2O_3	Al_2O_3	MgO	K_2O+ Na_2O	CO_2	SO_3	F	In.res.
Khibiny apatite concentrate	39.4[a]	51.5–52.0	–	–	0.2–0.4	0.8–1.0	–	–	3.0–3.1	–
Phosphate raw materials of brand P (ordinary), Karatau	24.5	36–39	1.0–1.7	0.8–1.2	3.5	0.3–0.7	8.0	–	2.2–2.4	15–20
Verkhnekamsk phosphate raw materials, primary concentrate	21.0–24.0	34.0–38.1	3.7–4.8	3.3–4.7	1.8–2.3	1.6	4.7–5.4	1.0	2.5–3.0	13–18
Egorievsk phosphate raw materials, primary concentrate	20.4–23.3	32.6–36.0	4.5–7.0	4.0–4.6	1.1–1.4	–	4.1–7.4	1.0	2.0–3.0	18–19
Kingisepp phosphorite, flotation concentrate	28.0	39.0–42.0	3		2.5	–	1.5–2.0	0.4	2–2.2	5–20
Polpinsky phosphorite, primary concentrate	19.5–21.5	30.5–33.3	2.6–4.0		0.5–2.3	–	3.4–5.0	1.1	2.1–2.8	31–37

[a] in GOST 22275-90, the content of P_2O_5 is not less than 39.0%. This is due to the gradual depletion of the field.

2. Experimental part

The object of the study was a representative sample of clod ore from the Polpinskoye field, which was subjected to grinding on a pilot plant, classification, and drying the powder to constant weight. Acid nitrate qualification "puriss." (GOST 4461-77) and distilled water (GOST 6709-72), SiO_2 qualification "puriss. spec." (GOST 9428-73, TU 6-09-3379-79) were used as reagents.

Physico-chemical studies were performed using a Nicole 380 FTIR spectrometer (ThermoScientific, USA), a Bruker D 8 Advance diffractometer (Bruker Corporation, USA), a KFK-3-01 photocolorimeter (ZOMZ, Sergiev Posad), a JEOL JSM-6510LV multifunction instrument (JEOL, Japan), ASAP 2020MP (Micromeritics, USA) adsorption analyzer of specific surface and porosity, MIN-8 polarization microscope (LOMO, Russia; 320 microscope magnification), IL-1 set of immersion liquids (Reakhim, Russia).

The initial sample with a particle size of 0.071–0.315 mm was subjected to decomposition with 0.1 M HNO_3 at a temperature of 25 ± 0.05°C in a reactor with an overhead stirrer. The process was completed upon reaching a decomposition rate of 99.7%, determined by the photometric method according to the ratio of the content of the extracted target component (P_2O_5water) to its initial content in the raw material (P_2O_5 total) [6–9].

The reaction of the target component of phosphate rock with nitric acid is described by the equation:

$$Ca_{10}(PO_4)_6CO_3 + 20HNO_3 = 6H_3PO_4 + 10Ca(NO_3)_2 + CO_2 + H_2O$$

The resulting suspension, consisting of the reaction products and the insoluble residue, was separated by vacuum filtration. The precipitate was washed with distilled water, dried for 1 hour at 200°C, and then sent for research. The filtrate is supposed to be used to obtain phosphorus-containing products for livestock and agriculture.

3. **Results and discussion**

According to the results of chemical analysis, the investigated ore from the Polpinsky deposit is a low-grade phosphate raw material. The share of the useful component (P_2O_5) is 15.3%, the share of the main impurity (SiO_2) is one third of the mass of the sample – 32.9%, the content of other impurities is: CaO – 27.4%, MgO – 0.23%, Fe_2O_3 – 1.4%, Al_2O_3 – 1.2%, F – 0.008%, CO_2 – 4.0%.

The results of IR spectroscopy of the analyzed sample in the region of 400–4000 cm^{-1} (Figure 1) show that each spectrum is an overlap of overlapping bands that are associated with different types of vibrations of the main functional groups: $H_2PO_4^-$, HPO_4^{2-}, PO_4^{3-} (569, 605, 1044 $см^{-1}$), SiO_2 (750–830 и 460–530 $см^{-1}$; it is important to note that the band at 694 cm^{-1} belongs to α-quartz), HCO_3^-, CO_3^{2-} (694, 866, 1428, 1458 $см^{-1}$), OH^- (3368, 3819 $см^{-1}$), and the absorption bands at 463, 516, 694 cm^{-1} and doublet 779–797 cm^{-1} correspond to the silica and glauconite components of phosphate [10, 11].

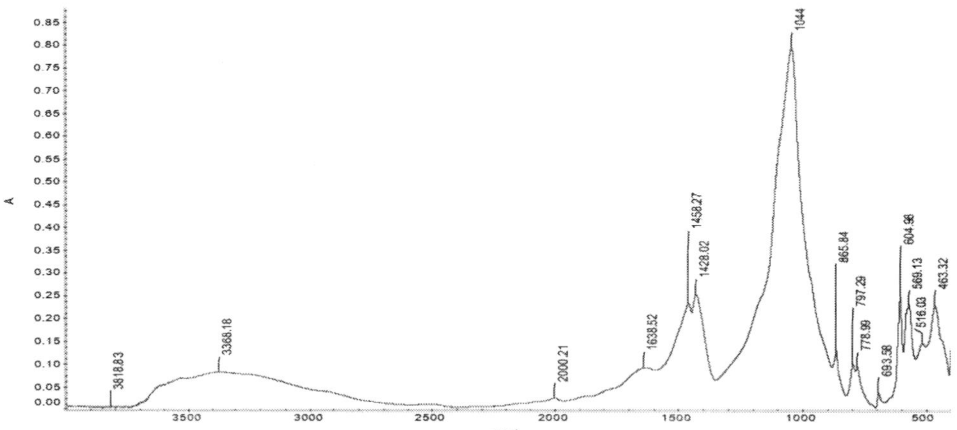

Figure 1. The IR spectrum of a sample of Polpinsk phosphate ore.

According to the results of X-ray diffraction, the phosphate mineral of the ore under investigation has the structure of a hydroxycarbonateapatite [01-074-2680] hexagonal syngony with the unit cell parameters a = b = 9.38Å, c = 6.89Å. The main accompanying impurity is alpha-quartz [00-046-1045] trigonal syngony. Parameters of the α-quartz unit cell: a = 4.9133Å, c = 5.4053Å. Alpha-quartz is a stable at low temperatures polymorphic modification of SiO_2.

The ore particles of the initial sample in the range of 0.071–0.315 mm have a porous structure (Figure 2). This is characteristic of sedimentary rocks with close mutual germination of minerals. For

IOP Conf. Series: Materials Science and Engineering **525** (2019) 012008 doi:10.1088/1757-899X/525/1/012008

the crystals of the main components of the ore, hydroxycarbonateapatite and α-quartz, are characterized by a cancerous fracture and imperfect cleavage [12, 13].

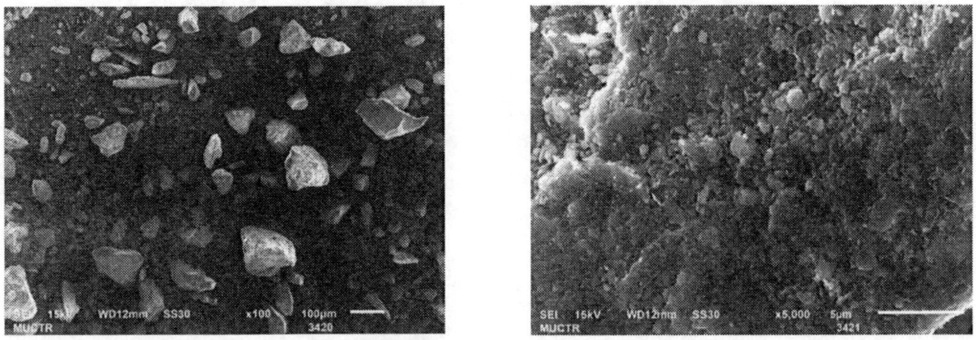

a) b)

Figure 2. Micrograph of polpinsky phosphorite particles at different magnifications.

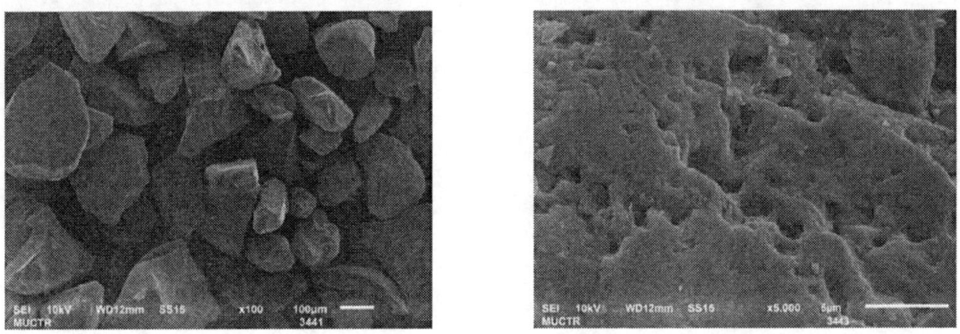

a) b)

Figure 3. Micrograph of particles of insoluble residue of polpinsky phosphorite at different magnifications.

Using the X-ray fluorescence method of analysis, the elemental composition of the insoluble residue of the Polpinsky phosphorite and the reference SiO_2 sample ("puriss. spec.") was investigated [14]. The results obtained indicate that the stoichiometric ratio of the elements in the sample under study (0.629) is preserved in comparison with the reference (0.621).

The optical properties of the insoluble residue (Figure 4) were estimated using the refractive index (GOST 9428-73, 15130-86); the obtained values of n_e and n_o were 1.555 and 1.538, respectively. According to the reference data, the optical identification characteristics of an alpha-quartz crystal are 1.553 and 1.544 [15, 16].

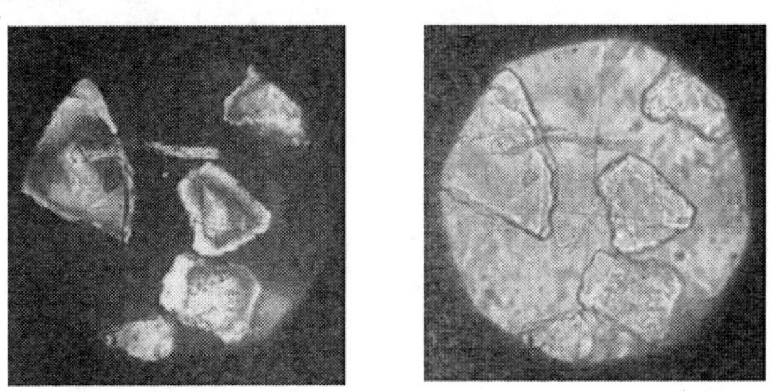

a) b)

Figure 4. Micrograph of α-quartz in determining the refractive index.

IOP Conf. Series: Materials Science and Engineering **525** (2019) 012008 doi:10.1088/1757-899X/525/1/012008

The isotherms of adsorption-desorption of phosphorite ore and its insoluble residue, obtained by the method of low-temperature (77 K) nitrogen adsorption (Figure 5 (a, b)) allowed us to calculate the values of specific surface area (S_{sa}) by the BET method, total pore volume (V_P) and volume distribution pore diameters (in the range from 1.7 to 300 nm) by the BJH method. For phosphate rock, S_{sa} and V_P were 8.03 m^2/g and 0.0474 cm^3/g, respectively, and for α-quartz – 11.48 m^2/g and 0.0455 cm^3/g [17]. An increase in S_{sa} and a redistribution of pore size in the sample of the insoluble residue relative to the original sample of phosphate rock, which is associated with the extraction of the target component and soluble impurities in the solution, are noted.

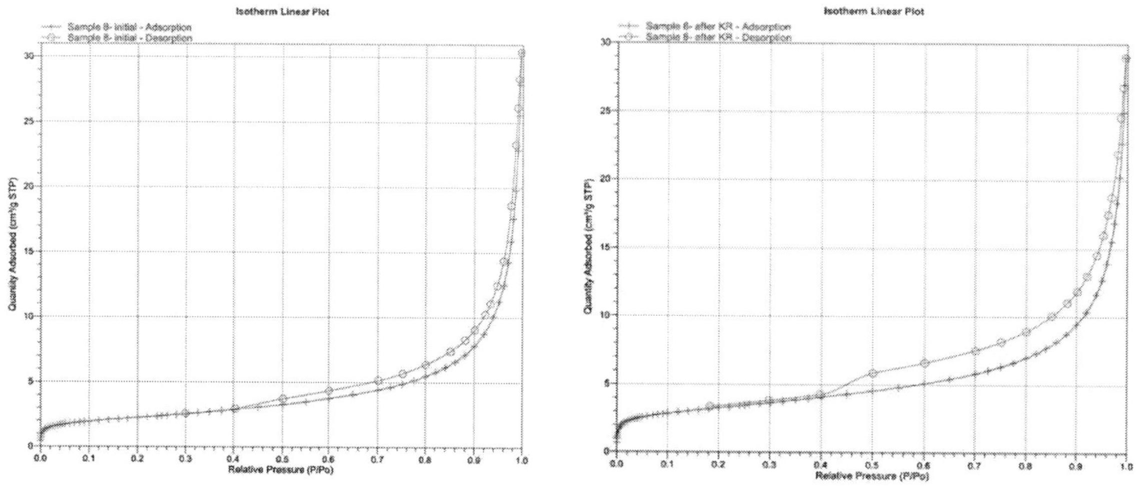

a) b)

Figure 5. Adsorption-desorption isotherm of the initial sample of the PF (a) and its insoluble residue (b).

For low-grade phosphate ore with a high proportion of silicon, complex processing seems appropriate, which consists in separating the suspension after the extraction stage by filtration into liquid and solid phases. The subsequent processing of the filtrate into soluble calcium and ammonium phosphates, and the precipitate (technical alpha-quartz) on glass is supposed to be carried out according to traditional technologies [18–20].

4. Findings

Specified literature data on the composition of high-silicon phosphate rock from the Polpino deposit. New data were obtained on the chemical composition, morphology, microstructure and optical properties of the siliceous component of the ore. The results of IR spectroscopy, X-ray diffraction, X-ray fluorescent analysis and polarization microscopy of the insoluble residue showed that it is alpha-quartz, its composition and optical parameters that determine its suitability for technological processing allow us to consider it as a raw material for the glass industry.

Acknowledgment

The work was supported by the Ministry of Education and Science of the Russian Federation within the framework of the implementation of the Federal Target Program "Research and Development in Priority Areas for the Development of the Russian Science and Technology Complex for 2014-2020." The unique identifier of the agreement RFMEFI58317X0068.

The authors thankful to the D I Mendeleev Center for Collective Use for assistance in conducting research.

IOP Conf. Series: Materials Science and Engineering **525** (2019) 012008 doi:10.1088/1757-899X/525/1/012008

References

[1] 1987 *Technology of phosphate and complex fertilizers* ed S D Evenchik and A A Brodsky (M: Chemistry) 464 p
[2] 2015 *State Report On the status and use of mineral resources of the Russian Federation in 2014* ed E A Kiselyov (M: Ministry of Environment of Russia) 319 p
[3] Klassen P V, Sushchev V S, Klados D K, Mironov V E et al 2010 Studying the possibility of using domestic phosphorites (using the example of Egorievsk) for the production of EPA and phosphorus-containing fertilizers *Chem. prom-ct today* **2** pp 24–31
[4] Filenko I A, Pochitalkina I A and Petropavlovsky I A 2016 Production of complex fertilizers from poor phosphate raw materials *Adv. in chem. and chem. tech.* vol 30 **3** pp 102–4
[5] 2007 Phosphate ores. Guidelines for the application of the Classification of reserves of deposits and forecast resources of solid minerals (M: Ministry of Environment of Russia) 42 p
[6] GOST 5716–74 *Phosphorite flour. Technical conditions* (M: Standards Publishing)
[7] GOST 20851.2–75 *Mineral fertilizers. Methods for the determination of phosphates* (M: Standards Publishing)
[8] Okrushko E Y, Pochitalkina I A and Petropavlovsky I A 2014 Kinetics of nitrophosphoric acid decomposition of polpinsky phosphorite *Adv. in chem. and chem. tech.* vol 28 **5** pp 133–5
[9] Fedotov P S, Ryashko A I, Kiselev V G, Pochitalkina I A et al. 2012 Study of the kinetics of hydrochloric acid decomposition of phosphorite from the Polpinsky field by the ionometric method. *Adv. in chem. and chem. tech.* vol 26 **8** pp 63–5
[10] Nakamoto K 1991 *IR spectra and spectra KR of inorganic and coordination compounds* (M: Mir) 536 p
[11] Plusnina I I 1976 *Infrared spectra of minerals* (M: Mos. University) 175 p
[12] 1990 *Chemical encyclopedia* in 5 tons v 2 Duffa-Medi ed I L Knunyants (M: Sov. encycl.) 671 p
[13] Eroshkina N A, Korovkin M O and Chamurliev M Y 2017 *Youth Research Herald* **5** pp 93-8
[14] Vlasov A I, Elsukov K A, Kosolapov I A 2011 *Electron microscopy: studies. Manual* (M: BMSTU) 168 p
[15] Wegman E F, Rufanov Y G 1990 *Crystallography, petrography and radiography* (M: Metallurgy) 256 p
[16] Vertushkov G N and Avdonin V N 1992 *Tables for the determination of minerals by physical and chemical properties* (M: Nedra) 489 p
[17] Pochitalkina I A, Kondakov D F, Artamonova O A and Vinokurova O V 2017 Textural and structural characteristics of phosphate Polpinskogo field *J. Inorg. Chem.* vol 62 **11** pp 1503–06 doi: 10.1134/S0036023617110146
[18] Isaev V A 2006 Structural impurities in quartz. Part I. Review and analysis of traditional methods for cleaning quartz from structural impurities *GIAB* **9** pp 11–23
[19] Zverev V A, Krivopustova E V and Tochilina 2013 T V *Optical materials Part 2 Tutorial* (SPb: SPb ITU ITMO) 248 p
[20] GOST 22551-77 *Quartz sand, ground sandstone, quartzite, and quartz vein for the glass industry* (M: Standards Publishing)

IOP Conf. Series: Materials Science and Engineering **525** (2019) 012009 doi:10.1088/1757-899X/525/1/012009

Nonuniform polarized states of ferroelectric nanoparticles in a dielectric matrix

V N Nechaev[1], **A V Viskovatykh**[2] **and A V Shuba**[1]

[1]Military Educational and Scientific Centre of the Air Force N.E. Zhukovsky and Y.A. Gagarin Air Force Academy, Starich Bolshevikov Str, 54a, 394064, Voronezh, Russia
[2]Russian Scientific Center of Roentgenoradiology (RSCRR), Profsoyuznaya Str., 86, 117997, Moscow, Russia

E-mail: ostrogvisk@mail.ru

Abstract. In the framework of the phenomenological theory of phase transitions (PT) Landau are showed that in nanoscale ferroelectrics (FE), a phase transition to a non-uniform polar state is possible. As a result of solving the eigenvalue problem for the equilibrium equation of polarization and electrostatic equations, a wave vector is defined, that characterizes a non-uniform phase in a thin ferroelectric film (FEF), temperature ranges of the «domain-like» phase are found, depending on the characteristic size of the FE inclusion and its surface properties. The possibility of controlling the domain structure in ferroelectric particles (FEP) using an external electric field is shown.

1. Introduction

High permittivity and polarization, and ability to control the operating parameters of the material by an external electric field or mechanical stresses, determine the wide use of nanocomposite materials (NCMs) based on FE in various devices and microwave technology, optoelectronics, information storage devices, pyro-receivers, humidity sensors and incite the researches, that needed to create a reliable model for predicting their properties [1–2]. In [3, 4] it was shown, that in dielectric (DE) matrix with FE inclusions, the formation of «domain-like» states in the polar phase is possible. In a thin FEF with a PT to the polar phase, the appearance of a spatially modulated state is also possible [5]. The use of the term «domain-like» state to characterize an inhomogeneous state is due to the fact that, unlike the usual domain structure, in this case there are no distinct boundaries separating the polar regions. Another type of inhomogeneous states was investigated when modeling the properties of relaxors [7].

We emphasize that a nonuniform state in the work means not only a state with a spatial dependence of the order parameter, that takes a place for any limited samples, but a state in which the polarization vector changes direction. The purpose of this work is to determine the temperature ranges of existence of inhomogeneous polarization states in thin FEF and in FE-DE nanocomposites, to study the conditions of their occurrence in a second-order phase transition depending on the size and shape of sample, its surface properties.

The sensitivity of the problem to consideration of electrostatic depolarizing fields makes it important to find the exact solution for polarization. As a rule, various approximations are used in studies of the properties of thin FEFs or NCMs with FE inclusions to consideration of electrostatic depolarizing fields [8–11]. For example, in the works [8–10], the Ritz direct variation method has been used, in

[11] the depolarizing field E_d is approximated by the formula $E_d = -(P - P_{mean})/\varepsilon$, where P_{mean} is the polarization average over the film thickness, ε is the permittivity. Note that a similar problem of magnetic fields consideration exists in the theories of magnetic films and domain structures [12,13].

An alternative to such approaches for NCM may be a numerical method, where the required accuracy can always be achieved by increasing the number of iterations. [3]. In the particular case for thin FEF it is possible, without any assumptions, to find the solution of the problem analytically, up to final formulas, which are analyzed numerically [5].

2. The features of PT in thin FEF

2.1. The system of relations, describing the PT

Consider a thin FEF with thickness l and perpendicular to the film plane ferroactive axis that is coincides with the 0z axis. The free energy of such a film is written in the form [14]:

$$F = \int_V \left(-\frac{\alpha}{2}P^2 + \frac{\beta}{4}P^4 + \frac{\kappa}{2}(\nabla P)^2 - \mathbf{E}\cdot\mathbf{P} \right) dV - \int_{V_\infty} \frac{\varepsilon}{8\pi}\mathbf{E}^2 dV + \int_S \frac{\alpha_s}{2}\mathbf{P}^2 dS\,, \tag{1}$$

where $\mathbf{P} = (0,\ 0,\ P)$ is the vector of polarization; $\alpha = \alpha_0(T_C - T)$ and β are the Landau coefficients in the expansion of free energy [15], T_C is the Curie temperature of homogeneous infinite crystal; $\kappa \approx a_0^2$, κ is the correlation constant, $a_0 \cong 3\cdot 10^{-8}$ cm is the lattice parameter; $\mathbf{E}$ is the electrostatic field strength, induced by a change of film polarization; ε is the contribution to the permittivity (mainly electronic), not considering the order parameter influence; α_s is the coefficient at the lowest degree $\mathbf{P}$ in the expansion of surface; V is the film volume; V_∞ is the volume around the film, included V; S is the film surface.

Passing in the functional (1) to independent functions – polarization $\mathbf{P(r)}$ and electric potential $\varphi(\mathbf{r})$, considering $\mathbf{E} = -\nabla\varphi$, we obtain a closed coupled system of equations [5]:

$$\begin{cases} -\kappa\Delta P - \alpha P + \beta P^3 = -\partial\varphi_i/\partial z \\ \varepsilon\Delta\varphi_i = 4\pi\cdot\partial P/\partial z \end{cases} \quad \text{for internal FEF region,} \tag{2}$$

$$\varepsilon_e \Delta\varphi_e = 0 \quad \text{for external FEF region} \tag{3}$$

and boundary conditions

$$\kappa\,\partial P/\partial z \mp \alpha_s P\big|_{z=0,l} = 0\,, \tag{4}$$

$$\varphi_i\big|_{z=0,l} = \varphi_e\big|_{z=0,l}\,, \tag{5}$$

$$\varepsilon\,\partial\varphi_i/\partial z - 4\pi P\big|_{z=0,l} = \varepsilon_e\,\partial\varphi_e/\partial z\big|_{z=0,l}\,, \tag{6}$$

$$\varepsilon_e\,\partial\varphi_e/\partial z\big|_{z=\pm\infty} = 0\,, \tag{7}$$

where φ_i, φ_e and ε, ε_e are the electric potential and permittivity for internal and external FEF region respectively.

2.2. Solving a boundary value eigenvalue problem

Considering that a second-order PT occurs by the mechanism of stability loss of the paraelectric phase, to determine PT temperature, it suffices to find the appearance point of a nonzero solution of the boundary value problem (2) – (7) in the linearized view – without βP^3 term in (2). The symmetry of the problem makes it possible to use the Fourier transform of x, y variables to find unknown functions $P(x,y,z)$ and $\varphi(x,y,z)$.

Equations for the Fourier transforms $\widetilde{P}(k_x,k_y,z)$ and $\widetilde{\varphi}(k_x,k_y,z)$ arising from equations (2), (3) and have the form

$$\begin{cases} \kappa \mathrm{d}^2 \widetilde{P}/\mathrm{d}z^2 + (\alpha - k_\perp \kappa)\widetilde{P} = \mathrm{d}\widetilde{\varphi}_i/\mathrm{d}z \\ \mathrm{d}^2 \widetilde{\varphi}_i/\mathrm{d}z^2 = k_\perp^2 \widetilde{\varphi}_i + 4\pi/\varepsilon \cdot \mathrm{d}\widetilde{P}/\mathrm{d}z \end{cases} \quad (8)$$

$$\mathrm{d}^2 \widetilde{\varphi}_e/\mathrm{d}z^2 = k_\perp^2 \widetilde{\varphi}_e \quad (9)$$

where $k_\perp^2 = k_x^2 + k_y^2$. The solution of equations (8), (9) must satisfy additional conditions arising from the boundary conditions (4) – (7) of the original problem.

Solving the system of equations (8), we obtain the equation

$$\kappa \mathrm{d}^4 \widetilde{P}/\mathrm{d}z^4 + \left(\alpha - 2k_\perp^2\kappa - 4\pi/\varepsilon\right)\cdot \mathrm{d}^2 \widetilde{P}/\mathrm{d}z^2 - k_\perp^2\left(\alpha - k_\perp^2\kappa\right)\widetilde{P} = 0, \quad (10)$$

which solution will be found in the form

$$\widetilde{P}(k_\perp, z) = C_1 \mathrm{e}^{\lambda_1 z} + C_2 \mathrm{e}^{\lambda_2 z} + C_3 \mathrm{e}^{-\lambda_1 z} + C_4 \mathrm{e}^{-\lambda_2 z}, \quad (11)$$

where C_1, C_2, C_3, C_4 are the arbitrary constants, which can be found from the boundary conditions, $\pm\lambda_1$, $\pm\lambda_2$ are the roots of characteristic equation:

$$\kappa\lambda^4 + \left(\alpha - 2k_\perp^2\kappa - 4\pi/\varepsilon\right)\cdot \lambda^2 - k_\perp^2\left(\alpha - k_\perp^2\kappa\right) = 0. \quad (12)$$

Then, we find the Fourier transform $\widetilde{\varphi}_i$:

$$\widetilde{\varphi}_i(k_\perp, z) = C_1\left(a\lambda_1^2 + b\right)\lambda_1 \mathrm{e}^{\lambda_1 z} + C_2\left(a\lambda_2^2 + b\right)\lambda_2 \mathrm{e}^{\lambda_2 z} - C_3\left(a\lambda_1^2 + b\right)\lambda_1 \mathrm{e}^{-\lambda_1 z} - C_4\left(a\lambda_2^2 + b\right)\lambda_2 \mathrm{e}^{-\lambda_2 z}, \quad (13)$$

where $a = \kappa/k_\perp^2$, $b = (\alpha - 4\pi/\varepsilon)/k_\perp^2 - \kappa$.

Equation (9) to determine $\widetilde{\varphi}_e$ is solved independently of the other system equations. For the region $z \le 0$ we get

$$\widetilde{\varphi}_e(k_\perp, z) = C_5 \mathrm{e}^{k_\perp z}. \quad (14)$$

Using boundary conditions for $\widetilde{P}$, $\widetilde{\varphi}_i$ and $\widetilde{\varphi}_e$, we obtain a system of five homogeneous algebraic equations to determine C_1, C_2, C_3, C_4, C_5 [5], which has nonzero solutions only if its determinant

$$\Delta = \frac{\alpha_s + \kappa\lambda_1 \,\mathrm{th}(\lambda_1 l/2)}{\alpha_s + \kappa\lambda_2 \,\mathrm{th}(\lambda_2 l/2)} - \frac{\lambda_1\left(a\lambda_1^2 + b\right)\left(\lambda_1 + k_\perp \,\mathrm{th}(\lambda_1 l/2)\varepsilon_e/\varepsilon\right) - 4\pi/\varepsilon}{\lambda_2\left(a\lambda_2^2 + b\right)\left(\lambda_2 + k_\perp \,\mathrm{th}(\lambda_2 l/2)\varepsilon_e/\varepsilon\right) - 4\pi/\varepsilon} \quad (15)$$

is equal zero. The equation $\Delta(T, k_\perp, l, \varepsilon, \varepsilon_e, \alpha_s) = 0$ is implicit dependence of the PT temperature T_f on the FEF parameters l, ε, ε_e, α_s and on wave vector $k_\perp$, characterizing inhomogeneous polarized state, to study which is really only numerically. The dependence of PT temperature T_f on wave vector $k_\perp$, determining the spatial modulation of the polar phase for different film thicknesses, is shown on Fig. 1. For draw graphics, the parameters of a triglycine sulfate (TGS) crystal were used: $T_C = 322\mathrm{K}$, $\alpha_0 = 3.92\cdot 10^{-3}\,\mathrm{K}^{-1}$, $\beta = 8\cdot 10^{-10}$ unitsESU [16].

2.3. Results analysis

It should be at once noted that the behavior of all film characteristics will be qualitatively different for different signs α_s (figure 1). This difference is described in detail in [17, 18] for the simplest situation when there is need not to consider electric or magnetic fields: if $\alpha_s > 0$ the temperature T_f decreases, polarization profile as a function of coordinate is a convex up function; if $\alpha_s < 0$ – converse situation. In both cases $T_f \to T_C$ at $l \to \infty$. When consider the depolarizing field in the FE phase, the PT can occur in a uniform (if the film parameters allow) or nonuniform state.

Figure 1 shows that in sufficiently thin films ($l < 30\cdot 10^{-8}$ cm) the PT takes place according to the «classical type»: the functions $T = T_f(k_\perp)$ steadily decrease with increasing $k_\perp$ from $k_\perp = 0$. If

$l \geq 30 \cdot 10^{-8}$ cm the PT occurs in spatially modulated nonuniform state with nonzero wave vector $k_{\perp}$, depending on parameters l, ε, ε_e, α_s. The PT to the nonuniform state takes place regardless of the sign parameter α_s.

The real phase transition temperature T_{fh} and the wave vector of the static polarization $k_{\perp}^{*}$ coincide with the point of maximum $T_{fh} = T_f^{(\max)}$ on the dependences $T = T_f(k_{\perp})$ (figure 1). This temperature as a function of l at various parameters α_s on the FEF surface is shown in figure 2.

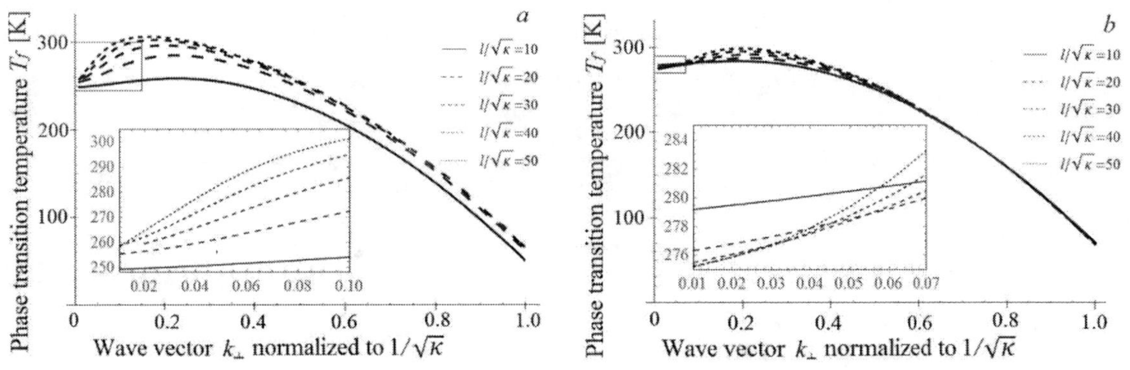

Figure 1. The dependences T_f on $k_{\perp}$ at different l : $a - \alpha_s = 0.75 \cdot 10^{-8}$ cm, $b - \alpha_s = -0.75 \cdot 10^{-8}$ cm.

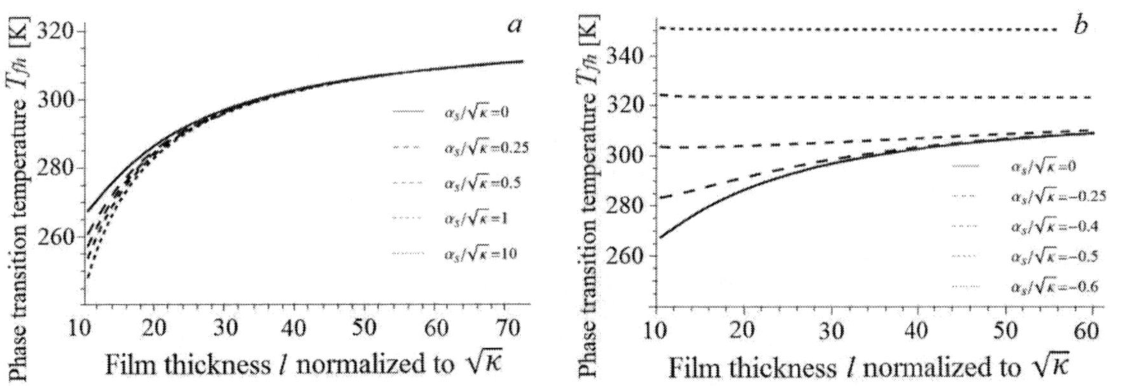

Figure 2. The dependences T_{fh} on l at different α_s : $a - \alpha_s > 0$, $b - \alpha_s < 0$.

It is shown, if $\alpha_s > 0$, then with increasing film thickness l the temperature $T_{fh}(l)$ increases and tends to T_C (figure 2a). The curve $T = T_{fh}(l)$ is clamped between the lines $T = T_f(l)$ and $T = T_C$, which are its asymptotes. The increase of $T_{fh}(l)$ with an enlarge l is explained by a decrease of the relative contributions of the correlation's and surface energies compared with other contributions to the free energy (1). The decrease of $T_{fh}(\alpha_s)$ with an enlarge α_s ($\alpha_s > 0$), naturally due to an increase in positive surface energy. Note that the dependence $T = T_{fh}(l)$ qualitatively coincides with the dependence of the PT temperature in a uniform state on the film thickness [18]. The dependence $T = T_{fh}(\alpha_s)$ be also traced in figure 2a: with increase of parameter α_s the PT temperature to nonuniform state deacreases, that is naturally, since in this case the surface works as a factor preventing the PT.

In the case the surface stimulates the transition to the FE phase, the parameter $\alpha_s < 0$ (figure 2b) and for small values of the parameter $|\alpha_s| < 1.2 \cdot 10^{-8}$ cm there is also a non-linear behavior of the function $T = T_{fh}(l)$, but its growth with increasing film thickness is weaker than with $\alpha_s > 0$, due to the negative contribution of the surface to the free energy (1). With decreasing parameter $\alpha_s < 0$ the influence of surface energy grows and already at $|\alpha_s| \geq 1.2 \cdot 10^{-8}$ cm the temperature T_{fh} practically does not depend on the film thickness (for the thicknesses used in the numerical calculations) – the surface tightens the PT into the up temperature region and fully controls the existence of the polar phase. The fundamental difference of behavior $T_{fh}(l)$ in this case is that $T_{fh} > T_f > T_C$ and deviation T_{fh} from the temperatures T_f, T_C increases with increasing parameter $|\alpha_s|$.

The width $\Delta T = T_{fh} - T_f(0)$ of the temperature existence of the nonuniform phase depends on l (figure 3). When $\alpha_s < 0$ (figure 3b) ΔT significantly lower than similar values for $\alpha_s > 0$ (figure 3a), what, apparently, to be related to the above features of the PT for this situation [19].

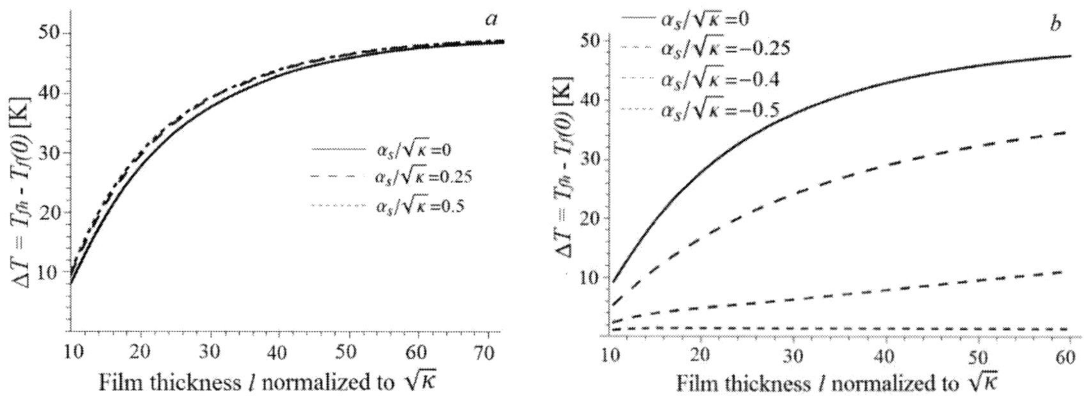

Figure 3. The dependences ΔT on l at different α_s : $a - \alpha_s > 0$, $b - \alpha_s < 0$.

2.4. Effect of shielding the surface charges

In a real situation, the environment of the film has a non-zero conductivity, and the larger it is, the stronger the surface charges are shielded. If we consider the characteristic shielding length L, then (14) and (15) will take the form

$$\tilde{\varphi}_e(k_\perp, z) = C_5\, e^{k_\perp z}\left(1 - e^{-2k_\perp (L+z)}\right). \tag{16}$$

$$\frac{\alpha_s + \kappa\lambda_1 \operatorname{th}(\lambda_1 l/2)}{\alpha_s + \kappa\lambda_2 \operatorname{th}(\lambda_2 l/2)} - \frac{\lambda_1\left(a\lambda_1^2 + b\right)\left(\lambda_1 + k_\perp \operatorname{th}(\lambda_1 l/2)\left(1 + e^{-2k_\perp L}\right)\varepsilon_e/\varepsilon\right) - 4\pi/\varepsilon}{\lambda_2\left(a\lambda_2^2 + b\right)\left(\lambda_2 + k_\perp \operatorname{th}(\lambda_2 l/2)\left(1 + e^{-2k_\perp L}\right)\varepsilon_e/\varepsilon\right) - 4\pi/\varepsilon} = 0. \tag{17}$$

The numerical solution of equation (17) is presented in figure 4. It shows that with increasing shielding (decreasing characteristic length L) the effect of the depolarizing field decreases and the PT temperature increases. Width ΔT in case $\alpha_s > 0$ about twice more (figure 4a), then in case $\alpha_s < 0$ (figure 4b). The shift of the temperature T_{fh} caused by electric fields in the second case is larger (2.6 K) than in the first one (2.3 K).

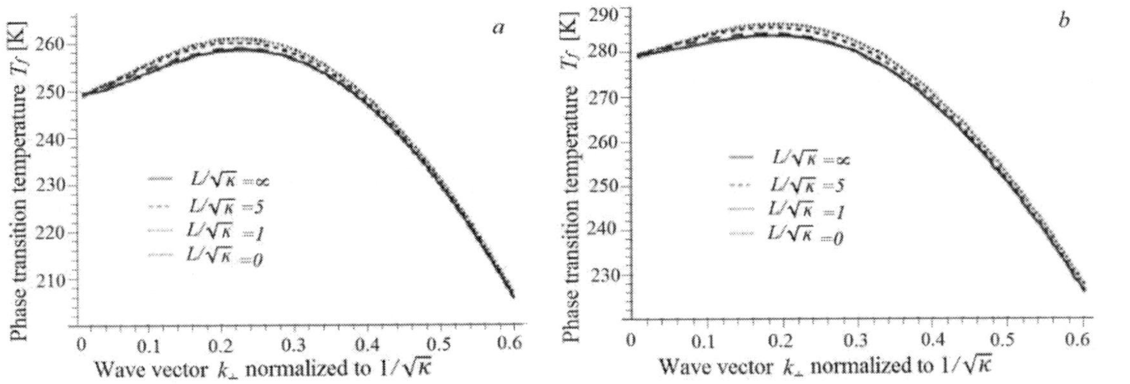

Figure 4. The dependences T_f on $k_\perp$ at different $L: a - \alpha_s = 0.75 \cdot 10^{-8}$ cm, $b - \alpha_s = -0.75 \cdot 10^{-8}$ cm.

3. Nonuniform polar states in FE nanoparticles

3.1. The system of relations, describing the PT

Consider the problem of the PT in the FE inclusion surrounded by a DE matrix. Let the FEP has the shape of a rotation ellipsoid with an elongation coefficient $\tau = b/a$ (b and a are major and minor semi-axes). DE matrix has the shape of a cube with c edge. The FE is uniaxial with the polar axis, parallel to the coordinate axis Oz and b. This problem can be solved numerically only. The system of equations for the FE-DE composite in an external electric field of a strength E_d similar to (2) – (3), except for the appearance E_d in the right-hand side of the first equation (2). The boundary conditions on the Γ surface of the FEP will take the form

$$\kappa \partial P/\partial \mathbf{n} - \alpha_s P|_\Gamma = 0,\ \varphi_i|_\Gamma = \varphi_e|_\Gamma,\ \varepsilon \partial \varphi_i/\partial \mathbf{n} - 4\pi P \cdot \mathbf{n}|_\Gamma = \varepsilon_e \partial \varphi_e/\partial \mathbf{n}|_\Gamma. \quad (18)$$

The boundary conditions, respectively, on the lower, upper, and lateral faces of the composite are

$$\varphi_e|_{D_1} = 0, D_1: \{|x|,|y| \le c/2, z = -c/2\};\ \varphi_e|_{D_2} = -E_d \cdot c, D_2: \{|x|,|y| \le c/2, z = c/2\};$$

$$\partial \varphi_e/\partial \mathbf{n}|_{D_3} = 0, D_3: \{x, y \le c/2, |z| = -c/2\}, \quad (19)$$

where $\mathbf{n}$ is the unit vector of the outer normal to the surface. The nonlinear boundary value problem (2), (3), (18), (19) has solved numerically by the finite element method in the software package Comsol Multiphysics.

3.2. Results analysis of the of numerical calculations

In figure 5 the polarization profiles in the FEP ($\tau = 5$) are shown. As in the case of the FEF (figure 1), for the FEP of size $a \le 30 \cdot 10^{-8}$ cm the PT occurs to polar phase in accordance with classical notions: at the definite temperature, that is different from T_c and depends on the FEP size, the polarization appears, which monotonously increases with distance from the PT point (figure 5a). For the FEP of size $a > 30 \cdot 10^{-8}$ cm the PT takes place to the polar nonuniform «domain-like» state, characterized by the regions with contrary polarization vector directions (figure 5b). Splitting into domains occurs in a direction perpendicular to the polar axis. In the direction of the polar axis, no splitting into domains is observed due to the increase in electrostatic energy. There is a central region (domain) with the maximum polarization at the center of the FEP decreasing in absolute value to the domain periphery. The dependence of the polarization on the coordinates, along with the behavior in external electric field, is one of the main differences between such regions and real domains. The next domain covers the central domain, and so on. The conditional boundaries of domains in the FEP are represented by concentric circles at the section by a plane perpendicular to the ferroactive axis. In each next region (domain) after the central one, the maximum polarization in absolute value is less, than that in the previous one.

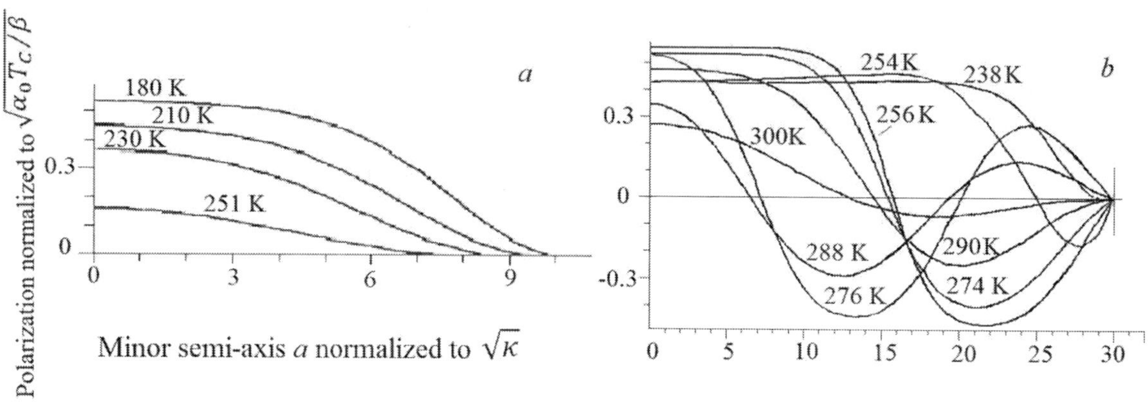

Figure 5. Distribution profiles P along a in the FEP with $\tau = 5$ at different temperatures:
a – $a = 3\text{nm}$, $a_s^{-1} = 0$; b – $a = 9\text{nm}$, $a_s^{-1} = 0$.

As mentioned above, the cause of the PT to nonuniform state is the same as the reason for splitting the infinite FE into domains: the tendency of the system to reduce a free energy by reducing the electrostatic energy of the depolarizing field. Firstly, this fact is proved by the strong dependence of the PT behavior on the size and shape of the FEP (figures 5a, b) [3]. Secondly, the direct proof of the role of the depolarizing field is the study of structural PT in which these is no the long-range electrostatic fields in the dielectric phase [20].

With an increase in the size of the FEP, the analysis of the temperature evolution of «domain-like» states becomes more complicated, the number of domains increases. Table 1 shows the temperature intervals for the existence of various polar states, which successively replace each other and for which different number of domains is characteristic.

Table 1. Temperature ranges of domains in the FE nanoparticle (K).

a (nm)	τ	a_s (nm)	T_f(K)	Number of domains in FE particle				
				1	2	4	3	2
3	5	∞	251	0-251				
6	5	∞	288	0-196	197-288			
6	3	∞	273	0-120	120-273			
9	5	∞	300	0-197	198-242	243-251	252-287	288-300
9	5	- 0.6	304	0-224	225-271	272-274	275-290	291-304

Figure 6 shows the polarization profiles for different a_s. In the general case, with decreasing $|a_s|$, the PT of the FE-DE system to the «monodomain» structure occurs through the several nonuniform states with different number of domains. For large absolute values of negative values a_s ($|a_s| \geq 6\text{nm}$ for $a = 9\text{nm}$, $T = 256\text{K}$ and $|a_s| \geq 1.5\text{nm}$ for $a = 9\text{nm}$, $T = 290\text{K}$) there is a deviation from the rule «from a large number of domains to a «monodomain» state with decreasing parameter $|a_s|$» (table 2).

Table 2. Influence of the parameter $|\alpha_s|$ on the domain structure (nm).

a (nm)	τ	T_f (K)	Number of domains in FE particle				
			1	2	4	3	2
6	5	240	≤0.75				0.75-3
9	5	256	≤0.3	0.33-1.5	3-3.75	≥6	
9	5	290	≤0.25	0.25-0.43		0.5-1.0	1.5-30

Note, when $|\alpha_s|$ decreases, the «multidomain» states, which the system undergoes by cooling from the dielectric phase, are repeated (figures 6a, b). For $a = 9\text{nm}$, $\tau = 5$ at $T = 290\text{K}$ failed to observe the polar state with four domains. It is possible due to the small interval of values $|\alpha_s|$.

The features of the polarization distribution, naturally, should influence the temperature dependences of the physical properties of researched materials. We shall discuss some possibilities of experimental observation of the manifestation of the nonuniform «domain-like» structure. At first, the new nonuniform state of the system should be reflected in its behavior in external electric field. The typical behavior of the temperature dependence of the dielectric permittivity for such systems is a sharp maximum at the PT temperature and its value strongly depends on the external field E_d, significantly decreasing with its growth. In the case of the PT in uniform state, the behavior of the dielectric permittivity differs radically (figure 7). In the temperature region of the nonuniform phase, there are several maxima, apparently related to the restructuring of the quasi-domain structure in a strong external field, and significant dielectric permittivity values between them, which is also associated with the presence of a nonuniform structure.

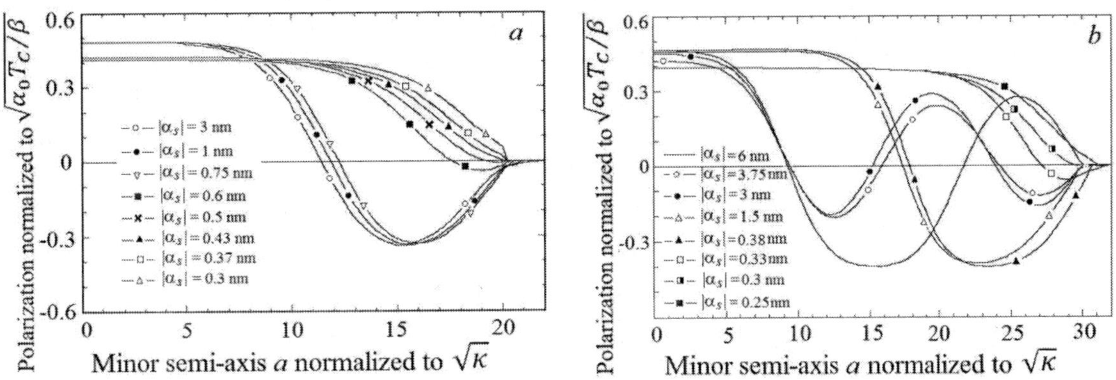

Figure 6. Distribution profiles P along a in the FEP ($\tau = 5$) for a different $|\alpha_s|$: a – $a = 6\text{nm}$, $T = 240\text{K}$; b – $a = 9\text{nm}$, $T = 256\text{K}$.

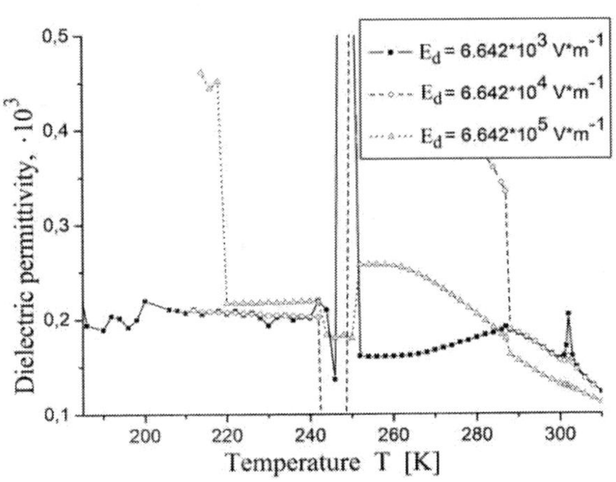

Figure 7. Temperature dependence of dielectric permittivity of the FEP ($a = 9\text{nm}$, $\tau = 5$, $\alpha_s^{-1} = 0$) for different E_d.

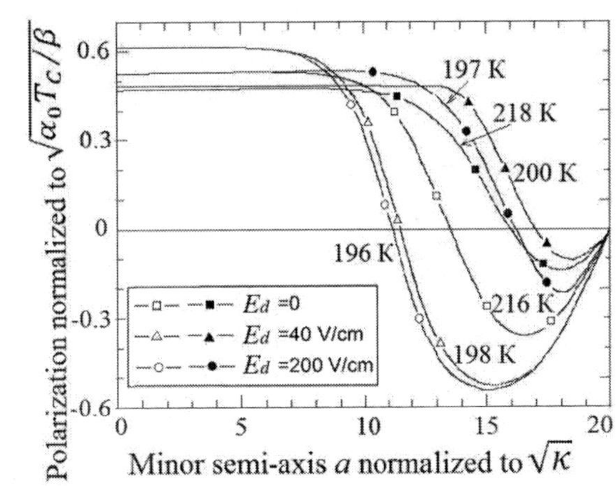

Figure 8. Distribution profiles P along a in the FEP ($a = 6\text{nm}$, $\tau = 5$, $\alpha_s^{-1} = 0$) for different E_d.

IOP Conf. Series: Materials Science and Engineering **525** (2019) 012009 doi:10.1088/1757-899X/525/1/012009

It is shown that the domain-like structure is strongly dependent on temperature and parameter α_s, but it is technically difficult to use these parameters for structure control. It is more convenient to control of the «nanodomain» structure with external electric field applied along the polar axis of the FEP (figure 8). With an increase in E_d, the region of the existence of the two-domain structure decreases, the transition to the single-domain state occurs at a higher temperature and a smaller absolute value of the polarization at the FEP center. In addition, at a direct transition from a two-domain structure to a single-domain structure, the width of the central domain in the presence of an electric field is larger, than that in the absence of a field.

3.3. Effect of screening of depolarizing field on the PT in the nanocomposite
The bound electric charges on the surface of the FEP are shielded to some extent. This effect can be consideration by imposing a boundary condition on the surface of the FEP with help a dielectric layer of thickness h. The thickness h of this layer plays the role of the shielding length, similar L in the FEF. Figure 9 shows the dependences of the polarization distribution in the FEP for different temperatures, h and boundary conditions. At $h < 3\text{nm}$ and $\alpha_S^{-1} = 0$, which corresponds to the Dirichlet boundary condition for P at the FEP surface, the PT takes place in the single-domain state at T = 306 K in according to the «classical type»; then with the temperature decreases the average polarization increases. At $h \geq 3\text{nm}$ and $\alpha_S^{-1} = 0$ the PT takes place to the nonuniform state, the temperature range ΔT of which expands with increasing h: $\Delta T = 304 - 302\text{K} = 2\text{K}$ at $h = 3\text{nm}$ (figure 9a) and $\Delta T = 302 - 292\text{K} = 10\text{K}$ at $h = 5\text{nm}$ (figure 9b). At $\alpha_S = -0.6\text{nm}$ and $h = 5\text{nm}$ (figure 9c), the PT also takes place to the FE phase on the «classical type».

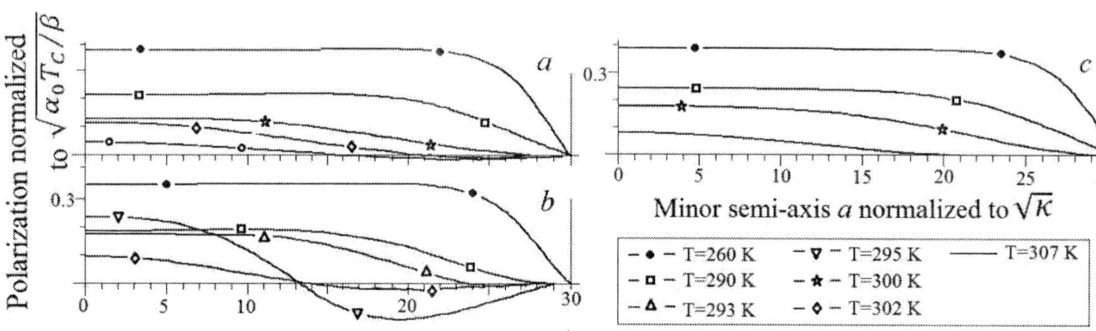

Figure 9. Distribution profiles P along a in the FEP ($a = 9\text{nm}$, $\tau = 5$) for various temperatures: $a - \alpha_S^{-1} = 0$, $h = 3\text{nm}$; $b - \alpha_S^{-1} = 0$, $h = 5\text{nm}$; $c - \alpha_S = -0.6\text{nm}$, $h = 5\text{nm}$.

The shielding of surface charges at the FE-DE interface leads to an increase in Curie temperature in the NCMs. Another mechanism of increasing the Curie temperature in the NCM due to the existence of an additional layer between the FE and the DE is considered in [21].

4. Conclusion

The temperature evolution of polar states can be investigated only in case of nonlinear approximation, which was done in this work with apply of numerical methods for an elliptical FE particle. Analysis of the results allows us to identify some general features typical of the PT in FEPs:

1. The presence of the following stages of the polarization profile evolution with distance from the PT point (if the particle parameter and the DE matrix parameters allow it):

a) the PT to the nonuniform state and the subsequent increase of polarization vector in domains with its contrary direction (it's possible fast restructuring of domains, changing its number with a changing of temperature);

b) growing of the central domain to the formation of a uniform single-domain state;

в) increase in the average polarization in the single-domain region with decreasing temperature.

2. Nonmonotonic oscillating behavior of the average polarization of unit volume as a function of temperature.

3. The existence of a certain temperature range for a nonuniform state near the PT point, the width of which depends on the shape and size of the FEPs, a concentration of the FEPs and their relative position [22], the parameters of the boundary between the FE and DE.

4. The conditions for the occurrence of a nonuniform state substantially depend on the existence in the material of free charge carriers that shield the electric field, arising at the PT.

The detected features of the PT will manifest themselves experimentally, for example, in the temperature anomalies of the heat capacity, the dielectric permittivity and the kinetic coefficients. In the case of massive samples, this effect will also appear as nonuniform states near the surface.

Note that the formation of the domain structure of the FE phase of the NCM, in addition to those considered in this article, should certainly be influenced by the following parameters. This is the volume concentration of the FE phase and its distribution in the NCM. In this case, it is necessary to consider the mutual influence of FEP on the formation of domains, taking into account changes of the depolarizing field. [22]. The preparation technology, the existence and concentration of defects, the cooling rate of FE material can also make a significant contribution to the formation of a «domain-like» structure. All this complicates the analysis of such states and is the subject of future research.

References

[1] Lallart M 2011 *Ferroelectrics – Applications* (London: InTechOpen) p 262
[2] Lallart M 2011 *Ferroelectrics – Material Aspects* (London: InTechOpen) p 530
[3] Nechaev V N, Viskovatykh A V 2015 *Phys. Stat. Sol.* **57** 704
[4] Nechaev V N, Viskovatykh A V 2018 *Russ. Phys. J.* **61** 1
[5] Nechaev V N, Shuba A V 2018 *Phys. Stat. Sol.* **60** 1322
[6] Strukov B A, Levanuk A P 1995 *Physical basis of ferroelectric phenomena in crystals* (Moscow: Nauka Phismatlit) p 229 (in Russian)
[7] Lahoche L, Luk'yanchuk I, Pascoli G 2008 *Integr. Ferroelectr.* **99** 60
[8] Glinchuk M D, Elisyeev E A, Stephanovich V A 2002 *Phys. B* **322** 356
[9] Morozovska A N, Eliseev E A, Glinchuk M D 2006 *Phys. Rev. B* **73** 214106
[10] Glinchuk M D, Rugulya A V 2010 *Nanoferroics* (Kiev: Naukova Dumka) p 313 (in Russian)
[11] Hua Z S, Tang M H, Wang J B, Zheng X J, Zhou Y C. 2008 *Phys. B* 3700
[12] Semenov V S 2012 *Thin film magnetic nanoelements. The structure of domain boundaries in thin magnetic films* (Moscow: Librocom) p 266 (in Russian)
[13] Belyaev B A, Tyurnev V V, Izotov A V Leksikov An A 2016 *Phys. Stat. Sol.* **58** 56 (in Russian)
[14] Landau L D, Lifshits E M 2005 *Theoretical physics* **8** (Moscow, Physmatlit) p 656 (in Russian)
[15] Landau L D, Lifshits E M 1976 *Theoretical physics* **5** (Moscow, Physmatlit) p 586 (in Russian)
[16] Smolensky G A, Bokov V A, Isupov V A, Krajnik N N, Pasynkov R E, Shur M S 1971 *Ferroelectrics and Antiferroelectrics* (Leningrad: Nauka) p 476 (in Russian)
[17] Tilley D R, Zeks B 1984 *Sol. State Com.* **49** 823
[18] Nechaev V N, Shuba A V 2016 *Ferroelectr.* **501** 32
[19] Lin S, Lu T, Cao W 2006 *Phys. Stat. Sol. (b)* **243** 2952
[20] Nechaev V N, Shuba A V, Viskovatykh A V 2015 *Russ. Phys. J.* **58** 711
[21] Nechaev V N, Viskovatykh A V 2016 *Phys. Stat. Sol.* **58** 2377
[22] Nechaev V N, Viskovatykh A V 2011 *Bull. Russ. Acad. Scien. Phys.* **75** 1378

IOP Conf. Series: Materials Science and Engineering **525** (2019) 012010 doi:10.1088/1757-899X/525/1/012010

Development of heat- and wear-resistant nanocomposite copper powder based material and technique of its obtaining used for plungers of die-casting machines

Y O Vladimirova and E P Shalunov

Joint Research and Development Laboratories of mechanically alloyed nanocomposite Materials, their Technology and Quality, Chuvash State University, Moscowskij Prospekt 15, Cheboksary, 428015, Russia

E-mail: yulivladimirova@mail.ru

Abstract. This paper presents the chemical composition, technology of manufacture and the basic properties of the nanocomposite Cu-Al-C-O copper powder based material for plungers of die - casting machines developed by the authors. The technology is based on the method of reactionary mechanical alloying in the attritor, which provides the formation of a dispersion-strengthened subgrain structure of the material containing strengthening particles γ-Al_2O_3 with an average size of 39±7 nm.
Its recrystallization temperature is 1000 ± 20 °C, and the hardness in the test temperature range of (20-800 °C) exceeds the hardness of standard bronzes, in particular, CuBe2NiTi, CuAl10Fe4Ni4 and CuAl10Fe3Mn2, widely used in plungers of such foundry equipment. The developed material has a low wear intensity $(3.20 \pm 0.85)\cdot 10^{-11}$ and, in contrast to the above bronzes, it can be used at an extremely high (up to 115 MPa) pressure between the plunger and the steel pressing chamber.

1. Introduction

Die-casting is one of the high-performance and high-precision methods of mold casting used for metal products made of aluminum, magnesium and copper alloys, as well as zinc, tin and lead and weighing from several grams to 50 kg [1, 2]. Application of the process in practice shows that the use of the new materials for die molds and other parts of the die-casting machine that come into the contact with the molten metal being pressed provides the increase in the quality of casting as well as its cost reduction [3, 4]. Plunger of a die-casting machine (Figure 1) is considered to be one of these critical and wear parts. It pushes a portion of the molten metal out of the pouring bush into the pressing chamber and then into the die casting mold.

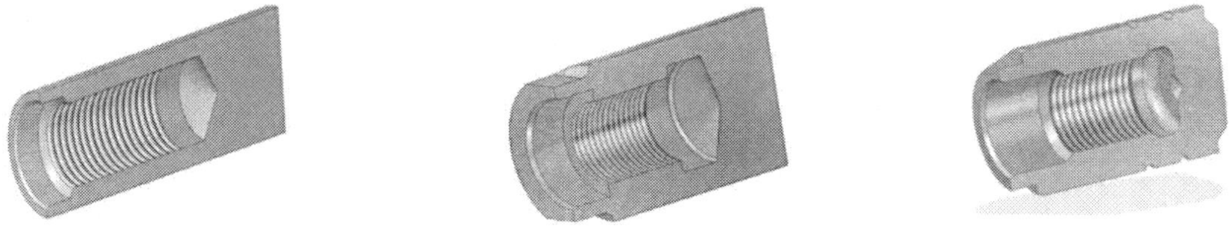

Figure 1. Types of plunger design for die-casting machine.

Along with this, moving inside the pressing chamber [1, 2, 5-7] the plunger simultaneously undergoes heavy specific loads, high temperature and significant friction forces. In particular, to obtain dense aluminum alloy castings, it is necessary to provide a pressing pressure of 35...85 MPa when the temperature of the molten metal reaches 720 °C [5]. The pressing pressure required to produce copper alloy castings increases up to 120 MPa (at a temperature of a molten metal, for example, brass - 960 °C) and in some cases it can reach 900 MPa [2, 6, 8, 9].

In case there is no «cold» space between the steel pressing chamber and the plate of a die-casting machine the contact pressure in the system «plunger-pressing chamber» can reach up to 250 MPa and if the «cold» space is 0.16 mm, the contact pressure falls down to 100 MPa. At such parameters of the contact pressure, according to [5], the coefficient of friction of the above tribosystem with a steel plunger is 0.39. Such a high value of the coefficient of friction indicates an extremely heavy interaction of tribosystem elements with each other that borders with possible seizure.

Based on the above investigation and the analysis of some published sources [2, 3-6, 10] we can make a conclusion that the promising material for a plunger must have high recrystallization temperature (more than 800 °C), high tensile strength (at least 800 MPa) and compressive resistance (at least 1000 MPa at standard temperature and at least 200 MPa at 800 °C), satisfactory thermal conductivity (at least 15% of the thermal conductivity of copper), low coefficient of linear thermal expansion (not more than $17 \cdot 10^{-6}$ 1/°C at 20 ... 100°C and C $23 \cdot 10^{-6}$ 1/°C at 700 – 800 °C), low wear intensity (not more than $5 \cdot 10^{-11}$). Along with this wear rate of the pressing chamber material must be much lower. Satisfying these requirements can provide the increase of the plunger service life by at least 1.5 times [11].

The absence of harmful impurities in the material and its complete technological reproducibility are also critical requirements.

Plungers are often made of various heat-resistant steels to be employed under these conditions. Since these steels have poor tribotechnical characteristics, the plungers made of them intensively wear out the pressing chamber of die-casting machines. At the same time, they wear out heavily themselves. To prevent wear of steel plungers in the source [12] it is proposed 1-3 rings made of ZrO_2-TZP based ceramic material and covered with Si-SiC material to be installed on them. In this case, the inner surface of a steel pressing chamber must be made of ZrO_2-TZP ceramic material containing TiO_2 and Al_2O_3 particles. MoS_2 , BN or graphite is recommended as a dry lubricant. Because of high cost of the casting tool of this type this engineering solution was not widely used.

Russian enterprises, rather often use gray cast-iron for example, EN GJL-240 brand [13] with a tensile strength of 240 MPa, for the manufacture of plungers for die-casting machines. Due to a small coefficient of linear thermal expansion, the plungers made of this cast-iron brand can work when the gaps are relatively small, providing a good sealing with the pressing chamber. However, due to the low thermal conductivity of cast-irons, the bottoms of the plungers made of them heat rapidly up to 350-400°C. Heating to such temperatures is highly undesirable, as it results in large abrasive wear of the plunger working part forming fatigue cracks and scuffing [14].

Copper alloys containing beryllium, cobalt, nickel, titanium, bronzes such as EN CuCo2Be (CW104), EN CuBe2NiTi, CuBe1,7NiTi etc. [4, 11] are most widely applied for the parts exploited under cyclic loads at high temperatures, including plungers for die-casting machines. They are highly resistant to destruction, wear and corrosion. Beryllium contained in these bronzes effectively reduces the coefficient of linear thermal expansion. Such materials are technological: they can be worked by pressure and do not cause serious problems when cut. The mechanical properties of beryllium bronzes can be compared with those of special steels: heat treatment increases their strength and they retain good thermal conductivity even at high temperatures. However, when the plungers made of these bronzes are heated up to 350-450 °C, intensive wear occurs [4, 5]. At the same time, this hard-to-find and expensive beryllium component makes the material brittle. High cost of beryllium bronzes, short supply, and toxicity of beryllium itself call for the search or development of alternative materials to replace the bronzes used for plungers of die-casting machines.

Aluminum-iron bronzes, in particular, bronzes such as CuAl10Fe4 EN, EN CuAl10Fe3Mn2 found wide application as materials for plungers of die-casting machines. Due to their composition, these bronzes have high mechanical and anti-friction properties and corrosion resistance. The oxide film formed on their surface allows these bronzes to resist corrosion in active wet or gas media. Alloying iron additives improve the structure of bronze alloys. They increase their wear resistance, strength and resistance to abrasion. In spite of the fact, that these materials do not contain environmentally harmful and expensive beryllium the properties of these materials are not inferior to those of beryllium bronzes mentioned above. Along with this EN CuAl10Fe4 and EN CuAl10Fe3Mn2 bronzes are among the least expensive and the most affordable brands, because they do not contain expensive tin either. However, with the temperature increase up to 350 °C and higher, hardness of these bronzes decreases significantly, and therefore the use of these materials for heat-stressed and abrasion subjected parts and, in particular, for die-casting machines plungers is not always effective.

The above proves that precipitation hardening technique used for producing the bronzes mentioned above does not provide the required degree of heat and wear –resistance. It can be explained by the fact that finely dispersed precipitation phases dissolve in the copper matrix at high temperatures.

Unlike the precipitation hardened alloys the dispersion- strengthened materials contain finely dispersed refractory particles (dispersoids) as strengthening phases that neither dissolve in the copper matrix nor react with it right up to their melt temperature [15].This phenomenon gives the highest strength to this type of materials at high temperatures. Moreover, the smaller the size of the dispersoids, the higher the recrystallization temperature [15].

The hardening coefficient (the ratio of the hardness of strengthened and non- strengthened copper) of copper dispersion- strengthened materials containing nanosized level dispersoids (less than 100 nm), is continuously increasing [15] with the temperature increase up to 0.90...0.95 of copper melt temperature. Besides the structure of these materials satisfies Georges Charpy` principle according to which antifriction materials must have ductile base with regularly arranged hard particles.

On the bases of above there are all the reasons to consider that it is this nanocomposite copper material with dispersion-strengthened structure that can be successfully used for plungers of die-casting machines as a promising material. The present article is devoted to development and obtaining of this exact material of the system Cu-Al-C-O.

2. Experimental materials and methods

To obtain Cu-Al-C-O nanocomposite copper material the initial mixture was prepared of electrolyte copper powder PMS-1 (at least 99.5 wt.% of copper, average particle size -38 μm), aluminium powder PP-1 (at least 97.0 wt.% of aluminium, 75 μm) an pencil lead powder GK-3(at least 92.0 wt.% of carbon, 40 μm). Carbon content varied depending on the quantity of aluminum in the powder mixture according to the following account: 1.0 wt.% fraction of aluminum should fit 0.1; 0.2; 0.3; 0.4 and 0.5 wt.% fraction of carbon, and aluminum content varied from 2.0 wt.% to 4.0 wt.%.

Reactionary mechanical alloying and certain operations of powder and granular metallurgy were used as methods for obtaining the required materials [16-18]. Reactionary mechanical alloying was performed in treating of 2 kg of powder mixture in the 15L attritor designed at the Chuvash State University. The mixture was worked for 1h in the chamber air medium at the atrittor rotor rotation frequency 600 min^{-1}. The granules obtained in the attritor were cold pressed at a pressure of 600 MPa into briquettes with of 25 mm to 95 mm in diameter. Then these briquettes were put into technological steel capsules filled wood carbonizer and heated at 850 °C for 30-90 min depending on the briquette diameter. At that temperature the capsules were transported one by one to the container of the hydraulic press heated up to 450 °C where the heated briquette was taken out of each capsule and put into the press container to be extruded through a tapered extrusion die (angle of taper 120° at a ring height of 3 mm) into a bar 13-52 mm in diameter.

These bars were used to produce some samples to be tested according to the standard techniques in order to determine physical and mechanical properties of the obtained materials. Fine structure of the material was examined by X-ray diffraction analysis and transmission electron microscopy applying

IOP Conf. Series: Materials Science and Engineering **525** (2019) 012010 doi:10.1088/1757-899X/525/1/012010

thin foils of this material and carbon replicas taken from its sections. Fine structure parameters (dispersoids, graines) were specified according to the technique mentioned in the source [19]. Methods of chemical analyses were also used.

Ware-resistance of the examined material was determined by friction test based on the system «pad-roller» using modernized friction machine SMZ-2, that provided contact pressure of 100 MPa and higher between the pad and the roller 100 mm in diameter made of hardened steel 42Cr4 with hardness of 52 HRC. The sliding speed of the roller over the pad was 5.3 m/s. Step loading of this simulative tribological system was performed with smooth increase in speed 0.03-0.09 MPa/s and specific load on the pad. Monitoring of the results of frictional torque and temperature of the pad was performed while testing. Load increment between the loading steps was 20 MPa. Testing was performed at every step of loading until the stable friction occurred or it was stopped in case the pad temperature was above 200 °C and / or some indication of system seizure appeared. For these reasons the testing was also stopped with load increasing. While rotating the roller was continuously lubricated with Isolat 128PP lubricant that is widely used for lubrication of plungers for die-casting machines.

Wear intensity of plunger material was determined as the ratio of linear wear of the pad to the length of its sliding path over the steel roller. Wear intensity of the roller material was determined as the ratio of the linear wear of the roller to the length of pad sliding path upon it. Along with this roller linear wear was determined by application of the well-known cratering method.

3. Test results and discussion

The test results of the Cu-Al-C-O material samples are shown in table 1.

Table 1. The properties of the tested Cu-Al-C-O system material.

Aluminum content in the powder mixture, wt.%	Carbon content in the powder mixture, wt. %	Tensile strength, MPa	Relative elongation, %	Hardness $HB_{5/250}$	Thermo-conductivity, % of copper termo-conductivity
2.0	0.0	772	5.5	206	22.9
	0.2	756	3.9	216	28.5
	0.4	767	3.1	225	25.3
	0.6	772	1.1	235	23.4
	0.8	838	1.7	243	19.6
	1.0	814	1.4	237	19.5
3.0	0.0	790	5.9	219	18.0
	0.3	816	3.5	230	22.9
	0.6	830	2.6	224	18.7
	0,9	836	2.5	238	17.4
	1.2[a]	–	–	254	15.6
	1.5[a]	–	–	244	15.4
4.0	0.0	782	6.2	223	16.4
	0.4	822	3.2	243	18.1
	0.8	850	2.0	260	15.2
	1.2	841	1.3	268	13,5
	1.6[a]	–	–	269	13.2
	2.0[a]	–	–	244	13.1

Note: [a] this carbon content did not allow testing the samples because of extremely high brittleness of the material.

Table1 shows that each given aluminum content has the best amount of carbon that gives the material highest thermoconductivity and hardness. In particular the maximum point of thermoconductivity occurs when the carbon content in the amount of 0.1 wt.% per 1.0 wt.% of aluminum, and the maximum hardness-at 0.4 wt.% of carbon per 1.0 wt.% of aluminum.

The reason for these maximum values formation as shown in [16-18, 20] is the reduction of copper oxides by carbon and aluminium. that occurred while working the powder mixture in the atrittor. Copper oxide occurs while working the powder mixture in the atrittor.

As it follows from table 2 carbon additives presented in the bars made of the tested material result in reduction of both oxygen content (caused by carbon reduction) and aluminum content spent for forming its own oxide as well as for reduction of copper from oxides. Moreover, the results of X-ray diffraction analysis show that aluminum content in the copper lattice stays unchangeable with the introduction of carbon.

Table 2. The results of chemical and X-ray diffraction analyses of the systems Cu-Al-O and Cu-Al-C-O materials.

Material system	Initial content in the powder mixture, wt.%		X-ray diffraction analysis		Chemical analysis	
	Al	C	Lattice period, Å	Aluminum content in the lattice, wt.%	Aluminum content out of the lattice, wt.%	Oxygen content in the material, wt.%
Cu-Al-O	0.5		3.6177	0.2	0.35	0.032
	1.0		3.6200	0.5	0.80	0.026
Cu-Al-C-O	0.5	0.15	3.6179	0.2	0.26	0.025
	0.5	0.25	3.6183	0.2	0.26	0.030
	1.0	0.30	3.6200	0.5	0.72	0.011
	3.0	0.90	3.6308	1.5	2.17	0.017

The analysis of the properties given in table1 shows that the material samples shown in table 3 possess the best combination of such characteristics as tensile strength, hardness, relative elongation (characterizes technological properties of the material – ductility) and thermoconductivity.

Table 3. The properties of the Cu-Al-C-O materials with different chemical properties.

Identity code of a material sample	Initial content in the powder mixture, wt.%		Tensile strength, MPa	Relative elongation, %	Hardness $HB_{5/250}$	Thermo-conductivity, % of copper thermo-conductivity
	Al	C				
CAG2/08	2.0	0.8	838	1.7	243	19.6
CAG 3/09	3.0	0.9	836	2.5	238	17.4
CAG 4/08	4.0	0.8	850	2.0	260	15.2

It goes without saying, that manufacturing a plunger of a glass shape (see figure 1) by turning the extruded bar will result in significant material loss. Therefore a work-piece for turning a plunger should resemble the final product in shape and in size as much as possible. Such a work-piece can be obtained by hot forming of the cylinder cut from the extruded bar. It`s clear that the higher the ductility of the material the easier the work-piece is hot formed.

Comparative analysis of the tested samples (see table 3) shows that the material with identity code CAG 3/09 possess the best ductility index. Because of the fact the further testing and investigations were performed over this material sample.

The performed tests showed that this material has an extremely high recrystallization temperature- 1000 ± 20 ° C, which is 0.92 of copper melt temperature and significantly exceeds all standard bronzes in this characteristic.

The analysis of the graphs (figure 2) shows, that the hardness the tested material is higher than that of the heat-resistant beryllium bronze EN CuBe2NiTi at the temperatures up to 420 °C and from 550 °C to 800 °C.

The hardness of the other two aluminum heat-resistant bronzes is lower than that of the tested material at the entire test temperature range. In this respect, the material CAG 3/09 can be classified as a heat-resistant material.

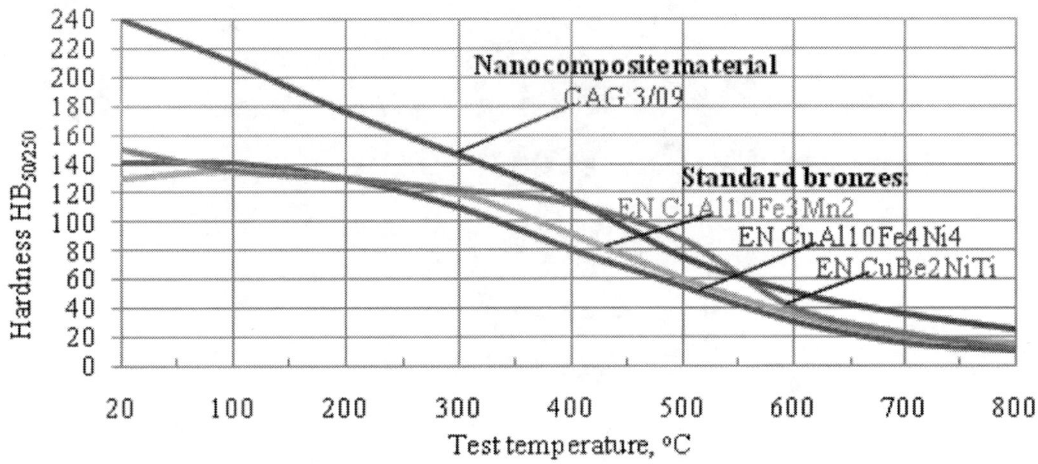

Figure 2. Diagrams of «hot» hardness of the tested material and standard heat-resistant bronzes depending on the test temperature.

High heat resistance of this material is primarily due to the peculiarities of its fine structure. The photo of the foil (figure 3) shows that the material CAG 3/09 has clear subgrain structure of the nanosized level, indicating the absence of mechanically alloyed copper recrystallization when its granules hot (850 °C) extruded into a bar.

The presence of the coresponding rings on a microelectronic diffraction patten taken of this foil, indicates that the fine structure of the material has a greater degree of polycrystallinity.

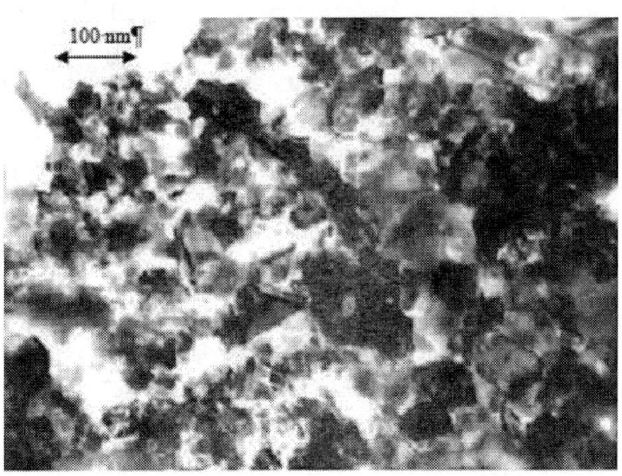

Figure 3. The photo of the copper composite material CAG 3/09 foil, taken by a transmission electron microscope, x57000.

The interpretation of foils microelectronic diffraction patterns and the analysis of X-ray diffraction patterns of this material anode deposit showed that the main strengthening phases in it is aluminium oxide γ-Al_2O_3, which was formed as a result of mechanochemical synthesis of aluminium powder and oxygen of the attritor chamber . Copper oxide is not seen neither in microelectronic diffraction patterns nor in X-ray diffraction patterns. This proves that the function of carbon in this system is to be a reducing agent for copper from its oxides. Diffraction rings and lines resulted from graphite also appear both in electronic diffraction patterns and in X-ray diffraction patterns.

Stereological analysis of replicas (figure 4) showed that the average particle size of the strengthening phases (γ-Al_2O_3) is 39±7 nm.

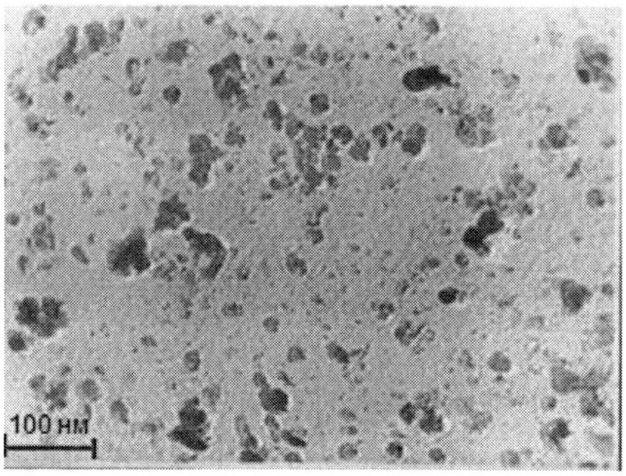

Figure 4. Tracks of inclusions (replica) in a copper composite material CAG 3 / 0.9, taken by a transmission electron microscope, x 20000.

The above features of the material fine structure and of its components parameters make it possible to specify the tested material as a nanocomposite material, and the presence of finely dispersed free (residual) carbon in the amount of 0.69 - 0.7 wt.% together with the above physico-mechanical properties, should also add good working properties to this material. This assumption was confirmed by the results of comparative tribological tests the main technological stages of which were given above. These tests showed that the phenomenon of beryllium bronze EN CuBe2NiTi pad seizure with a steel roller was observed when the contact pressure between the pad and the roller reached 85 MPa. For the pad of aluminum bronze EN CuAl10Fe4Ni4, the ultimate contact pressure was 90 MPa. During the test of the Cu-Al-C-O material pads, the first signs of seizure were recorded at a contact pressure of 115 - 120 MPa. Therefore, the tests were stoped at a contact pressure of 110 MPa. The linear wear of the pad and the steel roller were measured, and the path taken by the friction pair was also determined. The average values of the wear intensity of the tested material and the steel roller were $(3.20 \pm 0.85)\cdot 10^{-11}$ and $(1.14 \pm 0.67)\cdot 10^{-12}$ correspondently.

The obtained values of the ultimate contact pressure in a friction pair and the wear intensity of its elements give reasons for classifying the tested material as antifriction material for heavy-duty friction pairs.

4. Conclusions

Comparison of the obtained values of the physical-mechanical, technological and working properties of the developed nanocomposite material of the Cu-Al-CO system (CAG 3/09) with the requirements for promising materials for plungers of die casting machines given here allows us to recommend this material for foundry equipment parts and especially for plungers that undergo the high specific loads, high temperature and considerable friction forces simultaneously while moving inside the pressing chamber.

At the same time, the main reason to recommend the material for this field of application is that it is superior conventional heat-resistant beryllium and aluminum bronzes currently used for the plungers of die casting machines in heat resistance and wear resistance.

Acknowledgment
The authors of this paper thank the Russian Innovation Assistance Fund for financial support (competition UMNIK 17-12 (a), project No. 0038663) in carrying out the above studies.

References

[1] Bakemeyer H 2008 *Operating the die casting machine* Publication E-902 (North American Die Casting Association, Arlington Heights, Illinois) 116
[2] Becker M B 1990 *Pressure Die-Casting* Mashinostroenie, Moscow 400 (in Russian)
[3] Dańko R and Kowalczyk W 2015 *China Foundry* **12** (4) 305-309
[4] Murugarajan A and Raghunayagan P 2019 *Metalurgija* **58** 1-2 55-58
[5] Cheberyak O I 2017 *Proc. III Int. Conf. on Modern Technology in Engineering and the Foundry* (Cheboksary: Chuvash. State Uni.) 98-104 (in Russian)
[6] Bonollo F, Gramegna N and Timelli G 2015 *JOM* **67** 5 901-908
[7] Pasieka A and Konopka Z 2013 *Archives of Foundry Engineering* **13** 3 64 – 67
[8] Podprocká R, Malik J and Bolibruchová D 2016 *Manufacturing Technology* **15** 4 674-678
[9] Vinarcik E J 2003 High Integrity Die Casting Processes, 1st ed. (New York: Wiley) 3–25, 145–156
[10] Muhič M, Tušek J, Kosel F and Klobčar D 2010 *Strojniški vestnik – J. Mechan. Eng.* **56** 6351-356
[11] Vladimirova Y O and Shalunov E P 2018 *Proc. 4 Int. Forum on New Materials and Perspective Technologies* (Moscow) vol 1 (Moscow: Rus. Acad. Sci.) 106-110 (in Russian)
[12] Appl. EP 2620238A1, 26.01.2012
[13] Kolesnikov M S and Mukhametzyanova G F 2014 *Foundry Production* **08** 21–22 (in Russian)
[14] Mukhametzyanova G F, Kolesnikov M S, Mukhametzyanov I R and Astatshenko V I 2017 *IOP Conf. Ser.: Mater. Sci. Eng.* **240** 012053
[15] Suryanarayana C 2001 *Progress in Materials Science* **46** 1–184
[16] Dovydenkov V and Shalunov E 1998 *Proc. PM - 1998 World Congress* (Granada) vol 1 (Shrewsbury: EPMA) 372-376
[17] Dovydenkov V A, Simonov V A, Shalunov E P and Yarmolyk M V 2004 *Proc. PM-2004 World Congress (Vienna)* vol 1 (Shrewsbury: EPMA) 177-180
[18] Shalunov E P, Matrosov A L and Wendland S 2002 *Proc. Second Int. Conf. on Materials and Coatings for Performances: Investigations, Applications, Ecologically Safe Technologies for their Production and Utilization* (Katsiveli-town, Krimea) ed. V A Skorochod (Kiev: Institut Problems of Material Sciences NAS of Ukraine) 162-163
[19] Saltykov S A 1970 *Stereometric metallography* Metallurgia, Moscow 375 (in Russian)
[20] Shalunov E, Matrosov A and Chen L 2008 *Proc. Int. Congress & Exhibition on Powder Metallurgy – Euro PM – 2008* (Mannheim) vol 1 (Shrewsbury: EPMA) 113-119

IOP Conf. Series: Materials Science and Engineering **525** (2019) 012011 doi:10.1088/1757-899X/525/1/012011

Influence of heat treatment on the magnetic properties of Fe-28%Cr-7%Co-2%Mo-0.5%Si powder hard magnetic alloy

T A Vompe and I M Milyaev

A.A. Baikov Institute of Metallurgy and Materials Science, Russian Academy of Sciences, Leninskii pr. 49, 119334, Moscow, Russia

E-mail: tvompe@imet.ac.ru

Abstract. The hard magnetic alloy Fe-28%Cr-7%Co-2%Mo-0.5%Si (in wt. %) was prepared by the methods of powder metallurgy. The maximum magnetic properties on the alloy was received after the TMT temperature of 540 °C and was the following: residual induction B_r up to 1.3 T, coercive force H_c up to 40 kAm^{-1} and maximum energy product $(BH)_{max}$ up to 31 kJm^{-3}. The hardness of the alloy increases from 228 to 250 HV 30 in the temperature range of spinodal decomposition at 600 – 660 °C. The coefficient of thermal expansion of the alloy is $13{,}1*10^{-6}$ K^{-1}. It was found that the additional annealing of the alloy till 460 °C can increase the coercive force approximately for 6 kAm^{-1}.

1. Introduction

Permanent magnets based on the Fe-Cr-Co system are technologically important materials for electric power generation, transportation, power electronics and information technology [1]. They are widely used to manufacture permanent magnets due to their unique combination of magnetic and mechanical properties. Fe-Cr-Co magnetic materials have several advantages to be used as rotor ring for hysteresis motors [2]. The final cost of such magnets depends on amount of cobalt.

Fe-Cr-Co permanent magnet alloys achieve their properties after heat treatment due to decompose by a spinodal decomposition [3] of high temperature BCC α – solid solution into two isomorphic BCC α_1 and α_2 solid solutions. Besides, high magnetic α_1 phase is enriched by Fe and Co and weakly magnetic α_2 phase is enriched by Cr [4]. The significant changes in the microstructure and magnetic properties of Fe-Cr-Co system alloys occur during thermomagnetic treatment (TMT) [5]. Magnetic aging greatly improve the magnetic properties [6].

The production of the Fe-Cr-Co system for permanent magnets by the methods of powder metallurgy has advantages over the casting technology [7]. For example, the magnetic properties of the alloy can be increased due to the better compositional control offered by powder metallurgy processing as compared to melt practice [8].

Low cobalt Fe-Cr-Co alloys have better ductility and slower kinetics of phase transformations with comparisons with alloys with 15 % of cobalt and more. As cobalt content is reduced, the kinetics of spinodal decomposition slows down rapidly [9]. The spinodal temperatures of the low cobalt Fe-Cr-Co alloys are found to be higher than the Curie temperatures [10]. The important task is to investigate new low cobalt Fe-Cr-Co alloys for further commercial development.

In the present paper, we have investigated the influence of heat treatment on the magnetic properties of Fe-28%Cr-7%Co-2%Mo-0.5%Si powder hard magnetic alloy.

IOP Conf. Series: Materials Science and Engineering **525** (2019) 012011 doi:10.1088/1757-899X/525/1/012011

2. Materials and methods

2.1. Materials

The alloy Fe-28%Cr-7%Co-2%Mo-0.5%Si (in wt. %) was prepared by the methods of powder metallurgy. The commercial purity powders of iron, chrome, cobalt, molybdenum and silicon have been used.

2.2. Preparation of the samples

Particles of the powders were mixed in C 2.0 "Turbula" mixer (Russia). Single-ended dry pressing at 600 MPa in a zinc stearate lubricated hardened steel die on a manual press KNUTH-130042 (Germany) was used to compact the samples. The internal diameter of pressing die was 13,6 mm.

The density of the samples was determined by hydrostatic weighing. The density of raw samples was 78,8 – 79,1 %. Then compacted samples were sintered for 2.5 hours at the temperature of 1420 °C in a shaft furnace SshV-1,25/24-I (Russia) in 10^{-2} Pa vacuum. Finally, the sintered density of prepared samples was 98,6 – 99,1 %.

2.3. Characteristics of the samples

The samples of Ø12×20 mm in size were solution treated at the temperature of 1100 °C for 15 minutes, quenched in water, followed by the TMT (H = 4 kOe) and multiple step aging for various times. The TMT was carried out in the standard laboratory furnace equipped with armored electromagnet. The temperature was kept and monitored by proportional–integral–derivative controller. Standard laboratory muffle furnace equipped with the controller PROTAR 101 was used for heat treatment.

Magnetic hysteresis properties (residual induction B_r, coercive force H_c and maximum energy product $(BH)_{max}$) were measured by the Permagraph L EP-3 (Germany). The measurement error for coercive force H_c and residual induction B_r was 3 %, and it was 6 % for maximum energy product $(BH)_{max}$. The surface of sample was cleaned from slag using abrasive disk before measurements of the magnetic properties.

The samples were quenched at T_q = 1100 °C in water. The heat treatment was carried out by a two-stage continuous cooling starting from above the $\alpha_1+\alpha_2$ miscibility gap as shown schematically in figure 1. The initial cooling rates were equal 5 °C/h from T_2 = 580 °C and 3.4 °C/h from 540 °C. The second cooling rates were 2.3 °C/h and 3.6 °C/h from T_3 = 500 °C. After aging, there was also additional annealing till 460 °C to increase the magnetic properties. After that all the samples were air cooled from 460 °C to room temperature.

The dilatometric measurements were performed using a dilatometer NETZCH DIL 402 C72G (Germany). The samples were heated to the temperature 1300 °C and cooled to room temperature at the rate of 10 °C/sec with constant argon blow at the rate of 75 ml/min. The coefficient of thermal expansion (CTE) of the Fe-28%Cr-7%Co-2%Mo-0.5%Si alloy was also determined.

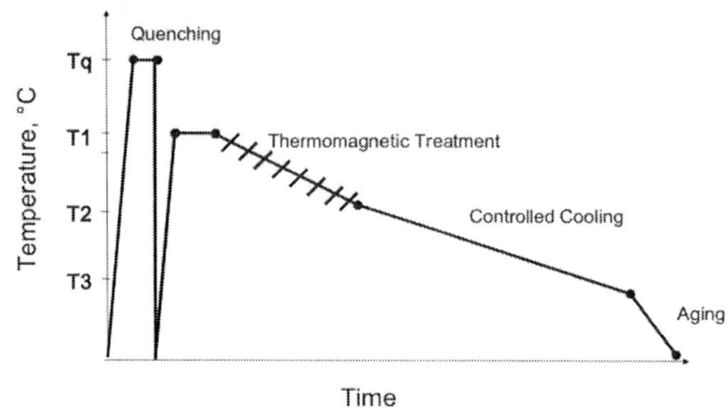

Figure 1. The heat treatment scheme for Fe-28%Cr-7%Co-2%Mo-0.5%Si alloy.

3. Results and discussion

3.1. Thermal expansion behavior of the alloy

Figure 2 shows the thermal expansion curve during heating and cooling of the Fe-28%Cr-7%Co-2%Mo-0.5%Si alloy at temperatures between 500 and 700 °C.

According to the phase diagrams and experimental data the phase transformation of $\alpha \rightarrow \alpha_1 + \alpha_2$ would be expected at the temperature rage 620 – 640 °C. The TMT parameters were selected based on this temperatures.

The average CTE of the alloy in the temperature between 20 °C and 900 °C is $13,1*10^{-6}$ K^{-1}.

Figure 2. Thermal expansion of the Fe-28%Cr-7%Co-2%Mo-0.5%Si alloy.

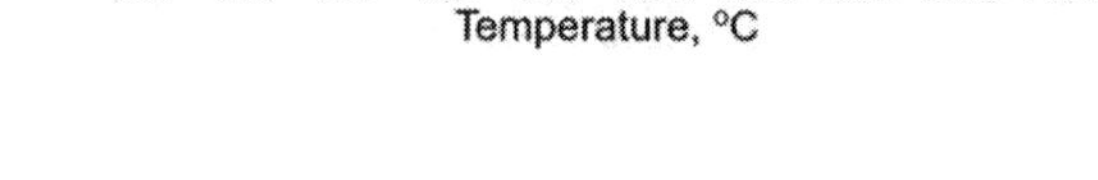

Figure 3. Temperature influence on the hardness of the Fe-28%Cr-7%Co-2%Mo-0.5%Si after 30 min annealing.

3.2. Hardness of the alloy after spinodal composition

The effect of 30 min annealing in the temperature range of 550 – 1300 °C on hardness of the alloy is shown at the figure 3. The hardness increases from 228 to 250 HV 30 in the temperature range of spinodal decomposition at 600-660 °C. In the temperatures from 900 °C to 1100 °C the hardness is approximately constant and after annealing at 660°C it is similar like after quenching.

3.3. Magnetic properties and additional annealing

Experimental data in table 1 shows magnetic properties that were received after different heat treatment regimes.

Table 1. Magnetic properties of the Fe-28%Cr-7%Co-2%Mo-0.5%Si alloy.

TMT	*Heat treatment regimes*		B_r, *T*	H_c, kAm^{-1}	$(BH)_{max}$, kJm^{-3}
A	660 → 580 °C 5 °C/h Field (4kOe)	580 → 500 °C 3.6 °C/h No field	1.22	28.4	18.6
B	660 → 540 °C 3.4 °C/h Field (4kOe)	540 → 500 °C 2.3 °C/h No field	1.24	34.0	26.9

The changing of the temperature at the first step to 540 °C and decreasing of the rate of the cooling to 3.4 °C/h leads to the increase of the magnetic properties of the alloy.

It was found that the additional annealing after 500 °C till 460 °C can increase the magnetic properties of the alloy. The temperature influence on the coercive force H_c of the Fe-28%Cr-7%Co-2%Mo-0.5%Si alloy after 20 hours annealing is shown at the figure 4 and 5.

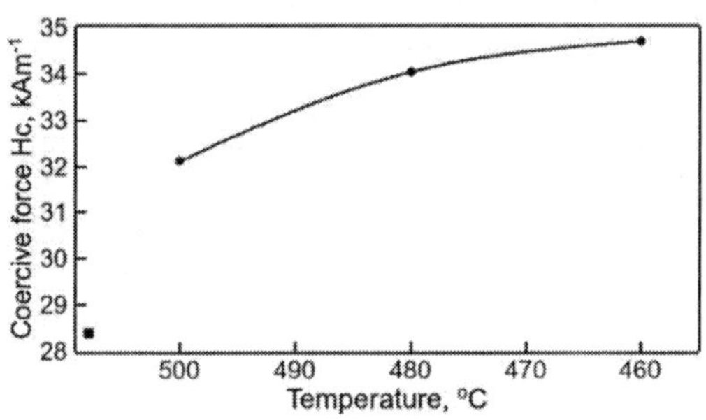

Figure 4. Temperature influence on the coercive force of the Fe-28%Cr-7%Co-2%Mo-0.5%Si alloy after A heat treatment and annealing.

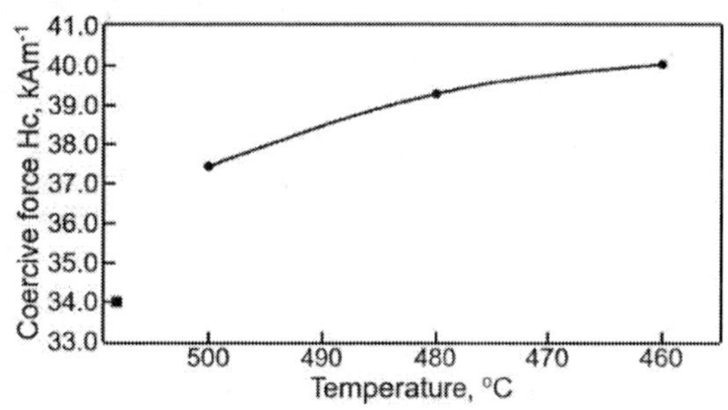

Figure 5. Temperature influence on the coercive force of the Fe-28%Cr-7%Co-2%Mo-0.5%Si alloy after B heat treatment and annealing.

Due to slower kinetics of phase transformations in low cobalt Fe-Cr-Co alloy the additional annealing was used to increase the value of the coercive force.

The curves of magnetization and demagnetization for Fe-28%Cr-7%Co-2%Mo-0.5%Si alloy after all heat treatments are shown in figure 6.

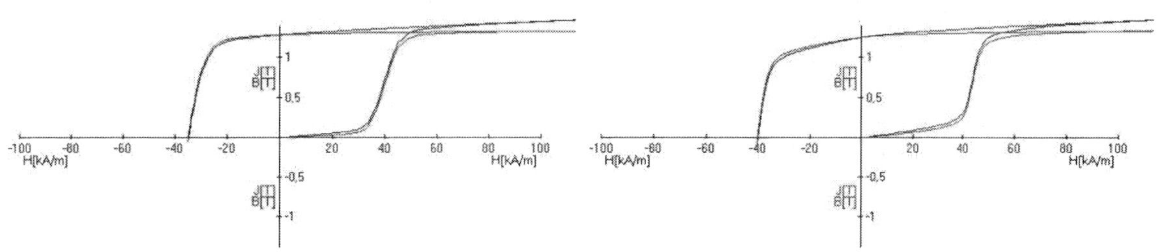

Figure 6. The curves of magnetization and demagnetization for Fe-28%Cr-7%Co-2%Mo-0.5%Si alloy after (a) – A heat treatment and additional annealing, (b) – B heat treatment and additional annealing.

The hysteresis loop squareness is about 62 % and was calculated by formula $\eta = (BH)_{max}/(B_r*H_c)$.

IOP Conf. Series: Materials Science and Engineering **525** (2019) 012011 doi:10.1088/1757-899X/525/1/012011

The following maximum magnetic properties can be obtained on the alloy: residual induction B_r up to 1.3 T, coercive force H_c up to 40 kAm^{-1} and maximum energy product $(BH)_{max}$ up to 31 kJm^{-3}.

4. Conclusion

The powder Fe-28%Cr-7%Co-2%Mo-0.5%Si hard magnetic alloy was investigated. The average coefficient of thermal expansion of the alloy is $13{,}1*10^{-6}$ K^{-1} in the temperature range between 20 °C and 900 °C, which is well-matched to that of the commercial Fe-Cr-Co alloy with 15 wt. % of cobalt.

The hardness of the alloy increases from 228 to 250 HV 30 in the temperature range of spinodal decomposition at 600 – 660 °C.

The maximum magnetic properties on the alloy was received after the TMT temperature of 540 °C and was the following: residual induction B_r up to 1.3 T, coercive force H_c up to 40 kAm^{-1} and maximum energy product $(BH)_{max}$ up to 31 kJm^{-3}.

It was found that the additional annealing of the alloy till 460 °C can increase the coercive force approximately for 6 kAm^{-1}.

Due to the level of the magnetic properties the studied powder magnetic alloy can be used as a material for permanent magnets.

Acknowledgments

The authors would like to thank V. A. Zelenskii and A. B. Ankudinov for the help in the receiving powder samples, D.D. Titov and A.S. Lusenkov for the help in studying thermal expansion.

The investigation of magnetic properties was carried out according to the State Task № 075-00746-19-00. Sintering regimes were chosen in accordance with the results of the study supported by RFBR project № 18-03-00666.

References

[1] H. Lewis L and Jimenez-Villacorta F 2013 Perspectives on Permanent Magnetic Materials for Energy Conversion and Power Generation *Metall. Mater. Trans. A* **44–Suppl.** 2

[2] Kubota T, Member A, Wakui G and Itagaki M 1998 Hysteresis motor using magnetically anisotropic Fe-Cr-Co magnet *IEEE Trans. Magn.* **34** 3888–96

[3] Cahn J W 1963 Magnetic Aging of Spinodal Alloys *J. Appl. Phys.* **34** 3581–6

[4] Kaneko H, Homma M and Nakamura K 1971 New Ductile Permanent Magnet of Fe-Cr-Co System *AIP Conf. Proceedings. Magn. Magn. Mater.* 1088–92

[5] Sugimoto S, Honda J, Ohtani Y, Okada M and Homma M 1987 Improvements of the magnetic properties of equiaxed Fe-Cr-Co-Mo hard magnets by two-step thermomagnetic treatment *IEEE Trans. Magn.* **MAG-23** 3193–5

[6] Mayo W E 1986 Non-linear theory of magnetic aging in spinodal systems *Scr. Metall.* **20** 1251–6

[7] Ustyukhin A S, Ankudinov A B, Zelenskii V A, Milyaev I M, Ashmarin A A and Alymov M I 2018 Phase Composition of Powder Hard Magnetic Fe–30Cr–(8–24)Co Alloys Sintered under Different Temperature Conditions *Dokl. Phys. Chem.* **482** 140–4

[8] Green M L, Sherwood R C and Wong C C 1982 Powder metallurgy processing of CrCoFe permanent magnet alloys containing 5–25 wt. % Co *J. Appl. Phys.* **53** 2398–400

[9] Milyaev I M, Milyaev A I and Yusupov V S 2009 Mechanism of formation of the high-coercivity state in nanostructured hard magnetic Fe-Cr-Co and Fe-Ni-Al-Co-Cu alloys *Russ. Metall.* **2009** 250–2

[10] Jin S and Gayle N 1980 Low-cobalt Cr-Co-Fe magnet alloys obtained by slow cooling under magnetic field *IEEE Trans. Magn.* **16** 526–9

IOP Conf. Series: Materials Science and Engineering **525** (2019) 012012 doi:10.1088/1757-899X/525/1/012012

Gold nanorods and a nanocomposite material based on them: analytical possibilities for spectrophotometric determination of total catecholamines

M V Gorbunova, A O Shlenova, R V Klimenko, S V Gutorova, V V Apyari and S G Dmitrienko

Department of Chemistry, Lomonosov Moscow State University, Leninskie gory, 1/3, Moscow 119991, Russia

E-mail: masha13_1992@mail.ru

Abstract. Gold nanorods and nanocomposite material based on them were synthesized and used as spectrophotometric reagents for determination of catecholamines. Approaches to the determination of total catecholamines content are discussed. Use of radar charts for evaluation of the error of catecholamines determination and for selection of the appropriate standard substance is proposed. Developed spectrophotometric techniques were used for determination of total catecholamines in a model mixture, urea samples and diluted blood serum. Accuracy of the techniques was proved by the standard addition method and by results of HPLC analysis.

1. Introduction

In recent years, there has been a steady tendency to study nanoobjects of non-spherical shape, for example, gold nanorods (AuNRs) characterized by several modes and several bands of surface plasmon resonance (SPR). The existing correlation between the composition, geometrical dimensions of AuNRs and their SPR-spectra provides a possibility of developing methods for spectrophotometric determination of various compounds that cause changes in these parameters. Thus, it was established that catecholamines reduce silver ions in the presence of gold nanorods in glycine buffer solution medium [1, 2]. Metallic silver covers longitudinal faces of AuNRs, changing particle shape. As a result, a hypsochromic shift of both maxima in the absorption spectrum of AuNRs is observed. This is a basis for spectrophotometric determination of this class of compounds.

As far as deposition of nanoparticles on a solid matrix contributes to their stabilization and, in some cases, improves their ease of use, synthesis and study of nanocomposite materials is also promising. Among various matrices for nanocomposite materials, polyurethane foam (PUF) has proven itself as a material of high strength, low cost and ease of use [3, 4].

Catecholamines (CAs) perform various vital functions in humans and animals: regulating stress, psychomotor activity, emotions, learning, sleep, memory [5, 6]. Epinephrine, norepinephrine and dopamine are of particular importance as far as these substances are directly produced by organisms. Abnormal concentration of these substances in the human body could serve as a marker of various diseases. Since biological objects, as a rule, contain mixture of catecholamines, the development of techniques for determining their total content is very topical.

Determination of CAs is mostly carried out by high-performance liquid chromatography with an electrochemical detector, fluorescence and chemiluminescence methods, mass spectrometry. A combined method for determination of CAs by HPLC or capillary electrophoresis in tandem with mass-spectrometric detection is widely used [7, 8]. It is advisable to develop methods that allow to increase analysis speed and reduce its cost; spectrophotometric methods meet these criteria. The analysis of literature data indicates that in most cases the detection limits of CAs using the spectrophotometric method are 0.1-1 μg mL^{-1}, which is insufficient for biological fluids. Use of metal nanoparticles instead of "classical" spectrophotometric reagents leads to significant decrease in the detection limits down to ~ 0.01 μg mL^{-1} [9]. Therefore, study of nanoparticles in relation to solving problems of the spectrophotometric determination of catecholamines seems relevant.

2. Experimental

2.1. Materials

Hydrogen tetrachloroaurate (chemically pure) and silver nitrate (pure) were used for preparation of AuNRs. Sodium borohydride (analytical grade) and ascorbic acid (chemically pure) were used as reducing agents, cetyltrimetylammonium bromide (CTAB, analytical grade) was used as a stabilizer of AuNRs. Epinephrine, norepinephrine hydrochloride and dopamine hydrochloride (all of analytical grade) were used as subjects of the study (figure 1). The substances stock solutions were prepared by dissolving their weighed portions in deionized water. Epinephrine was dissolved in 0.01 mol L^{-1} HCl. The stock solutions were frozen and kept in a refrigerator. Working 0.2 mmol L^{-1} solutions were prepared from the stock solutions immediately before use. Glycine buffer (pH 9.5) was prepared by addition of 74.8 mL of 0.1 mol L^{-1} glycine (7.51 g of glycine and 5.85 g of NaCl in 1 L) to 25.2 mL of 0.1 mol L^{-1} NaOH. Phosphoric acid (H_3PO_4), octane-1-sulfonic acid sodium salt, acetonitrile (all of analytical grade) were used for preparation of chromatographic eluent.

Commercially available polyether-type polyurethane foam (PUF) was used as a matrix for nanocomposites. Tablets of PUF of 16 mm in diameter (20 ± 2 mg) were cut with a metal punch from an industrial sheet of the polymer. To remove impurities, the PUF tablets were placed in acetone and shaken for 10 minutes. The procedure was repeated twice, after which the tablets were dried under stream of air.

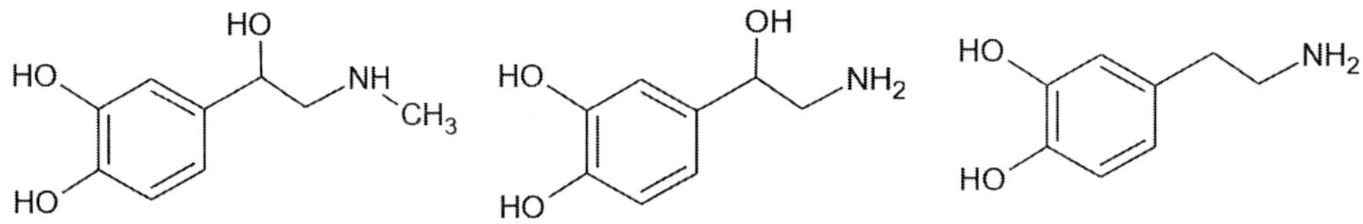

epinephrine norepinephrine dopamine

Figure 1. Molecular structures of the studied catecholamines.

2.2. Instruments

Absorption spectra of solutions were recorded by SF-103 spectrophotometer (Akvilon, Russia). Diffuse reflection in the visible region was recorded using an Eye-One Pro mini-spectrophotometer (X-Rite, USA). Chromatographic determination of catecholamines was performed using a Tsvett Yauza chromatograph (NPO Khimavtomatika, Russia) with an amperometric detector (E = +0.8 V). The chromatographic column "Eclipse XDB-C18" (Agilent) and guard column Security Guard C18 were used. Eluent contained acetonitrile and 0.1% H_3PO_4 in water (10:90, v/v) with addition of sodium 1-octanesulfonate (0.3 mmol L^{-1}). The flow rate was 0.4 mL min^{-1}. The eluent was degassed in a Bransonic 1510R-DTH ultrasonic bath (USA). Sample volume was 20 μL, injection was performed with a loop. Deionized water was obtained from the purification system Millipore Simplicity (Millipore, Germany). A mechanical shaker and a magnetic stirrer were also used.

2.3. Synthesis of AuNRs and the nanocomposite

AuNRs were synthesized according to the method previously reported by Nikoobakht and El-Sayed [10] with some modifications. A detailed scheme of synthesis is described in [2].

The nanocomposite material based on polyurethane foam and AuNRs was obtained by sorption modification of polyurethane foam [4]. Samples of the nanocomposite were modified with silver nitrate. To do this, the nanocomposite was placed into 5 mL of 0.001 mol L^{-1} $AgNO_3$, pressed with a glass rod, shaken for 10 minutes and dried between sheets of filter paper.

2.4. General procedure

To study the interaction of colloidal solutions of AuNRs with catecholamines, 0.3 mL of AuNRs, 20 μL of 0.01 mol L^{-1} $AgNO_3$, 0.1 mL of glycine buffer solution and 0 – 0.5 mL of 25 μmol L^{-1} catecholamines solution were added into a test-tube. The total volume of all components was 2.5 mL. The absorption spectra of the obtained mixtures were measured after 3 minutes.

Interaction of the nanocomposite modified with silver nitrate with catecholamines was studied by placing nanocomposite into 5 mL of solution containing 1 – 100 μmol L^{-1} catecholamines and 1 mL of glycine buffer solution. It was shaken on a mechanical shaker for 30 minutes. Then the nanocomposite was removed and dried between sheets of filter paper. The diffuse reflection coefficients (*R*) of the obtained nanocomposite were measured in a range of 400 – 730 nm with the 10 nm increment. Then, the values of the Kubelka-Munk function (*F*) were calculated from the formula: $F = \frac{(1-R)^2}{2R}$, here *R* is the diffuse reflection coefficient. The plotted spectra *F* versus λ were used to estimate changes in state of AuNRs in the phase of polyurethane foam.

3. Results and discussion

3.1. Interaction of catecholamines with $AgNO_3$ in AuNRs colloidal solution and AuNRs nanocomposite

The interaction of catecholamines with silver ions in the presence of AuNRs or the nanocomposite is accompanied by a hypsochromic shift of the SPR bands and an increase of their intensity (figure 2). These effects are visually detected as change of color of the AuNRs solution from pale purple to bright green and color of the nanocomposite from pale purple to grey. The shift of the maxima (Δλ) can be considered as a characteristic of completeness of the interaction.

A supposed scheme of the interaction includes oxidation of CAs catechol group by silver ions into ortho-quinonoid structure and subsequent deposition of metallic silver onto the surface of AuNRs (figure 3).

Dependencies of the shift of maxima on concentration of individual epinephrine, norepinephrine, dopamine and their equimolar mixture are represented in figure 4. Initial parts of these dependencies can be used as calibration graphs for determination of CAs. Some analytical characteristics of CA determination are listed in table 1. Determination of catecholamines with the use of colloidal solution of AuNRs is characterized by higher sensitivity in comparison with the use of nanocomposite material. Moreover, lower volumes of the analyzing samples are required. Nevertheless, nanocomposite material is characterized by higher selectivity to determination of CAs in the presence of inorganic ions that could be caused by its matrix influence. Ease of use could also serve as an argument for determination of CAs by nanocomposite material.

IOP Conf. Series: Materials Science and Engineering **525** (2019) 012012 doi:10.1088/1757-899X/525/1/012012

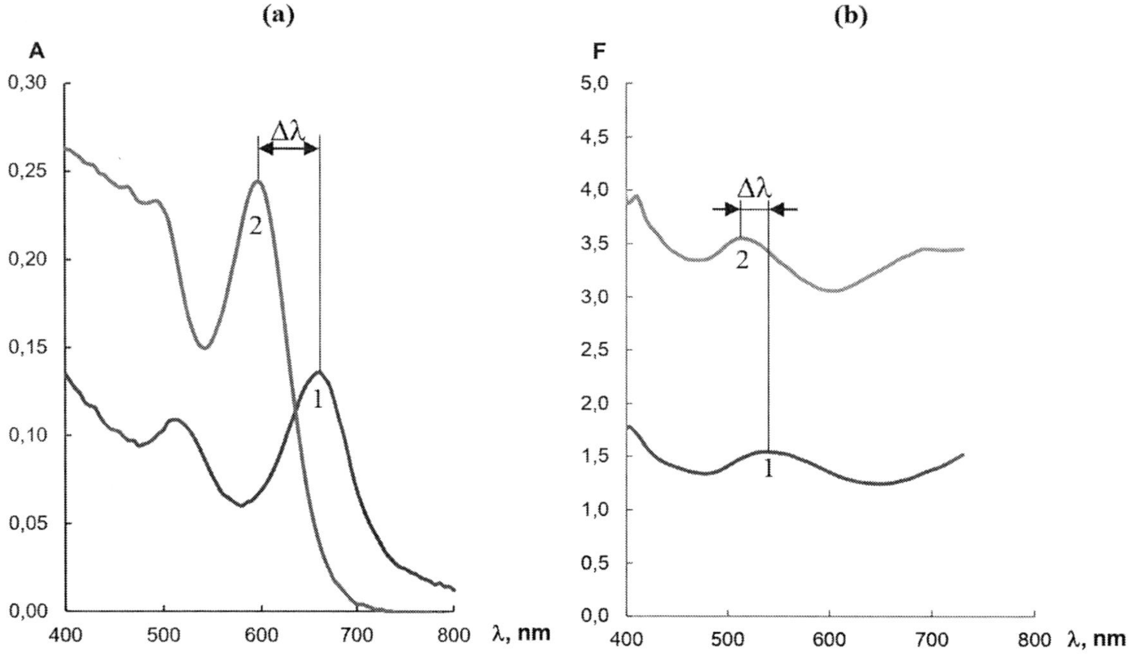

Figure 2. Absorption (a) and diffuse reflectance (b) spectra of AuNRs before (1) and after (2) interaction with catecholamines.

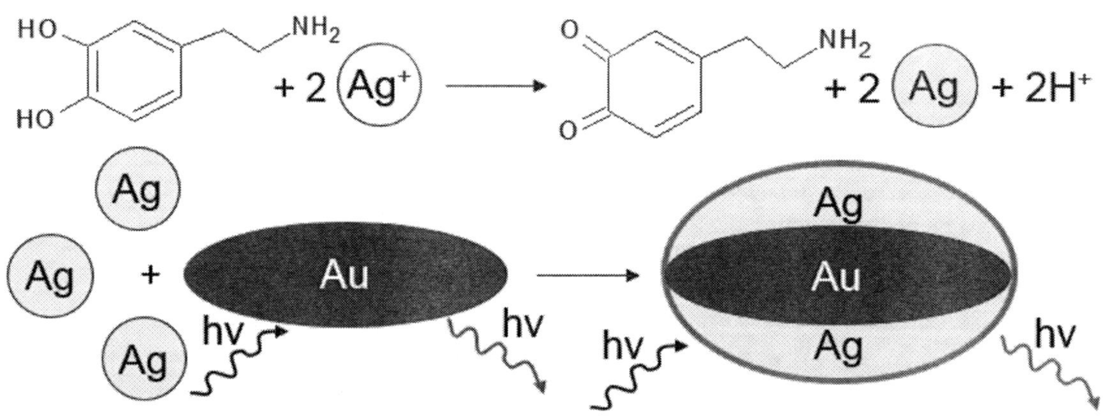

Figure 3. A supposed scheme of interaction.

IOP Conf. Series: Materials Science and Engineering **525** (2019) 012012 doi:10.1088/1757-899X/525/1/012012

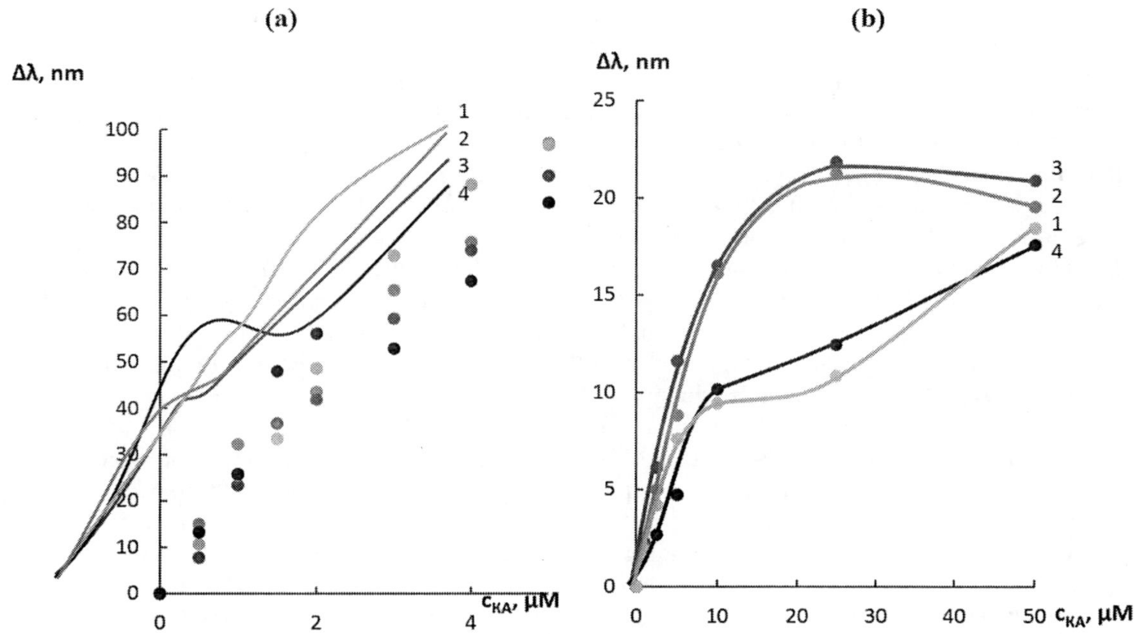

Figure 4. Dependencies of the shift of maxima of colloidal solution of AuNRs (a) and nanocomposite material (b) on concentration of individual epinephrine (1), dopamine (2), norepinephrine (3) and their equimolar mixture (4).

Table 1. Analytical characteristics of the determination of catecholamines (n = 3, P = 0.95).

		epinephrine	norepinephrine	dopamine	equimolar mixture
Using the colloidal solution of AuNRs	Calibration graph equation	y = 23.2x	y = 23.5x	y = 31.8x	y = 26.0x
	LOD, μg mL^{-1}	0.02	0.01	0.02	0.02
	Linear range, μg mL^{-1}	0.05 – 0.27	0.04 – 0.15	0.05 – 0.25	0.05 – 0.25
Using the nanocomposite	Calibration graph equation	y = 1.39x	y = 1.82x	y = 1.66x	y = 1.55x
	LOD, μg mL^{-1}	0.07	0.05	0.05	0.05
	Linear range, μg mL^{-1}	0.2 – 1.8	0.1 – 1.5	0.1 – 1.7	0.1 – 1.7

3.2. Determination of total catecholamines

In the case of some diseases progressions, a sharp change in the content of individual catecholamines is observed. Thus, in blood samples of patients with Alzheimer's disease, 10-fold decrease in the content of dopamine may be observed [11]. Biological fluids of patients with Addison's disease are characterized by 10-fold decrease in adrenaline levels [12]. For patients with pheochromocytoma, on the contrary, 10-fold rise of adrenaline production is peculiar [13]. Such changes in the content of individual catecholamines affect the overall content of catecholamines.

Since the sensitivity coefficients in the calibration graph equations for catecholamines, both in the case of colloidal solution of nanorods and nanocomposite, differ slightly (table 1), it is possible to use them to estimate the total content of catecholamines. Epinephrine, norepinephrine, dopamine, and their equimolar mixture were considered as possible standards for evaluation of CAs total content. Several mixtures containing these CAs in various proportions were analysed (tables A1, A2). The total content of catecholamines was calculated according to the equations of the calibration graphs represented in table 1 and compared with the added quantity.

In this paper, plotting a radar chart is proposed as a technique for estimation of the error in determining total content of CAs and selection of a standard system, in terms of which the total content should be expressed to minimize the error. The radar chart displays the module of relative error in the analysis of model mixtures. In this case, the mixtures were in order of increasing error by one of the standard substances (equimolar mixture). The resulting radar charts are shown in figure 5. Errors of the analysis are minimal for a standard (or standards) giving the smallest area of the diagram.

The advantages of the radar chart are ease of data interpretation and selection of the best standard; ease of estimating the minimum and the maximum error in determining sum of analytes within the studied samples (the closest and the furthest vertex of the obtained polyhedron) and the percentage of mixtures that can be analysed within the specified error (according to the angle of rotation of the radius vector, within of which the polyhedron does not exceed a circle characterizing this error). Moreover, the radar chart allows to evaluate compositions of real objects for which the smallest or the largest errors are achieved by using a particular standard.

As can be seen from figure 5, the standards giving the smallest error for the total content of CAs determined using colloidal solution of AuNRs are epinephrine and norepinephrine. Diagrams of these substances are practically equal and have the smallest area. In case of the nanocomposite material, dopamine or norepinephrine should be used as the standards: error in determination of the total amount does not exceed 0.2. The absence of obvious mutual influence of different catecholamines on each other allows to determine their total content.

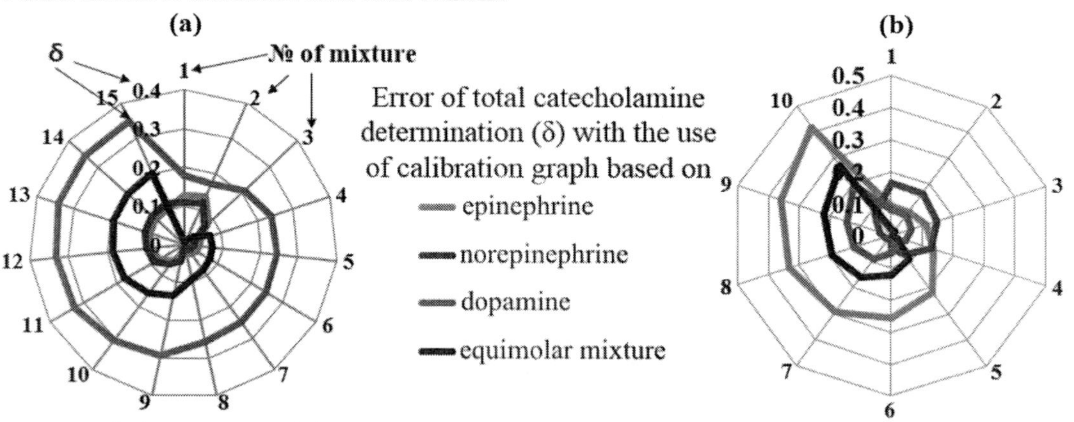

Figure 5. Relative error modulus (δ) of determination of the total amount of CAs with the use of colloidal solution of AuNRs (a) and nanocomposite material (b) in terms of epinephrine, norepinephrine, dopamine and the equimolar mixture.

To estimate a possibility of practical application of the proposed method, the analysis of a model mixture containing epinephrine (E), norepinephrine (NE), and dopamine (DA) in the ratio E:NE:DA = 1:7:32 (the average ratio of CAs in daily urine of a healthy adult) as well as inorganic ions in the following ratio "total content of CA: ion concentration": 1:100 Na^+, 1:100 K^+, 1:100 Mg^{2+}, 1:100 Ca^{2+}, 1:50 SO_4^{2-}, 1:50 NO_3^-, 1:450 Cl^- was performed. The results of the determination represented in table 2 indicate that the method can be used to determine the total content of CAs in mixtures containing 50 – 450-fold excess of common inorganic ions. As a standard substance for representing the total content of CAs, both individual catecholamines and their mixture can be considered.

Table 2. The results of determination of the total content of catecholamines in a model mixture with inorganic ions in terms of different standards (n = 3, P = 0.95).

		Standard substance			
		Epinephrine (E)	Norepinephrine (NE)	Dopamine (DA)	Mixture (E:NE:DA = 1:1:1)
Using the colloidal solution of AuNRs	Added, μM	1.00			
	Found, μM	1.05 ± 0.08	1.03 ± 0.07	0.76 ± 0.06	0.93 ± 0.07
	RSD, %	3			
Using the nanocomposite	Added, μM	5.0			
	Found, μM	6.2 ± 0.9	4.7 ± 0.7	5.2 ± 0.8	5.5 ± 0.8
	RSD, %	6			

3.3. Analysis of biological fluids

3.3.1. Determination of catecholamines in urine samples. The proposed approach was tested for determination of CAs in urine. Standard addition method was used for the analysis. For CAs determination by colloidal solution of AuNRs, 0.25 mL of urine and a mixture of catecholamines (E:NE:DA = 1: 7: 32) were added into the reaction mixture (0.30 mL of AuNRs colloidal solution, 0.10 mL of glycine buffer solution (pH 9.5), 0.02 mL of 0.01M $AgNO_3$, V = 2.50 mL). Total added CAs were 0/1/2/3 μM. For determination by the nanocomposite material, 3.00 / 0.30 mL of urine and a mixture of catecholamines (A:NA:DA = 1: 7: 32) were added into the reaction mixture (1.00 mL of glycine buffer solution (pH 9.5), V = 5.00 mL). Total added CAs were 0/5/10/15 μM). Then a tablet of the nanocomposite material modified with $AgNO_3$ was put, pressed with a glass rod, shaken on a mechanical shaker for 30 minutes and dried between sheets of filter paper.

As an independent method for estimating correctness of the determination, reversed-phase HPLC with amperometric detection was used. The mobile phase consisted of acetonitrile and 0.1% H_3PO_4 in water (10:90, v/v) with addition of sodium 1-octanesulfonate (0.3 mM), the flow rate was 0.4 mL min^{-1}. The results of CAs determination using gold nanorods and chromatography in urine samples are represented in table 3. Good agreement of the results obtained by various methods indicates their correctness.

Table 3. The results of determination of the total content of catecholamines in urine samples with the use of colloidal solution of AuNRs, nanocomposite material and HPLC.

	Sample 1, μM		Sample 2, μM	
Using the nanocomposite	10 ± 2 (RSD 9) 9.9 ± 0.4 (RSD 5)		14 ± 5 (RSD 5) 14 ± 1 (RSD 10)	
Using the colloidal solution of AuNRs	–		14 ± 0.7 (RSD 6)	
HPLC	10 ± 3 (RSD 20)	epinephrine: 7 ± 3 (RSD 20) norepinephrine: 1.4 ± 0.2 (RSD 8) dopamine: 1 ± 0.7 (RSD 40)	14 ± 5 (RSD 20)	epinephrine: 10.0 ± 0.7 (RSD 10) norepinephrine: 1.3 ± 1.1 (RSD 40) dopamine: 3 ± 4 (RSD 70)

3.3.2. Determination of catecholamines in blood serum. Standard addition method was used for evaluation of a possibility of CAs determination in serum samples. 0.10 and 0.25 mL of blood serum were added into reaction mixtures containing colloidal solution of AuNRs and nanocomposite respectively. The dependencies of $\Delta\lambda$ on the total concentration of catecholamines in the system were preliminarily measured. According to the results represented in table 4, the developed methods are applicable for the determination of CAs in serum at the level of 0.5 and 5 μM by using colloidal solution of AuNRs and nanocomposite material respectively.

Table 4. The results of catecholamines determination in blood serum with the use of colloidal solution of AuNRs and nanocomposite material ($n = 3$, $P = 0.95$).

	Using the colloidal solution of AuNRs	Using the nanocomposite
Added, μM	0.5	5
Found, μM	0.5 ± 0.1	5 ± 1
RSD	8	2

4. Conclusions

Interaction of catecholamines with silver nitrate in the presence of colloidal solution of AuNRs and a nanocomposite material has been examined. This interaction is accompanied by a hypsochromic shift of the maxima in the SPR-spectra of AuNRs, the magnitude of this shift can serve as a characteristic of completeness of the reaction. The possibility of using epinephrine, norepinephrine, dopamine and their equimolar mixture as standards in the determination of the total content of catecholamines has been studied. To estimate the errors in determination of catecholamines and select the best standard, the use of radar charts has been proposed. The polyhedrons area correlates with value of the error, and the angle of rotation of the radius vector, within which the diagram does not exceed a circle characterizing the given error, serves to determine the percentage of mixtures, the analysis of which can be made within this error. The standards giving the smallest error in determination of the total content of CAs with the use of colloidal solution of AuNRs are epinephrine and norepinephrine. For the determination by the nanocomposite material, dopamine or norepinephrine should be used as the standards. Analysis of model mixture confirmed applicability of the developed method for the determination of the total content of CAs. The developed method can be applied to the analysis of urine and blood serum.

Acknowledgments
This work was financially supported by the Russian Science Foundation (grant N 18-73-10001).

IOP Conf. Series: Materials Science and Engineering **525** (2019) 012012 doi:10.1088/1757-899X/525/1/012012

Appendices

Table A1. Mixture compositions and results of the determination of the total content of catecholamines in these mixtures using colloidal solution of gold nanorods in terms of individual compounds and their equimolar mixture.

Added (total): 1 μM		Total content of catecholamines (μM) in terms of:			
№ of mixture	Mixture composition (E:NE:DA), μM	Epinephrine (E)	Norepinephrine (NE)	Dopamine (DA)	Mixture (1:1:1)
		y = 23.2x	y = 23.5x	y = 31.8x	y = 26.0x
1	0.25:0.625:0.125	1.13	1.11	0.82	1.00
2	0.125:0.625:0.25	1.14	1.12	0.83	1.01
3	0.4:0.4:0.2	1.09	1.07	0.79	0.97
4	0.5:0:0.5	1.04	1.03	0.76	0.93
5	0.2:0.4:0.4	1.04	1.02	0.76	0.93
6	0.025:0.175:0.8	1.04	1.02	0.76	0.92
7	0.25:0.5:0.25	1.03	1.02	0.75	0.92
8	0:0.5:0.5	1.02	1.00	0.74	0.91
9	0.25:0.25:0.5	0.97	0.96	0.71	0.87
10	0.125:0.25:0.625	0.95	0.94	0.69	0.85
11	0.25:0.125:0.625	0.92	0.91	0.67	0.82
12	0.5:0.5:0	0.92	0.90	0.67	0.82
13	0.5:0.25:0.25	0.91	0.89	0.66	0.81
14	0.4:0.2:0.4	0.90	0.89	0.66	0.80
15	0.625:0.25:0.125	0.90	0.88	0.65	0.80
16	0.25:0.625:0.125	1.13	1.11	0.82	1.00

Table A2. Mixture compositions and results of the determination of the total content of catecholamines in these mixtures using nanocomposite in terms of individual compounds and their equimolar mixture.

Added (total): 5 μM		Total content of catecholamines (μM) in terms of:			
№ of mixture	Mixture composition (E:NE:DA), μM	Epinephrine (E)	Norepinephrine (NE)	Dopamine (DA)	Mixture (1:1:1)
		y = 1.39x	y = 1.82x	y = 1.66x	y = 1.55x
1	0.125 : 0.875 : 4	5.5	4.2	4.6	4.9
2	2 : 2 : 1	5.5	4.2	4.6	4.9
3	1.25 : 2.5 : 1.25	5.6	4.3	4.7	5.0
4	2.5 : 1.25 : 1.25	5.7	4.3	4.8	5.1

5	1.25 : 1.25 : 2.5	6.1	4.7	5.1	5.5
6	2.5 : 2.5 : 0	6.3	4.8	5.3	5.6
7	1 : 2 : 2	6.5	5.0	5.4	5.8
8	2.5 : 0 : 2.5	6.7	5.1	5.6	6.0
9	2 : 1 : 2	6.8	5.2	5.7	6.1
10	0 : 2.5 : 2.5	7.1	5.4	6.0	6.4

References

[1] Liu J M, Wang X X, Cui M L, Lin L P, Jiang S L, Jiao L and Zhang L H 2013 A promising non-aggregation colorimetric sensor of AuNRs–Ag^+ for determination of dopamine *Sens. Actuat.* B **176** 97 – 102

[2] Gorbunova M V, Apyari V V, Dmitrienko S G and Garshev A V 2016 Formation of core-shell Au@Ag nanorods induced by catecholamines: a comparative study and an analytical application *Anal. Chim. Acta* **936** 185 – 194

[3] Apyari V V, Arkhipova V V, Gorbunova M V, Volkov P A, Isachenko A I, Dmitrienko S G and Zolotov Yu A 2016 Towards the development of solid-state platform optical sensors: aggregation of gold nanoparticles on polyurethane foam *Talanta.* **161** 780 – 788

[4] Gorbunova M V, Matveeva M A, Apyari V V, Garshev A V, Volkov P A, Dmitrienko S G and Zolotov Yu A 2017 Sorption of gold nanorods on polyurethane foam as a way to obtain a nanocomposite material with a surface plasmon resonance for chemical analysis purposes *Nanotechnologies in Russia* **12** 185 – 192

[5] Pussard E, Neveux M, Guigueno N 2009 Reference intervals for urinary catecholamines and metabolites from birth to adulthood *Clin. Biochem.* **42** 536 – 539

[6] Whiting M J and Doogue M P 2009 Advances in biochemical screening for pheochromocytoma using biogenic amines *Clin. Biochem. Rev.* **30** 3 – 17

[7] Tsunoda M 2006 Recent advances in methods for the analysis of catecholamines and their metabolites *Anal. Bioanal. Chem.* **386** 506 – 514

[8] Bergquist J, Ściubisz A, Kaczor A and Silberring J 2002 Catecholamines and methods for their identification and quantitation in biological tissues and fluids *J. Neurosci. Methods* **113** 1 – 13

[9] Gorbunova M V, Gutorova S V, Berseneva D A, Apyari V V, Zaitsev V D, Dmitrienko S G and Zolotov Y A 2018 Spectroscopic methods for determination of catecholamines: a mini-review *Appl. Spectrosc. Rev.* DOI 10.1080/05704928.2018.1470980

[10] Nikoobakht B and El-Sayed M A 2003 Preparation and growth mechanism of gold nanorods (NRs) using seed-mediated growth method *Chem. Mater.* **15** 1957 – 1962

[11] Fonteh A N, Harrington R J, Tsai A, Liao P, Harrington M G 2007 Free amino acid and dipeptide changes in the body fluids from Alzheimer's disease subjects *Amino Acids* 32 213 – 24

[12] Bornstein S R, Breidert M, Ehrhart-Bornstein M, Kloos B, Scherbaum W A 1995 Plasma catecholamines in patients with Addison's disease *Clin. Endocrinol.* **42** 215 – 8

[13] Eisenhofer G, Keiser H, Friberg P, Mezey E, Huynh T T, Hiremagalur B, Ellingson T, Duddempudi S, Eijsbouts A, Lenders J W 1998 Plasma metanephrines are markers of pheochromocytoma produced by catechol-O-methyltransferase within tumors *J. Clin. Endocrinol. Metab.* **83** 2175 – 85

IOP Conf. Series: Materials Science and Engineering **525** (2019) 012013 doi:10.1088/1757-899X/525/1/012013

Radiation nanotechnology for selective modifications of atomic composition and properties of thin film materials

M M Dementyeva[1], K E Prikhodko[1,2], B A Gurovich[1], L V Kutuzov[1], B V Goncharov[1] and D A Komarov[1]

[1] National Research Centre "Kurchatov Institute", Moscow, Russia
[2] National Research Nuclear University MEPhI, Moscow, Russia

E-mail: dementyeva_MM@nrcki.ru

Abstract. We have demonstrated using of ion beam irradiation to control the atomic composition and properties of thin film materials by three different ways. Selective removal of atoms (SRA) technique allows us to transform Co_3O_4 to Co. By EELS on cross-section samples in STEM mode it was shown that target depth recovery profile has no monotonic character that proved the radiation nature of SRA process. Selective displacement of Atoms (SDA) technique under oxygen ion irradiation was used to control the critical current of ultrathin superconductive NbN film that can have an implementation during new cryogenic logic device design.

1. Introduction

In National Research Centre "Kurchatov Institute" over the past ten years, we were actively developing radiation nanotechnology to create functional elements of nanometer scale. Due to the impact of low-energy ion beams of different composition (protons, oxygen ions, nitrogen, etc.) at different doses, there is a controlled modification of the atomic composition and physical properties of the selected areas of thin-film materials (insulators, conductors, superconductors). This technology includes three methods of selective changes in the atomic composition of thin films: the method of selective removal of atoms (SRA), the method of selective association of atoms (SAA), and method of selective displacement of atoms (SDA) [1]. The main advantages of these methods are as follows. It is possible to create elements of specified shapes and sizes (minimum size from 3 – 10 nm). It is also possible of parallel transformation of the local composition and properties in different layers of multilayer thin-film structures.

In this paper we consider all three of the above mentioned methods of selective modification of the atomic composition and properties of thin-film materials. Radiation technology of selective removal of atoms (SRA) is presented by the example of removal of oxygen atoms from cobalt oxide Co_3O_4 [2], selective association of atoms by the example of alumina oxidation under oxygen ion irradiation [3], and selective displacement of atoms by the example of NbN thin superconductive film oxygen irradiation [4].

The reduction of cobalt oxide by radiation methods can be used to create a high-density pattered magnetic media (PMM). The formation of a media with a high recording density requires the creation of a special carrier consisting of separate magnetic granules (single-domain bits) with a size of 15 to 30 nanometers of the same size, shape and orientation (pattern magnetic media) [5]. In addition, when creating such structures, it is important to ensure the stability of the magnetization of individual bits

IOP Conf. Series: Materials Science and Engineering **525** (2019) 012013 doi:10.1088/1757-899X/525/1/012013

when the external magnetic field is switched off, which can be achieved by creating an antiferromagnetic layer, for example, from CoO along the ferromagnetic Co – bit boundary.

The application of the method of selective association of atoms (SAA) [1] is demonstrated to create a layer of aluminum oxide on the surface of a thin film of metal aluminum [3]. For the realization of radiation-induced oxidation of aluminum the irradiation with oxygen ions of low energy was used. Irradiation-induced high quality oxide can be used as of interlayer dielectric in the formation of the metallization and, on the other hand, to prevent unwanted sputtering of aluminum in the process of ion irradiation. The results obtained in this paper can be used in the creation of electrical insulation of metal conductors as an alternative to the methods of deposition of dielectric layers in traditional microelectronics and nanotechnology. The advantages of this method include precision oxidation of aluminum in depth, high dielectric properties of the resulting aluminum oxide and the ability to reduce the number of technological operations in the process of creating devices.

2. Instruments and experimental techniques

In this work, thin films of Co_3O_4, Al, NbN were irradiated with ion beams of various compositions extracted from a high-frequency (HF) plasma discharge (13.56 MHz). All experiments on irradiation in this work were carried out on a special installation "Copra Cube", the main part of which is the HF plasma source. The subsequent extraction of ions onto the sample was carried out by applying a pulsed HV bias to the table with the fixed sample. The ion beam consisted of protons with an energy of 1 keV and irradiation was carried out in the dose range (0.94-3.75)·1018 ions/cm2 in the SRA process; in the process of SAA samples were irradiated with oxygen ions with energy 0.2 and 0.5 keV in the dose range $(0.1\text{-}0.4)\cdot 10^{17}$ions/cm^2 at room temperature; SDA was carried out under irradiation with oxygen ions with energy (0.1–1) keV in the dose range $(0.62\text{-}6.25)\cdot 10^{15}$ ions/cm^2.

To analyze the degree of modification of films by depth after irradiation, the analytical transmission electron microscopy TEM and STEM techniques were used, the main advantage of the latter is the possibility of local analysis of the chemical composition of cross-section samples. The microstructure, phase composition of the initial and irradiated samples were studied using high-resolution TEM (HRTEM) bright-field images. The chemical composition of the samples was evaluated by electron energy loss spectroscopy (EELS) using GIF-2001 installed at "Titan 80–300ST" transmission electron microscope at an accelerating voltage of 200 kV. The phase composition of individual grains and irradiated samples was determined by diffraction patterns obtained by Fourier transform in the Digital Micrograph from the corresponding high-resolution images. The application of this technique is due to the fact that the analysis of SAD electron micro diffraction patterns obtained directly is difficult due to the blurring of the reciprocal lattice sites for nano-size grains.

The chemical composition of the experimental samples was determined by analyzing the profile spectra of electron energy losses by the method of relative concentrations:

$$\frac{N_A}{N_B} = \frac{I_A(\beta,\Delta)}{I_B(\beta,\Delta)} \cdot \frac{\sigma_B(\beta,\Delta)}{\sigma_A(\beta,\Delta)} \qquad (1),$$

where I_A, I_B — the integrated intensity of the peaks under the curve of energy loss after subtraction of the background, and σ_A and σ_B is the cross section of inelastic scattering of atoms A and B for a given collection angle of the spectrometer β [6].

For thin films cross section sample preparation we used the focused ion beam technique on the "Helios Nanolab 650" FIB facility with an accelerating voltage of the Ga ion gun of 30 keV and a current of 2.5 nA and a final thinning of the sample was held at a voltage of 2-5 keV and a current of 0.12 nA.

3. Results and discussion

3.1. Selective removal of atoms

After the proton irradiation of thin films of Co_3O_4 at different fluencies and temperatures the depth profiles of chemical composition of the film were measured out by EELS technique.

To find out the effect of the temperature during irradiation, figure 1 shows the distribution profiles of the elements in depth of the films irradiated to doses of $3.75\cdot 10^{18}$ ions/cm^2 at substrate temperatures of 20°C and 100°C. It is shown that at a temperature of 100°C film of cobalt oxide at a given dose is reduced to pure cobalt at a depth of 17 nm, in contrast to irradiation at 20°C. This is due to the faster release of displaced oxygen atoms from the sample during the selective removal of atoms from Co_3O_4 under proton irradiation. In addition, it was found that at both irradiation temperatures the target depth recovery profile is no monotonic, due to the no monotonic distribution of the damaging dose during irradiation. The direct correspondence of the cobalt reduction profile to the dose distribution profile at the target depth proves the radiation nature of the process of selective removal of atoms.

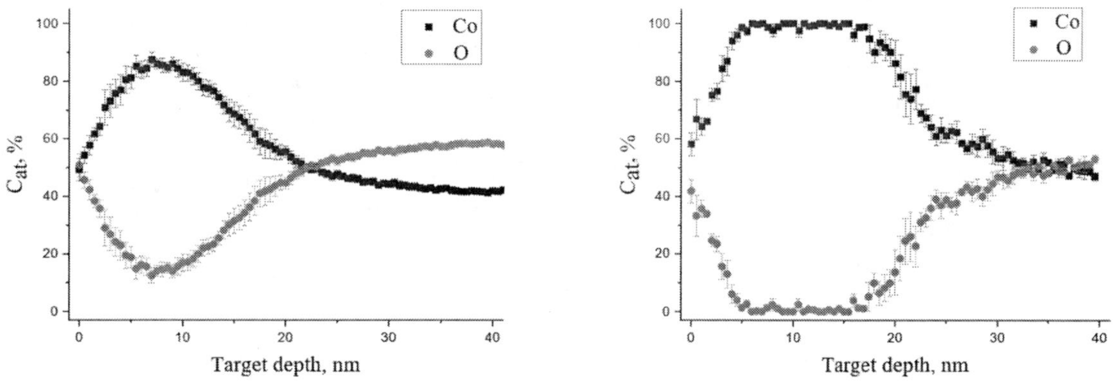

a b

Figure 1. Depth distribution profile of elements in the Co_3O_4 film after irradiation with 1 keV protons (calculated from EELS data) for dose of $3.75 \cdot 10^{18}$ions/cm^2 at different temperatures: (a)— 20°C; (b)— 100°C

The analysis of the diffraction patterns constructed by Fourier transform from the corresponding grains on high-resolution images is in full agreement with the EELS data. Figure 2 shows that the grains of the Co_3O_4 film irradiated at a dose of $3.75 \cdot 10^{18}$ ions/cm^2 at a temperature of 100°C correspond to the phase of pure Co hexagonal symmetry (P63/mmc) with the lattice parameter a=b=0.2514 nm, c=0.4105 nm.

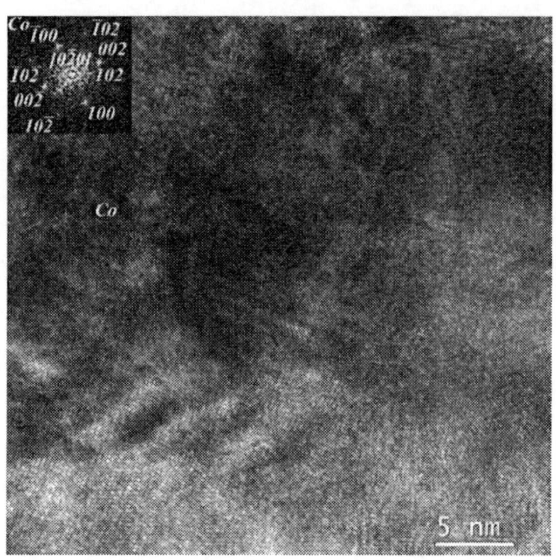

Figure 2. HRTEM image of the Co3O4 film cross section sample irradiated with 1 keV protons to dose of $3.75 \cdot 10^{18}$ions/cm^2 at a temperature of 100°C, FFT diffraction (inset)

IOP Conf. Series: Materials Science and Engineering **525** (2019) 012013 doi:10.1088/1757-899X/525/1/012013

3.2. Selective association of atoms

Figure 3 shows the depth distribution profiles of elements of the aluminum film irradiated with oxygen ions with energy 0.2 keV to a dose of $2.6 \cdot 10^{18}$ions/cm^2. As can be seen from figure 3, aluminum oxide is formed to a depth of ~20 nm, and a layer of metal aluminum remains below it. This result confirms the possibility of using the radiation method of aluminum oxidation to create an insulator layer on the metal surface. This is an example of ion beam irradiation implementation for forming an insulator layer at mask open areas on the surface of the sample.

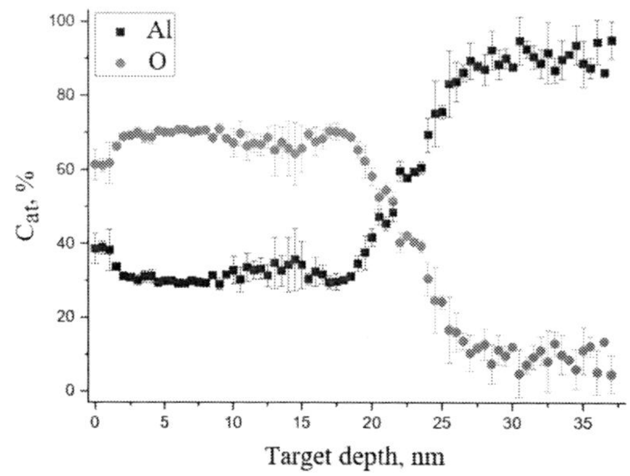

Figure 3. Depth-distribution profile of elements of aluminum film irradiated with oxygen ions with energy 0.2 keV to a dose of $2.6 \cdot 10^{18}$ions/cm^2[3].

3.3. Selective displacement of atoms

From the point of view of the development of a new element base to build low heat and high speed supercomputers, one of the most promising is the use of thin-film superconductors. In this regard, one of the examples of application of the method of selective displacement of atoms (SDA) is the development of a method of controlled reduction of the thickness of the superconducting ultra-thin (5 nm) film of niobium nitride under the low-energy oxygen ions irradiation, based on the replacement of nitrogen atoms with oxygen atoms in the irradiation zone [7]. The depth of radiation-induced replacement of nitrogen atoms with oxygen atoms is determined by the value of the total projective length of oxygen ions in the target material and reaches (2-3) nm at the particle energy (100-300) eV [7].

Figure 4 shows a controlled decrease in the thickness of the thin film of the superconducting NbN as a result of the process of selective displacement of atoms under the low-energy oxygen ions irradiation [7]. Reducing the thickness of the film allows you to change the superconducting properties and opens the possibility of manufacturing functional devices based on these changes.

The most important physical characteristic is the magnitude of the critical current of the superconductor, since the control of the critical current in the required places of the superconductor allows as to create different logic devices.

IOP Conf. Series: Materials Science and Engineering **525** (2019) 012013 doi:10.1088/1757-899X/525/1/012013

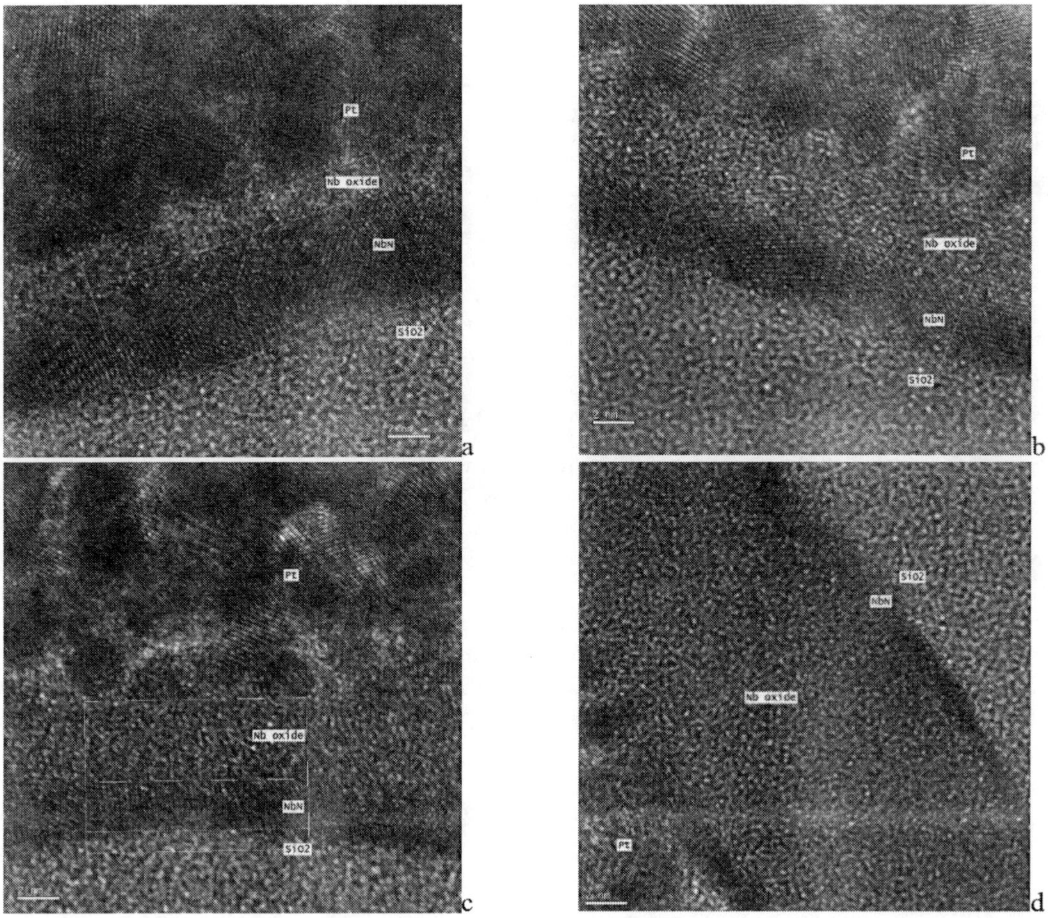

Figure 4. HRTEM images of ultra-thin NbN film irradiated with oxygen ions with low energy at different radiation doses [7].

With an SRA technique, it is possible to make a controlled local change in the critical parameters of NbN thin films: the critical current I_c and the superconducting transition temperature T_c, and to reduce the section of superconducting channel.

In this paper, the creation of areas with suppressed critical parameters (I_c and T_c) was made due to a controlled change in the composition and properties of the original NbN films under the irradiation with oxygen ions with an energy (0.02-0.1) keV in the dose range 0.7-1.8 d.p.a. (for nitrogen) at room temperature.

Forming of structures for electric measurements was carried out by the electron lithography and plasmo-chemical etching, in a manner similar to that described in [8]. As samples for research of controlled changing of I_c, the microbridges the sizes of 20x20 μm (see figure 5,a) made using a photolithography method with metal contacts were used.

Cooling was carried out by placing the measuring model, with the sample fixed on the measuring table, in the vessel with liquid helium "STG-40". The sample was mounted on the measuring table using clamping contacts. Figure 5,b shows the current-voltage characteristics of samples exposed to oxygen irradiation with various doses (for nitrogen).

From the figure 5,b it can be seen that with increasing dose the critical current of transition to the superconducting state drops. Thus, it was shown that oxygen irradiation can be used to control the change in the critical current value of thin-film niobium nitride.

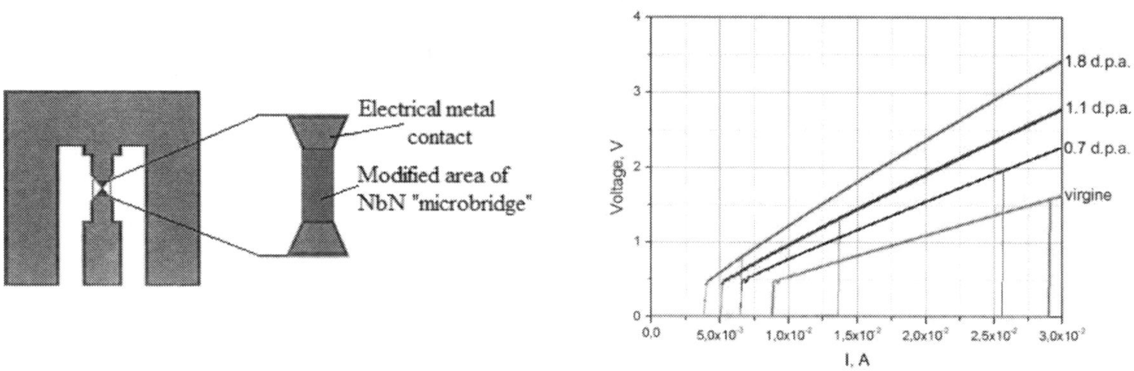

a b

Figure 5. Sample for electrical measurements (top view) (a) and Current-voltage characteristics of samples exposed to oxygen irradiation with various doses (for nitrogen) (b).

4. Conclusion

In this work we have shown the implementation of Selective Removal of Atoms, Selective Association of Atoms and Selective Displacement of Atoms techniques in thin film materials under ion beams irradiation to create different functional nano structures for various applications.

The most important results are as follows. We have shown that the transformation of cobalt oxide to metal cobalt under proton irradiation (SRA) characterized by no monotonic depth recovery character. This is the direct evidence that SRA process has radiation nature and begins at the depth with highest damage density production by ion beam.

Another important result is that we have found the possibility to control the critical current value of the ultrathin superconductive NbN film. This can be used during new generation of cryo- electronic device production.

Acknowledgments

The work was supported by the NRC “Kurchatov Institute” (№ 1639).

References

[1] Gurovich B A, Prikhod'ko K E, Kuleshova E A, Maslakov K I and Komarov D A, 2013 *JEPT* **116** 916

[2] Gurovich B A, Prikhodko K E, Kuleshova E A, Yakubovsky A Yu, Meilikhov E Z and Mosthenko M G 2010 *Journal of Magnetism and Magnetic Materials* **322** 3060

[3] Prikhodko K E, Dementyeva M M, Gurovich B A, Komarov D A and Kutuzov L V *Knowledge E Materials Science* 2018 1

[4] Gurovich B A, Tarkhov M A, Prikhod'ko K E, Kuleshova E A, Komarov D A, Stolyarov V L, Olshanskii E D, Goncharov B V, Goncharova D A, Kutuzov L V and Domantovskii A G 2014 *Nanotechnologies in Russia* **9** 383

[5] Meilikhov E Z and Farzetdinova R M 2014 *Physics of Solid State* **56** 2408

[6] Williams D and Carter A 2009 *Transmission electron microscopy* (New York: Springer) chapter 39 P.725

[7] Gurovich B A, Prikhodko K E, Komarov D A, Dement'eva M M and Kutuzov L V 2018 *Nuclear Inst. and Methods in Phys Research B* **435** 224

[8] Gurovich B A, Prikhod'ko K E, Tarkhov M A, Kuleshova E A, Komarov D A, Stolyarov V L, Ol'shanskii E D, Goncharov B V, Goncharova D A, Kutuzov L V, Domantovskii A G, Lavrukhina Z V and Dement'eva 2015 *Nanotechnologies in Russia* **10** 530

IOP Conf. Series: Materials Science and Engineering **525** (2019) 012014 doi:10.1088/1757-899X/525/1/012014

Effect of heat treatment on the structure and properties of the chromium-nickel alloy G-35

V A Ermishkin, D L Mikhailov, S P Kulagin, N A Minina, N A Palyi

A.A. Baikov Institute of Metallurgy and Materials Science, Russian Academy of Sciences, Leninskii pr. 49, 119334, Moscow, Russia

E-mail: Minina1951@rambler.ru

Abstract. This paper is devoted to the study of the nickel-chromium alloy G-35, which was intended for the manufacture of a chemical reactor vessel operating in molten chlorides in the temperature range of 550 - 650°C. A preliminary study of the structure and mechanical properties of the G-35 alloy was carried out in order to select the methods for their investigation. With the help of Photometric analysis of structural images it was found that the weld metal is in the austenitic state and the presence of an alloying of boron was also established. A class of functions has been established that describes the loading curves for three-point bending of samples.

1. Introduction

The choice of structural state of structural materials is associated with the search for answers to a number of important questions. Here we will not consider a complex of issues related to financial aspects of choice. The choice has already been made. It is dictated by the manufacturing conditions of the body of a chemical reactor operating in molten chlorides in the temperature range of 550 - 650°C. Among the mandatory requirements that must be met by the structural state of the nickel-chromium alloy G-35, should be mentioned: corrosion resistance in the working body of the reactor and, including, resistance to corrosion under voltage, heat resistance in a given temperature range, the necessary margin plasticity to perform rolling operations in the absence of heating, a sufficient level of technological strength in all operations for the manufacture of the reactor vessel, including weld ability and heat treatment of product. Taking into account the fact that most of the material properties that must meet the above requirements are structurally sensitive, special attention should be paid to the kinetics of the structural evolution of the alloy and phase transformations occurring in the alloy during the entire technological period of its manufacture and its production cycle. The properties of the alloy should be studied taking into account the dynamics of structural transformations, both in the manufacture of the reactor and in its operation.

2. Materials and Methods

Passport data alloy G-35 (USA) are shown in Table 1. Preliminary estimates of the phase composition of nickel-chromium alloys can be obtained from the Schaeffler diagram, so you need to find chromium and nickel equivalents(in mass.%), $[Cr]_{equiv}$ and nickel $[Ni]_{equiv}$ respectively, according to the formulae [1]:

$$[Cr]_{equiv} = (Cr) + 2(Si) + 1{,}5(Mo) + 5(V) + 1{,}75(Nb) + 1{,}5(Ti) + 5{,}5(Al) \text{ mass. \%} \quad (1)$$
$$[Ni]_{equiv} = (Ni) + (Co) + 0{,}5(Cu) + 0{,}5(Mn) + 25(N) + 30(C) \text{ mass. \%}$$

Table 1. The chemical composition of the G-35 alloy in mass. %

Mn	Al	C	Cr	Cu	Co	Fe
0,22-0,21	0,22-0,18	0,010-0,008	32,90-32,95	0,02-0,03	<0,05	0,14-1,22
W	**Mo**	**Ni**	**P**	**S**	**Si**	**V**
0,03	7,98-8,06	57,31	0,003	<0,002	0,14-0,16	<0,01

The calculations showed the following results: $[Cr]_{equiv}$ = 46.62 mass%, $[Ni]_{equiv}$ = 57.62 mass %. Thus, the G-35 alloy, according to the Schaeffler diagram, should be a single-phase solid solution of all elements, martensite stabilizers in austenite, in an equilibrium state, but due to non-equilibrium crystallization conditions, fluctuations of concentration clusters occur, which can have considerable sizes. Photometric analysis of structural images (PHASI) allows us to provide qualitative and quantitative information on the distribution of secondary phase precipitates on the observation surface and in size, as well as on their chemical composition. Figure 1 shows an example of such an analysis of a sample of alloy G-35 in the initial state and in the cast state in the weld.

PHASI, developed at IMET RAS [2], is a software-analytical complex based on computer analysis using a differential scheme comparing the reflection brightness spectra and structures of surface fragments of the reference and test samples with the output of the analysis results in a visual graphic form, in the form of tables and graphs. Its physical fundamentals and methodological possibilities were previously discussed in a number of publications [3, 4]. Here we will focus only on the features of the application of PHASI as applied to the study of diffusion processes. It begins with scanning the surfaces of samples from chemical elements that are alloying components of the alloy under study on the one hand and a fragment of the surface of the object under study on the other. Digitized images of scanned images are entered into a computer, in which, first of all, according to a special program included in PHASI, the luminance spectra of the reflection of visible light from the surfaces of compared objects are compared. These spectra are constructed in the coordinates "spectral density of the brightness of the reflection - P(I) - intensity of the brightness of the reflection - I". The spectral density was determined by the formula:

$$p(I) = \frac{n(I)}{N} \quad (2)$$

where: n (I) is the number of pixels with intensity of brightness - I, N is the total number of pixels into which the analyzed image is divided. The abscissa axis shows the values of the intensity of brightness in the divisions of the linear scale, varying from zero, corresponding to the total absorption of incident visible light, to one - to its full reflection from the surface under study. Above the spectra, mathematical operations are used, which make it possible to reveal the characteristic features of the structure of the materials under study. Using color staining of intervals corresponding to the spectra of the standards, it is possible to isolate these features

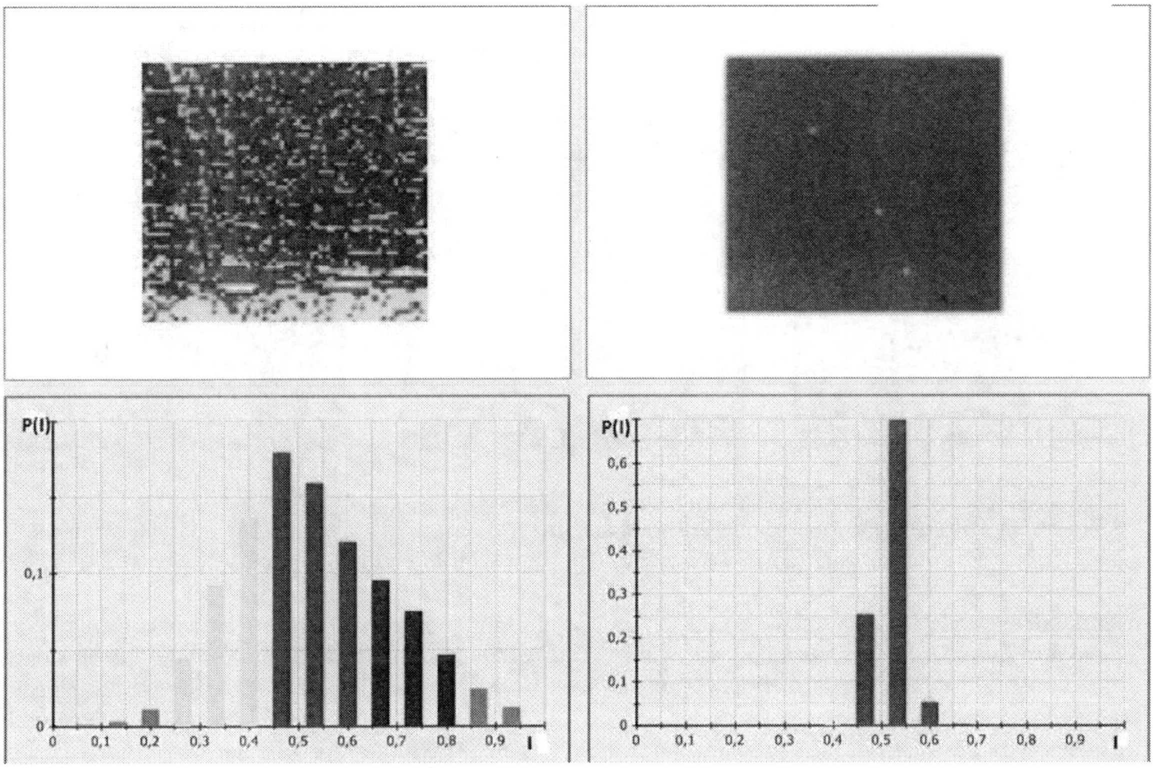

Figure 1. Comparison results of the base metal of the G-35 alloy. From above, the structure of the selected fragments, below them the spectra of the brightness of reflection of visible light from them: a) - with the weld metal (left panel) and initial state (right panel)

and then transfer this color to the images, i.e., identify those elements of the structure that contribute to the selected intervals of the spectrum. Using the alloying elements of the alloy under investigation as standards, one can study their concentrations on the observation surface, which are converted into volume concentrations by stereology methods. It can be seen from the figure that the base metal is homogeneous; its structural components are distributed evenly over the fragment, whereas the components painted in yellow and blue light have dropped out of the weld metal. Due to the fact that the material in the operating conditions will work for bending, the evaluation of its mechanical characteristics was also carried out during bending tests. Samples in the form of small beams with dimensions of $20 \times 4 \times 4$ mm^3 were loaded according to the three-point bending scheme on an Instron-3382 testing machine with a speed of 1 mm/min. Before and after deformation, the lateral surfaces of the samples were photographed by a digital camera with subsequent analysis by the PHASI method.

3. Results and discussion

The alloy samples were studied in three structural states: the base metal after cold rolling (samples 8), the main deformed (8-d), cast weld metal (4-1) and weld metal after annealing 800°C for 1 hour (4-2). To estimate the distribution of alloying elements, the PHASI method compared images of the structures of alloy fragments in the three listed states and the brightness spectra of the reflection of visible light from their surfaces with the standards of the alloying elements of the alloy under study. The peaks on the spectra of the standards were covered with intervals of 10 pixels wide, painted in the colors assigned to them: the peak on the chromium spectrum is colored magenta, nickel in blue, molybdenum in yellow and boron in red. Figure 2 shows the results of the analysis, the surface of the sample in the initial state has the layered structure with grains elongated along the rolling direction forming lines. The structure is predominantly colored blue, but due to the overlap of some spectra, it is difficult

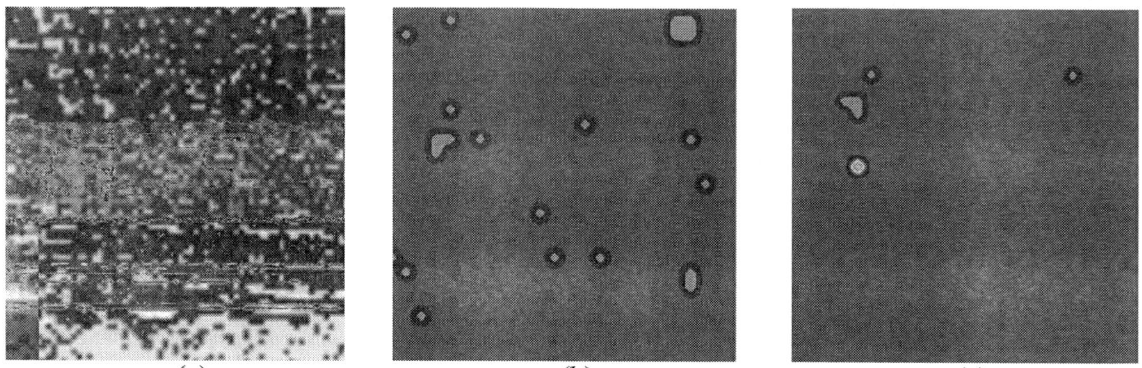

(a) (b) (c)

Figure 2. The structures of the studied states of the G-35 alloy at 125 times magnification: in the initial state (a); after welding (b); after welding and annealing (c).

to give an accurate estimate of the concentration of alloying elements on the surface of the fragments. The color of the surface fragments of welds, both after heat treatment and without it, is clearly dominated by the color of nickel. This can be seen by looking at the reflectance brightness spectra from nickel and chromium standards. It should be noted that the width of the spectra is greatly influenced by the quality of the preparation of their surface. The better it is, the narrower the spectrum. For rough surfaces are characterized by diffuse reflection and a wide range. Using the PHASI method, it was found that the G-35 alloy contains boron in an amount of 0.002 mass. %. Boron powder was used as a reference. Figure 3 shows the reflection spectra for nickel and chromium.

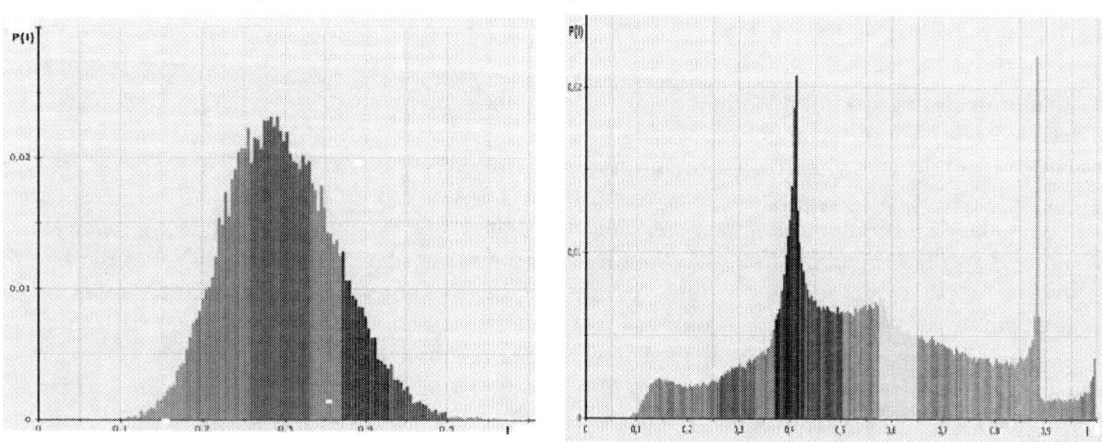

(a) (b)

Figure 3. The brightness spectra of the reflection of visible light from the surface of the standards: (a) - from nickel; (b) - from chrome

Figure 4 shows the boron distribution on the studied alloy fragments. All elements of the structure, except for the painted ones, were absorbed. The range of the boron spectrum was colored blue. It can be seen from the figure that in both cases boron is located along the grain boundaries and in the form of precipitates at the nodal points of the grain grid. In the heat-treated weld, the discharge is larger.

Inside the grains there is a fine discharge of the red colored phase. This color was colored characteristic peak, which does not belong to any of the peaks in the spectra of the

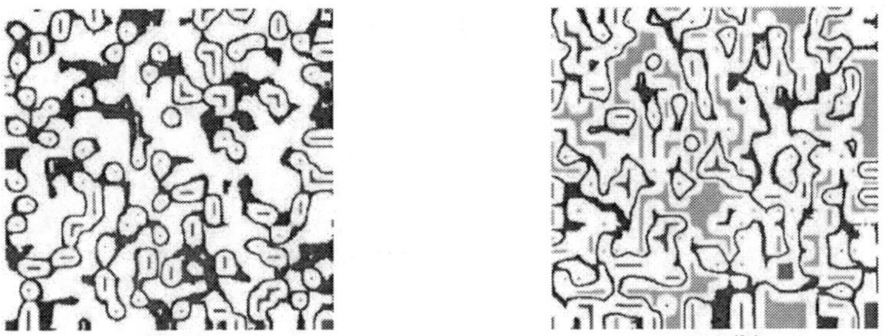

(a) (b)

Figure 4. Distribution of boron in the plane of observation of the surface fragments of the G-35 alloy: in the weld after heat treatment (a); in the base metal (b). × 970

main alloying elements of the alloy. The results of the analysis of mechanical tests showed that the deformation curves in the coordinates "load in kg P is a deflection arrow in mm f". well described by logarithmic functions of the form:

$$P(f) = a\ln(f) + b \quad (3)$$

The parameters of the function (2) for samples of all three states, together with the values of the confidence level (R^2), are given in Table 2.

Table 2. Parameters of the function (3) for the all studied states G-35 alloy samples

Alloy condition	a	b	R^2
Initial (base) state (8)	53,22	251,4	0,986
Welded seam (4-1)	61,34	341,2	0,957
Welded seam + heat treatment (4-2)	71,66	361,8	0,985

Figure 5 shows a typical loading curve for samples of alloy G-35 (weld metal). A summary of the mechanical test results is given in Table 3.

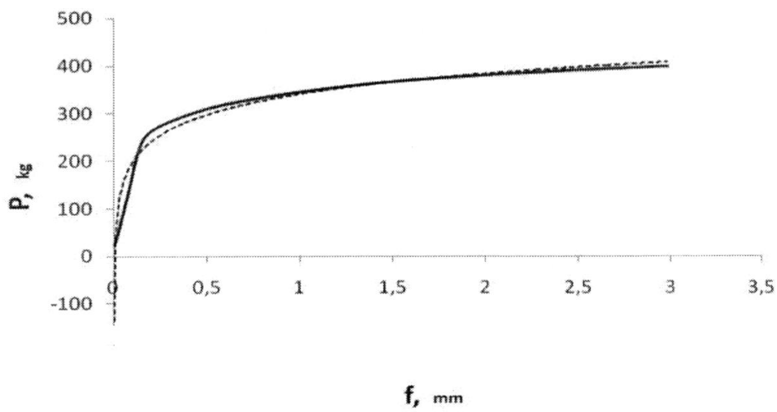

Figure 5. Typical loading curve of G-35 alloy sample with three-point bending (weld sample 4-2)

Table 3. The results of mechanical tests of the alloy G-35

№№	Sizes of samples, mm.			Mechanical properties				P, mm	ρ, mm.
	b, mm	h, mm	W, mm^3	σ_{02}, MPa	$\sigma_в$, MPa	δ, %	φ, %		
8-1д	3,56	3,78	8,477784	253,38	520,65	26,7	25,92	6,00	0,00112
8-2д	3,53	3,79	8,450879	244,36	477,75	24,5	27,17	7,20	0,00113
8-3д	3,47	3,72	8,003208	240,21	434,85	24,4	28,02	7,10	0,00115
8-4д	3,57	3,72	8,233848	285,42	391,95	20,1	23,76	3,43	0,00112
8-1	3,12	3,74	7,273552	292,03	657,15	33,7	27,87	4,63	0,00128
4-1	3,08	3,80	7,412533	279,28	649,35	33,3	28,90	4,62	0,00130
4-2	3,50	3,77	8,290858	320,77	639,60	32,8	29,78	5,33	0,00114

Conclusion

1. A preliminary study of the structure and mechanical properties of the G-35 alloy was carried out in order to select the methods for their investigation.
2. By the PHASI method, it was established that the weld metal is in the austenitic state.
3. Alloying of the alloy with boron, not specified in the alloy passport, has been established.
4. A class of functions has been established that describes the loading curves for three-point bending of samples.

Acknowledgments

The work was carried out according to the state task No. 007-00129-18-00 and with the financial support of the Russian Foundation for Basic Research (grant No. 17-08-00098a).

References

[1] Saito Y, Onay B and Maruyama T 2012, *High Temperature Corrosion of Advanced Materials and Protective Costings (Elsevier).*
[2] Ermishkin V.A., Murat D.P., Podbelsky V.V. Information technology of photometric analysis of fatigue damage of materials. *Information Technologies* 2007. **11**, p. 65-70.
[3] Ermishkin V.A., Murat D.P., Podbelsky V.V. The system of photometric analysis of structural images and its application to the study of materials in fatigue conditions. *Instruments and systems. Management, monitoring, diagnostics.* 2008. **10** p. 38-44
[4] Ermishkin V.A., Minina N.A., Fedotova N.L. "Method of photometric diagnosis of phase transformations in solids according to the analysis of the brightness spectra of the reflection of light from their surface", Patent No. 2387978, 2010. Bull. №12.

IOP Conf. Series: Materials Science and Engineering **525** (2019) 012015 doi:10.1088/1757-899X/525/1/012015

Identification of organic compounds using a multisensor detector based on SnO_2 films in gas chromatography

I I Efimov and V G Povarov

Saint-Petersburg Mining University, 199106, Saint-Petersburg, Russia.

E-mail: efimov.ignaty@gmail.com

Abstract. The possibilities of using a semiconductor multisensor device as a detector for gas chromatography are considered. Semiconductor films of various compositions were investigated for their performance as sensors, the optimal composition for inertia and sensitivity was established. The basic parameters of the detector were determined: sensitivity, linear dynamic range, detection limit. The applicability of the detector for solving typical problems of capillary gas chromatography is shown. Two methods for the identification of organic compounds using a detector based on three sensors have been proposed.

1. Introduction

The use of tin dioxide film sensors for analyzing various compounds in gaseous media began in the mid-1960s [1]. Currently, they are among the most common types of gas sensors [2]. The principle of their operation is based on the increase in the electrical conductivity of the film when the gaseous substance contacts the surface of the sensor. This effect is the result of several processes, among which the most important are gas adsorption and its effect on the surface Fermi levels of a semiconductor film. With increasing temperature, oxidation of volatile substances begins with sorption. This leads to the generation of new charge carriers in the surface layer of the film and opens up possibilities to the detection of organic compounds. Since the mid-1970s there studies in which sensors of this type are connected with simple gas chromatography [3]. It turned out that this type of sensor has a higher inertia compared to typical gas chromatography detectors (flame ionization, thermal conductivity detector). In addition to this drawback, there is the problem of the correct interpretation of the analytical signal of an individual sensor and the choice of its processing algorithm. This is necessary to create a universal and effective approach to the identification of organic compounds using this type of detector. An important task is the selection of such additives to tin dioxide and such film formation methods that would provide a set of selective, stably operating and low-inertia film sensors suitable for the analysis of multicomponent organic mixtures [4].

Proposed in [5, 6], a thermocatalytic detector based on tin dioxide with additions of transition metals can be used to perform typical analyzes in gas chromatography using both packed and capillary columns. As an illustration, Figure 1 shows the peak of hexane on two film sensors of different composition. The sensor signals differ in height and width, which can be explained by the different intensity of charge carrier generation processes and the difference in sorption / desorption times of organic compounds. Such a difference in the signals can be used to identify the components of the analyzed mixture.

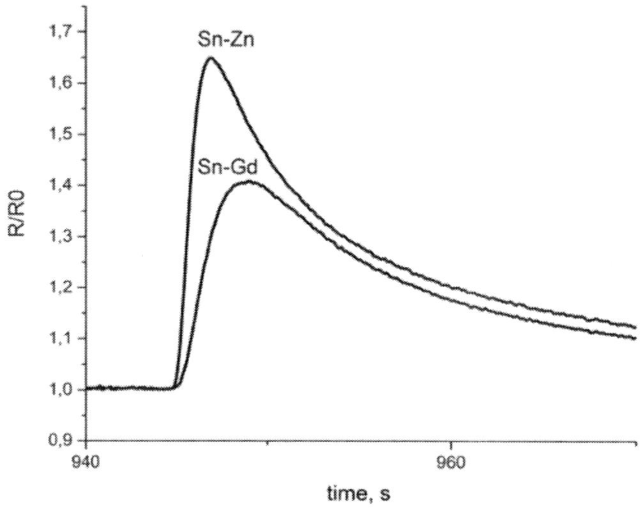

Figure 1. Analytical signal of 1 ml of saturated vapor of hexane on a pair of sensors (SnO_2-ZnO and SnO_2-Gd_2O_3)

2. Experimental part

SnO_2 films were deposited by pyrolytic decomposition. The glass tubes, with diameter of 2mm, were used as the substrate. Solutions for application were prepared by dissolving tin (IV) chloride in ethanol followed by the addition of nitrates of doping metals (Cu, Zn, Ni, Ag, Gd, Ce). The mass concentration of tin in the solution was 1% by weight. The ratio of tin to metal was 100:3. The process of film deposition occurred in two stages. In the first stage, the substrate was wetted with a solution and ethanol was evaporated at 120 ° C for 5 minutes. After that the substrate was calcined in a muffle furnace at a temperature of 720 ° C for 5 minutes and cooled to room temperature. The deposition process was repeated 3-5 times since, with a smaller number of depositions, the conductivity of the film remained too low for use as a sensor. In some cases, to improve the properties of the sensor, doping was carried out simultaneously with two oxides (Me_xO_y + Ag_2O), the amount of silver used was 10% of the amount of Me [7]. After the films were deposited, silver-coated steel wire was attached to opposite sides of the substrate and the film was incorporated into a measuring circuit.

Sensors (3 pieces) were placed in a quartz tube located in the heater. The output of the chromatographic column is connected to the bottom of the tube through a side entrance made of air flow (50 ml / min.). Oxygen is necessary for the rapid restoration of the working surface of the sensor, it also acts as a make-up gas. The temperature of the thermostat during the experiments was equal to 450 ° C, which ensured optimal selectivity and response of the sensors. The detector circuit is shown in Figure 2. The system was mounted on an Agilent serial gas chromatograph.

The measuring circuit of the detector is shown in Figure 3. It consists of a variable resistance, a sensor and a DC source with a voltage of 4.5V. All elements are connected in a series electric circuit. The magnitude of the voltage drop across the variable resistance was used as a signal, which was recorded using an ADC with a frequency of 10 Hz.

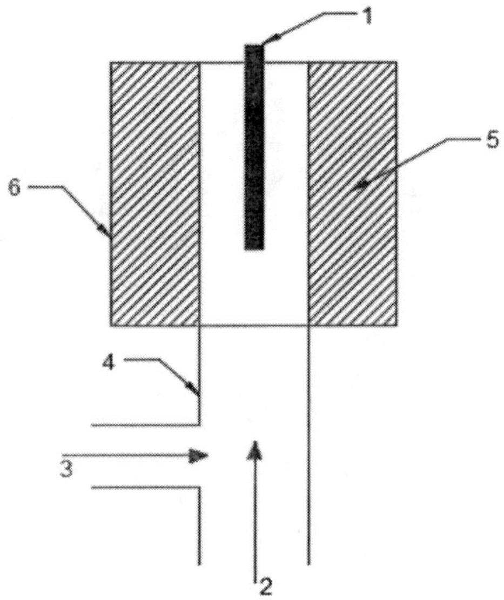

Figure 2. The detector diagram and its connection with a gas chromatograph. 1 - sensor, 2 - output end of the capillary column, 3 - air flow, 4 - quartz tube, 5 - heating element, 6 - heater body

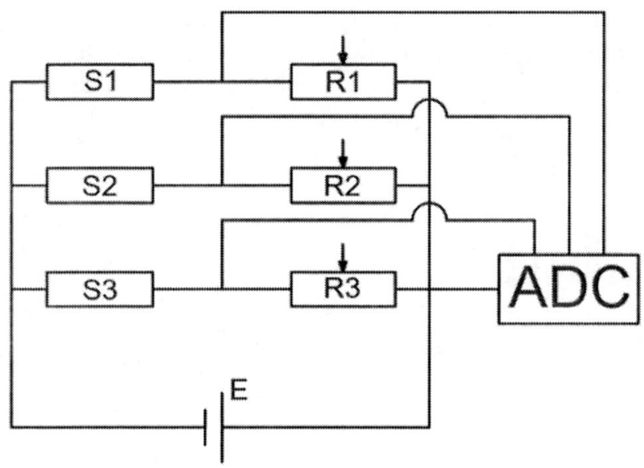

Figure 3. Measuring circuit S - sensor, E - EMF source, ADC – Analog-to-digital converter, R - variable resistor

Chromatogram processing was carried out in several main stages. At first, the chromatogram was manually labeled using a program written in C#. The primary dependence of the voltage drop on time was converted into the relative electrical conductivity of the sensor according to the formulae:

$$R/R_0 = (E/U_0 - 1)/(E/U_{max} - 1)$$

where E, U_0 и U_{max} - the EMF of the power source, the recorded voltage at the peak base and at the maximum of the peak, respectively. An example of conversion is shown in Figure 4.

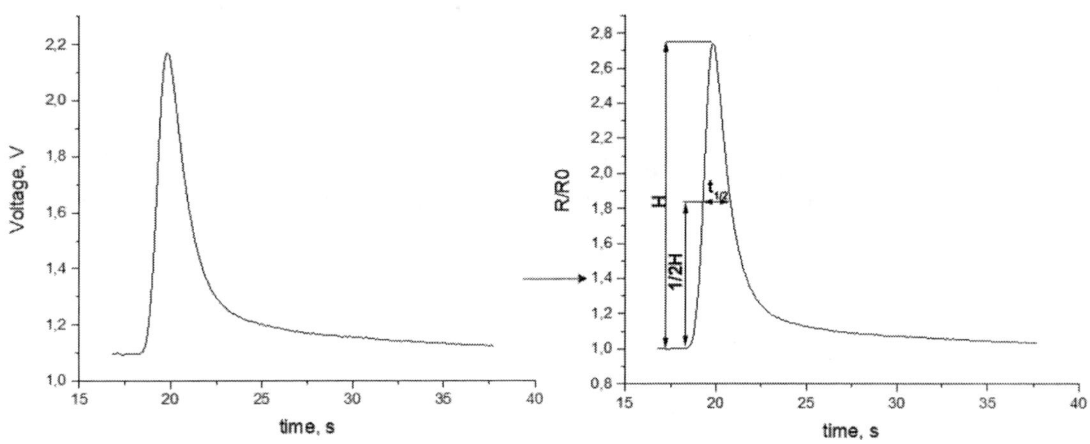

Figure 4. An example of transformation of the primary sensor signal to the dependence of the increase in electrical conductivity on time

After recalculation, the main parameters (height and width at half height) were extracted from the chromatogram for each peak. The text of the program is written in the programming language Python

IOP Conf. Series: Materials Science and Engineering **525** (2019) 012015 doi:10.1088/1757-899X/525/1/012015

3.7.1. The coordinate of the intersection point between half-width and peak was determined by interpolation at two adjacent points. Since the electrical conductivity of the sensor in the absence of the signal was minimal, y_{min} was taken as 1.

3. Results and discussion

Mixed alcohols and n-alkanes C_9-C_{14}, as well as solutions of essential oils, were studied as objects. Figure 5 shows an example of a chromatogram of fir essential oil in hexane, obtained on a SnO_2-Gd_2O_3 film sensor. As previously mentioned, the sensors have different responses, namely the heights and widths of the peaks.

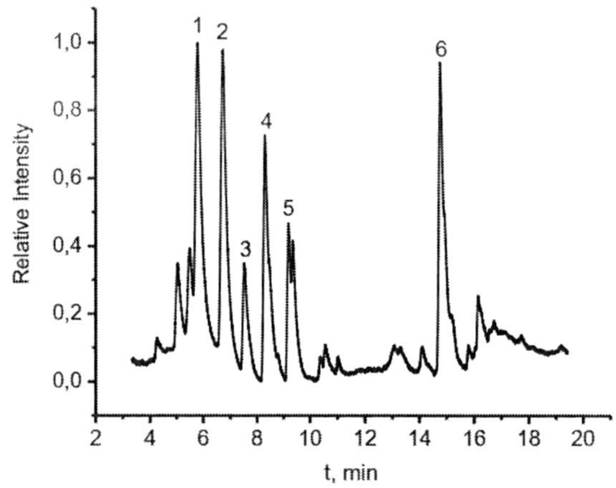

Figure 5. Chromatogram of fir essential oil in hexane. 1 - α-pinene, 2 - camphene, 3 - β-felllandrene, 4 - γ-terpene, 5 - limonene, 6 - bornyl acetate

The choice of the optimal composition of the films was made according to their sensitivity and inertia. For this, the sensors were calibrated for three substances (ethanol, hexane, acetone). The detection limit was calculated as a triple noise value multiplied by the sensitivity of the sensor. Linear dynamic range (LDR) was determined graphically from calibration curves. Films with the addition of ZnO, CuO, NiO turned out to be the most sensitive. It was also found that the addition of Ag_2O as the second dopant improves the inertial properties of the SnO_2-CuO and SnO_2-NiO films. The measurement results are presented in Table 1. It may be noted that, in terms of the LDR, film detector is similar to an electron capture, but at the moment gives way to serial detector sensitivity. Insufficient sensitivity can be corrected by the signal amplifier, but even in this form, the system is applicable for quantitative analysis

Table 1. Parameters of film sensors.

Sensor	Sensitivity, conventional units of area / mg	Detection limit, g	Linear range
Sn-Zn	512	$1{,}7 \cdot 10^{-5}$	10^2
Sn-Cu(Ag)	335	$0{,}5 \cdot 10^{-5}$	10^2
Sn-Ni(Ag)	423	$0{,}4 \cdot 10^{-5}$	10^2

To identify compounds in the mixture, it was proposed to use the difference in peak heights and their half widths, as well as additional chromatographic information in the form of retention indices. In order to eliminate the influence of the amount of substance and the velocity of the carrier gas on the identification features, we did not use the heights and half widths themselves, but their relations for two different sensors. As it turned out, within the linearity domain, the ratio of heights remains constant. Similarly, when the carrier gas speed changes, the ratio of half widths of peaks remains constant. Thus, if there are three sensors, we can calculate two independent height ratios and two independent half-width ratios. Considering the various combinations of such relationships, we sought to select such variants in which the entire set of the studied substances differs in the best way. In other words, the image of the figurative points of the compounds in the coordinates of the corresponding set should be as far as possible from each other, not depend on the quantities taken, and demonstrate a logical change in the homologous series.

In total, 5 different dependencies were investigated as potentially interesting variants: H_1/H_3-H_2/H_3, H_1/H_3-Retention index, W_1/W_3-W_2/W_3, H_1/W_1-H_2/W_2, H_1/H_3-W_1/W_3, where H_n and W_n are the height and half width of the signal of the n-th detector. As a result, it was found that the dependencies W_1/W_3-W_2/W_3 and H_1/W_1-H_2/W_2 do not allow the identification of substances due to the imposition of the figurative points of different classes of organic substances. The dependences of H_1/H_3-H_2/H_3 and H_1/H_3-Retention index (Fig. 6) showed themselves in the best way. As you can see, the identification signs of each substance are grouped in certain areas on the plane, and with increase in the molecular weight of alcohols, the figurative points of the compounds are gradually removed from the origin.

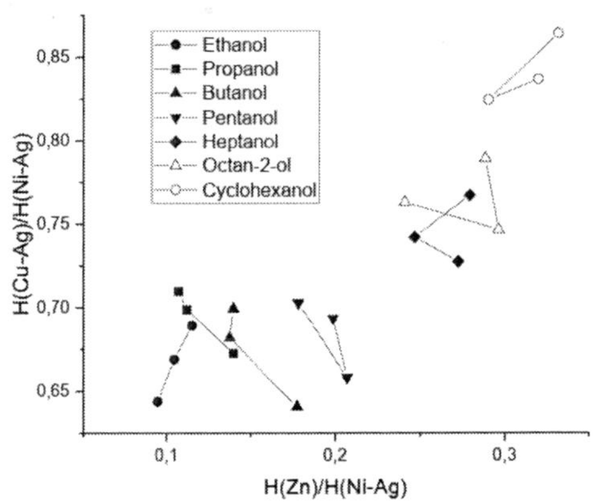

Figure 6. Figurative points of alcohols on the signal relationship plane.

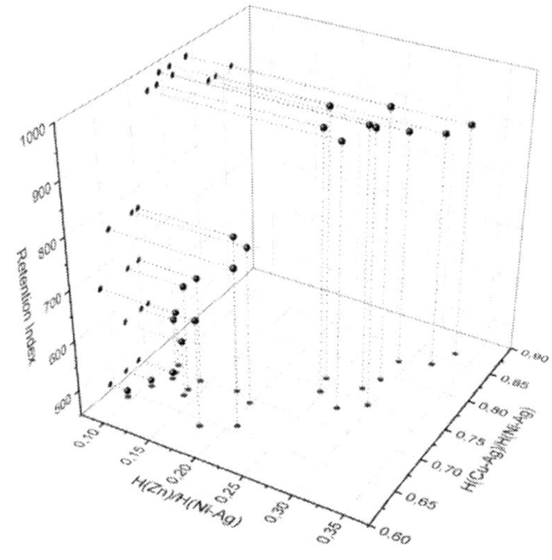

Figure 7. Figurative points of alcohols in the three-dimensional coordinate system.

The accuracy of identification can be significantly increased if retention indexes are used as the third parameter (Fig. 7). This technique is quite common in gas chromatography.

4. Conclusions

Thus, on the basis of the proposed data, we can draw a conclusion about the applicability of multisensor systems for solving typical problems of capillary gas chromatography. Its main advantage is the ability to identify compounds in mixtures. The proposed approach allows for the most qualitative separation of the figurative points of various compounds. However, the current design of the detector is not sensitive enough, which can be corrected through rational selection of the film composition, as well as the use of amplifying circuits.

IOP Conf. Series: Materials Science and Engineering **525** (2019) 012015 doi:10.1088/1757-899X/525/1/012015

References

[1] Seyama T, Kato A, Fujishi K and Nagatani M *1962 Anal. Chem. 34 1502*
[2] Barsan N, Koziej D and Weimar U 2007 Sens. Actuators B **121** 18–35
[3] Beling S, Bläser G, Bock J, Heinert L, Traxler M and Kohl D. *1998 Sens. Actuat. B. 52 15*
[4] Eranna G, Joshi B C, Runthala D P and Gupta R P 2004 Solid State Mater. Sci. **29** 111–188
[5] Povarov V G and Lopatnikov A I 2016 Journal of Analytical Chemistry **71** 901-906
[6] Povarov V G , Lopatnikov A I, Geibo D S , Bulatova A V and Razgonyaev A O 2013 Journal of Analytical Chemistry **68** 300-304
[7] Madler L, Roessler A, Pratsinis S E, Sahm T, Gurlo A, Barsan N and Weimar U 2006 Sens. Actuators B **114** 283–295

IOP Conf. Series: Materials Science and Engineering **525** (2019) 012016 doi:10.1088/1757-899X/525/1/012016

Development of functional materials with specific activities for degradation of toxins

E Efremenko[1,2], R Ahundov[1], A Aslanli[1], I Lyagin[1,2], O Senko[1,2], O Maslova[1] and N Stepanov[1,2]

[1]Lomonosov Moscow State University, Moscow, Russia
[2] Emanuel Institute of Biochemical Physics RAS, Moscow, Russia

E-mail: elena_efremenko@list.ru

Abstract. It was shown that it is possible to use different samples of hexahistidine-containing organophosphate hydrolase (His_6-OPH)-based enzyme preparations, including those obtained in the frame of functionalized materials, for the successful detoxification of organophosphorus compound such as Paraoxon and different mycotoxins (Zearalenone, Patulin, Sterigmatocystin) in their present in contaminated agricultural commodities. For the first time it was demonstrated that different toxins (xenobiotics and toxic compounds naturally produced by microscopic fungus) can be hydrolysed by the same enzyme and various biocatalysts prepared on the base of it use.

1. Introduction

Agricultural products can be contaminated with various xenobiotics, in particular, organophosphate pesticides, which are actively used at different stages of production and preservation of the crop, as well as natural toxicants, such as mycotoxins (McT), which produce various microscopic fungous cells that affect agricultural products [1]. These natural and synthetic toxins (organophosphorus compounds, OPC) cause significant harm to human and animal health, consuming agricultural products that contain them or having regular contact with various raw materials containing these substances to be stored or processed. The main problem is not the once consumed doses of these toxicants, causing acute poisoning, but regular consumption of these substances in small quantities, which leads to the formation of a cumulative effect and the development of chronic neurodegenerative diseases, nephropathy, hepatopathy, oncology.

To neutralize these toxicants, various means are being developed and applied in practice, but the most interesting and promising are combined functional materials that allow not only to ad(b)sorb toxic compounds from the mass (or surface) of agricultural raw materials or products, but also to protect them and catalyze their degradation. Since then, these materials should contain, first of all, various sorbents, including those presented in the form of fibrous materials, possessing enzymatic activity aimed at the destruction of McT or organophosphorus pesticides widely used in agriculture. Moreover, such materials are functionalized, as a result of imparting to them certain special catalytic characteristics.

It should be noted that the microbiologically confirmed absence of agricultural raw materials or the production of microscopic fungi does not mean that the object under study does not contain McT. So filamentous fungi can die during the processing of grain or other agricultural raw materials, and McT can retain their presence, since they are more resistant to chemical and heat treatments (melting point of zearalenone is 165°C, ochratoxins - 169–221°C, aflatoxins - 244–299°C, trichothecenes - 150–190°C). However, for people servicing storage and production facilities in which they come in contact with contaminated agricultural raw materials containing McT, protection from exposure to these substances is necessary, which can be implemented at the transdermal and inhalation levels, as well as *in vivo*.

There are no safe concentrations of MCT, since even a small amount of them has a negative effect on the animal's body while in cereal feeds, while they can accumulate in animal products (meat, milk, eggs), creating a trophic threat to human health. At the same time, even very low concentrations of McT can affect the state of health, since they "cooperate" with bacterial infections and the effects of xenobiotics, in particular, organophosphate pesticides, and thereby increase the overall negative effect. In addition, toxins reduce the body's resistance (suppress immunity), which increases its susceptibility to various diseases. Therefore, early and effective control of McT as well as defense against McT as well as against OPC is essential for animal health and the profitability of agricultural production.

However, there is another important problem associated with the accumulation of McT in agricultural products, which is that often in the feed there is the presence of several types of McT, which is due to the fact that each microscopic fungus can produce several toxins simultaneously. In this regard, in order to create effective means of neutralizing McT, it is necessary to select "maximally universal" enzymes that allow them to be used for the destruction of the widest possible range of McT.

As for the enzymatic and therefore environmentally friendly decomposition of organophosphorus pesticides in the same agricultural raw materials or products, here the selection of enzymes for such processing also involves the use of biocatalysts with the broadest substrate specificity [2]. The most interesting are the development of functional materials that allow decomposition of both McT and organophosphate pesticides.

The purpose of this work was to show that such biocatalytic functional materials can be obtained with respect to degradation of McT and OPC happening within contact or exposure of various agricultural raw materials with such biocatalytically active samples of materials. It was nesessary to show how it can lead to removal of McT and OPC from raw materials. It was decided to add additional functionality to these materials due to the presence of specially introduced stabilized forms of hexahistidine-containing organophosphate hydrolase (His_6-OPH), catalyzing not only the hydrolysis of various OPC, but also the McT [3-5]. The most effective way to stabilize this enzyme to neutralize these toxins by obtaining of non-covalent polyelectrolyte complexes of the His_6-OPH with polyaspartic acid was chosen previously [6], and was used in this work, taking into account the previously obtained positive results.

2. Materials and Methods

2.1. Contaminated objects for investigation

Agricultural raw materials (haylage, silage, feed grain, cake, pulp, meal) and feed (grain mixture, feed, extrudates), with the exception of flour (meat and bone, feather, fish) were prepared for the study as follows: dry substrates and substrates with low humidity thoroughly mixed and ground to 1 mm, silage, pulp and meal were ground with a knife and thoroughly mixed, flour was used without preliminary preparation. A sample of each of the feedstock was mixed with 1 mM Paraoxon solution to a final moisture content of 75 ± 5%. The obtained samples of raw materials were brought into contact with samples of functionalized materials containing enzyme or enzyme complexes, which were then exposed for 24 hours at room temperature.

2.2. Obtaining of biocatalytically functionalized materials

Recombinant cells of *Escherichia coli* SG13009[pREP4] were cultivated, and the enzyme was isolated and purified by methods published previously [7]. The purified preparation of His_6-OPH was characterized by enzymatic activity as described earlier [8]. The concentration of protein was determined by Bradford assay, and protein purity was analyzed by SDS-PAGE in 12% polyacrylamide gel using Mini-PROTEAN II cell (Bio-Rad, Hercules, CA, USA) followed by Coomassie Blue (R-250) staining. According to SDS-PAGE (data not shown), the purity of His$_6$-OPH dimer preparation obtained (MW ≈ 73 kg·mol^{-1}) was ca. 98%. The specific organophosphorus hydrolase activity of the purified enzyme was 4,300 U·mg^{-1}. One unit of enzymatic activity (U) was defined as the quantity of the enzyme necessary to hydrolyze 1 μmol of Paraoxon per min at 25°C.

Enzyme-polyelectrolyte complexes (EPC) were produced similarly to the known procedure [9, 10]. In brief, to produce an EPC of the enzyme with polyasparaginic acid (His$_6$-OPH/PLD$_{50}$), an aliquot of 20 g·L^{-1} PLD$_{50}$ water solution was added to the purified His$_6$-OPH (1.6 ± 0.1 g·L^{-1}) in 50 mM phosphate buffer (pH 7.5) containing 150 mM NaCl. The aliquot volume was calculated so that "enzyme:polymer" molar ratio was 1:5. Next, the mixture was held for 30 min at 8°C.

The organophosphorus hydrolase activity was determined as described earlier [8] with 7.8 mM aqueous Paraoxon stock solution at 405 nm using the Agilent 8453 UV-visible spectroscopy system (Agilent Technology, Waldbronn, Germany) equipped with a thermostatted analytical cell. The reaction was carried out in a 0.1 M carbonate buffer (pH 10.5).

Diatomite (Kvant, Russia), DVUM-carbon-containing fibrous material (Russia) and bacterial cellulose obtained according to a previously developed and published procedure [11] were chosen for biocatalytic fictionalization using the enzyme and its EPC.

3. Results and discussion

To demonstrate the feasibility of detoxification of OPC in agricultural raw materials and feed, specially prepared samples of various materials containing His$_6$-OPH or its EPC as stabilized forms of the enzyme were used.

To determine the effectiveness of detoxification of OPC, which was carried out for 1 day at room temperature under the action of biocatalytically functionalized materials and different active forms of the enzyme, extraction of OPC (as which Paraoxon was applied) from the obtained samples was carried out and the residual amounts of the contaminant were determined (Table 1).

Table 1. The effectiveness of the destruction of OPC in contact with powdered feed and agricultural raw materials under the action of the created samples of biocatalytically functionalized materials

Raw material	Residual concentration of Paraoxon, %				
	His$_6$-OPH	His$_6$-OPH/PLD$_{50}$	His$_6$-OPH with diatomite	His$_6$-OPH/PLD$_{50}$ with bacterial cellulose	His$_6$-OPH/PLD$_{50}$ with carbon containing fibrous material
Haylage	29.5 ± 1.1	12.2 ± 0.5	16.7 ± 0.2	19.7 ± 0.4	11.7 ± 0.6
Silage	22.2 ± 0.8	13.7 ± 0.5	17.2 ± 0.3	18.6 ± 0.6	10.3 ± 0.2
Coarse grains	24.6 ± 0.9	14.6 ± 0.4	16.4 ± 0.2	17.4 ±0.3	10.2 ±0.5
Bagasse	19.8 ± 0.8	12.8 ± 0.3	19.8 ± 0.3	16.8± 0.5	9.6 ± 0.5
Beet pulp	29.4 ± 1.1	13.2 ± 0.3	19.7 ± 0.3	15.6 ± 0.5	9.8 ± 0.6
Grist	22.5 ± 0.8	12.5 ± 0.3	15.5 ± 0.2	18.3 ± 0.3	9.3 ± 0.4
Grain mixture	24.2± 0.9	12.7 ± 0.3	14.6± 0.2	18.8 ± 0.4	8.8 ± 0.4
Mixed feed	29.1 ± 1.1	10.1 ± 0.3	19.1 ± 0.3	19.0 ± 0.6	9.0 ± 0.5
Extrudate	27.3 ± 0.9	12.0 ± 0.3	17.0 ± 0.3	18.2 ± 0.3	12.0 ± 0.4
Meet and bone meal	25.0 ± 0.8	17.8 ± 0.6	15.8 ± 0.2	21.3 ± 0.7	8.8 ± 0.3
Feather meal	26.8 ± 0.9	16.8 ± 0.5	18.6 ± 0.3	20.2 ± 0.4	9.3 ± 0.5
Fish meal	20.8 ± 0.7	10.6 ± 0.3	19.1 ± 0.3	19.3 ± 0.3	9.8 ± 0.4

It was found that even taking into account the obtained initial results of the experiments without any optimization of the conditions for the detoxification of OPC, as well as the applied enzyme activity, we can say that the enzyme, as well as its EPC with polyaspartic acid, has a high catalytic activity, which allows to count on the creation of new functionalized materials with special properties that form protection from exposure to toxicants.

Further, various biocatalytically active samples prepared on the basis of His_6-OPH, allowing to maintain the stability of the enzyme action, were used to remove McT from different types of raw materials and types of animal feeds (Table 2).

It should be noted that there is a known method for bioprotection of zearalenone (43.4 mg L^{-1}) in Tris-HCl buffer (pH 7.5) under the action of highly purified polyhistidine-containing recombinant lactonase, taken at a concentration of 1.2 mg L^{-1}, which allows to reach 100% efficiency of the mycotoxin decomposition within 38 hours [12]. The use of recombinant lactonase for the application of this method in practice demands the production of the enzyme in the required quantity and purification degree to the maximum level of catalytic activity, but the procedure is not optimized yet. However, a significant disadvantage of this method is the low rate of hydrolysis of Zearalenone (1.142 mg L^{-1} h^{-1}) and the possibility of using this enzyme method only for the hydrolysis of one mycotoxin, Zearalenone.

Table 2. Destruction of McT under the action of various forms of His_6-OPH

Raw material or fodder	Mycotoxin (concentration)	Catalytically active sample tested in the work	Period of 100%-decomposition, h
Wheat	Zearalenone (300 mg kg $^{-1}$)	His_6-OPH	3
Meet and bone meal	Sterigmatocystin (100 mg kg^{-1})	His_6-OPH/PLD_{50} with diatomite	2
Grain mixture	Patulin (2 mg kg $^{-1}$)	His_6-OPH/PLD_{50} with bacterial cellulose	1

In comparison with this method, the results achieved here are unique because:

- the possibility of carrying out the enzymatic biological neutralization of McT present in media with a complex chemical composition, and not only in the form of pure substances, was shown,
- a significant increase in the rate of effective decomposition of McT is shown while reducing the time required for their 100% decomposition (only in comparison with a pure substance - Zearalenone - an increase in the decomposition rate by an order of magnitude).

It should be noted that oxidative enzymes tested for the decomposition of McT, in particular, oxidoreductases (laccase, cytochrome P-450, F420-dependent oxidoreductase, Mn-containing peroxidase) [13, 14] provided a rather high rate of neutralization McT. For example, a maximum (90%) degradation of McT was achieved in 48 hours at a concentration of 312 mg L^{-1}, and the rate of decomposition of McT was 5.85 mg L^{-1} h^{-1}[15]. However, these methods of enzymatic oxidative decomposition of McT have a number of significant drawbacks: instability of the used enzymes, requirements for the special conditions of their storage and use in methods of McT detoxifying (use of additional expensive redox-mediators, introduction of hydrogen peroxide to the environment, the need to ensure elevated oxygen concentrations in media containing McT, etc.), and others. However, their main disadvantage is that as a result of the use of these oxidative enzymes, products are not less, but sometimes more toxic than the original McT [16, 17].

The results achieved herein enable a highly efficient biocatalytic detoxification of not one, but several different McT having a lactone ring in their molecular structures, using the same enzyme His_6-OPH, due to which the solution of the problem of decomposition and detoxification of McT produced by different filamentous fungi can be significantly simplified in practice.

4. Conclusion

Thus, it was shown that it is possible in principle to use different enzyme preparations based on His6-OPH, including in the composition of fibrous functionalized materials, for the detoxification of OPC using the example of paraoxon and several McT, which are often present in agricultural raw materials. In addition, for the first time, the possibility of decomposition of individual McT, in particular, Patulin and Sterigmatocystin, for which similar enzymatic decomposition has never been described anywhere else, has been shown. These results can solve the significant environmental and economic problem associated with the disposal of McT and an increase in the level of safe contact for the storage and processing of such raw materials.

Acknowledgment

This work was financially supported by RFBR (Project 18-29-17069).

References

[1] Jard G, Liboz T, Mathieu F, Guyonvarc'h A, and Lebrihi A 2011. Review of mycotoxin reduction in food and feed: from prevention in the field to detoxification by adsorption or transformation. *Food Addit Contam Part A*, **28(11)** 1590-1609.

[2] Efremenko E N, Lyagin I V, Aslanli A G and Makhlis T A 2017 The method of bioneutralization of mycotoxins. Ru Patent 2634914

[3] Efremenko E N, Senko O V and Maslova O V 2017 Optimization of the use of His_6-OPH-based enzymatic biocatalysts for the destruction of chlorpyrifos in soil. *Int. J. Environ. Res. Public Health.* **14** 1438.

[4] Efremenko E N, Lyagin I V, Makhlis T A, Maslova O V, Senko O.V., Stepanov N A 2017 Enzymatic detoxification of mycotoxins: known and promising solutions. *Acta Naturae* **special issue** 124-125.

[5] Maslova O V, Stepanov N A, Grigoryeva A I, Bruyako M G, Efremenko E N 2016 New effective His6-OPH-containing mineral carriers pretreated by low-temperature plasma for destruction of organophosphates in different types of soil. *IJPT* **8(4)** 27317-27333.

[6] Maslova O V, Senko O V and Efremenko E N 2018 The influence of enzymatic removal of chlorpyrifos from feed grain mixes on biochemical parameters of rat blood. *Biochem (Mocs) Suppl Ser B* **12(2)** 181–185.

[7] Efremenko E, Votchitseva Y, Plieva F, Galaev I and Mattiasson B 2006 Purification of His_6-organophosphate hydrolase using monolithic supermacroporous polyacrylamide cryogels developed for immobilized metal affinity chromatography. *Appl. Microbiol. Biotechnol.* **70**, 558–563.

[8] Votchitseva Y A, Efremenko E N, Aliev T K and Varfolomeyev S D. 2006 Properties of hexahistidine-tagged organophosphate hydrolase. *Biochemistry (Moscow)* **71** 167–172.

[9] Efremenko E N, Lyagin I V, Klyachko N L, Bronich T, Zavyalova N V, Jiang Y and Kabanov A V 2017 A simple and highly effective catalytic nanozyme scavenger for organophosphorus neurotoxins. *J. Control. Release* **247** 175–181.

[10] Lyagin I V and Efremenko E N 2018. Biomolecular engineering of biocatalysts hydrolyzing neurotoxic organophosphates. *Biochimie* **144** 115–121.

[11] Stepanov N and Efremenko E 2018 "Deceived" concentrated immobilized cells as biocatalyst for intensive bacterial cellulose production from various sources *Catalysts* **8** 1–16.

[12] Vekiru E, Fruhauf S, Hametner C, Schatzmayr G, Krskal R, Moll W D and Schuhmacher R 2016 Isolation and characterisation of enzymatic zearalenone hydrolysis reaction products *World Mycotoxin J.* **9(3)** 353-363.

[13] Wu Y Z, Lu F P, Jiang H L, Tan C P, Yao D S, Xie C F and Liu D L 2015 The furofuran-ring selectivity, hydrogen peroxide production and low Km value are the three elements for highly effective detoxification of aflatoxin oxidase. *Food Chem. Toxicol.* **76** 125-131

[14] Alberts J F, Gelderblom W C A, Botha A and van Zyl W H 2009 Degradation of aflatoxin B1 by fungal laccase enzymes Int. J. Food Microbiol **135** 47–52.

IOP Conf. Series: Materials Science and Engineering **525** (2019) 012016 doi:10.1088/1757-899X/525/1/012016

[15] Yehia R S 2014 Aflatoxin detoxification by manganese peroxidase purified from *Pleurotus ostreatus Braz. J. Microbiol.* **45(1)** 127-133.
[16] Zeinvand-Lorestani H, Sabzevari O, Setayesh N, Amini M, Nili-Ahmadabad A and Aramarzi M 2015 Comparative study of in vitro prooxidative properties and genotoxicity induced by aflatoxin B1 and its laccase-mediated detoxification products *Chemosphere* **135** 1–6.
[17] Karlovsky P 1999 Biological detoxification of fungal toxins and its use in plant breeding, feed and food production *Nat. Toxins* **7(1)** 1-23.

IOP Conf. Series: Materials Science and Engineering **525** (2019) 012017 doi:10.1088/1757-899X/525/1/012017

The prospects for hydrothermal processing leucoxene concentrates for the production of synthetic wollastonite

Y V Zablotskaya, G B Sadyhov, K. G Anisonyan and T V Olyunina

A. A. Baikov Institute of Metallurgy and Materials Science, Russian Academy of Sciences, 119334 Leninskii pr. 49, Moscow, Russia.

E-mail: Nboxclear@gmail.com

Abstract: The paper presents the results of a study of the simultaneously receive β-wollastonite in autoclave beskrainiye leucoxene ore Yarega field. The parameters of thermal treatment of calcium metasilicate hydrate with the production of β-wollastonite were determined by thermogravimetric and x-ray phase analyses. With the help of electron microscopy it was found that the resulting CaSiO3 has a needle shape. The possibility of using the obtained wollastonite in the production of ceramic composites was studied and the compositions of ceramic masses and the modes of its heat treatment were determined.

1. Introduction

World production of wollastonite (calcium metasilicate $CaSiO_3$) is about 1 million tons per year [1]. According to the data of the U.S. Geological Survey, natural wollastonite is produced by a relatively small number of mining enterprises located in the following countries: China, India, USA, Mexico, Finland [2]. Synthetic wollastonite is produced mainly in European countries that do not have large reserves of natural wollastonite (UK, Denmark, France, Germany). In Russia, wollastonite is rather an unconventional type of raw material and the volume of its consumption is very small.

The main areas of application of wollastonite are the production of ceramics 30-40%, polymers (plastics and rubber) 30-35 %, paints 10-15%, as well as used in the construction industry, metallurgy. The world market of wollastonite can be divided into two segments – needle wollastonite with a high ratio of grain length (L) to diameter (d) (L/d>3) (high-aspect-ratio – HAR) and powder wollastonite with L/d<3 (low-aspect-ratio – LAR). Grades HAR are mainly used as micro-reinforcing fillers in polymers, elastomers and coatings to provide thermal resistance, corrosion resistance and abrasion resistance. Wollastonite of the grades LAR is used in those areas where the chemical composition is more important than the physical (for example, reinforcing) properties [1].

Small proven reserves of natural wollastonite in the world and its price, as well as the increasing demand for this raw material stimulate interest in the development of methods for producing synthetic wollastonite. The cost of synthetic wollastonite is almost two times lower than the commodity price of natural. The properties and structure of synthetic wollastonite will be determined by the nature of the raw materials, the ratio of calcium and silicon - containing components, as well as the synthesis conditions. Unlike natural wollastonite, synthetic material is characterized by a higher degree of uniformity in composition and low content of impurities, as well as adjustable dispersion.

The main methods of synthesis of wollastonite are: solid-phase synthesis (sintering of calcium and silicon - containing components at 1000-1400 ° C), hydrochemical and hydrothermal synthesis of calcium hydrosilicates at temperatures from 80 ° C to 350 ° C and their subsequent dehydration and calcination at temperatures above 800 ° C [3]. The hydrothermal method is more promising and manageable for industrial use, since it allows to simultaneously vary many parameters of the process (temperature and duration of synthesis, chemical composition of hydrothermal solutions and the ratio of components) to obtain calcium hydrosilicates of a certain stoichiometric composition. The raw materials for the synthesis of wollastonite are a variety of compounds of calcium and silicon of natural (diatomite, flask, cristobalite, tridimite, gypsum, chalk, calcite, marble, limestone) or man-made origin (for example, phosphogypsum, borohypse, quartz glass, blast furnace slag).

In Russia, the Republic of Komi is one of the largest titanium deposits - Yareg silicon-titanium sandstones. Main titanium mineral is leucoxene, which is a mineral the structure of thin intergrowths of rutile and quartz (about 40%). Processing of such high-silicon raw materials is associated with significant difficulties. In IMET RAS proposed a new complex approach to the issue of processing ore leucoxene, which provides for the use of silica as raw materials for the simultaneous synthesis of wollastonite. As a result, not only titanium - containing product-artificial rutile, but also synthetic wollastonite is obtained as final products [4].

2. Experimental

Method of desiliconization of a leucoxene concentrate flows in the process of autoclave leaching at a temperature of 200-220°C , pre-reduced in terms of magnetizing roasting, with milk of lime in the presence of NaOH, serving only as the initiator of the process. The result is the dissolution of silica with soluble compounds Na_2SiO_3, followed by its interaction with $CaO{\cdot}H_2O$ and deposition of the hydrate calcium silicate (HCS). The process is carried out at the stoichiometric ratio CaO/SiO_2, which allows to obtain calcium metasilicate hydrate $CaO{\cdot}SiO_2{\cdot}nH_2O$. With the help of wet magnetic separation, the torsional concentrate is separated from the HSC, which is then dried and calcined at 1050-1100°C.

The phase composition of the material obtained in the process of desiliconization of a leucoxene and further heat treatment, were determined by x-ray diffraction, which was performed on the diffractometer DRON-3M in CuKα-radiation. Transcript of diffraction patterns was carried out using the PDF (POWDER DIFFRACTION FILE).

The specific surface area and porosity of the obtained synthetic wollastonite powders are determined using the Tristar 3000 analyzer. The morphology of the samples of calcium silicate hydrate and wollastonite were investigated using scanning electron microscopy on the microscope LeO – 430i. The determination of the compressive strength was performed on a universal Electromechanical Instron machine 3382.

In order to study the prospects of using the obtained synthetic wollastonite in the production of ceramics, the performance properties of ceramic samples based on it were studied, the compositions of ceramic masses and the modes of its heat treatment were determined. Based on the analysis of literature data, two variants of mixtures were chosen: wollastonite and kaolin clay; on the basis of wollastonite with the addition of oxalic acid [5, 6]. Silicate mixtures (ceramic masses) were prepared from the resulting material $CaSiO_3$, then composite billets of cylindrical shape were formed, which were fired at temperatures of 1100-1300°c with an isothermal exposure of 1 hour in the Nabertherm GMbH HTCT 03/16 furnace.

3. Results and discussion

In the process of desiliconization of a leucoxene concentrate was obtained, the calcium metasilicate, which is a white finely dispersed substance with a specific surface equal to 11,2 m^2/g and the volume of the inner pores is 1,1 m^3/g. the Dried material has the following composition: 50,2% of SiO_2, 46,9% of CaO, up to 0,9% of TiO_2 and 1,6% Na_2O. Using scanning electron microscopy established that in conditions of autoclave leaching the formation of calcium metasilicate, consisting of needle-like crystals to form separate agglomerates with sizes from 4 up to 20-30 μm (Figure 1 a, b).

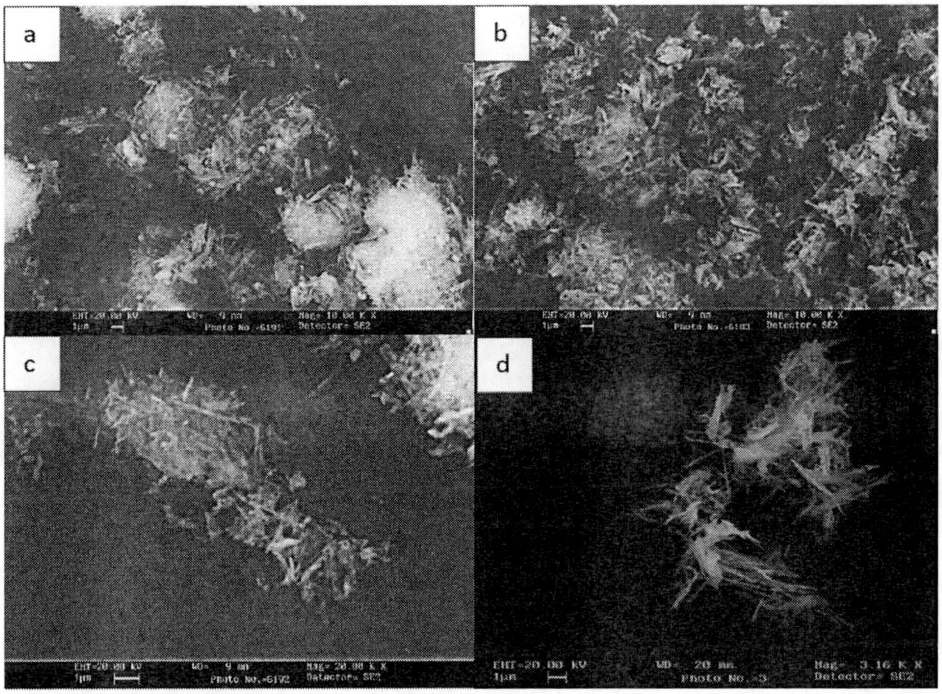

Figure 1. Photomicrographs of wollastonite, obtained by autoclave leaching of leucoxene concentrate with milk of lime: a, b – dried at 105°C; c, d – calcined at 1050°C.

As shown in figure 1, the morphology of the material dried at 105°C (a, b) or calcined at 1050°C (c, d) does not change. This suggests that the formation of "needles" occurs in the autoclave at the stage of hydrothermal treatment of silicon-titanium concentrate. It should be emphasized that the associated production of wollastonite needle modification in the processing of leucoxene concentrates Yarega field is one of the advantages of the proposed method.

The method of dry pressing at a pressure of 10 MPa formed ceramic samples of variable composition of wollastonite (W.) and kaolin clay (C.). In the process of firing clay forms a glass phase, which contributes to the sintering of the material. Samples after heat treatment, carried out to 1150°C, have a low degree of hardening, due to the insufficient amount of binder formed under these conditions. Samples of composition 50%W+50C%, 70%W+30%C. and 87.5%W.+12,5C%, annealed at a temperature above 1200°C, characterized by an optimal compressive strength of 6.4 ÷14,1 MPA, the linear shrinkage of the material is 6,9 to 17,6 per cent. The average density is 2,4 g/cm^3. It should be noted that the samples prepared from pure wollastonite without additional additives are characterized by a decrease in strength characteristics from 16 MPa to 5 MPa with an increase in temperature from 1250°C to 1300°C.

In the course of the study, comparative tests of ceramic samples based on wollastonite, additionally containing oxalic acid and water at the following ratio of components (wt. %): wollastonite 36-50, oxalic acid 4-5, H_2O – the rest. Preparation of the ceramic mass was carried out by the method of molding the mixture in a mold with pre-drying at room temperature and at 105°C. Further, the samples were fired at temperatures of 1200-1300°C with an isothermal exposure of 1 hour. In this version of the ceramic mass oxalic acid creates an acidic environment in which the dissolution of the wollastonite and its transformation into silica gel. During the subsequent high-temperature firing of the samples, the silica gel turns into a wollastonite bond, leading to the regeneration of the dissolved wollastonite.

As a result, it was determined that the optimal composition of the ceramic mass is a composition containing wollastonite 40%, oxalic acid 4%, at a firing temperature of 1250-1300°C. Figure 2 shows the ratio of strain to compression stress applied to this sample after high-temperature firing at 1300°C.

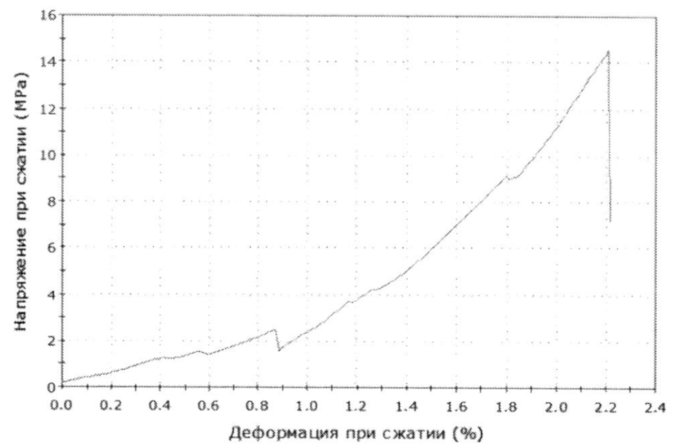

Figure 2. Deformation diagram of a ceramic sample based on wollastonite (40%) with the addition of oxalic acid.

4. Conclusion

The strength characteristics and density of the obtained calcined ceramic composites, consisting of a mixture of calcium metasilicate and clay, based on wollastonite with oxalic acid, correspond to the performance characteristics of high-temperature ceramics, can be recommended for use in industry, for example, as a lining and refractory material in the metallurgy of aluminum alloys. It should also be noted that according to the latest literature data, wollastonite with a needle structure is a promising and high-quality filler in cement, acts as a reinforcing component that increases the strength of cement stone.

Acknowledgements
The work was carried out according to the state task № 007-00129-18-00.

References

[1] Review of the wound of wollastonite in the CIS **(6)** 07.04.2017 Access mode: http://www.infomine.ru/files/catalog/58/file_58.pdf
[2] Wollastonite. Prepared by Daniel M. Flanagan U.S. Geological Survey, Mineral Commodity Summaries January 2017 Access mode: https://minerals.usgs.gov/minerals/pubs/mcs/2017/mcs2017.pdf
[3] Gladun V D, Akatyeva L V, Andreyeva N N and Kholkin A I (2004) Obtaining and application of synthetic wollastonite from natural and industrial row. *Chemical technology* **9** pp. 4 – 11.
[4] Sadykhov G B, Zablotskaya Yu V, Anisonyan K G, Kopyev D Yu, Olyunina T V and Goncharenko T V (2015) Obtaining of needle wollastonite under catalytic pressure leaching of leucoxene concentrate by lime milk *Journal of Advanced Materials* **1** pp 65-72.
[5] Suzdalytsev E, Vikulin V and Rusin M The method for producing ceramic products based on wollastonite *Patent RF, no. 2298537* publ. 10.05.2007.
[6] Simonova N, Samoylo A and Veresthagin V Ceramic mass on the basis of wollastonite *Patent RF, no. 2422402* publ. 27.06.2003.

IOP Conf. Series: Materials Science and Engineering **525** (2019) 012018 doi:10.1088/1757-899X/525/1/012018

One-step hydrothermal surface oxidation of copper foil for photocatalytic water splitting

D S Zimbovskii and A N Baranov

Lomonosov Moscow State University, Moscow, 119991, Russia

E-mail: ZimbovskiyDS@gmail.com

Abstract. In this work a hydrothermal treatment of copper foil in alkaline solution without addition of other oxidizing agents was suggested for surface copper oxidation producing Cu_2O layer. Obtained Cu_2O/Cu heterostructures absorb light in a visible range and demonstrate a photocatalytic activity in a process of water splitting. Maximum value of photocurrent density was reached up to -0.7 mA/cm^2 (-0.6 V vs Ag/AgCl), which corresponds to quantum efficiency of $\eta = 0.9\%$.

1. Introduction

Photocatalytic water splitting on semiconductor photocatalyst is a method for the hydrogen production under light irradiation. This technique is environmentally friendly, cheap and doesn't require complex equipment.

Photodegradation of water occurs in the presence of photocatalyst, which can be used in a form of powder or as a part of photoelectrochemical cells. In photocatalyst electron-hall pairs are generated under light irradiation (if energy of incident photon is higher than band gap of semiconductor). After that, charges are separated and transferred to surface of photocatalyst where they are involved in the reduction and oxidation reactions with water (Figure 3, right part).

$$H_2O + 2h^+ \rightarrow 2H^+ + {}^1/_2\,O_2$$

$$2H^+ + 2\bar{e} \rightarrow H_2$$

At this moment a lot of photocatalysts has been developed (TiO_2, Fe_2O_3, $BiVO_4$, TaON, g-C_3N_4, etc.), but none of them meets all the requirements for photocatalysts: optimal band gap, as well as the position of the valence and conduction bands, stability in aqueous solutions, low cost and non-toxicity [1, 2].

Copper (I) oxide corresponds to most of the parameters imposed on semiconductor photocatalysts and can absorb light in a visible range of the solar spectrum. It is a p-type semiconductor with band gap of 2.0 – 2.2 eV. Theoretically calculated value of photocurrent density for Cu_2O photoelectrodes is -15 mA/cm^2, which corresponds to a quantum efficiency of 18% [3, 4].

Hydrothermal method has many advantages in comparison with other approaches. It is a relatively simple method for synthesis of copper (I) oxide, which does not require high-purity reagents. This method allows, by varying pressure, temperature, time and pH of the solution, to obtain a well-crystallized product with a given composition and morphology [5-7].

The purpose of this work is to develop an approach for obtaining Cu_2O/Cu heterostructures via hydrothermal method and also to study their photocatalytic properties in the process of water splitting.

2. Experimental

2.1. Materials

Cu_2O copper foils with (10x15x0.5 mm sizes, grade M1, purity 99.9%) were used as the substrate. The plates were previously cleaned in an ultrasonic bath for 5 minutes in a mixture of isopropyl alcohol and acetone (1:1), then in 10% solution of hydrochloric acid, and finally in distilled water. The cleaned plates were fixed between PTFE supports (Figure 3, left part).

2.2. Synthesis

Hydrothermal synthesis was carried out at 180°C temperature for 1 hour in a NaOH solutions (analytical-grade) with concentrations from 0.05 to 1 mol/l, cell volume was 35 ml, the filling degree of the cell was 70% (Figure 3, upper part).

2.3. Characterization

The phase composition of the samples was determined by X-ray diffraction (Rigaku D/MAX 2500 diffractometer $\lambda CuK\alpha$ = 1.54178Å) and Raman spectroscopy (Renishaw InVia Raman microscope, laser wavelength - 514 nm, 20 mW, power output – 50%). The morphology of the samples was examined by scanning electron microscopy (SEM) (LEO Supra 50 VP).

The absorption spectra of the samples were recorded with Perkin Elmer Lambda 950 spectrophotometer in the wavelength range from 350 to 800 nm in the diffuse reflection mode. A sample of Labsphere's $BaSO_4$ WS-1-SL with known diffuse scattering characteristics was used as a standard.

The photocatalytic properties of the samples were examined using a three-electrode cell. The working electrode was Cu_2O/Cu sample, the counter electrode was a platinum coated titanium plate, the reference electrode was an Ag/AgCl electrode (3.5 mol/l KCl). The electrodes were immersed in a solution of 0.5M Na_2SO_4 (previously purged with a nitrogen for 30 minutes). A 20 W white LED 6300 K with a collimator lens was used as the light source. The measurement was performed with a potential sweep from 0 to -0.6 V at a speed of 10 mV/s in the chop-light mode (2 seconds light on, 2 seconds off, etc.). Intensity of the illumination at the sample position was determined at a level of 100 mW/cm^2 (Oriel PV Reference Cell System model 91150V).

3. Results and discussion

The SEM images show copper plates after hydrothermal treatment in 0.05 mol/l, 0.3 mol/l, 0.5 mol/l and 1 mol/l NaOH (Figure 1*a-d*). The surface of the substrate is coated with Cu_2O crystals, the size of which increases from 1 (0.05 mol/l) to 4 μm (1 mol/l) as a result of the recrystallization process (Figure 3 lower part) [8].

Figure 1. SEM images of the Cu_2O crystals on the surface of copper foil after hydrothermal treatment in 0.05 mol/l (*a*); 0.3 mol/l (*b*); 0.5 mol/l (*c*); 1 mol/l NaOH (*d*).

IOP Conf. Series: Materials Science and Engineering **525** (2019) 012018 doi:10.1088/1757-899X/525/1/012018

In accordance with XRD patterns (Figure 2), the intensities of the Cu_2O phase peaks increase with an increase of NaOH concentration from 0.05 mol/l to 0.3 mol/l, reaching a maximum value and, further, decrease at concentrations from 0.5 mol/l to 1 mol/l. The results obtained by X-ray diffraction are consistent with data obtained by Raman spectroscopy. The presence of CuO phase was not detected by XRD or Raman spectroscopy.

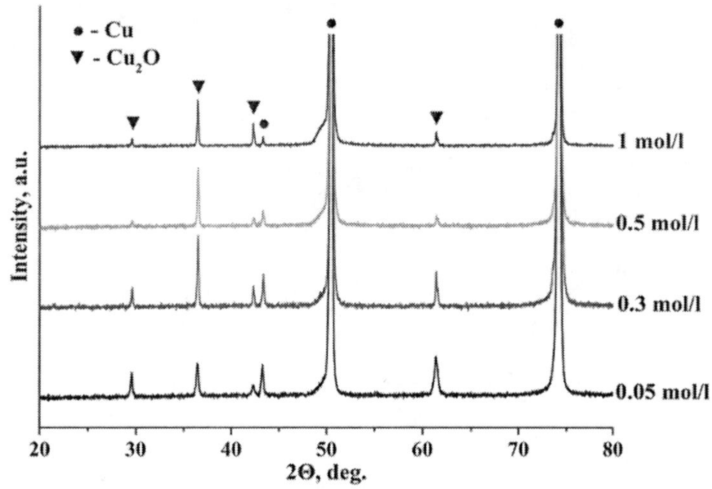

Figure 2. XRD patterns of Cu_2O/Cu samples obtained after hydrothermal treatment of copper foils in NaOH solutions with various concentrations.

According to [9, 10], two parallel processes take place during hydrothermal treatment of copper: 1*a* – oxidation of copper foil by oxygen, 1*b* – oxidation by hydroxide anions (Figure 3).

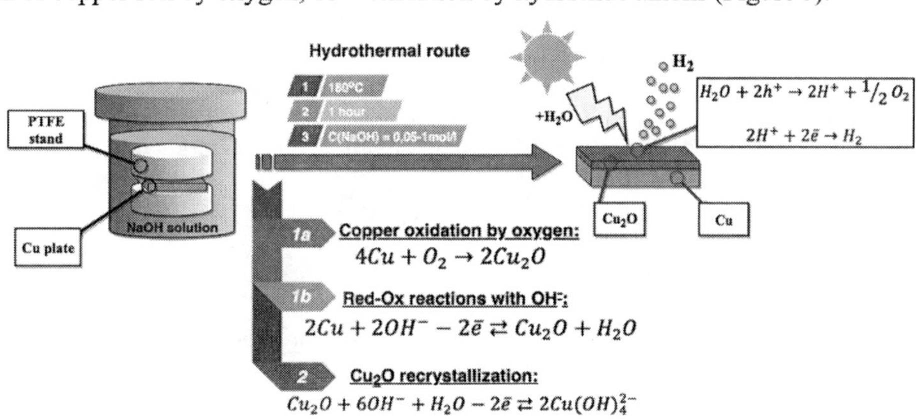

Figure 3. Scheme of the hydrothermal synthesis of the Cu_2O/Cu heterostructure and following photocatalysis of water.

During the synthesis of samples in NaOH solutions with different concentrations, a nonlinear dependence of the amount of copper (I) oxide on the alkali concentration takes place, which is caused by two competing processes: growth and dissolution of the copper (I) oxide layer. The amount of the Cu_2O phase increases with the increase of NaOH concentration to 0.3 mol/l. A further increase of the NaOH concentration leads to a decrease in the amount of oxide, due to the large impact of dissolution of the Cu_2O phase due to formation of a soluble complex $Cu(OH)_4^{2-}$ [11].

IOP Conf. Series: Materials Science and Engineering **525** (2019) 012018 doi:10.1088/1757-899X/525/1/012018

In accordance with the data obtained, the optimal conditions for the synthesis of the Cu_2O/Cu heterostructure are hydrothermal treatment of the copper foil in a 0.3 mol/l NaOH solution at 180°C within 1 hour. Samples synthesized under these conditions were used for further studies.

To determine the light absorption for the Cu_2O/Cu heterostructure, diffuse reflectance spectrum was obtained (Figure 4).

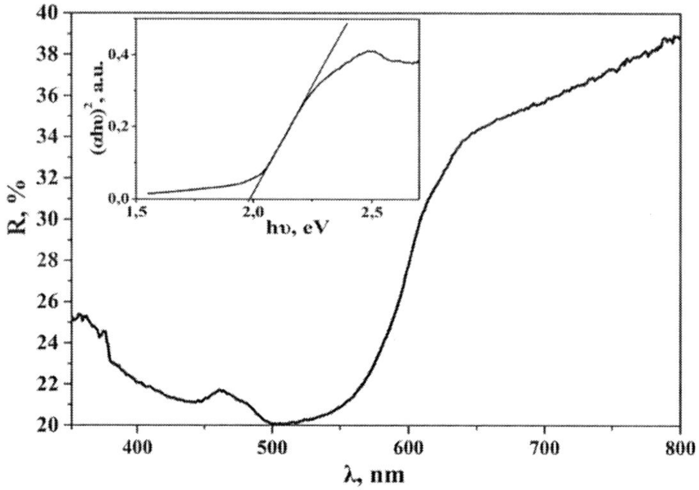

Figure 4. The diffuse reflectance spectrum of a sample obtained by hydrothermal treatment of a copper foil in a 0.3 mol/l NaOH solution at 180°C within 1 hour. The inset shows a graph of the dependence of the Kubelka-Munk function on photon energies.

According to the spectrum, a sample obtained absorbs light over the entire range from 800 to 350 nm. A sharp increase in absorption transition starts at 640 nm and reaches a maximum near bandgap at 500 nm.

To determine the band gap of the material, the absorption on the wavelength was rearranged in coordinates $(\alpha h\nu)^m$ from hν (inset in Figure 4), where α is the absorption coefficient, hν is the quantum energy, m = 2 (Cu_2O is a direct band gap semiconductor).

The optical absorption coefficient α was calculated by the Kubelka-Munk equation [12]:

$$\alpha = \frac{(1-R)^2}{2R} \quad (1)$$

where R is the diffuse reflectance of the sample.

In accordance with the dependence of the Kubelka-Munk function on photons energy, the value of the band gap of copper (I) oxide was determined as 1.982 ± 0.001 eV, which corresponds to the literature data [13].

To estimate the photocatalytic activity of the heterostructure, the dependence of the photocurrent density on the applied potential was measured (Figure 5).

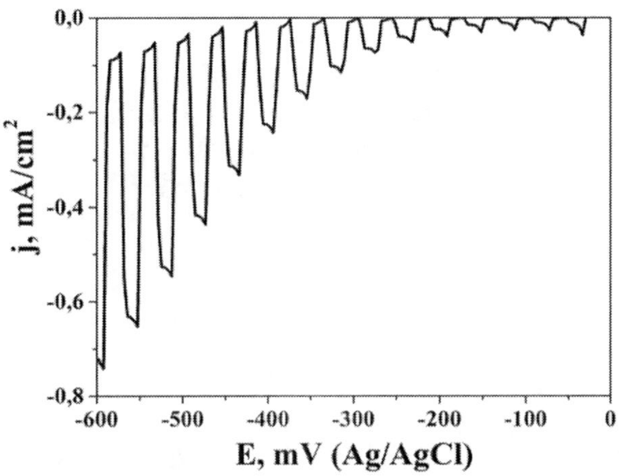

Figure 5. Dependence of the photocurrent density on the applied potential for a sample obtained by hydrothermal treatment of a copper foil in 0.3 mol/l NaOH solution at 180°C within 1 hour.

According to figure 5, heterostructure demonstrates photocatalytic activity: the maximum value of the current density corresponds to -0.7 mA/cm^2 at a potential of -0.6V. The quantum efficiency of the process of photocatalytic water splitting was estimated as η = 0.9% by the equation 2 [12]:

$$\eta(\%) = \frac{(1.23-E_{RHE})\cdot I}{P}\cdot 100\% \qquad (2)$$

where I – current density (mA/cm^2), E_{RHE} – applied potential (V vs RHE), P – power of the illumination (mW/cm^2). $E_{RHE} = E_{AgCl} + E^0_{AgCl} + 0.059pH$ $(E^0_{AgCl} = 0.197V\ (25^oC))$.

The lower value of the photocurrent density than the theoretical (-15 mA/cm^2) could be explained by the low electron mobility in copper (I) oxide and relatively low specific surface area (the oxide film consists of large crystallites with an average size of 2 μm).

4. Conclusions

The method for obtaining Cu_2O/Cu heterostructures by hydrothermal oxidation of copper foil in an alkaline solution without the addition of oxidizing agents is proposed. The optimal conditions for the synthesis of the Cu_2O/Cu heterostructure (0.3 mol/l NaOH at 180°C within 1 hour) were determined. This material absorbs in the visible range of the solar spectrum and is active in the process of photocatalytic water splitting: the maximum value of the photocurrent density is -0.7 mA/cm^2 at an applied potential of -0.6V (Ag/AgCl), which corresponds to a quantum efficiency η = 0.9%.

Acknowledgements

The equipment used in this work was acquired using the resources of the program of advancement of Moscow University. This work was funded by RFBR, grant № **16-03-00488.**

References

[1] Saraswat S K, Rodene D D and Gupta R B 2018 Recent advancements in semiconductor materials for photoelectrochemical water splitting for hydrogen production using visible light *Renew. Sustain. Energy Rev.* **89** 228–48

[2] Navarro Yerga R M, Consuelo Álvarez Galván M, del Valle F, Villoria de la Mano J A and Fierro J L G 2009 Water splitting on semiconductor catalysts under visiblelight irradiation *ChemSusChem* **2** 471–85

IOP Conf. Series: Materials Science and Engineering **525** (2019) 012018 doi:10.1088/1757-899X/525/1/012018

[3] Hara M, Kondo T, Komoda M, Ikeda S, Shinohara K, Tanaka A, Kondo J N, Domen K, Hara M, Shinohara K and Tanaka A 1998 Cu2O as a photocatalyst for overall water splitting under visible light irradiation *Chem. Commun.* **2** 357–8
[4] Wick R and Tilley S D 2015 Photovoltaic and Photoelectrochemical Solar Energy Conversion with Cu2O *J. Phys. Chem. C* **119** 26243–57
[5] Pan L, Zou J J, Zhang T, Wang S, Li Z, Wang L and Zhang X 2014 Cu2O film via hydrothermal redox approach: Morphology and photocatalytic performance *J. Phys. Chem. C* **118** 16335–43
[6] Dong X, Wang K, Zhao C, Qian X, Chen S, Li Z, Liu H and Dou S 2014 Direct synthesis of RGO/Cu2O composite films on Cu foil for supercapacitors *J. Alloys Compd.* **586** 745–53
[7] Ji R, Sun W and Chu Y 2013 One-step hydrothermal synthesis of a porous Cu2O film and its photoelectrochemical properties *ChemPhysChem* **14** 3971–6
[8] Byrappa K and Masahiro Y 2001 *Handbook of hydrothermal technology* (New Jersey: Noyes Publications)
[9] Ganzha S V, Maksimova S N, Grushevskaya S N and Vvedenskii A V 2011 Formation of Oxides on Copper in Alkaline Solution and Their Photoelectrochemical Properties **47** 191–202
[10] Vvedenskii A, Grushevskaya S, Ganzha S and Eliseev D 2014 Copper oxides: kinetics of formation and semiconducting properties. Part I. Polycrystalline copper and copper-gold alloys *J. Solid State Electrochem.* **18** 2755–70
[11] Giri S D and Sarkar A 2016 Electrochemical Study of Bulk and Monolayer Copper in Alkaline Solution *J. Electrochem. Soc.* **163** 252–9
[12] Chen Z, Dinh H N and Miller E 2013 *Photoelectrochemical Water Splitting - Standards, Experimental Methods, and Protocols* (New York: Springer-Verlag)
[13] Li C, Li Y and Delaunay J J 2014 A novel method to synthesize highly photoactive Cu2O microcrystalline films for use in photoelectrochemical cells *ACS Appl. Mater. Interfaces* **6** 480–6

IOP Conf. Series: Materials Science and Engineering **525** (2019) 012019 doi:10.1088/1757-899X/525/1/012019

The electrochemical behavior of nanostructured binary systems based on transition metals

N Ivanova[1], E Ivanova[1], A Lobanov[1], T Mikhailik[1], V Pugachev[2], Yu Zakharov[1,2] and A Valnyukova[2]

[1] Institute of Fundamental Sciences, Kemerovo State University, Kemerovo 650000, Krasnaya, 6, Russia
[2] The Federal Research Center of Coal and Coal Chemistry SB RAS, Kemerovo 650000, Sovetsky, 18, Russia

E-mail: sayganta@mail.ru

Abstract. By means cyclic and anodic linear sweep/stripping voltammetry methods the electrochemical behavior of bimetallic electrolytic nanoalloys and nanostructured powders Cu–Cd, Ni–Cd, Fe–Co, Fe–Ni was investigated in acidic (chloride, sulfate) and ammonia buffer media with the glassy carbon and carbon paste electrodes. The conditions (supporting electrolytes, concentrations of metal ions, voltammetric parameters) of the electrochemical synthesis and voltammetric characterization of the binary systems were established.

1.Introduction

Investigation of physicochemical properties of nanostructured (NS) bimetallic systems, finding of interrelation these properties with morphology and phase composition is an urgent problem in the field of new materials synthesis, as well as in the development of fundamental concepts of materials science. The binary systems based on the transition metals (Cu, Cd, Ni, Fe, Co) are widely used in the manufacturing of chemical and electrochemical reactions catalysts, magnetic devices, biomedical nanoparticles [1-5].

Electrochemical methods are known to be economical and available for the synthesis of the bimetallic systems, including Cu(Ni)-Cd и Fe-Ni(Co) [6-10]. In addition, the electrochemical analysis, for example, a combination of cyclic (CV) and anodic sweep/stripping linear voltammetry (ASLV), allows to obtain the unique information not only about the elemental and phase composition [10-13], but also about the spatial arrangement of the phases [14, 15]. In this respect voltammetry is able to compete with the known structure analysis techniques. The voltammetric methods are characterized by the highest sensitivity and provide the ability to detect the response of single nanoparticles placed on the surface of the electrode or present in the electrolyte. At the same time, an implementation of the voltammetric analysis requires thorough selection of experimental conditions, such as background electrolyte, cell design, working electrode material, sequence of stages during registration of VA curves. The above-mentioned transition metals are electrochemically active; however, the occurrence of anodic oxidation – cathodic reduction processes with their participation is accompanied by the formation of oxide-hydroxide compounds and catalytic hydrogen evolution, which significantly complicates the electrochemical experiment. In this paper we report and discuss the electrochemical behavior of the bimetallic systems Cu-Cd, Ni-Cd, Fe-Co, Fe-Ni in the form of electrolytic nanoalloys and nanostructured powders to identify suitable parameters for their electrochemical synthesis characterization.

IOP Conf. Series: Materials Science and Engineering **525** (2019) 012019 doi:10.1088/1757-899X/525/1/012019

2. Experimental

The CV and ASLV electrochemical measurements were performed in a three-electrode cell using potentiostat/galvanostat Parstat 4000. The working and counter electrodes were the glassy carbon cylinders ($l = 15$ mm, $d = 2$ mm) and the reference electrode – Ag/AgCl in 0.1 M KCl. Before experiment working and counter electrode surfaces were polished down to polishing cloths, etching in 5 M H_2SO_4, thoroughly washed with pure water and transferred to the electrochemical cell. Chemicals used were $CuSO_4$, $NiSO_4$, $CoSO_4$, $FeSO_4$, supporting electrolytes: 0.1 M 0.1 M NH_4Cl + 0.1 M NH_4OH (pH 9.8), 0.1 M Na_2SO_4 + H_2SO_4 (pH 1.5–2.0); 0.1 M KCl + HCl (pH 1.5–2.0); 0.075 M Na_2SO_4 + 0.025 M KCl + H_2SO_4 (pH 1.5) all of analytical grade. All solutions were prepared with double-distilled water. pH measurements were carried out using pH-meter Expert 001.

The electrodeposition of the nanoalloys Cu–Cd, Ni–Cd, Fe–Ni, Fe–Co was carried out in the potentiostatic conditions from the supporting electrolytes with various ratios of the metal ions concentrations ($1 \cdot 10^{-5} - 1 \cdot 10^{-4}$ M) at –1.6 V, time of the deposition was 2 – 5 min. During this process the solution stirred with magnetic stirrer. Anodic oxidation was carried out in the same electrolytic cell in the potentiodynamic mode, scan rate – 100 mV/s. The Cu–Cd powders were prepared by means electrolysis with glassy carbon cathode and graphite anode, electrolyte was 0,08 M $CuSO_4$ + 0.4 M $CdSO_4$ in various ratio with additives of 0.1 M H_2SO_4 at $i = 0{,}1$ A/cm^2, E = 4 V.

Nanostructured Ni–Cd, Fe–Ni, Fe–Co powders synthesized according to the procedure of chemical reduction given in [3, 19] were cleaned by water or etching agents and then were immobilized on the surface of the working electrode (glassy carbon or carbon paste, fabricated from chemical pure graphite powder and paraffine oil).

Powder X-ray diffraction (XRD) data were recorded by diffractometer DRON 2 (Cu Kα).

3. Result and Discussion

Preliminary the conditions for the formation of a reproducible CV and/or ASLV signal linearly dependent on the concentration of the metal ions into the electrolyzer were chosen for each metal.

The simple acidic (sulfate and chloride) or complexing and having a buffer effect media (ammonia buffer solution) were used as background electrolytes. It is shown that all the mentioned electrolytes can be used for the electrodeposition and characterization of the Cu-Cd, Ni-Cd systems. The investigation of Fe-Ni, Fe-Co systems is possible only in the sulphate electrolyte with pH not more than 2. Under these conditions the parameters of the anodic peaks and the relative electrodeposition rates of each metal were determined.

The considered binary systems can be divided into two types, in accordance with a predominant phase organization: the systems forming intermetallic compounds (IMC) – Cu-Cd, Ni-Cd, or solid solutions – Fe-Co, Fe-Ni.

By means voltammetry method for the NS electrolytic system *copper - cadmium* in the chloride media was found that the binary nanoalloy on the electrode surface consists of several phases. Anodic dissolution of this alloy is accompanied by the formation of up to four additional current peaks on the voltammogramms in the potential region between the oxidation peaks of pure copper and pure cadmium phases (Figure 1A). Table 1 shows the peak potentials for the chloride electrolyte. Most probably, each additional peak corresponds to a single IMC phase of a constant composition, since the peak potential does not depend on the ratio of the metal concentrations (excluding peak at –186 mV). Similar conclusions were reached by V.D. Jovič et al [12], who investigated the behavior of this system in 0.01 – 0.1 M $CuSO_4$, 0.2M $CdSO_4$, 10^{-3} M H_2SO_4, 0.5 M Na_2SO_4.

Table 1.The anodic peak potentials for the Cu-Cd electrolytic nanoalloys in comparison with the partial Gibbs energy of the IMC formation; 0.1 KCl + 0.1M HCl (pH 2), $c(Cu^{2+}) = 1\cdot10^{-5}$ M, $c(Cd^{2+}) = 4\cdot10^{-5}$ M.

Ep, mV	Cd –670	I –640	II –485	III –386	IV[a] –186 ÷ –150	Cu +23; +370
IMC		**Cd_3Cu**	**Cd_8Cu_5**	**Cd_3Cu_4**	**$CdCu_2$**	
ΔG_f^{298}(Cd), kJ/mol [16]		–0,84	–8,49	–6,98	–11,69	

[a] Peak potential depends on the cadmium concentration

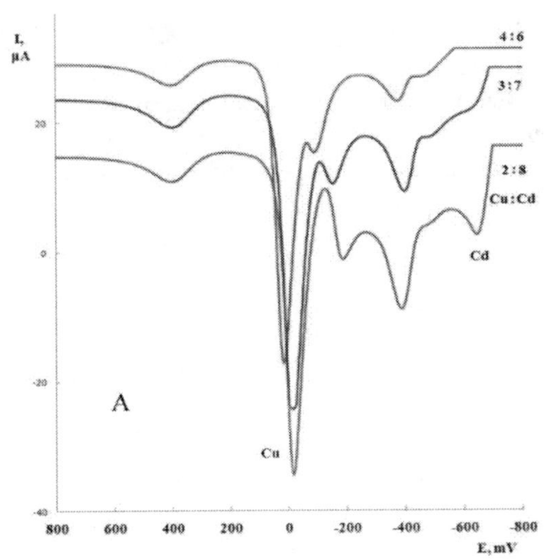

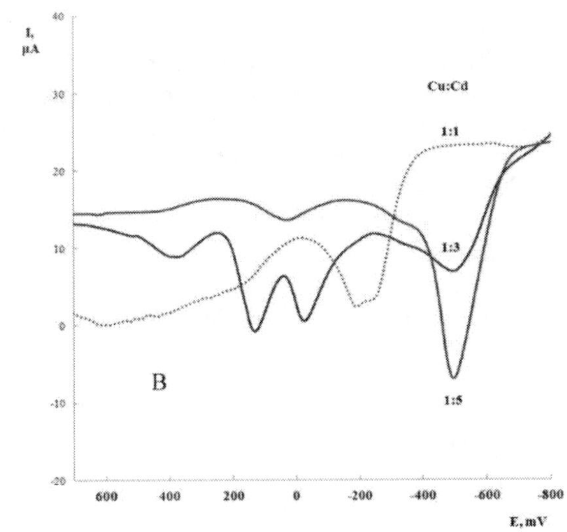

Figure 1. Anodic voltammograms obtained for the Cu-Cd electrolytic nanoalloys on glassy carbon (A) and NC Cu-Cd powders (B) on carbon paste electrode in the chloride media: 0.1 M KCl + HCl (pH 2). $c(Cu^{2+}) = 1\cdot10^{-5}$ M (A). The ion concentration ratios are shown in the graphs.

Using a mercury film electrode an ASLV analysis of the electrolyte after partial anodic dissolution of the copper-cadmium nanoalloys was carried out. It is established that anodic dissolution of the IMCs occurs through the selective oxidation of a more electroactive component (cadmium), because copper ions are not detected in the solution up to a scanning potential of -100 mV. The values of partial Gibbs energy of the IMC formation [16] suggest that the selective dissolution of cadmium from the IMC in the process of the anodic potential sweep and, accordingly, the formation of the IMC peaks occurs in the sequence: $Cd_3Cu \rightarrow Cd_8Cu_5 \rightarrow Cd_3Cu_4 \rightarrow CdCu_2$.

IOP Conf. Series: Materials Science and Engineering **525** (2019) 012019 doi:10.1088/1757-899X/525/1/012019

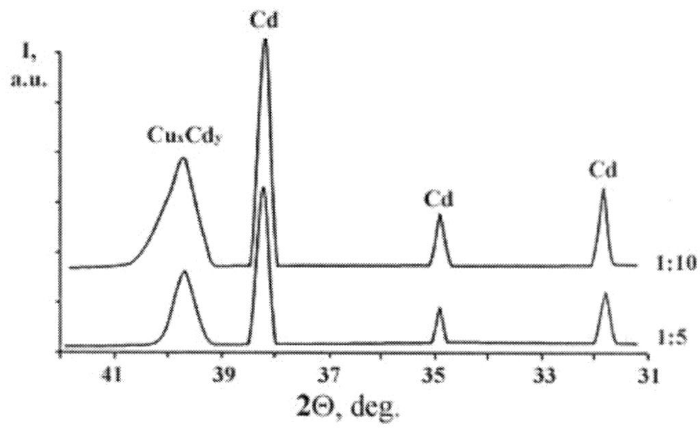

Figure 2. XRD data for the NS Cu-Cd powders with the different c(Cu):c(Cd) ratios.

The total amount of the IMCs in the NS electrolytic alloy is linearly related to the concentrations of copper and cadmium ions in the electrochemical cell. Each phase is characterized by the intrinsic existence region in the ion concentrations; however, it is not possible to bring these data into correlation with the fields of the phase diagram, apparently due to the non-equilibrium conditions of the alloy formation.

The XRD analysis of the NS powders Cu-Cd obtained by "massive" electrodeposition of the components confirmed the presence of the intermetallic compounds Cu_xCd_y (Figure 2) [17]. However, the overlapping of the reflexes did not allow to establish the exact composition of the intermetallides in the powder samples, since the observed in the region 39.2 – 40.2 deg. peak corresponds to the several IMCs ($CdCu_2$, Cd_8Cu_5 and Cd_3Cu_4) at once. The peak asymmetry at large angles (40.5 deg), arising from an increase in the amount of cadmium, implies the formation of an IMC Cd_3Cu.

The anodic linear voltammetry shows a commonality of the electrochemical properties of the electrolytic and nanopowder bimetallic alloys, which consists in the formation of oxidation peaks of intermetallic compounds Cu_xCd_y in the same potential fields (Figure..1b). A distinctive feature of the powder systems is the wider peaks on the anodic voltammogramms, probably, due to a diffusion limitation and, consequently, their greater overlapping. Furthermore, based on the number of peaks it must be taken into account, that the phase composition of powders is not so rich.

A qualitatively similar situation is observed during the investigation of the *nickel-cadmium* electrolytic nanoalloys and NS powders. According to the phase diagram [21] IMCs NiCd, $NiCd_5$ and solid solutions riched in nickel are formed in the Ni-Cd system, it should be noted that IMC Ni_3Cd was detected [3] in the NS powders produced via chemical co-reduction of salts. In this case, the ALSV data show, that at the electrodeposition in the sulfate-chloride electrolyte several binary phases of variable composition are formed, since the potentials of their anodic peaks (E_{Cd} = –710 mV, E_I = –550 mV, E_{II} = –450 mV, E_{III} = –310 mV, E_{IV} = –200 mV, E_{Ni} = –110 mV) depend on the ratio of the concentrations of nickel and cadmium (Figure3A). There are no individual phases of nickel and cadmium in the binary samples; it can be explained by the formation of solid solutions in a wider concentration range than in the Cu-Cd electrolytic system.

The phase composition of thin nickel-cadmium films depends on the type of supporting electrolyte; in ammonia buffer the electrodeposition occurs at a faster rate than in a sulphate-chloride one; apparently, this determines a larger amounts of phases in the first case [7].

It was proved that the voltammetric study (CV or ASLV) of the nanostructured nickel-cadmium powders is not realized without ammonium etching of the samples (2M NH_4OH water solution was used). Oxide-hydroxide layers present as impurities in the nanopowders reduce electrical conductivity and prevent obtaining of the correct results (Figure3C). In the other side the voltammetric analysis can be used for the detection the oxygen-containing compounds in the nanostructured alloys.

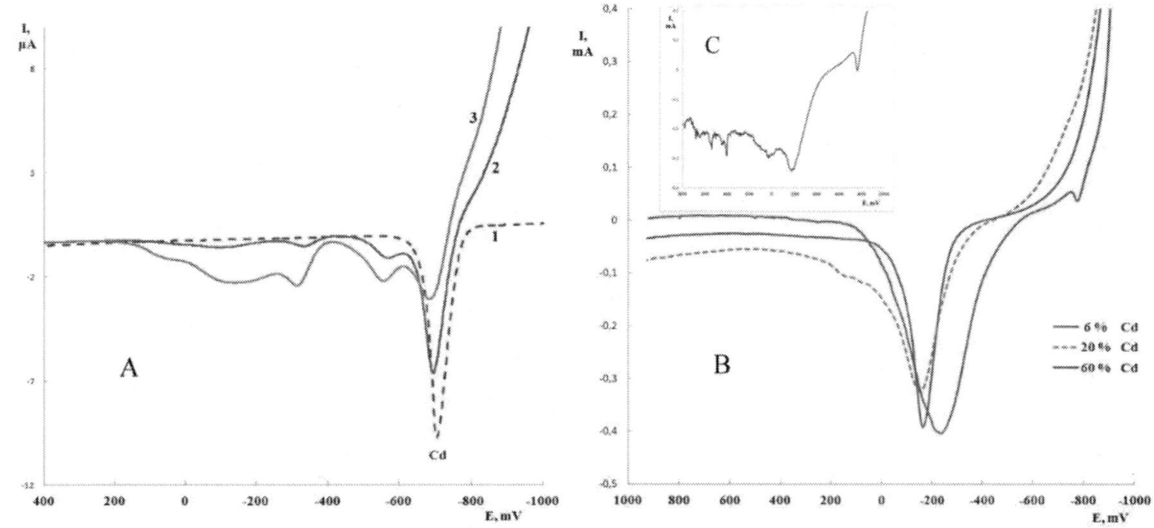

Figure 3. Anodic voltammograms obtained for the Ni-Cd electrolytic nanoalloys (A) and the NS Ni-Cd powders before (C) and after etching (B) on glassy carbon electrode; 0.075 M Na_2SO_4 + 0.025 M KCl + H_2SO_4 (pH 1.5). $c(Cd^{2+})$: 1– $5\cdot10^{-5}$ M + $c(Ni^{2+})$: 2 – $3\cdot10^{-4}$; 3 – $6\cdot10^{-4}$ M.

The electrochemical response of metal phases in the oxygen pure nanopowders (Figure 3B) correlates with the X-ray diffraction data [3].With an increasing in cadmium content from 6 to 60%, the total peak of the binary phase oxidation (E = –0.17 ÷ –0.25 V) is shifted to the field of more negative potentials, and then an appearance of a peak at –0.77 V supposedly corresponding to the IMC $NiCd_5$ oxidation is observed. The change in the phase composition of the NS powder was studied during successive CV scans, as well as when performing cathodic reduction, which lead to degradation of the initial powder structure and the appearance of a cadmium peak at the potential of –0.85 V. The voltammetric technique we developed has been used in the investigation of the chemical reduction mechanism which realized during the synthesis of the nanostructured powders Ni-Cd [18].

The electrochemical response of the NS electrolytic alloys *Fe-Ni* and *Fe-Co* in the sulfate electrolyte has a number of common features, consisting in the regular shift of an anodic peak of the first metal when the second metal is added into the system and in the increasing of the single anodic peak current (Fig.4). It should be noted that the direction of the peak shift is determined by the electrochemical properties of metals: if the more active metal is added, the peak potential decreases, if the more "noble" one is increased.

According to the phase diagrams and the results obtained by X-ray diffraction for the NS powders [19, 20], the components of the Fe-Ni and Fe-Co systems form solid solutions in a wide range of concentrations, therefore, the changes observed on the voltammetric curves can be interpreted as a transformation of the pure components phases to the phases of solid solutions having the variable composition. As can be seen in Figure 4 C, when the solid solution is enriched with one of the components, a regular change in the peak potential takes place. The obtained dependence can be used to determine the relative content of metals in the nanostructured systems using the voltammetric data. The presence of a single anodic peak on the VA curves indicates that the oxidation of the solid solution phase in each case occurs according to a simultaneous mechanism. The dependences of an electricity amount under the peaks on the concentrations of metal ions in the binary systems leads to the same conclusions. Additionally, it was found that in the sulfate medium during the deposition of nickel (or cobalt) in the presence of iron, so-called "induction period" needed for the formation of metal crystallization centers on the inert electrode surface is excluded.

The nanostructured Fe–Co, Fe–Ni powders are characterized an increased tendency to oxidation; therefore, only the peaks of oxide-hydroxide phases were observed on the voltammetric curves.

Preliminary preparation of the powders did not lead to the desired results; further experiments are meant to be used in an argon glove box.

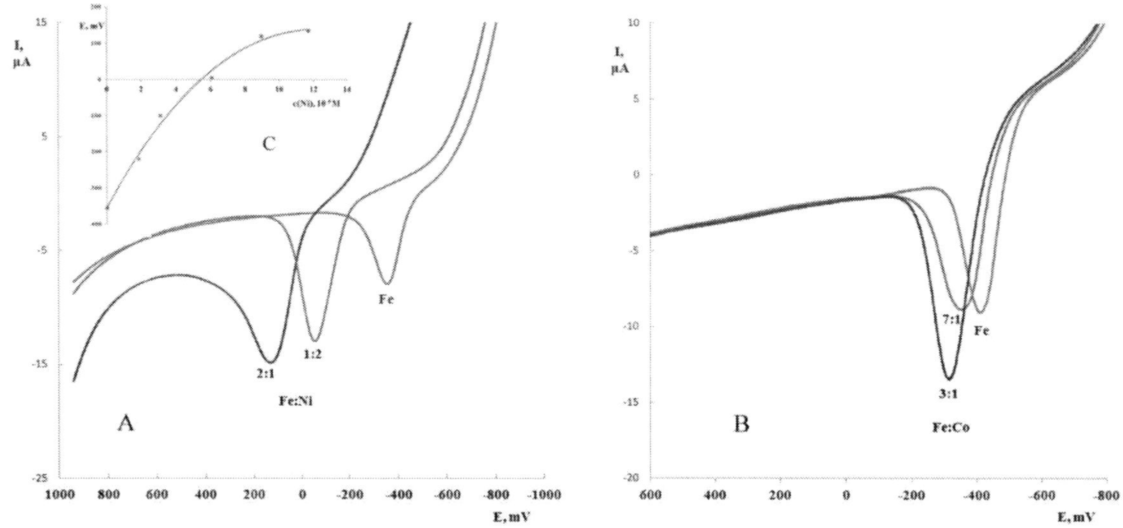

Figure 4. Anodic voltammograms obtained for NS Fe-Ni (A) and Fe-Co (B) electrolytic nanoalloys on glassy carbon electrode in the sulfate media: 0.1 M Na_2SO_4 + H_2SO_4 (pH 2), the dependence of the peak potential on the Ni^{2+} concentration (C), c(Fe^{3+}): $8 \cdot 10^{-5}$ M (A, C); $12 \cdot 10^{-5}$ M (B).

4. Conclusion

The electrochemical response of nanostructured bimetallic systems Cu– Cd, Ni–Cd, Fe–Co and Fe–Ni by means voltammetric methods (CV and ASLV) can be used in the investigation of their physico-chemical characteristics such as phase and elemental composition, spatial phase arrangement, stability to the oxidation, etc. The acidic sulfate (for Fe–Co, Fe–Ni, Cu–Cd, Ni–Cd), acidic chloride (pH 1.5 – 2.0) and ammonia buffer electrolytes (for Cu–Cd, Ni–Cd) are suitable for electrochemical synthesis and voltammetric characterization of the bimetallic electrolytic nanoalloys and nanostructured powders.

References

[1] Kishimoto M, Latiff H, Kita E and Yanagihara H, *J. Magn. Magn. Mater.* 2019 **476**, 229-233
[2] Abdelkareem M A, Al Haj Y, Alajami M, Alawadhi H and Barakat N A M 2018 *J. Environ. Chem. Eng.* **6** 332
[3] ugachev V M, Zaharov Yu A, Valnyukova A S, Dodonov V G and Datiy K A 2018 *Russ. Chem. Bull. Int. Ed.* **67** 1018
[4] Tabrizian P, Ma W, Bakr A and Rahaman M S 2019 *J. Colloid Interface Sci.* **534** 549
[5] Lu W, Guo X, Luo Y, Li Q, Zhu R, Pang H 2019 *Chem. Eng. J.* **355** 208
[6] Chaudhari A K and Singh V B 2018 *J. Alloys Compd.* **751** 194
[7] Ivanova N V, Zakharov Yu A, Palashkova T I and Voropay A A 2017 *Chem. Sustain. Develop.* **25** 627
[8] Bai A, Hu C-C 2003 *Electrochem. Commun.* **5** 78
[9] Torabinejad V, Aliofkhazraei M, Assareh S, Allahyarzadeh M H and Sabour Rouhaghdam A 2017 *J. Alloys Compd.* **691** 841
[10] Ivanova N 2017 *Electrochem. Commun.* **80** 48

IOP Conf. Series: Materials Science and Engineering **525** (2019) 012019 doi:10.1088/1757-899X/525/1/012019

[11] Dobrovolska Ts, Jović V D, Jović B M and Krastev I 2007 *J. Electroanal. Chem.* **611** 232

[12] Jović V D, Despić A R, Stevanović J S and Spaić S 1989 *Electrochim. Acta* **34** 1093

[13] Jović V D and Tošić N 1998 *J. Electroanal.* Chem. **441** 69

[14] Tschulik K, Ngamchuea K, Ziegler C, Beier M G, Damm C, Eychmueller A and Compton R G 2015 *Adv. Funct. Mater.* **25** 5149

[15] Zakharov Yu, Pugachev V, Kolmykov R, Russakov D, Dodonov V, Obraztsova I, Prosvirin I, Ivanova N and Ivanov N 2017 *Gold Bulletin* **50** 225

[16] Borg R 1961 *Trans. metallurg. soc. AIME* **221** 527

[17] X–ray diffraction date cards, ASTM

[18] Zakharov Yu A, Pugachev, V M Valnukova A S, Russakov D M, Ivanova N V, Dodonov V G and Kolmykov R P 2019 *Mater. Res. Express* **6** 036508

[19] Zakharov Yu A, Pugachev, Dodonov V G, Popova A N, Kolmykov R P, Postovtsev G A, Vasiljeva O V, Zyuzyukina E N, Ivanov A V and Prosvirin I P 2011 IOP J. Phys. Conf. Ser. **345** 012024

[20] Lyakishev N P 1996 Diagrammy sostoyaniya dvoinykh metallicheskikh sistem [The Diagrams of Binary Metallic Systems](Moscow: Mashinostroenie)

IOP Conf. Series: Materials Science and Engineering **525** (2019) 012020 doi:10.1088/1757-899X/525/1/012020

Technological features of receiving and research of high density ceramic materials on the basis of silicon carbide

G D Kardashova and Sh Sh Shabanov

Department of Engineering Physics, Dagestan State University, 43-a Gadzhiev Street, Makhachkala, the Republic of Dagestan, 367000, Russia

E-mail: gulya-ka11@yandex.ru

Abstract. The article is devoted to the establishment of regularities of synthesis and technological aspects of the formation of composite ceramics based on silicon carbide and aluminum nitride by the method of hot pressing, as well as the study of its structural properties. The paper presents the results of the study of the microstructure and elastic properties of ceramics based on silicon carbide, obtained by hot pressing at temperatures up to 2170 K and pressures up to 35 MPa, of various compositions (0.9SiC - 0.1AlN; 0.7SiC - 0.3AlN ; 0.5SiC - 0.5AlN; 0.3SiC - 0.7AlN; 0.1SiC - 0.9AlN), the average density of which is 3.21 g/cm^3. The microstructure of the obtained samples was studied with a scanning electron microscopy. Elastic moduli of SiC-AlN hot-pressed ceramic materials were determined by the method of resonant ultrasonic spectroscopy (RUS) depending on the composition. Our data on Poisson's ratio are in good agreement with the literature data and are well described by linear approximation. The values of Young's modulus are noticeably lower than the literature ones. The values of the elastic moduli lie below the additive right line of ceramic SiC and AlN with zero porosity, which is related to the influence of the porosity of our samples.

The rapid development of basic and applied researches of ceramic materials is based on the needs of all modern industries in brand new materials. It is ceramic materials based on silicon carbide and aluminum nitride that have unique functional and practical design properties (high-wearing feature, chemical resistance, heat resistance, special electrophysical properties, etc.). This determines the most promising market segments, where they gradually penetrate, displacing counterparts made of traditional materials.

The problem of the formation of new functional and structural ceramic materials based on silicon carbide and aluminum nitride with predictable properties highlights the monitoring of their properties.

The traditional method of forming products of silicon carbide powder is the method of hot pressing [1]. Upon obtaining of ceramic carbide silicon materials, the use of various technological methods, which provide for obtaining ceramics of a given composition, structure and set of useful properties, is required. The sintering process of SiC-AlN ceramics by hot pressing was carried out in N2 medium for 1 hour, at a temperature of less than 2170 K and pressure up to 35 MPa. Another important process operation in the production of ceramics is the uniform distribution of additives (AlN) and the maximum density of the initial raw materials. Samples with the following composition were obtained: 0.9SiC – 0.1AlN; 0.7SiC – 0.3AlN; 0.5 SiC – 0.5 AlN; 0.3SiC – 0.7AlN; 0.1SiC – 0.9AlN, the average density of which was 3.21 g/cm^3.

To study the microstructure of ceramics in this work, we used the method of scanning electron microscopy in the SE mode (secondary electron mode). Figures 1 and 2 show the morphology of the ceramic sample under study at an accelerating voltage of 12.5 and 15 kV. The average grain size according to the data obtained is 2-5 μm. There are also nanograins with sizes less than 100 nm.

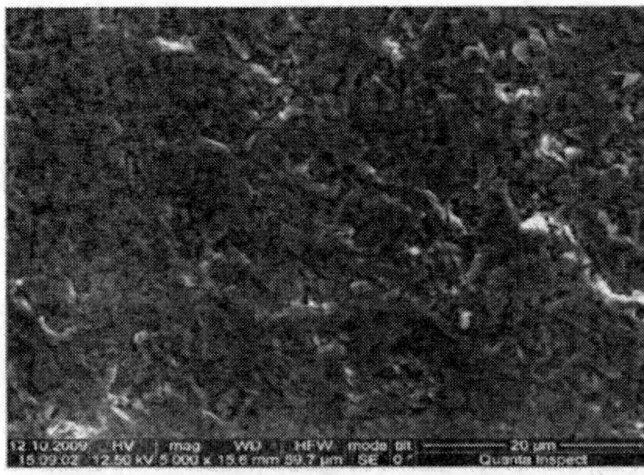

Figure 1. The morphology of the SiC-AlN ceramics sample (30% weight of AlN) at U = 15 kV.

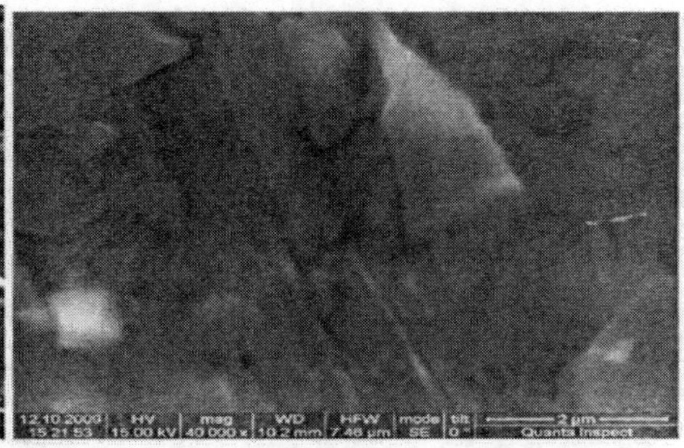

Figure 2. The morphology of the SiC-AlN ceramics sample (30% weight of AlN) at U = 12.5 kV.

A qualitive chemical analysis of a Sic-AlN ceramic sample for Si, Al, N, C elements was performed using scanning electron microscopy methods with a sensor resolution of 127 eV. The radiation was carried out at the brightest spectra in the Kα mode. The largest grain was selected for analysis (Figure 3).

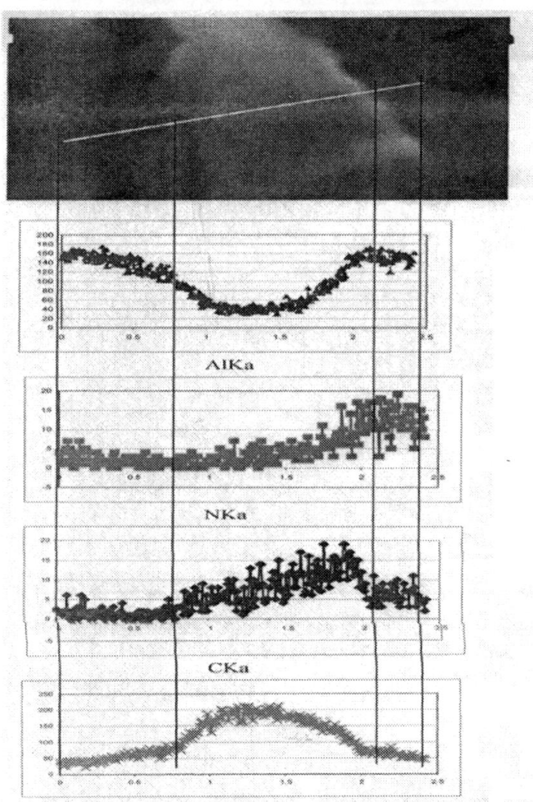

Figure 3. The results of SiC-AlN ceramics elemental analysis (30% weight of AlN).

As is clear from the image presented in Figure 3, the particle under study with a size of ~ 1 μm contains a significant amount of silicon and carbon, and a small amount of aluminum and nitrogen. Along the rim of this grain, an increase in the concentration of aluminum and nitrogen is observed,

IOP Conf. Series: Materials Science and Engineering **525** (2019) 012020 doi:10.1088/1757-899X/525/1/012020

which indicates the formation of a solid solution of silicon carbide with aluminum nitride in the boundary layer.

Of special interest is phase formation in hot-pressed silicon carbide. The structure and composition of SiC ceramics is affected by various media and sintering temperatures, as well as the proportion of the content of mineralizers and modifiers.

The results of phase analysis of the SiC-AlN ceramic sample (70% weight of SiC) are given below. Using the SEM compositional contrast ratio method, it is possible to identify phases, grain rims and to determine the nature of the distribution of elements over the grain section, the chemical composition of various inclusions. The observed grain contrast ratio is used to detect various dispersed phases. As a result of the studies, photographs of the morphology of the sample with the distribution of phases throughout the sample were obtained (Figures 4 and 5).

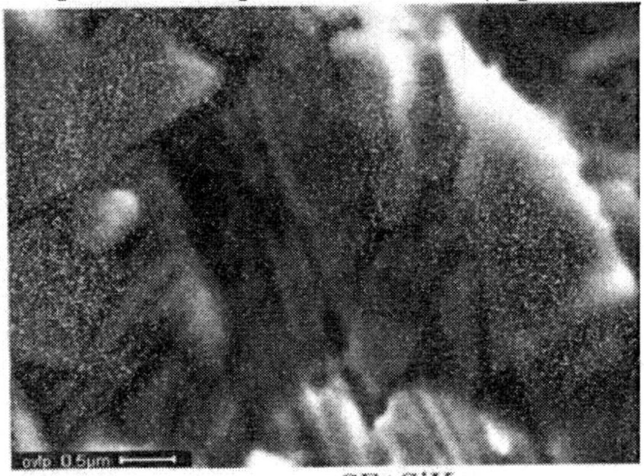
SE+SiKa

Figure 4. The results of SiC-AlN ceramics phase-shift analysis (30% weight of AlN). Phase Si distribution.

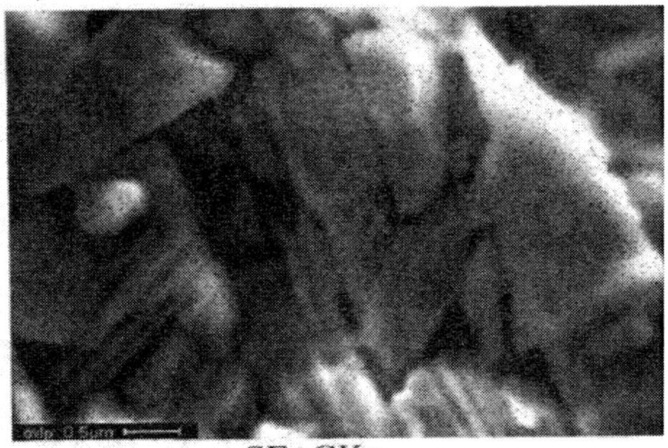
SE+CKa

Figure 5. The results of SiC-AlN ceramics phase-shift analysis (30% weight of AlN). Phase C distribution.

The figure shows that carbon is distributed evenly throughout the sample, and silicon is distributed in the form of carbide chains.

Table 1. Elastic properties of SiC-AlN ceramics.

AlN, %	C11	stdC11	C44	stdC44	std E	μ	std μ
10	439.20	6.7	175.30	4.4	409.45	10.2	0.1679
30	434.58	8.2	165.40	4.3	394.57	8.2	0.1928
50	403.18	8.1	148.38	3.1	358.73	9.3	0.2088
70	375.71	7.8	132.09	2.7	324.65	7.8	0.2289
90	334.94	8.5	114.47	2.9	283.98	9.7	0.2404

The elastic moduli of SiC-AlN hot-pressed ceramic materials are determined depending on the composition. The elastic properties of the samples at room temperature were measured by the method of resonant ultrasonic spectroscopy (RUS) on the apparatus described in Ref. 2 [2].

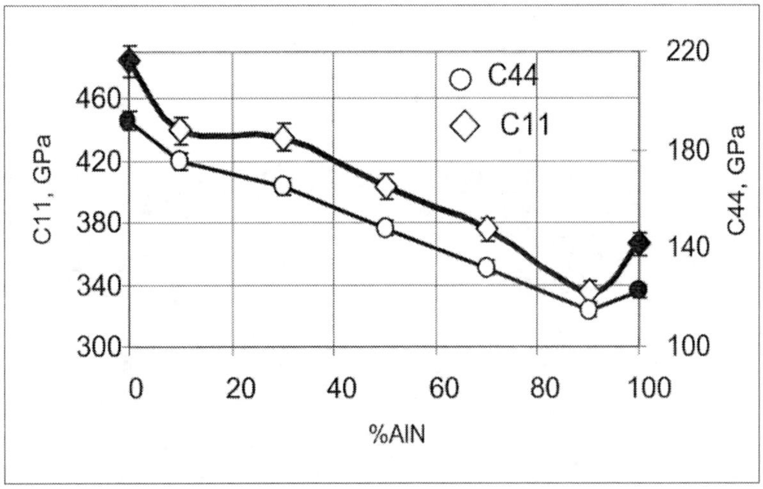

Figure 6. Concentration dependence of C11 and C44 elastic moduli

In our measurements, we used the point method of exciting and recording sample oscillations in the frequency range of 30 ... 1500 kHz. The array of digitized data "frequency-amplitude" was processed on a computer to search for the frequency and quality of the response. The Gauss-Lagrange approximation was used. The study results are presented in Table 1 and in Figures 6 and 7. The total error of measurement of elastic moduli does not exceed 2%.

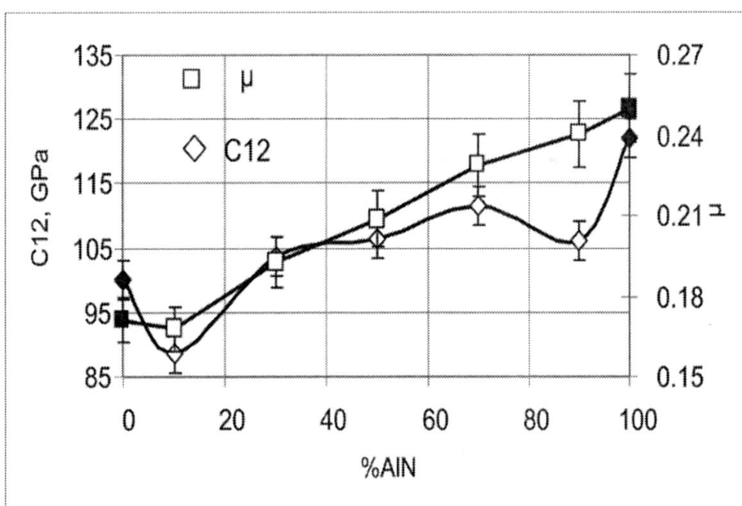

Figure 7. Concentration dependence of C12 elastic modulus and Poisson's ratio

The propagation speeds of Cl and Ct ultrasonic waves are calculated from the C11 and C44 elastic moduli, respectively. Dark icons (Figures 6-8) show literature data on pure ceramic SiC and AlN.

IOP Conf. Series: Materials Science and Engineering **525** (2019) 012020 doi:10.1088/1757-899X/525/1/012020

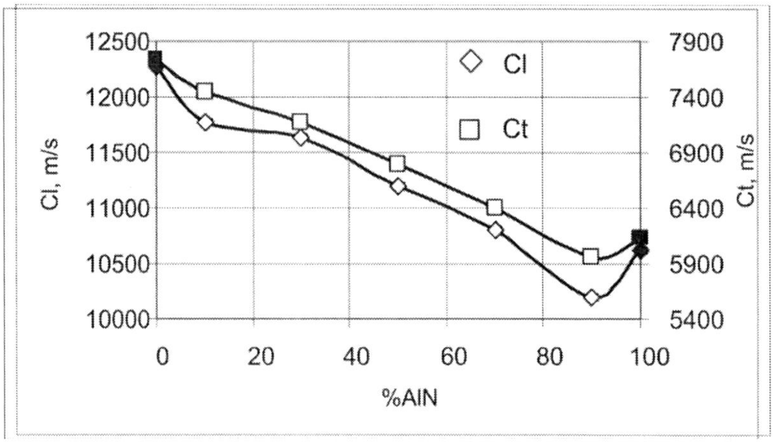

Figure 8. Concentration dependence of the propagation velocity of longitudinal Cl and transverse Ct waves.

Our data on Poisson's ratio are in good agreement with the literature data and are well described by linear approximation. The values of Young's modulus are noticeably lower than the literature ones. The values of the elastic moduli lie below the additive right line of ceramic SiC and AlN with zero porosity, which is related to the influence of the porosity of our samples.

The results obtained make it possible to establish regularities of composite ceramics synthesis based on silicon carbide and aluminum nitride by the method of hot pressing, as well as to optimize the technology of its production.

References

[1] Kardashova G D 2004 Sintering processes and electrophysical properties of ceramics based on silicon carbide with activating additives (Makhachkala: PhD dissertation)

[2] Pashuk E G and Khalilov Sh A 2006 *Proc. All-Russian Conf. on Physical Electronics (Makhachkala)* (Makhachkala: Dagestan State University) p 225-228

IOP Conf. Series: Materials Science and Engineering **525** (2019) 012021 doi:10.1088/1757-899X/525/1/012021

Composite oxide fibers and brittle matrix composites based on them

V M Kiiko, A A Kolchin, S T Mileiko and N I Novokhatskaya

Institute of Solid State Physics of the Russian Academy of Sciences, Chernogolovka Moscow district, 142432 Russia

E-mail: kiiko@issp.ac.ru

Abstract. Composite oxide fibers were obtained on the basis of various initial mixtures of oxides of aluminium, calcium, barium and yttrium. The fibers contain components that provide the necessary strength and fracture toughness. High strength is provided by alumina. Fracture toughness is mainly determined by the formation of layered components of calcium hexaaluminate or barium hexaaluminate in the fiber structure. Relatively weak planes in the structure of the hexaalumiates play the role of brakes of cracks in composites consisting of non-plastic components. Composites with brittle matrices were made on the basis of these fibers, tests of which showed their "quasiplastic" behavior under load. The strength values in the temperature range of 20–1400 °C and fracture toughness correspond to the requirements for structural materials.

1. Introduction

The growing requirements for structural materials, in particular, heat-resistant composite materials are developed for work in a wide range of temperatures, lead to the need to include in the structure of chemical compounds that were not previously used, which can provide work in new conditions. They can be strength and exhibit significant creep resistance at high temperatures. But many of them are brittle. This raises the problem of ensuring the overall non-brittle behavior of a material containing brittle components. It is known that fibrous composites with brittle components can exhibit quasi-plastic behavior. This is usually achieved by organizing relatively weak interfaces between the fibers and the matrix, which inhibit the propagation of cracks in the composite structure [1]. Such an opportunity, as a rule, is realized due to special coatings on the fiber, which provide the necessary characteristics of the fiber – matrix interface. As coatings, for example, lanthanum orthophosphate, which is capable of plastically deforming at sufficiently low temperatures or calcium hexaaluminate, which can exfoliate [2, 3]. As a rule, these coatings are performed on polycrystalline oxide fibers with limited creep resistance at high temperatures, and, in addition, the technological procedure for applying coatings leads to complication of composites manufacture and an increase in their cost. To obtain the fibers required for the reinforcement of brittle matrices, it is proposed to use composite fibers that include, for example, high-strength Al_2O_3 sapphire and calcium hexaaluminate $CaAl_{12}O_{19}$, the presence of weak planes in the crystal structure of which can provide the "quasi-plastic" behavior of the composite as a whole. The barium hexaaluminate $BaAl_{12}O_{19}$ can perform a similar role. The "weak" planes in these layered compounds in the destruction of the composite structure play the role of crack brakes. In addition, the proposed fibers are aimed at using them at high temperatures. Therefore, the creep resistance of such fibers is an important parameter. The possibility of a higher creep resistance at high temperatures of $Y_3Al_5O_{12}$ yttrium aluminum garnet based fibers, than sapphire-based Al_2O_3 fibers, is known, that is, such fibers can be of use for high-temperature composites [4]. Therefore, yttrium oxide Y_2O_3 as an additional initial component of the initial compositions of the mixtures of oxides of some batches of fibers was used in this work. The use of the internal crystallization method [5] to obtain fibers allows us to significantly solve the tasks. The internal crystallization method in brief is that a molybdenum carcass with straight hollow coaxial channels is made. A mixture of the necessary oxides is made also. In a high-temperature furnace, the mixture of oxides is melted in a crucible, the end of the molybdenum carcass is immersed in the melt, and the melt fills the channels due to capillary forces. Then the specimen with the channels rises into the cold zone, the melt crystallizes in the channels and forms composite oxide fibers. As a result of the technological procedure of the internal crystallization method associated with heating the specimen with a molybdenum carcass to about 2000 °C, followed by cooling to room temperature, molybdenum recrystallizes and becomes embrittled. Specimens with brittle molybdenum matrix and oxide fibers are obtained. As a result, a hierarchically organized composite system with composite oxide fibers (for example, Al_2O_3 – $CaAl_{12}O_{19}$) and with a brittle molybdenum matrix (Mo) are automatically formed using the initial mixtures of two oxides Al_2O_3 – CaO of a selected composition. The specimens of composites can be tested for strength and fracture toughness and studied their structure. The molybdenum carcass can also be removed by etching with a mixture of nitric and hydrochloric acids, and the fibers thus extracted are placed in another matrix (for example, the intermetallic NiAl). Using constitution diagrams of Al_2O_3 - CaO, Al_2O_3 – BaO and Al_2O_3 – Y_2O_3 systems as the basis, the choice of initial compositions of mixtures of oxides and technological regimes can be used to obtain various composite oxide fibers intended for use as a reinforcing component, first of all, brittle matrices.

2. Experimental procedures

The characteristics of the composite oxide fibers obtained by the internal crystallization method are determined, firstly, by the initial composition of the mixtures of oxides and, secondly, by the technological regimes, an important parameter of which is the pulling rate V of specimens into the cold zone during the crystallization of the fibers. These two parameters will later be indicated when presenting fibers (for example, Al_2O_3 – 4.2 wt.% CaO fibers, V = 50 mm/min were made from a charge of the initial mixtures of two oxides Al_2O_3 – CaO of the indicated weight composition and crystallized when the molybdenum carcass containing the fibers was drawn into cold zone at a rate of 50 mm / min). The fabrication of fibers based on the internal crystallization method presented in this work was based on the preparation of powders mixtures of various initial weight compositions of Al_2O_3 – CaO oxides, aimed primarily at obtaining composite fibers of the composition Al_2O_3 – $CaAl_{12}O_{19}$. Experimental variations of mixtures of oxides Al_2O_3 – CaO included from 0 wt.% CaO to 8.0 wt.% CaO. The composition of Al_2O_3 – 8.0 wt.% CaO corresponds nominally to calcium hexaaluminate $CaAl_{12}O_{19}$. The composition of Al_2O_3 – 20.0 wt.% BaO corresponds nominally to calcium hexaaluminate $BaAl_{12}O_{19}$. The initial composition of the mixtures of oxides Al_2O_3 – 39.0 wt.% Y_2O_3 – 2.8 wt.% CaO is aimed at obtaining fibers of the composition Al_2O_3 – $Y_3Al_5O_{12}$ – $CaAl_{12}O_{19}$, and other set of oxides Al_2O_3 – 24.0 wt.% Y_2O_3 – 2.4 wt.% CaO is aimed at obtaining fibers of the composition (Al_2O_3 – $Y_3Al_5O_{12}$) – $CaAl_{12}O_{19}$, – the eutectic is indicated in parentheses. Taking into account the complex conditions of crystallization of melts of mixtures of oxides in the channels of the molybdenum carcass, the actual final composition of the fibers could differ from that mentioned and contain other compounds, for example, $CaAl_4O_7$, $YAlO_3$ [6, 7]. It should be noted that the issues of the formation of the structure of oxide fibers in molybdenum carcasses require separate special studies. In any case, the composite structure of the fibers obtained contains high strong components and less strong components with relatively weak planes, which provide an increase in the fracture toughness of the whole composite material. The presented work is aimed at solving the problems of providing the necessary balance of strength and fracture toughness.

Composites with composite fibers and brittle matrix of recrystallized molybdenum, and fibers for their use in composites with other matrices were made by the internal crystallization method. The obtained fibers were tested for strength, and composite specimens for strength and fracture toughness. As an example of a composite with a different matrix, pilot specimens with an NiAl intermetallic matrix were made. Specimens were made by hot pressing in vacuum in graphite molds. The molds were filled with a powder of the matrix material with unidirectional fibers distributed in the powder. The resulting specimen preparation was pressed in a vacuum chamber in the regime of 1420 °C – 30 min – 20 MPa.

The fibers obtained in different regimes and with different compositions were tested for strength at room temperature using a technique specially developed for fibers of this type [8]. Briefly, it is as follows. The fiber is fixed on a metal foil substrate with natural rubber glue. The fiber with foil substrate are wound sequentially step by step on a series of rigid cylinders of decreasing diameters. After each step, breaks appear in the fiber due to bending. Then, the average distance between breaks at each step is measured ("fiber length") and the "ultimate strain" ε is determined at which these breaks occurred: $\varepsilon = d / 2R$, where d is the fiber "diameter" and R is the cylinder radius. Considering the brittle nature of the fracture of the fibers and the linear of strain curve of the fiber material, the bending strength of the fiber can be determined by multiplying the magnitude of the ultimate strain by the Young's modulus of the fiber material (up to 400 GPa).

The strength and fracture toughness of composites was determined under three-point bending testing by known methods on specimens in the form of rods of rectangular cross section. Moreover, the specimens for determining fracture toughness had side notch, and the values of fracture toughness, as critical stress intensity factors, were calculated using the formulas given in [9]. Testing of composite specimens was carried out at room temperature as well as at high temperatures (the latter is specifically indicated). At high temperatures, strength tests were carried out in a vacuum chamber.

Studies of the microstructure of fibres and composites were carried out in the digital scanning electron microscopes Tescan VEGA-II and CamScan MV230. Both microscopes have W-cathodes and are equipped with detectors of secondary and reflected electrons and energy dispersive X-ray microanalyzer.

3. Results and discussions

The side surface and fracture surface of the fibers (Figure 1) clearly demonstrate the layered structures of the calcium hexaaluminate $CaAl_{12}O_{19}$ and barium hexaaluminate $BaAl_{12}O_{19}$ fibers, the purpose of which is to inhibit cracks by the "weak" planes. Exfoliating in front of the crack tip, these materials blunt it, reducing the stress concentration, and inhibit its propagation, increasing the fracture toughness of the entire composite structure.

Figure 2 shows as an example the results of testing one of the batches of composite fibers made from the initial mixture of oxides Al_2O_3 – CaO. It can be seen that, over a fiber length of 1 mm, the strength may exceed more than 1000 MPa (Figure 2a), which is quite enough for their use in composite materials. Falling regression dependence is characteristic of brittle objects of this kind and is determined by the distribution of relatively weak "defects" in the bulk and on the surface of the fibers. The fiber fracture surface (Figure 2b) has a "non-brittle appearance", and surface fractures pass through calcium containing areas (of light color). The structure and composition of fibers in cross section (Figure 2c) has a completely composite appearance.

Figure 1. The side surface of calcium hexaaluminate fiber $CaAl_{12}O_{19}$, V = 50 mm/min (a), the fracture surface of the tested fiber of barium hexaaluminate $BaAl_{12}O_{19}$, V = 800 mm/min (b).

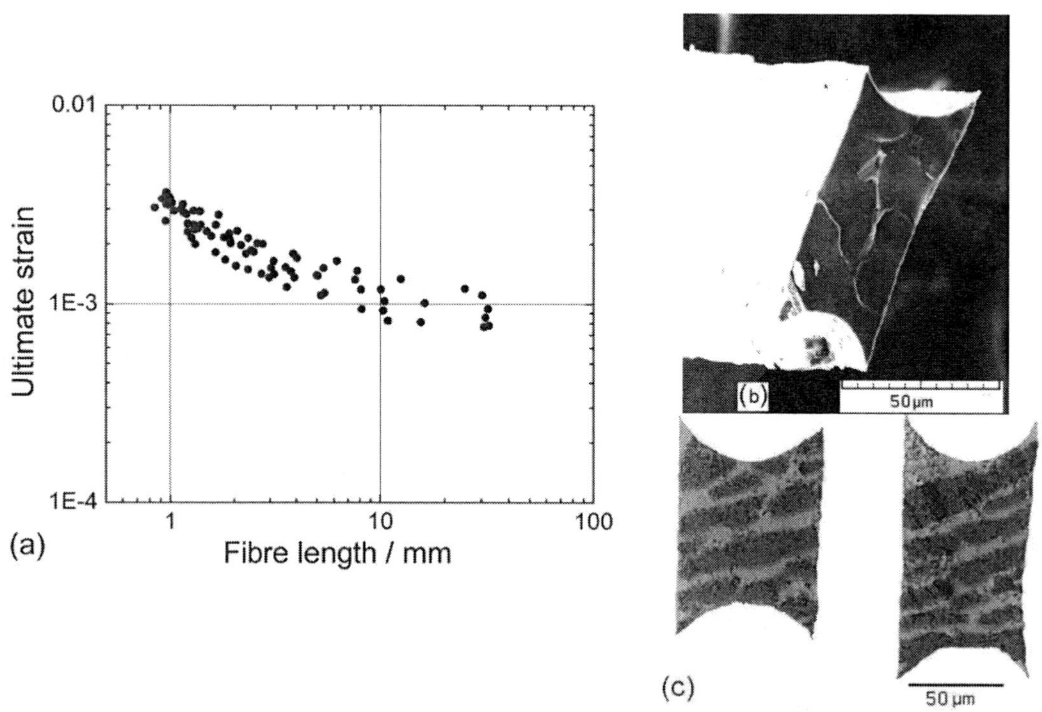

Figure 2. The dependence of the ultimate strain on the length of the fibers of the batch of initial composition Al_2O_3 – 4.2 wt.% CaO, V = 50 mm/min (a); the fracture surface of the tested fiber (b); fibers in cross section: dark areas – Al_2O_3, light areas – calcium containing compounds (c).

When tested, composite specimens with oxide composite fibers and brittle molybdenum matrix exhibit "quasi-plastic" deformation curves, as shown for one of the specimens in Figure 3a. The specimen was tested for fracture toughness, the test scheme is presented in the inset of the figure. The

IOP Conf. Series: Materials Science and Engineering **525** (2019) 012021 doi:10.1088/1757-899X/525/1/012021

value of the critical stress intensity factor for this specimen was 22 MPa·m$^{1/2}$, which is quite acceptable for construction materials of this type. The dependence of the values of the critical stress intensity factor of similar specimens with fibers of the same initial composition (Al_2O_3 – 3.2 wt.% CaO) on the pulling rate of them into the cold zone during crystallization of the fibers (Figure 3b) was also obtained. Dependence has a maximum, which opens up the possibility of optimizing the parameters of technological regimes. It is also seen that the fracture toughness of all composite specimens substantially exceeds the fracture toughness of the specimen from the matrix material without fibers (Mo carcass).

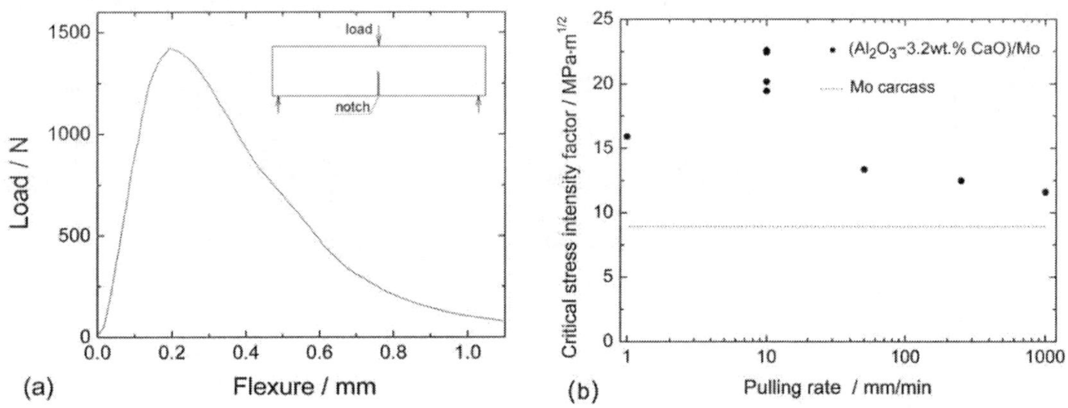

Figure 3. Test results of composite (Al_2O_3 – 3.2 wt.% CaO) / Mo, V = 10 mm/min of specimens with a notch for fracture toughness (specimen sizes ≈5 × 15 × 60 mm): the flexure dependence on the load of one of the specimens when tested, the arrows indicate the direction of load application (a); the dependence of the critical intensity factor of the specimens on the pulling rate them into the cold zone during the crystallization of the fibers; the dotted line is the value of the critical intensity factor of the molybdenum carcass without fibers (b).

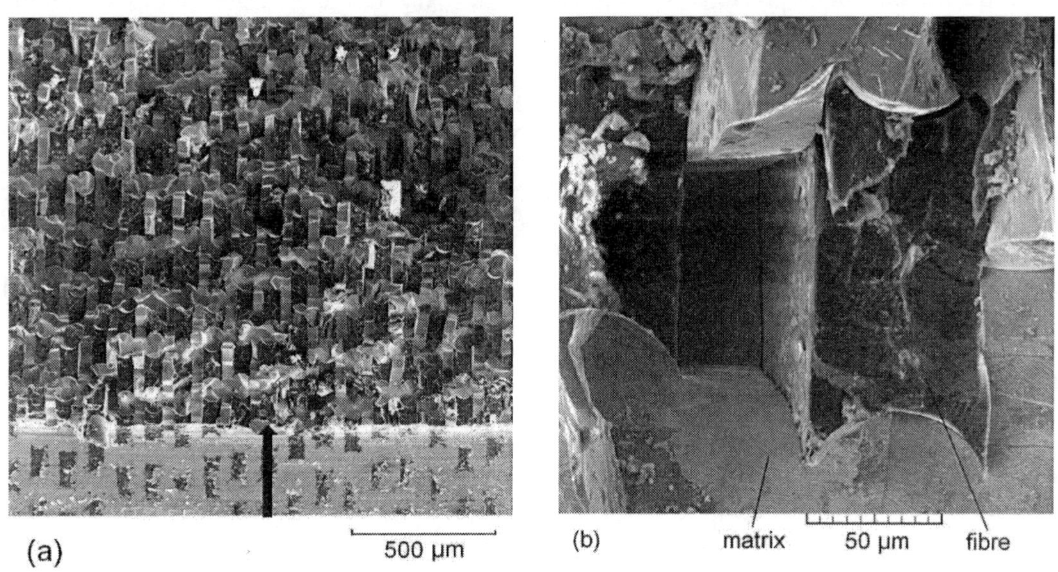

Figure 4. Fracture surface of the composite specimen (Al_2O_3 – 4.2 wt.% CaO) / Mo, V = 50 mm/min with a notch after tests for fracture toughness: the area of the fracture surface, the arrow shows the depth of the initial notch (a); one of the composite fibers in the molybdenum matrix after fracture (b).

The "non-brittle" of fracture is also confirmed by observations of the fracture surfaces of composite specimens. In Figure 4 one can clearly see delaminations and shifts along the fiber–matrix interfaces, pulling out the fibers from the matrix, the fibers crushing, the uneven relief of the fracture surface of the composite fibers themselves, which makes a significant contribution to the overall fracture toughness of the composite.

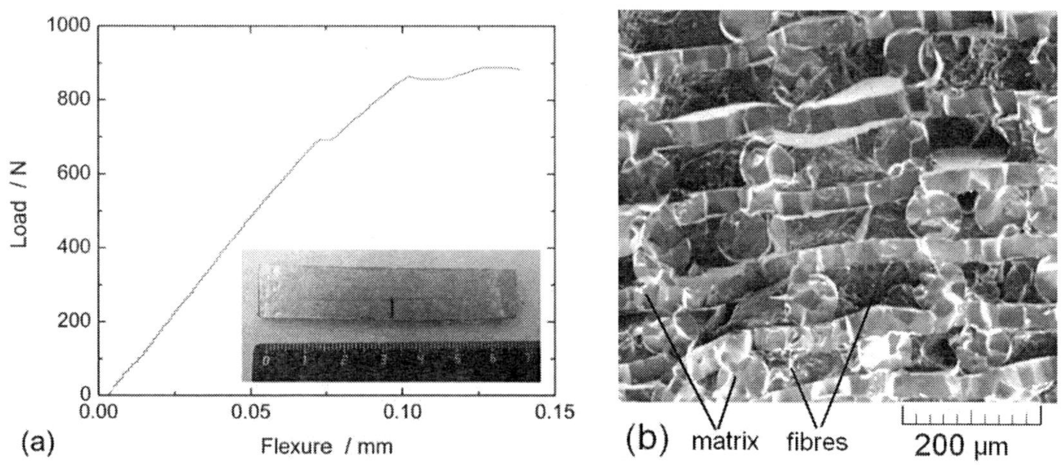

Figure 5. View of the curve of deformation of a composite specimen with a notch (Al_2O_3 – 39.0 wt.% Y_2O_3 – 2.8 wt.% CaO) / Mo, V = 50 mm/min during fracture toughness test, inset view of the specimen after test (a); the surface area of the fracture of the specimen (b).

In addition to composites with molybdenum matrix and fibers obtained from various mixtures of the initial oxides Al_2O_3 – CaO, composites with a molybdenum matrix and fibers from mixtures of the initial oxides Al_2O_3 – Y_2O_3 – CaO were obtained, tested and investigated. The results for one of such composite specimens are shown in Figure 5. The fracture toughness (critical stress intensity factor) of the specimen shown in the figure is an acceptable value of 19 $MPa{\cdot}m^{1/2}$. The strain curve is non-linear, the macrocracks path in the specimen is also non-linear (Figure 5a). The fracture surface has a non-brittle appearance (Figure 5b) with a set of microfractures (fiber splitting, delamination and shifts along the fiber–matrix boundaries), increasing the overall fracture toughness.

Fibers and composites are designed to work in a wide range of temperatures. Therefore, tests on the strength of composites with brittle molybdenum matrix and composite fibers of different composition were carried out in the temperature range of 20÷1400 °C. The results are shown in Figure 6. The strength over the entire temperature range is quite satisfactory for structural materials of this kind.

In addition to composites with composite fibers and brittle molybdenum matrix, the structure of which was formed in a single technological cycle, composite specimens were made with an intermetallic matrix with fibers extracted from the auxiliary molybdenum matrix. Specimens, as already mentioned, were made by hot pressing in a vacuum. Figure 7a shows a bunch of fibers prepared to be placed in a mold with a powder of matrix material (NiAl intermetallic compound). Figure 7b shows the fracture surface of one of the composite specimens with the specified fibers and matrix tested for strength. The strength of the presented composite specimen was 320 MPa. The surface of fracture has a "non-brittle" character: an uneven relief of the fibers and the composite as a whole are seen, traces of shifts along the fiber-matrix boundaries are observed, as well as crushing of the fibers.

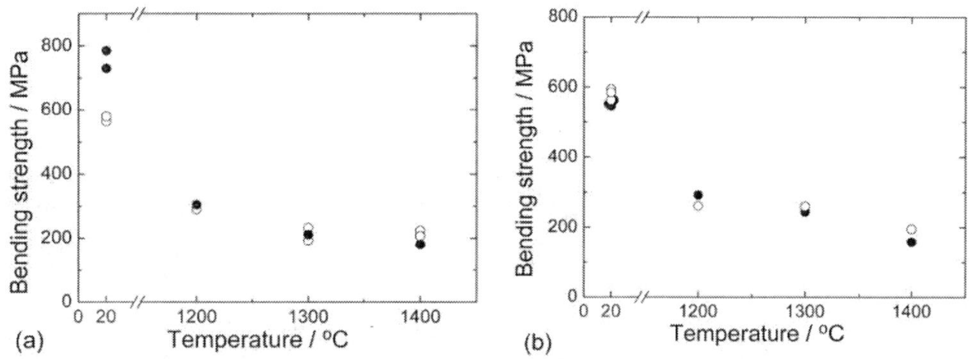

Figure 6. Dependence of the strength of composite specimen on the test temperature of various compositions and conditions of fiber crystallization: composites (Al_2O_3 – 3.2 wt.% CaO) / Mo, V = 50 /min (black dots on the graph) and (Al_2O_3 – 4.2 wt.% CaO) / Mo, V = 250 mm/min (a); composites (Al_2O_3 – 39.0 wt.% wt.Y_2O_3 – 2.8 wt. % CaO) / Mo, V = 50 mm/min (black dots) and (Al_2O_3 – 24.0 wt.% Y_2O_3– 2.4 wt.% CaO) / Mo, V = 50 mm/min (b).

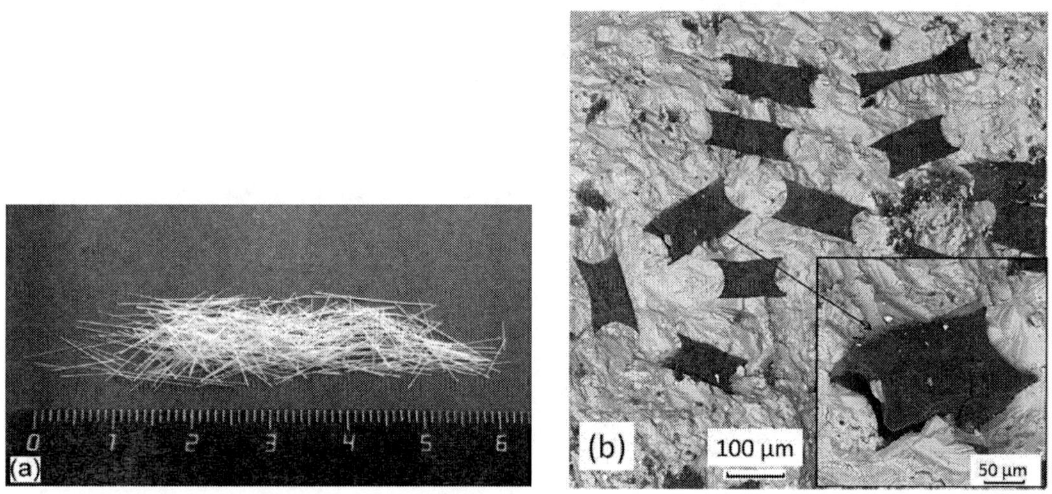

Figure 7. A bunch of composite fibers of a batch (Al_2O_3 – 4.2 wt.% CaO), V = 50 mm/min (a); a fracture surface of the composite with the indicated fibers and intermetallic matrix (Al_2O_3 – 4.2 wt.% CaO) / NiAl after the strength test (b).

4. Conclusion

Composite oxide fibers based on initial mixtures of aluminium and calcium or barium oxides, as well as yttrium, were obtained by the internal crystallization method. Composites with brittle matrices of recrystallized molybdenum and nickel-aluminium intermetallic and with these fibers were also made. Studies of the structure showed that the composite fibers contain both high-strength compounds and relatively weak ones with a specific layered structure braking cracks that provides acceptable fracture toughness in the whole of composites with non-plastic components. Thus, fibrous composites with "quasi-plastic" behavior were obtained from brittle components. Tests of composite specimens showed that the strength values in a wide range of temperatures and fracture toughness correspond to the conditions of use of materials of this kind.

Acknowledgements
The work was performed in accordance with State Assignment (No. 00-2016-0003) and with the financial support of the Russian Foundation for Fundamental Research (Project No. 17-03-01136).

References
[1] Cook J and Gordon J E 1964 A mechanism for the control of crack propagation in all-brittle systems *Proc. of Roy. Soc.* A **282** 508–20
[2] Popova N A and Tolstun A N 2007 A monazite imterphase in oxide/oxide composites with ICM-fibres *Sci. to Industry* **2** 58–61 (in Russian)
[3] Cinibulk M K 2000 Hexaluminates as a cleavable fiber-matrix interphase: synthesis, texture development, and phase compatibility *J. Europ. Ceram. Soc.* **20** 569–82
[4] Mileiko S T, Kurlov V N, Kolchin A A and Kiiko V M 2002 Fabrication, properties and usage of single-crystalline YAG fibers *J. Europ. Ceram. Soc.* **22** 1831–37
[5] Mileiko S T and Kazmin V I 1992 Crystallization of fibres inside a matrix: a new way of fabrication of composites *J. Mater. Sci.* **27** 2165–72
[6] Mileiko S, Kolchin A, Kiiko V and Novokhatskaya N 2017 Composite oxide fibres grown by internal crystallisation method *Fibers* **5** 1–10
[7] Mileiko S T, Kiiko V M, Novokhatskaya N I, Tolstun A N and Kolchin A A 2008 Single crystalline fibers of yttrium-aluminum garnet: microstructure and strength at room temperature *Deform. and Fracture of Mater.* **5** 2–7 (in Russian)
[8] Kiiko V M and Mileiko S T 1999 Evaluation of room temperature strength of oxide fibres produced by the internal crystallization method *Compos. Sci. and Technol.* **59** 1977–81
[9] Broek D 1974 *Elementary Engineering Fracture Mechanics* Leyden 74

IOP Conf. Series: Materials Science and Engineering **525** (2019) 012022 doi:10.1088/1757-899X/525/1/012022

Synthesis and study of the properties of thermoset oligoimides with propargyl fragment

T I Kolesnikov[1,2], A Y Tsegelskaya[2], M D Dutov[3], A M Orlova[1,2] and A A Kuznetsov[1,2]

[1] MIREA - Russian Technological University, Prospekt Vernadskogo, 86, Moscow 119571, Russia.
[2] Enikolopov's Institute of Synthetic Polymer Materials RAS, Profsoyuznaya st. 70, Moscow 117393, Russia.
[3] N.D. Zelinsky Institute of Organic Chemistry RAS, Leninsky Prospect, 47, Moscow 119334, Russia

E-mail: tikitaket@gmail.com

Abstract. A series of thermoplastic oligoimides with a propargyl substituent in the side chain was synthesized by the method of high-temperature catalytic copolycondensation in the melt of benzoic acid based on the monomer 5- (2-propyn-1-yloxy) benzene-1,3-diamine (PBD), 2,2-bis-[(3,4-di-carboxyphenoxy)phenyl]propane dianhydride (BPADA), 2,2-bis-[(4-aminophenoxy)phenyl]propane (BAPP) and phthalic anhydride (PA). The uncured imide oligomers showed good solubility (>20 wt% in N-methyl-2-pyrrolidone), wide temperature processing window (>50°C) and also flow well at 220°C. These imide oligomers were successfully converted to cross-linked structures after curing at 300°C. Cross-linked oligomers have a temperature of 5% weight loss >500°C and T_g >200°C.

1. Introduction

Aromatic polyimides (PI) have a complex of valuable properties: high thermal and heat resistance, good mechanical properties, chemical resistance, low moisture absorption, radiation resistance and performance in a wide temperature range. For the use of PI as matrix for polymer composites materials (PCM), prepolymer solutions are used — polyamic acids or imide-forming polymerization compositions of the PMR-15 type (polymerization of monomer reactants). In case of PMR-15, the solvent removal, the chain growth, imidization, and the formation of a three-dimensional network occur in a course of stepwise heating of the impregnated tissue to 350 °C for several hours. The disadvantage of these matrix for PCM is the difficulty of technological control during processing. It is necessary to carefully remove residual solvent and water, otherwise defects are formed, and the mechanical properties of PCM deteriorate. In this regard, the idea seems attractive of using the fully imidized oligomers with reaction groups capable to cure with formation of a three-dimensional network without release of volatile compounds [1].

In the 90s, Hughes (USA) [2] developed a series of Therimid oligoimides containing ethynyl end groups capable of cyclotrimerization with the formation of an aromatic cycle. Alkynyl groups are introduced using monofunctional alkyne-containing amines or anhydrides [3-5]. Also, oligoimide compositions based on 4-phenylethynylphthalic anhydride (PEPA) [6-8] are used, with the curing temperatures of about 380 °C. In oligomers of this type, the content of cross-linkable groups and the chain length, which determines the rheological behavior, are rigidly connected to each other, which makes it difficult to tune properties. In this connection, another approach is promising of using (co)oligoimides containing monomer moieties with cross-linkable groups in a side chain, the content of the latter in chain and the main chain length chain can be varied independently.

In this work, we have prepared a new type of cross-linkable thermoplastic oligoimides (OI) based on a new monomer 5-(2-propyn-1-yloxy) benzene-1,3-diamine (PBD). Due to presence of alkynyl group in the side chain, these OI can crosslink via cyclotrimerization to form 3D network. To obtain OI, an advanced method of high-temperature catalytic polycondensation in molten benzoic acid (BA) [9], BA playing the role of a solvent, catalyst and reactivity regulator in this process.

2. Materials and methods

2.1. Materials

2,2-Propylidenebis(1,4-phenyleneoxy)diphthalic anhydride (BPADA) (Aldrich) was crystallized from acetic anhydride, T_m=189°C. 2,2-bis-[(4-aminophenoxy)phenyl]propane (BAPP) was crystallized from isopropanol, T_m= 130 °C. Phthalic anhydride (PA) (Aldrich) was sublimed in vacuo, T_m=131°C. Benzoic acid (BA) was recrystallized from the mixture (ethanol/water) and dried in vacuo, T_m=122°C. Diethyl ether, acetone and chloroform (Chimmed, Russia) were used without additional purification.

2.2. Measurements

IR spectra were recorded on FTIR spectrometer Bruker, mod. Equinox 55/S (Germany) in a range of 4000-400 cm^{-1}. Samples were prepared in a form of CsI tablets with the substance content of 1%. NMR ^1H and ^{13}C were registered on Bruker AM-300 spectrometer at 300.13 and 75.47 MHz respectively. Chemical shifts are given with respect to signal of $Si(Me)_4$.

DSC thermograms were recorded on DSC-500 device (Russia) at heating rate of 16°C min^{-1}. GPC measurements were executed on the device "Stayer" (Akvilon) at 60°C, with Phenogel coloumn (2-70 kDa) and DMF as eluent (1 ml*min^{-1}). Polystyrene standards were used.

Thermomechanical analysis was (the stock penetration mode) performed using the UIP-70 instrument (Sputnik, Russia) at a load of 2.8 kg*cm^{-2} and a heating rate of 10°C/min. Polymer samples for TMA were obtained by hot pressing at 220–300°C.

Data of thermogravimetric analysis (TGA) were obtained using a PerkinElmer Pyris 1 TGA analyzer under a nitrogen flow of 100 mL/min at a heating rate 10°C/min in dynamic mode.

Determination of logarithmic viscosity was performed using a Ubbelode capillary viscometer for solutions of oligomers and polymers in chloroform with concentration 0,5 g/100 ml at 25°C.

2.3. Synthesis of 5- (2-propyn-1-yloxy) benzene-1,3-diamine (PBD)

To a solution of 2.22 g (0.01 m) 1,3-dinitro-5-(prop-2-en-1-yloxy)benzene [10] in 50 ml of dioxane at 10-15°C, a solution of 22.6 g (0.1 m) $SnCl_2{}^*2H_2O$ in 55 ml of concentrated HCl was added dropwise and maintained for 3 hours. Then, it was neutralized with a concentrated aqueous solution of NaOH until the yellowish color disappears (pH 9-10), and extracted with CH_2Cl_2 (4x100 ml), dried with Na_2SO_4, filtered through a layer of silica gel, and evaporated. To the residue 40 ml water and about 1 ml of concentrated HCl was added until positive acidic test, activated charcoal was added. Mixture was filtered and the filtrate basified to pH 8-9 with aqueous NaOH. The precipitate was filtered off, washed with water and dried to obtain 0,356 g. (22%:). T_m= 88-89°C, NMR ^1H: 3.41 (s. 4h); 3.49 (s, 1H); 4.55 (s, 1H);4.72 (s, 2H-); 5.46 -5.49 (m. 2H); MS (m/z) = 162 M+; IR spectrum (CsI): 3440, 3350 cm^{-1} (N-H), 3260 cm^{-1} (C-H acetyl), 3040 cm^{-1} (C-H ar), 2924 and 2870 cm^{-1} (CH_2), 2124 cm^{-1} (C≡C), 1620 cm^{-1} (N-H), 1600 and 1490 cm^{-1} (C-Car), 1170 cm^{-1} (C-O-C). 1H NMR (300 MHz, DMSO-d6): d 5.48 (d 1H), 5.43 (d 2H), 4.72 (s 4H), 4.52 (d 2H), 3.49 (t 1H). Elemental analysis: found, %: C, 66.48; H, 6.41; N, 17.11; calculated for $C_9H_{10}N_2O$, % C, 66.65; H, 6.21; N, 17.27.

IOP Conf. Series: Materials Science and Engineering **525** (2019) 012022 doi:10.1088/1757-899X/525/1/012022

2.4. Synthesis of polyimide and oligoimides with propargyl groups (PI 9-10, OI 4-5, OI 9-10) in benzoic acid

Crystalline BA (9.0 g) and 0.57 g (0.20 mmol) of BPADA were loaded into a glass reactor equipped with a stirrer and an inert gas inlet. After inert gas was blown into the reactor, the mixture was heated up to 140°C, and 0.404 g (0.18 mmol) BAPP was added. The reaction mixture was stirred for half an hour at 140°C with slow blowing of inert gas. Then, to the reaction mixture, 0.026 g (0.02 mol) of PBD was added and the synthesis was carried out for another 1.5 hours. The final molar ratio of monomers for polyimide-1 (PI-1) was BPADA:BAPP:PBD:PA 10:9:1:0. After cooling, the solidified mixture was extracted with diethyl ether. Polymer residue was filtered and dried in vacuo.

Similarly, oligoimides were obtained (OI-1 and OI-2). For obtaining OI-1 BA (9.0 g) and 0.57 g (0.20 mmol) of BPADA were loaded. After addition of BAPP (0.404 g (0.18 mmol)), the reaction mixture was stirred for half an hour at 140°C with slow blowing of inert gas. Then, 0.026 g (0.04 mol) of PBD was added to the reaction mixture, and after 20 minutes 0.029 g (0.04 mol) of PA was added. Synthesis was carried out for one hour. The final molar ratio of monomers for OI-1 was BPADA:BAPP:PBD:PA 5:4:2:2 and for OI-2 was BPADA:BAPP:PBD:PA 10:9:2:2. After extracrion with diethyl ether, was filtered and dried in vacuo.

3. Results and discussion

The syntheses of PI and OI-1, 2 were carried out according Figure 1. During the synthesis of OI-1,2, PA was introduced into the reaction system at the end of the process to close the terminal amino groups. To obtain propargyl-containing oligoimides with different degrees of polymerization, the ratio of initial monomers was changed (BPADA:BAPP:PBD:PA 5:4:2:2 for OI-1 and BPADA:BAPP:PBD:PA 10:9:2:2 for OI-2). Syntheses of all compounds were carried out in a homogeneous solution.

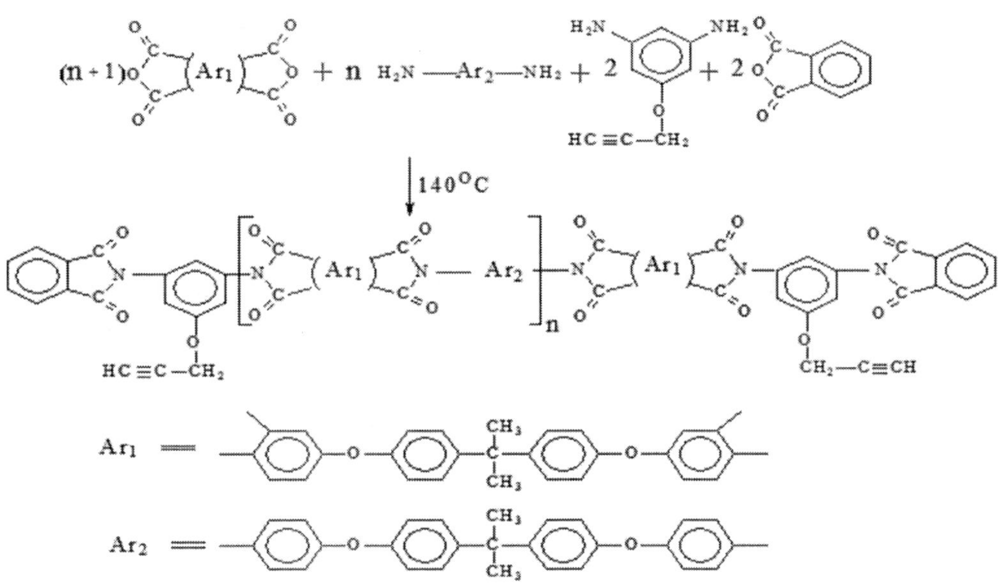

Figure 1. Scheme of synthesis of oligoimides (n = 4, 9 for OI-1, OI-2).

The resulting products were white powders or flakes, soluble in dimethyl sulfoxide (DMSO), dimethylformamide (DMF), tetrahydrofuran (THF) and chloroform. By increasing the molar ratio of the BPADA / BAPP monomers in the initial mixture (OI-1, OI-2), the logarithmic viscosity increases 0.345, 0.641 dl/g, respectively. Molecular weight characteristics of OI-1 and OI-2 samples are present in Table 1.

IOP Conf. Series: Materials Science and Engineering **525** (2019) 012022 doi:10.1088/1757-899X/525/1/012022

Table 1. Molecular Mass Characteristics of OI-1, OI-2 (GPC).

Sample	M_w	M_n	M_w/M_n
OI-2	79700	36500	2.2
OI-1	54100	34200	1.6

For copolyimide (PI-1), obtained by stoichiometry 10-9-1-0, the logarithmic viscosity is ήlog =1.2 dl/g. This value practically coincides with ήlog =1.22 dl/g for homo polymer (10:10:0:0) obtained from BPADA and BAPP at the same conditions. So, it could be concluded that alkynyl fragment in PI-1 is stable at conditions of the synthesis.

All products of copolycondensation possess a film-forming ability. Transparent films were obtained from their chloroform solutions: brittle for OI-1, durable for OI-2 and PI-1. The presence of "polymeric behavior" of oligomers OI-1 and OI-2 is associated with the structure of the initial monomers (a large length of the elementary unit).

Figure 2 shows the IR spectrum of propargyl-containing polyimide PI-1. Absorption bands at 1720 and 1780 cm^{-1} were attributed respectively to the symmetric and asymmetric C=O vibrations in imide cycle. The intensity of absorption of 3288 cm^{-1} corresponds to the vibrations of the C–H bond of the ethynyl group. Such a spectrum indicates that copolycondensation proceeds to high conversion, and that the final reaction products contain ethynyl groups.

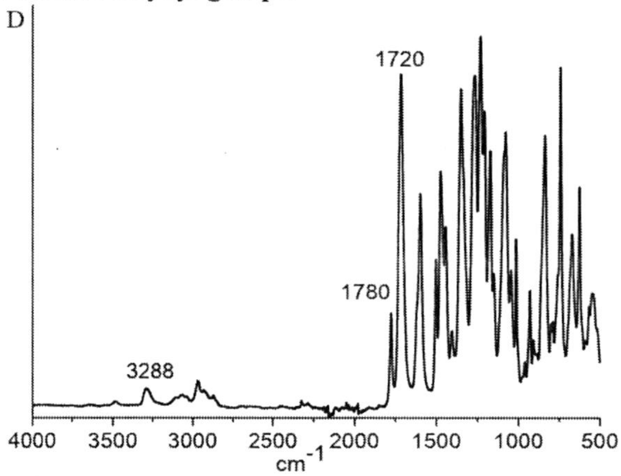

Figure 2. IR spectra of products of polycondensation PI-1.

Figure 3 shows the DSC thermograms of OI-1, OI-2 and PI-1. It is seen that all products demonstrate a jump in heat capacity in the region of 190-200°C corresponding to the glass transition temperature. In the higher-temperature region (270–330°C), an exothermic effect is observed corresponding to the cross-linking reaction of propargyl groups. The extremum of exoterrmic effect shifts to higher temperatures after heating which is apparently due to an increase in the effective activation energy of the cross-linking reaction (decrease in molecular mobility).

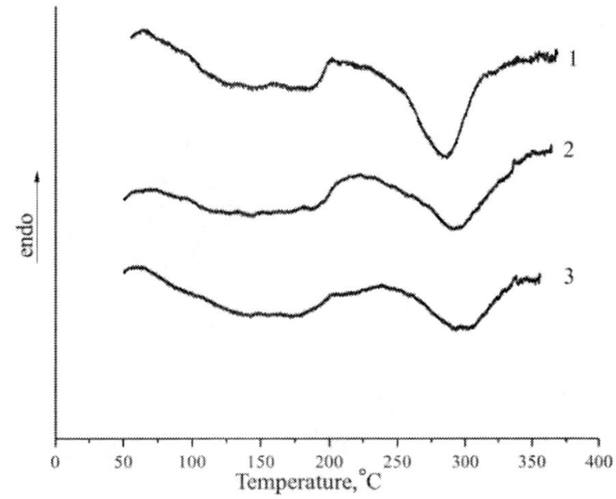

Figure 3. Thermograms DSC OI-1 (1), OI-2 (2) and PI-1 (3).

Curing of the synthesized samples was carried out in a hot press at different temperatures. Figure 4 presents the OI-1 thermograms with different degrees of curing. In the case of oligoimide curing at a temperature of 230°C (curve 3), the glass transition temperature does not change as compared with that of the uncured sample (curve 1). However, the shape of the exothermic peak of the compared samples is markedly different. Apparently, this is due to the partial cross-linking of the treated sample. The DSC data, as well as the fact that the treated sample remains thermoplastic and does not lose solubility, indicate that at 230°C only chain length increases, and a three-dimensional structure is not formed. Treatment at 300°C leads to loss of the solubility (curve 2) and disappearance of exothermic effect. Obviously, such changes are associated with the formation of a cross-linked polymer.

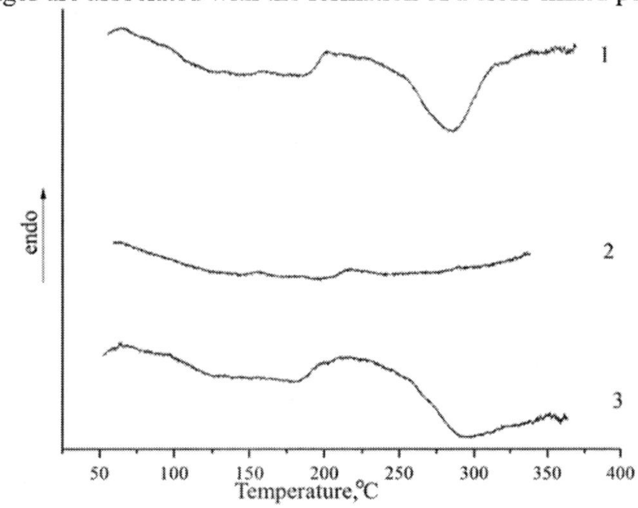

Figure 4. DSC thermograms of the initial oligoimide OI-1 (1), cured at 300 °C (2) and at 230 °C (3).

Cured propargyl containing compounds were investigated by IR spectroscopy. When comparing the IR spectra for the initial and for the cured polyimide PI-1 at a temperature of 300°C, it is shown Figure 5, you can see that characteristic absorption band of alkynyl group in the region of 3288 cm^{-1} disappears, indicating consumption of propargyl unit.

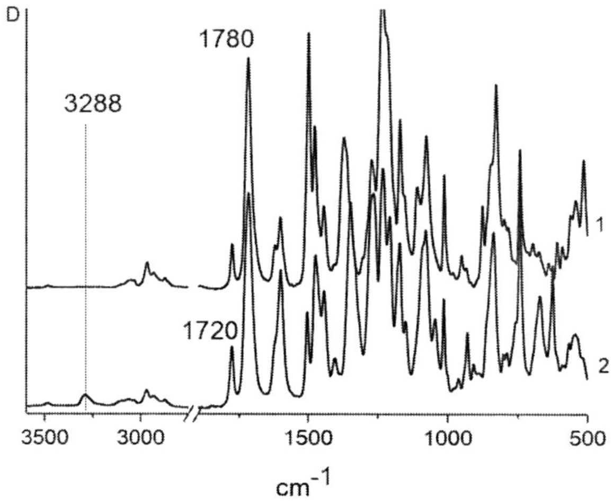

Figure 5. IR spectra of the initial PI-1 (2) and cured at a temperature of 300 °C for 20 minutes (1).

The results of the thermogravimetric analysis (TGA) of cured samples are shown in Figure 6 and Table 2. It can be seen that the obtained cross-linked products have excellent stability to thermal and thermal-oxidative degradation: $T_{5\%}$ are in the range of 500-518 °C in air, and the coke residue is about 50% in nitrogen atmosphere.

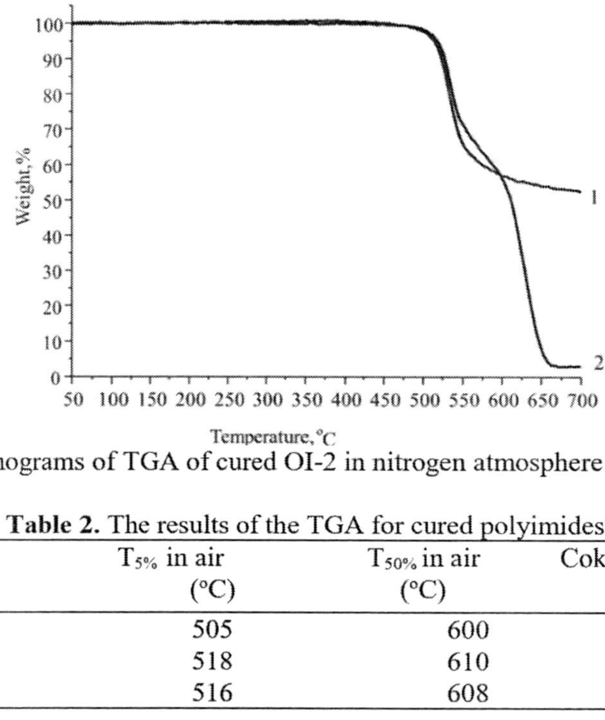

Figure 6. Thermograms of TGA of cured OI-2 in nitrogen atmosphere (1) and in air (2).

Table 2. The results of the TGA for cured polyimides.

Sample	$T_{5\%}$ in air (°C)	$T_{50\%}$ in air (°C)	Coke residue at 700 °C in N_2, (%)
OI-1	505	600	50
OI-2	518	610	53
PI-1	516	608	50

The products were investigated by the method of thermomechanical analysis (TMA). In Fig. 7, TMA curves are shown of OI-1 sample heat treated for 20 minutes in hot press at 230°C (1) and at 300°C (2). From fig.7, it can be concluded that the sample heated at 300°C is cross-linked system, whereas samples heated at 230°C retains a viscous flow behavior at temperatures higher than 230°C. Similar behaviors was observed for OI-2 sample. Both samples heated at 230°C retain solubility in organic solved.

IOP Conf. Series: Materials Science and Engineering **525** (2019) 012022 doi:10.1088/1757-899X/525/1/012022

In Fig. 8, analogous TMA curves for high molecular polyimide sample PI-1 are shown. It is seen that in contrast to lower molecular weight samples, after heat treatment both at 300°C and at 230°C samples lose ability to flow at temperature above 230°C. This result indicates that some cross-linking *via* reaction of propargyl groups takes place even at 230°C. In a case of low lower molecular weight samples this effect is small, and this stage can be considered rather as chain elongation than cross-linking. But for a high molecular weight product, even small amount of crosslinks in a sample is enough to lose its ability to viscous flow and ability to melt processing.

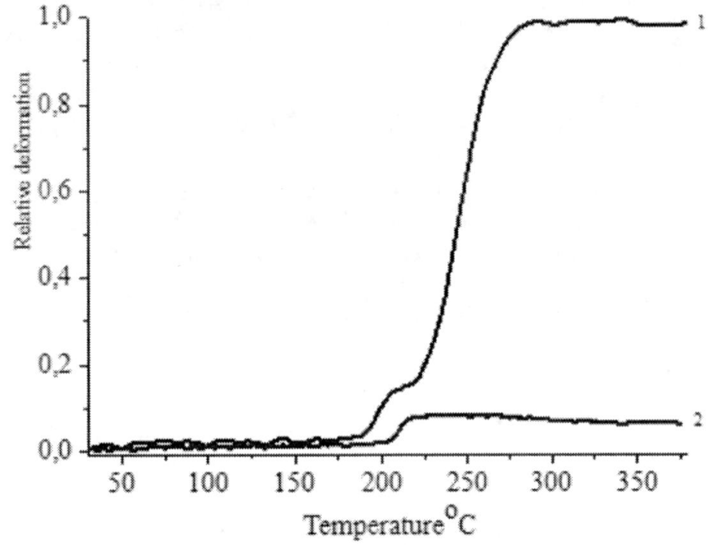

Figure 7. Thermograms TMA of OI-1 cured to 230 °C (1) and to 300 °C (2).

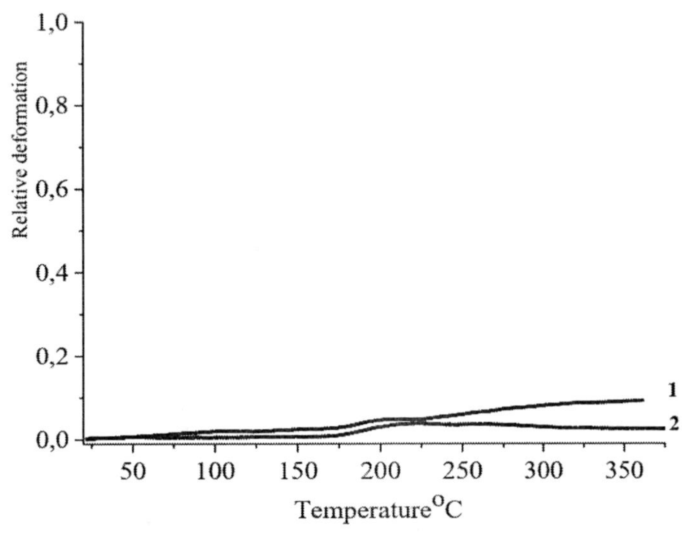

Figure 8. Thermograms TMA of PI-1 cured to 230 °C (1) and to 300 °C (2).

Thus, it was demonstrated that the obtained oligoimides have a "process window", the range of which is about 100°C and is sufficient for molding PCM based on it.

IOP Conf. Series: Materials Science and Engineering **525** (2019) 012022 doi:10.1088/1757-899X/525/1/012022

4. Conclusions

In this work, new thermoplastic oligoimides of different molecular weights containing propargyl groups in the side chain were synthesized. It was shown that the synthesized oligomers are able to transfer into viscous flow state at 200–230°C and cross-linked with subsequent heating to 300°C. Such materials are of potential interest for use as thermoplastic binders.

References

[1] Wilson D 1998 *British Polymer Journal* **20** 405 - 416
[2] Hanky A O and St Clair T L 1983 *Adhesion and adhesives* **3** 181-187
[3] Shoji Y, Ishige R, Higashihara T, Morikawa J, Hashimoto T, Takahara A, Watanabe J and Ueda M 2013 *Macromolecules* **46** 747-755
[4] Shengli C, Jianhua H, Xin W, Kuanyu Y, Xigao J and Jinyan W 2017 *Polymer* **115** 96-105
[5] Ping Y, Yan W, Junrong Y, Jing Z and Zuming H 2015 *Journal of Applied Polymer Science* **132** 42537
[6] Wei C, Mian J and Shi-yong Y 2016 *Chinese Journal of Polymer Science* **34** 933-948
[7] Seurer B, Vij V, Haddad T, Mabry J M and Lee A 2010 *Macromolecules* **43** 9337–47
[8] Hanyuan L, Wei W, Guofei C, Yinfeng L and Xingzhong F 2018 *Journal of Polymer Research* **25** 32
[9] Kuznetsov A A 2000 *High Perform Polymer* **12** 445–460
[10] Dutov M D, Serushkina O V, Nesterova E N, Shevelev S A and Aleksanyan D R 2018 *Russian Chemical Bulletin* **67** 822-825.

IOP Conf. Series: Materials Science and Engineering **525** (2019) 012023 doi:10.1088/1757-899X/525/1/012023

Lignosulfonate, anionic surfactants and their mixtures influence on water solutions surface tension and zinc concentrate pressure leaching

E B Kolmachikhina[1], E A Ryzhkova[1], D V Dmitrieva[1], V V Sviridov[2] and S S Naboichenko [1]

[1] Ural Federal University, 17, Mira St., Ekaterinburg, 620002, Russia
[2] Ural State Forest Engineering University, 37, Sibirskii trakt, Ekaterinburg, 620032, Russia

E-mail: e.b.khazieva@urfu.ru

Abstract. The formation of elemental sulfur during sulfide concentrate pressure leaching leads to the appearance of sulfur-sulfide granules, which significantly reduces zinc dissolution rate. The solution for this problem was the use of surfactants, which prevents the negative effect of molten sulfur, avoiding the granule formation during leaching. Lignosulfonates (waste from the woodworking industry) are most often used as reagents for sulfide concentrate pressure leaching. Influence of individual and mixtures of surfactants on surface tension of aqueous solutions, zinc extraction and size-grade distribution of cakes after sulfide zinc concentrate pressure leaching. Surface tension of aqueous solutions was analyzed by stalagmometric method. Leaching was performed in titanium autoclave in presence of lignosulfonate, sodium dodecylbenzenesulfonate and sodium dodecylsulfate. Solutions after leaching were analyzed on zinc by atomic absorption spectroscopy. Size-grade analysis of cakes after leaching was carried out by laser diffraction particle size analyzer. In this work, synergetic influence was observed of anionic surfactants and lignosulfonate on decreasing of aqueous solutions surface tension. The best results were obtained when LS-SDBS mixture was used, namely in the range of sodium dodecylbenzenesulfonate concentrations 400-600 mg / l. Usage of combined surfactants allowed to achieve high rates of zinc extraction and optimal particle size of cakes after leaching. At Ls and SDBS using zinc extraction was increased from 77,2 % up to 82,8 %, with the bulk of the cake (96,7%) having a particle size of -150 μm. Combined surfactants usage (lignosulfonate and sodium dodecylbenzenesulfonate) can be recommended for applying at pressure leaching of sulfide zinc concentrates.The first section in your paper

1. Introduction

Autoclave technologies for production of non-ferrous metals from sulfide raw materials became commonly used due to their environmental friendliness, low grade raw material treatment opportunity and associated elements recovery [1-3].Elemental sulfur formation during sulfide concentrates pressure leaching leads to appearing sulfur-sulfide pellets that results to emergency shutting-down of autoclave [4-5].Lignosulfonates usage became solution for this problem.

However, lignosulfonates composition (molecular weight, content of hydroxyl, carboxyl, sulfonic groups) is inconstant and ranged depending on processed timber species, delignification conditions, metals ions addition [6].In works of Lugovitskaya T.N. [7, 8] electrolytes composition influence on surface activity of lignosulfonates with variety of molecular weight were investigated in detail.It was shown that high molecular weight fraction of lignosulfonates had a maximum effect on surface tension decreasing that can be connected with their less hydrolysis and dissociation in comparison with medium molecular and low molecular samples.

Besides that, surfactants mixture creation, which can enhance individual reagents impact is very important task for non-ferrous metals hydrometallurgy.It is known that combined addition of lignosulfonate and sodium dodecylsulfate leads to micelle formation critical concentration decreasing [9].Interaction of dodecylbenzene sulfonate with polymethacrylic acid can be followed by formation of intermolecular aggregates [10] that leads to important changing of surface properties.

Combined surfactants (lignosulfonate and water soluble sulfonates) usage at pressure leaching of nickel pyrrhotite concentrates allowed to increase pyrrhotite decomposition degree, enhance of platinum group metal extraction from leaching cakes [4, 11]. Combined surfactants synergistic action allows to decrease solutions surface tension and this property is useful for ore and concentrates leaching intensification [12].

This work is devoted to investigation of simultaneous action of lignosulfonate and anionic surfactants (sodium dodecyl sulfate (SDS), sodium dodecylbenzene sulfonate (SDBS)) on solutions surface tension and zinc concentrates pressure leaching effectiveness.

2. Methods

Sodium lignosulfonate, sodium dodecylbenzene sulfonate and sodium dodecylsulfate were used for tests.Surface tension of individual and their mixtures solutions was determined by stalagmometric method at 298 K.Water solutions and sulfuric acid solutions (100 $g \cdot l^{-1}$) were used for tests; surfactants concentration in solutions was varied from 200 to 800 $mg \cdot l^{-1}$.

Pressure leaching was conducted in presence of individual surfactants and their mixtures. Zinc concentrates pressure oxidative leaching tests were carried out in 1l titanium autoclave at 418 K and oxygen pressure 5 MPa during 60 minutes. Zinc concentrate from Uchaly deposit had following composition, %: 44,2 Zn; 8,16 Fe, 1,22 Cu; 29,2 S.

Filter effluent and washwater were analyzed to find out concentrations of Zn, Fetotal, Cu and H2SO4 by atomic absorption spectrometer (novAA 300, Analytik Jena) and titration analysis. Mesh-size distribution of leaching residues was measured by laser light scattering particle size analyzer HELOS&RODOS (Sympatec GmbH).

3. Experiments results and discussions

Decreasing of surface tension of water solutions in presence of individual and mixed surfactants is indirect measure of reagents effectiveness at pressure leaching.Tensiometric investigations (figures 1, 2) testify about low surface activity of lignosulfonates, surface tension of this solutions is within the range 70-72 mN/m.Anionic surfactants, SDS and SDBS, decreased water solutions surface tension intensively from 72 mN/m to 20.6 mN/m and 38 mN/m respectively at 800 $mg \cdot l^{-1}$.

Lignosulfonate action during pressure leaching appears to consist in stabilization of sulfur particles due to a structural mechanical barrier formation. Whereas high surface activity of anionic surfactants promote wettability increasing of sulfides by water solutions and by that wettability decreasing by molten sulfur.

According to fig. 1b in the range of SDBS concentration 350-450 $mg \cdot l^{-1}$ a sharp change of slope corresponds critical concentration of micelle formation (CCM) adjusting (420 $mg \cdot l^{-1}$ [13]), similar curve shift on figure 1a was not observed because CCM for SDS equals to 2.4 $mg \cdot l^{-1}$ [14]. At sulfuric acid addition to SDBS solution (figure 2b) surface tension remained as before, but sulfuric acid addition to SDS solution (figure 2a) led to significant decreasing of surface tension and CCM reaching as early as at $\approx$200 $mg \cdot l^{-1}$.

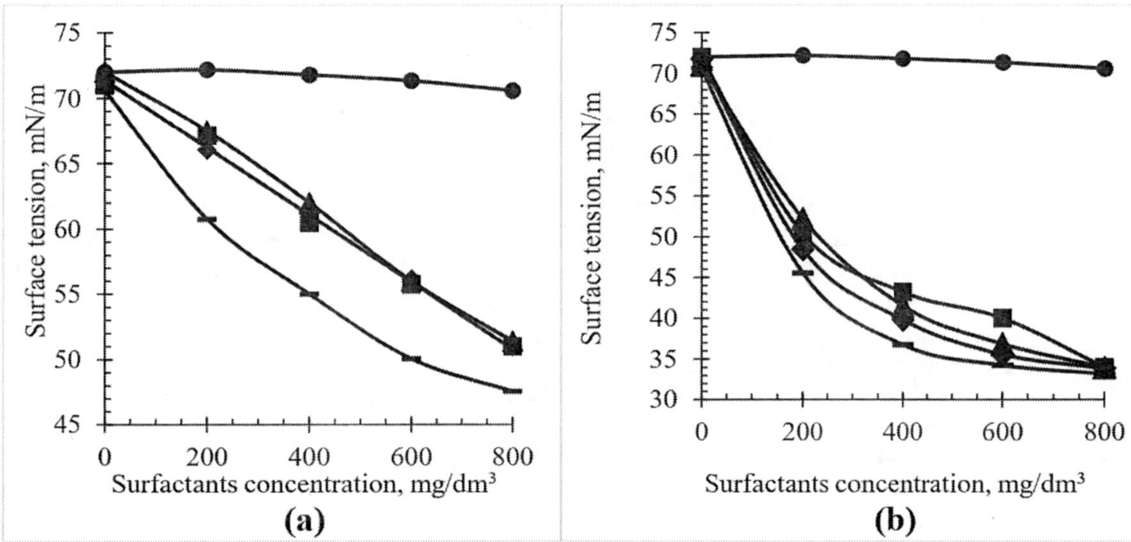

Figure 1. Surface tension changing of surfactants and their mixtures water solutions:
a) Concentration influence of Ls, SDS and their mixtures: ● – Ls; ▲ – SDS; ■ - 200 mg/dm³ Ls+SDS; ♦ - 400 mg/dm³ Ls+SDS; ▬ - 800 mg/dm³ Ls+SDS;
b) Concentration influence of Ls, SDBS and their mixtures:
● – Ls; ■ – SDBS; ▲ - 200 mg/dm³ Ls+SDBS; ♦ - 400 mg/dm³ Ls+SDBS; ▬ - 800 mg/dm³ Ls+SDBS.

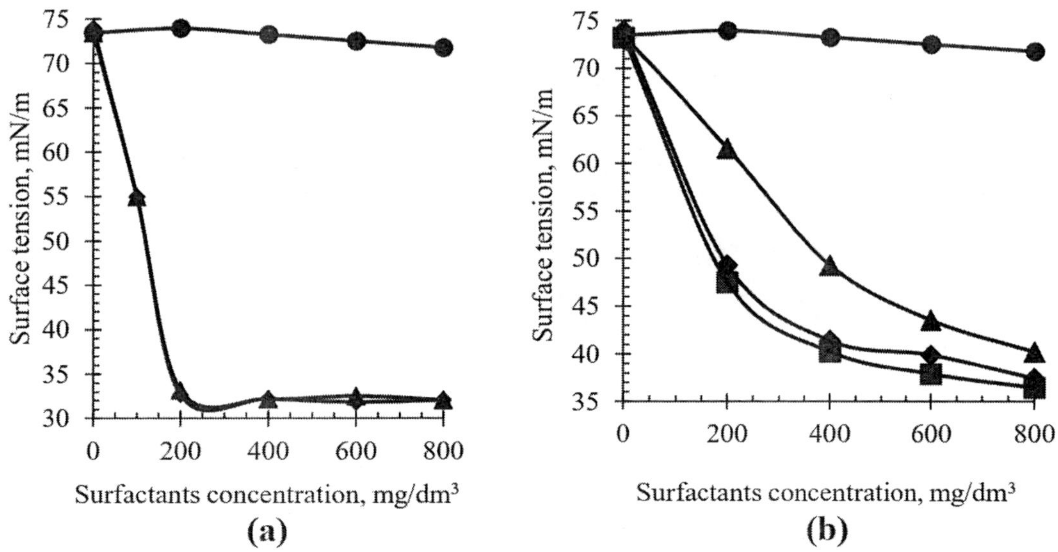

Figure 2. Surface tension changing of surfactants and their mixtures sulfuric acid solutions:
a) Concentration influence of Ls, SDS and their mixtures:
● – Ls; ▲ – SDS; ♦ - 200 mg/dm³ Ls+SDS;
b) Concentration influence of Ls, SDBS and their mixtures:
● – Ls; ▲ – SDBS; ♦- 200 mg/dm³ Ls+SDBS; ■ - 400 mg/dm³ Ls+SDBS.

Addition of Ls small quantity (200-400 mg·l^{-1}) to SDS solutions didn't lead to decreasing of surface tension (figure 1a), only at higher LS concentration (800 mg·l^{-1}) surface tension lowering on 10.0-11.3 % in comparison with individual SDS solutions was attained. In acidic media (figure 2a) synergetic effect of surface tension decreasing in presence mixture Ls and SDS was disappeared.

Significant synergetic effect was fixed in presence of mixture Ls and SDBS, where surface tension of water and sulfuric acid solutions was decreased at Ls concentration increasing in mixture. More noticeable effect appeared in the range of SDBS concentration 400-600 mg·l^{-1}. Moreover, CCM of mixture Ls-SDBS solutions was decreased at growing Ls concentration. For example, CCM of sulfuric acid solution of SDBS is equal to ≈450 mg·l^{-1}, and at the presence of 200 mg·l^{-1} of Ls in the mixture CCM was decreased up to 250-300 mg·l^{-1}. It is suggested [15] that reduction of surface tension of solutions in presence of surfactants can be related to hydrophobic interactions between surfactants hydrocarbon radicals and hydrophobic polymer chains, there is adsorption of low molecular surfactants on polymeric chain of Ls that causes surfactant CCM decreasing.

Results of leaching cake particle size analysis (figure 3) testify to significant stabilization action of Ls. At Ls presence content of fraction -150 μm increased from 80,2 to 91,5 % at the absence reagent and at presence of 400 mg·l^{-1} of Ls, respectively. Zinc extraction was enhanced from 44.9 % to 77.2 % (table 1). Optimal cake for flotation was received at addition 800 mg·l^{-1} of Ls, 97.5 % of cake particles had fineness -150 μm.

However, treatment of zinc cake with size less than 10 μm can be complicated. It is known that Ls can lead to downsizing of sulfur particles during leaching. Obtained results confirm this strong dispersing action of Ls: content of fraction -10 μm was increased from 8.39 % in the absence of reagents to 14.11 % and 26.93 % at Ls addition 400 mg·l^{-1} and 800 mg·l^{-1}, respectively.

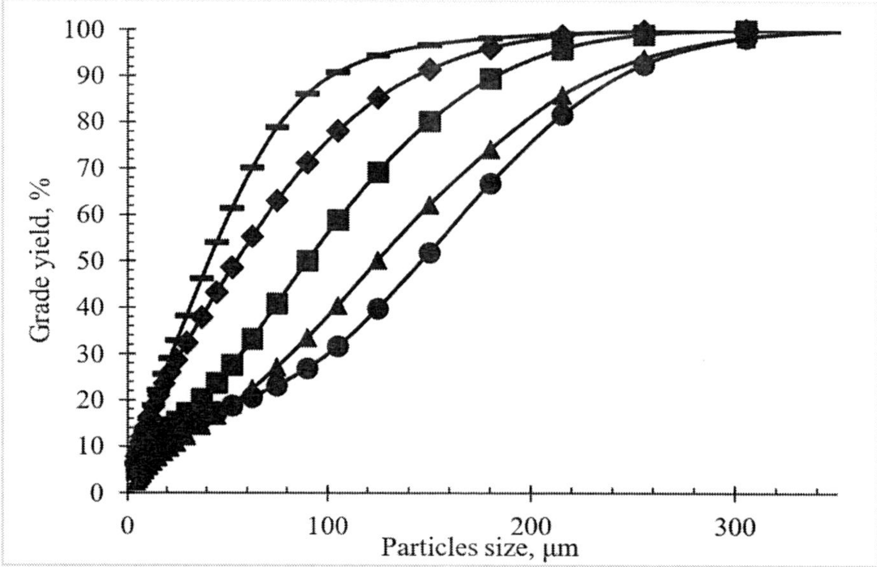

Figure 3. Size-grade distribution of cakes after leaching of zinc concentrate at presence of individual and composite surfactants: ■ – surfactants absence, ♦ – Ls 400 mg/dm^3; • – SDS 200 mg/dm^3, SDBS 200 mg/dm^3; ▲ – Ls 400 mg/dm^3+SDS 200 mg/dm^3; ▬ - Ls 400 mg/dm^3 + SDBS 200 mg/dm^3.

Table 1. Surfactants and their mixtures influence on zinc extraction at sulfide zinc concentrate pressure leaching.

Reagents concentration, mg/dm^3			Zinc extraction, %
Ls	SDS	SDBS	
-	-	-	49,2
400	-	-	77,2
-	200	-	52,4
-	-	200	58,8
400	200	-	76,6
400	-	200	82,8

In the context of autoclave technology of nickel pyrrhotite concentrates it is known [4, 11] that simultaneously addition of Ls with high molecular oil-products allows to receive cake with optimal fineness (10-150 μm) and increase velocity of pyrrhotite decomposition. Due to remarkable decrease of water solutions surface tension in presence of Ls and anionic surfactants mixtures, we have studiedinfluence of these reagent of zinc extraction and fineness of leaching cake particles.

At addition of 200 $mg \cdot l^{-1}$ SDBS and SDS increasing of +150 μm particles size up to 48.13 % was fixed, content of fraction of -10 μm was equal to 9 %. Zinc extraction at addition of anionic surfactants was in the range 52-59 % (table 1). Simultaneous addition of SDBS and Ls (figure 3) allowed to receive disperse cake: 96.7 % and 13.2 % of -150 μm and -10 μm fractions, respectively; zinc extraction was increased by 5.6 %. Mixture Ls-SDS have not significant influence on cake particles fineness, -150 μm fraction content was equal to 62.16 %, -10 μm – 5 %. Zin extraction was not changed in comparison with tests without Ls addition.

Consequently, surfactants mixture SDBS-Ls allows to enhance zinc extraction, receive optimal cake fineness due to decreasing of zinc sulfides wetting by molten sulfur (SDBS), stabilization of sulfur drops (Ls).

4. Conclusion

The work is devoted to influence investigation of addition of anionic surfactants and Ls mixtures on pressure leaching zinc concentrates effectiveness. For the purpose of supplementary definition of mixtures effectiveness regularities of solutions surface tension changing was studied.

On the base of obtained results a following conclusions and recommendations was made:

1) simultaneous addition of Ls with SDS or SDBS leads to remarkable reducing of surface tension due to decreasing CCM;

2) employment of surfactants mixture allows to enhance zinc extraction from 77.2 % to 82.8 % and obtain cakes with optimal fineness (10-150 μm).

Acknowledgment

Investigation of surfactants influence on solutions surface tension was funded by State Assignment No. 10.7347.2017/8.9.

Leaching experiments and grade size analysis were conducted at the expense of the Russian Science Foundation grant (project No. 18-19-00186).

References

[1] Sadeghi N, Moghaddam J, Ilkhchi M J 2017 *Physicochemical Problems of Mineral Processing* **53** 601-16
[2] Xu B, Yang Y, Li Q, Jiang T, Li G 2016 *Hydrometallurgy* **159** 87-94
[3] Zhukov V V, Laari A, Lampinen M, Koiranen T 2017 *Chemical Engineerin Research and Design* **118** 131-141

[4] Naftal M N, Naboychenko S S, Sharkiy R Yu, Petrov A F, Lapshina N A 2013 *Tsvetnye Metally* **9** 100-13
[5] Jorjani E, Ghahreman A 2017 *Hydrometallurgy* **171** 333-43
[6] Ouyang X, Qiu X, Chen P 2006 *Colloid and Surfaces A: Physicochemical and Engineering Aspects* **282-283** 489-97
[7] Lugovitskaya T N, Naboichenko S S, Bolatbaev K N, Mamyachenkov S V 2007 *Russian Journal of Non-Ferrous Metals* **48** 327-30
[8] Lugovitskaya T N, Naboichenko S S, Bolatbaev K N 2016 *Russian Journal of Applied Chemistry* **89** 1831-37
[9] Rana D, Neale G N, Hornof V 2002 *Colloid and Polymer Science* **280** 7750-78
[10] Pu B, Chen D 2001 *Journal of Colloid and Interface Science* **235** 262-72
[11] Naftal' M N , Naboichenko S S, Salimzhanova E V, Bol'shakova O V, Saverskaya T P 2015 *Russian Metallurgy (Metally)* **2015** 171-84
[12] Tan K, Li C, Liu J, Qu H, Xia L, Hu Y, Li Y 2014 *Hydrometallurgy* **150** 99-106
[13] Duan M, Ding Z, Wang H, Xiong Y, Fang S, Shi P, Liu S 2018 *Applied Surface Science* **427** 917-26
[14] Niraula T P, Shah S K, Chatterjee S K, Bhattarai A 2018 *Karbala International Journal of Modern Science* **4** 26-34
[15] Lugovitskaya T N, Naboichenko SS 2018 *Russian Journal of Applied Chemistry* **01** 82-89

IOP Conf. Series: Materials Science and Engineering **525** (2019) 012024 doi:10.1088/1757-899X/525/1/012024

Influence of trehalose additives on the properties of poly(vinyl alcohol) cryogels formed in aqueous as well as in organic media

O Yu Kolosova and V I Lozinsky

A.N. Nesmeyanov Institute of Organoelement Compounds of Russian Academy of Sciences

Corresponding author: kolosova@ineos.ac.ru

Abstract. Physical (non-covalent) macroporous cryogels of poly(vinyl alcohol) (PVACGs) are formed as a result of the consecutive freezing of concentrated PVA solutions, their incubation in a frozen state and then thawing. In this study the influence of the additives of trehalose and the solvent used on the PVA cryotropic gel-formation, as well as on the properties of the respective PVACGs was explored. It was shown that both the rigidity and the heat endurance of cryogels formed in the DMSO medium in the presence of trehalose progressively increased with increasing the additive concentration. It turns, in the case of cryogels prepared in the aqueous medium in the presence of trehalose the dependences of the physicomechanical characteristics on the additive concentration were of a bell-like (concave) type, while the heat endurance of the samples grew monotonically.

1. Introduction.

Polymeric cryogels are macroporous gel materials that are formed as a result of the consecutive freezing of molecular or colloidal solutions of certain precursors, subsequent incubation of the system in a frozen state, and then thawing [1–4]. The macroporosity of cryogels is generated by the polycrystals of the frozen solvent, while the gel formation itself proceeds within the volume of the remaining unfrozen inclusions (so-called unfrozen liquid microphase) [5] where the precursors are concentrated [1,4]. One of the most attractive cryogels are those based on poly (vinyl alcohol) (PVA), since these cryogenically-structured gel materials are of great applied interest in many fields, especially in biomedicine and biotechnology [6]. PVA cryogels possess by various favorable qualities, including excellent physico-mechanical and thermal characteristics, interconnected macroporosity, biocompatibility and non-toxicity PVACGs were successfully examined as artificial cartilages [7], wound and burn dressings, cardiovascular devices and drug vehicles [8].The physicochemical properties and porous structure of PVA cryogels depend on many factors, in particular, on the characteristics of the polyvinyl alcohol itself (the molecular weight, the amount of residual O-acyl groups, the chain tacticity), the PVA concentration in its solution prior to the freeze–thaw gelation, the solvent used, the conditions of cryogenic processing, namely, the rate of cooling upon the freezing of the polymer solutions, the temperature and duration of the freezing and frozen storage, the rate of heating upon defrosting the samples, and the number of freeze–thaw cycles [1]. And, certainly, the presence of low-weight additives in the initial solution of PVA influences on the properties of the resultant cryogels. Among such additives the kosmotropes are of especial significance because of their ability to promote hydrogen bonding [9] thus affecting positively on the efficience of PVA cryotropic gel-formation, and therefore natural approved for biomedical application kosmotropes can potentially be used upon the elaboration of various PVA-cryogel-based biomedical materials. Trehalose is one of such natural kosmotropes; it is a disaccharide formed by a 1,1-glycosidic bond between two α-glucose units (Figure 1). Trehalose was named after its source, trehalamanna, a sugar solution obtained from the brood and / or cocoon of some insects (Larinus species). Trehalose has found application in cosmetology as the face skin care agent [10]

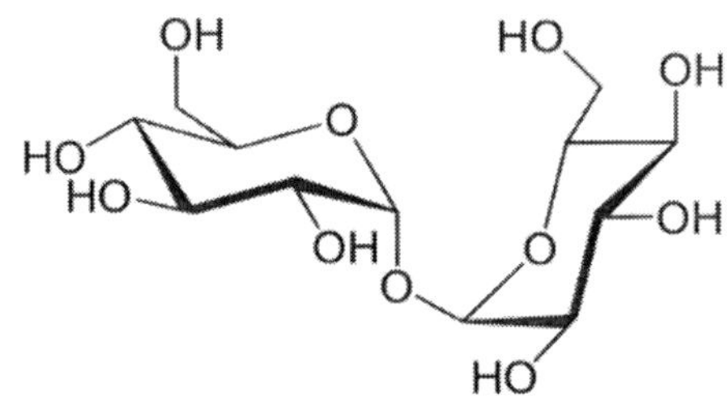

Figure 1. Structure of trehalose.

Most often, aqueous media are used as a solvent in the preparation of the initial PVA solutions for further cryogenic processing. However, there are few studies where PVA cryogels were formed from polymer solutions in organic solvents, in particular, in dimethyl sulfoxide (DMSO). Since DMSO is a thermodynamically better solvent than water for high-deacylated PVA samples, the PVA gelation in DMSO occurs poorly. Therefore, the PVA/DMSO cryogels are always weaker and have lower fusion temperatures than the PVACGs prepared from aqueous solutions of the same polymer concentration [11].

In this regard, it was reasonable to improve the mechanical properties and heat endurance of PVA cryogels formed in the DMSO medium using the trehalose additives, as well as to compare their influence on the PVA cryotropic gel-formation in both aqueous and DMSO milieus.

2. Materials and Methods

2.1. *Materials*

The following compounds were used: poly(vinyl alcohol) (molecular weight of ca. 86 kDa, the deacetylation degree of 100%; Acros Organics, USA), dimethyl sulfoxide (>99.8%; Komponent-Reaktiv, Russian Federation), trehalose (>98%; Panreac, Spain).

2.2. *Preparation of Feed PVA Solutions*

The dissolution of PVA in DMSO or in water was performed as described elsewhere [12]. Briefly, a known amount of dry polymer was dispersed in a calculated volume of DMSO to reach a PVA concentration of 100 g/L. The mixture was incubated for 18 h at room temperature for swelling of the polymer, and then, the system was heated for 1 h on a boiling water bath under stirring until the completion of PVA dissolution. When preparing the feed solutions that should contain trehalose the required amounts of additive were dissolved in the polymer solution at room temperature. The final solutions were sonicated for 20 min at room temperature in an UNITRA ultrasonic bath (Unitra, Poland) to remove air bubbles.

2.3. *Preparation of PVA Cryogels*

The samples of PVA cryogels were prepared analogously to the earlier described techniques [12]. In these experiments, samples for physicomechanical tests were formed in sectional duralumin molds (inner diameter 15 mm, height 10 mm). To determine T_f values, cryogels were formed in transparent polyethylene test tubes (inner diameter 10 mm), 3-mL portions of the polymer solution were poured, and a stainless steel ball (diameter 3.5 mm, weight 0.275 ± 0.005 g) was placed on the bottom of each tube. The containers and the tubes were put into the chamber of an FP 45 HP precision programmable cryostat (Julabo, Germany), and the samples were frozen and incubated at a preset temperature for 12 h. Then, the temperature was raised to 20°C at a rate of 0.03 °C/min controlled by the cryostat microprocessor.

2.4. Characterization of PVA cryogels samples

The compression Young's modulus (E) and the fusion temperature (T_f) of PVACG samples were evaluated in accordance with the protocols described elsewhere [11,12]. The E and T_f values were measured for three samples; the samples were examined in three to five independent experiments. The results obtained were averaged.

In brief, the E modulus of the PVACG samples was determined from the linear portion of the stress–strain dependence that was found using a TA-Plus automatic texture analyzer (Lloyd Instruments, UK) at a loading rate of 0.2 mm/min. The tests were performed until reaching a 30% deformation. The measurements were accomplished either for PVA cryogels prepared in water medium, either for the PVACGs prepared in DMSO and also for the samples in which the organic liquid phase (that is, DMSO and the dissolved solutes) was replaced by pure water. In the latter case, the cryogel samples were incubated for seven days at room temperature in glass beakers that each contained deionized water (100-fold excess relative to the volume of the PVACG sample); the water was replaced with a fresh portion every day. The fusion temperatures of the PVACG samples were measured by placing the tightly corked polyethylene tube containing cryogel with the stainless steel ball at the bottom upside down into the water bath. The bath temperature was increased at a rate of 0.4 °C/min. The gel fusion point was determined as the temperature when the ball fell down onto the stopper of the test tube after passing through the fused gel.

3. Discussion.

3.1. Physicomechanical properties of PVACGs prepared in aqueous medium with the addition of trehalose

Trehalose is a typical kosmotropic agent which promotes hydrogen bonding. Such effect on the physicochemical characteristics of PVA cryogels is similar to the influence of other kosmotropic agents, for example, sulfate or phosphate anions. So, we could expect that the higher concentration of trahalose is the stronger cryogels will be formed. But it was observed experimentally that with increasing in the trehalose concentration the rigidity of PVA cryogels initially lowered (Figure 2a), passed through a some minimum, and only then commenced to grow. When the concentration of trehalose in initial polymer solution reached 1 mol/L, the value of the Young's modulus for the resultant cryogel was of 19.5±0.5 kPa. At the same time, the fusion temperature of cryogels monotonically heightened with an increase in the trehalose concentration, and the value of fusion temperature raised by 7 °C for a sample with a trehalose concentration of 1 mol/L as compared with a sample formed without additives.

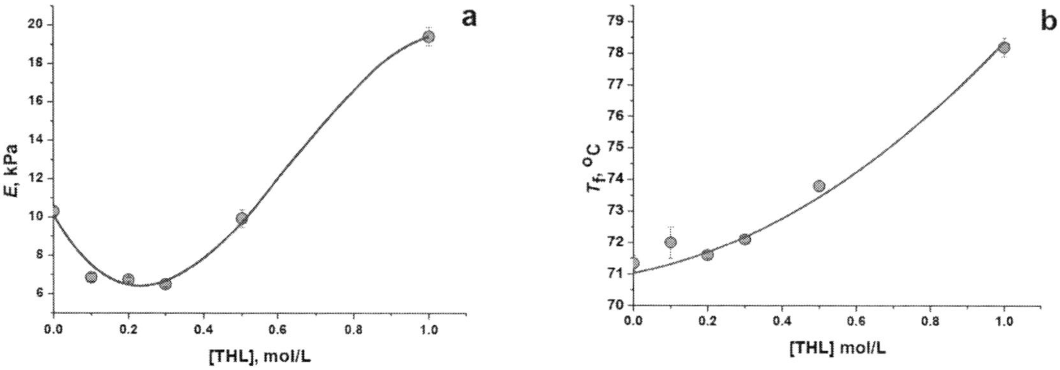

Figure 2. The dependence of the Young's modulus (a) and the fusion temperature (b) of PVA cryogels formed in aqueous medium on the trehalose concentration in the samples.

3.2. Physicomechanical properties of PVACGs prepared in DMSO medium with the addition of trehalose

It should be noted that the physicomechanical as well as the thermophysical properties of PVA cryogels formed from DMSO polymer solutions are very different from the properties of PVA cryogels obtained in aqueous medium under similar cryogenic processing conditions. Thus, PVA cryogels formed from the initial solutions of PVA in DMSO turned out to be less rigid and less heat resistant than the samples formed from aqueous solutions of the polymer [13]. Since DMSO is a thermodynamically better solvent for polyvinyl alcohol than water, that is, the affinity of the polymer for DMSO is higher, this reduces the efficiency of gelation due to the competition between interactions of the polymer-solvent and the polymer-polymer.

Since the crystallization temperature of DMSO is T_0 = 18.4 °C, describing the cryogenic treatment mode of such systems, it is more convenient to use not the absolute value of the temperature, but the difference value $\Delta T = T_i - T_0$, where T_0 and T_i are, respectively, the crystallization temperature of the pure solvent and the temperature of a specific experiment (expressed in the Celsius degrees). In our case, the cryogenic treatment temperature was -21.6°C, thus ΔT = -40°. PVA cryogels were formed under the following conditions: a concentration of PVA of 10 g/dL, a freezing temperature of -21.6°C, a freezing time was 12 h, a heating rate upon the system thawing was 0.03°C/min.

As it was noted earlier, PVA cryogels formed in DMSO medium have lower elastic modulus values E compared to PVA cryogels equiconcentrated with respect to the polymer, but formed from aqueous solutions of PVA. Thus, the value of the Young's modulus of a sample formed without additives was 3.35 ± 0.3 kPa (3 times less as compared to the PVA cryogel formed in the aqueous medium). However, an increase in the trehalose concentration in the samples led to a significant growth in the values of elastic modulus and reached 75.5 ± 3 kPa (at trehalose concentration of 1 mol/L).

A similar pattern was observed in the case of the fusion temperature of the samples. Thus, the fusion temperature increased from 42.5±0.5°C (sample without trehalose) to 83.0±0.2°C (PVA cryogels with the concentration of trehalose 1 mol/L).

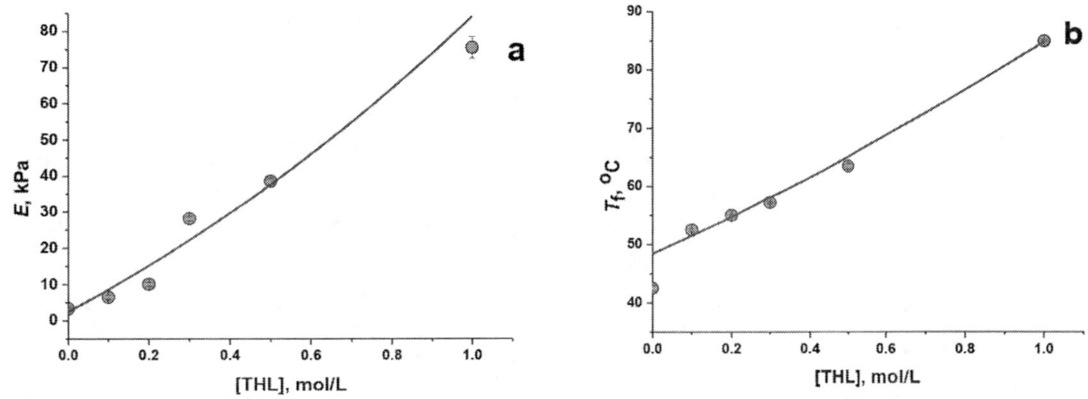

Figure 3. The dependence of the Young's modulus (a) and the fusion temperature (b) of PVA cryogels formed in DMSO medium on the trehalose concentration in the samples.

The next step of our study was to evaluate the changes in the physicomechanical and thermophysical properties of PVA cryogels after replacing an organic solvent for water. After rinsing with water of the cryogels formed in DMSO medium, the values of elastic moduli of cryogels and their fusion temperature were measured again.

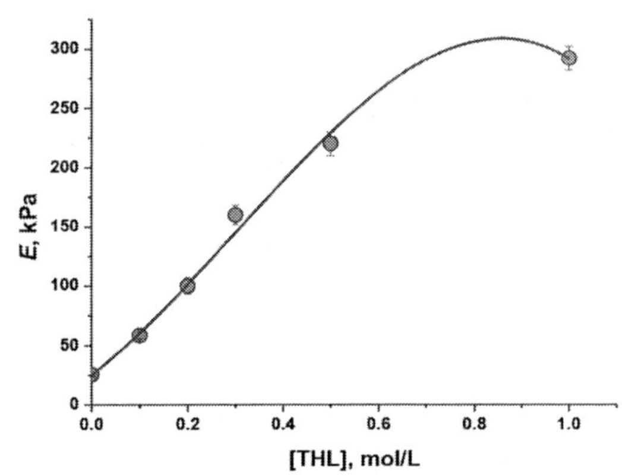

Figure 4. The dependence of the Young's modulus of PVA cryogels formed in DMSO medium and rinsed with water on the trehalose concentration in the samples.

The plots in the Figs. 4 and 5 show that the value of elastic modulus increased up to 292±5kPa, and the fusion temperature to 85±0.5°C. So, the change of a more thermodynamically good solvent (for PVA) DMSO with a less good one — water — resulted in the replacing of a part of the polymer-solvent interactions with polymer-polymer interactions, which, in turn, gave rise to a significant increase of the rigidity and fusion temperature of the cryogel samples.

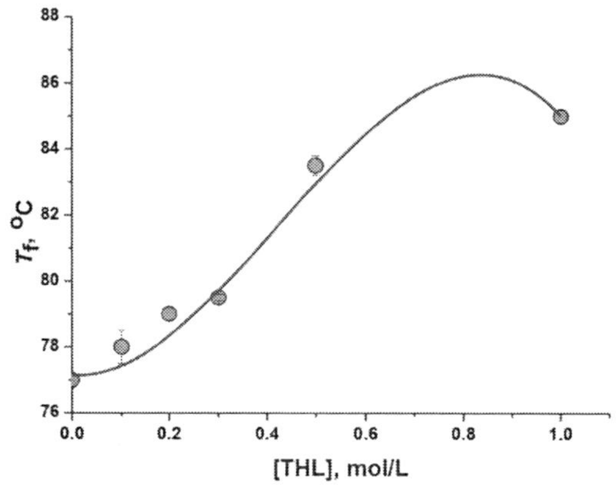

Figure 5. The dependence of fusion temperature of PVA cryogels formed in DMSO medium and rinsed with water on the trehalose concentration in the samples.

4. Conclusions

Taking in account a wide area of application of PVA cryogels [14] and, on the other hand, a rather simple technology of their preparation [15,16], it is reasonable to know well different approaches for varying and reliable control of the properties of these gel materials. So, the first, the introduction of the available low-molecular-weight additives (in our case trehalose) to the initial PVA solutions influence on the physicochemical characteristics of the PVA cryogels. And the second, the variation of the initial solvent for poly (vinyl alcohol) leads to the formation of cryogels with higher rigidity and heat resistance.

References

[1] Lozinsky V I 2002 *Rus. Chem. Rev.* **71** 489
[2] Gun'ko V M *et al* 2013 *Adv. Coll. Interface Sci. 187*
[3] Liu C et al 2014 *Prog. Chem* **26** 1190
[4] Lozinsky V I and Okay O 2014 *Adv. Polym. Sci.* **263** 49
[5] Singh B and Pal L 2008 *Eur. Polym. J.* **44** 3222
[6] Kim S J *et al* 2002 *J Appl. Polym. Sci.* **86** 1844
[7] Seal B L *et al* 2001 *Mater. Sci. Eng.* **4** 147
[8] Cascone M G *et al* 1995 *Biomaterials* **16** 569
[9] Kosmotropes and chaotropes. http://www1.lsbu.ac.uk/water/kosmotropes_chaotropes.html#r1567
[10] Matsumoto Y *et al* 2016 *Bioorg. Med. Chem. Lett.* **26** 301
[11] Trieu H H and Qutubuddin S *Polymer* **36** 2531
[12] Lozinsky V *et al* 1996 *J. Appl. Polym. Sci.* **61** 1991
[13] Sergeev G B and Batyuk V A 1976 *Rus. Chem. Rev.* **45** 391
[14] Peppas N A and Stauffer S R 1991 *J. Contr. Release* **16** 305
[15] Lozinsky V I. 2014 *Adv. Polym. Sci.* **263** 1
[16] Lozinsky V I and Okay O 2014 *Adv. Polym. Sci.* **263** 49

Influence of the content of n-butanol on the critical conversion of gel formation during "living" three-dimensional radical co-polymerization of styrene and divinylbenzene

R I Komendant, E O Perepelitsina and S A Kurochkin

Institute of Problems of Chemical Physics, RAS, Academician Semenov avenue 1, Chernogolovka, Moscow region, 142432 Russian Federation

E-mail komendant@icp.ac.ru, oligo@icp.ac.ru

Abstract. In this work, a study was made of the copolymerization of styrene and divinylbenzene in the presence of alkoxyamine synthesized "in situ" from benzoyl peroxide and 2,2,6,6-tetramethylpiperidine-1-oxyl in a solution of n-butanol (50-70 wt.%) at 120 °C. According to the experimental data obtained, the critical gelation conversion lies in the range of 50-75%. A decrease in the critical gelation conversion is also observed when the content of n-butanol 60%, which is accompanied by a separation of the system into two phases, that is, polymer is precipitated. With a lower content of precipitator, the separation of the system into two phases does not occur. It is assumed that in this case the removal of highly branched macromolecules from the polymerizing system occurs as a result of occlusion of "suspended" double bonds that become inaccessible to radicals, but "suspended" double bonds of linear and weakly branched macromolecules remain available for participation in the reaction with growth radicals.

1. Introduction

The synthesis of highly branched and network polymers is often carried out by the method of three-dimensional radical polymerization [1-14]. The fundamental aspects of the synthesis are described in sufficient detail in the reviews [15-20].

A promising way to obtain highly branched polymers of irregular structure is a three-dimensional radical polymerization proceeding in the mode of "living" chains [21]. The theoretical derivation of the dependencies that establish the relationship between the conditions of the process and the molecular mass distribution of macromolecules formed in the pre-gel period of a "living" three-dimensional radical polymerization was carried out in [5, 22]. In [23–26], three-dimensional radical polymerization of vinyl monomers with two or more double bonds was theoretically investigated, proceeding by the classical radical mechanism taking into account quadratic chain breakage or in the presence of inhibitors and chain transfer agents, as well as "living" three-dimensional radical polymerization with reversible inhibition. Using "living" three-dimensional radical polymerization, one can regulate the magnitude of the critical gelation conversion over a wide range by changing the ratio of a polyfunctional monomer with two or more double bonds and a "living" radical polymerization agent [3, 22].

It is known that a polymer product formed during three-dimensional radical polymerization prior to gelation is a mixture of linear, weakly branched and highly branched macromolecules. The theoretical content of highly branched macromolecules consisting of at least seven primary polymer chains does not exceed ~ 30 wt.%. That is, more than half of the polymeric substance is "ballast" that does not have unique properties [22]. To increase the content of the highly branched fraction, an approach was tested in [7], in which the "living" three-dimensional radical polymerization is carried out in an environment with a worsening thermodynamic quality, which is created as a result of adding a precipitating agent — n-butanol to the mixture. In particular, it is shown that the proportion of the highly branched fraction in the polymer product isolated near the critical gelation conversion increases with the addition of n-butanol up to 50 wt.%. At the same time, the value of the critical gelation conversion also increases. The styrene-divinylbenzene system, including the "living" radical polymerization of a mixture of these monomers, can be considered as a model due to comprehensive study [27–31].

In this work, the "living" polymerization was carried out in the presence of alkoxyamine, synthesized in situ from benzoyl peroxide and 2,2,6,6-tetramethylpiperidine-1-oxyl (TEMPO) according to the procedure [32]. The main objective of the work is to study the effect of the content of n-butanol, as a precipitator of polystyrene chains, on the copolymerization process of styrene and divinylbenzene in the presence of alkoxyamine. In particular, the task was to determine the dependence of the experimental critical gelation conversion with an n-butanol content of 50-70 wt.%, Provided that the value of the theoretical critical gelation conversion, calculated without taking into account the influence of the thermodynamic quality of the environment in which branched macromolecules are formed, in all experiments persisted and was equal to 0.5.

2. Materials and Methods

2.1. Materials

Styrene (ST) MM = 104.15 g/mol; $\rho = 0.906$ [33].

Styrene was purified from hydroquinone with a 10% aqueous solution of NaOH, washed with distilled water until neutral, dried over calcined CaCl2, distilled in vacuum, and immediately before the experiment further purified by freezing in high vacuum.

Divinylbenzene (DVB) MM = 130 g/mol; $\rho = 0.914$ [34].

Used divinylbenzene technical (80% mixture of isomers) of the company "Aldrich" without additional purification.

2,2,6,6-tetramethylpiperidine-1-oxyl MM = 156 g/mol [35] of the company "Sigma", 95–98% without further purification, were used as the "living" polymerization agent.

Benzoyl peroxide MM = 242 g/mol and dicumyl peroxide(PC) MM = 270 g/mol [34] were purified by recrystallization from ethanol.

N-butanol MM (N-but) = 84 g/mol; $\rho = 0.810$ [33] was purified by fractional distillation [36].

2.2Methods

The copolymer was obtained by the method of "living" radical copolymerization of styrene and divinylbenzene in the presence of alkoxyamine, synthesized "in situ" from benzoyl peroxide and TEMPO, and high-temperature initiator — dicumyl peroxide, in a solution of n-butanol in sealed glass ampoules with an internal diameter of 5 mm. The content of n-butanol ranged from 50 to 70 wt.%. The length of the primary polymer chains was regulated by the calculated concentration of TEMPO [37].

The reaction mixture with a known composition was poured into pre-weighed glass ampoules with an inner diameter of 5 mm. The oxygen dissolved in the monomer solution was removed on a vacuum unit with multiple repetitions of the freeze-thaw cycle in liquid nitrogen to a residual pressure of 0.01 mm Hg. The ampoules were unsoldered and placed in a WLM WS 100 heat treatment cabinet (T = 120 °C) and a DAK-1-1 microcalorimeter. The polymerization was carried out until the moment of gelation, the time of occurrence of which was determined by the moment of loss of the system fluidity, then the critical gelation conversion was determined using the kinetic curves obtained by the calorimetric method. The polymerization of the copolymers used in further studies was stopped for 15 min until the loss of system fluidity.

The contents of the ampoules of the obtained copolymer were removed in a Petri dish and dried from the volatile residual product at room temperature in a fume hood for 24 hours and in a vacuum oven at room temperature to constant weight.

The kinetics of radical polymerization were studied by the calorimetric method on a differential automatic microcalorimeter DAK-1-1 in the mode of direct recording of the heat release rate dQ/dt under isothermal conditions.

The molecular-mass characteristics of branched polymers obtained near the gelation, were determined by size exclusion chromatography on a gel chromatograph "Waters", equipped with a differential Refractometer, the eluent tetrahydrofuran, (T=35°C), elution rate 1 ml/min, consisted of two connected in series columns Plgel 5 μm MIXED-C.

3. Results and discussion

In all experiments, the ratio of styrene and divinylbenzene was the same $[ST]_0$: $[DVB]_0$ = 100: 1.17. The synthesis conditions are shown in table 1.

Table 1. - Initial conditions for the synthesis of copolymers. $[PC]_0$ = 0.3 wt.%. T = 120°C.

№ experience	$\omega_{n\text{-}but}$, wt.%	$[n\text{-}but]_0$, mol/l	$[TEMPO]_0$, mol/l	$[PB]_0$, mol/l
1	50	5.780	0.023	0.020
2	55	6.322	0.021	0.017
3	60	8.856	0.019	0.015
4	70	9.714	0.014	0.012

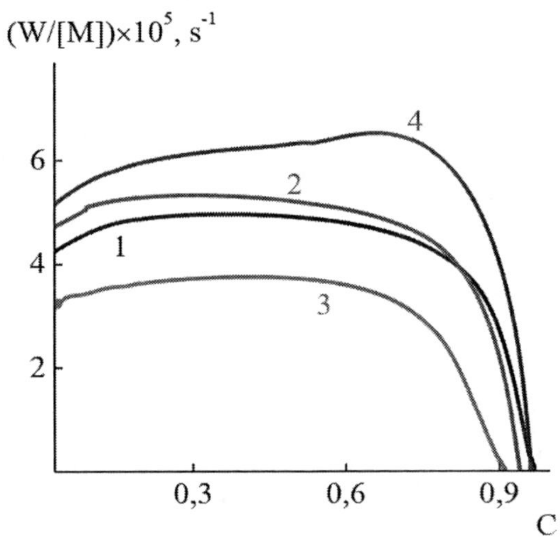

Figure 1. Kinetic curves of "living" radical copolymerization of styrene and divinylbenzene in the coordinates "reduced polymerization rate (W/[M]) - double bond conversion (C)" with n-butanol content 50 (1), 55 (2), 60 (3) and 70 wt.% (4). Conditions according to table 1.

Figure 1 shows the kinetic curves of "living" radical copolymerization of styrene and divinylbenzene. Due to the introduction of a high-temperature initiator (dicumyl peroxide) the polymerization rate reduced to the current concentration of monomers remains constant throughout almost the entire process and almost does not depend on the dilution of the system with n-butanol.

In figure 2 shows the dependence of the experimental critical gelation conversion on the n-butanol content in the mixture of polymerizing comonomers. It can be seen that there is a sharp drop in gelation at 60% of the precipitant content, which may be caused by the separation of the system into two

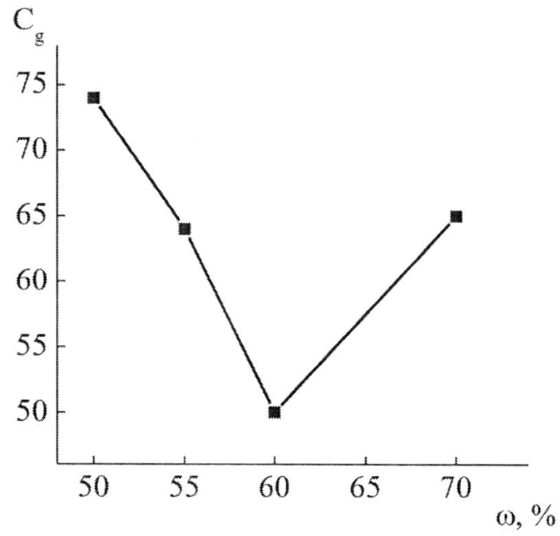

Figure 2. The dependence of critical conversion of gel formation from the concentration of n-butanol.

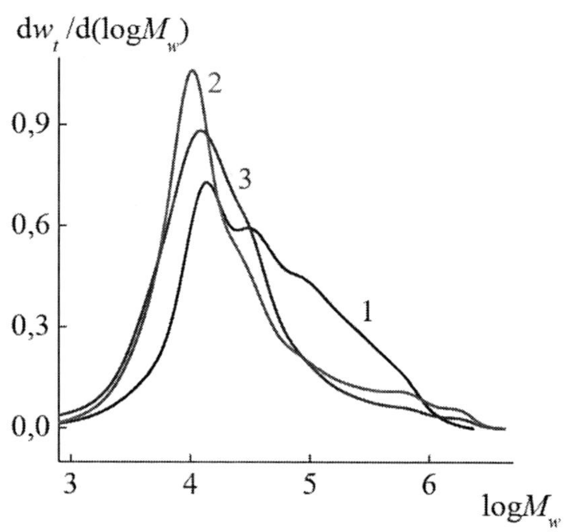

Figure 3. MMD of polymers obtained by "living" radical polymerization of styrene and divinylbenzene ($[ST]_0$: $[DVB]_0$ = 100: 1.17) in the presence of n-butanol – 50 (1) and 60 (2) and 70 (3) wt.%. The areas under the curves are normalized to unity.

phases, which is also observed at 70% of the content. Taking into account the fact that when the content of n-butanol is up to 50 wt.%, the critical gelation conversion, on the contrary, increased with an increase in the proportion of precipitator [7] and, accordingly, with the deterioration of the thermodynamic quality of the medium, this experimental fact is very interesting from the point of view of the mechanism of the "living" three-dimensional radical polymerization in the boundary conditions with the separation of the system into two phases.

When the precipitator content is less than 60%, the system is not divided into two phases. It is assumed that in this case the removal of highly branched macromolecules from the polymerizing system occurs as a result of occlusion of "suspended" double bonds that become inaccessible to growth radicals, but "suspended" double bonds of linear and weakly branched macromolecules remain available for participation in the reaction with growth radicals.

In figure 3 shows chromatograms of polymer products isolated near the gel point (for the conversion of monomers smaller by ~ 2% of the experimental critical conversion of gelation). From figure 3 shows that the highest content of the highly branched fraction with the highest molecular weight is formed in experiments in which the separation of the system into two phases is not observed.

4. Conclusion

Copolymers of styrene and divinylbenzene in the presence of 50, 55, 60 and 70 wt.% n-butanol were obtained by the method of three-dimensional radical polymerization in the mode of "living" chains. The values of the critical gelation conversion, which lies in the range of 50-75%, are determined.

Analyzing the results obtained, it can be said that highly branched macromolecules emerge from the polymerizing system due to the process of occlusion of "suspended" double bonds, which makes it impossible for the growth radicals to interact with these macromolecules. At the same time, this process does not interfere with the interaction of linear and weakly branched macromolecules with the radicals of other macromolecules, and allows them to continue to grow to a highly branched state, thereby increasing the yield of high-molecular product. In other words, there is a dependence of the rate constant of the reaction of radicals with "suspended" double bonds on the degree of branching of the macromolecule, which gives an advantage in the growth of molecular weight to weakly branched polymer chains. However, this process is possible only when no interphase separation occurs in the system, that is, the concentration of precipitant should not exceed 60% by mass.

Acknowledgments
This work was performed in accordance with the state task of IPCP RAS, the theme No 0089-2019-0010

References

[1] Kurochkin S A, Silant'ev M A, Perepelitsina E O Berezin M P, Baturina A A, Grachev V P and Korolev G V 2012 *Polymer Science* B **54** 223-233.
[2] Kurochkin S A, Silant'ev M A, Perepelitsina E O and Grachev V P 2013 *Polymer* **54** 31-42.
[3] Kurochkin S A and Grachev V P 2015 *Polymer Science* C **57** 20-31.
[4] Kurochkin S A, Grachev V P and Korolev G V 2008 *Polymer Science* A **50** 921-941.
[5] Kurochkin S A, Grachev V P and Korolev G V 2007 *Polymer Science* A **49** 227-233.
[6] Kurochkin S A, Silant'ev M A, Perepelitsina E O and Grachev V P 2014 *European polymer journal* **57** 202-212.
[7] Kurochkin S A, Makhonina L I, Vasil'ev S G, Perepelitsina E O, Zabrodin V A, Bubnova M L and Grachev V P 2017 *Polymer Science* A **59** 613-623.
[8] Kurmaz S V, Grachev V P, Kochneva I S, Perepelitsina E O and Estrina G A 2007 *Polymer Science* A **49** 884-895.
[9] Kurmaz S V, Kochneva I S, Ozhiganov V V, Baturina A A and Estrina G A 2008 *Russian Journal of Applied Chemistry* **81** 1831-1836.
[10] Kurmaz S V and Ozhiganov V V 2009 *Polymer Science* B **51** 157-165.
[11] Voit B I and Lederer A. 2009 *Chemical reviews* **109** 5924-5973.
[12] Wang Z, He J, Tao Y, Yang L, Jiang H and Yang Y 2003 *Macromolecules* **36** 7446-7452.
[13] O'brien N, McKee A, Sherrington D C, Slark A T and Titterton A 2000 *Polymer* **41** 6027-6031.
[14] Kurmaz S V and Pyryaev A N 2010 *Polymer Science* B **52** 1-8.
[15] Roshchupkin V P and Kurmaz S V 2004 *Russian Chemical Reviews* **73** 225-249.
[16] Korolev G V and Bubnova M L 2007 *Polymer Science* C **49** 262-284.
[17] Gao C and Yan D 2004 *Progress in polymer science* **29** 183-275.
[18] Yan D, Gao C and Frey H *Hyperbranched polymers: synthesis, properties, and applications* . (John Wiley and Sons, Vol. 8. 2011)
[19] Hult A, Johansson M and Malmström E 1999 *Branched Polymers II* 1-34
[20] Kim Y H 1998 *Journal of Polymer Science* A **36** 1685-1698.
[21] Korolev G V and Marchenko A P 2000 *Russian Chemical Reviews* **69** 409-434.
[22] Kurochkin S A 2010 *Polymer Science* B **52** 109-117.
[23] Korolev G V, Irzhak T F and Irzhak V I 2001 *Polymer Science* A **43** 594-598.
[24] Korolev G V, Irzhak T F and Irzhak V I 2001 *Polymer Science* A **43** 1236-1241.
[25] Korolev G V, Irzhak T F and Irzhak V I 2002 *Polymer Science* A **44** 1-4.
[26] Korolev G V, Irzhak T F and Irzhak V I 2002 *Journal of Advances in Chemical Physics* **21** 58.
[27] Tuinman E, McManus N T, Roa-Luna M, Vivaldo-Lima E, Lona L M and Penlidis A 2006 *Journal of Macromolecular Science* A **43** 995-1011.
[28] Alam M N, Zetterlund P B and Okubo M 2006 *Macromolecular Chemistry and Physics* **207** 1732-1741.
[29] Alam M N, Zetterlund P B and Okubo M 2009 *Polymer* **50** 1632-1636.
[30] Greszta D and Matyjaszewski 1997 *Polymer Science* A **35** 1857-1861.

[31] Scott A J, Nabifar A, Hernández-Ortiz J C, McManus N T, Vivaldo-Lima E and Penlidis A 2014 *European Polymer Journal* **51** 87-111.
[32] Zaremskii M Yu, Stoyachenko Yu I, Plutalova A V, Lachinov M B and Golubev V B 1999 *Polymer Science* A **41** 253-261.
[33] Goronovsky I T, Nazarenko Yu F and Nekryach E F *Brief guide to chemistry.* (Kiev, Naukova dumka, 1987) 833p.
[34] Aldrich: *directory of chemicals: a developer and manufacturer of Aldrich chemical company,* (Inc. – USA, Wisconsin, 1990) 2184p.
[35] Sigma: *directory of chemicals: a developer and manufacturer of Sigma chemical company,* (Inc. – USA, St. Louis, 1990) 2077p.
[36] Smyth C P and Stoops W N 1929 *Journal of the American Chemical Society.* **51** 3312-3329.
[37] Zaremski M Yu, Xin C, Orlova A P, Golubev V B, Kurochkin S A and Grachev V P 2010 *Polymer Science* B **52** 528-541.

IOP Conf. Series: Materials Science and Engineering **525** (2019) 012026 doi:10.1088/1757-899X/525/1/012026

Change of the structural organization in the polar elastomer film under the action of electrostatic field

N N Komova
Moscow Technological University (Institute of Fine Chemical Technologies), Vernadsky av., 78 Moscow, 119571 Russia

E-mail: komova_@mail.ru

Abstract The effect of the electrostatic field on the structural organization of macromolecules of polar elastomers was studied using polychloroprene (PCP) as an example. The impact of the electrostatic field on thin films of polychloroprene in the process of their formation leads to the formation of a supramolecular structure that is different from that in films formed under similar conditions, but in the absence of a field. An analysis of the dielectric relaxation spectra of samples in an electrostatic field shows an increase in order in the arrangement of macromolecules (degree of crystallinity). The results of the evaluation of the increase in the degree of crystallinity of the films studied by the dielectric method are in good agreement with the estimate of this parameter by analyzing the IR spectra. The results obtained give grounds to conclude that there is an increase in orderliness in the arrangement of polar molecules in thin films of elastomers when they are formed and kept in an electrostatic field. Increasing the order in the arrangement of polar macromolecules leads to an increase in electrical conductivity in thin films of such a material. The work substantiates and uses a method for assessing the degree of crystallinity of an elastomer by analyzing the temperature dependence of electrical capacity.

1. Introduction

Regulation of the physical properties of polymeric materials expands the prospect of their use in developing areas of science and technology. The creation of polymer materials of new quality is possible not only through synthesis, chemical modification, doping, doping, but also when traditional materials are exposed to external physical fields (electromagnetic, electrostatic, etc.). This type of modification is determined by the fact that the properties of the polymer material depend not only on the chemical structure of macromolecules, but also on their configuration (conformation of macromolecules and isomerism), the location of the macromolecules relative to each other, their packing density, which results in the supramolecular structure of the material and its physical properties. The impact of physical fields can change the conformation of macromolecules and their supramolecular organization, and in the most severe conditions change the chemical structure of macromolecules (oxidation, dehydrochlorination under the influence of hard gamma irradiation, etc.). Polymeric materials are widely demanded in various branches of technology due to a wide range of their mechanical and physical properties, as well as the possibility of their modification in various ways and imparting properties other than inorganic compounds [1-3]. Thus, polymer films are widely used not only as high-voltage insulation in various electrical engineering, but also as semiconductor elements [4]. Films with high electrical conductivity, pseudometallic and semiconductor properties are widely demanded in the manufacture of light-emitting diodes and thin-film field-effect transistors. As a result of studies of conducting polymers with a high degree of orientation of polymer chains in a bulk sample (which makes it possible to use the characteristics of the quasi-one-dimensional structure of macromolecules), polymeric quasicrystalline materials with high anisotropy of electrical characteristics and high carrier mobility were obtained [5,6].

Of particular interest in terms of sensitivity to electromagnetic fields and the possibility of using as organic semiconductors are polymers whose macromolecules have polar groups. For such polymers, the density of the electron clouds, forming a bond between atoms, is asymmetrically distributed, and the dipole moment of macromolecules depends on the polarity of the groups that make up the polymer, the frequency and symmetry of their arrangement along the molecule chain. Among the polar polymers is polychloroprene (PCP), the macromolecule of which includes the group C - Cl. PCP is an elastomer that is widely used for the manufacture of technical products.

The purpose of this work was to study the processes occurring in thin films of a polar polymer (polychloroprene) under the action of an electrostatic field and the influence of these processes on the electrophysical properties of this material.

2. Theoretical part

Polar polymers, being in an electrostatic field, can realize two types of polarization depending on their physical state [7]. In the glassy state, elastic-dipole and dipole-radical polarization occurs, and at temperatures when the polymer is in a highly elastic state, electron and dipole-elastic polarization is manifested. Both dipole-radical and dipole-elastic polarization are a kind of dipole-relaxation polarization. The difference lies in the fact that in the dipole-radical polarization there is a rotation not of the molecules as a whole, but of its individual segments or units (radicals). In the case of dipole-elastic polarization, entire segments (consisting of several molecules) are rotated [8], which is the case for elastomers. The result of the manifestation of polarization is a change in the position of individual parts of the macromolecules relative to each other with the appearance of local stresses that either relax under certain conditions or result in the destruction of the macromolecule [9].

The result of structural transformations in the film of polychloroprene under the influence of a constant electric field is reflected in changes in the temperature dependences of the dielectric constant and the tangent of the dielectric loss angle of samples compared to similar dependencies for samples that have not been subjected to such an impact. An analysis of the corresponding dependences makes it possible to determine the change in the degree of macromolecular ordering (quasicrystallinity) under the action of electrostatic fields on thin elastomer films.

Following the principle of superposition of relaxation processes in completely crystalline and fully amorphous polymer samples, formulas were obtained that establish the dependence of the dielectric properties of polymers on the degree of crystallinity (κ) [9]:

$$\varepsilon = \kappa(\varepsilon_{1\infty} + \int_0^\infty \frac{H_1(\tau)d\tau}{1+\omega^2\tau^2}) + (1-\kappa)[\varepsilon_{2\infty} + \int_0^\infty \frac{H_2(\tau)d\tau}{1+\omega^2\tau^2}], \quad (1)$$

where subscripts 1 and 2 refer to fully crystallized and fully amorphous samples, H (τ) is the density of the dielectric relaxation time spectrum, τ is the relaxation time, and ω is the circular frequency of the alternating electric field. Since most of the volume of the elastomer refers to the amorphous phase, the second term significantly exceeds the first term, then the expression takes the form

$$\varepsilon \approx (1-\kappa)[\varepsilon_{2\infty} + \int_0^\infty \frac{H_2(\tau)d\tau}{1+\omega^2\tau^2}] \quad (2)$$

From the presented relation it follows that the dielectric constant decreases with increasing degree of crystalline. If the crystalline phase is destroyed, then κ = 0, and the expression (2) takes the form

$$\varepsilon_{aм} \approx [\varepsilon_{2\infty} + \int_0^\infty \frac{H_2(\tau)d\tau}{1+\omega^2\tau^2}] \quad (3)$$

The ratio of the dielectric constant of the partially crystallized polymer (ε) to the dielectric constant of the completely amorphous polymer (ε) can be represented as $\varepsilon / \varepsilon_{am} = (1\text{-}\kappa)$, from where the degree of crystallinity can be estimated by the ratio

$$\kappa = 1 - \varepsilon/\varepsilon_{aм} \quad (4)$$

In the process of heating the PCP sample, the crystalline phase melts in the region of 333 K [10]. This argument was the basis for determining the degree of crystallinity and analyzing its change in the process of heating the sample by the dielectric method. Since the degree of crystallinity decreases with

IOP Conf. Series: Materials Science and Engineering **525** (2019) 012026 doi:10.1088/1757-899X/525/1/012026

increasing temperature, this should be reflected in an increase in dielectric constant. On the other hand, as the sample temperature changes, the dielectric constant also changes. The Debye equation for dielectric constant is

$$\varepsilon = \varepsilon_\infty + \frac{\varepsilon_0 - \varepsilon_\infty}{1+\omega^2\tau^2} \quad (5)$$

where ε_0 is the static low-frequency value of ε, ε∞ is the value of ε as $\omega \to \infty$. According to the Onsager, Fröhlich, Debye theory [9] for polar molecules, ε_0-ε_∞ = A / T, where A is a constant, T is temperature.

After substitution of this relation into equation (5)

$$\varepsilon - \varepsilon_\infty = \frac{A}{T(1+\omega^2\tau^2)} \quad (6)$$

Since the relaxation time τ depends exponentially on the temperature and activation energy [17]

$$\tau = \tau_0 \exp\frac{U}{kT} \quad (7)$$

the transformation of equations (6) and (7) leads to the dependence

$$\frac{1}{\varepsilon - \varepsilon_\infty} = \frac{T(1+\omega^2\tau_0^2)}{A} - \frac{2U}{Ak}\omega^2\tau_0^2 \quad (8)$$

If we take into account that ε <ε_∞ (with an increase in the frequency ω, the value of ε decreases [13]), then in the first approximation we can assume that ε-ε_∞ ≈ε. Then the ratio between the dielectric constant and temperature will take the form

$$\frac{1}{\varepsilon} = \frac{T(1+\omega^2\tau_0^2)}{A} - \frac{2U}{Ak}\omega^2\tau_0^2 \quad (9)$$

In the theories of Fröhlich and Buckingheim [8], the dependence of the dielectric constant relations under different conditions on temperature is also inversely proportional.

In (7), the parameter U represents the activation energy of the structural rearrangement process or the energy barrier of the transition of the system from one state to another. Since the applied electrostatic field contributes to the process of such a transition, reducing the energy barrier, the activation energy of the transition process will decrease due to the energy contribution of the electrostatic field: $W=U_{es}^2C/2$, where U_{es} is the voltage or potential difference of the electrostatic field, C is the electrical capacity. Then the activation energy of the process can be represented as

$$U=U_0-\gamma\, U_{es}^2C/2, \quad (10)$$

where γ is the coefficient characterizing the degree of reduction of the initial activation energy by the applied voltage of the electric field, U_0 is the initial potential barrier in the absence of an electric field. The joint solution (9) and (10), taking into account that $C = \varepsilon\varepsilon_0 S/d$ (S/d - geometric parameters of the film), with the necessary mathematical transformations make it possible to obtain a relation characterizing the influence of various factors on the dielectric constant of the film, having the form

$$\varepsilon = \frac{\left(b^2 + 2aAkU_{es}^2\right)^{1/2} - b}{aU_{es}^2}, \quad (11)$$

where $a = 2\frac{S}{d}\gamma\varepsilon_0\omega^2\tau_0^2$ is the parameter that is determined by the geometric dimensions of the sample; $b = k[T(1+\omega^2\tau_0^2) - 2\frac{U_0}{k}\omega^2\tau_0^2]$- temperature dependent parameter. Substitution of equation (11) into expression (4) makes it possible to obtain an expression for the degree of crystallinity of the form

$$\kappa = 1 - \frac{(b^2 + 2aAkU_{es}^2)^{1/2} - b}{a\varepsilon_{am}U_{es}^2} \tag{12}$$

From expression (12) it follows that with increasing voltage of the electrostatic field, the degree of crystallinity increases in a linear relationship, and with increasing temperature, the degree of crystallinity also decreases linearly. Thus, an idea of the influence of temperature and an electrostatic field on the ordering process of the structure of polar macromolecules has been obtained.

3. Materials and methods.

The object of the study was polychloroprene rubber brand Neopren (Neopren WD, DuPont). The proportion of the crystalline phase of this polymer at room temperature ranges from 8 to 30%. The dipole moment of the macromolecule PCP calculated on the monomer unit is 1.45 D (4.8 10^{-30} k • m). The viscosity average molecular weight, determined in a benzene solution, was 151.300.

A sample in the form of a film for the study was prepared by pouring a solution of polychloroprene in carbon tetrachloride on a polyamide substrate. The choice of substrate was determined by the absence of oxidative processes at the sites of contact of polychloroprene-polyamide and the possibility of separating the film of PVC from the surface of such a substrate without breaking down to form a fairly high-quality and uniform surface. The process of film formation during solvent evaporation was carried out at a temperature of 303 K for 24 hours. The control of solvent removal was carried out by weighing on an analytical balance (until a constant weight of the sample was reached). The thickness of the films was measured with a micrometer with an accuracy of 1 μm, they ranged from 150 to 250 μm.

The polychloroprene film formed during the drying of the solvent was removed from the substrate and placed on a copper cathode, to which a constant voltage in the range from 5 to 30 V was connected, for further morphology formation. The cathode was chosen as an electrode, since according to the data of [12], the greatest change in the properties of a polychloroprene film (change in conductivity) occurs precisely at the cathode.

Measurement of ac conductivity, dielectric loss tangent and electrical capacitance of the material between two steel electrodes was carried out on a Hewlett-Packard 4284A digital LCR meter in an equivalent circuit of parallel-connected resistor and capacitor at 1 kHz in a thermostat with a controlled heating temperature with an accuracy of 2 ° in the range of 290- 423 K. The temperature of the samples was monitored with a thermocouple chromel - copel. To determine the dielectric constant of the film, the measured electrical capacitance (C) of the measuring cell was recalculated from the relation

$$\varepsilon = \frac{Cm}{\varepsilon_0 \rho S^2}, \tag{13}$$

where m is the mass of the sample between the electrodes, the ρ-density of this sample, S is the area of the electrodes of the measuring cell. If the change in the sample density in the temperature range of the study between two electrodes (5×10^{-4} m in diameter) is neglected, then the relationship between the measured electrical capacity (C) and ε can be represented by the dependence ε = 0.2353 10^{14} C·m, $Ф^{-1}$ $кг^{-1}$.

Analysis of the degree of crystallinity and the content of cis and trans isomers in the polymer was evaluated by the IR spectra obtained on a Perkin Elmer Spectrum FTIR - FT spectrometer 100 at 295 ± 2K in the range of $4000 \leq n \leq 450$ cm^{-1} in transmitted light.

4. Experimental results and discussion.

Figure 1 shows the temperature dependence of the dielectric constant of the original polychloroprene film. This dependence has a stepped nature, the jumps of which reflect changes in the arrangement of macromolecules as a result of structural transitions [9]. An increase in ε with an increase in temperature characterizes the dipole orientational polarization. Moreover, an increase in ε is observed in the region of the maximum tangent of the dielectric loss angle (D).

Figure 2 shows the temperature dependence of the dielectric loss tangent of polychloroprene films at a frequency of 1 kHz. The kinks in the temperature dependence of the dielectric constant correspond to the temperatures of the maxima of the tangent of the dielectric loss angle (D) in Fig. 2 (curve 1).

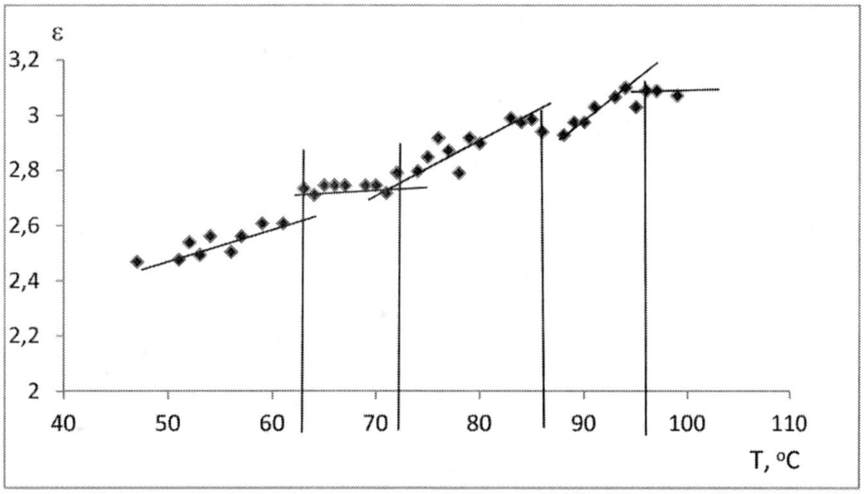

Figure 1.Temperature dependence of the dielectric constant of the original polychloroprene film.

In the range of melting points of the crystalline phase (60 ° C or 333 K), clear peaks are found in the samples studied, which should be associated with morphological transformations in the film. Comparison of the temperature of the dielectric loss of the sample after exposure to an electric field with the original sample (Fig. 2, curves 1 and 2) shows that the influence of an electric field leads to a shift of the maxima of the tangent values of the dielectric loss angle D in the temperature range from 45 ° C to 90 ° C higher temperatures and an increase in these values. The relaxation spectra reflect phase and structural transitions with a change in temperature at a constant frequency of an electric field or with a change in the frequency of a field at a constant temperature. The mechanism of phase transitions suggests a change in the cooperativity of thermal molecular mobility in polymer systems [11].

Upon annealing in amorphous-crystalline polymers, a liquid-liquid phase transition occurs, leading to the manifestation of mobility of various types of macromolecules of the intercrystallite layer and macromolecules of amorphous regions [13]. The melting of the crystal structure of polymers occurs in a finite temperature range and is associated with a decrease in crystallite size and a gradual transition into the amorphous phase of the intercrystalline layer [14]. For PCP, the temperature range for the melting of the crystalline phase coincides with the temperature of the transition from the highly elastic state to the viscous flow [10].

At temperatures above 310 K, an increase in the tangent of dielectric loss angle is due to the manifestation of the cooperativeness of the thermal molecular motion. The elimination of conformational restrictions imposed on parts of the macromolecules of the intercrystalline layer during the melting process leads to a decrease in D. The mobility of parts of the macromolecules of the intercrystalline layer is limited by fixing in crystals. This fact determines the cooperativeness of molecular motions and the growth of D at lower temperatures than amorphous elastomers [10]. As a result, in the melting temperature range of the crystalline phase, the tangent of the dielectric loss angle and the dielectric constant has the maximum value. The temperatures of their maxima coincide. In the range of melting points of the crystalline phase (310–353 K), the maxima of D appear. The more perfect the structure of crystallites, the higher the D value and less ε, but their maxima are shifted toward higher temperatures than for samples with a less perfect structure [9]. The presence of several maxima in the temperature range from 45 to 90 ° C is determined by the fact that the destruction of the crystalline (ordered) phase is accompanied by segmental movement of macromolecules (dipole-segmental polarization), as well as with an increase in the mobility of polar groups (dipole-group polarization). In the process of exposure to a constant electric field, the polar groups of the polymer acquire a certain orientation, contributing to greater orderliness in the arrangement of the macromolecules. Therefore, the observed change in the temperature dependence of the dielectric loss tangent occurs.

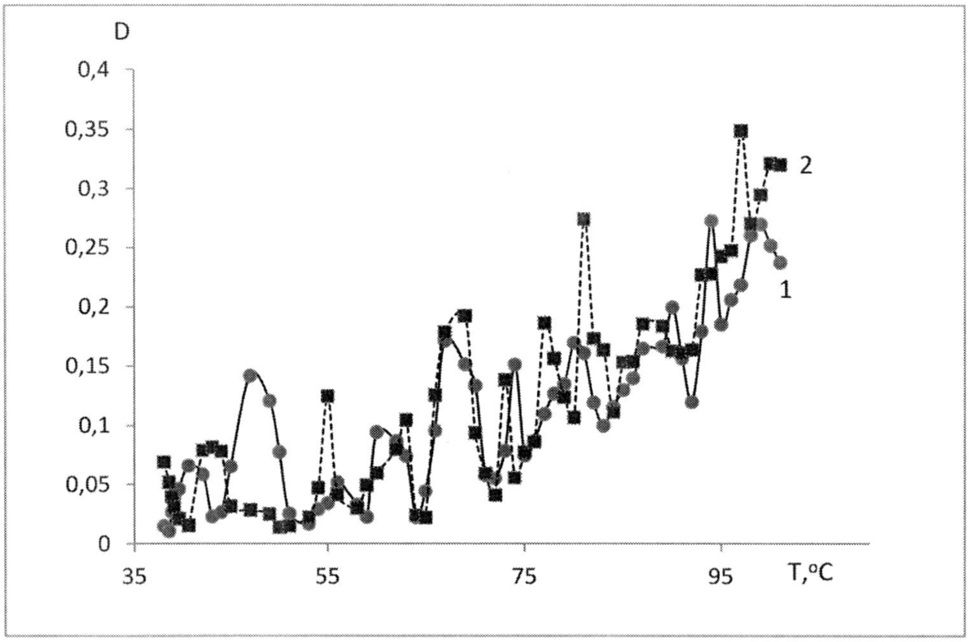

Figure 2. The dependence of the dielectric loss tangent of a polychloroprene film on temperature: 1- original sample; 2- the sample was held at the cathode at a voltage of 20 V for 120 minutes.

The ordering created by the field in the supramolecular structure of the elastomeric film can most likely be characterized as an increase in the degree of crystallinity or local ordering and can be estimated using relation (5). Processing the dependences of ε on temperature for samples aged for 6000 s in a constant electric field at different values of its magnitude makes it possible to obtain a dependence of the degree of ordering (crystallinity) of the system of macromolecules in a polychloroprene film. Figure 3 presents the dependence of the degree of ordering of macromolecules of polychloroprene, determined from the analysis of the temperature dependences of the dielectric constant of samples kept in an electric field of various sizes. The dependence is linear, the degree of order (local crystallinity) linearly depends on the voltage of the electric field, in which the structure of the elastomer film was formed by the ratio

$$\kappa = 0{,}0014U + 0{,}1376 \qquad (14)$$

Relation (14) indicates the structural changes occurring in the polychloroprene film under the action of an electric field. The dependence of the temperature of the maximum of the tangent of dielectric loss angle in the region of melting temperatures for samples sustained at a voltage higher than 10 V depends on the voltage applied to the electrode squarely (fig. 4), which corresponds to the expression

$$D = \frac{\alpha \tau U_{es}^2}{\delta^2}, \qquad (15)$$

where α is the value that determines the properties and structure of the polar molecule, U_{es} is the value

IOP Conf. Series: Materials Science and Engineering **525** (2019) 012026 doi:10.1088/1757-899X/525/1/012026

of the voltage of the electrostatic field in which the polymer film was formed, the δ-thickness of the film, the τ-time of stress relaxation in the polymer chain, depending on temperature.

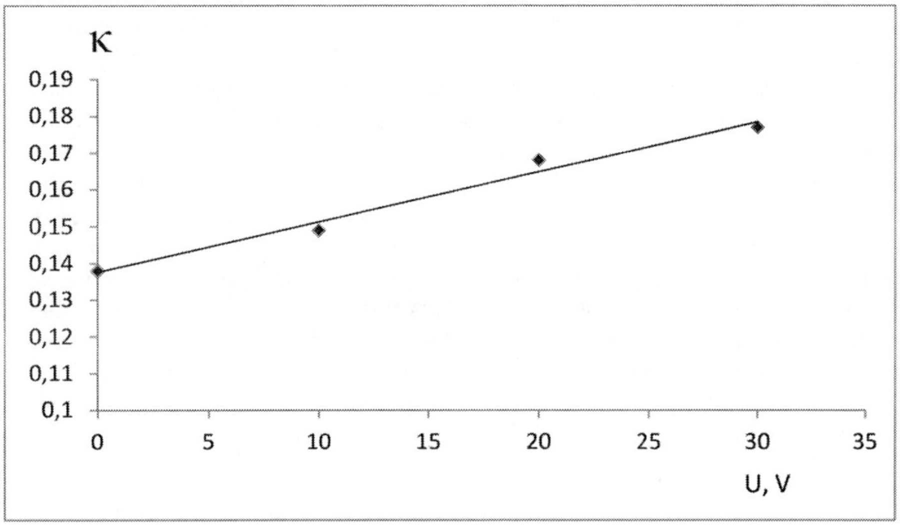

Figure 3. The dependence of the degree of crystallinity of films of PCP, aged in an electric field, on the magnitude of this field. The calculation of κ was carried out according to the relation (4).

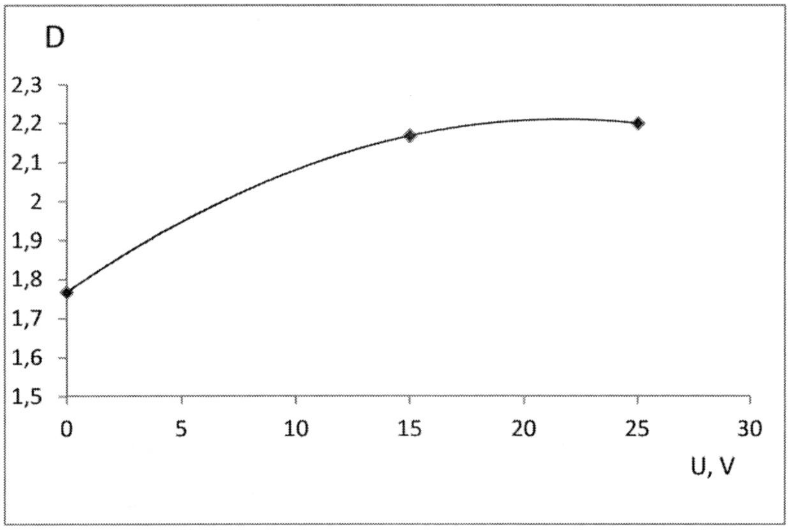

Figure 4. The dependence of the maximum of the tangent of the dielectric loss angle (D) in the region of melting temperatures of polychloroprene films on the voltage of the electric field. The exposure time of samples in an electrostatic field is 3600 s.

When comparing the IR spectra, a change in the intensity of absorption in the region of 780 cm-1 of PCP films held in an electric field for 3600 s at a voltage of 25 V and a temperature of 30 ° C with respect to films kept outside the field was established. The intensity of absorption in this region is associated with the presence of crystalline structures [15]. The change in the degree of crystallinity in these films is estimated from the ratio of the absorption intensities: the degree of crystallinity increases 1.2 times when exposed to an electric field under the specified conditions. Comparison of changes in the degree of crystallinity, determined by the dielectric method (fig. 3), when exposed to a field of 25

V also leads to similar results: the degree of crystallinity increases 1.22 times. This fact confirms the possibility of estimating the degree of crystallinity of PCP films from the temperature dependence of the electrical capacitance, which reflects the temperature dependence of the dielectric constant of materials of this type. Analysis of the IR spectra made it possible to estimate the change in the ratio of structural isomers in the PCP film as a result of an electric field. By absorption in the region of 1180 cm^{-1} [15], it was determined that when an electric field is applied to the film of polychloroprene, the proportion of trans-structures increases 1.2 times. This fact gives grounds to believe that an increase in the degree of crystallinity occurs due to an increase in the content of trans-isomers under the action of an electric field.

Changes in the structure of polychloroprene entail a change in the electrical properties of the films of this material. An increase in the degree of ordering and orientation of macromolecules under the action of an electric field leads to an increase in the conductivity of polar polymer films. Figure 5 shows the temperature dependence of the polychloroprene film formed without the participation of the field. Figure 6 presents a similar relationship for a film formed under the influence of an electric field of 25 V. It can be seen from the figures that during the formation of the film under different conditions, the nature of the temperature dependence of the conductivity changes. For a film formed under the influence of a field, conductivity in a wide range of temperatures is more stable and an order of magnitude higher than for a film obtained in the absence of a field.

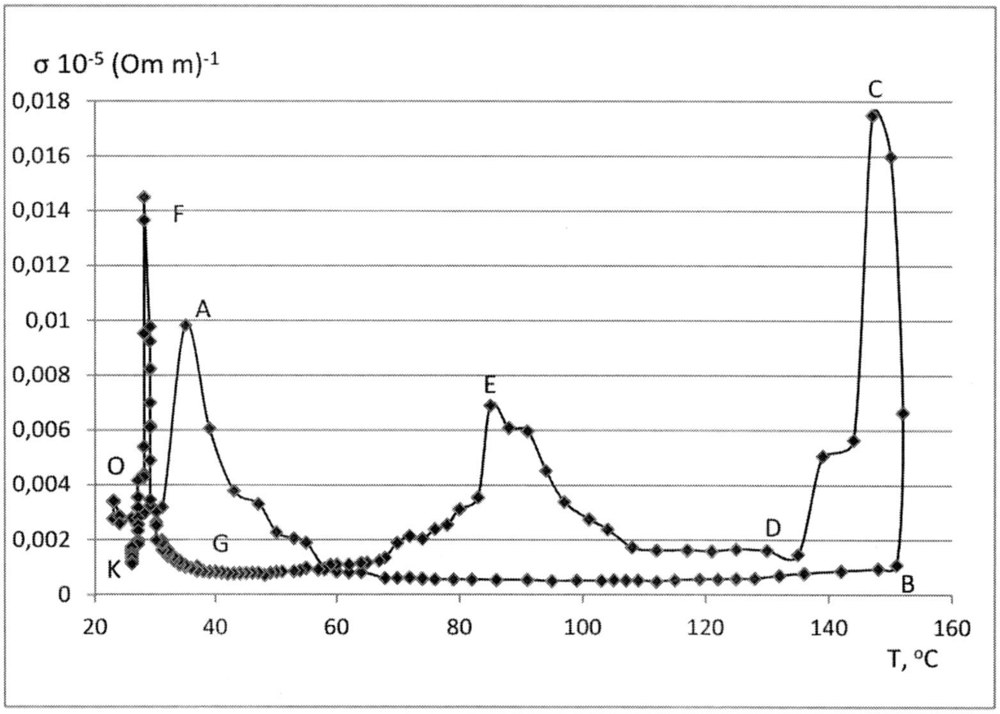

Figure 5. The dependence of the conductivity of the film of polychloroprene on temperature. O-A-B-C - heating at a rate of 6 deg. / min; C-D-E-G-F-K - cooling at a speed of 2.4 deg / min.

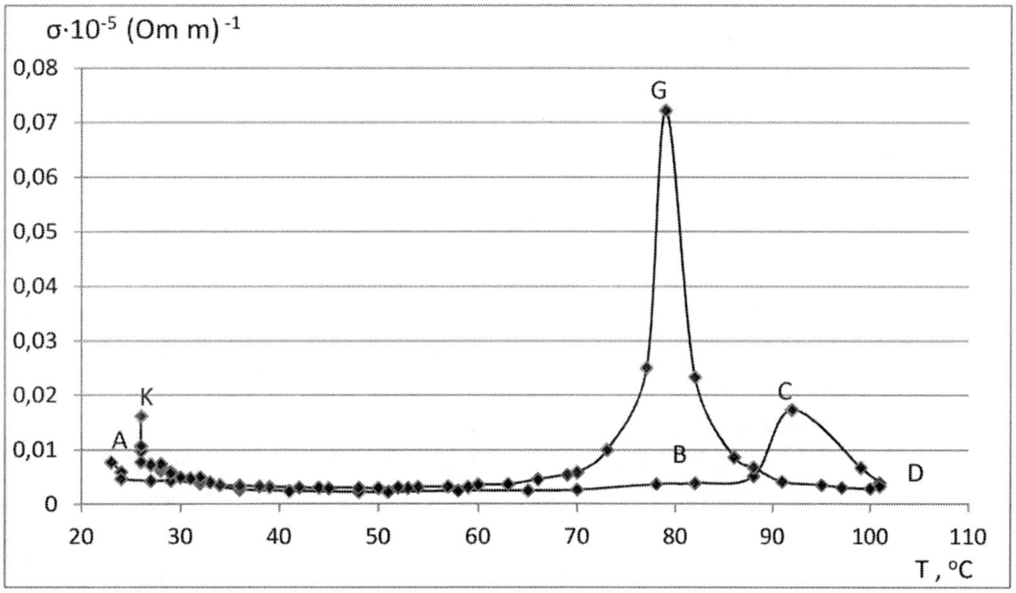

Figure 6. Temperature dependence of the resistivity of a polychlore film obtained from a solution by formation at the cathode. A-B-C-D –heating with a speed of 6 deg. / min; D-G-K - cooling at a speed of 2.4 deg. / min.

5. Conclusions

Based on the conducted research on the effect of the electrostatic field on the process of ordering the macromolecular structure of polar elastomers using the example of polychloroprene, we can conclude:

1. When finding and forming thin films of such a polar elastomer as polychloroprene, ordering occurs in the arrangement of the macromolecules due to the elastic-dipole and dipole-radical polarization. This fact is confirmed by the results of dielectric spectroscopy, the temperature dependences of the dielectric constant and dielectric loss tangent. This changes the ratio of the cis- and trans-conformation of macromolecules in favor of the latter. The degree change in the order of macromolecules (local crystallinity) corresponds to the change in the ratio of the cis and trans conformation of macromolecules PCP.
2. The obtained and theoretically justified dependences of the increase in the degree of local crystallinity (orderliness) in PCP films on the voltage of the electrostatic field and temperature, under the action of which the structure was ordered, were confirmed experimentally. An increase in the order (local crystallinity) in the films studied increases linearly with an increase in the electric field voltage in which the supermolecular structure of the films is formed.
3. With an increase in the order in the arrangement of macromolecules of polar polymers, an increase in the conductivity of thin films of this material occurs.
4. The dependence of the maximum temperature in the melting temperature region for samples aged at a voltage higher than 10 V depends on the voltage applied to the electrode, rather than linear, but quadratic, indicating several processes (changing the conformation of macromolecules, changing the arrangement of segments and parts of macromolecules, changing the size of macromolecules with an increase in the mobility of individual structural units), proceeding with the ordering of the structure of a polar elastomer under the action of an electrostatic field.

Thus, the electrostatic fields applied to the films of polar elastomers can modify the morphology of these films and change their properties.

IOP Conf. Series: Materials Science and Engineering **525** (2019) 012026 doi:10.1088/1757-899X/525/1/012026

Acknowledgements

The author is grateful to the Senior Researcher of the Laboratory of Diffusion Phenomena in Polymerization Systems of the Institute of Chemical Physics RAS Olkhov A.A. and senior researcher of the Russian University of Economics Mastalyginoy E.E. for providing the results of measurements of IR spectra of the studied samples and discussion of the results.

References

[1] Takeda Y, Andrew T L, Lobez J M, Mork A. J and Swager T M 2012 An Air-Stable Low-Bandgap n-Type Organic Polymer Semiconductor Exhibiting Selective Solubility in Perfluorinated Solvents *Angew. Chem.* **51** 36 9042–46.

[2] Hadziioannou G, Malliaras G G *Semiconducting Polymers: Chemistry, Physics and Engineering* 2007 2nd ed. (Weinheim: Wiley-VCH) p 321

[3] Wi H L and Yeong D P Organic Semiconductor Insulator Polymer Blends for High-Performance Organic Transistors 2014 *Polymers* **6** 1057-73

[4] Bässler H Charge transport in organic semiconductors 2012 *Top. Curr. Chem* **312** 1–66.

[5] Drobny J G 2012 *Polymers for Electricity and Electronics: Materials, Properties, and Applications* (New Jersey: John Wiley and Sons) p 332

[6] Hummel R E 2011 *Electronic Properties of Materials* (Springer) p 481

[7] Blythe A R, Blythe T and Bloor D 2005 *Electrical properties of polymers.* (Cambridge University Press) p 480.

[8] Nalwa H S 1997 *Handbook of Organic Conductive Molecules and Polymers: Conductive polymers : synthesis and electrical properties* (John Wiley & Sons Inc) p 888

[9] Perepechko I I 1978 *Introduction in physics of polymers* (Moscow: Khimiya) p 312

[10] Mashuryan A M , Gasparyan K A and Ovanesov G T Molecular Relaxation of Polychloroprene around its melting point 1985 *Pol. Sci. USSR* **27** 8 1862-65

[11] Bartenev G M and Barteneva A G *Relaxation Properties of Polymers* (Khimiya, Moscow, 1992) p 384

[12] Komova N N Conductivity at alternating current of thin films of polychloroprene formed in electric field 2018 *Fine Chemical Technologies* **13** 1 75-91.

[13] Mark J E *Physical Properties of Polymers Handbook* 2007 (New York Springer) p1073

[14] Gul' V E and Kuleznev V N 1992 *The Structure and Mechanical Properties of Polymers* (Moscow: Khimiya) p 248

[15] Dechant J 1972 *Ultrarotspektroskopische Untersuchungen an Polymeren* (Berlin: Akademie-Verlag) p 516

IOP Conf. Series: Materials Science and Engineering **525** (2019) 012027 doi:10.1088/1757-899X/525/1/012027

Palladium complex of octaphenyl-substituted pyrazinoporphyrazine: synthesis, photochemical and photophysical properties

A D Kosov[1], **T V Dubinina**[1,2] **and A N Volov**[1]

[1]Department of chemistry, Lomonosov Moscow State University, 119991 Moscow, Russian Federation.
[2]Institiute of physiologically Active Compounds, Russian Academy of Science, 142432 Chernogolovka, Moscow Region, Russian Federation.

e-mail: kosov.a.d@yandex.ru

Abstract. Novel palladium (II) tetrapyrazinoporphyrazinate, bearing eight peripheral phenyl groups, was obtained via metallation of the corresponding porphyrazine ligand. Target complex was identified using mass spectrometry, Fourier transform infrared and UV-Vis spectroscopy. Luminescence properties of palladium complex were investigated. Phosphorescence maximum of target compound was observed in the near IR region (828 nm). Singlet oxygen quantum yield (φ_Δ) was measured using the 1,3-diphenylisobenzofuran (DPBF) as the singlet oxygen trap. Value of φ_Δ (0.35) is in a good agreement with literature data for porphyrazine complexes.

1. Introduction

Palladium complexes of porphyrins and their analogues are of increasing importance due to their possible applications in optical oxygen sensors[1, 2]. They possess luminescence and absorption of visible light and can be applied in organic light-emitting diodes (OLEDs), chemical and biological sensing, photovoltaic devices, bioimaging[3]. Additionally, Pd (II) central ion enhances spin-orbit coupling, so that triplet state τ_0 is not so long-lived as to be totally quenched by environmental exposure or trace quenchers[2]. Interesting example of reversible oxygen binding by palladium complex of antracocyanine as a prospective agent for photodynamic therapy was described[4]. However, only little attention has been paid to tetrapyrazinoporphyrazine complexes of palladium[5-7].

The purpose of this work was to investigate spectral and photochemical properties of novel phenyl substituted tetrapyrazinoporphyrazine complex of palladium. Introduction of eight phenyl groups onto peripheral positions of porphyrazine macrocycle results in increasing of solubility of target complex comparing to unsubstituted or methylsubstituted[8] analogues.

2. Results and discussion

2.1. Synthesis and identification

Initial 5,6-diphenylpyrazine-2,3-dicarbonitrile **1** was obtained by the reaction between benzil and diaminomaleonitrile using ethanol as a solvent[9-11] (See Experimental section).

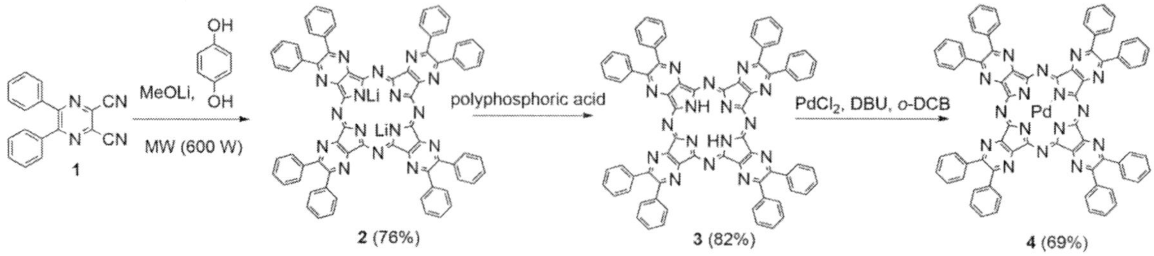

Scheme 1. Synthesis of phenyl-substituted tetrapyrazinoporphyrazine complex of palladium.

Palladium pyrazinoporphyrazinates, which were described earlier, were obtained by template cyclotetramerization of corresponding pyrazines. Pyridyl-substituted compound was obtained in a boiling DMSO in a high yield (80%)[5]. However, in the case of octapropyl-substituted complex this approach results in target palladium complex in a very low yield (11%)[6].

In the present work, alternative multi-step approach was realized (Scheme 1). Firstly, dilitium complex **2** was obtained by template synthesis in the presence of *p*-hydroquinone under MW irradiation. Compound **2** was further demetallated by freshly obtained polyphosphoric acid to give ligand **3**. The following synthesis of palladium complex **4** was performed in a boiling 1,2-dichlorobenzene (*o*-DCB) in the presence of 1,8-diazabicyclo(5.4.0)undec-7-ene (DBU) as a base and catalytic amounts of ammonium acetate. Total yield of three stages (dilithium complex formation, demetallation, and metallation) of the multi-step approach was 43% (Scheme 1).

All compounds were identified by mass spectrometry. Palladium complex was also characterized using high-resolution MALDI-TOF mass spectrometry (Table 1). The high resolution spectrum demonstrates good agreement between the observed mass and the monoisotopic mass calculated. The isotopic distribution of the molecular ion peak is almost equal to the theoretically calculated one (Figure 1). Besides the peak of molecular ions, peak of adducts with Na was also detected in the positive mode mass spectrum.

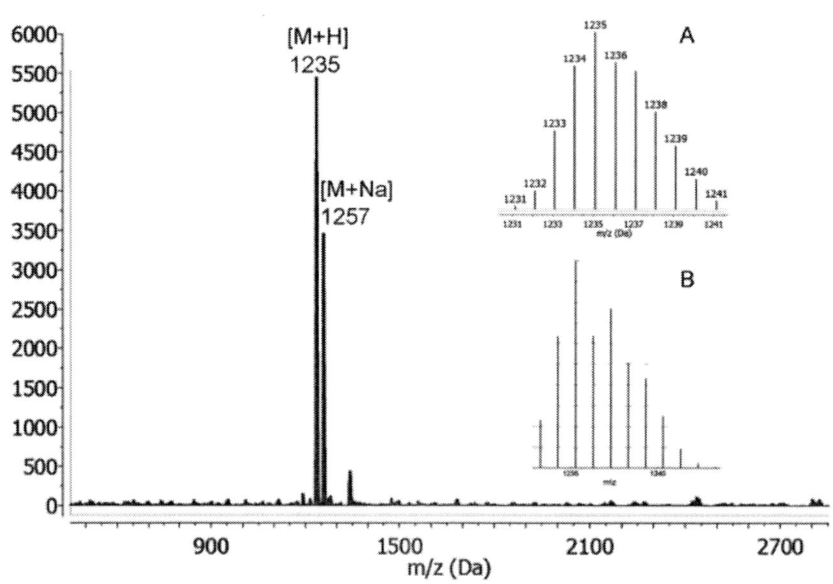

Figure 1. MALDI TOF of palladium complex **4,** isotopic pattern for the peak of $[M+H]^+$ ion (A) and simulated MS pattern (B).

Table 1. High-resolution MALDI TOF/TOF mass spectrometry data.

	Compound (Molecular formula)	Ion	Mass calculated	Mass found
4	$C_{70}H_{41}N_{16}Pd$	$[M+H]^+$	1235.2735	1235.2731

In FT-IR spectra of complex **4** (Figure 2), skeletal vibrations of the pyrrole and pyrazine fragments occupy the region from 1345 to 1635 cm^{-1}.

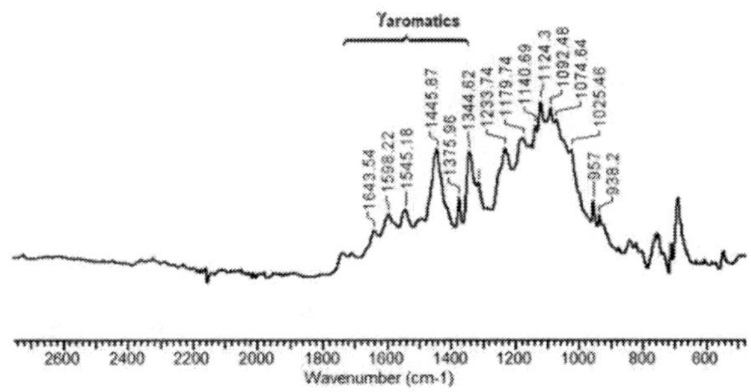

Figure 2. IR spectra of palladium complex **4** on the working surface of the internal reflection (ATR) diamond element.

2.2. UV-Vis spectroscopy and luminescence properties

In the UV-Vis spectrum of palladium complex **4** two main peculiarities were observed: *B* band (351 nm) in UV region and *Q* band (664 nm) in visible region (Figure 2). In comparison with phenyl-substituted phthalocyanine complexes, a hypsochromic shift of the *Q* band of about 30 nm is observed [12, 13].

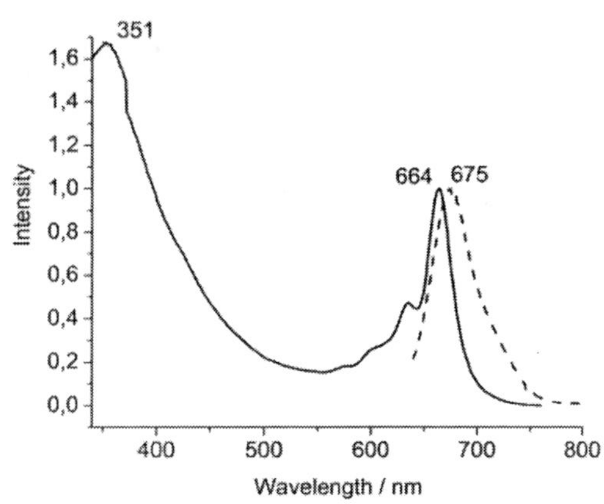

Figure 3. UV-Vis (solid line) and fluorescence (dashed line, λ_{ex}= 635 nm) spectra of palladium complex **4** in pyridine.

IOP Conf. Series: Materials Science and Engineering **525** (2019) 012027 doi:10.1088/1757-899X/525/1/012027

Fluorescence spectrum was also measured for target complex (Figure 3). The Stokes shift for *Q* band is very small (11 nm). This peculiarity is typical for porphyrazine complexes. For instance, in the case of methyl-substituted pyrazinoporphyrazines, Stokes shift is about 6 nm.
The phosphorescence spectrum of the palladium complex was measured for deoxygenated acetone solution at 77 K (Figure 4).
The main maximum of emission for compound **4** is observed at 828 nm. In comparison with *Q* band position in acetone solution (656 nm), the phosphorescence maximum is characterized by 172 nm bathochromic shift.

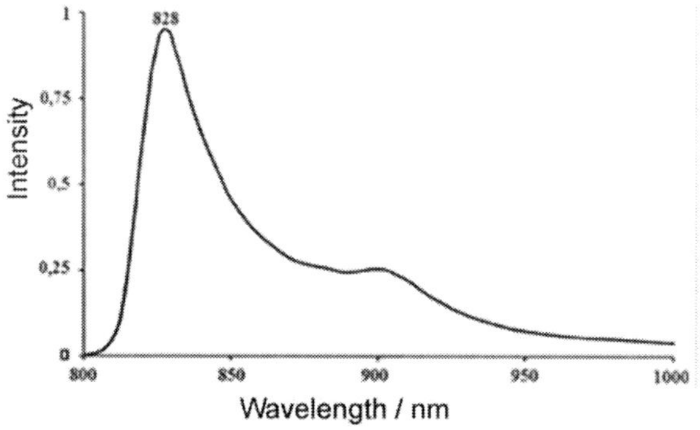

Figure 4. Phosphorescence spectrum of the complex **4** (77 K, acetone).

Similar large shift of the phosphorescence maximum to the near IR region was observed for platinum and palladium phthalocyaninates described in literature [1, 14, 15].

2.3. Study of the quantum yield of singlet oxygen generation

Singlet oxygen quantum yield (φ_Δ) was measured using the 1,3-diphenylisobenzofuran (DPBF) as the singlet oxygen trap. *meso*-Tetraphenylporphyrin (TPP) was chosen as a standard compound. According to the literature data, the value of the quantum yield of singlet oxygen generation by this porphyrin in acetone is 0.65 ± 0.05 [16, 17]. The interaction between the trap and singlet oxygen is determined by the chemical reaction, which is presented in the scheme 2 [18]. The reaction products do not absorb visible light. Therefore, the amount of singlet oxygen formed during the photoreaction can be judged by the loss of optical density in the region of the maximum absorption of DPBF.

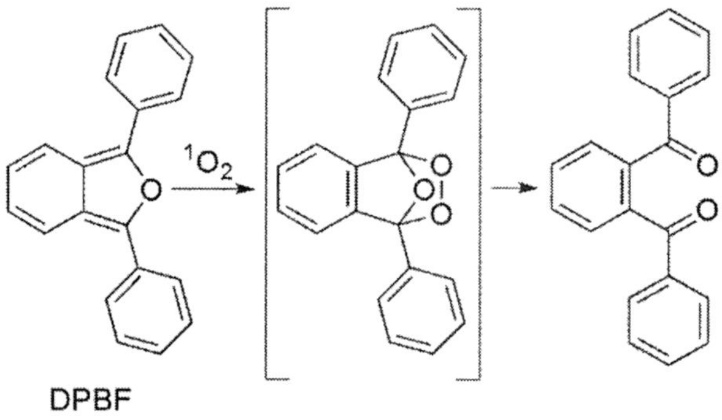

Scheme 2. The reaction between chemical trap and singlet oxygen.

IOP Conf. Series: Materials Science and Engineering **525** (2019) 012027 doi:10.1088/1757-899X/525/1/012027

To determine the quantum yields of singlet oxygen generation, solutions of the test compound and DPBF were mixed in a fixed volume of solvent. The wavelength of the light irradiated was chosen according to the position of the maximum of the most red shifted band for the compound investigated.

The calculation was made according to the following formula:

$$\varphi_\Delta = \varphi_{\Delta St} \cdot \frac{\Delta D_C \cdot t_{irrSt} \cdot I_{St} \cdot (1-10^{-D_{St}}) \cdot \lambda_{St}}{\Delta D_{St} \cdot t_{irrC} \cdot I_C \cdot (1-10^{-D_C}) \cdot \lambda_C},$$

Where $\varphi_{\Delta St}$ - is the quantum yield of singlet oxygen generation for the standard compound; ΔD_C - the average loss of optical density of DPBF in experiments with the test compound, ΔD_{St} - the average decrease in optical density of DPBF in experiments with a standard, t_{irrC} - the duration of irradiation of a mixture of DPBF with a test compound in s, t_{irrSt} - the duration of irradiation of a mixture of DPBF with a standard compound in s, I_C and I_{St} flux from a xenon lamp in experiments with a test compound and a standard compound in mW, D_C and D_{St} the optical densities of the compound studied and a standard compound corresponding to the wavelengths of booming light λ_C and λ_{St} (in nm).

Data obtained are summarized in Table 2. The value of φ_Δ for target palladium complex is comparable with substituted phthalocyanine analogs, described in literature. For e.g. for zinc tetrakis (benzylmercapto)phthalocyanine and for thiopheneethoxy-substituted zinc phthalocyanine φ_Δ– 0.33 [19, 20]. Thus the palladium phenyl-substituted tetrapyrazinoporphyrazine **4** is a prospective photosensitizer.

Table 2. Experimental data for photochemical measurements.

Compound	λ_{irr} (nm)	I (mW)	t_{irr} (s)	Optical densities	ΔD (DPBF)	φ_Δ
4	660	98	120	0.134	0.180	0.35±0.05
TPP	512	105	600	0.026	0.113	0.65±0.05

3. Experimental

3.1. Materials and methods

All reagents and solvents were obtained or distilled according to standard procedures. All reactions were TLC and UV/Vis controlled until complete disappearance of the starting reagents if not additionally specified. Thin-layer chromatography was performed using Merck Aluminium Oxide F_{254} neutral flexible plates.

The synthesis under microwave irradiation was performed using Samsung microwave oven. UV-Vis absorption spectra were recorded on a ThermoSpectronic Helios-α spectrophotometer using quartz cells (1×1 cm). Matrix assisted laser desorption/ionization time-of-flight (MALDI-TOF) mass-spectra were taken on a Bruker Autoflex II mass spectrometer with 2,5-dihydroxybenzoic acid (DHB) as the matrix. High-resolution MALDI mass spectra were registered on a Bruker ULTRAFLEX II TOF/TOF instrument.

IR spectra were recorded on IR spectrometer with Fourier transform IR 200 produced by Thermo Nicolet. Registration of spectra was carried out at a resolution of 4 cm^{-1}, the number of scans 20. Samples were placed on the working surface of the internal reflection (ATR) diamond element with the angle of incidence of 45°. Phosphorescence was excited by a focused light of a xenon lamp of 1 kW through a red border filter KS-11 ($\lambda \geq 660$ nm). The phosphorescence spectrum was measured using a monochromator with a replica of a diffraction grating. The gap width corresponded to 8 nm. The fluorescence spectra were recorded on a Hitachi F-2700 spectrometer in quartz cells (1×1 cm). The quantum yield of singlet oxygen generation by this compound was determined by the relative method using a chemical trap. 1,3-Diphenylisobenzofuran (DPBF) was used as a chemical trap. The measurements were carried out in square quartz fluorescent cells with an optical path length of 1 cm. The solutions were irradiated for a certain time with light passing through a monochromator from a xenon lamp of a Perkin Elmer MPF-44B fluorimeter, its power was measured using a ThorLabs PM-100D instrument with a S120VC sensor head. Optical density was determined using a Hitachi U-3400 dual-beam spectrophotometer.

3.2. Synthetic procedures

*3.2.1 5,6-Diphenylpyrazine-2,3-dicarbonitrile (1).*Benzil (1.06 g, 5.00 mmol) and diaminomaleonitrile (0.54 g, 5.00 mmol) were refluxed in 100 mL ethanol for 6h (TLC-control: SiO_2, ethyl acetate:hexane (1:2)). Then the reaction mixture was cooled to room temperature. The resulting residue was filtered off, washed with cold C_2H_5OH (3×30 mL) and dried at room temperature to give compound **1** (1.24 g, 88%). m.p. 244-246°C (lit. 246°C[21]). ^1H NMR (400.13 MHz, [D_5]Py) δH ppm: 7.35-7.39, 7.43-7.47 and 7.65-7.67 (m, Ph).

3.2.2 [2,3,9,10,16,17,23,24-Octaphenyl-1,4,8,11,15,18,22,25-octaazaphthalocyaninato]dilithium (2). A mixture of compound **1** (0.40 g, 1.42 mmol), MeOLi (0.11 g, 2.89 mmol) and p-hydroquinone (0.080 g, 0.73 mmol) were placed into a glass tube. The resulting mixture was irradiated in a microwave oven (600 W) during 15 min. Then the reaction mixture was cooled to room temperature and a MeOH-H_2O (20:1 v/v) mixture was added. The precipitate was filtered and washed with H2O, MeOH, and acetone. This yielded complex **3** (0.31 g, 76%). MS (MALDI-TOF) m/z: 1132 ([M-2Li+3H]$^+$, 30%), 1154 ([M-2Li+Na+2H]$^+$, 100%), 1176 ([M-2Li+2Na+H]$^+$. UV-Vis λ_{max} (Py)/nm 370 (lgε 4.56) and 660 (4.66). IR (ZnSe): ν (cm^{-1}): 1377-1612 (γ pyrrole and pyrazine).

3.2.3 [2,3,9,10,16,17,23,24-Octaphenyl-1,4,8,11,15,18,22,25-octaazaphthalocyanine (3). Complex **2** (0.31 g. 0.27 mmol) was dissolved in polyphosphoric acid (30ml) and stirred at 110°C for 2 h. Then the mixture was cooled to room temperature, and distilled H_2O was added. Ammonium carbonate was added until neutral medium of the reaction mixture was reached. The precipitate was filtered and washed with H_2O, MeOH, and acetone. This yielded compound **3** (0.25 g, 82%). MS (MALDI-TOF) m/z: 1131 ([M+H]$^+$, 100%). UV-Vis λ_{max} ($CHCl_3$)/nm 363 (lgε 3.98), 639 (4.56) and 672 (4.63). IR (ZnSe): ν (cm^{-1}): 1346-1600 (γ pyrrole and pyrazine), 3288 (st NH).

3.2.4 [2,3,9,10,16,17,23,24-Octaphenyl-1,4,8,11,15,18,22,25-octaazaphthalocyaninato]palladium (4). Compound **3** (0.050 g, 0.045 mmol), $PdCl_2$ (0.020 g, 0.135 mmol) were refluxed in *o*-dichlorobenzene (3 mL) (after complete dissolution catalytic amounts of NH_4OAc were added) in the presence of DBU (0.10 g, 6.00 mmol) for 2 h. Then the reaction mixture was cooled to room temperature and a MeOH-H_2O (20:1 v/v) mixture was added. The precipitate was filtered and washed with H_2O, MeOH, and acetone. This yielded complex **4** (0.039 g, 69%). MS (MALDI-TOF) m/z: 1235 ([M+H]$^+$, 100%), 1257 ([M+Na]$^+$, 60%). IR (diamond): ν (cm^{-1}): 1345-1635 (γ pyrrole and pyrazine). UV-Vis λ_{max} (Py)/nm 351 (I/I_{max} 1.00) and 664 (0.65).

4. Conclusions

Novel octaphenyl-substituted palladium tetrapyrazinoporphyrazinate was obtained in a high yield using the modified multi-step synthesis including the stage of formation of the porphyrazine ligand. Target complex was identified by high resolution MALDI-TOF mass-spectrometry, FTIR spectroscopy. In its UV-Vis spectrum, two main peculiarities were observed: *B* band (351 nm) in UV region and *Q* band (664 nm) in visible region. In comparison with phenyl-substituted phthalocyanine complexes, a hypsochromic shift of the *Q* band of about 30 nm is observed. Target complex possess phosphorescence at 828 nm. The value of singlet oxygen quantum yield φ_Δ=0.35±0.05 for title compound is comparable with substituted phthalocyanine analogs, described in literature. Thus, the investigated palladium tetrapyrazinoporphyrazine can be considered as a prospective photosensitizer.

Acknowledgements
We are grateful for financial support from the Russian Foundation for Basic Research 18-33-00519. We also thank the Council under the President of the Russian Federation for State Support of Young Scientists and Leading Scientific Schools (Grants MK-3115.2018.3). The work was performed within the framework of the State Assignment of 2019 (Theme 45.5 Creation of compounds with given physicochemical properties).

References

[1] S. M. Borisov, G. Zenkl and I. Klimant, *ACS Applied Materials & Interfaces*, 2010, **2**, 366-374.
[2] X.-d. Wang, H.-x. Chen, Y. Zhao, X. Chen, X.-r. Wang and X. Chen, *TrAC Trends in Analytical Chemistry*, 2010, **29**, 319-338.
[3] Y. Liu, G. Zhan, Z.-W. Liu, Z.-Q. Bian and C.-H. Huang, *Chinese Chemical Letters*, 2016, **27**, 1231-1240.
[4] W. Freyer and S. Flatau, *Tetrahedron Letters*, 1996, **37**, 5083-5086.
[5] M. P. Donzello, E. Viola, X. Cai, L. Mannina, C. Rizzoli, G. Ricciardi, C. Ercolani, K. M. Kadish and A. Rosa, *Inorganic Chemistry*, 2008, **47**, 3903-3919.
[6] K. Takahashi, Y. Aoki, T. Sugitani, F. Moriyama, Y. Tomita, M. Handa, K. Kasuga and K. Sogabe, *Inorganica Chimica Acta*, 1992, **201**, 247–249.
[7] W. Freyer, *Journal für Praktische Chemie/Chemiker-Zeitung*, 1994, **336**, 693-694.
[8] A. D. Kosov, T. V. Dubinina, M. Y. Seliverstov, L. G. Tomilova and N. S. Zefirov, *Macroheterocycles*, 2016, **9**, 201-205.
[9] E. H. Mørkved, N. K. Afseth and P. Zimcik, *Journal of Porphyrins and Phthalocyanines*, 2007, **11**, 130-138.
[10] E. M. Bauer, C. Ercolani, P. Galli, I. A. Popkova and P. A. Stuzhin, *Journal of Porphyrins and Phthalocyanines*, 1999, **3**, 371-379.
[11] H. W. Rothkopf, D. Wöhrle, R. Müller and G. Koßmehl, *Chemische Berichte*, 1975, **108**, 875-886.
[12] T. V. Dubinina, S. A. Trashin, N. E. Borisova, I. A. Boginskaya, L. G. Tomilova and N. S. Zefirov, *Dyes and Pigments*, 2012, **93**, 1471-1480.
[13] T. V. Dubinina, K. V. Paramonova, S. A. Trashin, N. E. Borisova, L. G. Tomilova and N. S. Zefirov, *Dalton Transactions*, 2014, **43**, 2799-2809.
[14] Y. Che, W. Yang, G. Tang, F. Dumoulin, J. Zhao, L. Liu and Ü. İşci, *Journal of Materials Chemistry C*, 2018, **6**, 5785-5793.
[15] W. Freyer, H. Stiel, M. Hild, K. Teuchner and D. Leupold, *Photochemistry and Photobiology*, 1997, **66**, 596-604.
[16] A. A. Krasnovsky, A. S. Kozlov and Y. V. Roumbal, *Photochemical & Photobiological Sciences*, 2012, **11**, 988-997.
[17] F. Wilkinson, W. P. Helman and A. B. Ross, *Journal of Physical and Chemical Reference Data*, 1993, **22**, 113-262.
[18] R. H. Young, D. Brewer and R. A. Keller, *Journal of the American Chemical Society*, 1973, **95**, 375-379.
[19] S. Tombe, W. Chidawanyika, E. Antunes, G. Priniotakis, P. Westbroek and T. Nyokong, *Journal of Photochemistry and Photobiology A: Chemistry*, 2012, **240**, 50-58.
[20] E. Dube, D. O. Oluwole and T. Nyokong, *Photochemistry and Photobiology*, 2018, **94**, 521-531.

[21] L. E. Hinkel, G. O. Richards and O. Thomas, *Journal of the Chemical Society (Resumed)*, 1937, DOI: 10.1039/JR9370001432, 1432-1437.

IOP Conf. Series: Materials Science and Engineering **525** (2019) 012028 doi:10.1088/1757-899X/525/1/012028

Synthesis of upconversion zirconia nanoparticles for bioimaging

A Paramonova, G Kiselev, A Fakhardo, P Krivoshapkin and E Krivoshapkina

Group of Ceramic and Natural Nanomaterials, Laboratory "Solution Chemistry of Advanced Materials and Technologies", ITMO University, Saint Petersburg, 197101, Russia

E-mail: kef@scamt-itmo.ru

Abstract. In this work, the upconversion Er^{3+}/Yb^{3+} co-doped ZrO_2 nanoparticles were synthesized and characterized. The nanoparticles were obtained via sol-gel method. To achieve upconversion activity ZrO_2 nanoparticles were doped with lanthanide ion pair of the Yb^{3+} and Er^{3+} and further calcined at 800°C. According to X-ray powder and high-resolution transmission electron microscopy the well-crystallized monoclinic nanoparticles were formed, which size ranges from 20 to 30 nm for undoped ZrO_2 and from 30 to 40 nm for doped ZrO_2. The calcined Er^{3+}/Yb^{3+} co-doped ZrO_2 nanoparticles demonstrated an upconversion luminescence visible by the bare eye. Namely, when excited with NIR-laser, the nanoparticles emit visible green light. Moreover, the Er^{3+}/Yb^{3+} co-doped ZrO_2 nanoparticles showed red (650-673 nm) and green (520–570 nm) upconversion luminescence due to the $^4F_{9/2} \rightarrow {}^4I_{15/2}$ of and ($^2H_{11/2}$ and $^4S_{3/2}$) $\rightarrow {}^4I_{15/2}$ transitions of Er^{3+} ion, respectively. The obtained nanoparticles exhibited intense luminescence alongside with low toxicity according to colorimetric assay for cellular growth and survival. It was concluded that Er^{3+}/Yb^{3+} co-doped ZrO_2 nanoparticles are perfect prospective bioimaging agents for using as extremely sensitive detectors of biomarkers or excessive contrast deep tissue indicators.

1. Introduction

The Biomedicine has significantly influenced the successful treatment of many human illnesses. Nevertheless, the clinical development of nanotechnology is only starting to emerge and there are numerous new methods of nanotechnology-based diagnostics and therapeutics being still in the process of clinical development [1]. Up to date, massive preclinical studies have provided vital information on essential design principles for the development of nanomedicine [2]. Although the future of nanomedicine looks bright and widespread, the clinical formulation of nanomedicine results faces the main challenge - it appears to be very complicated to engineer innovative nanoformulations of the drugs providing the ability to track the drag and its behaviour in the organism while not harming its clinical effectiveness [3]. Besides, new drugs are more and more often formulated with nanoparticles (NPs) to address drug delivery challenge such as solubility and therefore agents with any enhanced properties should be formulated in the form of nanoobjects [4].

While nanomedicine develops extremely rapidly its challenges and opportunities are only to be fully realized. Primarily, it is important to highlight that most of our present understanding of NPs behaviour in vivo is grounded on animal data, as a consequence its transformation to NP behaviour in human organism remains mostly unknown. For these purpose we offer optically active NPs intended for modification of biomedicine objects opening approaches for sensing and imaging of biomaterials.

IOP Conf. Series: Materials Science and Engineering **525** (2019) 012028 doi:10.1088/1757-899X/525/1/012028

Primarily, NPs with enhanced optical properties are required to identify the effect of nanotherapy, since NPs can influence NP–protein interaction or blood circulation. Inorganic NPs as optical agents suggest the benefit of being exceptionally robust, and hence highly resistant to degradation in the body liquids, which makes them really suitable for imaging and diagnostics [5].

One of the most essential and, simultaneously, the most advanced applications of luminescence is in the medical diagnostics and therapeutics. However, there are some unavoidable difficulties in the employment of photoluminescence based on the UV activation in biomedicine [6]. Human blood is known to absorb intensely both UV radiation and the emission of the optical agent [7]. The solution of this problem can be found by means of employing the upconversion luminescence where the absorption of two or more low-energy photons is followed by the emission of a higher energy photon. Additionally, one of the most reasonable advantages of using near infra-red (NIR) light in biomedicine is that it penetrates deep into tissues causing their minimal autoluminescence [8]. Moreover, there is practically no absorption by the whole-blood in the NIR region. Compared to conventional organic fluorophores upconversion NPs possess such advantages as: low toxicity, high photostability, nonphotobleaching, etc [9,10]. In fact, the upconversion nanoobjects have found a variety of applications in biomedicine as extremely sensitive detectors of biomarkers, excessive contrast deep tissue or single molecule bioimaging agents and many more [11].

Besides, it is worth noting that upconversion nanoobjects can be simply engineered for a multifunctional imaging purposes by integrating some appropriate lanthanide elements into the synthesising system [12]. Therefore, we utilized a well-known lanthanide ion pair of the Yb^{3+} (sensitizer) and Er^{3+} (activator) to produce upconversion bioimaging ability since this pair of ions demonstrates good UCL properties [13].

The zirconium oxide matrix seems to be an ideal medium for producing highly luminescent nanoobjects since it is photochemically and chemically very stable, possess low phonon energy and high refractive index [14]. The use of lanthanide dopants was rationalized to low concentrations for bio-applications because their toxicity is insignificant but the optical response is very high. In this work, we synthesized Er^{3+}/Yb^{3+} co-doped zirconium oxide aiming at obtaining a perfect bioimaging agent and studied its optical properties.

2. Synthesis and methods

NPs were prepared using sol-gel method. Sol-gel of zirconium oxide was produced by hydrolysis-polycondensation of zirconium (IV) propoxide solution (70%, Aldrich). To accomplish a controlled hydrolysis-condensation, an esterification reaction of acetic acid (98%, Chemmed) was a source of water and propanol-2 (Chemmed) served as a medium [15]. To achieve upconversion, the rare-earth acetate hydrates were added into the solution of precursors. The ratio of reagents for upconversion NPs was as follows: zirconium propoxide solution 1 ml, isopropanol 13 ml, acetic acid 0.6 ml. First, the desired amounts of erbium (III) acetate hydrate (99.9%, Aldrich) (1 mol%) and ytterbium (III) acetate hydrate (99.95%, Aldrich) (1 mol%) were dissolved in propanol-2 with stirring at 65°C. For undoped NPs, the above described step of adding rare-earth salts was skipped. After that, zirconium (IV) propoxide solution was added to the precursor's solution in inert atmosphere while stirring for 1 hour at 70°C. Then, acids were added to the solution with stirring for 6 hours at 65°C. Afterwards, the subsequent solution was left for gelation for 48 hours. To reach an upconversion activity zirconium oxide NPs doped with a pair of Er^{3+}/Yb^{3+} ions of the NPs were calcined so that dopants embedded in the crystal lattice of the oxide matrix. For an accurate comparison, undoped zirconium oxide NPs were calcined in a similar way. So, the resulting gel was calcined in muffle furnace LOIP-LF at 800°C at heating speed 5° per minute and cooled down to the room temperature at cooling speed 5° per minute.

3. Instrumentation for characterization

Hydrodynamic radius was measured via dynamic light scattering technique using a Photocor Compact-Z analyser. Zeta potential of the hydrosols was distinguished using a Photocor Compact-Z analyser. High-resolution Transmission Electron Microscopy (HRTEM) analysis was performed using

JEOL-2010. The X-ray diffraction (XRD) powder analysis of the calcined NPs was carried out using a D8 Advance Bruker powder X-ray diffractometer with Cu Kα radiation (λ = 1.5406Å) from 5° to 90° at a step of 0.01° per second and speed of 5° per minute. Photoluminescence and transmittance measurements of the NPs were conducted using Ocean Optics QE Pro spectrometer (the excitation was accomplished by HAL 100 and HBO 100). The NIR (980 nm) excitation was provided using a diode laser coupled to a 100 μm (core) optical fiber LASER MASTER PRO LSI980-2000 as a source of light. MTT assay was used to evaluate cytotoxic effects of pure and doped ZrO_2 NPs. The amount of surviving HeLa and human embryonic lung fibroblasts (HELF) (Biolot) cells was determined indirectly by measuring the MTT dye reduction by NAD(P)H dependent cellular oxidoreductase enzyme [16]. The absorption was measured at 570 nm using a plate reading spectrophotometer. HELF and HeLa cells were seeded at density 6×103 or 10×103 cells/well, respectively, in 96-well plates.

4. Results and discussion

4.1. Characterization of the NPs in hydrosols

A hydrosol of calcined NPs showed the average zeta potential of 12.3 ± 1.5 mV for pure ZrO_2 and 11.9 ± 1.2 mV for doped zirconia NPs. The average hydrodynamic radius of the calcined NPs in hydrosols (figure 1) observed via dynamic laser scattering analysis lies in the range of 35-40 nm for both pure and doped zirconia NPs. The collected data are in quite a good agreement with the high-resolution transmission electron microscopy (HRTEM) findings, which was performed to study the crystalline structures of the synthesized and calcined NPs.

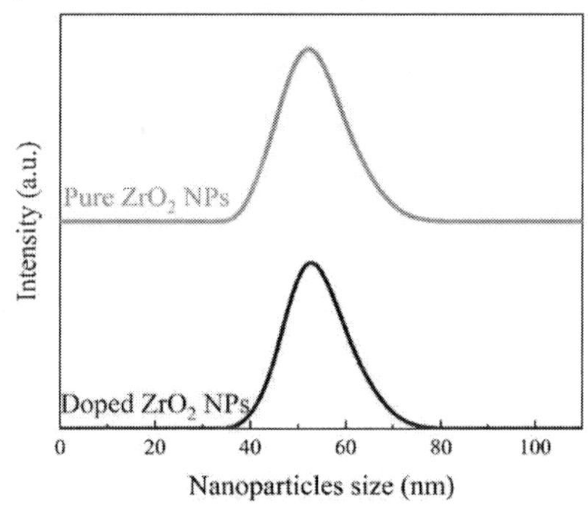

Figure 1. Hydrodynamic NPs size in hydrosols: calcined undoped ZrO_2 (orange line) and calcined Er^{3+}/Yb^{3+} co-doped ZrO_2 NPs (blue line).

4.2. The crystalline structures of the NPs via HRTEM

Consistent with HRTEM study the NPs size ranges 20-30 nm for undoped ZrO_2 (figure 2 A) and 30-40 nm for Er^{3+}/Yb^{3+} co-doped ZrO_2 (figure 2 B). The HRTEM data disclose well-defined crystalline structures of the obtained calcined NPs. Figure 2 (A) reveals distinct lattice fringes of pure ZrO_2 NPs with a P21/a space group and values of 0.3119 nm accurately matching the (-111) planes and 0.3658 nm matching the (110) planes of the monoclinic ZrO_2. The calcined Er^{3+}/Yb^{3+} co-doped ZrO_2 NPs demonstrate the similar tendency shown in figure 2 (B). A P21/a space group values of 0.5150 nm matching the (001) planes and 0.3577 nm matching the (011) planes of the monoclinic structure of ZrO_2.

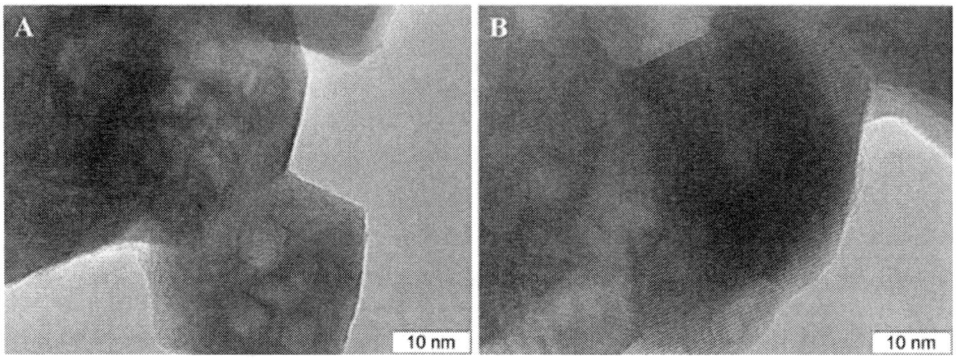

Figure 2. High-resolution TEM images of calcined undoped ZrO_2 NPs (A) and calcined Er^{3+}/Yb^{3+} co-doped ZrO_2 NPs (B).

4.3. The phase composition of the NPs via XRD

As seen from the XRD spectra of the both calcined pure and Er^{3+}/Yb^{3+} co-doped ZrO_2 in figure 3, well-crystallized monoclinic NPs were formed, matching correctly with the pattern of standard joint committee of powder diffraction standards (JCPDS) card No. 37–1484, as shown in figure 3 with black bars.

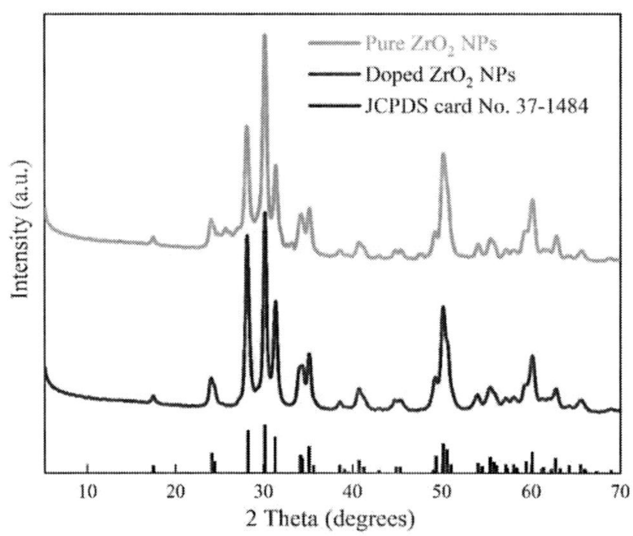

Figure 3. XRD pattern of calcined undoped ZrO_2 NPs (orange line) and calcined Er^{3+}/Yb^{3+} co-doped ZrO_2 NPs (blue line), black bars indicate JCPDS-37-1484 card data.

4.4. Study of the optical properties of the NPs

The room-temperature photoluminescence and transmittance spectra in the range of 400-1000 nm of the pure and Er^{3+}/Yb^{3+} co-doped ZrO_2 NPs were measured and presented in figure 4 (A) and figure 5 (A). In the case of calcined undoped ZrO_2 NPs, in photoluminescence spectrum, which is shown in figure 4 (A), we observe one broad peak in the range of 420–700 nm, which indicates that no energy transitions take place and, hence, upconversion luminescence of these undoped NPs is impossible to achieve. Moreover, the transmittance spectrum in inset of figure 4 (A) is clear without any peaks. This fact suggests that undoped NPs are not capable of absorbing the NIR light. Additionally, pure ZrO_2

IOP Conf. Series: Materials Science and Engineering **525** (2019) 012028 doi:10.1088/1757-899X/525/1/012028

NPs demonstrate blue luminescence under UV-excitation as seen in figure 4 (B), which correlates well with photoluminescence and transmittance measurements.

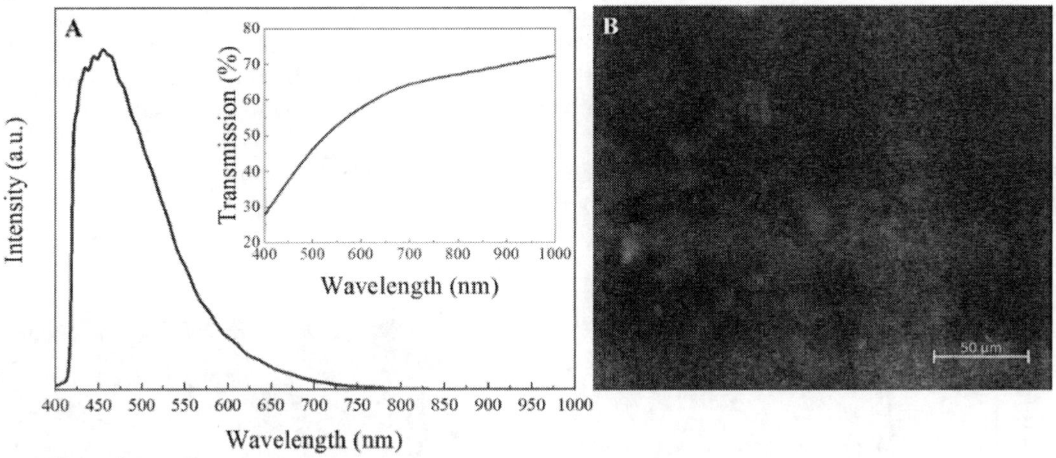

Figure 4. Photoluminescence spectrum of calcined undoped ZrO_2 NPs (A), the inset shows the transmission spectrum of calcined undoped ZrO_2 NPs; an optical micrograph of calcined undoped ZrO_2 NPs in photoluminescence microscope (B) under UV excitation.

In the case of calcined Er^{3+}/Yb^{3+} co-doped ZrO_2 NPs, we observe the photoluminescence peaks, which are exposed in figure 5 (A), indicating the transitions from the $^4I_{15/2}$ ground state to the excited states of Er^{3+} ion. This fact implies that Er^{3+} ions may absorb the pump light. The transmittance spectrum in inset of figure 5 (A) reveals an evident peak located at 975 nm, resembling the $^2F_{7/2} \rightarrow {}^2F_{5/2}$ transition of Yb^{3+} ion, which shows that Yb^{3+} ion can absorb the pump light. In addition, the peaks related to the transitions from the $^4I_{15/2}$ ground state to the excited states of the Er^{3+} ion are correspondingly appointed in inset of figure 5 (A) demonstrating the process of energy transfer from Yb^{3+} to Er^{3+} ion. Furthermore, crystals of calcined Er^{3+}/Yb^{3+} co-doped ZrO_2 NPs produce visible green light under UV-excitation as seen in figure 5 (B). The obtained data are in a good agreement with peaks in transmittance and photoluminescence spectra.

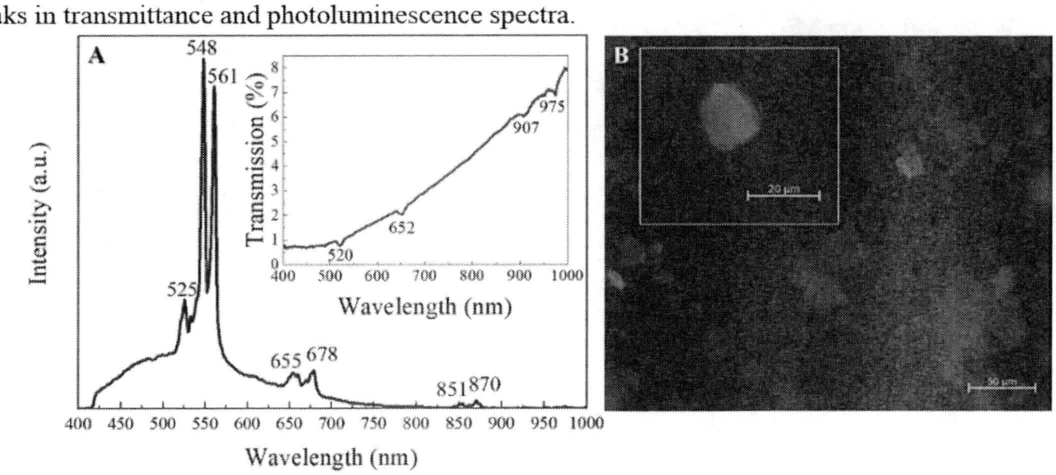

Figure 5. Photoluminescence spectrum of calcined Er^{3+}/Yb^{3+} co-doped ZrO_2 NPs (A), the inset shows the transmission spectrum of calcined Er^{3+}/Yb^{3+} co-doped ZrO_2 NPs; an optical micrograph of calcined Er^{3+}/Yb^{3+} co-doped ZrO_2 NPs in photoluminescence microscope (B) under UV excitation, the inset shows the optical micrograph of calcined Er^{3+}/Yb^{3+} co-doped ZrO_2 NPs in photoluminescence microscope under UV excitation at a higher magnification.

IOP Conf. Series: Materials Science and Engineering **525** (2019) 012028 doi:10.1088/1757-899X/525/1/012028

4.4.1. Upconversion luminescence of the NPs. The room-temperature upconversion luminescence spectrum of calcined Er^{3+}/Yb^{3+} co-doped ZrO_2 NPs under NIR (980 nm) excitation was obtained and presented in figure 6 (A). The obtained data reveal two obvious bands in the range of 500-700 nm where two green emissions about 523 nm and 546 nm can be observed. These emissions are correspondingly related to the radiative transitions of $^2H_{11/2} \rightarrow {}^4I_{15/2}$ and $^4S_{3/2} \rightarrow {}^4I_{15/2}$ of Er^{3+} ion. The red emission band lying in the range of 650-673 nm is assigned to the radiative transition of $^4F_{9/2} \rightarrow {}^4I_{15/2}$ of Er^{3+} ion. The obtained data correlate precisely and well with the well-known model of upconversion luminescence and the energy transfer process in Er^{3+}/Yb^{3+} co-doped samples [17]. The calcined Er^{3+}/Yb^{3+} co-doped ZrO_2 NPs demonstrate an upconversion luminescence visible by the bare eye. Namely, when excited with NIR-laser, the NPs emit visible green light as seen in figure 6 (B).

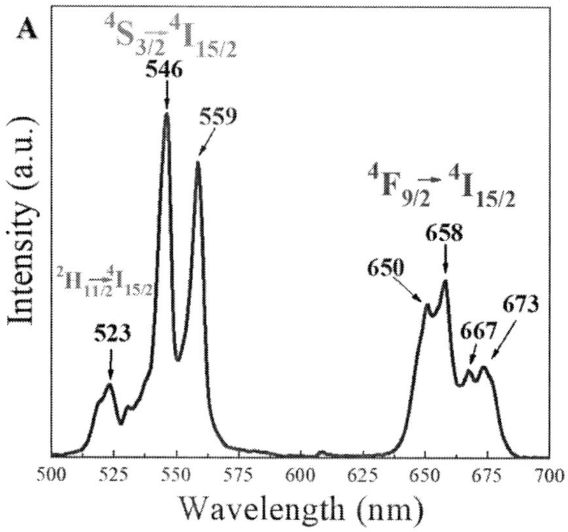

Figure 6. Upconversion emission spectrum of calcined Er^{3+}/Yb^{3+} co-doped ZrO_2 NPs under excitation with NIR laser (980 nm) (A); Digital photo of upconversion luminescence emission of calcined Er^{3+}/Yb^{3+} co-doped ZrO_2 NPs under the excitation with NIR laser (980 nm) (B).

4.5. Cytotoxicity of the NPs
The results of MTT assays concentration-dependent viability of HELF and HeLa cell lines with pure ZrO_2 and Er^{3+}/Yb^{3+} co-doped ZrO_2 NPs are presented in figure 7. The HELF cells were less sensitive to the NPs exposure than HeLa cells. It worth noting, that the death of the studied cell lines does not exceed 20%, from which it can be concluded that the obtained NPs are low cytotoxic. Moreover, Er^{3+}/Yb^{3+} co-doped ZrO_2 NPs show almost identical toxicity as pure ZrO_2 NPs, which means that release of the lanthanide ions doesn't occur. The obtained data identify that in investigated concentrations the obtained upconversion ZrO_2 NPs are inert to human cell lines and simultaneously they reveal upconversion characteristics making them a promising tool for in vivo diagnostics and therapeutics.

IOP Conf. Series: Materials Science and Engineering **525** (2019) 012028 doi:10.1088/1757-899X/525/1/012028

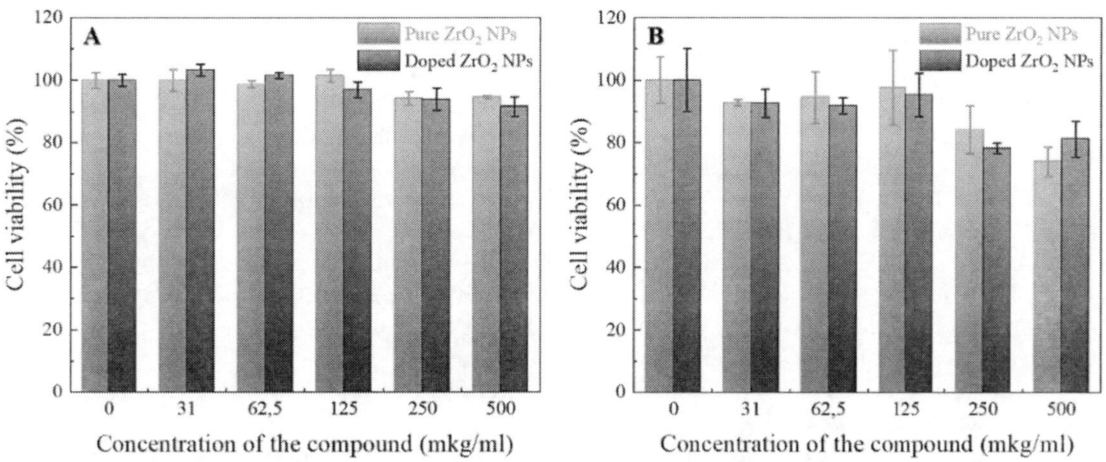

Figure 7. The results of MTT assays of pure ZrO_2 and calcined Er^{3+}/Yb^{3+} co-doped ZrO_2 NPs after 72 hours of exposure on HELF cells (A) and HeLa cells (B).

5. Conclusions

It can be concluded that the offered method is applicable for the obtaining of nanoparticles with upconversion activity. The exciting near-infrared light is located in the biological transparency window it can deeply penetrate into living tissues without causing autoluminescence, which is essential for biological imaging, for example, during tissue regeneration processes. The obtained upconversion materials possess well-crystallinity at nanoscale. Furthermore, the developed method allows to obtain stable Er^{3+}/Yb^{3+} co-doped ZrO_2 NPs hydrosols. This form is very convenient for the inject process for medical applications requiring biovisualization.

Acknowledgements

This work was financially supported by ITMO University, Project No. 418236. The X-ray powder diffraction studies were performed in the engineering center of the Saint-Petersburg State Technological Institute (Technical University).

References

[1] Chang E H, Harford J B, Eaton M A W, Boisseau P M, Dube A, Hayeshi R, Swai H and Lee D S 2015 Nanomedicine: Past, present and future - A global perspective *Biochem. Biophys. Res. Commun.* **468** 511–7

[2] Bhardwaj V and Nikkhah-Moshaie R 2017 Nanomedicine *Advances in Personalized Nanotherapeutics*

[3] Riehemann K, Schneider S W, Luger T A, Godin B, Ferrari M and Fuchs H 2009 Nanomedicine - Challenge and perspectives *Angew. Chemie - Int. Ed.*

[4] Kovalenko M V., Manna L, Cabot A, Hens Z, Talapin D V., Kagan C R, Klimov V I, Rogach A L, Reiss P, Milliron D J, Guyot-Sionnnest P, Konstantatos G, Parak W J, Hyeon T, Korgel B A, Murray C B and Heiss W 2015 Prospects of nanoscience with nanocrystals *ACS Nano*

[5] Sharma P, Brown S, Walter G, Santra S and Moudgil B 2006 Nanoparticles for bioimaging *Adv. Colloid Interface Sci.*

[6] Mathis G 1993 Rare Earth Cryptates with Human Sera *Clin. Chem.*

[7] Zijlstra W G, Buursma A and Meeuwsen-van der Roest W P 1991 Absorption spectra of human fetal and adult oxyhemoglobin, de-oxyhemoglobin, carboxyhemoglobin, and methemoglobin *Clin. Chem.*

[8] Misra S N, Gagnani M A, Indira D M and Shukla R S 2004 Biological and clinical aspects of lanthanide coordination compounds *Bioinorg. Chem. Appl.*

IOP Conf. Series: Materials Science and Engineering **525** (2019) 012028 doi:10.1088/1757-899X/525/1/012028

[9] Escobedo J O, Rusin O, Lim S and Strongin R M 2010 NIR dyes for bioimaging applications *Curr. Opin. Chem. Biol.*

[10] Zhou N, Ni J and He R 2013 Advances of Upconversion Nanoparticles for Molecular Imaging *Nano Biomed. Eng.*

[11] Suyver J F, Aebischer A, Biner D, Gerner P, Grimm J, Heer S, Krämer K W, Reinhard C and Güdel H U 2005 Novel materials doped with trivalent lanthanides and transition metal ions showing near-infrared to visible photon upconversion *Opt. Mater. (Amst).*

[12] Zhou J, Liu Q, Feng W, Sun Y and Li F 2015 Upconversion luminescent materials: Advances and applications *Chem. Rev.*

[13] Auzel F 2004 Upconversion and Anti-Stokes Processes with f and d Ions in Solids *Chem. Rev.*

[14] Saraidarov T, Reisfeld R and Pietraszkiewicz M 2000 Luminescent properties of silica and zirconia xerogels doped with europium(III) salts and europium(III) cryptate incorporating 3,3′-biisoquinoline- 2,2′-dioxide *Chem. Phys. Lett.*

[15] Agoudjil N, Kermadi S and Larbot A 2008 Synthesis of inorganic membrane by sol-gel process *Desalination* **223** 417–24

[16] Mosmann T 1983 Rapid colorimetric assay for cellular growth and survival: Application to proliferation and cytotoxicity assays *J. Immunol. Methods*

[17] Auzel F 1990 Upconversion processes in coupled ion systems *J. Lumin.*

IOP Conf. Series: Materials Science and Engineering **525** (2019) 012029 doi:10.1088/1757-899X/525/1/012029

Development of the technology of obtaining the composite based on the "magnesium-bone substance" for biodegradable implants by the method of powder metallurgy

P Krokhicheva, D Shornikov and M Savelyev

National Research Nuclear University "MEPhI", Russia.

E-mail: polinariakroh@gmail.com

Abstract. This work reports on the development of a composite "magnesium (Mg)- bone substance (HA)" for a biodegradable implant using powder metallurgy methods. In the course of the work, experiments were performed in two different types of a powder compaction: sintering this composite by standard compaction method and the spark-plasma sintering method (SPS). The optimum parameters for compacting the powder mixture were selected, the initial percentage of open porosity after sintering samples was calculated, X-ray microanalysis was performed, and the microstructures of the samples, which had been obtained by using various powder metallurgy methods, were compared.

1. Introduction

Sometimes, surgical intervention leads to extensive postoperative defects and, as a result, to functional insufficiency of the musculoskeletal system. A traditional approach is to replace a bone defect with an implant. The main purpose of this developing direction is the creation of an implant material, which must appropriate the following requirements:

- biocompatibility;
- bioactivity;
- corrosion resistance;
- strength.

Magnesium (Mg) exhibits excellent biocompatibility with the human body, the strength properties are most similar to the human bone strength properties, that is way, Mg is a perspective material for creating biodegradable implants. However, its corrosion performance at the current stage of development is still not good enough for increasingly diverse practical applications [1]. Thus, in order to reduce the corrosion rate and increase the time of implant degradation until the bone tissue is fully restored, various approaches are being discussed [2-7]. On the one hand, the authors of [2-6] propose to increase the corrosion resistance of magnesium implants by creating alloys - adding to magnesium alloying elements such as: Al, Zn, Mn, Zr, Gd, Sr. On the other hand, to prolong the time of implant degradation helps purification of magnesium from harmful impurities and anodizing, proposed [6]. In addition, to temporarily slow down the decomposition, appropriate types of coatings can be applied to the surface of a magnesium implant [7].

All approaches are promising, but at the present moment, none of them has yet led to the creation of commercially available products, the reason for this is not the full compliance of the implant material with the requirements. Regardless of the efforts undertaken, there is a real demand for solutions that allow realizing the production of biodegradable magnesium implants for the needs of traumatology, orthopedics and surgery.

Considerable efforts were directed to the development of ceramic materials based on Calcium hydroxyapatite (HA, $Ca_{10}(PO_4)_6(OH)_2$)- it is an analog on the phase and chemical composition of the mineral component of human bone tissue [8-11]. Materials based on HA do not have a negative impact in the human body, unlike a number of metals and polymers, and biologically active in relation to the formation of bone apatite. But the known hydroxyapatite ceramic materials have rather low mechanical characteristics in order to perceive physical exertion in many necessary situations with a high degree of reliability.

We have attempted to create a porous composite based on HA with reinforcing substance- magnesium (Mg) in the matrix of the material. To retain the initial properties- bioactivity of hydroxyapatite and biodegradability of magnesium, and close values of the strength characteristics with bone tissue, this composite can only be obtained by using powder metallurgy methods.

Some authors [12-14] reported the presence of magnesium (Mg) in calcined living tissues, which would indicate that Mg-ions could improve the biocompatibility and bioactivity of CaP ceramics [14-15]. Magnesium is one of the essential elements for all living organisms, approximately 60-65% of the total amount of Mg is in the human body- in the teeth and bones, and the remaining 35-40% is in the soft tissues, organs and blood. Mg is involved in important functions of the body, for example, it stimulates the muscle activity of the heart muscle, thereby helping to maintain a stable heart rate [16]. Mg, as well as Ca, provides strong and healthy bones, reducing the risk of osteoporosis. Osteoporosis of beneficial micronutrients from bone tissue is caused by a deficiency of Mg [17]. It is expected that implants based on the composite "magnesium (Mg)- bone substance (HA) will contribute to restore the bone, provide faster and more efficient regeneration of the damaged bones area. In the process of degradation of the implants in body medium gradually releasing Mg^{2+} cations, stimulating the appearance of osteocyte bone cells [18]. Physical and chemical properties of the material guarantee progressive process of bone regeneration until it is completely will be replaced by new bone.

In the conditions of the modern world and rapidly developing technologies, the goals of this development are not just to reduce the corrosion rate of implants and increase the strength of the composite obtained by powder metallurgy methods, but to create such a material that will increase the rate of bone splicing, the time of bone tissue regeneration, contributing to the formation of healthy callus by replacing the implant with live bone tissue.

2. Experimental procedure

2.1 Sample preparation

The initial powders of magnesium and hydroxyapatite were investigated; the particle size distribution was carried out on a laser particle analyzer ANALYSIS22. Visually, magnesium powder has a silver color, and HA powder has a pure white color. The analysis showed the average particle size of magnesium is 100μm, the size of agglomerates of synthetic HA powder obtained by the sol-gel method was 80μm, but the powder itself is fine dust with nanoscale particles. The pycnometric density of these powders was measured at the Helium pycnometer- AccuPyc II 1340. The density of magnesium powder is 1,729g/cm^3, which is consistent with the theoretical density, the density of HA is 3,472 g/cm^3. Using a Jeol JSM scanning electron microscope, an image of the particles was obtained, which could be used to determine their shape (figure 1-2):

Figure 1. The size and shape of magnesium powder particles (magnification × 500).

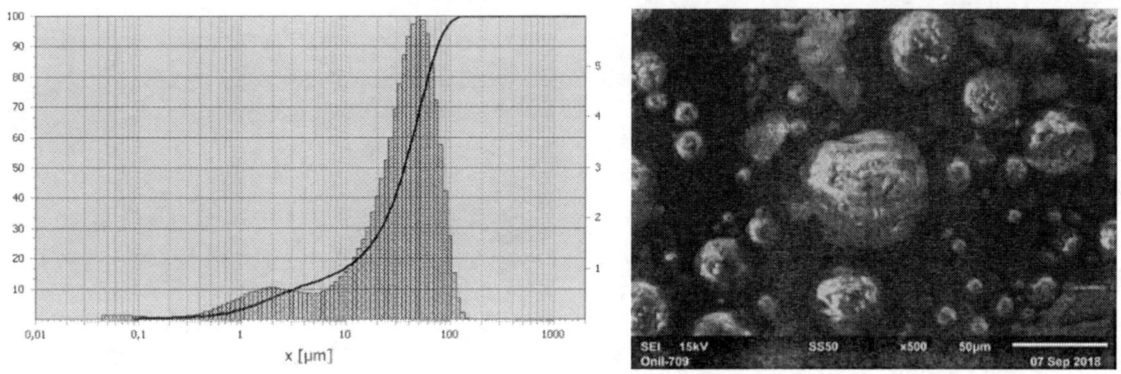

Figure 2. The size and shape of the particles of hydroxyapatite powder (magnification × 500).

Pure powders of the following components were prepared: HA and Mg; the mixture was mixed by a mechanical activation method; grinding of powders in a ML-1 ball vibratory mill. The mixture consisted of 9g HA and 1g Mg (volume ratio of 70% HA and 30% Mg), steel cups with steel balls were used; grinding was carried out in an Ar atmosphere in order to exclude the formation of magnesium oxide. Duration of grinding was 15 minutes. The size and shape of the powder particles after grinding is presented in figure 5.

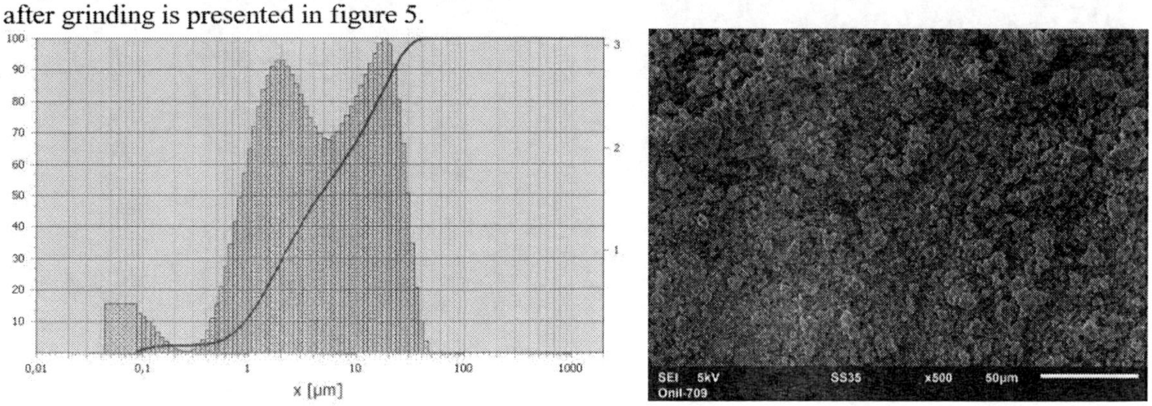

Figure 3. The size and shape of the particles of the mixture of the starting components after grinding in an argon atmosphere (magnification × 500).

IOP Conf. Series: Materials Science and Engineering **525** (2019) 012029 doi:10.1088/1757-899X/525/1/012029

As can be seen from the above figure, the size of the mixture after grinding average 8μm, but there is also an area with nanoscale particles.

2.2 Sintering of the composite by the traditional powder method

The first step in the experiment was to check the possibility of obtaining the composite "Mg-HA" using the standard powder metallurgy method — pressing and free sintering. Two batches of tablets were prepared: the ratio of the volume fractions of the components of 70% HA and 30% Mg. The sample sintering experiment was carried out in an argon (Ar) atmosphere and in a high vacuum in a furnace.

Previously, preliminary sintering tests were carried out on Mg- it was found that the optimal sintering temperature is about 500°C, at this temperature the samples would not crack, not form cracks, there would no gradient from the tablet to the center, sintering occurred throughout the volume evenly. The tablets were compressed in an Ar atmosphere, the applied pressure was 25MPa. The electronic image of the structure of the obtained samples is shown in figure 4.

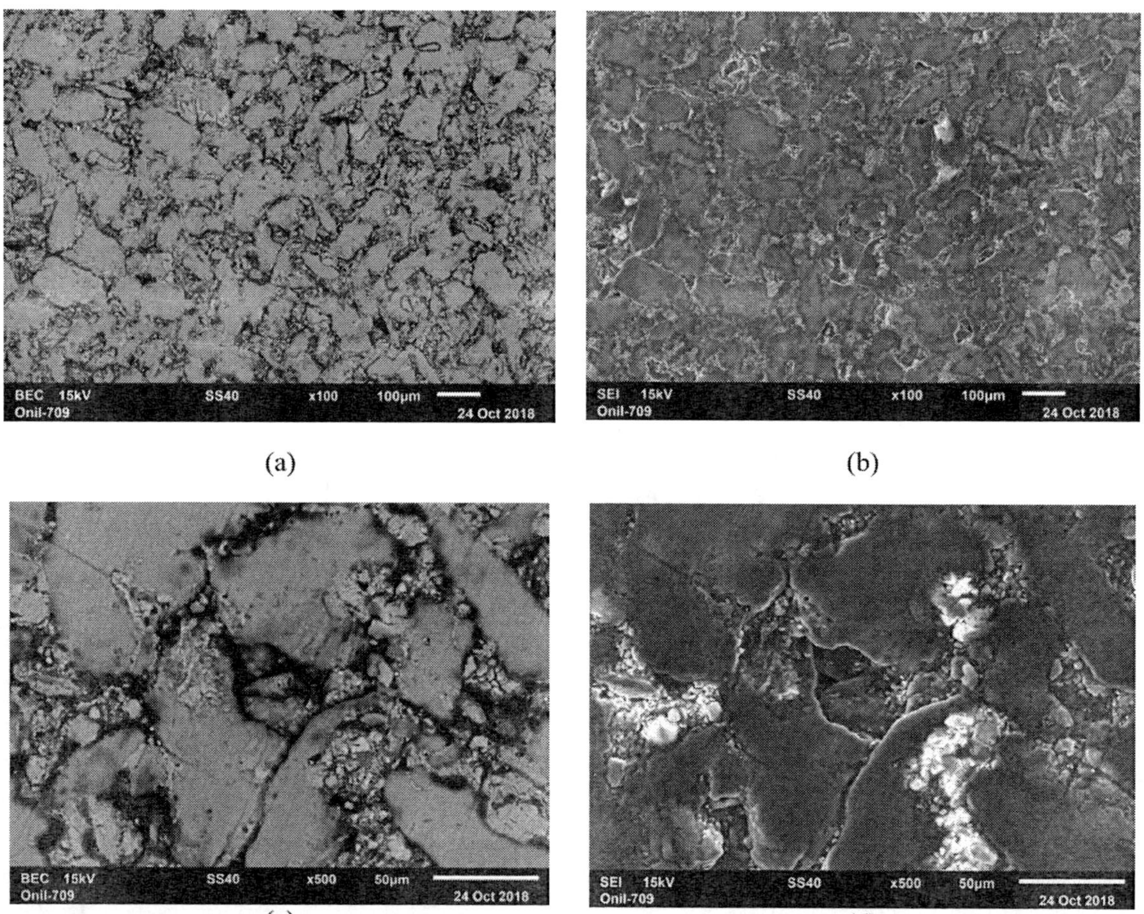

Figure 4. Electron image of the microstructure of the sample of composite "Mg-HA" obtained by standard powder method: (a) obtained in the mode of back-scattered electrons with magnification × 100, (c) with magnification × 500, (b) obtained in the mode of secondary electrons with magnification × 100, (d) at magnification × 500.

The microstructure of the composite, obtained by the standard powder metallurgy method, has a different morphology and grain size, the interphase boundaries are poorly defined, there are cracks passing both along the grain boundaries and intersecting them. The magnesium particles are unevenly arranged in the ceramic matrix. Therefore, in practice, the problem arises of preserving the structure of compacts before sintering, i.e. prevent intensive agglomeration of particles at high pressing pressures in order to preserve nucleation centers and prevent grain growth during sintering. The dispersion of powders has a much greater effect on their compactibility than their physical and mechanical properties. Table 1 presents data on the density of the samples before sintering and after.

Table 1. The porosity values of composite samples of 70% HA-30% Mg, obtained by the standard method of compaction in high vacuum

	$\Sigma\rho$, g/cm^3	$\rho_{geo(before)}$, g/cm^3	$\rho_{geo(after)}$, g/cm^3	ρ_{pic}, g/cm^3	% $_{open.por.}$
Sample 1	2,95	1,82	1,72	2,2	16,3
Sample 2	2,95	1,94	1,6	2,1	16,9

Traditional methods of static pressing lead to the formation of an inhomogeneous structure, grain growth in the compaction process and a not sufficiently high density of compacts. A promising method that can offer an alternative solution to these problems is electropulse compacting of powders - spark-plasma sintering (SPS).

2.3 Sintering of the composite by the SPS method

The second step in the experiment was to test the possibility of obtaining a composite "Mg-HA" using the spark-plasma sintering method. For this, a mechanically activated mixture was prepared; it was poured into a graphite matrix of 10mm in size, with graphite paper and tungsten foil, in order to eliminate interaction with graphite. Sintering parameters at the LABOX-152VHD unit: -500°C heating 20°C per minute, pressure-25MPa, holding time-10 min.At this stage, two mixtures of this composite with 70% HA-30% Mg by weight and 70% HA-30% Mg by volume were sintering to determine the effect of the amount of reinforcing substance (Mg) on the structure of the composite as a whole and the effect of the relative amount of HA on porosity of the obtained samples after sintering. The microstructure of the samples is shown in figure 5. Magnesium particles are more elongated and rounded grains, evenly spaced in a porous ceramic matrix, the interphase boundary is clearly visible:

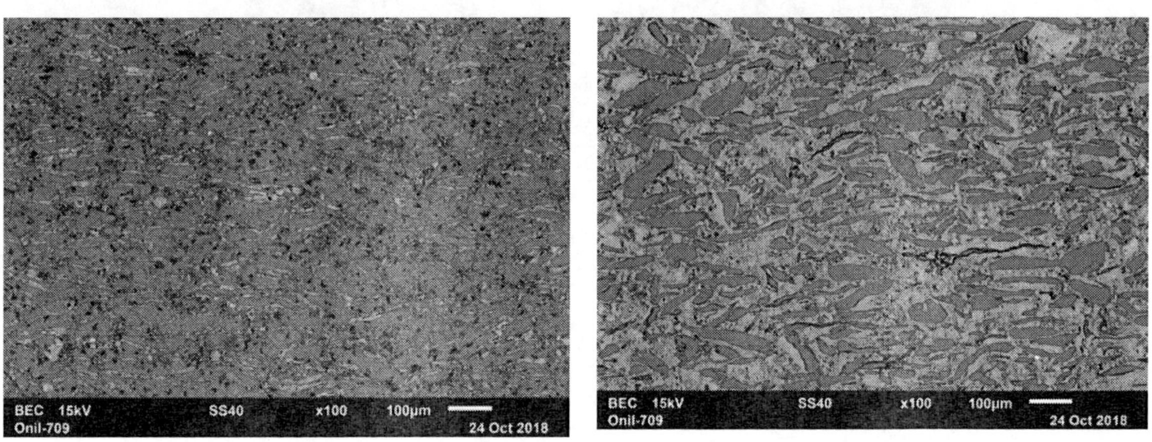

(a) (b)

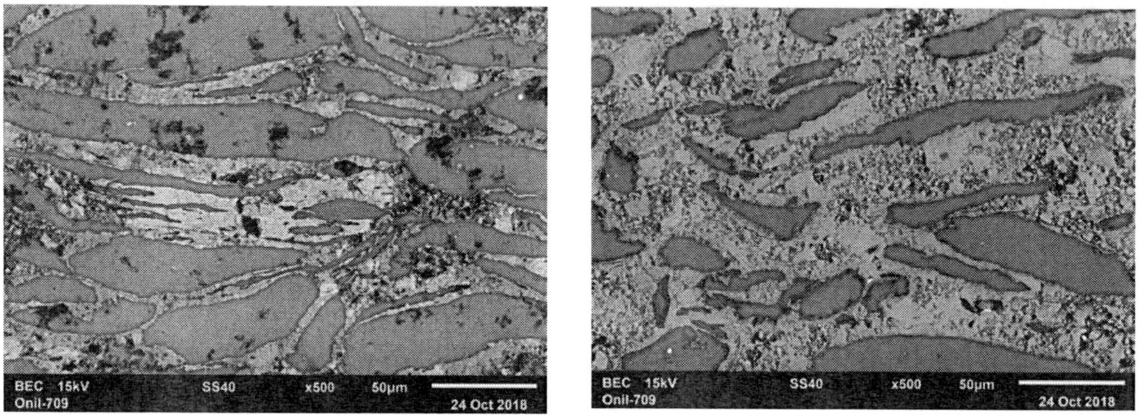

(c) (d)

Figure 5. Electron image in the back-scattered electrons of the microstructure of the sample of the composite "Mg-HA" obtained by the SPS method: a) the composition of the composite is 70% HA-30% Mg by mass, magnification × 100; b) the composition of the composite is 70% HA-30% Mg by volume, magnification × 100; c) 70% HA-30% Mg by weight with magnification × 500; d) 70% HA-30% Mg by volume with magnification × 500.

3. Results and discussion

During the experiment on sintering the composite "Mg-HA" by the standard powder metallurgy method, it was found that a batch of compressed tablets after sintering in an Ar atmosphere, did not sinter, due to the long duration of the process, the decomposition of HA and the color change of the tablets is observed on thr figure 6:

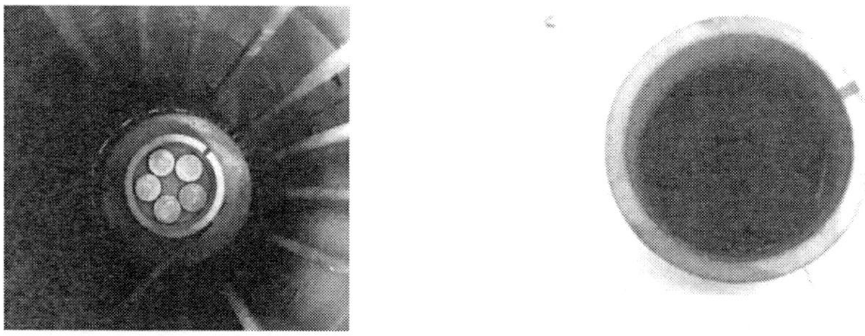

(a) (b)

Figure 6. Batch of composite samples: (a) prior to sintering in an Ar atmosphere, (b) after sintering in an Ar atmosphere.

One of the important stages in the technology for producing a composite is the study of its structural phase state; for this purpose, an X-ray microanalysis was carried out on a Jeol JSM microscope of composite samples of a "Mg-HA" obtained by various powder compaction methods.

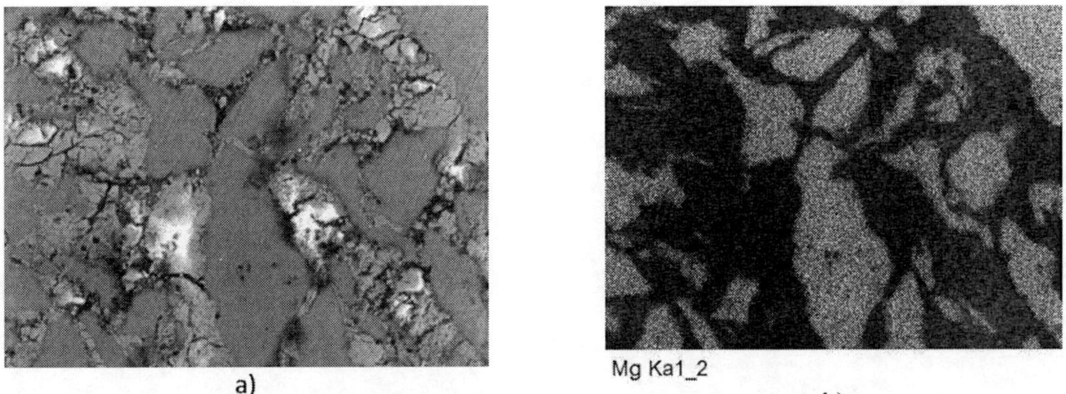

Figure 7. Map of elements on the plot of the composite 70% HA-30% Mg by volume: a) section of the surface of the ground section chosen for distribution of the elements; b) magnesium excretion.

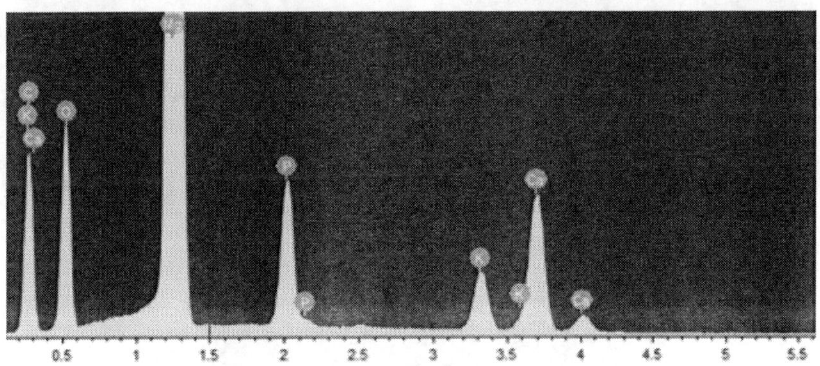

Figure 8. The total spectrum of chemical elements on the plot of the composite 70% HA-30% Mg by volume, obtained by the SPS method.

Figure 7 (a) illustrates a, section of a 70% HA-30% Mg composite by volume obtained by the standard powder metallurgy method; from the chemical element maps (b), it can be seen that the particles are distributed in the composite matrix are magnesium particles. The total distribution spectrum of the chemical elements shown in figure 8; the spectrum was detected insignificant potassium content.

According to the results of micro X-ray spectral analysis of the composite section sintered by the SPS method (figure 9), it can be seen that this composite is represented by two phases, large grains are magnesium- evenly spaced in a matrix consisting of elements such as phosphorus, calcium, oxygen. The boundaries of the diffusion interaction of magnesium and hydroxyapatite have not been identified, it can be concluded that at this sintering temperature does not the interaction between components, new phases are not formed. Thus, the mapping of the selected surface area can be very informative and widely used in the analysis of surface areas of both metal and composite samples.

IOP Conf. Series: Materials Science and Engineering **525** (2019) 012029 doi:10.1088/1757-899X/525/1/012029

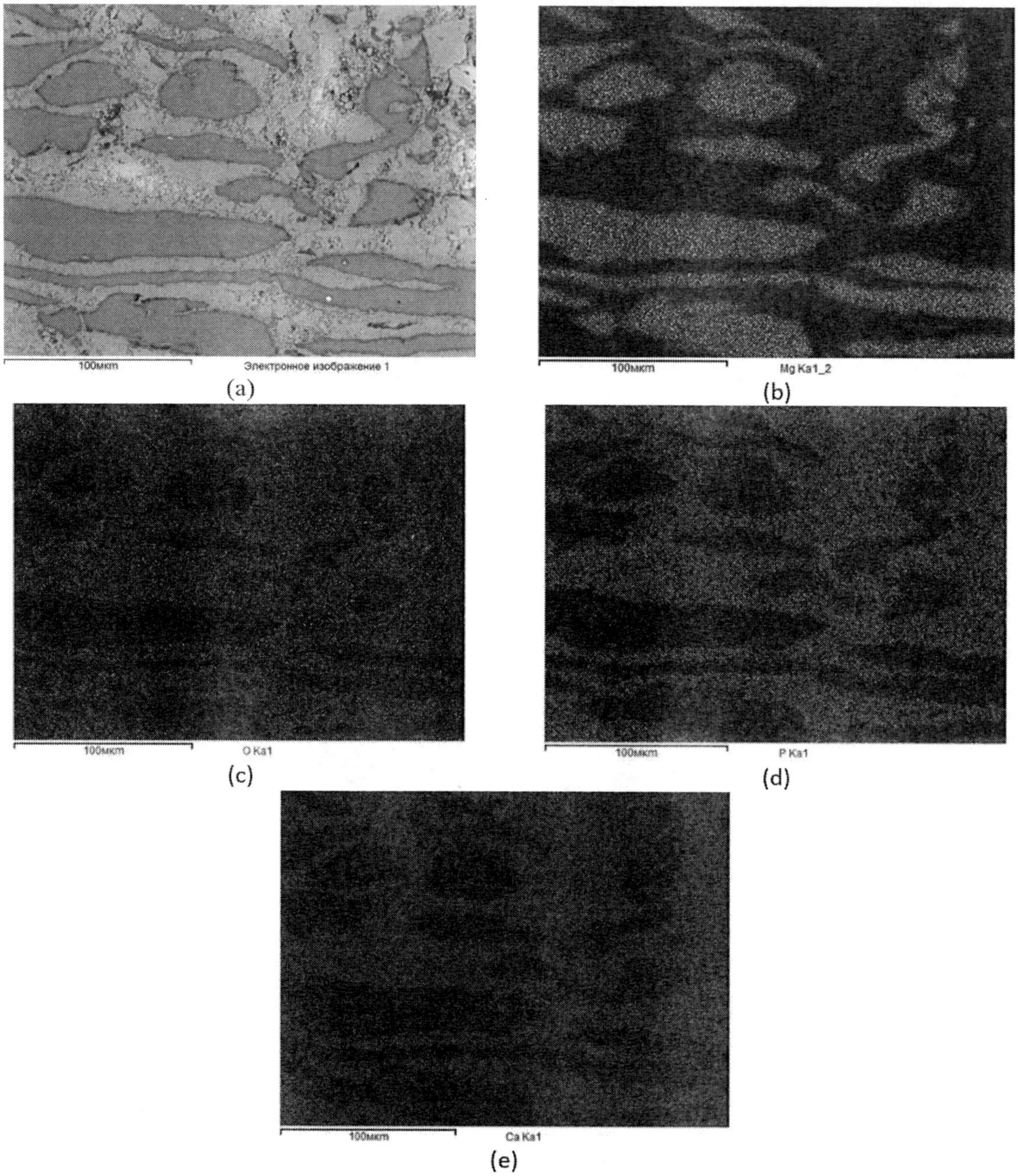

Figure 9. Map of elements on the plot of the composite 70% HA-30% Mg by volume, obtained by the SPS method: (a) the plot of the surface selected for distribution of the elements; (b) Magnesium distribution; (c) Oxygen distribution; (d) Phosphorus distribution; (e) Calcium distribution.

4. Conclusions

This study focused on the development of a composite "magnesium (Mg)- bone substance (HA)" for a biodegradable implant using powder metallurgy methods. During the experiments, the following conclusion was drown: it is necessary to significantly speed up the sintering process of the composite in order to exclude its interaction at all stages of the technology with the atmosphere and prevent decomposition of HA. A perspective option is the use of electro-pulsed methods of compaction, namely the use of SPS method.

Based on the experiments performed on sintering the composite using various powder metallurgy methods, the optimum compaction parameters were established and the initial open porosity was calculated after sintering the samples (table 2-3).

Table 2. The porosity values of composites obtained by various methods of compaction

Method	$\Sigma\rho$, g/cm^3	ρ_{geo}, g/cm^3	$\rho_{pic.}$ g/cm^3	$\%_{geo}$	$\%_{relative}$	$\%_{open.por.}$
SPS	2,9	1,72	2,1	58,4	71,9	13,5
Free sintering	2,9	1,71	2,2	58,2	75	16,8

Bone tissue is one of the types of connective tissue consisting of three types of cells and the extracellular matrix. The cells make up 1-2% of the total volume of bone tissue, the rest of the volume is occupied by pores and channels (for compact bone tissue, the porosity is 13-18%, for spongy it is higher) and the solid phase is organic and mineral components of bone plates. The organic component (40-50% of the solid phase) is represented by collagen. The mineral component (50-60% of the solid phase) is predominantly the crystals of hydroxyapatite $Ca_{10}(PO_4)_6(OH)_2$ and other calcium salts. Thus, it can be concluded that the composite "Mg- HA" obtained by powder compaction methods has a close initial porosity to the porosity of compact bone tissue, which is a good indicator for increasing the speed of bone growth in the composite and, therefore, accelerating regeneration.

Table 3. The optimal parameters of compaction of the composite

Parameters of compaction	Values
Pressing pressure, MPa	25
Sintering temperature, ^{0}C	500
Holding time, min.	10

References

[1] Song, Guangling. "Recent progress in corrosion and protection of magnesium alloys." *Advanced Engineering Materials*, 2005 , **7.7**: 563-586

[2] Zander, Daniela, and Naemi A. Zumdick. "Influence of Ca and Zn on the microstructure and corrosion of biodegradable Mg–Ca–Zn alloys." *Corrosion Science*, 2015, **93**: 222-233

[3] Homayun, Bahman, and Abdollah Afshar. "Microstructure, mechanical properties, corrosion behavior and cytotoxicity of Mg–Zn–Al–Ca alloys as biodegradable materials." *Journal of Alloys and Compounds*, 2014, **607**: 1-10

[4] Li, Huafang, Yufeng Zheng, and Ling Qin. "Progress of biodegradable metals." *Progress in Natural Science: Materials International*, 2014, **24.5**: 414-422

[5] Brar, Harpreet S., Joey Wong, and Michele V. Manuel. "Investigation of the mechanical and degradation properties of Mg–Sr and Mg–Zn–Sr alloys for use as potential biodegradable implant materials." *Journal of the mechanical behavior of biomedical materials*, 2012, **7**: 87-95

[6] Barchiche, C-E., et al. "Corrosion resistance of plasma-anodized AZ91D magnesium alloy by electrochemical methods." *Electrochimica Acta*, 2007, **53.2**: 417-425
[7] Hornberger, H., S. Virtanen, and A. R. Boccaccini. "Biomedical coatings on magnesium alloys–a review." *Acta biomaterialia*, 2012, **8.7**: 2442-2455
[8] Oonishi, H. "Orthopaedic applications of hydroxyapatite." *Biomaterials*, 1991, **12.2**: 171-178
[9] Kivrak, Nezahat, and A. Cuneyt Taş. "Synthesis of calcium hydroxyapatite- tricalcium phosphate (HA- TCP) composite bioceramic powders and their sintering behavior." *Journal of the American Ceramic Society*, 1998, **81.9**: 2245-2252
[10] Xiong, Guangyao, et al. "Characterization of biomedical hydroxyapatite/magnesium composites prepared by powder metallurgy assisted with microwave sintering." *Current Applied Physics*, 2016, **16.8**: 830-836
[11] Cuccu, Alessio, et al. "Optimization of the spark plasma sintering conditions for the consolidation of hydroxyapatite powders and characterization of the obtained products." *Chemical Engineering Transactions*, 2015, **43**: 1771-1776
[12] Brar, Harpreet S., et al. "Magnesium as a biodegradable and bioabsorbable material for medical implants." *Jom*, 2009, **61.9**: 31-34
[13] Staiger, Mark P., et al. "Magnesium and its alloys as orthopedic biomaterials: a review." *Biomaterials*, 2006, **27.9**: 1728-1734.
[14] Havaldar, Raviraj, S. Pilli, and B. Putti. "Effects of magnesium on mechanical properties of human bone." IOSR J. Pharm. Biol. Sci, 2013, **7.3**: 8-14
[15] Radha, R., and D. Sreekanth. "Insight of magnesium alloys and composites for orthopedic implant applications–a review." *Journal of magnesium and alloys*, 2017, **5.3**: 286-312
[16] Polimeni, Philip I., and Ernest Page. "Brief reviews: magnesium in heart muscle." *Circulation Research*, 1973, **33.4**: 367-374
[17] Kanis, John A., et al. "The diagnosis of osteoporosis." *Journal of bone and mineral research*, 1994, **9.8**: 1137-1141
[18] Zheng, Yufeng, Bin LIU, and Xuenan GU. "Research Progress in Biodegradable Metallic Materials for Medical Application [J]." *Materials Review*, 2009, **1**: 003

IOP Conf. Series: Materials Science and Engineering **525** (2019) 012030 doi:10.1088/1757-899X/525/1/012030

Study of electromagnetic characteristics of polymer materials based on single-walled and multi-walled carbon nanotubes

G E Kuleshov[1,2] and A V Badin[1,2]
[1] Laboratory of radiophysical and optical methods of exploring the surrounding environment, National Research Tomsk State University, Tomsk, Russia
[2] Department of Radioelectronics, Faculty of Radiophysics, National Research Tomsk State University, Tomsk, Russia

E-mail: grigorij-kge@sibmail.com

Abstract. The results of the study of the electromagnetic characteristics of polymer composite materials containing various carbon fillers are presented. Samples based on epoxy resin with single-walled and multi-walled carbon nanotubes, as well as conductive ABS plastic with carbon fibers were made. The frequency dependences of the complex dielectric constant on the frequency for these materials are shown. Based on the data obtained, the electromagnetic response from samples of composites located in free space and on the metal, taking into account the thickness, was calculated. The influence of the material thickness on the electromagnetic response from it was simulated. A composite material containing 4 wt.% SWCNTs is effective narrowband absorber at a frequency of 5 GHz, at a thickness of 3.0 mm. Or it can be used as a shielding high-frequency radiation material at frequencies from 7 GHz with a thickness of more than 2.5 mm.

1. Introduction

In recent years, the development of modern science and technology is ahead of theoretical research in obtaining the latest materials for the industrial complex. In structural and functional composites, reinforcing elements usually provide the necessary mechanical characteristics of the material (strength, stiffness, etc.), and the matrix ensures that the reinforcing elements work together and protect them from mechanical damage and aggressive chemical environment. These materials are used in all areas of industrial production. The use of protective materials and coatings to reduce the mutual influence of radio components, to improve the electromagnetic compatibility of various systems, as well as to improve the environmental situation is the most important direction of technological development [1]. As fillers of modern high range radio composite materials, ferrites with different crystal structures [2], the carbon material (amorphous carbon, graphite, fullerenes, carbon nanostructure compound) [3], carbonyl iron and other metal compounds [4] are commonly used. At present, we are getting closer to the limit values of the electromagnetic characteristics of materials that can be achieved using traditional fillers. Further development prospects in this area are associated with the development and use of new nanoscale high-tech materials. A promising class of materials for use in electronic devices are carbon nanotubes (CNTs) [5]. CNTs have a low bulk density, mechanical strength, ductility, and widely varying conductivity. They allow the development of high-strength and lightweight coatings based on them. By varying the concentration of CNTs, it is possible to produce materials both shielding and absorbing electromagnetic radiation [6, 7]. Therefore, it is of interest to investigate the difference in the electromagnetic properties of composites based on single-walled carbon nanotubes (SWCNT) and multi-layered carbon nanotubes (MWCNT) in the microwave radiation range.

At present, the manufacture of composite materials and structures based on additive technology is actively developing. Recently, the manufacture of plastic-based composite materials for 3D printing with the addition of fillers that effectively interact with electromagnetic radiation has become very popular [8]. Most often it is conductive plastics. This plastic consists of polymer resin and an appropriate filler. As fillers, either powders of well-conducting metals or various carbon structures are used [9]. It is the presence of carbon fiber in the composite that allows this type of ABS plastic to effectively conduct electric current. However, most conductive plastics we have today still are of high resistance. They are promising as functional and constructive radio materials.

In this paper, we study the electromagnetic characteristics of polymer radio materials based on carbon nanotubes and compare them with the results obtained for industrially produced carbon-containing ABS plastic of the "Conductive" brand.

2. Materials and Methods

The structure of the materials was determined by X-ray analysis. Images were obtained on X-Ray Diffractometer Shimadzu XRD 6000 with copper radiation (CuKα) and Wavelength Dispersive X-Ray Fluorescence Spectrometer XRF-1800. Powder samples and composites samples were studied. Shooting modes are tube voltage of 40 kV, anode current of 30 mA, goniometer speed when shooting is 2 deg/min, X-ray diffraction is $2\theta=(20\div60)^{\circ}$.

The structure of material was studied by optical microscope and by scanning electron microscopy. Electron-microscopic measurements were carried out with a scanning electron microscope with a focused ion beam QUANTA 3D.

2.1. Materials

Composite materials consist of a binder (matrix) and a filler. The epoxy resin EDP-20 with hardener PEPA was used as the polymer base of the composite material. Single-walled powders (SWCNT) and multi-walled carbon nanotubes (MWCNT) were used as fillers. Their main characteristics are presented in Table 1. The SWCNT was a Tuball material produced by Plasma-Chemical Technologies. It contains more than 70% of the main phase. The average diameter of carbon nanotubes is $d = 1.8$ nm. Used MWCNTs were synthesized by catalytic gas-phase deposition of ethylene [10] at the Institute of Catalysis G.K. Boreskov SB RAS. They contain more than 97.5% of the main phase with an average tube diameter $d = 9.4$ nm. MWCNT and SWCNT in concentrations from 2 to 4 wt.% Were mixed with epoxy resin.

Table 1. Comparative characteristics of MWCNT and SWCNT

Characteristics	MWCNT	SWCNT
CNT content	More than 97.5 wt.%	More than 70 wt.%
The average diameter of the nanotubes	9.4 nm (4 – 21 nm)	1.7 nm (1.66 – 2.28 nm)
Impurities of metals	Less than 1.7 wt.% (FeCo, Al_2O_3)	Less than 15 wt.% (FeCo)
External specific surface	320 m^2/g	> 300 m^2/g
Number of layers	7-9	1-2

TEM images of nanotubes	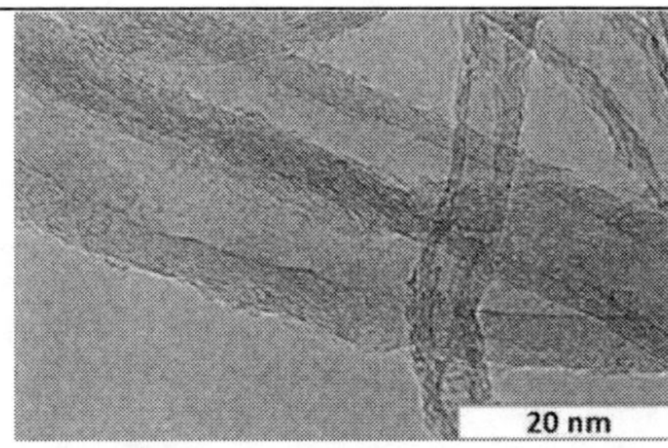	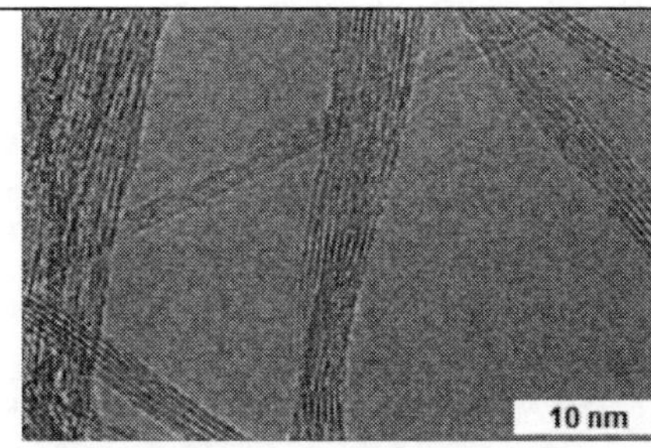

2.2. Obtaining experimental samples

For the manufacture of the samples the following scheme was used. At the first stage, the initial components were selected (epoxy resin with hardener, MWCNT and SWCNT). We carefully weighed the filler and the binder on an electronic scale (accuracy ~ 1 mg). After that, the components of the composite were combined in appropriate proportions (by weight) and thoroughly mixed until homogeneous state (using ultrasonic dispersers for 5 minutes at a power of 80 watts). Composite mixtures were obtained with appropriate concentrations of fillers by weight. The resulting mixture was poured into a coaxial form of fluoroplast (this material has a low adhesion to epoxy resin). Polymerization of samples occurred at room temperature for several hours. Next, the polymerized samples were processed to their full compliance with the dimensions of the selected measuring cell. This processing was performed by grinders devices. The samples are gradually ground off surplus from all sides, and also plane-parallel faces are formed. The fabricated samples are washers with the following dimensions: outer diameter d_{out} = 7 mm, internal diameter d_{in} = 3 mm, thickness h = 1.6 mm.

Thus, a series of experimental samples of composites for measurements were made. Their main characteristics are presented in table 2.

Table 2. Manufactured samples

Sample No	Content of epoxy resin, wt.%	Content of the filler, wt.%	
		MWCNTT	SWCNT
1	98	2	–
2	97	3	–
3	96	4	–
4	98	-	2
5	97	-	3
6	96	-	4

Also, using additive technology (printing on a 3D printer), a sample of the shape of a conductive plastic washer was made. Its size is fully consistent with the previously obtained samples. As a conductive plastic ABS plastic brand "Conductive" was used. It contains up to 20 wt.% carbon nanofibres and has a specific electrical resistance $\rho = (40 \div 80)\ 10^3$ Om / cm.

Thus, 7 experimental samples of composites for measurements were made.

2.3. Measuring equipment

Microwave measurements of electromagnetic characteristics were carried out in coaxial cells on a vector analyzer of P4M-18 circuits manufactured by "Mikran". The "to pass" measurement scheme was used when the experimental sample was installed in the center of the cell, and the elements of the dispersion matrix were measured at frequencies from 2 to 18 GHz. This method characterizes the interaction of electromagnetic radiation with a sample located in free space. A block diagram of the measurement setup is shown in Figure 1.

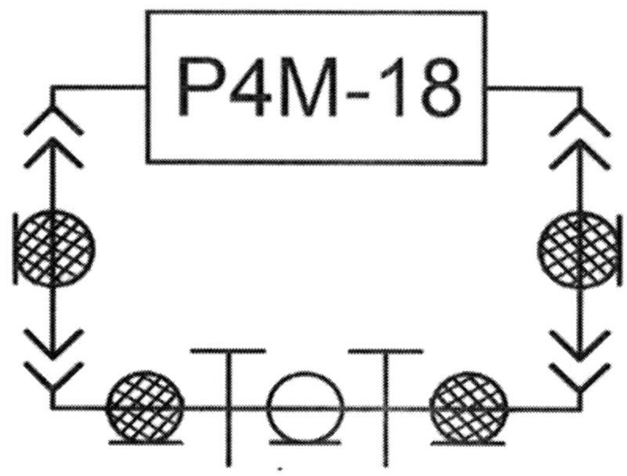

Figure 1. Block diagram of measurements "to pass"

The "to pass" scheme is used to measure *S*-parameters (*S*11, *S*12, *S*21 and *S*22). They correspond to the radiation power transmitted through the sample (*T*) and reflected (*R*) from the front face of the sample during the passage of an electromagnetic wave in the forward and backward directions. After that, according to the well-known formula, you can find the absorption coefficient: $A=1-T-R$. On the basis of the S-parameters obtained, taking into account the change in their phase, it is possible to calculate the spectra of the complex permeability and permittivity. For this purpose, the Becker-Jarvis technique [11] was used in this work. In the Mathcad programming environment, the corresponding software module was implemented.

The obtained values allow us to calculate the electromagnetic response from the materials at different thicknesses in further research. Two new modules have been created in the Mathcad programming environment. They allow to calculate the electromagnetic response from the material in free space (*T*, *R*, *A*) and on the metal (*RL*). To do this, they need a file with the data of values of electromagnetic parameters, calculated earlier. All calculation formulas are obtained in the approximation of plane waves.

In order to reduce the calculation error, an additional study of the effect of metallic impurities in CNT on the magnetic permeability of composites was conducted. This effect is most pronounced in the low-frequency region of microwave radiation. Therefore, a measurement setup based on the Keysight E4991B impedance analyzer was used. It has a test signal frequency range from 1 MHz to 3 GHz.

3. Results and discussion

The following are the results obtained for composites with MWCNTs (Figure 2) and for composites with SWCNTs (Figure 3). In addition, the results of the sample printed from carbon-containing ABS plastic "Conductive" were added to the graphs.

From the graph in Figure 2 it can be seen that with a change in the concentration of MCNTs, the dielectric constant increases nonlinearly. This is especially noticeable at lower frequencies. With increasing concentration of multi-walled carbon nanotubes, a significant increase in values of real (from 5 to 7.5 relative units) and imaginary parts of the permittivity (from 0.4 to 2 relative units) is observed. However, at frequencies above 5 GHz, a noticeable decrease in the real part of the complex permittivity can be noted. This type of behavior is associated with both the polarization characteristics of carbon nanotubes and the significant contribution of the conductive properties of the composite to the value of the complex permittivity. With increasing frequency, the contribution of conductivity to ε decreases, which we see in the graphs (decrease in permittivity with increasing frequency). The

IOP Conf. Series: Materials Science and Engineering **525** (2019) 012030 doi:10.1088/1757-899X/525/1/012030

carbon-based ABS plastic "Conductive" has higher complex permittivity values than a composite with 4 wt.% MWCNT.

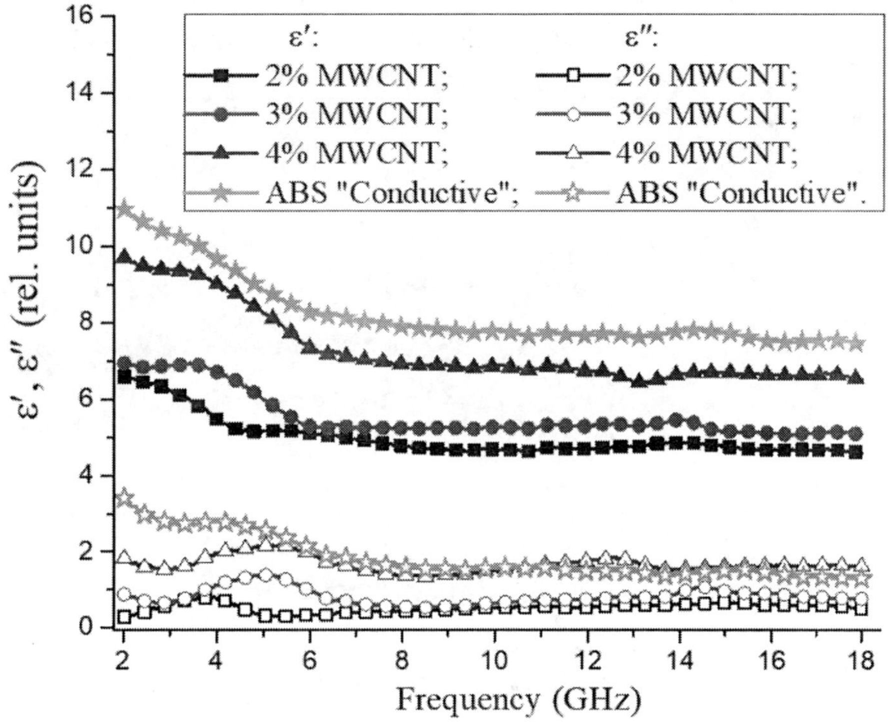

Figure 2. Frequency dependence of the complex permittivity of composites with MWCNT and ABS "Conductive" plastic

For composites with single-walled carbon nanotubes, the values of complex permittivity are several times greater than those for composites with MWCNT. When the concentration of the filler is 2 wt.% ε' varies within 22 ÷ 7 rel.units for the SWCNT and 7 ÷ 4 rel.units for MWCNT. This is probably due to high conductivity values of the material with a single-walled carbon nanotube, since only a few first layers of carbon nanotubes make a contribution to the electrical properties (i.e., for the MWCNTs, most of the layers make no contribution to the electrical properties and permittivity). A composite containing 4 wt.% of a single-walled carbon nanotube has already proved to be partially conductive. Resonance phenomena are observed in the frequency dependence of the permittivity in the range from 4.5 GHz to 7.5 GHz, which is probably due to the conductive properties of the composite. The material containing 4% by weight of single-walled carbon nanotubes has two times higher values of complex dielectric constant than carbon-containing ABS plastic.

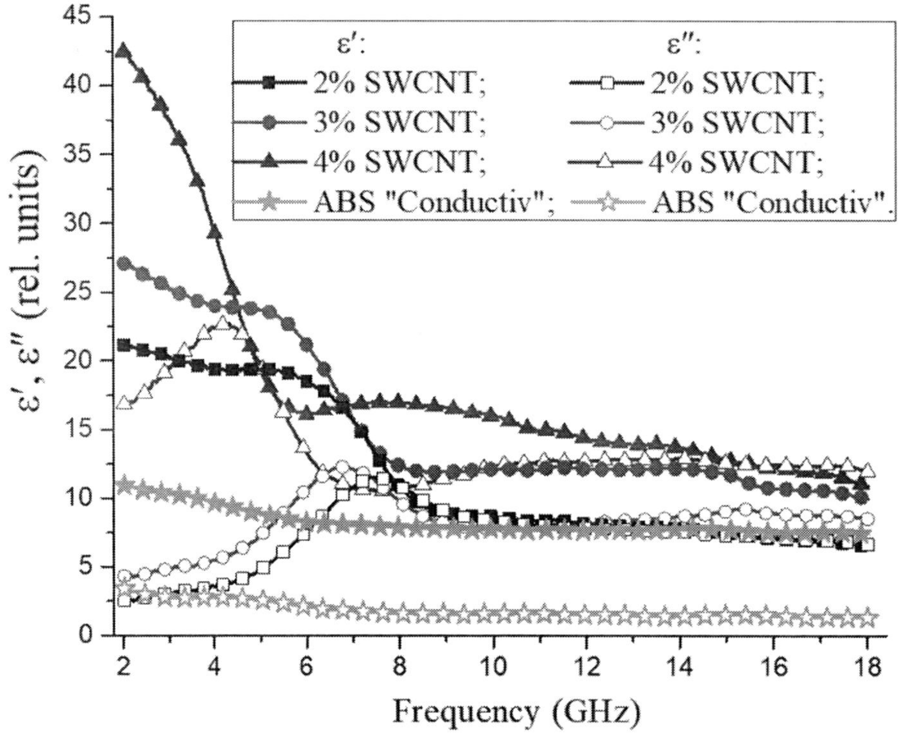

Figure 3. Frequency dependence of the complex permittivity of composites with single-walled carbon nanotubes and ABS "Conductive" plastic

To ensure that the features on the resulting frequency dependencies are not related to the presence of magnetic properties (due to the impurities of metals in carbon nanotubes), we conducted a study of complex magnetic permeability. It can be noted that composites based on carbon nanotubes are practically non-magnetic ($\mu' = 1$, and $\mu'' = 0.01 \div 0.05$).

On the basis of the obtained data about the complex dielectric constant, and in the absence of actual magnetic properties, the electromagnetic response was simulated from a layer of material located in the free space (*T*, *R*, *A*) and on the metal (*RL*). Distribution plots were obtained for the corresponding coefficients depending on the thickness and frequency of the layer.

The best shielding and absorbing properties are possessed by composite containing 4 wt.% SWCNT. Figure 4 shows the simulation results of the electromagnetic response from a layer of material in free space. This material has a good shielding ability. With increasing frequency and thickness of the material, the transmission coefficient *(T)* decreases. With a layer thickness of more than 2.5 mm at frequencies greater than 8 GHz, less than 10% of the radiation penetrates through the composite. In this case, more than 50% of the radiation is absorbed, and the rest is reflected. With increasing thickness and frequency, more and more radiation is shielded and at the maximum considered parameters less than 1% of radiation passes. Reflection is maximal (from 70 to 80%) at frequencies from 4 to 6 GHz with a material thickness of more than 3 mm. Composites based on 4 wt.% MWCNTs and ABS «Conductive» behave similarly. However, they interact with radiation two times weaker. At the same time, in a wide frequency band (from 4 to 18 GHz), they reflect more than 50% of the radiation. Only at very high frequencies with large thickness, they exhibit absorbing properties. Composites based on single-walled carbon nanotubes are several times better at shielding radiation (*T* ranges from 5% to 20% at frequencies higher than 5 GHz) than composites based on MWCNT (*T* ranges from 40% to 90% at frequencies above 5 GHz).

IOP Conf. Series: Materials Science and Engineering **525** (2019) 012030 doi:10.1088/1757-899X/525/1/012030

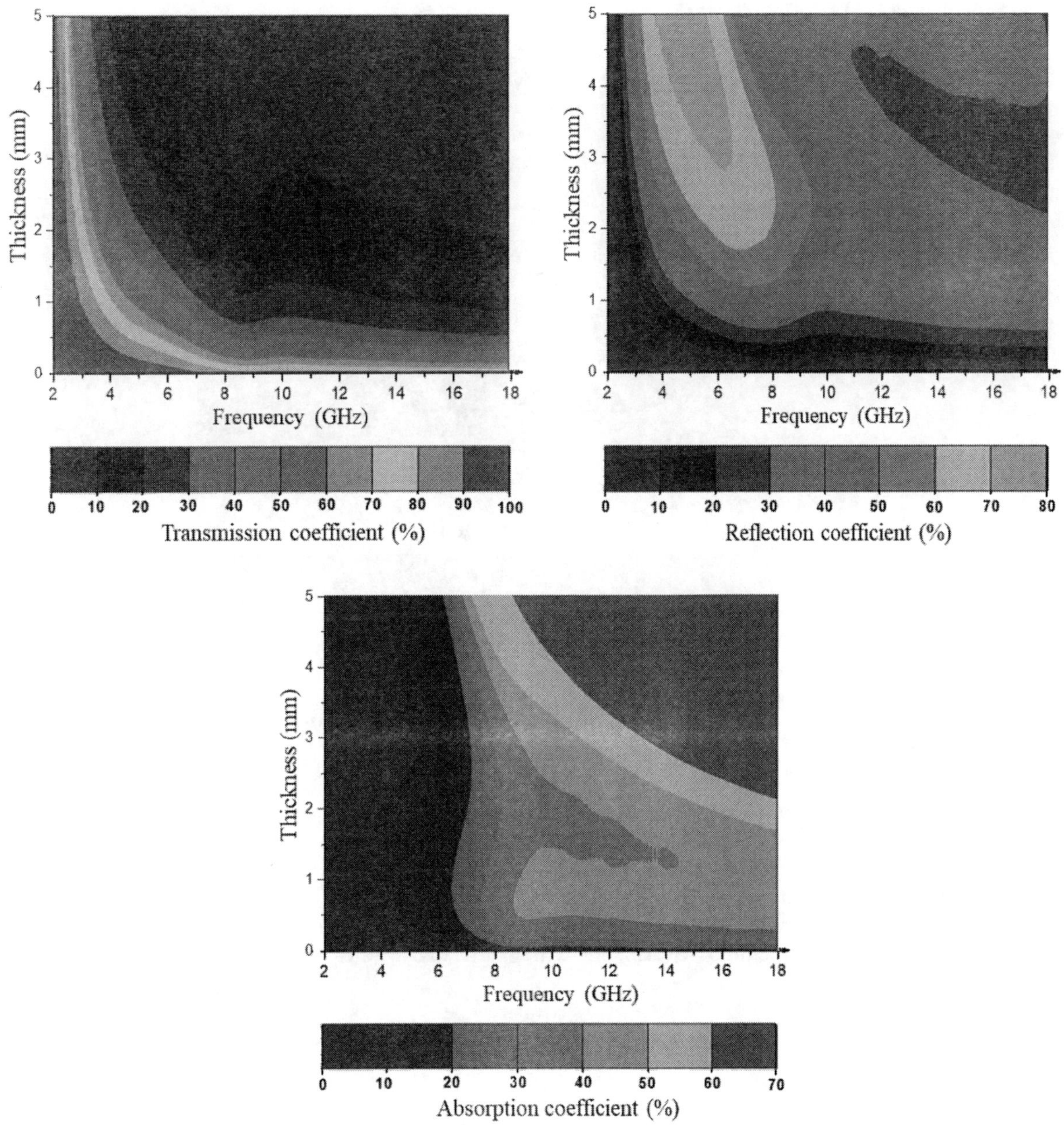

Figure 4. Dependence of electromagnetic response (*T*, *R*, *A*) from frequency and layer thickness for a composite material containing 4 wt.% SWCNTs in free space

Next, we simulated the electromagnetic response from the composite layer on the metal. Surface plots with level lines for reflection losses (*RL*) were obtained. Figure 8 shows the distribution of reflection loss from the frequency and layer thickness for a composite material containing 4 wt.% of a single-walled carbon nanotube in metal. It can be seen from the graph that the coating of this composite is a very narrowband absorber. The maximum attenuation of the radiation can reach -35 dB (more than 3160 times) at a frequency of 5 GHz with a thickness of 3.0 mm to 3.2 mm. In general, the level of attenuation of radiation is much less. In this case, the optimum thickness is 2-3 mm. With this thickness at frequencies above 6 GHz, the reflected power will be reduced by 3-10 times.

Other composites samples attenuate the level of the reflected radiation much weaker.

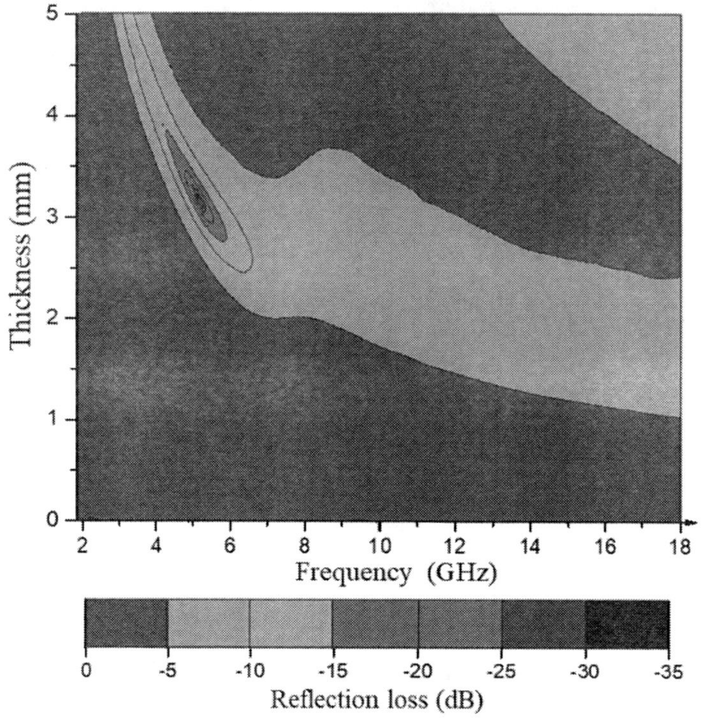

Figure 5. The dependence of the reflection loss on the frequency of radiation and layer thickness for a composite material containing 4 wt.% SWCNT

4. Conclusions

Electromagnetic response from materials with SWCNTs, MWCNTs and ABS «Conductive» has been investigated. It is shown that a composite material containing 4% by weight of single-walled carbon nanotubes is an effective narrowband absorber at a frequency of 5 GHz with a thickness of 3.0 mm. Or it can be used as a shielding high-frequency radiation material at frequencies from 7 GHz with a thickness of more than 2.5 mm. Investigated MWCNT and "Conductive" ABS plastic are more suitable for creating screening coatings, since they have high values of reflection coefficients.

Acknowledgment

The study was sponsored by RFBR in the framework of a research project №18-32-00810.

References

[1] Petzold J 2002 Advantages of softmagnetic nanocrystalline materials for modern electronic applications *J. Magn. Magn. Mat.* **242-245** 84–9

[2] Zhuravlev V, Minin R, Itin V, Lopushnyak Y, Zhuravlev A, Lilenko E 2017 Study of the magnetic anisotropy of the multiphase samples of the ferrimagnets with hexagonal crystal structure by the method of ferromagnetic resonance *IOP Conf. Series: Materials Science and Engineering* **168** 012080

[3] Tran H, Le T, Pejcinovic B, Brown J, Doneker R, Thompson K 2018 Characterization of Novel Magnetically Loaded Flocked Carbon Fiber Microwave Absorber *IEEE Symposium on Electromagnetic Compatibility, Signal Integrity and Power Integrity (EMC, SI & PI)* 41–6

[4] Wangchang L 2017 Enhanced and broadband microwave absorption of flake-shaped Fe and FeNi composite with Ba ferrites *J. Magn. Magn. Mat.* **426** 504–9

[5] Peng Q, Dearden A, Crean J, et al. 2014 New materials graphyne, graphdiyne, graphone, and graphane: review of properties, synthesis, and application in nanotechnology *Nanotechnology, Science and Applications* **7** 1–29.

[6] Liu L, Kong L, Yin W, Matitsine S 2011 Characterization of single- and multiwalled carbon nanotube composites for electromagnetic shielding and tunable applications *IEEE Transactions on Electromagnetic Compatibility* **53** 943–9

[7] Kuleshov G, Zhuravlyova Y, Dotsenko O 2015 Electromagnetic response from composite radiomaterials based on multiwall carbon nanotubes at microwave frequencies *SIBCON 2015 Proc.* 7147115

[8] Magisetty R, Shukla A, Kandasubramanian B 2019 Terpolymer (ABS) cermet (Ni-NiFe2O4) hybrid nanocomposite engineered 3D-carbon fabric mat as a X-band electromagnetic interference shielding material *Mater. Lett.* **238** 214–7

[9] Mishra S, Katti P, Kumar S, Bose S 2019 Macroporous epoxy-carbon fiber structures with a sacrificial 3D printed polymeric mesh suppresses electromagnetic radiation *Chemical Engineering Journal,* **357** 384–94

[10] Mazov I, et al. 2010 Electrophysical and Electromagnetic Properties of Pure MWNTs and MWNT / PMMA Composite Materials Depending on Their Structure *Fullerenes Nanotubes and Carbon Nanostructures* **18** 505–15

[11] Chalapat K, Sarvala K, Li J, Paraoanu G 2009 Wideband Reference-Plane Invariant Method for Measuring Electromagnetic Parameters of Materials *Microwave Theory and Techniques, IEEE Transactions on* **57** 2257–67

IOP Conf. Series: Materials Science and Engineering **525** (2019) 012031 doi:10.1088/1757-899X/525/1/012031

Outstanding problems of numerical simulation of the process of injection molding of PIM-feedstocks and component quality

A A Kutsbakh, A N Muranov, A B Semenov and B I Semenov

Laboratory of New Casting Methods, Bauman Moscow State Technical University, 2nd Baumanskaya st., 5, Moscow, 105005 Russia

E-mail: akutsbakh@mail.ru, semenovbi@bmstu.ru

Abstract. Based on the experience of foreign colleagues studied in publications and own exploratory work, the authors offer their own vision of the current knowledge and problems of numerical simulation of the injection molding process of feedstoks in PIM-technology. The results of the simulation of the injection molding of challenging component are presented in the Autodesk Moldflow software package.

1. Introduction

Three branches of modern metallurgy, Powder Injection Molding (PIM, MIM), Thixoforming and Thixomolding, united into the established metallurgy of thixotropic metals, are attributed to both new special casting methods and new powder metallurgy (PM) methods. During molding, the materials used, which have previously been given thixotropic properties, are molded in a two-phase (solid-liquid) state, which is characterized by the non-Newtonian behavior of the slurry flow. The flow behavior is influenced by numerous factors of the process; in particular, the behavior of the flow depends on the loading history and time. According to the data obtained from the literature, it is easy to establish that theoretical models for such deformable solid-liquid materials are still in the development stage, and the used constitutive models are built on the basis of experiments. Combining the advantages of injection molding of plastics and the strengths of PM techniques, PIM technology makes it possible to obtain, almost without waste, exact products of complex geometric configuration from metal and ceramic powders, possessing the required set of properties and optimal cost in serial production. The versatility of the technological methods and techniques used to work with powder fillers of different nature stimulated the rapid growth in the use and development of the scientific bases of PIM technology in industrialized world, and attracts more and more attention of Russian engineers in recent years [1–4].

In PIM, only one among the main ones but, according to experts, the determining stage of the technological process is the stage of injection molding of a polymer melt filled with powder – when the slurry is spontaneously structured during the molding process. A portion of the slurry, prepared according to certain criteria of suitability for fluidity and molding under conditions of intensive heat exchange, is injected into the metal mold. For the successful implementation of such a method of forming a shaped casting from powder, a proper assessment of the properties of a solid-liquid material during deformation is necessary. The obtained semi-finished piece at this stage is called the "green" part; it is its quality that largely determines the quality of the final part. In a high-pressure casting machine, specially designed for the injection of slurry, the feedstock, a mixture of a multicomponent polymer bond and powder filler [1], strictly controlled by composition, is heated in the barrel to the melting temperature of the binder and is brought by the screw to a condition of a viscous suspension with evenly distributed filler in it, and capable under the influence of excessive pressure of smoothly moving through all the cavities of the mold.

The metal tooling used for the manufacture of parts is an expensive mold with the shaping cavities of varying degrees of complexity, run channels, supplemented by various heating and cooling systems for the elements of the mold, a casting ejecting system, etc. Under these conditions, it is extremely important to choose the geometrical configuration of the casting cavity, consistent with the drawing of the part, and the casting conditions corresponding to the design of the mold, ensuring the production of defect-free castings. Compressibility and complex rheological behavior of the cast material with non-Newtonian properties under conditions of intensive heat exchange and solidification in channels of variable cross-section greatly complicate the design of tooling.

PIM manufacturers often have to find the right design and engineering solutions (DES) only by hit-and-miss method, as a result of which, when the mold is put into operation, they experience considerable difficulties in manufacturing defect-free castings [2–4]. The quest for the most efficient and rational DES, which allows to obtain defect-free final parts, the quality of which, as established, is 75–80% dependent on the quality of the shaped components, necessitates the use of numerical simulation of the injection molding process at the mold design stage.

2. Theoretical background

Since one of the phases of the polymer-powder mixture in the casting process is in the liquid state, and during injection, it is the slurry flow (feedstock melt) that is considered, the modeling of the injection molding process is the subject of hydrodynamic study of multiphase, spontaneously structured, and therefore thixotropic at the stage casting, media.

The main tasks of hydrodynamics as the science that deals with the motion of fluids is to determine the fields of distribution of velocities, pressure and temperature of the "fluid" in the studied space at any time. To determine these parameters, the following equations are traditionally used: the continuity equation expressing the law of conservation of mass of a fluid; the Navier-Stokes equation, expressing the law of conservation of the amount of the motion of fluid; the energy equation expressing the law of conservation of energy of a fluid; the equation of state (the constitutive model of the processed material), or as it is also called, the "closing" equation connecting the temperature, pressure, and density of the material being molded (PVT behavior).

The feedstock melt is a suspension, one of the phases of which is a dispersed filler in the form of solid powder particles, and the other is a liquid multicomponent polymeric binder. Based on this, there are 3 main approaches today to modeling the process of casting PIM-feedstocks [5]:

1) The feedstock melt is considered as a single-phase medium possessing effective macro properties of the suspension: thermal and physical properties, viscosity, equation of state, etc.

2) The particulate filler is considered as a pseudo fluid, and the flow of a two-phase medium, in which both phases are liquid, is simulated.

3) The simulation of a multiphase medium in which both the powder filler and the components of the polymeric binder are represented by fluids.

The second and third approaches, converting the suspension into an emulsion, require a larger amount of input data and the use of more complex mathematical models in comparison with the first method. Considering the flow of multiphase fluids, including suspensions, the mathematical description of the process uses the conservation equations for each phase, as well as adding new variables and equations describing the interaction of the phases with each other and characterizing the content of each component in the mixture, for example, concentrations or volume fractions of components.

The works [6–8], performed by modeling a multiphase medium, are focused on predicting the redistribution (liquation) of the feedstock components in the "green" part. The local excess of one component of the compound and, accordingly, the lack of another not only make it difficult to predict the change in the viscosity of the powder slurry during the injection molding process, but also lead to the formation of pores or uneven shrinkage (warpage) of the part at the stages of debinding and sintering, and local decrease in physical and mechanical properties of the product.

According to some foreign experts [5], the simulation of multiphase flow is consuming computing resources and time and, therefore, in practice it is not advisable for everyday use in the design of molds. However, in the matter of compound components redistribution evaluation during the casting process, the following solution is proposed: adding additional models to the existing mathematical models of single phase flow motion, which allow predicting the distribution of the components of a polymer powder mixture based on an assessment of the results of a single-phase flow simulation. Such models include the diffusion model [5], [V. Bilovol (2003), Ph.D. thesis], which is used in the software package SIGMASOFT®; "suspension balance model" (SBM) [9], which is used in the Moldflow software [Autodesk knowledge network. Suspension balance model from www.help.autodesk.com].

A team of authors [10] presented one of the basic works on simulation of the process of injection molding powder slurries. In this work, the feedstock melt is considered as a single-phase medium, endowed with effective suspension properties. The main emphasis is on the description of the rheological behavior of the feedstock. In the mathematical models used, the authors consider the presence of a yield strength and the presence of a so-called "slip" layer ("slip phenomena"), a characteristic phenomenon during the flow of highly filled materials, in which a thin layer of pure bond appears, which is formed under the action of shifts at the contact interface of the melt and walls of the mold. The 2.5D modeling approach for thin-walled PIM products used by the authors of [10] simplifies the solution of the proposed mathematical model with Hele-Shaw approximation, but at the same time limits the use of the developed PIMSolver® software for products that are not thin-walled and have bulky elements.

For the above works and a number of other publications in the field of simulation of the injection molding of polymer-powder compositions, attention is paid to the rheological behavior of the suspension. Most foreign colleagues [10–12] tend to use the Cross-WLF model, supplemented by the Herschel-Bulkley (HB) model and its various modifications. The use of the Cross-WLF+HB allows describing reversible transitions in the behavior of the thixotropic material in low shear rates most adequately, which leads to an increase in the accuracy of calculations. In the description of the rheological behavior of the dispersed systems under consideration, it is necessary to take into account the effect on viscosity not only of thermodynamic criterion and shear rate, but also of the time. In concentrated suspensions under the action of stationary shear, the effect of thixotropy appears, which is associated with structural changes in the dispersed system, leading to a change in viscosity during relaxation with a change in shear pull rate [1–3].

Components of the polymer bond experience a significant thermal expansion upon heating and shrinkage during cooling. For the description of PVT data of the feedstock, 2-domain Tait PVT-model is most often used [4, 10, 11, V. Bilovol (2003, Ph.D. thesis), K. Kate (2012, master thesis), V. Raymond (2012, master thesis)]. The change in the specific volume of the feedstock must be considered when selecting the parameters for the filling and the packing stages of the injection molding and when designing the mold.

An important issue for numerical simulation is to provide the mathematical models used with correct data on the properties of the polymer-powder compound. There is a significant difference between the values of the parameters in the process and in the research of the mixture with modern techniques. For example, the heating and cooling rates when studying with the DSC method and when determining PVT data, as well as changing the derivative of the shear rate when studying viscosity on the rheometers, can differ significantly from the real ones. The correctness of the obtained data on the properties of the feedstock [13–17] and, as a result, the correctness of the results of numerical simulation significantly depend on these parameters. At the moment, a clear boundary between the sufficient accuracy of the measured data on the properties of the polymer-powder compound and the acceptable accuracy of the computational experiment has not yet been established.

For the computational experiment, the value of the heat transfer coefficient between the melt and the mold, which will depend largely on the composition and properties of the compound and the surface quality of the mold cavities, is extremely important. The publications [18] and [V. Raymond, 2012, master thesis] thus give significantly different values of the heat transfer coefficient, which confirms the existence of this problem and the need to solve it.

3. Experimental procedure

In one of the challenging products, it was possible to successfully apply the obtained knowledge in the field of numerical simulation of the technological process under consideration [1]. The study object is presented in Figure 1. No problems with the liquation of the components of the compound were found in the study object; the method of simulation of a single-phase medium endowed with effective macro parameters of a polymer-powder compound was chosen as the approach used. The preparation and conduct of the computational experiment are described in more detail in publication [1].

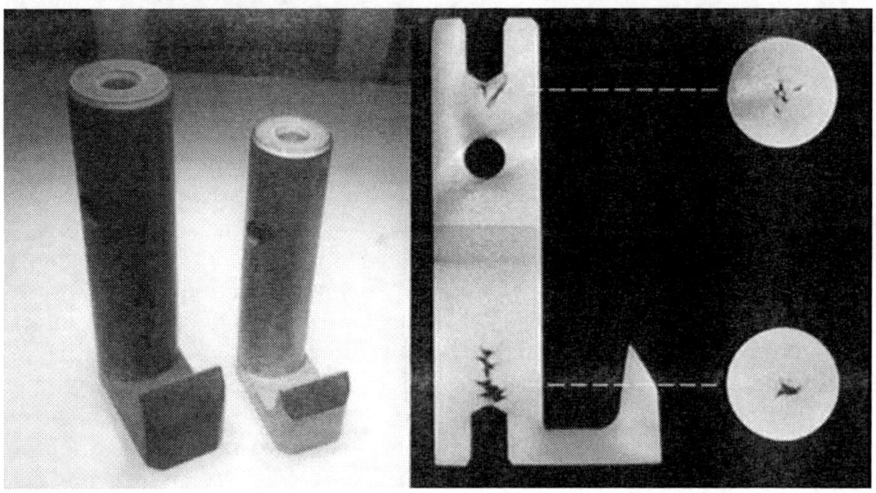

Figure 1. On the left: a general view of the casting and sintered part, respectively. On the right: images of defects of the "green" part, obtained on the computer tomography unit [1]

Using the method of differential scanning calorimetry (DSC) and the laser flash method (LFA), the effective thermal and physical parameters of the feedstock used for the manufacture of this part were investigated [19]. Based on the type of polymer binder used, on the morphological characteristics and particle size distribution of the powder filler, empirical coefficients of the most representative for this type of feedstock viscosity model (Cross-WLF) and coefficients of the 2-domain Tait model were selected from publications. Based on the feedstock processing parameters provided by the manufacturer, the injection molding process was simulated using the Moldflow software package. As Figure 2 shows, the suggested technique allowed obtaining results that rather well correlate with the observed defects in the real product. The selected area of maximum volumetric shrinkage coincides with the zone of defects in the cylindrical part of the component. The value of the maximum volumetric shrinkage of ~ 11% indicates a low efficiency of the compressing, which requires an adjustment of the parameters of the filling process at the packing stage. Since the compound has a massive part and is non-moldable for MIM, it may be necessary to change the geometry of the runner channels and the gate to prevent the run blocking before the end of the packing stage (or to use a hot runners mold).

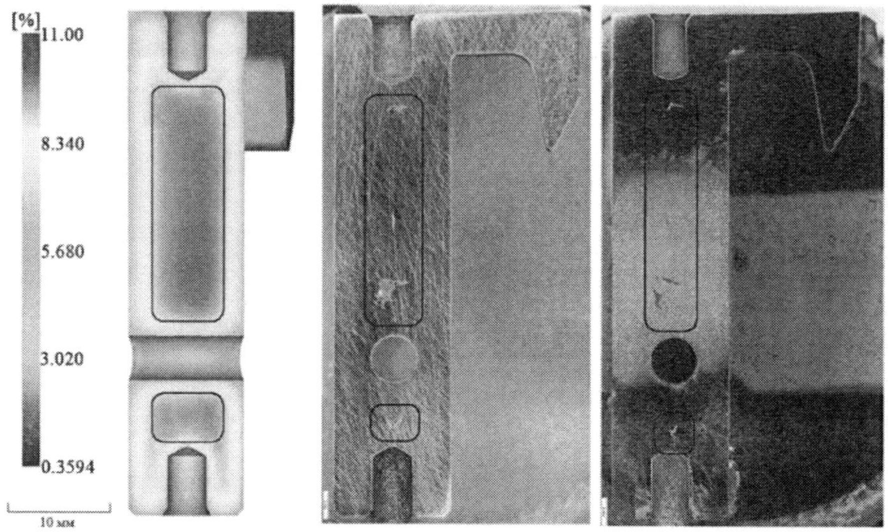

Figure 2. Comparison of the results of the calculation of volumetric shrinkage with the results of metallographic study of the part. The contour marks the zone of maximum volumetric shrinkage for the calculation results (on the left) and the defect location zones in the sintered part (on the right) [1]

4. Conclusion

Currently, the issue of numerical simulation of the injection molding of polymer-powder compounds in PIM technology is adequately treated, a lot of research has been done and software packages have developed for simulation of the technological process (SIGMASOFT®, PIM Solver®, modules in Autodesk Moldflow®, Moldex 3D®, etc.). The provision of the used mathematical models with correct data on the properties of the polymer-powder compound remains an important problem.

As shown above, even in the first approximation, using relatively simple mathematical models (approaches) with an incomplete amount of experimentally measured data on the properties of the feedstock, it is possible to simulate the process of injection molding of polymer-powder slurries for the purposes of analysis and evaluation of the DES taken at the stage of the mold design.

References

[1] Semenov A B, Kutsbakh A A, Muranov A N and Semenov B I 2019 Development of methodology for numerical simulation injection molding process in powder injection molding *Technologiya Metallov* **1** 9-16 DOI: 10.31044/1684-2499-2019-1-0-9-16

[2] Semenov A B, Gavrilenko A E and Semenov B I 2016 Next-generation casting technologies and their adaptation and development in Russia: I. At the beginning of a new technological paradigm. *Russian metallurgy (Metally)* **13** 1231-40 DOI: 10.1134/S0036029516130152

[3] Semenov A B, Muranov A N, Kutsbakh A A and Semenov B I 2017 Injection molding of structured multiphase materials. *RUDN J. of Engineering Researches* **18**(4) 407-25 DOI: 10.22363/2312-8143-2017-18-4-407-425

[4] Heaney D F 2012 Handbook of metal injection molding (Woodhead Publishing Limited) 601 p

[5] Thornagel M 2009 MIM-Simulation: A Virtual Study on Phase Separation *Euro PM2009 Proc. (Powder Injection Moulding - Quality & Simulation* vol 2*)* (Copenhagen: Bella Center) 6 p

[6] Barriere T, Gelin J C and Liu B 2002 Improving mould gesign and injection parameters in metal injection moulding by accurate 3D finite element simulation *J. of Materials Processing Technology* **125-126** 518-24 DOI: 10.1016/S0924-0136(02)00307-2

[7] Fang W, He X B, Zhang R J, You X M and Qu X H 2014 Effects of particle characteristics on homogeneity of green bodies in powder injection moulding *Powder Metallurgy* **57**(4) 274-82 DOI: 10.1179/1743290114Y.0000000101

[8] Samanta S K, Chattopadhyay H and Godkhindi M M 2011 Modelling the powder-binder separation in injection stage of PIM *Progress in Computational Fluid Dynamics* **11**(5) 292-304 DOI: 10.1504/PCFD.2011.042181
[9] Nott P R, Guazzelli E and Pouliquen O 2011 The suspension balance model revisited *Physics of Fluids* **23** 043304 DOI: 10.1063/1.3570921
[10] Ahn S, Chung S T, Atre S V, Park S J and German R M 2008 Integrated filling, packing and cooling CAE analysis of powder injection moulding parts *Powder Metallurgy* **51**(4) 318-26 DOI: 10.1179/174329008X284903
[11] Kukla C, Friesenbichler W, Duretek I and Thornagel M 2008 New Insights into Feedstock Behaviour and Injection Moulding Simulation for PIM *Euro PM2008 Proc. (Powder Injection Moulding* vol 2*)* (Mannheim) p. 287-392
[12] Thomas Y, Baril E, Ilinca F and Hetu J-F 2009 Development of titanium dental implant by MIM: experiments and simulation *Advanced in Powder Metallurgy & Particulate Materials* **4** 81-94
[13] Friesenbichler W, Duretek I, Rajganesh J and Kumar S R 2011 Measuring the pressure dependent viscosity at high shear rates using a new rheological injection mould *Polimery* **56**(1) 58-62
[14] Hayward A T J 1967 Compressibility equations for liquids: a comparative study *British J. Appl. Phys.* **18** 965-77
[15] Barvinskij I A and Barvinskaya I E 2014 Problemy` lit`ya pod davleniem izdelij iz polimerny`x materialov: uplotnenie *Polimerny`e materialy`*3-13
[16] Kowalski L, Duszcyk J and Katgerman L 1999 Thermal conductivity of metal powder-polymer feedstock for powder injection moulding *J. of materials science* **34** 1-5
[17] Kowalski L and Duszczyk J 1999 Specific heat of metal powder-polymer feedstock for powder injection molding *J MAT SCI L* **18**(17) 1417-20
[18] Thornagel M 2012 Injection moulding simulation: new developments offer rewards for the PIM industry *PIM International* **6**(1) 65-8
[19] Muranov A N, Semenov A B, Marakhovskiy P S, Chutskova E Yu and Semenov B I 2017 Teplofizicheskie svojstva polimerno-poroshkovoj smesi dlya izgotovleniya detalej iz stali 42CrMo4 metodom inzhekcionnogo lit'ya *Perspektivnye Materialy — Advanced Materials (in Russ)* 24-32 DOI: 10.30791/1028-978X-2018-9-24-32

Study of the interaction kinetics of apatite ore with nitric acid by independent analytical methods

I A Pochitalkina[1], P H Le[1], T T Vu[1] and I A Petropavlovskii[1]

[1] Department of Technology of Inorganic Substances and Electrochemical Processes, D I Mendeleev University of Chemical Technology of Russia, Moscow, 125480, Russia

E-mail: pochitalkina@list.ru

Abstract. With the help of ionometric and photometric methods of analysis, the kinetics of the interaction of a low-grade enrichment product of apatite ore with nitric acid was investigated. Presents the results of changes in the concentration of the ion H^+ and P_2O_5 in the reaction mixture during the reaction. The influence of physicochemical factors on the decomposition rate is shown.

1. Introduction

Apatite and phosphate ores are a source of phosphorus, which is extracted from mineral raw materials by acid extraction methods. The speed of interaction of components in a heterogeneous system depends on many factors: the chemical composition of the ore, hydrodynamic regime, acid concentration, particle dispersion, process temperature, time etc. The kinetics of acid decomposition of phosphate ores of various classes, deposits and genesis [1–4] is also determined by different energy values destruction of their crystal lattices. Therefore, the study of the kinetic regularities of certain types of phosphate ores with mineral acids is of great importance in the development of the scientific basis of the process of extraction of phosphoric acid from mineral raw materials and its implementation in industry. Identifying the factors that have the maximum effect on the kinetics of acid ore breakdown allows you to find effective technologies for its processing.

The relevance of work aimed at studying the kinetics of decomposition of low-quality phosphate raw materials, due to the resulting shortage of high-quality raw materials. This is due to the long-term exploitation of deposits, which leads to their depletion.

Objective: to study the effect of mixing, the initial concentration of nitric acid and temperature on the rate of extraction of phosphorus from poor apatite.

2. Experimental part

The object of the study was a representative sample of poor siliceous apatite ore from the Laokai deposit, containing: 21.23% P_2O_5, 28.08% CaO, 6.03% MgO, 0.98% Fe_2O_3, 1.38% Al_2O_3, 13% SiO_2, 1.5% F. Based on the X-ray diffraction pattern of the Laokai apatite ore, three main phases were identified [5]: phosphate substance with the structure of hydroxyfluoroapatite $Ca_5(PO_4)_3(OH)_{0.2}F_{0.8}$, admixture of carbonates with the structure of dolomite $CaMg(CO_3)_2$ and silicates – α-quartz SiO_2 [6]. We used the ion meter "Econix-Expert-001" (RF) and the spectrophotometer "SPECOL 1500/1" (Germany). The decomposition of apatite was performed with nitric acid with an excess of 40% of stoichiometry.

The decomposition of the phosphate substance and impurities present in the apatite of the Laokai deposit with nitric acid is described by the equations:

$$Ca_5(PO_4)_3F + 10HNO_3 = 3H_3PO_4 + 5Ca(NO_3)_2 + HF\uparrow$$

IOP Conf. Series: Materials Science and Engineering **525** (2019) 012032 doi:10.1088/1757-899X/525/1/012032

$$CaMg(CO_3)_2 + 2HNO_3 = (Ca, Mg)(NO_3)_2 + 2CO_2\uparrow + H_2O$$
$$SiO_2 + 4HF\uparrow = SiF_4 + 2H_2O$$

Experimental data were obtained by two independent methods. The process of interaction of nitric acid with apatite by the ionometric method was studied directly in the reactor. The analytical concentration of hydrogen ions was determined in 3 s steps. It was previously shown [7] that in the process of nitric acid decomposition of apatite, the potential of the glass electrode depends linearly on the logarithm of the calculated values of the concentrations of hydrogen ions. Consequently, the obtained concentration dependences can be used for the quantitative determination of H^+ ions in the studied range pH = 1–3. It should be noted that the decomposition process was carried out under conditions of constant ionic strength of the reaction mixture.

In parallel, in the same reactor using the traditional differential photocolorimetric method, the current concentration of extracted phosphorus (in terms of P_2O_5) was determined according to GOST [8] by sampling a volume of 1 ml. For braking the reaction, dilution of the sample with distilled water was used.

When studying the influence of physicochemical factors on the kinetics of decomposition of apatite, they varied: mixing speed in the range of 60–120 turn/min, HNO_3 concentration in the range of 0.05–1.0 M, and temperature in the range of 20–50°C.

Analysis of the kinetic dependences showed that the efficiency of decomposition of apatite increases with increasing mixing and reaches a maximum of 99.9% at a stirrer rotation speed of 120 turn/min (Fifure 1).

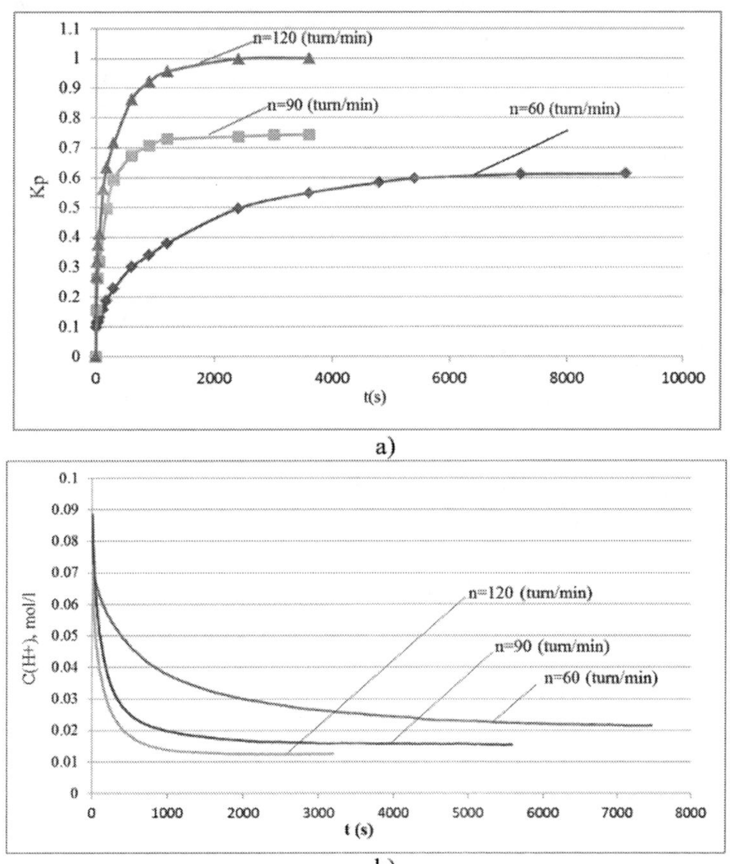

Figure 1. The effect of mixing on the concentration of H^+ in the resulting suspension (a) and the decomposition coefficient of Laokai apatite (b).

The completeness of decomposition of apatite was evaluated by the value of the decomposition coefficient (K_p). The K_p value of the photometric experiment was calculated as the ratio of the concentration of extracted phosphorus (P_2O_5) to the nitric acid solution to its concentration in apatite (Figure 1 a): $K_p = C(P_2O_5)water/C\ (P_2O_5)total$, and the ionometric experiment – using the ratio of current ion consumption hydrogen to its full consumption for the decomposition of apatite (Figure 1b).

An increase in the initial concentration of nitric acid contributes to an increase in the rate of the process; the maximum value of 0.04 (mol/l·s) is achieved when decomposition of apatite with nitric acid with a concentration of 1.0 M. In Figure 2, it can be seen that all the dependences presented have the same character.

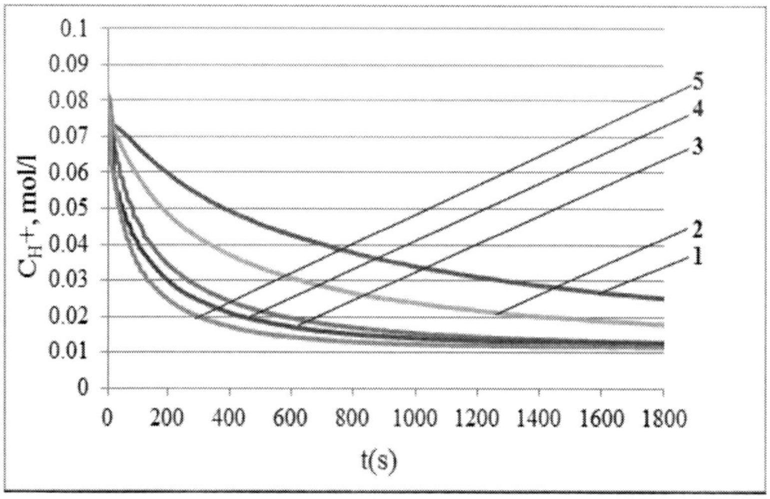

Figure 2. The change in the concentration of the H^+ ion with time at temperatures 20 (1), 30 (2), 40 (3), 45 (4) и 50°C (5); ([HNO_3] = 0.1M; υ=120 tur/min).

Using the graphical method, the order of reaction (n) is determined. It is shown that the approximation of the dependence of concentration on time in the coordinates $1/C-1/C_o = f\ (\tau)$ is expressed by a straight line, which corresponds to the second order of the reaction (Figure 3).

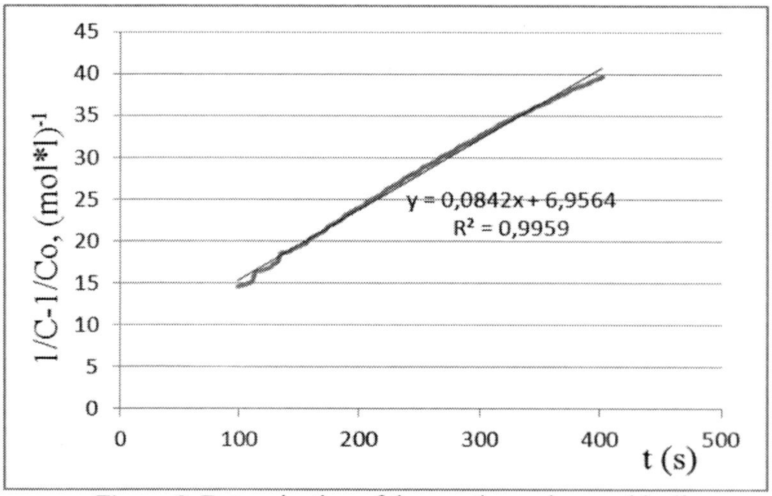

Figure 3. Determination of the reaction order graphically.

According to the Arrhenius equation, the apparent activation energy is determined: E_a = 48.75 kJ/mol (Figure 4).

IOP Conf. Series: Materials Science and Engineering **525** (2019) 012032 doi:10.1088/1757-899X/525/1/012032

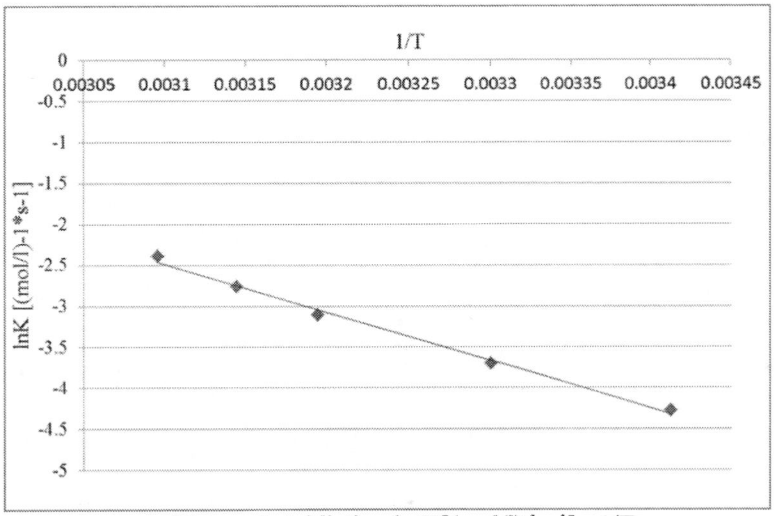

Figure 4. Addiction ln $K[(mol/l)^{-1}{\cdot}s^{-1}]$ - 1/T.

3. Results and discussion

As follows from the results of the kinetic experiment, with an increase in the concentration of nitric acid, there is a reduction in the half-transformation period calculated by the formula $\tau_{1/2} = (kC_0(CaCO_3))^{-1}$ and, as a result, the time to reach complete decomposition (Table 1).

Table 1. Determination of the half-transformation time by the kinetic equation (n = 2).

n	2				
T(K)	293	303	313	318	323
$lnK[(mol/l)^{-1}{\cdot}s^{-1}$	0.0139	0.0246	0.0449	0.0634	0.0912
$C_o(mol/l)\ [H^+]$	0.1	0.1	0.1	0.1	0.1
$t_{\frac{1}{2}} = \frac{1}{K \times C_o}$	719	407	223	158	110

Comparison of the results of ionometric and photometric analysis indicates their convergence. It is important to note that the half-transformation time obtained by intersecting the experimental kinetic dependences of the ionometric and photometric determinations – 150 s (Figure 5) is consistent with the calculated value $\tau_{1/2} = 158$ s, which indicates the correctness of the data.

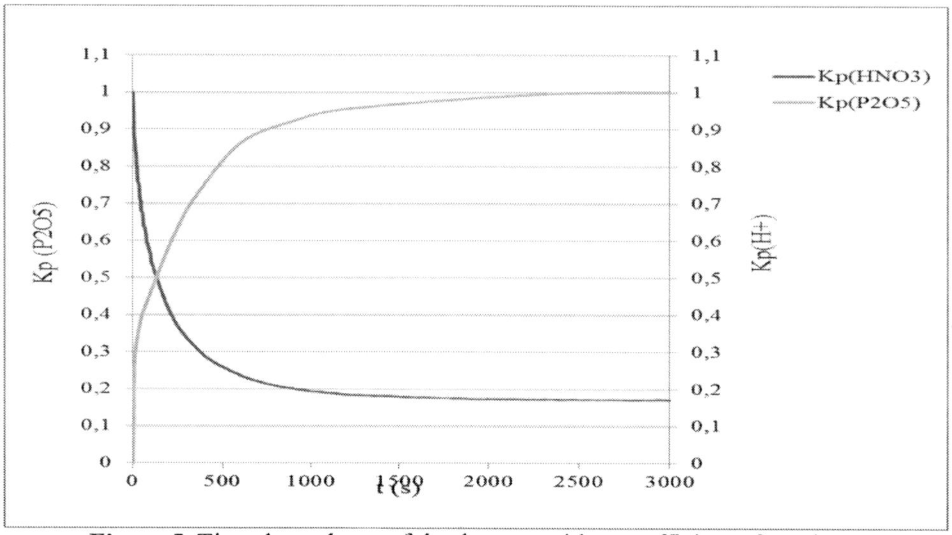

Figure 5. Time dependence of the decomposition coefficient of apatite.

At the initial stage of decomposition of apatite, the interaction of carbonate impurities contained in the composition of the ore with nitric acid is intensive. Over time, acid diffusion to phosphate particles becomes the limiting factor. The transition to the diffusion region is due to the formation on the phosphorite particles of a dense shell of a precipitate of SiO_2, which in the first stage was destroyed by the liberated CO_2, and hardly soluble compounds of iron and aluminum phosphates, which crystallize from a saturated solution. The low temperature coefficient (Table 2) of the decomposition reaction also indicates diffusion inhibition.

Table 2. Temperature coefficient of Van't-Goff.

n	2					
T (K)	293	303	303	313	313	323
lnK[(mol/l)$^{-1}$·s^{-1}	0.0139	0.0246	0.0246	0.0449	0.0449	0.0912
$k_2/k_1 = \gamma^{\Delta t/10}$	1.77		1.83		2.03	

4. Conclusion:

The optimal decomposition conditions in the temperature-concentration range studied are nitric acid concentrations of 1.0 M and a temperature of 50°C, with a decomposition coefficient of 99.7% achieved in 1800 s.

Acknowledgment

The authors thankful to the D I Mendeleev Center for Collective Use for assistance in conducting research.

References

[1] Lavrinenko A A, Shrader E A, Kharchikov A N and Kunilova I V 2013 *J. Mineral. Sci.* vol **49** p 811
[2] Seinazarov A, Namazov Sh and Beglov B J 2014 *Chem. Technol. Metall.* vol **49** p 383
[3] Abouzeid A Z M 2008 *J.of Mineral Proc.* vol 85 p 59
[4] Shaikh A M H and Dixit S G 1993 *Int. J. Mineral Proc.* vol **37** p 149
[5] Pochitalkina I A, Kondakov D F, Le P H and Vu T T 2018 *J. Inorg. Chem.* vol 63 **8** p 1046

[6] Pochitalkina I A, Le P H and Vu T T 2018 The fourth interdisciplinary scientific forum with international participation "New technologies and promising technologies" (M: Buki Vedi) vol 3 p 492

[7] Dobrydnev S V, Pochitalkina I A, Beskov V S and Bogach V V 2001 *J. of Applied Chem.*vol 74 **10** p 1579

[8] GOST 20851.2–75 *Mineral fertilizers. Methods for the determination of phosphates* (M: Standards Publishing)

IOP Conf. Series: Materials Science and Engineering **525** (2019) 012033 doi:10.1088/1757-899X/525/1/012033

Molecular modelling of fullerene C_{60} and its amino acid derivatives in aqueous medium

V B Krapivin [1,2] **and V B Luzhkov** [1,2*]

[1] Institute of Problems of Chemical Physics, Russian Academy of Sciences, Acad. Semenov av., 1, Chernogolovka, Moscow Region, 142432, Russian Federation
[2] Lomonosov Moscow State University, GSP-1, Moscow, 119991, Russian Federation

E-mail: vbl@icp.ac.ru

Abstract. The hydration free energy ΔG_s of fullerene C_{60} has been determined using the quantum mechanical methods and continuum solvent models, microscopic free energy molecular dynamics techniques, and compilation of known experimental data. Theoretical calculations using standard parameterization schemes of solvent models have predicted negative values of $\Delta G_s(C_{60})$ whereas application of the revised parameters, that take into account special features of carbon structures, shifts the computed $\Delta G_s(C_{60})$ to positive values. Reasonably accurate estimates of the hydration free energies of the neutral and anionic states of functionalized fullerenes were obtained from the generalized Born method. The described data have shown that the calculations of hydration thermodynamics of fullerenes apparently require separate parameterization for the carbon atoms met in these nanomolecules.

1. Introduction

Fullerenes are a widely studied carbon nanomolecules that find numerous chemical and biological applications as separate species and as functional fragments in complex molecular constructs [1−3]. The numerous practical applications of fullerenes determine significant interest in the rational understanding and theoretical prediction of the properties of these molecules in condensed phase and particularly in the aqueous biomolecular medium. The carbon atoms in fullerenes are interconnected in the networks of fused rings and form closed strained spheroid cages lined with the π-electrons of quasi-aromatic centers. Such electron structure and large molecular volume determine specific properties of these molecules in solution. Pristine fullerene have good solubility in the aromatic hydrocarbon, much lower solubility in the aliphatic hydrocarbons and polar organic solvents, and zero solubility in water under usual conditions [4−6]. However, the functionalized derivatives of fullerenes can be soluble in water [3]. In view of the active studies of the biologically active compounds containing pharmacophore of fullerene C_{60}, it is important to have good understanding of the mechanisms of fullerene–water interactions.

Modern theoretical chemistry methods can provide useful and accurate information about different physico-chemical characteristics of molecular systems. The computational techniques can especially be helpful in the studies of molecular properties that are difficult to measure experimentally. Exactly this situation exists for fullerenes where experimental data on hydration thermodynamics are insufficient or not available. The computational modelling of the molecules in solution generally involve either quantum mechanical (QM) calculations with the polarizable continuum model (PCM) of solvent or molecular dynamics (MD) force field (FF) free-energy simulations of microscopic systems.

IOP Conf. Series: Materials Science and Engineering **525** (2019) 012033 doi:10.1088/1757-899X/525/1/012033

In this work we consider the calculations of the thermodynamic potentials of fullerene C_{60} in water using the quantum mechanical PCM and free energy MD FF computational approaches. The influence of the special separate parameterization for the fullerene nanomolecules on the results of the modelling is reviewed. In addition, the hydration free energies of selected pentasubstituted amino acid derivatives of fullerene in neutral and multi-charged states are examined using the density functional theory (DFT) calculations with the polarizable continuum and generalized Born models of solvent.

2. Free-energy molecular dynamics simulations of fullerene C_{60} in water

The MD simulations of hydrated fullerene were conducted using the microscopic models with the minimal image periodic boundary conditions (PBC) and spherical boundary conditions (SBC) (Figure1). The potential energies of the system were computed using the OPLS force field with the OPLS-AA [7] and Bojan–Steele (B-S) [8] parameters for carbons of C_{60} and the TIP3P FF parameters for explicit water molecules. The OPLS-AA FF parameters are focused on reproduction of the properties of liquids while the B-S carbon–water FF parameters are based on the data for graphite–oxygen interactions. The fullerene atoms were not restrained in the PBC MD simulations whereas the treatment of solute–solvent system in the SBC simulations involved using the artificial spherical cavity for fullerene. The harmonic restraining potential applied to water molecules in order to create empty spherical cavity was derived using the radial distribution function of water molecules around fullerene [9]. The more detailed description of the FF for fullerene and simulation setup can be found in [9,10].

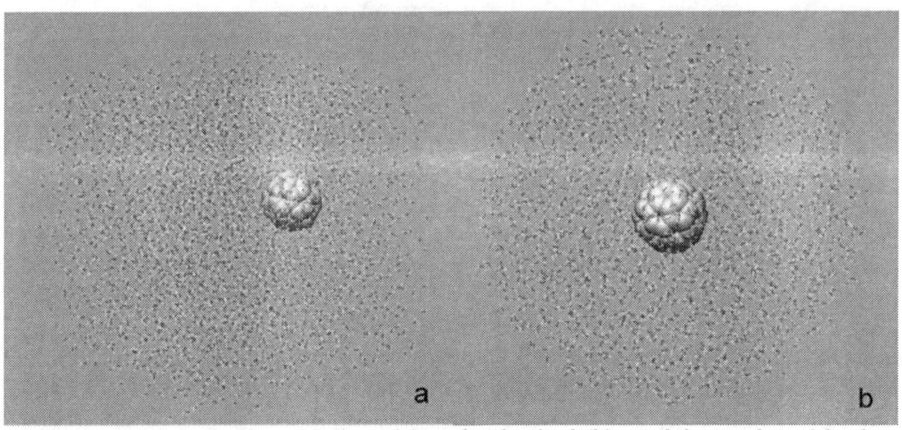

Figure 1. The periodic cubic box (a) and spherical (b) models employed in the MD simulations.

The hydration free energy of fullerene $\Delta G_s(C_{60})$ was determined from the alchemical transformation of fullerene into a dummy cage molecule $C_{60} \rightarrow d_{60}$ in water. The end-point d_{60} particle does not interact with the surrounding medium. The resulting free energy of mutation was calculated using the statistical mechanical free energy perturbation (FEP) method [11]

$$\Delta G_{\mathrm{mut,s}}\left(C_{60} \rightarrow d_{60}\right) = -RT \ln \left\langle e^{-\Delta U(\mathbf{r})/RT} \right\rangle_{A}. \quad (1)$$

Here, $\Delta U(\mathbf{r})$ is the perturbation potential, $\mathbf{r}$ denote the vector of atom coordinates, T is the absolute temperature, R is the gas constant , $\langle ... \rangle_A$ is the ensemble average over configurational space of the reference state A. Only parameters for the non-bonded interactions in the solute topology were changed in the perturbed state. In case of the spherical model, the calculation of $\Delta G_s(C_{60})$ involved a correction ΔG_{cav} for the free energy of creation of the spherical cavity for C_{60}

$$\Delta G_{s}\left(C_{60}\right) = \Delta G_{cav} - \Delta G_{\mathrm{mut,s}}\left(C_{60} \rightarrow d_{60}\right). \quad (2)$$

The cavitation free energy ΔG_{cav} of C_{60} was calculated using scaled particle theory [12]. For calculation of $\Delta G_s(C_{60})$ we used the cavity radius R_{cav}=5.414 Å which renders ΔG_{cav} =144.6 kJ mol^{-1}.

IOP Conf. Series: Materials Science and Engineering **525** (2019) 012033 doi:10.1088/1757-899X/525/1/012033

R_{cav} was determined as the sum of the distance of fullerene carbons to the center of C_{60} and the calibrated atom radius of atoms C [10].

The FEP/MD calculations using the OPLS-AA and B-S FF parameters for carbons predict sufficiently differing values of $\Delta G_s(C_{60})$ [10]. The $\Delta G_s(C_{60})$ free energies calculated using OPLS-AA have negative values of –50.2±2.6 kJ mol^{-1} for the SBC and –50.3±1.1 kJ mol^{-1} – for the PBC model. In case of B-S FF, the computed free energies $\Delta G_s(C_{60})$ are positive and equal 33.6±1.7 and 20.1±0.7 kJ mol^{-1} for the for SBC and PBC models, respectively. We can see in this connection that the 7.5-ns PBC multi-step FEP and 3-ns SBC single-step FEP calculations yield quite coherent results. The calculated with the B-S FF free energies well agree with the data from other MD simulations [13,14]. The FEP/MD simulations have also shown that $\Delta G_s(C_{60})$ is determined by the balance between large negative enthalpy and entropy terms [10].

3. Quantum mechanical modelling of fullerene C_{60} and its derivatives in water

3.1. QM PCM calculations of the hydration free energy of fullerene

The quantum mechanical calculations of hydration free energies ΔG_s were carried out using the 'united atom' (UA) [15] and 'standard model density' [16] versions of PCM in Gaussian. The computed ΔG_s for the solute in water cavity include electrostatic ΔG_{el} and non-polar ΔG_{np} contributions

$$\Delta G_s(C_{60}) = \Delta G_{el}(C_{60}) + \Delta G_{np}(C_{60}) . \tag{3}$$

The electrostatic contribution ΔG_{el} was obtained from the quantum mechanical calculation using integral equation formalism with the solvent reaction field included in solute Hamiltonian. The non-polar term ΔG_{np} was determined using empirical schemes. In the PCM(UA) method, ΔG_{np} includes the solvent cavity ΔG_{cav}, electron dispersion ΔG_{disp} and repulsion ΔG_{rep} contributions [15]. In the PCM(SMD) method, calculation of ΔG_{np} is based on the solvent-accessible surface area and surface tension parameters [16]. The calculated ΔG_s depend on the parameters of PCM for the solute–solvent interactions and on the level of theory employed. The cavity polarizable dielectric is assembled from the overlapping spheres around the atomic centers of the solute. The atomic radii, that define the solvent cavity in PCM, are of crucial importance for obtaining accurate ΔG_s values. For instance, the computed $\Delta G_s(C_{60})$ changes from negative to positive values upon increase of the atomic radius R(C) (Fig. 2a). It should be mentioned in this respect that the original UA and SMD models do not contain separate parameters for the atom types met in these nanomolecules.

The rational approach to estimate atomic cavity radius of the fullerene carbons R(C) in PCM considered several possible ways to derive R(C) for fullerene C60 [17]. This involved quantitative description of pyramidal hybridization of carbons in fullerene C_{60} and derivation of atomic radii from the constant electron density surfaces (Figure 2b) around the solute. The computed isodensity surfaces qualitatively support using spherical cavity in the MD SBC simulations of C_{60} (Figure 2b). The atomic radii were also calibrated using the thermodynamic cycle of the pK_a changes for dihydrofullerene and saccharin in DMSO and in a toluene–DMSO mixture. Overall, all three applied criteria have shown that the R(C) value in C_{60} exceeds the standard value of 1.725 Å for the aromatic carbon in the UA model [17]. The derived R(C) values from the calibration for pK_a are 1.842 Å for the HF method and 1.864 Å for DFT B3LYP [16].

The calculated values of $\Delta G_s(C_{60})$ are displayed in Table 1. The DFT B3LYP PCM(UA) calculations predict negative ΔG_s of C_{60} of –3.7 kJ mol^{-1} for standard value of R(C) in the united atom model and positive free energy ΔG_s = 14.8 kJ mol^{-1} for the recalibrated for fullerene R(C). The PCM(SMD) method produces negative free energy ΔG_s = –31.0 kJ mol^{-1}. The largest contribution to $\Delta G_s(C_{60})$ comes from the non-polar term although there is also present a small polar contribution (Table 1).

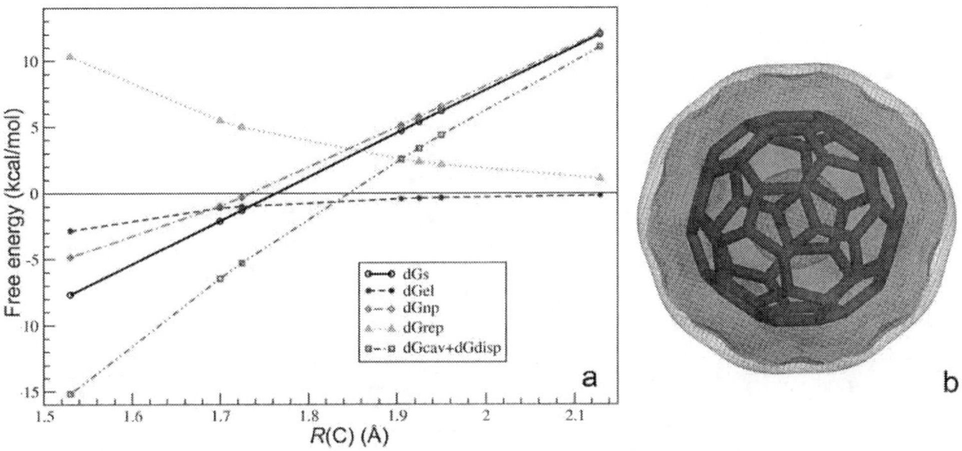

Figure 2. a) The hydration free energy $\Delta G_s(C_{60})$ and its contributions as a function of the carbon radius R(C) in the PCM(UA) scheme, b) the 0.002 e/a_0^3 contour isodensity (mesh) and solvent accessible (solid) surfaces of fullerene C_{60}. The calculations at the HF/6–31G(d) level of theory.

Table 1. The DFT PCM and GB–SA calculations of the hydration free energy of fullerene C_{60}. Calculations at the B3LYP/6–31G(d)//B3LYP/6–31G(d) level.

Model	Energy contributions, kJ mol^{-1}		
	ΔG_{el}	ΔG_{np}	$\Delta G_s(C_{60})$
PCM(UA, R(C)=1.725 Å)	–2.8	–0.9	–3.7
PCM(UA, R(C)=1.864 Å)	–1.3	16.0	14.8
PCM(SMD)	–8.9	–22.1	–31.0
GB–SA	0.0	17.3	17.3

3.2. Generalized Born calculations of hydration of fullerenes

The free energies of solvation were also determined from the calculations of the solute interactions with a continuum water dielectric using generalized Born (GB) treatment. The GB approach finds broad applications in the FF and QM studies of solvation of small molecules and proteins [18,19]. In this work, the calculations of ΔG_s of fullerenes were performed by means of the GB method with the fixed atomic charges and with linear dependence of the non-polar ΔG_{np} contrinution to the solvent-accessible surface area (SA) method [18]. We consider Eq.(3) where the polar ΔG_{el} and non-polar ΔG_{np} terms are expressed as the sums over solute atoms

$$\Delta G_{el} = -166(1-1/\varepsilon)\sum_i \sum_j \frac{q_i q_j}{f_{ij}^{GB}} , \quad (4)$$

$$\Delta G_{np} = \sum_i \sigma_i SA_i \quad . \quad (5)$$

In these equations, q_i are the partial atomic charges, f_{ij}^{GB} is the function of interatomic distances and the Born radii of atoms, SA_i is the solvent-accessible surface area of atom i, σ_i is an atomic solvation parameter, ε is the dielectric constant of water [18]. In the calculations of ΔG_s with Eq. (3–5), we have used the Amber-95 atomic radii and σ_i =30 J mol^{-1} (for all atom types). The employed Mulliken charges q_i were determined in gas phase at the B3LYP/6–31G(d)//B3LYP/6–31+G(d,p) level. The side groups in the studied compounds **1**,**2** (Figure 3a) were considered in a single extended conformation optimized in gas phase for uncharged state. The anionic states correspond to deprotonation of the carboxylic acid of side groups. In **2**, the oxidation of the more distant from the fullerene core –COOH group was considered. In computations of hydration free energies we used the partial atomic charges from the QM calculations, but the partial charges from the standard FFs can also be used in GB–SA to assess ΔG_s of the model structures.

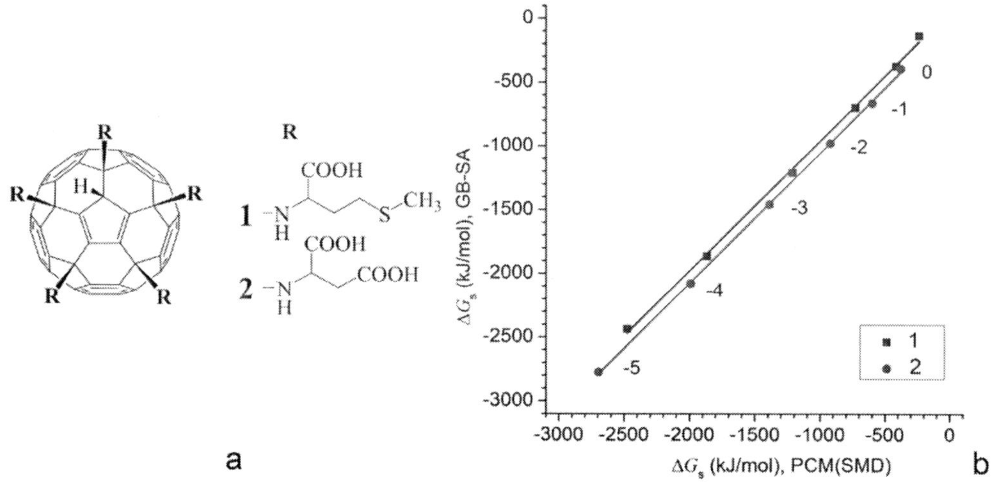

Figure 3. a) Chemical formulas of the studied fullerene derivatives, b) ΔG_s in water of the neutral and multi-anionic states of **1**, **2** from the GB–SA and PCM(SMD) calculations. The numbers at the plots indicate the total electronic charge of the molecules at the correspoinding points.

The computed GB–SA value of $\Delta G_s(C_{60})$ is 17.3 kJ mol^{-1} (Table 1) that is close to the result from the PCM(UA) model with the re-calibrated carbon radius. The hydration free energy of C_{60} in the GB–SA method totally comes from non-polar contribution (Table 1). We have also explored energetics in water for pentasubstituted amino acid fullerene derivatives **1**, **2** (Figure 3a). Hydration thermodynamics of these compounds are determined by the solute–water interactions with the central hydrophobic cage of fullerene and the peripheral hydrophilic groups. Step by step deprotonation of the side groups is accompanied by a steep reduction of ΔG_s (Figure 3b). Good liner correlations (with the slopes of the plot and correlation coefficients about unity) are observed between the data from the GB–SA model and more accurate PCM(SMD) method (Figure 3b). These results show that despite the simplified treatment of the solute–solvent system the GB–SA method can be useful in the studies of hydration of complex nanomolecules containing fullerene fragments.

4. Compilation of experimental data

The hydration free energy of fullerene C_{60} has not been measured experimentally (to our knowledge) due to extremely low solubility of this compound in water [5]. However, there are estimates of the aqueous solubility $S_{aq}(C_{60})$ of fullerene C_{60} based on the Hildebrand–Scatchard model and parabolic extrapolation of the solubilities in normal alcohols [5]. The reported value of $S_{aq}(C_{60})$ is 1.3×10^{-11} ng ml^{-1} [5,6]. There are also experimental data for the vapour pressure $P_{vap}(C_{60})$ of pure fullerene in equilibrium between the pure solid and gas phase ($\log P_{vap}(C_{60})=-22.6$) [6]. Consideration of the triple system 'solid C_{60} – gas state of C_{60} – water solution of C_{60}' allows to determine $\Delta G_s(C_{60})$ from the thermodynamic data for transitions solid↔gas and solid↔water. The standard hydration free energy ΔG_s° of fullerene was determined [10] from the equation

$$\Delta G_s^{\circ}(C_{60}) = 2.303RT\times\left(\log\left(P_{vap}(C_{60})/P^{\circ}\right)-\log S_{aq}(C_{60})\right). \qquad (6)$$

In this equation we use standard conditions of T=298 K and 1 M concentration and the pressure P° of an ideal gas of 24.45 atm. The described treatment predicts negative value of $\Delta G_s°(C_{60})= -24.5$ kJ

IOP Conf. Series: Materials Science and Engineering **525** (2019) 012033 doi:10.1088/1757-899X/525/1/012033

mol^{-1} which follows from the difference of large negative values for $\log S_{aq}(C_{60})$ and $\log(P_{vap}(C_{60})/P°)$ in Eq. (6). However, it was noted that additional consideration of the fullerene size effect via the Flory–Huggins approach reduces the calculated aqueous solubility $S_{aq}(C_{60})$ of fullerene by several orders of magnitude [5]. The estimated from the data of [5] value of $\log S_{aq}(C_{60})$ equals in that case −28.7. The calculation using this value of $\log S_{aq}(C_{60})$ in Eq. (9) shifts $\Delta G_s°(C_{60})$ to a positive value of 26.8 kJ mol^{-1}. Reverse calculations by Eq. (6) using the computed $\Delta G_s°(C_{60})$ and experimental $P_{vap}(C_{60})$ predict extremely low values of $S_{aq}(C_{60})$ in accordance with known practical experience.

5. Conclusions

We discuss in this presentation the calculations of the hydration free energy ΔG_s of fullerene C_{60} using the 'QM/implicit solvent model' approach and the FF FEP/MD technique with explicit water molecules. The reliable and conclusive experimental data on $\Delta G_s(C_{60})$ are not available. Judging by the hydrophobic properties of this nanomolecule, it would be safe to expect a positive value of $\Delta G_s(C_{60})$. However, theoretical computations using standard parameterization schemes predict negative values of $\Delta G_s(C_{60})$. The application of the revised parameters, that were adjusted to account for some special features of carbon structures, have shifted the computed $\Delta G_s(C_{60})$ to the range of positive values. The found variations of $\Delta G_s(C_{60})$ arise mainly from the non-polar contributions in the models considered. The re-calibrated parameters of the PCM(UA) model had been successfully employed in the calculations of the absolute pK_a values of substituted hydrofullerenes [20]. It was also found in this work that reasonably accurate estimates of the hydration free energies of the neutral and charged states of functionalized fullerenes can be obtained from the generalized Born method. In total, the described calculations have shown that correct calculations of hydration thermodynamics require separate parameterization for the atom types met in nanomolecules.

Acknowledgements

The authors gratefully acknowledge support from the Ministry of Science state theme № 0089-2019-0014 in IPCP RAS, and the assistance from the computer center at IPCP RAS.

References

[1] Kroto H W, Fischer J E and Cox D E 1993 *The Fullerenes* (Oxford: Pergamon Press)
[2] Bühl M and Hirsch A 2001 *Chem. Rev.* **101** 1153
[3] Nakamura E and Isobe H 2003 *Acc. Chem. Res.* **36** 807
[4] Ruoff R S, Tse D S, Malhotra R and Lorents D C 1993 *J. Phys. Chem.* **97** 3379
[5] Heymann D 1996 Fullerene Science and Technology **4** 509
[6] Abraham M H, Green C E and Acree W E 2000 *J. Chem. Soc. Perkin Trans.* **2** 281
[7] McDonald N A and Jorgensen W 1998 *J. Phys. Chem.* **102** 8049
[8] Bojan M J and Steele W A 1987 *Langmuir* **3** 1123
[9] Luzhkov V B 2012 *Molecular Simulation* **38** 326
[10] Luzhkov V B, Volokhov V M and Pokatovich G A 2017 *Chem. Phys. Lett.* **676** 95
[11] Zwanzig R W 1954 *J. Chem. Phys.* **22** 1420
[12] Pierotti R A 1976 *Chem. Rev.* **76** 717
[13] Redmill P S, Capps S L, Cummings P T and McCabe C 2009 *Carbon* **47** 2865
[14] Colherinhas G, Fonseca T L and Fileti E E 2011 *Carbon* **49** 187
[15] Cossi M, Scalmani G, Rega N and Barone V 2002 *J. Chem. Phys.* **117** 43
[16] Marenich V, Cramer C J and Truhlar D G 2009 *J. Phys. Chem.* B **113** 6378
[17] Luzhkov V B 2014 *Russ. Chem. Bulletin, Int. Edition* **63** 561
[18] Still W C, Tempczyk A, Hawley R C and Hendrickson T 1990 *J. Am. Chem. Soc.* ***112*** 6127
[19] Bashford D and Case D A 2000 *Annu. Rev. Phys. Chem.* **51** 129
[20] Luzhkov V B and Kotelnikov A I 2015 *Tetrahedron Lett.* **56** 6008

Effect of high-pressure treatment temperature on the structure of carbon phases formed from C_{60} fullerites under pressure

I N Lukina[1], O P Chernogorova[1], E I Drozdova[1], V A Stupnikov[2] and A V Soldatov[3,4]

[1] A.A. Baikov Institute of Metallurgy and Materials Science, Russian Academy of Sciences, Leninskii pr., 49, 119334, Moscow, Russia
[2] MSU Faculty of Chemistry, Lomonosov Moscow State University, Leninskiye Gory, 1-3, 119991, Moscow, Russia
[3] Department of Engineering Sciences and Mathematics, Luleå University of Technology, SE-97187 Luleå, Sweden
[4] Department of Physics, Harvard University, Cambridge, MA 02138, USA

E-mail: lukina.i.n@yandex.ru

Abstract. The synthesis of metal-matrix composite materials reinforced with superelastic hard carbon particles formed from C_{60} fullerites includes heating of the metal-fullerite powder mixture to temperatures above 800C under pressure. The structure evolution of the carbon particles upon heating from 700 to 800 C at a pressure of 5 GPa has been studied in detail by Raman spectroscopy and indentation hardness measurements. It is shown that the structure of the carbon particle passes the stage of a nanoscale mixture of one- and two-dimensional polymers with the resulting atomic superelastic solid phase. The determining properties of the carbon particles such as superplasticity and high hardness are attained before the complete disappearance of polymerized fullerites, the remnants of which are responsible for the disintegration of the material upon scratching.our abstract here...

1. Introduction

The study of the fullerite transformations upon heating under pressure is important for the search of the production methods for bulk samples of superelastic superhard carbon materials, which now are attainable mostly as thin coatings. It has been established that fcc C_{60} fullerites upon heating from 200 to 800°C at a pressure of not more than 8 GPa undergo two-dimensional polymerization [1]. At temperatures above ~800°C, the C_{60} molecules collapse to the atomic state both under pressure and without applied pressure [2, 3], and this transformation is accompanied by the disappearance of fullerene peaks in the XRD and Raman spectra and the appearance of lines characteristic of amorphous carbon or disordered graphite. The nanoclustered graphene phase (NGP) obtained by the high-pressure high-temperature treatment is characterized by superelasticity and high hardness [4,5]. Such material is advantageous with allowance for its wear resistance and tribological properties, and the particles of a superelastic hard phase obtained from fullerenes under pressure can be used to reinforce metal matrix composite materials. New data on the structure of this carbon material can be obtained by the examination using high-resolution Raman spectrometer CRM 200. The aim of the

IOP Conf. Series: Materials Science and Engineering **525** (2019) 012034 doi:10.1088/1757-899X/525/1/012034

work was to trace the sequence of the formation of the superelastic hard carbon phase upon heating at a pressure of 5 GPa.

2. Experimental

Iron-based composite material (CM) samples of 10 mm in diameter and 5-6 mm in height were prepared from a powder mixture of Fe and 10% wt % fullerites C_{60} (99.5%) obtained from the G.A. Razuvaev Institute of Organometallic Chemistry (Nizhny Novgorod, Russia). The CM samples were synthesized with a hydraulic press in standard quasi-hydrostatic high pressure chambers used for the synthesis of superhard materials. The pre-pressed pellet was heated at a pressure of about 5 GPa to 700-800 C and held for 0.5 min. The microstructure of the CM was examined with an Olympus optical microscope. The surface and fractographic features of the carbon particles were examined using a "Quanta 3D FEG" high resolution two-beam scanning electron microscope with a resolution of 1.2 nm. The identification of carbon phases obtained from fullerites under pressure was performed using Raman spectra obtained with a CRM 200 high-resolution Raman microscope at 532.0 nm argon laser excitation in the microsample investigation mode. A Shimadzu dynamic ultra-micro hardness tester at a load of 0.5 N was used to measure indentation hardness (H_{IT}), indentation modulus (E_{IT}), and elastic recovery ($\eta_{IT} = W_{elast}/W_{total}$, where W_{elast} and W_{total} are the elastic and total indentation works, respectively).

3. Results and discussion

For the study of the fullerite evolution leading to the formation of superelastic hard carbon phase directly on the sample, a temperature gradient of 20 K/mm or 0.02 K/μm was realized over the cross section of the CM sample held for 30 seconds at a temperature of 700-800 C and then this state was quenched. In this case, the particles in the center of the sample remained untransformed, while the superelastic hard carbon phase was formed closer to the edge. The color contrast between the transformed and untransformed particles allowed us to establish the transformation front in the temperature field on the sample polished surface, at which the non-transformed particles look darker (Figure 1).

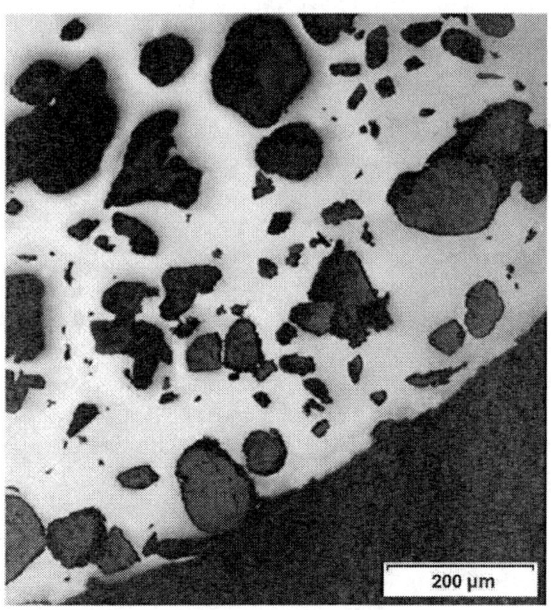

Figure 1. Microstructure of the CM obtained from Fe and C_{60} at a pressure of 5 GPa at temperature varied from ~700°C in the center of the sample to ~800°C closer to the edge of the sample.

IOP Conf. Series: Materials Science and Engineering **525** (2019) 012034 doi:10.1088/1757-899X/525/1/012034

The Raman spectra of the carbon particles corresponding to their positions in the sample are shown in Figure 2. The spectrum taken from a soft black particle, which lies 4000 μm away from the transformation front and exhibits a deep cavity upon indentation, shows that it consists of a mixture of C_{60} dimers and polymerized C_{60} phases such as one-dimensional orthorhombic phase and two-dimensional rhombohedral and tetragonal phases (Figure 2, line 1).

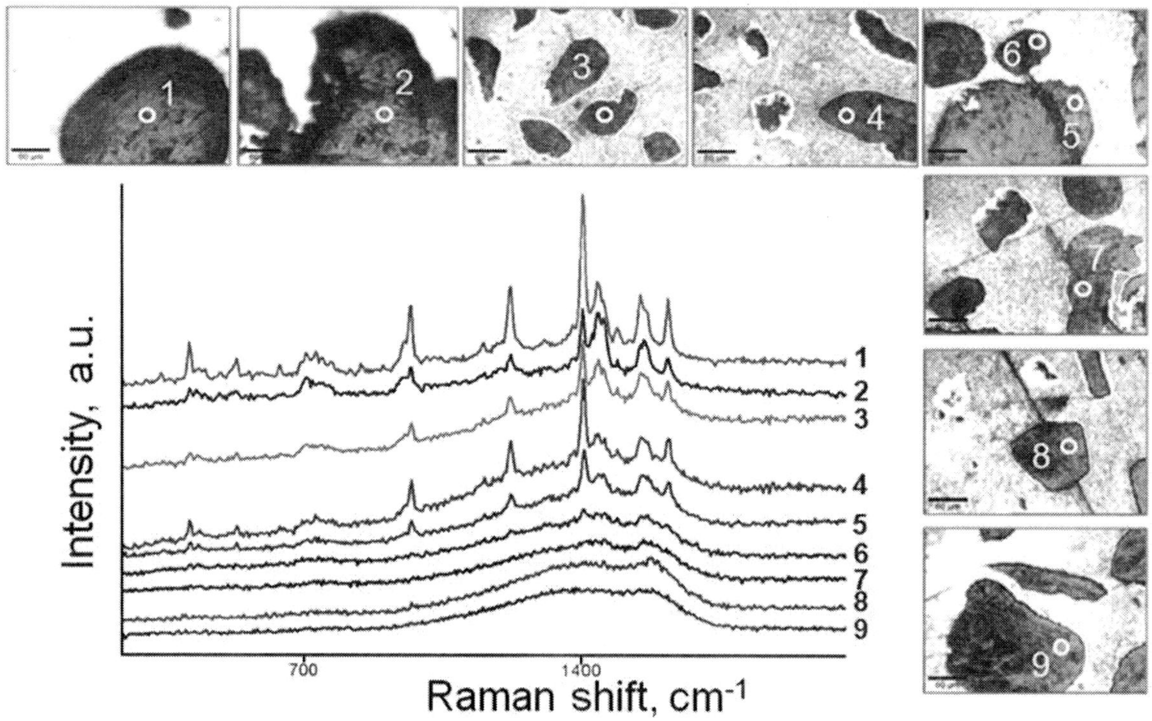

Figure 2. Changes in microstructure of the carbon particles in CM from the center to the edge of the sample and the corresponding Raman spectra.

As we trace the particle structure evolution from the center to the edge of the CM sample, that is, as the front of final transformation of the carbon particle to NGP is approached, the intensity of the peaks of two-dimensionally polymerized fullerites decreases, and the rise of the background line in a range of 1000–1700 cm^{-1} increases (Figure 2, lines 1 -7). Lines 8 and 9 in Figure 2 are characteristic of the superelastic hard phase, on which there is no trace of scratching by diamond pyramid. Since the Raman spectrum is recorded from a spot of about 1 μm in diameter, we can conclude that the nucleation of the superelastic hard carbon phase occurs in the interior of polymerized fullerite at a submicron level. At the same time, different types of polymers and the superelastic hard carbon phase coexist in a micron volume of the material. A more detailed examination of the Raman spectra shows that the rhombohedral C_{60} polymers are most "persistent". Thus, the structure of fullerite heated to ~ 800°C at a pressure of 5 GPa is a nanoscale mixture of one- and two-dimensional polymers, as well as the resulting atomic superelastic hard phase. The superelasticity and high hardness are attained by the carbon particles still containing insignificant amount of residual fullerites. The transformation ends with the complete disappearance of the peaks of polymerized fullerites. The correspondence of indentation curves to the structure of the carbon parcels is shown in Figure 3.

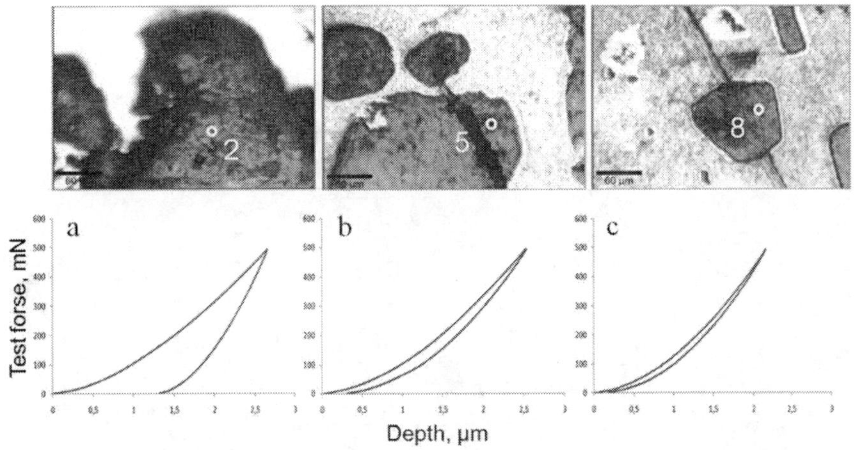

Figure 3. Indentation curves of the particles corresponding to Raman spectra (a) 2, (b) 5, and (c) 8 in Figure 2

The polymerized fullerites are characterized by relatively small elastic recovery (n = 48%) at moderate indentation hardness (H_{IT} = 5 GPa) and modulus (E_{IT} = 55 GPa). In the intermediate state characterized by the presence of retained fullerites, the particle upon microindentation becomes superelastic (η_{IT} = 83%), its hardness increases to H_{IT} = 8 GPa, but the indentation modulus somewhat decreases to E_{IT} = 46 GPa. This is caused by a decrease in the slope of unloading indentation curve as the carbon particle becomes superelastic and the loading-unloading loop becomes narrow. Despite the fact that the particles at the intermediate transformation stage become relatively hard and superelastic, their scratching by diamond pyramid under load leaves a pronounced trace, which can be considered in detail in Figure 4.

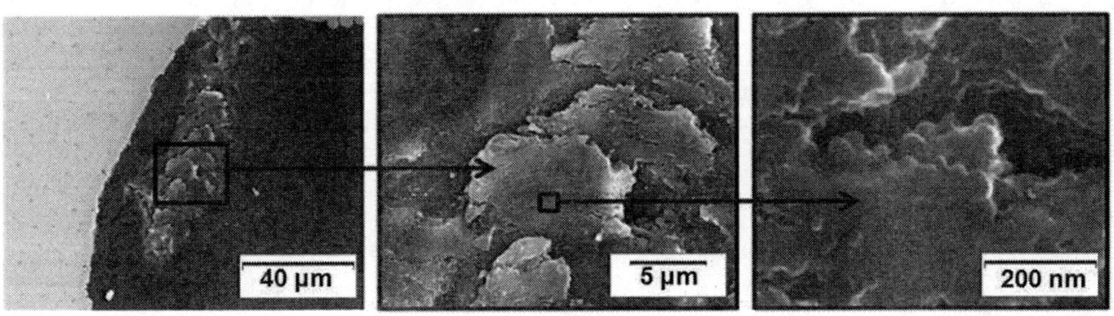

Figure 4. Scratch on the carbon particle in the area of unfinished transformation to NGP.

A detailed examination of the scratch on the particle, in which the transformation into NGP is not completed and the polymerized structure is retained, shows that the destruction occurs between hard globular "grains", which are weakly connected to each other. Such hard grains are 20-40 nm size, which corresponds to the average level of the fractal hierarchy of the granular structure of hard carbon [6]. The hard grains are still poorly connected to each other due to the presence of the interlayers of retained fullerites between them. The NGP particle, on which scratching leaves a thin even dent without torn edges is characterized by high hardness (H_{IT} = 12 GPa) typical of the C_{60} transformation product at a pressure of 5 GPa, a high elastic recovery (90%), and its indentation modulus reaches 62 GPa. The Raman spectra recorded from superelastic hard particles free from microscopically prominent traces of scratching by diamond pyramid are characterized by a wide unresolvable band typical of amorphous carbon (Figure 2, line 9).

A characteristic feature of the NGP structure is its inheritance link with the deformed structure of the fcc C_{60} fullerite. It is detected both by the relief on the polished surface, which reproduces the slip lines (planes) in fullerite, and by the characteristic deformed fcc structure observed in polarized light in an optical microscope (Figure 5).

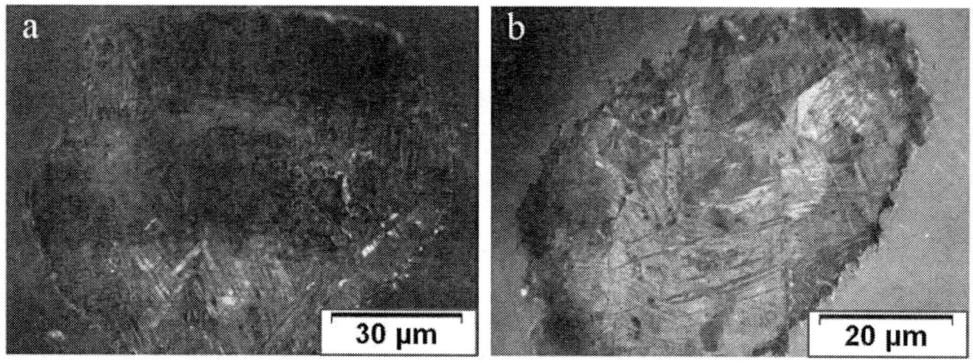

Figure 5. Optical micrographs of the (a) incompletely and (b) completely transformed carbon particles.

The presence of a picture of a deformed structure in polarized light is another characteristic sign of the NGP formation, since this effect is not detected in the case of the retention of polymerized fullerites in the structure (the center of the particle in Figure 5a).

4. Conclusion

With increasing temperature in a range of 700–800°C at a pressure of 5 GPa, the C_{60} fullerite particles in the metal composite material (CM) undergo changes from a nanoscale mixture of one- and two-dimensional polymers to an atomic nanocluster graphene phase. Such structure changes are accompanied first of all by an increase in indentation hardness and then by an increase in elastic recovery in the presence of an insignificant fraction of retained fullerites. However, even at a sufficiently high level of both characteristics, the scratching of a particle leads to disintegration along the boundaries between nanoscale grains of the hard phase. After the complete disappearance of polymerized fullerites, the carbon particles formed under pressure exhibit superplasticity and high hardness in the absence of noticeable scratches.

Acknowledgment

The work was carried out according to the state assignment number 075-00746-19-00.

References

[1] Davydov V A, Kashevarova L S, Rakhmanina A V, Dzyabchenko A V, Senyavin V M and Agafonov V N 2001 Polymeric high-pressure phases of fullerene C_{60}: synthesis, identification, research, properties *Ross. Khim. Zh* 45 **4** 25-34.

[2] Mochida I, Egashira M, Korai Y and Yokogawa K 1997 Structural changes of heat treatment for heat treatment, to graphitization temperature *Carbon* 35 **12** 1707-12.

[3] Nikonova R M, Pozdeeva M A, Ladyanov V I and Aksenova V V 2007 On the effect of temperature on the stability of fullerenes / fullerites *Bulletin of the Udmurt University, Physics* **4** 58-66.

[4] Wood R A, Lewis M H, West G, Bennington S M, Cain M G and Kitamura N 2000 Transmission electron microscopy, electron diffraction and hardness studies of high-pressure and high-temperature treated C_{60} *J. Phys. Condens. Matter.* ***12*** 10411−21

[5] Chernogorova O, Drozdova E, Ovchinnikova I, Soldatov A and Ekimov, E 2012 Structure and

properties of superelastic hard carbon phase created in fullerene-metal composites by high temperature-high pressure treatment. *Journal of Applied Physics* **111** 112601-1–5.
[6] Starešinić D, Dominko D, Rakić I S, Milat O, Ristić D, Ivanda M, RadićT M, Clément A, Saint-Paul M, Kozlov M E and Biljaković K 2017 Fractal nature of hard carbon prepared from C_{60} fullerene *Carbon* **124** 708-21.

IOP Conf. Series: Materials Science and Engineering **525** (2019) 012035 doi:10.1088/1757-899X/525/1/012035

Effect of the sintering aids on optical and luminescence properties of Ce:YAG ceramics

K E Lukyashin[1], A V Ishchenko[2], V A Shitov[1], V S Shevelev[2] and L V Victorov[2]

[1] Institute of Electrophysics UrB RAS, Yekaterinburg, Russia
[2] Ural Federal University, Yekaterinburg, Russia

E-mail: kostya@iep.uran.ru

Abstract. This reports on experimental results of a synthesis of Ce:YAG ceramics with a thickness of 2 and 8 mm and Ce^{3+} concentrations of 0.1, 0.5, 2 at% and a study of their optical and scintillation characteristics with and without using of a SiO_2 sintering aid. It is shown that using of the sintering aid (0.14 wt% SiO_2) in this ceramics production method results in deterioration of optical and scintillation characteristics. An effect of Si^{4+} ions on luminescence properties of Ce:YAG ceramics is revealed. The maximum light output has been achieved in the ceramics without the sintering aid (21900 photon/MeV – 0.5% Ce:YAG, h=8 mm) compared with ceramics synthesized by us previously (20800 photon/MeV – 1% Ce:YAG, h=2 mm).

1. Introduction

The most scintillation materials used in ionization radiation detectors are applied in a monocristalline condition. It is well known that growing of single crystals with the high melting point is a painstaking and costly process. In some cases, an alternative to difficult-to-make crystal scintillation materials may be similar transparent ceramic materials with comparable light output, optical quality, and luminescence decay properties [1, 2]. Currently, new transparent ceramics are intensively searched for, and their synthesis methods are constantly improved.

An yttrium aluminium garnet scintillator Ce^{3+}:$Y_3Al_5O_{12}$ (Ce:YAG) is well known among these materials. As of today, various methods for production of YAG-based scintillators, both monocrystalline and optically transparent materials, are developed. One of Ce:YAG scintillation ceramics production methods is the solid-phase reaction method. The materials to be synthesized have high melting points (1930°C for YAG). Therefore, sintering aids, such as SiO_2, MgO [3,4], are usually added to reduce the synthesis temperature and accelerate the material sintering process [5]. However, at the present time it is demonstrated that these aids may have an effect on optical and luminescence characteristics of Ce:YAG [4,6], but this question is sufficiently underexplored. Furthermore, attempts to synthesize transparent ceramics with the thickness over 5 mm are mostly unsuccessful. These unsolved issues lead to significant restrictions in industrial application of ceramics.

In this paper, we attempt to determine the effect of sintering aids on optical and scintillation characteristics of Ce:YAG ceramics and to study a possibility to synthesize large ceramics with the thickness over 5 mm.

IOP Conf. Series: Materials Science and Engineering **525** (2019) 012035 doi:10.1088/1757-899X/525/1/012035

2. Experimental

2.1 Samples synthesis

The samples were synthesized by a method published in [7]. The samples of Ce^{3+}:YAG ceramics were synthesized from nanopowders Al_2O_3, Y_2O_3, 1 at%Ce^{3+}:Y_2O_3 with the diameter of about 10-15 nm; these powders were produced by the laser ablation method [8]. Commercial powders with the purity of >99.99% were taken as raw materials. Firstly, the initial substances were annealed in the air in the electric muffle furnaces with SiC heating elements for 3 hours at 1200°C and 900°C for Al_2O_3 and Y_2O_3, 1 at% Ce^{3+}:Y_2O_3, respectively, to remove metastable phases.

The ceramics samples were synthesized by different methods. Samples synthesis details and stages are shown in table 1.

Synthesis of samples with sintering aids (samples 2587, 2421, 2427, 2409, 2520, 2519). The initial materials for samples were nanopowders Al_2O_3, Y_2O_3, 1 at% Ce^{3+}:Y_2O_3 (or 2 at% Ce^{3+}:YAG), which were mixed in the required stoichiometric ratio. The nanopowders were mixed for 48 hours in ethanol in a plastic container using an inclined-axis ball mill with yttria-stabilized zirconia (YSZ) balls. During mixing of the powders, 0.5 wt% TEOS (tetraethoxysilane $(C_2H_5O)_4Si$) was added as a sintering aid, which is equivalent to 0.14 wt% SiO_2. Then the mixture was vaporised in a rotary vacuum vaporizer and annealed in the atmosphere at 600 °C for 3 hours to remove alcohol residues. Additionally, the mixture of the initial powders for samples 2421, 2427, and 2409 was sieved using a 200 mesh sieve and annealed in the air at 600 °C for 3 hours to remove organic impurities.

Pressing of the powder mixtures for furnishing of the samples with the thickness of 2 mm was performed by the single-axis static pressing method under the pressure of 200 MPa in metal moulds with the effective diameter of 14 mm to the density equal to 40-50% of the full YAG density (4.55 g/cm^3). Compacts for furnishing of the samples with the thickness of 8 mm were initially pressed in metal moulds under the pressure of 16 MPa for compaction, and then isostatically pressed in a hydrostat under the pressure of 200 MPa in latex shells after vacuum treatment (2519 and 2524 samples). For removal of organic impurities, the ceramics were annealed in the air at 600-800 °C for 3 hours. The compacts were sintered in a high-vacuum electric furnace with graphite heating elements under $5 \cdot 10^{-5}$ mbar and at 1700 °C for 20 hours.

The samples without sintering aids (2524, 2525) were synthesized under an alternative procedure. The difference from the above mentioned procedures lay in that, before sintering, the powder mixture was compacted to the density equal to 20% of the full YAG density and annealed at 1200 °C for 3 hours in an atmospheric furnace to create YAG structures. Further, the samples were crushed and milled in the same way as for mixture making using an inclined-axis ball mill with yttria-stabilized zirconia (YSZ) balls. Then, after the pressing (see above) the samples were sintered in a vacuum furnace at 1780 °C for 20 hours. This sample synthesis procedure enabled to synthesize YAG beforehand in contrast to the procedures described above, where YAG was synthesized from the initial powders during the vacuum-sintering of the ceramics.

After the vacuum-sintering all samples were black due to oxygen deficiency in their structures. In order to replenish this deficiency the samples were annealed in the air atmosphere at 1300 °C for 15 hours. Then, the ceramics were ground and polished. The final dimensions of the ceramics were Ø10×2 and Ø 10×8 mm.

2.2 Equipment

The structures of the sintered ceramics were determined by X-ray diffraction (XRD) on copper radiation (CuK_α, λ=1.542Å) with the graphite monochromator on a secondary beam (D8 Discover GADDS, Bruker AXS, Germany). Processing of results of XRD was executed in the program of the full-height analysis TOPAS 3.

The optical transmittance and absorption spectra of the obtained Ce^{3+}:YAG ceramics samples were measured using a UV-2450 (Shimadzu, Japan) optical dual-beam spectrophotometer (200-900 nm). Photoluminescence (PL) and photoluminescence excitation (PLE) spectra were measured using a

Perkin Elmer LS55 spectrophotometer (NANOTECH Centre, Ural Federal University). The X-ray luminescence (XRL) spectra were measured using a monochromator MDR-23 equipped by an x-ray source of URS-1.0 (40 kV, 10 mA, W-anode) and Hamamatsu R928 and FEU-106 photo multipliers (experimental physics department, UrFU, Russia).

Table 1. Ceramic general flow chart

	Samples				
Own numbering	2421 2427 2409	2520	2519	2525	2524
Ce^{3+}, at%	0.1 0.5 2	0.5	0.5	0.5	0.5
Shape	Cylinder				
Diameter, mm	10				
Height, mm	2	2	8	2	8
Stage	**Synthesis conditions**				
1.	Raw materials of yttrium oxide, cerium oxide and aluminum oxide with purity >99,99%				
2.	Nanopowders producing (Al_2O_3 and Y_2O_3, 1-2 at.% Ce^{3+}:Y_2O_3 (10-15 nm) [8])				
3.	Nanopowders annealing (1200 and 900 °C/3 h)				
4.	Mixing (Y_2O_3+Ce:Y_2O_3):(Al_2O_3) = 3:5 (48 h) in ethyl alcohol (C_2H_5OH)				
5.	Addition 0.5 wt% TEOS (0.14 wt% SiO_2)			-	
6.	Mixture evaporation				
7.	Annealing (600 °C/3 h)				
8.	Sifting (200 mesh)	-		Densification (20%)	
9.	-			YAG synthesis (1200 °C/3 h)	
10.	-			Milling in C_2H_5OH	
11.	-			Mixture evaporation	
12.	Annealing (600 °C/3 h)	-		Annealing (600 °C/3 h)	
13.	Uniaxial static pressing (200 MPa)		Uniaxial static pressing (16 MPa)	Uniaxial static pressing (200 MPa)	Uniaxial static pressing (16 MPa)
14.	-		Cold isostatic pressing (200 MPa)	-	Cold isostatic pressing (200 MPa)
15.	Annealing (600-800 °C/3 h)				
16.	Vacuum sintering (1700 °C ~ 20 h)			(1780 °C ~ 20 h)	
17.	Annealing 1300 °C/15 h				
18.	Optical polishing				

3. Results and Discussion

The images of the several samples of synthesized ceramics are given on figure 1. Upon the synthesis results, it is found that using of sintering aids in the above-described synthesis procedure can lead to deterioration of optical characteristics in some cases. An example for this is Sample 2519, 8 mm thick, SiO_2 sintering aid. There has been observed a cluster of pores and second phases in a form of white

IOP Conf. Series: Materials Science and Engineering **525** (2019) 012035 doi:10.1088/1757-899X/525/1/012035

opaque areas in this sample. However, sample 2520, 2 mm thick, synthesized in the same way with the same sintering aid, has shown the monophasic structure and high optical quality (Figure 1b). Sample 2524, 8mm thick, synthesized without sintering aids, has also shown no additional phases. Furthermore, we had to extend the annealing period to 15 hours, i.e. 3 times longer than for the thinner samples (2 mm), in order to bleach the larger samples (to remove F-centres), which made total sense, since the thicknesses of the samples were also quite different (4 times), and the oxygen diffusion rate in the YAG lattice is constant.
There was made a microsection of Sample 2519 for phase composition analysis. According to XRD data the central part of the ceramics contain two crystalline phases: YAG – 97.46% and YAM (monoclinic yttrium aluminium oxide, $Y_4Al_2O_9$) – 2.54%. Presence of YAM phase in the ceramics indicates the shortage of aluminium, which could be attributed to the irregular shrinkage and YAG synthesis process rate due to presence of the sintering aid and to possible aluminium evaporation. The sample contains no additional phase in its periphery (YAG – 100%).

a b c

Figure 1. The images of several samples of synthesized Ce:YAG ceramics: a – samples 2519 (Ø10×2 mm) and 2520 (Ø10×8 mm) with 0.14 wt% SiO_2; b and c – samples 2525 (Ø10×2 mm) and 2524 (Ø10×8 mm) without sintering aid

For the analysis of optical and luminescence properties we have taken only monophasic samples 2524, 2421, 2427 and 2409, differing from each other in presence/absence of the sintering aid SiO_2 and Ce^{3+} dopant concentration.

The optical transmission and absorption spectra of the synthesized samples are given in figure 2. There are observed absorption bands in these spectra, which are typical for Ce^{3+} ions in YAG with the maximums in the range of 450, 340, and 250 nm, corresponding to $4f \rightarrow 5d$ transitions in Ce^{3+}. The optical transmission and absorption coefficients measured from the spectra are given in table 2. It is evident from the results that sample 2427 (0.5% Ce:YAG with the sintering aid SiO_2, figure 2a, table 2) has the best optical characteristics among the samples of the same thickness (2 mm). Notwithstanding the fact that the transmission coefficient of the sample with the thickness of 8 mm is the lowest among all the samples (57 %), in comparison with those of the samples with thickness of 2 mm, its absorption coefficient has the best value (see table 2). This fact allows for the conclusion that a sintering aid used for YAG synthesis acceleration and temperature reduction could degrade optical characteristics of YAG-based transparent ceramics. However, since the samples with and without SiO_2 aids were produced under different conditions, a significant effect of the synthesis conditions on the optical characteristics of the samples cannot be ruled out.

Variations observed in the absorption spectra should also be noticed (figure 2b). The spectra of the samples with the sintering aids show an absorption band on the area of 305 nm, which is not available in the absorption spectra of the samples without the sintering aids. Occurrence of this band is of unknown nature. Nevertheless, it might be safely assumed that this band occurs due to addition of the sintering aid SiO_2.

IOP Conf. Series: Materials Science and Engineering **525** (2019) 012035 doi:10.1088/1757-899X/525/1/012035

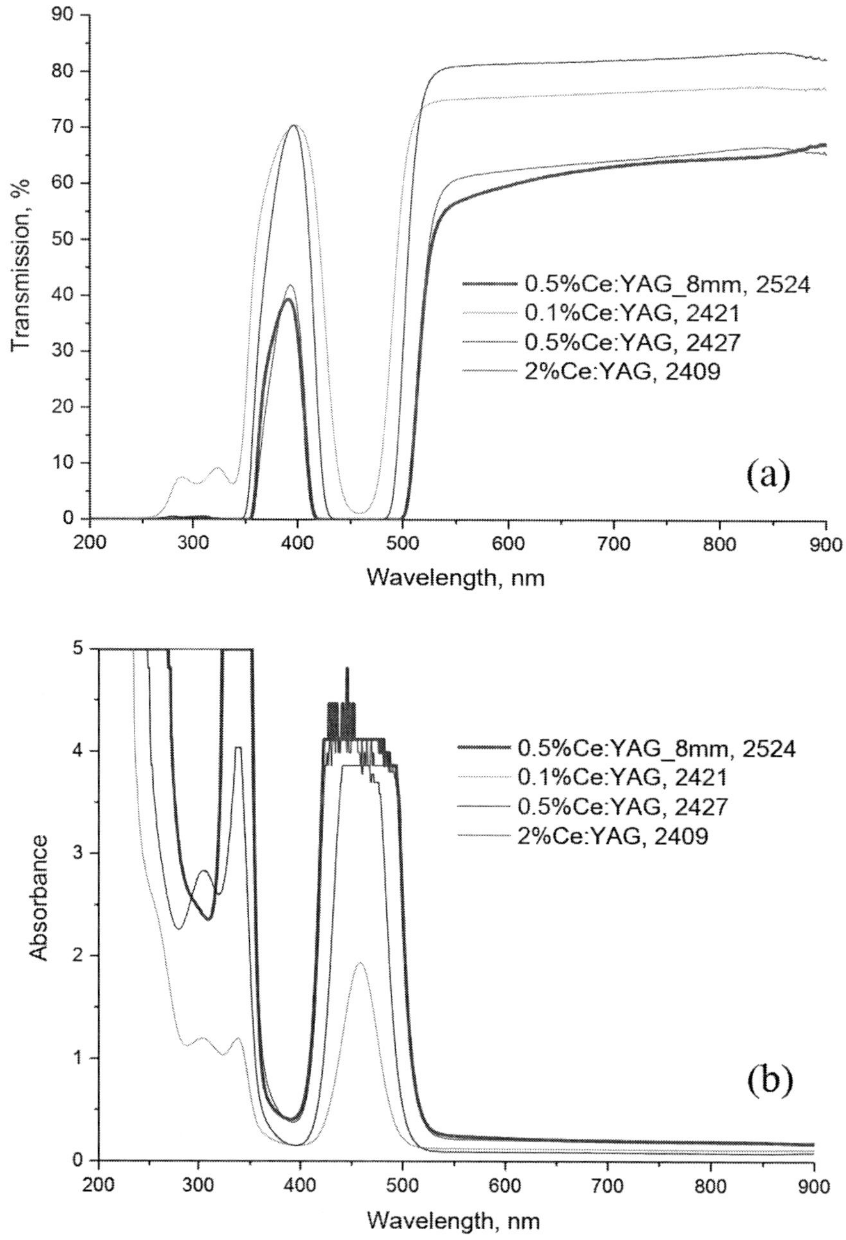

Figure 2. Optical transmission (a) and absorption (b) spectra of Ce:YAG synthesized ceramic samples

The XRL spectra of the synthesized samples are shown on figure 3. All the spectra contain an intensive luminescence band in the range of 525-540 nm, which corresponds to intracenter transitions $5d \rightarrow 4f$ in Ce^{3+} ions in the YAG host. The wide band in the range of 250-450 nm can be associated with light emission by self-trapped excitons (STE) in the YAG. It should be noticed that the higher the concentration of Ce^{3+} activator in the YAG is, the weaker the light emission of the STE band is. A series of narrow bands in this area could be attributed to luminescence of uncontrolled rare-earth impurities, Gd^{3+} and Tb^{3+}. It may be stated that the XRL spectra of all the examined samples are similar to each other, irrespective to presence or absence of the sintering aids or to synthesis methods.

The shift of the maximum of the Ce^{3+} luminescence band (525-540 nm) is caused only by the activator concentration.

Table 2. Optical and luminescence characteristics of the ceramics

Sample	Thickness, mm	T, % at 550 nm	α, cm^{-1} at 550 nm	Luminescence decay parameters	
				Amp. of fast comp; τ (ns)	Amp. of slow comp; τ (ns)
0.5%Ce, 2524	8	57	0.31	1; 90	0.011; 773
0.1%, 2421	2	75	0.62	1; 120	0.073; 966
0.5%, 2427	2	81	0.45	1; 104	0.058; 667
2%, 2409	2	61	1.08	1; 87	0.018; 680

For a more detailed study of effects of the sintering aids on the luminescence and optical properties of Ce:YAG ceramics the PL and PLE spectra were measured (figure 3b). There are observed luminescence excitation and Ce^{3+} light emission bands (transitions $4f \leftrightarrow 5d$), typical for Ce:YAG, in the PL and PLE spectra. However, it is noteworthy that photoluminescence properties of Ce:YAG ceramics respond unexpectedly to variations in the activator concentration and presence/absence of the sintering aid SiO_2. When the activator concentration increases from 0.1 to 2% (samples with the sintering aid SiO_2 – 2421, 2427, and 2409), a significant broadening of the luminescence excitation band in the range of 450 nm (transition $4f \leftrightarrow 5d_1$, figure 3b) is observed. This broadening is attributed to an occurrence of two additional shoulders with the maximums in the range of 410 and 510 nm. Interestingly, the most width of this band is observed in the PLE spectrum of sample 2524 with the thickness of 8 mm, 0.5% Ce:YAG, without the sintering aid. The highest intensity is observed in the shoulders in the 450 nm band on sample 2524 in comparison with other samples synthesized without using of the sintering aid. The similar behaviour can be observed in the luminescence excitation band with the maximum in the range of 340 nm. It is evident that the broadening band assumes a sophisticated irregular shape (figure 3b).

It is also noteworthy that the short wave part of the PL spectrum for sample 2524 (figure 3b) is significantly shifted. It could be attributed to occurrence of an absorption shoulder in the range of 510 nm. The shift of the maximum in the PL spectra of samples 2421, 2427, and 2409 from 526 to 535 nm in case of Ce^{3+} concentration increase from 0.1 to 2% could also be attributed to the intensity growth of the absorption band in a form of a shoulder with the maximum in the range of 510 nm.

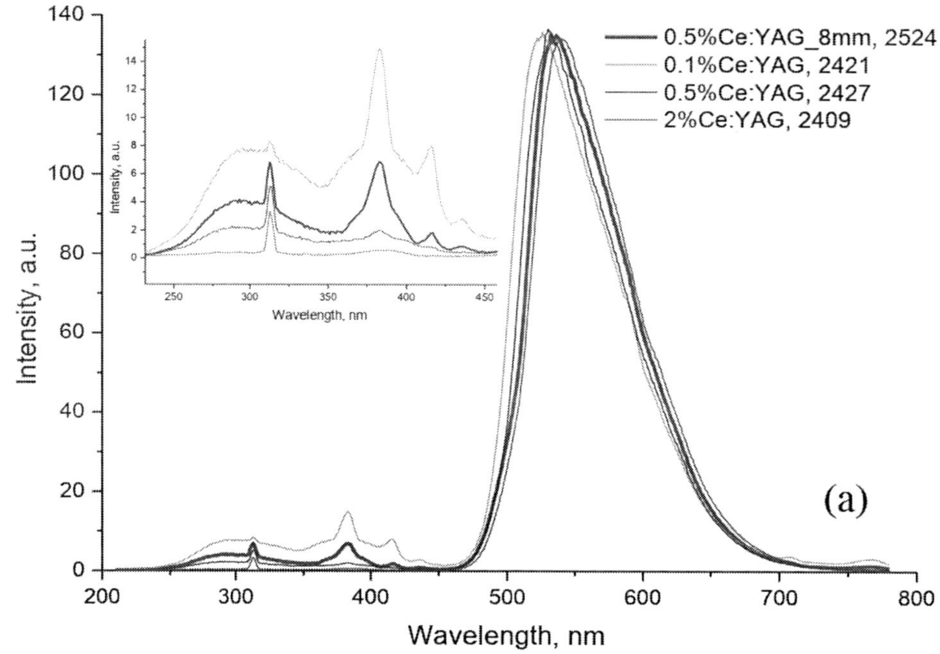
0.5%Ce:YAG_8mm, 2524
0.1%Ce:YAG, 2421
0.5%Ce:YAG, 2427
2%Ce:YAG, 2409
Intensity, a.u.
Wavelength, nm
(a)

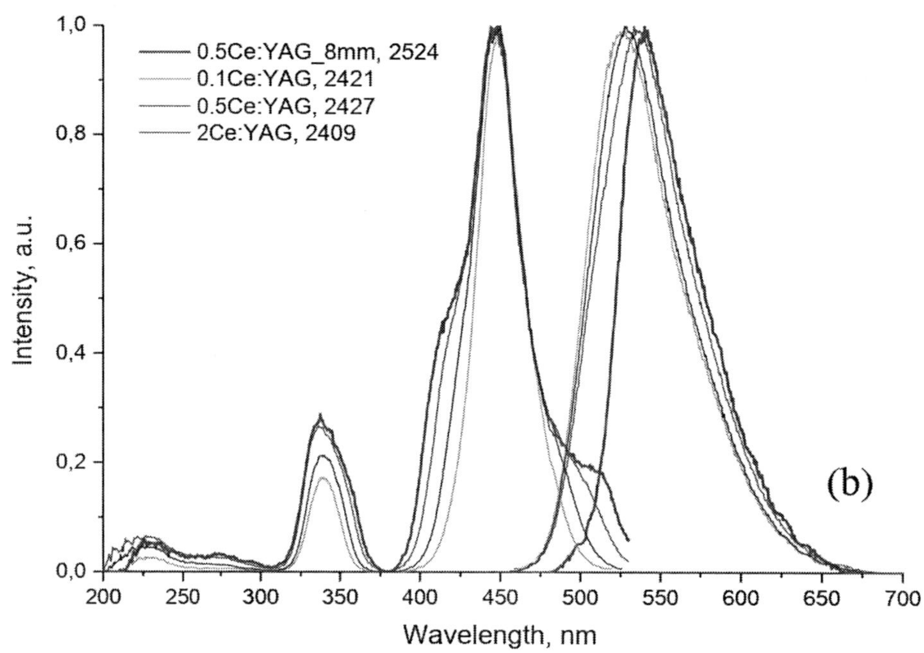
0.5Ce:YAG_8mm, 2524
0.1Ce:YAG, 2421
0.5Ce:YAG, 2427
2Ce:YAG, 2409
Intensity, a.u.
Wavelength, nm
(b)

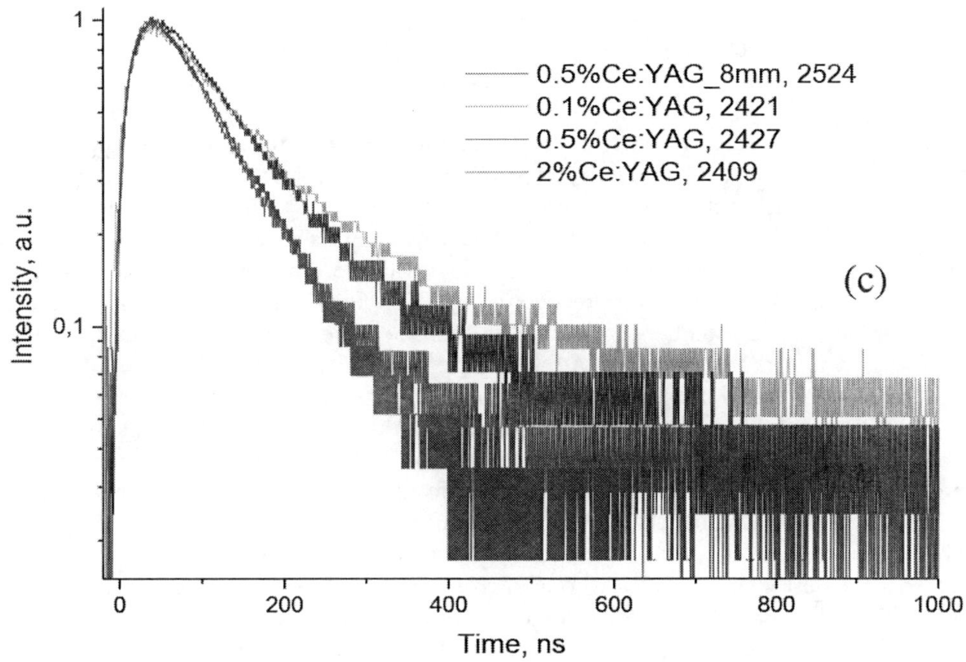

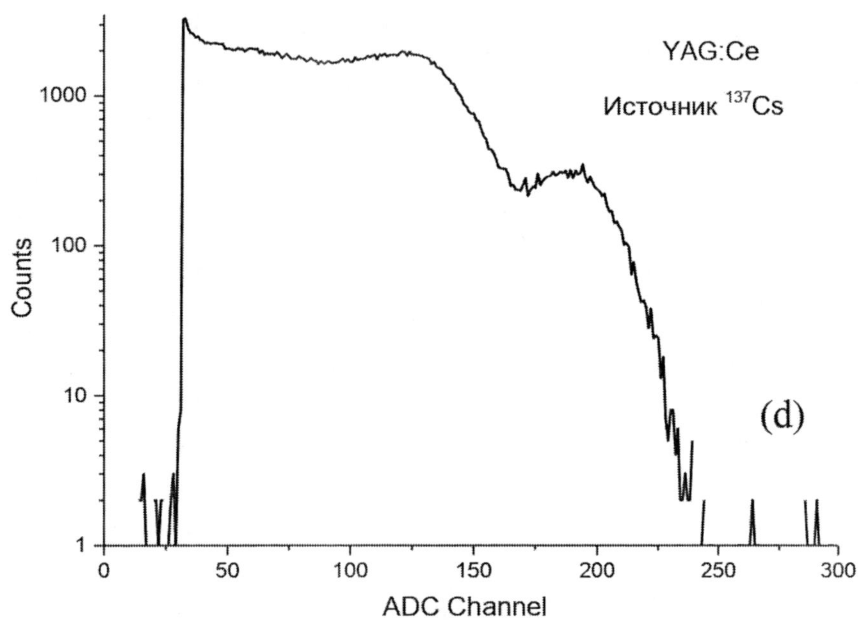

Figure 3. XRL spectra (a), PL and PLE spectra (b), decay curves (c) and pulse-height spectra (d) of Ce:YAG ceramic samples

Occurrence of additional shoulders in the excitation bands with the maximums in the range of 340 and 450 nm can be connected with changes in the symmetry of the impurity Ce^{3+} luminescence center. When introduced into the YAG host, Ce^{3+} ions replace Y^{3+} ions. However, due to the difference in the ion radius between Ce^{3+} and Y^{3+} ions in eightfold coordination (0.128 and 0.116 nm, respectively) a local distortion of YAG crystalline structure occurs, when Ce^{3+} ions are embedded. As the concentration of Ce^{3+} ions in YAG increases, a distortion degree also increases, which may lead to non-equivalent crystallographic positions of Ce^{3+} ions, particularly, to occurrence of some groups of luminescence centers with different symmetries. Energy levels $5d_1$ and $5d_2$ of cerium are very sensitive to their local environment, and changes in environment configuration may lead to changes in their energy, which becomes obvious in the excitation spectra, when Ce^{3+} concentration increases (figure 3b). Embedding of Si^{4+} ions into the YAG structure enables to compensate distortion of CeO_8 polyhedra to the certain extent by the effect of their less ion radius. Si^{4+} ions replace Al^{3+} ions occupying fourfold and six-fold coordination in the YAG. In this case the ion radius of Si^{4+} ions in any coordination is less than the radius of Al^{3+} ions (fourfold coordination 0.040 and 0.053 nm, and six-fold coordination 0.054 and 0.067 nm, respectively). Without the sintering aid SiO_2 the compensation effect does not occur, and shoulders appear in the excitation spectrum in the bands with the maximums at 340 ($4f \rightarrow 5d_2$) and 450 nm ($4f \rightarrow 5d_1$) with the highest intensity among all the samples studied.

For study of dynamic luminescence characteristics of Ce:YAG ceramics there were measured luminescence decay curves in case of pulsed electron beam excitation. Measurement results are shown on figure 3 and in table 2. It is revealed that the increase of Ce^{3+} ion concentration in the samples of transparent YAG-based ceramics leads to a significant shortening of luminescence decay kinetics (figure 3c). This effect was observed by us previously [2] and could be attributed to luminescence quenching in case of sufficiently high concentration of the activator in the host. It is interesting, however, that the best characteristics were shown in the sample synthesized without the sintering aid SiO_2 and with Ce^{3+} concentration equal to 0.5% (figure 3c and table 2). Particularly, near-complete absence of the slow component is observed (table 2). Presence of the slow component in the luminescence of the samples with SiO_2 can be attributed to occurrence of shallow traps for charge carriers during replacement of Si^{4+} ions with Al^{3+} ions. Due to the difference in charge states between Si^{4+} and Al^{3+} ions an electron trap occurs on Si^{4+} ions; this trap can catch an electron during the luminescence excitation and energy transmission processes. During the excited state relaxation periods the electron can leave the trap and recombine on the cerium luminescence center, resulting in occurrence of the slow component.

For study of dynamic luminescence characteristics of Ce:YAG ceramics there were measured luminescence decay curves in case of pulsed electron beam excitation. Measurement results are shown on figure 3 and in table 2. It is revealed that the increase of Ce^{3+} ion concentration in the samples of transparent YAG-based ceramics leads to a significant shortening of luminescence decay kinetics (figure 3c). This effect was observed by us previously [2] and could be attributed to luminescence quenching in case of sufficiently high concentration of the activator in the host. It is interesting, however, that the best characteristics were shown in the sample synthesized without the sintering aid SiO_2 and with Ce^{3+} concentration equal to 0.5% (figure 3c and table 2). Particularly, near-complete absence of the slow component is observed (table 2). Presence of the slow component in the luminescence of the samples with SiO_2 can be attributed to occurrence of shallow traps for charge carriers during replacement of Si^{4+} ions with Al^{3+} ions. Due to the difference in charge states between Si^{4+} and Al^{3+} ions an electron trap occurs on Si^{4+} ions; this trap can catch an electron during the luminescence excitation and energy transmission processes. During the excited state relaxation periods the electron can leave the trap and recombine on the cerium luminescence center, resulting in occurrence of the slow component.

In order to evaluate scintillation efficiency the amplitude luminescence spectrum was measured, when excited using a gamma-ray source with ^{137}Cs (figure 3d, table 3). Comparing the measured characteristics with our previous results [2], we reveal that sample 2524 (0.5%Ce:YAG, 8 mm thick, synthesized without the sintering aid) has the best characteristics among all the examined samples.

Table 3. Scintillation characteristics of the ceramics

Sample	Dimensions, mm	Resolution in our exp., %	Decay time, ns	ALY, photon/MeV
0.5%Ce, 2524	**Ø 10×8**	**20**	**90**	**21900**
1%Ce, 1851 [2]	Ø 10×2	25	99	20800
CsI-Tl	Ø 10×2	11.2	1000	65000
NaI-Tl	Ø 10×18	9.3	250	43000

Upon the results of the above-described performed measurements and published data [9] we managed to establish an energy diagram (figure 4), which illustrates energy transmission processes in Ce:YAG ceramics synthesized using the sintering aid SiO_2 and without it. The luminescence excitation and decay processes may develop as follows. Depending on the excitation energy level, there can be observed either light emission by STE in connection with interband transfer of electrons O2*p*↔Y4*d*, or light emission by impurity Ce^{3+} ions due to 4*f*↔5*d* transitions in Ce^{3+} ions. Due to availability of an energy transmission channel and increased possibility of non-radiative relaxation by the effect of phonons in case of the increase of Ce^{3+} activator concentration, STE band luminescence intensity decreases. The effect of the sintering aid SiO_2 on the energy dissipation processes in Ce:YAG is illustrated in a form of changed splitting of $5d_1$ and $5d_2$ levels and a shallow electron trap is occurred.

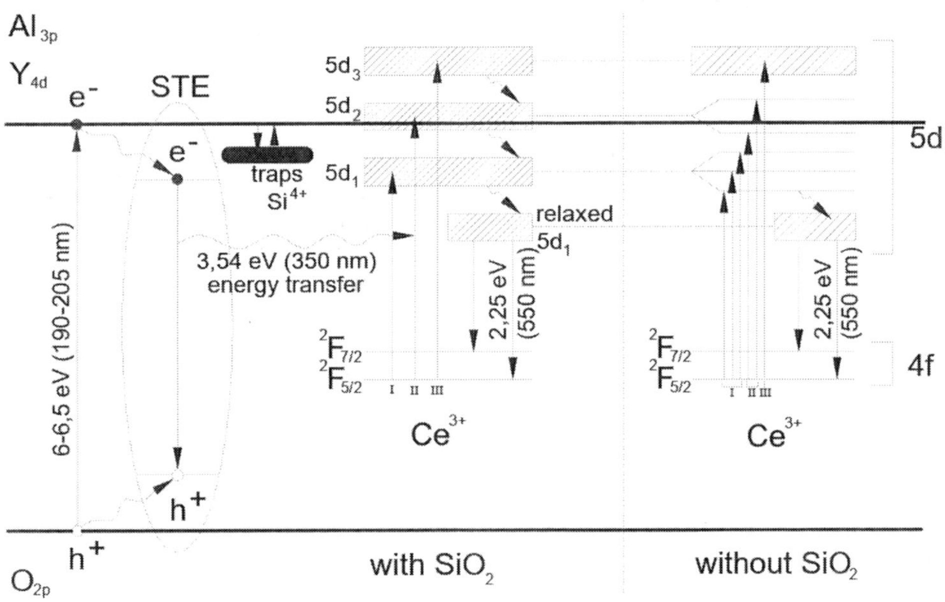

Figure 4. Energy diagram of luminescence processes in Ce:YAG with and without SiO_2

4. Conclusions

Scintillation Ce:YAG ceramics with the thicknesses of 2 and 8 mm and Ce^{3+} concentrations of 0.1, 0.5, 2 at% was synthesized by the solid-phase reaction method from nanopowders produced by the laser ablation method using the sintering aid SiO_2 and without it and without hot isostatic pressing. It is shown that using of the sintering aid (0.14 wt% SiO_2) in this synthesis method leads to degradation of optical and scintillation characteristics.

It is determined that the bleaching annealing duration for removal of F-centers in the ceramics for the samples with the thickness of 8 mm is 3 times longer than for those with the thickness of 2 mm. The absorption coefficient on the wavelength of 550 nm for the sample with 0.5% Ce:YAG (without the sintering aid) is the lowest (0.31 cm^{-1}) among all the ceramics being previously synthesized by us. It is revealed that the PLE spectrum of thin samples (2 mm thick) broadens with the growth of the concentration. The bands in the excitation spectrum of the sample with the thickness of 8 mm and 0.5% Ce have the maximum width.

The decay constant of the fast component in the sample with 0.5% Ce:YAG (8 mm thick) synthesized without the sintering aid is 90 ns, which corresponds to those (87 ns) in the ceramics with 2% Ce:YAG and 2 mm thickness, synthesized using the sintering aid (0.14 wt% SiO_2). The contribution of the slow component in the ceramic sample (8 mm thick) synthesized without the sintering aid is more than 5 times lower than in the sample (2 mm thick) synthesized using the sintering aid. The maximum light output has been achieved in the ceramics without the sintering aid (21900 photon/MeV – 0.5% Ce:YAG, h = 8 mm) compared with ceramics synthesized by us previously (20800 photon/MeV – 1% Ce:YAG, h = 2 mm).

Acknowledgments
This work was supported in part by RFBR grant № 19-08-00117. The authors gratefully acknowledge Dr. V.V. Platonov, R.N. Maksimov and Prof. V.I. Solomonov from Institute of Electrophysics UrB RAS for laser synthesis of nanopowders, optical polishing and fruitful discussion, respectively.

References
[1] Greskovich C, Duclos S 1997 *Annu. Rev. Mater. Sci.* **27** (1) 69–88
[2] Osipov V, Ishchenko A, Shitov V et al 2017 *Opt.Mat.* **71** 98-102
[3] Kochawattana S, Stevenson A et al 2008 *Journal of the European Ceramic Society* **28** 1527–1534
[4] With G and Dijk H 1984 *Mat. Res. Bull.* **19** 1669-1674
[5] Ikesue A, Aung Y et al 2006 *Annu. Rev. Mater. Res.* **36** (1) 397 – 429
[6] Zhang H, Chen Y, Zhu X, Zhou H, Yao Y and Li X 2019 *J. Lumin.* **207** 477–81
[7] Lukyashin K, Shitov V et al 2018 *IOP Conf. Ser.: Mater. Sci. Eng.* **347** 012013
[8] Osipov V, Kotov Yu et al 2006 *Las. Phys.* **16** 116-125
[9] You F, Bos A J J, Shi Q, Huang S and Dorenbos P 2012 *Phys. Rev. B - Condens. Matter Mater. Phys.* **85** 1–7

IOP Conf. Series: Materials Science and Engineering **525** (2019) 012036 doi:10.1088/1757-899X/525/1/012036

Carbazole-based donor-acceptor small molecules with hexyldicyanovinyl electron-withdrawing groups: synthesis and properties

A N Solodukhin[1], Yu N Luponosov[1,2]* and S A Ponomarenko[1,2]

[1]Enikolopov Institute of Synthetic Polymeric Materials of the Russian Academy of Sciences, Profsoyuznaya st. 70, Moscow 117393, Russia
[2]Chemistry Department, Moscow State University, Leninskie Gory 1-3, Moscow 119991, Russia

E-mail: luponosov@ispm.ru

Abstract. In this work, synthesis of two novel donor-acceptor small molecules based on donor 9-phenyl-9*H*-carbazole and acceptor hexyl-dicyanovinyl units was carried out. A comparative analysis of these molecules differing in the position of bithiophene π-spacer at the carbazole block only showed that changing the position allows to fine-tune absorption spectra, HOMO value, bandgap and phase behaviour of the materials.

1. Introduction

Donor-acceptor (D-A) small molecules are one of the most interesting and promising class of materials among various π-conjugated molecules for application in organic electronics, such as organic solar cells [1] and organic light emitting diodes [2] and organic field effect transistors. Binding of donor and acceptor moieties within one molecular structure allows fine-tuning properties of target materials: absorption and emission spectra and energy value of frontier orbitals - the highest occupied molecular orbital (HOMO) and lowest unoccupied molecular orbital (LUMO) [3]. Carbazole is a donor block which combines low redox potential and high chemical stability and uses actively to design functional materials for organic electronics [4-9]. In this work, we designed and synthesized novel D-A molecules by attaching bithiophene arms end-capped with electron-withdrawing hexyl-dicyanovinyl groups to the carbazole block at either 2 and 7 or 3 and 6 positions (Figure 1, molecules I and II, respectively). The optical and electrochemical properties and phase behavior of the obtained materials were studied and compared to each other.

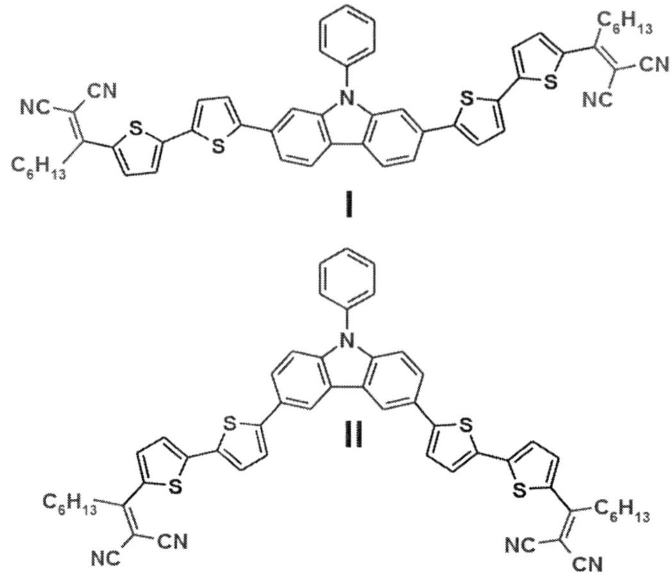

Figure 1. Structural formulas of I and II.

2. Experimental part

2.1. Materials

Malononitrile and tetrakis(triphenylphosphine)palladium(0) ($Pd(PPh_3)_4$) were purchased from Sigma-Aldrich Co and used as received. THF, toluene, hexane, and pyridine were dried and purified according to known procedures. All reactions, unless otherwise specified, were performed in an argon atmosphere and dry solvents. 2,7-dibromo-9-phenyl-9*H*-carbazole (1) [10], 3,6-dibromo-9-phenyl-9*H*-carbazole (2) [11], 2-[5'-(2-hexyl-1,3-dioxolan-2-yl)-2,2'-bithien-5-yl]-4,4,5,5-tetramethyl-1,3,2-dioxaborolane (3) [12] and 2-hexyl-5,5-dimethyl-2-[5'-(4,4,5,5-tetramethyl-1,3,2-dioxoborolane-2-yl)-2,2'-bithien-5-yl]-1,3-dioxane (4) [13] were prepared as described earlier.

2.2. Synthesis of molecules

Synthesis of 2,7-bis[5'-(2-hexyl-5,5-dimethyl-1,3-dioxan-2-yl)-2,2'-bithien-5-yl]-9-phenyl-9*H*-carbazole, 5a. In an inert atmosphere, $Pd(PPh_3)_4$ (0.08 g, 0.07 mmol) was added to a degassed solution of 2,7-dibromo-9-phenyl-9*H*-carbazole (1) (0.40 g, 1.00 mmol) and 2-hexyl-5,5-dimethyl-2-[5'-(4,4,5,5-tetramethyl1,3,2-dioxoborolan-2-yl)-2,2'-bithien-5-yl]-1,3-dioxane (3) (1.07 g, 2.39 mmol) in a mixture of toluene/ethanol (20/2 mL) and 2 M aqueous solution of sodium carbonate (3.6 mL). The reaction mixture was refluxed for 10 h, cooled to room temperature, and poured into a dropping funnel containing distilled water (100 mL) and toluene (100 mL). The organic phase was separated, washed with water, dried over sodium sulfate, and filtered. The solvent was evaporated in vacuum and the residue was dried at 1 Torr. The product was purified by column chromatography on silica gel using mixture of toluene–hexane (1 : 1) as the eluent to obtain a pure compound 5a (0.8 g, 90%) as a light brown powder. ^1H NMR (250 MHz, $CDCl_3$), δ: 0.79–0.93 (overlapping signals, 6H), 1.21–1.45 (overlapping signals, 16H), 1.94–2.05 (overlapping signals, 4H), 3.92–4.09 (overlapping signals, 8H), 6.89 (d, 2H, $J = 3.7$ Hz), 7.03 (d, 2H, $J = 3.7$ Hz), 7.08 (d, 2H, $J = 3.7$ Hz), 7.22 (d, 2H, $J = 3.7$ Hz), 7.50–7.74 (overlapping signals, 9H), 8.07 (d, 2H, $J = 8.9$ Hz). Calculated (%) for $C_{52}H_{53}NO_4S_4$: C, 70.63; H, 6.04; N, 1.58; S, 14.50. Found: C, 70.74; H, 6.24; N, 1.53; S, 14.39.

Synthesis of 3,6-bis[5'-(2-hexyl-5,5-dimethyl-1,3-dioxan-2-yl)-2,2'-bithien-5-yl]-9-phenyl-9*H*-carbazole, 5b. This compound was obtained according to the procedure described for compound 5a using 3,6-dibromo-9-phenyl-9*H*-carbazole (2) (0.70 g, 1.75 mmol), 2-hexyl-5,5-dimethyl-2-[5'-(4,4,5,5-tetramethyl-1,3,2-dioxaborolan-2-yl)-2,2'-bithien-5-yl]-1,3-dioxane (4) (2.06 g, 4.20 mmol), 2 M sodium carbonate (6.3 mL), toluene/ethanol (40/4 mL), and $Pd(PPh_3)_4$ (0.15 g, 0.13 mmol). The product was purified by column chromatography on silica gel (eluent - toluene : hexane 1:1) to give a pure compound 5b (1.2 g, 71%) as a light brown powder. ^{1}H NMR (250 MHz, $CDCl_3$), δ: 0.66 (s, 6H), 0.80–0.91 (overlapping signals, 6H), 1.24 (s, 18H), 1.35–1.51 (overlapping signals, 4H), 1.79–1.97 (overlapping signals, 4H), 3.41 (d, 4H, J = 11.3 Hz), 3.74 (d, 4H, J = 11.3 Hz), 6.88 (d, 2H, J = 3.7 Hz), 7.08– 7.17 (overlapping signals, 4H), 7.28 (d, 2H, J = 3.7 Hz), 7.38 (d, 2H, J = 8.3 Hz), 7.50– 7.71 (overlapping signals, 7H), 8.34–8.42 (overlapping signals, 2H). Calculated (%) for $C_{58}H_{65}NO_4S_4$: C, 71.94; H, 6.77; N, 1.45; S, 13.24. Found: C, 71.97; H, 6.82; N, 1.44; S, 13.16.

Synthesis of 1,1'-[(9-phenyl-9*H*-carbazole-2,7-diyl)bis(2,2'-bithiene-5',5-diyl)]diheptan-1-one, 6a. 1M hydrochloric acid (2 mL) was added to a solution of compound 5a (0.65 g, 0.7 mmol) in THF (30 mL) and the reaction mixture was refluxed with stirring for 4 h. Upon completion of the reaction, the product formed as yellow precipitate. The organic phase was separated using diethyl ether, washed with water, and filtered off to give a pure compound 6a (0.58 g, 98%) as yellow powder. ^{1}H NMR (250 MHz, $CDCl_3$), δ: 0.84–0.94 (overlapping signals, 6H), 1.27–1.42 (overlapping signals, 12H), 1.67–1.84 (overlapping signals, 4H), 2.85 (t, 3H, J = 7.6 Hz), 7.15 (d, 2H, J = 4.0 Hz), 7.25–7.31 (overlapping signals, 5H), 7.52– 7.64 (overlapping signals, 9H), 7.65–7.72 (overlapping signals, 2H), 8.08 (d, 2H, J = 8.9 Hz). Calculated (%) for $C_{48}H_{45}NO2S_4$: C, 72.41; H, 5.70; N, 1.76; S, 16.11. Found: C, 72.52; H, 5.79; N, 1.74; S, 15.99.

Synthesis of 1,1'-[(9-phenyl-9*H*-carbazole-3,6-diyl)bis(2,2'-bithiene-5',5-diyl)]diheptan-1-one, 6b. This compound was obtained according to the procedure described for compound 4a using 1M hydrochloric acid (4 mL), compound 5b (1.18 g, 1.2 mmol), and THF (18 mL), which yielded a pure compound 6b (0.79 g, 80%) as a yellow powder. ^{1}H NMR (250 MHz, $CDCl_3$), δ: 0.85–0.94 (overlapping signals, 6H), 1.28–1.43 (overlapping signals, 12H), 1.69–1.82 (overlapping signals, 4H), 2.87 (t, 4H, J = 7.3 Hz), 7.08–7.22 (overlapping signals, 2H), 7.29–7.73 (overlapping signals, 15H), 8.86 (s, 2H). Calculated (%) for $C_{48}H_{45}NO2S_4$: C, 72.41; H, 5.70; N, 1.76; S, 16.11. Found: C, 72.54; H, 5.81; N, 1.73; S, 15.96.

Synthesis of 2,2'-[(9-phenyl-9*H*-carbazole-2,7-diyl)bis(2,2'-bithiene-5',5-diylhept-1-yl-1-ylidene)]dimalononitrile, I. Compound 6a (0.52 g, 0.7 mmol), malononitrile (0.26 g, 3.9 mmol), and dry pyridine (12 mL) were placed in a reaction vessel and stirred under argon atmosphere for 8 hours at reflux using microwave radiation. Upon completion of the reaction, pyridine was evaporated in vacuum and the residue was dried at 1 Torr. The resulting residue was purified by column chromatography on silica gel using dichloromethane as the eluent. Further purification included the precipitation of the product with hexane from its solution in THF to give pure product as a black solid (0.47 g, 81%). ^{1}H NMR (250 MHz, $CDCl_3$), δ: 0.85–0.96 (overlapping signals, 6H), 1.26–1.49 (overlapping signals, 12H), 1.63–1.76 (overlapping signals, 4H), 2.91 (t, 3H, J = 7.6 Hz), 7.25–7.35 (overlapping signals, 6H), 7.52–7.62 (overlapping signals, 7H), 7.67–7.75 (overlapping signals, 2H), 7.93 (d, 2H, J = 4.3 Hz), 8.09 (d, 2H, J = 7.9 Hz. ^{13}C NMR (125 MHz, $CDCl_3$), δ: 13.88, 22.41, 29.19, 30.40, 31.28, 37.56, 76.88, 106.94, 113.79, 114.62, 118.87, 121.00, 123.30, 124.52, 124.82, 127.35, 127.50, 128.32, 130.37, 131.75, 134.30, 134.96, 135.34, 137.06, 142.30, 146.68, 147.65, 166.28. Calculated (%) for $C_{54}H_{45}N_5S_4$: C, 72.69; H, 5.08; N, 7.85; S, 14.37. Found: C, 72.78; H, 5.13; N, 7.88; S, 14.28.

Synthesis of 2,2'-[(9-phenyl-9*H*-carbazole-3,6-diyl)bis(2,2'-bithiene-5',5-diylhept-1-yl-1-ylidene)]dimalononitrile, II. This compound was obtained according to the procedure described for I using compound 6b (0.73 g, 0.9 mmol), malonitrile (0.36 g, 5.4 mmol), and pyridine (15 mL), which yielded a pure product (0.49 g, 60%) as a black solid. ^{1}H NMR (250 MHz, $CDCl_3$), δ: 0.77–0.97 (broad signal, 6H), 1.19–1.53 (overlapping signals, 12H), 1.58–1.79 (broad signal, 4H), 7.24 (d, 2H, J = 4.1 Hz), 7.29–7.43 (overlapping signals, 6H), 7.48–7.58 (overlapping signals, 3H), 7.59–7.71 (overlapping signals, 4H), 7.91 (d, 2H, J = 4.7 Hz), 8.28–8.38 (overlapping signals, 2H). ^{13}C NMR (125 MHz, $CDCl_3$), δ: 13.98, 22.45, 29.23, 30.49, 31.26, 37.42, 76.03, 110.67, 113.94, 114.78, 117.73, 123.49, 123.64, 124.47, 124.61, 125.97, 126.81, 127.68, 128.10, 130.13, 133.31, 134.81, 135.26, 136.77, 141.28, 147.08, 147.76, 166.21. Calculated (%) for $C_{54}H_{45}N_5S_4$: C, 72.69; H, 5.08; N, 7.85; S, 14.37. Found: C, 72.75; H, 5.11; N, 7.86; S, 14.31.

IOP Conf. Series: Materials Science and Engineering **525** (2019) 012036 doi:10.1088/1757-899X/525/1/012036

2.3. *Characterization*

^{1}H NMR spectra were recorded at a "Bruker WP-250 SY "spectrometer, working at a frequency of 250.13 MHz and utilising $CDCl_3$ signal (7.25 ppm) as the internal standard. ^{13}C NMR spectra were recorded using a "Bruker Avance II 300" spectrometer at 75 MHz. In the case of ^{1}H NMR spectroscopy, the compounds to be analyzed were taken in the form of 1% solutions in $CDCl_3$. In the case of ^{13}C NMR spectroscopy, the compounds to be analyzed were taken in the form of 5% solutions in $CDCl_3$. The spectra were then processed on the computer using the ACD Labs software.

Elemental analysis of C, N and H elements was carried out using CHN automatic analyzer CE 1106 (Italy). The settling titration using $BaCl_2$ was applied to analyze sulfur. Experimental error for elemental analysis is 0.30-0.50%. The Knövenagel condensation was carried out in the microwave "Discovery", (CEM corporation, USA), using a standard method with the open vessel option, 50 watts. In the case of column chromatography, silica gel 60 ("Merck") was taken. Absorption profiles were recorded with a Perkin Elmer Lambda-35 absorption spectrometer from 350 to 1100. All measurements were carried out at room temperature in diluted solutions (10^{-5} M) of THF. Films were cast from THF solutions on quartz substrate.

Cyclic voltammetry measurements were carried out using solid compact layers of the oligomers, which in turn were made by electrostatically rubbing the materials onto a glassy carbon electrode. Measurements were made in CH_3CN:o-$C_6H_4Cl_2$=1:4 solution using 0.1 M Bu_4NPF_6 as supporting electrolyte and using IPC-Pro M potentiostat. The scan rate was 200 mVs^{-1}. The glassy carbon electrode was used as a work electrode. Potentials were measured relative to a saturated calomel electrode (SCE). The HOMO and LUMO energy levels were calculated using the first standard formal oxidation and reduction potentials obtained from CV experiments in films according to the following equations: LUMO=-e(φ_{red}+4.40) (eV) and HOMO=-e(φ_{ox}+4.40) (eV) [14, 15].

DSC scans were obtained with Mettler Toledo DSC30 system with 20°C/min heating/cooling rate in temperature range of +20–290 °C for all compounds. N_2 flow of 50 mL/min was used.

3. Results and discussion

The molecules I and II were synthesized in three consecutive steps (Fig. 2). At the first step, precursors 5 a,b with protective dioxane groups were obtained by Suzuki cross coupling between branching center either (2,7-dibromo-9-phenyl-9*H*-carbazole (1) or 3,6-dibromo-9-phenyl-9*H*-carbazole (2)) and 2-[5'-(2-hexyl-1,3-dioxolan-2-yl)-2,2'-bithien-5-yl]-4,4,5,5-tetramethyl-1,3,2-dioxaborolane (3) or 2-hexyl-5,5-dimethyl-2-[5'-(4,4,5,5-tetramethyl-1,3,2-dioxaborolan-2-yl)-2,2'-bithien-5-yl]-1,3-dioxane (4) in yields of 90% and 71%, respectively. At the second step, the protective group was removed by refluxing of compounds 5a, 5b in THF at the presence of 1M hydrochloric acid as a catalyst. The yields of compounds 6a, 6b with carbonyl groups were 98% and 80%, respectively. At the last step, the Knoevenegal reaction between compounds 6a, 6b and malononitrile in pyridine resulted in the substitution of the carbonyl group for dicyanovinyl one. The crude products were purified by column chromatography to give D-A molecules I and II in yields of 81 and 60%, respectively. The chemical structures and purity of the molecules were determined by a set of modern physicochemical methods, such as ^{1}H and ^{13}C NMR spectroscopy, in combination with analytical GPC and elemental analysis. The NMR spectra displayed only typical signals for such compounds and the integral intensities of the signals agreed to the calculated ones. The determined contents of elements (C, H, and N) coincided well with the calculated values.

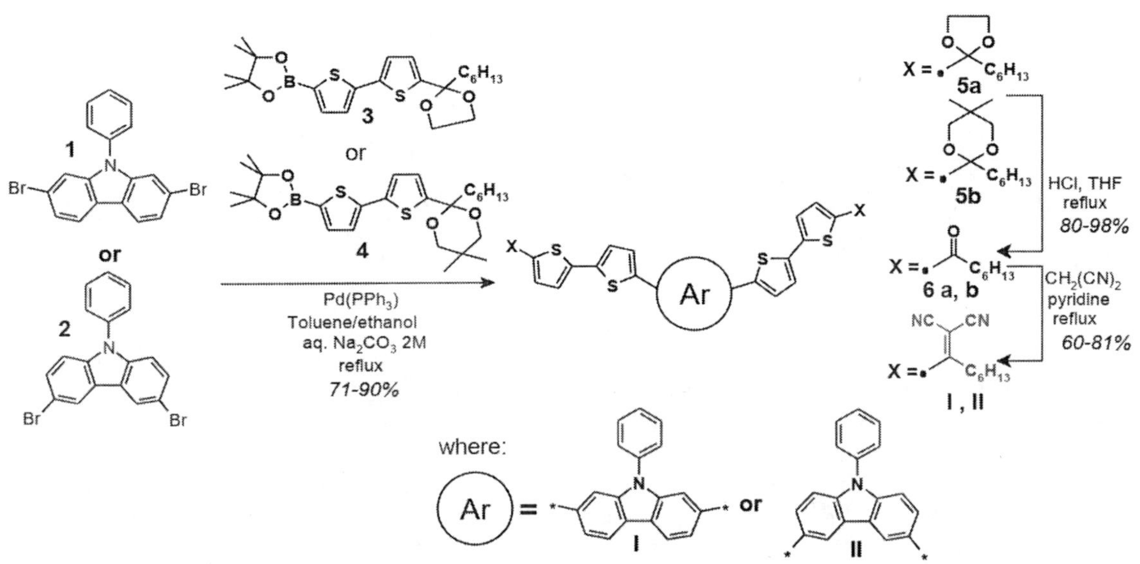

Figure 2. Synthesis of compounds I and II.

The optical properties of the synthesized molecules based on carbazole were studied by absorption spectroscopy in both solutions and thin films. The absorption spectra are shown in Fig. 3a.

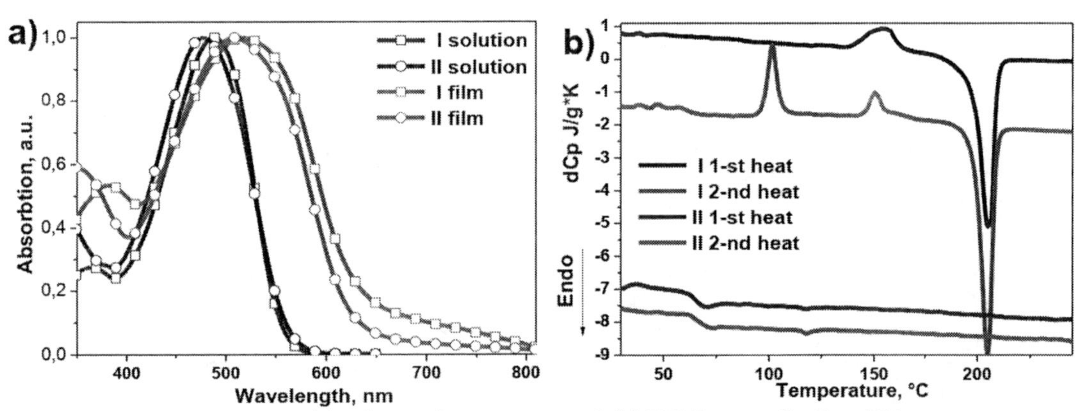

Figure 3. (a) Absorption spectra and (b) DSC scans for I and II.

The compounds efficiently absorb sunlight in the range from 350 nm to 650 nm. The spectra and absorption maxima (λ_{max}) of the molecules in diluted THF solutions have similar shape and peak values (Table 1). The spectra display two absorption bands: the first absorption band in a short-wavelength region is typically assigned to the π–π* transition and the second more intense band in a long-wavelength region is assigned to the intramolecular charge transfer between electron donating carbazole blocks with bithiophene fragments and electron-withdrawing dicyanovinyl substituents [16]. However, it has been shown that in fact both electronic excitations have a mixed character and the intensity of the high-energy absorption feature is a direct measure of the degree of conformational disorder in the molecule [17]. The absorption spectra in thin films obtained from molecules I and II by the solution method are broadened when compared to solutions and the absorption peaks are red-shifted by 30–34 nm (Table 1). Comparing the λ_{max} of the molecules one can see that the compound I has a bit red-shifted absorption both in solution and films.

Table 1. Optical and electrochemical properties of molecules I and II

	UV-vis spectroscopy		Cyclic voltammetry		
Molecule	Solution[a)] λ_{max}, nm	Film[b)] λ_{max}, nm	HOMO, eV	LUMO, eV	E_g, eV
I	488	518	-5.49	-3.30	2.19
II	477	511	-5.34	-3.30	2.04

[a] Measured in a THF solution.
[b] Film cast from THF solution.

Thus, the position of the bithiophene π-spacer relatively to the carbazole influence the λ_{max} of the compounds I and II, which can be explained by the fact that π-conjugation of the arms with carbazole through position 2 and 7 is more efficient as compared to that of through 3 and 6 positions. Therefore absorption spectra of compound I are slightly red-shifted both in solution and in the film as compared to that of compound II.

The electrochemical properties of the molecules were investigated by the cyclic voltammetry (CV). The reduction of the oligomers proceeds at the same values, since in these compounds the electron-withdrawing hexyl-dicyanovinyl group is mainly responsible for the reduction process and the position of the π-bridge should have no effect on that. However, the oxidation potentials were found to be depended significantly on the position of the thiophene π-bridge, since the oxidation occurs mainly through the carbazole-thiophene fragment. A facilitation of the oxidation process was observed when going from compound II to I. The values of HOMO and LUMO energies were calculated using the first standard formal oxidation and reduction potentials (Table 1). Values of the LUMO energy levels for this series of molecules are the same, whereas values of the HOMO energy levels tend to increase from I to II. The latter leads to a corresponding decrease of the electrochemical band gap from 2.19 eV to 2.04 eV.

The phase behavior of synthesized molecules were studied by differential scanning calorimetry (DSC). Figure 3b shows the thermograms for the first and second heats of the samples. It is seen that the compound II does not form a crystalline phase. The sample shows only a glass transition at 63–66°C region. In contrast, I has a pronounced tendency to crystallization. On the first DSC heating scan it can crystalize in region of 140 – 160 °C followed by melting at 205°C with enthalpy (ΔH_m) of 46 J/g. However, crystallization from the melt is hindered and on the second heating scan we observe a glass transition (60-65 °C) followed by two cold crystallization processes peaked at 101°C and 150 °C and then a compete melting at 205°C (ΔH_m = 46 J/g). Thus, the position of bithiophene arms at the carbazole block results in significant changes in the mechanism of intermolecular interaction and the appearance of an ordered phase.

4. Conclusions

To sum up, two donor–acceptor molecules based on the electron-donating 9-phenyl-9*H*-carbazole derivatives and electron-withdrawing hexyldicyanovinyl groups were designed, synthesized and characterized. The influence of the position of the bithiophene π-spacer at the carbazole block on the properties of donor-acceptor molecules was studied by optical absorption spectroscopy, CV and DSC techniques. It was shown that the attachment of bithiophene π-spacers to the carbazole at positions of 2 and 7 (I) leads to a red shift of the absorption spectra, decrease of HOMO energy values and appearance of the ability to form crystalline phases as compared to compound II, having the bithiophene arms at 3 and 6 positions of carbazole. The both materials have a potential to be used as donor component of active layer in bulk heterojunction organic solar cells.

Acknowledgments

The project has been implemented with the support of the Global Energy Association. Authors thank P. V. Dmitryakov (National Research Centre "Kurchatov Institute) for DSC experiments, E.A. Svidchenko

(ISPM RAS) for recording of absorption spectra and S.A. Peregudova (INEOS RAS) for CV experiments. The work was performed in the framework of leading science school NSh-5698.2018.3.

References

[1] Ilmi R, Haque A and Khan M S 2018 *Organic Electronics* **58** pp 53-62
[2] Duan L, Qiao J, Sun Y and Qiu Y 2011 *Advanced Materials* **23**(9) pp 1137-1144
[3] Ponomarenko S A, Luponosov Y N, Min J, Solodukhin A N, Surin N M, Shcherbina M A, Chvalun S N, Ameri T and Brabec C 2014 *Faraday Discuss.* **174** pp 313-339
[4] Wex B and Kaafarani B R 2017 *J. Mater. Chem. C.* **5**(34) pp 8622-8653
[5] Jiang H, Sun J and Zhang J 2012 **16**(17) pp 2014-2025
[6] Kato S, Yamada Y, Hiyoshi H, Umezu K and Nakamura Y 2015 *J. Org. Chem.* **80**(18) pp 9076–9090
[7] Damit E F, Nordin N, Ariffin A and Sulaiman K 2016 *Journal of Chemistry* **2016** pp 1–14
[8] Higginbotham H, Karon K, Ledwon P and Data P 2016 *Display and Imaging* **2** pp 207–216
[9] Bakirov A V, Solodukhin A N, Luponosov Yu N, Svidchenko E A, Obrezkova M A, Peregudova SM, Shcherbina MA, Ponomarenko SA and Chvalun SN 2017 *Nanotechnologies in Russia* **12** (7–8) pp 385–394.
[10] Fell V H K, Mikosch A, Steppert A-K, Ogieglo W, Senol E, Canneson D and Kuehne A J C 2017 *Macromolecules* **50**(6) pp 2338–2343
[11] Liu Y, Wu S, Wang G, Yu G, Guan J, Pan C and Wang Z 2014 *J. Mater. Chem. A* **2**(21) pp 7795–780
[12] Min J, Luponosov Y N, Gerl A, Polinskaya M S, Peregudova S M, Dmitryakov P V, Bakirov AV, Shcherbina M A, Chvalun S N, Grigorian S, Kaush-Busies N, Ponomarenko S A, Brabec C J 2013 Adv. Energy Mater **4**(5) p 1301234
[13] Luponosov Y N, Min J, Solodukhin, A N, Bakirov, A V, Dmitryakov, P V, Shcherbina, M A, Peregudova S M, Cherkaev G V, Chvalun S N, Brabec C J, Ponomarenko, S A 2016 *Journal of Materials Chemistry C* **4**(29) pp 7061–7076
[14] Ponomarenko S A, Kirchmeyer S, Elschner A, Alpatova N M, Halik M, Klauk H, Zschieschang U and Schmid G 2006 *Chem. Mater.* **18** p 579
[15 Ponomarenko S A, Rasulova N N, Luponosov Y N, Surin N M, Buzin M I, Leshchiner I, Peregudova S M and Muzafarov A M 2012 *Macromolecules* **45** p 2014
[16] Roquet S, Cravino A, Leriche P, Alévêque O, Frère P and Roncali J. 2006 *Journal of the American Chemical Society* **128**(10) pp 3459–3466
[17] Kozlov O V, Luponosov Y N, Ponomarenko S A, Kausch-Busies N, Paraschuk D Y, Olivier Y, Beljonne D, Cornil J and Pshenichnikov M S 2015 *Adv. Energy Mater.* **5**(7) p 1401657

Perspective approaches with the use of biocatalysts for improving the processes of polyaspartic acid production from oil benzene fraction after oxidative desulfurization

O Maslova[1], O Senko[1,2], N Stepanov[1,2] and E Efremenko[1,2]*

[1]Lomonosov Moscow State University, Moscow, Russia
[2] Emanuel Institute of Biochemical Physics RAS, Moscow, Russia

E-mail: elena_efremenko@list.ru

Abstract. Two promising approaches to greening the production of polyaspartic acid were demonstrated. The mixture of dibenzothiophene-sulfone ($DBTO_2$) and benzothiophene-sulfone (BTO_2) as potential products of oxidative desulfurization of petroleum products during benezene production was used for the first time for biocatalytic recovery to sulphide ions. The process was carrying out at the presence of biogenic H_2 produced during cultivation of anaerobic cells *Clostridium acetobutylicum* and *Desulfovibrio vulgaris*. It was shown that after 240 h conversion of sulfur, contained in mixture of 0.03 mM $DBTO_2$ and 0.03 mM BTO_2 was as minimum 32.2±2.1% using ethanol as extractant. This approach can be recommended for intensification and ecologization of oxidative desulfurization of various petroleum products. For production of stereoregular polymer of polyaspartic acid the correction of native process of chemical synthesis was done by addition of biocatalytic stage during conversion of ammonium fumarate into aspartic acid. The obtained immobilized biocatalyst on a base of *E. coli* BL-21 cell included into poly(vinyl) alcohol cryogel make it possible to accumulate 129.1±3.4 g/L aspartic acid during biotransformation of ammonium fumarate.

1. Introduction

More than forty states worldwide currently pursue explicit political strategies to expand and promote their bioeconomies. In connection with global environmental problems and the gradual depletion of resources the modern chemical production, as it is organized now, requires significant modernization. The main accent made on introduction of new ecoefficient technological decisions and realization of principles of bioeconomy [1]. Complete and deeper conversion of resources, minimization of wastes, the use of renewable raw materials, selective catalysts and biocatalysts, while the overall production efficiency should not be lower than that of existing analogues. Obviously, that achievement of all these targets is possible only through a comprehensive review of all stages, starting from the raw materials and ending with the production of a wide range of needed products that can be obtained during of a multistage conversion practically without waste formation. Increasing the market potential of such industries and processes is mainly possible if at least some of the products will be potentially with high value-added products or components for production such products.

Today such products are presented in innovative directions of medicine and pharmaceutical industry. Polymers based on lysine, aspartic (AA) and glutamic (GA) amino acids containing side functional groups that can be easily chemically modified are highly demand in dynamically developing innovative industries [2-3]. On their basis already obtained inexpensive biocompatible, non-toxic, biodegradable polymers, composite materials with desired properties, new systems of targeted drug delivery, therapeutic peptides, enzymes, combined systems for the therapy and diagnosis of a number of diseases. A number of copolymers-ampholytes on the basis of polyAA, polyGA or polylysine can show coassociative behavior with the formation of nanoparticles, which is especially valuable for the development of stable functioning systems of targeted delivery of pharmaceutical ingredients, monitoring of their prolonged releases and operations [2, 4].

Previously the authors of this work were demonstrate new effective approaches to the use of non-toxic polyelectrolyte complexes of polyAA and polyGA with hexahistidine-tagged organophosphorus hydrolase (His_6-OPH) in the agricultural, pharmaceutical and medical areas: for 100% degradation of pesticides in soils, [5], detoxification of animal feed [6], and increase the effectiveness of antibiotics [7-9].

Thereby, polyAA, polyGA and polylysine are products with potentially high added value, and modernization of their production, from the point of view of implementing the principles of green chemistry, is perspective.

Today the wide range of biocatalytic production processes of polyGA and polylysine from renewable sources are known. PolyGA can be obtained using biocatalytic method in one stage. [10]. In industry PolyAA is obtained by chemical synthesis from maleic anhydride during of multistage conversion of oil-and-gas substrates with production a non-stereo-regular product [2].

The absence of stereoregularity practically does not affect on biodegradability index. However, at the presence of calcium ions the chelating ability of non-stereoregular polyAA polymers was decrease on relation to cations [11]. This fact overlay restrictions on possible use of such polymer, in particular, in innovative areas. For example, when developing a dendriplexes based on electrostatic interactions for gene delivery systems, it is preferable to use stereoregular polymers of polyacids [12].

Thereby, the search of opportunities to modernize the scheme of the process for stereoregular polyAA production with a glance of principles of green chemistry and the needs of actively developing innovative areas is perspective and promising. The purpose of this work is demonstrate two promising approaches to greening the production of polyAA: (I) bioutilization of wastes, obtained after oxidative desulfurization of oil Benzene fraction, using *D. vulgaris* and *C. acetobutilycum* as biocatalysts, and (ii) correction of the initial chemical synthesis (Figure 1A), including into the process the biocatalytic stage with *E. coli* cells with aspartase (aspartase ammonia lyase, EC 4.3.1.1) activity immobilized in the polyvinyl alcohol (PVA) cryogel as biocatalyst for conversion ammonium fumarate to AA – monomer for synthesis of α-polyAA or β- polyAA (Figure 1B).

2. Materials and methods

2.1 Microorganisms and cultivation conditions

The bacterial strains *Clostridium acetobutylicum* B1787 and *Desulfovibrio* were obtained from the Russian National Collection of Industrial Microorganisms (www.genetika.ru). The *Clostridium acetobutylicum* strain B1787 was cultivated in the following medium (g/L): glucose – 20; triptone – 10; yeast extract – 5 (pH 6.8). The *Desulfovibrio vulgaris* strain B4053 was cultivated in the Postgate medium [13]. Cultivation of bacterial cells was performed under anaerobic conditions in an argon atmosphere at 37°C for 20–24 h and 48 h, respectively.

For conversion of ammonium fumarate to AA bacterial cells of *E.coli* SG13009[pREP4] (Qiagen, Hilden, Germany), BL-21 (Novagen, Darmstadt, Germany) and DH5α (Thermo Fisher Scientific, MA, USA) were used.

2.2 Methods.

2.2.1. Preparation of immobilized biocatalyst on base of E.coli cells for AA synthesis

Different samples of granules of immobilized biocatalyst (IBC) containing bacterial *E.coli* cells immobilized in polyvinyl alcohol cryogel were prepared by cryoimmobilization using a previously published technique [14]. IBC formation was carried out at -18°C during 30 h. After that, the granules (cylinders, d = 5.9 ± 0.1mm, h = 5.0 ± 0.1mm) were thawed at 4°C for 12 h, and the obtained samples of IBC were used in subsequent investigations.

For AA production in continuous and repeated modes at pH 8,5 and 37°C the following medium

IOP Conf. Series: Materials Science and Engineering **525** (2019) 012037 doi:10.1088/1757-899X/525/1/012037

was used (g/L): ammonium fumarate - 150, $MgSO_4 \times 7H_2O$ – 0,25.

2.2.2. Determination of concentrations sulphide ions and AA

Sulphide ions in liquid phase was monitored spectrophotometrically at 660 nm using Shimadzu UV-1202 (Japan) [15]. For the assay of AA concentration we used assay kit: Amplite™ Colorimetric (575±5 nm) L-Aspartate (Aspartic Acid) Assay Kit (AAT Bioquest, Sunnyvale, CA, USA).

The data were shown as means of at least three independent experiments ± standard deviation (± SD). Statistical analysis was realized using SigmaPlot 12.5 (ver. 12.5, Systat Software Inc., San Jose, CA, USA).

3. Results and Discussion

3.1. New approaches to the polyAA production scheme

Today, the most common process for polyAA production is chemical synthesis through the hydrolysis of polysuccinimide with formation of non-stereoregular polymer containing a mixture of α and β forms (Figure 1A) [2, 16].

A modified hybrid scheme of polyAA synthesis was proposed in this work using AA as initial monomer for stereoregular polyAA production (Figure 1B). This solution (Figure 2B) contains greater number of stages, but, despite of this, it's competitive and reasonable in terms of implementing the principles of green chemistry. All additional chemical stages were tested and actively used on an industrial scale in technological schemes for the production of other products. The production scheme of polyAA after the proposed modernization contains a stage of biocatalytic transformation of chemically obtained ammonium fumarate into a non-toxic monomer – AA (Figure 2B, gray area with *E.coli* immobilized in the PVA cryogel). For the first time for carrying out this reaction the use of a biocatalyst immobilized in PVA cryogel has been proposed. This carrier has been previously successfully used for production of biocatalysts based on cells of various microorganisms [17, 18].

The AA can be then transformed into a stereoregular polymer by known chemical methods. So the synthesis of α-polyAA can be realized using *N*-carboxyanhydride (NCA) method with protected pendant group [19, 20].

For this purpose AA is preliminarily transformed into aspartic acid β-benzyl ester, which then is convert by cyclization (the Leuchs method) method into the N-carboxyanhydride [21]. Further by the ring-opening polymerization of *N*-carboxyanhydride is obtained poly(β-benzyl-aspartate) and then hydrolyzed for conversion into α-polyAA [22]. By the ring-opening anionic assisted polymerization, poly (α-benzyl-aspartate) can be synthesized from the previously obtained aspartic β-lactam benzyl ester (3(R)-benzoxycarbonyl-2-azetidinone) by using solvents with low polarity. Poly(α-benzyl-aspartate) may be hydrolyzed for conversion into β-polyAA [23]. The price of stereoregular polyAA forms today is 6 times higher than the price of non-stereoregular forms (www.sigmaaldrich.com), in this regard, their production is more economically attractive.

The proposed process modernization is partially affects the production of benzene from hydrocarbons - the initial chemical compound in the polyAA synthesis scheme (Figure 1). Today, the production of benzene is based on the conversion of a number of raw materials: oil, toluene, the heavy pyrolysis fraction, and the coking coal tar. The main amount of benzene is obtained by catalytic reforming of a heavy oil gasoline fraction (naphtha), which is present itself a mixture of paraffins, naphthenes and aromatic hydrocarbons of the C6-C9 fraction, boiling away at 62-180°C. The reforming process is carried out in the presence of a catalysts (Al_2O_3/Pt/Re and other) at a temperature of 500-530°C and at pressure of 18-35 atm (or at 2-3 atm using machineries with continuous regeneration). The resulting product (stable reformate containing benzene) is cooled and removed. Crude benzene is mainly presented as a mixture of low-boiling aromatic hydrocarbons (80-95%), unsaturated hydrocarbons (5-15%), sulfur-containing compounds (0.2-2.0%) and others. The sulfur compounds are presented in raw benzene as carbon disulfide (CS_2), thiophene (C_4H_4S) and its homologues. Even minor impurities of sulfur in benzene, used for organic synthesis, can cause rapid deactivation of metal catalysts.

IOP Conf. Series: Materials Science and Engineering **525** (2019) 012037 doi:10.1088/1757-899X/525/1/012037

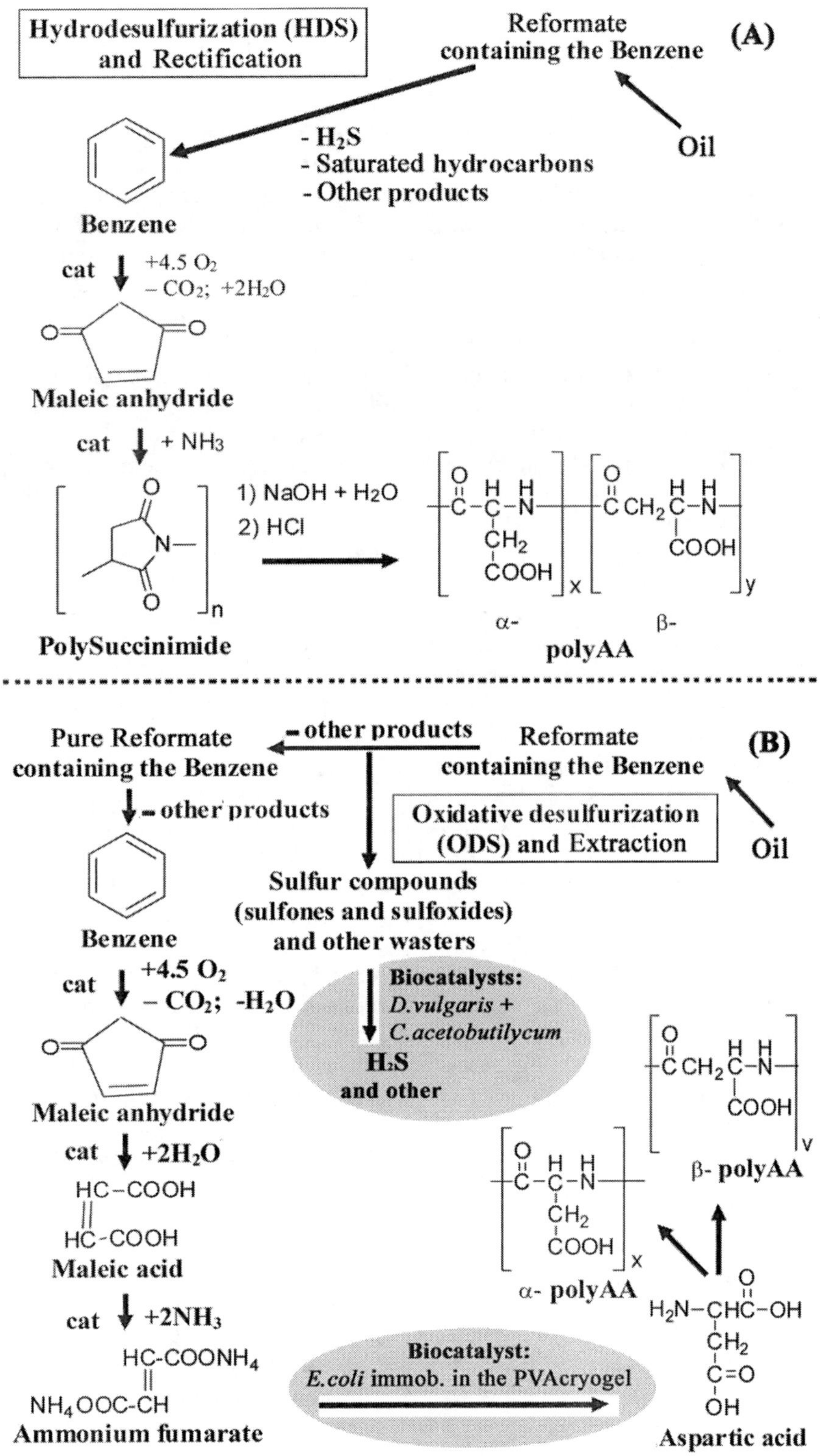

Figure 1. Scheme of non-stereoregular polyAA production from benzene, obtained from oil reformate and purified from sulfur compounds by hydrodesulfurization (HDS) (A) and Scheme of stereoregular polyAA forms production from benzene obtained, from oil reformate and purified from sulfur compounds by oxidative desulfurisation (ODS).

IOP Conf. Series: Materials Science and Engineering **525** (2019) 012037 doi:10.1088/1757-899X/525/1/012037

In this regard, it is necessary to purified the reformate from sulfur-containing and unsaturated compounds before distillation of reformate (when the main part of benzene boils at a temperature below 180°C) in order to isolate benzene, toluene, xylenes and solvent (solvent naphtha). For this purpose a low-waste method of catalytic hydrotreatment is recommended, which consists in treating of crude benzene and its fractions with hydrogen or coke oven gas. The target purification reactions are hydrodesulfurization (HDS) and hydrogenation of unsaturated hydrocarbons. Under preparation of high purity benzene the hydrodesulfurization reactions has a key role, especially the hydrogenolysis (destruction) of the most thermally stable compound-thiophene. The hydrodesulfurization catalysts are presented as sulfides or oxides of molybdenum, cobalt, tungsten, nickel, vanadium.

The process of hydrogenation is carried out under pressure in the gas phase over the catalyst; for this purified products are completely transformed into a vapor state and separate from the non-evaporating residue. The resulting vapors are mixed with pure hydrogen or with coke oven gas (about 57–60% of hydrogen) and go through the catalytic hydrotreating. Under the influence of temperature and pressure, various reactions take place, including the following: hydrodesulfurization reactions, hydrogenation of unsaturated compounds, hydrogenation of aromatic hydrocarbons, hydrocracking of saturated hydrocarbons, demethylation of benzene homologues. The chemistry of the hydrotreating process is consist in the sulfur-containing impurities are hydrogenated to form the corresponding hydrocarbons and hydrogen sulfide.

Until recently, hydrodesulfurization was considered the most effective way to purify hydrocarbon feedstock from sulfur compounds, however, in connection with hardening requirements for sulfur content, hydrofining capabilities almost reached the limit, and bringing the sulfur content in petroleum products to the required standards of 0.001 wt.% and below using this method is not economically rational [24]. As an alternative to hydrotreatment, hydrogenless methods for the removal of sulfur compounds are activity considered, of which the most perspective and low-cost from an economic point of view, are presented a catalytic oxidative desulfurization (ODS) [25].

In contrast to hydrodesulfurization, this approach is involves the catalytic treatment of the sulfur-containing fraction under mild conditions with a low-cost oxidant (for example, H_2O_2) with formation of dibenzothiophene, benzothiophene, etc., which are further separated by an extraction method. Catalytic systems have been developed for the efficient oxidative desulfurization of various petroleum fractions using H_2O_2 and O_3 and the most effective methods were founded for extracting of the oxidation products of sulfur compounds from hydrocarbon fractions using the extraction method [26, 27].

This approach is generally promising, including the possibilities of modernizing the technological scheme for polyAA production at the stage of obtaining of highly purified benzene. However, at the present time the question of environmentally safe and economically viable utilization of extracts containing sulfur compounds is still not resolved. During the modernization of the scheme presented in this work for the first time was proposed to recover such extracts by anaerobic biocatalytic way with *D.vulgaris* and *C.acetobutilycum* (Figure 1B gray area). If successful, this approach to the anaerobic degradation of waste oxidative desulfurization of petroleum products can be further combined with the production of commercial products, such as biogas.

3.2. Biotransformation of model waste oxidative desulfurization of hydrocarbons used to produce benzene

As a model extract containing oxidized forms of organic sulfur-containing compounds, which can be obtained after the oxidative desulfurization of hydrocarbons during the production of benzene, solutions containing a mixture of 0.03 mM dibenzothiophene-sulfone ($DBTO_2$) and 0.03 mM benzothiophene- sulfone (BTO_2) in ethanol or N,N-dimethylformamide (DMFA) were used. The cells of *D. vulgaris* and *C. acetobutilycum* were tested as biocatalysts for biotransformation of these compounds. To support the effectiveness of the action of biocatalysts, glucose (3 g /L) was used as a substrate.

From the obtained data (Figure 2), it is follow that after 240 h at least 30% of sulfur contained in a mixture of 0.03 mM BTO_2 and 0.03 mM $DBTO_2$ can be transformed into sulfide ions using ethanol as a potential extractant.

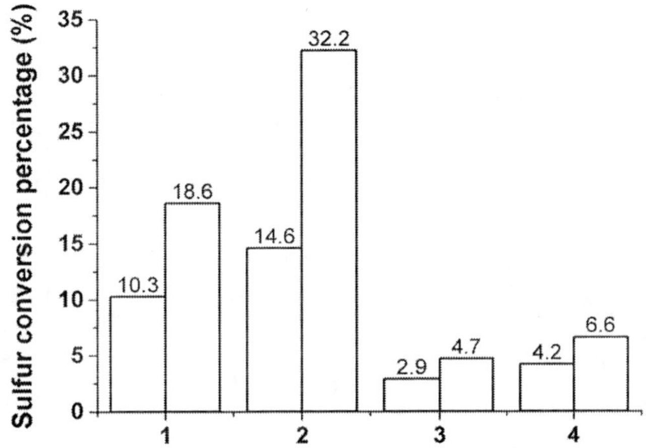

Figure 2. Biotransformation of sulphur from 0.03 mM $DBTO_2$ + 0.03 mM BTO_2 to S^{2-} with *D.vulgaris* (white bars) or *D.vulgaris* (95%) + *C.acetobutilycum* (5%) (shaded gray bars) after 120 h (1) and 240 h (2) with ethanol as extractant, after 120 h (3) and 240 h (4) with DMFA as extractant, respectively.

The use of DMFA as an extractant in the proposed process was not rational, since it's characterized minimal degrees of conversion of sulfur-containing compounds. This is apparently connected with toxicity of DMFA, which is showed inhibitory effect on the activity of biocatalysts.

For the most efficient run of bioconversion process of sulfones, it is preferred to additionally introduce of *C. acetobutilycum* cells to medium with sulfate reducers.

Thereby, for the fist time was shown the principal possibility of biocatalytic reduction of BTO_2 and $DBTO_2$ mixture as potential products of oxidative desulfurization of petroleum products, during the production of benzene to hydrogen sulfide. The process is run in the presence of biogenic hydrogen produced by anaerobic bacterial cells. This approach can be proposed for the intensification and ecologization of the processes of oxidative desulfurization of petroleum products.

3.3. Investigation of the possibility of transformation of fumaric acid into aspartic acid under the action of biocatalysts based on E. coli bacterial cells immobilized into poly(vinyl) alcohol cryogel

The main chemical method of industrial production of AA is the condensation of maleic anhydride with ammonia under high pressure and followed hydrolysis of the resulting ammonium salt of aspartic acid and separation of the free acid. In this case, a racemic mixture of AA isomers is formed with a yield about 72-80%. Currently, pure L-AA or D-AA is obtained by biotechnological methods. It was proved that using of cells with aspartate ammonia-lyase (aspartase) activity as biocatalysts is more efficient as compared with technologies used pure enzyme. The process of enzyme separation and purification is long, laborious and accompanied by decreasing of enzyme activity.

After a comparative analysis of the possibilities of biocatalytic production of L-AA by biotransformation of ammonium fumarate using bacteria of the genus *Serratia, Pseudomonas, Bacillus, Brevibacterium, E. coli*, etc. [28] it was determined that using of *E. coli* cells is more reasonable from a perspective of process productivity and product yield [29]. At the same time, it was determined that cells immobilized into polymer matrix (polyacrylamides, alginates, carrageenans, polyurethanes, polyazetidine, etc.) were characterized enhanced resistance and good productivity [30]. The activity of free cells was reduced when used substrate concentration above 1 M, and the immobilized cells was remained active up to 1.5 M of ammonium fumarate [31].

Since in natural *E.coli* strains aspartic activity was reduced, the processes of AA production using various genetically modified bacterial strains have been proposed [32].

However, the high cost of these recombinant strains, the hardness of their cultivation and the problems in utilization of biomass are limits use of such producers in practice. Thus, the development of a new efficient and stable biocatalyst based on natural *E. coli* strains is highly urgent.

In this regard, screening among strains of *E. coli* was done to identify the possibility of carrying out biotransformation of ammonium fumarate to AA (Table 1). Ammonium fumarate can be obtained by known chemical methods from benzene according to the proposed in Figure 1B. Also, ammonium fumarate can be obtained from renewable raw materials through the furfural and fumaric acid [33, 34].

The highest transformation efficiency of ammonium fumarate into AA was observed when *E. coli* BL-21 strain was used (Table 1). It is known that the best results of obtaining different products using *E. coli* cells can be achieved by using producers in immobilized form [35] and when PVA cryogel used as a carrier can be obtained stable and high effective biocatalysts [36].

Table 1. Efficiency of transformation of ammonium fumarate (1 M) into AA by *E. coli* cells.

E.coli strain	Concentration of AA, g/L	Conversion, %
SG13009[pREP4]	34.2±1.7	77.2±3.9
DH5α	38.7±1.9	87.4±4.3
BL-21	41.2±1.9	93.1±4.6

So, cells of the *E. coli* BL-21 were immobilized into PVA cryogel. During the formation of immobilized biocatalyst (IBC), the concentration of bacteria biomass introduced into the mixture with PVA solution was varied from 10 to 30 wt%. Maximal concentration of AA was accumulated when IBC with a cell concentration of 25–30 wt% was used (Table 2).

Under conditions of a flow reactor of column type, the period of IBC semi-inactivation with 25 wt% cell biomass was exceeded 180 days. Under periodic conditions such biocatalyst can be used more than 10 times. The characteristics of development IBCs were better or comparable with similar characteristics of known analogues [37, 38].

Table 2. Influence of cell concentration in the composition of IBC on conversion of ammonium fumatare (1 M) into AA.

Cell concentration, wt %	AA, g/L	Conversion of ammonium fumatare, %
10	119.1±3.9	89.5±1.5
15	123.5±3.1	92.9±1.3
20	125.2±3.2	94.1±1.4
25	127.3±3.3	98.2±1.5
30	129.1±3.4	97.1±1.5

Thus, the use of cells with aspartase activity immobilized into PVA cryogel can be perspective from the point of view of possibilities of further use in the processes of AA production and it's polymers.

4. Conclusion

Two promising approaches to greening the production of polyAA were demonstrated. The mixture of dibenzothiophene-sulfone ($DBTO_2$) and benzothiophene-sulfone (BTO_2) as potential products of oxidative desulfurization of petroleum products during benzene production was used for the first time for biocatalytic recovery to S^{2-} in anaerobic conditions. This approach can be proposed for the intensification and ecologization of oxidative desulfurization processes of various petroleum products. To obtain stereoregular polyAA was suggest a correction of the polymer initial chemical synthesis by introduction of biocatalytic stage during conversion of ammonium fumarate to AA - monomer. The high conversion (98,2±1,5%) of 1M ammonium fumarate to AA was shown under action of biocatalyst with *E. coli* BL-21 cells immobilized into PVA cryogel. Immobilization of cells with aspartic activity in PVA cryogel is promising way from the point of view of the development of IBC for the production of AA and its polymers.

Acknowledgment
The authors are sincerely grateful to the staff of the Department of Petroleum Chemistry and Organic Catalysis Faculty of Chemistry Lomonosov Moscow State University in the person of the head of the department Professor A.V. Anisimov for providing for research $DBTO_2$ и BTO_2.
This work was supported by the Russian Foundation for Basic Research (18-29-05064).

References

[1] de Besi, M., McCormick, K. 2015 Towards a bioeconomy in Europe: National, regional and industrial strategies. *Sustainability 7*(**8)** 10461-10478.
[2] Maslova O V, Senko O V and Efremenko E N 2018 Aspartic and glutamic acids polymers: preparation and applications in medicinal chemistry and pharmaceutics. *Russ. Chem. B* **67(4)** 614-623.
[3] de la Torre C, Domínguez-Berrocal L, Murguía J R, Marcos M D, Martínez-Máñez R, Bravo J and Sancenón F 2018 ϵ-Polylysine-capped mesoporous silica nanoparticles as carrier of the C9h peptide to induce apoptosis in cancer cells *Chem. Eur. J.* **24(8)** 1890-1897.
[4] Haque S, McLeod V M, Jones S, Fung S, Whittaker M, McIntosh M and Kaminskas L M 2017 Effect of increased surface hydrophobicity via drug conjugation on the clearance of inhaled PEGylated polylysine dendrimers *Eur. J. Pharm. Biopharm.* **119** 408-418.
[5] Senko O, Maslova O, Efremenko E 2017 Optimization of the use of His_6-OPH-based enzymatic biocatalysts for the destruction of chlorpyrifos in soil. *Int. J. Environ. Res. Public Health* **14(12)** 1438.
[6] Maslova O V, Senko O V and Efremenko E N 2018 The influence of enzymatic removal of chlorpyrifos from feed grain mixes on biochemical parameters of rat blood. *Biochem. (Mosc.), Suppl., Ser. B Biomed. chem.* **12(2)** 181-185.
[7] Maslova O V, Senko O V, Stepanov N A, Aslanli A G Q and Efremenko E N 2017 His_6-OPH and its stabilized forms combating quorum sensing molecules of gram-negative bacteria in combination with antibiotics. *JJNPP* **12(3)**.
[8] Maslova O, Aslanli A, Stepanov N, Lyagin I and Efremenko E. 2017 Catalytic characteristics of new antibacterials based on hexahistidine-containing organophosphorus hydrolase. *Catalysts* **7(9)** 271.
[9] Maslova O V, Aslanli A G, Senko O V and Efremenko E N 2018 The possibilities of reducing the minimal inhibitory concentration of puromycin and ceftiofur with their combination with His_6-OPH-based. *Mosc. Univ. Chem. Bull.* **73(6)** 298–302.
[10] Zhang C, Wu D and Qiu X 2018 Stimulatory effects of amino acids on γ-polyglutamic acid production by *Bacillus subtilis*. *Sci. Rep.* **8(1)** 17934
[11] Nakato T, Yoshitake M, Matsubara K, Tomida M and Kakuchi T 1998 Relationships between structure and properties of poly (asparticacid)s *Macromolecules* **31(7)** 2107-2113.
[12] Kurosaki T, Yamashita Y, Aki K, Harasawa H, Nakagawa H, Kodama Y and Sasaki H 2011 Secure and effective gene vector of polyamidoamine dendrimer pharmaceutically modified with anionic polymer. *J. Pharm. Sci.* **100(11)** 4855-4863.
[13] Ismayilov I T, Stepanov N A, Efremenko E N and Abbasov V M 2015 Evaluation of biocidal properties of vegetable oil-based corrosion inhibitors using bioluminescent enzymatic method. *Mosc. Univ. Chem. Bull.* **70(4)** 197–201.
[14] Maslova O, Stepanov N, Senko O and Efremenko E 2019 Production of various organic acids from different renewable sources by immobilized cells in the regimes of separate hydrolysis and fermentation (SHF) and simultaneous saccharification and fermentation (SFF). *Bioresour. Technol.* **272** 1-9.
[15] Trukhina A I, Gladchenko M A, Kalyuzhnyi S V 2011 Optimizations of sulphide and organic modifications of the DEAMOX process. *Appl. Biochem. Microbiol.* **47(9)** 841-845.
[16] Jalalvandi E, Shavandi A 2018. Polysuccinimide and its derivatives: degradable and water soluble polymers. *Eur. Polymer J.* **109** 43-54.

IOP Conf. Series: Materials Science and Engineering **525** (2019) 012037 doi:10.1088/1757-899X/525/1/012037

[17] *Immobilized cells: biocatalysts and processes*. Ed. E. Efremenko (Moscow RIOR) 2018, 500 p.

[18] Efremenko E, Stepanov N, Nikolskaya A, Senko O, Gudkov D, Spiricheva O and Varfolomeev S 2010 Immobilized cells of various microorganisms in production of liquid biofuels *Proceedings of the 18th European Biomass Conference and Exhibition*. ETA-Florence Renewable Energies 1753–1758.

[19] González-Henríquez C M, Sarabia-Vallejos M A and Rodríguez-Hernández J 2017 Strategies to fabricate polypeptide-based structures via ring-opening polymerization of N-carboxyanhydrides. *Polymers* **9(11)** 551.

[20] Deng C, Wu J, Cheng R, Meng F, Klok H A and Zhong Z 2014 Functional polypeptide and hybrid materials: Precision synthesis via α-amino acid N-carboxyanhydride polymerization and emerging biomedical applications. *Prog Polym. Sci.* **39(2)** 330-364

[21] Leuchs H and Geiger W 1908. Über die Anhydride von α-Amino-N-carbonsäuren und die von α-Aminosäuren. *Berichte der deutschen chemischen Gesellschaft* **41(2)** 1721-1726.

[22] Nakato T, Yoshitake M, Matsubara K, Tomida M and Kakuchi T. 1998 Relationships between structure and properties of poly(aspartic acid)s. *Macromolecules* **31(7)** 2107-2113.

[23] Rodriguez Galan A, Muñoz Guerra S, Subirana J A, Chuong B, Sekiguchi H. 1986 Synthesis of poly-β-(α-benzyl-L-aspartate) from L-aspartic β-lactam benzyl ester (3(R)-benzoxycarbonyl-2-azetidinone). In *Makromolekulare Chemie. Macromolecular Symposia* **6(1)** 277-284.

[24] Bösmann A, Datsevich L, Jess A, Lauter A, Schmitz C, Wasserscheid P 2001. Deep desulfurization of diesel fuel by extraction with ionic liquids. *Chem. Commun.* **23** 2494-2495.

[25] Craven M, Xiao D, Kunstmann-Olsen C, Kozhevnikova E F, Blanc F, Steiner A, Kozhevnikov I V 2018. Oxidative desulfurization of diesel fuel catalyzed by polyoxometalate immobilized on phosphazene-functionalized silica. *Appl. Catal. B-Environ.* **231** 82-91.

[26] Polikarpova P, Akopyan A, Shigapova A, Glotov A, Anisimov A, Karakhanov E 2018 Oxidative Desulfurization of Fuels Using Heterogeneous Catalysts Based on MCM-41. *Energy & Fuels* **32(10)** 10898-10903.

[27] Kairbekov Z K, Myltykbaeva Z K, Muktaly D, Nysanova B, Anisimov A V, Akopyan A V 2018 Peroxide Oxidative Desulfurization of a Diesel Fuel. *Theor. Found. Chem. Eng* **52(4)** 677-680.

[28] Rehm B H (Ed.). 2009 *Microbial production of biopolymers and polymer precursors: applications and perspectives*. Horizon Scientific Press.

[29] Chibata I, Tosa T, Sato T 1976 Production of L-aspartic acid by microbial cells entrapped in polyacrylamidegels. *Methods Enzymol.* **44** 739-746.

[30] Zajkoska P, Rebroš M, Rosenberg M 2013. Biocatalysis with immobilized *Escherichia coli*. *Appl. Microbiol. Biotechnol.* **97(4)** 1441-1455.

[31] Liese A, Seelbach K., Wandrey C (Eds.). 2006 *Industrial biotransformations*. JohnWiley&Sons.

[32] Szymańska G, Sobierajski B, Chmiel A. 2011 Immobilized cells of recombinant *Escherichia coli* strain for continuous production of L-aspartic acid Pol. J. Microbiol. **60(2)** 105–112.

[33] Li X, Jia P, Wang T 2016. Furfural: A promising platform compound for sustainable production of C4 and C5 chemicals. *ACS catalysis* **6(11)** 7621-7640.

[34] Araji N, Madjinza D D, Chatel G, Moores A, Jérôme F, Vigier K D O 2017 Synthesis of maleic and fumaric acids from furfural in the presence of betaine hydrochloride and hydrogen peroxide. *Green Chem.* **19(1)** 98-101.

[35] Zajkoska P, Rebroš M, Rosenberg M 2013 Biocatalysis with immobilized *Escherichia coli*. *Appl. Microbiol. Biotechnol.* **97 (4)** 1441-1455.

[36] Maslova O V, Sen'ko O V, Stepanov N A, Efremenko E N 2016 Lactic acid production using free cells of bacteria and filamentous fungi and cells immobilized in polyvinyl alcohol cryogel: A comparative analysis of the characteristics of biocatalysts and processes. *Catalysis in Industry* **8(3)** 280-285.

[37] Tajima T, Hamada M, Nakashimada Y, Kato J. 2015 Efficient aspartic acid production by a psychrophile-based simple biocatalyst. *J. Ind. Microbiol. Biotechnol.* **42(10)** 1319-1324.

[38] Derbikov D D, Novikov A D, Gubanova T A, Tarutina M G, Gvilava I T, Bubnov D M,

Yanenko A S 2017 Aspartic Acid Synthesis by Escherichia coli Strains with Deleted Fumarase Genes as Biocatalysts. *Appl. Biochem. Microbiol.* **53(9)** 859-866.

IOP Conf. Series: Materials Science and Engineering **525** (2019) 012038 doi:10.1088/1757-899X/525/1/012038

The high-temperature pseudo-invar effect in multilayer steel materials

V E Kabantseva, M D Safonov, and A I Plokhikh

Department of Material Science, Bauman Moscow State Technical University (BMSTU), 105005, 2nd Baumanskaya street, 5-1, Moscow, Russian Federation

E-mail: plokhikh@bmstu.ru

Abstract. The paper presents the results of research into Coefficients of Thermal Expansion (CTE) in multilayer steel material produced by hot pack rolling. It has been discovered that the material comprised of interleaving layers of steel features abnormally low CTE values in the direction normal to the rolling plane. A hypothesis has been proposed to explain the phenomenon by an interrelation between normal and tangent stresses generated in the interlayer boundaries within the material. A leading role of reactive stresses is confirmed by the results of the metallographic analysis. It has been shown that cyclic heating at a high temperature provokes recrystallization on the interlayer boundaries due to the difference between the CTE values of the adjacent layers and the high-temperature impact.

1. Introduction

Today, great efforts of the material scientists in research and development are focused on making the methods to improve operational characteristics of conventional materials by implementing the special types of structure in them. These structures can demonstrate adaptive behavior under an external impact due to non-linear strain or thermal characteristics. These effects are generally observed in composite materials and are caused by an interaction of local microscopic fields of elastic stresses, which is not the case for the uniform structured materials.

Many scientists worked on the theoretical aspects of production of such composite materials, which resulted in discovery of some interesting effects within the composite materials, and induced a search for the materials with a negative coefficient of thermal expansion (CTE) [1-5]. There is a mathematic model according to which CTE could have negative values within the materials with a layered structure based on binary (Iridium+Invar) or triplex (Iridium+Tungsten+Invar) metal compositions [6, 7]. The main criterion for selection of the composite components is the high difference in the normal elasticity modulus E and CTE values that may be 4-5 times different for the selected materials.

However, technological aspects concerning the compatibility of these materials, interlayer diffusion of alloy elements and the methods of making a continuous link between the layers that would make it possible to reveal this effect, have not been considered. This is why it is relevant to implement the approaches developed in the production of structural metal materials with layered structure, based, however, on the available and processable materials, such as steel.

2. Materials and Methods

Based on the considerations given above two model compositions made of AISI304+1008 and AISI304+AISIW108 steel have been studied. The primary composite blanks were 100 interleaving

IOP Conf. Series: Materials Science and Engineering **525** (2019) 012038 doi:10.1088/1757-899X/525/1/012038

sheets of the mentioned steels with the thickness of 0.5 mm, i.e. by 50 sheets per each grade. The output of the experimental process flow (Figure 1) was the flat steel blanks with a width of 100 mm and thickness of 10 mm [8-10]. The first cycle output bands with the thickness of 10 mm were rolled to the thickness of 2 mm, cleaned out and assembled into a pack, then, according to the described process flow, the blanks underwent the second hot pack rolling cycle to the thickness of 10 mm.

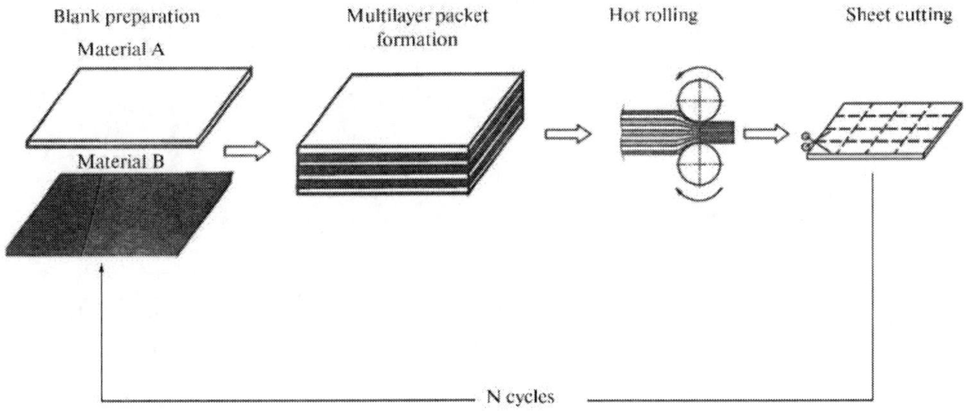

Figure 1. Experimental process flow chart of the multilayer metal materials production.

The samples were cut out in three directions to perform dilatometric studies: along the rolling direction (RD), in the transversal direction (TD) and in the normal direction (ND).

3. Results and discussion

The dilatometric study has shown (Figure 2) that the studied multilayer material AISI304+AISIW108 produced after the first and the second process cycles features the anomaly of thermal expansion (ΔL/L) along the ND. The transformation observed along the ND at ~900°C corresponds to the $\alpha \rightarrow \gamma$ transformation within the carbon steel which is a component of the studied composition. The studies of AISI304+1008 composition have shown the presence of the thermal expansion anomaly as well, that means the multilayer structure influences revealing of the discovered effect.

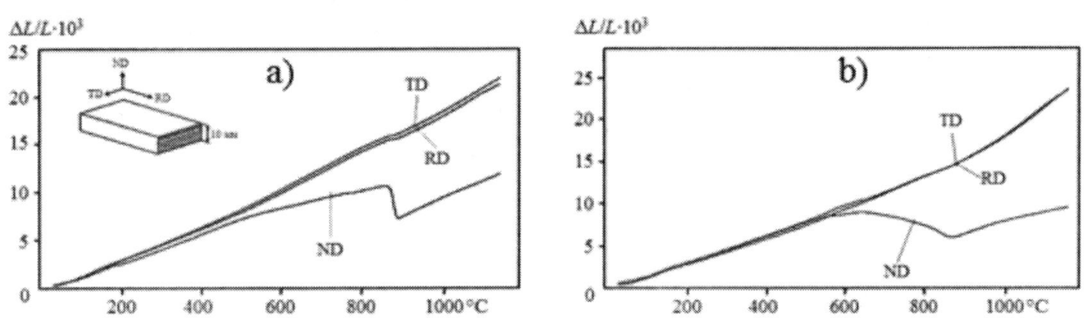

Figure 2. The dependence of relative elongation of the multilayer material in RD, TD, and ND within the samples with the thickness of 10 mm and the number of layers of 100 (a) and 1500 (b).

As we can see, the discovered invar-type effect increases after completion of the second process cycle and shifts into the area with lower temperatures. A hypothesis on the interaction between the reactive tangent and normal stresses during heating has been proposed to explain this effect. According to the hypothesis, the CTE values decrease is observed in RD and TD within the multilayer material due to mutual compensation of the thermal expansion of heterogeneous layers (Figure 3, a). This effect is highly visible within the interval $\alpha \rightarrow \gamma$ of carbon steel transformation. Thus, a decrease in the matrix parameters during the reverse transformation is compensated completely by the expansion of the AISI304 steel, which is demonstrated by the absence of the anomaly on the dilatograms in RD and TD. On the other hand, the compression effect is enhanced in ND, which is related to the expansion of the AISI304 steel overlapping the compressed layers undergoing $\alpha \rightarrow \gamma$ transformation (Figure 3, b). It is the compression which results in the CTE values in ND, corresponding to the values within conventional invar-type alloys.

a) RD, TD b) ND

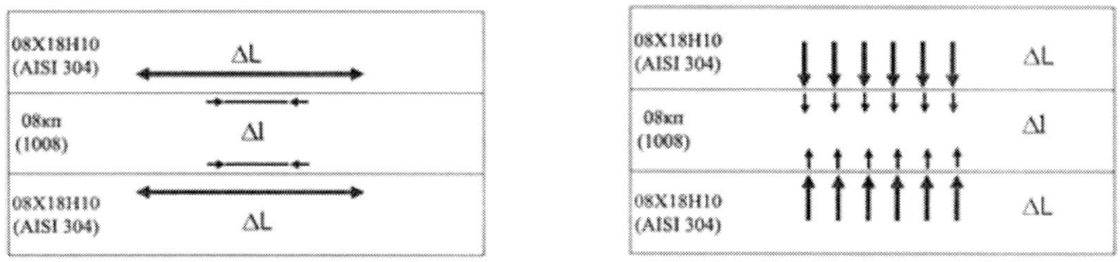

Figure 3. Tangent (a) and normal (b) stresses impact diagram exemplified by the 08X18H10+1008 composition.

Analysis of the hundred-layered samples microstructure, originally comprised of the AISI304 and AISIW108 steels, has shown that after completion of the first process cycle the transformed structure represents interleaving layers of low carbon steel and the layers of former AISI304 steel (Figure 4, a). Presence of high volume ratio of a finely-dispersed carbide constituent generated from the directed carbon diffusion is specific for the layers of AISI304 steel (Figure 4, b) [11].

a) b)

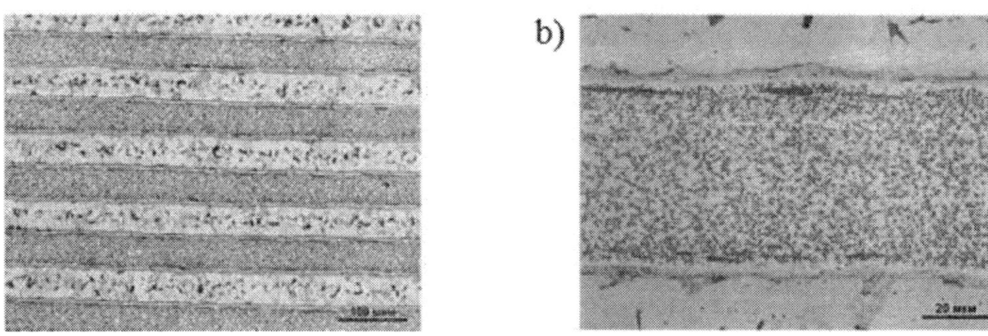

Figure 4. The microstructure of the AISI304+ AISIW108 (a) composition and AISI304 (b) steel layer.

To check this proposition hundred-layered samples of the AISI304+AISIW108 composition were repeatedly heated to the temperature of 1200°C, which knowingly is higher than the solution temperature of contained chromium carbides. As we can see (Figure 5), there are transformations within the layers of the studied multilayer material. Thus, carbide phase dissolved completely in the layers that formerly constituted the AISI304 steel, which resulted in active recrystallization with generation of a coarse-grained structure. At the same time the thickness of layers that formerly constituted the AISIW108 steel decreased from the original value of 100 μm in hot rolled condition at average down to 85–90 μm. Therefore, the increase of the thickness of layers that formerly constituted the AISI304 steel occurred due to the migration of the interlayer boundaries in direction of the layers that formerly constituted the AISIW108 steel.

Based on the foregoing we may conclude that this process is driven by interlayer stresses occurring due to the difference of the CTE values in the adjacent layers, which is different for the alloys with the similar chemical composition at given temperatures by more than 1.5 times. The mechanism of relaxation of these stresses is a dynamic recrystallization occurring on the boundary between the layers and shown by the metallography as interlayers. At the same time, the front of the moving boundaries is directed to the most plastic layer, which is explained by the least energy required for the microplastic deformation preceding the beginning of the recrystallization.

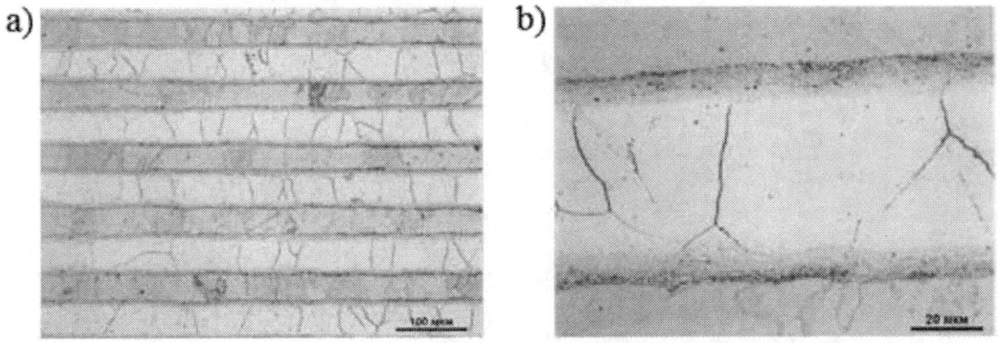

Figure 5. The microstructure of the AISI304+ AISIW108 composition after heating to the temperature of 1200°C.

In order to exclude the impact of diffusion processes on the mechanisms of interlayer stresses relaxation, similar studies have been performed on the AISI304+1008 compositions. It has been established (Figure 6, b), that repeated one-time heating to the temperature of 1200°C, also results in generation of an additional interlayer at the boundary of the layer facing the steel grade 1008. Three-time heating to the temperature of 1200°C has shown, that the thickness of this interlayer increases, which indicates recrystallization processes and confirms the proposition on a leading role of interlayer stresses in the structure generation in multilayer metal materials (Figure 6, c).

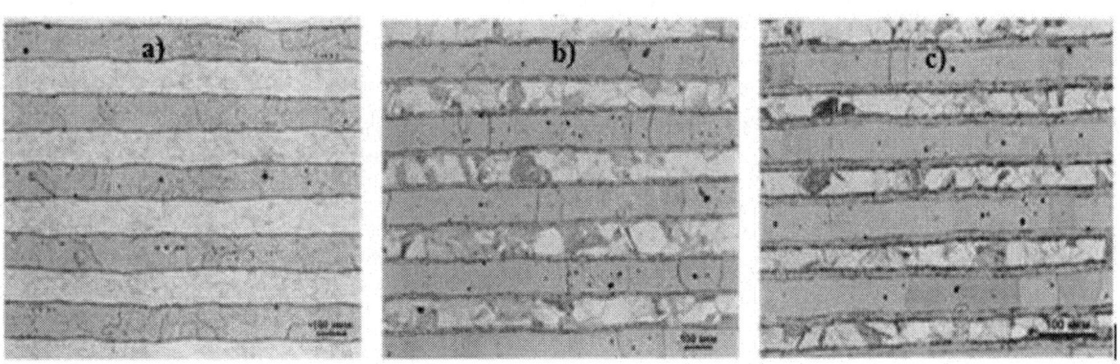

Figure 6. The microstructure of the AISI304+1008 composition in the original condition (a) and after repeated one-time (b) and three-time (c) heating to the temperature of 1200°C.

4. Conclusion

Therefore, the results of performed studies have confirmed the existence of the steel structural material which is capable of demonstrating flexible behavior under an external impact, as shown by distinct anisotropy of thermal expansion.

It has been established, that increase in the number of layers within the multilayer material, causes enhancement of the anomaly of thermal expansion in the direction normal to the rolling plane with a shift to the area with lower temperatures. It is demonstrated that such anisotropy of the properties is caused by the stresses occurring due to the difference of CTE values of steel grades contained n the

IOP Conf. Series: Materials Science and Engineering **525** (2019) 012038 doi:10.1088/1757-899X/525/1/012038

original multilayer material composition. The relaxation mechanism for such stresses is associated with a dynamic recrystallization that occurs on the boundary between the layers. The interlayer boundary migration is directed to the most plastic layer, which is explained by the least energy required for microplastic deformation preceding the beginning of recrystallization.

References

[1] Khomenko O A Origin and specific features of invar anomalies of physical properties: Fe-Ni alloys with an FCC lattice The Physics of Metals and Metallography. 2007. V. 104. № 2. pp. 146-156.

[2] Mary T A, Evans J S O. Vogt T & Sleight A W Negative thermal expansion from 0.3 K to 1050 K in ZrW_2O_8 Science *272* 90–92 (1996)

[3] Bodryakov V Yu, Povzner A A Invar and covar behavior of simple ferromagnets: Thermodynamic simulation Technical Physics. The Russian Journal of Applied Physics. 2007. V. 52. № 2. pp. 209-215.

[4] Monroe J, Gehring D, Karaman I, Arroyave R, Brown D and Clausen B, Tailored thermal expansion alloys Acta Mater 102 (2016) 333 –341

[5] Bensoussan A, Lions J-L and Papanicolaou G C, Asymptotic Analysis for Periodic Structures, North Holland Amsterdam 1978

[6] Kolpakov A G, Rakin S I To the task of synthesis of the composite material of a single-dimensional structure with specified characteristics, Journal of Applied Mechanics and Technical Physics 1986 V. 27, № 6 pp. 143-150

[7] Kolpakov A G, Rakin S I Deformation characteristics of layered composites with nonlinear deformations, Journal of Applied Mechanics and Technical Physics 2004 V. 45, № 5 pp. 157-166

[8] Kolesnikov A G., Plokhikh A I, Komisarchuk Yu S, Mikhaltsevich I Yu. A study of special features of formation of submicro- and nanosize structure in multilayer materials by the method of hot rolling. Metal Science and Heat Treatment. V.52, (2010), N. 5–6, pp. 273–278.

[9] Tabatchikova T I, Plokhikh A I, Yakovlev I L, Klyueva S Yu, Structure and properties of a steel-based multilayer material produced by hot pack rolling, The Physics of Metals and Metallography. 2013 V. 114, № 7 pp. 580-592.

[10] Tabatchikova T I, Yakovlev I L, Plokhikh A I, Del'gado Reina S Yu. Studying a multilayer material based on stainless steels and produced by hot pack rolling. The Physics of Metals and Metallography. 2014. V. 115 № 4 pp. 403-412.

[11] Vlasova D V, Plokhikh A I, Semenov M Y, Polyanskii V M, Safonov M D, Effect of diffusion processes during pack rolling on multilayer material stability, Metallurgist. 2018. V. 62. № 5-6. pp. 432-439.

IOP Conf. Series: Materials Science and Engineering **525** (2019) 012039 doi:10.1088/1757-899X/525/1/012039

Ultrasonication-assisted ultrafast solvothermal reduction of graphene oxide

P A Mikhaylov[1], I S Levin[1], A V Lubenchenko[2], G N Bondarenko[1] and V G Kulichihin[1]

[1] A.V. Topchiev Institute of Petrochemical Synthesis, Moscow Leninsky prospect, 29, 119991, Russia

[2] National Research University MPEI, Krasnokazarmennaya, 14, Moscow 111250, Russia

E-mail: pmih@ips.ac.ru

Abstract. Solvothermal reduction of graphene oxide (GO) is one of the suitable way to prepare a dispersion of reduced graphene oxide (RGO) in organic solvents that is stable at high concentration without surfactants. However, it is a long-term process and often requires elevated pressure. Ultrasonic irradiation of GO dispersion in N-methylpyrollidone (NMP) at high temperature promotes full reduction of GO in ten minutes, which is several decade faster than conventional solvothermal reduction of GO. Synthesized RGO was characterized by XRD, Raman, FTIR and XPS methods.

1. Introduction

Grapene is recently discovered allotrope of carbone, and it represents a monolayer of graphite. It was first obtained by Novoselov and Geim by mechanical exfoliation of highly oriented graphite [1]. Unique thermal, optical, electrical, optoelectrical and mechanical properties of graphene define the wide applications of this material [2,3].

Chemical transformation of graphite is one of the best ways for large-scale production of graphene [4]. For this purpose graphite is first converted to graphite oxide (GtO) by strong oxidative reagent ($KMnO_4$, $NaNO_3$, $KClO_3$, etc). As result, oxygen disrupts double bonds in graphite network with the formation of carboxylic, lactone, phenolic, anhydride functional groups on the edge of graphite basal planes [3]. The interplanar space in GtO significantly longer then in graphite due to decrease of interplanar Van-der-Vaals interactions. Unlike graphite, GtO is hydrophilic and can be dispersed in water as a few-layer or even a single-layer structure which is called graphene oxide (GO). Reduction of GO by various method allows forms a graphene-like material – reduced graphene oxide (RGO). Despite the fact that RGO has a defective structure, it is applicable for optoelectronic and electronic devices, as well as for energy storage and composites [3].

Firstly, in 1859 Brodie obtained GtO by oxidizing graphite with fuming nitric acid and $KClO_3$ [5]. The active stage of GtO investigation started after discovering the safe method by Hummers in 1958 [6]. Nowadays, the Hummers method and its modifications are widely used for preparation of GtO.

GO can be reduced chemically, thermally, solvothermally, electrochemically [7]. Hydrazine, ascorbic acid, and sodium borhydrite are commonly used as chemical reducers for GO [4,8,9]. Hydrazine is an effective and inexpensive GO reducing reagent, however its high toxicity and flammability are limitations for large-scale production. Moreover, the low stability of the dispersions of

IOP Conf. Series: Materials Science and Engineering **525** (2019) 012039 doi:10.1088/1757-899X/525/1/012039

hydrazine reduced GO is an obstacle for preparation RGO-polymer composites. By contrast, the solvothermal reduced GO (SRGO) often forms stable dispersion in organic solvents without surfactant. This allows one to produce polymer composites by solution process [10-13]. Electrochemical conductivity of SRGO is comparable or even higher of electroconductivity of hydrazine RGO.

Essentially, solvothermal reduction of GO involves heating a dispersion of GO in an appropriate solvent for a period of time. In the case of a low-boiling solvent, an autoclave is used. Various solvents have been employed as reduction media: water (hydrothermal reduction) [14], sulfuric acid [15], dimethylformamide (DMF) [16], DMF + hydrazine [13], N-methylpirollidone (NMP) [10,11], DMF or DMSO + sulfuric acid [17], alcohols [8,18,19]. Evidently, organic solvents are more preferable as reducing media due to the functionalization of graphene planes by solvent which improves the stability of the RGO dispersion. The main disadvantage of the solvothermal reduction process is the long duration. Is there any way to accelerate solvothermal reduction? One appropriated way is to use microwave to promote solvothermal reduction [20]. Another way is to use ultrasound irradiation. It was shown that ultrasound energy accelerates a reduction of GO in aqueous media with such reducing agent as zinc dust [21], ascorbic acid [22], and 2-aminobenzoxazole [23].

In the present work, it was shown that ultrasound irradiation while solvothermal reduction of GO can significantly decrease the duration of the process and it is a promising way to obtain high quality product. Firstly, we researched dispersions of GO in N-methylmorpholine-N-oxide monohydrate (NMMO). As previously reported [24], the dispersion GO in NMMO could be obtained by sonication of GO powder in melted NMMO. However, when we tried to do this using a horn-type sonotrobe applied to melted NMMO at 90 °C, the media became dark after less than one minute, which indicates reduction of GO.

The same kind of treatment was applied to the dispersions of GO in other solvents. The results of such treatment will be studied in this paper.

2. Sample preparation

2.1. Materials and Equipment

All used reagents were "reagent grade purity". The neat graphite (Zaval'e deposit, Ukraine) was ground by a domestic coffee mill and sifted through standard 0.08 mm sieves. Deionized water (DI) was less than 1 Sm/cm. A UZDN-A horn-type sonotrobe (130W, 22kHz) was used to prepare dispersions and for reduction of GO. Heidolph Hei-VAP vacuum-rotary evaporator with Vacumbrand MZ 2C NT vacuum pump (12 mbar) was used to remove/replace solvents.

2.2. Oxidation of graphite

GO was synthesized using the improved Hummers method [25]. In brief, 240 ml H_2SO_4, 27 ml H_3PO_4 and 2 g graphite were added into a 500 ml round-bottom flask with a Teflon-coated magnetic stirrer cooled with an ice bath. Then ice bath was removed and 12 g $KMnO_4$ was added to the vigorously stirred reaction mixture which was then slowly heated and kept for 12 h at 50 °C. The reaction mixture was poured into glass containing about 500 g ice and 6 ml 37% H_2O_2 was added. After that, the reaction mixture was repeatedly washed and decanted with DI water (2-3 l), 36 % HCl, (300 ml), ethanol (700-800 ml), then centrifuged for 20 minutes at 5000 rpm, and finally dried in vacuum (< 1 Torr) 12 h at 50-60 °C. GO aqueous dispersion prepared by sonication of 0.75 g GO in 74.25 ml DI water with a horn-type sonotrobe. The viscous dark-brown dispersion was filtered through 450 nm regenerated cellulose filter. If stored for a long time, GO dispersions was stirred for about 10 minutes with IKA UltraTurax before application. A dispersion of GO in NMP was obtained from dispersion of GO in water using rotary evaporator.

2.3. Reduction of GO

Series of sample was prepared according to the next protocol. Three ml of aqueous GO dispersion and 15 ml of solvent were added to 50-ml round-bottom flask and water was removed in vacuum using a rotary evaporator processing at gradually increasing temperature up to 70-80 °C. The GO dispersion in

IOP Conf. Series: Materials Science and Engineering **525** (2019) 012039 doi:10.1088/1757-899X/525/1/012039

organic solvent (~12 ml) was putted into a 20 ml vial and sonicated by a horn-type sonotrobe. During sonication, the temperature of the dispersion increased up to 175-185 °C and remained stable. The temperature of reaction mixture was controlled by thermocouple. All RGO samples was repeatedly centrifuged (30-60 min at 6000-8000 rpm) and washed with fresh solvent. Purified products were stored as slurries.

3. Results and discussions

GO was synthesized from natural graphite by an improved Hummers method [25]. This gave a product that was more oxidized and easily dispersed product in comparison with the original Hummers method.

As mentioned above, we observed the effect of fast GO reduction by ultrasound in NMMO solutions. Firstly, we try to replace NMMO by another with high boiling point solvent which is a liquid at ambient temperature. The most promising candidates were dimethylsulfoxide (DMSO), NMP and DMF (dimethylformamide). The blackening of a dispersion of GO in DMSO and NMP also was observed during sonication for few minutes, if the temperature of the reaction mixture was above 100-120 °C, while no significant darkening was observed with GO in DMF even after 40 minutes of sonication. As was mentioned above, Tien et al. [17], successfully reduced GO solvothermally when they used a mixture of sulfuric acid and DMF. We also observed the reduction of GO in DMF when catalytic amount (< 1 %) of sulfuric acid was loaded into reaction mixture. The precipitation of dark brown partially reduced GO flakes was observed. The XRD pattern of this product confirms presence of a GO phase. A reasonable explanation for this is that the coagulation of partially reduced GO occurs with sulfuric acid as coagulant in reaction mixture.

The sonication of GO dispersion in DMSO for 40 minutes resulted in partially reduced GO. The product color was black, but XRD shows the presence of a GO phase.

The sonication of a GO dispersion in NMP for only 10 minutes gave a product completely free of GO phase. The series of GO dispersions in NMP were prepared with 10, 15, 20 and 30 minutes of ultrasound exposure. All prepared dispersions demonstrated exceptional stability for several months.

The RGO samples were studied by X-ray diffraction method (XRD), Raman spectroscopy, FTIR and X-ray photoelectron spectroscopy (XPS). Firstly, the RGO samples were characterized by XRD to prove an absence GO phase.

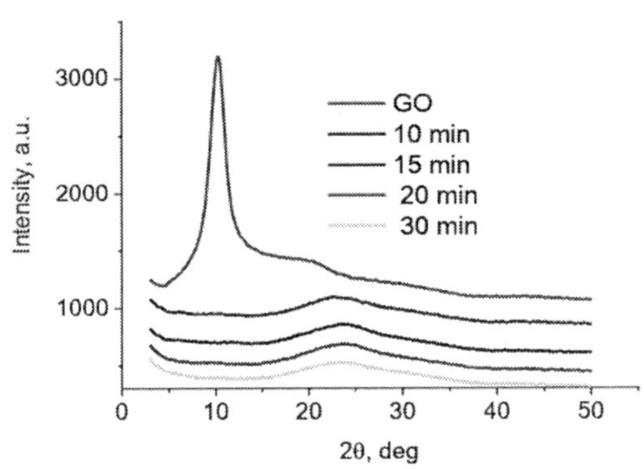

Figure 1. XRD patterns of sonicated dispersions.

3.1. X-Ray diffraction (XRD)

X-ray diffraction patterns were obtained using a Rigaku Rotaflex RU-200 X-ray source with a rotating copper anode (Cu K_α–radiation with characteristic wavelength λ=1.542 Å, a secondary curved graphite monochromator was used) at apparatus operating conditions 50 kV – 160 mA. The unit was equipped with a horizontal wide-angle goniometer Rigaku D/MAX-RC using Bragg-Brentano scheme in θ-2θ geometry. Scanning was carried out in the 3-50 angle range at 2θ scale with a speed of 2°/min and 0,04° increment. A scintillation counter was used as a detector of diffracted X-ray emission. XRD studies were carried out at room temperature in normal conditions.

IOP Conf. Series: Materials Science and Engineering **525** (2019) 012039 doi:10.1088/1757-899X/525/1/012039

Experimental XRD patterns of studied samples are shown at Figure 1**Figure 1**. The main peak with angle position $2\theta \approx 12$ degrees is intrinsic of a GO sample. As it seen, the reflex corresponding to GO disappears completely after 10 minutes of sonication.

3.2. Raman spectroscopy

Raman spectra were obtained using Horeba LabRam HR Evolution Raman spectrometer with excitation energy of 532 nm. The laser power was adjusted to prevent overheating of samples. Intensities of D (I_D) and G (I_G) bands were calculated using the Lorentz function approximation after baseline correction.

The time of sonication influences on of I_D/I_G ratio. According to the Tuinstra-Koening equation, the size of the sp^2-domain depends on I_G/I_D and can be calculated [26, 27] using the formula:

$$L_a(nm) = 2{,}4x10^{-10}\lambda^4(I_G/I_D), \quad (1)$$

where λ – is wavelength of the excitation laser.
These data are collected in Table 1.

Table 1. Ratio of intensity I_D/I_G, height, and size of sp^2-domains, for GO and RGO samples.

Sample	I_D/I_G	L_a, nm	H_D/H_G
GO	2.69	7.1	1.13
10 min	2.87	6.7	1.27
15 min	2.82	6.8	1.26
20 min	2.98	6.4	1.21
30 min	3.06	6.3	1.27

According to Table 1, the highest average size of sp^2-domains demonstrates GO, which gradually decreases with increasing the sonication time. A reasonable explanation for this behavior is the formation of a new sp^2-domain during reduction [28]. On other hand, a possible cause is a mechanical wrinkling of graphene sheets by ultrasound. As previously reported, such mechanical treatment of graphite as ball-milling is a cause for increasing a D-peak with an increase in ball-milling time [29].

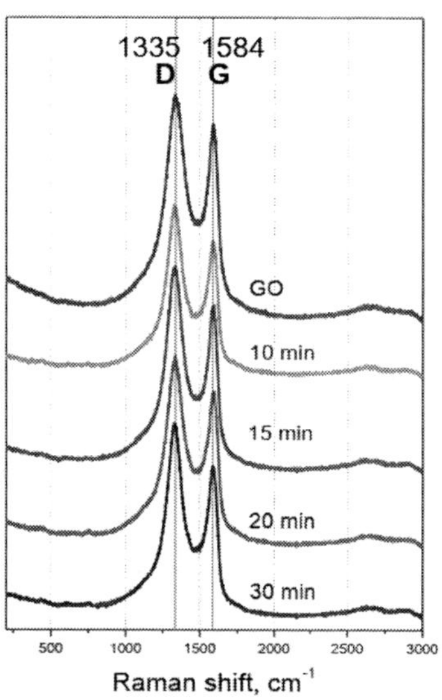

Figure 2. Raman spectra of GO and RGO.

3.3. Fourier-transform IR spectroscopy

IR spectra for all GO and dried RGO dispersions were registered in reflection mode with a HYPERION-2000 IR microscope (Germany), connected to a IFS-66 v/s Bruker IR-Fourier spectrometer (Germany)

IOP Conf. Series: Materials Science and Engineering **525** (2019) 012039 doi:10.1088/1757-899X/525/1/012039

in the frequency range of 600–4000 cm^{-1} (Ge-crystal, 100 scans, resolution 2 cm^{-1}). Spectral treatment was done using the software package OPUS-7.

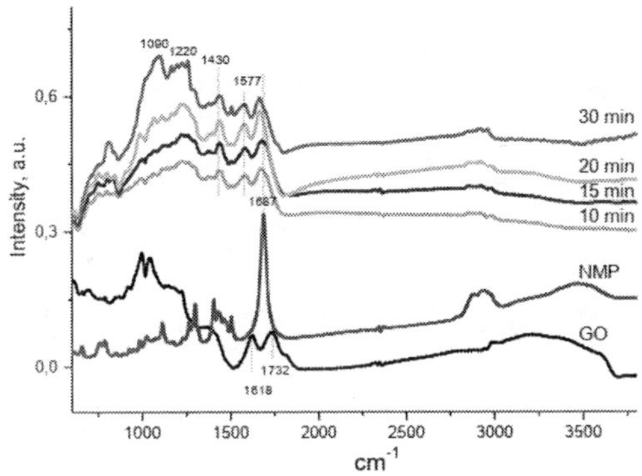

Figure 3. FTIR spectra of GO and RGOs.

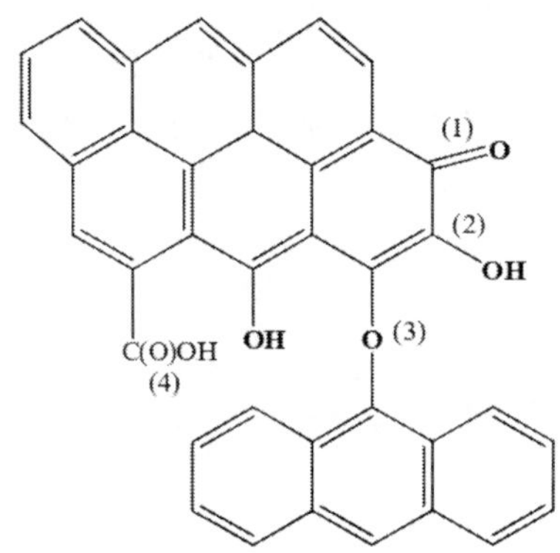

Figure 4. Functional groups on the edges of GO and RGOs.

The FTIR spectrum of GO (Fig. 3) demonstrates presence of a wide band, corresponding to sp^2 C-C (1618 cm^{-1}) and sp^3 C-C (1360 cm^{-1}) oscillations, similar to Raman spectrum bands G and D. C=O absorption ($\nu_{C=O}$) revealed as wide split band at 1732 cm^{-1} and C-O at 1200-990 cm^{-1}. Among them, bands above 1100 cm^{-1} usually belong to a C-O bond between a phenolic ring and bands around 1100 cm^{-1} belong to a single C-O bond in a carboxylic, alkylphenolic or dialkyl functional group in an ether or alcohols. Thus, the FTIR spectrum of GO presents bands of various functional groups, corresponding to the oxidized graphite-like structure, as is shown on Fig. 4. C=O oscillation ($\nu_{C=O}$) corresponding to functional groups 1 and 4 were detected at 1732 cm^{-1}. Apparently, bending and stretching vibrations of functional groups 1 and 4 were revealed at 998 cm^{-1} and 1040 cm^{-1}. Wide bands at 1200-1220 cm^{-1} correspond to the same vibrations of 3, while in 3300-3400 cm^{-1} region to stretched OH (ν_{OH}) in functional groups 2 and 4, which are associated via hydrogen bonds.

Consequently, the structure of GO represents a distorted graphite structure with highly oxidized edges of graphene honeycombs. Spectra of RGO after 10 and 15 min of ultrasound treatment are very similar (Fig. 3). C=O oscillations almost completely disappeared in these spectra, and this means that content of functional groups 1 and 4 (Fig. 4) were dramatically decreased in RGO. Intensities of bands at 998 cm^{-1} and 1040 cm^{-1} were also decreased, while intensity of wide band at 1200-1220 cm^{-1} was increased. The OH stretching bands (3300-3400 cm^{-1}) were not observed in RGO spectra. Intensities of bands at about 1090 cm^{-1} were tended to increase in the case of 20 and 30 min of ultrasound treatment.

Thus, RGOs were lost the most of functional group 1, 2 and 4 (Fig. 4). Long-term ultrasound treatment of GO (20 and 30 min) promoted the increase of intensity of C-O-C band (1090 cm^{-1}), corresponding to an alkyl-phenolic bond. Apparently, the solvent (NMP) took part in formation of these bonds. The bands of NMP present in all RGO spectra. It was remarkable, that the amide group band of NMP was blue shifted in RGOs from 1687 cm^{-1} to 1675 cm^{-1}, which is evidence of sufficiently stable complexing between NMP and functional groups of RGO. Formation of covalent bonds between graphene nanosheets and NMP and/or products of its degradation is possible according to FTIR.

3.4. *X-Ray photoelectron spectroscopy (XPS)*

In this work, The X-ray photoelectron spectra were recorded with the help of the electron-ion spectroscopy module on the base of the platform Nanofab 25 (NT-MDT). In the analytical chamber an ultra-high oil less vacuum about 10^{-7} Pa was achieved. The spectra were recorded by an electrostatic hemi-spherical energy analyzer SPECS Phoibos 225. The energy analyzer was calibrated by samples of Cu, Ag and Au. The energy resolution of the spectrometer by the line Ag 3d5/2 was 0.78 eV for non-monochromatic X-rays Mg Kα. All spectra were recorded using a Mg anode while the energy analyzer was operated in the FAT (Fixed Analyzer Transmission) mode. For survey spectra, the passing energy E_{pass} = 80 eV was set up in the lens of the energy analyzer, for detailed spectra it was E_{pass} = 20 eV.

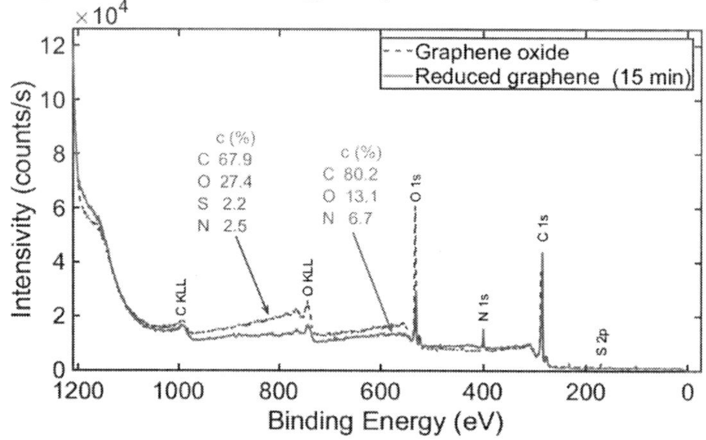

Figure 5. Survey XPS spectrum of GO and RGO (15 min).

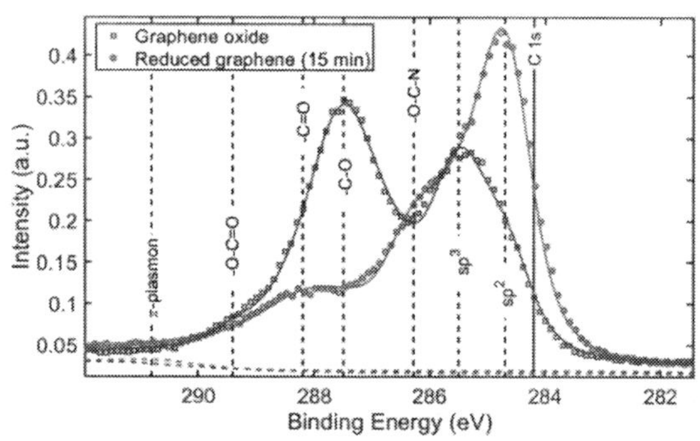

Figure 6. A detailed XPS spectrum of line of C 1s GO and RGO (15 min). Dots: experimental data; solid lines: calculation.

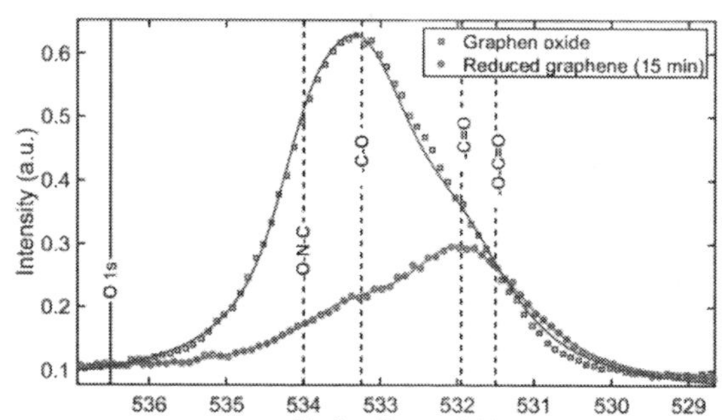

Figure 7. A detailed XPS spectrum of line of O 1s GO and RGO (15 min). Dots: experimental data; solid lines: calculation.

The line decomposition is based on the method described in [30]. The method includes: a new method for background subtraction accounting for depth non-uniformity of electron energy losses; a new method for photoelectron line decomposition into elementary peaks, which accounts for physical nature of the decomposition parameters; joint solving of both background subtraction and photoelectron line decomposition problems; criterion for assessing of line decomposition accuracy. The results are given in Table 2. Peaks at binding energies (BE) shown in Table 2 correspond to various bonds [10,31-32].

Table 2. Composition based on the XPS results

	BE, eV	*XPS deconvolution peaks (% of total peak area)*				
		GO	*RGO* (10 min)	*RGO* (15 min)	*RGO* (20 min)	*RGO* (30 min)
sp^2	284.7	14	50	49	48	47
sp^3	285.5	24	14	15	16	19
-C-N-O	286.6	6	14	14	14	14
-C-O	287.5	50	7	9	7	8
-C=O	288.2	5	14	10	13	11
-O-C=O	289.2	1	1	3	2	1

According to Table 2 the reduction of GO mainly occurs first 10 min of sonication and part of sp^2 carbon reaches a maximum. Extending of sonication time increases amount of sp^3-carbon as result of the functionalization graphene sheets. Consequently, rather functionalization than reduction occurred after 10 minutes of sonication. The lowest oxygen contents in RGO was observed after 30 min of sonication. Results of XPS were in good agreement with FTIR results.

4. Conclusion

A new method for solvothermal reduction of GO by simple sonication was demonstrated. It allows one to obtain the stable dispersion in organic solvents with high concentration due to functionalization of RGO nanosheets. It was shown that as little as 10 minutes of ultrasound exposure of a GO dispersion in NMP is enough to produce RGO nanosheets. This is up to several decades faster than previously described [10,11]. The solvent (NMP) plays a crucial in developed GO reduction method, functionalizing graphene nanosheets. The functional groups attached to RGO nanosheets can be easily controlled by increasing or decreasing sonication time. We hope that approach will be more energy and cost efficient in comparison with previously described methods.

Acknowledgments

This work was carried out in frames of Russian Science Foundation project # 17-79-30108. The authors also thank to Dr. Obraztsova E.D and Obraztsova E.A. (Prokhorov Institute of General Physics, of the Russian Academy of Sciences, Moscow, Russia) for registration of Raman spectra.

References

[1] Novoselov K S *et al* 2004 Electric Field Effect in Atomically Thin Carbon Films *Science* **306** 666–9

[2] Georgakilas V *et al* 2016 Noncovalent Functionalization of Graphene and Graphene Oxide for Energy Materials, Biosensing, Catalytic, and Biomedical Applications *Chemical Reviews* 116 5464–519

[3] Ferrari A C *et al* 2015 Science and technology roadmap for graphene, related two-dimensional crystals, and hybrid systems *Nanoscale* **7** 4598–810

[4] Chen H, Müller M B, Gilmore K J, Wallace G G and Li D 2008 Mechanically Strong, Electrically Conductive, and Biocompatible Graphene Paper *Advanced Materials* **20** 3557–61

[5] Brodie B C 1860 Sur le poids atomique du graphite *Ann. Chim. Phys.* **59** 466–72

[6] Hummers Jr W S and Offeman R E 1958 Preparation of graphitic oxide *Journal of the American Chemical Society* **80** 1339–1339

[7] Dang T T *et al* 2012 Superior dispersion of highly reduced graphene oxide in N,N-dimethylformamide *Journal Of Colloid And Interface Science* **376** 91–6

[8] Nethravathi C and Rajamathi M 2008 Chemically modified graphene sheets produced by the solvothermal reduction of colloidal dispersions of graphite oxide *Carbon* **46** 1994–8

[9] Fernández-Merino M J *et al* 2010 Vitamin C Is an Ideal Substitute for Hydrazine in the Reduction

of Graphene Oxide Suspensions *The Journal of Physical Chemistry C* **114** 6426–32
[10] Dubin S *et al* 2010 A One-Step, Solvothermal Reduction Method for Producing Reduced Graphene Oxide Dispersions in Organic Solvents *ACS NANO* **4** 3845–52
[11] Pham V H, Cuong T V, Hur S H, Oh E, Kim E J, Shin E W and Chung J S 2011 Chemical functionalization of graphene sheets by solvothermal reduction of a graphene oxide suspension in N-methyl-2-pyrrolidone *J. Mater. Chem.* **21** 3371–7
[12] Zhou D, Cheng Q-Y and Han B-H 2011 Solvothermal synthesis of homogeneous graphene dispersion with high concentration *Carbon* **49** 3920–7
[13] Wang H *et al* H 2009 Solvothermal Reduction of Chemically Exfoliated Graphene Sheets *Journal of the American Chemical Society* **131** 9910–1
[14] Zhou Y *et al* 2009 Hydrothermal Dehydration for the "Green" Reduction of Exfoliated Graphene Oxide to Graphene and Demonstration of Tunable Optical Limiting Properties *Chemistry of Materials* **21** 2950–6
[15] Kim D *et al* and Park C R 2012 Simple and cost-effective reduction of graphite oxide by sulfuric acid *Carbon* **50** 3229–32
[16] Kim S *et al* 2016 Solvothermal reduction of graphene oxide in dimethylformamide *Solid State Sciences* **61** 40–3
[17] Tien H N *et al* 2012 Enhanced solvothermal reduction of graphene oxide in a mixed solution of sulfuric acid and organic solvent *Chemical Engineering Journal* **211–212** 97–103
[18] Tkachev S V, Buslaeva E Y, Naumkin A V, Kotova S L, Laure I V and Gubin S P 2012 Reduced graphene oxide *Inorganic Materials* **48** 796–802
[19] Soares C P P, *et al* 2017 Solvothermal Reduction of Graphite Oxide Using Alcohols *Materials Research* **21**
[20] Murugan A V *et al* 2009 Rapid, Facile Microwave-Solvothermal Synthesis of Graphene Nanosheets and Their Polyaniline Nanocomposites for Energy Strorage *Chemistry of Materials* **21** 5004–6
[21] Mei X and Ouyang J 2011 Ultrasonication-assisted ultrafast reduction of graphene oxide by zinc powder at room temperature *Carbon* **49** 5389–97
[22] Abulizi A *et al* 2014 Ultrasound assisted reduction of graphene oxide to graphene in l-ascorbic acid aqueous solutions: Kinetics and effects of various factors on the rate of graphene formation *Ultrasonics Sonochemistry* **21** 1174–81
[23] Maktedar S S *et al* 2017 Ultrasound assisted simultaneous reduction and direct functionalization of graphene oxide with thermal and cytotoxicity profile *Ultrasonics Sonochemistry* **34** 856–64
[24] Kim C-J *et al* 2011 Graphene oxide/cellulose composite using NMMO monohydrate *Carbohydrate Polymers* **86** 903–9
[25] Marcano D C *et al* 2010 Improved Synthesis of Graphene Oxide *ACS Nano* **4** 4806–14
[26] Tuinstra F and Koenig J L 1970 Raman Spectrum of Graphite *The Journal of Chemical Physics* **53** 1126–30
[27] Cançado L G *et al* 2006 General equation for the determination of the crystallite size La of nanographite by Raman spectroscopy *Applied Physics Letters* **88** 163106
[28] Stankovich S *et al* 2007 Synthesis of graphene-based nanosheets via chemical reduction of exfoliated graphite oxide *Carbon* **45** 1558–65
[29] Kaniyoor A and Ramaprabhu S 2012 A Raman spectroscopic investigation of graphite oxide derived graphene *AIP Advances* **2**
[30] Lubenchenko A V, Batrakov A A, Pavolotsky A B, Lubenchenko O I and Ivanov D A 2018 XPS study of multilayer multicomponent films *Applied Surface Science* **427** 711–21
[31] Smith M, Scudiero L, Espinal J, McEwen J-S and Garcia-Perez M 2016 Improving the deconvolution and interpretation of XPS spectra from chars by ab initio calculations *Carbon* **110** 155–71
[32] Pham D M, Miyata Y, Awata T, Nakatake M, Zhang C F, Kanda K, Ogawa S, Ohta S, Yagi S

and Katayama A 2018 Development of sample preparation technique to characterize chemical structure of humin by synchrotron-radiation-based X-ray photoelectron spectroscopy *Surface and Interface Analysis*

IOP Conf. Series: Materials Science and Engineering **525** (2019) 012040 doi:10.1088/1757-899X/525/1/012040

Laser writing of optical structures in quartz glass

E S Oparin, K S Khorkov, D A Kochuev, R V Chkalov and A S Chernikov

Vladimir State University, Gorky str. 87, Vladimir 600000, Russia

E-mail: oparinvlsu@yandex.ru

Abstract. In this paper the interaction of ultrashort laser pulses with a quartz sample are shown. The effect of redistribution of the passing laser radiation intensity after sample treatment in the mode of multiple filamentation is shown. Examples of the formation of ordered structures in bulk transparent solids due to the refractive index modification of the regions exposed to femtosecond laser radiation are given.

1. Introduction

The processes of the ultrashort laser pulses interaction with a transparent dielectrics are known and were observing by different groups simultaneously with the appearance of these laser systems [1-5]. Nevertheless, the evolution of processes caused by the interaction of such radiation with the dielectric medium has not been sufficiently studied. Detailed study of the dynamics of changes in spatio-temporal and spectral characteristics of high-intensity light radiation in dielectrics is of great interest because of the wide range of possible applications.

When femtosecond pulses modify the materials, new physical processes and properties are revealed. The modification of transparent materials has been an active area of research for the past four decades. In the case of transparent materials, the absorption of incident laser radiation is nonlinear nature [1,6].

A nonlinear absorption mechanism must deposit laser energy into the material by promoting electrons from the valence band to the conduction band. The main types of nonlinear excitation mechanisms that affect the absorption process are photoionization (multiphoton ionization, tunnel ionization) and avalanche ionization. It should be borne in mind that the probability of multiphoton absorption in the case of ultrashort laser pulses is much higher, compared with pulses of longer duration, which is associated with a greater by several orders intensity of radiation.

When ultrashort laser pulses interact with a medium, in our case with quartz glass, filamentation can be observed. Laser filaments are beams, which are focused by a lens and then, due to high intensity, begin to self-focus into a long light channel. The indispensable conditions for filamentation are high pulse energy and short pulse width. The medium must be optically transparent. In the process of filamentation, annular structures may form around the filament, resulting from the interference of diverging radiation from the filament axis after defocusing and the light field surrounding the filament. Losses in defocusing are extremely small and almost all of the energy is again tightened to the filament axis, and if the power is many times greater than the self-focusing critical power, multiple filamentation can occur with the formation of daughter filaments. The existence of these structures is explained by the phenomenon of self-focusing [7-11].

The physical cause of self-focusing is the cubic optical polarizability of the medium (Kerr nonlinearity), which causes a self-induced increase of the refractive index in areas with high intensity

IOP Conf. Series: Materials Science and Engineering **525** (2019) 012040 doi:10.1088/1757-899X/525/1/012040

of radiation and the subsequent compression of the laser beam in the transverse direction. Pulsed radiation, the peak power of which is tens or more times greater than the critical self-focusing power, forms a large number of filaments. This is an inevitable consequence of the spatially modulated instability of an intense light field in a medium with cubic nonlinearity.

It should be noted that during filamentation, the plasma is formed mainly at the front of the pulse, while the back of the pulse is strongly defocused by this plasma, that is, these processes are fundamentally dynamic (non-stationary). The concentration of free electrons in plasma channels, which are formed during the filamentation of collimated or weakly ascending beams, according to estimates carried out in different works, is Ne ~ 10^{11} -10^{17} cm^{-3}. Such a spread of experimental data (six orders of magnitude or more) is due to the use of various measurement techniques and different experimental conditions.

During the filamentation of laser radiation in a quartz glass sample, the refractive index changes in the region where filamentation occurs. This change is irreversible, even after cessation of filamentation in the medium, the channel remains, which changes the intensity distribution of any radiation passing through the modified region of the sample. Being irradiated with ultraviolet radiation, the modified areas luminesce in the red wavelength range, while a quartz glass (KU-1) is transparent in ultraviolet spectral region. The prevailing plasma-forming processes are multiphoton and tunneling ionization, which do not lead to structural changes in the dielectric [12].

The filamentation of high-power laser radiation over extended paths is often random. This applies primarily to the longitudinal and transverse position of the filament formation point, as well as the number and dynamics of the interaction of multiple filaments in case of a significant excess of peak power. The main factors determining the stochastic nature of the process include inhomogeneities of the initial field distribution in a laser pulse (they can be random, noise-induced or artificially created, for example, in segmented beams), as well as fluctuations in the refractive index of the medium.

Additional complexity is the longitudinal and transverse instabilities of the beam and filaments, associated both with the instabilities of laser systems and with the fundamental stochasticity of the propagation process. For pulses whose power is hundreds of times greater than the critical self-focusing power, methods of forming an ordered array of a large number of filaments are investigated. It was experimentally and numerically shown that periodic intensity modulation can suppress the stochastic nature of filamentation caused by random fluctuations [13].

2. Optical structures in bulk transparent glass

2.1. Formation of optical channels in quartz glass

For inscription optical channels femtosecond Yb:KGW-laser system was used. The parameters of laser system: wavelength λ=1029 nm, pulse width τ=280 fs, pulse repetition rate of 10 kHz, pulse energy Emax=150 μJ [14-15]. The laser radiation was focused in the quartz glass sample by single lens with a focal length of 150 mm. The beam diameter in the focusing region was about 50 μm. The focused laser beam passed into the bulk of the quartz parallelepiped with dimension 5 mm x 5 mm x 20 mm. The sample is mounted on a two-axis translation stage.

Directly next to the sample two devices have been installed – a CCD-camera and a system for analyzing the cross section profile of the beam. With the aid of CCD-camera formed filaments were visualized, their geometrical dimensions relative to the exposure time and laser power were registered.

The filamentation of laser radiation in a quartz glass sample leads to a change in the refractive index of the exposure area. In this case, the intensity distribution in the transmitted beam also changes. Registration of the transverse profile of the distribution of laser radiation was carried out using the BeamStar system. A change in the beam profile as a function of power of laser radiation was registered.

At the beginning of the experiment, the area of the sample in which the beam intensity distribution corresponds to the Gaussian was chosen. As a source of transmitted laser radiation a DPSS-system with a wavelength of 532 nm was used. The radiation passing through the sample with inscribed structures thereby redistributing its intensity.

With the aid of laser beam analysis system the images of the intensity distribution at each point of the sample were obtained and then combined using a graphical editor into a panorama corresponding to the actual size of the sample (figure 1). Due to the small coverage of the camera, it was impossible to cover the entire sample in one continuous image, therefore this method of fixing data was applied: the images were combined according to a measurement matrix having 10 columns and 8 rows, each 5 millimeters in width across the sample.

Structural changes causing redistribution of the transmitted laser radiation are observed on obtained image. The green circles are normal Gaussian distribution; in these zones there is no modification. Red vertical arcs in columns 3 and 7 indicate increased intensity; the absence of color in column 6 in lines 2-6 indicates the absence of intense radiation in this zone. Both of these effects are caused by the redistribution of energy through the plasma channels created previously in the sample using filamentation.

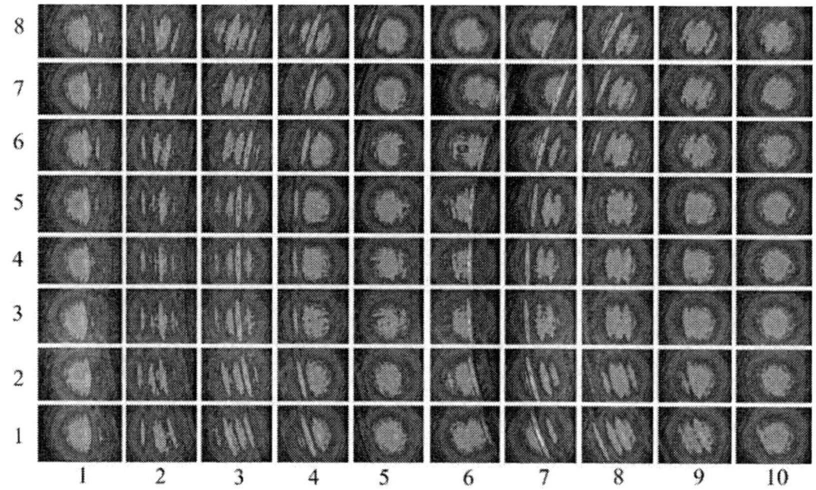

Figure 1. Panoramic image combined from images of the intensity distribution in various places of the modified quartz glass sample.

The formation of local maxima of laser radiation, in the presence of such kind of structures in the medium, occurs on the formed tracks. The propagation of laser radiation through the structures formed is independent of the power of the laser radiation. With a significant decrease in the intensity of laser radiation below the filamentation threshold, laser radiation retains the distribution structure characteristic of structures inscribed in the bulk of the medium. Thus, this property can be successfully used to form a given distribution profile of laser radiation in accordance with the requirements of material processing tasks.

To confirm this fact, the following experiment was conducted. A sample of KU-1 glass containing previously modified regions was irradiated with ultraviolet laser radiation with a wavelength of 257 nm. This spectral range was used for clarity of the experiment. By using a laser radiation with other wavelength, the nature of the propagation of laser radiation in the structure of the material is similar. The convenience of using ultraviolet laser radiation lies in the convenience of detecting the region of transmitted laser radiation (parallel beam) through the sample (figure 2).

IOP Conf. Series: Materials Science and Engineering **525** (2019) 012040 doi:10.1088/1757-899X/525/1/012040

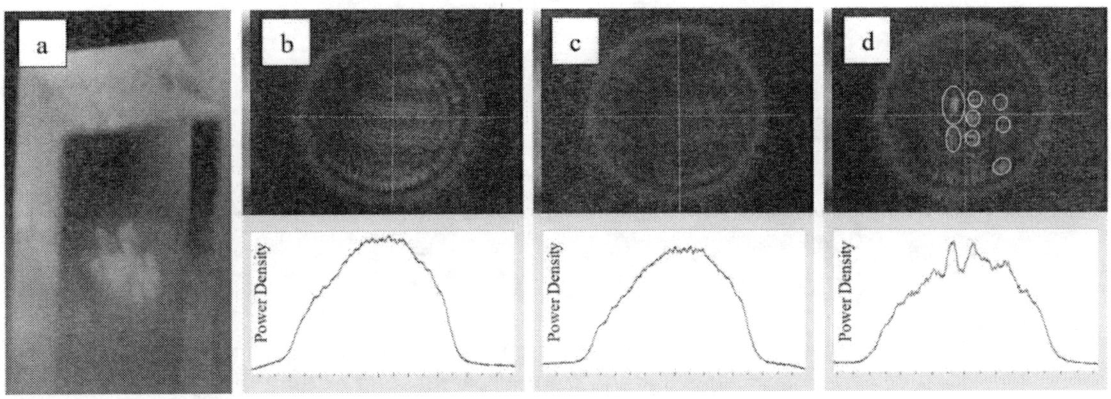

Figure 2. Registration of the laser radiation spatial distribution: a – the sample containing written structures; b – beam distribution at the exit of the laser system; c – laser beam passing through the unprocessed quartz sample; d – the formation of local radiation maxima in the beam cross section (outlined by ellipses).

2.2. Laser writing of information in transparent bulk samples

Another promising area of application of the presented laser method is controlled formation of structures in the volume of transparent medias. The scheme of the experiment is shown in figure 3. Structures were obtained using a 100x micro objective with a NA of 0.72, an average laser radiation power was 4 mW. Selective formation of structural entities was carried out by moving the sample relative to a stationary beam in coordination with the laser activation.

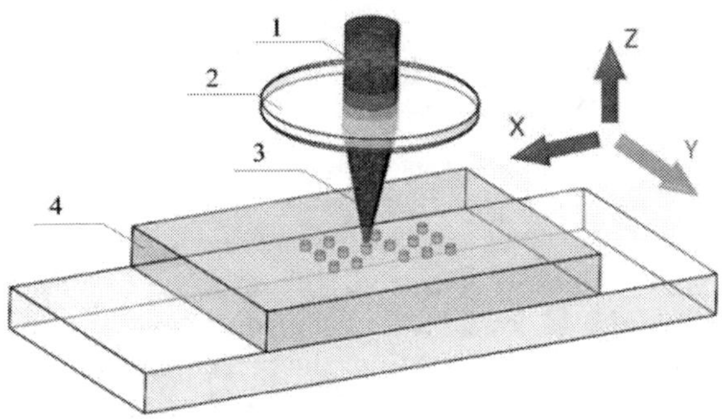

Figure 3. Schematic drawing of the processing area: 1 – femtosecond laser radiation, 2 – focusing optics, 3 – focused laser beam, 4 – transparent glass sample.

For the layer-by-layer formation of the topological pattern of the structures, specialized software was developed, allowing to set the parameters of sample processing and generate the executable command code for the laser complex (figure 4). Such a processing procedure involves the sequential impact by femtosecond laser pulse trains on a given area of a transparent solid medium with subsequent sample displacement in the X-Y plane, or along the Z-axis. The structures are written layer-by-layer from bottom to top. To provide additional protection for the encoded information, the algorithm can be extended by the data masking methods.

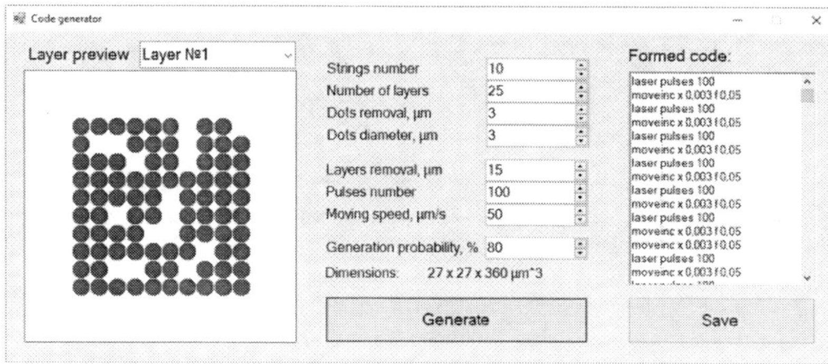

Figure 4. Interface of the software for the formation of the optical structures topology in transparent solid media.

Figure 5 illustrates an example of information encoded in the bulk glass in a form of dots array in accordance with the configuration of the generating program.

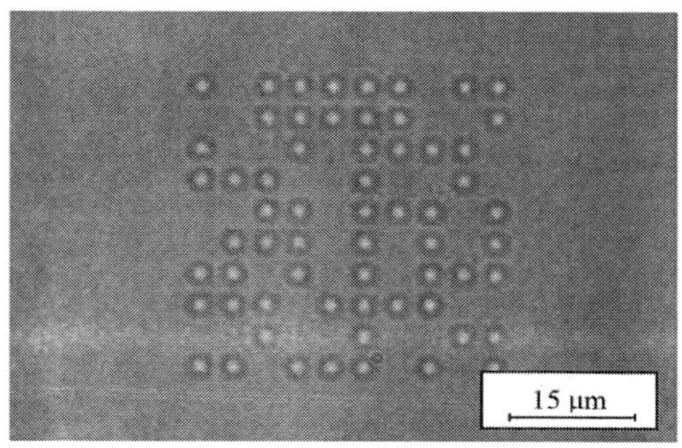

Figure 5. Dots array formed in a volume of the quartz sample.

The formed structures have a close to spherical shape that represents a hollow region in the volume of quartz glass. The minimum size of the formed dots can achieve up to 0.8 µm depending on the laser radiation power. The method allows for layer-by-layer direct writing and reading of structures in the sample volume. The optimal interlayer distance for the presented recording parameters was 15 µm.

This type of processing can be used as a method of originality authentication for substances packaged in thin-walled glass ampoules. This approach of formation of coded data in the volume of glass along with other methods of protection based on laser processing, such as recording of 3D-holographic images on a photosensitive emulsion, also meets the requirements for operation speed, cost and reproducibility of process.

3. Conclusion

The dependence of the formation of optical structures in bulk transparent samples on the intensity of laser radiation and the focal length of the optical system is studied. Lengthened optical channels were formed due to filamentation of femtosecond laser radiation. The program of forming dots arrays is described and the results of processing in accordance with the specified parameters are presented.

Acknowledgements

This work was performed as a part of Vladimir State University's State Task 3.5531.2017/8.9 GB-1106/17.

IOP Conf. Series: Materials Science and Engineering **525** (2019) 012040 doi:10.1088/1757-899X/525/1/012040

References

[1] Schaffer C B, Brodeur A, Mazur E 2001 Laser-induced breakdown and damage in bulk transparent materials induced by tightly focused femtosecond laser pulses *Meas Sci Technol* 12(11):1784-1794

[2] Anisimov S I and Luk'yanchuk B S 2002 Selected problems of laser ablation theory *Uspekhi Fiz Nauk* 45(3):293-324

[3] Mao S S, Quéré F, Guizard S, et al 2004 Dynamics of femtosecond laser interactions with dielectrics *Appl Phys A* 79(7):1695-1709

[4] Kondo T, Yamasaki K, Juodkazis S, Matsuo S, Mizeikis V, Misawa H 2004 Three-dimensional microfabrication by femtosecond pulses in dielectrics *Thin Solid Films* 453-454:550-556

[5] Shugaev M V, Wu C, Armbruster O, et al 2016 Fundamentals of ultrafast laser–material interaction *MRS Bull* 41(12):960-968

[6] Delone N Band Krainov V P 1998 Tunneling and barrier-suppression ionization of atoms and ions in a laser radiation field *Physics-Uspekhi* 41(5):469-485

[7] Apeksimov D V, Bukin O A, Bykova E E, et al 2013 Filamentation of the focused Ti: Sapphire laser pulse in air at two harmonics *Plasma Phys Reports* 39(13):1074-1081

[8] Couairon A and Mysyrowicz A 2007 Femtosecond filamentation in transparent media *Phys Rep* 441(2-4):47-189

[9] Kandidov V P, Shlenov S A, Kosareva O G 2009 Filamentation of high-power femtosecond laser radiation *Quantum Electron* 39(3):205-228

[10] Tarasova M A, Khorkov K S, Kochuev D A, Ivaschenko A V 2018 Study of the Filamentation Phenomenon of Femtosecond Laser Radiation Bull Lebedev Phys Inst 45(8):246-250

[11] Khorkov K S, Kochuev D A, Chkalov R V, et al 2018 Experimental study of the filaments parameters at the focusing with cylindrical lens *International Conference Laser Optics (ICLO)* IEEE 357

[12] Vislobokov N Yu 2014 Numerical simulation of the influence of dispersion on spectral continuatio powerful femtosecond pulses in silica glass *Journal of VDU* 2(80):23-24

[13] Kandidov V P, Akozbek N, Scalora M, et al 2004 A method for spatial regularisation of a bunch of filaments in a femtosecond laser pulse *Quantum Electron* 34(10):879-880

[14] Chkalov R V., Khorkov K S, Kochuev D A, et al 2018 Computerized laser complex for monitoring and controlling of the precision micromachining processes *Proceedings of the International Conferences on WWW/INTERNET 2018 and APPLIED COMPUTING 2018. - Budapest: IADIS* 395-399

[15] Chkalov R V, Khorkov K S, Kochuev D A, et al 2018 Thin film elements design: software and possibilities of femtosecond laser techniques *J Phys Conf Ser* 1109:012029

IOP Conf. Series: Materials Science and Engineering **525** (2019) 012041 doi:10.1088/1757-899X/525/1/012041

Sorption of cesium by aluminosilicate sorbents from rice straw

S B Yarusova[1,2], A E Panasenko[1,3], L A Zemnukhova[1,3], P S Gordienko[1], O D Arefieva[1,3] and A A Narbutovich[2]

[1] Institute of Chemistry, Far-Eastern Branch, Russian Academy of Sciences, Vladivostok, Russia
[2] Vladivostok State University of Economix and Service, Vladivostok, Russia
[3] Far Eastern Federal University, Vladivostok, Russia

E-mail: panasenko@ich.dvo.ru

Abstract. Using rice straw of various varieties of the Far Eastern selection as a source of silicon, a number of samples of sodium aluminosilicates with a Si:Al ratio from 1.02 to 1.56 were synthesized. For the obtained samples, the elemental and phase composition was found, the thermal properties were investigated. The morphology of aluminosilicates was studied: the particle size varies from 0.2 to 50 μm with a maximum distribution of about 10 μm, the specific surface area is 364 m2/g. The acid-base properties of the surface of aluminosilicates are characterized by pH-metry methods and the Hammett indicator method, the nature of the distribution of adsorption centers on the surface is established. The sorption capacity of the obtained samples, determined with respect to methylene blue (MB), is from 31.9 to 103.9 mg/g. By varying the synthesis conditions, it was shown that the greatest sorption capacity with respect to MB is achieved at the time of hydrolysis of the original rice straw for 60 minutes. It is shown that the sorption of Cs+ ions in the obtained samples is described by the Langmuir model. The sorption properties with respect to Cs+ ions from aqueous solutions with different salt background were studied.

1. Introduction

The spectrum of the chemical composition of silicates and aluminosilicates is quite wide [1]. They are mainly used as fillers for construction materials, sorbents, catalyst carriers and ion exchange materials [2]. An important characteristic of its surface is the acid-base properties [3], which make it possible to predict the reactivity and sorption capacity of a surface with respect to pollutants of various nature (metal cations and organic dyes).

Synthetic aluminosilicates surpass natural analogues in a number of indicators. The most significant advantages of synthetic aluminosilicates are their constant composition and the absence of impurities. The chemical and phase composition of synthetic aluminosilicates is largely determined by the method of preparation and the feedstock [4].

In addition to traditional mineral raw materials and wastes from their processing, silicophilous plants, in particular, rice, are an alternative source of silicon for the chemical industry. The by-products of rice cereal production – straw and fruit casings – are renewable raw materials, each year around 800 million tons are produced around the world. It is known to use rice husk ash as a source of silicon in the synthesis of zeolites [5–10]. Works on the synthesis and study of the functional properties of aluminosilicates from rice straw are much less [11–13]. A promising area of application of synthetic aluminosilicates is their use for the purification of aqueous solutions from cesium ions, which is confirmed by a number of studies [14–16].

Earlier, sodium and potassium aluminosilicates with a specific surface area of 69–134 m^2/g and a sorption capacity relative to Cs^+ ions up to 1.3 mmol/g were obtained from alkaline rice straw hydrolysates at the Institute of Chemistry of the Far East Branch of the Russian Academy of Sciences [17]. During the sorption of Cs^+ ions, the kinetic curve reaches a plateau already after 5 – 10 minutes and with an increase in temperature from 20 to 60 °C, the sorption capacity does not change, the degree of extraction of Cs^+ ions is 82.6–88.6% [18]. When choosing sorbents, it is also necessary to have some data on the effect of the ionic composition of the salt background on the degree of extraction of the corresponding ions from aqueous solutions. Of practical interest are the values of the sorption parameters in the sorption of cesium ions from natural sea water containing Cs^+ ions in various concentrations.

The purpose of this work is to obtain aluminosilicates from rice straw (RS), to study the acid-base properties of their surface, to study the sorption capacity of samples with respect to methylene blue and cesium ions, including sorption from solutions of complex salt composition.

2. Experimental

Rice straw *(Oryza sativa)* of the Far Eastern selection varieties harvested in 2017 was used as a source of silicon. Straw was ground into pieces 5–10 mm long, washed with water and air-dried. For the extraction of silicon, the straw was subjected to alkaline hydrolysis in a 1M solution of NaOH with a ratio solid-to-liquid 1:13 at a temperature of 90°C for a fixed time. A saturated aqueous solution of aluminum sulfate $Al_2(SO_4)_3{\cdot}18H_2O$, taken in a molar ratio of $Al_2(SO_4)_3{\cdot}18H_2O$: Si = 1 : 3, was added to the hydrolyzate obtained at room temperature, then the reaction mixture was neutralized with HCl solution. The precipitation was kept for at least 5 hours, separated with a filter and washed with water. The precipitate was dried at 105 °C [17].

An elemental analysis of aluminosilicate samples was performed by the method of energy dispersive X-ray fluorescence spectroscopy on a Shimadzu EDX 800 HS spectrometer (Japan). X-ray diffractograms were recorded on a Bruker D8 Advance diffractometer (Germany) in Cu K_α-radiation. The specific surface of the samples was found by the method of low-temperature nitrogen adsorption using a Sorbtometer-M device (Russia). A thermal analysis was performed on an MOM Q-1000 derivatograph (heating rate 5 K/min). Particle size distribution was determined on a Fritsch ANALYSETTE 22 MicroTec plus laser particle size analyzer.

The acid-base properties of the surface of sorbents were studied by the method of adsorption of indicators (Hammett's method) using 13 indicators with pKa values in the range from −0.29 to +16.80 [19]. According to the obtained values, the distribution curves of the centers of adsorption of indicators on the surface of aluminosilicates were constructed.

The sorption properties of the obtained aluminosilicates were studied on model solutions of methylene blue (MB) [20, 21] and solutions of cesium chloride. A study of the sorption of methylene blue from aqueous solutions was carried out under static conditions at a temperature of 20°C, with a sorbent : solution ratio of 1:200, with MB concentration of 80.8–841.6 mg/L. After stirring for 40 min, the suspension was centrifuged, the optical density of the solution was measured with a ZOMZ KFK-3-01 spectrophotometer (Russia) at a wavelength of 657 nm. Experiments on cesium sorption were carried out under static conditions at a ratio of sorbent : solution 1: 400 and temperature of 20 °C with stirring from 1 to 60 minutes. Aqueous solutions of CsCl with different initial concentrations of Cs^+ ions in the range from 0.16 to 3.2 mmol/L based on distilled sea water and 0.01 M of $NaNO_3$, KNO_3, NH_4NO_3 were used. The solutions were separated from the sorbent by filtration. Sea water, on the basis of which solutions with different contents of Cs^+ ions were prepared, was collected in the Sea of Japan, Amursky Bay (Vladivostok), filtered through a "white ribbon" filter.

The content of Cs^+ and Na^+ ions in solutions was determined by atomic absorption spectrometry (AAS) with a Solaar M6 spectrometer (Thermo Scientific, USA).

The sorption capacity (A_c, mmol/g) of the samples was calculated by the formula (1)

$$A_C = \frac{(C_{ini} - C_e)}{m} \cdot V, \quad (1)$$

where C_{ini} is the initial concentration of Cs^+ ions in the solution, mmol/L; C_e is the equilibrium concentration of Cs^+ ions in the solution, mmol/L; V is the volume of solution, L; m is the mass of sorbent, g.

The degree of extraction of Cs^+ ions (α, %) was calculated by the formula (2).

$$\alpha = \frac{(C_{ini} - C_e)}{C_{ini}} \cdot 100\%, \quad (2)$$

3. Results and Discussion

3.1. Characteristics of the samples

The samples synthesized in this work are X-ray amorphous, and a halo with a maximum corresponding to an interplanar distance of 3.21–3.28 Å is observed in the diffraction pattern. The composition of the samples corresponds to sodium aluminosilicate (Table 1).

Table 1. Characteristics of the samples.

Sample	Straw strain	Content of elements according to EDX, mass. %				M:Al:Si[a]	LI[b], %
		K	Na	Al	Si		
1	Khankaiskiy-429	3.16	16.36	39.05	41.43	0.21:1:1.02	24.98
2	Dariy	2.94	18.76	33.82	44.46	0.28:1:1.26	30.68
3	Lugovoy	2.15	17.66	31.08	50.22	0.27:1:1.56	31.74
4	Dubrava	3.11	17.28	31.08	47.94	0.30:1:1.47	43.35

[a] – molar ratio
[b] LI – loss on ignition, 1000 °C

The influence of the processing conditions of rice straw on the sorption properties of samples depending on the time of alkaline hydrolysis was studied. From straw strain *Lugovoy*, samples were obtained with a duration of hydrolysis of 60, 40 and 20 minutes (Table 2). As one can see, all samples have a similar composition.

Table 2. Characteristics of samples obtained at different times of hydrolysis.

Sample	Hydrolysis time	Content of elements according to EDX, mass. %				M:Al:Si	LI, %
		K	Na	Al	Si		
3	60 minutes	2.15	17.66	31.08	50.22	0.27:1:1.56	31.74
5	40 minutes	3.49	4.65	35.05	50.75	0.22:1:1.39	23.50
6	20 minutes	2.14	8.88	33.52	35.86	0.35:1:1.03	29.18

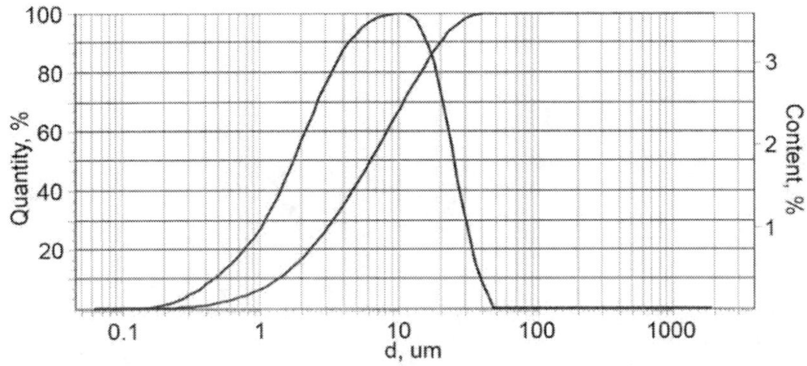

Figure. 1. Size distribution of sodium aluminosilicate (sample 3) particles.

The specific surface of sample 3 is 364 m^2/g, and the particle size has a distribution maximum of about 10 μm (figure 1). A thermogravimetric analysis of sample 3 (figure 2) showed that the removal of water begins at about 40 °C and ends at 220 °C. Above 220 °C, oxidation of the organic component begins, followed by an exoeffect on the DTA curve with a maximum at 280 °C, which corresponds to the oxidation of non-crystalline or slightly crystallized cellulose. A small maximum on the DTA curve at about 600–620 °C is associated with the oxidation of the carbonated residue. The total mass loss on ignition is 31.7%.

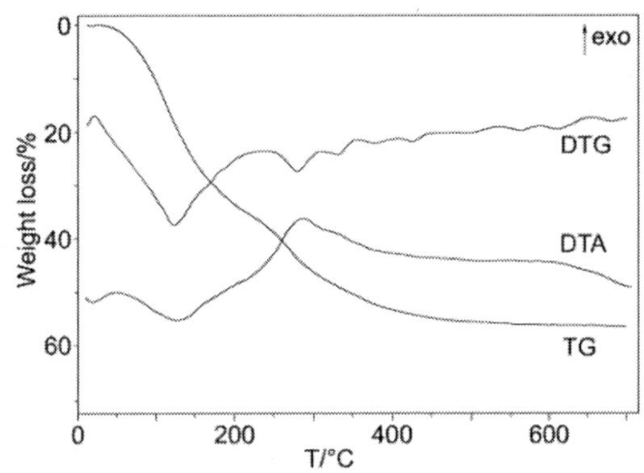

Figure 2. Thermogravigram of sodium aluminosilicate (sample 3).

The study of the acid-base properties of the surface of samples and the distribution of active centers by the indicator method showed that on the surface there is a set of Lewis and Brønsted sites of both acidic and basic types [19]. In the spectra of the distribution of adsorption centers, there are 3 main bands describing active sites (figure 3): Brønsted acidic (pKa +2.5), weakly acidic (pKa +6.4), and Lewis acidic (pKa +16.8) sites.

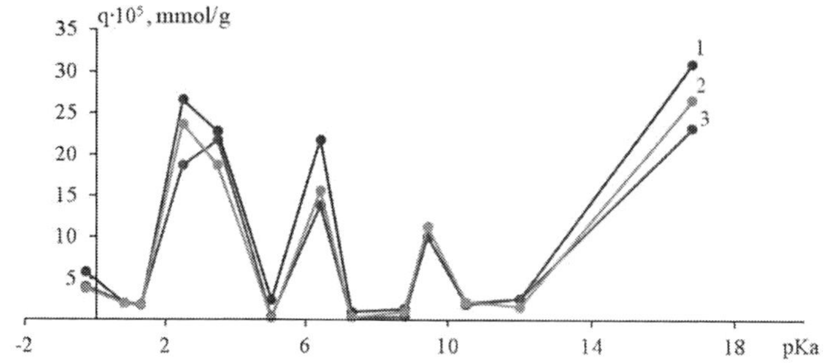

Figure 3. Distribution of adsorption centers of indicators on the surface of aluminosilicates (numbering as per table 2).

Brønsted acid sites with a pKa of +2.5 most likely correspond to the silanol groups ≡Si–OH [22]. The centers of adsorption with pKa +6.4 are characterized by the fact that they show neither basic nor acidic properties, due to the equality of E–O and O–H bonds [23]. The presence of weak Brønsted sites on the surface corresponds to OH groups of a different nature (for example, $=Si(OH)_2$, $-Al(OH)_2$) [22]. Lewis acid sites with pKa +16.8 are coordinatively unsaturated surface aluminum atoms with a localized positive charge.

3.2. Sorption characteristics

The methylene blue (MB) sorption on the obtained aluminosilicate samples occurs via a monolayer mechanism, as evidenced by the isotherm type (figure 4). The values of the sorption capacity for MB are shown in table 3.

Table 3. Value of the sorption capacity of aluminosilicates.

Raw materials and conditions for obtaining aluminosilicate	Sorption capacity with respect to MB, mg/g
Khankaiskiy-429, 60 minutes	32.9
Dariy, 60 minutes	31.9
Lugovoy, 60 minutes	103.9
Dubrava, 60 minutes	46.6
Lugovoy, 40 minutes	47.1
Lugovoy, 20 minutes	43.6

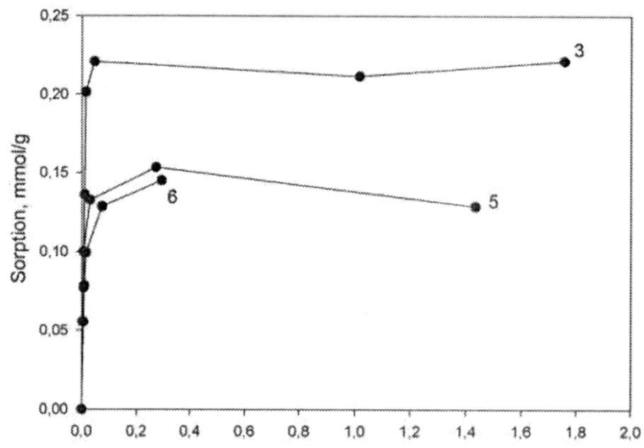

Figure 4. Sorption isotherms of methylene blue by samples of aluminosilicates (numbering as per table 2).

From table 3 it can be seen that the sorption capacity of the samples depends both on the rice variety, the straw of which was used as a raw material, and on the time of alkaline hydrolysis of the straw. Of all the varieties of rice, the use of RS of *Lugovoy* variety as a source of silicon allows obtaining aluminosilicate with the highest sorption capacity. In this case, the best result is achieved with an alkaline hydrolysis time of 60 minutes. Reducing the time of hydrolysis to 40 or 20 minutes dramatically reduces the sorption properties of the product.

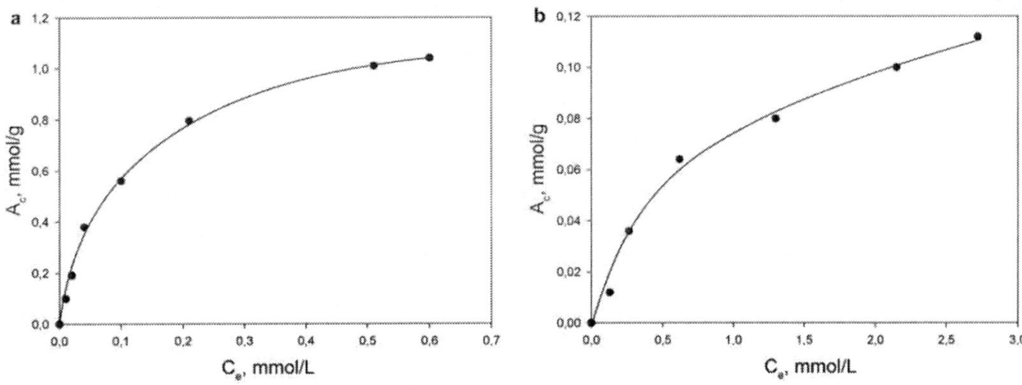

Figure 5. Sorption isotherms of Cs^+ ions with sodium aluminosilicate from solutions based on distilled (a) and sea (b) water.

Based on the obtained experimental data on the sorption of Cs^+ ions, sorption isotherms were built (figure 5). As can be seen from the figure, during the sorption of Cs^+ ions from sea water, a significant decrease in the sorption capacity is observed. The degree of extraction of Cs^+ ions from solutions without salt background reaches 96.2%, and from sea water it does not exceed 25.7%.

To assess the sorption properties, the isotherms obtained were analyzed in the coordinates of the Langmuir equation and the Freundlich empirical equation. Langmuir equation:

$$\frac{C_p}{A_c} = \frac{1}{A_m \cdot k} + \frac{C_p}{A_m},$$

where C_e is the equilibrium concentration of Cs^+ ions in the solution, A_m is the maximum sorption capacity, k is the Langmuir constant.

The constants of the equation were calculated from the slope and intersection of the lines on the graph in the corresponding coordinates of the linear C_e/A_e equation from C_e (figure 6).

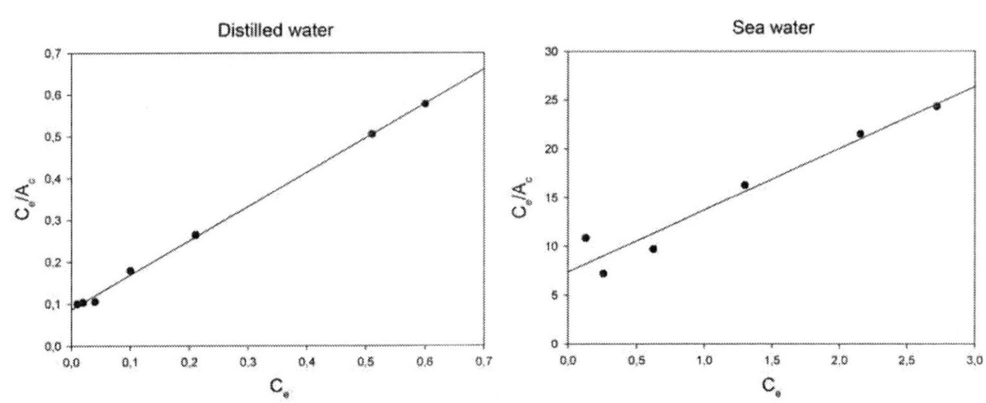

Figure 6. Sorption isotherms of Cs^+ ions with synthetic sodium aluminosilicate in the respective coordinates of the Langmuir linear equation.

IOP Conf. Series: Materials Science and Engineering **525** (2019) 012041 doi:10.1088/1757-899X/525/1/012041

The logarithmic form of the Freundlich equation was used to construct a linear dependence and ln A_e – ln C_e and a graphic definition of the parameters K_F and n:

$$\ln a = \ln K_F + \frac{1}{n}\ln C_e ,$$

where K_F is the equilibrium constant of the Freundlich equation relating to the sorption capacity; 1/n is a parameter indicating the intensity of the sorbent-sorbate interaction. The parameters of the Langmuir and Freundlich equations found graphically are presented in table 4.

Table 4. Parameters of the Langmuir and Freundlich equations obtained by analyzing the isotherms of sorption of Cs^+ ions with synthetic sodium aluminosilicate.

Parameters of equation	Type of solution containing Cs^+ ions	
	Distilled water	Sea water
	Langmuir equation	
k, L/mmol	9.5	0.88
A_m, mmol/g	1.2	0.16
R^2	0.9985	0.94
	Freundlich equation	
K_F, (mmol/g)·(L/mmol)$^{1/n}$	1.7	14.90
1/n	0.5521	0.66
R^2	0.9452	0.90

The table shows that for describing the sorption of Cs^+ ions with synthetic sodium aluminosilicate, the Langmuir model is most suitable, as evidenced by the corresponding correlation coefficients. The maximum sorption capacity of A_m during the sorption of Cs^+ ions from solutions without salt background and from sea water is 1.2 and 0.16 mmol/g, respectively.

Figure 7 (a, b) displays the integral kinetic curves of the sorption of Cs^+ ions from solutions of different salt composition with the studied sorbent.

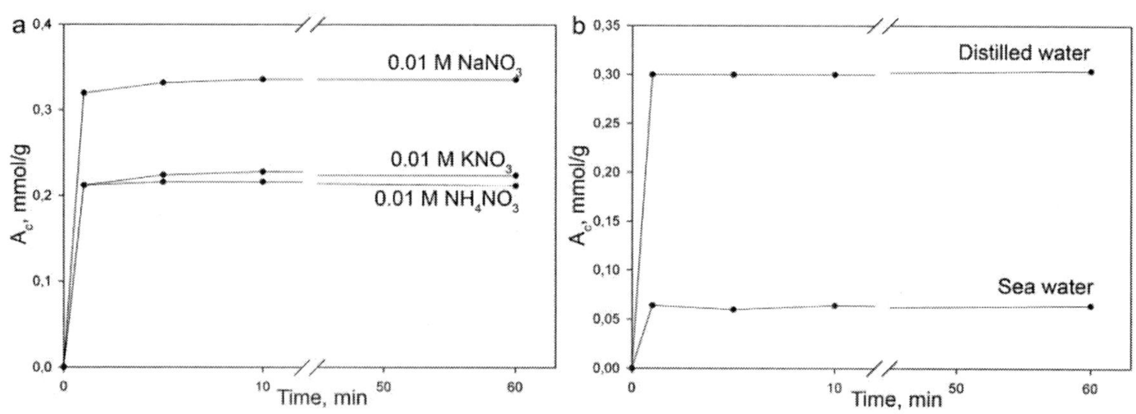

Figure 7. Integral kinetic curves of sorption of cesium by sodium aluminosilicate from solutions of different salt composition: a – 0.01 M solutions of nitrates; b – distilled and sea water.

In all the cases under consideration, the kinetic curves have already reached the plateau within 5 minutes, while the degree of extraction of Cs^+ ions in the case of distilled water is 96.4%; sea water – 19.7%; $NaNO_3$ – 87.4%; KNO_3 – 70.0%; NH_4NO_3 – 69.2%. Thus, the sorption of cesium in sea water is significantly reduced compared with all the studied types of solutions. During sorption of Cs^+ ions from 0.01 M solutions of nitrates, the highest values of the degree of extraction of Cs^+ ions are observed during sorption from a 0.01 M solution of $NaNO_3$. Similar data were obtained by the authors [24] in the study of the sorption properties of synthetic potassium aluminosilicate in the extraction of ^{137}Cs from solutions of nitrates-containing salts.

Acknowledgments
The work was partially supported within the frames of the State Order of the Institute of Chemistry FEBRAS, project no. 265-2019-0002, and partially supported by the Russian Federation President's Grants Council (State Support to Young Russian Candidates of Science Program, grant no. MK-2884.2017.3).

References

1. Shulz M M 1997 Silicaty v prirode i praktike cheloveka (Saint-Petersburg: Nauka) p 197 [in Russian]
2. Shoumkova A 2011 Austral. Inst. High Energ. Mat. Bull. 1
3. Tanabe K 1970 Solid Acids and Bases (New York: Academic Press)
4. Sembiring S, Simanjuntak W, Manurung P, Asmi D and Low I M 2014 *Ceram. Int.* **40** 7067
5. Melo C R, Francisco A C, Kuhnen N C, Rocha M R, Melo A R, Riella H G and Angioletto E 2014 *Mat. Sci. Forum* **798–799** 617
6. Mohamed R M, Mkhalid I A and Barakat M A 2015 *Arab. J. Chem.* **8** 48
7. Ng E P, Awala H, Tan K-H, Adama F, Retoux R and Mintova S 2015 *Microporous and Mesoporous Mat.* **204** 204
8. Mallapur V P and Oubagaranadin J U K 2017 *Transactions Ind. Ceram. Soc.* **76** 1
9. Takahashi M, Shibata K and Shiono T 2018 *Key Eng. Mat.* **766** 105
10. Ahmedzeki N S, Joodee A M, Abbas M N and Jaed Y M 2018 *J. Chem. Tech.* Met. **53** 11
11. Ali I O, Thabet M S, El-Nasser K S, Hassan A M and Salama T M 2012 *Microporous and Mesoporous Mat.* **160** 97
12. Ali I O, Salama T M, Thabet M S, El-Nasser K S and Hassan A M 2013 *Mat. Chem. Phys.* **140** 81
13. Jesudoss S K, Vijaya J J, Kaviyarasu K, Kennedy L J, Ramalingam R J and Al-Lohedan H A 2018 *RSC Advances* **8** 481
14. El-Naggar M R, El-Kamash A M, El-Dessouky M I and Ghonaim A K 2008 J. *Hazard. Mat.* **154** 963
15. Borai E H, Harjula R, Malinen L and Paajanen A 2009 *J. Hazard. Mat.* **172** 416
16. Singh B K, Tomar R, Kumar S, Jain A, Tomar B S and Manchanda V K 2010 *J. Hazard. Mat.* **178** 771
17. Zemnukhova L A, Fedorishcheva G A, Tsoy E A and Arefieva O D 2015 Patent RU 2557607
18. Panasenko A E, Yarusova S B, Zemnukhova L A and Gordienko P S 2017 *Book of abstracts of the International conference «Renewable plant resources: chemistry, technology, medicine»*, (Saint-Petersburg) p 63
19. Pakhnutova E A and Slizhov Y G 2014 Russ. *J. Phys. Chem.* A **88** 1408
20. Sheng J, Xie Y and Zhou Y 2009 *Appl. Clay Sci.* **46** 422
21. Bestani B, Benderdouche N, Benstaali B, Belhakem M and Addou A 2008 *Biores. Tech.* **99** 8441
22. Golubeva O Y, Ul'yanova N Y, Kostyreva T G, Drozdova I A and Mokeev M V 2013 *Glass Phys. and Chem.* **39** 533
23. Slizhov Y G, Matveev T N and Minakova T S 2012 *Russ. J. Phys. Chem.* A **86** 463
24. Gordienko P S, Yarusova S B, Shabalin I A, Zheleznov V V, Bulanova S B and Zarubina N V 2014 *Radiochemistry* **56** 518

Synthesis and electrochemical performance of Li-rich $xLi_2MnO_3{\bullet}(1\text{-}x)LiMn_{1/3}Ni_{1/3}Co_{1/3}O_2$ (x=0.2–0.5) cathode materials for lithium-ion batteries

L S Pechen[1], E V Makhonina[1], A M Rumyantsev[2,3], Yu M Koshtyal[2,3], A S Goloveshkin[4], V V Volkov[1], Yu A Politov[1] and I L Eremenko[1]

[1]N.S. Kurnakov Institute of General and Inorganic Chemistry, Russian Academy of Sciences, 31 Leninsky pr., Moscow, 119991 Russia
[2]Ioffe Institute, Russian Academy of Sciences, 26 Politekhnicheskaya, St Petersburg, 194021 Russia
[3]Peter the Great St.Petersburg Polytechnic University, 29 Polytechnicheskaya, St Petersburg, 195251 Russia
[4]Nesmeyanov Institute of Organoelement Compounds Russian Academy of Sciences, Russia, 28 Vavilova, Moscow, 119334 Russia

E-mail: lidia.s.maslennikova@gmail.com

Abstract. Li-excess transition metal oxides of the general composition $xLi_2MnO_3{\bullet}(1\text{-}x)LiMO_2$ ($M{=}Mn_aNi_bCo_c$, a+b+c=1) are synthesized by coprecipitation method. Synthesis conditions for obtaining spherical particles with porous structure were determined. The influence of the oxide phase composition on their electrochemical performance as cathode materials is studied in lithium half-cells. All the samples show very high first-cycle coulombic efficiencies (more than 90%) in the voltage range of 2.5–4.8 V at a current density of 0.5*C*. It is reported that the Li-rich oxide of the composition $0.35Li_2MnO_3{\bullet}0.65LiMn_{1/3}Ni_{1/3}Co_{1/3}O_2$ shows the better cycling performance in terms both discharge capacity and energy density and the best rate capability compared with the samples with larger and smaller contents of Li_2MnO_3.

1. Introduction

Currently, $LiCoO_2$, $LiNi_xCo_yMn_zO_2$, $LiMn_2O_4$, $LiNi_xCo_yAl_zO_2$, and $LiFePO_4$ are mainly used as positive electrode (cathode) materials for lithium–ion batteries (LIBs). However, the highest energy densities of these materials do not exceed 180–200 mA h/g. The development of high-energy cathode materials can significantly widen the LIB application. In addition, the cathode materials should have a good cyclability, be safe and not expensive.

Li-rich transition metal oxides have attracted attention as promising candidates for the new generation of cathode materials because of their high specific capacities [1, 2]. They can be written in two notations, $xLi_2MnO_3{\bullet}(1\text{-}x)LiMO_2$ ($M{=}Mn_aNi_bCo_c$, a+b+c=1) or in the form with excessive lithium $Li_{(1+y)}M'_{(1-y)}O_2$ [1-3]. The first notation shows nanoscale inhomogeneity of these materials, because $LiMO_2$ (trigonal, R-3m) and Li_2MnO_3 (monoclinic, C2/m) have compatible oxygen lattices, allowing integration of the two structures. Up to now, the atomic structure of these materials is debated in the literature [4-8].

Li_2MnO_3 monoclinic phase is electrochemically inactive when cycled to 4.4 V. After activation to 4.5-4.8 V, there is observed a significant capacity increase that can be attributed to activation of the monoclinic phase. The activation involves delithiation and deoxidation processes and formation of electrochemically active $LiMnO_2$ phase [9-11]. The discharge capacities of Li-rich oxides can reach up to 250-300 mA h/g. However, the materials have a large first-cycle capacity loss and suffer from capacity fade and voltage decay during cycling. Despite the large body of research, Li-rich materials without the above drawbacks are not obtained till now because of the complexity of these structures.

In this work, we studied effect of the $Li_2MnO_3/LiMO_2$ phase composition in $xLi_2MnO_3 \bullet (1-x)LiMO_2$ Li-rich structures on their electrochemical characteristics. We choose the $LiMn_{1/3}Ni_{1/3}Co_{1/3}O_2$ composition of the trigonal phase as optimal that in terms of the electrochemical performance [12]. The oxides with 50, 35 and 20% of Li_2MnO_3 were synthesized.

2. Experimental

2.1. Synthesis procedure of Li-rich oxides

Li-rich oxides were synthesized by coprecipitation method followed by a solid-state reaction with $LiOH \cdot H_2O$. The starting reagents were manganese nitrate tetrahydrate (98%, Alfa Aesar), nickel nitrate hexahydrate (99.9%, ABCR), cobalt nitrate hexahydrate (99% Acros Organics), sodium carbonate anhydrous (≥99%, Chimmed), ammonium hydroxide (special purity grade, Chimmed), and lithium hydroxide monohydrate (≥99%, Sigma-Aldrich). Solutions of metal nitrates (2 mol/L) and sodium carbonate (2 mol/L) with ammonium hydroxide (0.35 mol/L) were added dropwise in a quartz reactor partially filled with deionized water. Coprecipitation of carbonates MCO_3 ($M=Mn_xNi_yCo_z$) was carried out under CO_2 atmosphere.

The synthesis conditions of the carbonate precursors were optimized. The temperature was maintained at 60°C; solution pH was adjusted to 7.5, and the stirring speed fixed at 1000 rpm. Metal nitrates solution was added at a constant rate of about 120 mL/h. A Na_2CO_3/NH_3 solution was automatically added to maintain the solution pH at 7.5. The mature time was 12 h. Carbonate precursors prepared were thoroughly washed with deionized water, ethanol, and diethyl ether, and then dried at 120 °C at an argon atmosphere.

For solid-phase synthesis, appropriate amounts of lithium hydroxide monohydrate (with a 3% excess) and a precursor were thoroughly ground in an agate mortar and annealed at 480 (6 h) and 900°C (12 h). The mixture was repeatedly homogenized in the course of annealing. The samples containing 50, 35 and 20 % of Li_2MnO_3 phase ($0.5Li_2MnO_3 \bullet 0.5LiMn_{1/3}Ni_{1/3}Co_{1/3}O_2$, $0.35Li_2MnO_3 \bullet 0.65LiMn_{1/3}Ni_{1/3}Co_{1/3}O_2$ and $0.2Li_2MnO_3 \bullet 0.8LiMn_{1/3}Ni_{1/3}Co_{1/3}O_2$) were prepared.

2.2. Material characterization

The powder X-ray diffraction (XRD) patterns were recorded using a Bruker D8 Advance diffractometer (Cu Kα radiation, λ=0.15418 nm, 45 kV/250 mA) at room temperature. The data were collected in the Θ/2Θ scan mode, 0.02° scan step in the 2 Θ range of 5°–80°. The precise X-ray analysis was carried out with a Bruker D8 Advance Vario (Cu Kα1) equipped with a Ge monochromator and position-sensitive detector LynxEye with rotating sample stage (the Θ/2Θ geometry). The data were collected with the use of the Bruker DIFFRACplus program package, the analysis was performed with the EVA and TOPAS programs.

Compositions of the resulting compounds were determined by inductively coupled plasma mass-spectrometry (ICP-MS) with an Agilent 7500ce system ("Agilent Technologies Inc.").

The morphology and microstructure of the samples were studied by scanning electron microscopy (SEM) using a NVision-40 (Carl Zeiss) microscope.

Particle size analysis was performed with the use of a laser particle sizer Analysette 22 (Fritsch). We used the Fraunhofer model for data analysis.

2.3. Electrochemical tests

Electrochemical tests of the samples as cathode materials were carried out in CR2032 coin-type cells, using not less than 3 cells for each sample. The positive electrodes were prepared by mixing 92 wt% of an active material, 5 wt% of the conductive additive (Super C65, Timcal) and 3 wt% of polyvinylidene difluoride (Solef 5130, Solvay) in N-methyl-2-pyrrolidone. A mass load of active material in a cell was on average 8-12 mg. A lithium foil was used as anode and TC-E918 (Tinci) as the electrolyte.

The cells were cycled galvanostatically in different voltage ranges (50–300 cycles on average). Before the main electrochemical tests, all the cells were subjected to an activation stage. The activation stage included cycling in the voltage ranges of 2.5–4.2 V, 2.5–4.4 V, 2.5–4.8 V and again 2.5–4.4 V (two cycles for each range). Hereafter, if the sample showed an acceptable result, further studies in the required voltage ranges were performed. Rate capability of the Li-rich materials was also investigated in different voltage ranges. The charge current density was 10-100 mA/g, discharge current density was 10-1000 mA/g.

3. Results and discussions

Li-rich oxides containing 50, 35 and 20 % of Li_2MnO_3 phase ($0.5Li_2MnO_3{\bullet}0.5LiMn_{1/3}Ni_{1/3}Co_{1/3}O_2$, $0.35Li_2MnO_3{\bullet}0.65LiMn_{1/3}Ni_{1/3}Co_{1/3}O_2$ and $0.2Li_2MnO_3{\bullet}0.8LiMn_{1/3}Ni_{1/3}Co_{1/3}O_2$) prepared by the above procedure hereafter are denoted LRC50, LRC35, LRC20, respectively. The samples are black powders consisting of spherical agglomerates with narrow size distribution (Figure 1, Table 1).

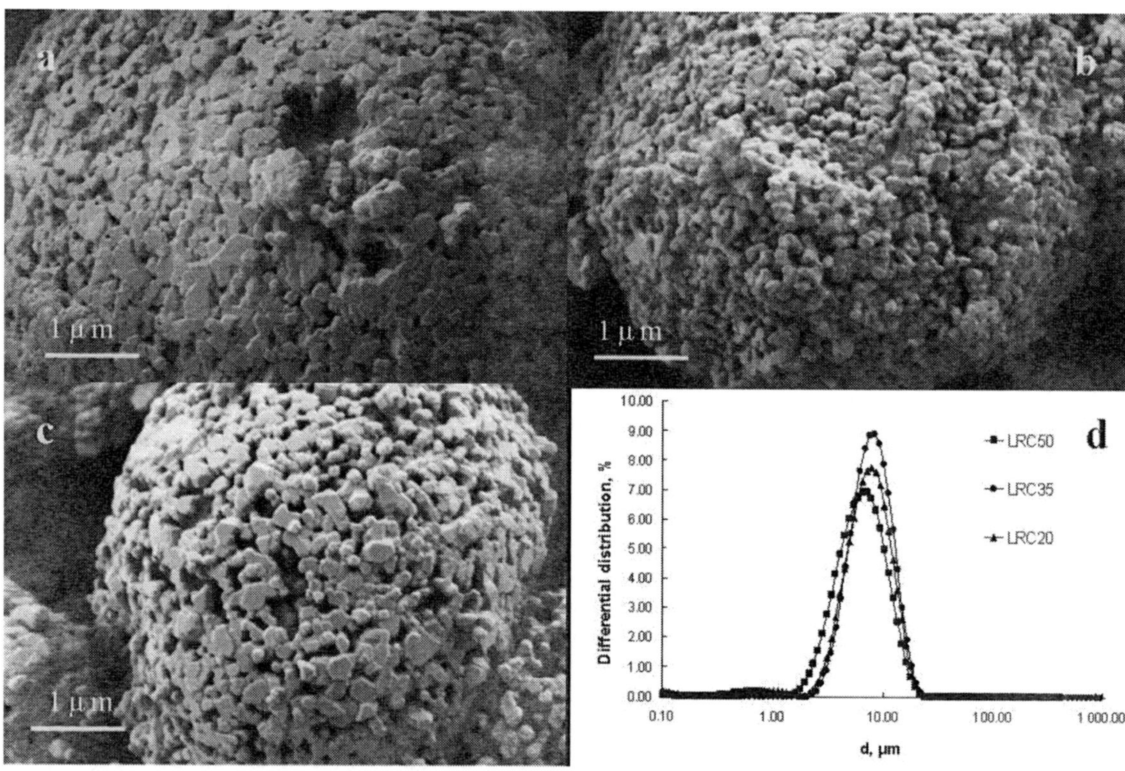

Figure 1. SEM images of samples (a) LRC50, (b) LRC35, and(c) LRC20; and (d) differential particle size distribution of the samples.

The agglomerate and primary particle sizes are close to each other for all the samples. The chemical compositions of the samples determined by ICP–MS analysis were found to be very close the target compositions (Table 1).

Table 1. Morphological characteristics of pristine Li-rich materials and data of ICP-MS analysis.

Sample	Primary particle size range, nm	Agglomerate size distribution, μm			ICP-MS data, at % (at.)			
		D10	D50	D90	Li	Mn	Ni	Co
LRC50	10-400	2.8	5.7	11.0	1.226	0.537	0.131	0.131
LRC35	100-300	4.1	7.3	12.4	1.138	0.483	0.186	0.193
LRC20	100-300	3.6	6.9	12.3	1.09	0.423	0.250	0.237

The XRD patterns of the samples with different $Li_2MnO_3/LiMO_2$ ratio are shown in Figure 2.

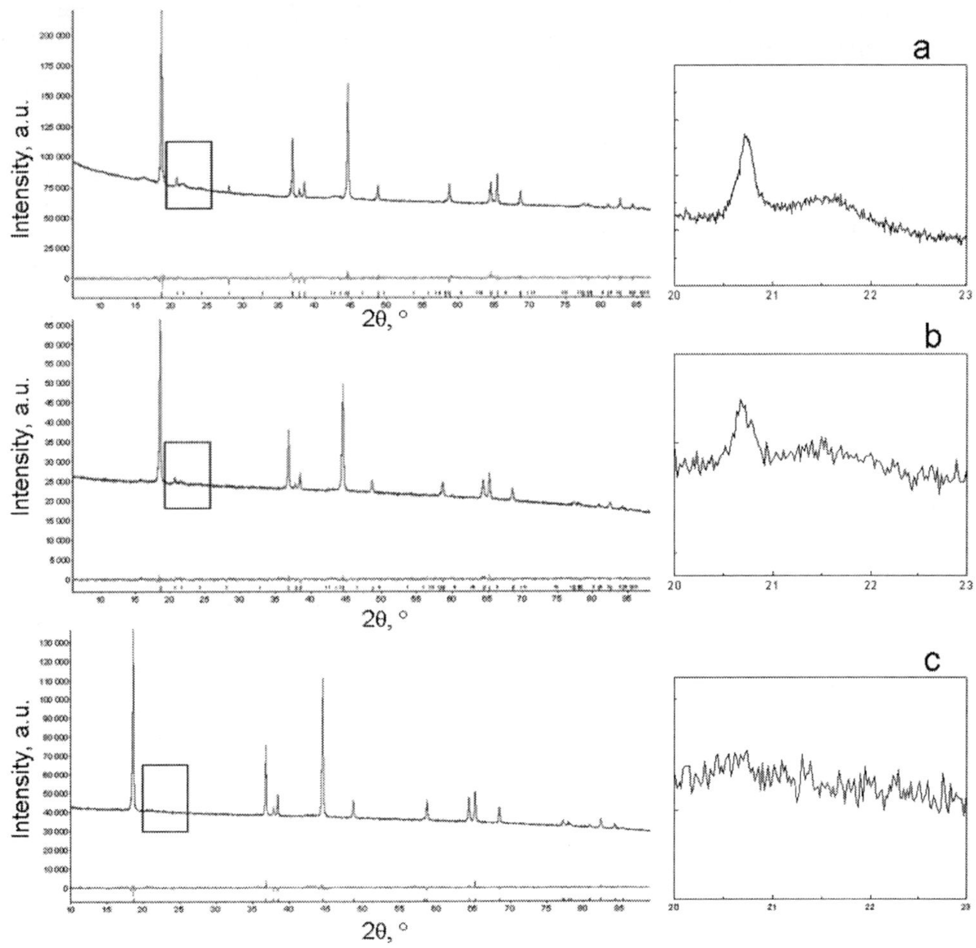

Figure 2. The Rietveld profile fittings of X-ray diffraction patterns of experimental (blue lines) and fitting (red lines) X-ray diffraction patterns of (a) LRC50, (b) LRC35, (c) LRC20 samples. The positions of Bragg reflections are denoted by vertical bars. The difference (experiment minus calculation) curve is shown by a gray line at the bottoms.

Most of the reflections can be indexed based on the trigonal R-3m structure. The small peaks between 20° and 23° (Figure 2, inserts) can be indexed in a monoclinic superstructure *C2/m* indicating

IOP Conf. Series: Materials Science and Engineering **525** (2019) 012042 doi:10.1088/1757-899X/525/1/012042

an ordering in the transition metal layer. Predictably, the intensities of these peaks decrease with decreasing Li_2MnO_3 content in the sample and practically vanish in the diffraction pattern of LRC20/80 sample. The crystallite sizes were estimated by Williamson-Hall method to be 50–70 nm for all the samples. The lattice parameters of the samples and coherent scattering lengths are summarized in Table 2. These values are lower than primary particle sizes observed by SEM (Table 1). Presumably, the particles can contain several crystallites or only a part of particle has a crystal structure.

Table 2. Lattice parameters of Li-rich samples.

Sample	a, Å	c, Å	V, Å3	Coherent scattering length, nm
LRC50	2.85177(3)	14.2371(3)	100.273(3)	56(1)
LRC35	2.85534(5)	14.2531(6)	100.634(6)	63(1)
LRC20	2.85839(3)	14.2453(3)	100.796(3)	78(2)

Figure 3a,b shows discharge voltage profiles of the samples in 3-rd and 50-th cycles in the range of 2.5-4.8 V at a current density of 100 mA/g (C/2). The discharge capacities of all the samples in the 3rd cycle are close to each other, at the same time we observe a tendency to decreasing the discharge capacity in the row LRC50–LRC35–LRC20. The samples have high values of the first-cycle coulombic efficiency in this voltage range, namely, 90% for LRC50 and LRC20 and 92% for LRC35. These values of the first-cycle irreversible capacity loss are considerably lower than those usually reported in the literature [13, 14].

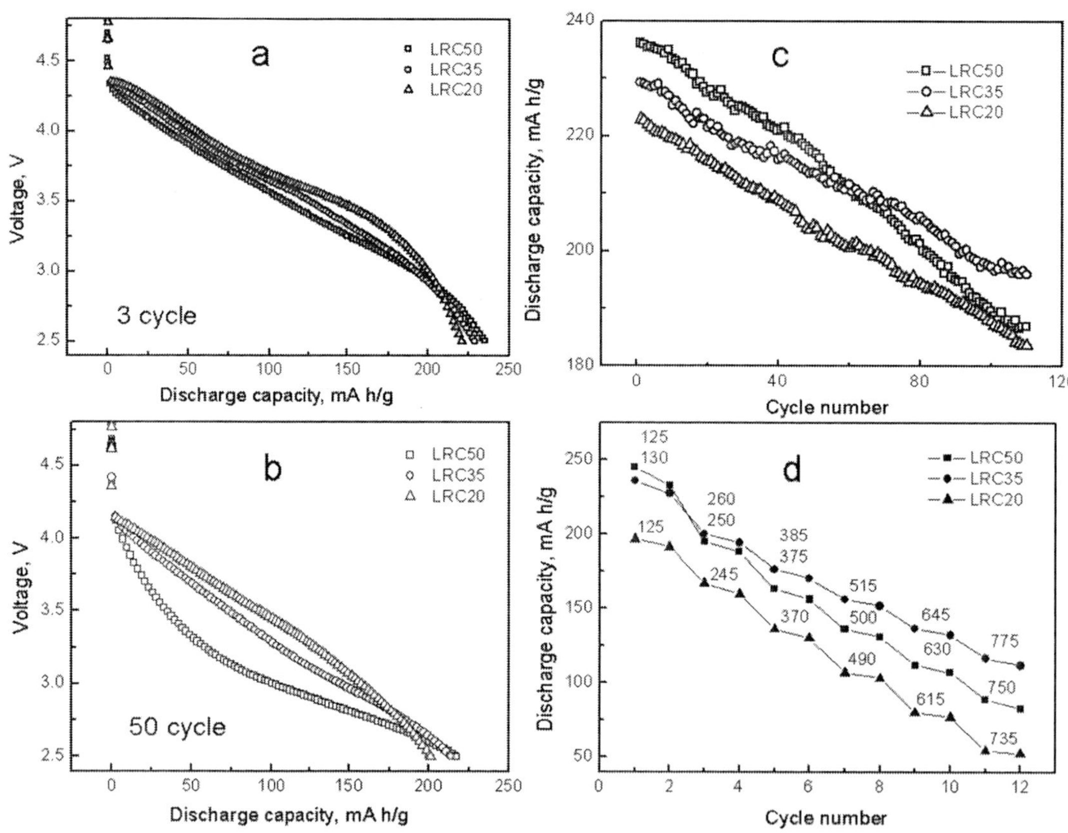

Figure 3. (a, b) Discharge voltage profiles for the 3-rd and 50-th cycles, respectively, for the samples in the voltage range of 2.5-4.8 V at a current density of 100 mA/g; (c) cycling behavior of the samples at a current density of 100 mA/g; and (d) the rate capability of the samples in the range of 2.5-4.8 V (current densities (mA/g) at the each step are shown in the figure).

In Figure 3c, we demonstrate cycling behavior of all the samples in the same voltage range of 2.5-4.8 V. As is seen, LRC35 sample is cycling better than other two samples. The capacity retention for this sample is 85% at the 110 cycle. The capacity retentions for LRC50 and LRC20 samples are 79% and 81%, respectively. LRC35 sample also shows the best rate capability (Figure 3d). The rate capabilities of the samples differ more considerably. The discharge capacity of LRC35 at the discharge current density of 775 mA/g is 47% from the initial discharge capacity at the current density of 125 mA/g, while the analogous values for LRC50 and LRC20 samples are 33 and 26%, respectively.

In the course of cycling, the average voltage of cells continuously decreases with each successive charge and discharge cycle. A serious voltage fade for these materials is generally considered to originate from the gradual structural transformation during cycling, mostly to a spinel-like structure [15-17] and appearing new low-voltage redox potentials. A new structure gradually formed undergoes, at least partially, reversible redox transformations [18, 19]. This process is related to the transition-metal (TM) ions migration from octahedral sites in the TM slabs to octahedral sites in the Li slabs through intermediary Li-slab tetrahedral sites in the course of lithiation/delithiation [18, 19].

The voltage fade is characteristic of all the samples synthesized, at the same time, the more the Li_2MnO_3 content, the more the voltage fade (Figure 3a). The values of the voltage fade at the 50th cycle are equal to 11.3, 7.3, and 6.2%, for LRC50, LRC35, and LRC20, respectively. To take into account this voltage decay, we compared the energy densities of the samples that characterize both specific discharge capacities and the discharge voltage. The energy densities for each cycle (E_n) were calculated as the integration areas under the corresponding discharge curves. The energy retentions for the 110-th cycle (E_{110}/E_3) are 63.1, 75.8, and 71.5% for samples LRC50, LRC35, and LRC20, respectively.

Therefore, the sample of the composition $0.35Li_2MnO_3 \cdot 0.65LiMn_{1/3}Ni_{1/3}Co_{1/3}O_2$ demonstrates a better electrochemical performance than the samples with both larger and smaller contents of the monoclinic phase.

4. Conclusion

In this work, we studied the effect of the phase composition in Li-rich complex oxides $xLi_2MnO_3 \cdot (1-x)LiMn_{1/3}Ni_{1/3}Co_{1/3}O_2$ obtained by coprecipitation method on their electrochemical behavior. All the samples show very high first-cycle coulombic efficiencies (equal to and more than 90%) in the voltage range of 2.5–4.8 V at a current density of 0.5*C*.

The Li-rich oxide of the composition $0.35Li_2MnO_3 \cdot 0.65LiMn_{1/3}Ni_{1/3}Co_{1/3}O_2$ (LRC35) shows the better cycling performance in terms both discharge capacity and energy density and the best rate capability compared with the samples with larger and smaller contents of Li_2MnO_3.

Acknowledgments

This work was supported by the Russian Science Foundation, Grant # 17-13-01424 (2017).

References

[1] Thackeray M M, Kang S H, Johnson C S, Vaughey J T, Benedek R and Hackney S A. 2007 *J. Mater. Chem.* **17** 3112–25
[2] Koga H, Croguennec L, Menetrier M, Douhil K, Belin S, Bourgeois L, Suard E, Weill F and Delmas C 2013 *J. Electrochem. Soc.* **160** A786–92
[3] Johnson C, Kim J, Lefief C, Vaughey J T and Thackeray M M 2004 *Electrochem. Commun.* **6** 1085–91
[4] Thackeray M M, Kang S H, Johnson C S, Vaughey J T and Hackney S A 2006 *Electrochem. Commun.* **8** 1531–38
[5] Croguennec L, Polacin M. R 2015 *J. Am. Chem. Soc.* **137** 3140–56
[6] Zheng J, Xu P, Gu M, Xiao J, Browning N D, Yan P, Wang C, Zhang J-G 2015 *Chem. Mater* **27** 1381–90

IOP Conf. Series: Materials Science and Engineering **525** (2019) 012042 doi:10.1088/1757-899X/525/1/012042

[7] Ohzuku T, Nagayama M, Tsuji K, Ariyoshi K 2011 *J. Mater. Chem.* **21** 10179–88
[8] Shukla A K, Ramasse Q M, Ophus C, Duncan H, Hage F and Chen G 2015 *Nature Comm.* **6** 8711
[9] Sathiya M, et al. 2013 *Nature Materials* **12** 827–35
[10] Roziera P and Tarascon J M 2015 *J. Electrochem. Soc.* **162** A2490–99
[11] Hy S, Liu H, Zhang M, Qian D, Hwang B-J and Meng Y S 2016 *Energy Environ. Sci.* **9** 1931
[12] Koyama Y, Makimura Y, Tanaka I, Adachi H and Ohzuku T 2004 *J. Electrochem. Soc.* **151** A1499–1506
[13] Mohanty D, Sefat A S, Li J, Meisner R A, Rondinone A J, Payzant E A, Abraham D P, Wood D L, Daniel C 2013 *Phys.Chem. Chem. Phys.***15** 19496-509
[14] Erikson E M, Schipper F, Penki T R, Shin J-Y, Erk C, Chesneau F-F, Markovsky B and Aurbach D 2017 *J. Electrochem. Soc.* **164** A6341–48
[15] Bettge M, Li Y, Gallagher K G, Zhu Y, Wu Q, Lu W, Bloom I and Abraham D P 2013 *J. Electrochem. Soc.* **160** A2046–55
[16] Yan P, et al 2015 *Nano Lett.* **15** 514–22
[17] Gallagher K G, Croy J R, Balasubramanian M, Bettge M, Abraham D P, Burell A K, Thackeray M M 2013 *Electrochem.Comm.* **33** 96–98
[18] Croy J R, Balasubramanian M, Gallagher K G and Burell A K 2015 *Acc. Chem. Res.* **48** 2813–21
[19] Croy J R, Gallagher K G Balasubramanian M, Long B R and Thackeray M M 2014 *J. Electrochem. Soc* **25** A318–25

IOP Conf. Series: Materials Science and Engineering **525** (2019) 012043 doi:10.1088/1757-899X/525/1/012043

Impact of UV treatment on polylactide–polyethylene film properties

M V Podzorova[1,3], Yu V Tertyshnaya[2,3], S G Karpova[2] and A A Popov[1,2]

[1] Plekhanov Russian University of Economics, 36 Stremyanny per., Moscow, 117997, Russia
[2] Emanuel Institute of Biochemical Physics, Russian Academy of Sciences, 4 Kosygina str., Moscow, 119991, Russia
[3] Federal Research Agro-Engineering Center VIM, 1st Institutskiy proezd, 5, Moscow, 109428, Russia

E-mail: mariapdz@mail.ru

Abstract. Influence of different environmental factors on the degradation of film samples based on polylactic acid and low density polyethylene with the addition of aged polyethylene was studied in this work. The article presents data after 100 hours influence of UV irradiation (λ=254 nm) obtained by differential scanning calorimetry (DSC) and electronic paramagnetic resonance (EPR). It was found that polylactide is more prone to UV degradation than polyethylene. It is shown that the effect of UV radiation for 100 hours reduces the melting point of polylactide from 165 to 135 °C, while the thermal characteristics of polyethylene vary slightly. There is a decrease in the melting temperature of the polylactide by 10-30 °C in all compositions. In the compositions with the content of PLA is more than 50 wt. % correlation time after UV exposure is significantly reduced.

Introduction

Creation of blends based on polylactide (PLA) with other materials is perspective for adjusting its properties and extend the range of possible applications [1]. Blends based on biodegradable polymers with the addition of synthetic polymers are the most perspective and in the focus of the researchers all over the world [2-4].

Biodegradable materials do not pollute the environment after using. Under the influence of ultraviolet, heat, air and soil microorganisms they degrade to low molecular compounds and easily assimilate in soil, being included in closed ecological circuit [2]. The most popular biodegradable polymers: polylactide (PLA), polycaprolactone (PCL), polyhydroxyalcanoates (PHB, PHV) [5-10]. The problem is the high price of these polymers. That is why they are widely used in medicine but still unavailable for packaging industry.

In [11] it is noted that blends of polylactide – polyethylene in the first stage has a significant impact of moisture and ultraviolet light. Photodegradation of polymers is influenced by light rays that can be absorbed by the polymer. The absorbing centres are most often carbonyl and other oxygen-containing groups. In the presence of oxygen, UV light initiates a chain oxidation reaction. The number of CO-groups in the polymer increases sharply during photo-oxidation.

IOP Conf. Series: Materials Science and Engineering **525** (2019) 012043 doi:10.1088/1757-899X/525/1/012043

It is possible to produce thin transparent films with sufficient mechanical properties from the blends of biodegradable polymers with polyolefins. Such blends are cheaper than pure biodegradable polymers, that is why they may be used in a wide range of applications including packaging.

Experimental

Blends of LDPE (TM 15803-020 from Neftekhimsevilen, OJSC, Russia) with PLA (4032D from Nature works, USA) in a wide range of compositions were studied. The PLA content in the polymer matrix was 20, 30, 50, 70, 80 wt% and pure polymers PLA and LDPE. There were also made three triple blends with 10, 20, 30 wt% of LDPEag (LDPE held in air, in an open vessel, at a temperature of 90°C for 250 hours, TM 15803-020 from Neftekhimsevilen OJSC, Russia) to neat LDPE respectively and 30 wt% of PLA. The blends were prepared by mixing for 6 minutes at Plasticorder PLD-651 (Brabender, Germany) with a temperature of 180°C. After cooling the samples of the obtained material were milled using a knife mill RM-120 (Vibrotechnik, Russia) and then compressed in a manual hydraulic press PRG-10 (VNIR, Russia) at 180°C under a load of 7 kN followed by rapid cooling. As a result, film round samples with a diameter of 8 cm and thickness of 100±10 μm were obtained.

The degree of crystallinity and melting temperature of the samples were studied with differential scanning calorimeter DSC 214 Polyma (Netzsch, Germany) at a heating rate of 10K/min and a sample weight of 5±0.1 mg. The temperature scale was calibrated against an indium standard (Tm = 156.6°C, ΔH = 28.44 J/g). The influence of ultraviolet on the obtained samples was studied with a lamp Vilber Lourmat VL-6.LC at λ=254 nm for 100 hours. Electron paramagnetic resonance (EPR) was studied by EPR spectrometer EPR-V (N.N. Semenov Institute of Chemical Physics, Russia). Stable nitroxide radical (2,2,6,6-Tetramethylpiperidin-1-yl)oxyl (TEMPO-1) was impregnated into films from the fumes at 30°C.

The infrared spectra of the samples were recorded on the FT-IR spectrometer Perkin Elmer Spectrum 100 at T = 23 ± 2 ° C in the wave number range $4600 \leqslant \nu \leqslant 450$ cm^{-1} in transmitted light. The software ACDlabs was applied for the mathematical processing of the obtained data.

Results and discussion

The research demonstrated that in binary blends of PLA and LDPE the ratio of polymers did not have an influence to their melting temperatures for the initial blend (Table 1). But, after UV irradiation melting temperature of PLA decreased more in the blends with low amount of PLA. Evidently, if PLA does not form its own phase, it is more prone to UV degradation. The blend 50_PLA:50_LDPE has the highest crystallinity degree for both polymers, apparently it would have the best mechanical and barrier properties. The crystallinity degree of both polymers fell significantly after UV irradiation, but in the blend 50:50 χ_{cr} of LDPE had the highest value, it looks that in this blend both polymers are the most stable to UV degradation.

It was found that the addition of LDPEag to the blend of PLA and LDPE does not influence their melting temperatures (method DSC), but decreases the crystallinity degree of LDPE and PLA, if to compare with binary blend 30_PLA:70_LDPE (Table 1). In triple blends, containing LDPEag, melting temperature and crystallinity degree of PLA reduced after the influence of UV irradiation as well. It was stated that the addition of 30 mas. % of LDPEag led to dramatic decline of these parameters in PLA: PLA melting temperature decreased from 165 to 135 °C, and PLA crystallinity degree fell from 45 to 16%. Three initial blends with 10, 20, 30% of LDPEag had the same melting temperatures and similar crystallinity, but after UV irradiation their characteristics differed from each other strongly.

The study of amorphous phase of PLA and LDPE in blends was made by EPR-spectroscopy with paramagnetic sonde. Nitroxide radical – tetramethylpiperidine oxide, has dual nature (polar and nonpolar parts), for that reason, it is usually concentrated in interphase regions of polymer blends. Segmental mobility in amorphous regions of blends was evaluated by rotation velocity of radical-sonde (rotational correlation time) and shape of EPR-spectra.

IOP Conf. Series: Materials Science and Engineering **525** (2019) 012043 doi:10.1088/1757-899X/525/1/012043

Table 1. Thermophysical characteristics of PLA/LDPE and PLA/LDPE/LDPEag blends.

Blends PLA:LDPE, wt.%	T_m, °C	T_m, °C after 100h UV	χ, %	χ, % after 100h UV
100:0	165	135	45	20
80:20	164:104	135:103	40:17	24:15
70:30	163:105	137:102	39:15	24:14
50:50	164:103	149:102	53:19	23:17
30:70	164:104	151:103	52:16	24:17
20:80	163:104	155:103	40:18	22:18
0:100	103	102	23	18
Blends PLA:LDPE:LDPEag, wt.%	**T_m, °C**	**T_m, °C after 100h UV**	**χ, %**	**χ, % after 100h UV**
30:60:10	165:104	150:102	50:14	34:12
30:50:20	165:104	148:103	47:14	20:13
30:40:30	165:104	135:102	45:16	16:13

Figure 1 presents EPR spectra of PLA:LDPE blends. Blends with high content of PLA have complicated shape of spectra. For the blends with the low (<30%) and high (>50%) content of PLA shape of triplets are different. Apparently, PLA form its own phase in blends just from the concentration of 50 %.

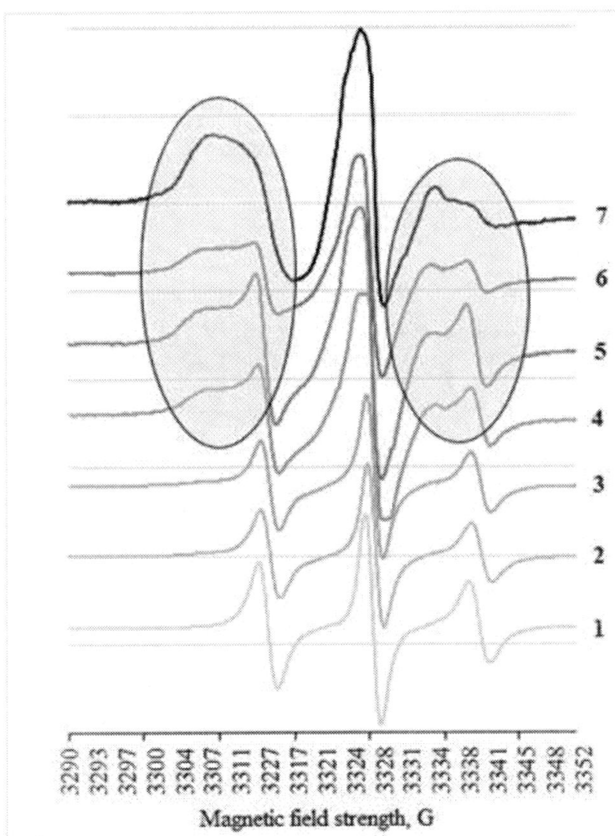

Figure 1. EPR spectra of nitroxide radical–sonde of double blends PLA:LDPE with the content of LDPE 100 (1), 80 (2), 70 (3), 50 (4), 30 (5), 20 (6), 0 (7) mass.%.

The correlation time (τ_c) for compositions of various compositions is calculated. The values of the rotation correlation time of the radical probe in PLA are much higher than in HDPE. First of all, this is due to the supramolecular structure of the blend components. In PLA, the degree of crystallinity is almost 2 times higher than that of LDPE, therefore, the PLA is less permeable for radicals.

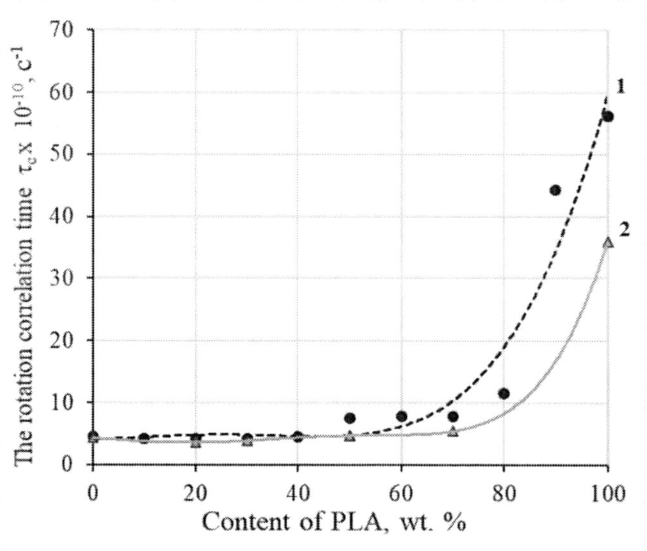

Figure 2. Change in correlation time 1- initial, 2-after 100 hours of UV.

From figure 2 shows that in the compositions of PLA:LDPE with a PLA content of more than 50 wt. % correlation time decreases after UV exposure within 100 hours. The observed regularities can be explained from the position of inhomogeneity of mixtures.

Table 2 presents correlation time data of compositions containing aged polyethylene in the matrix.

In these compositions, the correlation time is reduced by twice as much compared to the original composition 30PLA:70LDPE.

Table 2. Correlation time before and after exposure to UV radiation compositions containing fraction of aged LDPE.

Blends PLA:LDPE:LDPEag, wt.%	$\tau_c \times 10^{-10}$, c^{-1}	$\tau_c \times 10^{-10}$, c^{-1} after 100h UV
30:60:10	4,7	3,7
30:50:20	4,8	4,4
30:40:30	5,1	4,6

It can be seen from figures 3 and 4 that after exposure to UV radiation there is a change in the spectra, therefore, changes in segmental mobility and changes in the structure of the samples. No significant changes were found in the EPR spectra of the samples with aged polyethylene after UV.

The characteristic region for PLA is 1400-1000 cm^{-1}, which characterizes the stretching vibrations of C-O in esters. Also a band of 870 cm^{-1} is responsible for valence symmetrical and asymmetric oscillations in the C-O-C group.

As it can be seen at the figure 5, after 50 hours, the changes occur in the characteristic areas and bands of the PLA. Thus, the optical density of the 870 cm^{-1} band increases from 1,34 to 1,62, which indicates the oxidation and destruction of PLA under the influence of UV. Absorption at a frequency of 1160 cm^{-1} that appears after radiation, refers to the vibrations of C–O groups in carboxylic acid anhydrides formed during the oxidation of the polymer.

IOP Conf. Series: Materials Science and Engineering **525** (2019) 012043 doi:10.1088/1757-899X/525/1/012043

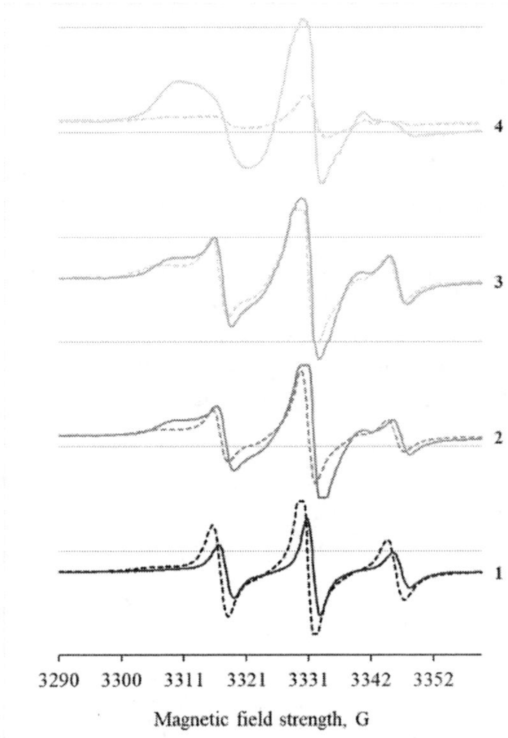

Figure 3. EPR spectra before (solid) and after (stroke) UV irradiation (100 hours λ=254 nm) PLA:LDPE: 30PLA:70LDPE (1), 50PLA:50LDPE (2), 70PLA:30LDPE (3), 100PLA (4) mass. %.

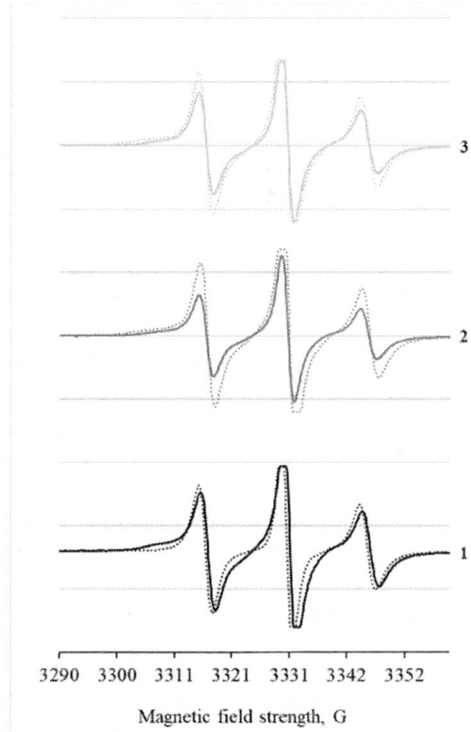

Figure 4. EPR spectra before (solid) and after (stroke) UV irradiation (100 hours λ=254 nm) 30PLA:60LDPE:10LDPEag (1), 30PLA:50LDPE:20LDPEag (2), 30PLA:40LDPE:30LDPEag (3) mass. %.

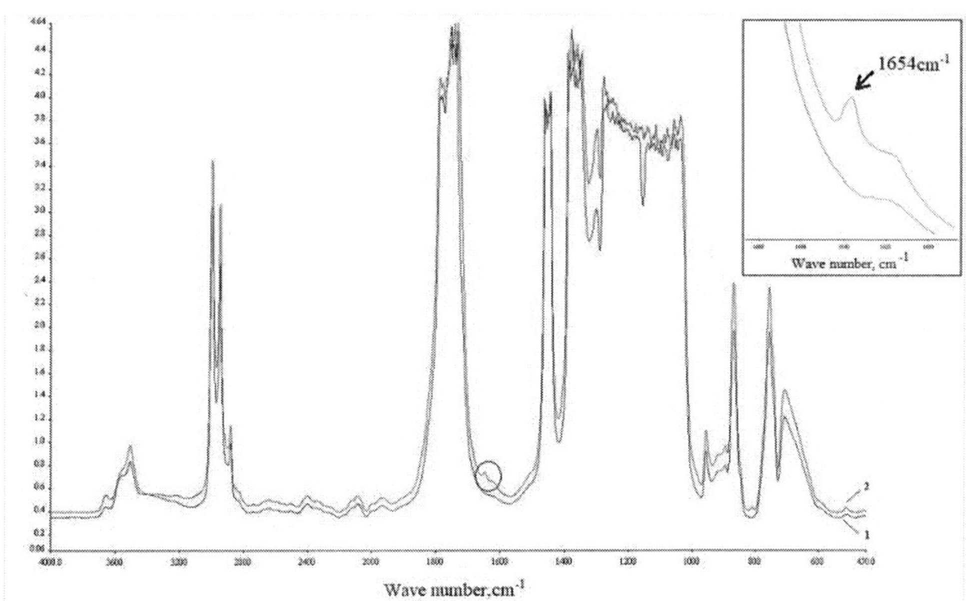

Figure 5. IR spectrum of the initial PLA (1) and after (2) 50 hours of UV radiation.

IOP Conf. Series: Materials Science and Engineering **525** (2019) 012043 doi:10.1088/1757-899X/525/1/012043

In our study, after 50 hours of exposure to UV radiation there is a peak of low intensity at 1654cm^{-1}, which refers to the oscillations of the vinyl group [12], so the degradation of PLA occurs with the formation of C=C bond. In mixtures UV absorption center is PLA, which due to occurs fragmentation of PLA: LDPE blends.

Conclusion

In blends made of polyethylene and polylactide more stable to UV degradation is polyethylene. Melting temperature and crystallinity degree of polyethylene was more stable after degradation than polylactide. Under the influence of UV irradiation aged polyethylene accelerates physical degradation of polylactide in triple blends. Therefore, polyethylene waste is a good additive to PLA:LDPE blends focused on rapid biodegradation. By EPR spectroscopy it was found that PLA do not form its own phase when its content in blend is less than 50%. PLA parts are encapsulated in polyethylene matrix and nitroxide radical is unable to penetrate into PLA in that case. When studying the effect of ultraviolet radiation at λ=254 nm on the above samples by IR spectroscopy method was found that the PLA decay proceeds with the formation of C=C bond.

References

[1] Imre B and Pukanszky B 2013 *Eur. Polym. J.* **49(6)**, 1215–1233.
[2] Carrasco F, Pagès P, Gámez-Pérez J, Santana O O and Maspoch M L 2010 *Polym. Degrad. Stab.* **95(2)**, 116–125.
[3] Sutivisedsak N, Cheng H N, Dowd M K, Selling G W and Biswas A 2012 *Ind Crops Prod.* **36(1)**, 127–134.
[4] Bajpai P K, Singh I and Madaan J 2012 *J Reinf Plast Compos.* **31(24)**, 1712–1724.
[5] Tertyshnaya Yu V, Khvatov A V and Lobanov A V 2017 *Rus J Phys Chem B.* **11(5)**, 828–832.
[6] Arrieta M P, López J, López D, Kenny J M and Peponi L 2016 *Ind Crops Prod.* **93**, 290–301.
[7] Arrieta M P, López J, López D, Hernández A and Ternary E R 2014 *Eur Polym J.* **50**, 255–270.
[8] Tertyshnaya Yu V and Podzorova M V 2018 *Rus J Appl Chem.* **91(3)**, 417–423.
[9] Bittmann B, Bouza R, Barral L, Diez J and Ramirez C 2013 *Polym Compos.* **34(7)**,1033–1040.
[10] Peponi L, Sessinia V, Arrieta M P, Navarro-Baena I, Sonseca A, Dominici F, Gimenez E, Torre L, Tercjak A, López D and Kenny J M 2018 *Polym Degrad Stab.* **151**, 36–51.
[11] Podzorova M V, Popov A A, Tertyshnaya Yu V 2018 *IOP Conf. Ser. Mater. Sci. Eng.* **347**, 12015.
[12] Nakayama N and Hayashi T 2007 *Polym. Degrad. Stab.* **92(7)**, 1255–1264.

IOP Conf. Series: Materials Science and Engineering **525** (2019) 012044 doi:10.1088/1757-899X/525/1/012044

Study of the energy band structure of $Lu_2SiO_5:Ce^{3+}$ single crystals by thermally stimulated luminescence method

V A Tedzhetov[1], A V Podkopaev[2] and A A Sysoev[2]

[1] Baikov Institute of Metallurgy and Materials Science, Russian Academy of Sciences, Moscow, Russia

[2] Joint Stock Company POLYUS Research Institute of M.F. Stelmakh, Moscow, Russia

E-mail: velmurad@yandex.ru

Abstract. Results of optical absorption spectroscopy and photoluminescence for $Lu_2SiO_5:Ce^{3+}$ (LSO) single crystals produced by modified Musatov method are reported. The energy structure of electron traps in LSO is studied by thermally stimulated luminescence method when the crystals are exposed to UV sources with different spectral and energy characteristics. The thermoluminescence glow curves allowed us to find a luminescence maximum corresponding to 300 K, which was not observed previously. The obtained results are interpreted taking into account the anisotropy of LSO and different energy of exciting photons.

1. Introduction

In recent two decades, owing to the rapid development of the high-energy physics and medicine, heavy scintillation materials, which are characterized by high both gamma-quanta absorption cross-section and quantum efficiency, and good temporal resolution as well, became necessary. The cerium-doped lutetium oxyorthosilicate $Lu_2SiO_5:Ce^{3+}$ (LSO) is a crystalline scintillator, which exhibits extremely well combined physical and luminescence properties. The LSO has the high density, and both effective atomic number and quantum efficiency, which result in the low radiation length and high conversion efficiency [1]. The luminescence maximum of LSO corresponds to the wavelength range of 400-450 nm; this fact determines the adequate coincidence with the spectral response of photomultiplier tube (PMT) applied now in scintillation detectors. Potentially, this material can substitute for $NaI:Tl^{+}$, which at present dominates among scintillation crystals. The substantial disadvantage of LSO consists in the afterglow, which worsens the time resolution of detectors and limits the application of the material. In accordance with the luminescence theory of crystalline materials, the afterglow phenomenon is related to the participation of electron traps (ET) in the luminescence process, which are defects that form metastable levels in the forbidden band and can radiationlessly trap electrons and release them for a long time [2]. There are phosphorescence models for the LSO in the literature, which take into account the conduction band [3] and ignore its [4]. However, at present, no comprehensive mechanism of phosphorescent transitions in LSO is available.

The subject of the present study is a LSO single crystal, the structure of which corresponds to the monoclinic (low-symmetry) crystal system (syngony) (point group symmetry 2/m, space group C2/c). There are two sites for Lu^{3+} ions in the LSO crystal lattice: Lu^{I} ion is locating in seven-oxygen-coordinated site (seven-apical coordination polyhedron) and Lu^{II} ion is locating in six-oxygen-coordinated site (oxygen ions form a distorted octahedron). The pure Lu_2SiO_5 compound exhibits

IOP Conf. Series: Materials Science and Engineering **525** (2019) 012044 doi:10.1088/1757-899X/525/1/012044

common luminescence parameters; because of this, to reach the intense and rapid luminescence used in detecting high-energy particles, the material is doped with Ce^{3+} activator. When doping, Ce^{3+} ions substitute for Lu^{3+} ions and, therefore, Ce^{3+} ions can occupy the two aforementioned sites. The Lu and Ce elements correspond two opposite ends of the lanthanide row; therefore, their ionic radii are substantially different, namely, 0.86Å (Lu^{3+}) and 1.02Å (Ce^{3+}). Doped Ce^{3+} ions are more likely to substitute for Lu ions in seven-oxygen-coordinated sites [5].

This work is a stage of investigations performed by us, which is aimed at the study of mechanisms of the efficient energy storage in LSO single crystals and at the finding of effects, which allow one to control the luminescence parameters of dielectric materials activated with rare-earth and transition metal ions. To study the luminescence properties and determine the luminescence mechanism, we apply the absorption spectroscopy and photoluminescence (PL) methods and thermally stimulated luminescence (TSL) using the exposition to different UV sources.

2. Experimental details

The optical-grade LSO single crystals studied in terms of the present work were grown along the direction perpendicular to the $(31\bar{1})$ plane using the modified Musatov method without rotation; this ensures the preparation of perfect crystals [6]. The studies were performed using plate-like samples 2x10x10 mm in size (optical absorption spectroscopy) and cubic samples 10x10x10 mm in size (PL and TSL experiments), the faces of which correspond to the $(31\bar{1})$ and (001) crystallographic planes. The optical absorption spectroscopy was performed at room temperature in a wavelength range of 180-800 nm using a UV-Vis-NIR Cary 5000 Varian (AgilentTech.) spectrometer.

Emission spectra were measured at 300 K with an original set-up [7] based on a MDR-23 monochromator and a FEU-100 PMT, which were mounted perpendicular to the exciting laser beam. The luminescence was excited using a YAG:Nd LTI-345 UV laser, which allows one to generate light pulses with the photon energy $h\nu_0$=3.49 eV.

TSL glow curves (intensity as a function of temperature) were measured in a temperature range of 125-570 K at heating rates of 4.5÷10 K/min using a VUP-5-based vacuum setup. This setup is equipped with a SZS-22 light filter (the transmission band is 400-600 nm) and a FEU100 PMT. Before the heating, the crystals were exposed with UV light from either a DDS30 low pressure deuterium lamp or a DRL400 high pressure mercury lamp at exposition times of 10, 15, 20, and 30 min.

3. Results and discussion

3.1. *PL parameters*

Optical transmission spectra of LSO plate-like samples (figure 1) demonstrate the stable high transmittance in the visible wavelength range (400÷800 nm), which confirms the optical-grade quality of the single crystals, and the decrease in the transmittance in the UV wavelength range, which is typical of the absorption of crystalline Lu_2SiO_5 matrix with Ce^{3+} activator. The linear approximation of the characteristic absorption edge (figure 1, see inset) allowed us to calculated the band gap energy E_g, which, according to the sample orientation, is 6.22 eV (001), 6.19 eV $(31\bar{1})$, and 6.26 eV (undirected samples) at $\Delta E_g = \pm 0.005$ eV; these results agree well with the literature data [8,9].

To analyze the Ce^{+3} activator absorption bands at 200÷400 nm, the transmission spectra were transformed into absorption spectra (figure 2). In this case, we use the well-known Bouguer-Lambert-Beer law (1) and the relationship (2) between the wavelength λ and photon energy:

$$\frac{I}{I_0} = \exp[-\alpha \cdot l] \quad (1)$$

$$h\nu = \frac{hc}{\lambda} \quad (2)$$

where I_0 and I are the intensities of incident and transmitted light, respectively; α is the absorbance of a material; and l is the sample thickness.

Emission spectra (figure 2) measured for the cubic LSO samples differing in the orientation (from herein, excitation face/ PL response face) were obtained by transformation of the dependences $I(\lambda)$ into $I(h\nu)$ by equation (3) to satisfy the luminous flux conservation condition $dF=Idh\nu=Id\lambda$.

$$I(h\nu)=I(\lambda)\frac{\lambda^2}{c} \qquad (3)$$

The approximation of absorption and emission spectra with the Gaussian function allowed us to identify four absorption and thee emission bands (figure 2), the position of which is almost unchanged (within the measurement error) with changing orientation of samples (table 1). The absorption bands correspond to optical transitions in Ce^{3+} ions [8,9] occupied two nonequivalent crystallographic sites Ce^{I} (at 3.47, 4.2, and 4.7 eV) and Ce^{II} (at 3.74 eV). The emission spectra demonstrate bands at ~2.96 and ~3.13 eV (related to Ce^{I}) and at ~2.70 eV (related to Ce^{II}), which were reported in [9,10].

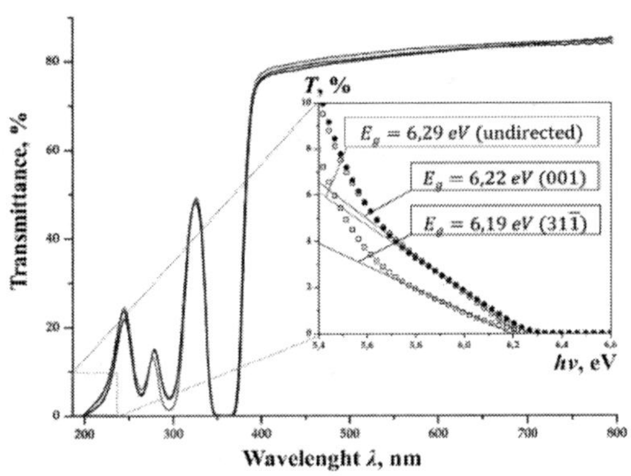

Figure 1. Transmission spectra measured for the LSO plates; inset shows the linear approximation of the characteristic absorption band edge.

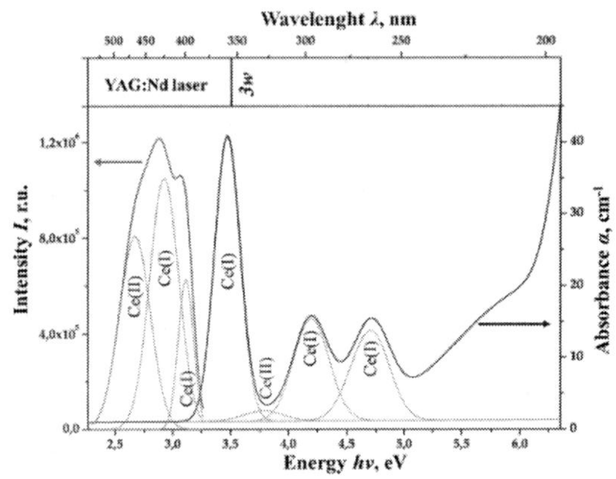

Figure 2. Absorption (blue line) and emission (red line) spectra measured for plate (001) and cubic (001)/(31$\bar{1}$) LSO samples, respectively. Green lines are results obtained with Gauss peak fitting of the spectra.

Table 1. Results of Gauss peak fitting of absorption and emission bands for LSO samples.

Absorption face	Absorption maximum[a] (eV)				PL response face	Emission maximum[b] (eV)		
	$h\nu_{a1}$	$h\nu_{a2}$	$h\nu_{a3}$	$h\nu_{a4}$		$h\nu_{e1}$	$h\nu_{e2}$	$h\nu_{e3}$
(001)	3.47	3.72	4.20	4.70	(31$\bar{1}$)	2.69	2.95	3.13
					undirected	2.72	2.97	3.12
(31$\bar{1}$)	3.47	3.74	4.20	4.70	(001)	2.68	2.95	3.12
					undirected	2.71	2.98	3.13
undirected	3.47	3.78	4.20	4.70	(001)	2.72	2.97	3.12
					(31$\bar{1}$)	2.69	2.96	3.13

[a] $\Delta h\nu_{ai}$~0.02 eV.
[b] $\Delta h\nu_{ei}$~0.03 eV.

3.2. Thermally stimulated luminescence

It follows from absorption bands parameters of Ce^{3+} in LSO that, the efficient excitation energy storage is possible in using gas-discharge UV-radiation lamps; the position of luminescence bands allows us to record the TSL glow curves in using sufficiently simple equipment, namely, optical glasses and a PMT.

When using the TSL method, the heating rate of sample affects substantially the recorded curves. Because of this, we performed the measurements at different heating rates (μ). Data obtained at a heating rate of 4 K/min are characterized by a low signal-to-noise ratio that impedes the data processing. At heating rates of above 10 K/min, the thermal destruction of samples takes place, which manifests itself in the appearance of triboluminescence flashes. The heating rate μ=7 K/min allows us to obtain intense curves without the triboluminescence.

The TSL glow curves obtained after exposition to the deuterium lamp radiation (figure 3) at different exposition times τ_{exp} demonstrate two TSL maxima at ~403 K (I) and ~345 K (II), which are well understood [4,9]. As in previously reported studies, no TSL peak was found in the low-temperature range (100÷300 K). The ratio of intensities of peaks I and II is 1:4; the intensities increase as the exposition time increases from 4 to 20 min (figure 3, see inset). In this case, the structure and positions of maxima change insignificantly. The use of the mercury lamp leads to the appearance of an additional peak in the TSL glow curves (figure 4), which corresponds to 300 K (III) and was not previously reported in the literature.

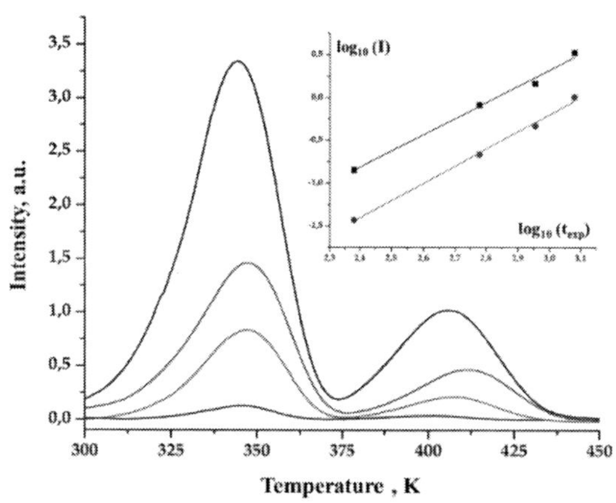

Figure 3. TSL glow curves of LSO, in increasing order of intensity, UV (the deuterium lamp) exposure times of 4, 10, 15, and 20 min. The curves were recorded at a heating rate of 7 K/min. Inset shows the proportional increase in the TSL peak intensity with increasing exposure time.

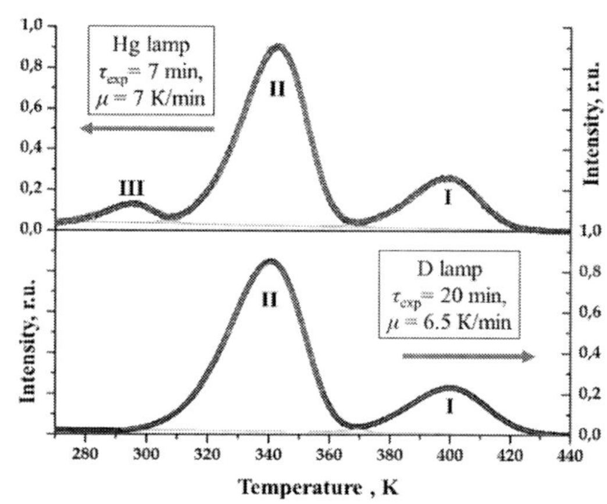

Figure 4. TSL glow curves measured for LSO after exposure to Hg lamp light (top curve) and deuterium lamp light (bottom curve). Green lines are results obtained with asymmetric Gauss peak fitting.

Table 2 – Parameters of electron traps

UV source,		Experimental data			Literature data [4,9]	
τ_{exp}, μ	Peak	T_{max} (K)	E_i (eV)	ln [s (Hz)]	E_i (eV)	ln [s (Hz)]
D lamp, 20 min, 6.5 K/min	I	403±4	1.18±0.04	30.9±0.7	1.17, 1.45	30.5, 29.5
	II	344±3	0.92±0.02	28.4±1.0	0.98, 1.21	29.3, 32.4
Hg lamp 4÷7 min, 7 K/min	I	400±1	1.12±0.03	30.7±0.2	1.17, 1.45	30.5, 29.5
	II	345±2	0.96±0.04	30.0±1.6	0.98, 1.21	29.3, 32.4
	III	298±3	0.88±0.01	32.2±0.3	–	–

IOP Conf. Series: Materials Science and Engineering **525** (2019) 012044 doi:10.1088/1757-899X/525/1/012044

The obtained curves were approximated with the asymmetric Gaussian function:

$$y = y_0 + \frac{\sqrt{2}A}{\sqrt{\pi}(w_1 + w_2)}\exp\left[-\frac{(x-x_c)^2}{2w_{1,2}^{\ \ 2}}\right]. \tag{4}$$

To calculate parameter of ETs, we used peak shape method, which is based on the linear approximation (equation (5)) of the temperature dependence of the intensity of emission *I(T)* due to the trap emptying on coordinates ln[*I(T)/S(T)*] – *1/T*:

$$\ln\frac{I(T)}{n^l} = -\frac{E_i}{kT} + \ln\frac{s}{\mu} \tag{5}$$

where n^l is the number of filled ETs, which is proportional to the area *S(T)* under the high-temperature portion of TSL peak; E_i is the depth or activation energy of ET; and *s* is the frequency factor. Results of the calculation are given in table 2 in comparison with the literature data.

4. Discussion

It is known that the electron states of the top of the valence band and the bottom of the conduction band in the LSO belong to the *2p*- and *5d*-electron states of O^{2-} and Lu^{2+} ions, respectively. Therefore, the variations of the E_g value obtained from the position of characteristic absorption edge mainly are due to the anisotropy of the density of electron states of anion-cation pairs O^{2-}-Lu^{3+} in the crystal matrix, in which the distance between ions, depending on the direction, is from 2.16 to 2.62 Å [11]. According to intracenter transition theory [2], the anisotropy of absorbed $h\nu_{a1\text{-}a4}$ and emitted $h\nu_{e1\text{-}e3}$ quanta energy also is determined by the crystal environment of Ce^{3+} ion. However, to determine such an interrelation, experiments should be performed at low temperatures (below 100 K), which ensure the low error of determination of spectral band parameters.

The characteristics of UV radiation of sources used in the present study for TSL experiments differ substantially. The low-pressure deuterium lamp gives the continuous UV radiation spectrum; the radiation intensity decreases as the wavelength decreases, and the radiant intensity is 1.8 mW/sr. The radiation of high-pressure mercury lamp is characterized by intense UV-range lines with the photon energies of up to 6.5 eV and the radiant intensity of 310 mW/sr. Thus, based on the absorption spectroscopy data for the LSO, the mercury lamp radiation not only can ensure the excitation of Ce^{3+} activator but also leads to the ionizing of it accompanied by the electron transfer to the conduction band ($h\nu_a + Ce^{3+} = Ce^{4+} + e^-$) and to the band-band transitions accompanied by the formation of electron-hole pairs. The analysis of the TSL data obtained in using the aforementioned UV radiation sources allows us to assume that traps responsible for the observed maxima are different in origin. Indeed, traps I and II, which take place in all experiments, are related to either the Ce^{3+} luminescence centre or objects associated with it, whereas traps III, the luminescence of which takes place only in using the mercury lamp, are related to states localized near the bottom of the conduction band of LSO. In any case, the study of the nature of found ET's is a matter of further investigations. No maxima at temperatures of above 400 K, which were indicated previously in [4,9], were found because the observation of them is possible only in using the high-energy radiation (X-ray or γ radiation).

The comparison of the obtained results of the absorption spectroscopy and photoluminescence with the TSL data allows us, in using data of [3,11], to construct schematic diagram of energy levels, which determine the luminescence mechanism in the single-crystal LSO (figure 5). The absorption of quanta with energies of 3.47-4.70 eV are realized by Ce^{3+} ion at the expense of electron transitions between doublets of ground *4f* and excited *5d* states. As the exciting quantum energy increases up to 5.83 eV, the ionizing of the luminescence centre to the Ce^{4+} state and the electron transfer to the conduction band take place. Further, the electron trapping with traps both localized near the conduction band bottom (trap III) and associated with Ce^{3+} (traps I, II) is possible. The final stage of luminescence process in the LSO is preceded by the electron release from traps and radiationless electron transitions from the conduction band and top *5d* levels of Ce^{3+} to the bottom *5d* level. The photon emission corresponds to the electron transitions from *5d* levels to the *4f* ground-state doublet with energies of 2.96, 3.12 eV (Ce^{I}) and 2.70 eV (Ce^{II}). Thus, we can conclude that the photoluminescence in the LSO has not only the intra-centre mechanism; the conduction band and trap states localized near the conduction band participate in the excitation energy storage.

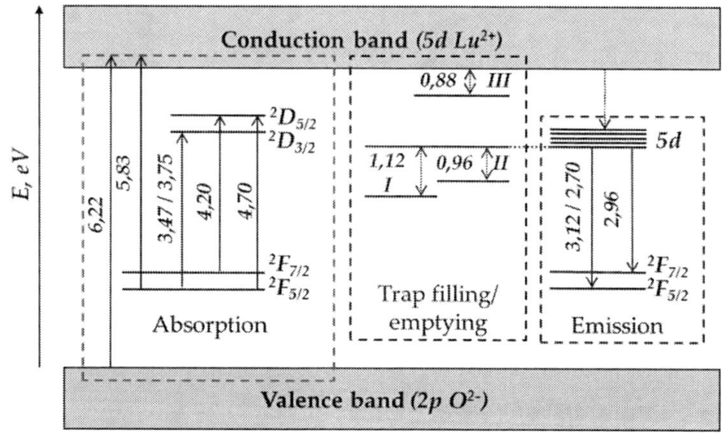

Figure 5. Simplified energy diagram of luminescence transitions in the single-crystal LSO. Vertical arrows correspond to radiation (solid) and radiationless (dotted) electron transitions.

5. Conclusions

The spectral characteristics of Ce^{3+} doping-activator in the LSO single crystals grown by the modified Musatov method, have been studied. The band gap energy was found to depend on the optical beam direction in the crystal; the dependence is due to the crystal anisotropy. No anisotropy was found for the centroids of absorption and luminescence bands of Ce^{3+}. The parameters of electron traps participating in the luminescence were determined by TSL method. Traps characterized by a depth of 0.88 eV were found for the first time; they make a contribution to the TSL process when the crystal is exposed to the more intense radiation produced by the high-pressure mercury lamp. Using the obtained data, a model of the energy structure levels determining the luminescence process in the LSO has been constructed.

Acknowledgments

This study was supported by the Russian Foundation for Basic Research (project. no 18-32-00656_mol_a "Investigation of the relationship between optical-luminescent and mechanical phenomena caused by reversible ionization of the Ce^{3+} activator in $Lu_2SiO_5:Ce^{3+}$ single crystals").

References

[1] Huber J S, Moses W W, Andreaco M S, Petterson O An LSO scintillator array for a pet detector module with depth of interaction measurement *IEEE Trans. Nucl. Sci.* 2001 **48** 684-688
[2] Curie D 1960 *Luminescence crystalline* (Paris: Dunod)
[3] Znamenskiĭ N V, Manykin E A, Petrenko E A, Yukina T G, Malyukin Yu V, Zhmurin P N, Grinev B V, Masalov A A, Shpak A P The nature and mechanism of charging of electron traps in $Lu_2SiO_5:Ce^{3+}$ *Journal of Experimental and Theoretical Physics* 2004 **99** 386-393
[4] Dorenbos P, Van Eijk C W E, Bos A J J, Melcher C L Afterglow and thermoluminescence properties of Lu_2SiO_5:Ce scintillation crystals *J.Phys.: Condens. Matter* 1994 **6** 4167-4180
[5] Varaksin A N, Sobolev A B, Kuznetsov A Yu, Keda O A Molecular-statics simulation of the cerium impurity in LSO crystals *Physics of the Solid State* 1997 **39** 426-427
[6] Bessonova L O, Garmash V M, Garmash M V, Tedzhetov V A Growth and study of the effect of

growth conditions on the structural perfection and morphological properties of cerium doped lutetium silicate crystals *Materials of Electronics Engineering* 2007 **1** 40-44 [In Russian]

[7] Garmash V M, Tedzhetov V A, Yakimova I O Correlation between the luminescent properties and the melting point in second column tungstate crystals *Materials of Electronics Engineering* 2009 **3** 26-32 [In Russian]

[8] Cooke D, Bennett B, McClellan K, Roper J, Whittaker M, Portis A Electron-lattice coupling parameters and oscillator strengths of cerium-doped lutetium oxyorthosilicate *Phys. Rev. B – Cond. Mat.* 2000 **61** 11973-11978

[9] Chen Y, Liu B, Shi Ch, Ren G, Zimmerer G The temperature effect of $Lu_2SiO_5:Ce^{3+}$ luminescence *Nucl. Ins. and Meth. in Phys. Res. A* 2005 **537** 31-35

[10] Yukihara E G, Jacobsohn L J, Blair M W, Bennet B L, Tornga S C, Muenchausen R E Luminescence properties of Ce-doped oxyorthosilicate nanophosphors and single crystals *Journal of Luminescence* 2010 **130** 2309-2316

[11] Pidol L, Guillot-Noël O, Kahn-Harari A, Viana B, Pelenc D, Gourier D EPR study of Ce^{3+} ions in lutetium silicate scintillators Lu_2Si2O_7 and Lu_2SiO_5 *Journal of Physics and Chemistry of Solids* 2006 **67** 643-650

IOP Conf. Series: Materials Science and Engineering **525** (2019) 012045 doi:10.1088/1757-899X/525/1/012045

Mechanical properties of multilayer materials

A I Plokhikh, D V Vlasova, K B Polikevich and A A Minakov

Department of Material Science, Bauman Moscow State Technical University (BMSTU), 105005, 2nd Baumanskaya street, 5-1, Moscow, Russian Federation

E-mail: plokhikh@bmstu.ru

Abstract. This article presents a study of mechanical properties of multilayer steel materials after the static, impact and cyclic tests, and evaluation of stress and strain behavior of such materials. The studies have shown an increase in the strength, impact strength and the fatigue life with the increase of the number of layers. Also, it has been shown, that high impact strength is maintained down to cryogenic temperatures. That indicates an absence of a definite point for the ductile-brittle transition within such materials. The studies of the stress and strain behavior of multilayer steel materials have shown that the rolling forces increase effect is specific for the production of such materials.

1. Introduction

The multilayer steel materials represent a new class of promising structural materials, in which the modern approach to creation of a structure that provides an improved set of mechanical properties is utilized. Mechanical properties of the produced multilayer steel materials outperform the set of properties of the structural polycrystal materials used in the machinery engineering for now. The mentioned approach provides for making an oriented multilayer structure in the submicron range by preserving a high-angle misorientation on the interlayer boundaries. The use of hot pack rolling method seems to be the most efficient and highly productive as compared with ARB-rolling, for instance [1].

2. Materials and Methods

This work uses the multilayer steel materials manufactured under the experimental process flow. A pre-cleaned multilayer pack was formed, comprising of 100 interleaving steel sheets ordered alternatively, represented by two steel grades (50 sheets per each grade) with the thickness of 0.5 mm each [2, 3]. Then, the multilayer pack was vacuumed and hot rolled at the temperature of 1000 °C. The first process cycle output is a multilayer steel material with a thickness of 2 mm comprising of 100 layers with the thickness of a single layer of 20 μm. The multilayer steel material sheets produced as a result of the first cycle were used in the second process cycle. The completed second process cycle resulted in the increase of the number of 2mm thick layers within the multilayer steel material up to 1400 layers, and single layer thickness is decreased to 3 μm.

The following multilayer steel material compositions were studied – AISI430+AISI304, W108+AISI304, which were used to make samples for static, dynamic and fatigue tests.

The mechanical tensile tests were carried out by 100 kN Zwick/Roell Z100 universal testing machine. Flat samples with the dimensions of 2x10x180 mm were cut out from the central part of multilayer steel bands along the rolling direction. The impact strength tests were carried out at the temperature of 20 °C, –70 °C and –196 °C on the samples cut out from the hot rolled band with the thickness of 10 mm in the transversal direction relative to rolling. The impact strength was evaluated in normal (⊥) and parallel (//) directions relative to the rolling plane.

The fatigue life was determined using the AISI430+AISI304 composition. The tests were carried out at the pure plane bending under symmetrical reverse loading by means of Schenk-Erlinger machine. The test load was 560 MPa (above the yield point of that composition after the first cycle, which equals 520 MPa, and lower than the second cycle yield point which equals 660 MPa).

A model experiment was carried out to evaluate the stress and strain behavior of the W108+AISI 304 multilayer steel material, during which the samples cut out from the hot rolled sheet in the normal direction relative to rolling were impacted by the Brinell machine ball with the diameter of 10 mm under 30 kN. After which microhardness of the first and the second process cycle samples was measured layer-by-layer by using longitudinal polished sections using the EmcoTest DuraScan 20 microhardness tester. The fractographic analysis of samples was carried out by Vega Tescan 5130 scanning microscope at the acceleration voltage of 20 kV.

3. Results and discussion

3.1. Study of the static properties

The mechanical properties of the studied composition samples are given in table 1. A low value of the elasticity modulus in comparison with the original steel grades, increase of the hardness and strength characteristics as layers thickness decreases, as well as a considerable decrease of the plasticity characteristics is specific for all compositions.

Table 1. Mechanical properties of the AISI 430+AISI 304 and W108+AISI 304 composition samples.

Composition	Number of layers	E, GPa	HB	$\sigma_{0,2}$ MPa	σ_B MPa	δ, %	ψ, %
AISI 430+AISI 304	100	160	180	520	780	7,0	43
	1400	170	370	660	800	4,5	48
W108+AISI 304	100	150	110	560	960	7,0	44
	1400	180	450	880	1470	4,0	18

3.2. Study of the impact strength

The studies have shown that the samples of AISI430+AISI304 and W108+AISI304 compositions tested in the direction normal to the rolling plane are not destructed by a drop hammer with the operational margin of 300 J, and demonstrate small breaks in the stress concentration area, after which they are affected by bending strain. With the decrease in the test temperatures, the sample values after the first process cycle, both in longitudinal and transversal directions, consistently decrease down to the minimum at the test temperature of –196 °C (table 2).

Table 2. Values of the impact strength for the studied compositions.

Composition	Concentrator position	KCU, J/cm² 20°C	–70°C	–196°C
AISI 430+AISI 304	⊥	360[a]	350[a]	160
		350[a]	360[a]	360[a]
	//	360[a]	150	30
		83	79	70
W108+AISI 304	⊥	80	47	13
		350[a]	350[a]	370[a]
	//	57	43	25
		80	60	30

[a] – sample not destructed
numerator – samples of (after) the first process cycle (100 layers)
nominator – samples of (after) the second process cycle (1400 layers)

The samples after two process cycles with1500 layers of the cross-section (the nominator values) behave differently. In the direction normal to the rolling plane the AISI430+AISI304 and W108+AISI304 composition samples are not destructed down to the temperature of –196 °C, which indicates an absence of the ductile-brittle transition point within the studied temperature range.

The increasing trend remains as to the destruction energy intensity for the multilayer steel materials and for the ones with the V-shaped concentrator. Thus, for the samples of AISI 430+AISI 304 composition tested at the temperature of 20 °C there is an almost double increase of Astr, and almost four times increase at the temperature of 196 °C below zero (table 3). This behavior within multilayer steel materials under impact loads nominally conflicts the common concepts of the ductile-brittle transition point within the polycrystal materials. The results of studies show that while decreasing the test temperature the work intensity of destruction of the impact low-alloy samples treated in the described way is increased, which indicates an absence of a definite ductile-brittle transition point (DBT) [4]. The absence of a definite ductile-brittle transition point in the multilayer steel materials can be achieved due to the oriented layered structure formation that increases the impact strength values in the direction normal to the rolling plane and improves the material strength at low temperatures.

Table 3. The impact strength values of the multilayer composition samples with V-shaped concentrator in the rolling direction (⊥).

Composition	Number of layers	Thickness of layer, μm	Astr (KCV), J/cm²		Sample condition (–196 °C)
			20 °C	–196 °C	
AISI 430+AISI 304	100	100	140	80	destructed
	1 400	7,0	280	260	not destructed
W108+AISI 304	100	100	80	15	destructed
	1 400	7,0	230	290	not destructed

The analysis of breaks within multilayer steel materials shows that high values of the impact strength of the samples tested in the direction normal to the rolling depend on a partial breakdown of layers and further strain of a sample part which remains integral (Figure 1).

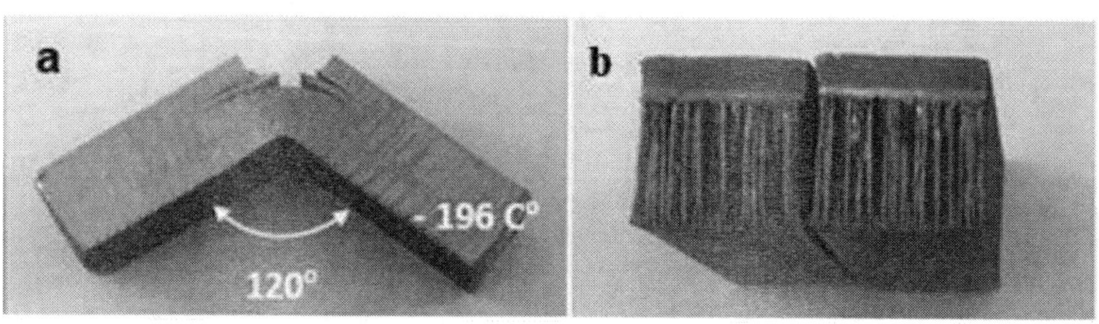

Figure 1. The samples of 08X18+08X18H10 composition with 1 400 layers after the impact bending test: a) ⊥ to rolling plane, b) // to rolling plane.

3.3. Study of fatigue properties (fatigue life)

The results of the fatigue life studies have shown that the fatigue life of the sample of multilayer steel material AISI 430+AISI 304 comprising of 100 layers is $2.8 \cdot 10^4$ cycles before destruction. For the sample after the second process cycle with the number of layers about 1 500 it is $4.6 \cdot 10^4$ cycles before destruction (Figure 2). Therefore, the fatigue life has increased 1.8 times with the increase of the number of layers within the sample from 100 to 1 500 (with a single layer thickness decrease from 20 μm down to 3 μm, accordingly). The obtained results have a good correlation with the tests carried out under cyclic tensile conditions [5]. The increase of the fatigue life can be explained by the following.

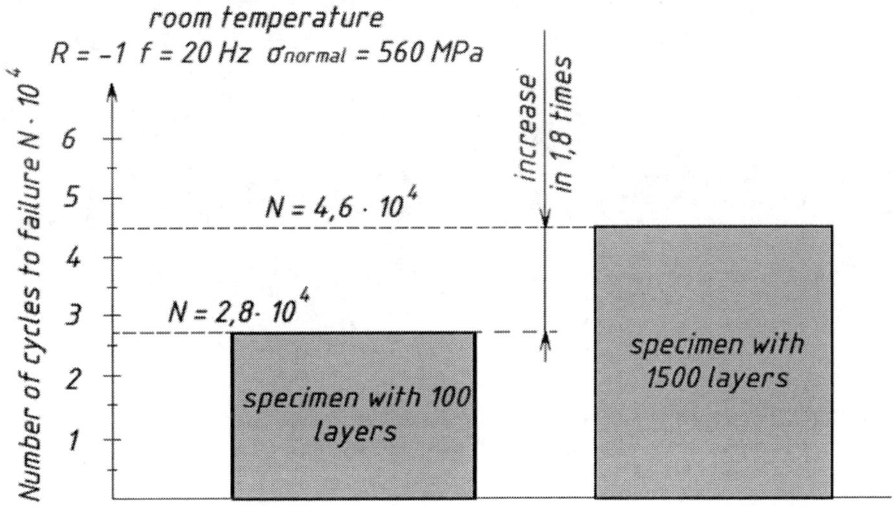

Figure 2. The results of fatigue life tests of the multilayer steel material AISI 430+AISI 304 under cyclic loading at the pure plane bending by means of Schenk-Erlinger machine.

It is known, that the fatigue life of a polycrystal material under cyclic loading will be high, provided that it has higher strength [6]. On the other hand, material fatigue greatly depends on its structural condition. Based on the Hall-Petch relationship, $\sigma \sim h^{1/2}$ (where σ – plastic strain, h – grain size), we may assert, that the polycrystal material fatigue life will be high if it has a greater range of the structural grain fineness.

The multilayer metal materials demonstrate different specific stress and strain behavior as compared with the polycrystal materials. The strain mechanisms are impossible to describe using the Hall-Petch model and for those materials, it shall be considered in a view of single dislocation loops slippage parallel to the section surfaces [7].

3.4. Evaluation of the stress and strain behavior of the multilayer steel materials

The results of studies on evaluation of the stress and strain behavior of W108+AISI 304 composition show that the strain depth after the first process cycle, with 100 layers in total, is about 50 % of the sample thickness (Figure 3a). In the samples after the second process cycle, with 1 500 layers in total, the strain depth did not exceed 15 % of the sample thickness (about 1.5 mm) (Figure 3b).

Such behavior of the multilayer steel materials can be explained by the following. Earlier it has been discovered, that one of the features of the multilayer materials production is the effect of considerable rolling forces increase in comparison with the values obtained on the blanks with polyhedral structure [8]. It has been established, that at the same temperature, speed and strain parameters during the process, the rolling forces on the multilayer materials comprising of 100 interleaving layers of AISI 1008 and W108 steel grades, exceed the rolling forces obtained on the samples comprising of the interleaving layers of W108 steel grade by 50% and exceed the rolling forces on the solid W108 steel sample by more than 100%. This effect is caused by the formation of special type structure, which is known as Bamboo-type, as well as by high interlayer stresses.

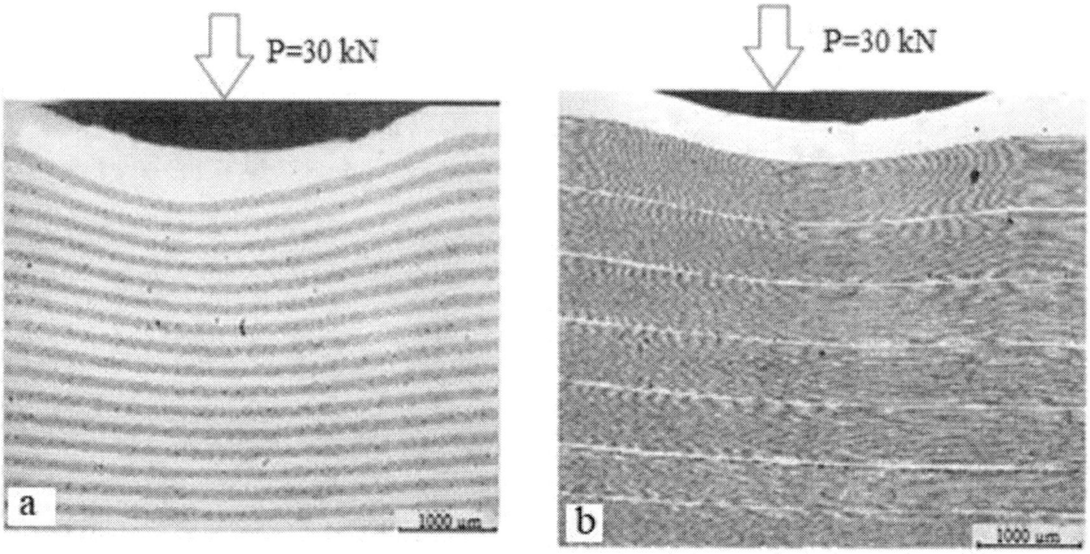

Figure 3. Microhardness distribution within W108+AISI 304 composition after the first (a) and the second (b) process cycle.

The study of the structure of layers within the area of strain, performed by an electronic scanning microscope on the samples of 100layers has shown, that no visible strain is observed within the layers of AISI 304 steel grade, while the layers of W108 steel grade are impacted highly, which results in slippage strips formation in them (Figure 4).

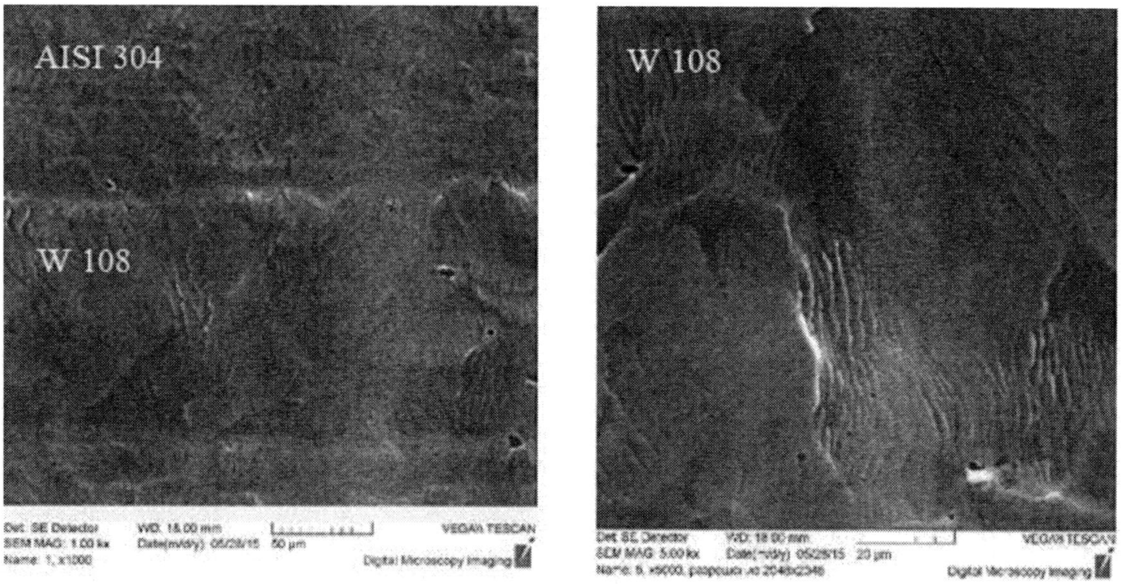

Figure 4. Electronic image of the structure of W108 steel grade layers affected by strain.

IOP Conf. Series: Materials Science and Engineering **525** (2019) 012045 doi:10.1088/1757-899X/525/1/012045

4. Conclusions

The results of performed studies have shown that due to the creation of a laminar structure within the multilayer steel materials a unique set of mechanical properties is observed. These materials have high strength in combination with high impact strength which is preserved down to cryogenic temperatures, thus allowing improving the strength of structural material. These materials are promising for applications requiring cyclic loads due to their high fatigue life, which increases with the increase of the number of the laminar structured layers.

References

[1] Saito Y, Utsunomiya H, Tsuji N and Sakai T 1999 Novel ultra-high straining process for bulk materials – development of the accumulative roll-bonding (ARB) process *Acta Materialia* **47** 579-583

[2] Kolesnikov A G, Plokhikh A I, Komisarchuk Yu S and Mikhaltsevich I Yu 2010 A study of special features of formation of submicro- and nanosize structure in multilayer materials by the method of hot rolling *Metal Science and Heat Treatment* **52** N5–6 273–278

[3] Tabatchikova T I, Plokhikh A I, Yakovlev I L and Klyueva S Yu 2013 Structure and properties of a steel-based multilayer material produced by hot pack rolling *The Physics of Metals and Metallography* **114** N7 580-592

[4] Kimura Y, Inoue T, Yin F and Tsuzaki K 2008 Inverse Temperature Dependence of Toughness in an Ultrafine Grain-Structure Steel *Science* **320** 1057-60

[5] Hertzberg R V 1983 Deformation and fracture mechanics of engineering materials, second edition (UK: John Wiley, Chichester) p. 715

[6] Schmiedt A, Luecker L, Kolesnikov A, Plokhikh A I and Walther F 2017 Production- and microstructure-based fatigue assessment of metallic AISI 304/430 multilayer materials produced by hot pack rolling *Materialpruefung* **59** N2 123-129

[7] Wang J and Misra A 2011 An overview of interface-dominated deformation mechanisms in metallic multilayers *Current Opinion in Solid State and Materials Science* **15** 20-28

[8] Rudskoi A I, Kodzhaspirov G E and Dobatkin S V 2012 Advanced technologies for manufacturing sheet products with an ultrafine-grained structure Russian metallurgy *Metally* **2012** N1 72-75

Elaboration of aluminium based metalmatrix composite manufacturing

E A Chernyshov [1], E A Romanova[1], A D Romanov[1], I D Romanov[1] and V V Myl'nikov[2]

[1] Nizhny Novgorod State Technical University n.a. R.E. Alekseev
603950, Nizhny Novgorod, Minin st. 24, Russia
[2] Nizegorodskij Gosudarstvennyj Arhitekturno-stroitel'nyj Universitet,
603950, Nizhni Novgorod, Ilyinskaja 65, Russia

E-mail: nil_st@nntu.ru

Abstract. The proposed technology provides for the formation of new hardening phases directly in the aluminum melt, thus enabling to produce MMC by a single-stage process and to achieve the high homogeneous distribution of dispersed particles and their sizes and also ensuring the economical efficiency of the process. The performed studies have shown the possibility for the significant variation of mechanical properties of the material

One of the effective solutions for reducing the vehicle weight is the development of new metal matrix composites with a reinforcing phase. In particular the use of metalmatrix composite materials (MMC) enables to reduce the weight of parts and structural elements by 15–50%, to increase by one-and-a-half times or twice as much their rigidity as compared to similar metal materials (for example, titanium and aluminum alloys), thus enhancing the efficiency, the design reliability and the considerable reduction of operating costs as compared to polymer composites [1 - 5].
Particularly among other composites aluminum alloy-based MMC have gained the widespread use. While maintaining aluminum alloy advantages (low density, high manufacturability, etc.) the reinforcement with dispersed particles significantly enhances the strength characteristics scarcely affecting the density and hence the weight of a product [6 - 13].
Notwithstanding such MMC advantages there should be noted their main drawbacks such as energy intensity, the complexity and duration of the process cycle and oftentimes environmental harm. Therefore, the scope of the MMC commercial use is yet inadequate to their technical and operational capabilities. The high cost of raw materials is a considerable deterrent to the widespread implementation of products made of such composites.
One of the developing manufacturing technologies is the "internal oxidation method", which has such advantages as obtaining a uniform stable dispersed phase in the volume of finished products. However, since the internal oxidation rate decreases with increasing the oxygen penetration depth and therefore, the enlargement of formed oxides takes place in the line therewith, the internal oxidation method is effective for articles or blanks small in thickness. Therefore, methods for manufacturing MMC directly of liquid melt are currently of interest, moreover that this is a single-stage production process, which ensures the optimum adhesion of the matrix-particle system [14 - 16]. The proposed technology provides for the formation of new hardening phases directly in the aluminum melt, thus enabling to produce

MMC by a single-stage process and to achieve the high homogeneous distribution of dispersed particles and their sizes and also ensuring the economical efficiency of the process.

This technology is based on the outcome of experiments in manufacturing an air-independent power plant on the basis of highly metalized gas-free fuel [17]. The literary and patent search in the course of preliminary works has revealed no direct analogs for the process elaborated.

To carry out the MMC manufacturing experiments, a test-bench was made for producing and casting of a certain alloy composition. During experimental works primary aluminum A6 (admixtures, mainly Fe and Si) was used as a matrix material to prevent the effect of alloying additives and to study the composite hardening only due to strengthening phase particles. Tests have been performed under the preliminary testing schedule for cast non-heat-treated MMC.

It has been established by metallographic and X-ray diffraction studies that the matrix of the manufactured sintered metal-powder material is aluminum and the main implementation phase is aluminum oxide. The particle size and the oxidation reaction completeness are regulated by the process duration and by the change in structural elements of a purge recovery unit as well as crystallization conditions. The obtained dimensions of the solid implementation phase are within the range from 2 μm to 2 mm. The sample microstructure is shown in Figures 1 through 8. The measurement was carried out with the aid of the software of microscopes RV-21 and KEYENCE VHX-1000 using the function of "3 points lying on a circle" and "the measurement of a distance between two points". Figure 4 shows the hardness measurement using PMT-3. Figure 8 depicts the sample fracture surface.

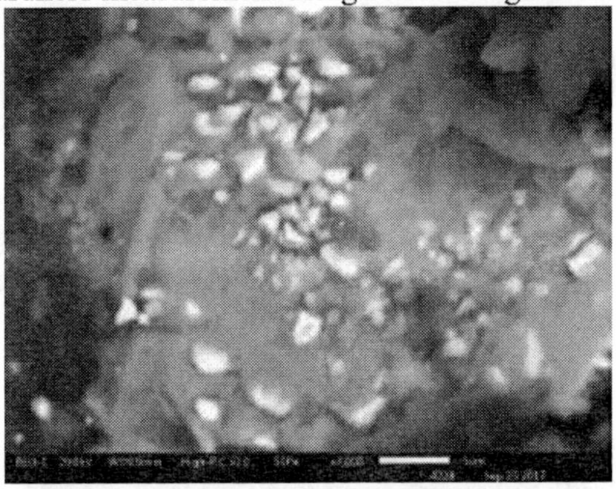

Figure 1. Sample microstructure, the mean particle size is 3 um (x3000)

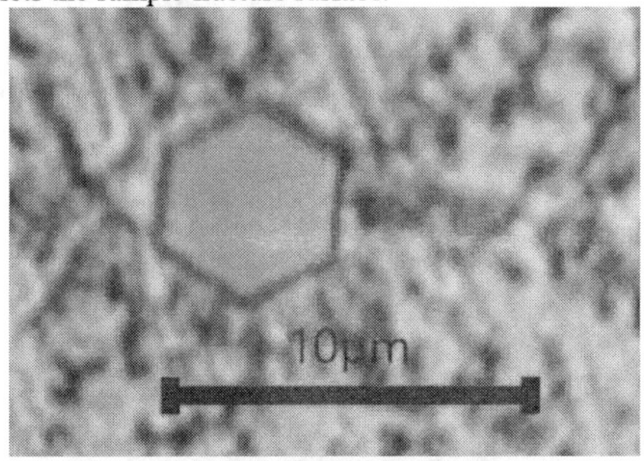

Figure 2. Sample microstructure, the particle size is about 6 um

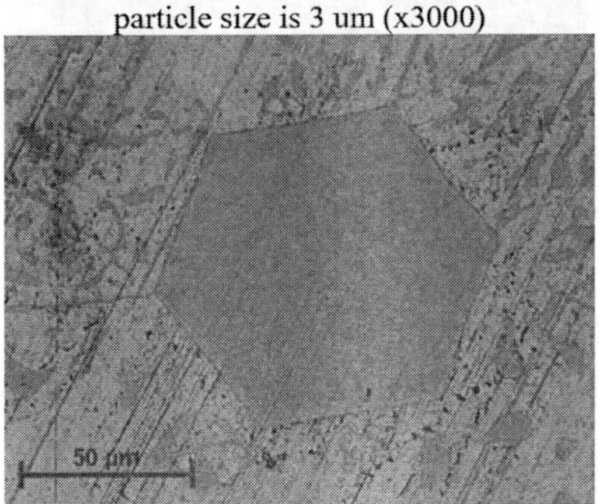

Figure 3. Sample microstructure, the particle size is about 60 um

Figure 4. Sample microstructure after the hardness measurement

IOP Conf. Series: Materials Science and Engineering **525** (2019) 012046 doi:10.1088/1757-899X/525/1/012046

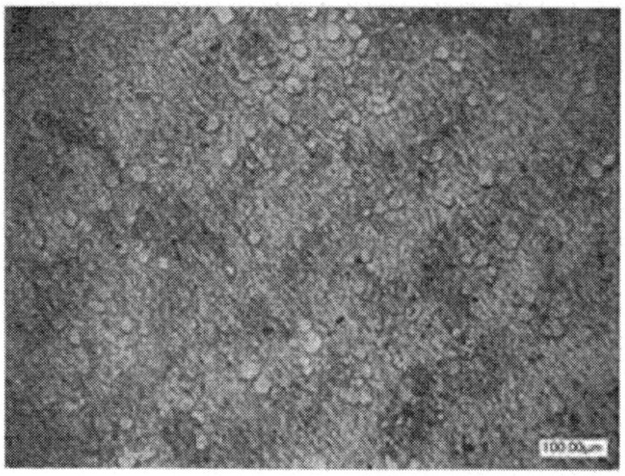

Figure 5. Sample microstructure with the particle size of about 60 um

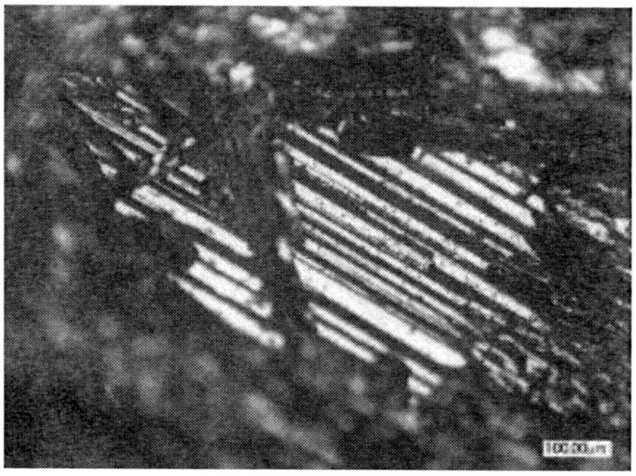

Figure 6. Sample fracture surface with the particle size of over 1 mm. Corundum particle fracture [18]

The purposeful variation of the size and content of the strengthening phase under this MMC manufacturing method shall enable to significantly change mechanical properties of the resultant product.

The study was carried out with a dispersion-strengthened aluminum-based composite material A6. The alloy was strengthened by the oxide solid phase Al_2O_3 (Figure 2-4), in most cases, of a hexagonal shape according to the type of the crystal structure, as was well known, for corundum it is a hexagonal lattice. For static-strength testing ingots have been cast and samples have been made therefrom. The mean particle size for the strengthening phase of predominantly prismatic morphology was 60–80 um. The uniaxial tensile tests of flat samples were performed using the Shimadzu AG-Xplus-0.5 universal testing machine at a room temperature. The fracture surface tests were performed using the KEYENCE VHX-1000 optical microscope with enhanced capabilities due to an improved long-focus optical system and the digital image processing. To study typical fracture areas, the JEOL JSM-IT300LV scanning electron microscope was used with energy and wave dispersed elemental analyzers.

Figure 7. Sample fracture surface - optical macrostructure x40

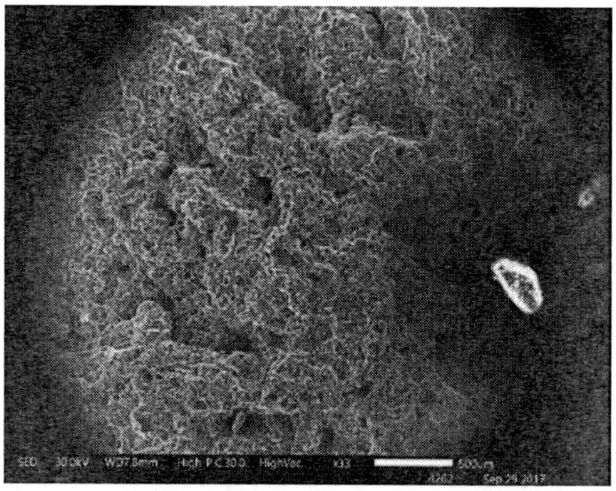

Figure 8. Sample fracture surface - macrostructure in the Optimal Image mode

Figure 7 shows the macroscopic view of the flat sample fracture surface. The fracture is of a mixed macrogeometrally inhomogeneous nature, which may be characterized as a dry fibrous fracture with some observed crystalline pimples and tear-outs.

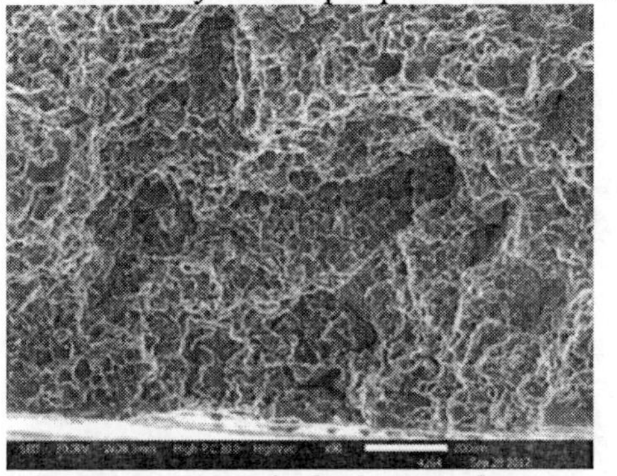

Figure 9. Sample fracture surface - macrostructure, scanning electron microscope

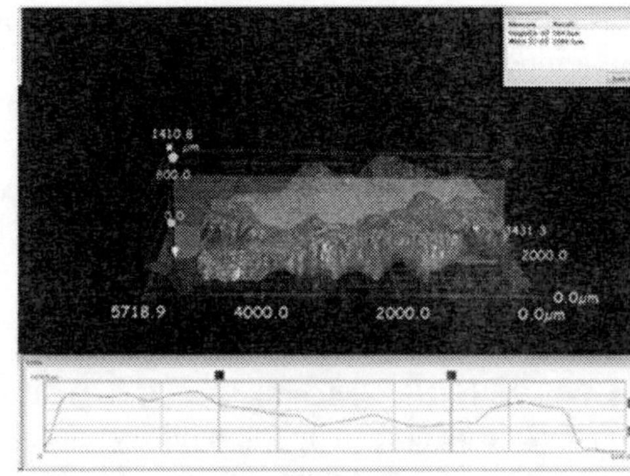

Figure 10. 3D-structure with detail section profilogram

The fracture nucleus and the crack birth are clearly visible in Fig. 8 in the area of conglomeration of dispersed phase inclusions. The studies of the macrostructure in polarized light in the Optimal Image mode (Figure 8) and the microstructure detect secondary cracking, which may be caused by high normal tensile stresses as a result of the separation of the matrix-hard phase interface along the maximum shear plane depending on each single grain orientation. The heterogeneity of the general fracture surface macrogeometry combines therein single areas characterized by the morphologically integral fracture surface, i.e. macrogeometrically homogeneous without geometric zones significantly differing in relief with available similar fracture elements. It is shown in Figure 9-10 as a characteristic of the degree of the fracture surface coarseness in the direction perpendicular to the plane of loading; that is associated with the stability of the high-energy crack propagation process during fracture on these sections.

The available amount of dispersed inclusions in the matrix of the material under study, which are located at considerable distances, causes the formation of low ridges in some directions. It may be explained by the fact that a crack growth front bypasses or bends around them as such propagation is easier than through the body of high-hard phases. Furthermore, the profilogram clearly shows a rather sharp one-time profile change directed at about 45 ° to the axis of tension (Figure 10); it may be interpreted as a shift on this area of the material typical for a shear zone. Microstructural studies (Fig. 9) show the alternation of crystalline areas and cleavage facets around oxides under a transcrystalline mechanism followed by thin fibrous streaks. A dendritic structure with intercrystalline fracture and quasishear are sometimes observed hereat.

The stated reduction in the cost of raw materials results from the disuse of powdered components. Moreover, the hardware, the purge recovery unit is also far cheaper than sintering, self-propagating synthesis or mechanical alloying plants. For comparison we shall give the calculation of costs for producing an aluminum alloy saturated with 30% Al_2O_3. (Table 1)

The aluminum oxidation is carried out by the following reaction

Al (1 g) + O_2 (0.9 g) = Al_2O_3 (1.9 g)

Table 1. The comparison of the cost of 1 kg of alloy containing 30 % Al2O3

Version 1 Adding powdered Al_2O_3	**Version 2 Internal oxidation**
0.7 kg of Al (100 Rub/kg) and 0.3 kg of Al_2O_3 (1000 Rub/kg)	0.86 kg of Al (100 Rub /kg) and 0.15 kg of O_2 (100 Rub/m3 or 140 Rub/kg)
the cost of 1 kg of alloy is 370 Rub/kg	the cost of 1 kg of alloy is 107 Rub/kg

The price of aluminum depends on a brand and as-received condition and ranges from 80 to 200 Rub/kg, the price of Al_2O_3 depends on the particle purity and size and ranges from 340 to 5000 Rub/kg, the price of oxygen depends on the purity and ranges from 60 to 150 Rub/ м^3.

IOP Conf. Series: Materials Science and Engineering **525** (2019) 012046 doi:10.1088/1757-899X/525/1/012046

Conclusions

The application of the elaborated MMC manufacturing technology based on the oxidation of aluminum with oxygen shall enable to significantly reduce the time and economic costs. Furthermore, depending on a technical assignment and any required operational characteristics quite a wide variability of the strengthening phase is possible without changing the general plant design. Reinforcing particles are synthesized thereat directly in the melt. It makes possible the single-stage manufacturing of composites and achieving the thermodynamic stability, the tight contact and adequate adhesion between the matrix and the strengthening phase. The degree of the melt saturation with ceramic elements is regulated by the reaction time. The performed studies have shown the possibility for the significant variation of mechanical properties of the material.

References

[1] Kablov, E. N., Ospennikova, O. G., and Lomberg, B. S. (2013). Strategic trends of development of structural materials and technologies of their processing for modern and future aircraft engines. *The Paton Welding,* **11,** 23–32.

[2] Kurganova Y. A., Chernyshova T. A., Kobeleva L. I., Kurganov S. V. (2011). Service properties of aluminum-matrix precipitation-hardenet composite materials and the prospects of their use on the modern structural material market. *Russian metallurgy*, **7**, 663-666.

[3] Belov N. A., Belov V. D., Alabin A. N., Mishurov S. S. (2010). New generation of economically alloyed aluminum alloys. *Metallurgist,* **54 (5-6),** 311-316.

[4] Shavnev A. A., Berezovskij V. V., Kurganova Yu. A. (2015). Osobennosti primeneniya konstrukcionnogo metallicheskogo kompozicionnogo materiala na osnove alyuminievogo splava, armirovannogo chasticami SiC. *Novosti materialovedeniya. Nauki I Tehniki,* **15**, 3 -10

[5] Adebisi A. A., Maleque M. A., Rahman M. M. (2011). Metal matrix composite brake rotor: historical development and product life cycle analysis. *International Journal of Automotive and Mechanical Engineering,* **4**, 471-480

[6] Pradeep Rohatgi, IUNIDO's State-of-the-art series: "Advances in Materials Technology: MONITOR", Compiled by the Industrial Technology Development Division Department of Industrial Promotion, Vienna, Austria, No.17, Feb. 1990

[7] Mortensen A., Cornie J. A., Flemings M. C. (1988). Solidification processing of Metal Matrix Composites *JOM*, **40(2)**, 12-19

[8] Geiger A. L., Walker J. A. (1991). The processing and properties of discontinuously reinforced Aluminium Composite. *JOM,* **43**, 8-15

[9] Liu Yao-Hui, Du Jun, Yu Si-rong, Wang Wei (2004). High temperature friction and Wear behaviour of Al2O3 and /or Carbon short fibre reinforced Al-12Si alloy Composites. *Wear*, **256**, 275-285

[10] Lloyd D.J. /In: Mallick PK, editor. Composite engineering handbook. New York: Marcel Dekker. 1997. 63–70

[11] Beffort O. Metal Matrix composites (MMCs) from Space to Earth /In.: Werkstoffe für Transport und Verkehr Materials Day. ETH-Zürich. 2001

[12] Chawla, N. Metal Matrix Composites / N. Chawla, K.K. Chawla //In: Springer Sience+Business Media, Inc. 2006. 401 p

[13] Rijesh M., James Valder, Jithin D., Dileep C. R., Abin R. N., Shibin F., Havila M. M. (2016). Production of Al-Al2O3 MMC by P/M Route and to Study the Feasibility of Fusion Welding. *American Journal of Materials Science,* **6(4A)**, 99-101

[14] Chien Chon Chen, Chih Yuan Chen, Hsi Wen Yang, Yang Kuao Kuo, Jin Shyong Lin (2014) Phase Equilibrium in Carbothermal Reduction Al2O3 → AlN Studied by Thermodynamic Calculations Atlas *Journal of Materials Science,* **1(2)**, 30–37

[15] Vlasova M., Kakazey N., Rosales I., Krushinskaya L., Bykov A., Tomila T., Voitsehovskaya E., Vinokurov V. (2010) Synthesis of Composite AlN–AlON–Al2O3 Powders and Ceramics

IOP Conf. Series: Materials Science and Engineering **525** (2019) 012046 doi:10.1088/1757-899X/525/1/012046

Prepared by High-Pressure Sintering *Science of Sintering,* **42,** 283-295

[16] Afanas'ev V. K., Gercen V. V., Dolgova S. V., Musohranov Yu. M., Popova M. V. (2015). O vliyanii vodyanogo para na formirovanie svojstv vysokokremnistyh Al-splavov *Metallurgiya mashinostroeniya,* **5,** 17-22

[17] Chernyshov E. A., Romanova E. A., Romanov A. D. (2015). Development of a heat generating element based on highly metallized gasless fuel *Vest. MGTU Im. N. É. Bauman, Ser. Mashinostroenie,* **6 (105),** 74-81.

[18] Chernyshov, E.A. Romanov, A.D., Romanova, E.A. (2018). Production of Highly Reinforced Dispersion-Strengthened Composite Material Based on Aluminum by Internal Oxidation *Metallurgist,* **7-8,** 815-819

IOP Conf. Series: Materials Science and Engineering **525** (2019) 012047 doi:10.1088/1757-899X/525/1/012047

Ultra-fast chip calorimetry accessories for in operando structural studies of nanogram-sized samples

A A Rychkov[1,2], A P Melnikov[1,2,4], A I Rodygin[1,2], D V Anokhin[1,2,4], M Rosenthal[3] and D A Ivanov[1,2,4,5]

[1] Lomonosov Moscow State University, Faculty of Fundamental Physical and Chemical Engineering, Leninskie Gory 1/51, 119991 Moscow, Russian Federation
[2] Moscow Institute of Physics and Technology (State University), Institutskiy per. 9, Dolgoprudny, Russian Federation
[3] European Synchrotron Radiation Facility, F-38043 Grenoble, France
[4] Institute of Problems of Chemical Physics, Russian Academy of Sciences, Chernogolovka, Moscow oblast, 142432, Russian Federation
[5] Institut de Sciences de Matériaux de Mulhouse-IS2M, CNRS UMR 7361, Jean Starcky, 15, F-68057, Mulhouse, France

E-mail: a.rychkov@inenergy.ru, dimitri.ivanov@uha.fr

Abstract. Ultra-fast chip calorimetry is a modern thermal analysis technique based on MEMS-type sensors, which allows performing quantitative measurements on samples of only a few nanograms. However, calorimetry alone is sometimes insufficient for in-depth analysis of complex phase transitions occurring during the experiment. Therefore, in-operando structural analysis was found to be helpful in deciphering the nature of the corresponding structural transitions. Here we report on design of accessories for in-operando synchrotron-based X-ray scattering experiments at low temperatures as well as under different atmospheres and humidities. The examples of applications in polymer physics include analysis of the double-melting behavior of poly(trimethylene terephthalate) and structure formation processes in isotactic polypropylene.

1. Introduction.

Since its introduction in the beginning of the 90s, the technique of ultrafast chip calorimetry, or nanocalorimetry,[1-3] has been attracting a steadily growing interest of a broad scientific community working in various fields materials' science. The main technological breakthrough of ultrafast chip calorimetry compared to the classical methods of thermal analysis is in its use of a MEMS-based measuring cell, which provides the possibility of carrying out experiments at extremely fast heating and cooling rates. The achievable rates can be up to one million times faster than the ones used in the classical DSC. This feature imparts to the novel technique the sensitivity to very small sample sizes of about several nanograms. The new applications of ultrafast chip calorimetry include but are not limited to ultra-thin organic and inorganic films and individual polymer single crystals [4–6]. However, quite often the complexity of the processes occurring during thermal experiments require additional information in order to be fully comprehended, as calorimetry alone can be insufficient to fully clarify the situation. Therefore, it is beneficial to couple the technique of ultrafast chip calorimetry to other in-situ methods of physical-chemical characterization. To this end, we designed a custom-built nanocalorimeter allowing for in-operando combination with synchrotron-based X-ray scattering experiments [7]. The nanocalorimetric device is compatible with various MEMS chips produced by the Xensor company [8]. The combination of nanocalorimetry and X-ray scattering with fast detection allows simultaneously addressing the thermal and structural parameters of the samples, which simplifies the analysis of the thermal events observable in the nanocalorimetric traces. For example, we revisited the so-called double-melting behavior of semirigid-chain semicrystalline polymers [9-10]. It is well documented that the semirigid-chain semicrystalline polymers can exhibit two and more melting peaks in the heating traces. This phenomenon is not always related to polymorphism and can be observed even for polymers possessing only one crystal modification. In the ongoing debate in the literature, the presence of several melting peaks was necessarily associated with reorganizaton during heating. However, in order to understand the details of the structure formation processes a combination of nanocalorimetry with in-situ X-ray scattering was badly needed.

In our recent publication, we explored the reorganization behavior of a typical semirigid-chain semicrystalline polymer using such combination of techniques [11]. We showed that the critical heating rate above which all reorganization processes cease to exist can be higher than 1000 °C/s, depending on the crystallization conditions. Therefore, the researcher has to be aware of the critical heating rate corresponding to each set of sample preparation conditions in order to completely exclude the reorganization processes during the experiment. This is obviously a prerequisite for acquiring meaningful thermoanalytical data that reflects the initial sample state and not just the interplay of the reorganization processes (i.e., recrystallization and subsequent melting) occurring during heating ramps. In the present work, we report on development of environmental variable-temperature stages for nanocalorimetric sensors which are compatible with optical microscopy operating in reflection and transmission and synchrotron-based X-ray scattering.

2. Devices designed for experiment.

To extend the operating temperature range of the the custom-designed nanocalorimeter, a low-temperature sensor stage and a prototype of the device were developed. The operation of the stage is based on liquid cooling and uses the thermoelectric Peltier converters. Due to the use of special clamps, a rigid fixation of the nanocalorimeter chip and immobility during thermal expansion of its structural elements is achieved, which is especially important for structural studies. The low-temperature stage also allows using an additional (reference) nanocalorimeter sensor, i.e. a differential setup, to subtract the addendum heat capacity (baseline), which is necessary for quantitative processing of the experimental data.

Figure 1 shows the scheme of the device with indication of the main structural elements. The left panel provides the view with spaced parts, while the right panel is the view of the stage assembly.

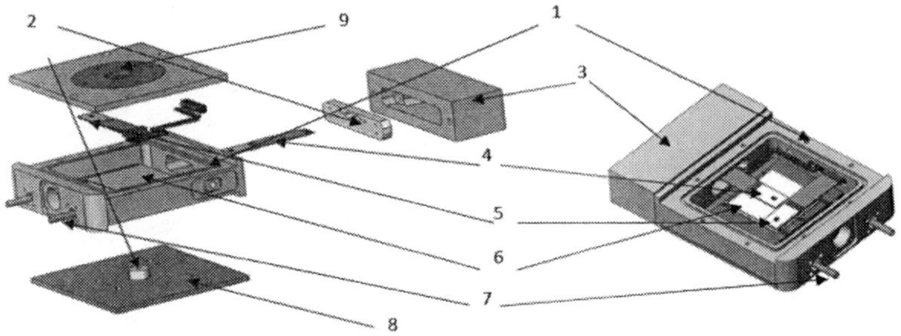

Figure 1. General view of the low-temperature stage for the nanocalorimetric sensor: 1 - housing of the stage, 2 - ceramic thermal screens, 3 - flange with a block of nanocalorimetric sensors, 4 - chip with the sample under study, 5 - reference chip, 6 - Peltier element, 7 - liquid cooling tubes, 8 - invar alloy cover, 9 - silicon nitride window.

The low-temperature stage was combined with X-ray scattering at the micro-focus ID13 beamline of the European Synchrotron Radiation Facility (ESRF) in Grenoble (France). An additional gaseous nitrogen blowing system was implemented for some elements of the setup, eliminating the possibility of condensation of moisture from the atmosphere. Figure 2 shows the scheme and photo of the experimental setup installed at the ID13 beamline. The implementation of the low-temperature stage made it possible to conduct combined nanocalorimetric and structural experiments in the temperature range from -25 to 400 °C, as well as using the modes of ultra-fast controlled quenching of samples up to 10 000 °C/s.

Figure 2. Schematics and photo of the environmental variable-temperature stage for the nanocalorimetric sensor integrated in the ID13 microfocus beamline of the ESRF (Grenoble, France).

The small size of the nano-calorimeter sensor makes it rather straightforward to create special environmental conditions around the chip. In particular, we developed a sealed cell with controlled humidity and atmosphere surrounding the sensor (Figure 3). The cell is connected to the electronic interface of the nanocalorimeter. The humidity inside the cell is controlled by regulating the flow of dry and wet gases into the housing of the chip independently and by mixing them inside the housing. The relative humidity values are measured by the Sensirion SHT31 electronic humidity and temperature sensor. The humidity sensor provides high accuracy and rate of relative humidity measurements and is equipped with a thermocouple, which makes it possible to synchronize the sensor's readings with the signals of the nanocalorimetric sensor. To optimize the cell for X-ray scattering experiments the cell is made prismatic, with two side windows of thin silicon nitride, which is an X-ray transparent material. The model of the assembled cell is given in Figure 3.

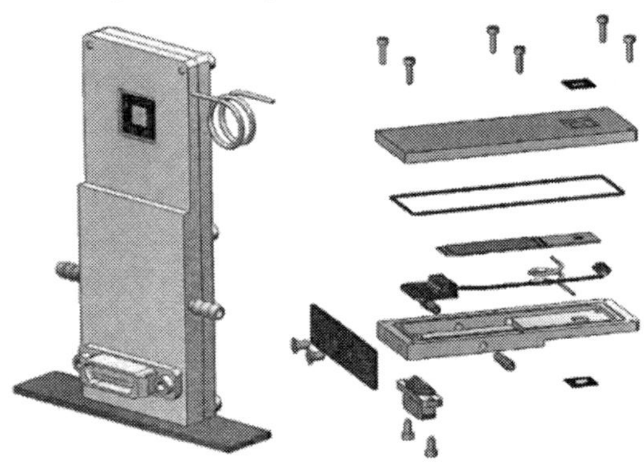

Figure 3. 3D model of the nanocalorimetric chip housing with controlled humidity.

The developed nanocalorimetric accessories make it possible to carry out a wide range of in-operando experiments using different sample geometries such as thin films, fiber microfragments, powder microparticles, etc. In the following of the article we will provide two examples of combined calorimetric and structural studies performed on poly(trimethylene terephthalate) (PTT) and on isotactic

IOP Conf. Series: Materials Science and Engineering **525** (2019) 012047 doi:10.1088/1757-899X/525/1/012047

polypropylene (iPP).

3. Experiments and results. The processes of structure formation and melting of semi-rigid chain polymers was elucidated using the example of a typical aromatic polyester, PTT. In the past, low-heating-rate synchrotron experiments and modelling of the semicrystalline structure of PTT [12] allowed concluding that the double-melting behavior is intrinsic to the thermodynamics of this and similar systems. Using modelling of the variable-temperature small-angle X-ray scattering curves with a generalized paracrystalline model it was concluded that the thermodynalic stability of PTT crystals depends not only on their thickness but also on the size of the amorphous gaps with the neighboring crystalline lamellae. The crystallization process at its later stage results in a build-up of negative pressure [12] imposed on the dominant crystals by the later grown subsidiary crystals. This in turn lowers down the melting point of the crystals according to the Clapeyron-Clausius equation. During heating, the melting process starts from the crystals squeezed in the smallest amorphous gaps where the negative pressure effect is the strongest. After melting of the least stable crystal fraction, the remaining crystals acquire additional stability and melt at a higher temperature. Therefore in this case the double-melting behavior can be accounted for by the complex interaction between the crystalline and amorphous regions and by the metastabile nature of the polymer crystals.

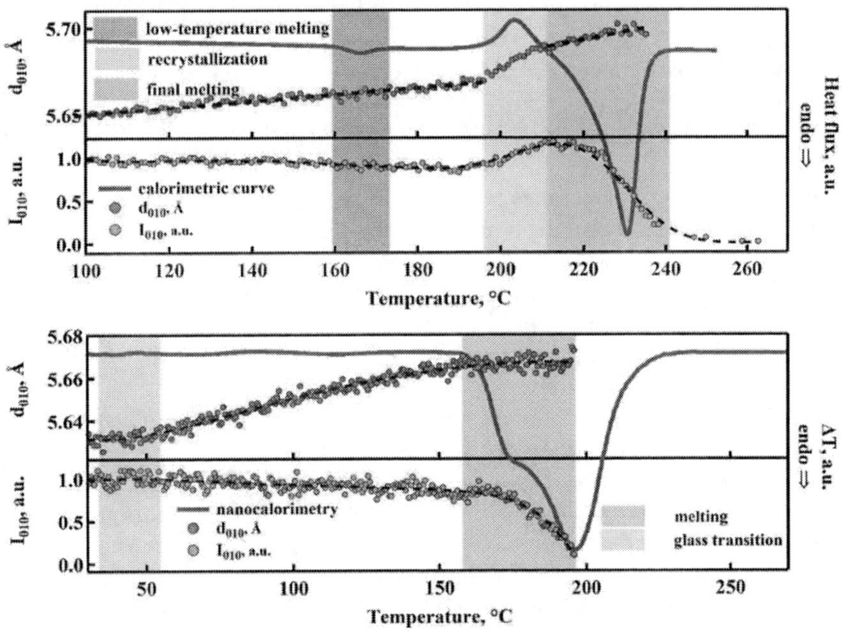

Figure 4. Temperature dependence of the 010 peak position and intensity and the corresponding calorimetric curves for PTT melt-crystallized at 150 °C. The heating rates employed are 1 °C/s (top) и 500 °C/s (bottom). For the sake of clarity the temperature regions of glass transition, as well as low-temperature melting, recrystallization and final melting are highlighted. The calorimetric curve shown in the top panel was measured with a Perkin Elmer 8000 differential scanning calorimeter.

It is noteworthy that in the literature, the double-melting behavior is often interpreted as a sign of reorganization on heating, which is in contrast to our interpretation. In order to clarify this issue, we performed combined in-situ thermal and structural studies of a melt-crystallized PTT at heating rates ranging from 1 °C/s to 3000 °C/s. As an example, figure 4 shows the evolution of position and intensity of the strongest 010 peak of the PTT lattice. The experiment shown in the top panel of figure 4 was conducted at a relatively low heating rate of 1 °C/s. It is noteworthy that at this slow heating rate our custom-built nanocalorimeter is not sensitive to the small sample and was therefore run in the so-called temperature-modulation mode (AC-mode). For the sake of simplicity, we present on the figure a

IOP Conf. Series: Materials Science and Engineering **525** (2019) 012047 doi:10.1088/1757-899X/525/1/012047

calorimetric curve measured in the usual DC mode on a conventional Perkin Elmer differential scanning calorimeter (DSC). The displayed DSC curve exhibits low-temperature melting at about 170 °C followed by exothermic recrystallization at ca. 200 °C and eventually the final melting. The structural data confirms that the sample reorganizes on heating. Indeed, simultaneously with the recrystallization peak in the DSC trace one observes a shift of the 010 peak position and an abrupt increase of the peak intensity. This clearly shows that a fraction of new crystals appears at this stage of heating.

The experiment presented in the bottom panel of figure 4 was conducted at a much higher heating rate of 500 °C/s. In this case, the nanocalorimetric curve shows a complex-shaped peak characteristic of the double-melting behavior. At the same time the structural data does not present any evidence for the reorganization processes. Indeed, the 010 peak position gradually shifts upwards due to the thermal expansion whereas the peak intensity reflects melting of the sample. This proves that the double-melting phenomenon is not necessarily coupled with the reorganization of the structure on heating but is inherent to the melting process in such system.

The complex phase transitions in polymer systems can be further illustrated using the example of iPP for which the heating and cooling rate can dramatically affect formation of different phases [cf. e.g., 13]. The left panel of figure 5 shows the nanocalorimetric curves measured during heating of the same iPP sample at 1000 °C/s. The difference between the curves is due to the different initial structures of the sample formed by cooling from the melt at the cooling rates indicated on the figure. It can be seen that for the slowest previous heating rate (i.e. 100 °C/s) the heating curve exhibits just one endothermic event, which can be assigned to melting. For the higher cooling rates the subsequent heating curve becomes more complex: it exhibits a thermal event at ca. 0 °C, as well as the ones at about 90 and 140 °C. The nature of this transitions can be clarified using the wide-angle X-ray scattering (WAXS) profiles measured before each of the heating ramps (cf. the right panel of fig. 5). It can be seen that upon relatively slow cooling from the melt the sample exhibits the semicrystalline state, while for faster cooling rates the crystalline peaks becomes weaker and eventually disappear leaving first the so-called mesophase state and then a largely amorphous structure. The combination of nanocalorimetry and in-situ WAXS makes it possible to evaluate the critical cooling rates at which the different sample states will be generated. This information is also helpful for understanding of the sample transformation on heating. Indeed, the fully crystallized samples, i.e the ones corresponding to the slowest previous cooling rates, will undergo melting. For the higher cooling rates, the initial state of the sample is partly mesomorphic, which is visible for example for the WAXS profile corresponding to the previous cooling rate of 300 °C/s. This would explain a more prominent glass transition (the T_g of amorphous iPP equals -7 °C). The thermal event in this region includes a step in the heat capacity due to devitrification followed by an endothermic peak of the so-called thermal ageing. The second endothermic peak in the curves at about 90 °C, which becomes visible already above the previous cooling rate of 100 °C/s stands for the mesophase crystallization, whereas the final melting temperature stays invariable. The enhancement of the glass transition at the highest previous cooling rates reflects the formation of a largely amorphous sample, i.e. at these cooling rates the formation of the mesophase is bypassed.

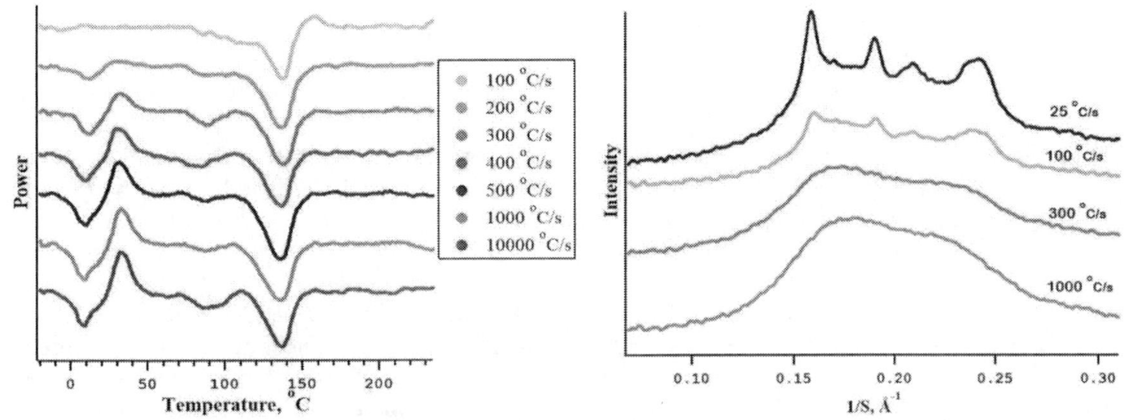

Figure 5. Left: Nanocalorimetric curves corresponding to iPP samples cooled down at different rates (indicated). The measurements were performed at 1000 °C/s. For the sake of clarity, all the curves are shifted vertically. Right: Wide-angle X-ray scattering profiles measured on the samples before starting the heating ramps. For the sake of clarity, the WAXS curves are shifted vertically.

In conclusion, the developed accessories for in-operando nanocalorimetry open a wide range of possibilities for conducting experiments under different atmospheres (including vacuum) and at different humidities. The compact design of the sensor stages equipped with high-quality transparent windows for visible light and X-ray beams make them compatible with the environment of the micro- and nano-focus synchrotron X-ray beamlines. The examples of the measurements provided in the paper show that this technique is useful for studies of polymer samples exhibiting complex thermal transitions on heating and cooling. The combination of such in-situ methods provides unique opportunities for advanced thermal and structural characterization of organic and inorganic samples under various atmorspheric conditions.

Acknowledgements
The authors acknowledge the Ministry of Science and High Education of the Russian Federation [contract No.14.616.21.0072 from 28 July 2016 (RFMEFI61616X0072)] for financial support. This work was performed in accordance with the state task no. 0074-2019-0014. The authors are grateful to Manfred Burghammer (ESRF) for excellent technical support at the beamline and fruitful discussions.

References

[1] Allen L H, Ramanath G, Lai S L, Ma Z, Lee S, Allman D D J, Fuchs K P 1994 1,000,000 °C/s thin film electrical heater: In situ resistivity measurements of Al and Ti/Si thin films during ultra rapid thermal annealing *Appl Phys Lett* **64(4):** 417–419

[2] Lai S L, Ramanath G, Allen L H, Infante P, Ma Z 1995 High-speed (10^4 °C/s) scanning microcalorimetry with monolayer sensitivity (J/m^2) *Appl Phys Lett* **67(9)**: 1229-1231

[3] Lai S L, Allen L H 1998 MEMS-based scanning calorimeter for thermodynamic properties of nanostructures *Microscale Thermophys Eng* **2(1):**11-19

[4] Kwan A T, Efremov M Y, Olson E A, Schiettekatte F, Zhang M, Geil P H, Allen L H 2001 Nanoscale calorimetry of isolated polyethylene single crystals *J Polymer Sci*: Part B 39 **622(11)**:1237-1245

[5] Olson E A, Efremov M Y, Zhang M, Zhang Z S, Allen L H 2003 The design and operation of a mems differential scanning nanocalorimeter for high-speed heat capacity measurements of ultrathin films *J Microelectromech Syst* **12(3):**355-364

[6] Efremov M Y, Olson E A, Zhang M, Allen L H 2003 Glass transition of thin films of poly(2-vinyl pyridine) and poly(methyl methacrylate): nanocalorimetry measurements *Thermochim*

Acta **403(1):**37-41
[7] Rosenthal M, Melnikov A P, Rychkov A A, Doblas D, Anokhin D V, Burghammer M and Ivanov D A 2016 Design of an In Situ Setup Combining Nanocalorimetry and Nano- or Micro-focus X-Ray Scattering to Address Fast Structure Formation Processes *In: Fast Scanning Calorimetry Springer International Publishing, Switzerland*, ISBN 978-3-319-31327-6, Chapter 9
[8] http:// www.xensor.nl/index.php/company
[9] Melnikov A P, Rosenthal M, Rodygin A I, Doblas D, Anokhin D V, Burghammer M and Ivanov D A 2016 *Re-exploring the double-melting behavior of semirigid-chain polymers with an in-situ combination of synchrotron nano-focus x-ray scattering and nanocalorimetry European Polymer Journal* **81**:508-606
[10] Rosenthal M, Melnikov A P, Burghammer M, and Ivanov D A 2017 Reorganization of semicrystalline polymers on heating: analyzing common misconceptions in the interpretation of calorimetric data. Response on the "Comment on "Re-exploring the double-melting behavior of semirigid-chain polymers with an in-situ combination of synchrotron nanofocus X-ray scattering and nanocalorimetry" by Dimitri A. Ivanov et al. [European Polymer Journal 81 (2016) 598–606.] *European Polymer Journal* **94**:517-523
[11] Melnikov A P, Rosenthal M, and Ivanov D A 2018 Thermal Analysis of Semicrystalline Polymers: Exploring the General Validity of the Technique ACS *MacroLetters* **7**:1426–1431
[12] Ivanov D A, Bar G, Dosière M, Koch M H J 2008 A Novel View on Crystallization and Melting of Semirigid Chain Polymers: *The Case of Poly(trimethylene terephthalate) Macromolecules* **41**:9224-9231
[13] Santonja-Blasco L, Zhang Z, Alamo R G 2015 Crystallization of Precision Ethylene Copolymers Advanced *Polymer Science* **276**:133-182

IOP Conf. Series: Materials Science and Engineering **525** (2019) 012048 doi:10.1088/1757-899X/525/1/012048

Ferro-gallium borate single crystals for nuclear resonance synchrotron experiments

N Snegirev[1], Yu Mogilenec[1], K Seleznyova[1], I Nauhatsky[1], M Strugatsky[1], S Yagupov[1], A Kulikov[2,3], D Zolotov[2], N Marchenkov[2,3], K Frolov[2] and I Lyubutin[2]

[1]Physics and Technology Institute, V.I. Vernadsky Crimean Federal University, 295007, Simferopol, Russia
[2]Shubnikov Institute of Crystallography of FSRC "Crystallography and Photonics" RAS, 119333, Moscow, Russia
[3]National Research Center "Kurchatov Institute", 123098, Moscow, Russia

E-mail: niksnegir@yandex.ru

Abstract. A series of ferro-gallium borate $Fe_{1-x}Ga_xBO_3$ single crystals was synthesized by a flux growth technique. The ratio of elements in crystals was determined by X-ray fluorescence analysis. The degree of structural perfection was studied by high-resolution X-ray diffraction techniques. These crystals are of a great interest for modern synchrotron experiments. It is established that diamagnetic impurity affects the degree of structural perfection of the samples.

1. Introduction

In traditional Mössbauer spectroscopy used in a laboratory, the γ-radiation from a radioactive source is uniformly distributed over the sphere (in 4π) and only a small part of it falls on the sample under study. Therefore, it takes a long time to accumulate a high-quality spectrum with a good signal-to-noise ratio. In this situation, the possibility of using high-intensity focused synchrotron radiation seems to be the most effective and optimal solution. Especially, it is critical for the experiments under extreme condition, such as high pressure studies in diamond anvil cells, where the sample size is very small [1]. This technique is called Synchrotron Mössbauer Spectroscopy (SMS) and it has recently been implemented at the ESRF (Grenoble, France) [2] and Spring-8 (Tsukuba, Japan) [3] synchrotron radiation facilities. The key element of the SMS scheme is an iron borate $FeBO_3$ single crystal, designed for the final monochromatisation of the synchrotron radiation with energy corresponding to Mössbauer resonance at the ^{57}Fe nuclei. In this case extremely high requirements are imposed on the structural perfection of single crystals [4].

The energy spectrum of the radiation reflected from $FeBO_3$ consists of several lines due to magnetic hyperfine splitting of the ^{57}Fe nuclear levels, which considerably complicates spectroscopic measurements [5]. This make it necessary to heat the crystal near the Néel temperature (~348.5 K), where the energy spectrum of the reflected radiation collapses into a single line [6]. However, heating, combined with highly ionized radiation, can lead to a degradation of the structural perfection of the crystal- monochromator. One of the possible solution to this problem could be a controlled decrease of the Néel temperature by isomorphous substitution of a part of paramagnetic iron ions in $FeBO_3$ with diamagnetic gallium ions [7-9]. Pure iron borate single crystals can be synthesized by vapor deposition

IOP Conf. Series: Materials Science and Engineering **525** (2019) 012048 doi:10.1088/1757-899X/525/1/012048

and flux growth techniques [10]. Using vapor deposition technique, bulk single crystals of iron borate can be obtained [11]. The flux growth technique allows obtaining single crystals with the shape of hexagonal basal plates. Such crystals have high structural perfection [12]. Thus, for the purposes of the present work the flux growth technique has proved to be the most appropriate. However, it is required to determine the degree of structural perfection of ferro-gallium borate ($Fe_{1-x}Ga_xBO_3$) single crystals.

The purpose of the present work is to synthesize the series of $Fe_{1-x}Ga_xBO_3$ single crystals and determine the degree of their structural perfection by high-resolution X-ray diffraction (HRXRD) techniques.

2. Synthesis of $Fe_{1-x}Ga_xBO_3$ single crystals

Single crystals $Fe_{1-x}Ga_xBO_3$ were synthesized by the flux growth technique. Crystallizations were carried out in the Ga_2O_3–Fe_2O_3–B_2O_3–PbO–PbF_2 system. The crystal-forming reagents are Fe_2O_3, Ga_2O_3 and B_2O_3, whereas PbO, PbF_2, and B_2O_3 serve as solvents. A typical composition of the reagents (in wt %) is as follows: Fe_2O_3 + Ga_2O_3 (5.73), B_2O_3 (51.23), PbO (29.31), and PbF_2 (13.73). Technological steps of the flux growth technique are: preparing of the charge, obtaining a homogeneous solution melt in a platinum crucible, slow cooling of the solution melt and extracting of synthesized crystals [13]. The synthesized crystals have the shape of hexagonal plates with the dimensions up to 5 mm in the basal plane and about 60 µm in thickness. Examples of the synthesized single crystals are shown in Figure 1.

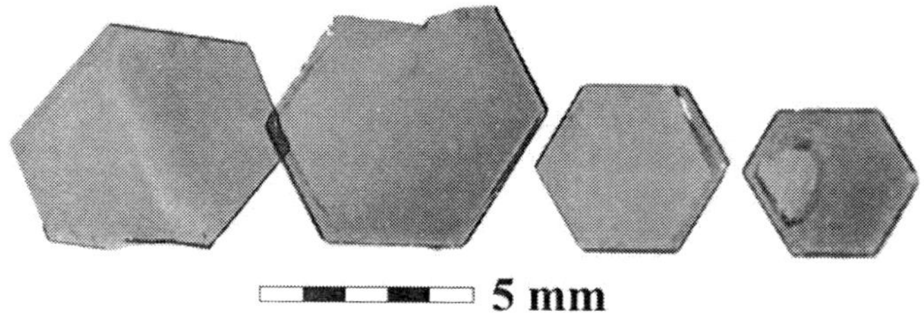

Figure 1. Examples of synthesized single crystals.

3. Characterization of the synthesized $Fe_{1-x}Ga_xBO_3$ single crystals.

The exact contents of Fe and Ga in the synthesized $Fe_{1-x}Ga_xBO_3$ single crystals were determined by X-ray fluorescence analysis (XRF) using Rigaku Supermini200 spectrometer. Studied crystals were fixed using a platinum diaphragm with a 4 mm hole. This made it possible to study crystals with different surface areas. The results of measurements are shown in Table 1.

Table 1. Gallium contents in $Fe_{1-x}Ga_xBO_3$ single crystals *vs.* gallium contents in charge.

x in charge	x in crystal
0.05	0.03
0.4	0.32
0.75	0.69
0.8	0.73
0.9	0.82

IOP Conf. Series: Materials Science and Engineering **525** (2019) 012048 doi:10.1088/1757-899X/525/1/012048

To determine the degree of structural perfection of $Fe_{1-x}Ga_xBO_3$ single crystals, the HRXRD techniques were used.

The double-crystal diffraction rocking curves (DRCs) were measured using an in-house diffractometer, the experimental scheme of which was described in detail in [14]. The scans were carried out in Bragg (for X-ray reflection) geometry for reflection 0012 with a Bragg angle in the range of 17.08° - 17.47° and extinction length L_{ext} in the range of 10.5 - 9.0 μm depending on the Ga contents (for x between 0 and 1). The local area of single crystals was illuminated by an X-ray beam, limited by an incidence slit with dimensions of 0.1×4 mm^2. A Mo X-ray tube ($K\alpha_1$ characteristic line with λ = 0.70932 Å) and a symmetric Si (110) monochromator tuned to the 220 reflection were used.

The X-ray topography studies of $Fe_{1-x}Ga_xBO_3$ single crystals were carried out using an in-house experimental setup «DITOM-M» [15]. The scheme of this setup is shown in Fig. 2. Radiation from conventional X-ray tube with a Mo anode was monochromatised using an asymmetrically cut Si (111) crystal (asymmetry coefficient $\beta \sim 10$), which also provided a beam of sufficient width to measure the whole sample in Laue (for X-ray transmission) geometry. Further adjustment of the beam size was achieved by slits. The acquisition was performed using a CCD camera with a 9 μm pixel size. The monochromator–sample and the sample–detector distances were about 1000 mm and 30 mm, respectively. Exposures were at least 45 minutes. The reflection 300 with a Bragg angle in the range of 15.40° - 15.61° and L_{ext} = 99.9 - 82.3 μm (for x between 0 and 1) was used.

The 2D-topograms were obtained at two various modes:

- Integrate mode is the scan along DRC (ω - scan). It is sensitive to structural features and defects in the sample volume [16].
- Static mode is the intensity acquisition at a fixed angular position at the DRC slope. This mode is sensitive to Bragg angle variation and deformations in the crystal lattice [16].

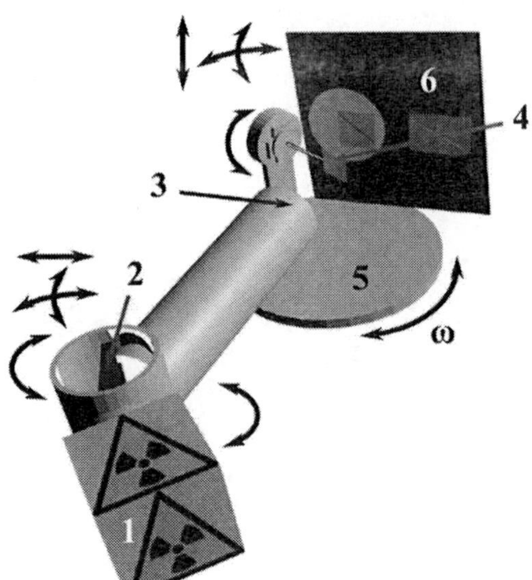

Figure 2. The scheme of «DITOM-M» setup: (1) X-ray source tube, (2) monochromator crystal, (3) tube collimator with the system of slits, (4) crystal under study, (5) goniometer and (6) CCD.

Figure 3 and Figure 4 shows double-crystal DRCs of of $Fe_{1-x}Ga_xBO_3$ single crystals with x = 0.0, 0.03, 0.32 and 1.0. The values of the half-width (FWHM) of DRCs are about 10 angle seconds for pure $FeBO_3$ and $GaBO_3$ crystals (Figure 3a,b). The DRCs for perfect pure crystals were calculated in accordance with the dynamic theory of diffraction taking into account the structural factor of the crystals, the instrumental function of the experimental scheme with allowance for the dispersion effect

[17, 18]. The calculated curves are in good agreement with the experimental data, which testifies the high perfection of the pure crystals.

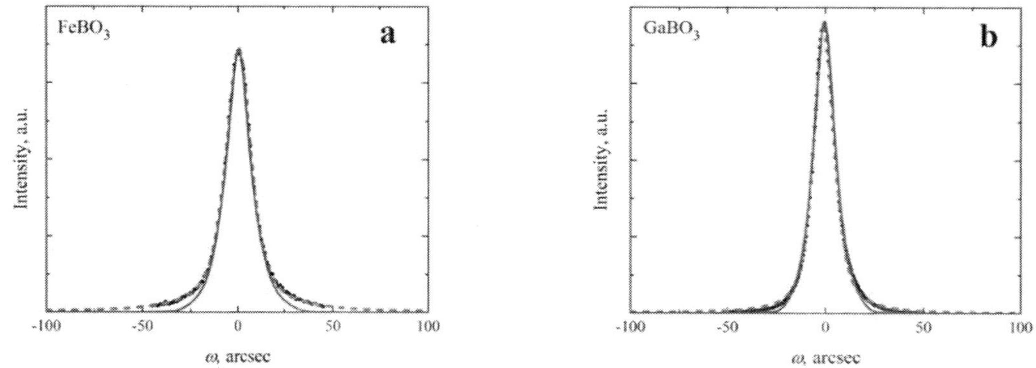

Figure 3. (Color online) Experimental (dotted), calculated (solid green lines) and fitted using pseudo-Voigt profile (dashed red lines) double-crystal DRCs of $FeBO_3$ and $GaBO_3$ single crystals. Reflection 0012 in Bragg geometry was used.

The DRCs of ferro-gallium borate $Fe_{1-x}Ga_xBO_3$ single crystals consist of several broadened peaks (Figure 4a,b). The existence of several peaks can be explained by defects in the crystal structure, for example, crystalline blocks. Due to different ionic radii of Ga and Fe, isomorphous substitution should lead to a distortion of crystal structure, which leads to a broadening of the rocking curves with respect to pure crystals.

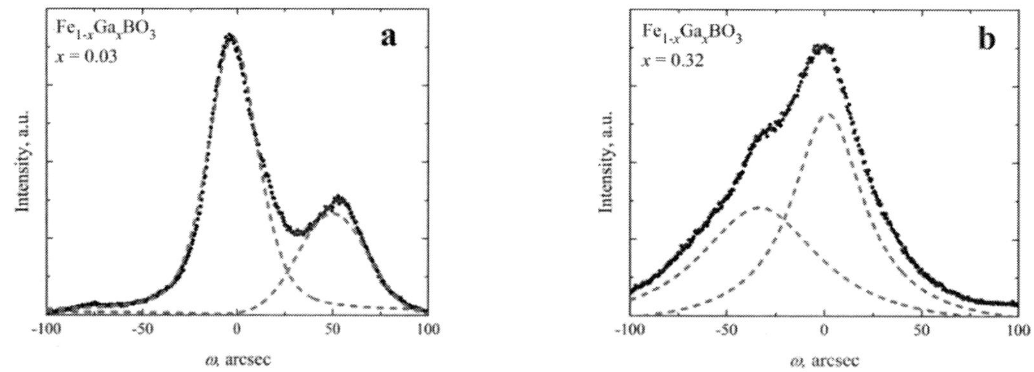

Figure 4. (Color online) Experimental (dotted) and fitted using pseudo-Voigt profile (dashed red lines) double-crystal DRCs of $Fe_{1-x}Ga_xBO_3$ single crystals. Reflection 0012 in Bragg geometry was used.

The X-ray topograms of $FeBO_3$ and $Fe_{1-x}Ga_xBO_3$ with $x = 0.73$ single crystals are shown in Figure 5. The absence of intensity contrast in the integrate scanning mode indicates the absence of growth steps and macroscopic structural defects, such as cracks and dislocations in both crystals [16]. However, the intensity contrast in the static mode for $Fe_{1-x}Ga_xBO_3$ single crystal with $x = 0.73$ is associated with mechanical strains [16]. The distribution of the mechanical strains can be caused by a non-uniform distribution of the iron/gallium contents over the crystal. It should be noted that the distribution of the mechanical strains does not exactly follow the distribution of the impurity contents.

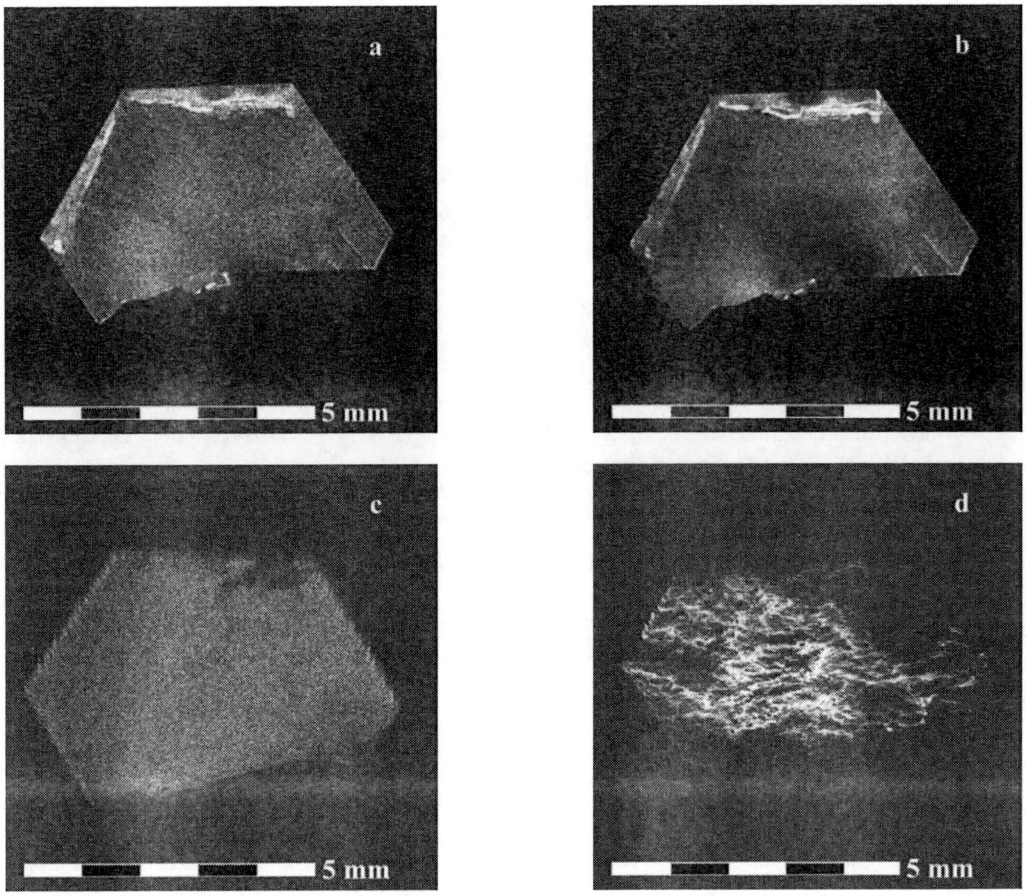

Figure 5. X-ray topograms of $FeBO_3$ (a, b) and $Fe_{1-x}Ga_xBO_3$ with $x = 0.73$ (c, d) single crystals, taken with integrate (a, c) and static (b, d) scanning modes. Reflection 300 in Laue geometry was used.

4. Conclusion

The series of iron borate based single crystals $Fe_{1-x}Ga_xBO_3$ was synthesized and studied by XRF and HDXRD techniques. Double-crystal diffraction rocking curves of pure $FeBO_3$ and $GaBO_3$ crystals are in excellent agreement with the calculated curves for perfect crystals. However, the DRCs of the crystals with mixed compositions $Fe_{1-x}Ga_xBO_3$ are broadened with respect to pure crystals. This can be explained by different ionic radii of Ga and Fe, and random distribution of gallium in the crystal lattice. Isomorphous substitution should lead to a distortion of crystal structure and to appearance of mechanical strains in ferro-gallium borate crystals.

Acknowledgment
This work was partially supported by the Russian Foundation for Basic Research (RFBR) and the Ministry of Education, Science and Youth of the Republic of Crimea in the framework of scientific projects Grants no. 17-42-92015 "p_a" and 18-42-910008 "p_a", by the RFBR in the framework of scientific project Grant no. 18-32-00210 "mol_a" (synthesis of experimental samples) and by the V.I. Vernadsky Crimean Federal University Development Program for 2015–2024.

X-ray diffraction studies were performed with the support of the Ministry of Science and Higher Education within the State assignment FSRC «Crystallography and Photonics» RAS.

References

[1] Lyubutin I and Gavriliuk A 2009 *Physics – Uspekhi* **52 (10)** 989
[2] Potapkin V, Chumakov A, Smirnov G, Celse J-P, Rüffer R, McCammon C and Dubrovinsky L 2012 *J. Synchrotron Radiat.* **19** 559–569
[3] Mitsui T, Hirao N, Ohishi Y, Masuda R, Nakamura Y, Enoki H, Sakaki K and Seto M, 2009 *J.Synchrotron Radiat.*, **16** 723
[4] Mitsui T 2005 *SPring-8/SACLA Research Frontiers* **8** 152
[5] Smirnov G, Zelepukhin M and van Bürck U 1986 *JETP Lett.* **43** 352–355.
[6] Smirnov G, van Bürc U, Chumakov A, Baron A and Rüffer R 1997 *Phys. Rev. B: Condens. Mater. Phys.* **55** 5811
[7] Seleznyova K, Sergeev N, Olszewski M, Stepien P, Yagupov S, Strugatsky M and Kliava J 2015 *Solid State Nucl. Magn. Reson.* **70** 38
[8] Seleznyova K, Strugatsky M, Yagupov S, Postivey N, Artemenko A and Kliava J 2014 *Phys. Stat. Sol. B* **251** 1393
[9] Yagupov S, Maksimova E, Nayhatsky I, Yagupov V, Milyukova E, Seleznyova K, Strugatsky M 2014 *Proc. Int. conference OMEE* 207
[10] Diehl R, Jantz W, Nolang B and Wettling W 1984 *Current Topics in Materials Science* **11** 241
[11] Pankratov A, Strugatsky M and Yagupov S 2007 *Sci. Notes of Taurida National University Ser. Phys.* **20(59)** 64
[12] Yagupov S, Strugatsky M, Seleznyova K, Mogilenec Yu, Snegirev N, Marchenkov N, Kulikov A, Eliovich Y, Frolov K, Ogarkova Yu and Lyubutin I 2018 *Cryst. Growth Des.* **18(12)** 7435
[13] Yagupov S, Strugatsky M, Seleznyova K, Maksimova E, Nauhatsky I, Yagupov V, Milyukova E and Kliava J 2015 *Appl. Phys. A: Mater. Sci. Process.* **121** 179
[14] Marchenkov N, Kulikov A, Petrenko A, Pisarevsky Yu and Blagov A 2018 *Review of Scientific Instruments* **89(9)** 095105
[15] Zolotov D, Buzmakov A, Elfimov D, Asadchikov E and Chukhovskii F 2017 *Crystallogr. Rep.* **62** 20
[16] Kulikov A, Marchenkov N, Blagov A, Kozhemyakin K, Nasonov M, Pashkov S, Pisarevskii Yu and Cherpukhina G 2016 *Acoustical Physics* **62(6)** 694
[17] Marchenkov N, Chukhovskii F and Blagov A 2015 *Crystallogr. Rep* **60(2)** 172
[18] Atknin I, Marchenkov N, Chukhovskii F et al 2018 *Crystallogr. Rep* **63(4)** 521

IOP Conf. Series: Materials Science and Engineering **525** (2019) 012049 doi:10.1088/1757-899X/525/1/012049

Effect of tempering on microstructure and mechanical properties of a Ta-added 9%Cr steel with high B and low N contents

E Tkachev[1], A Belyakov[2]

[1,2]Belgorod State University, Belgorod 308015, Russia

E-mail: tkachev_e@bsu.edu.ru

Abstract. Ta was added to a modified 9%Cr steel with high B and low N content to achieve an optimal dispersion of second phase particles. The effect of tempering on the microstructure and mechanical properties of an Fe-0.1C-9Cr-1.8W-0.6Mo-3Co-Nb-V-0,0013B-0,007N-0,085Ta steel was investigated. Typical martensite lath structure was observed after normalization at 1323 K for 0.5 h. The $M_{23}(C,B)_6$ and complex (Nb,Ta)(C,N) particles formed during tempering at 1023-1053 K. An increase in the hardness after tempering at 1038 K was attributed to the precipitation of the fine (Nb,Ta)(C,N) particles. The impact toughness depended significantly on tempering temperature. The impact toughness increased from 25 to 181 J/cm^2 with increasing the tempering temperature from 1038 to 1053 K.

1. Introduction

9%Cr martensitic steels are an important class of structural materials for elevated temperature applications because of combination of good corrosion resistance and high creep strength. Several concepts have been proposed for the development of new creep-resistant steels with increased creep resistance [1-5]. Recently, considerable interest was paid to boron containing 9-10%Cr steels. A small amount of B was found useful to improve the stability of tempered martensite lath structure (TMLS) under creep conditions especially due to fine dispersion of $M_{23}(C,B)_6$ particles [6-8]. However, the nitrogen content in B-containing steels should be controlled to avoid the formation of BN particles [9,10]. In addition, Co is known as substitutional element, which is important to suppress the δ-ferrite formation during high temperature normalization. Since the newest B-containing steels contain 9-10%Cr and 2-2,5%(W+Mo), the optimum Co content required to avoid the δ-ferrite formation was found to be ~3% [1,11,12]. Following this alloying design, several 9-10%Cr steels with increased creep strength were developed [13-15].

It is well known that MX-type carbonitride particles play an important role in creep resistance due to their fine dispersion and increased stability. Ta is a strong carbide/nitride forming element and is widely used to replace the Nb in reduced activation ferritic-martensitic (RAFM) steels, which are used in fusion power plant components. It was reported that in 9%Cr steel Ta formed fine TaC particles, which are smaller than those in steels without Ta [16]. Thus, in order to further improve the high temperature strength the combined addition of B and Ta was suggested.

The precipitation behavior and mechanical properties of high-chromium martensitic steels are closely connected with tempering temperature. Thus, the present work describes the effect of tempering on microstructure and mechanical properties of a modified Ta-added 9% Cr steel with high B and low N contents.

IOP Conf. Series: Materials Science and Engineering **525** (2019) 012049 doi:10.1088/1757-899X/525/1/012049

2. Materials and method

A modified 9%Cr-3%Co-2%W steel was produced by vacuum induction melting. The chemical composition of the steel is given in table 1. The steel samples were normalized at 1323 K for 0.5 h and tempered at 1023; 1038 and 1053 K for 3 h.

Table 1. Chemical composition of the studied steel (wt.%).

C	Ni	Mn	Cr	W	Mo	Nb	V	Co	Ta	N	B	Fe
0.11	0.02	0.30	9.05	1.99	0.61	0.06	0.19	3.06	0.085	0.007	0.013	Bal.

The microstructures of the present steel were analysed by optical microscopy and transmission electron microscopy (TEM) using Jeol JEM-2100 microscope operating at 200kV equipped with an INCA energy-dispersive X-ray (EDX) spectrometer. The dislocation densities were estimated by counting the individual dislocations in the lath interiors per unit area on arbitrarily selected TEM images. Extraction carbon replicas were prepared to determine the size distribution of different types of precipitates. The fracture surface of the impact tested specimens has been studied by using the FEI Nova NanoSEM 250 scanning electron microscope (SEM).

The hardness of studied steel after normalization and tempering was measured using Brinell tests with a load of 750 N and a ball diameter of 5 mm. Impact tests were performed at room temperature on an Instron IMP460 impact testing machine using standard 10×10×55 mm^3 specimens with a 2 mm V-notch.

3. Results and discussion

3.1. Microstructure after normalization and tempering

3.1.1. Microstructure of normalized steel The optical metallography showed that after normalization at 1323 K for 0.5 h the studied steel had a fully martensite microstructure without delta ferrite (figure 1). The mean size of prior austenite grains in this condition was 54±3 μm.

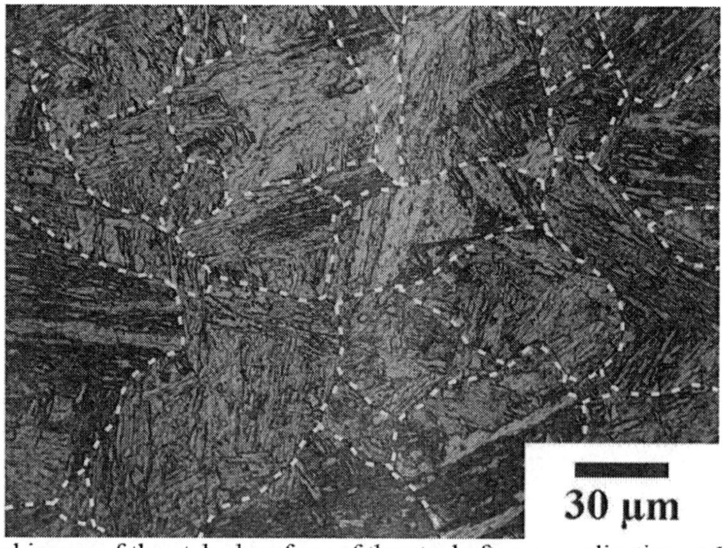

Figure 1. Optical image of the etched surface of the steel after normalization at 1323K for 0.5 h. The prior austenite grain boundaries are indicated by yellow dash lines.

IOP Conf. Series: Materials Science and Engineering **525** (2019) 012049 doi:10.1088/1757-899X/525/1/012049

The TEM observation revealed the lath martensite hierarchical substructure with an average lath thickness of 190±20 nm (figure 2). The high dislocation density of 4.5×10^{14} m^{-2} arisen from $\gamma\rightarrow\alpha$ transformation was maintained in the studied steel after normalization.

Similar to the previous studies on the effect of normalization and tempering on the structure of 9%Cr steels [17.18], the Fe_3C particles formed during cooling after austenization were observed in the studied steel (figure 2).

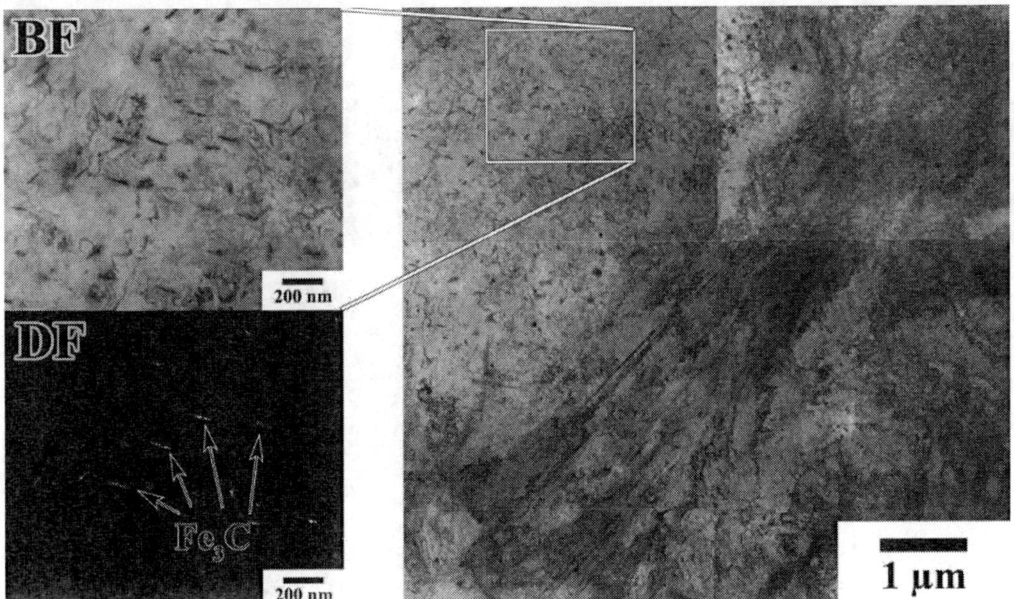

Figure 2. TEM microstructure of the steel after normalization at 1323 K for 0.5 h. The magnified bright field image (BF) and dark field image (DF) show the plate-like particles of cementite.

These particles has plate-like shape with mean size of 75×13 nm and arranged parallel to each other which suggests the presence of preferential crystallographic orientation relationship between these particles and α-martensite. In addition, the (Nb,Ta)(C,N) particles with round shape and mean size of 38 nm were observed.

3.1.2. Microstructure after tempering The tempering led to significant increase of the lath width (table 2). The lath widening was accompanied by the rearrangement of dislocation structure and the particle precipitation. Thus, the dislocation density gradually decreased to 0.9×10^{14} m^{-2} with increasing tempering temperature to 1053 K. Typical TEM micrographs of the studies steel after tempering at different temperatures are shown in figure 3.

IOP Conf. Series: Materials Science and Engineering **525** (2019) 012049 doi:10.1088/1757-899X/525/1/012049

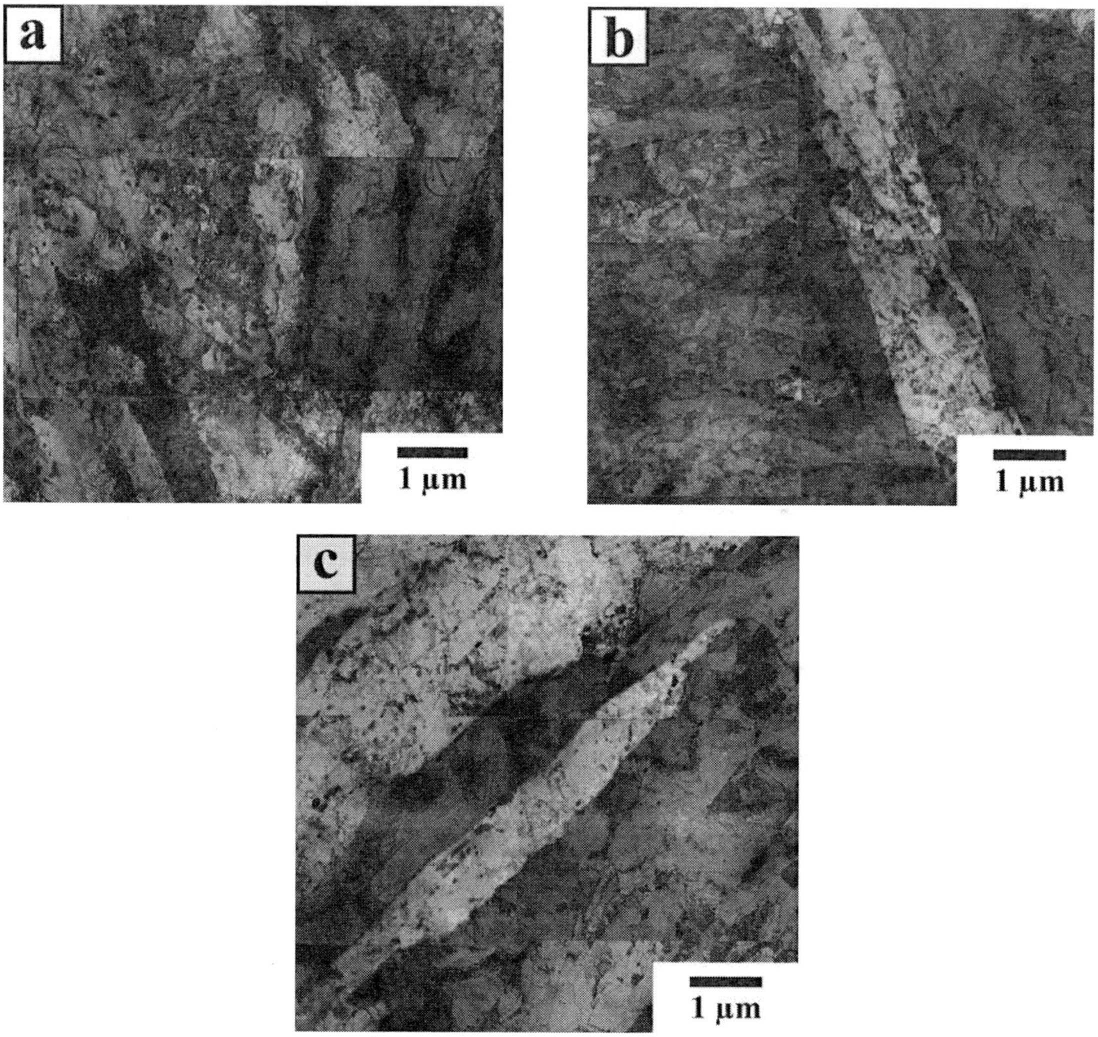

Figure 3. TEM micrographs of the steel after tempering at 1023 K (a); 1038 K (b) and 1053 K (c) showing the TMLS with decoration of $M_{23}(C,B)_6$ carbides on lath boundaries.

The main parameters of the tempered martensite lath structures (TMLS) and the mean diameters of each precipitate are represented in table 2. The tempering led to dissolution of Fe_3C particles. The major observed phase in the tempered conditions was $M_{23}(C,B)_6$ carbide which is more stable than cementite. The most of $M_{23}(C,B)_6$ particles are located along the lath and PAG boundaries and there is no significant difference in the size of these particles among the three steel specimens tempered at 1023; 1038 and 1053 K.

Table 2. Parameters of the microstructure after normalization and tempering at different temperatures.

Tempering Temperature	Normalized	1023 K	1038 K	1053 K
Lath width, nm	190±20	270±15	305±25	355±35
Dislocation density, $\times 10^{14}$ m^{-2}	4.5	3.5	2.4	0.9

IOP Conf. Series: Materials Science and Engineering **525** (2019) 012049 doi:10.1088/1757-899X/525/1/012049

Average size of Fe_3C, nm	75×13	-	-	-
Average size of $M_{23}(C,B)_6$, nm	-	71	75	80
Average size of (Nb,Ta)(C,N), nm	38	26	19	21

In order to clarify the precipitation behavior of (Nb,Ta)(C,N) particles, the size distribution histograms were analyzed (figure 4). It can be seen that undissolved primary particles of (Nb,Ta)(C,N) maintained in steel in normalized condition while additional fine particles of (Nb,Ta)(C,N) precipitate during subsequent tempering. The early-stage precipitation of the secondary (Nb,Ta)(C,N) particles was observed after tempering at 1023 K. The minimum size of these particles was found to be less than 10 nm. Increasing tempering temperature to 1038 K enhanced the precipitation of (Nb,Ta)(C,N) particles and, thereby, decreased the mean size of total population of these particles.

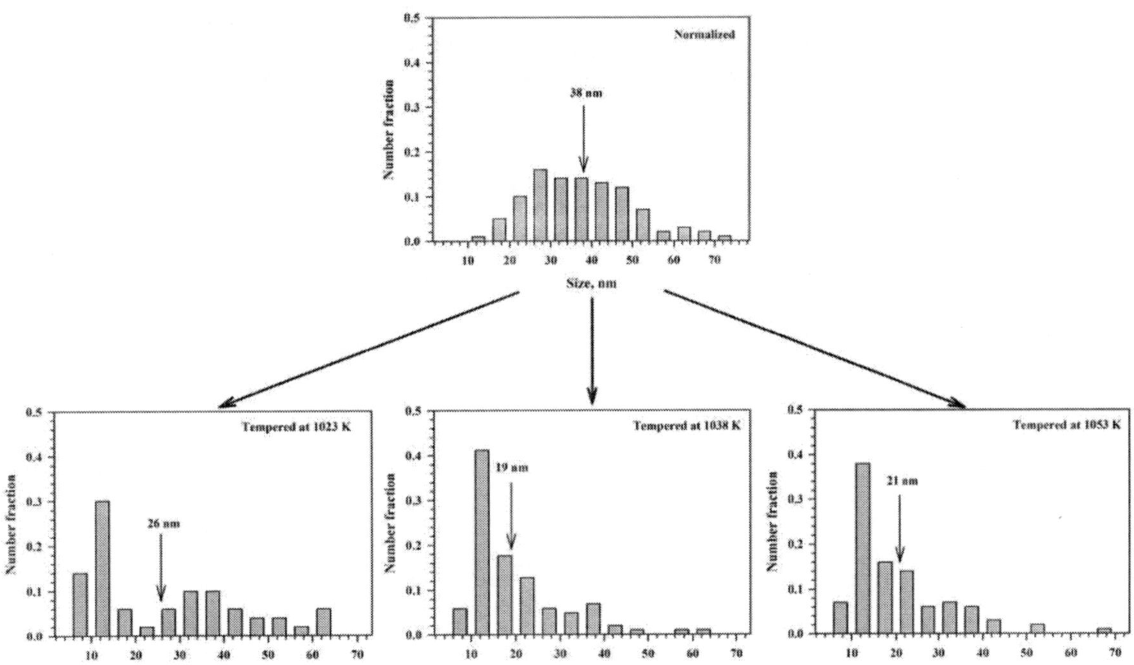

Figure 4. The size distribution of (Nb,Ta)(C,N) particles after tempering at 1023; 1038 and 1053 K. The mean size values are indicated by arrows.

EDX analysis showed that Nb/Ta ratio in (Nb,Ta)(C,N) particles was nearly equal to 50/50 at%. However, it should be noted that in contrast to the coarse primary (Nb,Ta)(C,N) particles, the fine secondary particles of this phase are characterized by increased Fe and Cr content as illustrated in figure 5.

IOP Conf. Series: Materials Science and Engineering **525** (2019) 012049 doi:10.1088/1757-899X/525/1/012049

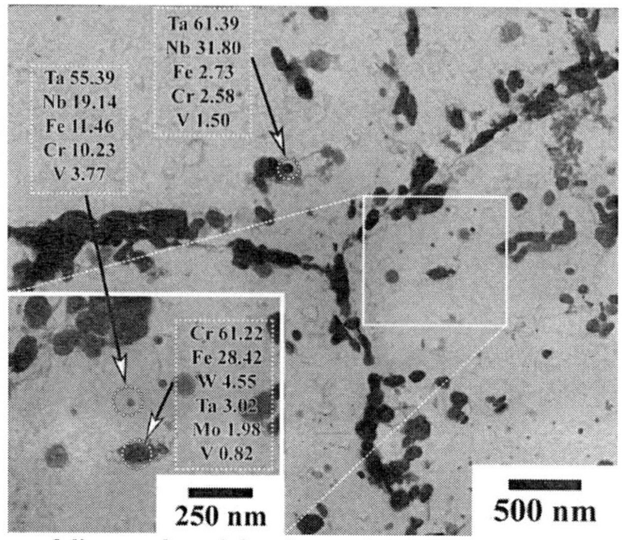

Figure 5. TEM image of dispersed particles that evolved after tempering at 1053 K for 3h. The compositions of particles are shown in wt%.

3.2. *Hardness, impact tests and fractography* The influence of tempering temperature on mechanical behaviour of the studied steel, including hardness and impact toughness, is represented in figure 6.

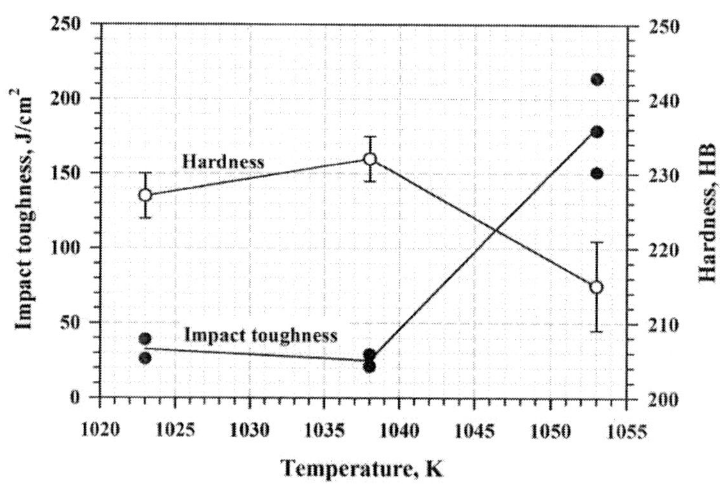

Figure 6. Impact toughness and Brinell hardness of the studied steel tempered at different temperatures.

The hardness of the studied steel after tempering at 1038 K was slightly increased as compared to that after tempering at 1023 K that might be caused by the secondary precipitations. A marked decrease of hardness occured after tempering at 1053 K. This can be attributed to the lath widening and a decrease in the dislocation density. In contrast to hardness, the toughness of the studied steel reached its minimum value of about 25 J/cm^2 after tempering at 1038 K. After tempering at 1053 K the studied steel displaied the high value of impact energy of 181 J/cm^2. Such a significant change in

IOP Conf. Series: Materials Science and Engineering **525** (2019) 012049 doi:10.1088/1757-899X/525/1/012049

the toughness led to quite different morphology of the fractured surfaces (figure 7), that is much similar to other studies [19,20].

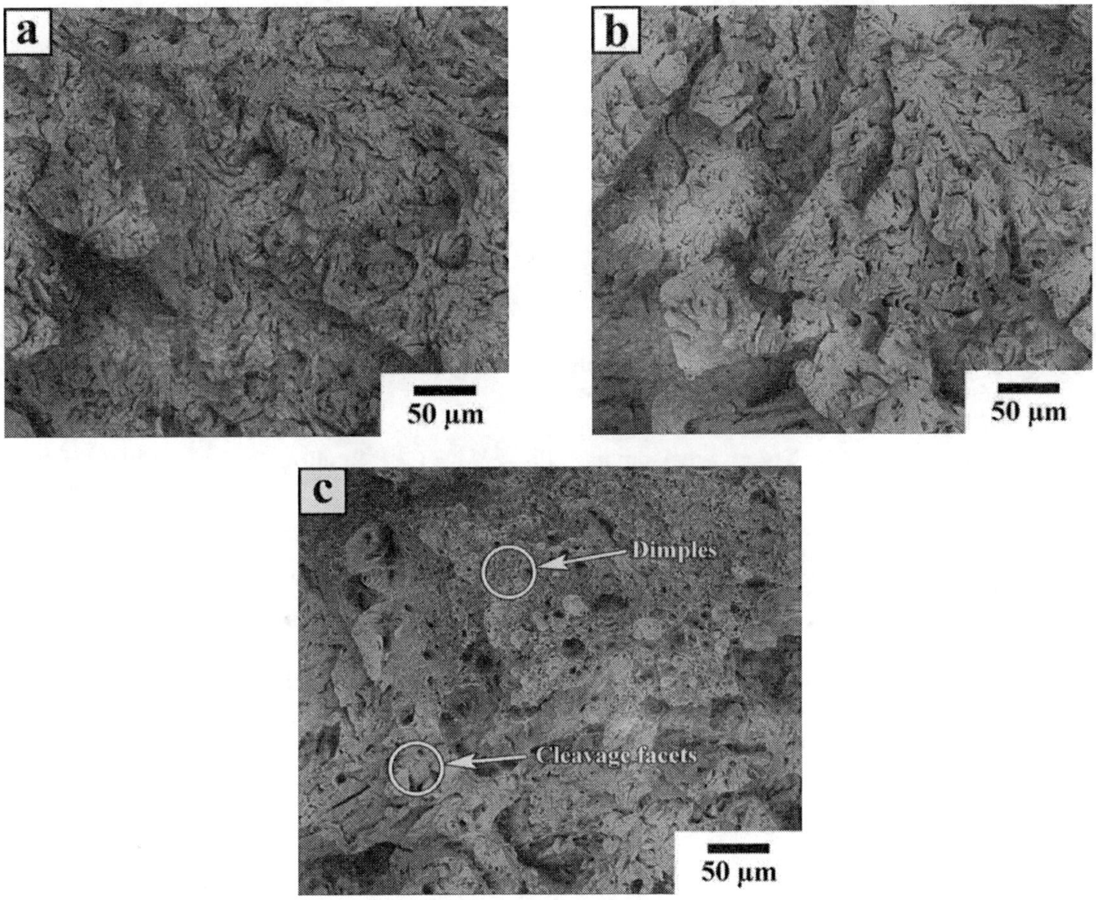

Figure 7. SEM fractographs in crack propagation zone of the studied steel after tempering at 1023 K (a); 1038 K (b) and 1053 K (c).

The SEM analysis of the fractured surfaces of broken Charpy impact specimens revealed the brittle cleavage fracture of the steel samples tempered at 1023 and 1038 K. However, the size of the cleavage facets of the sample tempered at 1023 K is smaller and more irregular, thus the propagation direction of the cleavage crack was deflected more frequently resulting in higher impact toughness. The fracture surface in crack propagation zone of the sample tempered at 1053 K is characterized by presence of ductile dimples tearing along with flat cleavage facets. This indicates the mixed mode of the fracture mechanism.

4. Summary and conclusions

The microstructure and mechanical behavior of a Ta-added 9Cr martensitic steel with high B and low N contents were investigated after normalization and tempering. The following key conclusions can be drawn:

1. The normalization resulted in full martensite structure with thin laths and high dislocation density. The observed secondary phase particles were identified as primary MX-type particles of (Nb,Ta)(C,N) with round shape and Fe_3C platelets.

2. Tempering at 1023-1053 K led to the precipitation of $M_{23}(C,B)_6$ and the secondary (Nb,Ta)(C,N) particles. The most pronounced precipitation of the secondary (Nb,Ta)(C,N) particles was observed at a temperature of 1038 K as revealed by an increase of hardness due to precipitation hardening.
3. Tempering at 1053 K is the most suitable regime for the studied steel because of optimum combination of hardness and toughness.

Acknowledgement
The reported study was funded by RFBR according to the research project № 18-38-00052.

References

[1] F. Abe, T.U. Kern, R. Viswanathan, Creep-Resistant Steels, 2008. doi:10.1533/9781845694012.
[2] F. Abe, M. Taneike, K. Sawada, Alloy design of creep resistant 9Cr steel using a dispersion of nano-sized carbonitrides, Int. J. Press. Vessel. Pip. **84** (2007) 3–12. doi:10.1016/j.ijpvp.2006.09.003.
[3] F. Abe, Precipitate design for creep strengthening of 9% Cr tempered martensitic steel for ultra-supercritical power plants, Sci. Technol. Adv. Mater. **9** (2008) 013002. doi:10.1088/1468-6996/9/1/013002.
[4] R.O. Kaybyshev, V.N. Skorobogatykh, I. a. Shchenkova, New martensitic steels for fossil power plant: Creep resistance, Phys. Met. Metallogr. **109** (2010) 186–200. doi:10.1134/S0031918X10020110.
[5] Z. Hong, X. Zhang, Q. Yan, Y. Chen, A new method for preparing 9Cr-ODS steel using elemental yttrium and Fe2O3oxygen carrier, J. Alloys Compd. **770** (2019) 831–839. doi:10.1016/j.jallcom.2018.08.196.
[6] R. Sahara, T. Matsunaga, H. Hongo, M. Tabuchi, Theoretical Investigation of Stabilizing Mechanism by Boron in Body-Centered Cubic Iron Through (Fe,Cr)23(C,B)6Precipitates, Metall. Mater. Trans. A Phys. Metall. Mater. Sci. **47** (2016) 2487–2497. doi:10.1007/s11661-016-3397-7.
[7] F. Abe, Effect of boron on creep deformation behavior and microstructure evolution in 9 % Cr steel at 650°C, Int. J. Mater. Res. **99** (2008) 387–394. doi:10.3139/146.101650.
[8] N. Dudova, R. Mishnev, R. Kaibyshev, Effect of Tempering on Microstructure and Mechanical Properties of Boron Containing 10%Cr Steel, ISIJ Int. **51** (2011) 1912–1918. doi:10.2355/isijinternational.51.1912.
[9] E. El-Kashif, K. Asakura, K. Shibata, Effects of Nitrogen in 9Cr-3W-3Co Ferritic Heat Resistant Steels Containing Boron., ISIJ Int. **42** (2002) 1468–1476. doi:10.2355/isijinternational.42.1468.
[10] L. Li, R. MacLachlan, M.A.E. Jepson, R. Thomson, Microstructural evolution of boron nitride particles in advanced 9Cr power plant steels, Metall. Mater. Trans. A Phys. Metall. Mater. Sci. **44** (2013) 3411–3418. doi:10.1007/s11661-013-1642-x.
[11] Å. Gustafson, J. Ågren, Possible Effect of Co on Coarsening of M23C6 Carbide and Orowan Stress in a 9% Cr Steel., ISIJ Int. **41** (2001) 356–360. doi:10.2355/isijinternational.41.356.
[12] L. Helis, Y. Toda, T. Hara, H. Miyazaki, F. Abe, Effect of cobalt on the microstructure of tempered martensitic 9Cr steel for ultra-supercritical power plants, Mater. Sci. Eng. A. **510-511** (2009) 88–94. doi:10.1016/j.msea.2008.04.131.
[13] E. Tkachev, A. Belyakov, R. Kaibyshev, Creep behavior and microstructural evolution of a 9%Cr steel with high B and low N contents, Mater. Sci. Eng. A. **725** (2018). doi:10.1016/j.msea.2018.04.032.
[14] A. Fedoseeva, I. Nikitin, N. Dudova, R. Kaibyshev, On effect of rhenium on mechanical properties of a high-Cr creep- resistant steel, Mater. Lett. **236** (2019) 81–84. doi:10.1016/j.matlet.2018.10.081.
[15] R. Mishnev, N. Dudova, R. Kaibyshev, On the origin of the superior long-term creep resistance of a 10% Cr steel, Mater. Sci. Eng. A. **713** (2018) 161–173. doi:10.1016/j.msea.2017.12.066.

[16] F. Abe, Bainitic and martensitic creep-resistant steels, Curr. Opin. Solid State Mater. Sci. **8** (2004) 305–311. doi:10.1016/j.cossms.2004.12.001.
[17] W.B. Jones, C.R. Hills, D.H. Polonis, Microstructural evolution of modified 9Cr-1Mo steel, Metall. Trans. A. **22** (1991) 1049–1058. doi:10.1007/BF02661098.
[18] I. Fedorova, A. Kostka, E. Tkachev, A. Belyakov, R. Kaibyshev, Tempering behavior of a low nitrogen boron-added 9%Cr steel, Mater. Sci. Eng. A. **662** (2016) 443–455. doi:10.1016/j.msea.2016.03.092.
[19] V. Dudko, A. Fedoseeva, R. Kaibyshev, Ductile-brittle transition in a 9% Cr heat-resistant steel, Mater. Sci. Eng. A. **682** (2017) 73–84. doi:10.1016/j.msea.2016.11.035.
[20] V. Dudko, J. Borisova, R. Kaibyshev, Ductile-Brittle Transition in Martensitic 12 % Cr Steel, Acta Phys. Pol. A. **134** (2018) 10–13. doi:10.12693/APhysPolA.134.649.

IOP Conf. Series: Materials Science and Engineering **525** (2019) 012050 doi:10.1088/1757-899X/525/1/012050

Carboxymethylcellulose sodium salt – effective “green” regent for management of calcium carbonate crystallization and natural gas hydrate formation

A V Fakhreeva[1,2], A I Voloshin[3], F F Musin[2], A G Telin[4] and V A Dokichev[1,2]

[1]Ufa State Aviation Technical University, Karl Marks st. 12, Ufa, Bashkortostan, 450077 Russia
[2]Ufa Institute of Chemistry, Russian Academy of Sciences, Oktyabrya pr. 71, Ufa, Bashkortostan, 450054 Russia
[3]OOO RN-UfaNIPIneft’, Sochinskaya st. 12, Ufa, Bashkortostan, 450103 Russia
[4]OOO Ufa-NTC, Aksakov st. 59, Ufa, Bashkortostan, 450076 Russia

E-mail: alsu.allagulova@mail.ru

Abstract. The effect of carboxymethylcellulose sodium salt (NaCMC) with molecular masses (Mw) of 90, 250 and 700 thousand on the crystallization of $CaCO_3$ polymorphs, their size and deposition on the surface of stainless steel from supersaturated aqueous solution at 80 ° C was studied. As a result of specific reversible adsorption, when forming crystalline nuclei formed, it is assumed that NaCMC changes the shape of the growing crystal, preferably forming vaterite — up to 95% in the presence of NaCMC with a molecular mass of 90 thousand during the interaction of $CaCl_2$ and $NaHCO_3$. In the presence of Mg^{2+} ions, $CaCO_3$ is formed mainly in the form of aragonite, and the influence of the molecular mass of NaCMC is significantly reduced. It was established that NaCMC with Mw 90 and 250 thousand inhibits the aggregation and agglomeration of $CaCO_3$ crystals, leading to a decrease in the average size of crystals from 24 to 10.6 and 16.5 microns, respectively. Under the action of NaCMC with a molecular mass of 700 thousand, on the contrary, an increase in the size of crystals to 37.2 μm is observed. It is shown that NaCMC with Mw 90 and 250 thousand slows down the speed and changes the conditions of gas hydrate formation, showing the properties of a thermodynamic and kinetic inhibitor with an efficiency that exceeds methanol by 300-450 times when used in the same dosages. The data obtained indicate the promise of creating new “green” reagents of complex action for oil and gas production on the basis of sodium salt carboxymethylcellulose.

1. Introduction

The formation of calcium carbonate from natural waters is one of the most common processes in nature. In technological processes, such as oil production, the formation of salts from produced water and their deposition on the surface of the well and oilfield equipment are among the most common complications [1-4].

In practice of oil production to prevent scale formation, the chemical method of inhibition using phosphorus-containing oil-field reagents and organic polymers is widely used [2-4], the shortcomings of which include low thermal stability and high toxicity.

Another significant complication that reduces the efficiency of oil production and transportation of hydrocarbons is the formation of gas hydrates [5,6]. Currently, thermodynamic and kinetic inhibitors of gas hydrate formation are used to control the formation of technogenic gas hydrates, among which methanol is most widely used [6-8]. The global trend in hydrate formation management is developing in the direction of replacing toxic methanol with low concentration gas hydrate formation inhibitors, that work effectively in dosages 100-500 times less than methanol (kinetic hydrate formation inhibitors, anti-agglomerates) [9].

The trends in the development of the processes of chemicalization of modern industries, including the extraction of oil and gas, dictate the need for the development of new "green" highly efficient reagents of a complex action. The of carboxymethylcellulose sodium salt (NaCMC) is one of the most accessible water-soluble polysaccharides, the physicochemical properties of which, as well as the availability, thermal stability, and biodegradability in aerobic and anaerobic conditions predetermine the search and creation on its basis of new "green" oil field reagents [10,11]. We have previously shown by gravimetric and electrochemical methods that NaCMC exhibits anticorrosion properties, is a corrosion inhibitor of mixed action - cathodic and anodic [12].

This paper presents the results of studies of the effect of carboxymethylcellulose sodium salt and its molecular weight on the crystallization processes of calcium carbonates, as well as the inhibition of gas hydrate formation of petroleum gases.

2. Experimental part

NaCMC produced by Aldrich with molecular masses of 90 thousand, degree of substitution 0.7 (NaCMC 90), 250 thousand, degree of substitution 0.7 (NaCMC 250), 700 thousand, degree of substitution 0.9 (NaCMC 700) were used.

As a model of saline water (MSW) used water following ion composition: Ca^{2+} - 750; Na^{+} - 863; Cl^{-} - 2662; HCO_3^{-} - 2290 mg/l (MSW 1) and Ca^{2+} - 250; Mg^{2+} - 85; Na^{+} - 3120.25; Cl^{-} - 4492.10; HCO_3^{-} - 1525.50 mg/l (MSW 2) prepared from salts of $CaCl_2$, $MgCl_2{\cdot}6H_2O$, NaCl, $NaHCO_3$, Na_2SO_4 qualification “Pro Analyse”.

The size of $CaCO_3$ crystals formed from the solutions were recorded on a laser diffraction particle size analyzer Analysette 22 NanoTec plus.

The morphology of calcium carbonate particles was studied using the Rigaku Ultima IV x-ray diffractometer. The analysis was performed in Cu-Ka radiation at a voltage of 20 kV and a current of 2 mA with a wavelength l = 1.5405 Å. Registration was carried out in the range of angles 2θ 15...115° with scanning step 0.02°. Processing of diffractograms and phase analysis were carried out in the program "Match!".

The microstructure of the samples was studied by scanning electron microscopy with field emission (FE-SEM) on a Hitachi SU8000 electron microscope. Images were taken in the registration mode of secondary electrons with an accelerating voltage of 2 kV and a working distance of 4-5 mm. The optimization of analytical measurements was carried out in the framework of the approach described earlier [13]. Before shooting, the samples were placed on the surface of an aluminum table with a diameter of 25 mm, fixed with a conductive glue, and a conductive metal layer (Pt/Pd, 80/20) 7 nm thick was deposited on them using the magnetron sputtering method previously described [12]. The morphology of the samples was studied taking into account the correction for the surface effects of deposition of the conducting layer [14].

The efficiency of NaCMC as a scaling inhibitor was determined at the polysaccharide concentration of 10, 20 and 50 mg/l by a method based on blocking the steel capillary formed by calcium carbonate deposits at a temperature of 80°C. A model of mineralized water (MMW) containing Ca^{2+}, Mg^{2+}, Na^{+}, Cl^{-} and HCO_3^{-} ions was pumped at a rate of 1 ml/min through a 2.5 m long stainless steel capillary with an internal diameter of 1 mm and the dynamics of pressure drop in the capillary caused by the deposition of $CaCO_3$ was recorded.

The effectiveness of inhibition (Table 1) was determined by the formula:

$$Z = 100{\cdot}(\Delta P_1 - \Delta P_2)/\Delta P_1$$

where ΔP_1 is the pressure drop on the capillary when pumping water without NaCMC,
ΔP_2-pressure drop on the capillary when pumping water from the NaCMC

The inhibition of gas hydrate formation was studied by rheological method in the high-pressure cell "Thermo" in the cooling mode from room temperature to -5 °C. The cell was pre-injected with natural gas, then the cell was cooled, simultaneously with the temperature decrease, the cell pressure and shear stress were recorded [15].

3. Results and discussion

It is known that NaCMC can form sufficiently stable complexes with Ca^{2+} ion [16-18]. Therefore, assuming that the NaCMC can prevent the precipitation of calcium salts from supersaturated aqueous solutions and inhibit scaling, we studied its effect on the crystallization of calcium carbonate.

The study of the effect of molecular weight of NaCMC on the process of inhibition of carboxymethylcellulose sodium salt scaling by capillary testing at a temperature of 80°C showed that at concentrations of 10, 30, 50 mg/l polysaccharides inhibit the process of $CaCO_3$ scaling (Figure 1, Table 1).

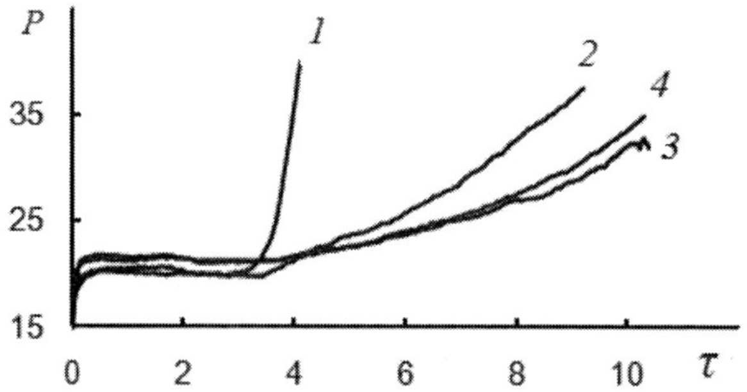

Figure 1. The influence of NaCMC 90 on the process of scaling $CaCO_3$ at concentrations of 0 (*1*), 10 (*2*), 30 (*3*) and 50 (*4*) mg/l at 80 °C. *P* - pressure (kPa), τ - time (min).

Table 1. The dependence of the efficiency of inhibition of calcium deposition on the NaCMC molecular weight.

Molecular mass NaCMC	NaCMC Concentration, mg/l		
	10	30	50
	Inhibition efficiency, %		
90	52	61	60
250	50	61	60
700	20	51	62

All NaCMC samples inhibit the formation of calcium carbonate. However, at low concentrations (10 mg/l), the efficiency of low molecular weight samples is 2.5 times higher.

The molecular mass of NaCMC plays an important role in the formation of calcium carbonate crystals. In the presence of NaCMC, there is a significant change in the size of the resulting $CaCO_3$ crystals (Fig.2). The average size of the formed particles in the presence of NaCMC 90 decreases from 24.0 to 10.6 μm, for NaCMC 250 to 16.5, and NaCMC 700, on the contrary, increases the resulting calcium carbonate crystals to 37.2 μm.

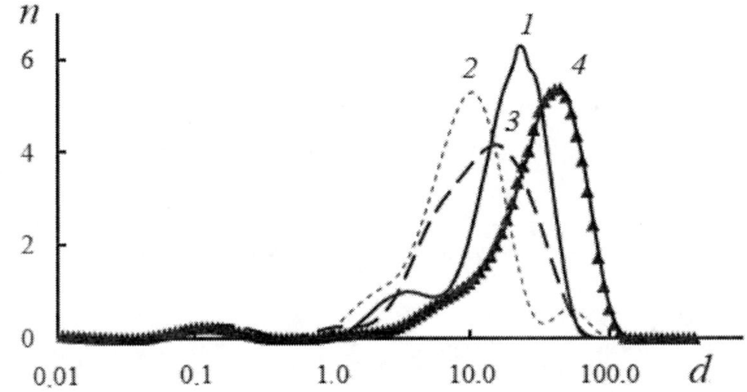

Figure 2. Influence of molecular weight of NaCMC on the size distribution of $CaCO_3$ crystals formed from the solution of MSW 2 in the presence of 30 mg/l of NaCMC: without NaCMC (*1*), NaCMC 90 (*2*), NaCMC 250 (*3*), NaCMC 700 (*4*). *n* - volume fraction of crystals (%), *d* – diameter (μm).

Studies on the x-ray diffractometer "Rigaku Ultima IV" showed a significant effect of the molecular weight of NaCMC and the composition of saline water on the phase composition of calcium carbonate crystals (Figure 3a, b; Table. 2.3).

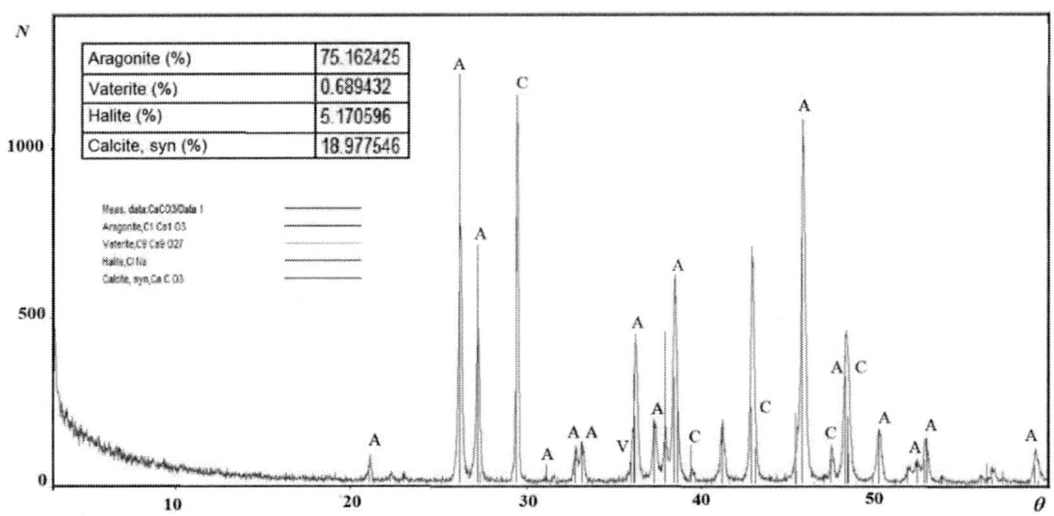

Figure 3a. Diffractogram of $CaCO_3$ crystals obtained without NaCMC in MSW 1. θ - deflection angle (deg), N – intensity (cps).

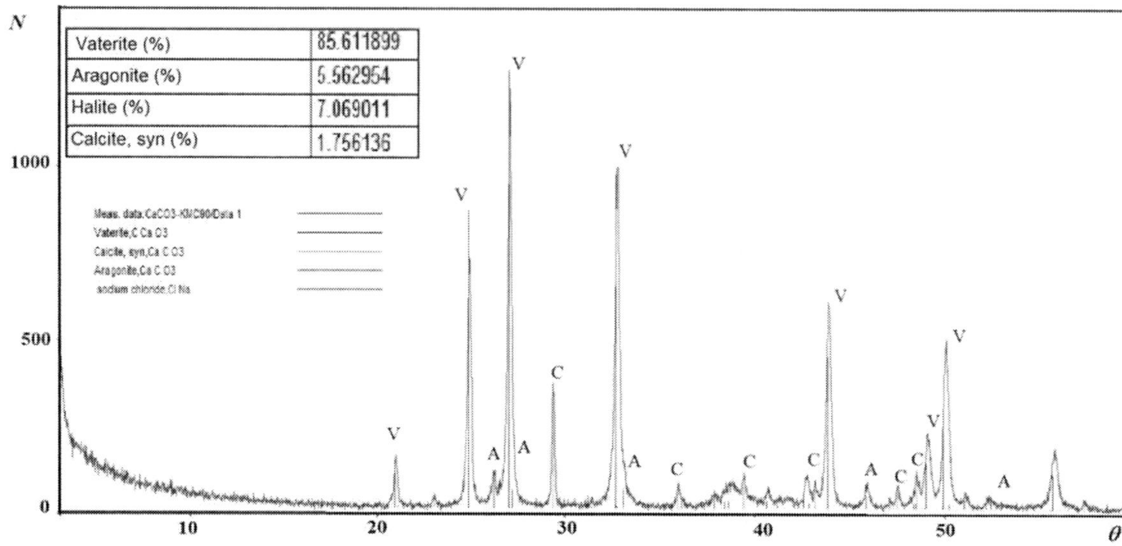

Figure 3b. Diffractogram of $CaCO_3$ crystals obtained in the presence of NaCMC 90 in MSW 1. θ - deflection angle (deg), N – intensity (cps).

In MSW 1 and MSW 2 without polysaccharides, mainly aragonite and calcite are formed. In the presence of NaCMC, the amount of aragonite increases from 3 to 7% depending on the molecular weight of NaCMC, and calcite is replaced by vaterite (Table 3). The addition of a low molecular weight sample of a polysaccharide to the solution during the interaction of $CaCl_2$ and $NaHCO_3$ makes it possible to obtain vaterite in a yield of up to 86%, while an increase in the molecular weight of NaCMC reduces the yield of vaterite to 59%. In the presence of Mg^{2+} ions, $CaCO_3$ it is formed mainly in the form of aragonite.

Table 2. Dependence of the phase composition of crystalline $CaCO_3$ on the molecular weight of NaCMC, obtained in MSW 1.

NaCMC	Calcite, %	Aragonite, %	Vaterite, %	Halite, %
without NaCMC	19	75	1	5
90	5	7	86	2
250	11	10	77	2
700	12	27	59	2

Table 3. Dependence of the phase composition of $CaCO_3$ on the molecular weight of NaCMC, obtained in the MSW 2.

NaCMC	Calcite, %	Aragonite, %	Vaterite, %	Halite, %
without NaCMC	10	88	1	1
90	2	95	3	-
250	2	94	4	-
700	6	91	2	1

The data obtained by electron microscopy are consistent with the results of x-ray diffraction analysis. On SEM images, calcium crystals formed during crystallization represent irregularly shaped structures in the form of threads and stars, while crystals obtained in the presence of a polysaccharide have a spherical shape characteristic of vaterite (Figure 4).

a *b*

10 µm 10 µm

Figure 4. Electron micrograph of $CaCO_3$ crystals, obtained without (*a*) and in the presence of (*b*) NaCMC 90.

A sample of $CaCO_3$ obtained from MSW 1 and MSW 2 in the absence of a polysaccharide contains elongated crystals of aragonite, which under selected conditions is the thermodynamically most stable phase (Figure 4). During the crystallization of calcium carbonate in the presence of all NaCMC samples, the particles of aragonite acquire an irregular shape with rounded edges. They cause low adsorption on the metal surface of pipes and equipment.

The mechanism of interaction between NaCMC and calcium salts is rather complicated. Probably, the main reason for the influence of NaCMC on crystal formation can be considered the specific adsorption of NaCMC on the forming faces of salt crystals, both due to the electrostatic interaction of ionized carboxyl groups with Ca^{2+} ions located on the crystal surface, and due to coordination and hydrogen bonds with oxygen atoms and OH- groups of D-glucose fragments [19]. Such an interaction leads to the formation of polymorphic forms without pronounced morphological signs and changes in the size of the particles of calcium carbonate.

4. Inhibition of gas hydrate formation

Inhibiting formation of hydrates the ability of the NaCMC investigated in terms of quasi-equilibrium thermodynamic experiment

As a hydrate-forming model medium, two gas compositions were used: 99.5% methane and 0.5% ethane – the composition of 1, 93% methane and 7% ethane – the composition of 2. Composition 1 was used to study the effect of methanol concentration, and composition 2 was used to study the effect of NaCMC concentration. Phase curves of gas-hydrate formation for composition 1 and 2 were calculated using the software package HWHYD (Heriot-Watt University). The efficiency of NaCMC as an inhibitor of gas hydrate formation was evaluated by the ratio of the concentrations of polysaccharide and methanol with the same effect of changing the pressure of the beginning of hydrate formation.

Among the studied samples of NaCMC with different molecular weight (90, 250 and 700 thousand) the highest inhibitory activity was demonstrated by polysaccharide with molecular weight of 90 000 at concentrations of 0.005, 0.01 and 0.05% (Figure 5). It should be noted the influence of the molecular weight of NaCMC on the process of gas hydrate formation: the compound with a mass of 700 thousand had no effect on the formation of hydrates. The of carboxymethylcellulose sodium salt alters the conditions of hydrate formation, showing the thermodynamic properties of the inhibitor, the efficiency of which is superior to methanol in 350 times. The reagent is not only an effective thermodynamic, but also a kinetic inhibitor of gas hydrate formation, since with the increase in the concentration of polysaccharide, the pressure drop of gas hydrate formation increases and the rate of gas hydrate formation decreases.

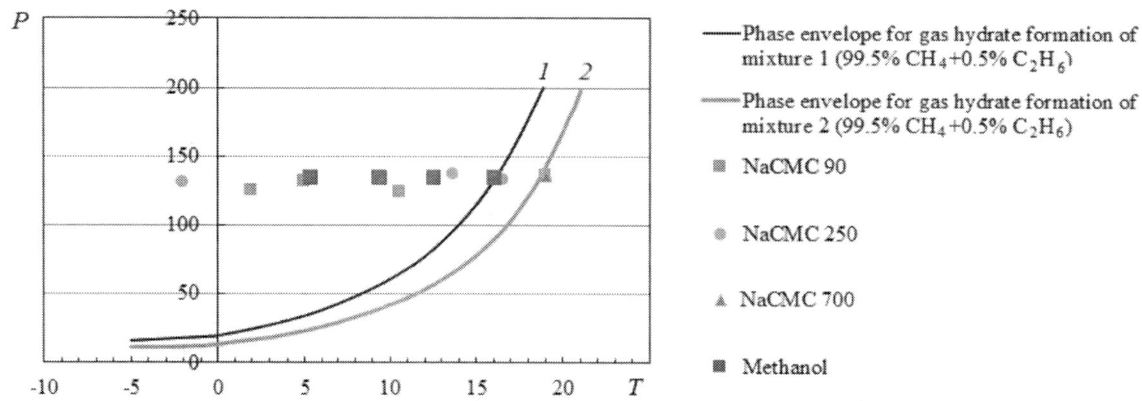

Figure 5. The effect of the NaCMC and methanol on the hydrate formation of gas composition of *1* (the effect of the NaCMC) and composition *2* (the effect of methanol). *P* - pressure (bar), *T* - temperature (°C)

Probably, inhibition of gas hydrate formation occurs as a result of electro-static interaction of ionized carboxyl groups of NaCMC and sodium cations with water molecules, and the formation of hydrogen bonds with oxygen atoms and OH-groups of D-glucose fragments [20].

5. Conclusion

Thus, the influence of molecular weight of NaCMC on the process of crystallization of calcium carbonate was investigated. A low molecular weight sample of NaCMC with a molecular weight of 90 000 allows obtaining a metastable form of calcium carbonate-waterite. At the same time, in the presence of magnesium ions, waterite is completely replaced by aragonite. The of carboxymethylcellulose sodium salt alters the conditions of hydrate formation showing the properties of a thermodynamic inhibitor, with efficacy superior to methanol in 350 times.

Acknowledgement

The results were obtained within the framework of the state task of the Russian Ministry of Education and Science (project No. 4.2703.2017/ПЧ) on No. AAAA-A17-1170011910021-8 and the financial support of the grant of the Russian Scientific Foundation (project 14-33-00022).

References

[1] Crabtree M, Eslinger D, Fletcher P et al. 2002 *J. Oilfield Rev.* **11 N3** 30–45
[2] Olajire A A 2015 *J. Pet. Sci. Eng.* **135** 723–737
[3] Pereyma A A 2015 *Oil Industry* **2** 84–87
[4] Voloshin A I, Gusakov V N, Fakhreeva A V et al. 2018 *Oilfield Engineering* **N11** 60-72
[5] Birchwood R., Dai J., Shelander D. et al. 2010 *Oilfield Rev.* **22 N1** 18-33
[6] Carroll J 2014 *Natural Gas Hydrates. Gulf Professional Publishing* p 340
[7] Kelland M A 2017 *Energy Fuels* **31 N5** 5046-5054
[8] Grunvald A V 2007 *Oil and Gas Business* **2** 1-25
[9] Perrin A, Musa O M and Steed J W 2013 *Chemical Society Reviews* **5** 1913-2204
[10] Fraser-Reid B O, Tatsuta K, Thiem J et al. 2008. *Glycoscience. Chemistry and Chemical Biology,* 2nd ed.; Springer-Verlag; Berlin-Heidelberg-New York
[11] Fakhreeva A V, Gusakov V N, Voloshin A I et al. 2016 *Russ J Appl Chem* **89 N12** 1541-1545
[12] Chernyaeva E Yu, Sayapova V V, Alimbekova S R et al. 2018 *Corrosion: materials, protection* **5** 32–36
[13] Kachala V V, Khemchyan L L, Kashin A S et al. 2013 *Russ. Chem. Rev.* **82 N7** 648–685
[14] Kashin A S and Ananikov V P 2011 *Russ. Chem. Bull., Int. Ed.* **60 N12** 2602 – 2607

[15] Dokichev V A, Fakhreeva A V, Voloshin A I et al. 2018 *Equipment and technologies for oil and gas complex* **N5** 43-48
[16] Rao A, Berg J K, Kellermeier M et al. 2014 *Eur. J. Mineral.* **26** 537-552
[17] Zakharov N A, Ezhova Zh A, Koval' E M. et al. 2005 *Inorg. Mater.* **41** 509-515
[18] Backfolka K, Lagergeb S, Rosenholmc J et al. 2002. *J. Colloid Interface Sci.* **248** 5-12
[19] Sawada K 1997 *Pure Appl. Chem.* **69** **N5** 921–928
[20] Altamash T, Qureshi M F, Aparicio S. et al. 2017 *J. Nat. Gas Sci. Eng.* **46** 873-883.

IOP Conf. Series: Materials Science and Engineering **525** (2019) 012051 doi:10.1088/1757-899X/525/1/012051

Investigation of multi-tip large area emitters using computerized field emission projector

S V Filippov[1], A G Kolosko[1], R M Ryazanov[2], E P Kitsyuk[2] and E O Popov[1]

[1]Ioffe Institute, ul. Polytechnitscheskaya 26, St.-Petersburg, 194021, Russia
[2]Scientific-Manufacturing Complex "Technological Centre", Shokin Square, Bld. 1/7, Zelenograd, 124498, Russia

E-mail: s.filippov@mail.ioffe.ru

Abstract. The paper presents an original approach to the numerical analysis of emission patterns of the flat field emission projector. A system for testing and numerical characterization of the emission properties of field cathodes was developed. A sample of the multi-tip large area field emitter based on non-oriented carbon nanotubes grown on a tungsten substrate was studied by the emission glow pattern analysis.

1. Introduction

The development of manufacturing technologies for conducting nanoscale structures (carbon nanotubes, graphene plates, nano-diamonds, etc.) made it possible to use such structures as field sources of free electrons for the needs of vacuum micro- and nanoelectronics. Multi-tip emitters based on carbon nanotubes (CNT) have proven themselves in the best way. This is due to the high aspect ratio of CNTs, their high thermal conductivity and the ability to withstand high current densities (up to 10^7 A/cm^2).

Today, prototypes of devices using CNT-based emitters exist: vacuum electrodynamic tethers [1], compact mass spectrometers [2], vacuum sensors [3], medical tomographs [4] and etc. However, a number of related factors affecting emissions limit the stability and the lifetime of this type of field emitters. The distribution of the current loading over the set of individual nanoscale emission sites increases their individual stability, however, the stochasticity of the geometric parameters and the mutual position of the sites leads to the occurrence of explosive emission.

Investigation of the instability of the total emission current and its fluctuations of multi-tip field emitters requires observation of the state and evolution of their surface.

The main methods for obtaining information about the surface of a nanostructured emitter are scanning electron microscopy (SEM) [5], field emission scanning electron microscopy (FE-SEM) [6], field emission microscope with a scanning anode (SAFEM) [7]. However, the most accessible in terms of ease of data analysis and technical implementation is the method of direct observation of emission sites on the emitter surface – field emission projector (FEP) [8]. Computer processing of the glow patterns registered with the FEP, together with the registration and online processing of the field emitter current-voltage characteristics (IVC), opens up broad possibilities for obtaining information on the emission ability and the interaction of individual emission sites.

For example, in [9], digitized FEP luminescence patterns were used to calculate the field emission area of individual emission sites. In [10], instead of the luminescent coating on the anode, resistive PMMA thin film was used, which is sensitive to the flux of electrons emitting from the surface of the field cathode. Using this original technique, histograms of the distribution of microscopic emission sites by heights were constructed. At the same time, SEM pictures were used for surface analysis and the IVC of the sample were recorded.

In [11], the YAG: Ce (yttrium aluminum garnet doped with cerium) coated with a Mo film was used as a fluorescent anode in FEP. Using FEP field emission properties of nanodiamond films were studied. Image analysis of individual emission sites made it possible to determine the nonlinear dependence of the effective emission area on the applied electric field. The effect of saturation of the emission current with increasing voltage for semiconductor emission structures was also demonstrated.

In [12], a computerized characterization system for multi-tip field emitters using FEP in conjunction with recording voltage and emission current levels is presented. It is proposed to use the total brightness of pixels corresponding to individual emission sites in the glow pattern as weighting factors when evaluating their individual emission current. The histograms of the pixel spread of individual emission sites over the current level were plotted.

In [13], ZnO nanowire circle pad arrays and coplanar-gate circle pad arrays were studied. From the luminescence patterns, the uniform distribution of the geometric characteristics of ZnO nanowires was determined. The macroscopic IVC in the Fowler-Nordheim coordinates made it possible to estimate the average field enhancement factor (FEF) of nanoscaled electron sources. A numerical estimate of the temporal stability of the emission current level was also made. The receipt of SEM pictures of the surfaces studied accompanied all measurements.

The purpose of this work was to study the emission properties of a multi-tip field emitter based on non-oriented CNTs using a computerized FEP.

2. Experimental

To study the emission properties of multi-tip field emitters, a unique experimental technique [14] was used. It is based on multichannel registration and online processing of field emission data. IVCs were recorded in the "fast" mode (one IVC for 20 ms) in two configurations: emitter with a metal anode (diode) and an anode coated with a conductive ITO layer on the glass and a layer of luminescent substance SVETOSOSTAV K-49. The latter configuration is equipped with a long-focus USB microscope mounted outside the vacuum chamber on a vacuum window and is called the FEP. The microscope records the glow patterns that appear on the luminescent coating and reflect the distribution of emission sites over the cathode surface. In fig. 1 schematically shows the device of the cathode systems of the measuring setup.

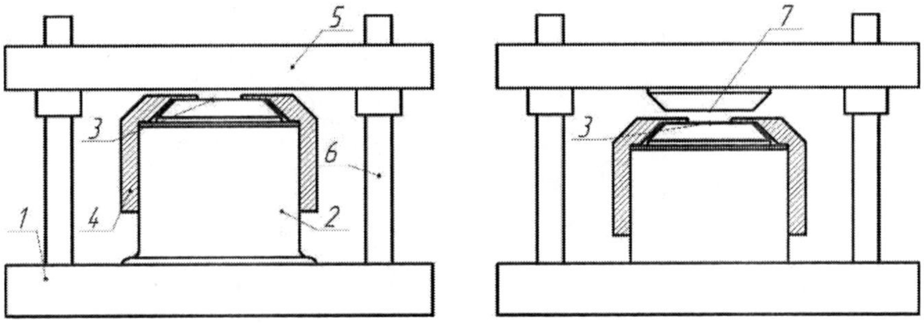

Figure 1. Scheme of experimental setup: (a) diode configuration and (b) FEP. 1 – Base, 2 – cathode holder, 3 – emitter on the substrate surface, 4 – guarding diaphragm, 5 – glass covered with ITO and luminescent powder, 6 – isolation stands, 7– tantalic anode.

The developed procedure for characterization of emitters is carried out in several stages:

1. Determination of the voltage of the emission current occurrence (above a given level, in our case it is 100 nA).
2. IVC registration with a maximum emission current at level of 10 μA or 100 μA, or 1 mA.

IOP Conf. Series: Materials Science and Engineering **525** (2019) 012051 doi:10.1088/1757-899X/525/1/012051

3. Recording the time dependence of the current and voltage levels during the so-called sample training (activation and aging of the emitter surface under the influence of a high electric field in diode configuration with a metal anode). Estimating of the maximum emission current achievable at sample training.

4. Determination of the voltage of the emission current occurrence of the trained sample.

5. IVC registration at stable level of the pulse amplitude of the emission current of 10 μA or 100 μA, or 1 mA.

6. Evaluation of the temporal stability of the emitter at current level of 1 mA.

7. Determination of the effective parameters of the sample - the emission area (A_{eff}) and the field enhancement factor (β_{eff}) with a stable current level of 1 mA, the construction of histograms of fluctuations of these parameters.

8. Characterization of the change in the emissivity of the sample with a sequential increase and decrease in the voltage level: registration of the change in the shape of the IVC with a smooth change in U and registration of the kinetics of a change in the current level I with a step change in U.

9. Replacing the metal anode with an anode with luminescent layer (FEP) and registering the pattern of emission site distribution over the cathode surface.

10. Determining the number and location of emission sites, their maximum emissivity and construction a histogram of the distribution of brightness sites.

11. The calculation of the local FEFs of the found sites and plotting corresponding histogram.

12. Construction of histograms of emission sites distribution over the cathode surface, numerical evaluation of its uniformity.

The studied sample of a multi-tip field cathode consisted of non-oriented CNTs. Tungsten was used as a substrate, on the surface of which thin metallic layers of Ti (10 nm thick) and Ni (2 nm thick) were deposited. These layers served as a catalyst for the synthesis of CNT by the PECVD method. The synthesis took place at the following parameters: temperature 680 °C, pressure 2 Torr, synthesis time 10 min. A mixture of Ar, H_2, C_2H_2, NH_3 was fed into the chamber. The diameter of the grown CNTs was 10–20 nm, the length was up to 3 μm. Figure 2 shows SEM images of a slice of the substrate with CNTs grown on it.

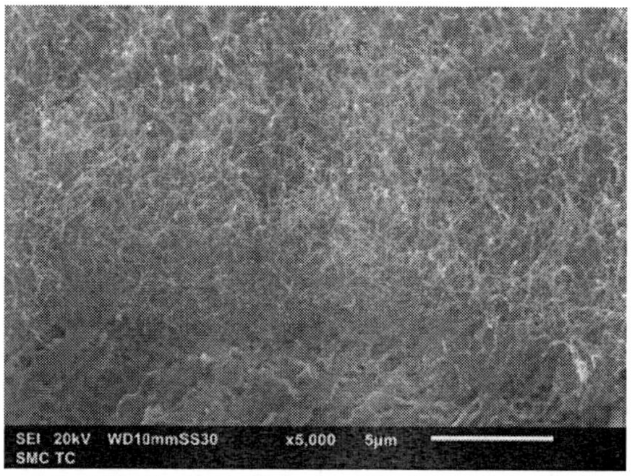

Figure 2. SEM image of the emitter surface at 45° angle.

3. Results

Before measuring the main emission characteristics of an as-prepared emitter, a high-voltage training procedure was required to activate and clean the surface of emission sites, as well as to shorten the highest and most unstable emission sites with a vacuum discharge. For this, a diode configuration with a flat metal anode and an interelectrode distance of 650 μm was used (taking into account the thickness of the metal diaphragm pressing the sample).

The change in the state of the emitter as a result of training is characterized by a change in the voltage value corresponding to the appearance of the emission current against the background of the noise of the measurement setup, as well as the IVC with a current amplitude of 1 mA (see Fig. 3a). The voltage of the emission of the as-prepared sample was U_{tho} = 350 V for current in I_{app} ~ 100 nA, and for the trained sample - U_{tho} = 900 V.

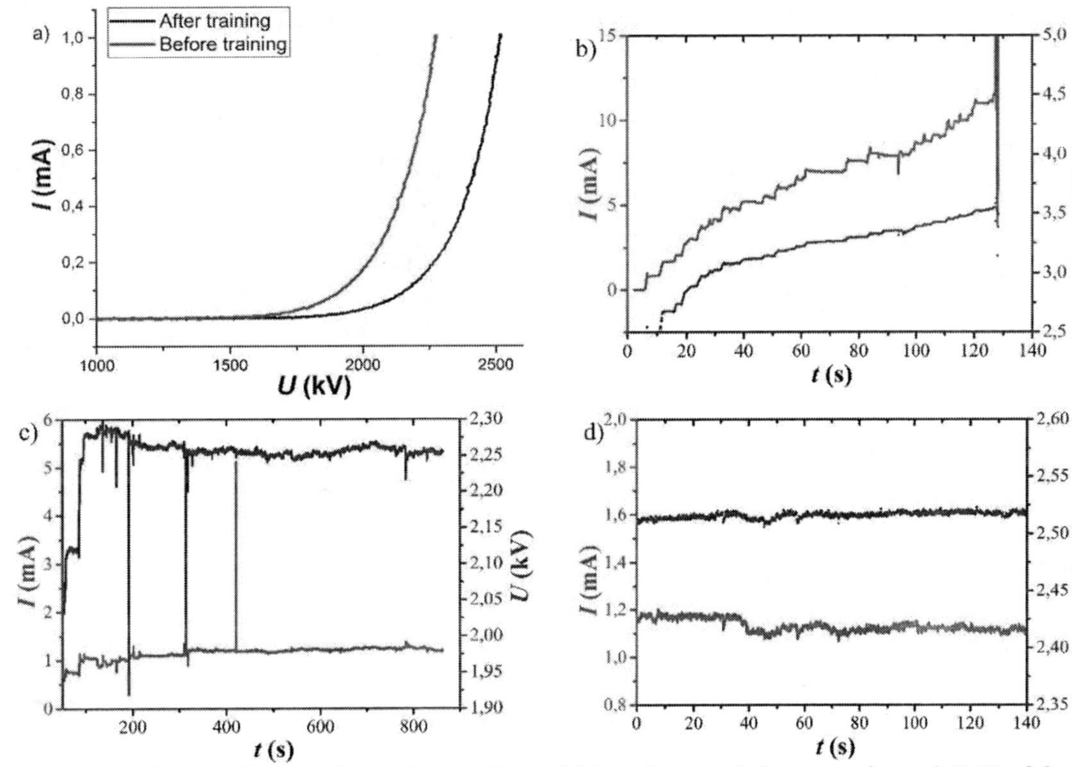

Figure 3. Emitter characteristics change due to high voltage training procedure: a) IVC of the emitter before and after training, b) changes in the amplitude of voltage pulses and corresponding emission current pulses during sample training, c), d) temporal stability of the emitter current at 1 mA before and after training.

Fig. 3b shows the process of training of the sample (the dependence of the current level on time). The maximum current reached during training was 11.5 mA. Above this level, discharges observed in the form of flashes in the interelectrode space steadily appeared, and the system could not maintain emission at a given current level.

Fig. 3 c, d represents the time dependences of the current and voltage levels recorded before and after training the sample at a current level of 1 mA. Sharp current level surges correspond to short-term vacuum discharges, because of which the highest emission sites were destroyed. The threshold voltage of the sample increases because of this, the total current decreases, but the probability of a vacuum discharge is reduced and the temporal stability of electron emission increases.

After activation of the emitter surface, the stability of its emission current over time was estimated (see Fig. 3d). Stability was assessed using the formula:

$$S = \frac{I_{max} - I_{min}}{I_{orig}} \cdot 100\%$$

where I_{max} and I_{min} are the maximum and minimum levels during the observation of current stability, respectively, and I_{orig} is the initial current level.

At amplitude of current pulses of 1 mA in 10 minutes, the stability S=14% was obtained.

The effective parameters of the emitter were estimated at stable current level from IVC in the Fowler – Nordheim coordinates (IVC-FN), [15]. Figure 4 presents the statistics of fluctuations of the effective enhancement factor β_{eff}, which indirectly characterizes the spread of emission characteristics over the sample surface [16]. As the voltage level increases, the β_{eff} values shift towards to lower values, which is explained by the inclusion of lower emission sites in the emission process.

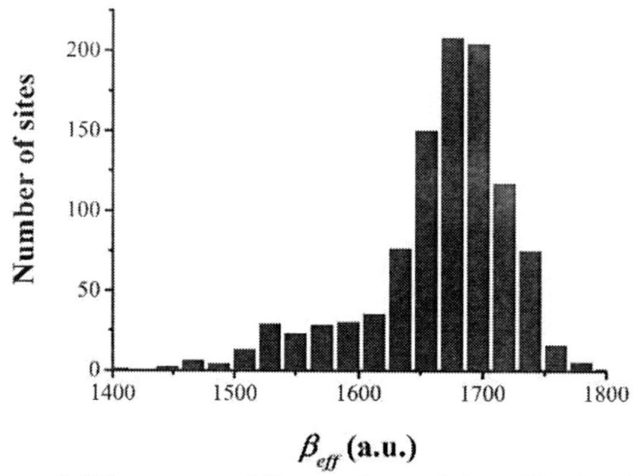

Figure 4. Histogram of fluctuations of the effective field enhancement factor of a CNT emitter at 200 μA current level.

The mean values of the parameters were as follows: for a current level of 1 mA: $S_{eff}\sim 5.0\cdot 10^4$ nm^2 and $\beta_{eff}\sim 1520$, and for a current level of 200 μA: $S_{eff}\sim 4.8\cdot 10^4$ nm^2 and $\beta_{eff}\sim 1650$.

In Fig. 5 shows the IVC for the current level of 200 μA and the corresponding trend line. Note that all registered in the fast mode IVC pass the Forbes test for compliance of the experimental data with "pure" field emission [17].

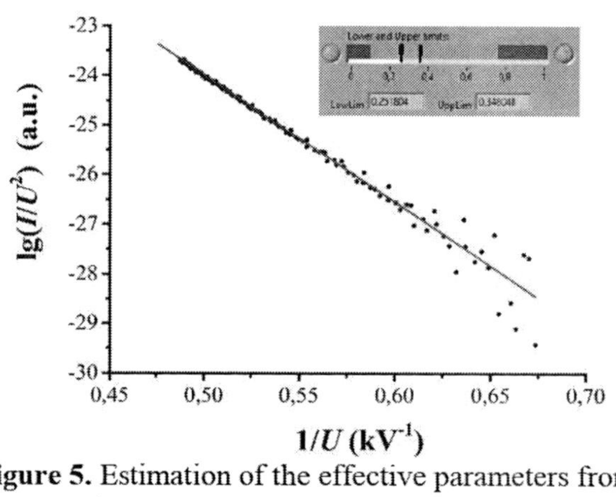

Figure 5. Estimation of the effective parameters from the IVC-FN (200 μA current level).

Figure 6 demonstrates the dynamic characteristics of the change in the emissivity of the sample with a stepwise increase and decrease in the voltage level. At the same time, there is an exponential increase and decrease in the level of the emission current when the amplitude of impulses of applied voltage reaches a new level. A cyclic change in the IVC shape is also observed, which is associated with a change in the concentration of volatile products in the interelectrode space and on the cathode surface [18].

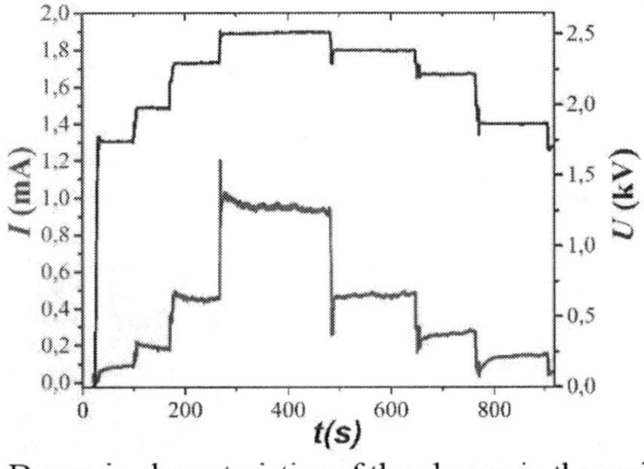

Figure 6. Dynamic characteristics of the change in the emissivity of the sample with a step increase and decrease in the voltage level.

The distribution of emission sites over the cathode surface and their local emission characteristics was studied in another diode system with a luminescent anode (FEP). Figure 7 presents the FEP emission pattern at different current levels (130, 250, 450 and 1000 μA).

Due to the instability of the luminescent coating to electron bombardment, obtaining high emission currents in such a configuration is impossible, moreover, at high currents, the image of emission sites begin to overlap and cease to be distinguishable. Therefore, the current levels in the study did not exceed 5 mA.

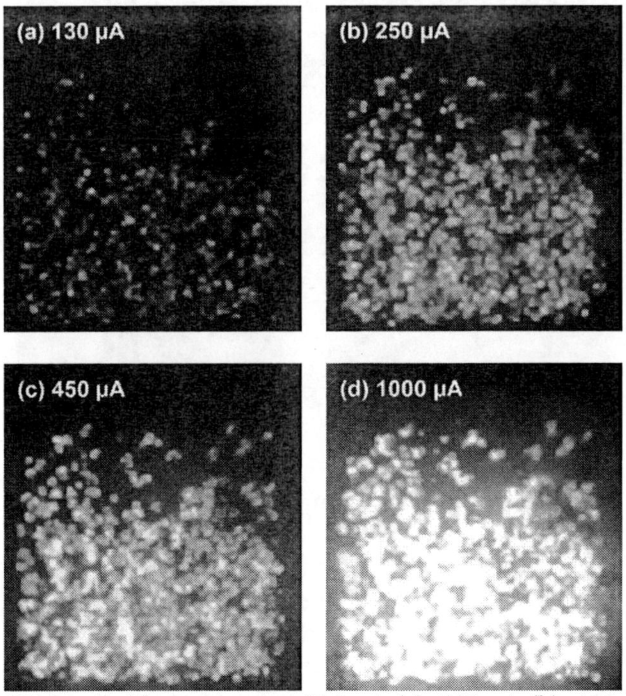

Figure 7. Glow pattern at 130 μA (a), 250 μA (b), 450 μA (c), 1000 μA (d) emission current levels.

Collection of emission sites was made for I = 220 μA. In fig. 8 the dependence of the number of registered emission sites on time was shown.

After the amount of registered emission sites reached a relatively stable level, we can assume that all the main emitting sites of the cathode were found. Then, each of the site was monitored to determine its maximum emissivity. After the output of the sum of the maximum brightness of the sites found at a relatively stable level, we can assume that all maximum brightness and currents are found.

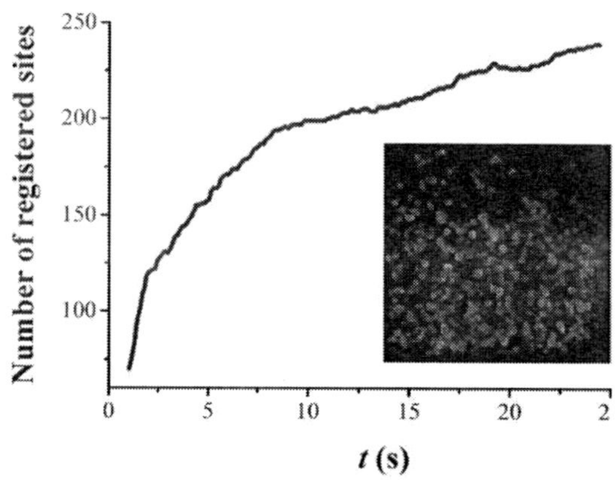

Figure 8. Dependence of the number of registered emission sites on time at current level of 220 μA.

The distribution of microscopic emission sites of the emitter is determined from these currents by the magnitude of the FEF (see Fig. 8b). The method of online calculation of local emission currents and FEFs of individual emission sites was described in detail in [19, 20]. It is based on the idea of the brightness of the emission site in the luminescence pattern as the weight coefficient of the current contribution of this site. Then, numerically solving the Fowler-Nordheim equation with respect to FEF for a given work function (for CNTs, it is 4.6 eV) and area emission of one emission center (was chosen equal to 200 nm^2 based on the ratio of the effective FEA and the number of centers found).

The values of the obtained local FEF correlate with the effective FEF (~ 1650).

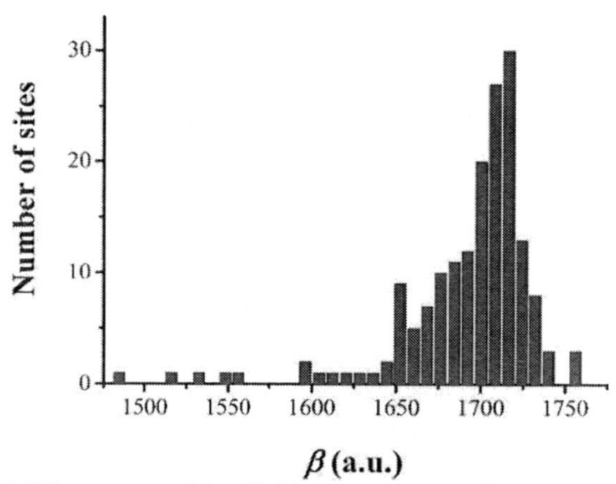

Figure 9. Histogram of the field enhancement factor distribution.

Figure 10 shows the result of the analysis of the emission site distribution over the cathode surface - the radial distribution diagram. Since the technique is designed for samples with circular substrates, and the sample holding diaphragm had a square hole, the analysis was performed for a circular zone inscribed in this square.

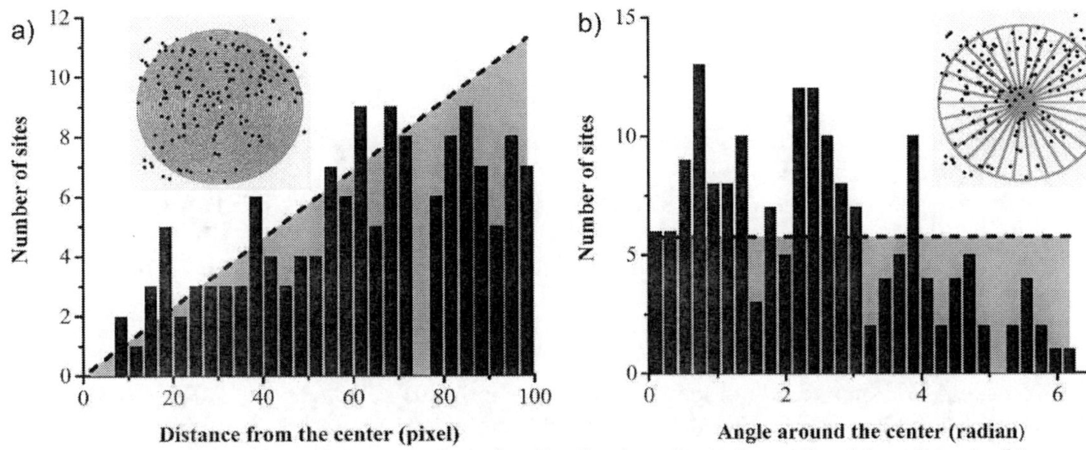

Figure 10. Chart of the emission site distribution along the radius (a) and angle (b).

Similarly, diagrams of the site distribution of brightness and a diagram of the distribution of total current loading of surface segments over the whole sample were obtained. Using these diagrams, the correlation coefficient R_{cor} and the covariance coefficient V_{cov} were calculated, which make it possible to estimate the ideality of the site distribution [19]. For this sample, they were: $R_{cor} = 53\%$, $V_{cov} = 63\%$. This numerical estimate allows directional optimization of the surface quality of the cathodes.

4. Conclusion

The paper presents a characterization of a sample of a multi-tip field cathode based on non-oriented CNTs grown on a multilayer substrate by the PECVD method. The developed system for testing and numerical characterization of the emission properties of field cathodes includes: a description of the sample activation process by high voltage, an assessment of the temporal stability of the emission current, as well as its dynamic characteristics when the voltage level changes, checking for compliance of the emission signal with the pure field emission mode, plotting fluctuation histograms effective parameters, the construction of the emission site distribution by the values of the local field enhancement factor and, finally, the estimate uniformity of emission sites distribution over the cathode surface and the level of their emission activity.

The constructed histograms of the distribution of microscopic emission sites by the magnitude of the local field enhancement factor showed good agreement with the effective value of this parameter. Despite the fact that the macroscopic effective parameters and local estimates of β are linked at the level of estimates of macroscopic and microscopic emission areas, the method of decomposing the effective β_{eff} into many local β has great prospects for understanding and analyzing the fundamental laws of operation of multi-tip field emitters.

References

[1] Ohkawa Y, Okumura T, Iki K, Okamoro H and Kawamoto S 2018 *31st Int. Vac. Nanoelectronic Conf.* Kyoto, Japan 09-13 july 40-41

[2] Yang Ch and Velásquez-García L F 2019 *J. Phys. D: Appl. Phys.* **52** 075301-1-9

[3] Rudyk N N, Il'in O I, Il'ina M V, Fedotov A A, Klimin V S and Ageev O A 2017 *J of Phys.: Conf. Series* **917** 082008-1-5

[4] Mavalankar A, Cameron J and Travish G 2017 *30th Int. Vac. Nanoelectronics Conf.* Regensburg, Germany 300-301

[5] Lin P-H, Sie C-L, Chen C-A, Chang H-C, Shih Y-T, Chang H-Y, Su W-J and Lee K-Y 2015 *Nanoscale Res. Lett.* **10** 297

[6] Chiu C-C, Tsai T-Y and Tai N-H 2006 *Nanotechnology* **17** 2840–4

[7] Bandurin D A, Kleshch V I, Smolnikova E A, Obronov I V, Nasibulin A G, Kauppinen E I and Obraztsov A N 2013 *J. of Nanoelectronics and Optoelectronics* **8** 114–118
[8] Pandey A, Prasad A, Moscatello J P, Engelhard M, Wang C and Yap Y K 2013 *ACS Nano* **7** 117–25
[9] Patra R, Singh A, Vankar V D, Ghosh S 2016 *Adv. Mater. Lett.* **7(10)** 771-776
[10] Li Y, Sun Y, Jaffray D A and Yeow J T W 2017 *Nanotechnology* **28** 155704 1-12
[11] Chubenko O, Baturin S S, Kovi K K, Sumant A V and Baryshev S V 2017 *ACS Appl. Mater. Interfaces* **9** 33229–3323
[12] Kopelvski M M, Galeazzo E, Peres H E M, Ramirez-Fernandez F J, Silva D A C and Dantas M O S 2016 *Measurement* **93** 208–214
[13] Zhao L, Chen Y, Zhang Z, Cao X, Zhang G, She J, Deng S, Xu N and Chen J 2018 *Sci. Reports* **8** 12294
[14] Popov E O, Kolosko A G, Filippov S V, Romanov P A and Fedichkin I L 2018 *Materials Today: Proceedings* **5** 13800–13806
[15] Kolosko A G, Popov E O, Filippov S V and Romanov 2015 *J of Vac.Sci. & Tech. B* **33** 03C104
[16] Kolosko A G, Popov E O, Filippov S V and Romanov P A 2014 *Technical Physics Letters* **40 5** 438–441
[17] Kolosko A G, Filippov S V, Romanov P A, Popov E O and Forbes R G 2016 *J of Vac.Sci. & Tech. B* **34** 041802
[18] Popov E O, Kolosko A G, Filippov S V, Fedichkin I L and Romanov P A 2015 *J of Vac.Sci. & Tech. B* **33**, 03C109
[19] Filippov S V, Popov E O, Kolosko A G and Vinnichek R N 2017 *Jour. of Phys.: Conf. Ser* **917** 092022
[20] Popov E O, Kolosko A G, Filippov S V and Terukov E I 2018 *J of Vac.Sci. & Tech. B* **36** 02C106

IOP Conf. Series: Materials Science and Engineering **525** (2019) 012052 doi:10.1088/1757-899X/525/1/012052

Laser synthesis of graphene in liquid nitrogen

K S Khorkov, D A Kochuev, V A Ilin, V G Prokoshev and S M Arakelian

Vladimir State University, Gorky str. 87, Vladimir 600000, Russia

E-mail: freeod@mail.ru

Abstract. In this paper discusses the properties and methods of obtaining graphene. The processes and mechanisms of graphene formation under the influence of femtosecond laser radiation in liquid nitrogen are considered. Experimental schemes of laser-induced graphene exfoliation in liquid nitrogen are presented.

1. Introduction

Carbon is one of the paramount substances necessary for life on our planet. Due to the ability of carbon atoms to connect in various ways, carbon-based systems exhibit various allotropic structures and properties. Since ancient times, certain allotropic states of carbon, such as diamond and graphite, have been known. With the emergence of the term "nanotechnology", quoted in a lecture by Nobel laureate R. Feynman called "There is plenty of room at the bottom" [1], new methods of imaging substances at small scales, leading to the discovery of fullerenes. Discovery of fullerenes by R.E. Smalley, H. Kroto, and R.F. Curl in 1985 sparked an interest in nanoscience, bringing it to the fore.

The results were the discovery of carbon nanotubes by S. Iijima in 1991. However, there is earlier evidence of the discovery of carbon nanotubes, for example, in 1974–1976 [2]. Works have been published that describe thin tubes with a diameter of less than 100 Å made by the method of vapor condensation, but a more detailed study of the structure has not been carried out.

Interestingly, the two-dimensional form (graphene) was obtained quite recently and immediately attracted much attention. The first attempts to obtain a two-dimensional carbon structure were made long before the discovery of "real graphene". For example, in 1962, the name graphene was born, when Hanns-Peter Boehm described the properties of carbon one atom thick [3]. In the works of Peierls and Landau (1935–1937), it was said that strictly two-dimensional (2D) carbon crystals are thermodynamically unstable and cannot exist [4, 5]. This was one of the reasons why no one tried to obtain two-dimensional crystals before their discovery. Thermodynamic stability turned out to be possible due to thermal fluctuations that create a "ripple" – a waviness of the graphene surface with a height of $\approx$5 nm. The discovery of real graphene occurred in 2004, after publication in the journal Science [6], wherein it was reported about getting carbon monolayer was on an oxidized silicon substrate by dint of mechanical splitting. This technique consists in peeling graphene scales as a result of friction of crystalline graphite against each other, or in "peeling off" them from the surface of graphite using a special adhesive tape with its subsequent dissolution.

Graphene is a single-layer two-dimensional carbon structure organized into a regular hexagonal crystal lattice with carbon atoms at the vertices arranged in plane. Graphene has a layer thickness of one atom, and the distance between them is 0.142 nm, with each atom associated with three neighboring atoms with covalent chemical σ-bonds with sp^2-hybridization, and the fourth valence

IOP Conf. Series: Materials Science and Engineering **525** (2019) 012052 doi:10.1088/1757-899X/525/1/012052

electron is included in the conjugated π-system of graphene. Thus, three bonds located in the plane define the geometric structure of graphene, and the fourth, its unique electronic properties.

2. Properties of graphene and methods of synthesis

The most attractive feature of graphene is its unique properties. This material is the first known two-dimensional atomic crystal [7]; the thinnest object ever received; the most durable material [8]; charge carriers in it are massless Dirac fermions [9–11]; it conducts electricity very well [12] and heat [13]; very elastic and impenetrable for many molecules. Graphene has a huge number of different properties, a detailed review of which is impossible to give in this work. For this reason, only certain properties will be considered.

2.1. Optical properties

The optical properties of graphene are very paradoxical, since graphene is one atom thick, and it absorbs most of the light. In the infrared limit, the absorption coefficient is $\pi\alpha \approx 2.3\%$, where $\alpha = e^2 / \hbar c$ is the fine structure constant, corrections to this value in the visible spectral range are less than 3%. Such a noticeable absorption coefficient makes it possible to see graphene with an unarmed eye. This is the paradox, since graphene is the smallest particle. At higher frequencies, the absorption increases, reaching 10% due to the Van Hove singularity at the edges of the zone. The optical absorption of graphene can be changed by shifting the positions of the Fermi level by adjusting the concentration of charge carriers [14].

2.2. Linear dispersion dependence and chirality

Graphene is a semimetal with a zero forbidden band, since the valence band and the conduction band touch at two points K and K'. Such an electronic structure is a consequence of the hexagonal symmetry of the lattice, and it does not apply to Bravais lattices, since it contains two atoms in each unit cell. The carbon atom orbitals hybridize to form zones, the intersection of which at points K and K' forms a gapless spectrum with a linear dispersion. Consequently, the absence of a gap between the valence band and the conduction band of graphene is very stable and is a consequence of the symmetry between the sublattices.

Linear dispersion itself makes graphene special, but not limited to this. The symmetry and charge conjugation of the electron-hole pairs present in graphene determine the chiral characteristics of pairs of charge carriers. The specificity of chirality is due to the pairwise inequivalent sublattices of the crystal structure of graphene. Such properties of graphene allow one to "see" the relativistic effects of quantum electrodynamics (QED).

2.3. Klein paradox

The Klein paradox, theoretically predicted in 1929, is the visual result of the chiral symmetry of quasiparticles in graphene [14]. The paradox is associated with an increase in the tunneling of a relativistic particle approaching unity, when the barrier height exceeds $2m_0c^2$ (where m_0 is the rest mass of the particle, c is the speed of light). This can be viewed as a consequence of backscatter suppression, since massless relativistic particles, such as photons, always move at a constant velocity – the speed of light, whereas backscattering requires that the speed at the turning point is zero. One can also consider the birth and annihilation of particle-antiparticle pairs, which in regions of a strong electric field are due to the Schwinger mechanism.

The Klein paradox for chiral quasiparticles in graphene leads to the transformation of an electron inside a potential barrier into a hole and to an equal unit probability of tunneling through such a barrier, at least during normal incidence. This, in turn, leads to the absence of localization and the final minimum conductivity even in graphene with moderate disorder, which in the limit of the formally zero concentration of charge carriers breaks up into electron-hole puddles. The absence of backscattering, leading to the Klein paradox, also provides a positive sign of quantum (interference) corrections to conductivity, leading to weak anti-localization.

IOP Conf. Series: Materials Science and Engineering **525** (2019) 012052 doi:10.1088/1757-899X/525/1/012052

2.4. Quantum Hall effect

Quantum Hall effect occurs in graphene under normal conditions, that is, at room temperature and magnetic fields [15]. Measurements of the transverse resistivity of the Hall demonstrate the ambipolarity of the current-voltage characteristic of graphene. It is manifested in the fact that electrons and holes carry current together, being chiral, have half-integer spin indices, different in signs, and are located in adjacent, adjacent valleys – between the points of Dirac. The mechanism of ambipolar electrical conductivity is due to the mutual conjugation of electrons and holes and the properties of graphene as a semimetal associated with this conjugation. Ambipolarity means the ability to control the type of carriers of another sign (by switching electrons and holes) and is a remarkable property of the current-voltage characteristics of the transistor inclusion of graphene.

2.5. Methods of graphene synthesis

A carbon monolayer was first obtained by micromechanical cleavage technique of crystalline graphite [6, 7]. This technique consists in separating graphene sheets from crystalline graphite, either as a result of friction of small graphite crystals against each other, or with the help of adhesive tape, the subsequent dissolution of which results in individual graphene layers. Studies show that this technique allows you to select single-layer graphene sheets, having an ordered structure and having about 10 microns in width and about 100 microns in length.

The main problem with this method is the difficulty of identifying the resulting graphene, since the number of layers varies from one to one hundred. To identify such objects, standard observation methods based on the use of scanning probe microscopy, which does not allow determining the number of layers of a multilayer structure, are unsuitable. The use of an optical microscope was a solution on the way to identifying and obtaining samples of single-layer graphene.

The chemical vapor deposition (CVD) method is widely used for the synthesis of carbon nanostructures. Before the appearance of nanocarbon materials, used to produce thin carbon filaments, later found application for the synthesis of carbon nanotubes in macroscopic quantities. The essence of this method is the thermocatalytic decomposition of gaseous hydrocarbons on the surface of some metals to form different nanocarbon structures [16].

The laser ablation method is based on the removal of matter from the surface by a laser pulse. Obtaining graphene by this method is based on the stratification of crystalline graphite under the influence of laser radiation. As a target, a crystal of highly ordered pyrolytic graphite is used, which is placed on a silicon substrate. Graphene synthesis is carried out in argon atmosphere. Carbon particles formed as a result of laser ablation of the graphite surface fall on a silicon substrate. When the energy density of the laser radiation on the target is from 1 to 10 j/cm^2, thin scales appear on the surface of the target, which spontaneously separate from the target and fall on the surface of the substrate.

3. Method of graphene laser-induced synthesis in liquid nitrogen

The main idea of method is the effect of femtosecond laser radiation on a graphite sample immersed in a liquid nitrogen medium [17]. The physical meaning lies in the heat-conducting property of graphite, as well as the change in its structure at low temperatures. The thermal expansion of the graphite lattice is anisotropic; it manifests itself more strongly if the lattice order is higher. In the direction perpendicular to the basic planes, the coefficient of thermal expansion is positive, practically independent of temperature, and more than 20 times higher than the average absolute value for the basal planes. Consequently, the temperature coefficient of linear expansion in the direction of the basal planes is much less than in the perpendicular direction, thereby contributing to the separation of the structure with intensive heating.

3.1. Experimental setup

To obtain graphene samples by the method of laser synthesis in liquid nitrogen an experimental setup was developed (figure 1). For the experiment, a double-circuit cryostat with vacuum thermal insulation was also developed. The design of the cryostat made it possible to eliminate such difficulties during the experiment as: boiling of liquid nitrogen during heat exchange with the external environment; cooling and maintaining a low sample temperature; icing glass windows. Liquid nitrogen was used as a source of low temperatures; it is also possible to use other cryoagents.

The cryostat design is a steel cylindrical vessel equipped with transparent windows through which observation or introduction of laser radiation into the vessel volume can be carried out (the glass used is transparent for the working wavelength of laser radiation). The product is equipped with two independent circuits, allowing to work in different temperature ranges, as well as to place samples in different environments (liquids, gases). The assembly of the product is carried out by means of threaded connections. At each butt section of the flanges a special polytetrafluoroethylene gasket is installed. The unit has an object table, on the surface of which there are threaded holes for reliable fastening of the sample. The object table is made of a material with high thermal conductivity to ensure good heat exchange with the sample. The design of the cryostat provides an additional scheme for the removal of combustion products.

The carbon sample is processed by a pulsed femtosecond Yb:KGW laser system TETA-10. A sample of carbon is placed in a cryostat. After creating the vacuum in chamber, liquid nitrogen is supplied to the cryostat and laser treatment is performed. Removal of products of combustion and vapors of liquid nitrogen is carried out through the ventilation system. Liquid nitrogen is removed from carbon materials in an evaporator equipped with a substrate for the deposition of graphene.

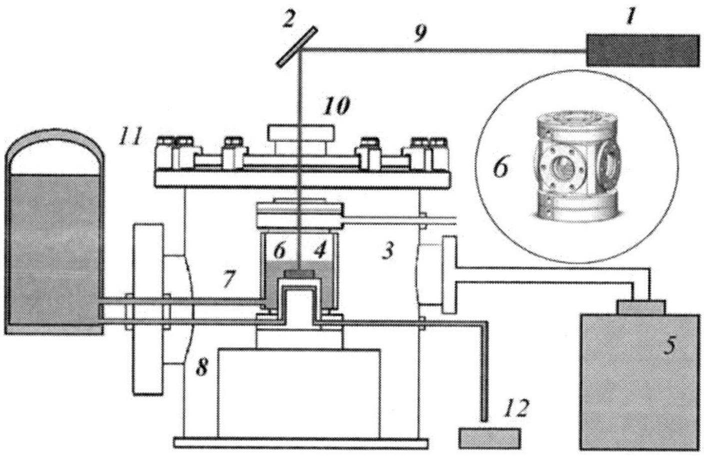

Figure 1. The experimental setup: 1 - laser, 2 - mirrors system, 3 - ventilation system, 4 - carbon sample, 5 - turbomolecular pump, 6 - cryostat, 7 - liquid nitrogen supply system, 8 - vacuum chamber, 9 - laser radiation, 10 - input window of laser radiation, 11 - insulating vessel, 12 - nitrogen evaporator.

3.2. Mechanisms of graphene exfoliation

Under normal conditions, the distance between the atoms of graphite is 1.42 Å, and the interplanar distance is 3.38 Å. When graphite is immersed in liquid nitrogen, it is cooled. Low temperature leads to a decrease in lattice vibrations of the source material of graphite; therefore, the interplanar distance decreases to 3.393 Å, and when heated increases to 3.465 Å [18]. The diameter of the liquid nitrogen molecule is approximately 0.32 Å, which is less than the interplanar distance of the cooled graphite. This dependence allows the nitrogen molecule to penetrate into the interplanar distance of graphite, through cracks, chips and delaminations.

The nitrogen molecule is heated at laser pulses, where the main process is the transfer of energy through a sample of graphite - a molecule, due to thermal conductivity. During a laser pulse, electrons transfer energy to the lattice of the substance, and it heats up in the laser-induced area. As a result of laser heating and, as a result, an increase in the amplitude of vibrations of nitrogen molecules, a volume expansion of the region occupied by nitrogen molecules occurs (transition from the liquid phase to the gas phase). Thus, interplanar carbon π-bonds (16-40 kJ / mol), weaker than σ-bonds (≈ 450 kJ/mol), are destroyed. Effective heating of nitrogen is due to the transfer of thermal energy from the graphite lattice due to the absence of nitrogen absorption bands at the working wavelength.

For successful application of laser exfoliation of graphene it is necessary to provide special conditions for cooling the laser exposure zone. Laser interaction on the target in the cryogenic medium forms a larger temperature gradient, which increases the thermoelastic stresses. A schematic representation of the process of laser separation of graphene is shown in Figure 2.

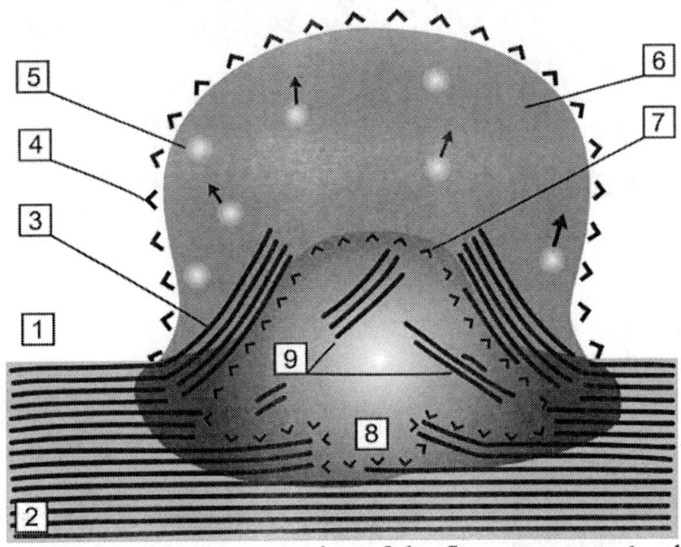

Figure 2. Schematic representation of the flow processes: 1 – liquid nitrogen, 2 – carbon target, 3 – exfoliating graphene, 4 – propagation front of vapor-gas region, 5 – evaporated nitrogen, 6 –vapor-gas area, 7 – front of high-pressure region, 8 – high-pressure area (the heating of intercalated nitrogen, laser ablation), 9 – exfoliated graphene

At the interaction of ultrashort laser pulses with a graphite sample in liquid nitrogen, several variants of the processes of structural changes in the material can be distinguished. When the laser radiation power is insufficient for laser ablation processes, the energy of photons is absorbed by the carbon lattice and intercalated nitrogen. Weak interplanar π-bonds are destroyed due to a sharp increase in lattice vibrations and nitrogen molecules.

When energy of laser radiation in excess of the ablation threshold is the destruction of the graphite surface in the affected area. As a result of the processes occurring during laser ablation, a high-pressure region is formed on the walls of the formed cavity. The stratification structure of the graphite to multilayer graphene in this mechanism is the result of a reaction effect of ablation products.

3.3. Results of experiments

Processing of carbon targets surface was carried out by femtosecond laser system TETA-10 with following parameters: wavelength $\lambda = 1030$ nm, radiation pulse duration $\tau=280$ fs, pulse repetition rate $f = 10$ kHz, pulse energy $E_{max}=150$ μJ. Surface processing of graphite was carried out in liquid nitrogen, which covered it with a layer 10 mm thick. The diameter of the laser spot on the target surface reached 100 μm. As the source samples to obtain graphene was used, the target of highly oriented pyrolytic graphite HOPG-1,7-10x10x1-1 and glassy carbon SU-2000.

After processing graphite in liquid nitrogen, the obtained samples were studied to confirm the result. The result of the separation of graphene on the surface of HOPG is demonstrated on Figure 3.

Obviously, the separation begins from the center of the crack in accordance with the mechanism described above. SEM-images show that some structures flake and bend in the direction from the beginning of the separation remain at the base, and in some places also flake. The next stage of using the obtained graphene sheets is the separation from the original sample using ultrasonic treatment, the formation of a solution and deposition on the required substrates.

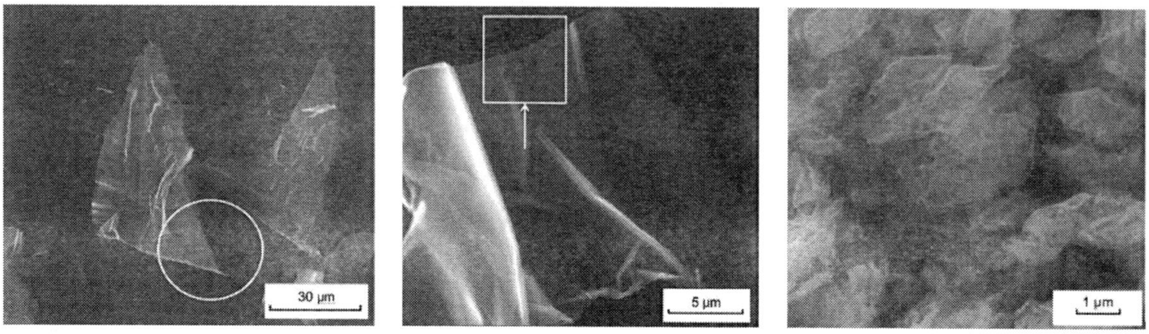

Figure 3. SEM-images of graphene layers

4. Graphene application

The most attractive feature of graphene is its properties, which are superior to those of competing materials. This makes it perceptible for many applications. One of the promising areas is the use of graphene as a porous electrode for supercapacitors. The use of graphene as an electrode material in electrochemical supercapacitors is shown in [19]. The authors used graphene obtained by three independent methods. The first method involved thermal exfoliation of graphitic oxide. In this method, graphitic oxide was prepared by reacting graphite with concentrated nitric acid and sulphuric acid with potassium chlorate at room temperature for 5 days. Thermal exfoliation of graphitic oxide was carried out in a long quartz tube at 1050°C. In the second method, graphene was obtained by heating nanodiamond at 1650°C in a helium atmosphere. The last method involved the decomposition of camphor over nickel nanoparticles. The reaction was carried out in a two-stage furnace and the camphor was slowly sublimed from the first furnace to the second furnace held at 770°C where the micron sized nickel particles were placed.

The results obtained in [19] showed that the use of graphene as an electrode for supercapacitors is possible and promising. Since the specific capacity of exfoliated graphene in an aqueous electrolyte is comparable with the capacity obtained with activated carbons, and exceeds the capacity of carbon nanotubes, while the value of the energy density of graphene capacitors is one of the highest.

5. Conclusion

The basic properties of graphene and methods of its production are considered. The experimental schemes and software for the complex of femtosecond laser processing of materials have been modernized to investigate the influence of processing conditions and raw materials on the nature of the formed nano- and microstructures. Specialized cryostat for laser synthesis of graphene has been developed. It is shown that common mechanism of graphene exfoliation under femtosecond laser action in liquid nitrogen is separation of carbon surface layers in samples as a result of the disturbance of interplanar bonds due to the expansion of intercalated nitrogen into the gas phase under laser heating.

Acknowledgements
This work was performed as a part of Vladimir State University's State Task 3.5531.2017/8.9 GB-1106/17.

References

[1] Feynman R P. There's plenty of room at the bottom: An invitation to enter a new field of physics //*Handbook of Nanoscience, Engineering, and Technology*, Third Edition. 2012, CRC Press: 26–35

[2] Oberlin A, Endo M, Koyama T. High resolution electron microscope observations of graphitized carbon fibers //*Carbon*. 1976; **14(2):**133–135

[3] Boehm H. The adsorption behavior of very thin carbon films //Z. *Anorg. Allg. Chem.* 1962;**316(3–4):**119–127

[4] Peierls R E. Quelques proprietes typiques des corpses solides. Ann. IH *Poincare*. 1935; **5**:177–222

[5] Landau L D. Zur Theorie der phasenumwandlungen. II. Phys. Z. Sowjetunion. 1937;11:26–35

[6] Novoselov K S. Electric field effect in atomically thin carbon films. *Science.* 2004;**306(5696):**666–669

[7] Novoselov K S. Two-dimensional atomic crystals. *Proceedings of the National Academy of Sciences of the United States of America*. 2005;**102(30)**:10451–10453

[8] Lee C. Measurement of the elastic properties and intrinsic strength of monolayer graphene. *Science.* 2008;**321(5887):**385–388

[9] Geim A K, Novoselov K S. The rise of graphene. *Nature materials.* 2007;**6(3)**:183

[10] Novoselov K S. Two-dimensional gas of massless Dirac fermions in graphene. *Nature.* 2005;**438(7065)**:197

[11] Zhang Y. Experimental observation of the quantum Hall effect and Berry's phase in graphene. *Nature.* 2005;**438(7065)**:201

[12] Meric I. Current saturation in zero-bandgap, top-gated graphene field-effect transistors. *Nature nanotechnology*. 2008;**3(11)**:654

[13] Balandin A A. Superior thermal conductivity of single-layer graphene. *Nano letters.* 2008;**8(3)**:902–907

[14] Novoselov K S. Graphene: Materials in the Flatland. *Reviews of Modern Physics.* 2011;**181(12):**1299–1311

[15] Alekseenko A G. Graphene. M. BINOM. Laboratory of Knowledge. 2014:168

[16] Li X. Large-area synthesis of high-quality and uniform graphene films on copper foils. *Science.* 2009;**324(5932):**1312–1314

[17] Khorkov K S. Mechanisms of graphene exfoliation under the action of femtosecond laser radiation in liquid nitrogen. *J. Phys. Conf. Ser.* 2018;**951**:012014

[18] Mortazavi S Z, Parvin P, Reyhani A. Fabrication of graphene based on Q-switched Nd: YAG laser ablation of graphite target in liquid nitrogen. *Laser Physics Letters.* 2012;**9(7):**547

[19] Vivekchand S R. C. Graphene-based electrochemical supercapacitors. *Journal of Chemical Sciences.* 2008;***120*(1)**:9–13

IOP Conf. Series: Materials Science and Engineering **525** (2019) 012053 doi:10.1088/1757-899X/525/1/012053

Sol-Gel synthesis of high-temperature aluminosilicate glass-ceramics and composite materials on its basis

A S Chaynikova, N E Shchegoleva, S Y Modin, A A Akopian, D V Grashchenkov and M L Vaganova

All-Russian Scientific Research Institute of Aviation Materials, 105275 Moscow, Russia

E-mail: anna-chanikova@mail.ru

Abstract. Powders of strontium aluminosilicate glass-ceramics were synthesized by a sol-gel method. Composite materials based on them with the addition of industrial powders α-Si_3N_4, β-Si_3N_4, Y_2O_3, ZrO_2 and HfO_2 were obtained by the method of spark plasma sintering. It was found that reinforcement with silicon nitride leads to an increase in K_{1C} of glass-ceramics by more than 2 times. The temperature of sintering and the properties of composite materials (CM) are complexly dependent on the nature of the refractory oxide. Maximum strength is typical for CM containing Y_2O_3 additive, and maximum microhardness is for material with HfO_2 additive. Modification by refractory oxides leads to an improvement in oxidation resistance of composite materials.

1. Introduction

One of the most important problem of modern materials science is the development of new materials that can be operated under the influence of high temperatures and various aggressive media. From this point of view, a great deal of attention is being devoted to glass-ceramics in the system SrO-Al_2O_3-SiO_2. A review of scientific-technical data and patents [1–10] shows that strontium aluminosilicate glass-ceramics are promising to use as matrices of high temperature composites for application in aerospace technology. This is because strontium aluminosilicate glasses have high softening temperatures T_g (700–800 °C) while the principal phase crystallizing in this system (monoclinic strontium anorthite $SrAl_2Si_2O_8$) possesses a high melting point of 1650 °C, good mechanical properties (E = 100 GPa, σ_{bend} = 100–120 MPa), low thermal coefficient of linear expansion (26-48 × 10^{-7} °C^{-1}) and phase stability up to the melting point [11, 12].

Strontium aluminosilicate glass-ceramics of stoichiometric composition ($SrO \cdot Al_2O_3 \cdot 2SiO_2$) is shows excellent service characteristics due to minimum content of residual glass phase in it. But it is necessary to use high temperatures for production of such glass-ceramics using conventional glass methods and developing ceramic (powder) technologies. Thus the preparation of strontium aluminosilicate glass-ceramics of stoichiometric composition by a sol-gel method is promising. The main advantage of the sol-gel method is the possibility of obtaining materials with a nanocrystalline structure at low temperatures.

Increases of fracture toughness and bending strength are the main goals of development of high temperature composites by reinforcement of strontium aluminosilicate glass-ceramics with various fillers. The most promising fillers are particles and whiskers of Si_3N_4 [13–17]. Recently, research in the field of synthesis of composites with barium aluminosilicate glass-ceramic matrices and dispersed

IOP Conf. Series: Materials Science and Engineering **525** (2019) 012053 doi:10.1088/1757-899X/525/1/012053

fillers based on Si_3N_4 has been actively carried out in China. For example, the authors of [16] demonstrated that the critical stress intensity factor (K_{IC}) of the barium aluminosilicate glass-ceramics increases to 8.9 MPa·m$^{1/2}$ upon introduction of silicon nitride grains in an amount of 70 wt %. This value of K_{IC} is maximally attained to the present moment when the glass-ceramic matrix is reinforced with a dispersed filler. In [17], use of 60 wt % Si_3N_4 led to an increase in the K_{IC} of the barium aluminosilicate glass-ceramics to 7.4 MPa·m$^{1/2}$, which exceeds by 270% that of the glass-ceramic matrix.

The sintering temperature of these materials is very high (~1800°C), while its operating temperature is low due to the high propensity of silicon nitride to oxidation when exposed to elevated temperatures. To lower the sintering temperature and reduce the sintering time, so far the relatively effective sintering method is spark plasma sintering (SPS). To increase the operating temperature modification of glass-ceramics by refractory oxides, such as Y_2O_3, ZrO_2 and HfO_2 is of great interest [14–16].

The goal of our study was sol-gel synthesis of high-temperature aluminosilicate glass-ceramics and synthesis of composite materials on its basis via introduction of finely dispersed powders of α-Si_3N_4 and β-Si_3N_4, Y_2O_3, ZrO_2 and HfO_2 using of spark plasma sintering and to examine the influence of the nature of refractory oxides on the sintering, phase composition, structure and properties of the materials obtained.

2. Experimental

The stoichiometric composition of strontium anorthite ($SrO·Al_2O_3·2SiO_2$, SA2S) was used to prepare sol-gel solutions. Solutions were prepared by sequentially adding components.

The composite materials were obtained by the method of spark plasma sintering (SPS) using equipment from FCT (Germany). As the main methods of study were used laser granulometry (Analysette 22 MicroTec, Fritsch), differential scanning calorimetry (STA 449 C Jupiter, Netzsch), X-ray phase analysis (XPA) (D2 Phaser, Bruker), and dilatometric method for determining the shrinkage during heating in argon (DIL 402 PC dilatometer, Netzsch). Four-point bending strength and Vickers microhardness were determined using a T1-FR010THW A50 testing machine (Zwick Germany) and an HV-1000 microhardness tester (TIME Group, China), respectively. The analysis of the microstructure of materials was carried out on a JSM-6490LV scanning electron microscope (Jeol, Japan). Apparent density was measured by using the Archimedes principle.

3. Results and discussion

After completion of the hydrolysis and polycondensation processes, the gels were dried. The study of dried gels by the DSC/TG method showed that in the temperature range of 100–900 °C, crystalline water and solvent are removed from them, as well as the decomposition of inorganic residues of salts of the starting elements. The ongoing physical and chemical transformations are accompanied by significant mass losses of more than 50%, which stabilize only in the temperature range of 900 °C. In the temperature range of 900–1400 °C, gels crystallize with the formation of monoclinic strontium anorthite by successive transformations of strontium silicates and hexagonal strontium anorthite. According to the results of DSC data analysis, the dried gels were subjected to high-temperature treatment until the formation of the monoclinic strontium anorthite phase was completed.

Industrial powders α-Si_3N_4, β-Si_3N_4, Y_2O_3, ZrO_2 and HfO_2 were used as fillers of composite materials. Intergrinding in a planetary mill of fillers and glass-ceramic powders, obtained after high-temperature treatment was carried out. The dominant particle size in the mixtures was 4 μm (Fig. 1).

The study of linear shrinkage curves obtained by dilatometric method showed that the introduction of silicon nitride in glass-ceramics leads to a shift in temperature range of composites sintering to higher temperatures (Fig. 2). The presence of refractory oxides in composites intensifies their sintering. The consolidation of composites is occurs by the viscous-flow mechanism and accompanied by a shrinkage of 8–15% in the temperature range from 1200 °C to >1800 °C. At a temperature of ~1600 °C, sintering slows down due to the phase transition of α-Si_3N_4 to β-Si_3N_4.

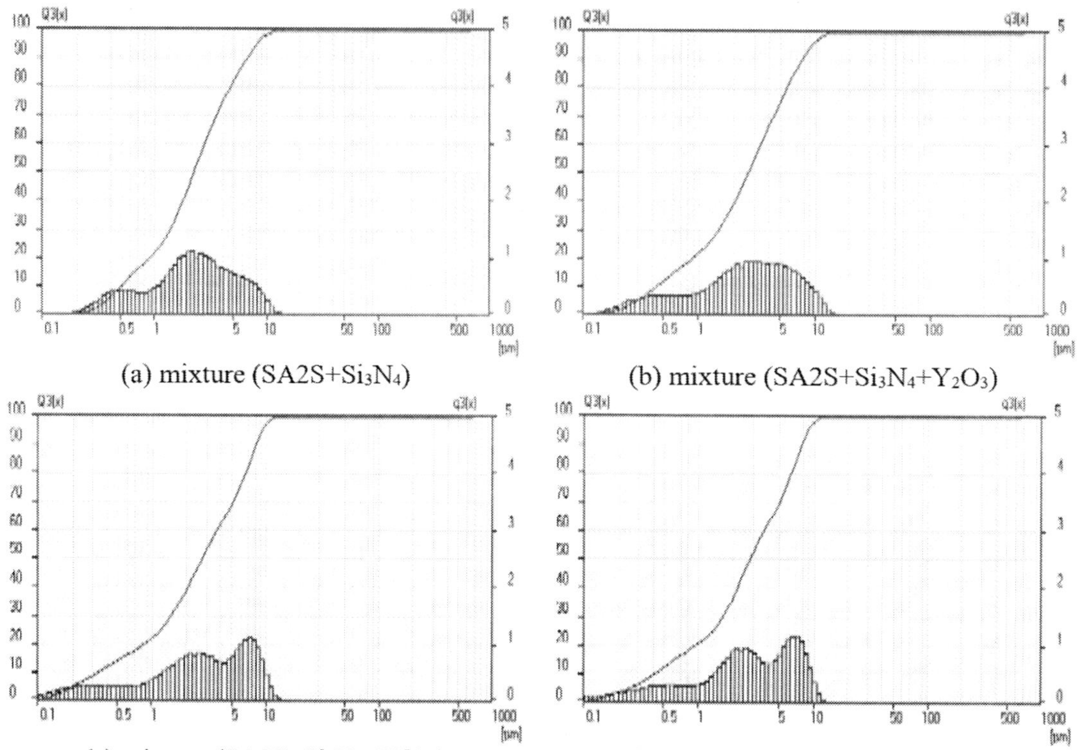

(a) mixture ($SA2S+Si_3N_4$) (b) mixture ($SA2S+Si_3N_4+Y_2O_3$)

(c) mixture ($SA2S+Si_3N_4+HfO_2$) (d) mixture ($SA2S+Si_3N_4+ZrO_2$)

Figure 1. Results of laser granulometry of powder mixtures of composite materials

The study of linear shrinkage curves during the SPS process showed that the initial temperature of sintering of CM was in the temperature range of 1100–1350 °C. The introduction of refractory oxides into glass-ceramics leads to decrease of initial temperature of sintering in the series: ($SA2S+Si_3N_4$) → ($SA2S+Si_3N_4+HfO_2$) → ($SA2S+Si_3N_4+ZrO_2$) → ($SA2S+Si_3N_4+Y_2O_3$) (Fig. 3). The dependence of the relative density of the synthesized composites on the nature of the oxides is of a similar nature. However, the shorter sintering range (Fig. 3) during isothermal holding at the sintering temperature of the composite with the addition of ZrO_2 led to its density being lower than the density of the sample with HfO_2 addition (table 1).

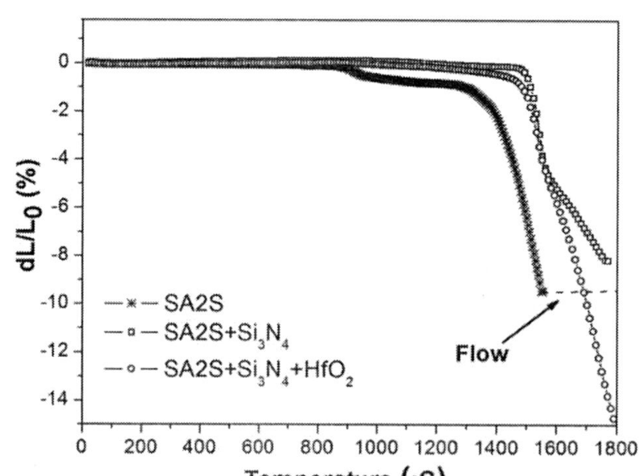

Figure 2. Curves of linear shrinkage dL/L_0 of the materials in their heating in argon

IOP Conf. Series: Materials Science and Engineering **525** (2019) 012053 doi:10.1088/1757-899X/525/1/012053

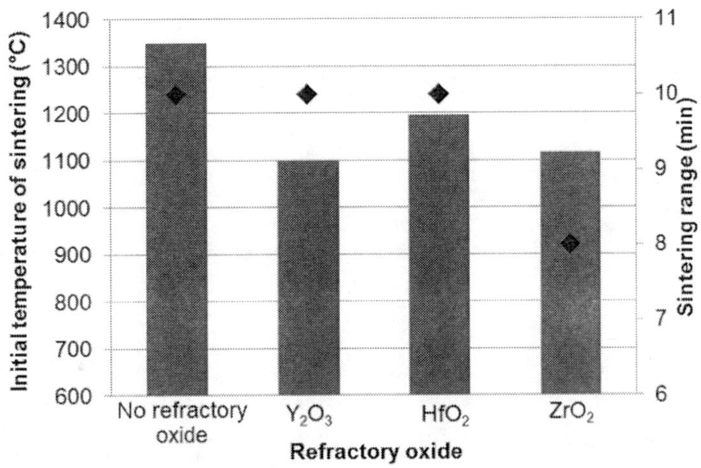

Figure 3. The dependence of the initial temperature of sintering (columns) and the sintering range during isothermal holding at the sintering temperature (symbol) on the nature of the refractory oxide

Figure 4 shows scanning electron micrograph of the fractured surface of synthesized composites. It was found that all composite materials are characterized by a fine-grained structure. The samples ($SA2S+Si_3N_4$) and ($SA2S+Si_3N_4+HfO_2$) contain intergranular pores with a size of about 1–2 μm. In CM, containing Y_2O_3 and ZrO_2, intergranular pores with a size of about tenths of a micrometer are found. The content of pores in the samples with refractory oxides is significantly lower compared with composite ($SA2S+Si_3N_4$).

(a) composite ($SA2S+Si_3N_4$) (b) composite ($SA2S+Si_3N_4+Y_2O_3$)

(c) composite ($SA2S+Si_3N_4+HfO_2$) (d) composite ($SA2S+Si_3N_4+ZrO_2$)

Figure 4. Scanning electron micrograph of the fractured surface of the composites

The results of the study of the mechanical properties of the synthesized composites are presented in table 1. It was found that samples containing Y_2O_3 are characterized by maximum strength. This may be due to their highest density. However, it should be noted that despite the higher density of samples containing additives of HfO_2 or ZrO_2 than CM without additives, their bending strength is lower than for the sample ($SA2S+Si_3N_4$).

Table 1. Physical and mechanical properties of composites

Composition of composites	Property				
	$\rho_{apparent}$ (kg/m^3)	$\rho_{relative}$ (%)	σ_{bend} (MPa)	H_v (GPa)	K_{IC} (MPa·m$^{1/2}$)
$SA2S+Si_3N_4$	2730	87 ± 0.5	174 ± 10	14 ± 1	5.3 ± 0.1
$SA2S+Si_3N_4+Y_2O_3$	3140	99 ± 0.5	186 ± 10	17 ± 1	5.3 ± 0.1
$SA2S+Si_3N_4+HfO_2$	3080	97 ± 0.5	128 ± 10	19 ± 1	5.3 ± 0.1
$SA2S+Si_3N_4+ZrO_2$	3000	95 ± 0.5	155 ± 10	17 ± 1	5.3 ± 0.1

The modification with refractory oxides did not affect the fracture toughness of composite materials. Regardless of the nature of the modifying oxide, the critical stress intensity factor for all samples were (5.3 ± 0.1) MPa·m$^{1/2}$, which is more than twice the K_{IC} value of strontium aluminosilicate glass-ceramics (2.9 MPa·m$^{1/2}$). This demonstrates the effectiveness of the use of selected fillers for the reinforcement of glass-ceramics.

The study of the thermal properties of samples showed that the modification by all refractory oxides leads to an improvement in oxidation resistance of composite materials.

4. Conclusions

Thus, as a result of the research, it has been established that the temperature range of sintering and properties of composite materials based on synthesized by a sol-gel method strontium aluminosilicate glass-ceramics with the addition of industrial powders α-Si_3N_4, β-Si_3N_4, Y_2O_3, ZrO_2 and HfO_2 obtained by the SPS are in complex dependence on the nature of the refractory oxide. The initial temperature of sintering during SPS is decreases in the series: ($SA2S+Si_3N_4$) → ($SA2S+Si_3N_4+HfO_2$) → ($SA2S+Si_3N_4+ZrO_2$) → ($SA2S+Si_3N_4+Y_2O_3$). Maximum strength is typical for CM containing Y2O3 additive, and maximum microhardness is for material with HfO_2 additive. The modification with refractory oxides did not affect the fracture toughness of composite materials. Regardless of the nature of the refractory oxides reinforcement with silicon nitride leads to an increase in K_{IC} of glass-ceramics by more than 2 times. The modification by refractory oxides leads to an improvement in oxidation resistance of composite materials.

Acknowledgements

The authors are grateful to Osin I.V. (All-Russian Scientific Research Institute of Aviation Materials, Moscow, Russia) for spark plasma sintering of samples.

References

[1] Beall G.H. 2009 *J. of the European Ceramic Society*. **29** 1211–1219
[2] Bansal N.P. 1997 *J. of Materials Research*. **12(3)** 745–753
[3] Roether J.A., Boccaccini A.R. 2005 *Handbook of ceramic composites*. Boston: Kluwer Academic Publishers 485–511
[4] Ye F., Liu L., Wang Y., Zhou Y., Peng B., Meng Q. 2006 *Scripta Materialia*. **55** 911–914
[5] Liu L., Ye F., Zhou Y., Zhang Z. 2010 *Scripta Materialia*. **63** 166–169
[6] Academician Sarkisov P.D., Paleari A., Chainikova A.S., Orlova L.A., Popovich N.V.,

Filippov V.V. 2012 *Doklady Chemistry.* **446(2)** 220–222.
[7] Kablov E.N., Grashchenkov D.V., Isaeva N.V., Solntsev S.S., Sevast'yanov V.G. 2012 *Glass and Ceramics.* **69(3-4)** 109–112
[8] Kablov E.N., Grashchenkov D.V., Isaeva N.V., Solntsev S.S. 2010 *Ros. Khim. Zh.* **LIV(1)** 20–24
[9] Grashchenkov D.V., Vaganova M.L., Schegoleva N.E., Chainikova A.S., Lebedeva Yu.E. 2017 *Aviats. Mater. Tekhnol.* **S** 290–305 DOI: 10.18577/2071-9140-2017-0-S-290-305.
[10] Chainikova A.S., Orlova L.A., Popovich N.V., Lebedeva Yu.E., Solntsev S.S. 2014 *Aviats. Mater. Tekhnol.* **3** 45–54 DOI: 10.18577/2071-9140-2014-0-3-45-54.
[11] Sung Y.M., Kim S. 2000 *J. of Materials Science.* **35** 4293–4299
[12] Chainikova A.S., Voropaeva M.V., Alekseeva L.A., Orlova L.A., Samsonov V.I. 2014 *Aviats. Mater. Tekhnol.* **S6** 45–51 DOI: 10.18577/2071-9140-2014-0-s6-45-51.
[13] Kablov, E.N., Shchetanov, B.V., Ivakhnenko, Yu.A., Balinova, Yu.A., 2013 *Trudy VIAM.* **2** 5 (viam-works.ru)
[14] Chainikova A.S., Schegoleva N.E., Grashchenkov D.V., Vaganova M.L., Modin S.Yu. 2016 *VIII Vseros. Conf. Po Isp. Issled. Sv. Mat. «Testmat»*, (Moskow)
[15] Chainikova A.S., Grashchenkov D.V., Vaganova M.L., Modin S.Yu., Kachaev A.A., Osin I.V., Akopian A.A. 2015 *Conf. «Sovr. Dost. V obl. Sozd. Persp. Nemet. Komp. Mat. I Pokr. Dlia aviac. I kosm. Teh.»* (Moscow)
[16] Ye F., Liu L., Zhang J., Meng Q. 2008 *Composites Science and Technology.* **68** 1073–1079
[17] Ye F., Liu L., Zhang J., Iwasa M., Su C.-L. 2005 *Composites Science and Technology.* **65** 2233–2239

IOP Conf. Series: Materials Science and Engineering **525** (2019) 012054 doi:10.1088/1757-899X/525/1/012054

$RBa_2Cu_3O_Y$ crystal structure and properties

L A Cherepanova, S G Titova and A V Fetisov

Institute of Metallurgy Urals Division of Russian Academy of Sciences, 620016, Ekaterinburg, Russia

E-mail: freefly88@mail.ru

Abstract. In this study, the effect of R (R= Y, Pr, Nd, Sm, Eu, Lu) substitution on crystal structure and properties HTS compositions $R_{1-x}Ca_xBa_2Cu_3O_y$. The XRD results showed that the substitution of Y / R in Y-123 leads to the disappearance of spikes in the linear and volumetric coefficients of thermal expansion. Mechanical activation showed that the composition of Nd-123 demonstrates enhanced chemical stability with respect to water vapor and CO_2 when exposed to air.

1. Introduction

High-temperature superconducting materials (HTS) are interesting from the fundamental physical effects variety point of view, and from their applications in superconducting electronics. Currently, there is an active introduction of electrical equipment, superconducting power cables, current limiters, transformers and other unique technical devices based on HTS in various fields of electrical engineering. With the improvement of technologies, properties of current-carrying elements are improved. For HTS by X-ray and neutron diffraction observed structural anomalies, which appear as spikes in thermal expansion coefficients and in primitive cell parameters, as well as decrease of the Cu – O apical bond length which is non-monotonic by temperature. Such changes during thermal cycling can lead to deformations and destruction of superconducting products. When selecting a superconducting composition for a particular product and typical operating conditions, it is necessary to take into account the chemical stability of HTS materials with respect to the components of air and water.

2. Experimental Procedure

The objects of our study are $R_{1-x}Ca_xBa_2Cu_3O_y$, where R = Y, Pr, Nd, Sm, Eu, Lu. The following materials were used as starting materials: R_2O_3, $BaCO_3$, CuO, CaO in quantities corresponding to the desired composition. To remove the adsorbed water, the starting components were calcined at 600 ° C for 2 hours. The powders were thoroughly mixed in planetary monomill the Fritsch Pulverisette using agate balls d = 15mm. Afterwards, sample formulation into tablets on hydraulic press at a pressure of 15 MPa. Thereafter, annealed in a Nabertherm N11 furnace at temperatures of 860^{0}C, 900^{0}C, 920^{0}C, 930^{0}C, 950^{0}C in during the 24 hours with intermediate grinding in a planetary mill with the addition of ethyl alcohol with agate balls d = 35 mm and pressing. After the final annealing, the samples were quenched on a copper plate. The phase composition was controlled using x-ray phase analysis. To obtain samples

IOP Conf. Series: Materials Science and Engineering **525** (2019) 012054 doi:10.1088/1757-899X/525/1/012054

with various oxygen treatment, the phases were treated in a quartz boat placed in a quartz tube, to which an external gas flow (oxygen) was connected under the following conditions:

• 24 hours at 550 ° C in an atmosphere of oxygen, quenched in a quartz tube under a stream of cold water;
• 24 hours at 500 ° C in the atmosphere of air, quenched in a quartz tube under a stream of cold water;
• 24 hours at 700 ° C in an air atmosphere with quenching on a copper plate in air.

Certification of chemical compounds was carried out on a XRD 7000 Maxima diffractometer by Shimadzu with CuKa radiation (wavelength = 1.5406 Å), graphite monochromator, scattering angle interval 2Θ = 15-80 degrees. According to X-ray phase analysis, samples of Y, Lu, Ca-123, Y, Eu-123, Nd, Eu-123 contained impurities of complex barium cuprates $YBa_3Cu_4O_x$ (00-041-0454), $Ba_2Cu_3O_x$ (00-047-0117), and also BaO (00-026-0178) in quantities not exceeding 3-5%. Sample Eu, Pr-123 had the same impurities, but in a larger amount ~ 10-12%. The list of investigated samples and their characteristics are given in table 1.

Using the powder X-ray diffraction method, we studied the effect of oxygen content and the degree of substitution on the crystal structure of HTS materials in the temperature range of 80–300 K. The temperature dependences of the primitive cell parameters, determined by the least squares method by 18–20 lines, are shown in Figure 1.

3. Results and Discussion

3.1. Crystal structure at 80-300 K

Substitution (Y^{3+}/Ca^{2+}) allows simplifying the technology for producing superconducting products, reduces the cost of electricity at the final stages of roasting, in some cases leads to an increase in the critical current [1-3]. The compositions of Nd, Eu-123; Sm, Ca-123; Eu-123; Y, Lu, Ca-123; Eu, Pr-123 with a high oxygen content (y> 6.8) demonstrate smooth monotonic dependences of lattice parameters on temperature, which are well described by quadratic dependences; that is, linear thermal expansion coefficients decrease with cooling. The Nd-123 and Y, Eu-123 compositions with low oxygen content (y <6.8) demonstrate a small maximum in the temperature dependence a (T) near 130 K, and the Sm-123 compositions with high oxygen content (y = 6.9) and Eu- 123 with a low oxygen content (y = 6.64) demonstrate a maximum on the temperature dependences of the lattice parameters near ~ 250 K. Samples of the remaining compositions did not have pronounced features on the temperature dependence of the lattice parameters.

Table 1. Characteristics of the $R_{1-x}Ca_xBa_2Cu_3O_y$ samples

Composition	Primitive cell parameters (T=293 K)			Obtaining conditions		Volumetric coefficient of thermal expansion, ×10^5, K^-1	
	a, Å	*b*, Å	*c* , Å	T, °C	lgP(O_2)	α_1 (93-173 K)	α_2 (182-293 K)
$NdBa_2Cu_3O_{6.75}$	3.8870	3.8937	11.6887	550	-0.7 (air)	1.9	3.1
$Nd_{0.5}Eu_{0.5}Ba_2Cu_3O_{6.65}$	3.8693	3.9123	11.7636	700	-0.7 (air)	1.99	4.34
$Nd_{0.5}Eu_{0.5}Ba_2Cu_3O_{6.76}$	3.8635	3.9096	11.7573	550	-0.7 (air)	2.48	2.97
$Nd_{0.5}Eu_{0.5}Ba_2Cu_3O_{6.83}$	3.8607	3.8607	11.7495	550	0 (O_2)	2.38	3.44
$SmBa_2Cu_3O_{6.9}$	3.8615	3.8936	11.7160	550	0 (O_2)	2.06	3.02
$Sm_{0.9}Ca_{0.1}Ba_2Cu_3O_{6.8}$	3.8585	3.9078	11.7398	550	-0.7 (air)	2.3	4.35
$EuBa_2Cu_3O_{6.64}$	3.8562	3.9036	11.7455	700	-0.7 (air)	1.87	2.79

$EuBa_2Cu_3O_{6.77}$	3.8528	3.9083	11.7465	550	-0.7 (air)	2.13	4.36
$EuBa_2Cu_3O_{6.85}$	3.8523	3.9058	11.7373	550	0 (O_2)	2.35	4.28
$Y_{0.5}Eu_{0.5}Ba_2Cu_3O_{6.6}$	3.8629	3.8859	11.7314	700	-0.7 (air)	0.63	3.99
$Y_{0.45}Lu_{0.45}Ca_{0.1}Ba_2Cu_3O_{6.8}$	3.8280	3.8829	11.6724	550	-0.7 (air)	2.6	4.1
$Eu_{0.45}Pr_{0.45}Ca_{0.1}Ba_2Cu_3O_{6.8}$	3.8580	3.8967	11.7242	550	-0.7 (air)	1.31	3.44

From the temperature dependences of the lattice parameters, the values of the thermal expansion coefficients are determined using the expression $\alpha_x = -\frac{1}{X}\frac{\Delta X}{\Delta T}$, where X is the volume or the primitive cell parameters. The obtained data are shown in Figure 1 for Eu-123 compounds with different oxygen content. For the remaining compositions, the results have a similar appearance.

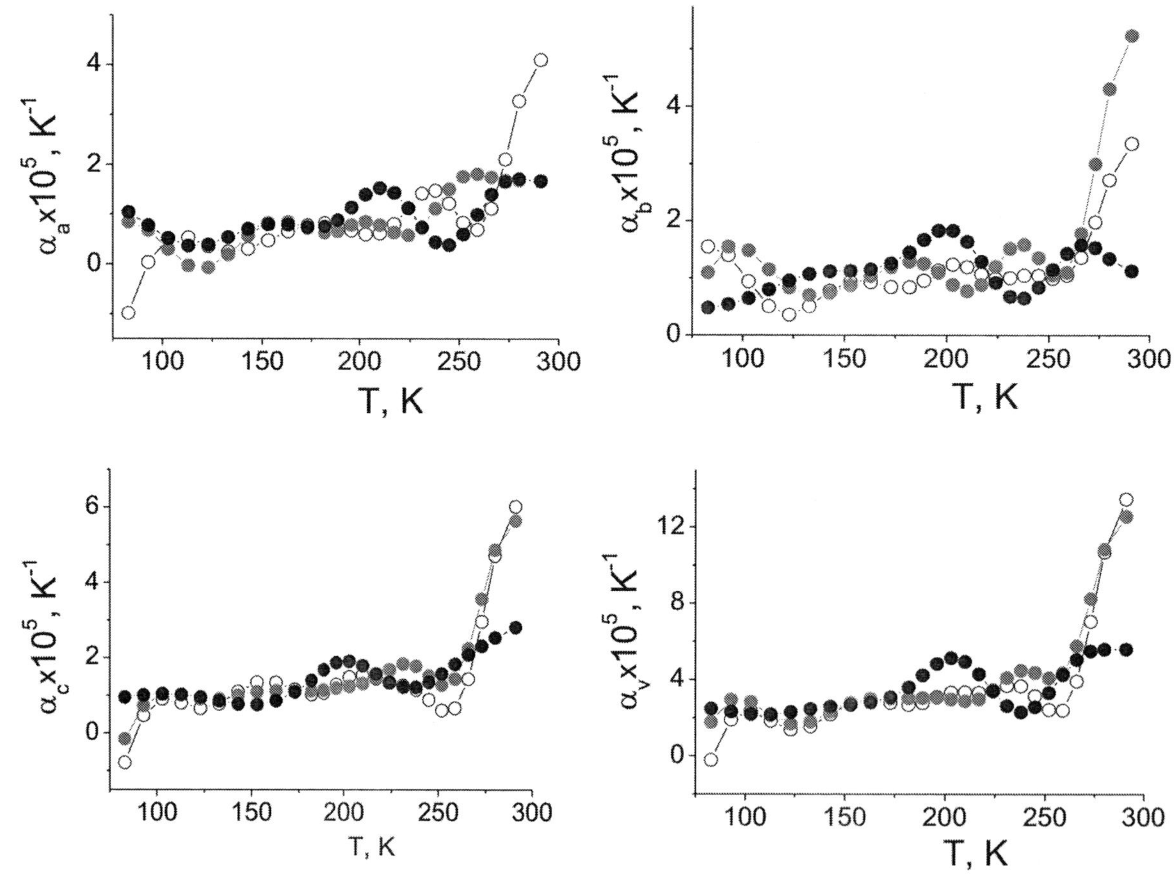

Figure 1. Temperature dependences of linear and volumetric coefficients of thermal expansion for samples $EuBa_2Cu_3O_y$; y=6.64 (white characters), y = 6.67 (gray characters) and y = 6.854 (black characters).

IOP Conf. Series: Materials Science and Engineering **525** (2019) 012054 doi:10.1088/1757-899X/525/1/012054

3.2. Mechanical activation

The task of studying the stability of $RBa_2Cu_3O_y$ under the water vapor and carbon dioxide was solved using mechanical activation, since during mechanical activation there is a significant increase and activation of the free surface that is in contact with the atmosphere, $RBa_2Cu_3O_y$ particles also cleavage predominantly along the base plane of the crystallites, which brings chemically active structural layer of BaO to the open surface. The particle size of the resulting material, determined by analyzing images of a Carl Zeiss EVO 40 scanning electron microscope, was 3–5 μm. The particles were agglomerates consisting of coalescent crystallites ~ 0.3 microns in size. Mechanical activation during the first 1-2 minutes in the AGO-2 mill leads to a significant reduction in the size of coherent scattering regions, with an increase in the duration of grinding to 10 minutes, there is practically no further change. In the initial single-phase Y-123, with an increase in the grinding time, the amount of the impurity phase Y_2BaCuO_5 continuously increases, and in the Nd-123, on the contrary, a small amount of the impurity phase $Ba_2Cu_3O_{5+x}$ present in the initial sample disappears in mechanically activated samples. This confirms the high chemical resistance of Nd-123 to the effects of atmospheric moisture and CO_2 compared to Y-123.

Thus, as a result of mechanical activation, the surface of Y-123 powders was covered with the products of hydrolysis and carbonization to a considerable depth. The process of chemical degradation occurs in the usual way — through the formation of the "green" phase of Y_2BaCuO_5. The study of chemical degradation for mechanically activated Nd-123 showed that the interaction of this oxide with CO_2 is not directly related to the hydrolysis process, as it happens in $YBa_2Cu_3O_y$. Carbonization of Nd-superconductor, in this case, is homogeneous, through the stage of carbon dioxide dissolution in the lattice. This mechanism now eliminates the usual hydrolysis reaction for $RBa_2Cu_3O_y$ with the formation of the characteristic "color" phase R_2BaCuO_5. As a result, the rate of chemical decomposition of mechanically activated $NdBa_2Cu_3O_y$ under the action of water is relatively low.

Thus, it was found that for Y-123 oxide, mechanical activation serves only as an accelerator of chemical degradation, while for Nd-123 it changes the mechanism of its interaction with moisture to less active, and superconducting characteristics close to the maximum.

3.3. Phase analysis

The stability of R-123 compounds in air at room temperature was investigated by comparing the initial X-ray diffraction data and after exposure on the air for several (3 to 7) years. The samples were kept in the form of powder or compressed tablets with a diameter of 1 cm and a thickness of 1-2 mm for Nd-123, as well as for all other R-123 studied, long exposure to air does not lead to the destruction of the superconducting phase, unlike the composition of Lu, Y, Ca-123, the tablet surface of which was found to be covered with heterogeneous formations. According to the XRF data, for the Lu, Y, Ca-123 superconducting phase completely disappeared, the decay products were determined using the PDF-2 database as $Y_{1.4}BaCu_{1.6}O_5$ (00–046–0895), $Ba_2Cu_3O_{5+x}$ (00-040-0312) and $CaCO_3$ (00-001-0628), are shown in Figure 2.

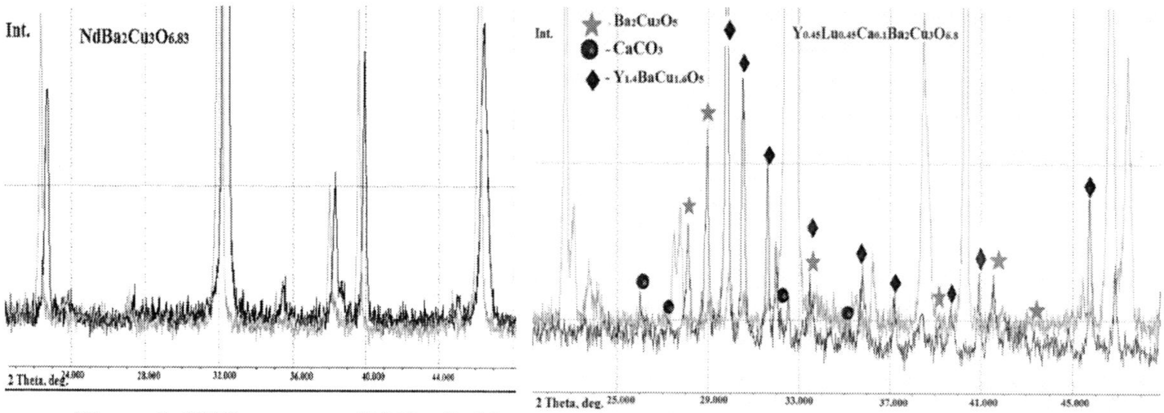

Figure 2. XRD patterns of $NdBa_2Cu3O_{6.83}$ samples (left) and $Y_{0.45}Lu_{0.45}Ca_{0.1}Ba_2Cu_3O_{6.8}$ (right) newly synthesized (grey line) and aged in air for 3 years (black lines).

4. Conclusion

The substitution of Y / R in Y-123 leads to the disappearance of spikes in the linear and volumetric coefficients of thermal expansion. The substitution (Y^{3+}/Ca^{2+}) is accompanied by an increase in the microhardness of the samples. The composition of Nd-123 demonstrates enhanced chemical stability with respect to water vapor and CO_2 when exposed to air.

Acknowledgement

The work was performed within the framework of the State task of the Institute of Metallurgy of the Ural Branch of the Russian Academy of Sciences, no. 0396-2015-0075 "Chemical equilibria in multicomponent manganites, ferrites and REM cuprates as the basis for creating new materials with promising functional properties"

References

[1] Sedky A, Bander Abu-Ziad., New investigation for Tc depression by Ca in $Y_{1-x}Ca_x$: 123 superconducting systems // Physica C, 2010, V.470. P. 659–668.

[2] Nazarova E., Zaleski A., Buchkov K., Doping dependence of irreversibility line in $Y_{1-x}Ca_xBa_2Cu_3O_{7-d}$ // Physica C, 2010, V.470. P.421–427

[3] Liyanawaduge N.P., Anuj Kumar, Shiva Kumar, Karunarathne B.S.B., Awana V.P.S., The Role of Ca in Superconducting and Magnetic Properties of $Y_{1-x}Ca_xBa_2Cu_3O_{7-\delta}$ (x = 0.0–0.30) //J.Supercond. Nov. Magn, 2012, V.25. P.31–37.

IOP Conf. Series: Materials Science and Engineering **525** (2019) 012055 doi:10.1088/1757-899X/525/1/012055

Investigation of viscosity of construction mixtures applied for 3D printing

A V Sharanova[1], A D Panfilova[1], A A Plahtiy[1] and M A Dmitrieva[1]

[1] Immanuel Kant Baltic Federal University, 14 Nevskogo St, Kaliningrad 236016, Russia

E-mail: a.sharanova@gmail.com

Abstract. Additive technologies meet the requirements of the modern technological stage, which applies to all areas industrial activities. Construction using additive technologies is actively developing and this technology has been introduced into production. All this makes 3D concrete printing a relevant and promising direction for the modernization of the construction industry. Concrete mixtures with certain characteristics are used as the "ink" for the construction 3D-printers. In this paper, we investigate the rheological properties of mortars for construction. We also determined the effectiveness of the mechanical activation of the initial components of the mixture, evaluated its effect on the rheological properties and the setting time of the mortar. The optimal values of the dynamic shear stress for the compositions for construction 3D printing are calculated.

1. Introduction

Additive technologies are actively used in many areas of production (mechanical engineering, metallurgy, etc.), as they correspond to the modern technological stage requirements aimed at ensuring the rational use of energy and material resources. In construction, additive technologies have also found wide application, and at the moment concrete 3D printing is successfully developing and being applied in practice. This method of construction has several advantages compared with the traditional method. The layer printing of structures can significantly reduce construction time, material consumption, and labor and machine costs. During construction using additive technologies, there is practically no construction waste left, since only as much material is used as necessary. In addition, it becomes possible to create complex geometry and make the objects architecturally expressive using construction 3D printing [1, 2]. All this allows one to achieve significant results of construction, and makes concrete 3D printing one of the most popular research subjects in this area.

For the construction 3D-printers, concrete mixtures are used as “ink”. Their deformative properties should make it possible to grow the building structures layer by layer. The curing rate of the formed layer is the defining characteristics of the concrete mixtures for a 3D printer. The deformative properties of concrete mixtures are characterized by the rheological parameters [1]. In the construction industry, instead of the real rheological characteristics (viscosity, ultimate shear stress, etc.), some generalized indicators are used: conditional viscosity, texture of the binder, workability of mortar or concrete mixes, etc.

It has been proved that the mechanical activation of cement leads to an increase in the activity of cement at the early stages of hydration, as well as a reduction in the setting time [3]. The mechanical activation of cement contributes to an increase in the strength of concrete in the initial periods of hardening. This is a necessary condition for the using of additive technologies in construction. Thereby, the development of the methods of dry building mix activation is the key to developing a technology for preparing concrete for additive technologies in construction.

In the present work, we carried out the mechanical activation of individual components of the dry building mixture. The influence of the mechanical activation of components on the rheology of the mortar was determined. We considered the flowability and consistency of cement paste as the rheological characteristics of concrete solutions that determine their applicability in 3D printers. Such properties as workability, i.e. the ability to fill the form, while maintaining its uniformity, viscosity, as well as the ultimate dynamic shear stress were investigated. Simultaneously with the rheological characteristics, the setting time of mortars was controlled.

2. Materials and Methods

In the experimental part of the work, we carried out the mechanical activation of the initial components of the concrete mixture. The following rheological parameters were determined: the flowability (using ring spreading, GOST 31356-2007) and the consistency of the cement paste by the Vicat apparatus with a plunger (GOST 310.3-76). When choosing the optimal composition of the initial building mix, it is necessary to take into account not only the rheological characteristics but also the speed of curing. For this purpose, the setting time of the concrete mixture was determined using the Vicat apparatus with a needle according to GOST 310.3-76. In another work, an additional calorimetric analysis was performed to assess the activity and rate of cement paste hydration [4].

2.1. Materials

Compositions of the initial dry building mixtures were selected based on the known experimental data [1, 5, 6, 7]. The following materials were used for the preparation of the mortar: 1) construction sand: 0–3 mm fraction; size modulus Ms = 3.15; origin location: Kaliningrad region; 2) Portland cement Eurocem super 500, CEM I 42.5H, Petersburg Cement; 3) polypropylene fiber: length of 12 mm, thickness of 35 μm, country of origin: Russia; 4) fly ash; 5) highly active metakaolin: white, country of origin: Russia, Chelyabinsk region, Synergo; 6) STACHEMENT 1267 hyperplasticizer based on polycarboxylates for the production of ready-mixed concrete: country of origin: Poland.

2.2. Methods

The initial components of the building mixtures were modified by mechanical activation in a Retsch EMax high-speed ball mill at speeds of 500 rpm, 750 rpm and 1000 rpm for 5 minutes. Corundum balls were used as the grinding media. A series of activations was carried out to study changes in the rheological properties of the mortars caused by the mechanical activation of the initial components. We carried out activations of the following components: 1) cement, 2) ash, 3) cement and ash, 4) cement and sand. Table 1 shows 13 testing compositions.

Table 1. Testing compositions.

Sample No.	Cement (g)	Sand (g)	Water (ml)	Fly ash (g)	Metakaolin (g)	Fiber (g)	Hyperplast. (ml)
1	150 (500 rpm)	321.7	114.2	43	21.5	0.33	3.22
2	150 (750 rpm)	321.7	99.2	43	21.5	0.33	3.22
3	150 (1000 rpm)	321.7	99.2	43	21.5	0.33	3.22
4	150	321.7	104.2	43 (500 rpm)	21.5	0.33	3.22
5	150	321.7	99.2	43 (750 rpm)	21.5	0.33	3.22
6	150	321.7	99.2	43 (1000 rpm)	21.5	0.33	3.22

IOP Conf. Series: Materials Science and Engineering **525** (2019) 012055 doi:10.1088/1757-899X/525/1/012055

7	150 (500 rpm)	321.7	99.2	43 (500 rpm)	21.5	0.33	3.22
8	150 (750 rpm)	321.7	99.2	43 (750 rpm)	21.5	0.33	3.22
9	150 (1000 rpm)	321.7	99.2	43 (1000 rpm)	21.5	0.33	3.22
10	150 (500 rpm)	321.7 (500 rpm)	99.2	43	21.5	0.33	3.22
11	150 (750 rpm)	321.7 (750 rpm)	99.2	43	21.5	0.33	3.22
12	150 (1000 rpm)	321.7 (1000 rpm)	106.7	43	21.5	0.33	3.22
13	150	321.7	114.2	43	21.5	0.33	3.22

2.2.1. The ring flowability test. The flowability of dispersed mixtures is determined using a ring test by the diameter of the mixture spread (mm) flowing out of the ring when the ring is removed (Figure 1). The auxiliary devices for testing are a ring of stainless steel with a polished inner surface with an inner diameter of 70 mm and a height of 50 mm, a 400x400 mm glass plate, and a metal ruler. The prepared mixture is placed within to the ring installed in the center of the glass plate. The glass and the ring are pre-wiped with a damp cloth. The surface of the mixture is struck of flush with the top edge of the ring. After 10-15 seconds, the ring is lifted vertically to a height of 10-15 cm and set aside. Immediately after lifting the ring, the diameter of the mixture spreading is measured with a ruler in two mutually perpendicular directions with an error of ± 5 mm. Then, the average value of the results of the two measurements of the spreading diameter is calculated. The average is rounded off to whole mm. The test is repeated using a different mortar sample. The average value of the spreading diameter of the two mixture samples, rounded off to whole mm, is taken as the final result.

2.2.2. The mortar consistency using the Vicat apparatus. To determine the consistency of the mixture, the Vicat apparatus is used. The apparatus is equipped with a metal plunger with a diameter of 10 mm and a ring with an internal diameter of 65 mm and a height of 40 mm (Figure 2). After mixing the solution, the ring is filled quickly in one step. Then, the ring is shaken five or six times, while tapping the plate against a solid base. The surface of the mixture is struck off flush with the edges of the ring. The plunger of the device is brought into contact with the surface of the mixture in the center of the ring and the rod is secured with a locking device. Then the rod is quickly released and the plunger sinks freely into the mortar. The depth of plunger immersion is measured by the instrument scale in 30 seconds.

2.2.3. The setting time using the Vicat apparatus. The Vicat apparatus with a metal needle and a ring is used to determine the setting time (Figure 3). The needle of the device is brought into contact with the surface of the mixture, which had been prepared and placed into the ring. In this position, the rod is first secured a stopper and later released. The needle.sinks freely into the mortar. The moment when the setting begins is determined with the free lowering of the needle. The needle is immersed in the mixture every 10 minutes. The ring is moved after each dive so that the needle does not fall into its original position. The time from the moment of water pouring to that when the needle is 2-4 mm above the plate is considered the initial setting time of the cement paste.

Figure 1. Ring flowability test.

Figure 2. Consistency test.

Figure 3. Initial setting time test.

3. Results and Discussion

The following table shows the results of the tests.

Table 2. Results.

Sample No.	Ring flowability (mm)	Vicat depth (mm)	Initial setting time (min)
1	70	37	190
2	72	35	195
3	72	33	210
4	70	36	220
5	71	37	200
6	71	37	180
7	71	31	250
8	71	36	230
9	70	36	260
10	70	36	180
11	70	37	160
12	70	35	160
13	73	31	240

The concrete mixtures are classified as non-Newtonian fluids. Therefore the value of concrete viscosity is not constant and depends at any time on the shear stress and the strain rate (the shear rate). In [8] the group of researchers derived a dependence that makes it possible to obtain the numerical characteristic of dynamic stress during shear on the basis of the experimental data obtained.

$$\tau_0=3/2\pi Rh$$

Based on the value of the Vicat plunger depth and the radius of the plunger (5 mm), the following data were obtained:

Table 3. Dynamic shear stress.

Sample No.	h (mm)	τ_0(MPa)
1	3	0,032
2	5	0,019
3	7	0,014
4	4	0,024
5	3	0,032
6	3	0,032
7	9	0,011
8	4	0,024
9	4	0,024
10	4	0,024
11	3	0,032
12	5	0,019
13	9	0,011

The best results were observed for mortars with the intensive joint mechanical activation of sand and cement. Mechanical activation of fly ash was ineffective.

4. Conclusion

As applied in civil engineering, additive technology is a new research-intensive method to construct buildings and structures. The use of 3D printing in construction has several significant advantages, such as saving time and labor or the rational use of raw materials and finished materials. Although the principles for the development of building mixtures for 3D printers exist already, this topic requires a large amount of work. This work includes the creation of a regulatory framework and the development of a technology for mixtures production on an industrial scale.

The mechanical activation of components is one of the possible modifiers of dry construction mixtures used in additive technologies. Additional grinding of the components reduces the setting time of the mixture and increases the initial strength of the mortar. This paper investigated the effect of the mechanical activation of components on the rheological properties of the mixture. The compositions modified by activation have good workability; the mixture holds its shape well and does not spread (the ring flowability test). The initial setting time is also reduced. It was established that the joint activation of cement and sand was the most effective, as it reduced the initial setting time 1.5-fold compared with the non-activated composition. The values of the dynamic shear stress are determined by the Vicat apparatus. The calculations showed that for the compositions suitable for 3D printing, the values of dynamic shear stress were within 0.019-0.032 MPa.

A study into the rheological properties of mortars and into the rate of hardening makes it possible to evaluate whether the mortars can be used as "ink" for a 3D construction printer. The method of mechanical activation of the original cement-sand mixture can be applied to control effectively the rheological characteristics of concrete mixtures.

References

[1] Vatin N, Chumadova L, Goncharov I, Zykova V, Karpenya A, Kim A and Finashenkov E 2017 3D printing in construction *Construction of Unique Buildings and Structures* **1(52)** 27–46 (rus)

[2] Batrakov V 2006 Concrete modifiers: new opportunities and prospects *Stroitelnye materialy* **10**

IOP Conf. Series: Materials Science and Engineering **525** (2019) 012055 doi:10.1088/1757-899X/525/1/012055

1-8 (rus)
[3] Sobolev K, Lin Z, Cao Y, Sun H, Flores-Vivian I, Rushing T, Cummins T and Weiss W J 2016 The influence of mechanical activation by vibro-milling on the early-age hydration and strength development of cement *Cement and Concrete Composites* **71** 53-62
[4] Sharanova A, Lenkova D and Panfilova A 2018 Study of strength kinetics of sand concrete system of accelerated hardening *IOP Conf. Series: Materials Science and Engineering* **347** 012014
[5] Koval Y 2017 The benefits of 3d printing technology with the use of fibre in low-rise construction *Synergy journal* **11** 960-964 (rus)
[6] Belyakova E, Moskvin R and Tarakanov O 2017 Cement on fly ash for modern building technologies *Regional architecture and engineering* **1** 5-10 (rus)
[7] Slenkov V, Minakov Y, Kononova O, Anisimov S, Smirnov A and Leshkanov A 2015 Heave concreate production plasticizing additives effectiveness *Sovremennye problemy nauki I obrazovaniya* **2** 1 (rus)
[8] Lootens D, Jousset P, Martinie L, Roussel N and Flatt R J 2009 Yield stress during setting of cement pastes from penetration tests *Cement and Concrete Research* **39(5)** 401-408

IOP Conf. Series: Materials Science and Engineering **525** (2019) 012056 doi:10.1088/1757-899X/525/1/012056

Low-temperature cementation – perspective technology of hardening of toothed parts

T V Shveyova and A I Shveyov

Kazan Federal University, Russia

E-mail: asttv@mail.ru

Abstract. The structure, properties and geometrical accuracy of gear wheels after chemical heat treatment which cementation was carried out at 930ºC and 870ºC are investigated. Advantage of low-temperature cementation of toothed parts in size of a radial runout, fluctuation of a side clearance, ovality of an internal opening and the area of a spot of contact between teeths of the strengthened gear wheels is shown. Influence of structure of the endothermic atmosphere on mechanical properties and intensity of saturation by carbon steel 15HGN2TA and 20HGNMTA is investigated. It is established that increase in content of hydrogen from 20% to 40% in the endoatmosphere increases saturation speed carbon steel by 10 – 34%, excludes their hydrogen saturation and keeps indicators of mechanical properties of the cemented parts. The technology of chemical heat treatment of toothed parts including cementation at 870ºC in the endothermic atmosphere containing 40% of hydrogen in the structure is developed and recommended for application.

1. Introduction.

The problem of increase in firmness of gear details in operation is considered one of most "ageless". Many experts solve it by application high-strength steel or due to use of the effective strengthening technologies. And in those and other cases indispensable condition is stability as on the accuracy of all geometrical parameters of detail, and on mechanical-and-physical properties of material of product. The special attention is deserved by the gear details high-loaded in operation which during the work at the same time are exposed to wear and test static, impact, flexural, contact and cyclic loads [1-5].

Results of experimental works and synthesis of production data show that in gear details it is inherent dispersion on macro – and to microgeometry, structure and mechanical-and-physical properties [4-7]. Receiving cases on the strengthened details of "spotty" cementation [8] are frequent that undoubtedly reduces the steel resilience to cyclic and other types of loadings. It is possible to achieve stabilization at the required level on deformation and buckling of details at high complex of mechanical and special properties of hardware due to development and implementation of innovative technologies in machine-building production. Considering that the center of collapse of details is in most cases observed on surface [5,9-10], the paramount value has to be given to blanket.

The work purpose – research and realization of the hidden reserves in technologies of steel redistribution at the machine-building enterprises directed to increase in accuracy of the geometrical sizes of gear details and stabilization at the high level of strength and plastic properties of hardware.

IOP Conf. Series: Materials Science and Engineering **525** (2019) 012056 doi:10.1088/1757-899X/525/1/012056

2. Methods of probe and results of work.

One of ways of complete solution of objective is the choice and justification of temperature of cementation of details. For confirmation of this hypothesis in work influence of technology of chemical heat treatment on structure, properties and geometrical accuracy of the conducted pinion gears (fig. 1) made of steel 20HGNMTA is investigated.

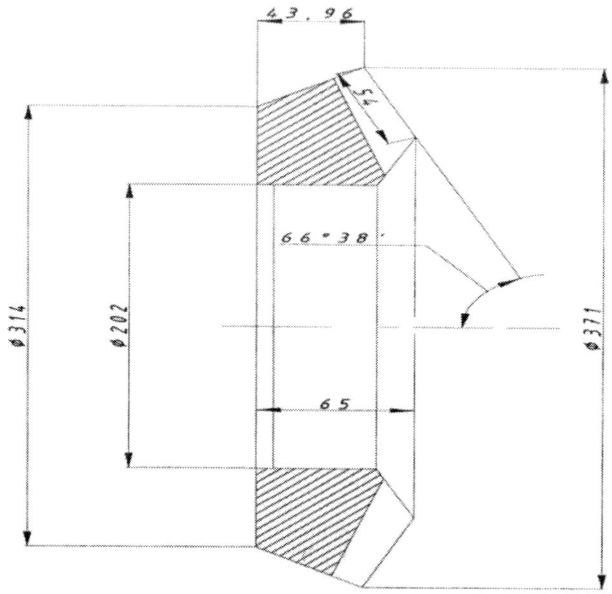

Figure 1: Form and the sizes of the conducted pinion gear from steel 20HGNMTA

The Chemical Heat Treatment (CHT) was executed in units of «Holkroft» and «Ipsen» with use of the chamber mechanized furnaces for cementation. On units of «Ipsen» «The Karbo-Prof» program allowing to steer the carbon potential of the atmosphere on duration of process of cementation of details is used. After cementation of detail chill to 850°C, tempered in the MZM hot oil – 26, washed out and subjected to holiday at temperature 180-200°C. Later CHT on details controlled the hardness, microhardness, depth and microstructure of the strengthened layer and core. Applied microscopes to metalgraphic probes «Neofot – 21», «Epitip – 2» and «IM 7200» with the system of the analysis of panoramic images «Video the Test — M» and the «Thixomet PRO» software product.

Took distance from surface to zone with the microhardness of HV of 550 determined by microhardness tester of «Dyurimet» at loading (0.1 N) for the effective depth of the cemented layer. Content of chemical elements in steel was determined on AH-7529 and AH-7560 analyzers, the AFS-51 spectrograph with the specialized software of SBP and NEXT and spectrometer «Spektrolab M-12» (Germany). Hydrogen content in steel was estimated according to GOST 17745 – 90 on the RH device – 2 «Leco».

Control of geometrical parameters of the strengthened details was carried out in laboratory and working conditions with use of instrumentation and the specialized equipment. The form, arrangements and the area of spot of contact piece between teeths of pinion gears in gearing were registered on control – the inspection machine. Relative areaof spot of contact piece was defined by the relation actual to the area of spot recommended by regulations.

For determination of ovality of bore the caliper of ShTsTs-2 and Mahr 16 EW (Germany) used. Nonflatness of butt, size of radial run-out and backlash variation were measured on the multiple-purpose center Mar Gear GMX-400 (Germany).

Results of comparative probes of geometrical accuracy of pinion gears after CHT, structure and properties of the case are given in table 1. At measuring control it is established that all details which

IOP Conf. Series: Materials Science and Engineering **525** (2019) 012056 doi:10.1088/1757-899X/525/1/012056

have undergone low-temperature cementation at CHT have undoubted advantage on stability of change of the analyzed dimensional indicators in comparison with details cemented on traditional technology – at temperature 930°C. Drop of temperature of cementation has allowed to reduce interval of dispersion of values on fluctuation of side clearance by 1.8 times, ovality of internal bore by 1.3 time, not planeness of butt by 2.0 times and the size of radial run-out by 3.5 times. Higher instability in dimensional parameters of details after cementation at temperature 930°C it is connected with disintegration of separate chemical compounds, austenite alloying with these elements and change of the grain building became which have impact on processes of phase and structural transformation when heating and cooling and, respectively, on change of volume and the sizes of hardware.

Table 1: Properties of the cemented* gear steel 20HGNMTA details

Tempe-rature (time) of cementation, °C	Thickness of the cemented layer, mm	Surface hardness, $HV_{0,1}$	Microstructure of the cemented layer	Change of parameters of geometrical accuracy of details, mm				Relative area PC, %
				KBZ	EWO	NT	VRB	
930°C (13 hours)	1,3	762	$M+A_{res}$	0,05 – 0,13	0,08 – 0,14	0,02 – 0,07	0,03 – 0,16	58,86 / 66,99
870°C (17,5 hours)	1,4	827	$M+A_{res}$	0,05 – 0,09	0,02 – 0,06	0,01 – 0,03	0,04 – 0,07	86,10 / 81,69

Note: * At cementation the gas atmosphere with the maintenance of 20% of H_2, 20% with and 60% of N_2 is used. Designation: KBZ – backlash variation; EWO – ovality of internal bore; NT – notflatness of butt; PK- contact patch: in numerator on bent – in denominator on the convex party of teeth; VRB – size of radial run-out. M – martensite; A_{res}. – residual austenite.

In manufacturing techniques of the majority of details of cars from chromium-nickel steel including tappets of the hinge of the KAMAZ car, thermal methods of processing – hot plastic deformation are used, thermal and chemical heat treatment. Temperature of heating of metal at these stages of processing usually exceeds 900 °C. For example, heating under hot volume stamping makes 1220-1250 °C, and at heat treatment ~930 °C. At such temperatures of heating in steel type 20H2N4A, 12HN3A and similar low resilience to growth of grain (fig. 2) is observed.

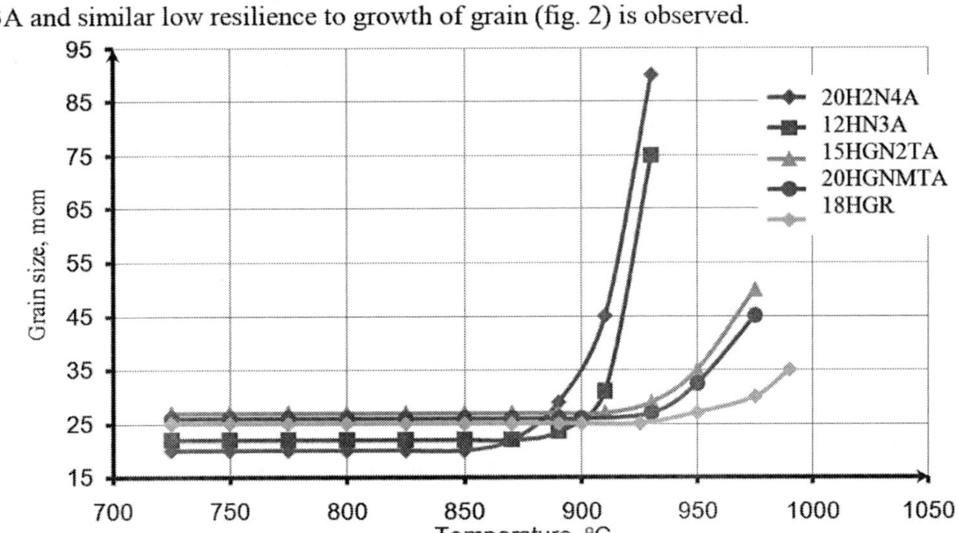

Figure 2: Temperature dependence of growth of grain in steel

Chemical heat treatment in the specialized unit for chromium-nickel l steel (tab. 2, scheme 1) have not allowed to reach demanded (200 thousand cycles) the level of cyclic durability of details.

Periodic tests of 118 details strengthened on the specialized unit (tab. 2, scheme 1) have shown that their cyclic firmness to be ranging from 58 up to 118 thousand cycles. The characteristic type and the place of collapse of detail and also the macrostructure of surface of break are presented in figure 4. As a result of metalgraphic probes of details after collapse it is established that the hardness of surface was in limits of 57-60 HRC, and cores changed from 36 HRC to 45 HRC. Depth of the cemented layer fluctuated from 1.20 mm to 1.60 mm. In microstructure of the strengthened cemented layer presence of martensite (4-7ball), residual austenite and carbides (1-4ball) is revealed. Existence so-so – and coarse needled martensite testifies to the coarse-grained structure of steel [2].

It is established that the reason of the coarse-grained building in steel 20H2N4A, over crystallizable at 830 °C (scheme 1, tab. 2) is inheritance of large grain from cementation heating (920°C) and preservation of not broken up austenite at isothermal endurance (580±20 °C) before heating under training. It is shown that such condition of steel reduces the impact strength and indicators of strength of steel products (tab. 2 and tab. 3).

One of technology solutions on providing at the high level of properties of the cemented details from chromium-nickel steel application of low-temperature gas cementation. Despite reduction in the rate of diffusive saturation carbon, in such steel does not observe growth of grain at offered (870 °C) temperature that will not demand additional heat treatment on correction of the coarse-grained building. Chemical heat treatment of details on the offered technology (tab. 2, scheme 5) has provided the high level of properties of steel and its fine-grained building (fig. 3). It is necessary to notice that the general time of thermal treatment of steel (apart from tempering) on this technology is 22 hours instead of 20-23chas by traditionally known options in which the steel over crystallizable is provided (tab. 2, scheme 1 and 2).

Many indicators of operational properties of gear details are connected with location, form and the area of spot of contact piece between teeths in gearing. The smoothness of gearing, noise level and size of contact tension depends on this criterion. The contact piece spot on teeths of pinion gears is normalized according to GOST 1643-81. It is revealed that CHT of pinion gears at cementation temperature 870°C allows to receive more favorable area and arrangement of spot of contact piece on teeths, than on teeths of the details which have undergone cementation at more high temperature.

On technical and economic indicators of drop of temperature of cementation can inevitably lead to increase in product cost. It is explained by reduction in the rate of saturation became carbon [12-13] and need of preservation on details of thickness of the strengthened layer. Therefore in thermal production are forced to increase duration of process of CHT.

The analysis of the CHT technologies used by various domestic and foreign enterprises demonstrate that at cementation in overwhelming number of cases the endothermic atmospheres as a part of which are present CO, H_2 and N_2 [11-13] are applied. Have essential difference of the atmosphere on the hydrogen content (20% and 40%) and nitrogen (60% and 40%). Not enough data on comparative characteristic of the endoatmospheres it is also impossible to give preference in information sources any of them. Therefore in work influence of hydrogen in such atmosphere on steel saturation speed by carbon and indicators of mechanical properties of the cemented products is estimated.

It is established that saturation speed carbon of details of various configuration and weight and made of steel 20HGNMTA, 15HGN2TA and 12HN3A at cementation with use in the endothermic atmosphere of 40% of hydrogen is higher, than when processing in the atmosphere from 20% of hydrogen. So, on steel 20HGNMTA the speed of saturation has increased from 0.073-0.090 mm/hour to 0.100-0.108 mm/hour, and on steel 12HN3A – from 0.087 to 0.108 mm/hour at temperature of the cementation of details corresponding 910°C in the mechanized furnaces of Holkroft. The know-how and statistical data have shown that due to change of content of hydrogen from 20 to 40% in the endothermic atmosphere the speed of saturation steel carbon increases by 10 – 34% (tab. 4).

IOP Conf. Series: Materials Science and Engineering **525** (2019) 012056 doi:10.1088/1757-899X/525/1/012056

Table 2: Structure and properties 20H2N4A after hardening of details according to various schemes CHT

№	Scheme CHT and duration of process	Microstructure of the strengthened layer	Hardness $\frac{HRC_{surf.}}{HRC_{core}}$	Durability	
				$\frac{M_{dest.}}{M_{prop}}, \kappa N \cdot m$	N, thousand cycles
1	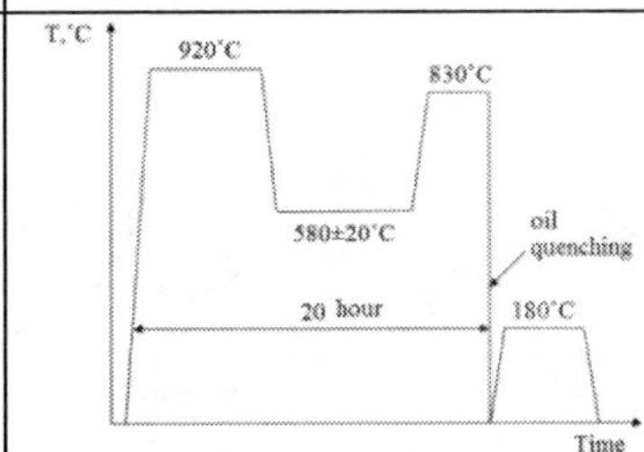	$M_{(5\delta)} + A_{res} + K$	$\frac{59}{43}$	$\frac{24{,}32}{17{,}18}$	90,6
2	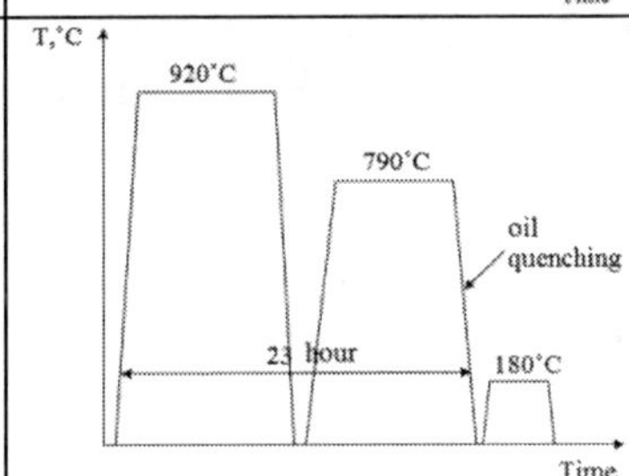	$M_{(7\delta)} + A_{res} + K$	$\frac{55}{43}$	$\frac{24{,}79}{17{,}60}$	77,7
3	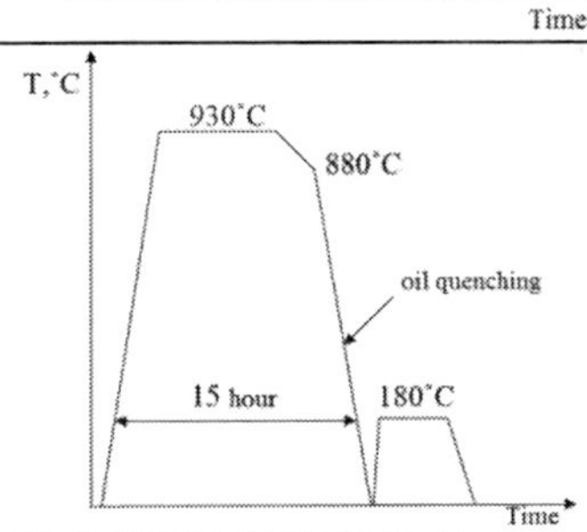	$M_{(6\delta)} + A_{res} + K$	$\frac{57}{42}$	$\frac{22{,}30}{15{,}90}$	66,6
4	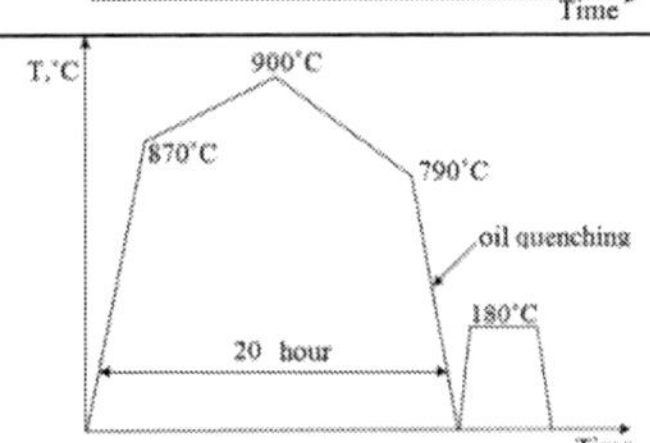	$M_{(5\delta)} + A_{res} + K$	$\frac{58}{44}$	$\frac{25{,}12}{17{,}80}$	127,1
5	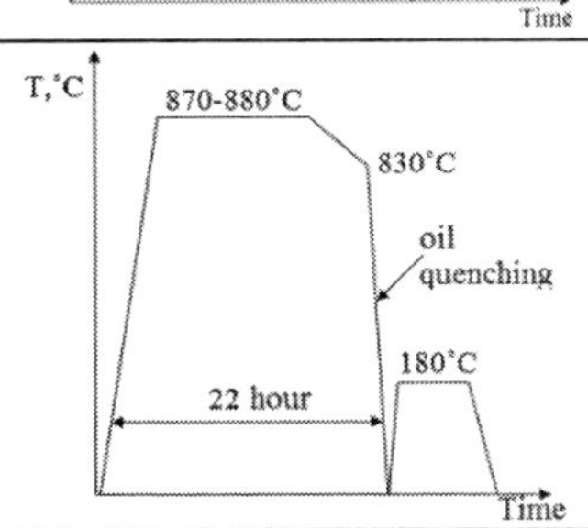	$M_{(3\delta)} + A_{res} + K$	$\frac{58}{42}$	$\frac{24{,}20}{18{,}40}$	314

Designations: M – martensite (point); A_{res}– residual austenite; K – carbides; M_{dest}– destroing moment; M_{pr}– proportionality moment; N – number of cycles before collapse; $HRC_{surf/core}$– hardness of surface (core).

Table 3: Properties of steel 20H2N4A after "false" cementation

Parameters of "false" cementation T,°C	Time, hour	Grain size, point	Impact elasticity (KCU), Dj/m^2
870	20	8-11	4,15-5,04
890	20	7-11	3,50-4,94
910	15	6-9	-
920	14	5-9	1,2-1,7
930	15	4-7	1,1-1,7

Note: At each option about 15 samples are tested for impact elasticity.

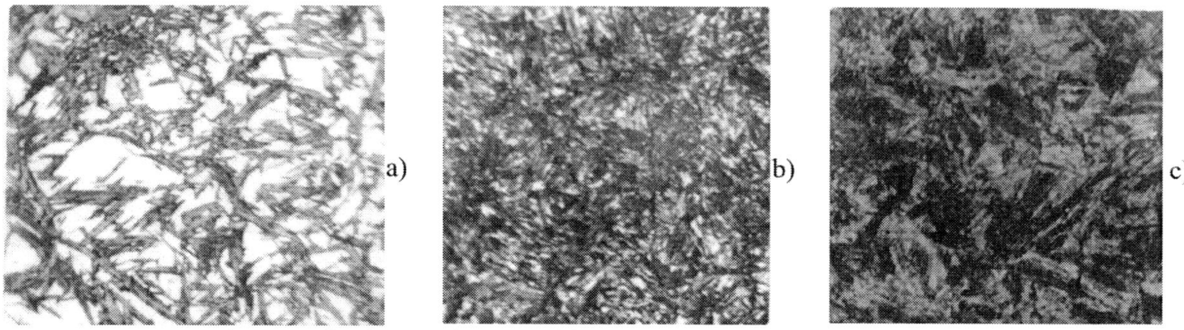

martensite + retained austenite + carbides troostomartensite

Figure 3: Microstructure of the strengthened layer at distance of 0.3 mm (a) and 0.5 mm (b) from surface and microstructure of core (c) of details from steel 20H2N4A after CHT. ×1000.

Table 4: Saturation speed carbon at cementation of details from variety of steel

Detail, material	Cementation time, h	Depth of layer, mm		Speed of saturation, mm/h	
		In the hollow	On the worker surfaces	In the hollow	On the worker surfaces
cylindrical gear, steel 20HGNMTA	14/17	1,29±0,14 1,20±0,16	1,40±0,07 1,25±0,15	0,092/0,070	0,100/0,073
gear, steel 20HGNMTA	14/17	1,33±0,16 1,28±0,13	1,43±0,20 1,29±0,13	0,095/0,076	0,102/0,082
the same	11/12	1,22±0,18 1,06±0,18	1,30±0,16 1,07±0,15	0,101/0,088	0,108/0,089
-//-	11/12	1,18±0,13 1,07±0,10	1,26±0,13 1,08±0,09	0,098/0,089	0,105/0,090
-//-	11/12	1,22±0,12 1,06±0,13	1,29±0,12 1,07±0,13	0,101/0,088	0,102/0,089
piston pin, steel 12HN3A	12/16,5	-	1,3/1,4	-	0,108/0,087

Note 1. In numerator sizes the corresponding cementations are specified in the atmosphere from 40% of hydrogen, in denominator - from 20%
2. Depth of layer corresponds to results of mathematical processing of 80 and more details

About negative impact of hydrogen in steel it is specified in many works of domestic and foreign scientists [14-16]. Therefore in work it was important to establish about possible hydrogen saturation of steel in the course of gas cementation in the water containing atmosphere.

To define how hydrogen content in the gas atmosphere affects the mechanical properties cemented steel, the samples which have undergone cementation in environments from 20 and 40% of hydrogen tempered in oil MZM– 16, washed out in KM-1 solution, released at 180°C within 120 minutes and then, those from them which on depth of the cemented layer differed no more than for 10%, were tested on torsion in accordance with GOST 3565-80. It is established that indicators of strength and the limit of proportionality cemented steel 15HGN2TA and 20HGNMTA in the atmosphere of the furnace practically do not depend on hydrogen content (tab. 5). At the same time, amount of hydrogen in processed steel also does not depend on its contents in the atmosphere: after processing in the atmosphere from 40% of hydrogen it is equal in steel 15HGN2TA to 1.25 ppm., from 20% of hydrogen – 1.41 ppm; in steel 20HGNMTA - 1,33 and 1,34 ppm respectively.

The fraktografichesky probes of breaks same cemented by steel it is established that in cementation zone the collapse displaced "fragile intergranular" and "visdisity calyceal". Quantity of fragile and visdisity components in break became identically in case of its processing in the gas atmosphere with 20 in 40% of hydrogen [17-18].

Thus, results of mechanical tests and fraktografichesky probes of breaks have shown that change of content of hydrogen in the endothermic atmosphere from 20 and 40% does not increase fragility investigated by steel.

Table 5: Mechanical properties of steel the endoatmospheres cemented in various structures

Steel	Depth of the cemented layer to HV550mm	Surface hardness HRC	Ultimate strength, MPa	Limit of proportionality, MPa
15HGN2TA	1,43/1,30	60/60	1450/1500	750/740
20HGNMTA	1,77/1,72	62/61	1440/1500	860/860

Note. In numerator sizes, the corresponding cementations are specified in the atmosphere from 20%, in denominator – from 40% of hydrogen.

As a result of complex probe of CHT technology of gear details it is revealed that drop of temperature of cementation with 930°C to 870°C allows to minimize change of the sizes of details after hardening, and application of the endothermic atmosphere from 40% of hydrogen instead of 20% will keep the capacity of thermal units in connection with bigger activity its sating ability.

3. Conclusions

1. Influence of temperature of cementation and structure of the endoatmosphere on geometrical accuracy is investigated, the speed of saturation there were carbon and properties of gear details after CHT. Advantage of cementation of gears is shown at 870°C.
2. Increase in content of hydrogen from 20% to 40% in the endothermic atmosphere increases steel saturation speed carbon by 10 – 34% and keeps indicators of mechanical properties of the cemented details at the high level.
3. It is recommended to apply cementation to gear details at 870°C with use of the endoatmosphere containing 40% of hydrogen.

References

[1] Kozlowski I.S. Chemical heat treatment of the gears. (1970)
[2] Sagaradze V.S. Increase in reliability of the cemented details. (1975)
[3] Kudryavtsev V.N. Details of cars. (1980)
[4] Kalner V. D. Quality control of heat treatment of semi-finished products and details: The reference book. (1984)
[5] Zinchenko V.M. Engineering of surface of tooth gears by methods of chemical heat treatment. (2001)
[6] Astashchenko V.I. Technological methods of management of structurization became by

production of details of cars. *Moscow, Akademiya Publishing house.* (2006)
[7] Tylkin M.A. Reference book by the heat-treater of repair service. *Moscow, Metallurgy.* (1981).
[8] Astashchenko, V.I., Karginova, L.A. Blotchy carburization of steel Metal Science and Heat Treatment (1982)
[9] A.M. Sulima, V.A. Shulov, Yu.D. Yagodkin. Blanket and operational properties of details of cars. *Moscow, Mechanical engineering.* (1988)
[10] Suslov A.G. Quality of blanket of details of cars. *Moscow, Mechanical engineering.* (2000)
[11] Firger I.V. Heat treatment of alloys: Reference book. *Mechanical engineering.* (1982)
[12] Borisenok G. V. Chemical heat treatment of metals and alloys: Reference book . *Moscow, Metallurgy.* (1981)
[13] Shmykov A.A. The controlled atmospheres. *Moscow, Mechanical engineering.* (1953)
[14] Morozov A.N. Hydrogen and nitrogen in steel. *Moscow, Metallurgy.* (1968)
[15] Kolachev B.A. Hydrogen fragility of metals. *Moscow, Metallurgy.* (1985)
[16] Shapovalov V.I. Influence of hydrogen on structure and property of iron carbon alloys. *Moscow, Metallurgy.* (1982)
[17] Zinchenko V.M. To question of hydrogen role at chemical heat. MEATH, No. 8, page 2-7. (1986)
[18] Zinchenko V.M., Astashchenko V.I., etc. Intensification of process of cementation there were. Automotive industry, No. 4 – page 30-31. (1986)

IOP Conf. Series: Materials Science and Engineering **525** (2019) 012057 doi:10.1088/1757-899X/525/1/012057

Nanometallocarbosilanes and organoelementoxanes as precursors of components of promising ceramic composites

G I Shcherbakova[1], P A Storozhenko[1], T L Apukhtina[1], D V Zhigalov[1], M S Varfolomeev[1,2], A I Drachev[1] and A A Ashmarin[3]

[1]SSC RF "State Research Institute for Chemistry and Technology of Organoelement Compounds", 105118, Moscow, Russia
[2]Moscow Aviation Institute (National Research University), 121552, Moscow, Russia
[3]Baikov Institute of Metallurgy and Material Science RAS, 119334 Moscow, Russia

E-mail: galina7479@mail.ru

Abstract. Nanometallocarbosilanes and organoelementoxanes represent a new class of unequaled precursors of components of promising ceramic composites. Ceramic-forming polymers (oligomers) have been synthesized in GNIICHTEOS. They can be used to obtain components of carbide and oxide CCM: nanometallocarbosilanes - precursors of SiC ceramics, modified by mixed carbides ($Ta_xZr_yC_z$ or $Ta_xHf_yC_z$). Organoelementoxane oligomers represent a new class of precursors for components (binders; coatings; fibers, etc.) of corundum (α-Al_2O_3), aluminosilicate ($xAl_2O_3{\cdot}ySiO_2$), yttrium aluminum ($xY_2O_3{\cdot}yAl_2O_3$), yttrium-aluminosilicate ($xY_2O_3{\cdot}yAl_2O_3{\cdot}zSiO_2$), magnesium aluminum ($xMgO{\cdot}yAl_2O_3$). yttrium-aluminum-magnesium ($xY_2O_3{\cdot}yAl_2O_3{\cdot}zMgO$) and high-temperature binary ceramics based on aluminum and yttrium oxides modified by oxides of refractory metals (zirconium, hafnium or chromium) compositions for the preparing high-strength high-temperature and oxidative-resistant ceramic composites.

1. Introduction

The development of new materials with a complex of properties that significantly surpass the physical and mechanical performance of existing materials for operation in extreme conditions becomes the primary task. Ceramic-forming organoelement polymers (oligomers) - nanometallocarbosilanes and organoelementoxanes represent a new class of unequaled precursors for producing components of high-temperature oxidation-resistant high-strength nanostructured ceramic materials: ceramic fibers, matrices, protective coatings, etc. [1, 2].

The active research aimed at the development and investigation of ceramic-forming poly(oligo)carbosilanes (PCS) is in progress in many countries [3-9].

The prospects for the development of modern oxide ceramic materials are determined by the enormous funds that the advanced nations invest in the creation of new highly heat-resistant and chemically inert ceramic materials based on aluminum oxide [9-11].

2. Experimental details

SiC fibers modified with refractory metal compounds were obtained by melt spinning polymer fibers from nanometallocarbosilanes with the following oxidative hardening and carbidization.

The nanometallocarbosilanes and polymer fibers on their basis were pyrolyzed in a Nabertherm 50/500/11 tubular furnace under an argon atmosphere during heating to 1100°C at a rate of 5 °C/min, followed by holding for 1 h and pyrolysis at temperatures up to 1600 °C [12-14].

The organoelementoxanes were heat-treated in a SNOL 12/16 resistance-heated electric furnace at 1300 °C.

The surface morphology and elemental composition of the ceramic samples of the nanometallocarbosilanes and organoelementoxanes were assessed by scanning electron microscopy (SEM) on a Philips SEM 505 equipped with a Sapphire Si(Li) SEM10 energy dispersive detector and a Micro Capture SEM3.0M image capture system and on a Quanta 250.

Phase compositions were determined by X-ray diffraction at room temperature on a Shimadzu XRD-6000 vertical X-ray diffractometer using monochromatized copper radiation with a wavelength of $\lambda_{K\alpha cp}=(2\lambda_{K\alpha 1}+\lambda_{K\alpha 2})/3=1.54178$ [Å]. The crystalline and X-ray amorphous phases present were identified using ICDD PDF Release 2003 data.

Using a previously described procedure, we prepared binders based on organoalumoxanes [15, 16] and organoyttriumoxanealumoxanes [17-19]; film-forming agents and binders based on organoyttriumoxanealumoxanesiloxanes [20-22]; high-purity alumosilicate [23], alumomagnesium [24] ceramics and high-purity ceramics based on oxides of aluminum, yttrium and magnesium [2].

3. Results and discussion

Nanometallocarbosilanes are precursors of SiC ceramics modified with nanoparticles of compounds of refractory metals (Zr, Hf, Ta) [1, 2, 12-14] and SiC ceramics modified by mixed carbides ($Ta_xZr_yC_z$ or $Ta_xHf_yC_z$) [2].

Below are micrographs of the SEM and X-ray elemental microanalysis of SiC ceramics modified with mixed carbides ($Ta_xZr_yC_z$ or $Ta_xHf_yC_z$) (figure 1).

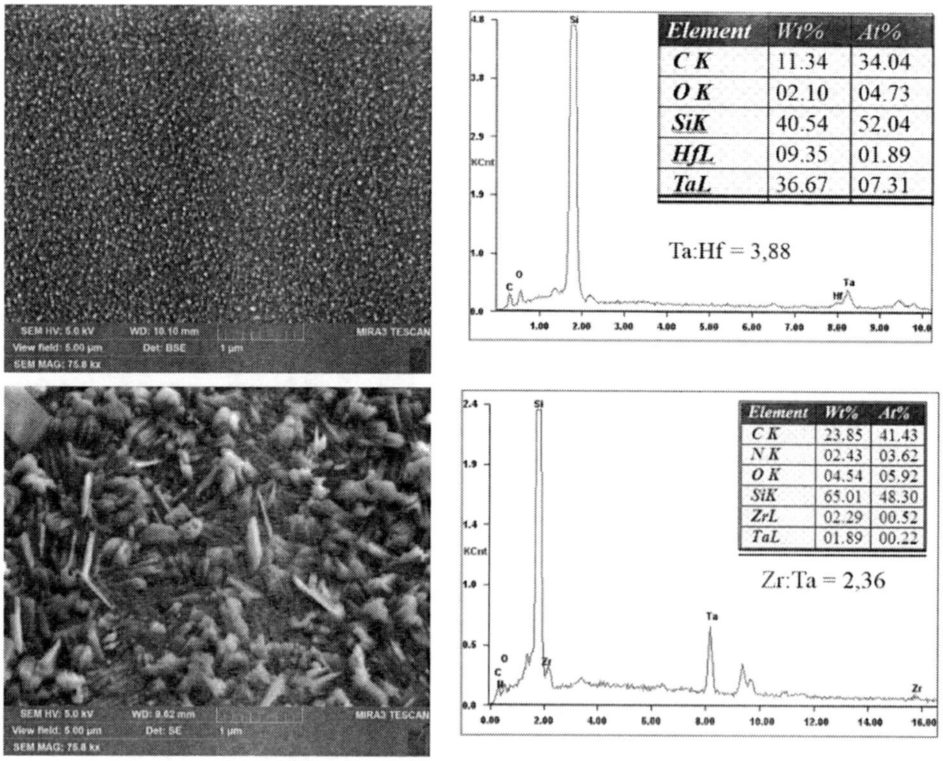

Figure 1. SEM and X-ray elemental microanalysis of modified SiC ceramics

IOP Conf. Series: Materials Science and Engineering **525** (2019) 012057 doi:10.1088/1757-899X/525/1/012057

Samples of SiC ceramics modified with mixed carbides (TaxZryCz or TaxHfyCz) were investigated by powder X-ray diffraction. The phase of nitrogen and carbon carbide was observed in the samples (Table 1). The nitrogen carbide phase was described by the structural Moissanite 3C. The third phase in the studied samples was a system similar to hafnium, tantalum and zirconium carbides. All above carbides have a cubic cell with "NaCl" structural type, the difference between these structures was observed only in the cell parameters. The lowest value of this parameter is inherent to tantalum carbide (4.43Å), in hafnium and zirconium carbides it amounts to 4.64 and 4.70Å, respectively. The “a” parameter of a cubic cell, determined using a refining diffractogram of Rietveld method, is shown in Table 1. In all samples it exceeds this value for tantalum carbide but is lower for hafnium and zirconium carbides. Thus, we can assume that the formation of carbide of a mixed structure takes place in the studied system.

The results are presented in Table 1.

Table 1. Maximum X-ray diffraction of CuK_α radiation on samples of modified SiC ceramics

Sample	HfC-ZrC-TaC		C	SiC	TaC
	wt %	a, Å			
$SiC(Ta_xZr_yC_z)$ 1600 °C N_2	4.3(6)	4.517(5)	1.34(13)	94(1)	0.42(11)%, a =4.476(6)Å, D=12(3)nm
$SiC(Ta_xHf_yC_z)$ 1500 °C Ar	5.2(19)	4.522(2)	0.48(17)	95 (3)	0.09(4)%, a =4.4839(9)Å, D=50(10)nm
$SiC(Ta_xHf_yC_z)$ 1600 °C N_2	7.1(4)	4.495(3)	1.46(6)	91(2)	0.14(2)%, a =4.4734(8)Å, D=46(10)nm
$SiC(Ta_xHf_yC_z)$ 1600 °C Ar	6.8(5)	4.501(2)	1.41(7)	91(1)	0.73(6)%, a =4.4867(4)Å, D=75(6)nm

It should be emphasized that nanometallocarbosilanes posses fiber-forming properties. Polymer fibers (figure 2, a) were produced from them by melt processing, after the polymer fibers were subject to oxidative curing (figure 2, b) and carbidization SiC fibers modified with refractory metal compounds (figure 2, c, d) were obtained. Figure 2, d shows the results of the SEM and X-ray elemental microanalysis of SiC fiber modified with Ta and Zr compounds.

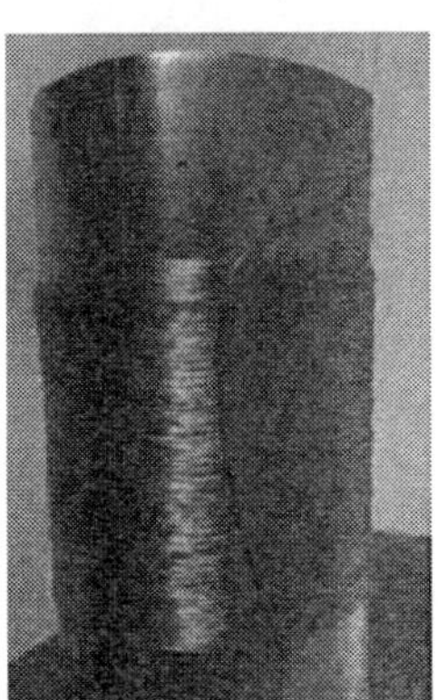

a

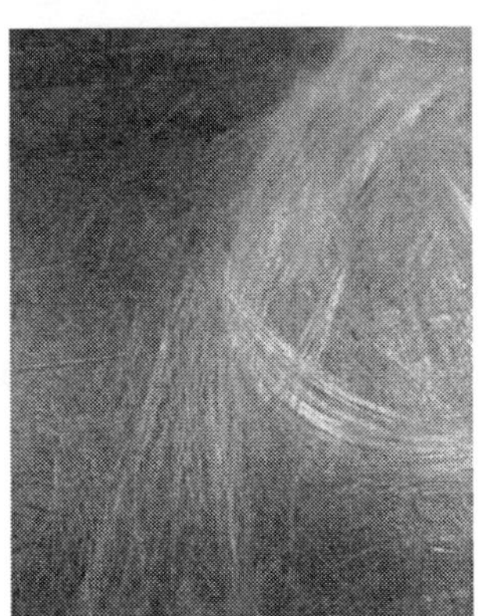

b

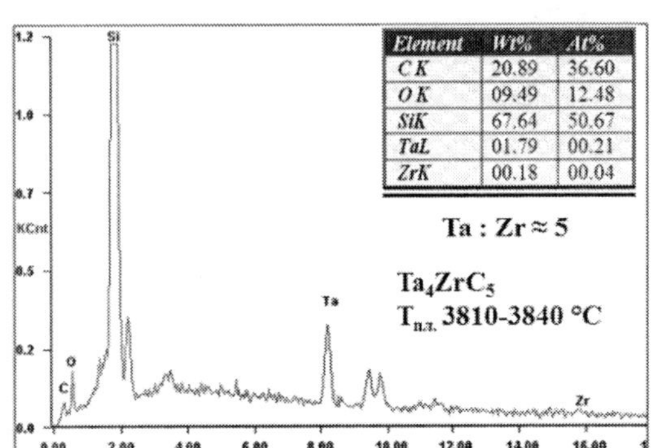

IOP Conf. Series: Materials Science and Engineering **525** (2019) 012057 doi:10.1088/1757-899X/525/1/012057

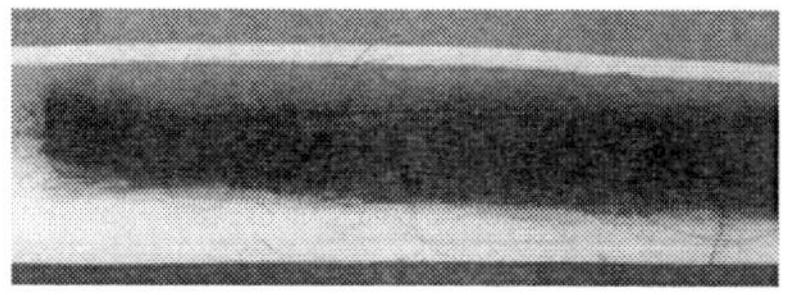
c

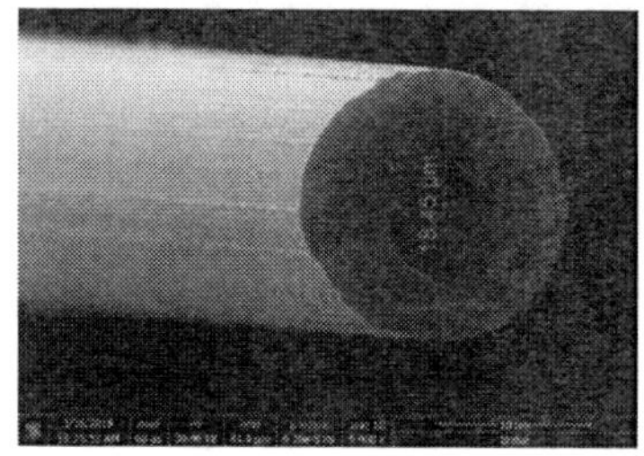
d

Figure 2. Photographs of nanometallocarbosilanes fibers: (a) polymer, (b) cured, (c) pyrolyzed, (d) micrograph of SEM and X-ray elemental microanalysis of $SiC(Ta_xZr_yC_z)$

We have synthesized organoelementoxane oligomers that can be used to obtain components of the CCM of oxide composition: chelated alkoxy alumoxanes — precursors of highly heat-resistant and chemically inert corundum ceramics [15, 16]; organoalumoxanesiloxanes - precursors of high-purity aluminosilicate ceramics with a given Al:Si molar ratio (in particular, mullite $3Al_2O_3 \cdot 2SiO_2$) [23]; organoyttriumoxanealumoxanes – precursors of alumoyttrium ceramics with a given Al:Y molar ratio (in particular, garnet $3Y_2O_3 \cdot 5Al_2O_3$) [17-19]; organoyttriumoxanealumoxansiloxanes – precursors of the components for chemically resistant, high-strength, clear or enamel protective coatings, as well as transparent and opaque glass ceramics based on yttrium, aluminum and silicon oxides with any desired molar ratio of $xY_2O_3 \cdot yAl_2O_3 \cdot zSiO_2$ [20-22]; organomagnesiumoxanealumoxanes – precursors of high-purity ceramics based on aluminum and magnesium oxides (alumomagnesium spinel $MgAl_2O_4$, in particular)[24]; organomagnesiumoxaneyttriumoxanealumoxanes – precursors of high-purity ceramics based on aluminum, yttrium and magnesium oxides, in particular, yttrium aluminum garnet ($Y_3Al_5O_{12}$) and perofskite ($YAlO_3$) modified by magnesium oxide, spinel ($MgAl_2O_4$) modified by yttrium oxide, as well as corundum (α-Al_2O_3) modified by yttrium and magnesium oxides [2]; organometalloxaneyttriumoxanealumoxanes – the precursors of high-purity, high-temperature binary ceramics based on aluminum and yttrium oxides modified by oxides of refractory metals (zirconium, hafnium or chromium) [2].

Micrographs of SEM, X-ray elemental microanalysis and diffractograms of oxide ceramics obtained in the result of pyrolysis of elementoxane oligomers at 1300°C follow (Figures 3 - 9)

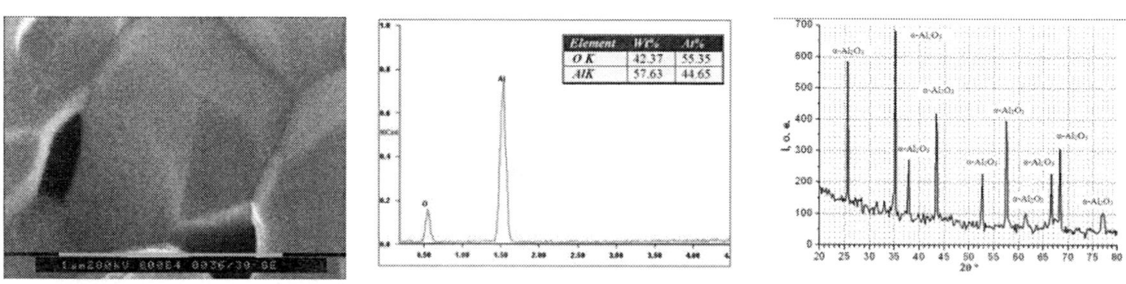

Figure 3. Micrograph of SEM, X-ray elemental microanalysis and diffractogram of alumina ceramics obtained in the result of organoalumoxane pyrolysis

IOP Conf. Series: Materials Science and Engineering **525** (2019) 012057 doi:10.1088/1757-899X/525/1/012057

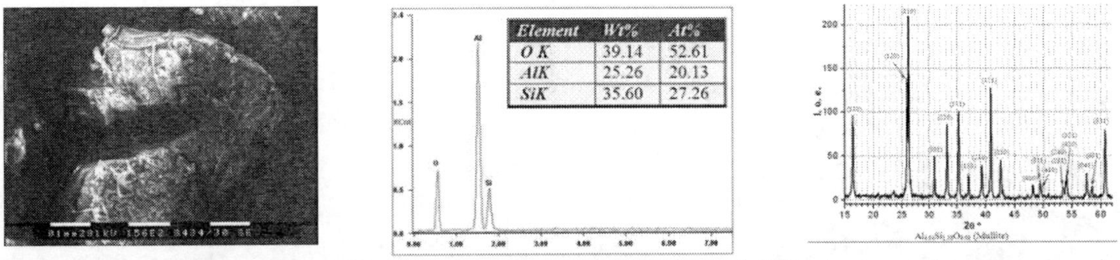

Figure 4. Micrograph of SEM, X-ray elemental microanalysis and diffractogram of mullite ceramics, obtained in the result of organoalumoxanesiloxanes pyrolysis

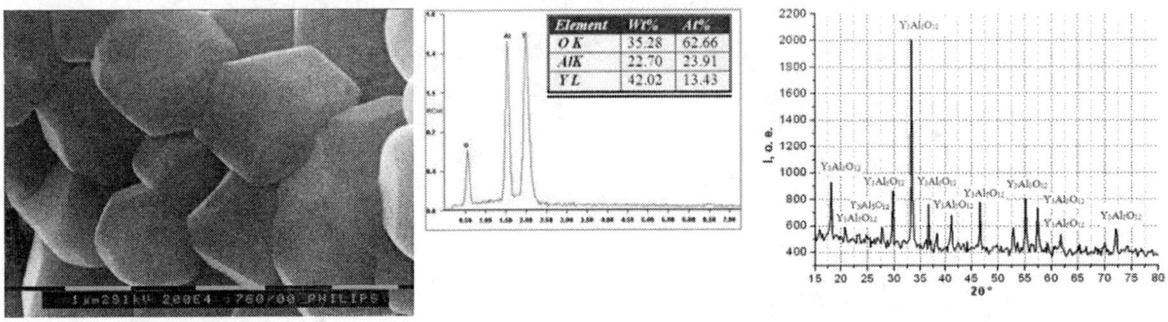

Figure 5. Micrograph of SEM, X-ray elemental microanalysis and diffractogram of aluminum-yttrium ceramics obtained in the result of organoyttriumoxanealumoxanes pyrolysis

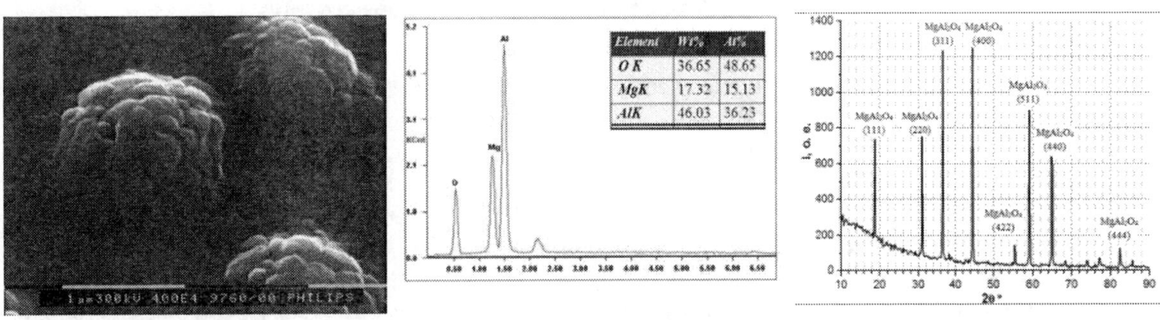

Figure 6. Micrograph of SEM, X-ray elemental microanalysis and diffractogram of aluminum-magnesium ceramics obtained in the result of organomagnesiumoxanealumoxanes pyrolysis

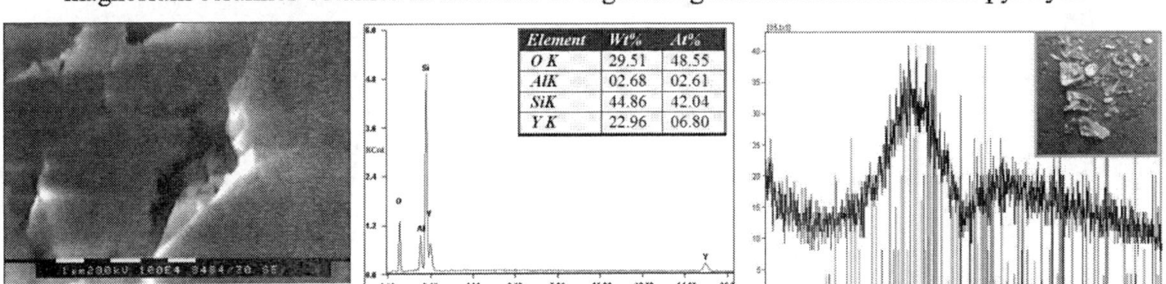

Figure 7. Micrograph of SEM, X-ray elemental microanalysis and diffractogram of glass obtained in the result of organoyttriumoxanealumoxanesiloxanes pyrolysis

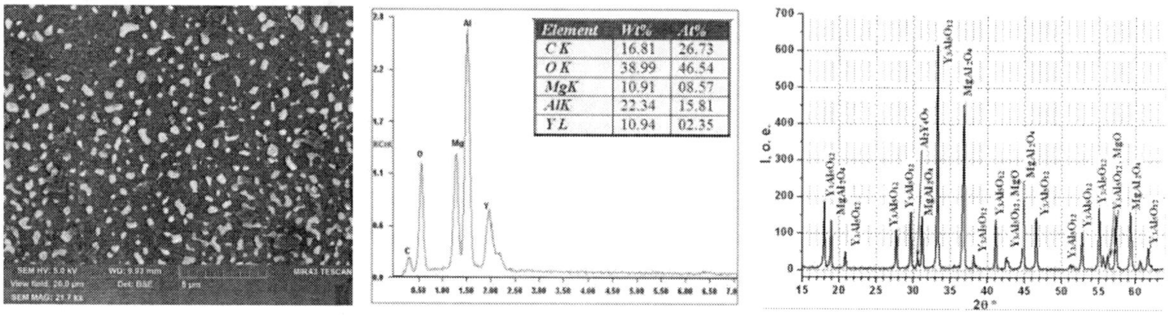

Figure 8. Micrograph of SEM, X-ray elemental microanalysis and diffractogram of ceramics based on aluminum, yttrium and magnesium oxides obtained in the result of organoyttriumoxanemagnesiumoxanealumoxanes pyrolysis

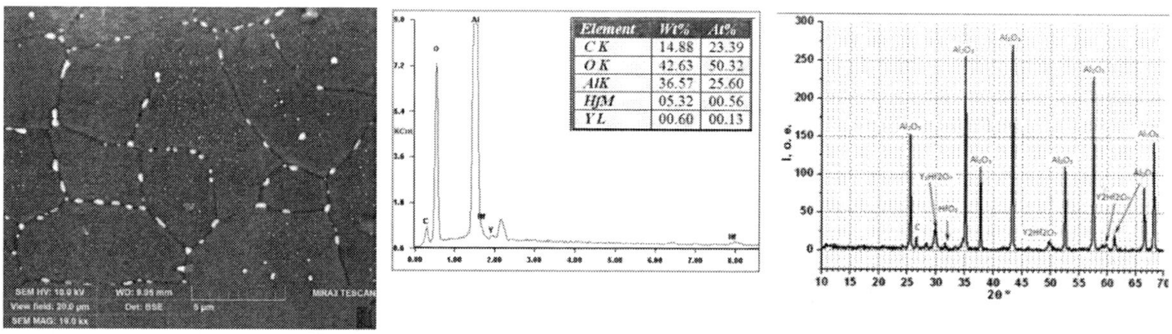

Figure 9. Micrograph of SEM, X-ray elemental microanalysis and diffractogram of ceramics based on aluminum, yttrium and hafnium oxides obtained in the result of organoyttriumoxanehafniumoxanealumoxanes pyrolysis

Conclusion

Effective methods for obtaining components of CCM (fibers, coatings, matrices, etc.) on the basis of ceramic-forming nanometallocarbosilanes and organoelementoxanes have been developed:

- SiC- fibers modified by compounds of refractory metals (Zr; Hf; Ta);
- binders, film-forming and impregnating compositions;
- sintering additives for SiC- composites;
- special and barrier coatings, in particular, glass-ceramic coatings of Y_2O_3–Al_2O_3–SiO_2 composition with a specified molar ratio of components;
- amorphous powders obtained during pyrolysis of organomagnesiumoxaneyttriumoxane-alumoxanes as a sintering additive for SiC composite.

Acknowledgements

The authors are grateful to A.S. Goloveshkin, postgraduate student of INEOS RAS, for his assistance in the analysis of samples SiC ceramics modified with mixed carbides ($Ta_xZr_yC_z$ or $Ta_xHf_yC_z$) by means of XRPD.

This work was supported by the Russian Foundation for Basic Research, project No. 17-03-00331 A), part of the work was performed within the framework of the government task No 007-00129-18-00.

References

[1] Abakumov G A et al. 2018 *Russ. Chem. Rev.* **87** (5) 393–507

[2] Shcherbakova G I, Storozhenko P A, Apukhtina T L, Varfolomeev M S, Zhigalov D V, Blokhina M Kh, Korolev A P, Kutinova N B and Riumina A A Components of ceramic

composite materials based on organoelement oligomers. FNM2018. IOP Conf. Series: Journal of Physics: Conf. Series **1134** (2018) 012054
[3] Flores O, Bordia R K, Nestler D, Krenkel W and Motz G 2014 *Adv. Eng. Mater.* **16** (6) 621–36
[4] Kaur S, Riedel R and Ionescu E 2014 *J. Eur. Ceram. Soc.* **34** (15) 3571–8
[5] Vijay V V, Nair S G, Sreejith K J and Devasia R 2016 *J. Inorg. Organomet. Polym.* **26** 302–11
[6] Chen S, Wang J and Wang H 2016 *Mater. Des.* **90** 84–90
[7] Wen Q, Xu Y, Xu B, Fasel C, Guillon O, Buntkowsky G, Yu Z, Riedel R and Ionescu E 2014 *Nanoscale* **6** (22) 13678–89
[8] Lu Y, Chen F, An P, Ye L, Qiu W and Zhao T 2016 *RSC Adv.* **6** (91) 88770–6
[9] Advanced Ceramics. US Industry Study with Forecasts for 2015-2020, September 2011 (2794), 290P https://www.freedoniagroup.com/brochure/27xx/2794smwe.pdf
[10] Chawla K K 2012 *Composite Materials: Science and Engineering* (Springer) p 542
[11] Fergus J W 2012 *J. Eur. Ceram. Soc.* **32** 525–40
[12] Shcherbakova G, Storozhenko P, Blokhina M, Shatunov V, Sidorov D, Sidorov D and Yurkov G 2014 *J. Chem. Chem. Eng.* **8** (3) 232–42
[13] Shcherbakova G I, Blokhina M Kh, Storozhenko P A, Zhigalov D V, Sidorov D G, Apukhtina T L, Varfolomeev M S, Sidorov D V, Kuznetsova M G and Yurkov G Yu 2014 *Inorg. Mater.* **50** (4) 423–30
[14] Shcherbakova G I, Blokhina M Kh, Storozhenko P A, Zhigalov D V, Varfolomeev M S, Drachev A I and Yurkov G Yu Thermal transformation of nanohafniumcarbosilanes. *Ceramics International.* **45** (2019) 122–130
[15] Shcherbakova G I, Apukhtina T L, Krivtsova N S, Varfolomeev M S, Sidorov D V and Storozhenko P A 2015 *Inorg. Mater.* **51** (3) 206–14
[16] Storozhenko P A, Shcherbakova G I, Tsirlin A M, Murkina A S, Varfolomeev M S, Kuznetsova M G, Polyakova M V and Trokhachenkova O P 2007 *Inorg. Mater.* **43** (3) 320–8
[17] Varfolomeev M S, Moiseev V S, Shcherbakova G I, Storozhenko P A and Shatunov V V 2015 *Inorg. Mater.* **51** (1) 76–81
[18] Shcherbakova G I, Storozhenko P A, Kutinova N B, Sidorov D V, Varfolomeev M S, Kuznetsova M G, Polyakova M V, Chernyshev A E, Drachev A I and Yurkov G Yu 2012 *Inorg. Mater.* **48** (10) 1058–63
[19] Varfolomeev M S, Moiseev V S, Shcherbakova G I, Krivtsova N S and Yurkov G Yu. 2015 *Inorg. Mater.* **51** (7) 722–7
[20] Shcherbakova G I et al. 2014 *Inorg. Mater.* **50** (3) 306–13
[21] Shcherbakova G I, Apukhtina T L, Varfolomeev M S, Sidorov D V, Drachev A I and Yurkov G Yu 2014 *Inorg. Mater.* **50** (6) 636–41
[22] Apukhtina T L, Shcherbakova G I, Sidorov D V, Varfolomeev M S, Sidorov D G and Drachev A I 2015 *Inorg. Mater.* **51** (8) 806–10
[23] Storozhenko P A, Shcherbakova G I 2011 *Inorg. Mater.* **47** (2) 167–71
[24] Shcherbakova G I, Storozhenko P A, Apukhtina T L, Varfolomeev M S, Kuznetsova M G, Drachev A I and Ashmarin A A 2017 *Polyhedron* **135** 144–152

IOP Conf. Series: Materials Science and Engineering **525** (2019) 012058 doi:10.1088/1757-899X/525/1/012058

The effect of polyvinylpyrrolidone nanowires on the metabolic activity of Lactobacillus acidophilus

O Y Berezina[1], A V Vasilyeva[2], N A Sidorova[3], A I Savushkin[4] and N P Marcova[5]
[1]K. f.-m. Sc., Associate Professor, Deputy Director of the Physico-Technical Institute, PetrSU, Petrozavodsk, Russia
[2]Student, Institute of Biology, Ecology and Agrotechnology, PetrSU, Petrozavodsk, Russia
[3]K. b. Sc., Associate Professor, Institute of Biology, Ecology and Agrotechnology, PetrSU, Director of the Small Innovative Enterprise «Microbiom», Petrozavodsk, Russia
[4]Deputy Director of the Small Innovative Enterprise «Microbiom», Petrozavodsk, Russia
[5]Senior Lecturer of the Physico-Technical Institute, PetrSU, Petrozavodsk, Russia

E-mail: kennard@inbox.ru, fagafon@yandex.ru

Abstract. The effectiveness of the use of polyvinylpyrrolidone (PVP) nanowires for increasing the metabolic activity of probiotic cultures of *Lactobacillus acidophilus* has been shown. PVP nanowires are obtained by electrospinning in two versions: PVPI based on a solution of high molecular weight polyvinylpyrrolidone and distilled water; PVPII based on a solution of two water zinc acetate in distilled water and a solution of high molecular weight PVP in ethanol. In the presence of PVP, there was an increase in viable lactobacillus cells up to 7.4×10^8 CFU/ml and accumulation of DL-lactic acid during acidification of the test solution to 3.10 and an increase in Eh to + 360.5 mV. PVPI is a more preferred carrier for lactobacilli than PVPII with ZnO. The zinc ions contained in the PVP matrix are capable of causing oxidative stress, followed by the formation of free radicals that damage membrane lipids, proteins and DNA of the prokaryotic cell. The results are aimed at developing new forms of immobilized drugs that can be used as alternative probiotic agents.

1. Introduction

Poly-(N-vinylpyrrolidone) refers to synthetic polymers-gamma-vinyl lactam N-aminobutyric acid, obtained by homopolymerization of an aqueous solution of N-vinylpyrrolidone at a temperature of 50-80 °C under the action of 0.05-2.5% H_2O_2 in the presence of NH_3. Due to the complex of physical, chemical and biological properties, polyvinylpyrrolidone (PVP) is a unique compound, has biocompatibility and the ability to interact with cell membranes, non-toxicity, chemical stability and solubility in water and organic solvents [1]. According to the degree of impact on the human body, PVP belongs to the 4th hazard class or low hazardous substances, has an affinity for complex hydrophobic and hydrophilic substances, and is capable of forming complexes with transition metal ions, drugs and toxic compounds of different origin. The listed PVP properties have found application in medical technologies, in optics and electrical engineering, on its basis created highly permeable composite membranes with a selective layer, adhesives, cetamines, various coatings and fibers [2; 3]. The pharmaceutical industry uses the ability of PVP to bind the functional components of a drug until the drug reaches the stomach, and then dissolve quickly to release the drug. Using studies in this direction, it has been proved that PVP can also have the opposite effect if it is used as a coating and delayed the release of a drug in order to have a prolonged effect and eliminate multiple dosages. Thus, the use of PVP as a coating for venlafaxine-montmorillonite antidepressants is described in order to slow the release of venlafaxine [4].

PVP nanofibres obtained by electrospinning are used in tissue engineering as a biocompatible extracellular matrix [5]; PVP and emodin membranes, as well as PVP and alginate hydrogel nanowires with silver nanoparticles in the form of AgNPs are successfully used for drug delivery and accelerated wound healing [6]. There is evidence of magnetic nanoparticles coated with silica and embedded in PVP fibers obtained by electrospinning. In the experiment, modified PVP nanoparticles showed unique fluorescent and magnetic properties [7]. There are data on the development of films based on chitosan and PVP doped with AgNPs with prolonged bactericidal activity, which is promising for solving the problems of device-associated infections [8]. Studies on the encapsulation of DNA molecules by PVP nanoparticles were performed [9]. When DNA was immobilized with PVP, it was found that 80% transfection efficiency is observed in vitro, which makes PVP an effective carrier of functional molecules for therapeutic delivery [9].

The purpose of the work performed is related to the study of the biological effect of PVP nanowires on *Lactobacillus acidophilus* microorganisms with a pronounced probiotic potential. Compared with other lactobacilli, *L. acidophilus* are the most recommended microbial cultures for use as probiotics. They are involved in the restoration of microintestinal balance [10; 11], modify the activity of bacterial enzymes, affect the permeability of the intestinal mucosa and the regulation of the immune system [12; 13]. The metabolic products of probiotic cultures exhibit antibiotic properties and inhibit the vital activity of pathogens of the genus *Klebsiella, Enterobacter, Pseudomonas, Salmonella, Serratia, Bacteroides*, and others [12]; Probiotics control the level of glucuronidase, β-glucuronidase, nitroreductase and azoreductase in the intestine. These enzymes catalyze the conversion of procarcinogens into carcinogens, such as nitrosamine and secondary bile acids [13; 14]. Low levels of these enzymes significantly reduce the risk of cancer in the colon. It has been suggested that lactobacilli immobilized by modified PVP nanofibres are capable of increasing their metabolic activity and long-lasting viability with prolonged probiotic action.

2. Method

The effect of polyvinylpyrrolidone nanowires was tested on cells of the *L. acidophilus* strain Ep. n. v. 317/402 in the composition of the commercial preparation «Narine», series 500051. The preparation contains lyophilisate *L. acidophilus* in the amount of 1×10^9 CFU / g, corn starch, sucrose and magnesium stearate (Mg $(C_{18}H_{35}O_2)_2$. Lactobacillus cultures were grown under experimental and control conditions on agar media of the recommended composition [15]. In the experiment, 0.075 g of PVPI (option 1) and 0.085 g of PVPII with ZnO (option 2) were added to the nutrient medium. In the control, lactobacilli were cultured without adding PVP. solution based on high molecular weight of polyvinylpyrrolidone (Mr = 1.3×10^6 g / mol) and distilled water at a rate of 0.13 g / ml; PVPII - when mixing a solution of two aqueous zinc acetate Zn $(CH_3COO)_2 \times 2H_2O$ in distilled water and a solution of high-molecular PVP (Mr = 1.3×10^6 g / mol) in ethanol. A medical syringe (needle diameter 0.7 mm) was used in an electrospinning unit. The PVPI and PVPII solutions were supplied from the syringe with a NE-300 syringe pump at a rate of 0, 5 ml / hour An electric field with a strength of 1.8 kV / cm was created between the needle and the metal substrate with the help of a high-voltage source «AHVS -30/5» (adjustable high voltage source). The diameter of the obtained filaments 200 - 300 nm. The photo of the obtained PVP nanowires is presented in Figure 1.

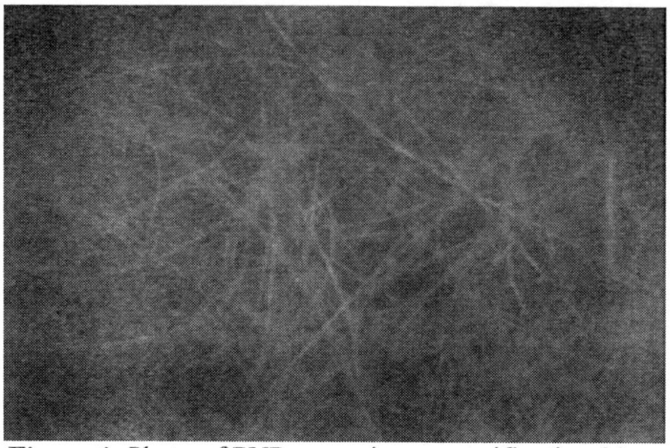

Figure 1. Photo of PVP nanowires, magnification 80 ×.

The metabolic activity of immobilized *L. acidophilus* strains was assessed by enzyme activity and the formation of DL-lactic acid. The presence of DL-lactic acid was determined by the results of a qualitative reaction with H_2SO_4, $CuSO_4 \times 5H_2O$ and thiophene solution. Additionally, we analyzed the pH and redox potential (Eh) of the medium, which are technological parameters and control the direction of biochemical processes; The experiment also studied the morphology and cultural properties of bacteria. The morphology of the lactobacillus colonies was analyzed using a «MOTIC DM-BA-30» biological microscope, the morphology and cell linear dimensions – using SEM microscopy using a «HITACHI SU1510» microscope with an energy dispersive attachment. To do this, microbial cells were applied to the surface of a microscope slide, dried, fixed in ethyl alcohol, followed by applying a layer of gold by ion sputtering using a magnetron sputtering system and mounted on a microscope stage. *Lactobacillus* microscopy was performed under a high vacuum condition. The viability of lactobacilli in the experiment and control was taken into account within 30 days of the experiment according to the results of planting the corresponding tenfold dilutions of the suspensions studied on agar media followed by counting the colony forming units after the incubation time. Statistical processing of the data was performed according to the methods used in microbiological studies [16].

3. Results and Discussion

In the experiment with PVPI and PVPII, lactobacilli actively adhered to the surface of the carrier, which led to an increase in the number of viable cells from $3{,}1 \times 10^3$ CFU / ml (PVPII, 24 h) to $7{,}4 \times 10^8$ CFU / ml (PVPI, 30 days) . In the control, the number of actively growing bacteria did not exceed $3{,}4 \times 10^5$ CFU / ml, and by the end of the experiment it decreased to $2{,}5 \times 10^5$ CFU / ml (Table 1). Based on the above data, it should be noted that the nutrient medium with 0.075 g of PVPI was preferable for activating the growth and reproduction of microbial cells than the medium with the addition of 0.085 g of PVPII with ZnO, in all cases an increase in the share of viable microorganisms from 8.8% (24 h) up to 44.6% (30 days).

Table 1. The number of viable cells (CFU / ml-1) of *L. acidophilus* in experiment and control (X ± I95)

The period of immobilization	PVPI	PVPII	Control
24 hours	$(3{,}4 \pm 0{,}4) \times 10^4$	$(3{,}1 \pm 0{,}4) \times 10^3$	$(2{,}1 \pm 0{,}7) \times 10^4$
10 days	$(4{,}3 \pm 0{,}6) \times 10^6$	$(3{,}7 \pm 0{,}4) \times 10^4$	$(2{,}8 \pm 0{,}7) \times 10^6$
20 days	$(5{,}9 \pm 0{,}5) \times 10^8$	$(4{,}7 \pm 0{,}6) \times 10^6$	$(3{,}4 \pm 0{,}6) \times 10^5$

30 days	$(7{,}4\pm0{,}6)\times10^8$	$(4{,}1\pm0{,}6)\times10^7$	$(2{,}5\pm0{,}7)\times10^5$

As a result of SEM microscopy, in the experiment and control, the presence of large straight rods with rounded ends, located singly or in chains of 2-3 cells (Fig. 2a) and large globular structures of exopolysaccharide nature (Fig. 2b), was established as a result of an increase in metabolic activity of immobilized cells. Lactobacilli are immersed in the semi-liquid matrix of exopolysaccharide and are practically not visible on electronic photographs. Despite the morphological uniformity of bacteria grown both in the presence of PVP and without it, the culture characteristics (size of colonies) of the control and experimental variants of lactobacilli on agar nutrient media were significantly different. The maximum diameter of colonies (0.29 ± 0.06 cm) reached the bacteria in the presence of PVPI (Fig. 3b), in the presence of PVPII, the diameter of the colonies was 0.34 ± 0.06 cm (Fig. 3c), and in the control it was not exceeded 0.20 ± 0.05 cm (Fig. 3d).

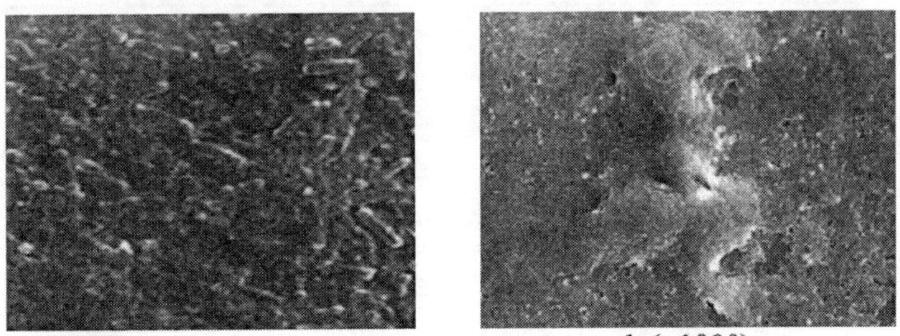

a (×3000) b (×1000)

Figure 2. Electronic photographs of immobilized PVP *L. acidophilus* (made using a «HITACHI SU1510» scanning electron microscope, gold plating)

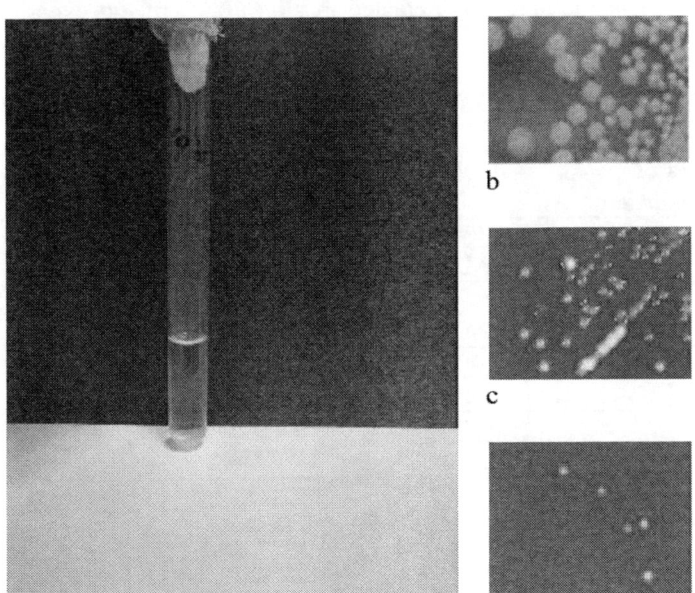

b
c
a d

Figure 3. The appearance of the tubes with the appropriate dilution of *L. acidophilus* (10-4) (a); photographs of the culture growth of *L. acidophilus* on agar media in the experiment with PVPI (b), PVPII (c) and in the control (d)

IOP Conf. Series: Materials Science and Engineering **525** (2019) 012058 doi:10.1088/1757-899X/525/1/012058

The effect that stimulates the metabolic activity of bacteria caused by the presence of polyvinylpyrrolidone is confirmed by the results of studying the dynamics of pH and Eh in experiment and control (Table 2 and 3). For 30 days of the experiment, the pH of the culture medium due to the accumulation of lactate decreased to 3.10 (PVPI) and to 3.80 (PVP II). In the control, the pH values were only 4.50. Apparently, such acidity was found to be a threshold for native lactobacilli, which, with the depletion of the nutrient substrate and the accumulation of extracellular toxic metabolites, caused the inhibition of the growth of microorganisms and a decrease in their viability. The values of Eh in the experimental conditions steadily increased from +102 mV to +360 mV (for PVPI) and from +90 mV to +342 mV (for PVPII). In the control, an increase in Eh was observed only during the 20 days of the experiment (from +72 mV to +203 mV), and by 30 days – Eh decreased to +169 mV (Table 2).

Table 2. Dynamics of pH and Eh of the culture medium in the experiment and control

Period of immobilization	PVPI	PVPII	Control
24 hours	4,92* 102,7**	5,20 90,4	6,10 72,8
10 days	3,70 258,8	4,62 210,7	5,89 107,1
20 days	3,24 311,4	4,10 260,1	4,43 203,2
30 days	3,10 360,5	3,80 342,4	4,50 169,7

Note: * - pH values, ** - Eh values

The presence of DL-lactic acid was detected in all variants of the experiment and in the control (Table 3).

Table 3. Titratable acidity of the culture medium in the experiment and control

Period of immobilization	PVPI	PVPII	Control
24 hours	80 DL MK +	70 DL MK +	50 DL MK +
10 days	110 DL MK +	90 DL MK +	100 DL MK +
20 days	140 DL MK +	180 DL MK +	130 DL MK +
30 days	310 DL MK +	220 DL MK +	130 DL MK +

However, the maximum increase in titrated acidity by 30 days of the experiment was recorded for PVPI, and the minimum - for control.

4. Conclusions

According to the results of studies of the biological effect of PVP nanowires on *L. acidophilus*, we can state the proven positive effect of polyvinylpyrrolidone nanowires on the metabolic activity of probiotic cultures. Immobilized PVP cells of lactobacilli are characterized by rapid expression of genes controlling the catabolism of lactose. In the presence of PVP, cells maintain a long-term viability level compared with lactobacilli growing in liquid suspension cultures. Probably, the growth of microorganisms in the presence of PVP nanowires determines the degree of cell aggregation, having an indirect effect on the synthesis of secondary metabolites. It should be noted that zinc ions contained in the PVP matrix can cause oxidative stress with the subsequent formation of free radicals, damaging lipid membranes, proteins and DNA of the prokaryotic cell. This, in turn, regulates a cascade of mechanisms that accelerate metabolic processes associated, in this case, with the splitting of lactose to lactic acid and carbon dioxide. Subsequent acidification of the culture mixture (to pH 3.1), apparently, triggers the expression of genes that ensure the formation of biofilm formation in lactobacilli cells immobilized on PVP. The results will be aimed at developing new forms of immobilized drugs that can be used as alternative probiotic agents. The work was performed within the framework of the state assignment of the Ministry of Education and Science of Russia № 16.5857.2017 / 8.9 and the implementation of the Development Program of the support university of FSBEI HE «Petrozavodsk State University» for the period 2017-2021.

References

[1] Teodorescu M and Bercea M 2015 Poly(vinylpyrrolidone) – a versatile polymer for biomedicaland beyond medical applications *Polymer-Plastics Technology and Engineering* 54 pp 923–943

[2] Osadchenko S et al 2013 New Composite Membranes with a Selective Layer Based on the Polyaniline-Poly- (N-Vinyl Pyrrolidone) System *Bulletin of the Nizhny Novgorod University* I Lobachevsky 2 (1) pp 79-85

[3] Zhi X *et al* 2013 The immunotoxicity of graphene oxides and the effect of PVP-coating *Biomaterials* 34 pp 5254–5261

[4] Patel H A *et al* 2011 Sustained release of venlafaxine from venlafaxine–montmorillonite polyvinylpyrrolidone composites *Appl. Clay Sci.* 51 pp 126–130

[5] Wang M *et al* 2014 Preparation of PVDF/PVP core–shell nanofibers mats via homogeneous electrospinning *Polymer (Guildf)* 55 (9) pp 2188–2196

[6] Dai X Y *et al* 2012 Electrospun emodin polyvinylpyrrolidone blended nanofibrous membrane: a novel medicated biomaterial for drug delivery and accelerated wound healing *J. Mater. Sci. Mater. Med.* 23 pp 2709–2716

[7] Chen M *et al* 2012 *Magnetic electrospun fluorescent polyvinylpyrrolidone nanocomposite fibers Polymer (Guildf)* 53 pp 4501–4511

[8] Wang B L *et al* 2012 Fast and long-acting antibacterial properties of chitosan Ag/polyvinylpyrrolidone nanocomposite films *Carbohydr. Polym.* 90 pp 8–15

[9] Saxena A *et al* 2006 Ultra-low sized cross-linked polyvinylpyrrolidone nanoparticles as non viral vectors for in vivo gene delivery *Biomaterials* 27 pp 5596–5602

[10] Betoret N *et al* 2003 Development of probiotic-enriched dried fruits by vacuum impregnation *J. of Food Engineering* 56 pp 273-277

[11] Sullivan A and Nord C E Probiotics and gastrointestinal diseases *J. of Internal Medicine* 257 pp 78-92

[12] Krasaekoopt W *et al* 2003 Evaluation of encapsulation techniques of probiotics for yoghurt *International Dairy Journal* 13 pp 3-13

[13] Adams M R and Moss M O 2000 Food Microbiology *Second Edition, The Royal Society of Biochemistry* (Cambridge, UK) pp 318-323

[14] Gorbach S L 1997 Health benefits of probiotics *IFT Annual Meeting, Orlando, FL, Abstract* pp 73-81

[15] Likhacheva A Yu et al 1992 The current state of the issue of the taxonomy of bacteria of the genus Lactobacillus Journal of Microbiology 9-10 pp 74-78

[16] Ashmarin I P and Vorobiev A A 1962 Statistical methods in microbiological research *L.: Medgiz* p 280

IOP Conf. Series: Materials Science and Engineering **525** (2019) 012059 doi:10.1088/1757-899X/525/1/012059

Siliciding of carbon fabrics with gaseous SiO

E I Istomina[1], P V Istomin[1], A V Nadutkin[1], V E Grass[1], M Y Presniakov[2], A S Lysenkov[3] and M G Frolova[3]

[1] Federal Research Centre "Komi Science Centre of the Ural Branch of the Russian Academy of Sciences", 24 Kommunisticheskaya ul., Syktyvkar, Komi Republic, 167982, Russian Federation
[2] National Research Centre "Kurchatov Institute", 1 Pl. Kurchatova, Moscow 123182, Russian Federation
[3] Baikov Institute of Metallurgy and Materials Science, Russian Academy of Sciences, 49 Leninskii pr., Moscow 119991, Russian Federation

istomina-ei@yandex.ru

Abstract. Silicon carbide fabrics were synthesized through siliciding of carbon fiber precursors with gaseous SiO using three original batch-type reactors of different design. The prepared samples were analyzed by XRD and SEM/EDX. It was found that the choice of SiO gas source affected the microstructure characteristics of SiC fiber materials obtained by the proposed method.

1. Introduction

Silicon carbide fibers and fabrics made thereof have attracted much attention due to their superior high temperature thermal stability and mechanical properties [1–4]. Two widespread methods now exist for producing SiC fibers on a commercial scale: i) chemical vapor deposition using methylchlorosilanes as a source gas and tungsten or carbon filaments as a substrate [5]; ii) pyrolysis of polycarbosilane polymer filaments [6]. An alternative cost-effective and convenient way to fabricate SiC fiber materials is through siliciding of carbon fiber precursors with gaseous SiO [7–11]. From a chemical standpoint, this method is based on the conversion of carbon into silicon carbide in a SiO gas environment according to the following reaction:

$$2C + SiO_{gas} = SiC + CO_{gas} \quad (1)$$

The rate of reaction (1) decreases drastically with growing SiC layer inward the carbon filament, therefore a rather long-term heat treatment is required for the complete conversion of the carbon fiber into SiC one. This is obviously a key limiting factor for productivity of the method, but not when using a high-load batch-type reactor which can produce a large amount of material in a single run. In the present work we aim to demonstrate our technical and technological solutions for siliciding of carbon fabrics with gaseous SiO using laboratory-scale batch-type reactors and also to consider issues concerning the choice of SiO gas source.

IOP Conf. Series: Materials Science and Engineering **525** (2019) 012059 doi:10.1088/1757-899X/525/1/012059

2. Materials and methods

Carbon fabric with a surface density of 200 g/m^2 made of Toho Tenax HTS40-E13-3K carbon fiber (Toho Tenax, Japan) consisting of 3000 filaments 7 μm in diameter was used as a carbon fiber precursor. Siliciding heat treatment procedures were carried out using three original batch-type reactors of different design shown in Fig. 1. The reactor was mounted in a vacuum furnace chamber and heated to 1350–1650 °C depending on the choice of SiO gas source. At temperature above 1300°C, the heating rate was maintained at less than 6 °C/min. The whole duration of heat treatment was not less than 2 hours. The gas pressure in the furnace chamber was maintained at less than 0.1 Pa by continuously pumping out the gaseous products over the entire time of heat treatment. In order to generate SiO gas during siliciding treatment, the following three reactive powder mixtures were used: i) an equimolar mixture of Si and SiO_2; ii) a mixture of SiC and SiO_2 taken in 1:2 molar ratio; iii) an equimolar mixture of activated charcoal and SiO_2. The first one was meant to be a base case, and two others were chosen as alternatives. To prevent deposition of SiO gas on cooled parts of vacuum furnace, the activated charcoal was optionally used as a SiO gas absorber.

The prepared samples were analyzed by X-ray diffraction (XRD) using Ni-filtered CuKα radiation, scanning electron microscopy with energy-dispersive X-ray analysis (SEM/EDX), and high-resolution scanning electron microscopy (HR-SEM).

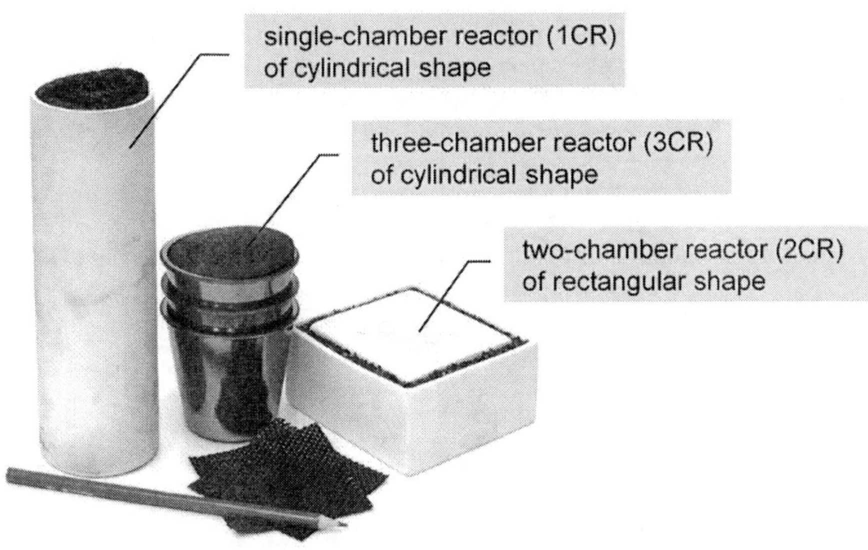

Figure 1. Laboratory-scale batch-type reactors for siliciding of carbon fabrics with gaseous SiO.

3. Results and discussion

3.1. Synthesis of SiC fabrics through siliciding of carbon fiber precursors with gaseous SiO in batch-type reactors

Schematics of the batch-type reactors we used for siliciding of carbon fabrics with gaseous SiO are sketched in Fig. 2. The single-chamber reactor (1CR) has been described in detail elsewhere [11]. A large volume of the reaction chamber and the simplicity of design are major advantages of 1CR. In other hand, when the volume of 1CR is too large, there is a distinct gradient in siliciding environment along the reaction chamber. To overcome this problem, we have designed the reactors provided with a separate semi-closed reaction chamber where the gas environment is assumed to be rather uniform. The two-chamber reactor (2CR) has an additional separate section for SiO gas absorber. The three-chamber reactor (3CR) is build of three crucibles where the fabrics to be silicided, SiO gas source and SiO gas absorber are placed separately from each other.

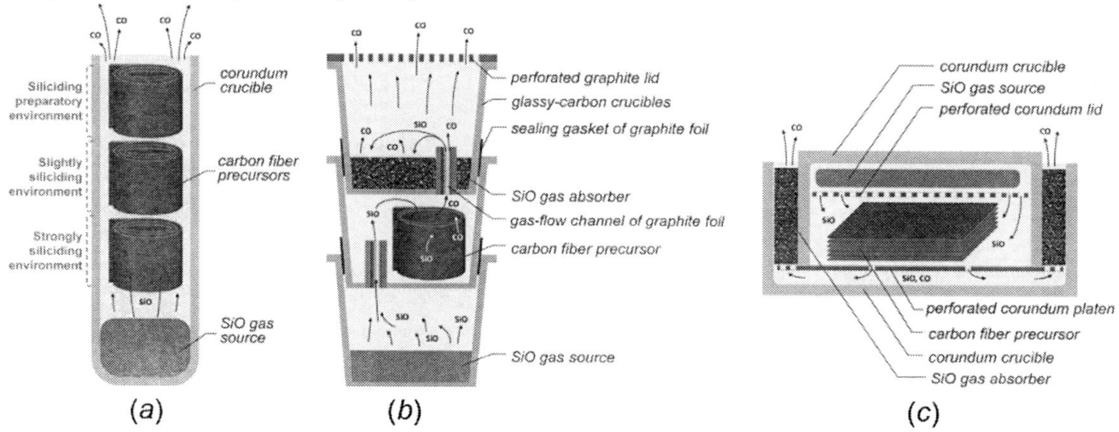

Figure 2. Schematics of the reactors: 1CR (*a*), 3CR (*b*) and 2CR (*c*).

Our experiments proved the efficiency and convenience of the above batch-type reactors. In all cases when the technological requirements have been met, the conversion of carbon fabrics into SiC one was fully completed and no carbon residue was present in the products. According to typical XRD pattern given in Fig. 3, siliciding of carbon fiber materials with gaseous SiO resulted in the formation of cubic polytype of silicon carbide (β-SiC). The content of hexagonal and rhombohedral polytypes of silicon carbide (α-SiC) did not commonly exceed 20 vol %.

SEM images of the SiC fibers prepared by siliciding of carbon fiber precursors with gaseous SiO using Si/SiO_2 reactive mixture as SiO gas source are shown in Fig. 4. The obtained SiC fibers have a smooth surface with multiple small longitudinal cracks and grooves but no crucial inner damage. The average crystallite size is found to be about 15–20 nm as determined by XRD and HR-SEM.

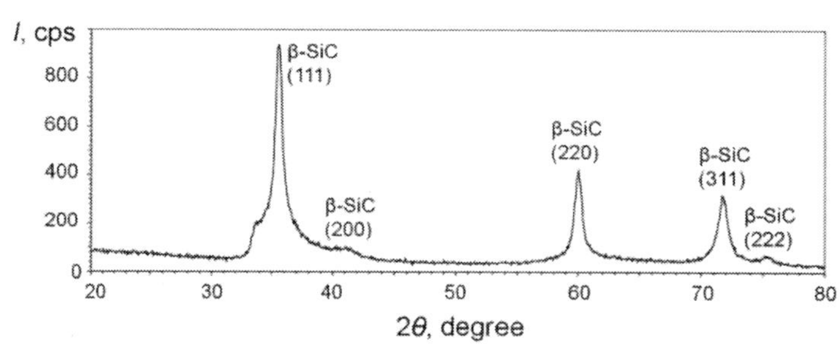

Figure 3. XRD pattern of the SiC fibers obtained by siliciding of carbon fiber precursors with gaseous SiO.

3.2. Influence of synthesis conditions on the characteristics of SiC fibers

A good way to improve the characteristics of SiC fibers is through providing much milder conditions for siliciding of carbon precursors. In particular, this could be achieved by decreasing SiO gas pressure in the reaction chamber during siliciding treatment. To confirm or deny this supposition, we have experimented with three different reactive powder mixtures to generate SiO gas by the following reactions:

$$SiO_2 + Si = 2SiO \quad (2)$$

$$2SiO_2 + SiC = 3SiO + CO \quad (3)$$

IOP Conf. Series: Materials Science and Engineering **525** (2019) 012059 doi:10.1088/1757-899X/525/1/012059

$$SiO_2 + C = SiO + CO \quad (4)$$

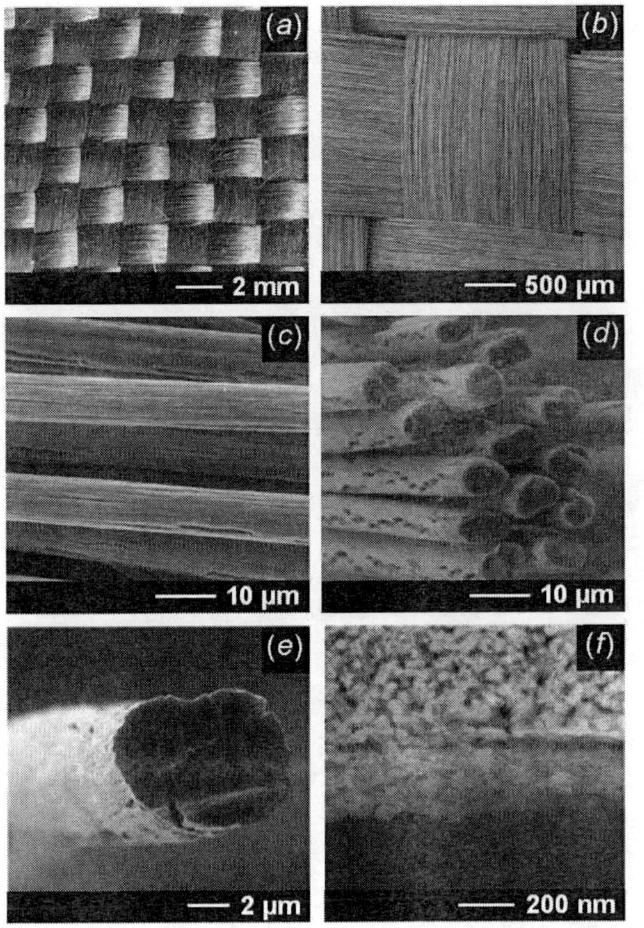

Figure 4. Close-up photograph (*a*), SEM (*b–e*) and HR-SEM (*f*) images of the SiC fibers obtained by siliciding of carbon fiber precursors with gaseous SiO using Si/SiO_2 reactive mixture as SiO gas source.

As seen from Fig. 5, SiO gas pressure over the systems $SiC–SiO_2$ and $C–SiO_2$ is significantly lower compared to that over the system Si–SiO2. It would seem that this is an excellent reason to use SiC/2SiO2 and C/SiO2 reactive mixtures as sources of SiO gas when "mild" siliciding conditions are needed. However, we have actually found that these alternatives cannot be seriously considered to be better choices to serve as SiO gas sources. Their major drawback is that they favor an enhanced crystal growth causing the surface of the obtained SiC fibers to be too rough.

SEM images of the SiC fibers prepared by siliciding of carbon fiber precursors with gaseous SiO using $SiC/2SiO_2$ and C/SiO_2 reactive mixtures as SiO gas sources are shown in Figs. 6 and 7, respectively.

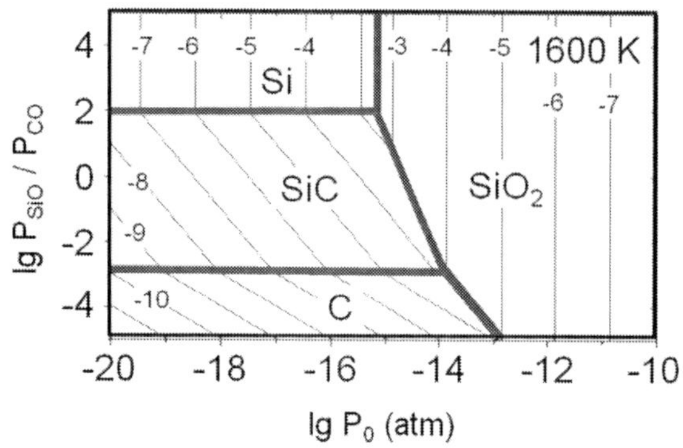

Figure 5. Calculated solid-gas phase equilibria in the Si-C-O system at 1600 K. Thin lines represent the isobars of SiO gas partial pressure logarithm over Si/SiO_2, $SiC/2SiO_2$ and C/SiO_2 reaction mixtures to be used as SiO gas sources.

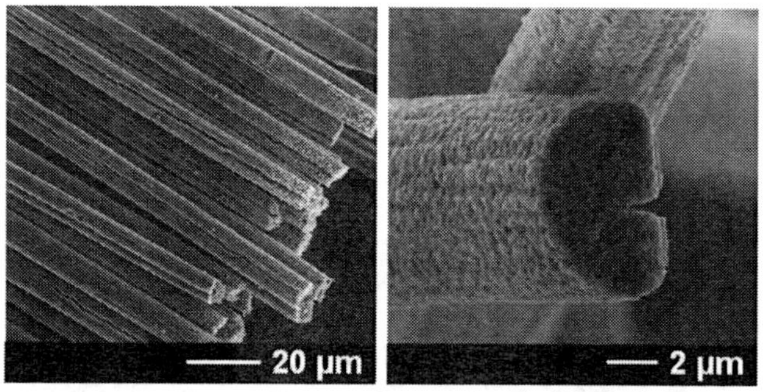

Figure 6. SEM images of the SiC fibers obtained by siliciding of carbon fiber precursors with gaseous SiO using $SiC/2SiO_2$ reactive mixture as SiO gas source.

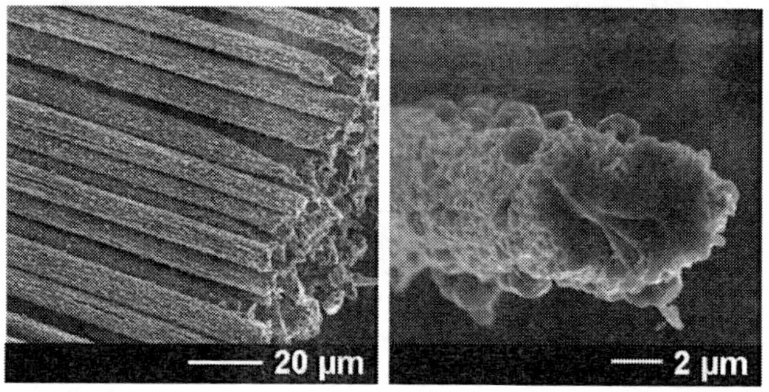

Figure 7. SEM images of the SiC fibers obtained by siliciding of carbon fiber precursors with gaseous SiO using C/SiO_2 reactive mixture as SiO gas source.

The observed effect can be explained as follows. The initial stage of the siliciding process is associated with the nucleation of SiC on the surface of the carbon fiber precursors. The nucleation site density depends on the SiO gas pressure in the reaction chamber. When the partial pressure of SiO gas is low, the number of the nucleation sites is small, and therefore the nucleated crystals can grow to large sizes without spatial constraints. Moreover, in the cases of $SiC/2SiO_2$ and C/SiO_2 a much higher temperature compared to that for the case of Si/SiO_2 is required for the reaction to proceed. This is also an important matter because crystal growth rate tends to increase with increasing temperatures. Thus, such results do not encourage the use of $SiC/2SiO_2$ and C/SiO_2 reactive mixtures to SiO gas generating. It is obvious that the strength characteristics of coarse-grained SiC fibers are much worse than those of fine-grained ones obtained using Si/SiO_2 reactive mixture as SiO gas source.

4. Conclusion

Being focused on designing and developing techniques for the synthesis of silicon carbide fabrics, we have demonstrated our technical and technological solutions for siliciding of carbon fiber precursors with gaseous SiO using three original batch-type reactors of different design. We have shown that the choice of reactive powder mixtures to be used to SiO gas generating strongly influences the synthesis conditions that in turn affects the microstructure characteristics of SiC fibers. The better results were achieved when using equimolar mixture of Si and SiO_2 as SiO gas source.

Acknowledgments
The authors are grateful to the Russian Foundation for Basic Research for grant #18-08-01460. The study was performed using the equipment of the Center for Shared Use of Scientific Equipment "Khimiya" of the Institute of Chemistry, Komi Science Center, Ural Branch of the Russian Academy of Sciences. The experimental part of the work partially done on the equipment of the Resource Center of Probe and Electron Microscopy (Kurchatov Complex of NBICS-Technologies, NRC "Kurchatov Institute").

References
[1] Bunsell A R and Piant A 2006 A review of the development of three generations of small diameter silicon carbide fibres *J. Mater. Sci.* **41** 823
[2] Schawaller D, Clau B and Buchmeiser M R 2012 Ceramic filament fibers – a review *Macromol. Mater. Eng.* **297** 502
[3] Flores O, Bordia R K., Nestler D, Krenkel W and Motz G 2014 Ceramic fibers based on SiC and SiCN systems: current research, development, and commercial status *Adv. Eng. Mater.* **16** 621
[4] Liu Y, Fu Q, Li L, Shen Q, Li H and Li K 2016 Microstructure and mechanical properties of SiC fibres after exposure at elevated temperature *Adv. Appl. Ceram.* **115** 449
[5] Galasso F, Basche M and Kuehl D 1966 Preparation, structure and properties of continuous silicon carbide filaments *Appl. Phys. Lett.* **9** 37
[6] Yajima S, Okamura K, Hayashi J and Omori M 1976 Synthesis of continuous silicon carbide fibers with high tensile strength *J. Am. Ceram. Soc.* **59** 324
[7] Okada K, Kato H and Nakajima K 1994 Preparation of silicon carbide fiber from activated carbon fiber and gaseous silicon monoxide *J. Am. Ceram. Soc.* **77** 1691
[8] Ryu Z, Zheng J, Wang M and Zhang B 2002 Preparation and characterization of silicon carbide fibers from activated carbon fibers *Carbon* **40** 715
[9] Lee J C and Park M J 1999 Effect of hold time on reaction of silicon monoxide with activated carbon fiber composites *Carbon* **37** 1075
[10] McHugh K M, Garnier J E, Griffith G W 2011 Synthesis and analysis of alpha silicon carbide components for encapsulation of fuel rods and pellets *ASME 2011 Small Modular Reactors Symp.* 165
[11] Istomina E I, Istomin P V, Nadutkin A V, Kargin Y F and Lysenkov A S 2018 Preparation of a SiC fiber textile material *Inorg. Mater.* **54** 787

IOP Conf. Series: Materials Science and Engineering **525** (2019) 012060 doi:10.1088/1757-899X/525/1/012060

The structure of the alloy Ti - (20-30) Nb - 5Zr after smelting and homogenizing annealing

S V Konushkin, M I Baskakova, A V Leonov, E O Nasakina, M A Sudarchikova, A A Kolmakova, A Bespamiatnova, K V Sergiyenko, Yu O Leonova and M A Sevostyanov

Baikov Institute of Metallurgy and Material Sciences, 49 Leninsky Avenue, Moscow, Russia

E-mail: nacakina@mail.ru

Abstract. The formation of the Ti– (20–30) Nb-5Zr alloys (at.%) structure was studied. The structure of the samples was determined using an light microscope. It is shown that after melting by the method of vacuum arc melting, the alloys have a developed dendritic structure. Homogenizing annealing at a temperature of 900 °C does not lead to the destruction of the dendritic structure and recrystallization. Annealing at a temperature of 1000 °C completely destroys the dendritic structure in the alloys.

1. Introduction

Currently, Ti-based alloys not containing Ni exhibiting the shape memory effect and superelasticity are actively studied. These alloys are considered as possible candidates for the replacement of the nitinol widely used in medicine, since nickel carcinogenic properties limit further medical use of this alloy [1-5]. The solution to this problem may be to create biocompatible corrosion-resistant surface layers on it [5-14]. However, in this case, the question of the mutual adhesion of the components arises, especially under alternating dynamic loads.

Alloys based on the Ti-Nb-Zr system are of interest for orthopedics and endovascular surgery. Alloy properties, incl. the magnitude of the shape memory effect and superelasticity, depend on the chemical composition, so for the accuracy of the research it is important to ensure the uniformity of the chemical composition of the alloy. For this it was important to work out the technology of smelting and homogenizing annealing of the selected alloys ingots [15-18].

In this paper, we investigated the structure of ingots of Ti-Nb-Zr system alloys after smelting and homogenizing annealing.

2. Materials and methods

The alloys of the Ti-Nb-Zr systems of the following compositions were studied: Ti-20Nb-5Zr, Ti-23Zr-5Zr, Ti-25Nb-5Zr, Ti-28Nb-5Zr, Ti-30Nb-5Zr (at.%). The alloys were smelted by the method of vacuum electric arc melting from pure charge materials. Ingots were obtained in the form of a biconvex lens with a diameter of 40 mm and a thickness in the center of 10-12 mm, weighing 30 g. The ingots were annealed in vacuum at a temperature of 900-1000 °C for 12 hours in a vacuum of 10^{-5} millimeters of mercury and cooled with the furnace. The temperature range of the homogenizing annealing was chosen, assuming that at 900 °C the processes of dendrite dissolution and recrystallization will begin, and after annealing at a temperature of 1000 °C a fully recrystallized β-phase will form.

Preparation of samples for metallographic studies was carried out by grinding on P320 abrasive paper, polishing on a diamond disk with a suspension with a size of polycrystalline diamond particles of 3 and 1 μm, and polishing on a felt disk with a suspension of colloidal quartz with particle sizes of 0.1 μm. The microstructure was detected by etching in an HF: 3HNO3: 5H2O solution and examined using a Neophot 2 light microscope.

3. Results and discussion

As shown by a light microscopy study in a state after smelting the structure of Ti-Nb-Zr alloys has 3 distinct zones over the ingot cross section (figure 1). The zone adjacent to the bottom of the copper crystallizer during smelting is consist of large homogeneous grains. The width of this zone was about 1 mm in the center of the ingot and decreased to its ends. Then there is a zone of dendritic grains, mainly oriented perpendicular to the crucible bottom. The grain size is about 1-1.5 mm wide and up to 5 mm long. Next the zone with smaller and multidirectional dendrites comes. There dendrites axis of the 1st and 2nd orders are visible. The microstructure of these ingot zones is presented in figure 2.

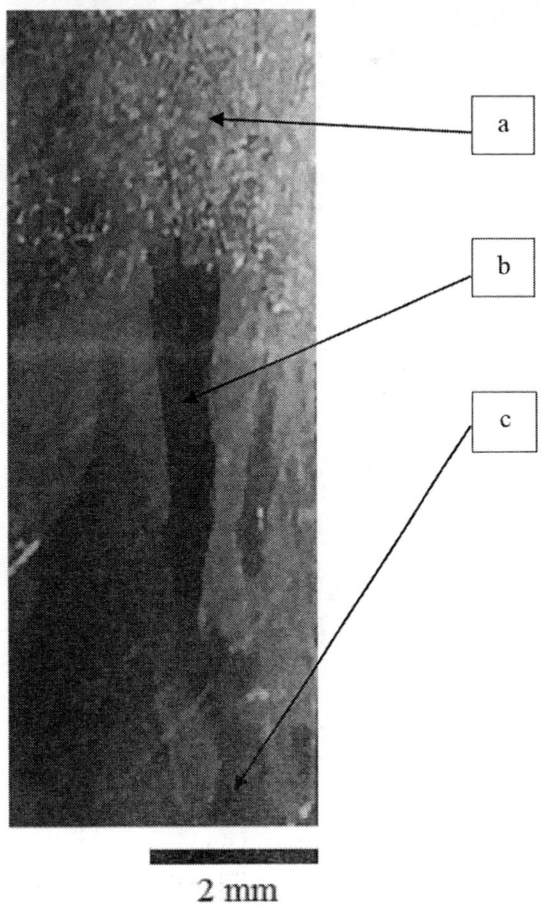

Figure 1. The microstructure of the Ti-Nb-Zr alloy ingot after smelting: a - the upper part of the ingot; b - the middle part of the ingot; c - the lower part of the ingot

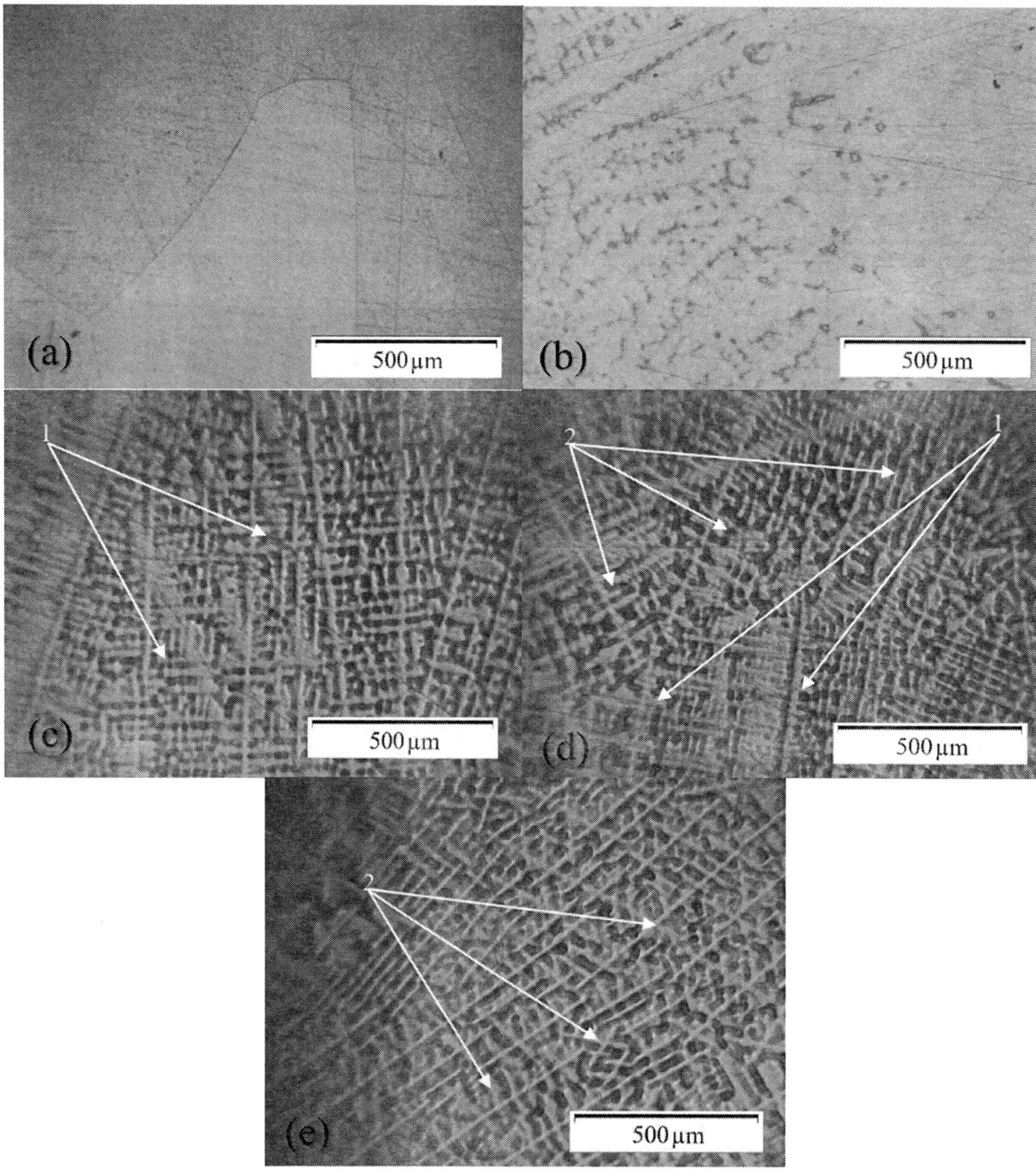

Figure 2. The microstructure of the ingot after smelting: (a) - in the lower part of the ingot; (b) - transition zone between the lower and middle parts of the ingot; (c) - in the middle part of the ingot; (d) - transition zone between the middle and upper parts of the ingot; (e)- in the upper part of the ingot. 1 - codirectional branches of dendrites, 2 - multidirectional branches of dendrites

Figure 2a shows the grain structure of the lower zone of the ingot. Grains have a size of up to 1 mm and are directed perpendicular to the bottom of the mold. Next comes the transition zone between the lower and middle parts of the ingot (figure 2b), in which the continuous, uniform structure is replaced by the dendritic one. The middle part of the ingot consists of large grains of co-directed dendrites elongated along the height of the ingot (figure 1, figure 2c). Figure 2d shows the transition zone between the middle and upper parts of the ingot, where the codirectional branches of the dendrites are replaced

IOP Conf. Series: Materials Science and Engineering **525** (2019) 012060 doi:10.1088/1757-899X/525/1/012060

by multidirectional. The upper part of the ingot is shown in Figure 2e and represents an area consisting of multidirectional dendrites.

The microstructure of ingots of different Ti-Nb-Zr compositions is shown in figure 3. As can be seen from the presented images, the microstructure of Ti-Nb-Zr alloys does not have significant differences in the shape and size of its elements. Characteristic features for all alloys are a developed dendritic structure in the upper part of the ingot and a coarse-grained structure in the lower part of the ingot.

The dendritic structure of the melted ingot is a marriage because is a chemical liquation. To eliminate the dendritic structure, long-term homogenizing annealing is used to activate the diffusion processes and equalize the chemical composition. The structure of the ingots after annealing at a temperature of 900 ° C for 12 hours is shown in figure 4.

As can be seen from the microstructure images annealing at a temperature of 900 °C led to a significant dissolution of dendrites and recrystallization in the Ti-20Nb-5Zr alloy. The boundaries of equiaxed grains began to be etched. The grains of dendrites in the ingots of the remaining alloys are still well etched.

For further dissolution of dendrites and recrystallization, alloy ingots were annealed at a temperature of 1000 °C for 12 hours. The structure of the ingots is shown in figure 5. As can be seen from it annealing at a temperature of 1000 °C led to the growth of grains in the Ti-20Nb-5Zr alloy and complete dissolution of the dendritic structure. In the Ti-23Nb-5Zr, Ti-25Nb-5Zr, Ti-28Nb-5Zr, and Ti-30Nb-5Zr alloys, the boundaries of polyhedral grains and subgrains also began to be etched.

Grains reached a size of 200-400 microns, which is quite suitable for further plastic deformation of melted ingots

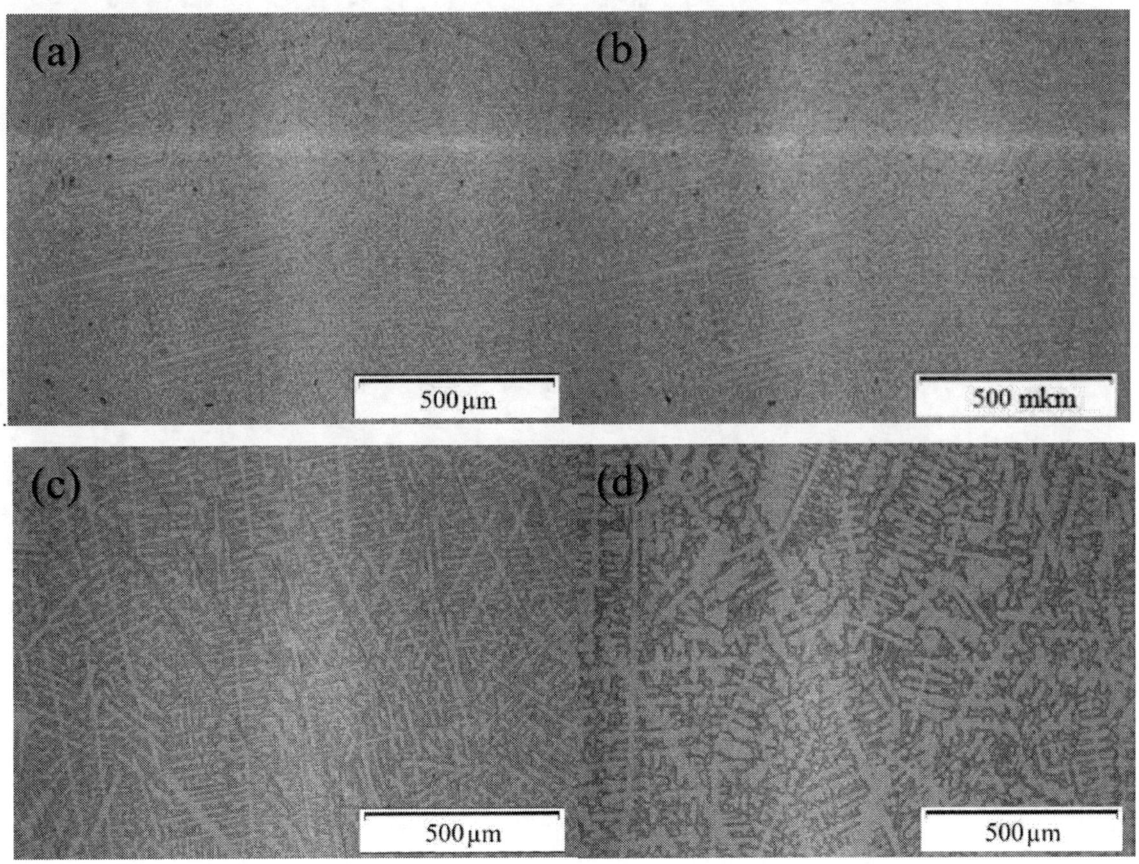

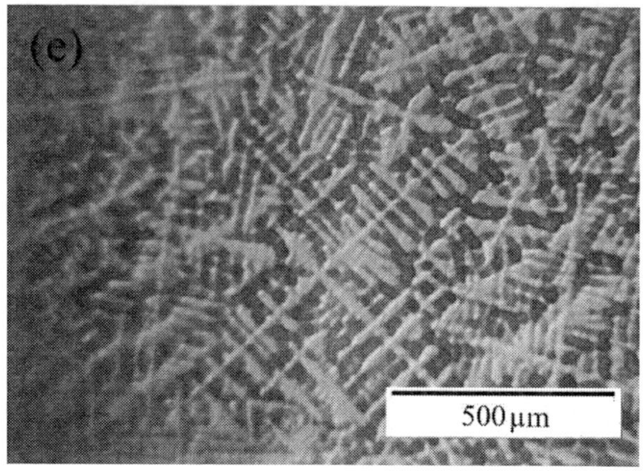

Figure 3. Dendritic structure of Ti-Nb-Zr alloy ingots after smelting: (a) - Ti-20Nb-5Zr; (b) - Ti-23Nb-5Zr; (c) –Ti-25Nb-5Zr; (d) - Ti-28Nb-5Zr; (e) - Ti-30Nb-5Zr

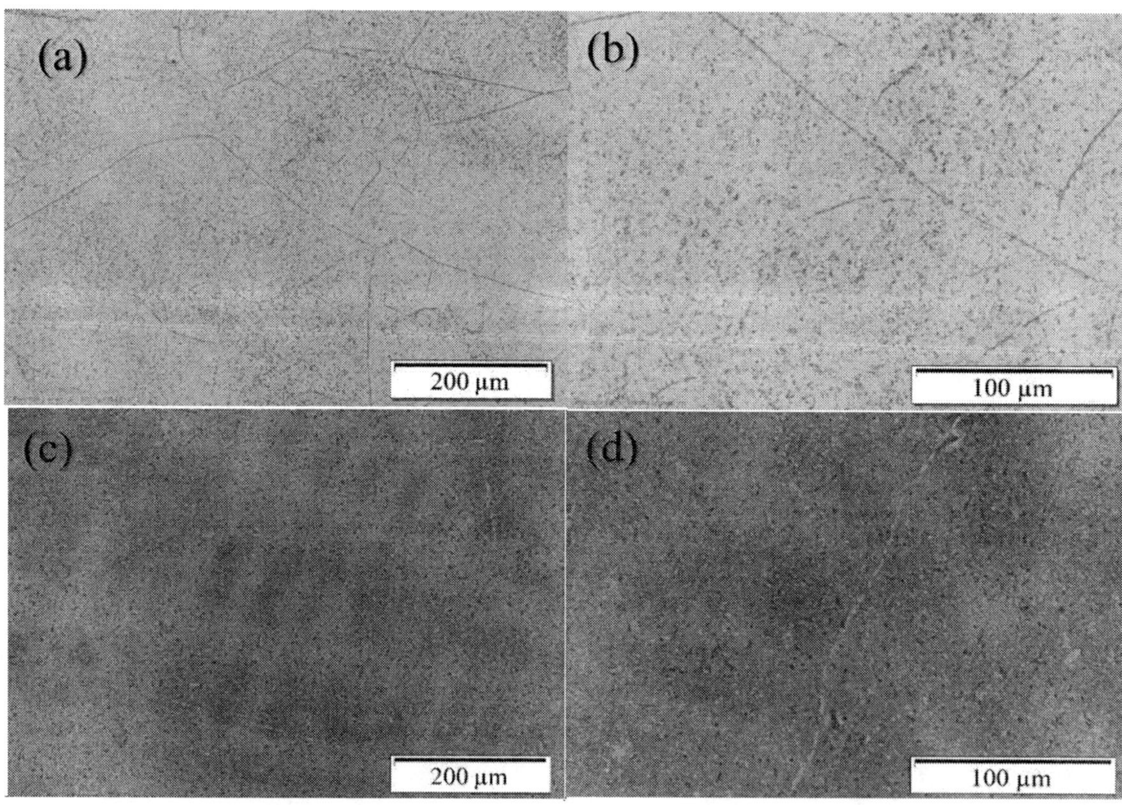

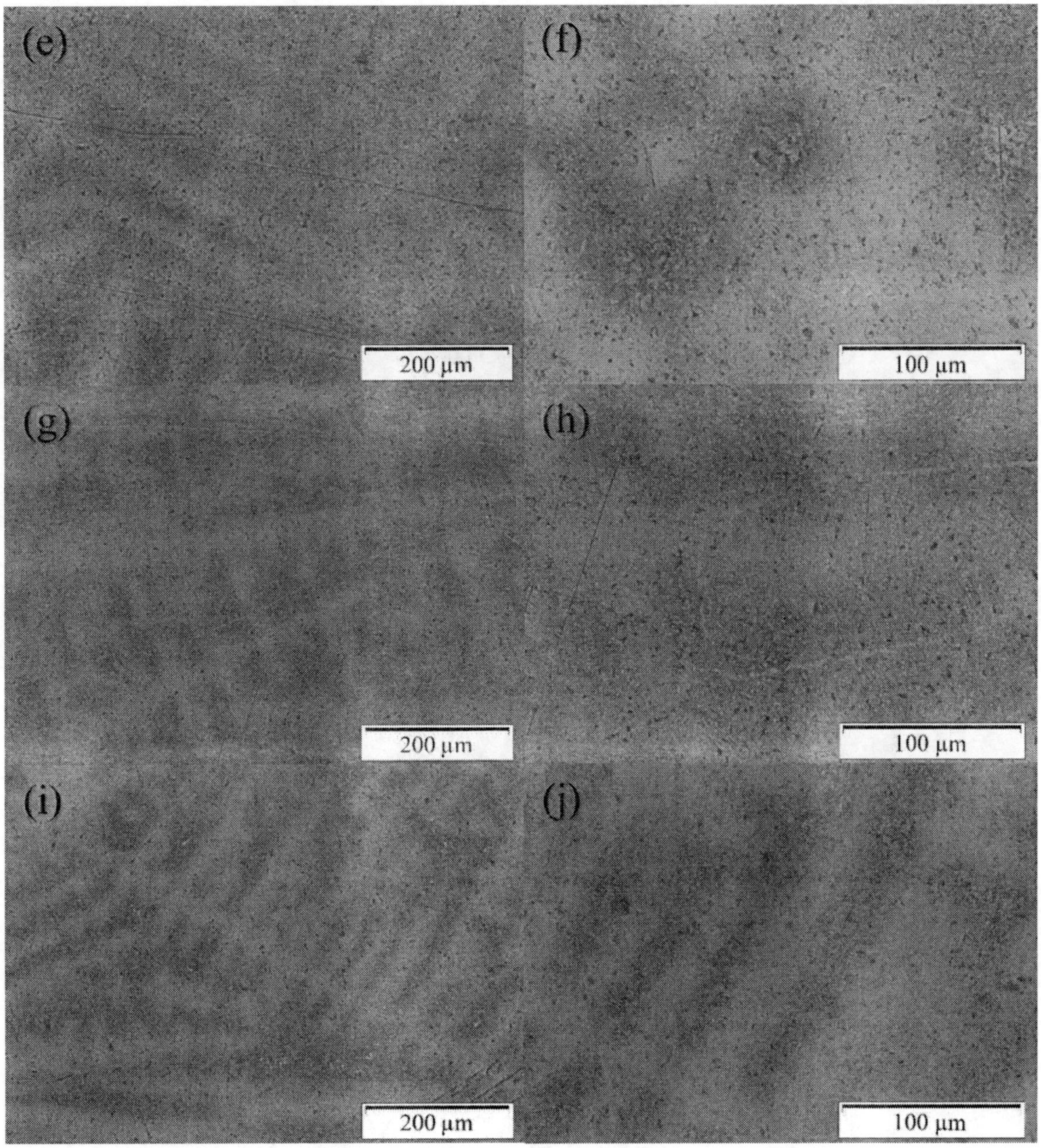

Figure 4. The structure of ingots of Ti-Nb-Zr alloys after annealing at 900 °C: (a), (b) - Ti-20Nb-5Zr; (c), (d) - Ti-23Nb-5Zr; (e), (f) - Ti-25Nb-5Zr; (g), (h) - Ti-28Nb-5Zr; (i), (j) - Ti-30Nb-5Zr

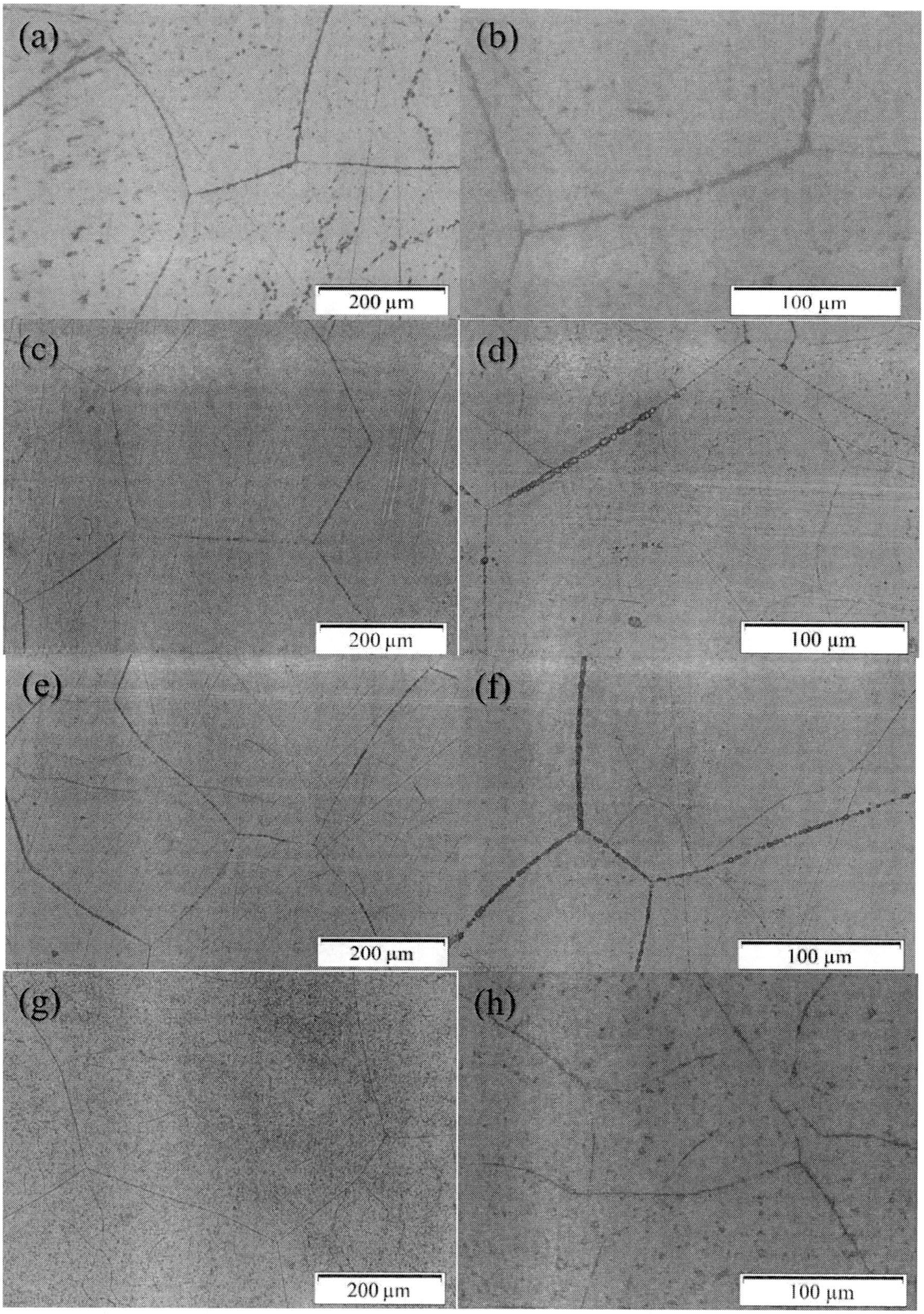
(a)
200 μm
(b)
100 μm
(c)
200 μm
(d)
100 μm
(e)
200 μm
(f)
100 μm
(g)
200 μm
(h)
100 μm

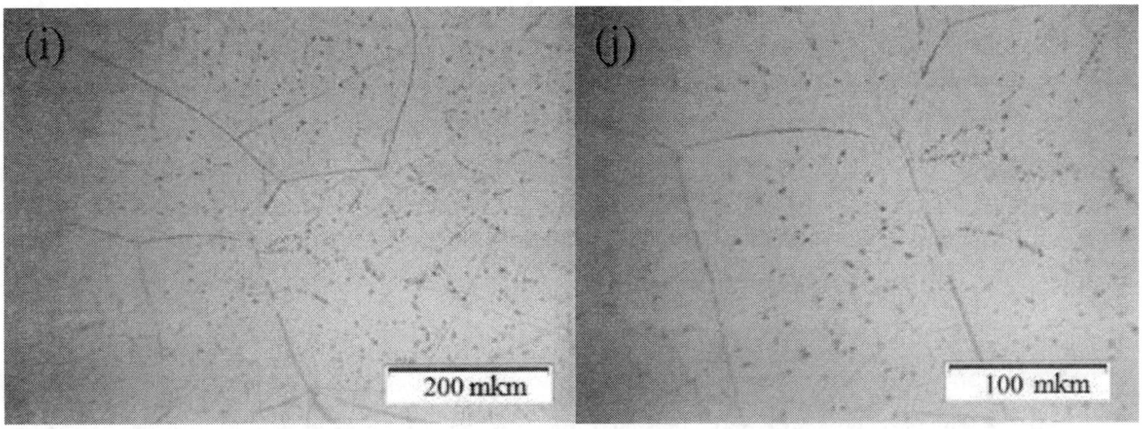

Figure 5. The structure of the ingots of alloys Ti-Nb-Zr after annealing at 1000 °C: (a), (b) - Ti-20Nb-5Zr; (c), (d) - Ti-23Nb-5Zr; (e), (f) - Ti-25Nb-5Zr; (g), (h) - Ti-28Nb-5Zr; (i), (j) - Ti-30Nb-5Zr

4. Conclusions

1. After vacuum electric arc melting the alloys of the Ti-Nb-Zr system have a developed dendritic structure, including elongated dendrite grains in the middle of the ingots and multidirectional grains in the upper part.

2. Homogenizing annealing at a temperature of 900 °C for 12 hours does not lead to the destruction of the dendritic structure and recrystallization. Annealing at a temperature of 1000 °C for 12 h completely destroys the dendritic structure in alloys Ti-20Nb-5Zr, Ti-23Zr-5Zr, Ti-25Nb-5Zr, Ti-28Nb-5Zr, Ti-30Nb-5Zr.

Acknowledgments

The work was supported by the Russian Science Foundation (grant № 18-79-00128).

References

[1] Davis J R. 2003 *Handbook of materials for medical devices* (ASM Int.) 350 p

[2] Sergienko K V, Sevost'yanov M A, Konushkin S V, Nasakina E O, Baikin A S, Shatova L A, Kolmakov A G 2018 Investigation of static properties of medical alloys Ti-(20-30)Nb-(10-13)Ta-5Zr *IOP Conf Ser: Mater Sci Eng* **347** 012049

[3] Nasakina E O, Sevostyanov M A,Golberg M A, Dyomin K Yu, Baikin A S, Goncharenko B A, et al. 2015 LongTerm Corrosion Tests of Nanostructural Nitinol of (55.91 wt % Ni, 44.03 wt % Ti) Composition under Static Conditions: Ion release *Inorg. Mater. Appl. Res.* **6(1)** 59–66

[4] Nasakina E O, Sevostyanov M A,Golberg M A, Dyomin K Yu, Baikin A S, Goncharenko B A, et al. 2015 LongTerm Corrosion Tests of Nanostructural Nitinol of (55.91 wt % Ni, 44.03 wt % Ti) Composition under Static Conditions: Composition and Structure before and after Corrosion *Inorg. Mater. Appl. Res.* **6(1)** 53–58

[5] Nasakina E O, Baikin A S, Sevost'yanov M A, Kolmakov A G, Zabolotnyi V T, Solntsev K A 2014 Properties of nanostructured titanium nickelide and composite based on it *Theor Found Chem Eng+* **48(4)** 477–486

[6] Nasakina E O, Sevost'yanov M A, Mikhailova A B, Gol'dberg M A, Demin K Yu, Kolmakov A G, et al. 2015 Preparation of a nanostructured shape memory composite material for biomedical applications *Inorg. Mater.* **51(4)** 400-404

[7] Nasakina E O, Sevostyanov M A, Dyomin K Yu, Mikhaylova A B, Goldberg M A, Kolmakov A G, Zabolotnyy V T 2015 The use of ion-atomic deposition in the fabrication of one-dimensional composites *Inorg. Mater. Appl. Res.* **6(4)** 293-297

[8] Nasakina E O, Seregin A V, Baikin A S, Kaplan M A, Konushkin S V, Sergiyenko K V, et al.

2017 Formation of biocompatible surface layers depending on the sputtering distance *IOP Conf. Series: J Physics* **857** 012032

[9] Sevost'yanov M A, Nasakina E O, Baikin A S, Sergienko K V, Konushkin S V, Kaplan M A et al. 2018 Biocompatibility of new materials based on nano-structured nitinol with titanium and tantalum composite surface layers: experimental analysis in vitro and in vivo *J. Mater. Sci. – Mater. Med.* **29** 33

[10] Nasakina E O, Baikin A S, Sergiyenko K V, Kaplan M A, Konushkin S V, Yakubov A D, et al. 2018 Formation of biodegradated polymers as components of future composite materials on the basis of shape memory alloy of medical appointment *IOP Conf Ser: Mater Sci Eng* **347** 012016

[11] Nasakina E O, Baikin A S, Sergienko K V, Sevost'yanov M A, Kolmakov A G, Goncharenko B A, et al. 2015 Biocompatibility of nanostructured nitinol with titanium or tantalum surface composite layers formed by magnetron sputtering *Doklady Chemistry* **461(1)** 86–88

[12] Nasakina E O, Baikin A S, Konushkin S V, Sergienko K V, Kaplan M A, Fedyuk I M, et al. 2018 Receiving of layered composite materials with shape memory effect of medical appointment *IOP Conf Ser: Mater Sci Eng* **411** 012051

[13] Nasakina E O, Baikin A S, Sergiyenko K V, Leonov A V, Kaplan M A, Seryogin A V, et al. 2017 Formation and Investigation of Composite Material Silver–Nitinol for Medical Purposes *Inorg. Mater. Appl. Res.* **8(1)** 112–117

[14] Nasakina E O, Sevostyanov M A, Mikhaylova A B, Baikin A S, Sergienko K V, Leonov A V et al. 2016 Formation of alpha and beta tantalum at the variation of magnetron sputtering conditions *IOP Conf Ser: Mater Sci Eng* **110** 012042

[15] Sevostyanov M A, Baikin A S, Shatova L A, Nasakina E O, Berezhnov A V, Gudkov S V, et al. 2018 Biocompatibility of the Ti81Nb13Ta3Zr3 Alloy *Doklady Chemistry* **482(1)** 204–206.

[16] Kim J I, Kim H Y, Inamura T, Hosoda H,Miyazaki S 2005 Shape memory characteristics of Ti–22Nb–(2–8)Zr(at.%) biomedical alloys *Mater. Sci.Eng. A* **403** 334–339.

[17] Kim H Y, Sasaki T, Okutsu K, Kim J I, Inamura T, Hosoda H, Miyazaki S 2006 Texture and shape memory behavior of Ti–22Nb–6Ta alloy *Acta Mater* **54** 423–433

[18] Dubinskiy S M, Prokoshkin S D, Brailovski V, Inaekyan K, Korotitskiy A V, Filonov M R, Petrzhik M I 2011 Structure Formation During Thermomechanical Processing of Ti-Nb-(Zr, Ta) Alloys and the Manifestation of the Shape-Memory Effect *PHYS MET METALLOGR* **112(5)** 529–542

Effect of mixing duration and sintering temperature on the magnetic properties of hard magnetic powder alloy Fe-30Cr-8Co (wt.%)

I M Milyaev, T A Vompe, M I Alymov, V S Yusupov, V A Zelenskii, A B Ankudinov and D M Abashev

Baikov Institute of Metallurgy & Materials Science (IMET RAS), Moscow, Russia

E-mail: tvompe@imet.ac.ru

Abstract. Effect of mixing duration and sintering temperature on the magnetic properties of hard magnetic powder alloy Fe-30Cr-8Co (wt.%) was studied. The obtained magnetic hysteresis properties of the hard magnetic powder alloy are better than as for industrial Fe-Cr-Co alloy which cobalt content is twice bigger. The optimum sintering temperature was 1350-1400 °C. At a sintering temperature of 1400 °C for 2.5 hours the mixing duration does not affect on the magnetic hysteresis properties. This technological operation is not critical.

1. Introduction

Hard magnetic materials of Fe-Cr-Co system have a special place among the hard magnetic materials due to their high mechanical (strength and plastic) characteristics along with a sufficiently high level of magnetic hysteresis properties (residual induction B_r, coercive force H_c, maximum energy product $(BH)_{max}$) [1].

Currently, special attention of researchers is attracted to low-cobalt (7-10 wt. % Co) hard magnetic Fe-Cr-Co alloys [2, 3] due to the fact that price of the cobalt in the global markets has been dramatically rise.

The previous investigations of the powder low cobalt hard magnetic Fe-Cr-Co alloys [4], that were held in the decade of 1980s, showed the fundamental possibility of obtaining very high magnetic hysteresis properties, nevertheless they have not received commercial application up to now.

The aim of the present work was to determine the effect of mixing duration and sintering temperature on the magnetic properties of hard magnetic powder alloy Fe-30Cr-8Co (wt.%).

Our findings open pathways towards fine production of Fe-Cr-Co hard magnetic alloys with controlled characteristics for distinctive applications.

2. Materials and methods

2.1. Materials

The samples of hard magnetic Fe-30Cr-8Co (wt. %) powder alloy were prepared by the methods of powder metallurgy. Powders of iron (VS), chrome (PHS-1), cobalt (PK-1N) of commercial purity have been used.

IOP Conf. Series: Materials Science and Engineering **525** (2019) 012061 doi:10.1088/1757-899X/525/1/012061

2.2. Preparation of the Samples

The powder particles were mixed in turbulence mixer C 2.0 "Turbula" in the interval 1-60 minutes. The powder was compacted by single-ended dry pressing at 600 MPa in a hardened steel die on a manual press KNUTH-130042. The pressing die had internal diameter 13,6 mm.

Then samples have been sintered for 2,5 hours at different temperatures in a shaft furnace SShV-1,25/24-I1 in 10^{-2} Pa vacuum. The height of the cylinder samples was 20 mm.

2.3. Characteristics of the samples

The sintered density of the hard magnetic Fe-30Cr-8Co (wt. %) powder alloy samples was determined by hydrostatic weighing. Then it has been correlated with the theoretical density of the alloy.

The treatment of samples was carried out in the standard laboratory furnace equipped with armored electromagnet. The magnetic field strength was 4000 Oe. The temperature was kept and monitored by proportional–integral–derivative controller. Heat treatment was carried out in standard laboratory muffle furnace equipped with the controller PROTAR 101.

Magnetic hysteresis properties including residual induction B_r, coercive force H_{cB} and maximum energy product $(BH)_{max}$ for the solution-treated alloys were determined from B-H loops automatically recorded by the Permagraph L EP-3 hysteresis graph. The measurement error for coercive force H_c and residual induction B_r was 3 %, and it was 6 % for maximum energy product $(BH)_{max}$.

The scheme of the heat treatment is shown in the figure 1.

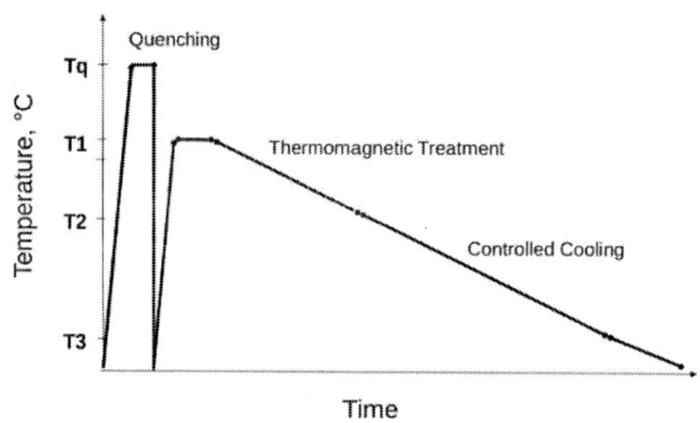

Figure 1. Schematic diagram of the heat treatment of the Fe-30Cr-8Co alloy.

The samples were quenched at 1100 °C in water and annealed from 670 °C till 580 °C at the rate of 4 °C/h in a 4000 Oe magnetic field. Further controlled cooling was conducted up to 500 °C without magnetic field at the rate of 4 °C/h. There was also additional annealing till 460 °C.

The x-ray diffraction studies were performed using a DRON-3M diffractometer, FeK_α radiation.

3. Results and discussion

3.1. Effect of Sintering Temperature

Density of samples of Fe-30Cr-8Co powder alloys and magnetic hysteresis properties of Fe-30Cr-8Co alloys versus sintering temperature in the range 1100 – 1400 °C are shown in figures 1 and 2, respectively.

One can see that there is a direct correlation between the increasement of the density of sintered samples with sintering temperature increasement and magnetic hysteresis properties increasement.

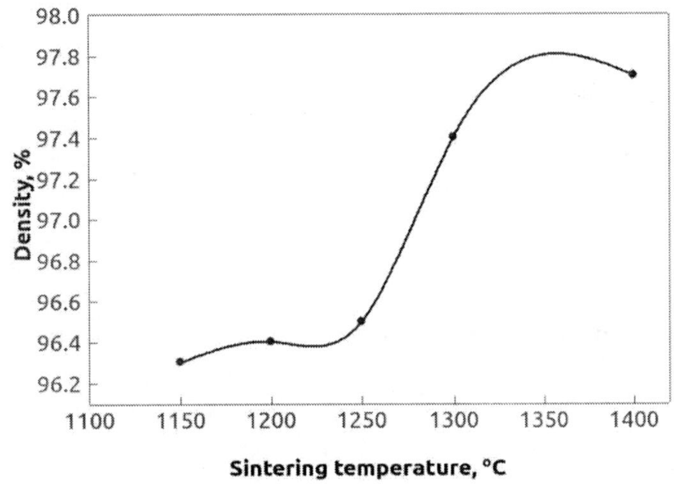

Figure 2. Density of samples of Fe-30Cr-8Co powder alloys versus sintering temperature

Relative increment of the residual induction B_r and coercive force H_c at sintering temperature at 1400 °C as compared to those obtained for 1100 °C are practically equal to the relative increment of density. All values increase by about 2%. Moreover, the curve of the temperature dependence of the coercive force has the same appearance as the curve for density change. These facts indicate that the structure of the investigated alloy as a functional magnetic material has been formed after sintering already at 1100 °C. Further sintering temperature increase affects only on the shape of the demagnetization curve. The evidence of this phenomenon is the temperature dependence of the maximum energy product, when the growth $(BH)_{max}$ is disproportionately high in comparison with the increase in density up to 1300 °C.

3.2. The Effect of Mixing Duration

Magnetic hysteresis properties of Fe-30Cr-8Co alloys versus mixing duration after sintering at 1200 °C during 2.5 hours is shown in the figure 4.

The figure 4 shows that the effect of the mixing duration on the magnetic hysteresis properties can be described by logarithmic function. Mixing for more than 15 minutes has practically no significant effect on the magnetic properties.

The x-ray diffraction studies of the samples, sintered in the entire studied range of sintering temperature, revealed that the obtained materials are single-phase materials with a ferrite structure, which is typical for low-cobalt alloys even at a low synthesis temperature. The trace of γ-phase was not detected. The X-ray diffraction pattern of the sample that was sintered at 1150 °C is typical (figure 5), similar x-ray patterns were obtained at other temperatures up to 1400 °C.

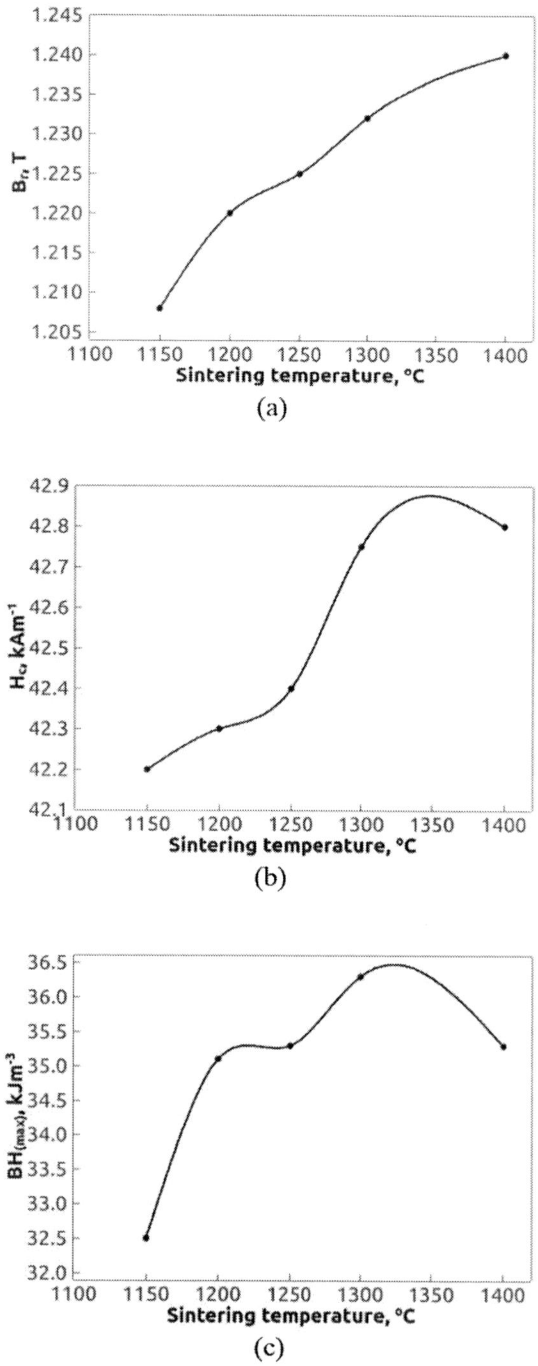

Figure 3. Magnetic hysteresis properties of Fe-30Cr-8Co alloys versus sintering temperature: (a) – B_r, (b) – H_{cB} and (c) – $(BH)_{max}$.

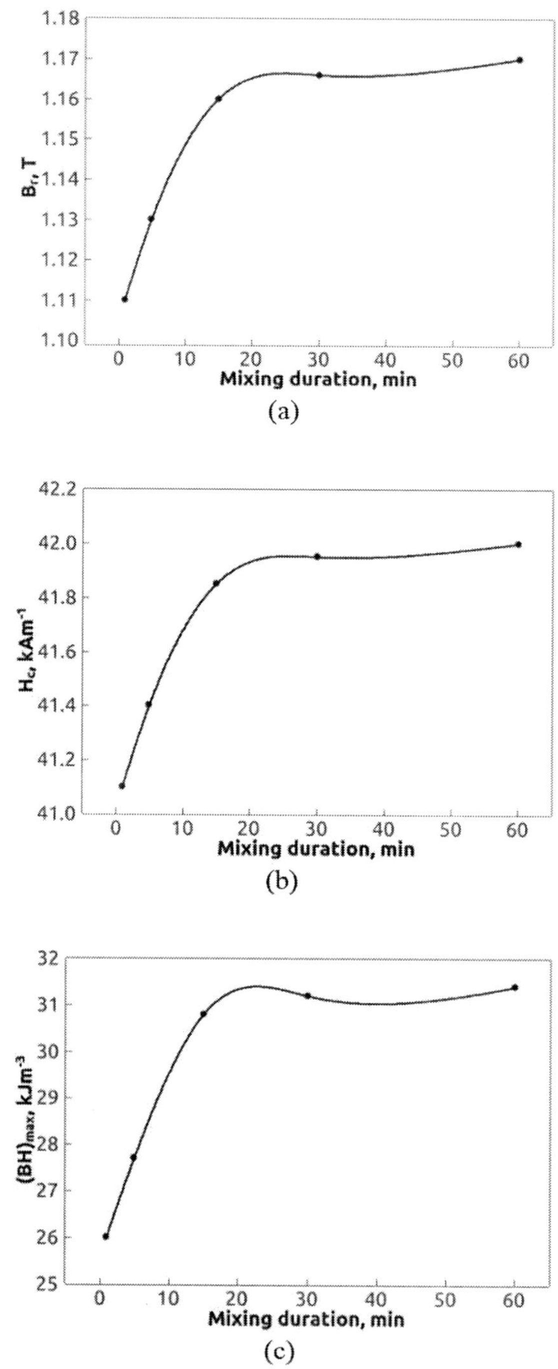

Figure 4. Magnetic hysteresis properties of Fe-30Cr-8Co alloys versus mixing duration of powder after sintering at 1200 °C: (a) – B_r, (b) – H_{cB} and (c) – $(BH)_{max}$.

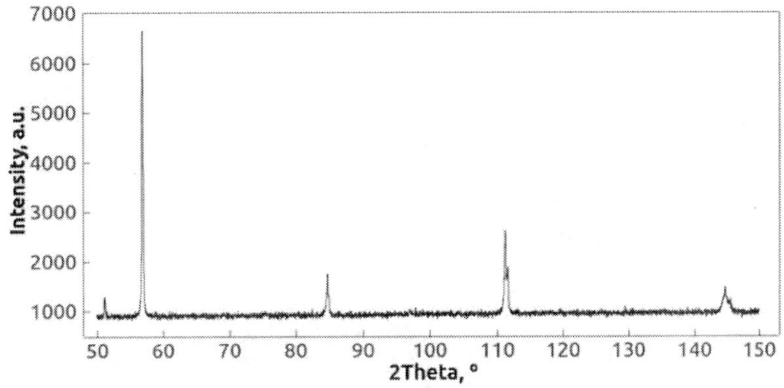

Figure 5. X-ray diffraction pattern of Fe-30Cr-8Co (after sintering at 1150 °C).

Magnetic hysteresis properties of Fe-30Cr-8Co alloys depending on the mixing duration after sintering at 1400 °C are shown in the table 1.

Table 1. Magnetic hysteresis properties of Fe-30Cr-8Co alloys depending on the time of mixing the powder mixture after sintering at 1400 °C.

Time of mixing, min	B_r, *T*	H_{cB}, *kAm*$^{-1}$	$(BH)_{max}$, *kJm*$^{-3}$
1	1.245	43.5	38.2
5	1.247	43.2	37.9
15	1.245	43.4	37.8
30	1.248	43.4	38.1
60	1.252	43.5	38.4

The mixing duration from 1 minute to 60 minutes does not affect on the magnetic properties after sintering at 1400 °C for 2.5 hours.

For samples sintered at 1200 °C and 1400 °C the difference in the behavior of the magnetic characteristics depending on mixing duration is quite explainable. With an increase in the sintering temperature increases by 200 °C, diffusion processes are activated significantly. In particular, the diffusion length increases significantly at a fixed sintering time, which determines the degree of completeness of the sintering process in terms of obtaining a homogeneous alloy. At a temperature of 1200 °C, after sintering samples with a short mixing time, the sintered material is not homogeneous. As a result, magnetic properties have smaller values. At this temperature, the sintering time of 2.5 hours is clearly not enough to create a homogeneous structure of the sintered material from insufficiently mixed powders with a short mixing time. Thus, it is possible to eliminate the heterogeneity in the sintered material and to obtain higher values of magnetic properties only after prolonged mixing.

4. Conclusion

The obtained magnetic hysteresis properties of the hard magnetic Fe-30Cr-8Co (wt. %) powder alloy are better than as for industrial Fe-25Cr-15Co-Al-Nb-V alloy which cobalt content is twice bigger.

The structure of the investigated alloy as a functional magnetic material turns out to be formed at 1100 °C temperature sintering. The resulting materials are single phase materials with a ferrite structure.

The optimum sintering temperature of the powder alloy Fe-30Cr-8Co should be recognized as 1350-1400 °C.

At a sintering temperature of 1400 °C for 2.5 hours the duration time does not affect on the magnetic hysteresis properties of the Fe-30Cr-8Co alloy. Thus, this technological operation is not critical.

Our findings can be used in synthesizing and achieving desired magnetic properties of the hard magnetic Fe-30Cr-8Co powder alloys.

Acknowledgments

The investigation of magnetic properties was carried out according to the State Task № 075-00746-19-00. Sintering regimes were chosen in accordance with the results of the study supported by RFBR project № 18-03-00666.

References

[1] Kaneko H, Homma M and Nakamura K 1971 New Ductile Permanent Magnet of Fe-Cr-Co System *AIP Conf. Proceedings. Magn. Magn. Mater.* 1088–92

[2] Vompe T A, Milyaev I M and Yusupov V S 2015 Magnetic hysteresis properties of the hard magnetic Fe-25%Cr-10%Co-2%Mo-0,5%Si powder alloy *Lett. Mater.* **5** 424–7

[3] Milyaev I M, Alymov M I, Bouryakov I N, Yusupov V S and Abashev D M 2018 Magnetic properties of powder hard magnetic Fe–27Cr–10Co–0.5Mo and Fe–27Cr–10Co–2Mo alloys *IOP Conf. Ser. Mater. Sci. Eng.* **347** 012053

[4] Green M L, Sherwood R C and Wong C C 1982 Powder metallurgy processing of CrCoFe permanent magnet alloys containing 5–25 wt. % Co *J. Appl. Phys.* **53** 2398–400

IOP Conf. Series: Materials Science and Engineering **525** (2019) 012062 doi:10.1088/1757-899X/525/1/012062

Investigation of the surface layer thickness uniformity at the magnetron sputtering depending on the geometry of the flow

E O Nasakina, M A Sudarchikova, G S Sprygin, M I Baskakova, I M Fedyuk, A Bespamiatnova, E A Danilova, S V Konushkin, A V Leonov, M A Sevostyanov and A G Kolmakov

Baikov Institute of Metallurgy and Material Sciences, 49 Leninsky Avenue, Moscow, Russia

E-mail: nacakina@mail.ru

Abstract. The processes of formation of Ti surface layers by magnetron sputtering depending on the geometry of the flow for the subsequent creation of a layered composite are studied. The structure and composition of the resulting layered composite materials were determined using SEM, atomic emission spectrometry and Auger electron spectrometry. The correspondence of morphology of new surfaces and the substrate itself is noted. High adhesion of the components is supposed. The inhomogeneity of the sprayed stream was noted, the main sputtering zone of the target was identified. The gradient of the distribution of the surface layer thickness within the limits of the most intense flux is found, depending on the angle of its incidence on the substrate.

1. Introduction

Modern science offers the possibility to create completely new materials with enhanced performance characteristics of several traditional ones at once - composites combining properties previously seen to be incompatible [1–19]. Including layered composites, which are widely used in electronics, optics, medicine, in the creation of structural and other functional objects and can be effectively obtained using vacuum ion-plasma methods. One of these methods is magnetron sputtering, which contributes to the production of thin films and surface layers of very different nature and thickness on substrates of almost any composition and geometrical configuration. The parameters of the resulting layers are directly related to the process parameters, which can vary in a fairly wide range. Among other things, the mutual arrangement in space of the sputtered target and the substrate, on which the layer is deposited, is essential, since it determines the intensity of the resultant flow along with the power of the process.

The purpose of this work was to study the uniformity of the thickness of the surface layer during magnetron sputtering depending on the flow geometry.

2. Materials and methods

As the basis of the formed composite materials, massive steel substrates with a diameter of 40 mm were used (which is economically more advantageous to study the effect of parameters and working out modes) after successive grinding of the surface with emery paper with a grain size of 180 to 1000 grit and by diamond paste (6-1 micron). For cleaning, activating and polishing the surface of the substrate it was bombarded by argon ions with discharge parameters Ue = 900 V, Ie = 80 mA —

IOP Conf. Series: Materials Science and Engineering **525** (2019) 012062 doi:10.1088/1757-899X/525/1/012062

preliminary ion etching. Chemically pure titanium was used as the material of the surface layer, regardless of the process parameters, forming a single phase, which facilitates the identification of results.

The creation of metal composite materials was carried out by forming surface layers using a magnetron at a constant current of ~ 760 mA at a voltage of ~ 400 V in an argon gas atmosphere at a working and residual pressure of ~ 0.4 and $4x10^{-4}$ Pa, respectively, during the sputtering time 12- 60 min at a sputtering distance of 140-210 mm with the substrate displaced relative to the longitudinal axis of the magnetron with the aim of maximum flow overlap. The analysis points on the sample surface facing the sprayed stream were arranged in an arbitrary order so that both the middle and the edges of the plane being formed were covered. The temperature on the surface of the substrate did not exceed 150 °C.

The morphology and the layered elemental composition (including using transverse thin sections) of the material surfaces were examined using a TESCAN VEGA II SBU scanning electron microscope (SEM) equipped with INCA Energy, an energy dispersive analysis device, JEOL Auger spectrometer, combined with ion etching when argon is bombarded at an angle of 30° and a GDS 850A glow discharge atomic emission spectrometer with a high-frequency alternating current source.

3. Results and discussion

Depending on the magnetron design, the flow of the sprayed substance has a different geometric configuration. However, the general is the heterogeneity of the flow of the substance sprayed. At the same time, due to the complexity of the target geometry in the sputtering zone, where gradual etching occurs, which is usually not straightforward (when the target is consumed, its relief changes with a gradual uneven curvature of the sprayed zone; and it is extremely difficult to determine from which point of the target surface specific atom/ion of a substance that forms one or the other flow zone will fly out; therefore this curvature was neglected, as if an untouched target were used at each moment of spraying), and due to the chaotic character of knockout of the target sputtering particles by ions of the working gas someone can not speak about the absolute angle of incidence of each individual atom/ion of sprayed material on a substrate. It is possible to analyze only the flow of the sprayed substance as a whole.

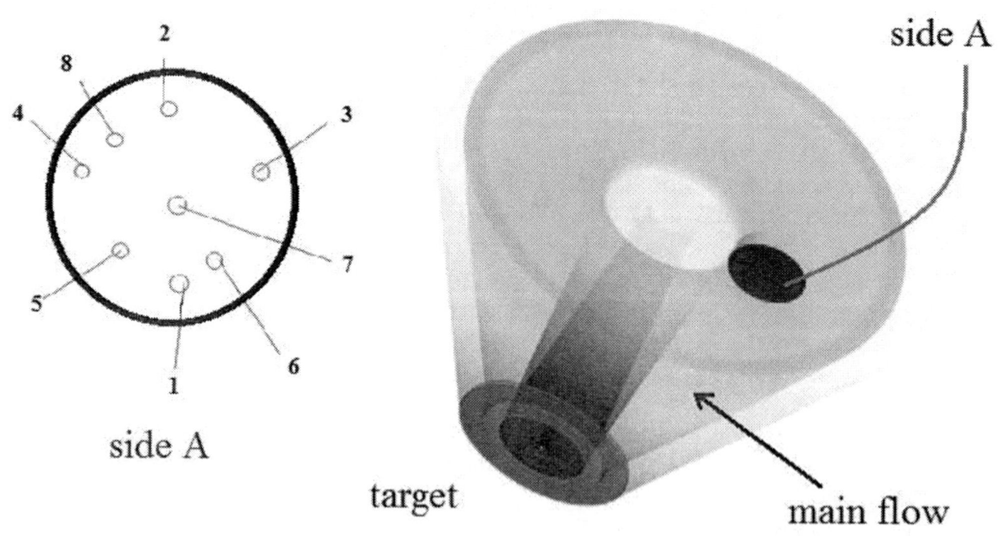

Figure 1. Distribution of sputtering flow and analysis points in space

In this work, the main (middle annular) target spraying zone and three more zones (2 internal and 1 external) of the stream with a much lower concentration of the sprayed substance, also differing from each other in intensity, were conditionally allocated. Figure 1 shows the location of the substrate in the flow of the sprayed substance, the target sputtering zones (the main is lighter one), the flow zones, and analysis points on the surface of the substrate facing the titanium flow.

Initially it was assumed that within the main flow zone the thickness of the layers formed would be fairly uniform. However, upon closer examination, a gradient of the distribution of the thickness of the surface layer within the entire flow was found, depending on the angle of its fall on the substrate.

Figure 2 shows the results of layer-by-layer analysis of samples with a conditional thickness of 0.4 μm (sprayed for 12 minutes), obtained at a distance of 140 mm in the most intense sputtering flow. In order not to overload the image, the distribution curves over the thickness of the concentrations of only two main elements are shown - iron and titanium. The depth of intersection of these curves was taken as a comparative thickness of the surface layer. It was shown that even on a site with a diameter of 40 mm, in the absence of substrate rotation, the thickness of the formed layer is significantly different, which means the divergence of the flow intensity even in the zone of the highest sputtering of the target. Similar results were obtained under other conditions. It can be noted that, in general, the thickness of the layer decreases within the limits of the most intense flow with distance from the longitudinal axis of rotation of the flow. The deviation of the thickness between the extreme points is about 30 nm, i.e. approximately 7% of the specified thickness (based on previous work).

At the same time, the same multilayer structure is preserved under all the listed conditions (with a conditional layer thickness of 0.4 μm, it is more difficult to visually distinguish the surface layer thickness than with a larger length), the thickness of the transition layer does not change, and the angle of incidence of the main flow does not affect adhesion.

Rotation of the substrate in a plane perpendicular to the longitudinal axis of the magnetron led to a smoothing of the thickness.

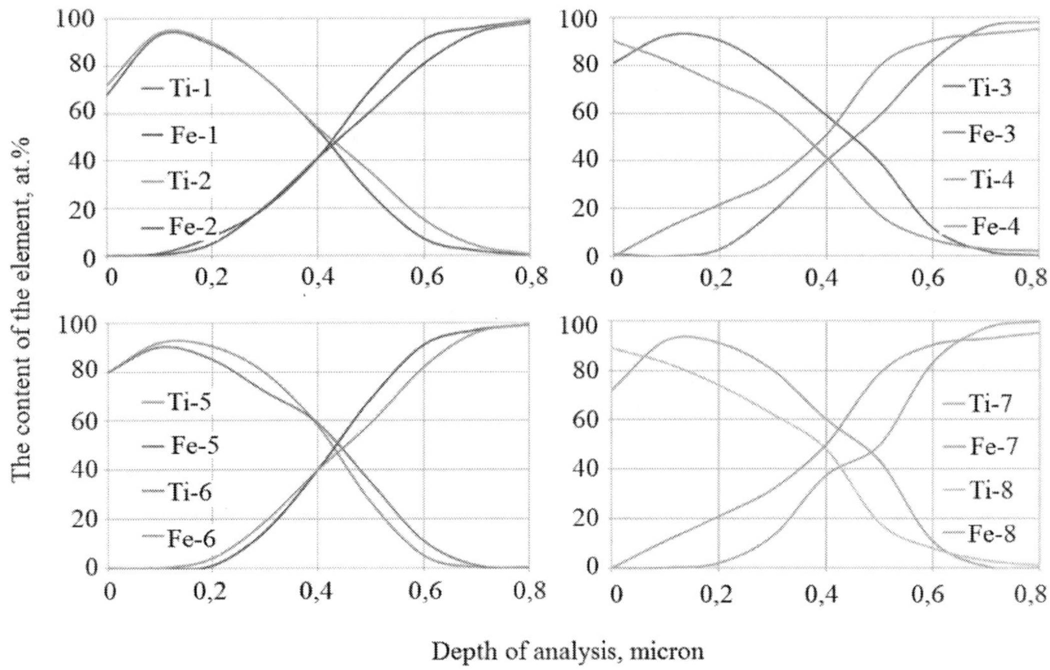

Figure 2. The dependence of the elements concentration on the depth of layer-by-layer analysis in sample No. 1 at analysis points 1-8

4. Conclusions

The processes of formation of Ti surface layers by magnetron sputtering depending on the geometry of the flow for the subsequent creation of a layered composite were studied.

The inhomogeneity of the sprayed stream was noted, the main sputtering zone of the target was identified.

The gradient of the distribution of the surface layer thickness within the limits of the most intense flux is found, depending on the angle of its incidence on the substrate.

The angle of incidence of the main stream does not affect the adhesion of the layers.

Acknowledgments

The work was supported by the Russian President Program for Young Scientists "MK-4521.2018.8".

References

[1] Arash Fattah-alhosseini, Hasan Elmkhah, Ghazaleh Ansari, Faridreza Attarzadeh, Omid Imantalab 2018 A comparison of electrochemical behavior of coated nanostructured Ta on Ti substrate with pure uncoated Ta in Ringer's physiological solution *J. Alloys Compd.* **739** 918-925

[2] Seidl W M, Bartosika M, Kolozsvári S, Bolvardi H, Mayrhofer P H 2018 Improved mechanical properties, thermal stabilities, and oxidation resistance of arc evaporated Ti-Al-N coatings through alloying with Ta *Surf. Coat. Technol.* **344** 244-249

[3] Sevost'yanov M A, Nasakina E O, Baikin A S, Sergienko K V, Konushkin S V, Kaplan M A et al. 2018 Biocompatibility of new materials based on nano-structured nitinol with titanium and tantalum composite surface layers: experimental analysis in vitro and in vivo *J. Mater. Sci. – Mater. Med.* **29** 33

[4] Li P, Zhang X, Xu R, Wang W, Liu X, Yeung K W K, Chu P K 2013 Electro-chemically deposited chitosan/Ag complex coatings on biomedical NiTi alloy for antibacterial application *Surf. Coat. Technol.* **232** 370–375

[5] Kononova S V, Romashkova K A, Kruchinina E V, Gusarov V V, Potokin I L, Korytkova E N, Maslennikova T P 2010 Polymer-inorganic nanocomposites based on aromatic polyamidoimides effective in the processes of liquids separation *Russian Journal of General Chemistry* **80(6)** 1136-1142

[6] Sevostyanov M A, Fedotov A Yu, Kolmakov A G, Zabolotnyi V T, Barinov S M, Goncharenko B A, et al. 2014 Mechanical Properties of Nanostructured Nitinol/Chitosan Composite Material *Inorg. Mater. Appl. Res.* **5(4)** 344–346

[7] Sevost'yanov M A, Fedotov A Yu, Nasakina E O, Teterina A Yu, Baikin A S, Sergienko K V, et al. 2015 Kinetics of the release of antibiotics from chitosan-based biodegradable biopolymer membranes *Doklady Chemistry* **465(1)** 278-280

[8] Cheng Y, Cai W, Li H T, Zheng Y F 2006 Surface modification of NiTi alloy with tantalum to improve its biocompatibility and radiopacity *J Mater Sci* **41** 4961–4964

[9] Krivoshapkin P V, Mikhaylov V I, Krivoshapkina E F, Zaikovskii V I, Melgunov M S, Stalugin V V 2015 Mesoporous Fe-alumina films prepared via sol–gel route *Microporous and Mesoporous Mater* **204** 276–281

[10] Dong-Won Lee, Yong-Nam Kim, Myung-Yeon Cho, Pil-Ju Ko, Daeseok Lee, Sang-Mo Koo, Kyoung-Sook Moon, Jong-Min Oh 2018 Reliability and characteristics of magnetron sputter deposited tantalum nitride for thin film resistors *Thin Solid Films* **660** 688-694

[11] Asa'ad Al-masha'al, Andrew Bunting, Rebecca Cheung 2016 Evaluation of residual stress in sputtered tantalum thin-film *Appl Surf Sci* **371** 571-575

[12] Jamil Siddiqui, Tousif Hussain, Riaz Ahmad, Zeeshan A Umar 2017 On the structural, morphological and electrical properties of tantalum oxy nitride thin films by varying oxygen percentage in reactive gases plasma *Chinese Journal of Physics* **55(4)** 1412-1422

[13] Mukesh Kumar, Neelam Kumari, VSRS Praveen Kumar, Vinod Karar, Amit L Sharma 2018

IOP Conf. Series: Materials Science and Engineering **525** (2019) 012062 doi:10.1088/1757-899X/525/1/012062

Determination of optical constants of tantalum oxide thin film deposited by electron beam evaporation *Materials Today: Proceedings* **5(2.1)** 3764-3769

[14] Colin J J, Abadias G, Michel A, Jaouen C 2017 On the origin of the metastable β-Ta phase stabilization in tantalum sputtered thin films *Acta Mater* **126** 481-493

[15] Cacucci A, Loffredo S, Potin V, Imhoff L, Martin N 2013 Interdependence of struc-tural and electrical properties in tantalum/tantalum oxide multilayers *Surf. Coat. Technol.* **227** 38–41

[16] Navid A A, Hodge A M 2012 Nanostructured alpha and beta tantalum formation—Relationship between plasma parameters and microstructure *Mater Sci Eng A* **536** 49– 56

[17] Bernoulli D, Müller U, Schwarzenberger M, Hauert R, Spolenak R 2013 Magnetron sputter deposited tantalum and tantalum nitride thin films: An analysis of phase, hardness and composition *Thin Solid Films* **548** 157–161

[18] Zhou Y M, Xie Z, Ma Y Z, Xia F J, Feng S L 2012 Growth and characterization of Ta/Ti bi-layer films on glass and Si (1 1 1) substrates by direct current magnetron sputtering *Appl. Surf. Sci* **258** 7314– 7321

[19] Bunshah R F at al. 1982 *Deposition technologies for films and coating* (Park Ridge, New Jersey (USA): Noyes Publikations) 489 p.

Influence of various welding methods on the structure and properties of welded austenite steel joints with nitrogen content ~ 0,5%

V S Kostina, M V Kostina and S O Muradian

A.A. Baikov Institute of Metallurgy and Materials Science, Russian academy of Sciences, Moscow, 119334, Russia

E-mail: vskostina@yandex.ru

Abstract. The analysis of literature data on the weldability of nitrogen-containing austenitic steels with high equilibrium nitrogen concentrations (0.4-0.6) is performed, and the results of studies of the structure and phase composition and mechanical properties are given for Cr-Mn-Ni-Mo-N welded joints with nitrogen concentration ~ 0.5% obtained by laser welding (for thin sheet metal) and argon arc and manual arc welding (for hot rolled metal with a thickness of up to 20 mm)

1. Introduction

According to the ISO standard: "Steels or metals can be considered weldable if the continuity of the welded joint is achieved during the welding process. At the same time, welding methods are used to obtain a joint that satisfies the requirements for their local properties and the effect on the structure of which they are part". An important indicator of the weldability of austenitic corrosion-resistant nitrogen-containing steels with high nitrogen concentrations is the absence in the welded joints (WJ) of overheating sites, gas pores and / or release of chromium nitrides in the heat affected zone (HAZ), reducing the level of performance properties of the welded product, including the reduction of mechanical strength, viscosity of the weld (W), and its corrosion resistance [1-4].

It is known that one of the causes of gas pores is a partial burnout of the main alloying elements, such as chromium and manganese (elements that contribute to the solubility of nitrogen) [5]. Due to the decrease in their concentration in the molten metal zone, the solubility of nitrogen is limited, withal, the crystallizing metal prevents the release of undissolved nitrogen, leaving it in the W as gas pores. M. Raffi et al. [6] investigated the microstructure and corrosion resistance of welded joints of high-nitrogen steel (~ 0.5% N), welded with Cr-Mn-N filler material by argon-arc welding in a protective gas. WJ showed a high level of properties due to the complete solubility of nitrogen in the metal of W. The authors recommend carefully choosing a filler material so that its composition contains enough chromium and manganese for nitrogen solubility. It is also necessary to achieve equal strength of the W in comparison with the base metal (BM). Equal strength of the WJ can be achieved by using as a filler material a metal corresponding in its chemical composition to the base metal. Such studies were carried out for steels Х22АГ16Н8М [7] and 05Х22АГ15Н8М2Ф [8]. WJ were obtained using argon arc welding and submerged arc welding ANK-67. It was found that, compared with BM, the content of nitrogen, chromium, and manganese in the metal of the W is lower. However, the WJ showed high strength, the metal of W remained austenitic, nitrogen was evenly distributed in W and HAZ, and there were no pores in the metal of W.

In general, the issues of weldability of austenitic steels with high equilibrium nitrogen concentrations are resolved; however, the development of welding process technology is required for each specific steel.

2. Materials and Methods

We studied the WJ of two grades of austenitic steels: 04Х20Н6Г11М2АФБ [9] and 05Х22АГ15Н8М2Ф [10] with an equilibrium nitrogen concentration ~ 0.5% (find chemical composition in Table 1) obtained by several methods of fusion welding. Taking into account the analysis of the literature data, for their welding the methods and modes given in Table 2 were selected, which allow to obtain high-quality non-porous WJ [11, 12]. Steels for which WJ have been investigated:

1) hot rolled sheet steel 04Kh20N6G11M2AFB (thickness 10 and 20 mm), obtained using filler materials S_{v1}-09Kh16N25M6AFS, S_{v2}-10Kh20N18M3APS and EA-868/20 (10Kh19N23G2M5FAT) [12] (WJ-1, WJ-2, WJ-3);

2) 05Kh22AG15N8M2F steel - hot rolled and cast (WJ-4, WJ-5), thickness - 3 mm (chemical composition is presented in Table 1), welded without the use of filler materials.

Welding modes are shown in Table 2.

Table 1. The chemical composition of the base metal and filler materials

Grade	Chemical composition, mass% (Fe - and impurities - the rest)								
	C	N	Si	Cr+Mo	Ni+Mn	V	Nb	S	P
04Kh20N6G11M2AFB	0,05	0,47	0,15	20,3	17,8	0,13	0,15	0,004	0,014
S_{v1}-09Kh16N25M6AFS	0,12	0,15	0,9	21,8	26,8	0,80	-	0,007	≤0,015
S_{v2}-10Kh20N18M3APS	0,12	0,15	0,9	22,8	19,3	0,80	-	0,007	≤0,015
EA-868/20 (10Kh19N23G2M5FAT)	0,12	0,25	0,47	23,1	25,3	1,85	-	0,005	0,015
05Х22АГ15Н8М2Ф	0,04	0,47	0,2	24,0	23,2	0,2	-	0,008	0,01

Table 2. Modes and methods of welding

Welded Joint	Structural condition /thickness	Filler material	Welding method	Edging
WJ-1	hot-rolled /20 mm	S_{v1}-09Kh16N25M6AFS	TIG: I_w=200-220 A, U_a=24-28 B	X-shaped
WJ-2		S_{v2}-10Kh20N18M3APS		
WJ-3	hot-rolled /10 mm	EA-868/20 (10Kh19N23G2M5FAT)	MMA: I_w = 110-140 A, U_a= 24 ÷ 28B.	
WJ-4	hot-rolled /3 mm	No	Laser welding: power P = 1 … 6, kW; welding rate V= 1…7 m/min, depth of focus from the surface +3 … -3 mm[a]	No
WJ-5	cast /3 mm			

[a] Development of IPG IRE-Polus (Ytterbium Fiber Laser)

The microstructure of WJ was examined on an Olympus light microscope and a LEO-1420 scanning electron microscope with an Oxford Instruments X-ray Microscope Analysis Unit (MRSA). The nitrogen content in samples from different zones of WJ was determined on the Eltra ONH-2000 gas analyzer ("Venta Lab"). Tensile tests were carried out according to GOST 1497-84 on Instron 3382 equipment; (test speed 1 mm / min). Impact bending tests were carried out according to GOST 9454-78 on Amsler RKP-450 (copra energy - 300 J). The microhardness of the various structural components of the WJ (fusion line — FL, weld — W, base metal — BM, and heat-affected zone of weld — HAZ) was determined according to GOST 9450-76 on Volpert 402MVD hardness meter (load 50g for 10 s). Ferritometry was performed using IMP-2M ferritometer.

3. Results and Discussions

As a result of welding according to the specified modes, defect-free (according to X-ray inspection data) WJ were obtained from the flat rolled steel 04Kh20N6G11M2AFB with thickness 10 and 20 mm.

Zones of noticeable porosity were found along the austenite grain boundaries (pore size 20-100 nm, length of zones up to 7-10 microns) in the WJ-1 (with the use of S_{v1}-09Kh16N25M6AFS). Sometimes the appearance of pores is accompanied by the appearance of particles of excess nitride or carbonitride of the type Cr2N. Porosity was absent in the WJ-2 (using S_{v2}-10Kh20N18M3APS).

There is no pronounced HAZ of the weld in WJ-3. The zone of relatively small grains (10-40 microns) having a width of 100-200 microns adjoins the FL. At WJ-1 (Figure 1, a) and WJ-2 (Figure 1, b), the HAZ of the weld is observed wherein primary recrystallization took place with the formation of small equiaxed grains, and directly near the FL a collective recrystallization took place. In this part of the HAZ larger equiaxed austenite grains were formed.

WJ-4 and WJ-5 (thickness - 3 mm) are characterized by the presence in the central part of W of a clear vertical line of joining dendrites with size from ~ 5 to 150 ± 30 μm of each of the two halves of WJ; this is especially noticeable in WJ-4 (Figure 1, c). WJ-5 is characterized by the presence of ferrite in BM in the amount of ~ 2.86% (Figure 1, d), ~ 0.3% of ferrite is contained in BM of welded joints of hot rolled steel. Ferrite and σ-phase precipitates in the weld metal of WJ-5 were not observed. In WJ-4 and WJ-5 the metal microstructure in the heat affected zone of the weld did not differ from the microstructure of the base metal.

The MRSA method was used to determine the content of basic metal alloying elements. In WJ-1 and WJ-2 the chemical composition of the base metal (BM) at a distance of 8–10 μm from the FL corresponds to the grade composition of steel 04Kh20N6G11M2AFB. In some areas of the metal of the melted grains on the FL small changes in the concentration of Cr, Mn, Ni are noted. In welds, at a distance of 10–30 μm from the FL, a gradual approximation of the chemical composition of the metal to the composition of the welding wire begins (the content of Mn decreases, the content of Ni and Mo increases, in WJ-1the content of Cr also decreases). At a distance of ~ 130 and ~ 240 μm from the FL, the compliance of the composition of the metal WJ with the chemical composition of the welding wire was recorded. In WJ-4 and WJ-5 in BM the content of the main alloying elements corresponds to the grade composition of steel 05Kh22AG15N8M2F. In the weld there is a slight decrease of the concentration of Mn (on average, 14.5% versus 15.3% in BM), the content of the remaining alloying elements is almost the same as in BM. According to the gas fractional analysis of the WJ metal of hot-rolled steel, the nitrogen concentration in BM is 0.47, and in the weld it is 0.39.

Measurements of microhardness in various zones of WJ-3 showed that the highest hardness is observed for the metal on the fusion line (HV_{50} = 345), then it decreases in the series: HAZ → BM (310 → 280). Metal of weld has the lowest hardness (260). The average values of microhardness in the WJ-4 are as follows: in BM - 329, in the W - 324. For the WJ-5 lower values were obtained: in BM - 312, in the weld - 318.

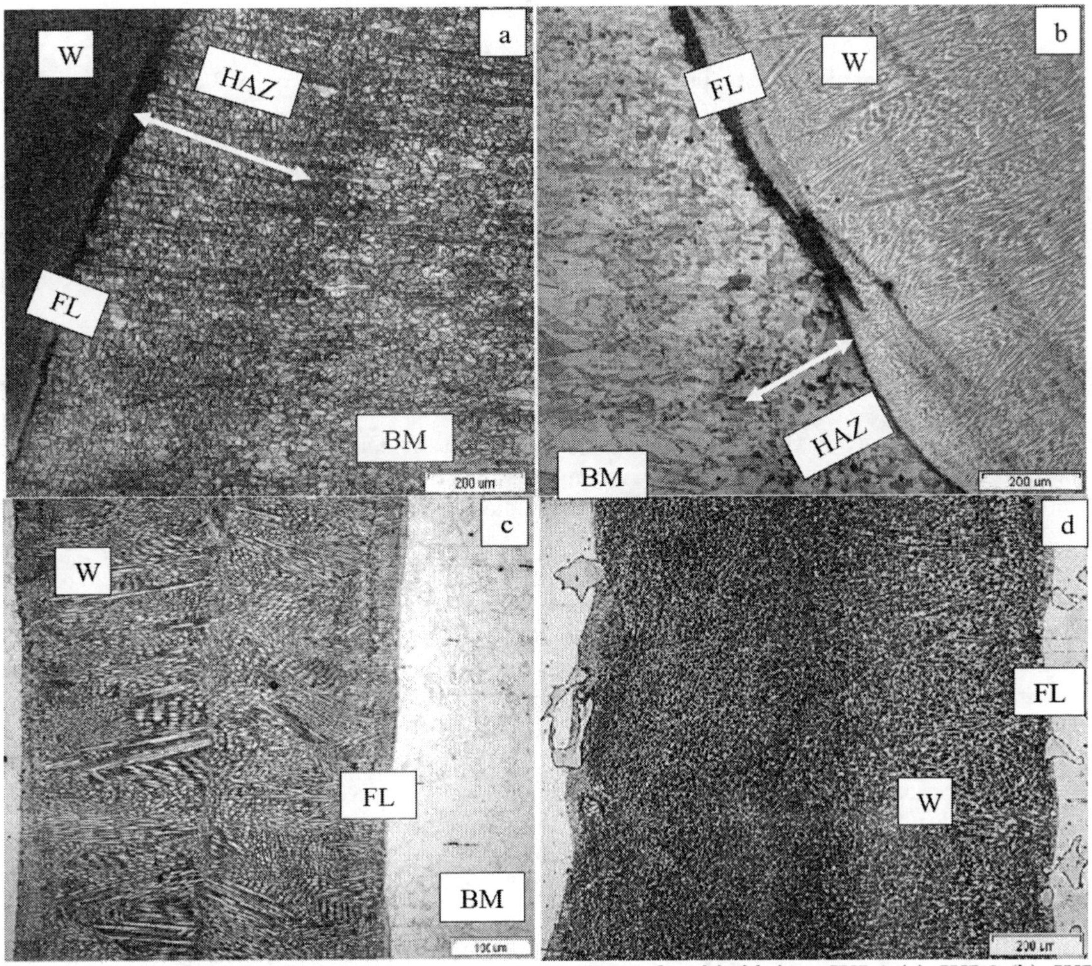

Figure 1. Microstructure of various zones of welded joints: WJ-1 (a), WJ-2 (b), WJ-4 (c) and WJ-5 (d).

Tests on the mechanical properties of WL-3 and base metal 04Kh20N6G11M2AFB were carried out on cylindrical samples. The yield strength of tensile samples WJ-3 is higher than the yield strength of BM (thickness 10 mm) and of the metal of the welding wire (Table 3). Higher strength of the metal WJ is probably achieved due to a significant refinement of the structure in the zone of fusion line (the size of the cast grains on the FL is 4 μm, whereas in BM it is from 50 to 100 μm). The ductility characteristics of the WJ-3 specimens are lower than the ductility characteristics, determined separately on the BM (thickness 10 mm), and the metal of the welding wire. This is concerned with the presence of inclusions, first of all - large particles in the metal of the weld.

Table 3. Mechanical properties of WJ-3, filler material and steel 04Kh20N6G11M2AFB at 20 °C.

Type of metal	$\sigma_{0,2}$, MPa	σ_u, MPa	δ, %	ψ, %	KCV, J/cm^2
WJ-3	597	800	21	29	≥ 200
EA-868/20 (10Kh19N23G2M5FAT)	530	770	33	55	107,5
BM, rolled 10 mm	540	854	57	71	250

IOP Conf. Series: Materials Science and Engineering **525** (2019) 012063 doi:10.1088/1757-899X/525/1/012063

Tensile tests were carried out on flat samples of WJ-4 and WJ-5 and base metal 05Kh22AG15N8M2F in a cast state, thickness - 3 mm (find in Table 4).

Table 4. Mechanical tensile properties at 20 °C of samples WJ-4 and WJ-5 and base metal 05Kh22AG15N8M2F (not welded samples of cast steel).

Type of metal	$\sigma_{0,2}$, MPa	σ_u, MPa	δ, %	ψ, %	Comments	
WJ-4	375	914	31	18	FL	Region of breakage
	336	920	36	17	W	
WJ-5	308	657	51	45	BM	
	268	625	39	35	FL	
05Kh22AG15N8M2F (cast steel without welding)	355	650	49	33	With edge treatment	
	321	648	47	45	Without edge treatment	

The WJ samples were flat, without smooth edges, and the sharp edges of the samples were stress concentrators. When testing the cast metal a comparison was made: flat samples of BM were tested with treated edges (smoothed out and not treated). On the tested specimens the WJ region in which the specimen was destroyed was detected by the metallographic method, identifying zones of the weld, FL, HAZ, BM.

The tensile strength of WJ-4 is about 900 MPa, which is typical for non-welded samples of this steel after hot plastic deformation and short-term annealing at 1000-1050 ° C with cooling in water [13]. It indicates an equal strength of WJ and BM is observed. This correlates with the results of microhardness measurements. At the same time, the yield strength of these WJs, especially those destroyed in the weld zone, is at the level inherent in the cast state of 05Kh22AG15N8M2F nitrogen-hardened steel.

The destruction of WJ-5 samples occurred both in BM and in FL. Samples destroyed in the BM zone naturally had a level of strength similar to that of the cast non-welded metal samples (~ 650 MPa). Slightly smaller values of strength and yield strength among WJ-5 samples had samples destroyed in FL. Tests of cast non-welded metal (BM) samples showed that flat BM samples with non-smoothed edges have a slightly lower yield strength than samples with treated edges.

4. Conclusion

1. High-quality welded joints obtained by various methods of fusion welding, using and without welding filler materials.
2. The influence of the modes and methods of welding on the structure and properties of welded joints of high-nitrogen austenitic stainless steels (~ 0.5% N) in hot-rolled and cast states was established.
3. For welded joints (WJ-1, WJ-2 and WJ-3) made using the filler material, it has been established: the higher the welding current, the larger the HAZ size at a comparable rolled thickness. In WJ-4 and WJ-5. thickness 3 mm, welded with an ytterbium fiber laser (developed by IPG IRE-Polus), the HAZ is absent.
4. The content of basic chemical elements corresponds to the grade composition of steels and filler materials, however, in the zone of fusion line, the content of Mn and Cr slightly decreased in some samples.
5. At WJ-1, WJ-2 and WJ-3, the metal on the fusion line has the highest hardness, then it decreases in the series: HAZ → BM → W.
6. WJ-1, WJ-2 and WJ-3 have high toughness and strength, including based on the release of nanoscale particles of chromium nitrides, which are an obstacle to the movement of dislocations.
7. WJ-4 and WJ-5 can be considered equal-strength to the base metal, which correlates with the data of microhardness measurements of BM and weld. The yield strength of the studied WJ as a whole is at the level inherent in the cast state of the 05Kh22AG15N8M2F steel strengthened by nitrogen; it also depends on the zone in which the destruction occurred.

IOP Conf. Series: Materials Science and Engineering **525** (2019) 012063 doi:10.1088/1757-899X/525/1/012063

Acknowledgement
The work was carried out according to the state assignment No. 075-00746-19-00.

References

[1] Zhao L, Tian Z and Peng Y 2007 High-Frequency Stainless Steel *ISIJ Int* **47** pp. 1772-1775
[2] Dong W, Kokawa H, Tsukamoto S and Sato Y S 2005 Nitrogen desorption of high-nitrogen steel weld metal during CO2 laser welding *Metallurgical and Materials Transactions B* **36B** pp. 677-681
[3] Harzenmoser M 2004 Welding of high nitrogen steels Materials and Manufacturing Processes **19** pp. 75-86
[4] Frolov V V, Vinokurov V A and Volchenko V N 1999 Theoretical foundations of welding (Moscow: Higher School) p 592
[5] Balitsky A I, Kostyuk I F and Krokhmalny O A 2003 Physico-mechanical inhomogeneity of welded joints of high-nitrogen chromium-manganese steels and their corrosion resistance *Automatic welding* **2** pp. 28-31
[6] Mohammed R, Madhusudhan Reddy G and Srinivasa Rao K 2015 Microstructure and corrosion free shielded metal arc welded high nitrogen nitrogen *Defense Technology* **11** pp. 237-243
[7] Bannykh O A, Yushchenko K A, Blinov V M, Chornamoryan S A, Solokha A M and Kazennov N P 1990 Non-magnetic welded stainless steel *Patent USSR* No. 1595936 C 22 *Discoveries of Invention* I36 S 120
[8] Blinov V M, Bannykh O A, Ilyin A A, Sokolov O G, Kostina M V, Blinov E V, Rigina L G and Zvereva T N 2005 High-strength and highly viscous non-magnetic welded steel *Patent RF* No. 2303648
[9] Bannykh O A, Blinov V M, Kostina M V, Lyakishev N P, Rigina L G, Gorynin I V, Rybin V V, Malyshevsky V A, Kalinin G Y, Yampolsky V D, Butsky E V, Rimkevich V S and Sidorina T N 2003 High-strength non-magnetic corrosion-resistant weldable steel *Patent RF* No. 2205889
[10] Bannykh O A, Blinov V M, Blinov E V, Kostina M V, Muradyan S O and Rigina L G 2010 High-strength foundry non-magnetic, corrosion-resistant steel and products maded from it *Patent RF* No. 2445397
[11] Kostina M V, Muradyan S O, Kalinin G Yu, Fomina O V, Blinova E N, Kostina V S and Shatalov A V 2015 Structure and properties of plate welded joints of new austenitic nitrogen-containing steel for work in conditions of high static and alternating loads, corrosive environment *Materials Science Issues* **81**, 1 pp. 95-107
[12] Bishokov R V, Avvakumov Yu V, Malyshevsky V A, Baranov A V and Baryshnikov A P 2006 Electrode of grade EA-868/20 for welding high-strength steels of pearlitic and austenitic class *Patent RF* No. 2268128
[13] Bannykh O A, Blinov V M, Kostina M V, Blinov E V and Zvereva T N 2007 Research on the weldability of high-nitrogen corrosion-resistant austenitic steels of type Kh22AG16N8M *Metals* **4** pp. 51-67

IOP Conf. Series: Materials Science and Engineering **525** (2019) 012064 doi:10.1088/1757-899X/525/1/012064

The corrosion resistance investigation of the austenitic nitrogen Cr-Mn-Ni-Mo-N steel in the cast and deform state

S O Muradyan, M V Kostina, V S Kostina and P Yu Polomoshnov

Institute of metallurgy and materials science by A.A. Baykov Russian Academy of sciences, Moscow, Russia

E-mail: muradianso@gmail.com

Abstract. In the article are given the analyzing corrosion resistance data of nitrogen containing steels. The resistance on the pitting, crevice corrosion with high nitrogen containing is estimated. The corrosion resistance of the steel 05Kh22AG15N8M2F in the condition after cold deforming in chloride containing substances is defined. Laboratory and field tests two types of steels 05Kh22AG15N8M2F and 12Kh18N10T are provided. It`s determined that the samples of the steel 05Kh22AG15N8M2F are demonstrating larger corrosion resistance at the laboratory and field investigations in comparison with the steel 12Kh18N10T. The properties of the steel 05Kh22AG15N8M2F with known AISI 316 are compared. It is established that the steel 05Kh22AG15N8M2F a top of AISI 316 on the resistance to the pitting corrosion and the critical pitting temperature. Pitting Resistance Equivalent (PREN) for the steel 05Kh22AG15N8M2F and the critical pitting temperature are determined.

1. Introduction

The intergranular (ICC), pitting and crevice corrosion present a problems for the stainless steels and one of the ways to solve this unacceptable phenomenon – the alloying by nitrogen. As known, nitrogen improve the resistance of the austenitic steels to the corrosion, especially for the local corrosion form.

In the highlowcarbonnitrogen steels ICC can develop from the formation not carbides, but a nitrides chromium, such as Me_2N. In this case, ICC slow down because of the less depletion border areas by chromium. Conditions beginning ICC in the result formation nitride Me_2N in the steel X24H20 were studied [1]. To explanation the influence of nitrogen on ICC resistance steel O3X19AГ3H10 studied to ICC after the temper in comparison with changes of condition of the grain boundary zone [2]. It`s shown that ICC regardless of the test method, was determined by release of carbides during the tempering, not nitrides. At the same time, the tendency to ICC decreased with the increasing nitrogen content, but the width of the chromium-depleted border zone did not depend on nitrogen content (the latter investigated using transmission electron microscopy). From the data obtained by the authors, it concluded that nitrogen does not affect diffusion processes, but changes in the energy state of the boundaries. Experiments have shown that with an increase in nitrogen content, the relative free energy of the boundaries is decreasing. The formation of the particles of a new phase should be accompanied by a decrease in free energy ΔF. During the formation of the particles at the grain boundaries:

$$\Delta F = -V \bullet \Delta fV + S\sigma + Z - \Delta F_b. \quad (1),$$

IOP Conf. Series: Materials Science and Engineering **525** (2019) 012064 doi:10.1088/1757-899X/525/1/012064

where V – volume; ΔfV – variation of free energy per unit of volume of the formed particle; S – summery quantity of the particle surface; σ – unit surface energy on border particle-matrix; *Z*- elastic energy arising due to the differences in a specific volumes of a particle and matrix; ΔF_b – difference of border energy. Because of nitrogen injection reduces ΔF_b, then ΔF increases. Since a probability carbide nucleus formation, than with increasing ΔF probability carbide formation decreases. The obtained ideas about the effect of nitrogen on the energy state of a grain boundaries, apparently, can also be used in explaining the effect of nitrogen on the other properties of the corrosion-resistant steels [3].

A summary of the above and previously known data indicates the positive effect of nitrogen on the increasing in the resistance against ICC. However, it should be keep in mind, that the stability chromium nitride Cr_2N lower the stability carbide $Cr_{23}C_6$ in the hardened chromium-nickel and chromium-nickel-manganese austenitic steels in the wide range of the potentials, as a result chromium nitrides can be etched [4]. Carbo-nitride, such as $Cr_2(C_xN_y)$ have significantly more higher resistance, than Cr_2N. Therefore, to ensure the maximum corrosion resistance of nitrogen-containing steels, it is advisable to choose the heat treatment that ensures the absence or minimum amount of the nitride precipitates [3].

The pitting corrosion is one of the well-known problems for the stainless steels. It can also be sensitized, especially any local Cr decreasing can provide the pitting corrosion. Thus, any chromium segregation caused by the welding or the heat treatment, etc., may have an effect on the pitting formation resistance.

The critical pitting temperature determines the temperature of the whole formation. The general range of this temperature for the corrosion resistant steels is 10–100 ° C and depends from the composition of the chemical composition of the steel. A source [5] reports, that the crevice corrosion critical temperature lies on 20°C below the pitting formation critical temperature (Figure 1) [6]. The conventional pitting resistance index for the corrosion resistance steels with nitrogen – Pitting resistance equivalent + N (PREN), is determined as:

$$PREN = Cr\,(\%) + 3{,}3\,Mo\,(\%) + x\,N\,(\%), \quad (2)$$

where x = 13…30.

The author [7] suppose, that x=16 using for the steels with a molybdenum containing to 4,5 % and x=30 for the steels with a molybdenum containing 4,5-7,0 %.

The authors [6] suggesting own version to the calculation pitting resistance critical temperature, instead of PREN index they suggesting MARC («Measure of Alloying for Resistance to Corrosion»):

$$MARC = \%Cr + 3{,}3\%\,Mo + 20\%\,N + 20\%C - 0{,}5\%Mn - 0{,}25\,\%\,Ni, \quad (3)$$

On figure 1 performed the pitting formation critical temperatures some of commercial alloys. The authors approve that chromium, molybdenum, carbon and nitrogen (in the solid solution) increasing the corrosion resistance, while Mn and Ni slightly reduce it.

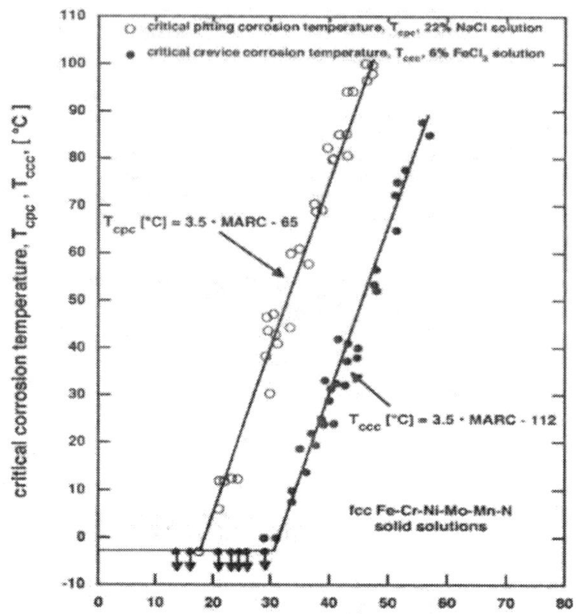

Figure 1. Critical temperature of the pitting and crevice corrosion commercial Fe-Cr-Ni-Mo-N steels in dependence from MARC index [6].

Crevice corrosion, in generally, follows the same principles (that shown by second dependence on figure 1), such as the pitting, but the conditions her significantly harder because of geometrical influence (electrolyte condensed in the crack) [8].

2. Corrosion resistance investigation of high nitrogen steels

The pitting, crevice and corrosion cracking in significantly interconnecting and typical for the substances, containing chloride-ions or other halide ions. The pitting and crevice corrosion resistance in the chromiumnickelmolybdenum steels increasing with nitrogen insertion to them [3]. So, the investigation of chromiumnickelmolybdenum austenitic steels with the different content of nitrogen (from 0,08 to 0,39 %) and manganese (0,9-4,8 %) let to determined, that even with a relatively high content of manganese and sulfur, the influence of 0,38% N provides high resistance to the pitting [9]. Increasing the pitting and crevice corrosion resistance in the high alloyed steels such as 20Cr-25Ni and 20C-25Ni-4,5Mo with the presence 0,21 % N [10]. The achieved effect is associated with the formation of an ammonium ion NH_{4+}, which increases the pH of the solution and facilitates the repassivation. A positive effect of a nitrogen influence was find in the steel as with molybdenum, so without him, and this showing on a possibility of the creation the economically alloyed steels without molybdenum and resistant to the pitting and crevice corrosion. For example, the results of the investigation of the austenitic steels without molybdenum with 20 % Cr and 10 % Ni with a super equilibrium nitrogen containing – 0,7 % N showed the pitting potential ≥0,90 V (for comparison – pitting potentials the same steel, but with 0,35 % N and also for the steel type X17H13M2 and X18H10 respectively consist 0,35; 0,30 and 0,18 V) [11].

Analyzing the given information on the corrosion resistance of the nitrogen-containing steels, we can conclude that to provide higher, which for the steels of the type 18-10 corrosion resistance, the steel should contain more than 18% Cr in its composition, should be alloyed with a molybdenum and nitrogen. The content of austenite-formation elements Ni and Mn should ensure the austenitic structure of the steel (including due to the influence of Mn on the solubility of nitrogen, a strong element - austenitizing agent in iron-based solid solutions).

In this work the next compounds of the cast nitrogen steels were studied – table 1.

Table 1. Chemical compound of the investigated experimental cast nitrogen steels

Chemical compound, %												
C	Mn	Si	Cr	Ni	Mo	V	S	P	N	N_{count}	B	Nb
0,03-0,05	14,0-16,0	≤0,45	21,0-22,0	7,5-8,5	1,0-1,5	0,15-0,25	≤0,02	≤0,02	0,45-0,55		-	-
0,04	14,4	0,24	22	7,6	1,12	0,22	0,01	0,011	0,47	0,55	0,004	-
0,02	15,5	0,23	20,3	7,1	0,9	0,17	0,008	0,01	0,47	0,49	-	0,12
0,03	14,8	0,2	20	7,8	0,83	0,2	0,008	0,01	0,62	0,46	-	0,01

2.1.Materials and methods

The steels melted in JSC "RPA "CNIITMASH". The steel melt №1 melted in the open induction furnace (weight of melting 70 kg) by the method of the smelting pure raw materials with using: an armco iron; Ni and Mn electrolytical; carbon-free ferrochrome; nitrated ferrochrome; metallic Mo and ferrovanadium. A final deoxidation was performed by silicocalcium in a ladle. Nitrogen introduced into the metal using nitrogenous ferrochrome (with ~ 10% N) in a final melting period.

The steel № 1was flooded into a specially made mold, which allows obtaining a multistage model casting. The stepped construction of the experimental casting was developed by taking into the account the analysis of the range of the parts for the cast reinforcement, which showed that the maximum absolute thickness of the assemblies and the walls of the cast parts does not exceed, as a rule, 30 - 50 mm [12]. Produced casting in the form of a slab with thickness of a steps from 50 to 1 mm, a length 725 mm, and a width of 286 mm made it possible to simulate the process of the solidification of the castings of the specified composition in the thickness range from 1 to 50 mm and to 120 mm.

2.2. Results and discussion

In this work the corrosion resistance of the steel 05Kh22AG15N8M2F in the condition after cold rolling (list, thickness 4 mm) in chloride-containing substances were determined. The laboratory and field tests of two types of the steel were carried out - 05Kh22AG15N8M2F and 12Kh18N10T. At the first stage, the laboratory studies carried out in 3% NaCl solution. At the second stage, field tests were performed in a sea water and atmosphere. The steel – 12Kh18N10T – the austenitic stainless steel – selected as a comparison material. According to the results of research, it founded that the samples of the steel 05Kh22AG15N8M2F demonstrate the greater corrosion resistance under the conditions of both laboratory and field tests, in comparison with the steel 12Kh18N10T. After carrying out the field tests for the atmospheric and marine corrosion for the samples from the steel 05Kh22AG15N8M2F, the current density of the corrosion is 15% and 25% less than for the samples from the steel 12Kh18N10T, respectively. The analysis of a gravimetric data confirmed the conclusions made from the results of the electrochemical measurements, indicating a lesser weight loss of the samples of 05Kh22AG15N8M2F steel by 1,6–1,8 times [13].

For the intergranular corrosion the samples of the cast steel melt 1 was tested in the thermally treated condition (1200 ^{o}C, 1 h, water).

The established absence of inclusions of nitrides and carbides at the grain boundaries of the cast steel and the steel after the heat treatment should have guaranteed the steel from the ICC. Really, inspecting tested ICC samples even in a cast, most a heterogeneous structural condition, downturn to the angle 90 ^{o}, showed the absence of a cracks not only on the sample, which was not exposed in a corrosion substance, but also on tested. The steel in condition after the annealing at 1200 ^{o}C more don't show the signs of ICC [14, 15].

With the tests on the pitting corrosion the cast samples after the heat treatment: (1) the annealing at 1200 °C, 1 h, the hardening in the water and (2) the annealing at 1200 °C, 1 h, the hardening in the water and the provocative aging at 700 °C, 100 h was compared.

The results of inspection of the surface of the test samples for the subject of absence and intensity the pitting corrosion damage in the dependence of the continuance of exposure in the solution of iron trichloride and the temperature of the test are given, respectively, in the tables 2 and 3 and on the figure 2. From the data presented in the table 2 it is obviously, that at 25 °C, with the increasing continuance of exposure from 1 to 10 h, there is no practical difference in the resistance against the pitting corrosion of the cast metal in the hardened state and the metal after the hardening and the additional provocative aging.

Table 2. The results of the visual inspection of the surface of the cast steel 05Kh22AG15N8M2F samples after the tests on the resistance to the pitting corrosion after the two types of the heat treatment: 1 – annealing at 1200 °C, 1 h, hardening in the water; 2 – annealing at 1200 °C, 1 h, hardening in the water + provocative aging at 700 °C, 100 h.

Heat treatment	Continuance of exposure in the solution, h			
	1	2	5	10
The annealing at 1200 °C, 1 h, hardening in water	No focuses of the pitting corrosion was detected	No focuses of the pitting corrosion in the concrete	No focuses of the pitting corrosion in the concrete. The microsites of the erosion of the surface	The single uneven distributed focuses of the pitting corrosion
The hardening in the water + the provocative aging at 700 °C, 100 h				

Table 3. The resistance of the steel 05Kh22AG15N8M2F after the annealing at 1200 °C, 1 h, the hardening in the water and annealing at 1200 °C, 1 h, the hardening in the water + the provocative aging at 700 °C, 100 h against pitting corrosion at 35-50 °C

t, °C	τ, h	Heat treatment	The view of the surface of the sample	RSD*, %/ The mark of the resistance
35	5	1200 °C, 1 h, water	The small focuses of the pitting corrosion	0,3 / 8
		1200 °C, 1 h, water + 700 °C, 100 h	The pitting, such as at the hardening structure	
	10	1200 °C, 1 h, water	The many smallest focuses of pitting corrosion	1,0 / 6
		1200 °C, 1 h, water + 700 °C, 100 h		
45	5	1200 °C, 1 h, water	The irregularly distributed focuses of pitting corrosion of different sizes	16 / 3
		1200 °C, 1 h, water + 700 °C, 100 h		
50	5	1200 °C, 1 h, water	The many focuses of the pitting corrosion	33 / 2
		1200 °C, 1 h, water + 700 °C, 100 h		

RSD* - The rate of the surface damage about by the scale ASTM DIN 610

According to the scale of the standard ASTM DIN 610 the corrosion resistance to the pitting corrosion is estimating by 10-rate scale of the sizes of the surface damage square (in %). The mark ~8 point is given at the damage square 0,1 %. On the surface in this case single linear focuses of the pitting corrosion are initiated. Such scene observed at the 10 h test. The provocative aging don't degrade the resistance of the steel to the pitting corrosion: the appearance of the hardening and the aging samples after tests is perfect uniform, the low damage rate evidence about the resistance of the cast steel to the pitting corrosion under the given test conditions.

With rising the temperature of the test solution to 35-50 °C the pitting corrosion intensity is increasing. It should be particularly noted that the difference in the corrosion behavior between the steel in the hardened and aged state was also not observed, the appearance of the samples was the same (table 3). The sizes of damage focuses at the temperature 35 °C is still the small at the test period to 5 h. With rising the exposure continuance period of the solution with the temperature 35 °C the intensity of the pitting corrosion increasing, and, according to scale ASTM DIN 610, she consist of 1 % (the rate of the resistance 6). At the test 5 h at 45 °C immunity to the pitting corrosion abruptly reduced (to rate 3), on the surface at the same time the smallest microfocuses the focuses of the pitting corrosion are initiating. Based on the figure 1, it's can considering that for the 05Kh22AG15N8M2F cast steel, which have PREN = 33,2 in the hardened and also in the aging condition, the value of the critical pitting formation temperature is up $t_{crit.P}$ = 43 °C (Figure 2). With increasing the test temperature to 70-80 °C it's take a place the intensive common solution of the samples surface, the pitting formation potential is going down because of prevailing the velocity of general corrosion over a pitting growth velocity. (It should be consider, that initiation of the focuses of the pitting corrosion determined not only by chemical composition, but also (in specific stage) by the presence of microdefects of the casting at the pitting resistance of the cast metal product estimation).

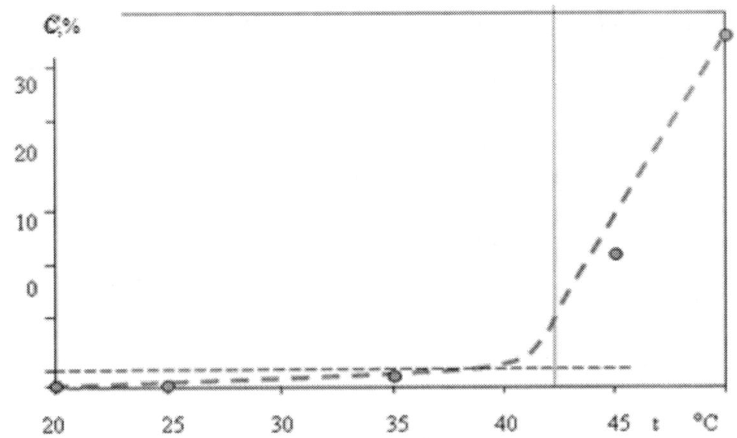

Figure 2. The influence of the test temperature in the 10 % $FeCl_3$ (5 h) on the damage pitting corrosion rate (RSD) of the samples of the steel 05Kh22AG15N8M2F

The comparison of the properties of the steel 05Kh22AG15N8M2F with the well-known steel AISI 316 (table 4) showed, that the new cast nitrogen steel significantly exceed the steel AISI 316 in the pitting resistance corrosion – the coefficient PREN – 33 against 24 (at steel AISI 316) and the pitting formation critical temperature [14].

Table 4. The comparison of the properties of the corrosion resistance of the steels 05Kh22AG15N8M2F and the steel AISI 316.

The mark	%Cr	%Mo	%N	PREN[1]	PCT[2] (°C)
AISI 316	16,5	2,1	0,05	24,2	15[2]
05Kh22AG15N8M2F	22	1,5	0,5	33	43[3]

[1] Pitting resistance equivalent, PREN = %Cr+3,3%Mo+16%N;
[2] Pitting formation critical temperature, PCT: by ASTM G-48A (at 72 h in the 6% $FeCl_3$ solution); [3] by the Russia state standards (at 5 h in the 10% $FeCl_3$ solution).

3. Conclusions

Thus, it can be conclude, that: (1) the steel 05Kh22AG15N8M2F in the cast state, such as after the partially homogenizing the structure annealing at 1200 °C (1 h, the water), is not inclined to ICC due to the absence the influences of the excess phases particles on the grain boundaries and the other cast metal structural elements (2). The investigated steel in the state after the hardening (1200 °C, 1 h) and the hardening with further aging 100 h at 700 °C under the accelerated tests in 10 % solution $FeCl_3$ by the State Standard 9.812-89 at the temperature 25 °C and the exposition time in the substance up to 10 h, and also at the 35 °C, 5 h tests, is not demonstrating the pitting corrosion tendency. (3) For the studied steel, in the state after the hardening, the hardening and aging, pitting formation critical temperature determined PCT = 45 °C - exceeding that for the traditional corrosion-resistant steels of the type 18Cr-8-12Ni.

References

[1] 1986. *Kovove materialy.* **V. 24, № 1.** pp. 3-24
[2] Shapiro M. B., Barsukova I. M. Vliyaniye azota na korrozionnuyu stoykost' nizkouglerodistoy austenitnoy stali // 1984 J. Zashchita metallov V. 22, **№ 2** pp. 250-254
[3] Feldgandler E. R., Savkina L. Ya. Azot v korrozionnostoykikh stalyakh 1990 *J* Byulleten' Chernaya metallurgiya **№ 11** pp. 24-34
[4] Babich S. G. Korrozionno-elektrokhimicheskiye svoystva karbidov i nitrida khroma i ikh vliyaniye v kachestve izbytochnykh faz na korrozionnoye povedeniye nerzhaveyushchikh staley 1988 *Abstract of the cand.sc. dissertation* Moscow 24 p.
[5] Pedrazzoli R., Speidel M. O. Korrosion und spannungsriss corrosion von stick offhaltigen stahlen 1991 *Ergebnisseder werkstofforschung, verlag thubal-kain* Schweiz, Zurich pp. 103-121
[6] Speidel M. O. Nitrogen containing austenitic stainless steels 2006 *Mat-wiss. U. Werkstoiftech* **V. 37, № 10** pp. 875-880
[7] Pleva J. Korrosion und spannungsriss corrosion von stick offhaltigen stahlen 1991 *Ergebnisseder werkstofforschung, verlag thubal-kain* Schweiz, Zurich pp. 153-165
[8] Roman Ritzenhoff, Andre Hahn Corrosion resistance of high nitrogen steels, corrosion resistance, Dr Shih (Ed.), ISBN: 978-953-51-0467-4, InTech, DOI: 10.5772/33037. Available from: http://www.intechopen.com/books/corrosion-resistance/corrosion-resistance-of-high-nitrogen-steels
[9] Kolchin G. G., Ermakov B. O. Ob ustoychivosti khromonikelevykh austenitnykh staley, legirovannykh azotom i molibdenom, k pittingovoy korrozii 1988 *Zashchita metallov* **V. 24 № 6** pp. 989-990
[10] Wallin T., Jargelius R. The effect of nitrogen alloying on the pitting crevice corrosion resistance of CrNi and CrNiMo austenitic stainless steels 1986 *Bulletin korrosionist* **V. 101** pp. 161-164
[11] Degallaix S., Fact I. Nitrogen in austenitic stainless steels 1988 *Memories et etudes scientifiques de la revue de metallurgie* **V. 2** pp. 111-123
[12] Nazaratin V. V., Rigina L. G., Kostina M. V., Muradyan S. O., Gorbach V. D., Kuzmin E. V., E. V. Statcukovsiy Issledovaniye vozmozhnosti primeneniya novoy vysokoazotistoy stali dlya proizvodstva litykh zagotovok 2009 *J. Liteynoye proizvodstvo* **V. 6** pp. 23-28
[13] Gnedenkov S. V., Sinebryukhov S. K., Egorkin V. S., Vualyi I. E., Imshinetskii I. M., Kostina M. V., Muradyan S. O., Sergienko V. I. Corrosion stability of austenitic steels 05Kh22AG15N8M2F and 12Kh18N10T in Chloride-Containing Media 2017 *Protection of Metals and Physical Chemistry of Surfaces* **V. 53, № 5** pp. 910-915
[14] Kostina M. V., Harina I. L., Muradyan S. O., Nemov V. V. Issledovaniye stoykosti novoy liteynoy vysokoazotistoy Cr-Mn-Ni-Mo-N stali k mezhkristallitnoy, pittingovoy i shchelevoy korrozii 2011 Saint-Petersburg *XVI mezhdunarodnaya nauchno-tekhnicheskaya konferentsiya «Problemy resursa i bezopasnoy ekspluatatsii materialov i konstruktsiy»* pp. 229-233
[15] Kostina M. V., Harina I. L., Muradyan S. O., Nemov V. V. Stoykost' k lokal'noy korrozii novoy liteynoy vysokoazotistoy vysokoprochnoy austenitnoy stali 18-20 may 2011 Moscow *Fundamental'nyye aspekty korrozionnogo materialovedeniya i zashchity metallov ot korrozii* pp. 144-145

IOP Conf. Series: Materials Science and Engineering **525** (2019) 012065 doi:10.1088/1757-899X/525/1/012065

Some features of behavior of solid dielectrics under high pressure combined with shear stress

I B Oparina and V V Schienok

Baikov Institute of metallurgy and materials Science, RAS, Moscow, Russian Federation

ibo@imet.ac.ru

Abstract. The paper presents the results of experimental research of behavior of solid dielectrics under high pressure combined with high shear stresses with emphasis on specificities of explosion of this solids under such conditions, which is known as reological explosion. Revealed the dependence of shear deformation, which is necessary for explosion on pressure, speed of product of explosion was measured, mechanism of appearance of hard radiation which arising during the explosion was established.

1. Introduction

Reological explosion of solid state bodies is the specific form of destruction of said bodies under very high pressure and shear stresses. This phenomenon described first by P.W. Bridgman in his classical work [1]. He considered the explosions of the samples, which occurred from time to time during experiments, as a kind of disturbing factors and paid no attention for study of this phenomenon. Moreover, his apparatus with two pairs of anvils gave no possibility for such studies. The investigators, which were followed him later [2-4], emphasized on study of some concrete occurrences of the explosions, such as radio radiation, chemical reactions, etc. Just all of them gave many propositions on the nature of explosion, without complete experimental study of phenomenon [5, 6]. Such study was done by V. Shienok and T.Gorazdovsky [7-9], which introduced first the term «reological explosion» in this works. M. Yaroslavsky participated in some experiments and reported about achievement of comprehensive pressure as high as hundreds of thousands kilobars and many other very interesting results [10].

2. Experimental Procedure

The studies of this phenomenon were carried out with the apparatus (figure1) included one pair of Bridgman's anvils one of which were fixed at the ram of 100 tons hydraulic press and another at the shaft of mechanical reductor, which were mounted at this press. The reductor allowed to rotate said anvil around the axis, passed through the centers of both anvils. The angle of rotation was measured with accuracy of $\pm 0{,}2\ ^{\circ}\mathrm{C}$ and was unlimited. Speed of rotation was constant with valuation about $1\ ^{\circ}\mathrm{C}/sec$.

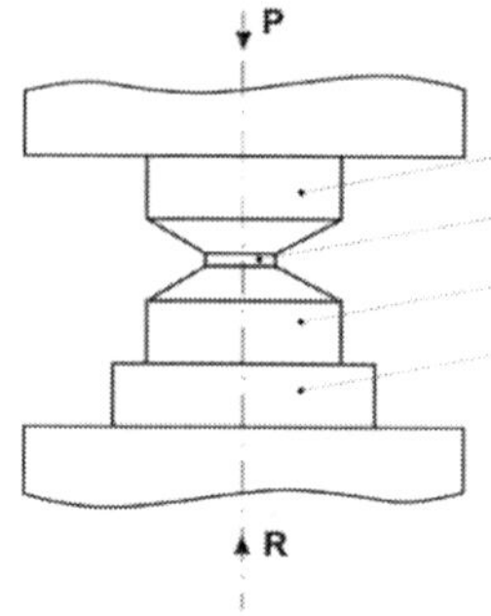

Figure 1. Scheme of the apparatus (1-upper anvil, 2-sample, 3-lower anvil, 4-shaft of redactor)

The anvils had a form of truncated cones made of alloy steel forced up HRC60, or of cemented carbide with 8 % of cobalt. The working surfaces of anvils had a square of 2-3 cm^2 and were grinded for roughness of a few of microns. The objects of investigations were polycrystalline dielectrics such as various forms of $CaCO_3$, coal, brick, salts, boric acid, etc. Chalk was the most convenient from them.

Pressure was valuated as average force to square of working surface of anvils because it was not hydrostatic with maximum in the center. Moreover, gradient of pressure was changing during deformation. Thus exact measuring of pressure was useless. Deformation valued as the angle of rotation of anvil because it was different in different points of the sample and exact measuring was useless too. The samples exploded when pressure and deformation achieved some values, individual for each material.

3. Results and Discussions

It was estimated, that explosion occurred only when pressure was higher, then some initial, which marked as P_0. The explosion never occurred below this value independently on deformation. The explosions, which were observed under pressures higher, then P_0 had a force depended on the value of pressure, namely: higher a pressure – strongly a force of explosion. The value of deformation, which was necessary for explosion (α), depended on the value of pressure. The form of this dependence one can see at the figure 2.

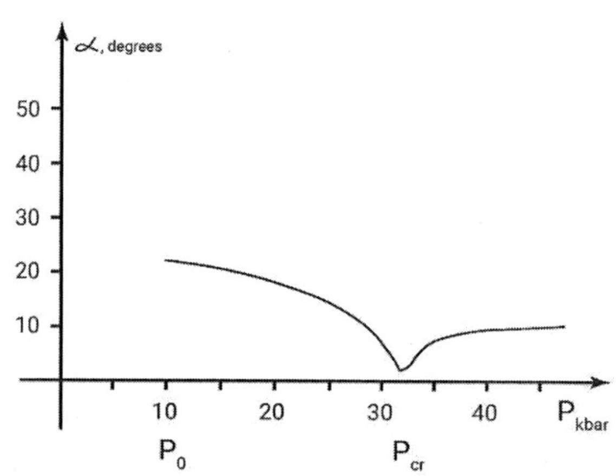

Figure 2. Dependence of rotation angle of anvil, which is necessary for explosion on pressure

The curve have a minimum, which is sharp enough and which is different for different materials. The value of pressure, corresponding to this minimum, is marked as P critical (P_{cr}). The value of α under P_{cr} is as small, as 1-3 degrees. The value of α is changing from tens of degrees at pressure P_0 to 1-3 degrees at P_{cr} and then it is increasing slowly with pressure to the value of about 10 degrees. The

IOP Conf. Series: Materials Science and Engineering **525** (2019) 012065 doi:10.1088/1757-899X/525/1/012065

scatter of values of α is changing from tens of percents at pressures near P_0 to some percents at pressure P_{cr} and higher. Finding of this dependence was very useful, because the field of experiments became wider.

A tablet there was formed of material, which had been pressed between the anvils, under the pressure about P_0 and higher. The tablets were not formed under lower pressures – material was cracked only. Residual tablet remained between the anvil after explosion. The thickness of tablets depends on the sort of material and on the value of pressure and was about a few millimeters. The thickness and mass of residual tablets was less than the thickness and mass of tablets before the explosion. The tablets lost from a few percents to about 50 % of mass after explosion. The value of this loss depended on pressure and changed from minimum at P_0 to 50 % at P_{cr}. This loss just not changed at pressures higher, than P_{cr}. The density of the tablets after explosion was some percents higher, than before explosion.

The structure of the tablets differed drastically after explosion and before it. There are a lot of layers, which are laying under ±45° to the axis of symmetry of the tablets. This layers were similar to the cleavage surfaces in crystals, because its were the surfaces of destruction, when the tablets were crashed. The metal flakes were formed in the residual tablets, when the material was doped with metal powder. The orientation of this flakes was same, as orientation of said layers.

The material, which is flied out at the explosion have a form of dust. This dust have very special spectrum of particles dimensions and forms. Most of the particles (about 90 %) have dimensions of 1-2 microns and are just isotropic. The fracture surfaces were covered with a lot of this particles, which could not be removed by a flow of air.

The working surface of anvils obtained specific damages after explosion. There were two types of damages: «ripple» and scratches. This damages were observed both at the steel and hard alloy anvils. The «ripple» had «wave length» about a few millimeters and amplitude about some tenths of millimeter. The «wave length» was just independent on material of anvils, but amplitude was greater at steel anvils than at hard alloy anvils. The orientation of «ripple» was similar to orientation of ordinary ripple at the surface of water after stone drop.

The scratches had a length from some tenths of millimeter up to 2-3 millimeters, depth and width up to some tenths of millimeter. The forms of a scratches were very manifold and it's width and depth were changed along one scratch. The directions of a scratches was just chaotic at the central part of working surface of anvils, but this directions became radial at the peripheral part. It is necessary to note, that a hardness of all exploded materials and hardness of inclusions was noticeably less, then that of materials of anvils.

One more specificity of reological explosion was discovered. This specificity is named «related explosion», and it's nature is following: the material under the pressure higher than P_{cr} which is subjected to the deformation less, that is necessary for explosion, can remain in this state indefinite time and explode when the pressure decreased. The explosion under such conditions occurred when the pressure dropped to a value about P_{cr} and had full force. Thus we are observing a metastable state of solid body, which is existing due to potential barrier of definite magnitude by way of the pressure and is rupturing when the barrier drops below this magnitude. This metastable state is rupturing in the time less than 10^{-5} sec. Discovering of this phenomenon allowed to do a lot of useful experiments, such as high-speed filming of explosion, which is shown at the figure 3 and, which were unreal without «related explosion» experimental technique.

Dropping of electrical resistance of material in the moment of explosion was observed, but detailed investigation of this effect was not done.

The reological explosion is following with some interesting effects. One of such effects is a hard radiation, which is observing during an explosion and after it. The radiation was researched with the apparatus which had the high sensitive scintillation detector and computing device. The radiation, which appeared in the moment of explosion, and first had been observed [11], was electromagnetic and had upper energetic threshold of 10-12 keV. The number of registered particles depended on the energy so, that it decreased quickly above this threshold and increased below it. The spectrum of energies of the particles was continuous below the threshold.

Figure 3. High-speed film of reological explosion speed of filming 10.000 (a, b) 5.000 (c) frames/sec. The curve at the frames 1-6 – dependence of speed of particles on time (a), frame 0 – last before explosion. The beginning of explosion: frame 1 (a), 3(b), 2(c).

The comparison of this results with the results of researching of other specificities of reological explosion made it possible to explain the nature of described radiation.

It was established with ballistic pendulum technique [10], that the speed of products of explosion was about 500-1000 m/sec at 5-10 cm from the center of exploding sample. The product of explosion of chalk under 30 kbar was dust with basic dimension of particles about 1,5 μm.

There was calculated the loss of speed of the particle, when it moves through air. The calculations, based at aerodynamics have many limitations of ranges of values of parameters, such as speed, dimensions, density, temperature etc. which are leading to incorrectness out of limitations. There was used method of direct calculation of loss of speed of the particle after collisions with air molecules to avoid possible mistakes. There was supposed, that air molecules have the mass, average between the mass of nitrogen and oxygen molecules, that molecules are motionless and collisions are elastic.

The particle with initial speed V_p^0 will have speed V_p^1 after one collision with air molecule in accordance to the laws of saving of energy and impulse:

$$V_p^1 = \left(\frac{m_p - m_a}{m_p + m_a}\right) \times V_p^0$$

where m_p – is the mass of particle and m_a – is the mass of air molecule.

After second collision the particle will have speed V_p^2:

$$V_p^2 = \left(\frac{m_p - m_a}{m_p + m_a}\right) \times V_p^1 = \left(\frac{m_p - m_a}{m_p + m_a}\right)^2 \times V_p^0$$

After i-collisions the particle will have speed V_p^i:

$$V_p^i = \left(\frac{m_p - m_a}{m_p + m_a}\right)^i \times V_p^0$$

Supposing that a particle have a spherical form, so it's mass is:

$$m_p = \rho\, 4/3\pi R^3$$

where ρ – density 3 g/cm^3 and R – is radius of particle 0,75μm.

So:

$m_p = 3 \times 4/3\pi(0{,}75 \times 10^{-4})^3 = 5{,}2 \times 10^{-12}\ g$

$m_a = m_{g-m}/N = 29/6 \times 10^{23} \approx 5 \times 10^{23}\ g$

where m_{g-m} – the mass of gram-molecule, N – Avogadro's number.

There was calculated the number of collisions of the particle and molecules of air at the distance of 1cm. This number (n) is equal to the number of molecules of air in the cylinder, having base area: $\pi R^2 = 1{,}75 \times 10^{-8}\ cm^2$ and generatrix – 1cm, i.e. in the volume $V_c = 1{,}75 \times 10^{-8}\ cm^3$

$$n = \frac{N}{V_0} \times V_c$$

where: N - is Avogadro's number, V_0 – is the volume of gram-molecule.

Thus:

$$n = \frac{6 \times 10^{23} \times 1{,}75 \times 10^{-8}}{22{,}4 \times 10^3} \cong 5{,}25 \times 10^{11}$$

The total mass of this molecules is:

$$m_a \times n = m_n = 5 \times 10^{-23} \times 5{,}25 \times 10^{11} \approx 2{,}610^{-11}\ (g),$$

i.e. at the distance of 1cm the particle impacts with a mass ≈ 20 times as much as it's mass and loses just all it's speed.

The speed of the products of explosion, as it is mentioned above, is 500 – 1000 m/sec at the distance of 5 – 10 cm from the center of exploding sample. It means, that even the particle loses only half of speed at 1 cm, V_p^0 have a value about 100-300 km/sec.

In accordance to this results, it became possible to make the assumption on the mechanism of appearance of radiation in process of reological explosion.

It is known from quantum mechanics, that excitation of atoms can occur as a result of collision of a bodies – «impact or shock excitation».

The probability of excitation (W) up the energy, which corresponds to the frequency ω is:

$$W = 2(\omega_{exc}/\omega_{km})^2$$

where:

$\omega_{exc} = 2\pi(V/d)$;
ω_{km} – is specific frequency of atom;
V – speed of impaction;
d – lattice constant of impacting material.

Calculation of a value of speed of collision of a nitrogen molecule with a particle of chalk, which is necessary for excitation and following radiation up energy of 10 keV gives (in assumption of full transformation of kinetic energy to energy of radiation):

$$mV^2/2 = 10\ keV = 1{,}6 \times 10^{-8}\ erg$$

$$V = \sqrt{\frac{2 \times 1{,}6 \times 10^{-8}}{5 \times 10^{-23}}}\ cm/sec \cong 250\ km/sec$$

Analysis of high-speed films of explosion gives the same results. There is a curve of speed dependence of particles on time at the figure 3,a. Extrapolation of this curve to the moment of origin of explosion leads to the value of initial speed of some hundreds km/sec/. Results of measuring of average speed of particles with ballistic pendulum, mentioned above may be used as independent reference point.

The comparison of this results with the results of experiments leads to conclusion: the hard radiation in process of reological explosion is a result of inhibition of high speed particles of exploding material in air. The upper energetic threshold of radiation corresponds to maximal speed of particles. The origin of some kind of damages of the surfaces of anvils became clear too – the objects, having speed of hundreds km/sec can damage the materials with hardness significally higher than that of such objects.

4. Conclusion

Revealed the dependence of shear deformation, which is necessary for explosion on pressure. It was established, that the speed of products of explosion can reach about 250 km/sec. The mechanism of appearance of hard radiation which arising during the explosion was established: it is arising due to collisions of high-speed products of explosion with the molecules of air.

References

[1] Bridgman PW 1935 *Phys. Rev.* **48** pp 825-847
[2] Aleksandrov A I, Aleksandov I A, Prokofyev A I 2013 *Russ. Dokl. Ac. Sci.* **451 pp** 50-52
[3] Aleksandrov A I, Aleksandov I A, Zezin S B, Degtyarev E N, Dubinskii A A, Abramchuk S S, Prokofyev A I 2016 *Russ Chem. Phys.* **35** pp 78-85
[4] Dubovik A V 2016 Russ Chem. Phys. **35** pp 37-43
[5] Dubovik A V, Matveev A A 2015 *Russ. Gorenie i Vzriv* **8** pp 99-104
[6] Fateev E G, Han V P 1991 *Russ. Journ. Tech. Phys. Lett.* **17** pp 50-56
[7] Gorazdovskii T Y, Shienok V V 1975 *Russ. Trudi Moskovskogo Gidromeliorativnogo Instituta* **44** pp 113-116
[8] Gorazdovskii T Y, Shienok V V 1976 *Russ. Trudi Moskovskogo Gidromeliorativnogo Instituta* **50** pp 89-95
[9] Shienok V V, Gorazdovskii T Y 1977 *Russ. Trudi Moskovskogo Gidromeliorativnogo Instituta* **54** pp 114-133
[10] Gorazdovskii T Y, Yaroslavskii M A 1975 *Russ. Trudi Moskovskogo Gidromeliorativnogo Instituta* **44** pp 101-106
[11] Gorazdovskii T Y 1967 *Russ. Journ. Exp. Theor. Phys. Lett.* **5** pp 78-82

Structure, hardness and fracture toughness of arc-melted LaB_6-TiB_2 eutectic alloy

D D Nesmelov[1], S V Vikhman[1], E S Novoselov[1], S N Perevislov[2] and S S Ordan'yan[1]

[1]Saint-Petersburg State Institute of Technology, Saint-Petersburg, Russia
[2]Institute of Silicate Chemistry of Russian Academy of Sciences, Saint Petersburg, Russia

E-mail: dnesmelov@yandex.ru

Abstract. Microstructural and indentation-techniques mechanical investigations have been performed on a LaB_6-TiB_2 eutectic composite, obtained by the electric arc-melting in argon flow. Structure of solidified eutectics consists of directionally crystallized TiB_2 fibers in LaB_6 matrix. XRD, SEM and EDX analysis of eutectics were performed. The median diameter of the TiB_2 fibers, calculated using statistical analysis of SEM micrographs, was 0.45 μm. The volume fraction of the TiB_2 phase in eutectics was 16% as calculated by area fractions from SEM micrographs. The Vickers hardness (25.5 GPa) and fracture toughness (6.7 $MPa \cdot m^{1/2}$) were measured on arc-melted LaB_6-TiB_2 alloy. Results were compared with the same ones, measured on floating-zone melted monocrystalline LaB_6.

1. Introduction

In addition to outstanding thermionic characteristics [1-3] and promising optical properties [4-7], lanthanum hexaboride exhibits high strength, hardness and Young's modulus. This fact is due to strong covalent bonds within the B_6 octahedron. One of the key characteristics of brittle materials, fracture-toughness, in case of LaB_6 single crystal is quite low - about 2.5-3.5 $MPa \cdot m^{1/2}$. Near full-density polycrystalline LaB_6 demonstrates the same values of fracture-toughness. However, mechanical properties of LaB_6 can be enhanced by creating heterophase structures. It is possible on the base of LaB_6-MeB_2 systems (where Me = Ti, Zr, Hf, Cr, V, Nb etc.), all of them are eutectic [8]. As shown in [8-10], the creation of composite ceramic materials based on LaB_6-MeB_2 systems makes it possible to increase the fracture-toughness of the material to 3.7-4.5 $MPa \cdot m^{1/2}$. This effect is associated with a more fine-grained structure of two-phase materials in comparison with single-phase ones due to the restriction of grain growth during sintering and a decrease in sintering temperature due to the presence of eutectic.

However, even more increase in mechanical properties can be achieved by creating a specific structure of directionally crystallized eutectics [11-17]. For example, Bogomol et al. described materials obtained by the floating-zone method, the crack resistance of which reaches 8.9 $MPa \cdot m^{1/2}$ [11]. In a number of other works, sufficiently high values of crack resistance in directionally crystallized eutectics in LaB_6-MeB_2 systems were also obtained [12-17].

In present study we investigated the microstructure, hardness and fracture-toughness of electric arc-melted LaB_6-TiB_2 eutectic composite. In contrast to floating-zone method, arc-melting does not allows to obtain unidirectional solidified alloy at macro-scale. The structure of arc-melted alloys can consist of differently oriented (according to different temperature gradients during crystallization) blocks of eutectics. It was of considerable interest to investigate the Vickers hardness and fracture-toughness of arc-melted LaB_6-TiB_2 composite in comparison with the same characteristics of obtained by floating-zone method monocrystalline LaB_6.

2. Materials and Method

Commercial powders of LaB_6 (99.0 wt. %) and TiB_2 (99.0 wt. %) were used in this study. The average size of starting powders (the 50th percentile particle size, d_{50}) determined by laser diffraction was 12 and 15 μm respectively. Mixing combined with milling of powders was provided on vibratory ball-mill in gasoline-solvent media with TiB_2 balls (balls to material weight ratio B:M = 10:1) during 40 hours. Powders were mixed in eutectic concentration ratio 66 mol. % LaB_6 – 34 mol. % TiB_2 [8] with taking into account the additional 1.3 mass. % TiB_2, which appears at mixture due to the wear of TiB_2 balls. After the ball-milling the average size of mixture was d_{50} = 3.5 μm.

Vacuum-dried mixture was compacted into the tablets with diameter of 20 mm and height of 12 mm under the pressure of 1 t/cm^2 and then sintered at 1850 °C in vacuum of 10^{-2} Pa during 1 hour.

Sintered tablets were put on a water-cooled cooper plate (anode) under argon gas flow and then electric arc-melted. After arc-melting and cooling of samples they were cut and polished for analysis.

Phase composition of alloy was determined using the X-ray diffraction (XRD) analysis (Rigaku SmartLab 3, Cu Kα radiation, λ = 1.5406 Å) in the 2θ range of 10–90° with 0.01 ° step. Microstructure of alloy was investigated using optical microscope (Meiji 7000 with Thixomet-Lite system for image analysis) and SEM Tescan Vega 3SBH. For EDX analysis SEM was used with Aztec Energy Advanced analyzer (Oxford Instruments).

Indentation tests were carried out with use of Vickers hardness tester PMT-3 under the load of 7 to 13.2 N. Direction of indentations were so that the indenter diagonals were oriented perpendicular to the TiB_2 fibers. For calculation of *Hv* (Vickers hardness, GPa) we use equation (1):

$$Hv = 1{,}854 \cdot \frac{P}{D^2}, \qquad (1),$$

where *P* – load, N; *D* – measured diagonal of the indentation, μm.

Fracture-toughness was K_{Ic} calculated according to equation (2):

$$K_{IC} = 0{,}0889\sqrt{\frac{Hv \cdot P}{4l}}, \qquad (2)$$

where *P* – load, N; *l* – measured length of the crack from corner of indentation, μm.

At each load, at least 10 indentations were made, from which the mean values of the lengths of the diagonals and cracks was calculated.

3. Results and discussion

After the arc-melting the samples externally had the form of distorted ellipses with a matte surface with a metallic luster and a purple shade, inherent to lanthanum hexaboride. The lower part of the alloy, which was in contact with the copper substrate, was removed.

As can be clearly seen in the SEM images of cross-sectional arc-melted LaB_6-TiB_2 sample (Figure 1) the structure of crystallized objects is typically eutectic with the second phase (TiB_2) distributed in matrix (LaB_6). The distribution of the TiB_2 fibers (rods) is uniform, the individual eutectic blocks are separated by areas of crystallization of pure lanthanum hexaboride (Figure 1b). The median diameter of the TiB_2 fibers, calculated using statistical analysis of SEM micrographs of different fields of alloy, was 0.45 μm. The volume fraction of the TiB_2 phase in eutectics was 16% as calculated by area fractions from SEM micrographs, which is consistent with the literature data [8]. Weight Ti : La = 1 : 5.7 ratio determined by means of EDX also corresponded to the eutectic concentration of related borides.

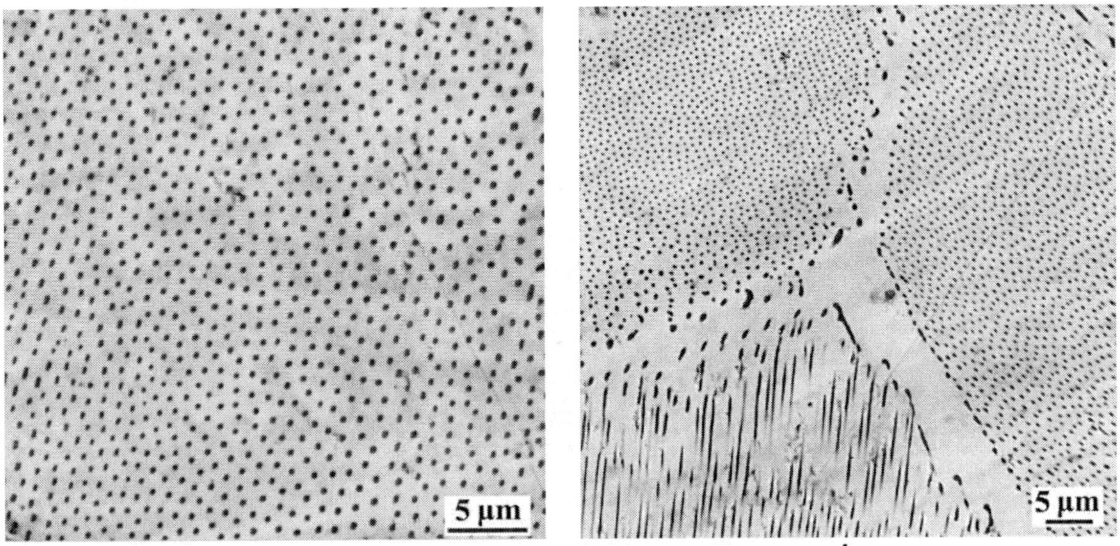

Figure 1. SEM images of cross-sectional arc-melted LaB_6-TiB_2 eutectic alloy

XRD-spectra of arc-melted alloy cross-section is presented at fig.2. In the spectra there are diffraction peaks of only two phases LaB_6 and TiB_2. Almost threefold excess of the standard intensity of the LaB_6 [300] peak at 67.5623° 2θ indicates the fixation of a predominantly oriented structure.

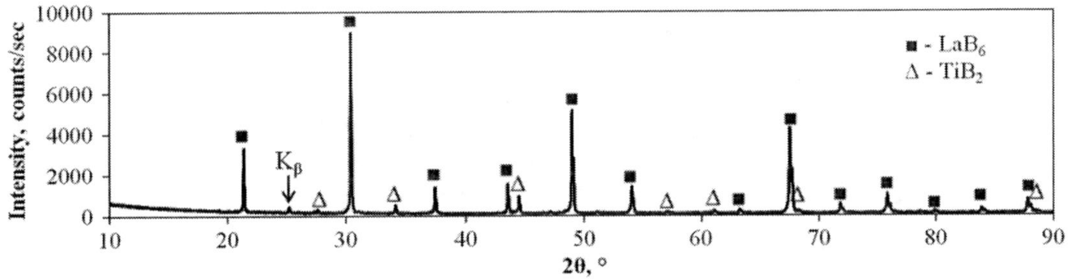

Figure 2. XRD patterns of arc-melted LaB_6-TiB_2 alloy cross-section

Mean hardness of LaB_6-TiB_2 alloy obtained by melting in electric arc (Hv = 25.5 GPa regardless of the load) was lower than the theoretical value calculated by the rule of mixtures (28.4 GPa). One of indentations is shown on fig. 3a. Hardness of single-crystalline LaB_6 (26 GPa) was correctly measured only at load of 7 N. Increasing the load led to the appearance of ring indentation-induced cracks and peeling of the material (Figure 3b), which did not allow to calculate the hardness and fracture-toughness of single crystal according to accepted methods. Unlike single-crystal, arc-melted alloy maintained without critical destruction all the loads applied to it, including 13.2 N. There no formation of ring cracks was observed.

Fracture toughness of LaB_6-TiB_2 alloy and LaB_6 single crystal was respectively 6.7 and 3.1 MPa·m$^{1/2}$. High fracture-toughness values of arc-melted alloy are due to its "fiber self-reinforced" structure. Main mechanism of reinforcement in this case is due to an increase in the length of the crack propagation path because of its deflection by TiB_2 fibers, as it is shown on fig. 3c.

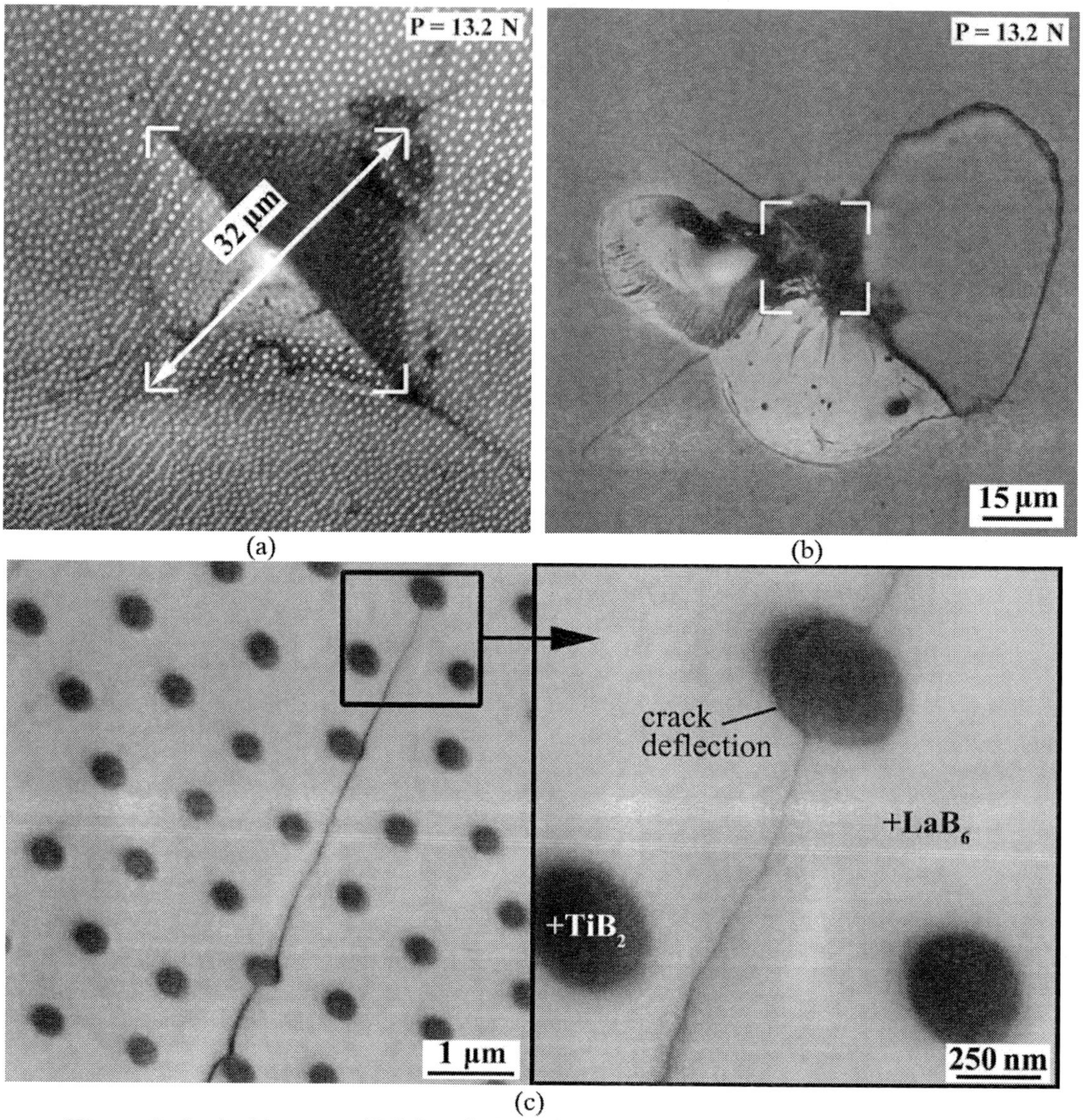

Figure 3. Optical images of Vickers indentations and indentation-induced cracks: **a** - arc-melted LaB_6-TiB_2 eutectics, **b** - single-crystalline LaB_6, and **c** – SEM micrograph of crack deflection in arc-melted LaB_6-TiB_2 eutectics

4. Conclusion

Directionally-solidified LaB_6-TiB_2 eutectic alloy was obtained by the melting at electric arc in argon flow. Structure of eutectics consists of directionally crystallized TiB_2 fibers (rods) in LaB_6 matrix. The median diameter of the TiB_2 fibers was 0.45 μm, the volume fraction of the TiB_2 phase in eutectics was 16%. Arc-melted LaB_6-TiB_2 alloy demonstrated high values of Vickers hardness (25.5 GPa) and fracture toughness (6.7 $MPa \cdot m^{1/2}$), which is due to mechanism of crack deflection by TiB_2 fibers.

Acknowledgments

The reported study was funded by RFBR project No 18-33-20221. Authors are grateful to the Engineering Centre of Saint-Petersburg State Institute of Technology for the equipment provided.

References

[1] Lafferty J M 1951 *Journal of Applied Physics* **3** 299
[2] Taran A, Voronovich D, Oranskaya D, Filipov V, Podshyvalova O 2013 *Functional materials* **4** 485
[3] Paderno Y B, Taran A A, Voronovich D A, Paderno V N, Filipov V B 2008 *Functional materials* **1** 63
[4] Kowalczyk J M, Hadmack M R, Szarmes E B, Madey J M 2014 *International Journal of Thermophysics* **8** 1538
[5] Mattox T M, Coffman D K, Roh I, Sims C, Urban J J 2018 *Materials* **2** 226
[6] Mattox T, Urban J 2018 *Materials* **12** 2473
[7] Sani E, Mercatelli L, Meucci M, Zoli L, Sciti D 2017 *Scientific Reports* **1** 718
[8] Ordanyan S S, Vikhman S V, Nesmelov D D, Danilovich D P, Panteleev I B 2014 *Advances in Science & Technology* **89** 47.
[9] Min G H, Gao R, Yu H S, Han J 2005 *Key Engineering Materials* **297** 1630
[10] Wang X., Zhang J X, Yang X Y, Hu K, Zhang J W 2017 *Advances in Applied Ceramics* **3** 132
[11] Bogomol I, Nishimura T, Nesterenko Y, Vasylkiv O, Sakka Y, Loboda P 2011 *Journal of Alloys and Compounds* **20** 6123
[12] Soloviova T O, Karasevska O P, Vleugels J, Loboda P I 2017 *Journal of Alloys and Compounds* **729** 749
[13] Chen C M, Zhang L T, Zhou W C 1998 *Journal of crystal growth* **4** 873
[14] Yang X, Wang P, Wang Z, Hu K, Cheng H, Li Z, Zhang J 2017 *Materials & Design* **133** 299
[15] Volkova H, Filipov V, Podrezov Y. 2014 *Journal of the European Ceramic Society* **14** 3399
[16] Jouanny I, Sennour M, Berger M H, Filipov V B, Ievdokymova A, Paderno V N, Sayir A 2014 *Journal of the European Ceramic Society* **9** 2101
[17] Loboda P I, Soloviova T O, Bogomol Y I, Remizov D O, Bilyi O I 2015 *Journal of Superhard Materials* **6** 394

IOP Conf. Series: Materials Science and Engineering **525** (2019) 012067 doi:10.1088/1757-899X/525/1/012067

Methods and techniques for producing ceramics from aluminum oxynitride

D V Prosvirnin, A G Kolmakov, M D Larionov, M E Prutskov and A V Levina

Baikov Institute of metallurgy and materials Science, RAS, Moscow, Russian Federation

E-mail: mail.imetran@gmail.com

Abstract. In this paper, we analyzed the literature on the subject of obtaining transparent ceramics from aluminum oxynitride (AlON). In the ceramic industry there are a large number of methods for the synthesis of raw materials and their sintering, in this paper we reviewed the main approaches given in the literature and analyzed the results.

1. Introduction

Oxinitride ceramics (AlON) is a very promising material today. AlON ceramics is used in many areas of technology: engineering, aviation, instrument making and others. Ceramics of aluminum oxynitride has a combination of properties that are not inherent in one material, durability, transparency and low density. However, the production technology of this material implies the exact execution of each production stage, since any deviation from the technology leads to a decrease in transparency. To obtain ceramics of the required quality, it is necessary to observe high purity of the raw materials at all stages prior to sintering, as well as to choose the optimal sintering technique and composition.
Aluminum oxynitride (ALON) is a solid solution in the pseudo-binary system Al_2O_3 - AlN. The composition of AlON is consistent with the formula Al $_{(64+x)/3}$ O $_{32-x}N_x$, where the composition with x = 5 is the main stable phase in the γ-AlON region. Thus, its stoichiometric formula is $Al_{23}O_{27}N_5$. The mechanical properties given by Surmet [1] are presented in table 1.

The ceramic material in the Al_2O_3-AlN system has a stable γ phase like cubic spinel [2]. For the first time, the possibility of the existence of this phase in the Al2O3-AlN system was announced by Japanese scientists Yamaguchi and Yanagida, this discovery occurred in 1959 [3]. Later, other independent research teams confirmed the existence of this phase. In 1964, the first phase diagram of this system was presented (figure 1a) [4] and then, following this diagram, other phase equilibria of the system [5] were found (figure 1b). Such data became a prerequisite for obtaining a dense material, based on the phase of cubic spinel Al-O-N in the system of aluminum oxide - aluminum nitride. In 1976, according to literary sources, the first samples were obtained with transparency (figure 2) [6]. In 1979, a refined state diagram for the γ phase of aluminum oxynitride obtained by MacCool (figure 3) [7] was presented.

The required γ phase (AlON) is in the range of 64–78 mole percent Al_2O_3, the beginning of the formation of this phase with a composition of 72–73 mole percent aluminum oxide in the region of 1640 °C, this concentration allows to achieve the lowest possible temperature in a double system without sintering additives.

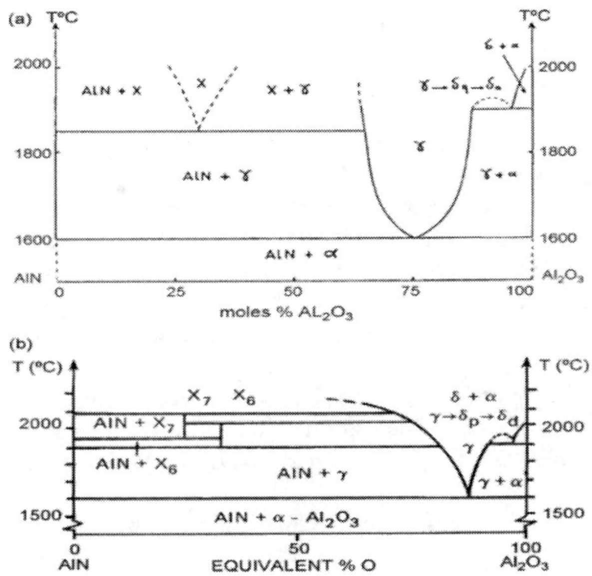

Figure 1. (a) Al2O3-AlN [4] state diagram, (b) Refined Al2O3-AlN state diagram from source [5]

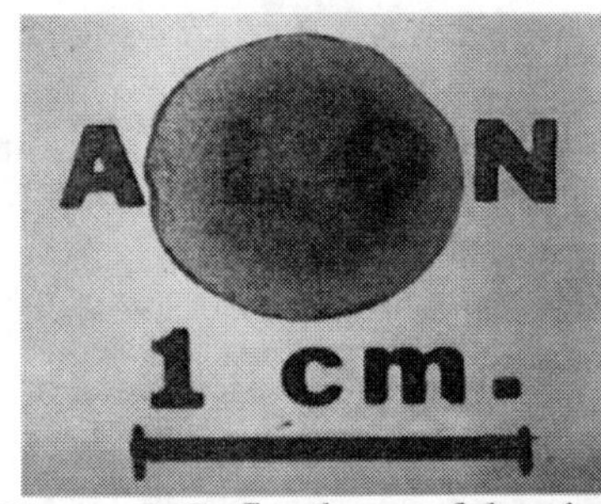

Figure 2. Refined part of the phase equilibrium diagram. [5]

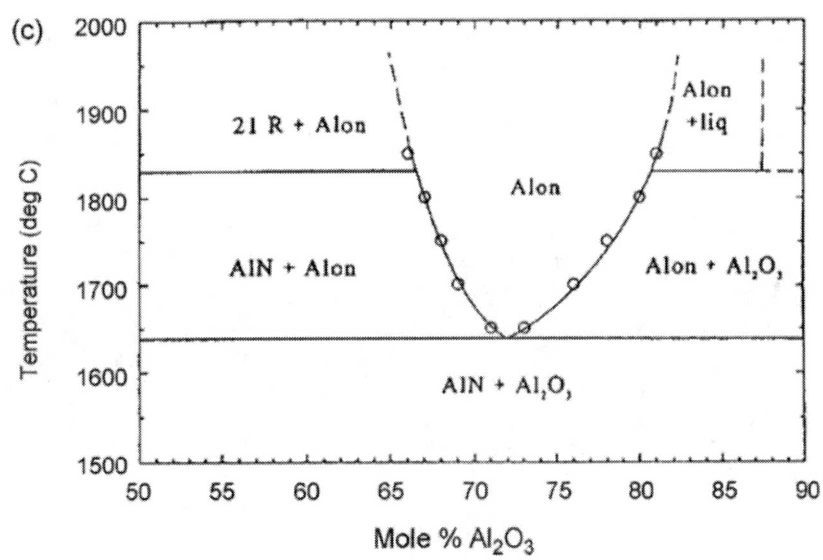

Figure 3. Refined part of the phase equilibrium diagram. [7]

Table 1. Characteristics of Al_2O_3 and AlN powders

Material	Flexural strength, MPa	Hardness according to Knoop, kg / mm^2	Young's modulus, GPa	Density, g / cm^3	Temperature Melting, °C
ALON	379 -700	1850	380	3,69	2140

IOP Conf. Series: Materials Science and Engineering **525** (2019) 012067 doi:10.1088/1757-899X/525/1/012067

2. Synthesis

The initial powders of aluminum oxide and nitride (Al_2O_3 and AlN) are subject to the strictest requirements for chemical purity and dispersion of particles [8,9]. For example, in the work of the researcher Kong [8], the initial powders had a chemical purity of 99.99% and an average particle size for Al_2O_3 - 1 μm, and for AlN - 0.5 μm.

It should be noted that in the production of oxynitride ceramics micro-additives of sintering components such as: Y_2O_3, Sc_2O_3, Lu_2O_3, La2O3, and others are used.

Y_2O_3, the main sintering additive used in oxynitride ceramics, as a promising optical material, has excellent physical and chemical properties, such as high melting point (2430 ° C), wide transparency range (0.2-8 microns), high corrosion resistance, a density of 5.01 g cm-3 and a boiling point of 4300°C. The sintering additive of yttrium oxide can cause the formation of a liquid phase during sintering, which is beneficial for compacting and removing pores. Y^{3+} ions increase the mobility of grain boundaries and accelerate grain growth.

Lanthanum oxide (La_2O_3) has the opposite effect. A small amount of La^{3+} ions can significantly slow down mass transfer and, therefore, suppress grain growth. Such additives can be used together to compensate for the negative effects of both additives and to reduce porosity and intergranular defects [10].

Self-propagating high-temperature synthesis. This process, being locally initiated, then propagates itself in the form of a reaction wave; “High-temperature” - because it is accompanied by the release of heat, which heats the substance in the reaction zone to a high temperature; and the word "synthesis" indicates that during this process valuable substances and materials can be synthesized [11]. According to the researcher Borovinskaya [12], experiments on the synthesis of AlON using the SHS method - the joint production of two or more target products were carried out. As a result of the conducted studies of the phase composition and structure of the combustion products of Al – Al_2O_3 mixtures at high pressures of gas-like nitrogen (10–100 MPa) under conditions of CBC-gas gassing, optimal parameters for the synthesis of powders (γ-ALON) were found. Experiments using highly exothermic coupled boron combustion reactions in gaseous nitrogen for additional heating of the initial mixtures showed the possibility of obtaining in the SHS mode not only powders with clearly formed layered particles collected in mosaic macrostructures, but also glassy polycrystalline mass with areas of transparency in the visible region (figure 4).

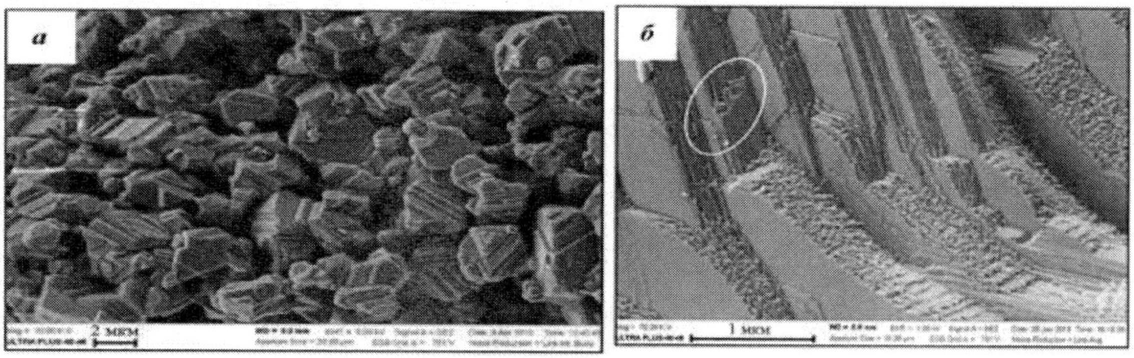

Figure 4. The microstructure of the particles of the powder γ-AlON (a); nanostructure of the layers of the particle γ-AlON (b) [13]

Plasma chemical synthesis. The synthesis of oxides in a plasma arc discharge is carried out by evaporation of the metal, followed by oxidation of the vapor or oxidation of the metal particles in an oxygen-containing plasma [13]. The characteristics of the obtained powders depend on the raw materials used, the synthesis technology and the type of plasmatron. The particles of plasma-chemical powders are single crystals and have sizes from 10 to 100-200 nm and more.

IOP Conf. Series: Materials Science and Engineering **525** (2019) 012067 doi:10.1088/1757-899X/525/1/012067

Carbothermic reduction method with nitriding. In the work of Chinese researchers [14], AlON powder was synthesized from nano-sized γ-Al_2O_3 of high chemical purity (99.99%) and carbon with a purity of 99.95%. The powders were mixed in a zirconium oxide microreducer for 12 hours, followed by drying at 80 ° C for 4 hours.

Then, the resulting mixture of γ-Al_2O_3 and carbon powder were placed in a crucible and heated under nitrogen atmosphere inside a graphite furnace. The mixture was first heated at a heating rate of 10 to 1550 ° C for 1 hour to obtain AlN and then heated at a heating rate of 3 ° C / min to 1650, 1750 and 1800 ° C for 2 hours to obtain AlON powders. The resulting powders were treated for 4 hours in air at 700 ° C to remove excess carbon or impurities.

However, in this work, the best results were obtained at 1750 ° C with a powder size of 42 μm. Such a particle size does not satisfy the sintering conditions, however, it should be noted that the required phase is formed as a result of the application of this method (figure 5).

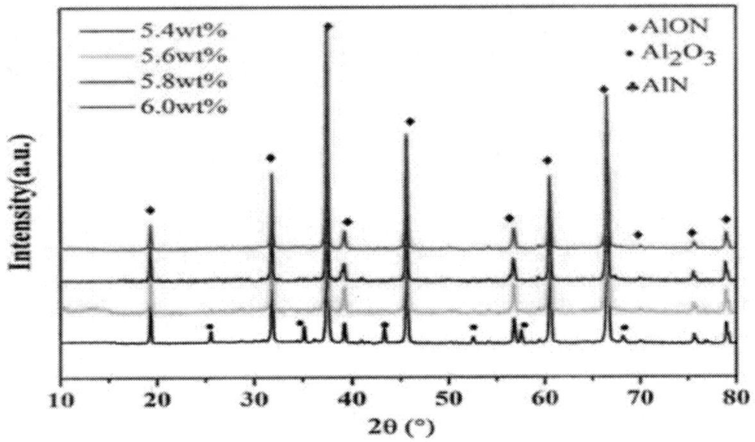

Figure 5. Diffraction patterns of the obtained powders with different mass percent carbon addition [14]

3. Sintering

Sintering is one of the key stages in the creation of ceramics from aluminum oxynitride, since the required (γ-phase AlON) and compaction of the product is formed during the sintering process. An important parameter is the density of the finished product, namely, 97 percent or more of the theoretical density, since below this threshold the optical properties drop to almost zero. Thus, the key points in this process are the balance between pressure, temperature and the environment in which sintering takes place. The difficulty lies in the choice of such parameters, which will allow to achieve the desired compaction at the lowest temperatures and pressure, since this directly affects the complexity of production and the cost of the finished product.

Reaction sintering. Sintered material is compacted by the formation of a new crystalline substance by the reaction between the solid-phase substance in the body of the product as one component and the vapor phase of the other substance. There is a kind of impregnation of the porous body with a vaporous substance, which reacts with the main or specially introduced phase to carry out the reaction. In this case, the formation of oxynitride is influenced by the medium, namely the chamber with the sintered compact is subjected to purging with nitrogen. Therefore, the denser and larger the product, the more difficult is the process of reaction sintering, what means that the force during compaction also affects the sintering process.

The reaction sintering is carried out at a temperature of 1700-1900 °C, consistent with the phase equilibrium diagram [7, 15]. The disadvantages of this method are the following factors: a sufficiently long exposure time (from 12 hours) and the difficulty of finding the optimal pressing force and porosity of a compact, for the passage of diffusion between the gas and solid phases.

The research group Revva IB and Ditz A.A. [16] investigated in their work the effect of pressing force on the synthesis of oxynitride using the method of reaction sintering. Samples were compressed with different forces, 20, 70, 140 and 210 MPa. Sintering was carried out at a temperature of 1750-1800 °C. According to X-ray phase analysis, samples pressed under a pressure of 70 MPa have the greatest amount of the target phase (figure 6).

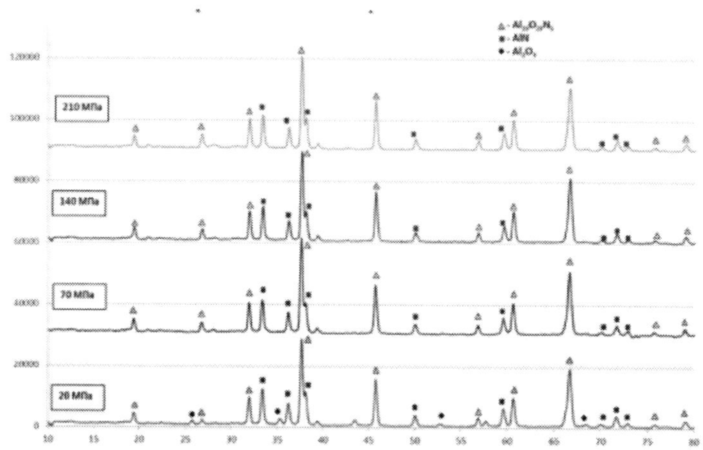

Figure 6. X-ray diffraction patterns of samples calcined at 1,800 ° C, molded at different pressures [17].

These data show that the reaction proceeds more intensively in samples compacted with a force of 70 MPa, since there remains sufficient porosity for the reaction of the solid phase with the gas.

Hot pressing method. The process is a simultaneous pressing and sintering. The mixture is placed in a mold, which walls are coated with an alcohol-based boron nitride coating to prevent the introduction of elements constituting the mold. During the sintering process, the vacuum chamber is subjected to "nitrogen flushing" cycles, the specimen is subjected to mechanical pressing during sintering. This method is more expensive than the reaction sintering method, since mechanical pressure is used in combination with vacuum and high temperature.

The advantage of hot pressing is the possibility of carrying out the process at temperatures lower than the sintering temperature in the free unloaded condition, since compaction is also achieved due to mechanical effort and the healing of the pores occurs at lower temperatures than with reaction sintering.

There is a variety of hot pressing, namely hot isostatic pressing. The process of simultaneous heating and forming of a powder billet, in which the powder is contained in a sealed sheath of flexible tin or glass, and thus the contained powder is subjected to equal pressure in all directions at a temperature high enough to cause plastic deformation and sintering.

In the work of a group of researchers from China, led by Junming X. [17], sintering was performed by hot isostatic pressing at various temperatures of 1800, 1850, and 1900 ° C. A mixture of powders with a composition of 67.5 mol% Al2O3, 33.5 mol% AlN and 0.5 wt% Y_2O_3 was sintered, the powders were mixed for 12 hours and dried at 80 ° C. The duration of the sintering process at different temperatures was the same and was 120 minutes. Sintering took place with a pressure of 20 MPa in a nitrogen atmosphere. The test results of sintered samples are presented in table 2.

Table 2. Mechanical properties of samples [18]

Sample	Sintering temperature ° C	Density g/cm^3	Hardness HV_{10} (GPa)	Crack resistance K_{IC} (MPa·m$^{1/2}$)
a	1800	3.68	14.67	2.45
b	1850	3.67	14.97	2.91

c	1900	3.65	15.66	3.37

According to the work, it was possible to achieve a high density of 3.65-3.68, (theoretical density of 3.69 g / cm^3). Although the density at a temperature of 1850-1900 °C is lower, but the mechanical characteristics are higher. The sample in Figure "c" has a more uniform grain size and an optimal structure, which has a positive effect on the light transmission of the sample.

Sintering using plasma sparks. This method allows to reduce the sintering temperature up to 1650, provided that the sintering additive Y_2O_3 is added, the exposure time is also shortened.

Spark plasma sintering (SPS - Spark plasma sintering), also known as field assisted sintering technique (FAST), is a process of rapid sintering to obtain a fully dense and fine-grained ceramic at a low temperature with a short holding time.

The driving force of the SPS method is the cumulative effect of force, heat, electric field and plasma generated by a pulsed current within the entire volume of the sample. It differs from the usual sintering force created by external heating elements, in which heating occurs through radiation and convection [18].

In the work of the group of researchers Patel P., Gilde G. and McCauley J. W. [19], sintering was performed with a temperature of 1600 ° C, a dwell time of 15 minutes and a pressure of 20 to 50 MPa. The compacts had a different composition, namely, the mass percentage of the Y_2O_3 additive differed. The research results showed that the optimal concentration for the lowest activation energy and the greatest compaction is 0.6 wt% Y_2O_3 (table 3). Figure 7 shows the porosity of the samples obtained, the pores are circled.

Table 3. Density of samples with the addition of Y_2O_3 [19]

Mas% Y_2O_3	0	0,3	0,6	1,2
Density from theoretical in%	96,0	98,3	99,2	97,2

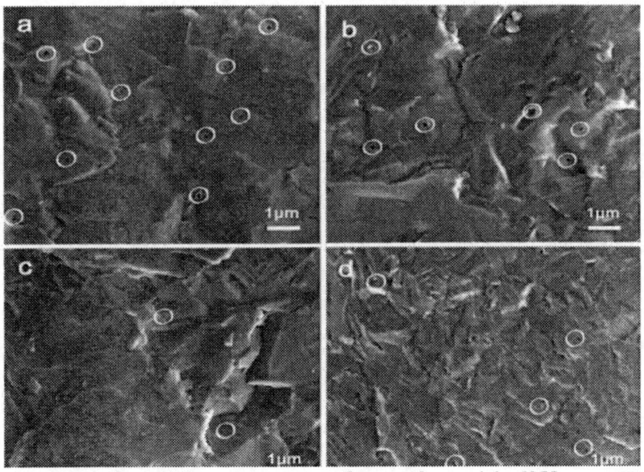

Figure 7. X photographs of the structure of samples with different Y_2O_3 contents: a - 0; b - 0.3%; c - 0.6%; d - 1.2% [19].

This result is due to the fact that when adding 1.2% Y_2O_3, the formation of the $Y_3Al_5O_{12}$ (YAG) phase occurs, this phase adversely affects the mechanical characteristics of the sintered samples, in turn, a concentration below 0.6% does not allow for the desired compaction and elimination of pores and defects. Figure 8 shows the dependence of the phase reflections on the concentration of the sintering additive Y_2O_3.

The obtained density of samples with the optimal concentration of the sintering additive Y_2O_3 (0.6%). In conjunction with not great sintering temperatures and holding time, it makes it possible to declare the viability of this method, however, the high complexity and high cost of the installation are an undoubted disadvantages.

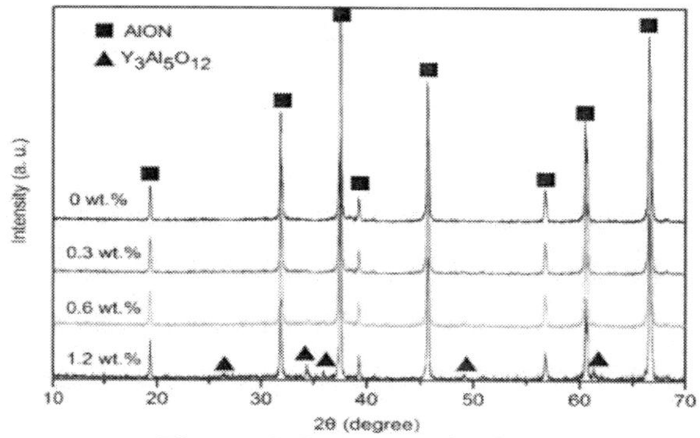

Figure 8. XRF samples [19]

Liquid sintering. Another sintering method used in the production process is transparent ceramics, in particular γ-AlON, to eliminate porosity and reduce secondary phases that are crucial for transparency. Dr. Patel investigated transient liquid phase sintering of AlON ceramics as a technology for the manufacture of dense transparent material [19].

This method consists in heating in two stages in a vacuum oven in a nitrogen atmosphere (N_2). In the first stage, the mixture of powders is heated to the temperature of the heterogeneous region of existence of the liquid and γ-AlON and held for a long time at a temperature of about 2150 ° C. In the second stage, the temperature is reduced until the homogeneous γ-AlON phase is reached(figure 9). The meaning of the temperature change is that in the first stage we obtain the liquid phase, which, when the temperature drops, allows the pores to heal in the specimen. Micro pores and intercrystalline defects are completely filled with the liquid phase, which makes it possible to obtain a dense product with light transmittance (transparency) during cooling (Figure 10).

As part of the work, the loss of specimen mass was detected at temperatures above 1950 ° C. So in the next work of Patel [20], the mass loss due to a change in the lattice parameter was explained. The loss of mass is due to a change in the relative composition determined by the lattice parameters. This mass loss and change in the lattice parameters are probably due to the formation of the vapor phase at elevated temperatures, either from a liquid or from a solid phase. It was found that at elevated temperatures, the composition migrates to the lattice parameters of the layer and, consequently, to a higher nitrogen content.

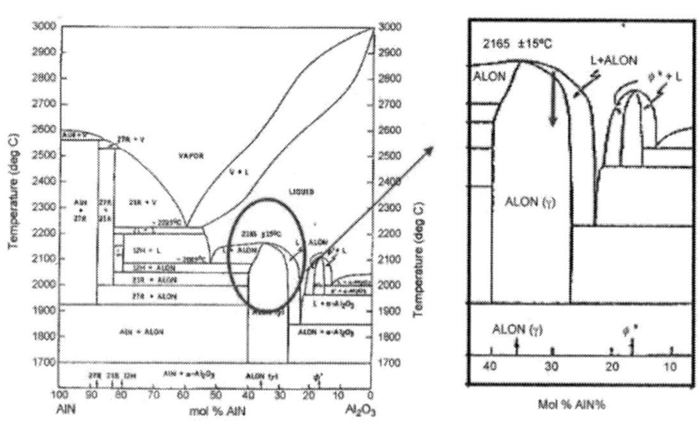

Figure 9. Liquid sintering method [19]

Figure 10. The sample obtained by the method of liquid-phase sintering [19]

Figure 11 shows the surprisingly strong dependence of the lattice parameter AlON, at a nitrogen pressure and a temperature of 2000 ° C, which leads to a loss of weight as the temperature rises to 2000 °C. A reduced lattice parameter will mean a loss of nitrogen or a shift in the composition towards Al_2O_3. An increase in nitrogen content is equivalent to a decrease in oxygen content, and this is a relative change in the ratio, which is reflected in the lattice parameter. However, the observation reveals a decrease in the percentage of nitrogen, in the ratio of nitrogen - oxygen, with increasing pressure of nitrogen. One explanation is the suppression of the oxygenated vapor phase in the form of Al_2O, [21] with increasing pressure. It is believed that nitrogen does not react under these conditions and there is no significant dissolution of nitrogen in AlON. It was determined that the lattice parameter AlON decreased from 0.7960 to 0.7926 nm, since the excess pressure was increased from 0.11 MPa to 0.2 MPa.

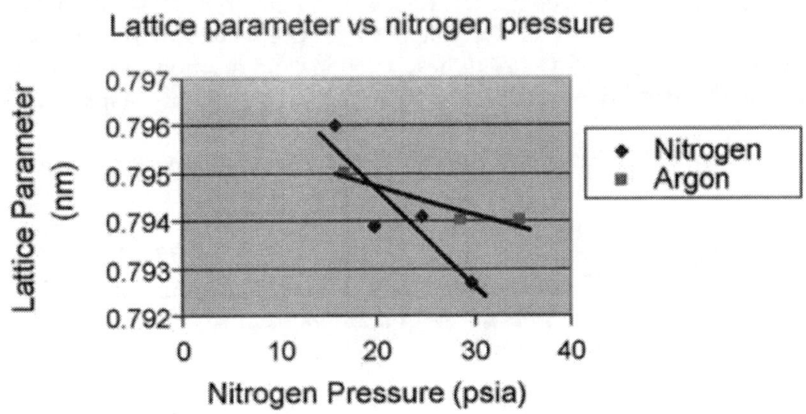

Figure 11. Dependence of gas pressure on the lattice parameter AlON [20]

The following conclusions were made: the lattice parameter increases with the nitrogen content, the lattice parameter decreases with increasing oxygen content, increasing the pressure of nitrogen or argon stabilizes the system, suppressing the loss of oxygen (Al_2O).

Liquid-phase sintering has positive sides: high compaction and defect elimination, due to the liquid phase, but the disadvantages of this method are that high temperatures up to 2200 °C and exact compliance with the nitrogen pressure parameter to control mass loss are required

4. Conclusions

There are quite a number of methods and approaches to the synthesis and sintering of ceramic material from aluminum oxynitride. Powders obtained by different methods have different characteristics of dispersion and purity, but in all methods they tend to obtain powders with particles of nano size and purity up to the second decimal place, i.e. 99.99%. The literature data under study are not straightforward, since in each individual case researchers resort to different methods of synthesis and sintering with their technological features and in many cases claim to obtain transparent samples, but not large.

Literary data is required to be reproduced experimentally, in order to evaluate the results and determine the optimal technological regimes, the compositions of powder mixtures, the modes of synthesis, pressing and sintering. It should be noted that today there is no proven technology for the production of oxynitride aluminum ceramics, both in Russia and in other countries. Accordingly, research in the field of obtaining a transparent ceramic material from aluminum oxynitride is necessary, since this material can replace a whole class of transparent structural materials.

Acknowledgments
The work was carried out according to the state assignment №007-00129-18-00

IOP Conf. Series: Materials Science and Engineering **525** (2019) 012067 doi:10.1088/1757-899X/525/1/012067

References

[1]. Goldman L M et al. 2010 ALON® optical ceramic transparencies for sensor and armor applications *Structure* **380** 700
[2]. Kingery W D et al. 1976 *Introduction to ceramics*. (New York: Wiley) p 500.
[3]. Yamaguchi G and Yanagida H 1959. Study on the reductive spinel—a new spinel formula AlN–Al2O3 instead of the previous one Al3O4. *Bulletin of the Chem. Soc. of Jap.* **32** №. 11, pp. 1264-65.
[4]. Lejus A 1964 Formation at high temperature of nonstoichiometric spinels and of derived phases in several oxide systems based on alumina and in the system alumina–aluminum nitride. *Rev. Int. Hautes Temp. Refract.* **1** pp 53-95
[5]. Gauckler L J and Petzow G 1977 In representation of multi-component silicon nitride based systems *Nitrogen ceramics* ed F L Riley (Leyden: Noordhoff) pp 41–62.
[6]. McCauley J W 1974 Viechnic. Dj. Reaction studies in SI3N4-AL2O3 system *Amer. Cer. Soc. Bull.* , (Westerville: Amer Ceramic Soc) **53** № 8 pp 620-620.
[7]. McCauley J W Corbin N D 1979 Phase relations and reaction sintering of transparent cubic aluminum oxynitride spinel (ALON) *Journ. of the Amer. Cer. Soc.* **62** № 9- 10 pp 476-479
[8]. Kong L B et al. 2015 Transparent Ceramic Materials *Transparent Ceramics* (Cham: Springer)pp 29-91.
[9]. Cheng J et al 2001 Microwave reactive sintering to fully transparent aluminum oxynitride (ALON) ceramics *Journ. of Mater. Sc. Let.* **20** №. 1 pp. 77-79.
[10]. Wang J et al 2015 Effect of Y2O3 and La2O3 on the sinterability of γ-AlON transparent ceramics *Journ. of the Eur. Cer. Soc.* **35** № 1 pp 23-28.
[11]. Merzhanov A G 2006 Self-propagating high-temperature synthesis: about the past, nowadays, and future *Izv. Vysh. Ucheb. Zav. Tsvet. Met.* **5** pp 5-8.
[12]. Borovinskaya I P et al. 2016 Structuro I fasoobrasovanie productov goreniia pri sinteze γ-ALON v rezhime svs-gazostatirovaniia *Izv. Vysh. Zav. Por. Met. I Funct. Pokr.* **3** pp 23-30
[13]. Gusev A I and Rempel A A 2001 *Nanocrystalline materials* (Moscow: Nauka) 224
[14]. Sahin F C, Kanbur H E and Apak B 2012 Preparation of AlON ceramics via reactive spark plasma sintering *Journ. of Eur. Cer. Soc.* **32** №4 pp 925-929
[15]. Prosvirnin D V et al. 2017 Effect of reaction sintering conditions on properties of ceramics based on alumina oxynitride *Materialovedenie* **10** pp 45-48
[16]. Revva I B, Ditz A A and Shulzhenko A S 2015 Vliyanie davleniia pressovaniia na sintez oxinirida aluminia *Vysokie tehnologii v sovremennoy nauke I tehnike: sbornik nauchnih trudov IV mezhdunarodnoi nauchno-tehnicheskoi konferencii molodyh uchenyh, aspirantov i studentov* (Tomsk: Izdatelstvo Topskogo polytehnicheskogo universiteta) pp 164-167
[17]. Junming X et al. 2008 Hot-pressed translucent aluminum oxynitride (AlON) ceramics *Key Eng. Mat.* **368-372** pp 450-452.
[18]. Li X, Luo J and Zhou Y 2015 Spark plasma sintering behavior of AlON ceramics doped with different concentrations of Y2O3 *Journ. of the Eur. Cer. Soc.* **35** № 7 pp 2027-32.
[19]. Patel P 2000 Phase equilibrium and kinetics in the multi-component non-oxide Al–O–N system Ph. D. (Baltimore: The Johns Hopkins University).
[20]. Patel P, Gilde G and McCauley J W 2006 Transient liquid phase reactive Sintering of Aluminum Oxynitride (AlON) Patent US-7,045,091 B May 16, 2006.

IOP Conf. Series: Materials Science and Engineering **525** (2019) 012068 doi:10.1088/1757-899X/525/1/012068

Corrosive researches of nonnickel shape memory alloy

M A Sevostyanov[1], M Losertova[2], E O Nasakina[1], O G Kuznetsova[1], A M Levin[1], A A Kirsankin[1], K V Sergiyenko[1], S V Konushkin[1], M I Baskakova[1], A V Leonov[1], M A Sudarchikova[1], I M Fedyuk[1], M A Kaplan[1], L A Shatova[3] and A G Kolmakov[1]

[1] Baikov Institute of Metallurgy and Material Sciences, 49 Leninsky Avenue, Moscow, Russia
[2] ŠB – Technical University of Ostrava, Faculty of Metallurgy and Materials Engineering, 17. listopadu 15, 708 33 Ostrava-Poruba, Czech Republic
[3] Voronezh State Technical University, st. 20-letiya Oktyabrya, 84/4, Voronezh, Russia

E-mail: nacakina@mail.ru

Abstract. Corrosion resistance of a nickel-free shape memory alloy TiNbTaZr of several compositions were studied. The structure and composition of the materials were determined using SEM, atomic emission spectrometry and Auger electron spectrometry. Electro-chemical parameters and alloy dissolution in physiological modeling media were investigated. It has been shown that the alloys are quite corrosion-resistant: no dissolution and high Ebd potential.

1. Introduction

Materials with shape memory effect (SME) are widely used in the modern world and are considered the best candidates for creating medical implants used in noninvasive endoscopic surgery, due to plastic deformation in a chilled state to an extremely compact form, contributing to easier and less traumatic delivery to the necessary part of the body without abdominal surgery, and independent adoption of a functional form in specified operating conditions without additional exposure [1–2]. The most famous medical material from this class is nitinol, endowed with mechanical characteristics similar to the behavior of living tissues, which helps it to adapt to physiological loads [1–7]. However, in addition to the positive mechanical characteristics, this alloy is endowed with a number of disadvantages: difficulty of processing in the manufacture of products, a high content of toxic element [8–10], a controversial level of biocompatibility and corrosion resistance [3, 11–22], which limits its applicability.

At the same time, SME and superelasticity are also found in other materials — alloys and polymers. These include titanium nickel-free alloys, which, due to the corrosive and biological properties of their elements, perfectly meet medical requirements. Unfortunately, at the moment, these alloys are inferior to nitinol in the manifestation of these unique mechanical characteristics. In addition, the technology for obtaining thin wire or other geometrical objects, which are suitable for the production of minimally invasive implants, has not yet been developed. At the same time, the configuration and the state of the surface of an object, defined by the production process, strongly influence the manifestation of all material properties.

IOP Conf. Series: Materials Science and Engineering **525** (2019) 012068 doi:10.1088/1757-899X/525/1/012068

This work is devoted to a long-term study of the dissolution of shape memory alloys of the composition Ti–(20–30)Nb-(10–13)Ta-5Zr in the form of a thin wire in media that simulate physiological, in static conditions.

2. Materials and methods

The object of the research was wire with a diameter of 280 μm and a length of 80 mm from Ti-20Nb-10Ta-5Zr; Ti-20Nb-13Ta-5Zr; Ti-25Nb-10Ta-5Zr; Ti-25Nb-13Ta-5Zr; Ti-30Nb-10Ta-5Zr, Ti-30Nb-13Ta-5Zr in the initial state (after drawing) and after grinding the surface. In order to improve the quality of the surface, wires in the initial state were subjected to successive grinding of the surface with emery paper of grit from 180 to 1000 and final processing with diamond paste to a mirror surface.

The morphology and layered elemental composition (including using transverse thin sections) of the surfaces of the materials were examined with a TESCAN VEGA II SBU scanning electron microscope (SEM) equipped with an INCA Energy dispersive energy analyzer, which also carried out fractographic studies of the samples, and electronic Auger spectrometer JAMP-9500F company JEOL in combination with ion etching when bombarded with argon at an angle of 30°.

The corrosion dissolution of the material was studied under static conditions by the method of immersion in solutions of various acidity and composition, since in the human body, the pH varies from almost 1 to 9. A neutral 0.9 mass % sodium chloride solution (NaCl, pH 6.31), artificial plasma (NaCl (92.3 mmol), NaHCO3 (26.3 mmol), K2HPO4 (0.9 mmol), KCl (2.7 mmol), NaH2PO4 (0.22 mM), CaCl2 (2.5 mM), MgSO4 • 7H2O (0.82 mm), Na2SO4 (1.48 mm), D-glucose C6H12O6 (1 g / l), pH 7.36) and standard buffer solution for reproducing at a given level of an acidic medium prepared from the corresponding mixture of the Merk company (potassium tetraoxalate KH3C4O8x2H2O, 0.05 M, pH 1.68) were used [13–15].

Samples of each type in the form of a 32.6 g wire were placed in polypropylene flat-bottomed flasks filled with 100 ml of the selected solution and kept for 30 days at 37 °C, making samples after 6, 13, 21 and 30 days, respectively.

For the experiments, Ti-Nb-Ta-Zr alloy wires of six compositions were used in 2 states: after delivery (after drawing, sample 1), after polishing (sample 2). After a selected period of time, samples were taken from the solutions for analysis. The analysis was carried out on a sequential atomic emission spectrometer with induction plasma with the aim of using the ICP-AES method (inductively coupled plasma atomic emission spectrometry) for direct simultaneous determination of titanium, niobium, zirconium and tantalum in solutions.

Electro-chemical corrosion indicators were investigated by the method of cyclic voltammetry (CVA) in a standard electrochemical cell using a universal potentiostat IPC-Pro. The potential sweep rate was 10 mV/s. Electrolyte was saline 0.9% NaCl at temperature 20 °C. Determination of corrosion parameters (stationary potential Ecorr, breakdown potential Ebd and repassivation potential Er) was carried out. The test samples were in the form of a wire with a diameter of 0.028 cm as the working electrodes. A glass graphite ring counter electrode was used as an auxiliary, and a saturated silver chloride silver electrode served as a reference electrode. The surface of the samples before conducting the experiments was treated with ethyl alcohol and washed with distilled water. Samples only after grinding the surface were investigated due to the need for surface homogeneity for analysis.

Scanning of CVA was started with a potential of -1.40 V and was carried out until a amperage of no more than 5 mA/cm^2 was achieved, based on the recommendations of the standard. Scanning in the opposite direction led to the value of the repassivation potential. The current density was measured in mA/cm^2.

3. Results and discussion

The surface morphology of the Ti-20Nb-10Ta-5Zr alloy wire is shown in Figure 1a. High roughness and heterogeneity are noted. It was of interest to study the structure of the surface in areas different from each other and at the same time regularly repeating. The composition of the wire surface in areas

IOP Conf. Series: Materials Science and Engineering **525** (2019) 012068 doi:10.1088/1757-899X/525/1/012068

differing in the SEM image is shown in Figure 1 b-c. It is noticeable that it differed in the analyzed areas - areas with high carbon content were presumed to remain after drawing (carbon-containing lubricant was used) and hardened during numerous anneals. In other areas, a high oxygen content was observed, which is quite natural - titanium and tantalum actively adsorb this element, forming oxides, and their formation is characteristic of almost any new surface.

The surface morphology of the Ti-20Nb-10Ta-5Zr alloy wire after grinding is shown in Figure 2a, and the surface composition of the wire is shown in Figure 2 b. It can be seen that after grinding the surface, its uniformity increased. The content of tantalum and niobium was reduced to zero on the surface, while titanium and zirconium remained in the amount of 5–10 atomic percent, and were most likely observed in oxides. The high carbon content on the surface is due to mechanical contamination of the surface.

Similar patterns were obtained for other compositions. It can be concluded that wires of any composition behave identically.

During the immersion tests, the release of metals in the used neutral solutions was absent or was at a level below the detection limit of the device (0.01 mg/l), which allows to call the alloys highly corrosion-resistant. A slight dissolution, slowing down with time due to surface repassivation, was noted in an acidic solution, which is presumably connected with the complexation of titanium and tetraoxalate. There were no significant differences in the behavior of the alloys.

It was shown that the potential of surface repassivation is within 1400 V for all the samples studied, and there is no hysteresis dependence, which is an indicator of the high corrosion resistance of materials (Fig. 3). The values of the breakdown potential of the passive film speak in favor of the same conclusion. For all compositions it was from 500 V and higher, and it can be clearly noted that an increase in the content of both niobium and tantalum in the alloy shifts this characteristic to an even more electrically charged direction.

a

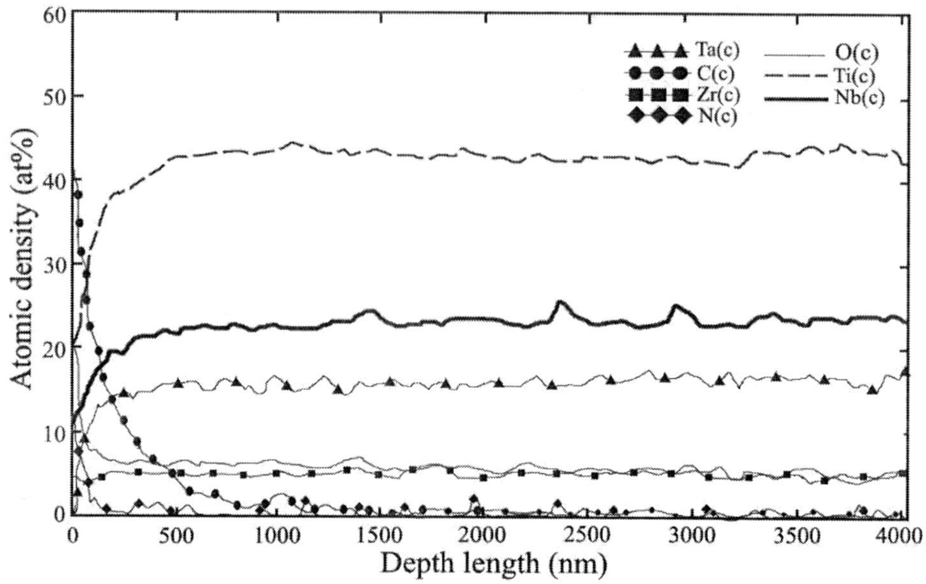

b

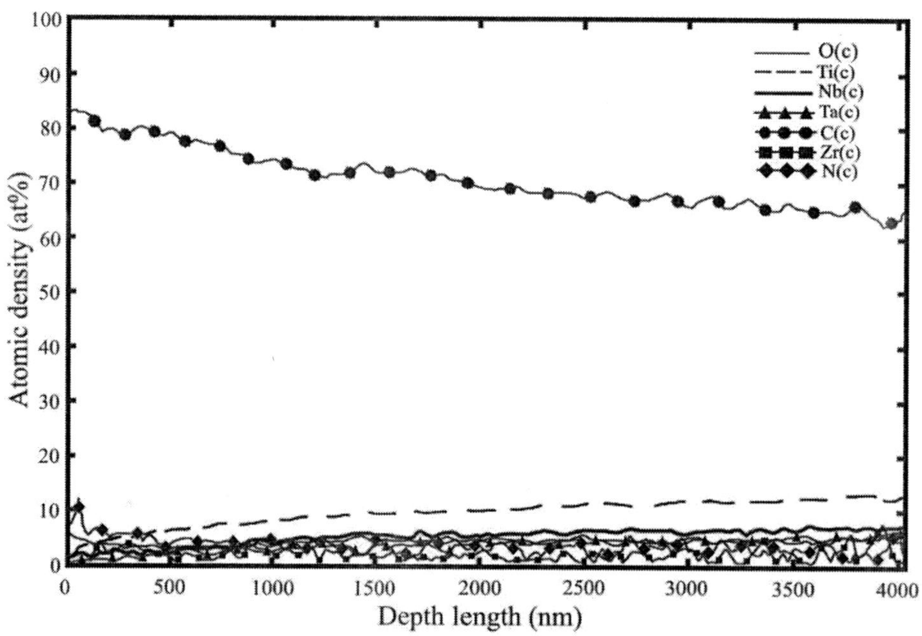

c

Figure 1. Morphology (a) and layered composition of light (b, on "a" denoted as "1") and dark (c, on "a" denoted as "2") surface areas of alloy in the initial state

IOP Conf. Series: Materials Science and Engineering 525 (2019) 012068 doi:10.1088/1757-899X/525/1/012068

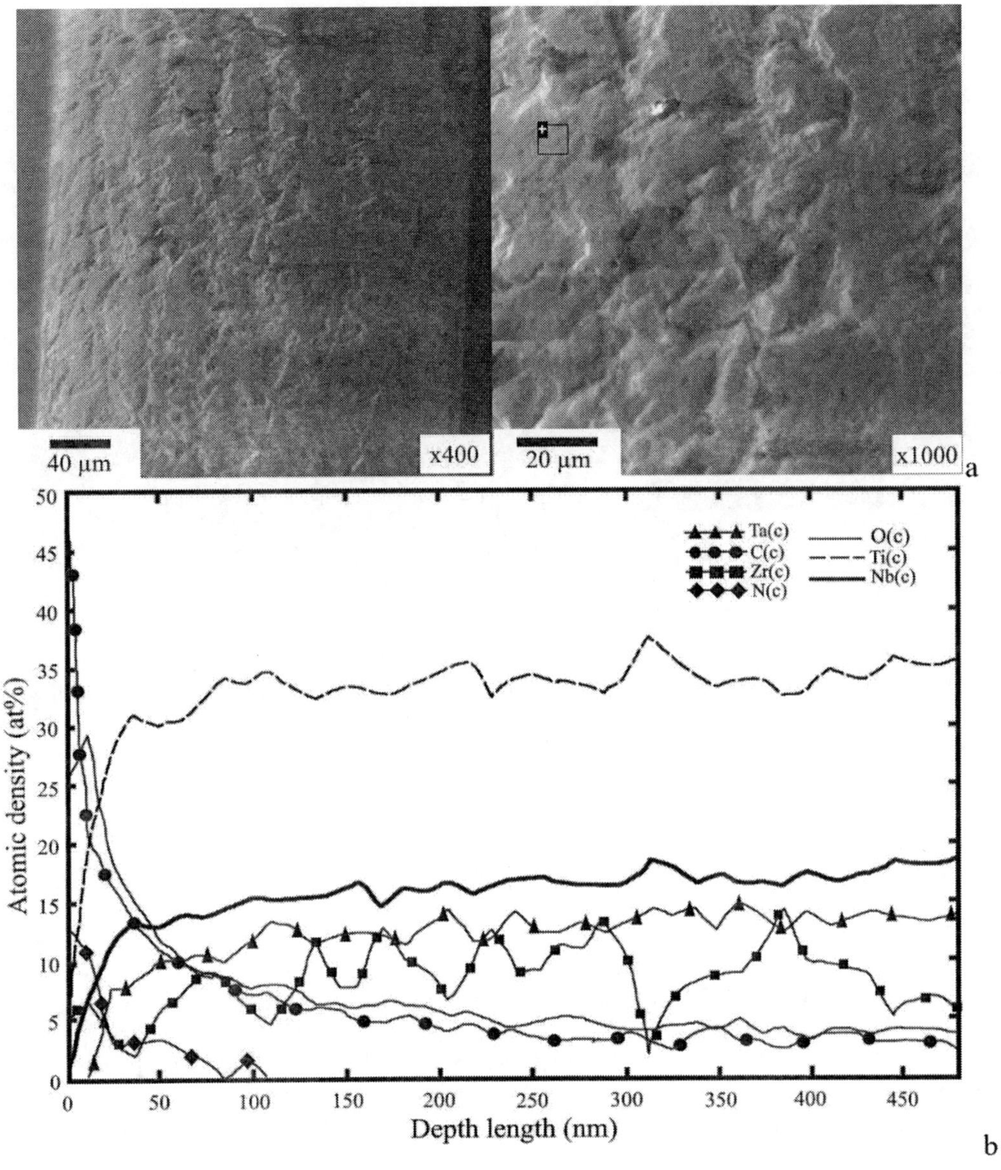

Figure 2. Morphology (a) and composition (b) of the alloy surface after grinding

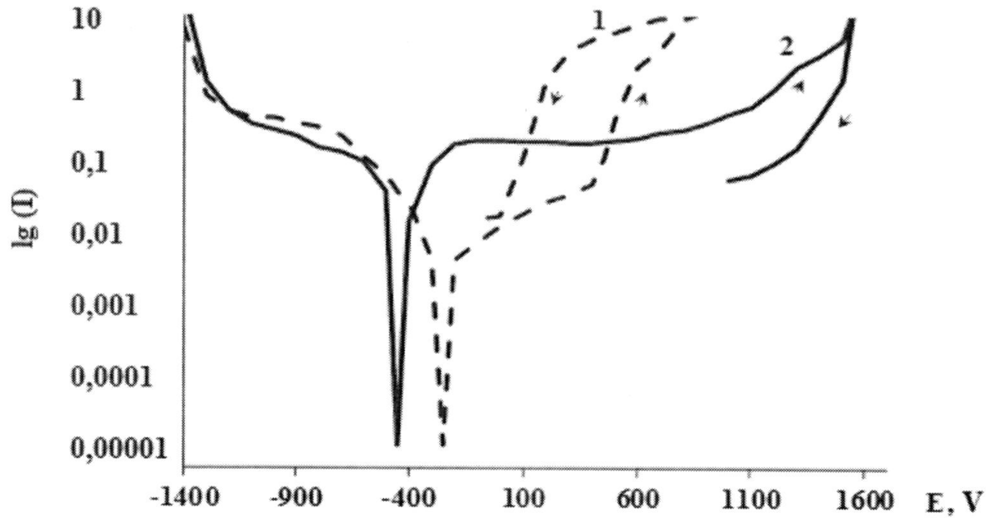

Figure 3. Cyclic voltammograms of TiNi (1) and Ti-Nb-Ta-Zr (2) alloys in 0.9% NaCl solution at a temperature of 20 °C

4. Conclusions

Ti-Nb-Ta-Zr of several compositions were prepared in the form of thin wire and investigated for its corrosion resistance. Electro-chemical parameters and alloy dissolution in physiological modeling media were studied. It has been shown that the alloys are quite corrosion-resistant: no dissolution and high Ebd and Ere potential.

Acknowledgments

The work was supported by the Ministry of Education and Science of Russia (grant identifier RFMEFI60417X0196).

References

[1] Petrini L, Migliavacca F 2011 Biomedical Applications of Shape Memory Alloys *Journal of Metallurgy* **2011** 1-15

[2] Duerig T W, Melton K N, Wayman C M, Stockel D 1990 *Engineering aspects of shape–memory alloys* (Oxford: Butterworth Heinemann Ltd) pp 181 – 194

[3] Shabalovskaya S 1996 On the nature of the biocompatibility and medical applications of NiTi shape memory and superelastic alloys *Bio Med Mater Eng* **6** 267 – 289

[4] Marjan Bahrami Nasab, Mohd Roshdi Hassan 2010 Metallic Biomaterials of Knee and Hip - A Review *Trends in Biomaterials and Artificial Organs* **24(1)** 69 - 82

[5] Surdell D, Shaibani A, Bendok B, Eskandari M K 2007 Fracture of a Nitinol Carotid Artery Stent That Caused Restenosis *J Vasc Interv Radiol* **18(10)** 1297–1299

[6] Bose A, Hartmann M, Henkes H A 2007 Novel, Self–Expanding Nitinol Stent in Medically Refractory Intracranial Atherosclerotic Stenosis: Wingspan Study *Stroke* **38** 1531–1537

[7] Stoeckel D, Pelton A, Duerig T 2004 Self-expanding nitinol stents: material and design considerations *Eur Radiol* **14** 292-301

[8] Xiaoying Lu, Xiang Bao, Yan Huang, Yinghua Qu, Huiqin Lu, Zuhong Lu 2009 Mechanisms of cytotoxicity of nickel ions based on gene expression profiles *Biomater* **30** 141–148

[9] Uo M, Watari F, Yokoyama A, Matsuno H, Kawasaki T 1999 Dissolution of nickel and tissue response observed by X-ray scanning analytical microscopy *Biomater* **20** 747–755

[10] Wataha J, O'Dell N, Singh B, Ghazi M, Whitford G, Lockwood P 2001 Relating nickel-induced

tissue inflammation to Ni release in vivo *J. Biomed. Mater. Res.* **58** 537–544

[11] Balazic M, Kopac J 2007 Improvements of medical implants based on modern materials and new technologies *J. Achiev. Mater. Manuf. Eng* **25(2)** 31-34

[12] Kim J H, Shin J H, Shin D H, Moon M W, Park K, Kim T H et al. 2011 Comparison of diamondlike carbon-coated nitinol stents with or without polyethylene glycol grafting and uncoated nitinol stents in a canine iliac artery mode *Br J Radiol* **84** 210-215

[13] Nasakina E O, Baikin A S, Sevost'yanov M A, Kolmakov A G, Zabolotnyi V T, Solntsev K A 2014 Properties of nanostructured titanium nickelide and composite based on it *Theor Found Chem Eng+* **48(4)** 477–486

[14] Nasakina E O, Sevostyanov M A,Golberg M A, Dyomin K Yu, Baikin A S, Goncharenko B A, et al. 2015 LongTerm Corrosion Tests of Nanostructural Nitinol of (55.91 wt % Ni, 44.03 wt % Ti) Composition under Static Conditions: Ion release *Inorg. Mater. Appl. Res.* **6(1)** 59–66

[15] Nasakina E O, Sevostyanov M A,Golberg M A, Dyomin K Yu, Baikin A S, Goncharenko B A, et al. 2015 LongTerm Corrosion Tests of Nanostructural Nitinol of (55.91 wt % Ni, 44.03 wt % Ti) Composition under Static Conditions: Composition and Structure before and after Corrosion *Inorg. Mater. Appl. Res.* **6(1)** 53–58

[16] Nasakina E O, Sevost'yanov M A, Mikhailova A B, Gol'dberg M A, Demin K Yu, Kolmakov A G, et al. 2015 Preparation of a nanostructured shape memory composite material for biomedical applications *Inorg. Mater.* **51(4)** 400-404

[17] Nasakina E O, Baikin A S, Sergienko K V, Sevost'yanov M A, Kolmakov A G, Goncharenko B A, et al. 2015 Biocompatibility of nanostructured nitinol with titanium or tantalum surface composite layers formed by magnetron sputtering *Doklady Chemistry* **461(1)** 86–88

[18] Nasakina E O, Baikin A S, Konushkin S V, Sergienko K V, Kaplan M A, Fedyuk I M, et al. 2018 Receiving of layered composite materials with shape memory effect of medical appointment *IOP Conf Ser: Mater Sci Eng* **411** 012051

[19] Nasakina E O, Baikin A S, Sergiyenko K V, Leonov A V, Kaplan M A, Seryogin A V, et al. 2017 Formation and Investigation of Composite Material Silver–Nitinol for Medical Purposes *Inorg. Mater. Appl. Res.* **8(1)** 112–117

[20] Nasakina E O, Sevostyanov M A, Mikhaylova A B, Baikin A S, Sergienko K V, Leonov A V et al. 2016 Formation of alpha and beta tantalum at the variation of magnetron sputtering conditions *IOP Conf Ser: Mater Sci Eng* **110** 012042

[21] Tomić S, Rudolf R, Brunčko M, Anžel I, Savić V, Čolić M 2012 Response of monocyte-derived dendritic cells to rapidly solidified nickel-titanium ribbons with shape memory properties *Eur. Cell. Mater.* **23** 58–81

[22] Chun-Che Shih, Shing-Jong Lin, Yuh-Lien Chen, Yea-Yang Su, Shiau-Ting Lai, Gaston J. Wu, Ching-Fai Kwok, Kwok-Hung Chung. 2000 The cytotoxicity of corrosion products of nitinol stent wire on cultured smooth muscle cells *J. Biomed. Mater. Res.* **52** 395–403

Investigation of amorphous alloys nanostructure and phase composition formed during integrated and split exposure of deformation and flash lamp annealing

N D Bakhteeva[1], **E V Todorova**[1] **and S V Kannykin**[2]

[1] Institution of Russian Academy of Sciences A.A. Baikov Institute of Metallurgy and Material Science RAS, Russia, Moscow, Leninsky av., 49
[2] Voronezh State University, Russia, Voronezh, Universitetskaya sq. 1

E-mail: nbakhteeva@imet.ac.ru

Abstract. Microstructure and phase composition of the $Al_{85}Ni_7Fe_4La_4$ amorphous alloy after separate and complex impacts of severe plastic deformation by shift under the pressure (SPD) with P=8 GPa, n=1-3 revolutions and flash lamp annealing (FLA) with the fluence F=10-25 J/cm^2 were investigated. By X-ray structural analysis and transmission electronic microscopy (TEM) was shown, that in initially amorphous structure partial multiphase crystallization with formation of amorphousnanocrystalline composite (solid solution of crystalline Al_{cr} and amorphous Al_{am} and $Al_{11}La_3$, $Al_9(Fe, Ni)_2$ in aluminum) is developing during the SPD. FLA leads to partial nanocrystallization at F=10 J/cm^2 and full crystallization after the FLA at 20-25 J/cm^2. Phase transformation and static recrystallization processes are developed along with the crystallization in amorphous structural component at SPD and FLA integrated exposure.

1. Introduction

Aluminum alloys obtained in amorphous condition demonstrate increased levels of strength, hardness, corrosion characteristics, high levels of tribological properties at low specific weight preservation – thus define interest in new compositions development and in treatment technologies. Research groups in Russia, USA, Japan and PRC are engaged in development of new technologies of creation of nanocrystalline materials and multiphase amorphous-nanocrystalline composites on the basis of amorphous alloys. The attention of researchers is focused on the problems of multicomponent aluminum alloys optimization – multicomponent aluminum alloys of "Al - transition metal - rare-earth element" system able to amorphizate under conditions of high-speed quenching from the liquid state [1, 2]. Multicomponent amorphous alloys of the Al-Ni-Fe-La system are included in this class of materials and are investigated in the scopes of this article. Scientific achievement results of interrelationship between amorphous alloys structure and properties study are widely presented in the monograph [3]. Phase transformations and nanocrystalline structure formation trends in amorphous alloys of the Al-Ni-La system with silicon additives at flash lamp annealing (FLA) are investigated in scientific papers of academician Ievlev V.M. and his school [4, 5]. FLA of gas-discharge lamps (IFO) method allows energy localizing in the material near-surface region and reducing the heat treatment time. Researchers in Russian Federation have accumulated wide experimental material on the structural transformations study, mechanical properties changes; light exposure mechanisms on the material, the treatment effect dependence on the emission wavelength, pulse duration and strength were discussed [6]. In open sources there are no publications about systematic studies of aluminum-based amorphous alloys nanocrystallization during simultaneous deformation and radiation treatment. Such studies are required to optimize amorphous alloys technological processing to obtain nanocomposite structure with advanced mechanical properties.

2. Material and techniques

Amorphous structure in $Al_{85}Ni_7Fe_4La_4$ alloy in the samples in the form of the ribbons was obtained after the melt spinning method with the speed of 10^6 K/s. Amorphous ribbons were exposed to severe plastic deformation (SPD) by the shift under the 8 GPa pressure on the Bridgeman anvils. Rotation angle of the moveable anvil was equal to one or three full revolutions (φ=360° and φ=360°×3). Amorphous ribbons after the high-speed quenching and SPD were exposed to flashed lamp annealing with the F=10, 20 and 25 J/cm² fluence in conditions of gas-discharge lamp radiation.

Alloy structure was investigated by X-Ray spectroscopy, DRON-3M diffractometer (Russia) with Cu-Kα radiation (exposure mode: U=38 kV, I=20 mA, 1x2x0.5 slit + Sollers slit system) and ARL X'TRA diffractometer, copper radiation, geometry of a parallel beam (parabolic mirror on the primary beam, thin film collimator – on the secondary beam) were used. X-Ray pattern phase analysis was performed using PDF-2 data. Mechanical tests were performed on an Instron 5848 MicroTester machine at room temperature in pneumatic grippers with 0.01 and 10 mm/min tensile rates. Pneumatic grippers allow quick and continuous adjusting of the ribbons clamping forces and ensure reliable clamping of the samples. Working area of the test specimens length was 20 mm, the width of the ribbon is 1.1 mm. Samples microhardness was measured on 402 MVD Wolpert Wilson Instruments hardness-testing machine (Germany) at a load of 10 g, 10 s.

3. Results and its discussion

Structure, phase transformations and microhardness in the amorphous state and after the severe plastic deformation by a shift under the pressure 8 GPa were previously investigated by the authors [7] in amorphous alloys of the Al-Ni-Fe-La system. It was shown that nanocrystallization develops during plastic deformation in the structure of aluminum alloys with the formation of a multiphase amorphous-nanocrystalline structure including nanoscale grains of intermetallic compounds of various types and aluminum.

$Al_{85}Ni_7Fe_4La_4$ alloy diffraction pattern obtained by the X-ray diffraction method after the high-speed quenching and subsequent shift under the 8 GPa pressure with 360° torsion angle is shown in Figure 1. The $Al_{85}Ni_7Fe_4La_4$ alloy after spinning has an X-Ray amorphous structure with a typical halo on the diffractogram (Figure 1, curve I) according to the data of X-ray structural analysis. SPD shift under the 8 GPa pressure with the deformation degree matching to one movable anvil full revolution (φ=360°) leads to partial crystallization development with formation of small amount of crystal aluminum grains and $Al_{11}La_3$, $Al_9(Fe,Ni)_2$, $Al_8Ni_xFe_{1-x}La$ intermetallic compounds. Intensity peaks corresponding to these phases appear on the diffractogram (Figure 1, curve II). An amorphous-nanocrystalline structure with a small volume fraction of crystalline phases is formed in the alloy after the SPD with φ=360°.

Quenched and deformed ribbons were subjected to pulsed photon processing with a fluence of 10, 20, 25 J/cm². Multi-phase crystallization develops in the entire fluence parameters range of the FLA process both in quenched (initially amorphous) and in deformed (with φ=360° and φ=360°×3) ribbons. Multiple intensity peaks are formed on diffractograms, corresponding to the phase composition $Al_{кр}$ + $Al_{11}La_3$ + $Al_9(Fe,Ni)_2$. Diffraction maximum presence at 2Θ=36,09° with an intensity of 100% does not exclude the presence of the $Al_8Fe_{2-x}Ni_xLa$ phase in the structure. Lesser intensity lines for the indicated intermetallide coincide with the diffraction maxima of the other phases. The intermetallic compound $Al_8Fe_{2-x}Ni_xLa$ (Al_8Fe_2Eu prototype) was first discovered and certified by the authors in [8]. This phase refers to a rhombic system (*Pbam* space group) with lattice parameters a=1.258(3) nm, b=1.448(3) nm, c=0.405(8).

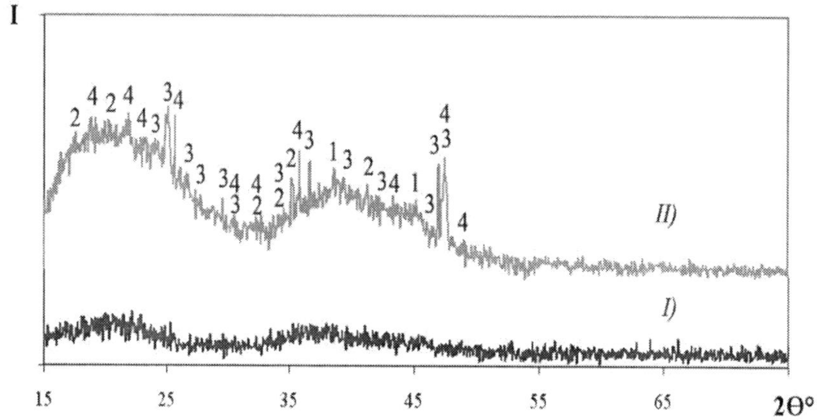

Figure 1. Alloy $Al_{85}Ni_7Fe_4La_4$ diffraction pattern: I) amorphous condition, II) SPD 8 GPa, φ=360°. Symbols: 1 – Al, 2 - $Al_{11}La_3$, 3 - $Al_9(Fe,Ni)_2$, 4 - $Al_8Ni_xFe_{2-x}La$.

SPD and FLA integrated exposure does not lead to noticeable changes in diffraction patterns. Halo is partially retained after the SPD with φ=360° and FLA with a fluence of 10 J/cm^2 on the diffractogram which indicates that crystallization is incomplete. Volume ratio of uncrystallized amorphous matrix decreases with fluence magnitude growth; the halo on the diffractograms disappears completely after the F=25 J/cm^2 fluence. Alloy phase composition is a mixture of crystalline phases: an aluminum based solid solution and $Al_{11}La_3$, $Al_9(Fe,Ni)_2$ intermetallic compounds. Deformation degree increase to φ=360°×3 under the integrated exposure of SPD and FLA does not significantly affect the phase transformations regular pattern. Reflections intensity from the crystalline aluminum decreases and reflexes intensity from intermetallic phases increases as the fluence increases at FLA. The effect of intensity redistribution between the different phases reflexes is associated with volume ratio of aluminum based crystalline solid solution decrease and intermetallic phases volume increase.

Mechanical tensile testing results of amorphous ribbons in the initial state at room temperature with tensile rates 0.01-10 mm/min are presented in Table 1. For comparison the mechanical properties of industrial cast aluminum alloys are shown in Table 2.

Table 1. $Al_{85}Ni_7Fe_4La_4$ amorphous alloy tensioning mechanical tests results

Alloy	Speed of the test, mm/min	Tensile strength, MPa
$Al_{85}Ni_7Fe_4La_4$	*0.1*	*384*
	10	*553*

In all test modes the amorphous alloys yield strength is equal to the tensile strength which is stipulated by the strain hardening absence during the strain process. It was established that with deformation rate increase from 0.1 to 10 mm/min the tensile strength increases from 384 to 553 MPa (Table 1). Similar results conserning the effect of deformation rate on the amorphous ribbons tensile strength were obtained and described earlier in [9]. As noted by the author this dependence is typical for the homogeneous deformation realization in amorphous alloys, which, apparently, is also observed in the ribbons studied after uniaxial tension. The obtained mechanical characteristics are close to the tensile strength values of Al-Cu based industrial high-strength alloys utilized in aircraft manufacturing (VAL10, VAL12, Table 2).

The tensile test method applied by the authors like most of the modern mechanical testing methods does not allow complete mechanical properties characterization of the new amorphous high-strength materials due to the small sizes of ribbon samples and especially of the samples exposed to shift under

IOP Conf. Series: Materials Science and Engineering **525** (2019) 012069 doi:10.1088/1757-899X/525/1/012069

the pressure. Various indentation methods are widely used to determine and characterize mechanical properties of such materials.

Table 2. Results of industrial casting alloys mechanical tests [10]

Alloys and their assignment	Grade	Tensile strength, MPa
Al (~85 %)-Si based high-technology alloys doped with Mg, Ti, Cu, Mn, Fe. For production of modular castings	AL9-1	*320*
	V 124	*400*
Al-Cu based high-strength alloys doped with Cu, Mn, Ti, Zn, Mg, Cd, Zr, Fe, Si. For castings production (aircraft industry)	VAL10	*420*
	VAL12	*550*
Al-Cu-Ni based high-temperature alloys doped with Cu, Ni, Mn, Ti, Zr, Ce, Mg, Fe, Si, works at 200-250 °C	AL33	*280*
	VAL18	*300*

Original amorphous structure hardness of the $Al_{85}Ni_7Fe_4La_4$ alloy of the amorphous ribbon is 406 HV. The maximum harness values 659 and 651 HV were obtained on an amorphous $Al_{85}Ni_7Fe_4La_4$ alloy after the FLA with a fluence of 10 and 20 J/cm^2 respectively.

The results of $Al_{85}Ni_7Fe_4La_4$ amorphous alloy samples microhardness measuring after the SPD and FLA integrated exposure depending on the radiation dose demonstrated that the sample hardness after deformation increased with twist angle rotation increasing from 406 HV in the amorphous state to 580 HV and 619 HV after the SPD with φ=360° and φ=360°×3 respectively. However, after irradiation this indicator decreased in samples with a large twist angle (φ=360°×3) reaching the minimum value of 495 HV after treatment with 20 J/cm^2 fluence. The most probable cause of hardness decrease is the static recrystallization development at thermal exposure on a sample during FLA as a result of FLA. Microhardness gradually increased and reached 653 HV value at energy dose increase to 20 J/cm^2 for the samples subjected to SPD exposure with P=8GPa, φ=360°. The microhardness values of both series of the samples turned out to be approximately the same (within the measurement error limits) after FLA with the maximum energy dose of 25 J/cm^2.

4. Conclusions

Separate and complex SPD and FLA exposure of amorphous ribbons processing methods leads to solid-phase crystallization processes development and phase transformations in the $Al_{85}Ni_7Fe_4La_4$ amorphous alloy with the multiphase composite structure formation. Volumetric fraction of crystallization development and the grain size depend on the FLA (fluence 10…25 J/cm^2) and SPD (φ=360°…360°×3) parameters. With these characteristics increase the crystallization volume fraction increases. Gradient structure is formed over the ribbon cross section at FLA as against to the deformation processing; this allows creating of high-strength coatings of the given thickness on the processed products surface.

The maximum microhardness of 659 and 651 HV was obtained after FLA with fluences 10-20 J/cm^2 resulted in multiphase nanocrystalline structure formation at the alloy which included aluminum-based solid solution (Al_{cr}) and $Al_{11}La_3$ + $Al_9(Fe,Ni)_2$ intermetallic compounds.

Integrated exposure including SPD (8 GPa with φ=360°) + FLA (with 20 J/cm^2 fluence) provides 653 HV microhardness and does not introduce additional hardening as compared to an amorphous alloy after FLA.

Acknowledgments

The investigations have been fulfilled according to the Federal order № 075-00746-19-00.

References

[1] Wang S H, Bian X F Structure relaxation and crystallization of $Al_{83}Ni_{10}Ce_7$ metallic glass *Journal of Alloys and Compounds* **450** Issues 1–2 14 February 2008 pp. 260-263

[2] Popov V V, Rassolov V V, Tkach V G et al. Effect of partial replacement of nickel by cobalt on the thermal stability of amorphous and nanocomposite structures in alloys *Physics and Technics of High Pressure* 2009 **19 (2)** pp. 76–86. (In Russ.)

[3] Inoue A, Suryanarayana C *Bulk Metallic Glasses* CRC Press 2011 523 p

[4] Ievlev V M Activation of solid-phase processes by radiation of gas-discharge lamps *Uspekhi Khimii — Russian Chemical Reviews* 2013 **82 (9)** pp. 815–834

[5] Belousov O K, Vavilova V V, Ievlev V M, et al. Nanocrystallization in amorphous $Al_{83}Ni_{10}La_7$ and $Al_{83.5}Ni_{9.5}La_{5.6}Si_{1.4}$ alloys during thermal annealing and flash lamp processing *Inorganic Materials* 2009 **45 (8)** pp. 873–879

[6] Sinelnikov A A *The influence of flash lamp processing on the processes of recrystallization of Au, Pt, Pd films and Ti oxidation* Abstract of the candidate's thesis Voronezh 2011

[7] Bakhteeva N D Nanocrystallization in Al-based amorphous Al-Ni-Fe-La alloys 2010 *Nanotech. in Russia* **5 (3-4)** pp. 235-249

[8] Vasil'ev A L, Ivanova A G, Bakhteeva N D, et al. Microstructure of the Al-La-Ni-Fe system 2015 *Crystallography Reports* **60 (1)** pp. 23-29.

[9] Zolotukhin I V Amorphous metallic materials *Soros educational journal, Physics* 1997 **3 (4)** pp. 73–78 (In Russ.)

[10] *Aviation materials* Ed. by academician Kablov E N Moscow: MISIS, 2002. 426 p.

IOP Conf. Series: Materials Science and Engineering **525** (2019) 012070 doi:10.1088/1757-899X/525/1/012070

The study of ceramic materials system SiC-YAG

A M Tsareva, A V Leonov, M A Sevostyanov, A S Lysenkov

Baikov Institute of Metallurgy & Materials Science (IMET RAS), Moscow, Leninskiy prospect 49, Russia

Email: alyo.tsaryowa2012@yandex.ru

Abstract. Dense ceramic materials based on SiC with an oxide activating additive YAG by hot pressing at a temperature of 1850 ° C were obtained. The oxides, melting, fill the space between the SiC particles, contributing to the compaction and mass transfer of the material in the process of hot pressing. It is shown that for a high compaction of the SiC – YAG material, a small amount of oxides is necessary - 9 wt. %

1. Introduction

SiC based ceramics find reaching use in technology and manufacturing applications. This is due to many of its key properties. In Structural applications, take advantage of its high hardness, high stiffness, low density, and low coefficient of thermal expansion. High temperature and thermal applications take advantage of the phase stability and high thermal conductivity. Since SiC is a semi-conductor, its electronic properties can be adjusted. Polycrystalline SiC based ceramics could find broader use if they could be machined easier by rendering them more electrically conductive. Higher conductively enables efficient electro-discharge machining (EDM) of complex shapes. Strength enhancement would allow for competition with structural Si3N4 based applications [1-3].

Solid-state junctions of carbide ceramics with platinum-group metals are relevant in a variety of areas of materials technology, including microelectronics, ceramic joining technology, metal–ceramic composites. Chemical interactions between the metal and ceramic layers in many applications can be an important consideration, acutely affecting the microstructure, properties, and performance of a particular system [4-5].

The functionality of the material is determined not only by its fundamental properties, but also by the correctness of the choice of initial components and technologies of their combination [6]. To obtain materials based on silicon carbide use a large number of sintering additives [7-11]. One of the most commonly used additives is Y2O3 - Al2O3, which at the sintering stage forms yttrium aluminum garnet [12].

Recently, in several papers [13-16], it was shown that α- and β-SiC powders, when adding Al2O3 and Y2O3, are compacted by the liquid-phase sintering mechanism at a relatively low temperature of 1850-2000 °C. In the Y2O3-Al2O3 system 3 modifications were recorded:

- 3Y2O3-5Al2O3 (garnet phase) of the cubic syngony;
- Y2O3-Al2O3 (perovskite phase) of the orthorhombic system;
- the monoclinic phase 2Y2O3-Al2O3.

In the present work, the physicomechanical properties of materials based on SiC, obtained by hot pressing in argon with sintering additive Y2O3-Al2O3, were investigated.

IOP Conf. Series: Materials Science and Engineering **525** (2019) 012070 doi:10.1088/1757-899X/525/1/012070

2. Experimental procedure

In this work, SiC powders were used to create the material:

- SiC-1, Chernogolovka, ISMAN RAS, grain size: 1 micron;
- SiC-2, St. Petersburg, mark M5, grain size: 1 micron.

To the initial powders of silicon carbide was added sintering additive in the amount of 9 wt.%, consisting of oxides Y2O3 and Al2O3, in a quantitative ratio of 3.5 wt.% and 5.5 wt.% respectively. This additive allows obtaining more dense materials with a high level of mechanical properties and high crack resistance [17].

Table 1. Sintering additives in the composition

Al2O3	Y2O3	Humidity	General content
5.50 %	3.50 %	1 %	6 %

Sintering additive was obtained from Y2O3, Al (OH) 3 * 6H2O. Powders mixed wet method in the environment of isopropyl alcohol in a planetary mill, with a ratio material: balls - 1: 5 for 30 minutes. Then the powder was dried in an oven at a temperature of 80°C, and then it was sieved through a sieve with a mesh size of 63 µm. After then the mixture was placed in a muffle furnace for 4 hours with a temperature of 1000°C. Received oxide additive was added in the amount of 9 wt.% powders SiC-1 and SiC-2.

Two different compositions were prepared that differ in the type of silicon carbide. The original components were mixed in a planetary mill according to the technology outlined above. The finished powders are preformed into a roasting billet. The raw material is a disk with a diameter of 25 mm, the weight of the powder was 8 g. A dense material based on SiC and YAG was obtained at the Thermal Technology Inc. hot-pressing facility. model HP20-3560-20. Calcination was carried out in a graphite mold at a temperature of 1800, with an exposure time of 60 minutes, in argon medium. Unlike nitrogen, sintering in an argon atmosphere can reduce mass loss and grain growth [18]. The specific pressing pressure was 30 MPa.

Table 2. Compositions of different samples and processing conditions

Designation	Composition, wt.%		Firing mode
	SiC	$Y_2O_3 + Al_2O_3$	
SiC-1	91	9	T = 1850 °C, 60 min shutter speed, p = 30 MPa, in the Ar environment.
SiC-2	91	9	

The density of the material was determined by hydrostatic weighing in water. The true density was calculated by the additivity rule. For bending tests, specimens were produced. The research were carried out on an Instron 3382 unit with a load cell of 100 kg. Microhardness was measured on a Wolpert Wilson Instruments 401/402-MVD unit with a load of 0.98 N.

3. Results and discussion

Investigations of the influence of the type of sintering additive in the composition on the mechanical properties of the material and the microstructure were carried out. Result are presented in Table 3.

Table 3. Mechanical properties of materials

Composition	Firing mode	Bending stress, MPa	Microhardness, GPa	Density, g cm-3
SiC-1	T = 1850 °C, 60 min shutter speed, p = 30 MPa, in the Ar environment.	495,7	29,4	3,23
SiC-2		509,5	47,53	3,27

Flexural strength A sample of SiC-2 (St. Petersburg, M5, granule size: 1 μm) showed the best result - 509.5 MPa.

The SiC-2 sample (St. Petersburg, M5, granule size: 1 μm) has the best microhardness indicators - 47.5 GPa.

The resulting material has high rates of physico-mechanical characteristics. The best result was shown by a sample based on SiC-2 (M-5 grade, with a granule size of 1 μm) with an oxide additive YAG in the amount of 9 wt.%: $\sigma = 509.5 \pm 10$ MPa, HV = $47.5 \pm 0,2$GPa, $\rho = 3.27 \pm 0.1$ g / cm3.

4. Conclusion

In the work, the physicomechanical properties of materials based on SiC, obtained by hot pressing in argon with sintering additive Y2O3 - Al2O3, were investigated. The optimal amount of the activating additive is 9% by mass, its decrease leads to incomplete wetting by oxides of the SiC surface and, as a result, to a decrease in mechanical properties. Increasing the concentration of the additive to 20% leads to a decrease in the hardness of the material.

Further studies will focus on studying the microstructure of the obtained samples, as well as the effect of temperature on the physical and mechanical properties of the samples.

References

1. Luchin V., Maltsev P., Malyakov E. // Electronics: Science, Technology, Business // Silicon carbide: strategic material of the electronics of the future 1997 №3-4 p.61-64
2. Silicon carbide: technology, properties, application / OA Ageev, A.E. Belyaev, N.S. Boltovets, V.S. Kiselev, R.V. Konakova. Kharkov.: Publishing house "ISMA", 2010 - 5323.
3. Tairov Yu.M. and Tsvetkov V.F. Investigation of growth processes of ingotsof silicon carbide single crystals. J. Crystal Growth. 1978 Vol. 43 P. 209
4. Borovinskaya I. P. et al. (ed.). Concise Encyclopedia of Self-propagating High-temperature Synthesis: History, Theory, Technology, and Products. – Elsevier, 2017. C. 214-218.
5. Abavare E. K. K. et al. Indirect phase transition of refractory nitrides compounds of: TiN, ZrN and HfN crystal structures //Computational Materials Science. – 2017. – T. 137. – C. 75-84.
6. Perevislov S.N., Chupov V.D., Tomkovich M.V. Influence of activating additives of yttrium aluminum garnet and magnesian spinel on compactibility and mechanical properties of SiC ceramics // Materials Science. 2011 № 1 P.123-129.
7. Mulla, M.A. Reaction-inhibition during sintering of SiC with Al2O3 additions / M.A. Mulla, W.T. Thompson, V.D. Krstic // Canadian Metallurgical Quarterly, 1995 V. 34, № 4 P. 357–362
8. Suzuki, K., Pressureless sintering of SiC with addition of Al2O3, in SiC Ceramics, vol. 2 / K.Suzu-ki, S. Somiya, Y. Inomata // London: Elsevier. 1991 P. 163–182.
9. Martin, H.–P. Design of electrical and thermal properties of liquid phase sintered silicon carbide / H.–P. Martin, J. Adler // Proc. Materials Week 2001, Symposium K5 – Multifunctional Ceramics, Munich. 2001 P. 38
10. Martin, H.–P. Design of Electrical Properties of LPS–SiC / H.–P. Martin, J. Adler // Proc. Materi-als Week 2001, BMBF–Statusseminar "MaTech Highlights", Munich 2001 P. 76

IOP Conf. Series: Materials Science and Engineering **525** (2019) 012070 doi:10.1088/1757-899X/525/1/012070

11. Kirianov, A. Sintering and oxidation resistance of compositions in the SiC–La2O3–Al2O3–Cr2O3 system: I. Pressureless sintering of compositions in the SiC–La2O3–Al2O3–Cr2O3 system at low temperatures / A. Kirianov, A. Yamaguchi // Ceramics International. 2000 V.26 P. 441–446.
12. Sigl, L.S. Core/Rim Structure of Liquid-Phase-Sintered Silicon Carbide / L.S. Sigl, H.J.Kleebe // J. Am. Ceram. Soc. 1993 V.76. P. 773–776.
13. Farries P.M., Bullock E., Rawlings R.D. – Quantitative assessment of sintering behavior of silicon carbide with additions of alumina and yttria // J. Mater. Sci. Lett., 1999. v. 18. № 21. P. 1727-1730
14. Sciti D., Bellosi A. – Production and characterization of silicon carbide by liquid-phase sintering // Int. Ceram. J., 2000. v.18. № 2. P. 35-39.
15. Wang G., Krstic V.D. – Effect of Y2O3 and total oxide addition on mechanical properties of pressureless sintered бета-SiC // J. Mater. Sci. and Technol., 2003. v. 19. – № 3. P. 193-196
16. Balbo A., Sciti D., Bellosi A. – Pressureless liquid phase sintering of silicon carbide in function of the powders characteristics // Int. Ceram. J., 2004. v. 22. – № 4. P. 33-37.
17. Sciti, D. Effects of additives on densification, microstructure and properties of liquid-phase sintered silicon arbide / D. Sciti, A. Bellosi // J. Mater. Sci. 2000 V. 35 P. 3849–3855.
18. de Carvalho, M.T. The influence of nitrogen and argon atmospheres on the liquid phase sintering of silicon carbide / M.T. de Carvalho, O.M.M. da Silva, K. Strecker, C.R.M. da Silva // Key Engineering Ma-terials. 2001 V. 189–191. P. 126–131.

IOP Conf. Series: Materials Science and Engineering **525** (2019) 012071 doi:10.1088/1757-899X/525/1/012071

Synthesis of alumomagnesian spinel by mechanical activation method

N A Popova[1], E S Lukin[1], L T Pavlyukova[1], M A Sevostyanov[2], A V Leonov[2]

[1] Dmitry Mendeleev University of Chemical Technology of Russia, Miusskaya sq. 9, 125047 Moscow, Russia

[2] Baikov Institute of Metallurgy & Materials Science (IMET RAS), Moscow, Leninskiy prospect 49, Russia

Email: a_leonov.imet@list.ru

Abstract. The influence of the mechanical activation of the initial alumina and magnesium oxide powders on the formation of alumomagnesian spinel has been studied. It has been established that the mechanical activation of the starting powders accelerates the formation of alumomagnesian spinel, significantly reducing the temperature of the subsequent heat treatment.

1. Introduction

Alumomagnesian spinel is used as a structural and refractory material for modern technology, due to such important technological properties as high melting point, increased mechanical strength, high resistance to neutron irradiation and low neutron capture cross section with the formation of radioactive products [1].

Much attention is paid in the literature to the synthesis of stoichiometric $MgAl_2O_4$. The difficulty of obtaining alumomagnesian spinel from aluminum and magnesium oxides during their solid-phase reaction is the formation of the $MgAl_2O_4$ phase at sufficiently high temperatures above 1400 ° C [2]. Mainly chemical methods based on the processes of wet chemistry (coprecipitation, sol-gel methods, SHS processes, hydrothermal method, Pechini method, etc.) are used to obtain a reactive and single-phase material based on $MgAl_2O_4$ [3,4]. However, these chemical methods of obtaining imply the use of chemicals that are sensitive to the environment (sol-gel process, Pechini method), the possibility of leaching cations during filtration (coprecipitation process), complex instrumentation (hydrothermal method). Mechanically activated synthesis exceeds the methods of wet chemistry, allows mixing and grinding of components, and solid-phase reactions of a multicomponent system in one process, significantly reducing the subsequent heat treatment temperature. Mechanically activated synthesis is actively used to obtain various ceramic materials of functional purpose [5].

The main objective of this study is to establish the effect of mechanical activation in the dry grinding of magnesium and aluminum oxides on the heat treatment temperature in order to synthesize alumomagnesian spinel. The production of $MgAl_2O_4$ is possible directly during grinding as a result of physicochemical processes associated with the effect of mechanical energy on powder particles.

2. Materials and methods

The initial components are γ-form of aluminum oxide (99.999%, produced by Nanocorund LLC, Sarov, the average size of the agglomerates is 5 μm) and magnesium oxide (the average size of the aggregates is 4 μm). Highly dispersed magnesium oxide was obtained by thermal decomposition of magnesium carbonate. To obtain stoichiometric alumomagnesian spinel, γ-alumina and magnesium oxide were taken in a molar ratio of 1: 1. Mechanical activation was performed in a planetary mill (Fritsch Pulverisette 5) at room temperature in air, in corundum drums (volume 250 ml) with grinding corundum bodies (diameter 15 mm), the rotation speed of the mill was 400 revolutions per minute. The duration of grinding varied from 30 to 120 minutes (composition M30, M60 and M120), the ratio of material and grinding media was 1:10. The obtained powders after mechanical activation were investigated by the methods of differential thermal (EXSTAR TG / DTA 7300 (SII)) and X-ray phase analyzes (D2 PHASER, Bruker). Heat treatment of the mixtures was carried out in the temperature range of 500-1400°C, holding at the maximum temperature was 1 hour. The crystallite sizes for all the studied powders were calculated from the broadening of the peaks in the diffractograms according to the Scherrer equation: $D = K \cdot \lambda / (\beta \cdot \cos\theta)$, where D is the average particle size, K is the Scherrer constant or form factor (0.89–0, 90), λ is the wavelength of the incident X-ray radiation (copper - 0.15406 nm), β is the full width of the peak at the level of half the intensity, θ is the angle of diffraction.

3. Results and discussion

Figure 1 shows the X-ray patterns of the obtained powders after dry grinding at different duration mechanical activation. Diffraction peaks correspond to the MgO (JCPDS No. 78-0430), gamma-Al_2O_3 (JCPDS No. 77-0403) and $MgAl_2O_4$ (JCPDS No. 73-1959) phases. As a result of mechanical action on the charge (γ-Al2O3: MgO) over time, the intensity of the diffraction peaks of γ-alumina and magnesium oxide is significantly reduced with simultaneous broadening of the peaks and is primarily due to a decrease in the particle size of the charge during the grinding process. The formation of the primary crystallites of alumomagnesian spinel phase on the surface of oxide particles occurs during the first 30 minutes of mechanical activation. An increase in the time of mechanical treatment of powders up to 120 minutes leads to an increase of aluminomagnesian spinel phase during mechanical action, as well as on the size and degree of crystallinity of the initial particles of the mixture.

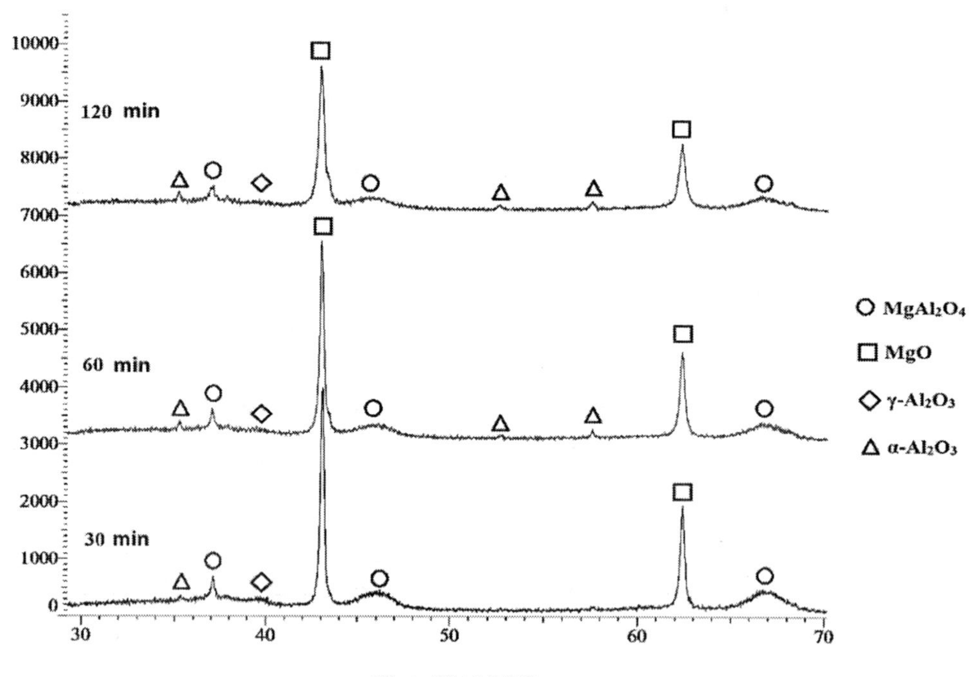

Figure 1. Radiographs of mechanically activated mixture γ-Al_2O_3: MgO in a molar ratio of 1:1 after 30, 60 and 120 minutes of grinding.

When the charge oxide is activated, structural destruction processes occur, causing amorphization of the surface layer of particles. By the nature of radiographs (Figure 1), the degree of crystallinity of the components of the mixture decreases with increasing mechanical exposure time of 30, 60, 120 minutes, respectively. The process of mechanical activation of the original nanosized powders can be described in three stages. At the first stage the initial oxide particles undergo a certain degree of amorphization during grinding, but a solid-phase reaction towards the formation of the product still does not occur. Further mechanical action, characterized as the second stage of activation, contributes to the formation of nuclei of a new phase and its crystallization. At the final stage, the grinding has a mechanical effect only on the synthesized powders, forming structural defects in the particles. Comparison of the curves of differential thermal analysis of mechanically activated powders showed the effect of mechanical activation time on the synthesis of $MgAl_2O_4$ (Figure 2). The endothermic effect observed up to 350 °C is caused by the loss of volatile surface water, as mechanically activated magnesium oxide has a higher rate of adsorption of water molecules and, possibly, chemisorption of gas molecules on the surface of particles. The DTA curves for M60 and M120 powders exhibit an exothermic effect in the temperature range of 755–930 ° C, which is caused by the solid-phase reaction of unreacted magnesium oxide and aluminum oxide and the residual redistribution of cations between the octahedral and tetrahedral positions in the spinel structure, because the formation of the MgAl2O4 phase has occurred at the start of machining. The appearance of crystallites of alumomagnesian spinel during machining significantly reduces the subsequent temperature of heat treatment of powders. Moreover, with an increase in the time of mechanical activation, the manifestation of the exo-effect becomes more pronounced. The subsequent heat release at ~ 1000°C, apparently, is the result of a change in the degree of defectiveness of particles towards their complete disappearance. On the derivatogram of M30 powder, there is one exothermic effect, which corresponds to the residual solid-phase reaction with the formation of defective particles of alumomagnesian spinel. Further removal of powder particle defects will either lead to an increase in the heat treatment temperature, or to an increase in the exposure time at 1000 ° C.

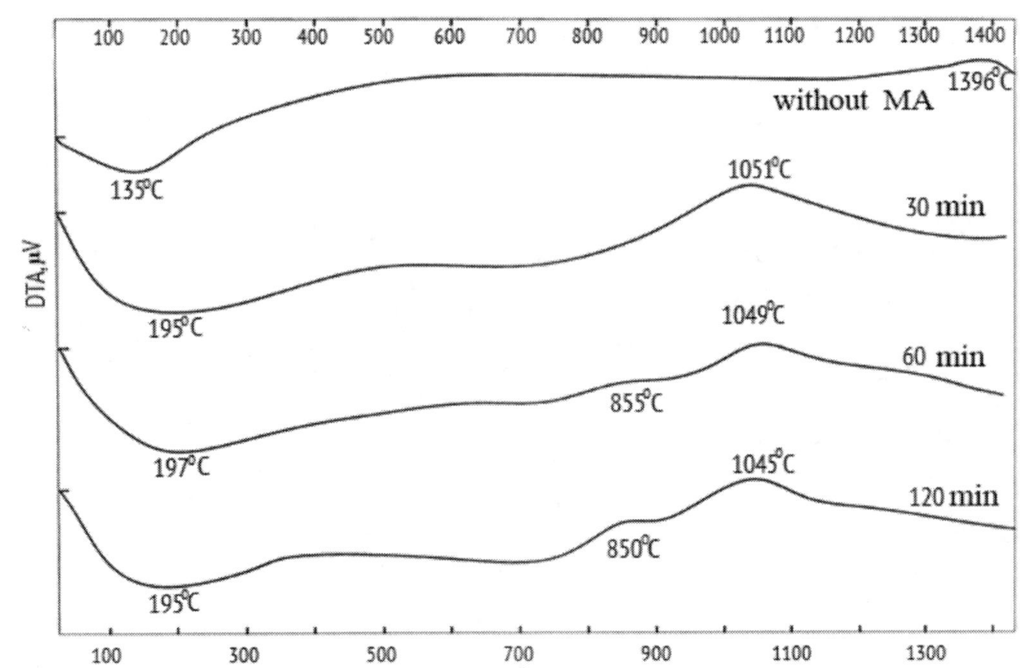

Figure 2. DTA curves of the studied powders.

In order to identify the effectiveness of mechanical activation on the powder, the composition of magnesium oxide and aluminum oxide was prepared, also taken in a 1: 1 molar ratio. The grinding of this composition was carried out on a roller mill for 6 hours. The speed of rotation of the mill was 48-50 revolutions per minute. The result of the DTA revealed the temperature of the synthesis of alumomagnesian spinel for the composition, not subjected to mechanical activation, as 1400 °C (Figure 2). Mechanical activation promotes subsequent thermal activation, reducing the heat treatment temperature of the powder by 400 ° C. The dependence of crystallite size (CSR) and the crystallization of alumomagnesian spinel on temperature was studied on the composition of M30 to exclude the contribution of free corundum to the process of formation of MgAl2O4. With an increase in the synthesis temperature, the crystals grow (Figure 3 (a)), which can be seen in the diffraction patterns as a decrease in the width of the peaks and an increase in their intensity (Figure 3 (b)).

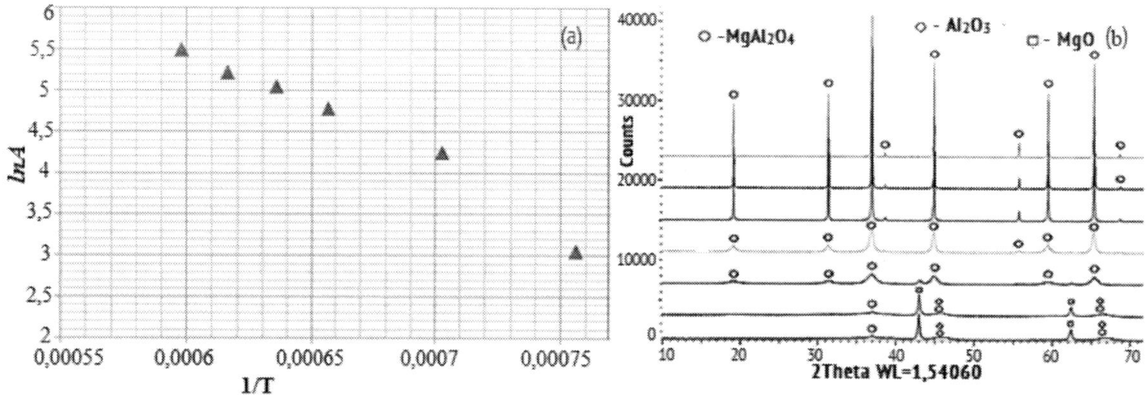

Figure 3. (a) Dependence of the average crystallite size on the synthesis temperature, (b) Radiographs of the heat-treated composition M30.

Mechanical action provides a sufficiently rigid impact on the structure of the initial oxides, which leads to excessive defects (disorder) in the crystal lattice and, accordingly, the longer the particles are exposed to the higher rate of formation of the new compound. The morphology of alumomagnesia spinel powders after mechanical activation is shown in Figure 4.

Figure 4. (a) Photographs of the microstructure of the powders (a) M30; (b) M60; (c) M120.

The powders are subject to a sufficiently strong agglomeration, increasing with an increase in the duration of the mechanical action. Regardless of the shape of the agglomerates, nanoparticles of the powder are clearly visible in the images. The average size of the aggregates for 30 minutes of grinding was ~ 50 nm, for 60 minutes ~ 120 nm, for 120 minutes ~ 150 nm. With an increase in the time of mechanical action on the powder, the density and size of the aggregates increases, in the SEM image (Fig. 4 (c)). For 120 minutes of mechanical activation, the morphology of the rounded agglomerates is characterized by a tighter fit of the particles to each other.

4. Conclusions

Thus, it can be concluded that the use of mechanical activation of the initial γ-alumina and magnesium oxide powders accelerates the formation of alumomagnesian spinel, significantly reducing the temperature of the subsequent heat treatment. Mechanical activation increases the surface of the reacting phases (the reaction zone) and thus facilitates the course of reactions towards the formation of the product. An increase in the time of mechanical action leads to a decrease in the particle size and the degree of crystallinity of the starting oxides.

References

[1] R.J. Bratton, Sintering and grain-growth kinetics of MgAl2O4 // *J. Am. Ceram. Soc.,* 2008, **52. № 8**. P.141-143.
[2] W. Bakkar, J.G. Lindsay, Reactive magnesia spinel, preparation and properties // *Am. Ceram. Soc. Bull.,* 2010, **46. № 11**. P. 1094 – 1097.
[3] Carmen Baudin, Rafael Martinez, and Pilar Pena, High-Temperature Mechanical Behavior of Stoichiometric Magnesium Spinel // *J. Am. Ceram. Soc.,* 2009, **42. № 4**. P.235-242.
[4] W. Bakkar, J.G. Lindsay, Combustion synthesis, characterization and sintering behavior of magnesium aluminate ($MgAl_2O_4$) powders// *J. Am. Ceram. Soc.,* 2011, **46. № 4**. P.111-123.
[5] Huang ZC, Zhang L, Pan W., Mechanochemical synthesis of barium titanate// *J. Eur. Ceram Soc.* 2012, **34**. P.783–790.

IOP Conf. Series: Materials Science and Engineering **525** (2019) 012072 doi:10.1088/1757-899X/525/1/012072

Reinforced composite materials based on silicon carbide and silicon nitride

S N Perevislov[1], A S Lysenkov[2], D D Titov[2], M G Frolova[2] and M.V. Tomkovich[3]

[1]Institute of Silicate Chemistry I.V. Grebenshchikov Russian Academy of Sciences, St. Petersburg, Russia
[2] Baikov Institute of Metallurgy and Materials Science, Russian Academy of Sciences, Moscow, Russia
[3]Physico-Technical Institute A.F. Ioffe Russian Academy of Sciences, Saint-Petersburg, Russia

E-mail: perevislov@mail.ru

Abstract. In this work, ceramic materials based on silicon carbide and silicon nitride with an additive were obtained to increase the mechanical and operational properties of fibers and whisker crystals of SiC and Si_3N_4. Studied the basic properties and applications of reinforced materials.

1. Introduction

Dense materials based on silicon carbide is produced by the reaction, liquid phase sintering and hot pressing. Despite the wide spread of a method of manufacturing articles for constructional purposes, each of the methods of producing SiC (Si_3N_4) – materials has both its advantages and disadvantages, which determine their great importance to obtaining products of a wide range of applications.

Reaction-sintered body (impregnation) silicon carbide (SiSiC) the most common form of ceramic materials [1, 2], is prepared by impregnating porous preforms from silicon carbide particles and carbon black technical silicon melt. For SiSiC-keramiki characteristic substantially nonporous structure ($P \leq 0{,}5$ %) material in which main phase grains are silicon carbide primary SiC^{I}, the phase between them - the secondary carbide SiC^{II}. As a part of the finished product is also present unreacted 5 to 15 % residual silicon. A positive feature of SiSiC-materials is sintering at a relatively low temperature (~ 1600°C) the use of large SiC starting powders, dimensionalty stability during sintering. The disadvantages in the material a high content residual silicon, which limits the temperature region of use (Table 1).

A liquid phase-sintered bodies (LPSSiC) obtained by sintering silicon carbide ($d_{0,5} \leq 1$ μm) with oxide additives (10–20 wt.%) At temperatures 1850–1950°C [3–5]. The resulting materials are more corrosion-resistant and temperature stable (working temperature up to 1450°C in an air atmosphere). However, high temperature sintering, unlike SiSiC and large shrinkage limit number of products manufactured by this method (table 1).

Hot-pressed silicon carbide (HPSiC) and silicon nitride (HPSN) prepared by sintering fine powder α-SiC or β-Si_3N_4 in the presence of small amounts (up to 5 wt.%) of oxides [6, 7]. The resulting material has a higher level of mechanical characteristics (Table 1). The disadvantages of this method can apply high energy and the inability to obtain articles of complex geometrical shape without additional machining, which limits the industrial manufacture of articles by hot pressing.

Table 1. Physical and mechanical properties and high temperature the ceramic

Material	$\rho\pm$ 0,02 (g/cm^3)	$\sigma_{fl}^{20}\pm$ 10 (MPa)	$\sigma_{fl}^{1200}\pm$ 10 (MPa)	$\sigma_{comp}^{20}\pm$ 20 (MPa)	$K_{IC}\pm$ 0,2 (MPa·m$^{1/2}$)	$HV\pm$ 0,2 (GPa)	Working temperature Protective atmosphere	Working temperature Air atmosphere
SiSiC	3.05	400	150	1200	3.5	22	1300	1200
LPSSiC	3.20	500	400	2000	5.5	21	1600	1450
HPSiC	3.15	750	500	2200	6.5	26	1700	1500
HPSN	3.20	900	500	2800	7.5	20	1600	1400

ρ – density; σ_{fl}^{20} – flexural strength, measured at a temperature of 20 °C; σ_{fl}^{1200} – flexural strength, measured at a temperature of 1200 °C; σ_{comp}^{20} – the compressive strength measured at 20 °C; K_{IC} – fracture toughness; HV – Vickers hardness

The purpose is to obtain reinforced structural fibers and whiskers of materials based on SiC and Si_3N_4 covalent compounds.

2. Materials and Method

The density and porosity of the samples was determined by hydrostatic weighing; Elastic modulus was determined method of measuring the resonant frequency of longitudinal vibrations on the "ZVUK–130"; strength at three-point bending and compression on the installation «Shimadzu AG-300kNX»; Vickers hardness – by pressing into the material of Vickers pyramid, on the hardness installation TR–7r 1; crack resistance coefficient - measuring the length of cracks emanating from Vickers pyramid base. The microstructure was examined using a scanning electron microscope «Quanta 200».

3. The results and discussion

Conventionally, structural materials consisting of covalent compounds reinforced structure can be divided into three types: 1) Reinforced fiber (SiC_f) SiSiC-materials; 2) hot-pressed materials based on SiC and Si_3N_4, reinforced whisker (SiC_w and Si_3N_{4w}); 3) Materials based on SiC and Si_3N_4 advanced samoarmirovannoy structure during sintering.

1) randomly distributed fibers SiC_f are in the «SiC–C»preform, which are impregnated with a silicon melt. The maximum density after impregnation of the porous billet can not be obtained, and in the contact areas SiC_f is the surface porosity (Figure 1). For reinforced SiSiC-materials used fibers SiC, as a continu-ous filament, obtained by pyrolysis polikarbosilanovoy thread (Japan) with a diameter d = 140 mm, which were crushed to size (length) l = 0,5–1,0 mm and introduced into the composition of the feedstock composition SiSiC-material.

With increasing content SiC_f composed material increases its porosity and strength properties are reduced, but fracture toughness value is increased (to 12 vol. % SiC_f). With further increase of the content SiC_f, due to increased porosity, reduced level of mechanical properties (Table 2).

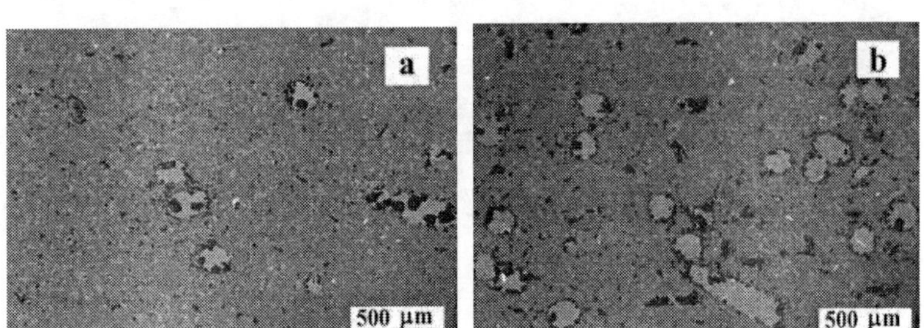

Figure 1. Microstructures SiSiC-material: **a** – 4 % vol. SiCf; **b** – 12 % vol. SiCf

Table 2. Physical-mechanical properties of the reinforced ceramic

Material	Fibers SiC_f	Whiskers	$\rho \pm 0{,}02$ (g/cm³)	$\sigma_{fl}^{20} \pm 10$ (MPa)	$\sigma_{comp}^{20} \pm 20$ (MPa)	$K_{IC} \pm 0{,}2$ ($MPa{\cdot}m^{1/2}$)	$HV \pm 0{,}2$ (GPa)
	4 %	–	3,00	400	1200	4,0	20,0
SiSiC	8 %	–	2,92	320	900	5,0	–
	12 %	–	2,85	280	750	5,5	–
	–	4 % SiC_w	3,12	850	2200	7,0	24,0
HPSiC	–	8 % SiC_w	3,10	950	2300	8,0	22,0
	–	12 % SiC_w	3,07	900	2200	7,5	20,5
	–	4 % Si_3N_{4w}	3,20	1000	2800	8,5	19,0
HPSN	–	8 % Si_3N_{4w}	3,15	1100	2800	9,0	18,0
	–	16 % Si_3N_{4w}	3,05	970	2550	8,2	–

2) In the hot-pressed materials based on silicon carbide and nitride are introduced up to 25 vol. % SiC_w and Si_3N_{4w}. Unlike reinforced fibers SiC_f materials SiSiC hot-pressed materials are formed to a higher density. With increasing quantities SiC_w and Si_3N_{4w}, increases the level of mechanical properties of materials, despite the slight decrease in density. When administered whiskers > 10 vol. % increased porosity materials and reduced level of mechanical properties (Table 2). The micrographs of the microstructure of whiskers and hot-pressed materials with addition SiC_w and Si_3N_{4w} shown in Figure 2 and Figure 3.

3) The widespread in the industry as a structural material was sintered silicon nitride. During the sintering process occurs growth of reinforcing structure SSN grains (Figure 4a), which gives increased strength properties of the sintered material [8]. The disadvantage can be considered SSN decrease in hardness compared with LPSSiC-materials. Sintered silicon nitride has been successful as a structural material working in extreme conditions of erosive environments (hot gas streams at elevated temperature – 1400 °C) (Table 3).

Figure 2. Micrographs whiskers: **a** – SiCw; **b** – Si_3N_{4w}

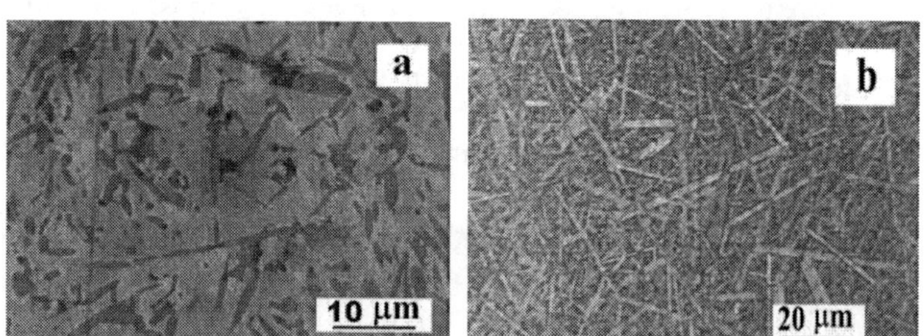

Figure 3. The microstructure of materials: **a** – HPSiC+8 vol. % SiC_w; **b** – HPSN + 16 vol. % Si_3N_{4w}

Figure 4. Photomicrographs of materials: **a** – SSN; **b** – RSSN

Reaction-sintered (nitrided), silicon nitride (RSSN) obtained sintered material consisting of a composition «Si_3N_4–Si» in N_2 atmosphere at a temperature ≥ 1600 °C. In the process of nitriding, a structure consisting of grains of silicon nitride bonded thin elongated grains $Si_3N_4^{II}$ (analogs whiskers) (Fig. 4b). RSSN-materials are resistant to high temperatures. Reinforced structure during sintering makes it lasting the material up at work to high temperatures (1400 °C) (Table 3). RBSN the advantages include lower sintering temperatures, little shrinkage, the use of coarse raw powders and the possibility of forming large products of complex geometric shape.

Table 3. Physical and mechanical and high-temperature properties of hot-pressed materials

Material	$\rho \pm 0{,}02$ (g/cm^3)	$\sigma_{fl}^{20} \pm 10$ (MPa)	$\sigma_{fl}^{1200} \pm 10$ (MPa)	$\sigma_{comp}^{20} \pm 20$ (MPa)	$K_{IC} \pm 0{,}2$ (MPa·m$^{1/2}$)	$HV \pm 0{,}2$ (GPa)	Working temperature	
							Protective atmosphere	Air atmosphere
SSN	3,18	650	400	1700	6,0	18	1450	1350
RSSN	2,85	300	250	1300	4,5	16	1600	1500

Adding to the initial particles consisting of powder α-SiC, a small amount of β-SiC powder leads at a temperature> 1600 ° C to activate the growth of elongated grains (Figure 5). Grain growth is carried out by β → α phase transition silicon carbide. The presence of α-SiC powder in the starting material increases the rate of β → α junction, which is fully completion at the extended sintering time (≥ 4 h).

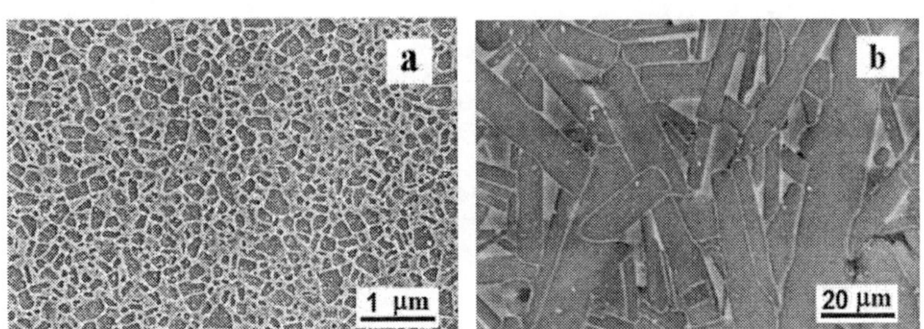

Figure 5. The microstructure of the sintered body 90 wt. % β-SiC + 10 wt. % YAG, at different sintering durations: **a** – 1 h; **b** – 8 h

SiC-grain growth speed is determined by particle size of the main phase and the amount of oxides. The growth of elongated grains in the process of β → α phase transition reduces the material strength. Fracture toughness is inversely related to strength (see Table 4). Materials with large grains can be destroyed on the mechanism of grain fracture, thereby defining a high mechanical properties self-reinforcing coarse grain SiC-ceramic.

Materials without the addition of β-SiC has low values K_{IC}. The reinforced ceramics crack occurs in contact with an elongated grain encloses it or pass through it, breaking the grain and loses speed, which is characteristic for the mechanism of grain fracture of ceramics [9].

Table 4. Physico-mechanical properties self-reinforcing ceramics based on SiC

Material	Contents YAG (wt. %)	Duration of sintering time (h)	$\rho \pm 0{,}02$ (g/cm^3)	$\sigma_{fl}^{20} \pm 20$ (MPa)	$K_{IC} \pm 0{,}2$ (MPa·m$^{1/2}$)
5 wt. % α-SiC + 85 wt. % β-SiC	10	1	3,23	505	4,5
		2	3,20	470	4,8
		4	3,15	420	5,4
		8	3,10	380	6,0

4. Conclusions

Applying the methods of reinforcing fibers carbon and silicon carbide and growth of elongated grains during sintering using different technology of dense ceramic materials, or adjusting the particle size distribution of the material or structure can receive new product with complex mechanical and performance properties.

References

[1] Nesmelov D D, Perevislov S N 2015 *Glass and Ceramics* **9-10** 313.
[2] Perevislov S N, Shcherbak P V, Tomkovich M V 2018 *Refractories and Industrial Ceramics* **1** 32.
[3] Perevislov S N, Nesmelov D D 2016 *Glass and ceramics* **7** 15.
[4] Perevislov S N, Panteleev I B, Shevchik A P, Tomkovich M V *Refractories and Industrial Ceramics* **5** 577.
[5] Perevislov S N, Bespalov I A, Tomkovich M V 2018 *Refractories and Industrial Ceramics* **4** 359.
[6] Perevislov S N, Lysenko A S, Titov D D, Tomkovich M V 2017 *Inorganic Materials* **2** 206.
[7] Lukianova O A, Khmara A N, Perevislov S N, Kolesnikov D A, Krasilnikov V V 2018 *Ceramics International* **9** 624.
[8] Lysenkov A S, Kim K A, Titov D D, Frolova M G et al. 2018 *J. Phys.: Conf. Ser.* **1134** 012036.
[9] Kim J-Y, An H-G, Kim Y-W, Mitomo M 2000 *J. Mater. Sci. Lett.* **15** 3693.

IOP Conf. Series: Materials Science and Engineering **525** (2019) 012073 doi:10.1088/1757-899X/525/1/012073

Liquid-sintered SiC based materials with additive low oxide oxides

S N Perevislov[1], A S Lysenkov[2], D D Titov[2], K A Kim[2], M V Tomkovich[3], D D Nesmelov[4] and M A Markov[4]

[1]Institute of Silicate Chemistry I.V. Grebenshchikov Russian Academy of Sciences, St. Petersburg, Russia
[2]Baikov Institute of Metallurgy and Materials Science, Russian Academy of Sciences, Moscow, Russia
[3]Physico-Technical Institute A.F. Ioffe Russian Academy of Sciences, St.-Petersburg, Russia
[4]St.-Petersburg State Institute of Technology (Technical University), St.-Petersburg, Russia

E-mail: perevislov@mail.ru

Abstract. By the method of free liquid-phase sintering, dense materials based on silicon carbide ($P \leq$ 0.5%) were obtained. In the preparation of charge powders, the method of co-precipitation of oxides on silicon carbide powder was used. As a sintering additive, a eutectic mixture of the components MgO, Y_2O_3, and Al_2O_3 was used, corresponding to the composition on the garnet-spinel line. The resulting materials have a high level of mechanical properties: E = 410 GPa; σ = 680 MPa; K_{IC} = 6.0 MPa*m$^{1/2}$; HV = 21.5 GPa, approaching the level of properties of hot-pressed silicon carbide materials.

1. Introduction

The classical method of liquid phase sintering of SiC with oxide additives includes mechanical mixing micron powders, molding the preforms and high temperature sintering materials, what It has several disadvantages: uneven mixing powders of oxide between the silicon carbide particles, the complexity of a homogeneous distribution of small amounts ($\leq$ 5 wt. %) of oxides, The impossibility of a uniform distribution of the oxide components in strictly eutectic ratio. Alternative mechanical components may be kneaded mortar mix techniques (sol-gel method, precipitation and co-precipitation from salt solutions, colloidal activation method, and others.). Which was investigated by some researchers [1–4].

The purpose of this paper is to obtain on the basis of materials SiC with additives of oxides MgO, Y_2O_3 and Al_2O_3, deposited on the silicon carbide surface by means of co-precipitation from salt solutions. Basic work task - to achieve a high uniformity of distribution of oxide between the SiC particles and, hence, a high density material SiC–MeO.

2. Starting materials and methods

The problem of obtaining SiC–MgO–Al_2O_3 homogeneous mixture compositions, SiC–Y_2O_3–Al_2O_3 and SiC–MgO–Y_2O_3–Al_2O_3 can be solved as follows: depositing on 1/3 of SiC powder of magnesium hydroxide as a precipitant using NaOH solution; depositing on 2/3 of SiC powder and aluminum hydroxides of yttrium, using as the precipitant NH_4OH solution. For the preparation of solutions of magnesium nitrate salt was used, the yttrium and aluminum.

As activators during sintering SiC-material using a mixture of oxides of yttrium and aluminum, forming the step sintering aluminum-yttrium garnet (YAG) and magnesium, yttrium and aluminum, forming a ternary mixture of oxides corresponding to the eutectic composition on the line garnet-spinel ternary system diagram of state $MgO–Y_2O_3–Al_2O_3$ (MYA) [5]. The solutions were prepared based on the weight percent based on the material composition 80–95 wt. %. SiC + 5–20 wt. % MeO.

Additional studies to obtain fine-grained oxide powders systems $Y_2O_3–Al_2O_3$ and $MgO–Y_2O_3–Al_2O_3$, held at heat treatment hydroxide at a temperature of 600–700 °C (for $Y_2O_3–Al_2O_3$ system) and 400–500 °C (for $MgO–Y_2O_3–Al_2O_3$ system). The obtained X-ray amorphous synthesized powders have a large specific surface area – 57±3 m^2/g (for $Y_2O_3–Al_2O_3$ system oxides) and 61±5 m^2/g (for $MgO–Y_2O_3–Al_2O_3$ system oxides), with increasing temperature and time of heat treatment specific surface area powders decreases.

The samples heat treated in a vacuum furnace at a temperature of 700 °C. To obtain a molding powder SiC+MeO agglomerates crushed pieces in the cone inertial crusher and ground in a planetary mill to size $d_{0.5}$ = 1 μm. Feedstock composition SiC+MeO, obtained by co-precipitation, kneaded 2 % solution of an organic binder and pelletized. From the powders were compressed samples of size of 6×6×60 mm at a pressure of 100 MPa, was dried and sintered in a vacuum furnace in the Ar atmosphere at 1850±10 °C (materials with additive YAG) and 1800±10 °C (MYA materials with additive).

X-ray phase analysis was performed on diffractometer «Rigaku Smartlab 3» with CuKα-radiation and a Ni-filter; density and porosity of the sintered samples was determined by a hydro-static weighing; Study of the microstructure of the sintered body and fraktogrammy, determining the size of silicon carbide grains and crystals synthesized oxides as well as X-ray fluorescence analysis was performed on an electron microscope «Quanta 200». The grain size distribution was studied using a program «Thixomet Pro» program. The flexural strength was studied in the installation «ShimadzuAG-300kNX». The equipment provided «Engineering Center» SPbSTI (TU).

3. Results of the study

The morphology of the composite powder composition of 85 wt %. SiC + 15 wt. % MYA obtained by co-precipitation, shown in Fig. 1, where larger particles (≈ 1 μm) is SiC, small (50–200 nm) oxides. The distribution of magnesium oxide, yttrium and aluminum in the composite powder under stirring from the solution is more uniform. Uniform distribution of MYA and YAG oxides over the surface of SiC powder helps to reduce the melting point of the oxide melt and reduce the liquid-phase sintering temperature of the ceramic while maintaining the high level of physical and mechanical properties of SiC-based composite materials.

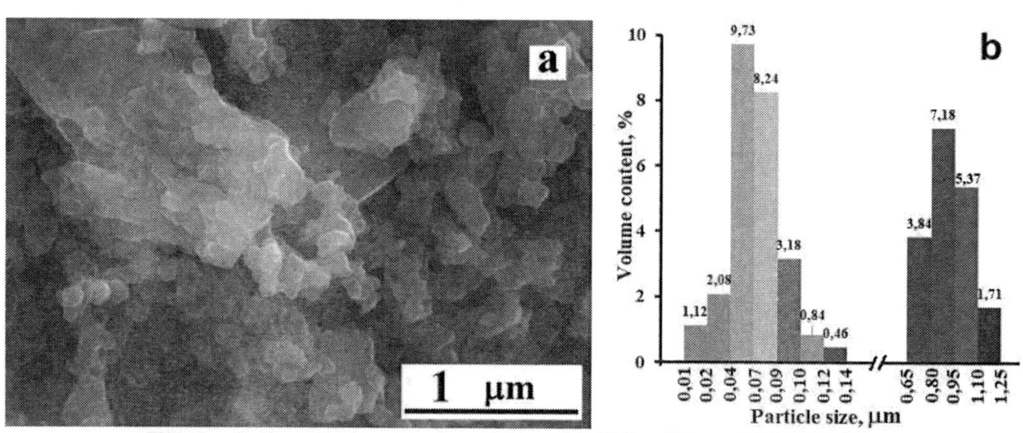

Figure 1. Composite powders 80 wt. % SiC + 20 wt. % MYA (**a**).
Histogram distribution of powder particle size (**b**)

IOP Conf. Series: Materials Science and Engineering **525** (2019) 012073 doi:10.1088/1757-899X/525/1/012073

Oxides of Y_2O_3 and Al_2O_3 obtained by co-precipitation, react to form aluminum yttrium garnet at a temperature significantly lower (≈1000–1050 °C) [4], than the temperature of fusion of micron-sized YAG powders (≈1500–1550 °C) [6]. For the system $MgO–Y_2O_3–Al_2O_3$ characteristic similar dependence reduce synthesis temperature to 900–950 °C ceramics based on the SiC deposited by co-precipitation oxides sintered to a dense state at a temperature on the 80–100 °C below the temperature of preparation of standard liquid-phase-sintered SiC-material powders of micron-sized – 1930 °C (with added YAG) [7, 8] and 1860 C (with addition MYA) [5]. μm

X-ray phase analysis of the sintered samples showed that the main phase is the α-SiC (6H modification), the secondary phase - $Y_3Al_5O_{12}$ and the mixture $Y_3Al_5O_{12} + MgAl_2O_4$. In this case there was no individual phases MgO, Y_2O_3 or Al_2O_3, which means that the phase constituents MYA and YAG completely synthesized.

Analyzing the mechanism of liquid phase sintering in accordance with the data [6], we may conclude that during sintering stoichiometric YAG and MYA synthesized at temperatures ≈ 1000 and 900 °C, a shrinkage of the material due to the sliding and rotation of particles SiC relative to each other (at a temperature ≈ 1500–1600 °C), the material is compacted due to mass transfer and contact sintering of SiC particles (≈ 1800–1850 °C).

Densities, porosity, grain size of the SiC and the bending strength of liquid-phase-sintered SiC-materials are given in Table 1. The density obtained SiC-materials increases with increasing content of the oxide additive. The minimal porosity of the material is fixed SiC + 20 wt. % MYA, sintered at (1800 ± 10 °C): P = (0,4 ± 0,1)%, at a density ρ = 3,36 ± 0,02 g/cm^3 (Table 1). By comparing the properties of standard materials SiC+ YAG (MYA), sintered at 1950 and 1860 °C, with new materials prepared at 1850 and 1800 °C, it can be noted that at the lower temperature sintering process is completed totally. At lower temperatures, mass transfer processes are slowed down more that defines a smaller grain size silicon carbide for material obtained using the co-precipitation method (Table 1). Materials consisting of small grains SiC between which thin layers of oxides are formed, it has a high level of strength properties (Table 1).

When increasing oxide content SiC-material flexural strength (σ_{fl}) increases monotonically. The maximum values of the flexural strength attained in a material of composition 80 wt. % SiC + 20 wt %. MYA, Sintered at 1800 °C – σ_{fl} = 750 ± 10 MPa (Table 1). By comparing the obtained result with values of standard materials [5, 6], may be mentioned improve-ment of σ_{fl} by 20–30 %.

Table 1. Physical and mechanical properties of liquid-phase-sintered SiC–MeO materials

Oxide content	Sintering temperature (°C)	Density ρ ± 0,02 (g/cm^3)	Porosity $P \pm 0,1$ (%)	Grains size of SiC $d_{0,5} \pm 0,1$ (μm)	Flexural strength $\sigma_{fl} \pm 10$ (MPa)
5 wt. % YAG	1850	3,19	2,7	1.4	450
10 wt. % YAG		3,27	2,0	1.6	510
15 wt. % YAG		3,35	1,4	2.2	600
20 wt. % YAG		3,43	0,9	2.7	660
5 wt. % MYA	1800	3,20	1,9	1.1	510
10 wt. % MYA		3,21	1,0	1.2	570
15 wt. % MYA		3,32	0,8	1.4	640
20 wt. % MYA		3,36	0,4	1.6	730

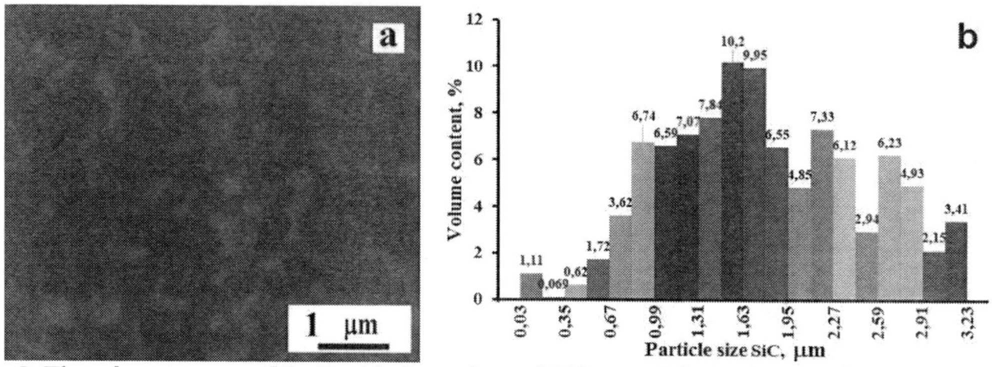

Figure 2. The microstructure of the liquid-phase-sintered SiC-material, supplemented 20 wt. % MYA (**a**), the histogram distribution of SiC grain size (**b**)

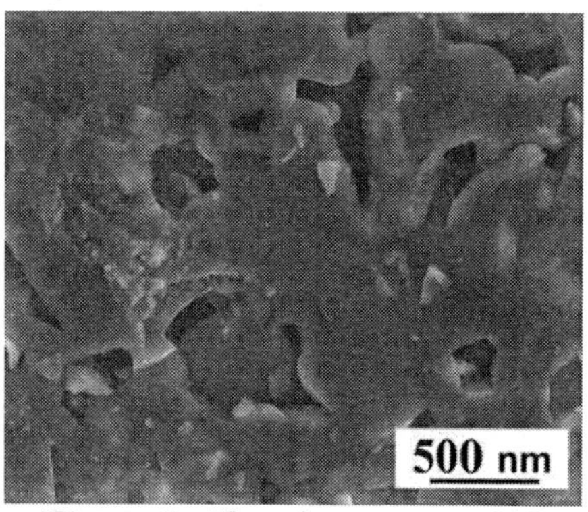

Figure 3. The microstructure SiC ceramics containing 5% by weight. MAY

Figure 2 shows the microstructure of a liquid phase sintered SiC-material comprising 20 wt. % MYA. Grains of SiC, mainly correct (round) form, with uniformly distributed between the oxide phase. Oxides well moisten the surface of the silicon carbide grains materials do not contain SiC particles in the sintered agglomerate.

Fraktogramma destruction SiC-ceramics, sintered with the addition of 5 wt. % MAY at 1800 °C is shown in Figure 3. The crystal size of the oxide phase is much smaller SiC grains, which contributes to a more uniform distribution between the grains of silicon carbide and has a positive effect on the compressibility of liquid-phase-sintered materials.

Elemental analysis of X-ray fluorescence (EDS) showed that silicon carbide grains include phase Si, Mg, Y, Al and O, which indicates that the SiC has a structure «core-shell» where «shell» is a solid oxide solution carbide silicon (containing small amounts of solute elements Mg, Y, Al and O), which confirmed earlier results obtained in [6]. Intergranular phase is characterized by the elements Mg, Y, Al, O and Si, appear due to the dissolution of the SiO_2 film, on the surface of the SiC particles, in the oxide melting with the formation of a siloxisilicate melt.

Application of the method of co-precipitation of the salt solution for a uniform deposition of oxides on the SiC powder to reduce their total number and to receive the high-density SiC-materials with high mechanical characteristics.

The results are a prerequisite for further research in the field of producing high density materials based on SiC, in particular, using as activating additives oxides fusible eutectic systems $CaO–Al_2O_3$, $CaO–TiO_2$, $MnO–TiO_2$ et al.

Acknowledgments

The reported study was funded by RFBR project No 18-38-00383.

References

[1] Guo X-Zh, Yang H 2005 *J. Zhejiang Univ SCI* **3** 213.
[2] Perevislov S N, Nesmelov D D 2016 *Glass and ceramics* **7** 15.
[3] Nesmelov DD, Kozhevnikov O A, Ordan'yan S S, Perevislov S N 2017 *Glass and Ceramics* **2** 43.
[4] Nien Y-T, Chen Y-L, Chen I-G, Hwang C S et al. 2005 *Mater. Chem. Phys* **93** 79.
[5] Perevislov S N, Panteleev I B, Shevchik A P, Tomkovich M V 2018 *Refractories and Industrial Ceramics* **5** 577.

IOP Conf. Series: Materials Science and Engineering **525** (2019) 012073 doi:10.1088/1757-899X/525/1/012073

[6] Perevislov S N, Lysenko A S, Titov D D, Tomkovich M V 2017 *Inorganic Materials* **2** 206.
[7] Perevislov S N 2013 *Glass and ceramics* **7-8** 265.
[8] Frolova M G, Leonov A V, Kargin Y F, Lysenkov A S et al. 2018 *Inorganic Materials: Applied Research* **9** 675.

Materials based on boron carbide obtained by reaction sintering

S N Perevislov[1], A S Lysenkov[2], D D Titov[2], M V Omkovich[3], D D Nesmelov[4] and M A Markov[4]

[1]Institute of Silicate Chemistry I.V. Grebenshchikov, Russian Academy of Sciences, St. Petersburg, Russia
[2] Baikov Institute of Metallurgy and Materials Science, Russian Academy of Sciences, Moscow, Russia
[3]Physico-Technical Institute A.F. Ioffe Russian Academy of Sciences, St.-Petersburg, Russia
[4]St.-Petersburg State Institute of Technology (Technical University), St.-Petersburg, Russia

E-mail: perevislov@mail.ru

Abstract. Porous boron-carbide billets containing different amounts of carbon (0-15 wt%) were infiltrated with molten silicon to produce dense materials (p = 99.6 % of theoretical) from starting B_4C powders with $d_{0.5}$ = 90 and 12.5 μm. The phase composition, microstructure and mechanical properties of the material were studied. A solid solution of boroncarbide in silicon was determined to be present.

1. Introduction

Ceramics from of boron carbide possesses a low density, high hardness and Young's modulus, which allows the use of these materials in extreme conditions of high dynamic loads [1, 2]. However, the covalent bond in boron carbide creates difficulties in the preparation of dense materials at its base. Typically, these materials are prepared by hot pressing [3].

An alternative is the method of reaction sintering B_4C (RBBC). According to this, molded a preforms of boron carbide and the soot, which were impregnated with a silicon melt. During infiltration by capillary action, the liquid fills the pores and silicon reacts with the carbon black to form secondary silicon carbide [4–6]. Also, silicon partially dissolves boron carbide to form a solid solution $B_{12}(C,Si,B)_3$ which is significantly embrittle the material and negatively affect its mechanical properties, particularly crack resistance coefficient. For decreasing the formation of solid solution can restricting contact of molten silicon during sintering to the surface of the workpiece based on boron carbide.

The aim work is to study new aspects for RBBC-material, consisting in the impregnation of porous materials through the channels of sacrificial porous workpieces, study of the structure, phase composition and physical properties of sintered materials.

2. Starting materials and methods

The original components used in the work - boron carbide $d_{0,5}$ = 42,0 μm (B_4C_{lg}) and $d_{0,5}$ = 2,5 μm (B_4C_{sm}), technical carbon black «K–354» and silicon «KR00» milled to size $d_{0,5}$ = 0,5–1,0 sm. The compositions of the materials were shown in Table 1.

Samples of the compositions №1 and №2 were prepared through impregnation sacrificial preform; №3 composition samples prepared according to standard techniques reaction sintering (with an excess of silicon during impregnation) and used for comparison of properties.

Table 1. The initial composition of the workpieces RBBC-materials

№ composition	Content of components in workpiece, wt. %				Density $\rho \pm 0,02$ (g/cm^3)	Porosity $P \pm 0,1$ (%)
	B_4C_{lg}	B_4C_{sm}	Carbon	Si		
1	60	25	15	-	2.70	0.9
2	56	24	15	5	2.72	1.2
3	60	25	15	-	2.55	6.4

3. Results of the study

The results of XRF sample 3 (Figure 1, c) confirm the presence of a phase of Si, $B_{12}(C,Si,B)_3$, β-SiC and a small amount of boron carbide. The material structure 1 shows the phase of silicon content due to minimal dosed (optimal) amount entering the silicon workpiece. For the same reason, fewer phase content $B_{12}(C,Si,B)_3$ (Figure 1, a). When introduced into the batch 5 wt. % Si (composition 2) of the sintered sample is also present silicon phase and the β-SiC (Figure 1b). The intensity of the peaks corresponding phase $B_{12}(C,Si,B)_3$ is reduced, the secondary phase silicon carbide - is increased.

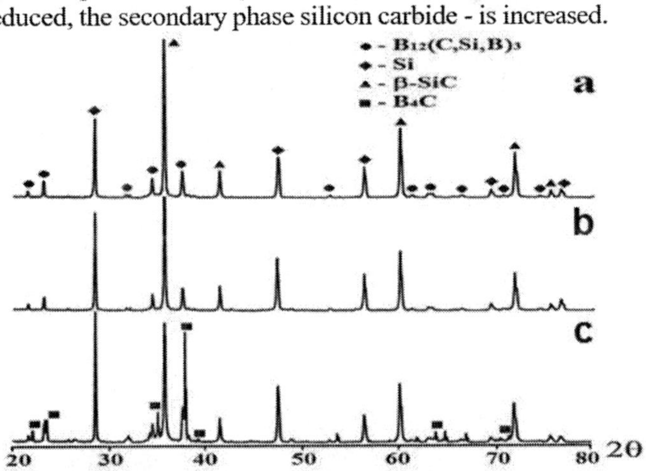

Figure 1. X-ray phase analysis of material structure 1 (**a**), composition 2 (**b**), composition 3 (**c**)

The microstructure of the material 1 is characterized by the presence of large grains of boron carbide $d_{0,5}$ = 39,5 μm (gray, Figure 2a), which proves small grains B_4C dissolution in the silicon melt. Introduction of additional Si in the composition of the material increases the solubility of grains B_4C (Figure 2b). In the field of weak contact with the silicon melt intergranular porosity is observed (Figure 2c).

The obtained SEM photograph RBBC microstructure (Figure 3) demonstrates the presence of B_4C, $B_{12}(C,Si,B)_3$, SiC^{II} (β-SiC), Si phases in the materials, which confirms the conclusions of the microstructure analysis made on an optical microscope (Figure 2).

Properties reactive sintered materials are shown in Table 2.

Dense materials (composition 1) have a high level of mechanical performance, far exceeding the level of properties RBBC-materials sintered under standard reaction sintering conditions (composition 3, Table 2).

When impregnate porous workpieces with silicon melt through the pores of sacrificial specimens, materials based on boron carbide with a high density (ρ = 99,0 %), which have a high level of mechanical characteristics, are obtained, compared to materials obtained by standard impregnation with an excess amount of silicon [4].

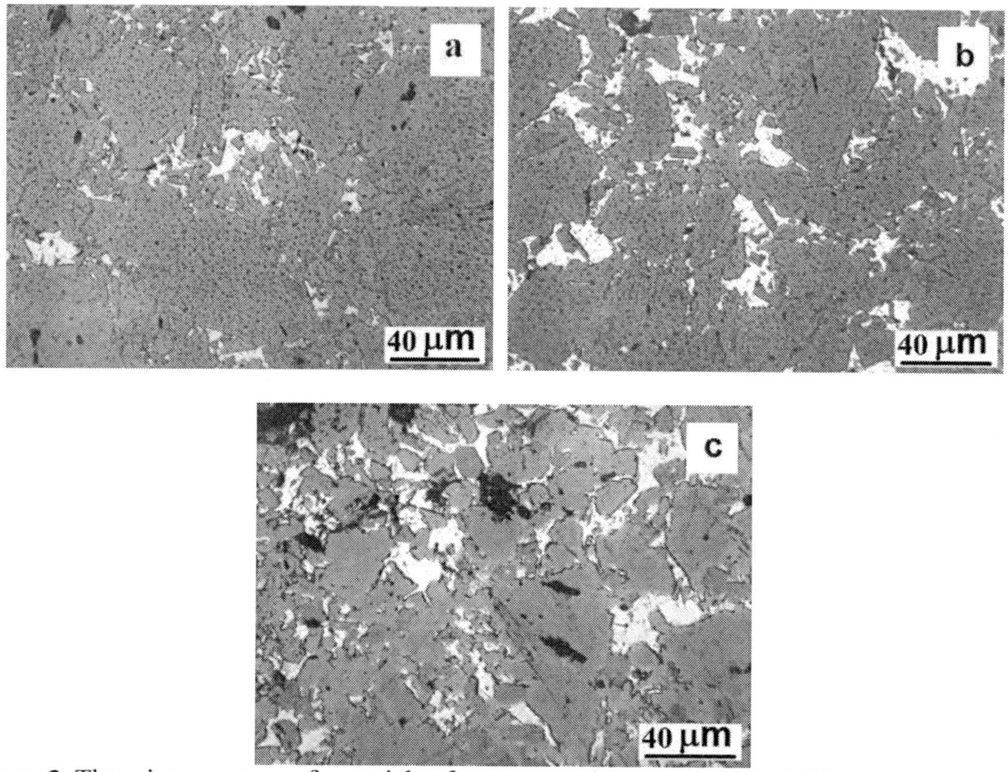

Figure 2. The microstructure of materials of structure 1 (**a**), composition 2 (**b**), composition 3 (**c**)

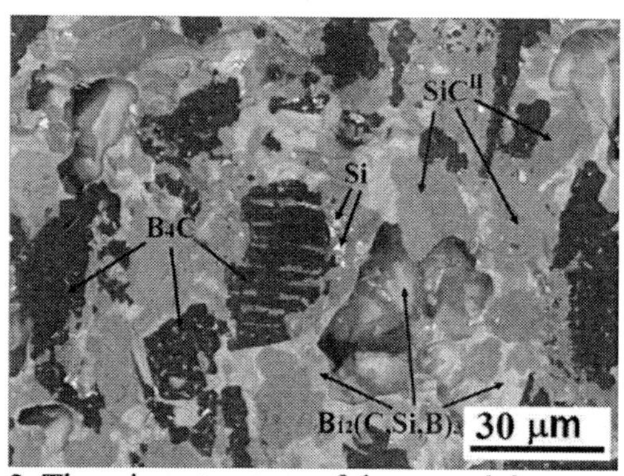

Figure 3. The microstructure of the material composition 3

Table 2. Mechanical properties RBBC-materials

№ composition	Elastic modulus $E_{elast} \pm 10$ (GPa)	Flexural strength $\sigma fl \pm 10$ (MPa)	Fracture toughness $K_{IC} \pm 0,2$ (MPa·m$^{1/2}$)	Vickers hardness $HV \pm 0,1$ (GPa)
1	360	450	4.8	30.5
2	340	420	4.2	28.5
3	320	380	3.8	27.0

From the comparative analysis based on boron and silicon carbide materials properties shows that obtained a reaction-sintered B_4C, the level of physico-mechanical characteristics superior reaction-sintered SiC and approaches the level of properties of hot-pressed materials based B_4C (Table 3).

Table 3. Comparative characteristics of materials

Material	ρ (g/cm³)	*P* (%)	σ_{fl} (MPa)	$K_{IC} \pm 0{,}2$ (MPa·m$^{1/2}$)	*HV* (GPa)
Reaction-sintered B_4C	2,70	0,9	450	4,8	30,5
A hot-pressed composite B_4C-Al [7]	1,67	28,0	125	-	-
Hot-pressed B_4C [3]	2,54	1,0	665	5,3	36,1
Reaction-sintered SiC [4]	3,10	0,5	400	4,5	21,0

References

[1] Hayun S, Rittel D, Frage N, Dariel M P 2008 *Mater. Sci. Eng. A* **487** 405.
[2] Suri A K, Subramanian C, Sonber J K, Murthy T C 2010 *Mater. Int. Rev* **1** 4.
[3] Perevislov S N, Lysenkov A S, Vikhman S V 2017 *Inorganic Materials* **4** 376.
[4] Nesmelov D D., Perevislov S N. Glass and Ceramics **9–10** 313.
[5] Perevislov S N, Shcherbak P V, Tomkovich M V 2018 Refractories and Industrial Ceramics **1** 32.
[6] Perevislov S N, Shcherbak P V, Tomkovich M V 2018 *Refractories and Industrial Ceramics* **2** 179.
[7] Rumyantsev I A, Perevislov S N 2017 *Refractories and Industrial Ceramics* **4** 405.

Properties of spherical stainless steel powders

M A Kaplan, A A Kirsankin, M A Smirnov, T A Kalaida, E E Baranov, Yo O Ustinova and M A Sevostyanov

Baikov Institute of Metallurgy and Materials Science RAS

E-mail: mishakaplan@yandex.ru

Abstract. The additive technologies are gaining popularity due to the possibility of creating custom products for complex applications in many industries. Spherical metal powders are initial components for metal additive manufacturing. Initial materials are very important part of manufacturing, because their quality has an influence on stability of production process and quality of final product. In this work properties of spherical powders including 304L, 310, 316L and 17-4PH grade were determined using scanning electron microscopy, laser diffraction particle size analysis, reductive melting method and X-ray analysis.

1. Introduction

Additive manufacturing methods are becoming more common as there is a demand for rapid prototyping and production of parts that require design features that cannot be manufactured by the conventional press and sinter process normally used in powder metallurgy (for example biomedical implants) [1-7]. Additive manufacturing is process which builds a part layer by layer out of particulate material, a chemical binder and a digital design.

The powder characteristics most desirable for additive manufacturing are spherical powders with a fine particle size. Having a distribution of particles that flows and forms dense or well packed beds is important in both powders for additive manufacturing methods mainly for selective laser melting [8-9].

Stainless steel, a class of ferrous alloys, is well-known for their resistance to corrosion, creep and high temperature applications. Due to increasing demand of powder metallurgy components made from stainless steel in variety of applications, including aerospace, automotive, chemical processing and biomedical field, it became of a great interest in the research domain. In this work we studied the properties of spherical stainless steel powders of grades 310, 304L, 316L and 17-4PH products from which are widely used in various industries.

2. Methods and materials

Spherical stainless steel powders were obtained by gas atomization method. During the gas atomization process, the molten steel is atomized through the use of inert gas jets into fine metal droplets which cool down during their fall in the atomizing tower. Metal powders obtained by gas-atomization offer a perfectly spherical shape combined with a high cleanliness level. The chemical composition of steels is presented in Table 1.

IOP Conf. Series: Materials Science and Engineering **525** (2019) 012075 doi:10.1088/1757-899X/525/1/012075

Table 1. Chemical composition (mass %) of stainless steel.

Steel grade powder	Cr, %	Ni, %	Cu, %	Nb, %	Mo, %	S, %	P, %	C, %	Si, %	Mn, %	Fe, %
310	24	19	0	0	0	0,03 max	0,04 max	0,08 max	1 max	2 max	Balance
304L	18	11	0	0	0	0,03 max	0,04 max	0,03 max	1 max	2 max	Balance
316L	16	15	0	0	3	0,03 max	0,04 max	0,03 max	1 max	2 max	Balance
17-4PH	16	4	4	0,3	0	0,03 max	0,04 max	0,030 max	1 max	1 max	Balance

The particle size distributions of powders were determined using an Analysette 22 NanoTec laser diffraction particle size analyzer from Fritsch. The particle size analyzer measuring cell contains the sample particles prepared by the dispersion unit which are irradiated with laser light. By changing the spacing between detector and measuring cell, a different angle range of the scattered light is detected. The particle size distribution is calculated from this data.

The phase composition was determined by an Ultima IV X-ray diffractometer (Rigaku, Cu$K\alpha$ radiation) with a graphite monochromator that is intended for the use of different polycrystalline inorganic and organic substances, metals, alloys, films, coatings, and composites with a vertical goniometer and D/teX high-rate semiconductor detector using Cu$K\alpha$ radiation according to the Bragg–Bretano method and by the oblique incidence technique with a fixed rotation angle of the X-ray tube.

The Leco TC-600 is a software-controlled instrument that measures both nitrogen and oxygen in a wide variety of metals, refractories, and other inorganic materials. The inert gas fusion principle is employed. A weighed sample, placed in a high-purity graphite crucible, is fused under a flowing helium gas stream at temperatures sufficient to release oxygen, nitrogen, and hydrogen. The oxygen in the sample combines with the carbon from the crucible forming primarily carbon monoxide (CO). In some instances, depending upon sample type and furnace temperature, some oxygen can be released directly as carbon dioxide (CO). The nitrogen present in the sample releases as molecular 2 nitrogen, and any hydrogen present is released as hydrogen gas. Hydrogen determination by inert gas fusion thermal conductivity detection of the Leco RHEN-602 gas analyzer. The Leco CS-600 system is designed for wide-range measurement of carbon and sulfur content of metals, ores, ceramics, and other inorganic materials.

Scanning electron microscope (SEM) Tescan Vega II SBU with an INCA Energy unit for energy dispersive analysis was used to study the surface morphology and to carry out the fractographic investigation of the specimens after the static tests.

3. Results and discussion

The particle size distributions of spherical stainless steel powders 304L, 310, 316L and 17-4PH is shown in Figure 1. 304L steel and 310 steel powders consist 10-30 μm size particles, 316L and 17-4PH – 40 and 120 μm respectively.

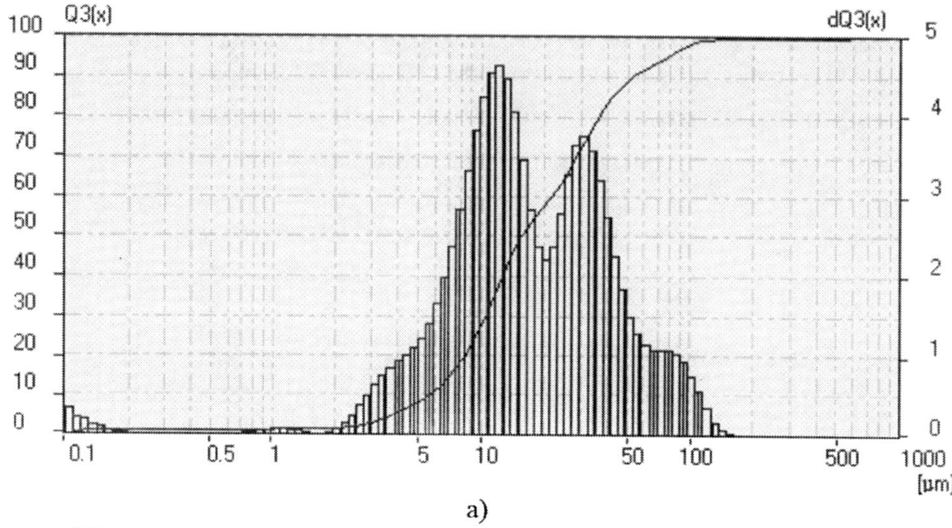
Q3(x)
dQ3(x)
100
90
80
70
60
50
40
30
20
10
0
5
4
3
2
1
0
0.1
0.5
1
5
10
50
100
500
1000
[µm]

a)

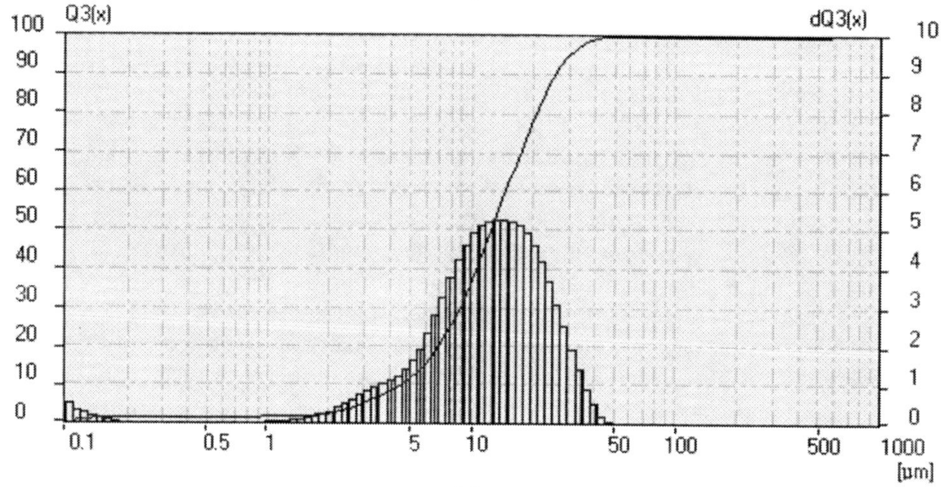
Q3(x)
dQ3(x)
100
90
80
70
60
50
40
30
20
10
0
10
9
8
7
6
5
4
3
2
1
0
0.1
0.5
1
5
10
50
100
500
1000
[µm]

b)

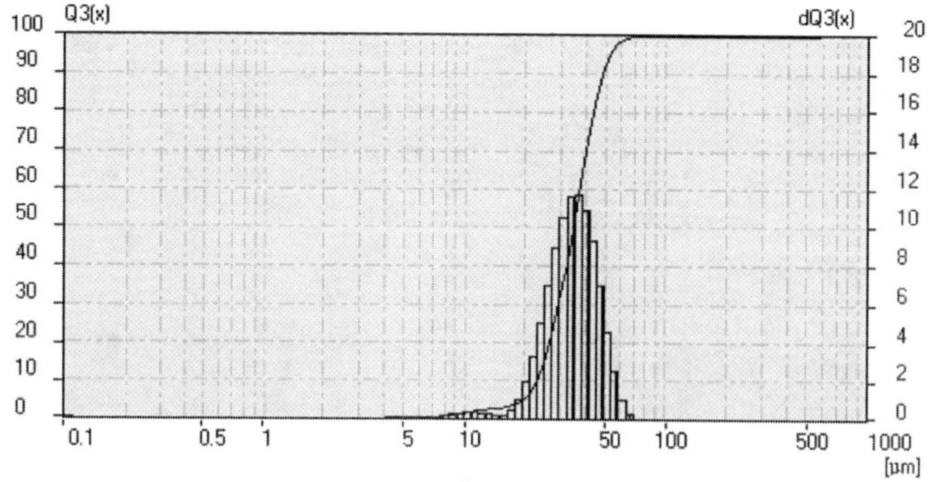
Q3(x)
dQ3(x)
100
90
80
70
60
50
40
30
20
10
0
20
18
16
14
12
10
8
6
4
2
0
0.1
0.5
1
5
10
50
100
500
1000
[µm]

c)

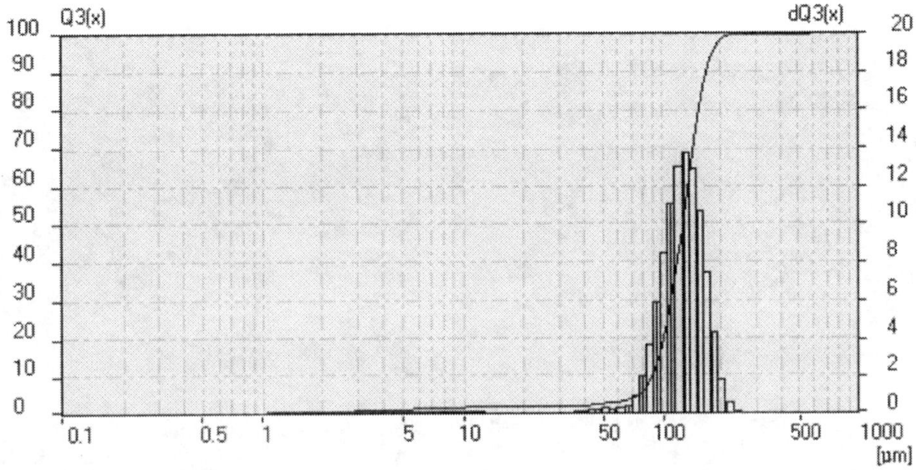

d)

Figure 1. Particle size distribution of the powders: a - steel 304L, b - steel 310, c - steel 316L, d - steel 17-4PH.

The phase composition was determined by an Ultima IV X-ray diffractometer. X-ray diffractometer showed that 304L and 17-4PH steel powders consist of both austenitic and ferritic phases, steel 310 and 316L powders – only austenitic phase.

Measuring of oxygen, nitrogen, carbon and hydrogen content was carried out using gas analyzers TC-600 from Leco, RHEN-602 from Leco, CS-600 from Leco. The results of the study are shown in table 2.

Table 2. Impurity content (mass %).

Steel grade powder	O, %	N, %	C, %	H, %
310	0,061±0,001	0,14±0,01	0,372±0,003	0,0016±0,0003
304L	0,066±0,003	0,21±0,01	0,030±0,001	0,0016±0,0004
316L	0,061±0,005	0,077±0,001	0,017±0,001	0,0008±0,0001
17-4PH	0,037±0,003	0,11±0,01	0,030±0,001	0,0008±0,0002

The carbon and hydrogen content in all powders is the same as in bulk stainless steel. Increased oxygen content in all samples may reduce the corrosion resistance of final product. In spherical stainless steel powders 304L nitrogen content also exceeded, that can cause the formation of defects (holes, cracks, etc.).

The study of the powder morphology was carried out using a Tescan Vega II SBU scanning electron microscope. SEM images (Figure 2) show that particles have a spherical shape. However, on the surface of some particles there are defects in the form of satellites. Satellites are the most characteristic and inevitable defect of powders obtained by gas atomization.

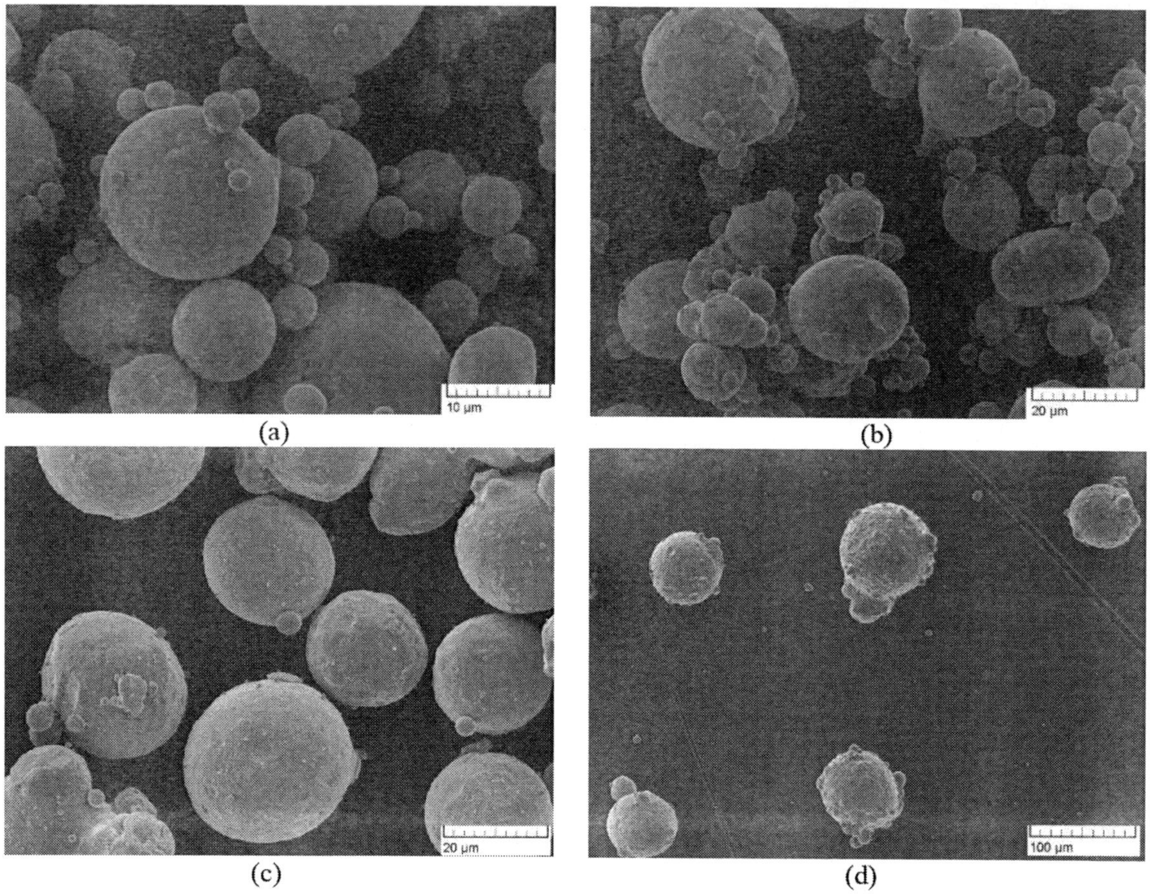

Figure 2. SEM images of powder: (a) - steel 304L, (b) - steel 310, (c) - steel 316L, (d) - steel 17-4PH

4. Conclusion

The spherical powder obtained by gas atomization has defects in the form of satellites, which can affect the fluidity of the powder. Increased oxygen and nitrogen content in all samples may reduce the corrosion resistance and formation various deffects in final product. Application spherical stainless powder will reqire determination the correlations between the properties of the powders and the properties of final products.

Acknowledgments

The work was carried out according to the state task No. 075-00746-19-00

References

[1] Nasakina E O, Baikin A S, Sergiyenko K V, Leonov A V, Kaplan M A, Seryogin A V, Konushkin S V, Myasnikova N V, Sevostyanova M A, Kolmakov A G, Simakov S V 2017 Formation and Investigation of Composite Material Silver–Nitinol for Medical Purposes *Inorganic materials: applied research* **8** №1 pp 112–117

[2] Nasakina E O, Baikin A S, Sevostyanov M A, Kolmakov A G, Zabolotnyi V T, Solntsev K A 2014 Properties of Nanostructured Titanium Nickelideand Composite Based on It *Theoretical Foundations of Chemical Engineering* **48** №4 pp 477–486

[3] Kaplan M A, Sevost'yanov M A, Nasakina E O, Baikin A S, Sergienko K V, Konushkin S V, Kolmakov A G 2018 Influence of the Surface Modification on the Mechanical Properties of

NiTi (55.8 wt % Ni) Alloy Wire for Medical Purposes *Inorganic Materials: Applied Research* **9** №4 pp 751–756

[4] Yap C Y, Chua C K, Dong Z L, Liu Z H, Zhang D Q, Loh LE, Sing S L 2015 Review of selective laser melting: Materials and applications *Appl. Phys. Rev.* **2** 041101

[5] Kruth J P, Froyen L, Vaerenbergh J V, Mercelis P, Rombouts M, Lauwers B 2004 Selective laser melting of iron-based powder *J. Mater. Process. Technol.* **149** pp 616-622

[6] Schleifenbaum H, Meiners W, Wissenbach K, Hinke C 2010 Individualized production by means of high power Selective Laser Melting *CIRP-JMST* **2** №3 pp 161-169

[7] Yadroitsev I, Gusarov A, Yadroitsava I, Smurov I 2010 Single track formation in selective laser melting of metal powders *J. Mater. Process. Technol* **210**, pp 1624–1631

[8] Wohlers T 2014 Additive Manufacturing and 3D Printing State of the Industry *Annual Worldwide Progress Report. Wohlers Associates*Inc 275 p

[9] Smirnov M A, Kaplan M A, Sevostyanov M A 2018 Receiving finely divided metal powder by inert gas atomization *IOP Conf Ser Mater Sci Eng.* **347** pp 1-5

IOP Conf. Series: Materials Science and Engineering **525** (2019) 012076 doi:10.1088/1757-899X/525/1/012076

Investigation of the properties of heat-resistant spherical powders

M A Smirnov, M A Kaplan, A A Kirsankin, T A Kalaida, E O Nasakina and M A Sevostyanov

A.A. Baikov Institute of Metallurgy and Materials Science RAS, Russia, Moscow

E-mail: mexfexpex@gmail.com

Abstract. Nowadays, additive manufacturing (AM) become very perspective method for fast fabrication of any details, parts or products. One of the main methods of AM is selective laser melting (SLM), which allows to obtain complex shape parts from various metal powders. However, this process requires high demands on the initial materials. To determine the quality of the initial EP648, EP741I, and Hastelloy X powders, such methods as scanning electron microscopy, laser diffraction, reductive melting and X-ray were used.

1. Introduction

Spherical metal powders are the main raw material to produce products by various additive methods. One of the most promising AM methods is selective laser melting, which is based on layer-by-layer sequential fusion of metal powder by laser. This method allows you to get complex shape parts, which is technologically difficult or impossible to obtain by traditional methods [1]. SLM products, are widely used in various fields.

One of the most important advantages of this method is the ability to create products with desired properties by selecting a powder with appropriate characteristics [2, 3]. In this paper the results of impurity content, fractional and X-ray phase analysis of spherical heat-resistant powders EP648, EP741 and Hastelloy X are presented.

2. Materials and methods

All spherical powders were obtained by inert gas atomization method. The fractional composition of the powders was determined using a Fritsch Analysette 22 NanoTec laser diffraction particle size analyzer and consist up to 60 microns for EP648, from 20 to 80 microns for EP741, and from 20-120 microns for Hastelloy X. The composition of the initial powders is presented in table 1

Table 1. Composition of the initial powder (weight %)

Steel grade powder	Ni	Cr	Mo	Fe	Co	W	C	Si	Al	Nb	Ti
EP648	Bal.	32-35	2.3-3.3	4 max	-	4.3-5.3	0.1 max	0.4 max	0.5-1.1	0.5-1.1	0.5-1.1
EP741	Bal.	8-10	3.5-4.2	0.5 max	15-16.5	5.2-5.9	0.02-0.06	0.5 max	4.8-5.3	2.4-2.8	1.6-2
Hastelloy X	Bal.	22	9	18	1,5	0,6	0,1	1 max	0,5 max	0,5 max	0,15 max

IOP Conf. Series: Materials Science and Engineering **525** (2019) 012076 doi:10.1088/1757-899X/525/1/012076

The principle of operation of a laser diffractometer is based on measuring the angular distribution of the intensity of scattered light during the passage of a laser beam through a dispersed sample. Large particles mainly scatter light at small angles to the laser beam, while small particles - at large angles. The Mi theory of light scattering is able to determine the size of the particles that form the scattering indicatrix, which coincides with the measured data on the angular dependence of the intensity of scattered light. Particle size is expressed as the sphere diameter of equivalent volume. The study of the morphology of the powder particles was carried out using a Tescan Vega II SBU scanning electron microscope.

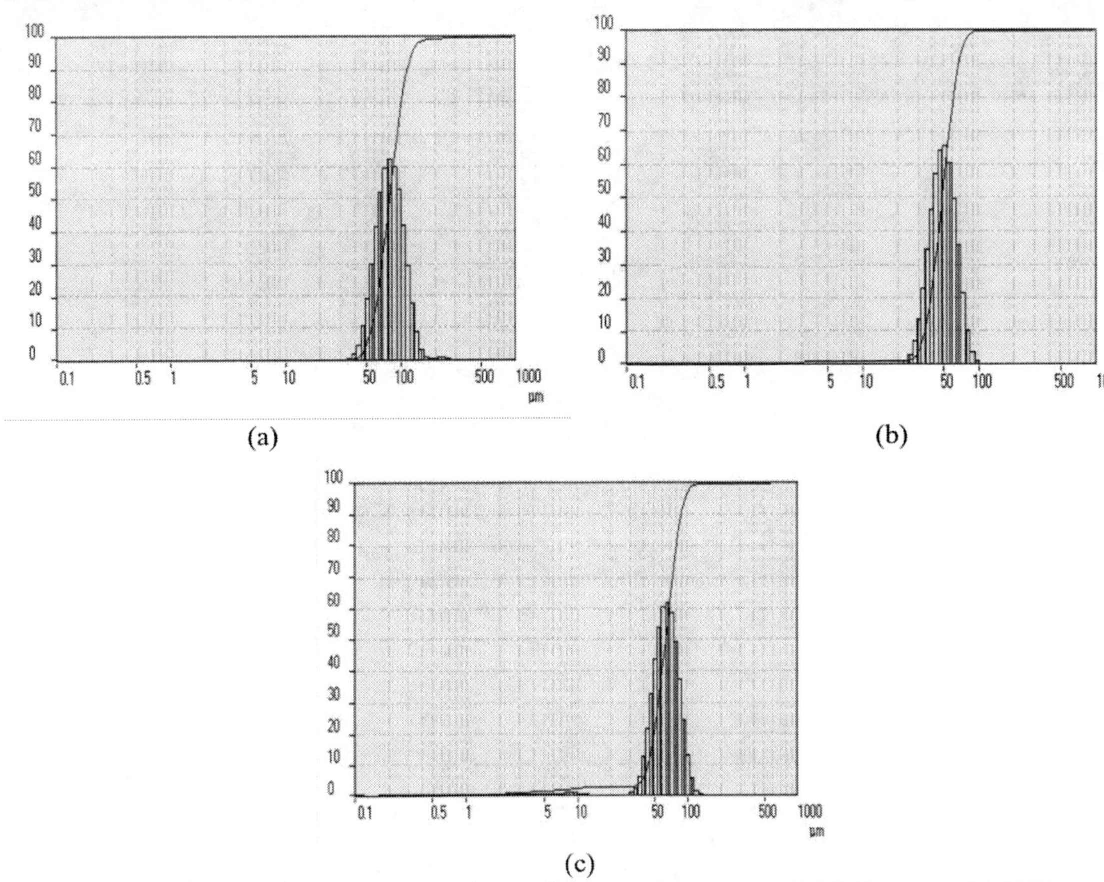

(a) (b) (c)

Figure 1. The granulometric composition of the powders: a - steel EP648, b - steel EP741, c - Hastelloy X alloy.

Figure 1 shows the particle size distribution of EP648, EP741 and Hastelloy X high-temperature alloy powders. All the studied powders have a rather narrow peak of particle size distribution, which is important for further use.

IOP Conf. Series: Materials Science and Engineering **525** (2019) 012076 doi:10.1088/1757-899X/525/1/012076

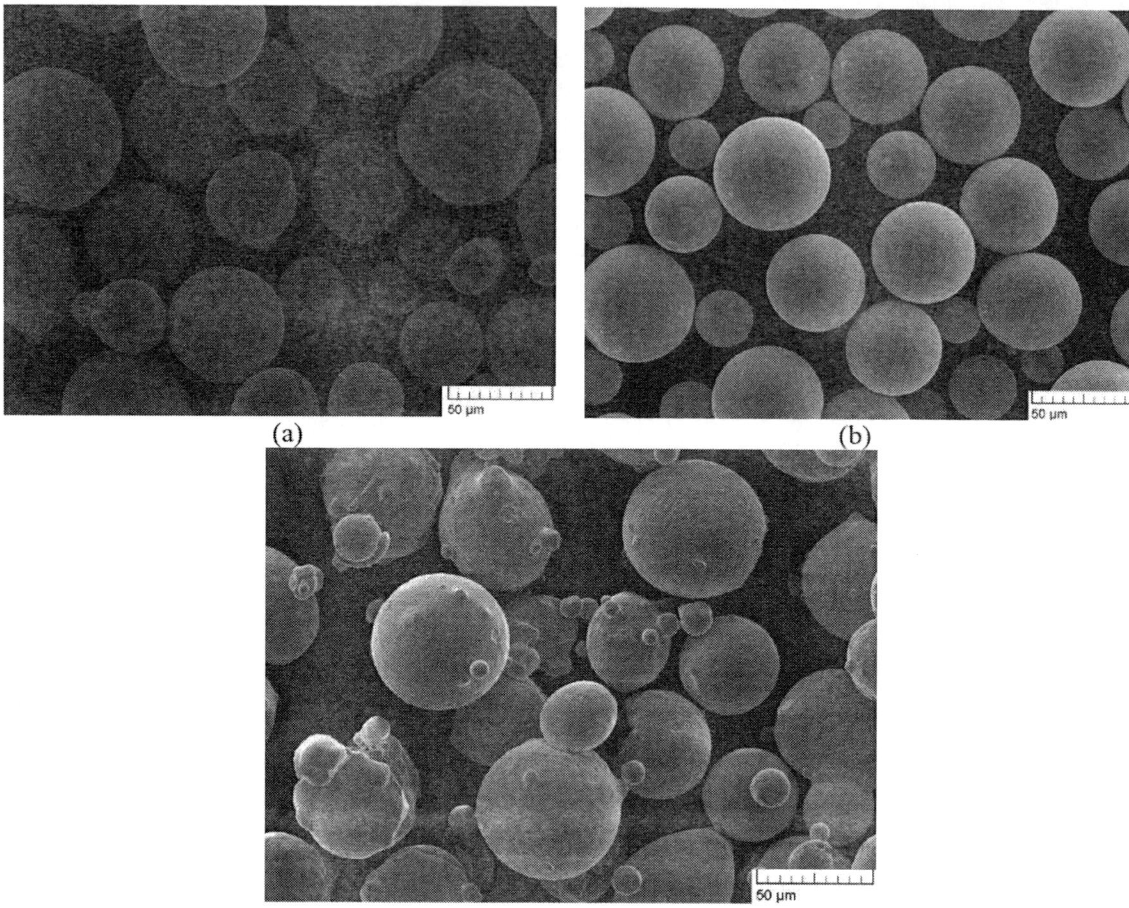

(a) (b)

(c)

Figure 2. SEM images of powders: a - EP648, b - EP741, c - Hastelloy X.

Figure 2 shows the SEM images of powder particles. As you can see, most particles have a regular spherical shape. However, there are defects in the form of satellites on the surface of some particles, especially at Hastelloy X surface. Satellites are the most characteristic and inevitable defect of powder particles obtained by gas atomization. At the exit from the flow, the particles collide with each other and leave traces of impacts (Figure 2a) with full crystallization or sticking of particles occurs (Figure 2) with temperature differences. This process usually occurs during the collision of large particles (from the liquid phase) with small particles (from the vapor phase).

3. Results and discussion

The content of oxygen and nitrogen in the samples was determined by the method of reductive melting in a stream of inert gas with subsequent detection of oxygen in the infrared cell and nitrogen in the conductometric cell of the Leco TC-600 gas analyzer. Helium was used as a inert gas. To determine the hydrogen content in the samples, the method of reductive melting was also used, but only with the high purity argon as the inert gas. The detection of hydrogen was carried out in a conductometric cell of a Leco RHEN-602 gas analyzer. Determination of carbon content was carried out by the method of reductive melting in a ceramic crucible in an induction furnace and detected by the amount of evolved gaseous CO2 in the infrared cell of the Leco CS-600 gas analyzer. The results of the study are shown in table 2.

IOP Conf. Series: Materials Science and Engineering **525** (2019) 012076 doi:10.1088/1757-899X/525/1/012076

Table 2. Mass fraction of oxygen, nitrogen, carbon and hydrogen in the samples of powders.

Steel grade powder	O, %	N, %	C, %	H, %
EP648	0,065±0,005	0,11±0,001	0,07	0,00027
EP741	0,024±0,003	0,0005 max	0,05	0,00096
Hastelloy X	0,016±0,003	0,022±0,001	0,075±0,001	0,0004±0,0002

As can be seen from the table, powders from steels EP648 and EP741 have a high oxygen content, which is caused, apparently, by the technology for their preparation. In the EP648 sample the nitrogen content is quite exceeded, which can lead to the formation of various defects (cavities, cracks, etc.) with the further fabrication of products by the method of selective laser melting. The content of carbon and hydrogen in all samples corresponds to the content in the monolithic material.

To obtain the phase composition of the powders, X-ray structural phase analysis was used. The diffraction spectra of the powders are presented in Figure 3. In all cases, the powder has a single-phase austenitic composition having an fcc lattice with a large grain size (> 1000 Å)

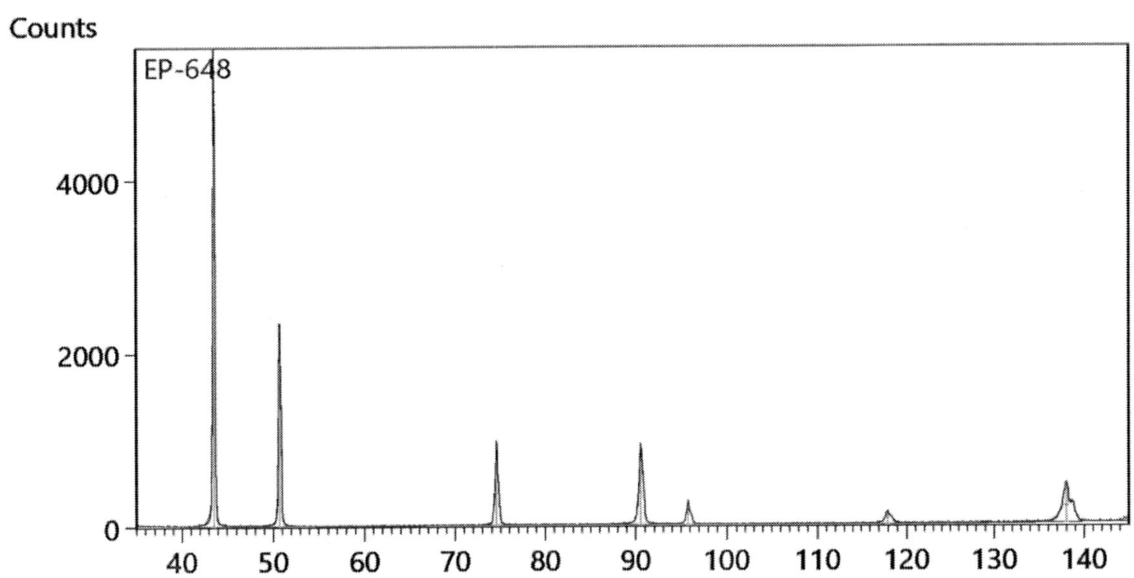

(a)

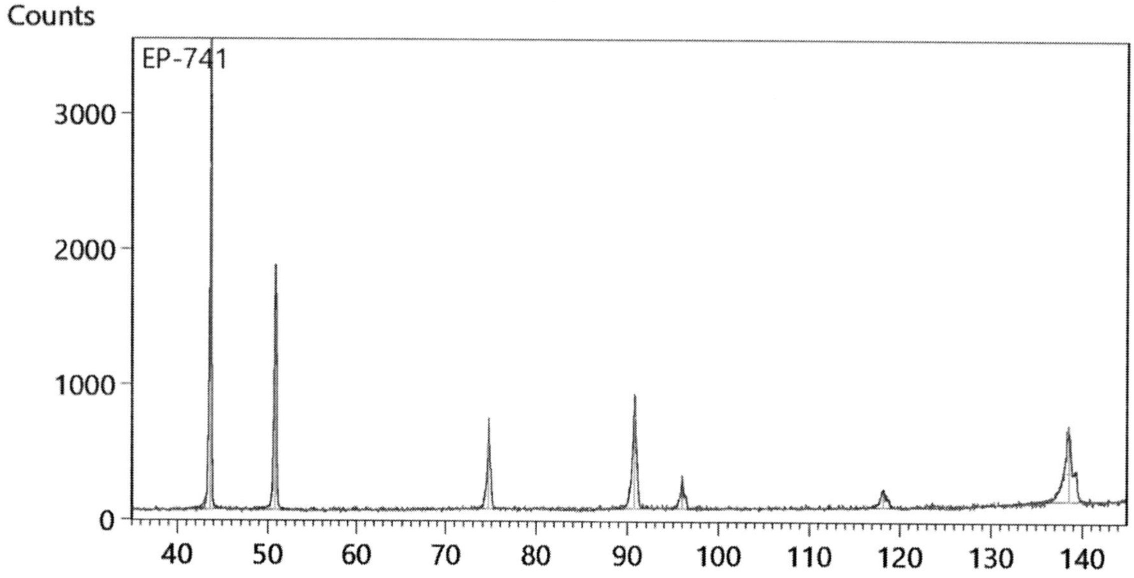

(b)

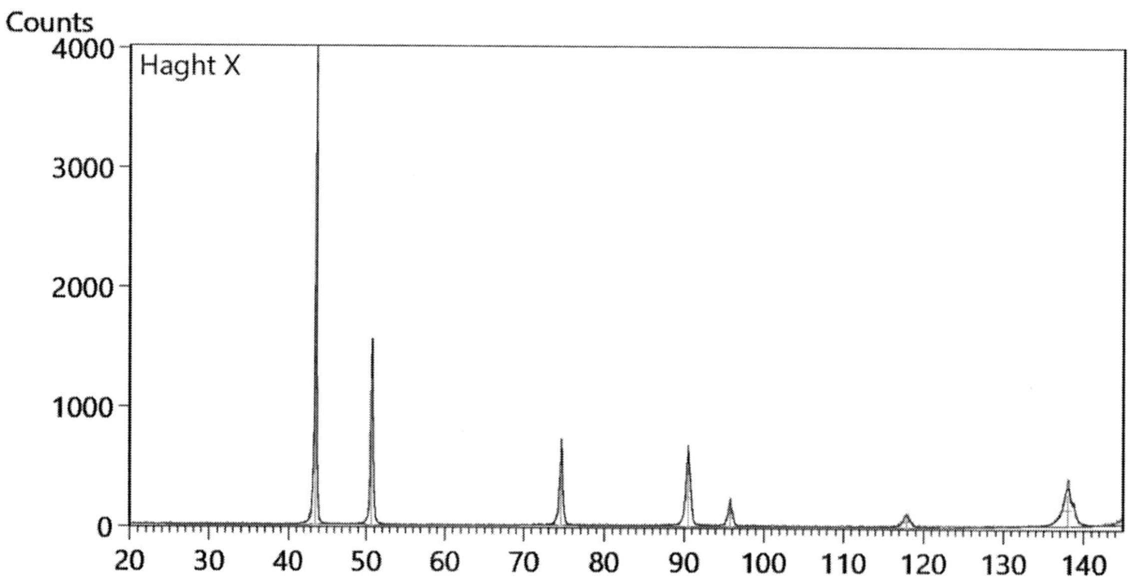

(c)

Figure 3. Diffraction spectra of powders: a - steel EP648, b - steel EP741, c - Hastelloy X

4. Conclusion

All studied powders have a single-phase austenitic composition with an fcc lattice and a large grain size. These powders also have a rather narrow peak in the particle size distribution and high sphericity. However, due to the peculiarities of the technology (gas atomization method), the satellites are found on the surface of some particles (especially in Hastelloy X powder), as well as EP648 and EP741 powders have an increased oxygen content, which can provoke cracking during SLM.

Thus, the studied powders can be recommended for fabricating products using the SLM method.

Acknowledgments

The work was carried out according to the state task No. 075-00746-19-00

References

[1] Smirnov M A, Kaplan M A, Sevostyanov M A 2018 Receiving finely divided metal powder by inert gas atomization *IOP Conf Ser Mater Sci Eng*. **347** pp 1-5

[2] Nasakina E O, Kaplan M A, Konushkin S V, Baskakova M I, Baikin A S, Fedyuk I M, Sudardchikova M A, Sergienko K V, Sevostyanov M A, Kolmakov A G 2018 Mechanical Characteristics of Non-Nickel Shape Memory Alloy of Medical Appointment *International Conference on Power, Energy and Environmental Engineering - DEStech Publication Inc. U.S.A.* pp 304-308

[3] Kaplan M A, Sevost'yanov M A, Nasakina E O, Baikin A S, Sergienko K V, Konushkin S V, Kolmakov A G 2018 Influence of the Surface Modification on the Mechanical Properties of NiTi (55.8 wt % Ni) Alloy Wire for Medical Purposes *Inorganic Materials: Applied Research* **9** №4 pp 751–756

Rheological properties of $MoSi_2$-$NbSi_2$ powders obtained by SHS-method and solid-phase mixture

D D Titov[1], P A Miloserdov[2], A S Lysenkov[1], M G Frolova[1], E A Gumennikova[1,3] and Yu F Kargin[1]

[1] Baykov Institute of Metallurgy & Material Science RAS, Moscow, Russia
[2] Institute of Structural Macrokinetic and Material Science RAS, Chernogolovka, Russia
[3] Mendeleev University of Chemical Technology of Russia, Moscow, Russia

e-mail: mitytitov@gmail.com

Abstract. $(Mo_{1-x}Nb_x)Si_2$ powders ($0.1 \leq x \leq 0.9$) were obtained by self-propagating high-temperature synthesis (SHS) in ISMAN RAS. The same mixtures were obtained by commercial powders $MoSi_2$ and $NbSi_2$ solid-phase mixing in a planetary mill. The commercial disilicides were synthesised by magnesium-thermal reaction in Plasmotherm Ltd. During the experiment on powder mixtures deformation an array of data was obtained as a result of loading in a constant-speed regime. A kinetic curves representing the pressure dependence on time was a result of the experiment, then the necessary pressure values were determined in the preliminary cold pressing to provide the selected density of the workpieces, the compressibility modulus and the compressibility factor for each composition were calculated.

1. Introduction

Recently interest has increased in two-component systems based on molybdenum disilicide and niobium disilicide for the production of new generation heating elements and coatings on their basis, which, due to high resistance and good adhesion to high-temperature niobium alloys [1] could increase the products working temperature and their lifespan at 1600°C in oxidising environments [2]. Such materials would increase the thermal efficiency of combustion systems such as jet engines for airplanes, and next-generation ultra-high temperature gas turbine systems in power plants such as the gas turbine combined cycle (GTCC).

The results demonstrated in research [3] that the simultaneous improvement of high-temperature creep strength and room temperature fracture toughness can be first accomplished by the development of unique cross-lamellar microstructure in $(Mo_{0.85}Nb_{0.15})Si_2$ composite, which opens a potential avenue for the development of novel ultrahigh-temperature (UHT) materials as alternatives to existing Ni-based superalloys. Koji Hagihara [3] and other researchers [4] shows that $(Mo_{0.85}Nb_{0.15})Si_2$ composite is very perspective future contracture material. The initial stage of raw preparation and pressing is no less important than the sintering process.

The most important question in the theory and practice of powders cold pressing consists in establishing the relationship between the applied pressure (P) and the compacts density. The presses-density dependence lets it possible to select the required value of P, which provides the given value of the density (ρ). A large number of experiments are carried out under conditions of static loading in the regime of constant pressure (P = const) [5]. In practice, powder blanks pre-pressing is carried out in special molds on hydraulic presses, which create a constant pressure. However, in this case there are methodological difficulties associated with some uncertainty of such an experiment. This is due to the fact that the compaction process is essentially non-stationary and the compaction degree is continuously changing though the time. Therefore, the density corresponding to the preset pressure depends also on the pressing process duration. If ρ is understood as its maximum value, it is also affected by the dwell time. For different materials, the conditions for obtaining holistic blanks having an optimum density are usually empirically. For various compositions of charge, the corresponding porosity value is 30-50% [6] and it is optimal to obtain a billet blank with a relative density of 60-70% to dilatometric studies. It is important all the compositions have the same relative density when determining the sintering process activation energy [7, 8].

2. Materials and methods of researches

The cast mixture $(Mo_{1-x},Nb_x)Si_2$ (where $0.1 \leq x \leq 0.9$) was obtained by the SHS method at the Institute of Structural Macrokinetic and Material Science Russian Academy of Sciences (ISMAN RAS) in Chernogolovka Russia [9] and the same mixtures were obtained by solid-phase mixing commercial $MoSi_2$ and $NbSi_2$ (Plazmoterm LLC). All powders were characterised by SEM (Supra 50 VP LEO, with microanalysis system INCA Energy + Oxford, Germany), X-ray (Shimadzu XRD-6000, Japan) and granulometric analyses (Fritch, Analysette 22, Germany). The dominant fraction is 10÷15 μm particles in both case. Rheological properties were tested on the mechanical testing machine (Instron 5581, Great Britain). The stain velocity was 1 mm/min.

The deformation was calculated by the Eq. (1):

$$\varepsilon = \frac{\Delta h}{h_0} \tag{1}$$

where Δh is the change in the height of the bulk layer over time; h_0 is the initial height of the bulk layer.

The compressibility modulus is numerically equal to the tangent of the linear section slope angle of the diagram strain against deformation, and is calculated by the Eq. (2):

$$G = \frac{\Delta p'}{\varepsilon *} \tag{2}$$

where $\Delta p'$ is the limiting change in the linear section pressure value; ε * is the final value of the linear section deformation.

The compressibility coefficient (k_c) characterises the reversible decrease in the sample height (volume) under the pressure action and is quantitatively determined by Eq. (3):

$$k_c = -\frac{1}{h_0} \cdot \frac{\Delta h'}{\Delta p'} \tag{3}$$

where $\Delta h'$and $\Delta p'$ are the limiting values of the height and the linear section pressure variation, h_0 is the initial height of the backfill.

The physical meaning of the compressibility coefficient is characterisation the material ability to compact at the initial (linear) stage, during which the compaction intensity is maximal.

3. Results and discussion

Figure 1 shows stress strain curves for the powder compositions $(Mo_{1-x}Nb_x)Si_2$ (where $0.1 \leq x \leq 0.9$) obtained by two methods: the SHS method and solid-phase mixing in a planetary mill. The first stage on the rheological curve is limited to deformation values up to 10-15% and is accompanied by a linear increase in stain and increasing ε, which occurs due to the powder particles movement into the pores.

The second stage corresponds to deformations from 15-20% to 30-35%. In the P-ε diagram, it corresponds to a very steep nonlinear rise in stain with increasing ε. The particles movement is due to the accommodation (adaptation) of them among themselves and from the other hand due to deformation of their volume [10].

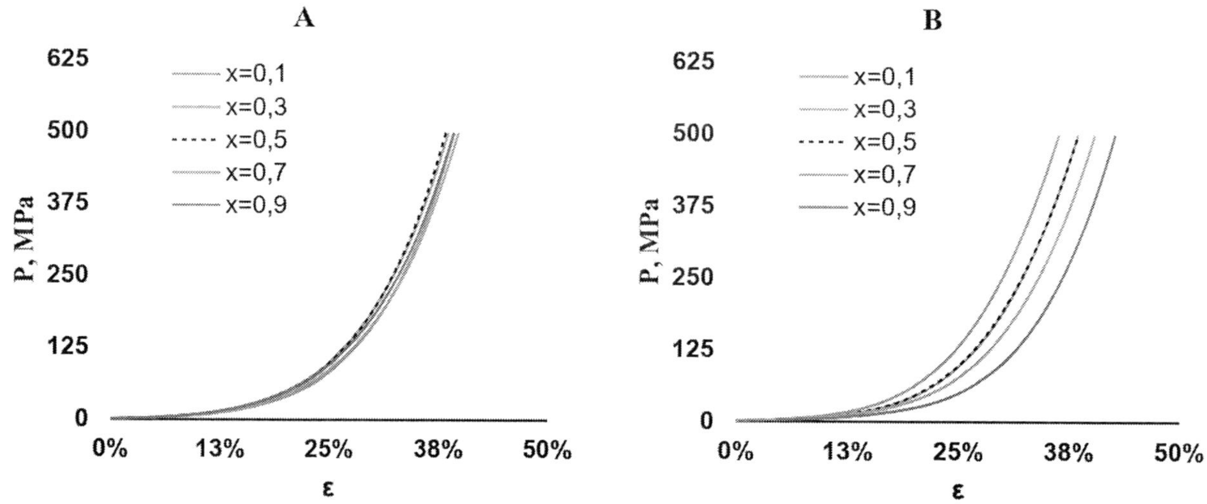

Figure 1. Diagram strain against deformation: A-SHS and B-solid-phases mixtures powder $(Mo_{1-x},Nb_x)Si_2$, where $0.1 \leq x \leq 0.9$

In the third stage, when the stain is increased over a wide range, a small change in the strain occurs. The sealing process is carried out mainly due to the growth of the contact surface during deformation of the particles.

We note that, other things being equal (the main fraction is 10-15 μm), the powders obtained by SHS method have a narrower divergence of the stress-strain curves of different compositions with respect to the curves for the powders obtained by solid-phase mixing (Fig. 1). Apparently this is due to the uniform distribution of niobium disilicide in molybdenum disilicide. X-ray analysis showed that in SHS synthesis, the crystal lattice deforms from tetragonal to hexagonal with an increase in the concentration of $NbSi_2$ (Table 1). That is, there is basically one hexagonal phase in the SHS powder, and there are simultaneously hexagonal crystals of $NbSi_2$ and tetragonal $MoSi_2$ in the solid-phase mixture, which makes their compatibility difficult when pressed.

Table 1. The cell parameters in $(Mo_{1-x}Nb_x)Si_2$ system, where $0.1 \leq x \leq 0.9$, obtained by the SHS method.

$(Mo_{1-x},Nb_x)Si_2$, x	a [Å]	c [Å]	V [Å^3]
$MoSi_2$ [80-0544]	3.200	7.850	80.4
0.1	-	-	-
0.3	4.666	6.532	123.2
0.5	4.709	6.536	125.5
0.7	4.749	6.557	128.1
0.9	4.780	6.587	130.3
$NbSi_2$	4.790	6.583	130.8

IOP Conf. Series: Materials Science and Engineering **525** (2019) 012077 doi:10.1088/1757-899X/525/1/012077

Determination of the necessary pressure in the preliminary cold pressing to ensure the selected density of the workpieces is an important task for further dilatometric studies. To compare the sintering processes of different sample compositions, it is required to obtain blanks of the same relative density. We plot the graphs in the "stress-relative density" coordinates (Figure 2).

Table 2. The rheological parameters of SHS powder $(Mo_{1-x}Nb_x)Si_2$, where $0.1 \leq x \leq 0.9$

$(Mo_{1-x},Nb_x)Si_2$, x	G, МПа (корреляция 0.980)	ε*	k_c, 10^{-6}Па
0.1	48.0	0.0642	13.0
0.3	49.8	0.0817	14.3
0.5	34.4	0.0498	13.6
0.7	31.7	0.0506	14.6
0.9	21.8	0.0043	4.16

Table 3. The rheological parameters of solid-phase powder $(Mo_{1-x}Nb_x)Si_2$, where $0.1 \leq x \leq 0.9$

$(Mo_{1-x},Nb_x)Si_2$, x	G, МПа (корреляция 0.980)	ε*	k_c, 10^{-6}Па
0.1	17.8	0.0234	10.7
0.3	35.6	0.0657	15.1
0.5	32.1	0.0551	15.3
0.7	59.1	0.0831	13.1
0.9	76.2	0.0899	19.0

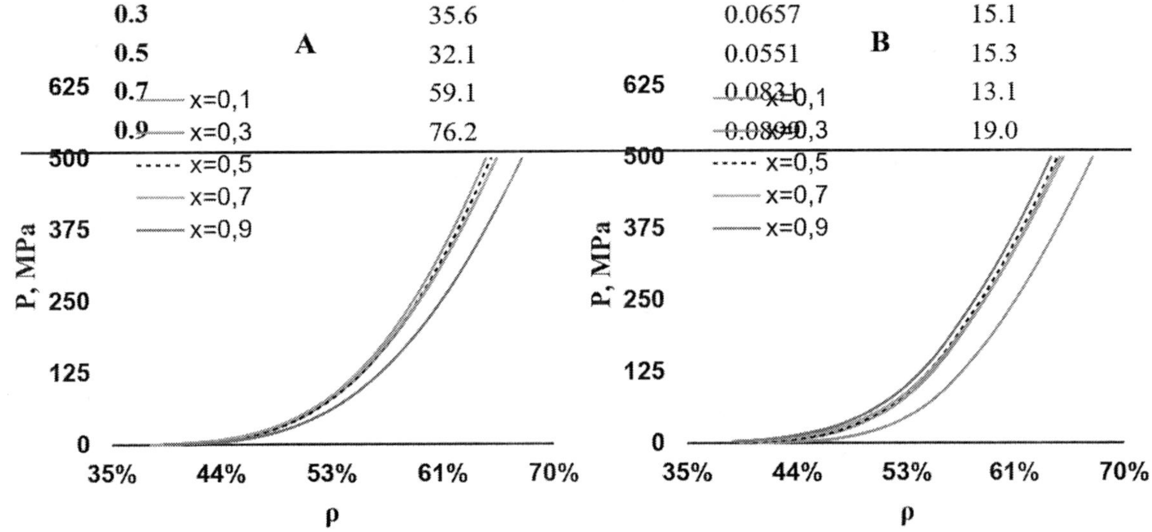

Figure 2. The diagram strain against relative density: A-SHS and B-solid-phases powder $(Mo_{1-x}Nb_x)Si_2$, where $0.1 \leq x \leq 0.9$

Analysis of the curves showed a direct relationship between the concentration of niobium disilicide and the relative density of the green body. In case of SHS powder density increases when the concentration of $NbSi_2$ increased and of solid-phase mixture of two disilicides it decreases. This behaviour can be explained by the difference in the phase composition, which leads to differences in the compaction of powders.

4. Conclusion

A rheological approach was used to study $(Mo_{1-x}Nb_x)Si_2$ powder materials, where $0.1 \leq x \leq 0.9$, obtained by SHS method and solid-phase mixing, based on the study of the kinetics of material deformation in the constant speed mode of the press plunger.

IOP Conf. Series: Materials Science and Engineering **525** (2019) 012077 doi:10.1088/1757-899X/525/1/012077

The rheological behaviour of powder materials in the system $(Mo_{1-x}Nb_x)Si_2$, obtained by SHS method and solid-phase mixing in a planetary mill, is investigated. The numerical characteristics of the materials under investigation have been experimentally obtained: the compressibility modulus G, the final value of the deformation of the linear section of the stress-strain curve ε*, the compressibility coefficient kc, at constant loading rates of 1 mm/s.

Acknowledgements
The reported study was funded by RFBR project № 18-38-00327. The methodical part of the work (X-ray and SEM) was carried out according to the state task № 075-00746-19-00. The authors are grateful to Plasmotherm LLC for providing the initial materials and for a fruitful discussion.

References

[1] Sun J, Fu Q and Guo L 2016 *Intermetallics* **72** 9
[2] Sun J et al. 2018 *Applied Surface Science* **439** 1111
[3] Hagihara K, Ikenishi T, Araki H and Nakano T 2017 *Sci. Rep.* 7 12
[4] Pan Y, Mao P, Jiang H, Wan Y and Guan W 2017 *Ceramics International* **43** 5274
[5] Andrievsky R A 1988 *Introduction to powder metallurgy* (Frunze: Ilim) p 174
[6] Stolin A M, Bazhin P M and Pugachev D V 2008 *PM and FM izvestiya vuzov* **4** 28
[7] Titov D D et al 2018 *IOP Conf. Ser.: Mater. Sci. Eng.* 347 012024
[8] Titov D D et al 2018 *Inorg. Mat.* **54** 1113
[9] Yukhvid V et al 2016 *Adv. Eng. Mater.* **18** 1930
[10] Murashova N A 2004 *Uspehi sovrem. estestvozn.* **4** 12

IOP Conf. Series: Materials Science and Engineering **525** (2019) 012078 doi:10.1088/1757-899X/525/1/012078

Synthesis of Ti–Cr–C composite materials from $CaCrO_4$ based mixtures

P A Miloserdov[1], V A Gorshkov[1], V I Yukhvid[1], O M Miloserdova[1], O A Golosova[1] and D D Titov[2]

[1]Merzhanov Institute of Structural Macrokinetics and Materials Science, Russian Academy of Sciences, Chernogolovka, 142432 Russia
[2]Baykov Institute of Metallurgy and Materials Science RAS, Moscow, 119334 Russia

E-mail: yu_group@ism.ac.ru

Abstract. Regularities of combustion and autowave chemical transformation of $CaCrO_4$/Al/C, $CaCrO_4/TiO_2$/Al/Ca/C, and $CaCrO_4$/NiO/TiO_2/Al/Ca/C highly exothermic mixtures was studied. It was shown that the mixtures can burn in a wide range of carbon content. The variation in mixture composition makes it possible to prepare cast refractory chromium compounds with different composition and microstructure. The addition of titanium oxide leads to a decrease in the combustion temperature and, accordingly, adversely affects the synthesis parameters and quality of target product. Highly exothermic CaO_2/Al additive significantly increases the combustion temperature of mixture, expands the limits of combustion and phase separation. The introduction of nickel oxide into mixture positive effects on synthesis parameters. Reduced nickel reacts with free aluminum and in doing so inhibits the formation of Cr_2AlC MAX phase. The product consisting of Cr_3C_2 and $Ti_{0.8}Cr_{0.2}C$ carbides and Ni_3Al intermetallic binder was prepared.

1. Introduction

Chromium and chromium–titanium carbides possessing high hardness and resistance to aggressive environment at elevated temperatures are widely used for producing composite materials and applying protective coatings on the working surfaces of parts operating under extreme conditions. As the matrix, nickel aluminides characterized by high heat resistance and hardness and increasing the adhesion strength of coatings with steel base and their resistance to impact, as well as reducing their porosity are most often used [1]. As a rule, carbide ceramics is produced by powder metallurgy methods using high-temperature set ups. The most effective approach to produce such materials in the cast form is the metallothermic SHS process using mixtures of chromium and titanium oxides with aluminum and carbon as a raw material. The temperature released during the redox reaction is sufficient to produce a high-temperature melt used for obtaining cast synthesized products. Refractory chromium compounds $Cr_{23}C_6$, Cr_7C_3, and Cr_3C_2 possess properties necessary for solving technical problems (high hardness, strength, and resistance to wear, heat, and corrosion) [2] and are widely used to create protective coatings [3, 4]. In the Ti–Cr–C system, TiC and Cr_3C_2 phases are the most important for practical application. The solubility of Cr_3C_2 in TiC at 1700°C is 30%. At the chromium carbide content of 30%, the microhardness of titanium carbide (3000 kg/mm^2) increases to 4000 kg/mm^2 [1].

2. Results and discussion

In the present paper, we studied the possibility of using $CaCrO_4$ calcium chromate as a chromium-containing agent instead of chemically unstable chromium oxide (VI) CrO_3.

The overall reaction schemes can be represented in the forms:

$$CaCrO_4 + Al + nC \Rightarrow Cr_xC_y + Al_2O_3 + CaO \quad (1)$$

$$TiO_2 + (70\%\ Al/30\%\ Ca) + C \Rightarrow TiC + Al_2O_3 + CaO \quad (2)$$

$$NiO + 2Al \Rightarrow Ni + Al_2O_3 \quad (3)$$

The content of mixtures (2) and (3) was varied in the basic mixture (1). In the mixture (2), a part of aluminum was replaced with calcium for more complete reduction of TiO_2 [5]. Thermodynamic analysis was carried out using a THERMO program [6].

In the system (1), the carbon content was varied to prepare various chromium carbides: $Cr_{23}C_6$, Cr_7C_3, and Cr_3C_2. The analysis showed that the adiabatic temperature of chemical transformation of the mixture (T_{ad}) exceeds 3000 K. The increase in the carbon content in the mixture (nC) from 0 to 3.7% leads to a decrease in T_{ad} and an increase in the content of metallic and gas phases (figure 1a).

Experiments on this system showed that within the range nC = 0–3.7%, the mixture retains the ability to burn. Combustion products have a cast form and are easily divided into two layers: metal (target) and oxide (slag). As the carbon content in the green mixture increases, the burning velocity and relative mass loss during combustion decrease, and the yield of the target product in the ingot increases (figure 1b).

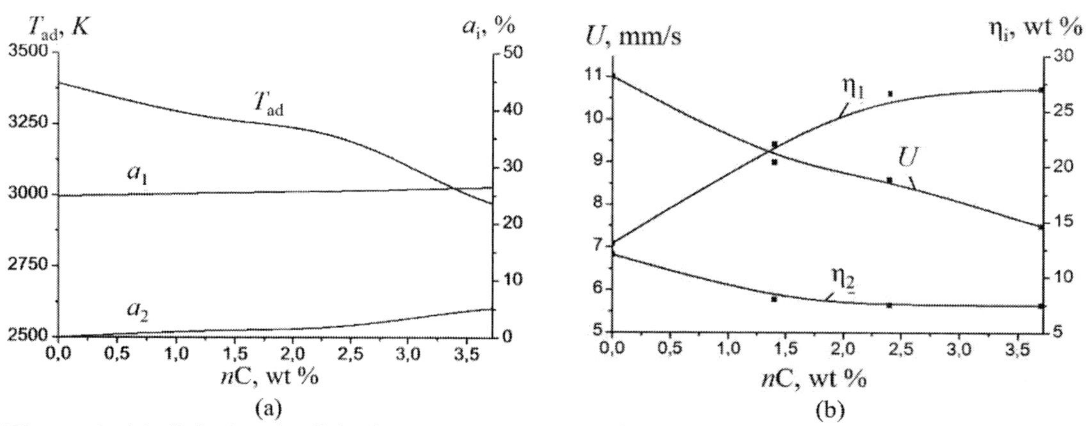

Figure 1. (a) Calculated adiabatic temperature T_{ad} and mass fractions of metallic (a_1) и gaseous (a_2) chemical conversion products as a function of nC; (b) burning velocity U, yield of metallic phase η_1, and spread of combustion products (dispersion) η_2 as a function of nC. $U = l/t$, where l is the height of the mixture, t is the time of burning; $\eta_1 = m/M_1$, $\eta_2 = [(M_1-M_2)/M_1] \times 100\%$, M_1 is the mass of the green mixture, M_2 is the mass of the final combustion products, and m is the mass of the metal ingot.

The XRD results show that target products consist of different chromium carbides including Cr_2AlC MAX phase. At nC = 2.4% (calculated carbon content to prepare Cr_7C_3), the final product is seen in figure 2 to contain mainly Cr_2AlC MAX phase.

Figure 3 shows the results of the thermodynamic analysis of mixtures, which were calculated from different ratios of mixtures (1) and (2) to produce titanium–chromium carbide. It can be seen that an increase in the fraction of mixture (2) (α) to 70% leads to a smooth decrease in the combustion temperature. Within the range α = 70–100%, the combustion temperature drops to 2600 K. The quantity of gaseous combustion products decreases to zero at α = 50%. The yield of the desired product (a_1) increases with increasing α. Experiments revealed that the mixture burns within the range α = 0–40%. With increasing α, the burning velocity U, yield of metallic phase η_1, and spread of combustion products η_2 decrease. At α = 10%, there is no phase separation. The introduction of highly exothermic additive CaO_2/Al made it possible to increase the phase-separation limit to 30% (figure 3b).

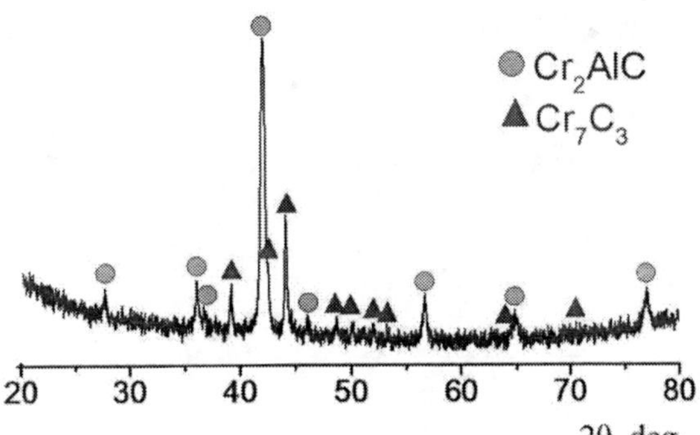

Figure 2. X-ray diffraction pattern of the product prepared at nC = 2.4%.

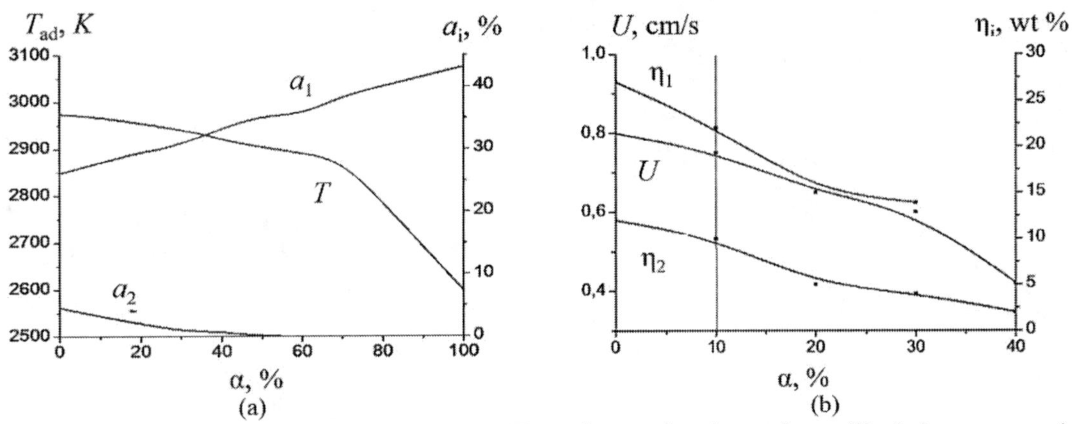

Figure 3. (a) Calculated adiabatic temperature T_{ad} and mass fractions of metallic (a_1) и gaseous (a_2) chemical conversion products as a function of α; (b) burning velocity U, yield of metallic phase η_1, and spread of combustion products (dispersion) η_2 as a function of α. $\alpha = [m_2/(m_1 + m_2)]\times 100\%$, where m_1 is the mass of mixture (1) and m_2 is the mass of mixture (2); $U = l/t$, where l is the height of the mixture, t is the time of burning; $\eta_1 = m/M_1$, $\eta_2 = [(M_1 - M_2)/M_1]\times 100\%$, M_1 is the mass of the green mixture, M_2 is the mass of the final combustion products, and m is the mass of the metal ingot.

The XRD data of prepared products show that an increase in the fraction of mixture (2) in the charge contributes to a decrease in the amount of the Cr_2AlC phase in the final product while $Ti_{0.8}Cr_{0.2}C$ phase content grows (figure 4).

In order to produce composite materials, a charge with α = 30% and mixture (3), the content of which was calculated to obtain 30% nickel in the final product, was chosen. The addition of a given amount of the mixture based on nickel oxide into the mixture makes it possible to increase the calculated combustion temperature to 2900 K that positively affects the synthesis parameters: overall yield of target phase and its phase composition. The microstructure and X-ray diffraction pattern of this composite material are presented in figure 5. As can be seen, the synthesized product consists of Cr_3C_2, intermetallic binder Ni_3Al, and $Ti_{0.8}Cr_{0.2}C$.

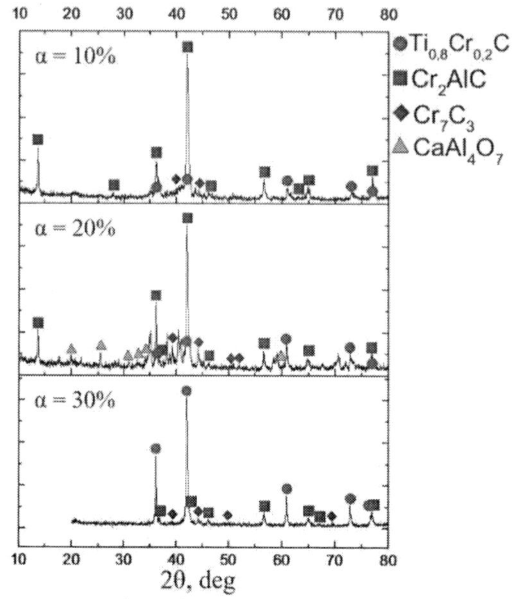

Figure 4. X-ray diffraction patterns of products prepared at different ratio of mixtures (1) and (2).

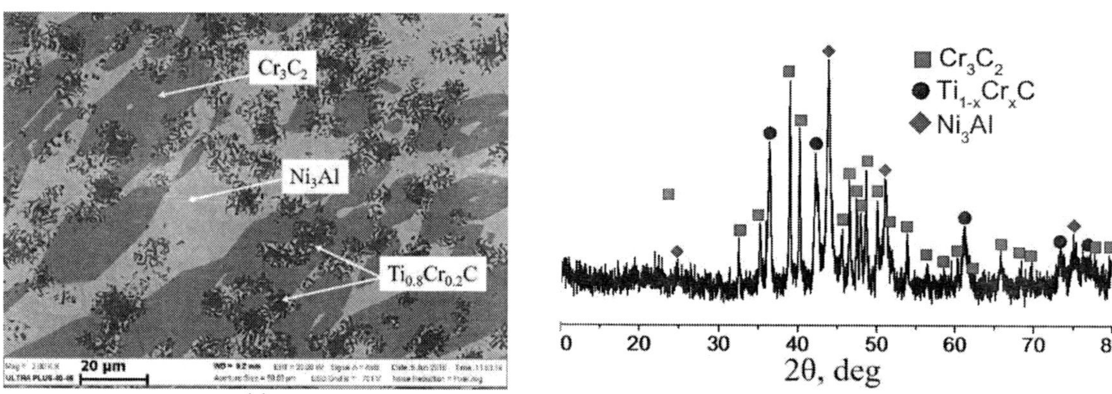

(a) (b)

Figure 5. (a) Microstructure and (b) X-ray diffraction patter of the composite material.

Conclusions

1) The regularities of combustion and autowave chemical transformation of highly exothermic mixtures $CaCrO_4/Al/C$, $CaCrO_4/TiO_2/Al/Ca/C$ and $CaCrO_4/NiO/TiO_2/Al/Ca/C$ are studied.

2) The introduction of TiO_2 leads to a decrease in the combustion temperature and adversely effects on synthesis parameters and quality of target product. The highly exothermic CaO_2/Al additive significantly increases the combustion temperature of the mixture and expands the limits of combustion and phase separation.

3) Composite material consisting of Cr_3C_2 and $Ti_{0.8}Cr_{0.2}C$ carbides and intermetallic binder Ni_3Al are obtained.

Acknowledgements

The work was supported by the Russian Foundation for Basic Research (project no. 18-08-00804).

References

[1] Kiffer R and Benezovskii F 1968 *Solid materials* (Moscow: Metallurgiya) 384 p
[2] Rudneva V V and Galevskii G V 2007 *Izv. vuzov. Cvetnaja Metal.* **2** 67–70

[3] Guilemagy J M, Espallargas N, Suegama P H and Benedetti A V 2006 *Corros. Sci.* **48** 2998–3013
[4] Korzhik V N, Borisova A L, Popov V V, Kolomytsev M V, Chaika A A, Tkachuk V I and Vigilyanskaya N V 2014 *Avtomat. Smarak* **12**
[5] Andreev D E, Yukhvid V I, Ikornikov D M and Sanin V N 2017 *Eng. Mater.* **746** 262–266
[6] Shiryaev A 1995 *Int. J. Self-Propag. High-Temp. Synth.* **4** 351–362

IOP Conf. Series: Materials Science and Engineering **525** (2019) 012079 doi:10.1088/1757-899X/525/1/012079

Research of evaluation criteria quality control of the worked oil, charged into the main gearbox of «VR-14»

D V Ratenko, I S Melnikova and A I Sergeev

Federal State Unitary Enterprise State Research Institute of Civil Aviation, Russia, Moscow

E-mail: 124otd@gosniiga.ru

Abstract. The main unit of the helicopter transforming torque from the motor shaft to the rotor and to the drives of the helicopter units is the main gearbox. The operating conditions of the reducer relate to the field of high-cycle fatigue. Oil is supplied by the gearbox lubrication system, to reduce the main gearbox mechanisms kinematic connection working parts friction coefficient and it necessary cooling. The reliability of the lubrication system largely depends on the quality of the oil used in accordance with the regulatory requirements for it, which regulate the properties of fresh oil. Regulatory requirements contain indicators that characterize its reliability, but do not reflect the requirements imposed on the properties of oils that are in use. The working oil requirements are presented by the product developer. It is used and indicated in the chemotological map of the design documentation with the reflection of these requirements and in the operational documentation. The quality control evaluation criteria of the «BP-14» main gearbox of Mi-8AMT / MTV (Mi-8) helicopters is discussed in the article. The measures aimed at improving the reliability, preserving the quality of the used oil and evaluating its parameters, as well as ensuring the durability of the whole mechanism. It was found in the course of studying the requirements for fresh oils and workers that the existing requirements do not allow us to estimate the real state of the oil and the technical condition of the system as a whole due to its partial replacement with fresh oil for different ones. The characteristics of the oil deteriorate, and at the during operation there are no requirements for oil having a working life in the design and operational documentation. The article proposes to define the following indicators for determining the quality of working oil and deciding on further operation: "determining the amount of mechanical impurities" according to GOST 10577 and "determining the purity class" according to GOST 17216

1. Introduction

A gearbox is provided in order to transfer torque from a high-speed gas turbine engine to the screws helicopters. The torque transmission to the screws is due to the transmission coupling. The Mi-8, Mi-8MTV-1, Mi-8AMT helicopters and their modifications main transmission units are the main gearbox BP-14, the intermediate gear PR-8, the tailing gear XP-8, the transmission shafts with the brake rotor brake.

The main gearbox is a separate unit it consists of bodies fastened together. There are inside: the main gearbox mechanism with a planetary stage, the drive units and the oil system of the gearbox [1].

The gearbox lubrication system purpose is to supply oil to all moving parts - gear wheels, bearing supports, splined connections to reduce the coefficient of friction and effective cooling [2]. In addition to ensuring sufficient heat removal and lubricating properties, lubricating substances have to retain their

IOP Conf. Series: Materials Science and Engineering **525** (2019) 012079 doi:10.1088/1757-899X/525/1/012079

properties for a sufficiently long operating time, considering the different operating modes and the downtime of the helicopter in changing meteorological conditions. However the quality of the oil deteriorates due to mechanical loads and chemical processes during operation: the destruction of viscous additives hydrocarbon chains under high pressure conditions, oxidation as a result of contact with air, emulsification and foaming with air and moisture. Thus, irreversible processes occur in the oil, which cause changes in its properties. For example, due to an increase in viscosity, the efficiency of the system is reduced, the frictional moment on the bearings and other nodes increases, and the heat sink intensity decreases. The corrosion occurs due to the increase in acidity.

To reduce the negative impact viability oils properties of the whole mechanism, it is replaced with a fresh one. There are two operation methods that determine the stage of oil change: the operation of the oil as it is, and the operation according to the operating time and service life. The oil change as a state is carried out on the oil observation basis according to the samples analysis results taken during operation, which is based on the establishment of limit values for the most important indicators (kinematic viscosity, acid number, additive concentration). Replacement on the appointed term (operating time, service life) is made regardless of the oil condition, design features and the aircraft equipment (AT) operating conditions. The frequency change oil according to the established regulations on the AT-based product is individual for each type of aircraft and oil grade [3].

2. Results and discussion

The working oil requirements are determined by the product designer and indicated in the chemotology map of the design documentation.

The following synthetic oils are used for the Mi-8 brand and its modifications serial helicopters in the oil systems of the main gearbox BP-14 and turbo-shaft engines TB3-117: B-3B according to the specifications TU 38.101295-85, LZ-240 according to the specifications TU 301-04 -010-92, TURBONYCOIL-98 (TN-98) to DEF STAN 91-98/2 specification.

Consider the main adverse points occurring during operation in the main gearbox BP-14. There are cases of oil flooding in BP-14 main gearbox operation process. It is due to the humid air vapor presence it has passed through the cooler, in violation of the gearbox storage rules or the oxidation of oil itself [4]. The oxidation process when the oil interacts with oxygen dissolved in the air is another source of water in the oil. It should be noted the use of synthetic oils for BP-14 gearbox have drawbacks, it is high hygroscopic and the ability to form corrosive compounds. Thus, under the catalytic effect of additives and various impurities, water promotes the hydrolysis of the oil to form alcohols and acids. When they interact with mechanical impurities, oil-insoluble compounds are formed, which precipitate and concentrate on the oil filter sections [5]. It should be noted on the working surfaces of moving parts a thin solid coating is formed, which can lead to parts jamming. All this leads to a decrease in the basic properties of the oil, which characterizes the ability to maintain its physicochemical parameters (e.g. thermal oxidative stability). Therefore, the indicators "water content" and "kinematic viscosity" in oil are subject to periodic monitoring.

Also, natural wear of friction pairs occurs in the process of product operation, as a result of metal particles enter the oily environment. In cases of critical or non-design loads, or deficiencies of a production nature it leads to further wear of friction pairs, changes in the geometric dimensions of the kinematic connection parts, distortions, the tracks and bearings occurs rolling elements deformations, which increased vibrations, clogged channels of the oil system, erosion wear of sliding surfaces or rolling (in this case bearings). These circumstances worsen the parameters of the main gearbox, and entail prerequisites for an accident. Therefore, oil from the gearbox is periodically monitored for the content of wear products to prevent the above negative consequences.

To solve the problems of controlling the performance properties of the oil and determining the limiting state during the operation of the main gearbox BP-14, the following measures were developed and implemented:

1. sampling for water availability (in accordance with the Maintenance Schedule (RO) of the Mi-8AMT / MTV helicopter every 2 months);

IOP Conf. Series: Materials Science and Engineering **525** (2019) 012079 doi:10.1088/1757-899X/525/1/012079

2. sampling for the maintenance of wear products in oil (in accordance with the Directive letter of the UNPLG GVS FSNST MT RF dated April 6, 2005 No. 5.10-21GA and the TN-98 Oil Sampling Control Methodology, considering the introduction of temporary standards for iron content in TB2-117, TVZ-117 engines oil and main gearboxes BP-8 (BP-8A) and BP-14 every 50 hours);

3. sampling for the purpose of controlling the quality of the oil (when working in accordance with the Work Programs for assessing the main gearboxes BP-14 technical condition when operating according to the technical condition for the regulated operating conditions, more than those established during the manufacture or repair of resources or service life) would);

4. oil change in the BP-14 main gearbox and in engines TV3-117 with a frequency of 300 hours, but at least 1 time per year (in accordance with the PO per helicopter). Also, oil change when changing the grade of oil (manual for technical operation (OM) on BP-14).

In the course of studying the requirements of the operational and regulatory technical documentation for the Mi-8 helicopter and its modifications, differences were found. E.g. Program No. 78-769-754ПР (works on the evaluation of the BP-14 main gearboxes technical condition with a phased increase in service life to the 1st repair (overhaul) over 15 years, after operation and storage in the original packaging, up to 19 years) is sampling oil for analysis only for the presence of water and mechanical impurities. After operating time more than 100 hours or the lifetime of 6 months the last quality control or after the last replacement, the kinematic viscosity is monitored. The same is indicated in Bulletin No. К7870-14 БЭ-Г (informing operating organizations about the resource establishment for the BP-14 main gearboxes produced by JSC Gearbox-PM before the 1st overhaul and overhaul of the technical condition up to 4000 hours for the regulated operating conditions in Civil Aviation (GA)), the addition is to perform an oil analysis, replace it, if it is impossible.

Thus, we obtained three controlled indicators that required current regulatory documents: kinematic viscosity, water mass fraction, mechanical impurities content.

The undissolved phase content in the oil can be assessed visually, by weighing, by determining the sieve composition of the particles. The contamination visual method is difficult to determine. This method has a low sensitivity and classifies the "absence" of mechanical impurities and water, when there are separate particles of contamination and water droplets with 15-20 μm average size in the oil. Therefore, they strive to use methods and devices capable of fixing particles smaller than 5 μm in size for a more accurate result.

The most common method for determining mechanical impurities is weighing. The basic principle of weighing is to determine the weight of the filter after passing a certain amount of oil through it.

GOST 6370 "Oil, petroleum products and additives. The determination of mechanical impurities method. The method essence consists in filtering the tested products with preliminary dissolving slowly filtered products in gasoline or toluene, washing the precipitate on a filter of the White Ribbon or Red Ribbon brands with a solvent, followed by drying and weighing. The mechanical impurities content is judged by the filter weight (in mass%). Mass fraction of mechanical impurities up to 0.005% inclusive is estimated as their absence.

GOST 10577 "Oil products. The mechanical impurities determining content method. The method essence consists in determining the mass of mechanical impurities that are retained by membrane filters with a pore size of 0.8-0.9 μm. when filtering the test oil through them. The content of mechanical impurities is judged by the filter weight (in wt.%).

The results obtained by determination the index of mechanical impurities according to GOST 6370 and GOST 10577 may differ due to the different structure of the filter material. Figure 1 shows the scanning electron microscopy (Tescan Vega II) of the filtration materials structure:

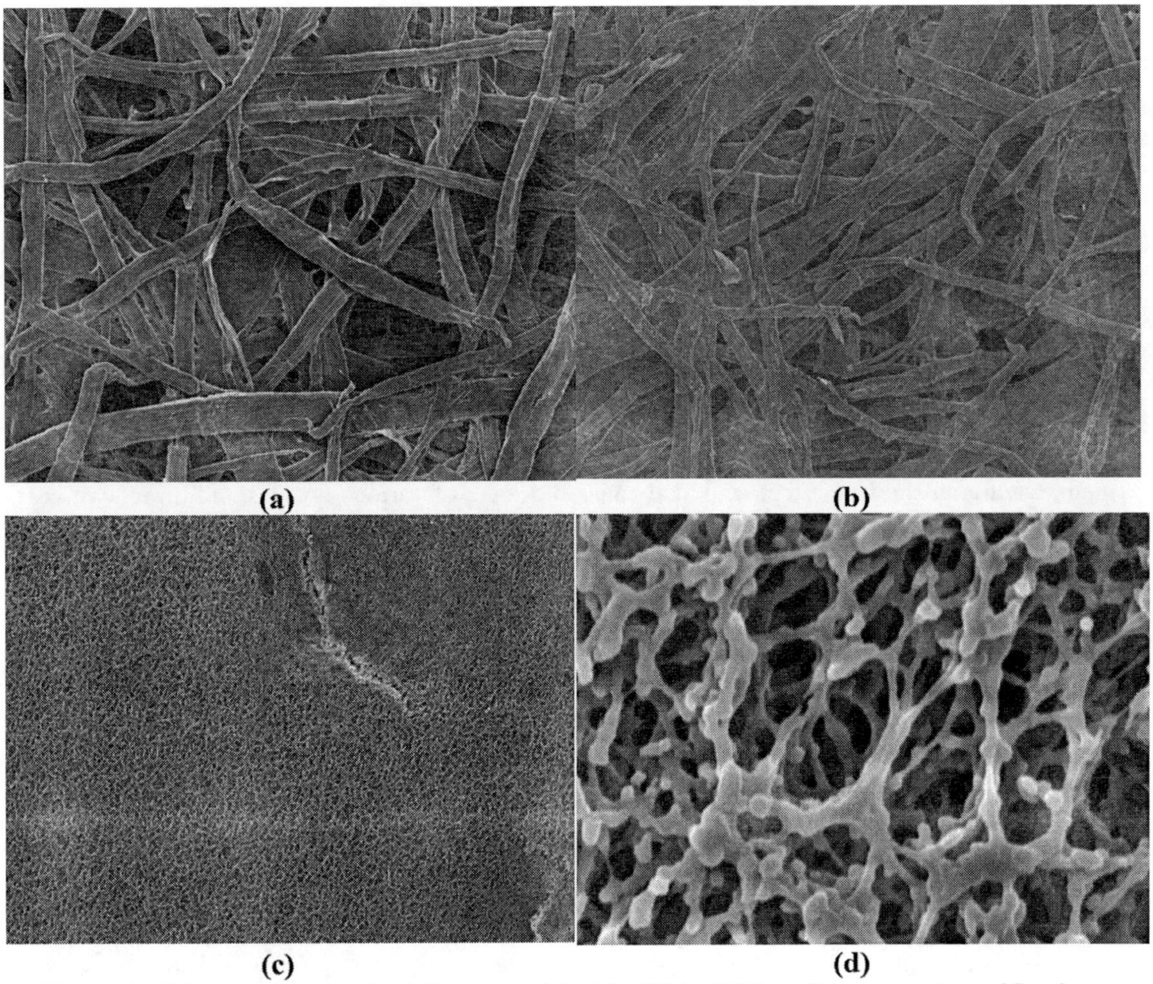

Figure 1. Electron micrograph of filter material: (a) - White Ribbon filter paper (magnification 500x), (b) - Red Ribbon filter paper (magnification 500x), (c) and (d) - Cellular cellulose membrane filter brand MFAS-HB produced by Vladipor (magnification 500x and 1000x)

A filter paper is obtained using a special technology from carefully cleaned and pressed, without sizing, long-fibrous paper pulp. The filter paper has a fibrous structure with overlapping fibers in different directions (Figure 1a, b). The filter paper porosity depends on its density and decreases with increasing density. Filter paper "White ribbon" - has a high filtration rate and the pore size is ~ 8-12 μm. Filter paper brand "Red Ribbon" - has a moderate filtration rate and the pore size is ~ 5-8 μm. Thus, the long fibers of the filter paper perform the role of a prefilter, e.g. the particles are retained at the intersections of the fibers, both on the surface and in the thickness of the filtering material [6].

Membrane filters are produced by making colloidal solutions from acetylcellulose, applied in a thin layer on glass plates and subjected to regular evaporation, during it gelation occurs, lead to the formation of open colloidal membrane structure. Membranes with porosity are obtained in very wide limits depending on the gelatinization conditions [7].

Membrane filters have a smooth and uniform surface. Such a surface creates a contrasting background that contributes to a better particles visibility. Membrane filters are sieves with medium pore sizes in a narrow range of 0.8-0.9 μm with a clear microporous structure and a stable pore configuration (Figure 1c, 1d). Thus, they play role of a surface filtration material, since particles almost completely linger on the filter surface.

IOP Conf. Series: Materials Science and Engineering **525** (2019) 012079 doi:10.1088/1757-899X/525/1/012079

The paper keeps impurities on the surface as well as inside the filter fibers, thereby creating a secondary filter layer. This layer does not allow to determine the nominal porosity of the filter paper the specified retention range of filter paper of the Red Ribbon brand and the White Ribbon brand can be used as recommendations.

Membrane filters have greater particle trapping efficiency. Particles larger than the nominal porosity remain on surface, while smaller particles pass through the filter if other interactions in the filter do not keep them in the filter matrix.

Based on the above, the membrane filter uniform pore size with a clear microporous structure and stable throughput provides greater accuracy in determining the content of mechanical impurities.

The gear oil system will be discussed further. The energy is transmitted and controlled by means of a pressurized fluid inside a closed circuit in the oil system of the gearbox. Fluid is simultaneously played a role a lubricant and means of energy transfer. The solid particles presence in the liquid reduces its lubricating properties, lead to wear parts. Contamination by particles affects the reliability and safety of the system. The degree of contamination should be monitored and should not exceed the level adopted for this system. For example, particles up to 5 μm have a significant effect on the wear of friction units, since getting into the gaps between the working surfaces of the engine parts and units, thereby jamming them, having an abrasive effect, and also cause blockage of various nozzles and throttle valves. The particles, getting into the oil system, contribute to the rupture of the oil film, worsening the lubrication mode and causing an increased oxidation of the oil.

The particle size distribution of oil pollution is determined by sedimentation and microscopic methods. Optical methods of sedimentation analysis are widely used, based on a photo colorimetric method for measuring the amount of settling particles, which are used to determine the size of wear particles present in oil. Microscopic method for determining the pollution particle size distribution is based on counting the number of particles, studying the appearance and particles shape. The particles nature can be determined using a special device SPECTROSCAN, based on X-ray fluorescent analysis of elements in a solid, powdered, dissolved state, as well as deposited on the surface or deposited on filters.

GOST 17216 "Industrial cleanliness. Liquids purity classes". The standard establishes a classification of industrial cleanliness of liquids used in the manufacture, operation and repair of machines and devices (working fluids of hydraulic drive systems and machine controls, tool drives; lubricating oils, liquid fuels, solvents), as well as coding industrial cleanliness of liquids used in hydraulic systems depending on the number and contamination particles size in 100 ± 0.5 cm3 of the analyzed liquid.

The contamination particles are all foreign particles, including sand grains, oxidation and wear products, bacteria colonies and their metabolic products. The pollution particles size, except fibers, is determined by the greatest measurement result. Fibers consider particles with a thickness of no more than 30 μm with a length to thickness ratio of at least 10 μm. GOST 17216 provides for working fluid purity classes: 00, 0, 1, 2 etc. to 17. The contamination particles larger than 200 μm (not counting fibers) are not allowed in oils. Aircraft engine oils are usually of 8 ... 13th grade purity. The 8th class of cleanliness usually has oil during the control flushing of a new engine, and the 13th class - before replacement in operation according to the rules established for each engine. Particular attention should be paid to the fibers that can pass through the filters grids together with the oil oxidation products to contribute to nozzle clogging. As practice shows, fibers appearing in oil are the nap of synthetic and cotton cleaning materials.

3. Conclusion

Investigating in practice the programs implementation to ensure the oil quality and vital components of the BP-14 main gearbox, it was found that testing samples in accordance with GOST 6370 in terms of "mechanical impurities" does not evaluate the contamination formed during operation in the gearbox, and controls only the quality of the oil itself for its compliance with the manufacturer's specifications. This circumstance can be solved by introducing an oil assessment procedure into the design and

IOP Conf. Series: Materials Science and Engineering **525** (2019) 012079 doi:10.1088/1757-899X/525/1/012079

operational documentation by the developer of the main gearbox BP-14 for compliance with the requirements of GOST 17216 and GOST 10577 not only in the manufacture and assembly of gearboxes with strict observance of industrial hygiene, but also during operation products, but also in the manufacture and supply of oil before refueling it in the main gearbox.

Thus, to determine the source of pollution, using GOST 10577 and GOST 17216 with a description of the appearance of the solid non-evolving phase in oil, the problem of ensuring the early detection of faults in main gearboxes will be solved.

References

[1] Mi-8,14,17 1995 Main gearbox BP-14. Manual operation. 7871RE, - p 79
[2] Vulgakova E B 1981 Aircraft gears and gearboxes: a Handbook (Moscow: Machine building) p 288
[3] Yanovsky A S, Kharin A A, Babkin V I 2016 Fundamentals of chemotology: a textbook (Moscow, Berlin: Direct Media) p 482
[4] Chernozhukov N I, Kerin S E 1955 Oxidability of mineral oils (Moscow: Gostoptekhizdat) p 398
[5] Syroedov N E, Galko S A, Sharykin F E, Muratkhanov N.A 2016 *Scientific Herald of MSTU GA* **225** 74
[6] Vurasko A V, Ageev A Ya, Ageev M A 2011 Technology of production, processing and processing of paper and cardboard: study guide (Ekaterinburg: Ural. state forestry. Univ.) p 272
[7] Brock T 1987 Membrane filtration, trans. from English (Moscow: Mir) p 464

Silicon nitride ceramics with light-melting sintering additive in CaO-TiO_2 system

A S Lysenkov[1], S N Ivicheva[1], D D Titov[1], Yu F Kargin[1], K A Kim[1,5], M G Frolova[1], N V Petrakova[1], V G Leontiev[1], M V Tomkovich[2], S N Perevislov[3], I S Melnikova[4] and K D Danilin[1,5]

[1]Baikov Institute of Metallurgy and Materials Science, Russian Academy of Sciences, Moscow, Russia
[2]Ioffe Physical-Technical Institute of the Russian Academy of Sciences, Saint Petersburg, Russia
[3]Institute of Silicate Chemistry of Russian Academy of Sciences, Saint Petersburg, Russia
[4]Federal State Unitary Enterprise State Research Institute of Civil Aviation, Moscow, Russia
[5]Mendeleev University of Chemical Technology of Russia, Moscow, Russia

E-mail: toxa55@bk.ru

Abstract. The aim of this work was to obtain Si_3N_4/SiC composites, where silicon nitride served as the matrix, and the silicon carbide content varied from 3 to 50wt.%. To reduce the sintering temperature (hot pressing) of the ceramic, it is used a sintering additive of calcium aluminate eutectic. The charge was prepared as follows: the silicon nitride powder was mixed with 10 wt.% sintering calcium aluminates, then the charge was added over 100 % of silicon carbide in an amount of 3, 5, 7, 10, 15, 20 and 50 wt.% in a planetary mill. Ceramic materials were hot pressed at 1660°C, the pressure was 30 MPa for 60 minutes in N_2 atmosphere. The 10 wt.% SiC composite has highest properties: 3.16 g/cm^3 density, 650 MPa bending strength, 22 GPa microhardness.

1. Introduction

Different types of structural ceramics compete with each other. In the case of silicon nitride for ball bearings, its combination of properties gives advantages over others. Silicon nitride has a unique combination of properties: high wear resistance, low density, low weight, wide operating temperature range, high resistance to oxidation or corrosion. This complex of properties allows to use silicon nitride products in many industries ranging from agriculture to electronics and aerospace [1-4].

Structural ceramics ensures the fulfillment of strict requirements that are imposed in the engineering, medical, environmental and manufacturing industries. Direct replacement of metal parts for ceramic is increasingly common in industry due to ceramics outstanding properties. Of course, ceramic products have several disadvantages due to the fragility and problems of fastening or mounting on the support. It is often necessary to consider various aspects of the system as a whole. Industrial ceramics cannot be welded or bolted to the right place, ceramics require more complex and careful fastening systems. Ceramics are highly heat resistant and hard due to these properties, engineers prefer ceramics with metals and polymers. The competition between these types of materials continues with varying success. There are a number of examples when ceramic products are replaced with metal ones, often for reasons of cheapness.

There is a search for ways to reduce the final cost of products made of silicon nitride ceramics continues. One of the ways to reduce costs is the use of sintering additives with a lower melting point than those have used in industry analogues. However, a decrease in sintering temperature may lead to a decrease in the main material properties. It is necessary to develop a sintering additive, one will reduce the sintering temperature up to 150-200°C while the properties of the resulting ceramics will not change.

The preparation of full density Si_3N_4 ceramics is difficult due to the high dissociation during sintering. Therefore, to obtain full density silicon nitride ceramic, additives are needed. It is contributed to the compaction of samples at below the temperature which silicon nitride starts to dissociate. Most often, alumina, yttrium and magnesium oxides, and their mixtures are used to densify the ceramics [5]. Unfortunately, the use of above refractory oxides increases the cost of the final products due to the high sintering temperature.

The studies were carried out to obtain low-melting sodium and yttrium oxides for the synthesis of dense and durable ceramics based on silicon nitride at 1600-1700°C [6, 7]. We have also previously conducted studies on the use of a sintering additive in the CaO-Al2O3 system of eutectic composition with a melting point of 1597°C [8]. The materials obtained had high strength characteristics: relative density up to 98%, bending strength up to 850 MPa at room temperature, bending strength up to 400 MPa at 1400°C, microhardness up to 19 GPa [9].

The aim of this research was to study the effect of a low-melting sintering additive of $CaO\text{-}TiO_2$ eutectic composition system (a melting point of 1454°C) and to analyze the properties of silicon nitride ceramics.

2. Results and discussion

The commercially available silicon nitride powder («Plasmotherm» LLC) is represented by well-crystallized fibrous particles of silicon nitride. An average size is about 3-4 μm in length and up to 200 nm thick (Figure 1). The powder contented $\alpha\text{-}Si_3N_4 \geq 95$ % and specific surface is 9.2 m^2/g.

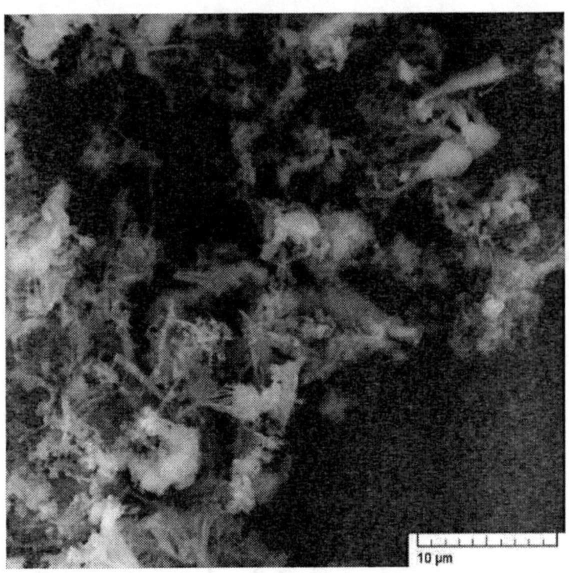

Figure 1. SEM micrographs $\alpha\text{-}Si_3N_4$ powder

IOP Conf. Series: Materials Science and Engineering **525** (2019) 012080 doi:10.1088/1757-899X/525/1/012080

Sintering additive was prepared by sol-gel method of calculating 25 mol% CaO : 75 mol% TiO_2 (Figure 2). Tetraethoxide $(C_2H_5O)_4Ti$ (> 98% Merck), isopropyl, ethyl alcohols and mineral acids were used to prepare the sol.

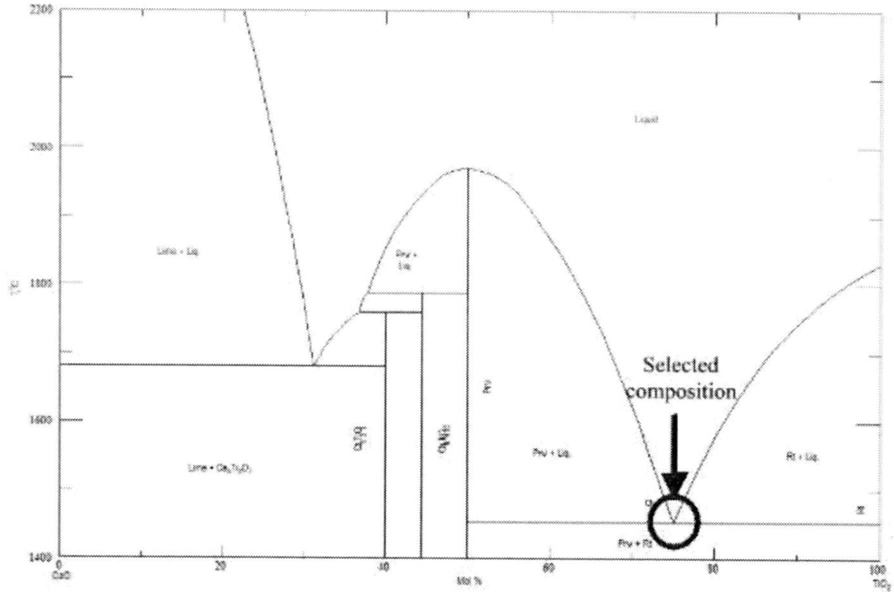

Figure 2. Diagram of the CaO-TiO_2 system.

The calculated amount of calcium oxide was introduced into the sol in the form of acidic solutions of calcium salts (chlorides or nitrates). The temperature of the additive synthesis was determined by DTA (STA 409 Luxx with mass spectrometer QMS 403 C Aëolos by Netzsch). The synthesis temperature additive lies in the range of 870-900°C (Figure 3).

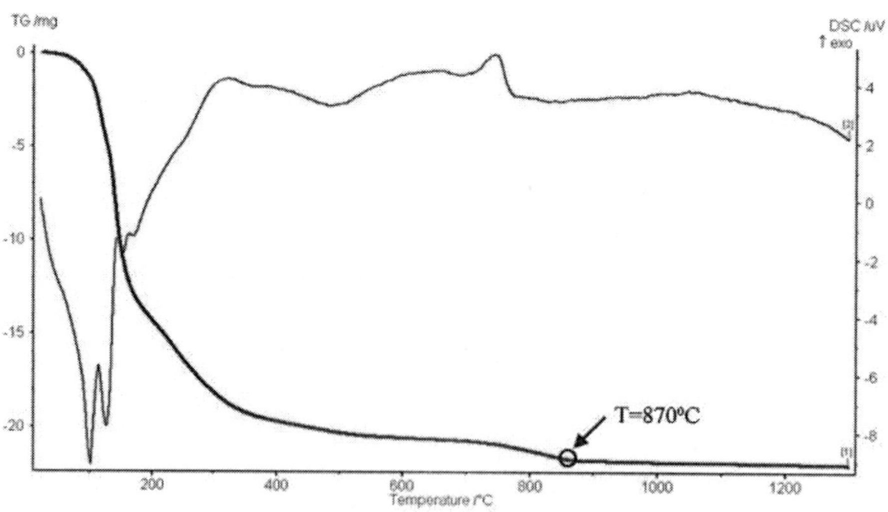

Figure 3. DTA of CaO-TiO_2 system

Si_3N_4 powder was introduced into the liquid sol with vigorous stirring on the basis of a given amount of sintering additive (7 and 10 wt.%). Gelation of a sol containing hydrated CaO – TiO_2 particles on silicon nitride powders was carried out in air, and the gels in the resulting mixtures were aged at room temperature. The removal of tetraethoxytitanium hydrolysis and calcium salts from a mixed sol, deposited on Si_3N_4 (water, alcohols and acids) was carried out by heat treatment in a muffle furnace at

IOP Conf. Series: Materials Science and Engineering **525** (2019) 012080 doi:10.1088/1757-899X/525/1/012080

900°C. In this case, the synthesized additive is evenly distributed over the volume of the charge. The additive grain size is 50-120 nm (Figure 4).

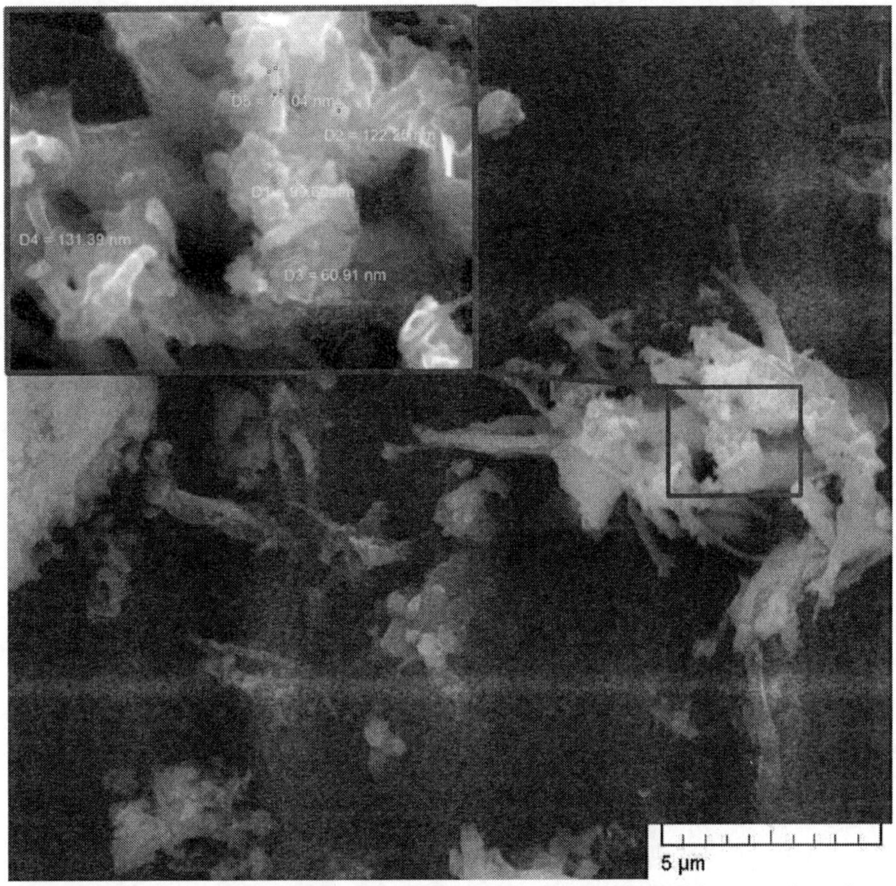

Figure 4. SEM micrographs of α-Si3N4 and sintering additive powder mix

Samples from the resulting mixture were molded blanks in the form of cylinders. The molding was performed by cold uniaxial two-sided pressing in a steel mold at 100 MPa with 3 wt.% addition of 10% aqueous solution of polyvinylpyrrolidone.

Ceramic materials were hot pressed at 1500-1600°C, a pressure was 30 MPa, for 60 minutes long in an N_2 atmosphere.

For dense ceramics used hot pressing furnace Thermal Technology Inc. high temperature experts, model HP20-3560-20. To identify the phase and chemical composition were used X-ray diffraction analysis (XRD 6000 SHIMADZU diffractometer, CuKα radiation, $\lambda = 1.5406$ Å, scanning speed $2\theta =$ 2 deg/min). The samples phase composition was identified using the PDF-2 database, JCPDS-ICDD (Set 1-2002). The surfaces of the samples were studied on a Tescan Vega II SBU scanning electron microscope with an embedded INCA Energy X-ray microanalysis system (SEM). The conductive coating (gold) application was performed using a Q150R Quorum Technologies sputtering unit. Density was measured by the Archimedes method in water. The flexural strength three bending test were studied by Instron 5581 test machine. The microhardness test was studied by Micro-hardness Tester 401/402 MVD.

The best properties had samples with 10 wt.% sintered at 1600°C. Mechanical strength was 350 MPa, density 3.06 g/cm3, microhardness 17.4 GPa.

Acknowledgments

The reported study was funded by RFBR project No 18-38-00383. The methodical part of the work (X-ray, SEM, flexural strength, density) was carried out according to the state task No 075-00746-19-00.

References

[1] Guzman I Ya 2003 Chemical Ceramics Technology: Study Guide (Moscow: Building Materials) p. 514
[2] Bocanegra-Bernal M H and Matovic B 2010 *Materials Science and Engineering: A* **6** 1314
[3] Katz R Nathan 1993 *Materials Technology* **7-8** 142
[4] Hampshire S 2007 Journal of Achievements in Materials and Manufacturing Engineering **1** 43
[5] Perevislov S N 2013 *Perspektivnye Materialy* **10** 47
[6] Tatli Zafer and Thompson Derek P 2007 *Journal of the European Ceramic Society* **27** 791
[7] Tatli Zafer and Thompson Derek P 2012 *Ceramics ŭ International* **38** 15
[8] Kargin Yu F, Lysenkov A S, Ivicheva S N, Zakorzhevskii V V, Borovinskaya I P, Kutsev S V, Solntsev K A 2012 *Inorganic Materials* **11** 1158
[9] Kargin Yu F, Lysenkov A S, Ivicheva S N, Solntsev K A, Zakharov A I, Popova N A 2010 Inorganic Materials **7** 799
[10] Lysenkov A S, Kargin Yu F, Titov D D, Petrakova N V, Ivicheva S N, Zakharov A I, Popova N A, Zakorzhevskii V V, Borovinskaya I P and Melnikova I S 2018 IOP Conf. Ser.: Mater. Sci. Eng. 347012040
[11] Toropov N A, Barzakovsky V P, Lapin V V, Kurtseva N N 1969 State Diagrams of Silicate Systems. Directory. Release the first. Binary systems (Leningrad: Nauka) p. 822

IOP Conf. Series: Materials Science and Engineering **525** (2019) 012081 doi:10.1088/1757-899X/525/1/012081

Transparent ceramic based on magnesium aluminate spinel for armor

D O Lemeshev, M O Senina, M S Pedchenko and A V Boyko

Dmitry Mendeleev University of Chemical Technology of Russia, Miusskaya sq. 9, 125047 Moscow, Russia

E-mail: diolem@muctr.ru

Abstract. Prospects of appliance of transparent ceramic based on magnesium aluminate spinel are considered. Sintering additive that makes it possible to synthesise transparent ceramic material from spinel was selected. Boron oxide was selected as such compound. The influence of concentration of this additive component as well as the influence of the method of its injection into the load on the properties of products based on magnesium aluminate spinel are considered.

1. Introduction

Optically transparent ceramic is the ceramic that is transpoarent to electromagnetic waves. This material is distinguished by fine optical transmission, density, durability, corrosion stability which make possible its usage in many fields [1]. Thus, transparent ceramic is popular in metallurgy, radioengeneering, chemical industry, protection and armor system and nuclear power industry being a prospective raw material for it. That is why optically transparent ceramic property investigation and synthetic methods from spinel is a highly topical issue of the research.

Scientists develop new processes which allow to enter manufacturing of products based on different forms of optical ceramic including details that may be used in airplane wings in future.

Global study report of world market in the field of polycrystalline transparent ceramic is serially made. It comprehends historical sources as well as the prognosis before 2025. The report contains information about leading global producers of this ceramic type, main types of products, tendencies and the field way forward. Thus, main companies which are producting optical ceramic were indicated: CoorsTek Inc. (USA), Surmet Corporation (USA), Schott AG (Germany), II-VI Optical Systems (USA), CILAS (France), Brightcrystals Technology Inc. (China), CeramTec-ETEC GmbH (Germany) [2].

Asian-Pacific Region is a large market of transparent ceramic due to its high demand in many technical fields. The growth of demand for electronics and arms industry products in major countries such as China and India stimulates development of markets in this part of the world. As estimated Asian-Pacific Region is the fastest growing market for transparent ceramic production. The region offers potential opportunity of development due to strong production basis and administrative norms which promote foreign investments receipt resulting in market growth in developing countries.

In North America the USA is the primary customer for transparent ceramic. It is associated with increase of transparent ceramic usage in the military field.

In Russia transparent ceramic production has worse development rates than in European, American and Asian countries.

One of the non-conventional materials for transparent ceramic production is magnesium aluminate spinel which is not studied enough in that field. Spinel may be used in the capacity of armour-coated glass which has the advantage by weight in comparison with sorts of glasses which are used nowadays, protecting masks for soldiers, laser optics and multispectral sensory glasses.

Magnesium aluminate spinel (MgAl2O4) is a synthetic material which crystallizes as cubes. The crystalline order of MgAl2O4 is characterized by tetrahedral coordinated magnesium ions and octahedral coordinated aluminum. Elementary cell contains 32 oxygen ions, 16 octahedral cations and 8 tetrahedral tetrahedral. Due to lack of polymorphic transformations spinel does not undergo thermic phase change. Spinel is distinguished by high hardness (13,5 hPa), thermal shock resistance, resistance to mineral acids, alkali melts, carbon and various metals, fine optical transmission in UV- (180-400 nm), visible (180-400 nm) and IR- (740-5000 nm) spectral band. This is attributable to cubic lattice structure which allows to avoid light scattering due to different refraction indexes along the axis of the crystal, and, therefore, permanent optical transmission til 92% in visible spectral band. Magnesium aluminate spinel melting point is high accounting for 2135oC [3].

Production of transparent ceramic from magnesium aluminate spinel comes amid some difficulties. Getting close texture material with a porosity slim to none is near-impossible without the use of special additives. The additive when interacting with powdery ceramic material under sintering may form interstitial solid solution or substitutional solid solution or a novel compound either fully from its own components or from a new blend composition which appears when interacting with a basic material. The additives that are forming liquid phase and are not interacting with spinel and thus, not forming second phase presence of which resulting lack of transparence in obtainable ceramic. Additives of such sort intensify the process of liquid-phase sintering and then vaporize due to temperature increase. Particularly, such additives are B2O3, LiF, CaF2 etc.

Addition of CaO observably increased the sintering rate and had an effect on MgAl2O4 grain growth. Addition of just 0,1 wt % of CaO allows to densify severely, maximum values of density became available at 0,5-1,0 wt %. In contrast the amount of additive which is more than 1,5 wt % had a negative effect on the process. When optimum quantity of CaO was used and temperature ranged from 1800 to 1900°C vacuum sintering allowed to get material with density value from 99,7 to ~ 100 % of theoretic density value. Porosity of material was almost reduced while CaO concentrations were of 0,5 and 1,0 wt % and the samples were semitransparent [4].

LiF also is very common additive for spinel ceramic sintering. It is possible to synthesize transparent ceramic based on magnesium aluminate spinel using 1 wt % LiF via pressure sintering at a temperature up to 1550°C during 2 hours. The samples are characterized with fine visible transmission coefficient ~ 85 % [5].

It was found [6, 7] that addition of a small amount of boron oxide B2O3 (0,05-0,5 wt %) promoted much more effective MgAl2O4 powder sintering which was exposed to hot isostatic pressing method. Experimental results made it clear that presence of boron oxide not only allows to decrease precursor-powder sintering temperature but inhibits ceramic grains growth as well that is very important to provide fine transparence. The 1 mm thick samples may have fine optical transmission of 80 % when the concentration of B2O3 was optimal: 0,15 wt %.

Though LiF is necessary for spinel packing to be transparent it may react with aluminum in spinel matrix producing regions enriched with magnesium. These regions decrease transparence in consolidated material due to poor sintering. Also it is possible that LiAlO2 crusts during the reaction that causes nontransparent regions formation [8].

2. Results and discussion

This work is focused on investigation of the influence of vaporized boron oxide sintering additive concentration as well as injection method of this additive on properties of ceramic based on magnesium aluminate spinel.

IOP Conf. Series: Materials Science and Engineering **525** (2019) 012081 doi:10.1088/1757-899X/525/1/012081

Aluminium hydroxide (R) and magnesium hydrocarbonate (R) were used as input materials for production of ceramic based on magnesium aluminate spinel. Aluminum oxide and magnesium oxide ratio was 50:50 mol. %. As temporary technological bond paraffin (P-2) was used.

Powder fusion temperature is an important technological parameter while producting spinel.

With a rise of temperature higher than 1000°C the composition of spinel chages greatly. So called γ-nonstoichiometry appears: magnesium aluminate spinel enriches with quite large excessive mass of Al_2O_3. Any stoichiometry change comes with an appearance of disorder in crystal – appearance of vacancies in sites of the lattice or voids colonization [9], that may affect material transparence and structural characteristics in a negative way. It is found that spinel formation process is reliant on many factors such as dispersiveness of primary components, their nature and impurity types, special additives, annealing conditions etc.

A large body of research of powder structure influence on sintering, microstructure and properties of ceramic allows to choose necessary requirements to initial powder:

- lack of aggregates with a size of more than 1 micron;
- controllable phase composition and chemical purity;
- high dispersion ability;
- powder sintering activity [10].

Magnesium aluminate spinel powder was synthesized via thermic method as it is the easiest and the most cost-efficient method in comparison with other techniques in operation.

To investigate the influence of sintering additive on the product properties 1,5; 3; 5; 7; 10 % wt of boron oxide were added into synthesized spinel.

For workpiece formation semidry molding was used. Molding pressure was equal to 100 MPa.

Sintering was carried out in two stages. At the first stage temporary technological bond was removed from the workpiece at a temperature of 800°C in air. At the second stage sintering was carried out at a temperature of 1650°C under vacuum. For the achievement of high-density state of the samples the second stage was carried out at exposures at a maximum temperature within 3 hours.

In this work the following research methods were used:

- Determination ignition loss (DIL) was determined in accordance with Russian government technical requirements 3594.15 – 93.
- Differential thermal analysis (DTA) was carried out per unit Q-1500D of Paulig-Paulig-Erdey system. Quantitive evaluation was carried out via programme "ECOKHROM".
- Scanning electron microscopy (SEM) investigation was carried out on Tescan MIRA III XMU scanning electron microscope.
- X-ray phase analysis was carried out on DRON-3M apparatus. Results evaluation was carried out via programme Crystallographica.

In order to determine magnesium aluminate spinel fusion temperature differential thermal analysis of aluminium hydroxide and magnesium hydrocarbonate mixture was carried out. The results are presented on figure 1.

IOP Conf. Series: Materials Science and Engineering **525** (2019) 012081 doi:10.1088/1757-899X/525/1/012081

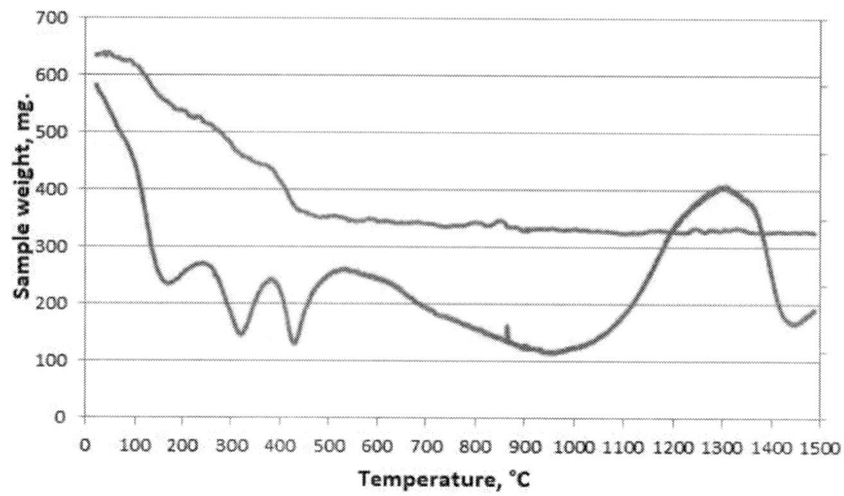

Figure 1. The results of differential thermal analysis of aluminium hydroxide and magnesium hydrocarbonate mixture

The synthesis of magnesium aluminate spinel started at a temperature of 550°C, active crystallization occured at a temperature of 1350°C. To complete the process of spinel phase formation the mixture was calcinated at a temperature of 1350°C during 1 hour. The results of X-ray phase analysis are presented on figure 2 below.

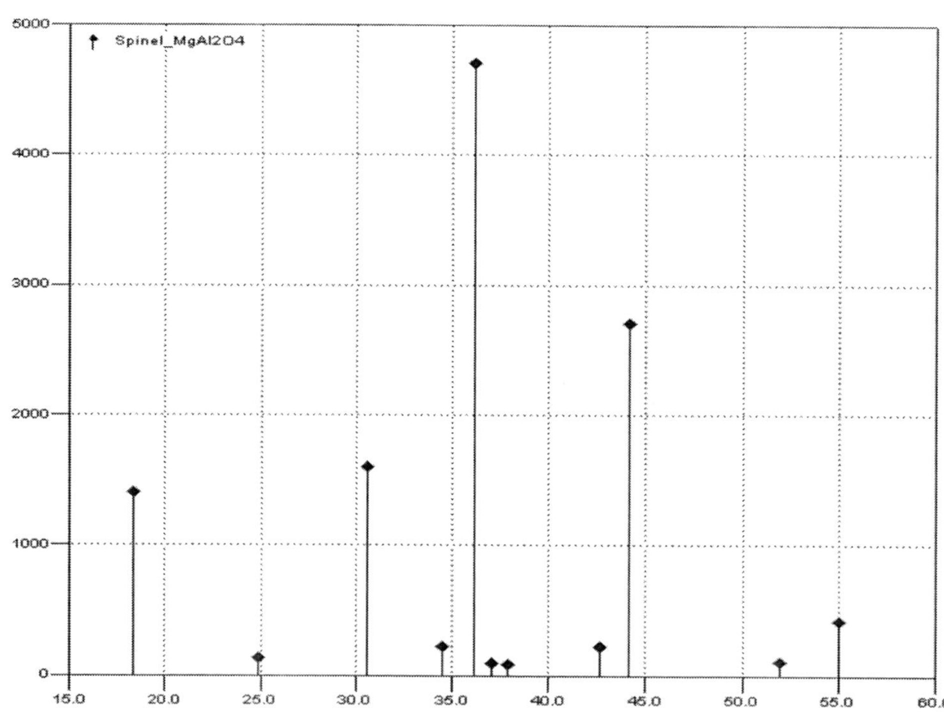

Figure 2. X-ray diagram of powder synthesized at a temperature of 1350°C

IOP Conf. Series: Materials Science and Engineering **525** (2019) 012081 doi:10.1088/1757-899X/525/1/012081

The results of X-ray phase analysis suggest that at a temperature of 1350°C pure phase of magnesium aluminate spinel is being formed as is evidenced by lack of extraneous phases.

Figure 3 shows microstructure of synthesized magnesium aluminate spinel powder.

According to SEM data magnesium aluminate spinel powder is agglomerated. There are agglomerates of spherical shape at rates of 5-15 microns and plate shaped agglomerates at rates of 50-70 microns.

Plate shape of the particles has a fine influence on sintering process helping to achieve condition of densely-sintered material. But the plates form bulky agglomerates that worsen optical indexes because of the structure in which pores are inside the material. Polydisperse structure of powder has a fine influence on sintering.

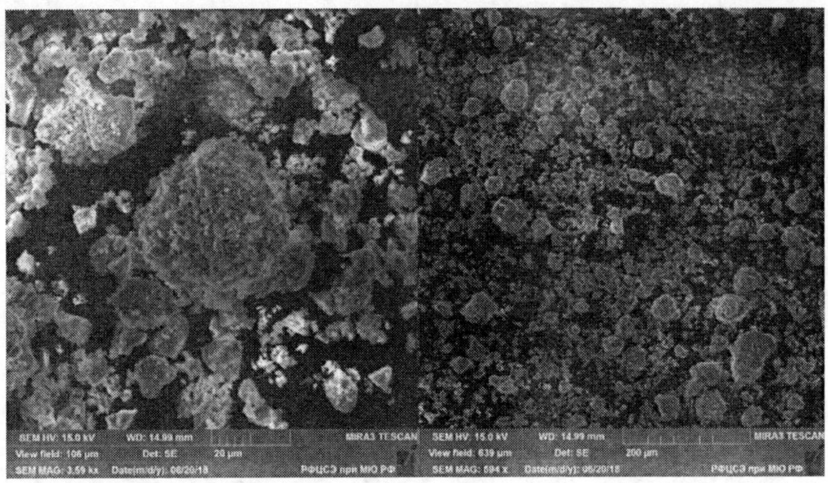

Figure 3. Microstructure of thermically synthesized magnesium aluminate spinel powder

After calcination middle-density and effective porosity of samples were determined. The results are presented on figure 4 (a, b).

Results of investigation suggest that increase of additive ratios decreases effective porosity and increases middle-density. But at larger concentration ratios (7 and 10 wt %) dramatic effective porosity increase and middle-density decrease occur due to evaporation of a large amount of additive and material cavitation. In this case optimum concentration is 5 wt %.

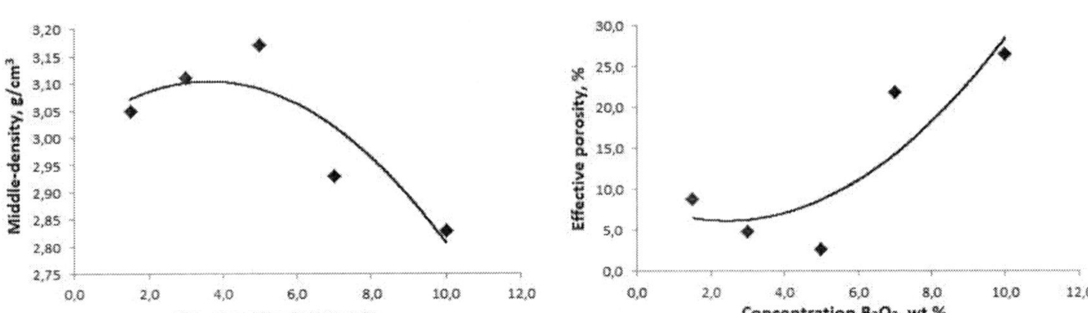

a) b)

Figure 4. Values of middle-density (a) and effective porosity (b) of samples, calcinated at a temperature of 1650°C.

An important technological stage is sintering additive injection method into a load. Boron oxide in amount of 5 wt % was added to spinel powder in two ways: dry and wet.

Values of ceramic middle-density and effective porosity are almost the same for dry and wet methods. During the experiment sample density values were 3,23 and 3,30 g/cm^3 and their porosity values were 2,5 and 2,6 % respectively. Wet method is preferable because in this case the additive dispenses homogeneously by powder volume and the method results higher values of ceramic middle-density. However, these values do not allow to get samples with fine optical transmission due to great number of pores (figure 5). It is possible to solve this problem with heat treatment conditions variation of products under vacuum.

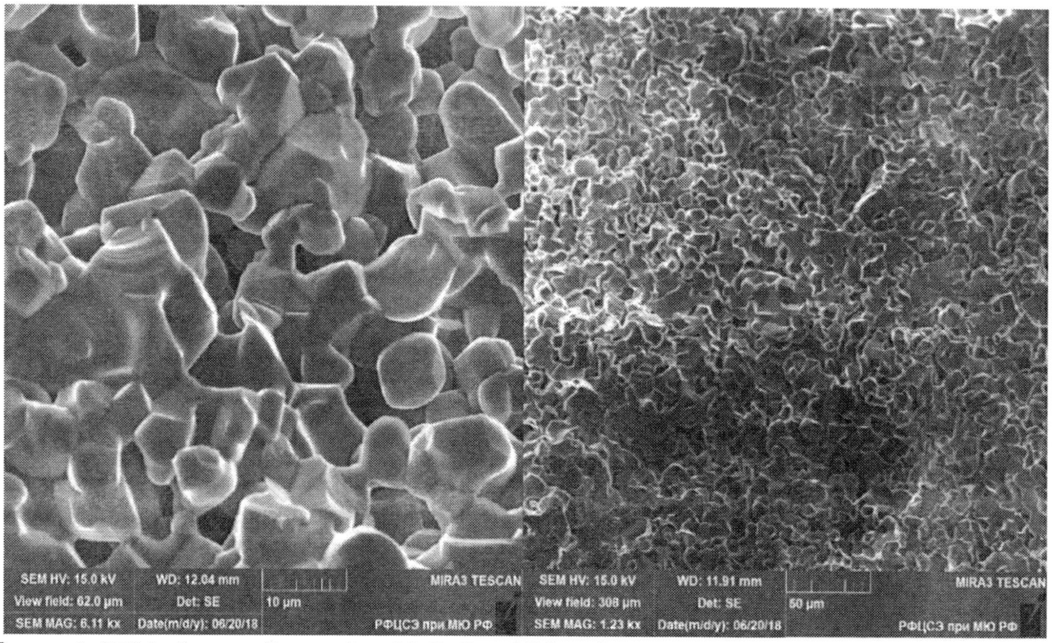

Figure 5. SEM-picture of ceramic with B_2O_3 additive that was synthesized via wet method $T_s = 1650°C$

According to the analysis data it can be said that sample material structure is nonhomogeneous with grain dimensions of 5-20 microns. Broad range of sizes is accounted for beginning of material recrystallization process. A quantity of pores has been observed as well.

Figure 6 shows synthesized ceramic samples.

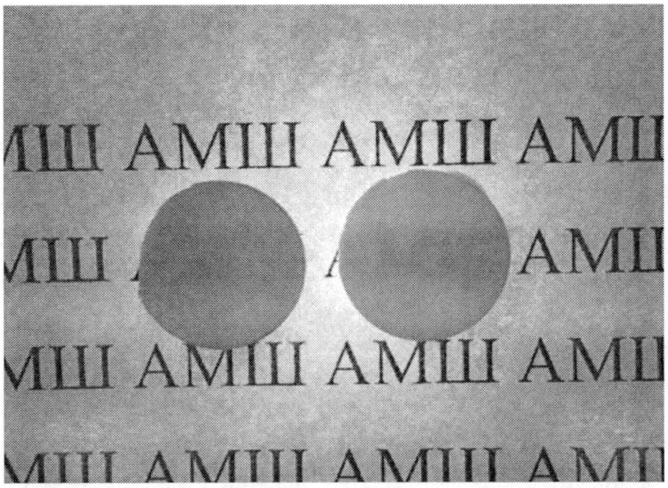

Figure 6. Samples with B_2O_3 additive that were synthesized via dry (on the left) and wet (on the right) methods, T_s = 1650°C

3. Conclusion

Thus, basing on the research findings it may be said that B2O3 addition into magnesium aluminate spinel helps to form more dense samples with low-grade of open pores in their structure. Optimum concentration should not be more than 5 wt %, density value was 3,30 g/cm3, porosity value was 2,5%. Larger concentration ratios (7 and 10 wt %) dramatic effective optical properties due to material cavitation.

After calcination material became monophasic and quite dense that allows to draw the conclusion that transparent ceramic production with the help of vaporized additives is possible.

Acknowledgments

The reported study was funded by RFBR according to the research project № 18-33-00507.

References

[1] Vydrik G A, Solov'eva T V, Kharitinov F Y 1980 *Transparent ceramic* (Moskva: Energiya) p 97
[2] URL: https://www.mordorintelligence.com/industry-reports/transparent-ceramics-market
[3] Strelov K K, Mamykin P S 1978 *Refractory engineering* (Moskva: Metallurgiya) p 376
[4] Bratton R J 1974 *Journal of the American Ceramic Society* **57** 283
[5] du Merac M R, Reimanis I E, Smith C, Kleebe H- J, Müller M M *2013 International Journal of Applied Ceramic Technology* **10** 33
[6] Tsukuma K 2006 *Journal of the Ceramic Society of Japan* **114** 802
[7] Misawa T, Moriyoshi Y, Yajima Y, Takenouchi S, Ikegami T 1999 *J Ceram Soc Jpn* **107** 343
[8] Villalobos G R, Sanghera J S, Aggarwal I D 2005 *Journal of the American Ceramic Society* **88** 1321
[9] Kovtunenko P V 1997 *Glass and ceramic* **8** 12
[10] Lukin E S 1996 *Refractories and technical ceramic* **9** 2

Composite ceramics based on silicon carbide with layered location of reinforcing SiC fibers

K A Kim [1,2], A S Lysenkov [2], D D Titov [2], Yu F Kargin [2], M G Frolova [2], A V Leonov [2], S N Perevislov [3], E I Istomina [4] and D O Lemeshev [1]

[1]Mendeleev University of Chemical Technology of Russia, Moscow, Russia
[2]Baikov Institute of Metallurgy and Materials Science, Russian Academy of Sciences, Moscow, Russia
[3]Institute of Silicate Chemistry of Russian Academy of Sciences, Saint Petersburg, Russia
[4]Komi Institute of Chemistry, Syktyvkar, Russia

E-mail: Const552@gmail.com

Abstract. The ceramic composites based on SiC with layered location of SiC fibers were obtained and studied. The samples were obtained by hot pressing. In this study the effect of layer quantity on mechanical strength was revealed.

1. Introduction

Silicon carbide is one of the few materials that includes high mechanical properties, stable in aggressive environments and extremely temperatures as well. The main disadvantage of silicon carbide, along with its positive properties, is very fragile, like any ceramic material, this fact is the main reason of properties improvement for example crack resistance and mechanical durability could by improved by reinforcing. As reinforcing material is used SiC nanotubes, SiC whiskers and SiC fibers (SiC_f). It will allow spread a scope of using ceramics.

Special attention is paid to the introduction reinforcing particles in ceramic matrix. High surface energy of particles provides compatibility with ceramic matrix, metals and polymers as well. Fibers durability is high despite small diameter 0.01 – 2 μm and up to 1600 °C can be used [1].

Using of highly effective SiC_f is promising direction in the high durable materials development. The main method of SiC_f obtaining based on SiC sedimentation on carbon fibers at a temperature 1300 °C. In this case SiC receives due to heat treatment of alkylsilane.

Authors of many scientific works, for example [2 – 4], homogenize composition of mixtures as much as possible to get uniform spreading of fibers in the matrix, however there are few study works about components organizing. In the scientific research [5] was determined that diameter increasing leads to composite durability decreasing. Optimal amount of SiC_f is up to 8 wt.% [6], if fibers concentration is more than 8 wt.%, then randomness of distribution is increased and it leads to packaging failure. In this study we considered a potential of obtaining and studying properties of composite ceramics with layered location of reinforcing SiC fibers.

IOP Conf. Series: Materials Science and Engineering **525** (2019) 012082 doi:10.1088/1757-899X/525/1/012082

2. Materials and Method

As initial raw materials we used SiC powder manufactured by «Saint-Gobain» (Figure 1a) and SiC fibers by Komi Institute of Chemistry production (Figure 1b). SiC powder included 9 wt.% sintering agent Y_2O_3-Al_2O_3 (YAG). The particles size was 0.5 – 1 μm.

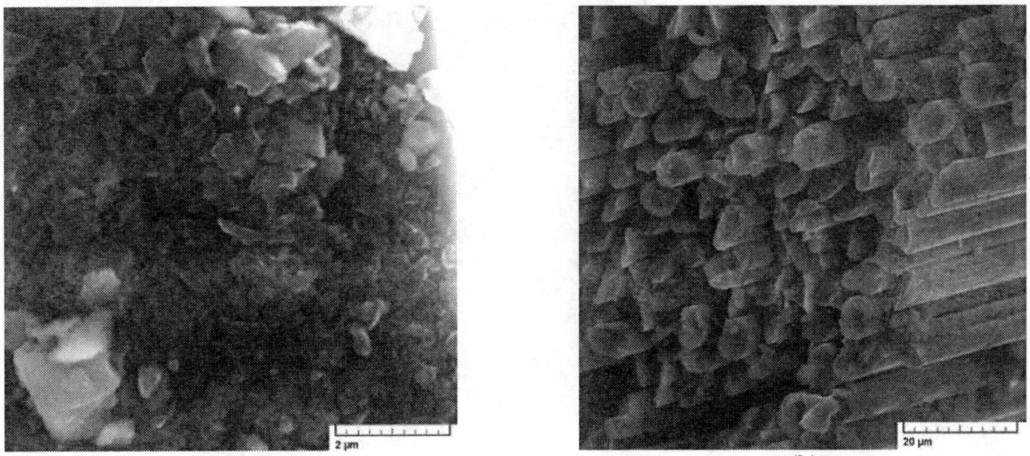

(a) (b)

Figure 1. SEM microstructure of (a) – SiC granule, (b) – SiC_f

3. Results and discussion

To obtaining layered ceramics we were guided by simplicity of technology. The main aim was to get ceramics of SiC-SiC_f 3, 5-layer sets. The total weigh of SiC powder was 7 g and weigh of SiC_f was depended on number of layers (≈7 wt.% in the sample). Low mass of SiC_f in the layer due to low bulk density. Obtained properties are shown in the table 1.

Table 1. Properties of ceramic samples

SiC	SiC_f quantity, mass. %	Temperature, °C	Density, g/cm^3	Porosity, %	Durability, MPa
Granilar SiC «Saint-Gobain»	7 (3 layers)	1850	3,20	0,95	363
Granular SiC «Saint-Gobain»	7 (5 layers)	1850	3,16	0,22	422

Receiving sequence of layered ceramics is simple but it takes high accuracy. The press form was filled in series with the layer of SiC and after with the layer of SiC_f (Figure 2). To spread reinforcing particles of the layer we used a double sided sticky tape. To obtain composite samples we used hot press Thermal Technology Inc. (HP20-3560-20) that allowed to get quite dense ceramic composite without open porosity.

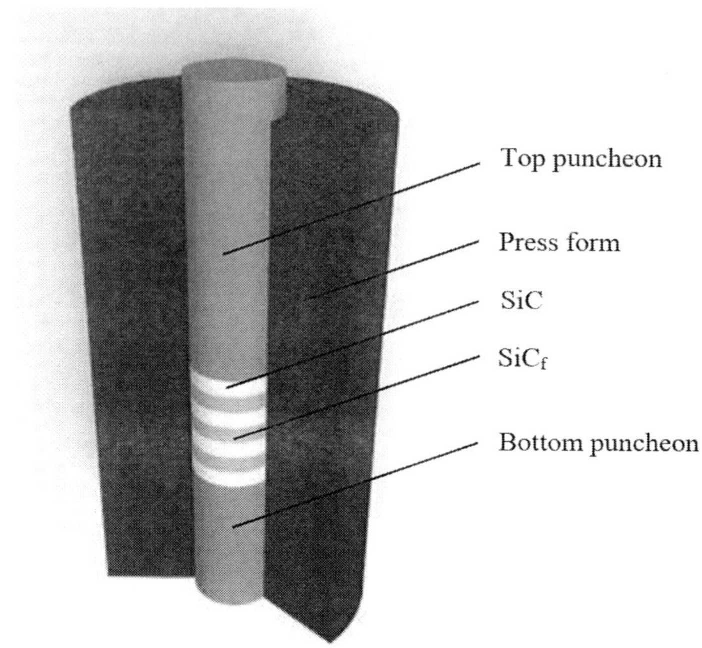

Figure 2. Schematic picture of the samples manufacture

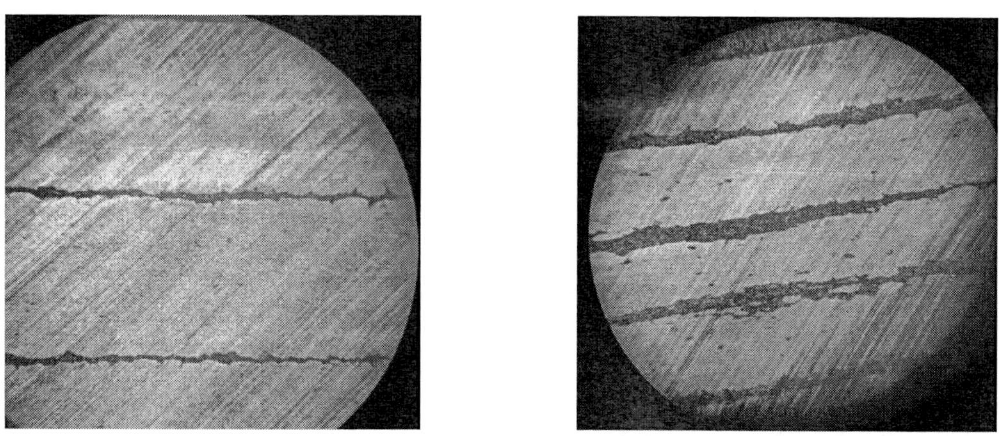

(a) (b)

Figure 3. SiC-composite structure with layered SiC_f (a – 3 sets, b – 5 sets)

4. Conclusions

A study was conducted to determine ceramic composites mechanical properties and results allow us to make conclusions:

- Mechanical properties depend on the number of SiC_f directly. Depending on this fact it is possible to vary final properties of the composite;
- Depending on the fibers length could be a difference between mechanical strength and resilience. The longer fibers in composite is greater composite resilience;
- To obtain ceramic composite we used 7 wt.% SiC_f independently on number of layers. The most durable sample had 7 layers.

Acknowledgments
The work was carried out according to the state task № 075-00746-19-00.

References
[1] Tel R 2012 *Refractories and technical ceramics* **11-12** 53
[2] Chai Y, Zhou X, Zhang H 2017 *Ceramics International* **13** 9934
[3] Silvestroni L et al 2015 *Journal of the European Ceramic Society* **16** 4377
[4] Zoli L. et al 2015 *Journal of the European Ceramic Society* **16**
[5] Grishina O I, Serpova V M 2017 *Proceedings of VIAM* **49** 1
[6] Perevislov S N, Tomkovich M V, Lysenkov A S, Frolova M G 2018 *NOVYE OGNEUPORY* **10** 37

IOP Conf. Series: Materials Science and Engineering **525** (2019) 012083 doi:10.1088/1757-899X/525/1/012083

Synthesis the composites Si_3N_4-TiN by hot pressing

Yu F Kargin[1], A S Lysenkov[1], K A Kim[1], A Yu Ivannikov[1], M G Frolova[1], D D Titov[1], S N Ivicheva[1], N A Ovsyannikov[1], A A Konovalov[1], S N Perevislov[2] and N V Petrakova[1]

[1]Baikov Institute of Metallurgy and Materials Science of Russian Academy of Sciences, Moscow, Russia
[2]Institute of Silicate Chemistry of Russian Academy of Sciences, Saint Petersburg, Russia

E-mail: yu.kargin@rambler.ru

Abstract. The Si_3N_4-TiN composites were obtained by hot pressing at 1600–1800 °C in a nitrogen atmosphere from mixtures of powders of silicon nitride and metallic titanium. The influence of particle morphology, Ti content in the initial charge and synthesis conditions on the microstructure, phase composition, mechanical strength of the samples of ceramic composites Si_3N_4-TiN was studied. It was established that in the process of sintering, complete nitriding of titanium occurs with the formation of a non-stoichiometric nitride of composition $TiN_{0.9}$. Samples of composites Si_3N_4-TiN, obtained from a mixture containing 5-30 % Ti, have a density of 3.02-3.41 g/cm3, water absorption 0.01-0.14 %, open porosity 0.03-0.44 % and flexural strength from 250 to 584 MPa. It is shown that ceramics Si_3N_4-TiN with a sintering additive of calcium aluminates is characterized by a dense intergrowth of silicon nitride crystals, which provides an increase in the strength of the samples.

1. Introduction

Composites based on silicon and titanium nitrides attract attention with the possibility of realizing high values of physicomechanical parameters and electrical conductivity [1-3]. The high hardness of Si_3N_4-TiN-based nanodispersed crystalline films is comparable to that of diamond. Composites in the Si_3N_4-TiN system can be obtained from highly dispersed powders at temperatures of 1500–1900 °C using sintering additives Y_2O_3-Al_2O_3. The literature contains information on various methods of producing composites Si_3N_4-TiN including using $TiSi_2$ [4] or modified grain sol-gel grains of powders Si_3N_4 with TiO_2 nanoparticles and their subsequent nitriding [5-8]. At the same time the use of metallic titanium as an initial component is presented in a single publication on spark plasma sintering of mixtures of powders of Si_3N_4 and Ti [9]. This paper presents the results of the production of composites in the Si_3N_4-TiN system by hot pressing mixtures of Si_3N_4 powders and highly dispersed metallic titanium in a nitrogen atmosphere combined with the nitriding of Ti with the formation of TiN.

2. Results and discussion

The starting components for the production of Si_3N_4-TiN composites were commercial silicon nitride powder («Plasmotherm» LLC, the content of α-Si_3N_4 $\geq$ 95 %, powder specific surface 9.2 m^2/g), titanium powders obtained by the plasma-chemical method at IMET RAS (designation I) and by the magnesium-thermal method at «Polema» JSC (designation P). It can be seen in Fig.1. that the powders

IOP Conf. Series: Materials Science and Engineering **525** (2019) 012083 doi:10.1088/1757-899X/525/1/012083

are significantly different in their morphology and particle size distribution. Particles of titanium powders synthesized by the plasma-chemical method are predominantly spherical in shape (3 to 35 μm in diameter), and those obtained by the magnesium-thermal method are splices of irregular shape (from 5 to 60 μm in size).

To identify the phase and chemical composition of the initial powders and the obtained ceramic samples, we used X-ray phase analysis (XRD, XRD 6000 Shimadzu diffractometer, CuKα radiation, $\lambda = 1.540598$Å), local X-ray spectral analysis (INCA Energy, Tescan Vega II). The microstructure and morphological features of the samples were investigated using scanning electron microscopy (SEM) electron microscopy (electron microscope Tescan Vega II SBU).

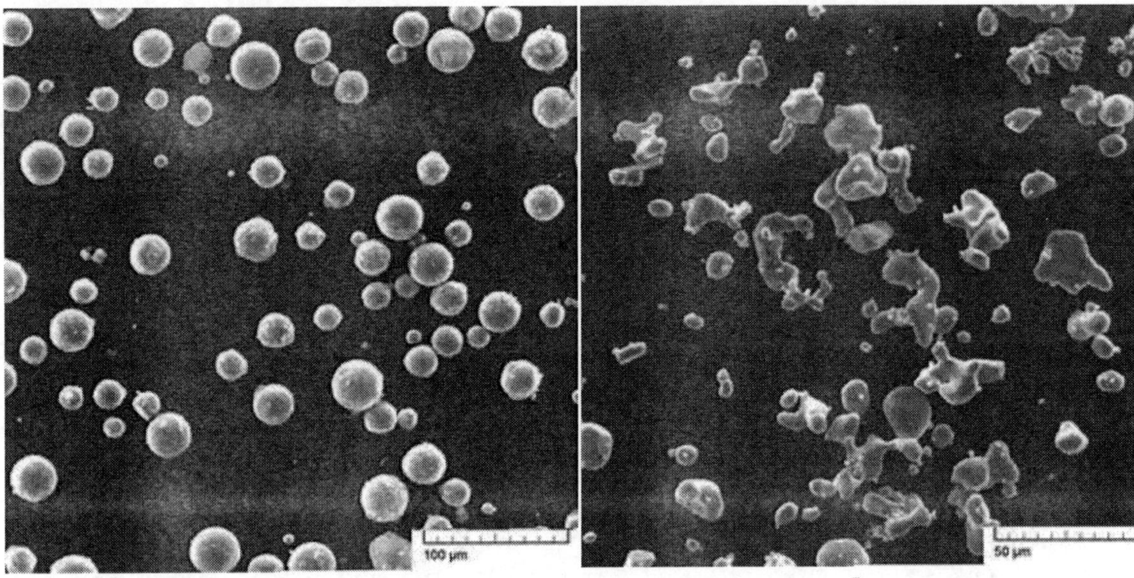

a б

Figure 1. SEM image of titanium powders obtained: a - plasma-chemical, b - by magnesium-thermal methods.

The peculiarities of the process and the sintering conditions were established on the basis of the dilatometric curves of the initial samples obtained using a DIL 402 C7G dilatometer (NETZSCH Geraetebau GmbH). On the continuous shrinkage curves of the initial mixtures of silicon nitride and titanium, the processes of active compaction of samples (with a decrease in linear dimensions from 5 to 20%) are observed in the temperature range 1400-1650 °C. Sintering of Si_3N_4-TiN ceramic composites samples was performed in a hot-pressing HP20-3560-20 press (Thermal Technology Inc.) at a temperature of 1650 °C for 1-2 hours in a nitrogen atmosphere at a uniaxial pressing pressure of up to 30 MPa. The sintering process was carried out in MPG-7 graphite molds. The original mixture was prepared as follows. Silicon nitride powder was mixed with 10 wt. % sintering calcium aluminate additive of eutectic composition (with melting temperature Tmelt. = 1597 °C) ensures maximum compaction of the composite material at relatively low firing temperatures. Then, to the mixture was added 5, 10, 15, 20 and 30 wt. % of the corresponding metal powder of titanium (P or I). The mixture was mixed in a planetary mill with the addition of isopropyl alcohol for 40 minutes, dried and rubbed through a sieve. From the resulting mixture was molded by the method of semi-dry pressing, disks with a diameter of 25 mm and a mass of 8 g. Then the disks (raw materials) were placed in a graphite mold and burned by hot pressing. In the process of heating, a specific pressure of 5 MPa was applied to the preform, which was maintained to an isothermal holding temperature. When the temperature reached 1650 °C, the pressure was increased to 30 MPa and held for 60 minutes in a stream of nitrogen. Thus,

IOP Conf. Series: Materials Science and Engineering **525** (2019) 012083 doi:10.1088/1757-899X/525/1/012083

the processes of compaction of the composite material and nitriding of titanium and flow together in the temperature range 1000-1650 °C.

The effect of titanium content in the initial charge and synthesis conditions on the microstructure, phase composition, mechanical strength of the samples of ceramic composites Si_3N_4-TiN was studied. According to the X-ray phase analysis, all the samples obtained consist of two components — silicon nitride and titanium nitride. According to the X-ray phase analysis all obtained samples consist of two components - silicon nitride and titanium nitride. At the same time both modifications of silicon nitride are present: α-Si_3N_4 (s.g. P31c, card PDF - [83-0700]) and β-Si_3N_4 (s.g. $P6_3/m$, card PDF - [83-0701]). The position of the diffraction maxima of titanium nitride on radiographs indicates the formation of a phase of nonstoichiometric composition TiN0.9 (s.g. Fm3m, card PDF - [71-0299]). The presence of calcium aluminates was not detected, since the melt formed during their melting actively interacts with silicon nitride crystals with the formation of Ca-sialon and contributes to the compaction of composites by the method of liquid-phase sintering. The remnants of the original metallic titanium were also not detected, which indicates the completeness of the nitriding reaction with the formation of the titanium nitride phase ($2Ti + N_2 \rightarrow 2TiN$). Composites Si_3N_4-TiN, obtained from mixtures containing 5-30% Ti (I or P), have a density of 3.14 - 3.41 g/cm^3 (I) or 3.02 - 3.38 g/cm^3 (P), water absorption 0.01 - 0.04% (I) or 0.02 - 0.04% (P), open porosity 0.03 - 0.44% (I) or 0.07 - 0.16% (P) and flexural strength from 365 to 524 MPa (I) or from 250 to 584 MPa (P). The best properties are observed in composites of silicon nitride with titanium nitride obtained from a mixture containing 10 wt. % titanium (I): flexural strength 524 MPa, density 3.27 g/cm^3, open porosity 0.03%. Testing samples for the presence of electrical conductivity found that composites obtained from the mixture with 5, 10, 15 wt.% Ti (I and P), have high resistance values (obviously, corresponding to the electrical conductivity of silicon nitride), samples with 20 wt.% Ti are characterized non-uniform volumetric conductivity (there are areas with both high and low conductivity). Samples of Si_3N_4-TiN composites obtained from a mixture containing 30 wt.% Ti (I and P) are conductors (i.e., a continuous conductive (percolation) cluster is formed from them in contacting metal particles).

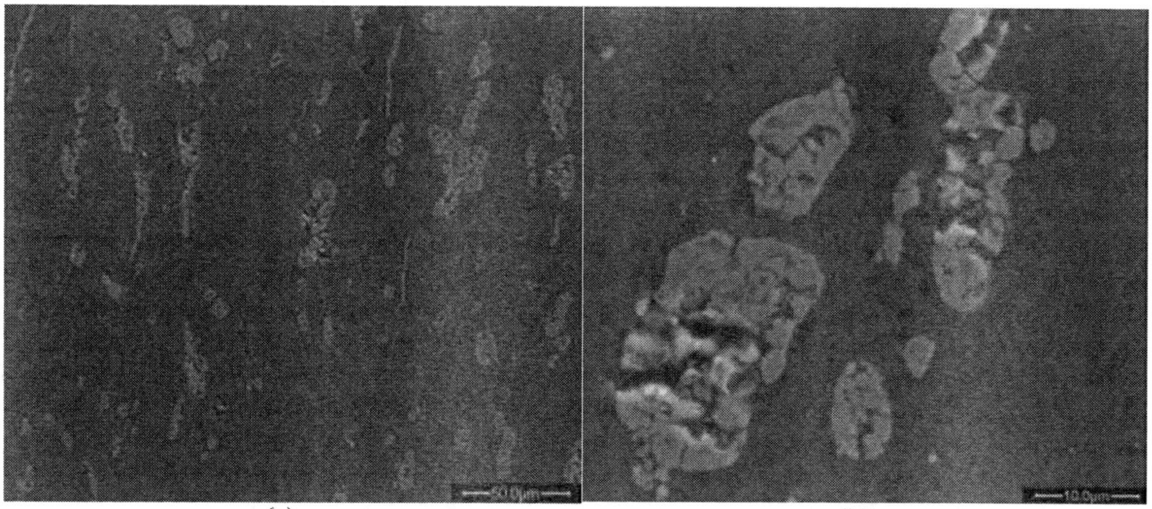

(a) (b)

Figure 2. The distribution of TiN particles in the Si_3N_4 matrix

In Figure 2 one can see a sufficiently uniform distribution of particles of titanium nitride formed in the silicon nitride matrix and the absence of their interaction with the matrix. In the areas of formation of titanium nitride voids (pores) are observed which apparently is a consequence of a higher density of TiN (5.44 g/cm^3) relative to Ti (4.54 g/cm^3). Thus, ceramics Si_3N_4-TiN with sintering calcium aluminates obtained by hot pressing at 1650 °C in a nitrogen atmosphere from mixtures of powders of silicon nitride and metallic titanium is characterized by dense accretion of silicon nitride crystals which provides rather high values of density and strength of samples.

Acknowledgments
The work was carried out according to the state task № 075-00746-19-00 with the financial support of the Basic Research Program of the Presidium of the Russian Academy of Sciences "Scientific basis for creating new functional materials".

References
[1] Bellosi A, Guicciardi S, Tampieri A 1992 *Journal of the European Ceramic Society* **9** 83
[2] Bracisiewicz M, Medri V, Bellosi A 2002 *Applied Surface Science* **202** 139
[3] Martin C, Zivkovic C Lj, Nikolic Z, Boskovic S, Miljkovic M 2004 *Journal of Alloys and Compounds* **373** 231
[4] Borodianska H, Krushinskaya L, Makarenko G, Sakka Y, Uvarova I, Vasylkiv O 2009 *Journal of Nanoscience and Nanotechnology* **11** 6381
[5] Shuichi Kawano, Junichi Takahashi, Shiro Shimada 2003 *Journal of the American Ceramic Society* **86** 701
[6] Shuichi Kawano, Kenichi Tsukurimichi, Junichi Takahashi and Shiro Shimada 2001 *Journal of Materials Chemistry* **11** 2625
[7] Zheng S, Gao L, Watanabe H, Meguro T 2007 *Ceramics International* **33** 355
[8] Kargin Yu F, Ivicheva S N, Lysenkov A S, Ovsyannikov N A, Shvorneva L I, Solntsev K A 2012 *Inorganic Materials* **48** *897*
[9] Norhayati Ahmad, Hidekazu Sueyoshi 2010 *Ceramics International* **36** 491

IOP Conf. Series: Materials Science and Engineering **525** (2019) 012084 doi:10.1088/1757-899X/525/1/012084

Zol-gel synthesis of SiAlON materials dopped by rare-earth elements

S N Ivicheva, A S Lysenkov, N A Ovsyannikov and Yu F Kargin

Baikov Institute of Metallurgy and Materials Science of Russian Academy of Sciences, 119334, Moscow, Leninskii pr., 49, Russia

E-mail: ivitcheva@mail.ru

Abstract. The phase composition and morphological features of сн-sialons doped with REE (Eu, Tb, Ce) were studied by nitriding the mixed sol prepared on the basis of silicon and aluminum alkoxides without introducing the corresponding nitrides. It was established that as a result of nitriding of a doped REE mixed Si-Al sol at 1650 °C, the β-SiAlON phases with a hexagonal crystal structure corresponding to β-$Si_3Al_3O_3N_5$ are formed. At a firing temperature of 1700 °C crystalline sialon types corresponding to the χ-phase with a monoclinic structure ($Si_2Al_3O_7N$ and $Si_6Al_{10}O_{21}N_4$) are formed. The effect of the addition of red-earth elements on the morphology of crystalline particles of the resulting sialon phases is shown.

1. Introduction

Complex nitride and oxonitride compounds, in particular, SIALON (SiALON) are promising matrices for doping with rare-earth ions and manufacturing phosphors and luminescent detectors of ionizing radiation [1-6]. Actual task is to study the influence of the method of synthesis of high-purity SiALON and the method of introducing doped REE ions on the luminescent properties of SIALONS.

Sol-gel methods for the synthesis of sialon materials make it possible to obtain high-purity and highly dispersed powders with a uniform introduction and distribution of sintering and dopants. Earlier [5, 6], the features of sialon synthesis with a nominal composition $Si_3Al_3O_3N_5$ were studied using silicon nitride or aluminum nitride and a mixture of silicon and aluminum nitrides as starting components using the corresponding silica or aluminum obtained by the sol – gel method [7].

The goal of this work was to obtain high-purity β-sialon (β-$Si_4Al_2O_2N_6$) doped REE (Eu, Tb, Ce, etc.) by nitriding the mixed sol prepared on the basis of silicon and aluminum alkoxides without introducing the corresponding nitrides.

2. Results and discussion

The silica obtained by its interaction with aluminate ions has a very strong specific interaction, which is proved by the extremely low solubility of aluminosilicate minerals, for example, such as alumina. It seems that the nitriding of highly dispersed mixed oxides will allow to obtain sialons with high functional characteristics. The special relationship between aluminum and silicon is probably explained by the fact that for both atoms the coordination number with respect to the oxygen atom can under appropriate circumstances have a value of 4 or 6, and also Al and Si have relatively close in magnitude ionic radii [8]. Therefore, the aluminate – ion $Al(OH)_4^{-1}$ is geometrically similar to $Si(OH)_4$ which promotes a closer interaction with the formation of a mixed oxide during subsequent nitriding

IOP Conf. Series: Materials Science and Engineering **525** (2019) 012084 doi:10.1088/1757-899X/525/1/012084

of which it is possible to obtain high purity and uniformity of SiALON.

Interest in the use of aluminum and silicon alcoholates in the process of sialon synthesis is primarily due to the use of their hydrolysis products to obtain pure (by chemical and phase composition) and active highly dispersed silicon and aluminum oxides and more uniform distribution of REE, which are injected in the required amount into sol in the form of homogeneous acid solutions. Mixed oxides of aluminum and silicon obtained using alkoxotechnologies differ in structural and textural characteristics from traditional crystalline varieties and amorphous silicon and aluminum hydroxides obtained, for example, by the coprecipitation method. By varying the conditions of hydrolysis, one can influence the basic characteristics of the final product, including its particle size and phase composition. In addition, when forming the synthesis products, sol-gel methods allow one to obtain narrow-fractional spherical powders of micro- and nano-dimensions, which is especially important since the maximum contribution to the excess surface energy of the system, which is the driving force in the synthesis of ceramic material, increases significantly with decreasing particle sizes.

To prepare sols, alkoxides of silicon and aluminum were used: tetraethoxysilane $(C_2H_5O)_4Si$ (TEOS), purity > 98% of Merck and aluminum sec-butoxide $Al(OCH(CH_3)CH_2CH_3)_3$, prepared by reacting metallic aluminum with butyl alcohol in the presence of catalyst (compounds gallium). In the synthesis process, ethyl and isopropyl alcohols, mineral acids, oxides or nitrate salts of REE were also used. Gelation of a sol containing mixed hydrated particles of mixed oxides of silicon and aluminum doped with REE was carried out in air, the gels were aged at room temperature. Preheating and heat treatment to remove the products of hydrolysis of alkoxides of silicon and aluminum were not performed. The obtained Al-Si-xerogel powders were burned in corundum crucibles without preliminary pressing in a furnace with a graphite heater in a nitrogen atmosphere at temperatures of 1600-1700 °C for 2 hours.

Figure 1. Photomicrograph SEM of the original x-ray amorphous mixed sol dried at room temperature

To identify the phase and chemical composition of the initial xerogel powders and their calcination products (SiAlON), X-ray phase analysis was used. The morphology and structural features of the samples were studied using scanning electron microscopy. The change in the morphology of the mixed particles of silica and aluminum oxides before and after calcination at 1600 °C in nitrogen atmosphere is shown in microphotographs (SEM), Figure 1 and 2. The initial sol dried at room temperature is represented by agglomerates 2-5 μm in size from spherical nanoparticles 20-50 nm (Figure 1). After calcination at 1600 °C of the sol doped with Ce, the resulting ceramic sample consists of intergrowths of prismatic crystals of elongated shape with an average cross section of 1 μm and a length of 2-4 μëm (Figure 2).

Figure 2. Micrograph of the SEM of a SiAlON sample doped with Ce after calcination at 1600 °C in a nitrogen atmosphere

According to X-ray analysis the sialon samples obtained by nitriding doped with REE mixed Si-Al sol at 1650°C are β-SiAlON powders with a hexagonal crystal structure corresponding to the $Si_3Al_3O_3N_5$ phase. However, on X-ray diffraction patterns, all X-ray reflections are shifted toward larger angles which is probably due to the introduction of rare-earth elements, which differ in size from silicon and aluminum ions. With an increase in the firing temperature to 1700°C according to XRD in addition to $Si_3Al_3O_3N_5$ other crystalline sialons $Si_2Al_3O_7N$ (35-0023) and $Si_6Al_{10}O_{21}N_4$ (36-0832) are observed [JCPDS-International Center for Diffraction Data. All rights reserved PCPDFWIN v.2.4, 2003], characterized as a χ-phase with a monoclinic structure in which there is some deficiency in Si probably due to the volatility of SiO which can be formed in the reducing environment of the experiment. According to SEM data the appearance of the χ-phase of sialon is manifested by the

IOP Conf. Series: Materials Science and Engineering **525** (2019) 012084 doi:10.1088/1757-899X/525/1/012084

elongation of crystals and the appearance of whiskers penetrating the structure of crystallites (Figure 3).

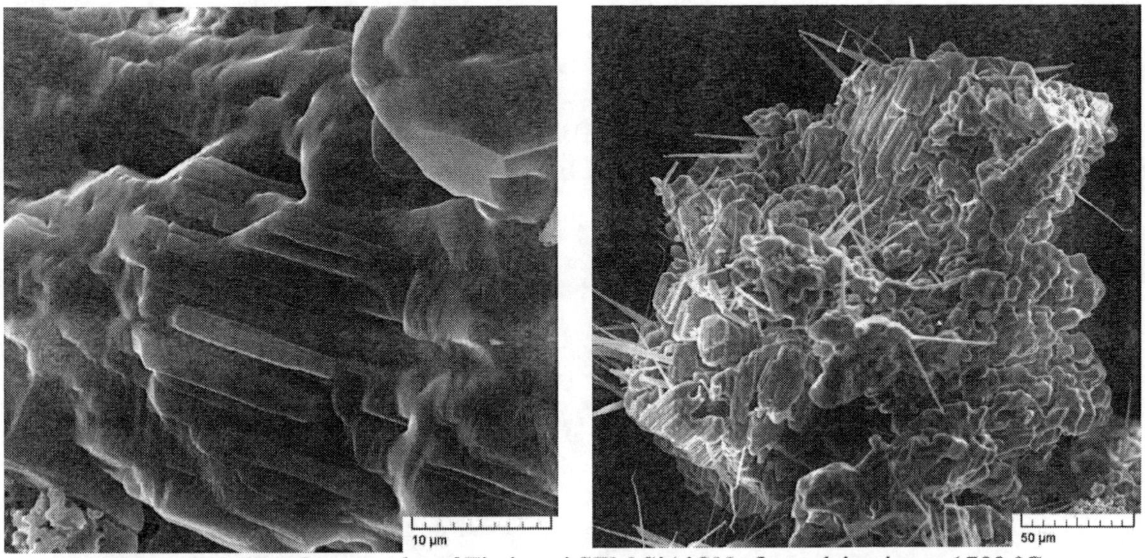

Figure 3. Microphotographs of Tb doped SEM SiAlON after calcination at 1700 °C.

Acknowledgments
The work was supported by the Russian Foundation for Basic Research, project No. 17-03-00630a.

References

[1] Shyan-Lung Chung, Shu-Chi Huang, Wei-Chi Chou and Wira Wibisono Tangguh 2014 *Current Opinion in Chemical Engineering* **3** 62
[2] Dhia A Hassan, Jian Xua, Yibin Chena, Langkai Lia, Renjie Zenga 2016 *Materials Research Bulletin* **79** 69
[3] Young Jun Yun, Jin Kyu Kim, Ji Young Ju, Seul Ki Choi, Woon Ik Park, Ha-kyun Jung, Yongseon Kim, and Sungho Choi 2016 *Inorganic Chemistry* **55** 8750
[4] Yuwaraj K Kshetri, Bhupendra Joshi, Soo Wohn Lee 2016 Journal of the European Ceramic Society **36** 4215
[5] Slesarev A I, Ivicheva S N, Lysenkov A S, Kargin Yu F, Solntsev K A, Ishchenko A V, Yagodin V V, Shul'gin B V 2017 *Problemy spektroskopii and spektrometrii* **38** 3
[6] Yagodin V V, Ishchenko A V, Ivicheva S N, Lysenkov A S, Ovsyannikov N A, Kargin Yu F, Baibalova G F 2018 *Problemy spektroskopii and spektrometrii* **40** 87
[7] S. N. Ivicheva, A. S. Lysenkov, N. A. Ovsyannikov, D. D. Titov and Yu. F. Kargin 2018 *IOP Conference Series: Materials Science and Engineering* **347** 012046
[8] Ailer R K 1982 *Chemistry of silica* (Moscow: MIR) p 712

Silicon carbide ceramics reinforced SiC fibers

M G Frolova[1], Yu F Kargin[1], A S Lysenkov[1], S N Perevislov[2], D D Titov[1], K A Kim[1], A V Leonov[1], E I Istomina[3], P V Istomin[3] and M V Tomkovich[4]

[1] A.A. Baikov Institute of Metallurgy and Materials Science, Russian Academy of Sciences
[2]Institute of Silicate Chemistry of Russian Academy of Sciences, Saint Petersburg, Russia
[3]Komi Institute of Chemistry, Syktyvkar, Russia
[4]Physico-Technical Institute A.F. Ioffe Russian Academy of Sciences, Saint-Petersburg, Russia

E-mail: frolovamarianna@bk.ru

Abstract. Ceramic materials based on silicon carbide, SiC fibers reinforced were densely obtained by hot pressed at 1850°C. SiC fibers were obtained by siliconizing. Yttrium aluminum garnet (YAG) was used as an activating additive in an amount of 8 wt.%. The highest level of physical and mechanical properties was achieved ρ = 3,19±0,01 g/cm^3; δ = 571±33 MPa; K_1C = 5,7±0,1 MPa·m$^{1/2}$ in the material reinforced with 5 wt.% SiC fibers.

1. Introduction

Silicon carbide (SiC) is an interesting material that has found application in a variety of industries. Carbides are a class of ceramic materials that has drawn great interest to the scientific community due to their exceptional mechanical, physical, chemical, and even biological properties. Therefore, carbides are the basis of advanced ceramics with numerous applications and a future prospect. Although they have been known for over one hundred years and being important materials in many technological fields such as industry tools, mechanical engineering, engine industry, optic, electronic, catalytic, nuclear technology, and chemical industry, most of their applications are recent.

There are two best known applications of this material an abrasive material and more recent use as a wide band gap semiconductor for high power and high temperature electronic devices. It is well-known for many years that SiC has a high hardness, so it is used as tools in machining and other structural applications [1].

There is the need for creating fundamentally new materials with a complex of high physicomechanical properties, capable of working under harsh conditions in connection with the development of science and technology. Attempts to create such materials have led to the emergence of composite ceramic materials reinforced with fibers and threadlike crystals. Ceramic composites are actively used for the manufacture of high-temperature protective structures, elements of gas turbines and engines, elements of optical measuring systems, sliding and rolling bearings, cutting tools, elements of armor, nozzles, friction units, etc. [5, 6]. However, their creation is associated with a number of problems, the most important are achieving uniform distribution of reinforcing additives, creating the necessary interface between components in composition and structure, and achieving a high density of material. [7]. The composite properties are derived from the combination of two or more materials properties. The composite components are not independent and can reveal strong interfacial interactions that might influence the composite's properties significantly. Specifically, mechanical characteristics depend on stress transfer from the matrix to the inclusions through the interface. Thus, the properties of the composite are affected by the compatibility of the two or more components. In some instances, the composite components can be incompatible, causing agglomerations, weak interfacial bonding, voids, or other possible defects in the composite material [2, 3].

It is therefore understandable that an overriding consideration in ceramic matrix composites (CMCs) is to toughen the ceramics by incorporating fibers in them and thus exploit the attractive high temperature strength and environmental resistance of ceramic materials without risking a catastrophic failure.

A ceramic composite is a ceramic material (matrix) carried out by introducing into its volume ultrafine micro- and nano-sized particles: whiskers, fibers, nanotubes, fullerenes, microspheres, powders of a different nature ceramic materials, etc. Al_2O_3, mullite, cordierite, Si3N4, AlN, SiC, carbon, etc. are used as the matrix material. Oxide, carbide, nitride, and boron fibers are used as the reinforcing phase. Fibers improve the physical and mechanical properties more than powders. However, if we are talking about a high temperature zone, then fibers have limitations up to 1500°C. SiC and carbon fibers are of greatest interest to developers [5-7].

Together with boron filaments and carbon fibers, SiC fibers are important continuous inorganic reinforcement materials with high modulus. Owing to their high-temperature properties and the resistance to oxidation they are particularly well suited for structural parts where high stiffness at high temperatures is required.

2. Materials and Method

The main aim of the research is to obtain a composite based on silicon carbide reinforced with SiC fiber, that has a combination of physicomechanical properties, namely, high strength and hardness, low thermal expansion coefficient and wear resistance.

The raw materials used in the work were granulated industrial silicon carbide powder (by Saint Gobain), an average particle size was 1 μm (Figure 1a), and silicon carbide fibers obtained by siliconizing SiO vapor (Figure 1b) in accordance the procedure [8].

(a)

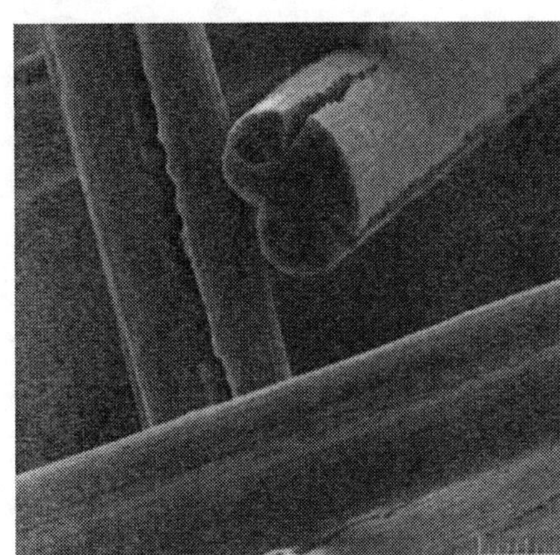

(b)

Figure 1. SEM (a) nitrogen carbide powder; (b) fiber-reinforced nitrogen carbide, obtained by siliconizing carbon fibers (the fibers are fully siliconized)

The mixture of silicon carbide powder and reinforcing fiber SiC (5 wt.%) with a sintering additive (8 wt.% yttrium-aluminum garnet), was obtained in an isopropanol environment in a planetary mill for

60 min. The resulting mixture was dried at the next stage. The blanks were molded by cold uniaxial two-sided pressing in a steel mold at 100 MPa pressure with the 3 wt.% addition of 10 % aqueous solution of polyvinylpyrrolidone. Sintering was carried out by hot pressing (Thermal Technology Inc. HP20-3560-20) in a graphite mold at 1850°C in an argon atmosphere. The pressure was 30 MPa.

The obtained samples in the form of tablets were and cut into beams polished in order to study the physicomechanical properties. The density and open porosity of the samples was measured according to GOST 15139-69. Flexural strength was determined on an Instron 5581 testing machine. Crack resistance measurements were carried out according to the indentation method on a TP-7p-1 testing machine. The method does not involve study of propagating cracks, which is typical of all other crack resistance testing methods. It is based on the study of stopped cracks after the introduction of the Vickers indenter into the sample surface.

3. Results and discussion

In this work, the microstructure and some physicochemical properties of hot-pressed silicon carbide-reinforced silicon carbide materials were studied (Figure 2). SEM analysis of the obtained ceramic samples showed no interaction between the matrix material and the reinforcing component. It may indicate the obtained ceramics is a composite material.

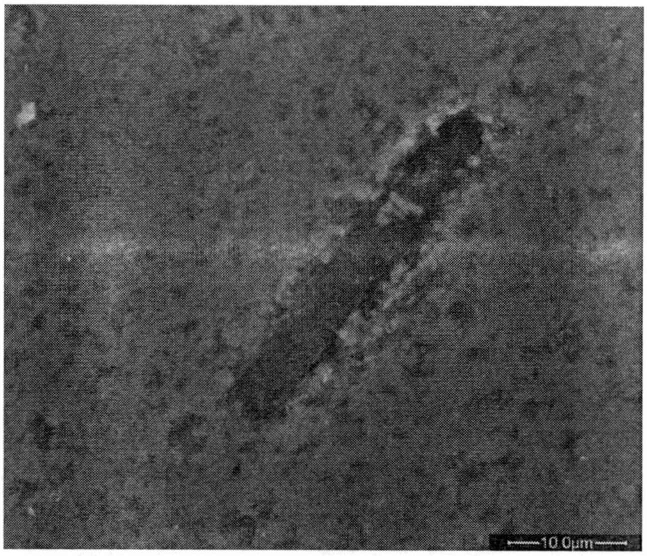

Figure 2. SEM ceramics reinforced with SiC fibers

The fibers concentration increases in SiC matrix leads to an increase in the mechanical properties of the composite. The maximum physicomechanical properties of ceramics based on silicon carbide, reinforced with SiC fibers, were the following. The samples density was obtained 3.17 g/cm^3, strength 430 MPa.

4. Conclusion

The possibility of using silicon carbide fibers as a reinforcing material causes a combination of properties such as high strength and electrical conductivity, low density, implementation of stiffness control, corrosion resistance and high-temperature creep, heat resistance, the ability to create intelligent materials, etc., as well as their compatibility with matrices of different chemical nature [1-3, 5]. In view of the listed features above, composite materials reinforced with SiC fibers can work for a long time in conditions of high temperatures and corrosive environments, while maintaining a high level of operational characteristics.

IOP Conf. Series: Materials Science and Engineering **525** (2019) 012085 doi:10.1088/1757-899X/525/1/012085

Acknowledgments
The study was carried out with the financial support of the project UMNIK.

References

[1] Gerhardt R 2011 *Properties and Applications of Silicon Carbide* (N.Y.: In Tech) p 546
[2] Hull D, Clyne T 1996 *An introduction to composite materials* (Cambridge University Press: Cambridge) p 326
[3] Barrera E V Lozano K 2000 *Journal of the Minerals, Metals & Materials Society* **11** 32
[4] Keçeli Z, Ögünç H, Boyraz T 2009 *J. Achievements in Materials and Manufacturing Engineering* **2** 423
[5] Simon G, Bunsell A R 1984 *J. Mater. Sci.* **11** 3649
[6] Iio S, Watanabe M, Matsubara M, Matsuo Y 1989 *Journal of the American Ceramic Society* **10** 1880
[7] Ye F, Lei T, Zhou Y 2000 *Materials Science and Engineering: A* **1-2** 305
[8] Istomina E I, Istomin P V, Nadutkin A V, Kargin Y F, Lysenkov A S 2018 *Inorganic Materials* **8** 787

Tree etude problems on composite fuel gas tank

A N Polilov[1], N A Tatus[1] and O Yu Sklemina[2]

[1] Blagonravov Mechanical Engineering Research Institute of Russian Academy of Science
[2] Moscow Polytechnic University
38 Bolshaya Semyonovskaya str., Moscow, 107023, Russia

E-mail: nikalet@mail.ru

Abstract. As an illustration of the design features of composite winding pressure gas tanks, the article provides three illustrative simple examples:
1. calculation of the critical pressure in composite cylinders according to the layer-by-layer analysis scheme for pairs of symmetrically wound layers;
2. assessment of the permissible wall thickness by a rational design based on the "thread analogy";
3. analysis of the effectiveness of multi-cavity spherical gas tank, which are located in each other - box in a box model ("nested doll").

1. Calculation of bearing capacity by criteria for pairs of layers

The cylindrical part of the gas tank is wind with a unidirectional tape with a symmetrical alternation of orientation angles, so that the reinforcement system can be broken down into orthotropic pairs of layers ($\pm\alpha$). If the stacking symmetry were not observed, under internal pressure, not only the axial and circumferential stresses would arise, but also shear ones, leading to a curvature of the cylinder axis, to its twisting, which is clearly undesirable.

We illustrate the use of a layer-by-layer method [1, 3] for pairs of layers using the example of a cylindrical section of thin-walled tank. Axial $\sigma_z = pR/(2h)$ and circumferential $\sigma_\theta = pR/h$ stresses are created by the internal pressure p. Where R, h are the average cylinder radius and wall thickness. If we denote E_{ij} – four constants of elasticity modules matrix of a monolayer, then for each pair of layers the Young's modulus in the axial direction is expressed in the following form:

$$E_z(\alpha) = \frac{\left(E_{11}E_{22} - E_{12}^2\right)\cos^2 2\alpha + E_{66}\left(E_{11} + E_{22} + 2E_{12}\right)\sin^2 2\alpha}{E_{11}\sin^4\alpha + E_{22}\cos^4\alpha + \left(2E_{12} + 4E_{66}\right)\sin^2\alpha\cos^2\alpha} \tag{1}$$

The Young's modulus $E_\theta(\alpha)$ is expressed by the Eq. (1) with *cosα* replaced by *sinα*. Under Voigt's assumption of equal deformation of all layers and the composite as a whole, the effective elasticity modules of the composite are found by averaging over all pairs of layers. Knowing the elastic modules and average stresses, we can determine the average circumferential and axial strains, multiplying which by the corresponding elastic modules (1), we find the circumferential and axial stresses for each pair of layers. Further, for each pair of layers under the conditions of the known biaxial stress state we apply the "tilting rhombus type" strength criteria [1, 2, 3] (Figure. 1):

$$\left|\frac{\sigma_z}{\sigma_z^*(45)}\operatorname{tg}\alpha - \frac{\sigma_\theta}{\sigma_\theta^*(45)}\operatorname{ctg}\alpha\right| = 1. \quad (2)$$

where $\sigma_z^*(45)$ and $\sigma_\theta^*(45)$ - axial and circumferential strengths of the composite tube wound with one pair of layers at angles ±45°.

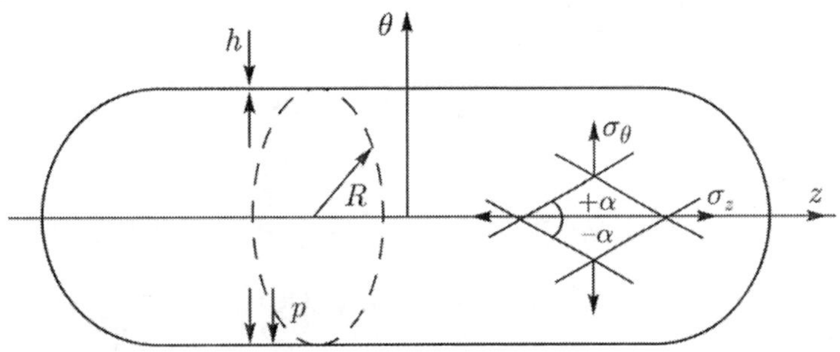

Figure 1. The cylinder scheme and the rhombus model of inextensible threads.

The fiber rupture condition (straight line 3 in Fig. 2) takes the form:

$$\sigma_z = \sigma_0 \cos^2\alpha;\ \sigma_\theta = \sigma_0 \sin^2\alpha, \quad (3)$$

where in rupture moment $\sigma_0 = \sigma_z^*(0)$ – critical tensile stress along the fibers.

The highest strength according to (2) corresponds to the optimal winding angle α^* (or the optimum ratio of applied stresses $tg\beta = \sigma_\theta/\sigma_z$) [4, 5, 6]:

$$\alpha^* = arctg\sqrt{tg\beta\ \frac{\sigma_z^*(45)}{\sigma_\theta^*(45)}}. \quad (4)$$

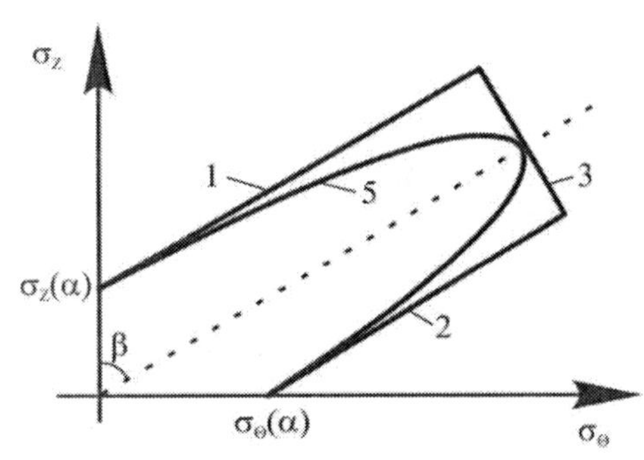

Figure 2. Limiting surfaces in the stress space for a tube with winding (±α) under biaxial tension: 1, 2 – the conditions for tilting rhombus type (2), 3 – fiber rupture condition (3), 5 – limiting ellipse (5).

The use of piecewise linear approximation (2) - (3) in computer calculations is more difficult than using a single expression for the strength criterion. In the biaxial tension of the tubes under stresses σ_z and σ_θ, the experimental data in the σ_z–σ_θ coordinates are well described by the ellipse section (*5* in Fig. 2). The ellipse passing through points on the axes ($\sigma_z(\alpha)$ and $\sigma_\theta(\alpha)$) corresponding to the strengths under uniaxial loading, and through a point on the optimal loading line (4) from the origin to the length associated with the strength (3) along the fibers $\sigma_z^*(0)$:

IOP Conf. Series: Materials Science and Engineering **525** (2019) 012086 doi:10.1088/1757-899X/525/1/012086

$$\sigma_z^2 t^4 + \sigma_\theta^2 - \sigma_z\sigma_\theta \left(2t^2 - \frac{\sigma_z^{*2}(45)\left(1+t^2\right)^2}{\sigma_z^{*2}(0)} \right) = t^2\sigma_z^{*2}(45), \tag{5}$$

where $\sigma_z^*(0)$, $\sigma_z^*(45)$ are the tensile strengths of tubes with windings (0) and (±45°); $t = tg\alpha$.

2. Rational design of the cylinder is a thread analogy

A simpler scheme for calculating the winding gas tanks is to use three assumptions, the so-called 'thread analogy' [5]:

1. The fibers work only on tension condition and carry the entire load, and the matrix is not loaded at all.
2. All fibers are equally stressed.
3. The structural failure occurs as a result of the limit stress $\sigma_0 = \sigma_z^*(0)$ reaching in all fibers (at the same time, according to the second assumption).

Assumption 2 is very 'strong', it restricts the considered winding systems only to the class of equally stressed fibers: there are no superfluous fibers, all fibers work "at the end of tether". Such constructions in reality do not exist, and should not exist. The failure of them is like an explosion – all fibers are break simultaneously, but this model can be used for engineering estimates. Using the example of the 'thread analogy' it is easy to explain the concept of '*rational design*'.

The often used term '*optimal design*' – in the broadest sense of the word means the creation of the best structure. In a narrower sense, the optimal design is traditionally called the process of finding in the design space bounded by force, kinematic, criterial, technological and other conditions, a multidimensional vector of design parameters that implements a minimum of some goal function (for example, weight, cost or some convolution of these or other criteria). Usually, optimization problems are solved by methods of non-linear programming with the use of 'penalty' functions.

Designing is called rational when certain parameter relationships are predetermined, for example, the equal strength and equal stress in all fibers. This greatly simplifies the search for an optimum. In fact, this finding of a conditional optimum is not the "top of the mountain", but the highest point on a certain section "of the mountain". In the simple case of equally stressed fibers, we get a certain maximum optimum: no fiber can be added – it will be underloaded and no fiber can be eliminated: the stresses in the remaining fibers will exceed the ultimate strength.

As an example, consider the scheme of reinforcement of the cylindrical part of a gas tank by two families of fibers: 1 – with orientation $\pm\alpha_1$ and with layer thickness h_1 and 2 – with orientation $\pm\alpha_2$, with thickness h_2. The total wall thickness $h = h_1 + h_2$. The average axial and circumferential stresses

$$\sigma_z = \frac{pR}{2h}, \quad \sigma_\theta = \frac{pR}{h} \tag{6}$$

in a critical state can be expressed through the limiting stress along the fibers σ_0, summing up the forces created in the two families of layers,

$$\frac{A_0}{2} = h_1 \cos^2\alpha_1 + h_2 \cos^2\alpha_2$$
$$A_0 = h_1 \sin^2\alpha_1 + h_2 \sin^2\alpha_2 \tag{7}$$

where $A_0 = pR/\sigma_0$.

Two equations (7) contain four design parameters, two of which can be found after setting the other two arbitrarily (?). The question-mark (?) means that we want to make sure that we can really specify any values of the angles. Suppose that $h_2 = 0$. There remain two parameters that are found from (7):

$$h = h_1 = \frac{3A_0}{2}; \; \alpha_1^* = \alpha^* = 54^0 44'; \; \left(\mathrm{tg}^2\alpha^* = 2\right).$$

Choosing angles α_1, α_2 equal to (±30/90), (±45/90), (0/±60), we will see that the total thickness of the wall remains the same, $3A_0/2$. The same will remain for the algebraic sum of the thicknesses and for a combination of angles (±60/90), (0/±45) or (0/±30), only one of the thicknesses will be negative, which means that it is impossible to provide equilibria in two stresses if two families of fibers have an orientation angle both smaller or both larger than α^*. In fact, not the thickness is negative, and the stresses in these, 'unsuccessfully' selected families, should become compressive to ensure the stress uniformity of the fibers (7).

This example shows that any rational project of equally stressed families of fibers provides the same weight of the structure under given loading conditions; choosing the best remains for technological or constructive reasons.

3. Multi-cavity spherical gas tank

Multi-cavity gas tanks (of the type "nested doll") - (Fig. 3) are of interest to reduce the overall dimensions, since for a given external radius of the increasing number of the inner tanks to increase their allowable pressure, and hence can increase the total mass of injected gas.

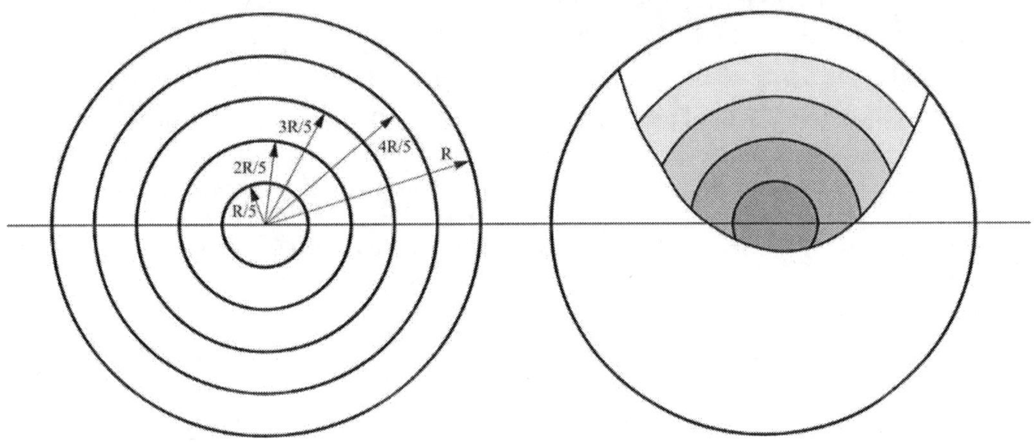

Figure 3. Multi-cavity gas tank in section.

As a model, consider ***n*** equidistant spherical gas tank put one into another (box in a box model or "russian nested doll") with a constant wall thickness h. The outer cylinder with the number n has a radius R, the inner, the smallest with the number 1 – R/n accordingly, the i-th cylinder has a radius iR/n.

1) The total mass of the gas tank, which naturally increases with the number of tanks:

$$M=\sum_{i=1}^{n} m_i = 4\pi\rho h R^2 \sum_{i=1}^{n}\frac{i^2}{n^2} = B\frac{n(n+1)(2n+1)}{6n^2} = \frac{B}{6}\left(1+\frac{1}{n}\right)(2n+1); \tag{8}$$

where $B=4\pi\rho hR^2$ is the mass of the outer (largest) sphere;
ρ - the density of the tank material.

2) The volume of the cavities between the spheres:

$$v_1=\frac{4}{3}\pi\left(\frac{R}{n}\right)^3=\frac{A}{n^3};\quad v_i = A\frac{i^3-(i-1)^3}{n^3};\ i=1,\ldots,n; \tag{9}$$

where $A=4\pi R^3/3$ - is the volume of the external spherical gas tank.

3) Calculate the allowable pressure in each cavity. This is analytically - the most complex, iterative procedure, not solvable in algebraic form, since there is no general formula for partial sums of a harmonic series. The permissible stress $[\sigma]$ for each cavity is expressed in terms of the external p_{i+1} and internal p_i pressure, which, for convenience, we refer to 0.1 MPa = 1 atm to estimate how many times the volume of accumulated gas exceeds the volume of the cavity v_i:

$$[\sigma]=(p_i-p_{i+1})\frac{R_i}{2h};\ N_i=\frac{p_i}{p^0};\ \sigma^*=\frac{[\sigma]}{p^0}\Rightarrow$$
$$N_n=C;\ N_i=C\frac{n}{i}+N_{i+1}=Cn\left(\frac{1}{n}+\frac{1}{n-1}+...+\frac{1}{i}\right)=Cn\sum_{k=0}^{n-i}\frac{1}{n-k}=Cn\sum_{j=i}^{n}\frac{1}{j},\qquad(10)$$

where $C=2h\sigma^*/R$ the permissible gas pressure (in atm.) in the outer (largest) spherical gas tank.

4) Calculate from (9), (10) the total volume of gas in all cavities in terms of the normal pressure of 1 atm. The volume of accumulated gas increases with the number of cavities with the former external dimensions, i.e. with the same radius of the external sphere:

$$V=\sum_{i=1}^{n}(v_iN_i)=\sum_{i=1}^{n}\left\{\frac{AC}{n^2}\left[i^3-(i-1)^3\right]\sum_{j=i}^{n}\frac{1}{j}\right\}.\qquad(11)$$

5) We find from (8), (11) the efficiency factor K, as the ratio of the largest possible volume of gas in a multi-cavity tank to the mass of the walls:

$$K=\frac{V}{M}=\frac{AC}{B}=\frac{2\sigma^*}{3\rho}.\qquad(12)$$

As can be seen from equality (12), obtained on the basis of numerical calculations, the efficiency factor K (in the 1st approximation, without taking into account the effect of wall thickness on cavity volumes) does not depend on the gas tank radius, nor on the wall thickness, nor on the number of embedded tanks. This is, in fact, a property of the gas tank material: metal, composite. You can compare, for example, the effectiveness of steel and aluminum cylinders. The result is not obvious and ambiguous. It is necessary to check the validity of formula (12) by direct arithmetic calculations for n = 1, 2, 3, 4, 5, etc.

Table 1. The results of calculations by formulas (8) - (12) for small number of cavities.

	1 cavity	2 cavities	3 cavities	4 cavities	5 cavities
The mass of gas tank M (Eq. 8) kg	$4\pi\rho R^2h\sigma^*=B$	$\frac{5}{4}B$	$\frac{14}{9}B$	$\frac{15}{8}B$	$\frac{11}{5}B$
The volume of gas tanks v_1 (Eq. 9) m³	$\frac{4}{3}\pi R^3=A$	$v_1=A/8$; $v_2=7A/8$	$v_1=A/27$; $v_2=7A/27$; $v_3=19A/27$.	$v_1=A/64$; $v_2=7A/64$; $v_3=19A/64$; $v_4=37A/64$.	$v_1=A/125$; $v_2=7A/125$; $v_3=19A/125$; $v_4=37A/125$; $v_5=61A/125$.
Permissible pressure in each tank N_n (Eq. 10)MPa	$\frac{2h\sigma^*}{R}=C$	$N_1=3C$; $N_2=C$;	$N_3=C$; $N_2=5C/2$; $N_1=11C/2$;	$N_4=C$; $N_3=7C/3$; $N_2=5C/2$; $N_1=11C/2$;	$N_5=C$; $N_4=9C/4$; $N_3=47C/12$; $N_2=77C/12$; $N_1=137C/12$.
Bulk-volume gas V (Eq. 11) m³	$v_iN=AC$	$\frac{5}{4}AC$	$\frac{14}{9}AC$	$\frac{15}{8}AC$	$\frac{11}{5}AC$
Efficiency factor $K=\frac{V}{M}$	$\frac{2\sigma^*}{3\rho}=\frac{AC}{B}$	$\frac{AC}{B}$	$\frac{AC}{B}$	$\frac{AC}{B}$	$\frac{AC}{B}$

To obtain an analytical expression for the partial sums of the harmonic series, no one else (since the time of Euler) has succeeded.

Two words about the features of the harmonic series. The name “harmonic” itself is connected with the fact that when the length of a string (violin) decreases by a factor of **i**, its tone corresponds to the **i**–th harmonic with respect to the total length of the string. Euler proposed the following approximation:

$$S_n = \sum_{i=1}^{n} \frac{1}{i} = \ln n + \gamma + \varepsilon_n; \quad \varepsilon_n \xrightarrow[n\to\infty]{} 0;$$

where γ=0.5772 - Euler-Macheroni constant.

Table 2. Partial sums of the harmonic series

n	1	2	3	4	5	6	7	8	∞
S_n	1	3/2	11/6	25/12	137/60	49/20	263/140	761/280	∞

Another surprising property of the harmonic series (apart from the absence of an analytical expression for partial sums) is that these sums grow very slowly. Despite the fact that the series diverges, that is, its partial sum tends to infinity, in order to achieve, for example, a value of 100, it is necessary to sum up an insanely large number of members of the order of 10^{43}. The computer with a speed of a billion operations per second will calculate over millions of years such a number members. But this fact, apart from curiosity, has nothing to do with the task at hand: it’s just that - a surprising number.

4. Conclusion

1. Using the cylindrical part of the gas tank as an example, the easiest way to illustrate the possibility of optimizing the structure is by choosing the winding angles.
2. Any rational project of equally stressed families of fibers provides the same mass of structure under given loading conditions; choose the best is from technological or design considerations.
3. In the first approximation, without taking into account the wall thickness, the ratio of the largest amount of gas stored in a multi-cavity gas tank to the mass of its walls does not depend on the tank radius, nor on the wall thickness, nor on the number of built-in tanks.

Acknowledgments

This work was financially supported by the Russian Foundation for Basic Research, grants No. 18-08-00372 and 18-58-53020.

References

[1] Polilov A N Tatus' N A 2018 *Biomekhanika prochnosti voloknistyh kompozitov* (M.: FIZMATLIT) p 328 ISBN 978-5-9221-1760-9

[2] Polilov AN Tatus' NA 2012 Eksperimental'noe obosnovanie kriteriev prochnosti voloknistyh kompozitov, proyavlyayushchih napravlennyj harakter razrusheniya. *Vestnik Permskogo nacional'nogo issledovatel'skogo politekhnicheskogo universiteta. Mekhanika* no 2 pp 140-166

[3] Polilov A N 1980 Method of calculating the strength of oblique-reinforced composites in plane stress *Mechanics of Composite Materials* vol 16 no 2 pp 154-158

[4] Polilov A N Tatus' N A 2008 Kriterii prochnosti polimernyh voloknistyh kompozitov, opisyvayushchie nekotorye ehksperimental'no nablyudaemye ehffekty *Problemy mashinostroeniya i avtomatizacii* no 3 pp 103 - 110

[5] Polilov A N 2001 Strength and Reliability Model of GFRP Fuel gas Tanks and tube Structures. *Proceedings of the International Conference on FRP Composites in Civil Engineering CICE 2001 12-15 December 2001* (Hong Kong China/Elsevier London) vol 2 pp 1421-1478

[6] Polilov A N Tatous N A 2007 Some new Strength Criteria for FRP Test Methods Substantiation *Proc. Int. Conf. Exp. Mech. ICEM 13 Grece 2007* pp 121-122

IOP Conf. Series: Materials Science and Engineering **525** (2019) 012087 doi:10.1088/1757-899X/525/1/012087

Creation and application of fluoropolymer photoconversion films for greenhouses: Concept.

S V Gudkov[1,2,3], A V Simakin[1], V E Ivanov[4], E V Barmina[1], I V Baimler[1,5], I I Rakov[1], L A Katicheva[2], V A Vodeneev[2] and G A Shafeev[1,6]

[1] Prokhorov General Physics Institute, Russian Academy of Sciences, 38 Vavilova st., Moscow, 119991 Russia
[2] The Institute of Biology and Biomedicine, Lobachevsky State University of Nizhni Novgorod, Nizhni Novgorod, 603950 Russia
[3] All-Russia Research Institute for Phytopathology, B. Vyazyomy, Moscow Region, 143050 Russia
[4] Institute of Theoretical and Experimental Biophysics, Russian Academy of Sciences, Institutskaya St. 3, Pushchino, Moscow Region 142290, Russia
[5] Moscow Institute of Physics and Technology, Institutsky lane 9, Dolgoprudny, Moscow region, 141700 Russia
[6] National Research Nuclear University MEPhI (Moscow Engineering Physics Institute), Kashirskoe sh. 31, Moscow, 115409 Russia

E-mail: S_makariy@rambler.ru

Abstract. The purpose of this investigation is to create a scientific basis and technology of the production of fluoropolymer photoconversion films for greenhouses to improve the performance of greenhouses in the area of risk farming. The aim basis of photoconversion technology in greenhouses is reduced to photoconversion of UV radiation into blue-violet, and green and yellow light into red light necessary for plants. In other words, when sunlight passes through fluoropolymer photoconversion films, the intensity of the blue-violet and red regions of the spectrum should increase, and the intensity of the UV, green and yellow regions of the spectrum should drop. The article presents the manufacturing technology of fluoropolymer photoconversion films and examples of its use in greenhouses.

1. Introduction

Visible light is electromagnetic radiation with wavelengths of 380–780 nm and quantum energy of 1.6–3.2 eV, perceived by the human eye. Visible light is very important for life on Earth, for most ecological communities it is the main source of energy [1]. The exception, perhaps, are only the deep-water communities of "black smokers", unable to use the energy of solar radiation and living at the expense of the energy of chemical compounds released on the fractures of lithospheric plates [2]. In general, visible light can modulate various biological processes [3,4], and applied in the diagnosis and treatment of various diseases [5,6]. With the participation of light, such important processes as, phototaxis, phototropism, photoperiodism, circadian rhythms and vision take place [1]. It is known that in order to obtain energy in the process of photosynthesis, plants use mainly quanta of the red and blue-violet part of the spectrum [7]. Most of the Russian Federation and other northern countries are located in the zone of risk farming, which is determined, among other things, by the insolation regime.

For many crops grown on greenhouse conditions in the zone of risk farming, the average daily intensity of the red and blue-violet part of the spectrum is usually sufficient only on clear summer days. The problem of insufficient illumination in greenhouses is usually solved with the help of artificial light sources [8]. This project involves the creation of scientific basis and technology for the production of polymer photoconversion coating for greenhouses. The coating produces photoconversion of UV light into blue-violet light, and green and yellow light into red light. In the future, the use of such a coating will significantly improve the performance of greenhouses and move on to their year-round use without "extra illumination" during the day in the area of risk agriculture.

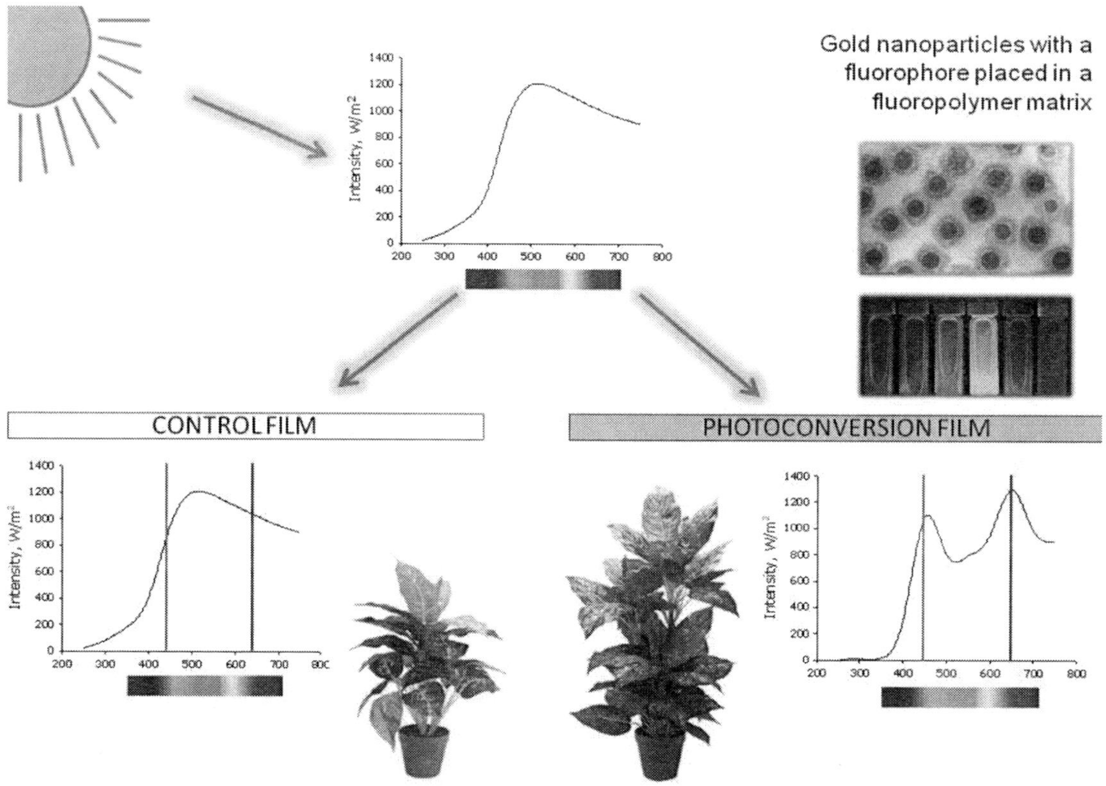

Figure 1. The concept of the presented work. The upper graph shows the energy spectrum of the sun on the surface of the Earth. When passing through the control film, it does not significantly change (lower left graph). When passing through fluoropolymer photoconversion films, its contribution to the blue and red components increases (lower right graph). As a result, in high latitude conditions, plants growing under a fluoropolymer photoconversion films have an advantage.

The task of creating photoconversion polymer coatings is long standing. The first coatings contained organic fluorophores, the main problem of such coatings was the rapid bleaching of the fluorophore [9]. Later, rare-earth metals and their compounds, often europium, were used as phosphors. The stability problem was partially solved, but the quantum yield was small [10]. At the present stage, nanoparticles with plasmon or exciton emission are used, cadmium-selenium, zinc-sulfur pairs, etc. are popular [11]. These fluorophores have significant problems with their inclusion in the polymer matrix and the stability of relatively active oxygen forms, which are always formed in vapor-permeable materials. Thus, there is the problem of quantum yield, inclusion in polymers and vapor permeability. In this article we will show how can solve the indicated problems. We use gold nanoparticles doped with fluorophores. Such particles are relatively low toxic and fairly stable.

IOP Conf. Series: Materials Science and Engineering **525** (2019) 012087 doi:10.1088/1757-899X/525/1/012087

Nanoparticles with fluorophore are able to effectively convert UV radiation, a part of green and yellow light into blue light, and red light needed by plants. The concept of the work is shown in Figure 1.

2. Methods

Gold nanoparticles were obtained by laser ablation of solid target in a deionized water. Au bulk was located at the bottom of a glass cuvette under a thin layer of working fluid (several millimeters) and irradiated with radiation of Ytterbium fiber laser (90 J/cm2 at 1060-1070 nm, pulse width of 90 ns, repletion rate of 20 kHz) (Figure 2). The optical density of the samples was measured using an Ocean Optics USB3000T spectrometer. Morphology of the generated nanoparticles was investigated by Libra 200 FE HR transmission electron microscope (Carl Zeiss). The size of the nanoparticles was determined using an analytical centrifuge DC24000 (CPS Instruments) [12].

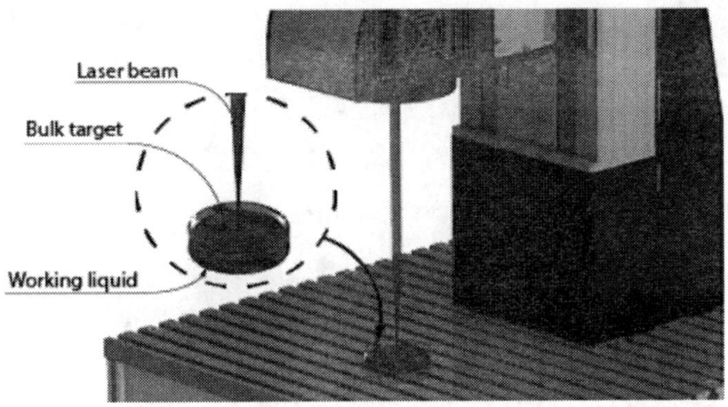

Figure 2. Experimental setup for laser ablation of solids in a liquid environment. An enlarged image of an experimental cuvette with sample for laser ablation .

The combination of gold nanoparticles and fluorophore was made using laser technology. Laser ablation of nanoparticles and fluorophore in the absence of the solid target was carried out using the radiation of the Nd:YAG laser at a wavelength of 1064 nm, pulse duration of 4 ns and fluence 140 J/cm^2 [13]. The transfer of nanoparticles to a fluoropolymer Functionalization of fluoropolymer by nanoparticles was carried out using the technology developed earlier [14]. Various routine procedures were used to study the spectral characteristics of fluoropolymer photoconversion films [3]. A CMP-3 pyronometer (Kipp & Zonen), a UV-BC spectrometer USB3000T (OceanOptics), an IR spectrometer Alf (Bruker), a measuring station of wide range Field Master (Coherent) were used as measuring stations. Practical tests of the developed fluoropolymer photoconversion films were carried out in greenhouses. The main indicator was the biomass of the studied plants.

3. Results

The generated gold nanoparticles are characterized by its size and morphology (Figure 3). As can be noticed absorption spectrum of Au nanoparticles has a maximum at 520 nm (Figure 3A) that corresponds the transverse plasmon resonance of spherical nanoparticles with size 10-20 nm [15]. This data is confirmed TEM imaging and disc centrifuge analysis (Figure 3B-D). As can be seen nanoparticles have a size of less than 30 nm. Remarkably, the portion of nanoparticles with average sizes of 15-20 nm amounts to more than 95% of all the gold weight in colloid while the number of nanoparticles with average sizes of 6-10 nm is less than 5% (Figure 3B). The data obtained by disc centrifuge were correlated with TEM image (Figure 3C). As can be seen all nanoparticles have a spherical shape (Figure 3D).

In this work, we attached fluorophores (Cd/ZnSe and Eu ions) to gold nanoparticles. Composites of gold nanoparticles and Cd/ZnSr showed a higher quantum yield compared with gold nanoparticles doped with Eu ions. With the help of the technology developed by us, nanoparticles were incorporated into the photoconversion fluoropolymer. In this case, nanoparticles are combined into aggregates of the "chain" type. Obviously, nanoparticles cannot be directly incorporated into polymer chains, and the matrix flows around them. This process is shown schematically in figure 4. This aggregation is

IOP Conf. Series: Materials Science and Engineering **525** (2019) 012087 doi:10.1088/1757-899X/525/1/012087

accompanied by a change in the plasmon resonance spectrum of metal nanoparticles and synergistic amplification of the optical wave field. It should be noted that the fluoropolymers are practically water vapor impermeable. This judgment supports the recent soak study of sulfonated tetrafluoroethylene-based fluoropolymer-copolymer [16-18].

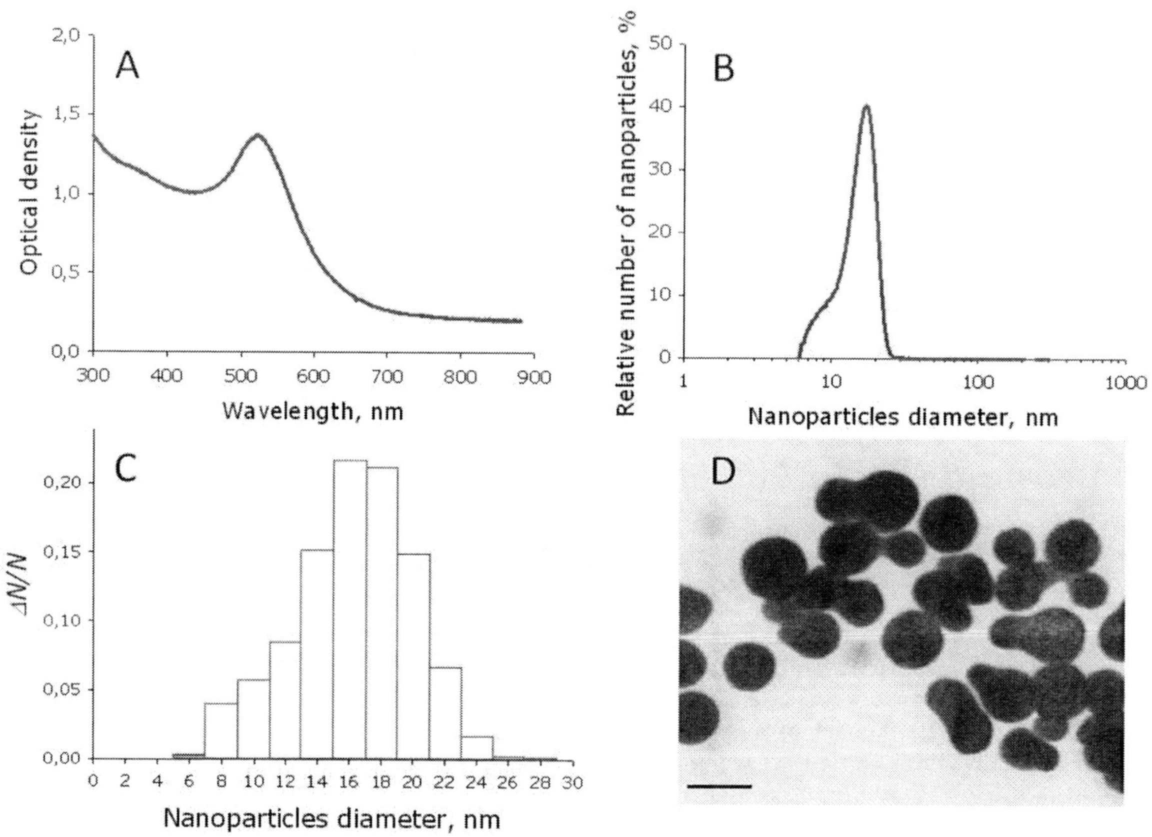

Figure 3. Characteristics of gold nanoparticles obtained by laser ablation. A – Absorption spectrum of of gold nanoparticles in water; B – Nanoparticles size distribution by weight determined by an analytical disc centrifuge; C – Size distribution of nanoparticles from TEM images; D – TEM image of gold nanoparticles. Scale bar, 20 nm.

With the help of artificial light sources simulating sunlight, primary tests of fluoropolymer photoconversion films were carried out. It was shown that when illuminated with model sunlight through a photoconversion polymer films, a noticeable increase in the biomass accumulation of lettuce plants (30%), cabbage (25%) and peas (15%) is observed. We assume that this is due to the more efficient operation of the photosynthetic apparatus. Of course, this assumption needs further verification.

4. Conclusions

The purpose of this study is to create a scientific basis and technology for the manufacture of fluoropolymer photoconversion films for greenhouses. Such films should increase the productivity of greenhouses in the zone of risk farming (high latitudes). The essence of the photoconversion technology in greenhouses is reduced to photoconversion of UV radiation, green and yellow light into blue and red light necessary for plants. Gold nanoparticles are used effectively as a fluorophore carrier. The technology of incorporating nanoparticles into fluoropolymer films in the form of chains has been used for supersharing enhancement of the optical wave field. The obtained fluoropolymer photoconversion films passed primary agricultural tests and showed their effectiveness. In the future, work is needed on consumer qualities (increase quantum yield, increased service life, cheaper prices, etc.).

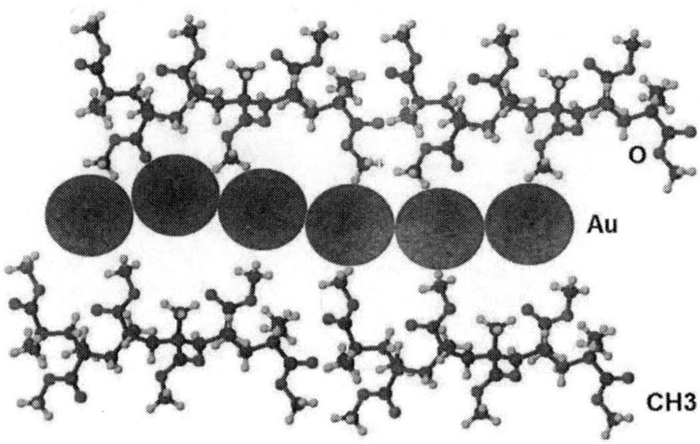

Figure 4. Polymer molecules align gold nanoparticles into chains in the nanocomposite. Gold nanoparticles are red.

Acknowledgements
The work was partially supported by the Grant MD-3811.2018.11 of the President of the Russian Federation for Governmental Support of Young Russian Scientists - Doctors of Sciences and RFBR Projects 19-02-00061_a, 18-52-70012 e-Asia_a. The optical part of the work was carried out with the support of R&D program (AAAA-A18-118021390190-1). The authors are grateful to the Center for Collective Use of the GPI RAS for the equipment provided.

References
[1] Gudkov SV, Andreev SN, Barmina EV, Bunkin NF, Kartabaeva BB, Nesvat AP, Stepanov EV, Taranda NI, Khramov RN and Glinushkin AP 2017 *Phys. Wave Phenom.* **25** 207–213.
[2] Weiss MC, Sousa FL, Mrnjavac N, Neukirchen S, Roettger M, Nelson-Sathi S and Martin WF 2016 *Nature Microbiol.* **1** 16116
[3] Chernov AS, Reshetnikov DA, Kovalitskaya YuA, Manokhin AA and Gudkov SV 2018 *J. Photochem. Photobiol. B.* **188** 77-86
[4] Ivanov VE, Usacheva AM, Chernikov AV, Bruskov VI and Gudkov SV 2017 *J. Photochem. Photobiol. B* **176** 36–43.
[5] Guryev EL, Volodina NO, Shilyagina NY, Gudkov SV, Balalaeva IV, Volovetskiy AB, Lyubeshkin AL, Sen' AV, Ermilov SA, Vodeneev VA, Petrov RV, Zvyagin AV, Alferov ZI and Deyev SM 2018 *PNAS* **115** 9690-9695
[6] Brilkina AA, Peskova NN, Dudenkova VV, Gorokhova AA, Sokolova EA and Balalaeva IV 2018 *J. Photochem. Photobiol. B* **178** 296–301.
[7] Shindy HA 2017 *Chem. Int.* **3** 97-105
[8] Nelson JA and Bugbee B 2014 *PLoS One* **9** e99010.
[9] Amin F, Yushchenko DA, Montenegro JM and Parak WJ. 2012 *Chemphyschem.* **13** 1030-1035
[10] Garciia-Murillo A, Le Luyer C, Garapon C, Dujardin C, Bernstein E, Pedrini C and Mugnier J 2002 *Opt. Mater.* **19** 161-168
[11] Reda SM 2008 *Acta Materialia* **56** 259-264
[12] Barmina EV, Mel'nik NN, Rakov II, Ivanov VE, Simakin AV, Gudkov SV and Shafeev GA 2018 *IOP Conf. Series: Materials Science and Engineering* **347** 012005
[13] Barmina EV, Gudkov SV, Simakin AV and Shafeev GA 2017 *J. Laser Micro/Nanoeng.* **12**,

254-257
[14] Barmina EV, Mel'nik NN, Rakov II and Shafeev GA 2017 *Phys. Wave Phenomen.* **25** 165–169
[15] Creighton JA and Eadon DJ 1991 *J. Chem. Soc., Faraday Trans.* **87** 3881-3891
[16] Bunkin NF, Kozlov VA, Shkirin AV, Ninham BW, Balashov AA and Gudkov SV 2018 *J. Chem. Phys.* **148** 124901
[17] Bunkin NF, Shkirin AV, Kozlov VA, Ninham BW, Uspenskaya EV and Gudkov SV 2018 *J. Chem. Phys.* **149** 164901.
[18] Bunkin NF, Kozlov VA, Aliev IN, Molchanov II, Abdullaev SA, Belosludtsev KN, Astashev ME, Gudkov SV 2015 *Phys. Wave Phenomen.* **23** 255-264.

IOP Conf. Series: Materials Science and Engineering **525** (2019) 012088 doi:10.1088/1757-899X/525/1/012088

Electrochemical recycling of nickel-cobalt-containing tungsten alloys

O G Kuznetsova, A M Levin, M A Sevostyanov and A O Bolshih

A.A. Baikov Institute of Metallurgy and Materials Science of Russian Academy of Sciences, 49 Leninsky pr., Moscow, Russian Federation

E-mail: okuznetsova@imet.ac.ru

Abstract. The potentiodynamic method of linear scanning voltamperometry and chronopotentiometry have been used to investigate the electrochemical dissolution of VNZhK alloy (wt.%: W 90, Ni 7,2, Fe 1,8, Co 1), VNZh alloy (wt.%: W 90, Ni 7, Fe 3), as well as their components in alkaline-ammonium solutions. The high anode current density of tungsten (~700 mA/cm^2), the passivity of iron and the similarity of nickel and cobalt anode behavior in the studied electrolytes were revealed. A higher electrochemical activity of VNZhK alloy has been established as compared with VNZh alloy, it is related with a decrease in the iron content in the alloy. Using the example of VNZh alloy, the possibility of electrochemical processing of tungsten heavy alloys with a degree of tungsten extraction into solution of more than 99% has been shown.

1. Introduction

The expansion of tungsten and its alloys application creates the necessity for the recirculation of this metal from waste, such as tungsten-based alloys [1, 2]. Nickel-cobalt-containing tungsten alloys are widely used in the manufacture of ammunition. Their production is based on the W-Ni-Fe-Co phasing diagram (VNZhK type alloys) and W-Ni-Fe phasing diagram (VNZh type alloys). The preparation of these heavy alloys usually involves the mixing of a tungsten powder with a small amount of metals with a lower melting point. It provides the formation of a liquid phase when heated to appropriate temperatures. Base domestic alloys usually contain 90% W, 7% Ni and 3% Fe. The substitution of 1–2% Fe by Co in the alloy composition has a positive effect on its sintering ability and mechanical properties [3, 4].

A significant amount of older armor-piercing ammunition results in the development of an environmentally safe hydrometallurgical process of valuable components regeneration to obtain pure tungsten compounds (ammonium paratungstate, tungsten acid, tungsten trioxide, etc.), which are the initial products for the production of metallic tungsten powder.

The processing of heavy tungsten alloys, including a binding phase from the metals of the iron subgroup (nickel, cobalt, iron) can be carried out by chemical dissolution methods [5, 6]. However, their effective application requires aggressive reagents and preliminary grinding of the processed material, which has high strength. At the same time, electrochemical oxidation is the most promising method for processing metal tungsten-containing wastes [7], since it provides high extraction of valuable components and quality of commercial products without the using of highly aggressive reagents.

Selective electrochemical dissolution of the iron-nickel binding phase of VNZh alloy was proposed in [2]. For this purpose, the using of acid electrolytes is most appropriate, however, tungsten is passivated in the acid media and the alloy dissolution stops due to the formation of a passivating film. It is also important to note, that according to [3], in case of electrolytic removal of the Ni – Fe – Co binding phase from a tungsten heavy alloy, the bond between the tungsten grains remains strong and causes significant diffusion difficulties with further dissolution of non-ferrous metals. Tungsten at high speed passes into solution when using alkaline electrolytes, but, in this case, nickel and iron are passivated and block the alloy dissolution [8].

A significant intensification of the VNZh and VNZhK alloys processing can be achieved through the dissolution of both the tungsten base and the binding phase. The using of a two-component electrolyte, containing sodium hydroxide and sodium chloride, where the chloride ion promotes the passing of non-ferrous metals into solution, and hydroxide ions - that of tungsten, was proposed in work [9]. The addition of chloride ions improves the electrochemical dissolution of non-ferrous metals, but leads to decrease in the rate of tungsten dissolution.

Currently, the main trends in the processing of tungsten-containing raw materials are aimed at introducing technologies, that provide the most efficient and careful environmental management. This is achieved, first of all, by using alkaline reagents for raw material processing [10]. At the same time, the using of mixed alkaline and ammonium solutions can improve the recirculation process due to the high dissolution rate of tungsten in alkaline electrolytes [9, 11], as well as the ability of nickel and cobalt to form soluble ammonium complexes [12]. In addition, the low electrical conductivity of ammonium solutions significantly increases due to the presence of sodium hydroxide in the electrolyte [13].

The purpose of this work is to study the anode behavior of the components of heavy tungsten-containing alloys (VNZh and VNZhK) in alkaline -ammonium electrolytes and to develop a method for the electrochemical extraction of tungsten from these alloys.

2. Experimental part

Investigations were carried out using linear voltammetry in potentiodynamic conditions in standard chemical cell with universal IPC-Pro potentiostat. Samples of tungsten with purity of 99,9 wt.%, N0 nickel, C0 cobalt, Armco iron, VNZh alloy, wt. %: W - 90, Ni - 7, Fe - 3, and VNZhK alloy, wt. %: W - 90, Ni - 7,2, Fe - 1,8, Co - 1 were analyzed. The electrolyte was a solution of sodium hydroxide with a concentration of 2M and ammonium hydroxide with a concentration of 0 ÷ 6M and a temperature of 20°C. The auxiliary electrode was made of platinum and the reference electrode was a saturated chlorine-silver electrode. The potential sweep speed was 1 mV/s. Investigated samples were preliminarily mechanically cleaned and washed with 4M hydrochloric acid solution and bidistilled water. In addition, anodic dissolution of VNZh alloy was investigated by the using of chronopotentiometry method in a solution, containing sodium hydroxide 2M and ammonium hydroxide 4M. Analytical determination of metals content in solutions was performed by atomic emission spectroscopy.

The study of the anode polarization of tungsten in a solution of NaOH 2M + NH_4OH with a concentration of NH_4OH from 0 to 6M showed that with an increase in the electrode potential from -0,4 to 0,2 V the anodic current density rises significantly. With a further increase in the potential to 1,0 V, the value of the anode current density reaches a plateau for electrolytes, containing NaOH + NH_4OH; however, for the solution of NaOH 2M, this value decreases. It was found that the introduction of ammonium hydroxide in the electrolyte increases the maximum current of tungsten dissolution by more than 30% (up to ~ 700 mA/cm^2), that is consistent with [14]. It should be noted that an increase in the concentration of NH_4OH from 4 to 6M does not result in significant intensification of the process, but is accompanied by some decrease in the electrical conductivity of the electrolyte [13].

Figure 1 shows the polarization curves of nickel in the studied electrolytes. It can be seen, that the current is spent mainly on the formation of passivating oxide film in case of nickel anode polarization in sodium hydroxide solution (figure 1, curve 2). When a potential reaches the value of ~ 0,5V and above, oxygen is also released.

The addition of ammonium hydroxide to the investigated solution leads to the appearance of a peak on the polarization curves, that apparently corresponds to the active oxidation of nickel. At the same time the potential of the oxygen evolution reaction shifts to the electropositive region to approximately 0,9V. The current at the peak of the anode nickel oxidation reaches a maximum at concentrations of NH_4OH in the electrolyte 4 ÷ 6M. However, deeper passivation of the metal occurs in the potential range of 0,7 ÷ 0,9V at 6M NH_4OH.

The cobalt polarization curves (figure 2), like those of nickel, show the anodic oxidation peak in the potential range of 0,6 ÷ 0,7V, when ammonium hydroxide is added to the sodium hydroxide solution. At the same time, the peak appears at lower concentrations of ammonium hydroxide - 2M. It may be associated with a greater tendency of cobalt to complexation in ammonium solutions, compared with nickel [12].

The polarization curves of iron indicate, that the metal in the studied electrolytes behaves passively and almost all the current, flowing through the electrode, is spent on the release of oxygen at a potential higher, than 0,6V. At the same time, with an increase in the NH_4OH concentration in the solution, a decrease in the polarization of this process is observed.

Thus, the data obtained show, that the using of an alkaline-ammonium electrolyte activates the electrochemical oxidation of tungsten (which constitutes VNZh and VNZhK alloy base), as well as nickel and cobalt. It allows us to recommend a solution of the composition NaOH 2M + NH4OH 4M for processing heavy tungsten-containing alloys.

Anode polarization curves of VNZh and VNZhK alloys (three uninterrupted cycles in the potential range from -0,5 to 0,6 V) are shown in figures 3 and 4, respectively, in a solution of NaOH 2M + NH_4OH 4M.

Anode polarization curves of VNZh alloy demonstrate the behavior characteristic of passivating materials (figure 3). The initial part of the alloy dissolution curves (in the potential range from -0,4 to 0,1 V) is determined, apparently, by the predominant dissolution of the tungsten base.

In addition, it can be seen, that in each subsequent cycle, the maximum anode current density

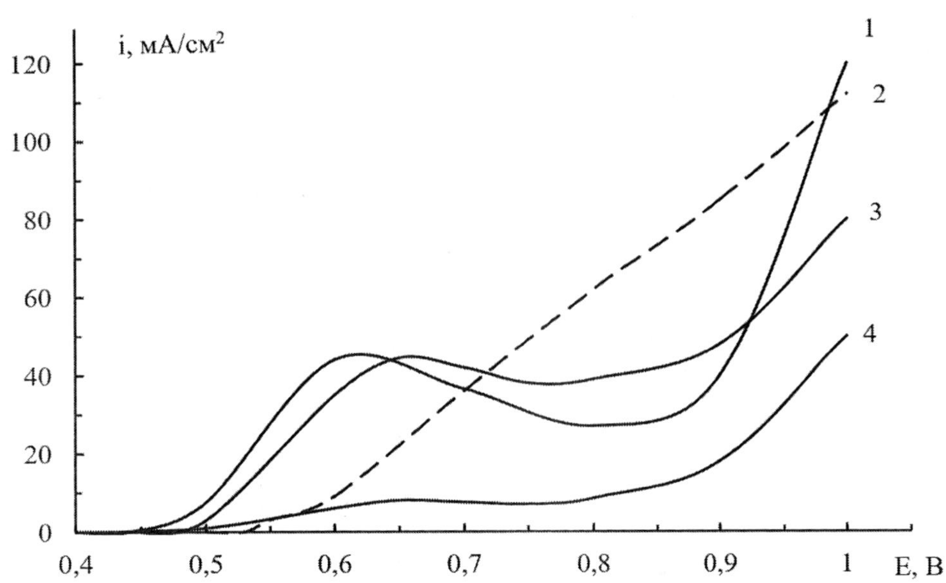

Figure 1. Anode polarization curves of nickel in NaOH 2M + NH_4OH solution. The concentration of NH_4OH, M: 1 – 6, 2 – 0, 3 – 4, 4 – 2.

IOP Conf. Series: Materials Science and Engineering **525** (2019) 012088 doi:10.1088/1757-899X/525/1/012088

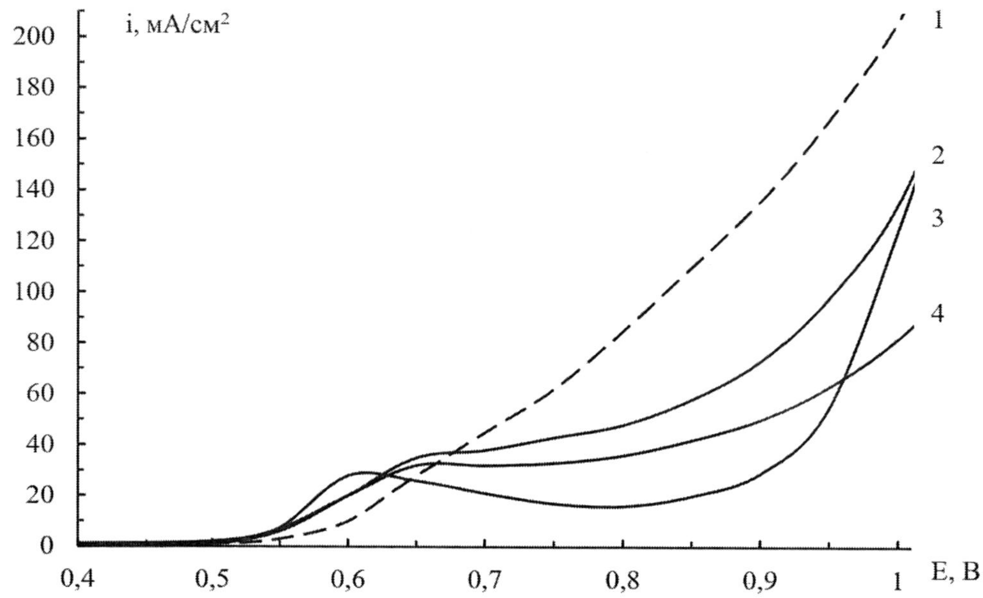

Figure 2. Anod polarization curves of cobalt in NaOH 2M + NH_4OH solution. The concentration of NH_4OH, M: 1 – 0, 2 – 4, 3 – 6, 4 –2.

of the alloys active dissolution decreases significantly. It may indicate a preferential passing of tungsten into solution with simultaneous accumulation of binder components on the anode surface.

It should be noted that the maximum current density of dissolution of fresh alloys of the VNZh and VNZhK is rather high (about 220 mA/cm^2 for VNZh alloy and 270 mA/cm^2 for VNZhK alloy). It shows their high electrochemical activity in the selected electrolyte (figures 3, 4). A higher maximum anode current density of VNZhK alloy, as compared with VNZh alloy, is associated with a decrease in the iron content in the alloy (the most passive component).

The electrochemical dissolution of VNZh alloy at the potential of -0,1V was studied. This potential corresponds to the maximum rate of alloy dissolution in the electrolyte of the composition NaOH 2M + NH_4OH 4M. The duration of process was 9 hours. It was established, that the dissolution is carried out at an average rate of 0,57 kg/m^2·h with a current efficiency close to 100% (based on tungsten ionization in oxidation level +6) and specific power consumption 1,51 kW·h/kg. The degree of tungsten extraction in solution amounted to 99,3%, and Ni+Fe - 0,7%. Thus, when the alloy is being dissolved in a alkaline-ammonium electrolyte, the tungsten base of the alloy preferentially passes into solution. The weak electrochemical activity of the iron-nickel binder alloy may be related with the prevailing influence of iron, that is deeply passivated in alkaline and ammonium media.

The obtained data suggest that the process of dissolution of the VNZhK alloy will proceed with no less high rates under similar conditions due to the more active electrochemical behavior of VNZhK alloy in comparison with VNZh alloy (figures 3, 4).

3. Conclusions

1. The electrochemical behavior of the individual components of VNZh (W-Ni-Fe) and VNZhK (W-Ni-Fe-Co) alloys has been investigated in solutions of sodium hydroxide with a concentration of 2M and ammonium hydroxide with a concentration of 2 ÷ 6M. It has been shown, that the dissolution rate of tungsten is 10 times higher, than that of other alloy components in these electrolytes.

IOP Conf. Series: Materials Science and Engineering **525** (2019) 012088 doi:10.1088/1757-899X/525/1/012088

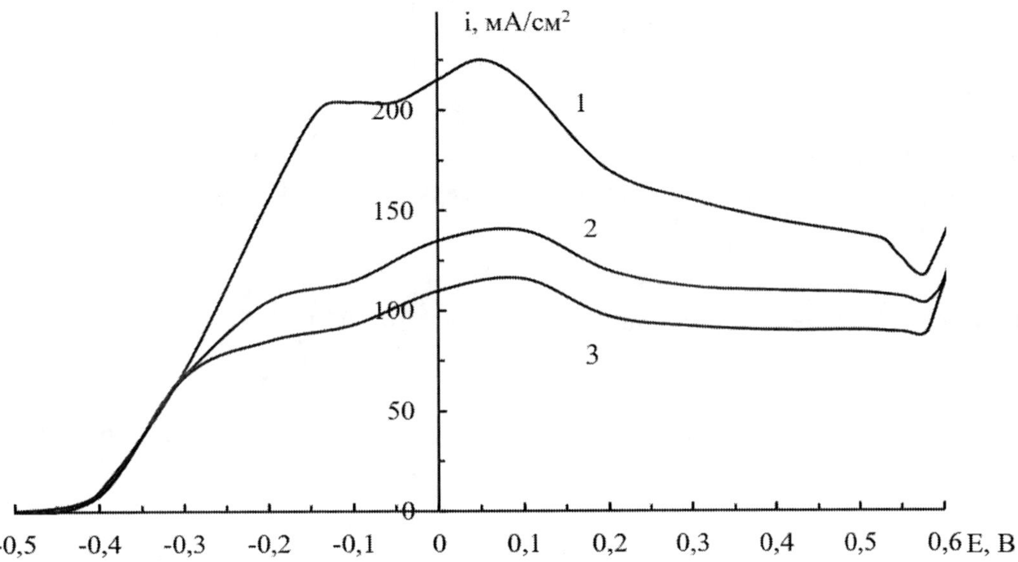

Figure 3. Anode polarization curves of VNZh alloy (W-Ni-Fe) in solution NaOH 2M + NH_4OH 4M. 1- the first cycle, 2 - the second cycle, 3 - the third cycle.

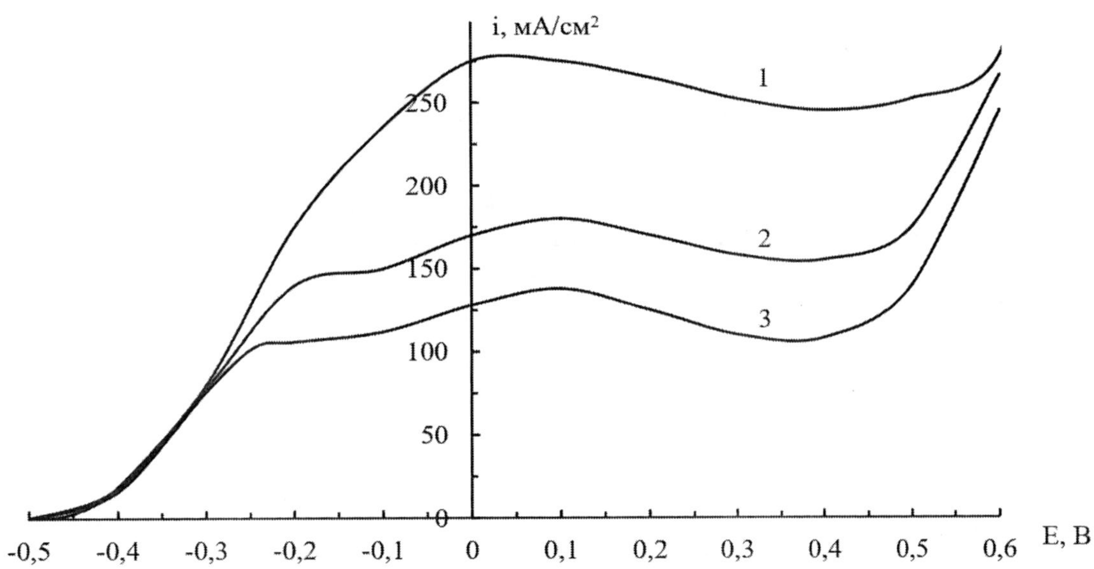

Figure 4. Anode polarization curves of VNZhK alloy (W-Ni-Fe-Co) in solution NaOH 2M + NH_4OH 4M. 1- the first cycle, 2 - the second cycle, 3 - the third cycle.

2. It has been revealed, that the addition of ammonium hydroxide to the sodium hydroxide solution: increases the electrochemical dissolution rate of tungsten by more than 30%, leads to the appearance of the peak on nickel and cobalt polarization curves, responding to their active anode oxidation and does not have a significant effect on the electrochemical behavior of iron. The optimal composition of the electrolyte is NaOH 2M + NH_4OH 4M in terms of the process rate and the specific conductivity of solution.
3. Electrochemical dissolution of VNZh alloy in the solution of NaOH 2M + NH_4OH 4M has been investigated. It has been shown, that a degree of tungsten extraction into solution exceeds 99% at the anode potential of 0,1V with a current efficiency close to 100% and a specific energy consumption ~ 1,5 kW·h/kg.

Acknowledgments
The work was carried out according to the state ask № 007-00129-18-00

References
[1] Cherniak G V, Povarova K B 2014 *Tungsten in ammunition* (Moscow: FGUP "TSNIIHM") p 360
[2] Kovalenko V 2017 Selective anodic treatment or W(WC)-based superalloy scrap *Eastern-European Journal of Enterprise Technologies* vol **85** 1/5 53-58
[3] Povarova K V, Makarov P V, Ratner A D 2002 Heavy alloys of type VNZh-90. I. Influence of doping and tungsten powder production regimes on their structure, microstructure and properties of sintered alloys *Metals* **4** 39-48
[4] Povarova K V, Alymov M I, Gavrilin O S 2007 Investigation of the influence of sintering modes of nanopowders of heavy alloys of the W-Ni-Fe-Co system on the structure and density of compact samples *Metals* **6** 65-72
[5] Reznichenko V.V, Butenko A M, Loboyko O.Y 2009 The nature of iron extraction from tungsten-containing instrumental alloys in the presence of ozone *Proceedings of the Odessa Polytechnic University* **2** 205-209
[6] Baranov G A., Gavrish M N, Sanikovich D D 2011 Production of nanosized powders during waste processing on the basis of tungsten-containing alloys, the study of their particle size distribution *Bulletin of the National Technical University of Ukraine "Kiev Polytechnic Institute". Machinery Series* **63** 42-46.
[7] Palant A A, Levin A M , Palant L A 2013 The method of electrochemical processing of metal tungsten-copper alloy *Patent RU2479652*
[8] Hairunisha S,. Sendil G K, Prabhakar-Rethinaraj J, Srinivasan G N, Adaikkalam P, Kulandaisamy S 2007 Studies on the preparation of pure ammonium tungstate from pure ammonium para tungstate from tungsten alloy scrap *Hydrometallurgy* **85** 67-71
[9] Parshutin V V 2008 Corrosion and electrochemical behavior of pseudo-alloys based on tungsten and their components *Electronic processing of materials* **6** 27-45
[10] Kraydenko R I, Perederin Y V, Filatov D S, Manucharyants A B 2015 Tungsten mining technology: state-of-the-art technology. *Polzunovsky Herald* vol **2** 4 135-139
[11] Suhotin A M 1981 *Handbook of electrochemistry*, ed A M Suhotin (Leningrad: Chemistry) p 214
[12] Dyatlov M N, Temkina V Y, Popov K I 1998 *Complexons and complexonates of metals* (Moscow: Chemistry) p 554
[13] Levin A M, Kuznetsova O G, Sevostyanov M A Electrical conductivity of aqueous solutions of the NaOH-NH_4OH system at 20 °C Conf. "In the world of science and innovation". 2017 April 20, Kazan (Ufa: AETERNA) part 5 pp.31-34

IOP Conf. Series: Materials Science and Engineering **525** (2019) 012088 doi:10.1088/1757-899X/525/1/012088

[14] Belov S F, Igumnov M S, Levin A M 1991 Anodic dissolution of tungsten in sodium hydroxide in the presence of ions NH^{4+}, SO_4^{2-}, CO_3^2 *Izv. Universities. Non-ferrous metallurgy* **1** 124–127

Biogas production from biomass of microalgae Chlorella vulgaris in the presence of benzothiophene sulfone

O Senko[1,2], O Maslova[1], M Gladchenko[1,2], S Gaydamaka[1,,2] and E Efremenko[1,2]

[1]Lomonosov Moscow State University, Moscow, Russia
[2] Emanuel Institute of Biochemical Physics RAS, Moscow, Russia

E-mail: elena_efremenko@list.ru

Abstract. The possibility of methanogenesis of microalgae *Chlorella vulgaris* biomass under the action of natural and artificially created anaerobic consortia in the presence of sulfur-containing xenobiotic benzotiophensulfone demonstrated. It has been shown that, despite a slight decrease in the concentration of intracellular ATP in the used biocatalysts, the process of methanogenesis proceeded quite effectively. It has been established that the components of hydrolyzate of microalgae *Chlorella vulgaris* biomass, remaining after the extraction of lipids, are excellent substrate for the production of biogas.

1. Introduction

Microalgae biomass has long established itself as a promising substrate for the production of various products under the influence of microbial biocatalysts [1-5]. Its use was also effective as a co-substrate (an additional source of proteins and carbon) in the processes of biocatalytic decomposition of xenobiotics [6].

Currently, the possibility of biocatalytic utilization of sulfur compounds using aerobic bacterial cells, in particular, obtained during the desulphurisation of petroleum products, has been shown. The research in the field of alternative technologies desulfurization during the last 30 years was mainly focused in the development of biotechnological strategies to remove sulfur via a serial of reactions realized under aerobic conditions, known as 4S path-way [7-8]. This path was first described for gram-positive bacteria *Rhodococcus erythropolis* IGTS8 [9], but later it was found in other bacteria [10-11]. This multistage process provides a biocatalytic transformation of heterocyclic sulfur-containing compounds dibenzothiophene through a 4-step sequential education dibenzothiophene-sulfoxide (catalyst - monooxygenase DszC), dibenzothiophene sulphone (catalyst - monooxygenase DszC) and 2-hydroxydiphenyl-sulfinate (catalyst - monooxygenase DszA) 2-hydroxybiphenyl and sulfite (the catalyst - desulfenasa DszB) [9]. Then sulfite is assimilating by its oxidation to sulfate under the action of sulfite oxidoreductase [12]. In addition, the reduced form of flavin mononucleotide is consumed at three stages of oxidation. Its presence in the medium is ensured by the functioning of flavin reductase (DszD) [13].

The main disadvantages of the 4S path-way are the duration of its implementation even at the stage of transition to dibenzothiophensulfone (up to 1 month) and low yields of each intermediate product due to the inhibitory effect of sulfur-containing compounds on the catalytic activity of cells. It is known that the formation and accumulation of sulfones leads to inhibition of whole biocatalytic oxidation of sulfur-containing compounds [14]. Genetic modification and immobilization of cells

IOP Conf. Series: Materials Science and Engineering **525** (2019) 012089 doi:10.1088/1757-899X/525/1/012089

allow intensifying 4S path-way, however the optimal solution is still not found. There are a few studies about the effect of such compounds on anaerobic cells metabolism.

Biocatalytic consortia containing sulfate-reducing bacterial strains that can utilize sulfur compounds under anaerobic conditions and its transformation to the same final product as in hydroconversion - to hydrogen sulfide is known [15]. However, such processes are characterized by long time-consuming costs. The speed of the process significantly slows down as hydrogen sulfide accumulates, that imposes restrictions on the possibility of increasing the process speed by increasing biocatalyst concentration.

There are studies describing the effect of other sulfurcontaining xenobiotics, most often dimethyl sulfoxide, on the metabolism of anaerobic cultures [16]. However, specially allocated or adapted anaerobic bacterial cells used at such studies.

As a promising and original solution associated with environmentally safely destruction of sulfones - waste of oxidative desulfurization of petroleum products, it would be possible to consider their biotransformation combined with the methanogenesis processes. However, no data have been presented about the effect of the presence of sulfur-containing xenobiotics (sulfones) at the metabolism of producers during methanogenesis.

So, the aim of this work was to assess the effect of sulfones presence at methanogenesis efficiency of *Chlorella vulgaris* microalgae biomass under action of natural and artificially created anaerobic consortia.

2. Materials and methods

2.1. Microorganisms

Green microalgae cells *Chlorella vulgaris [Beier.] rsemsu Chv-20/11-П3* were taken from IBCP RAS collections. Wastewater from Ostankino milk processing plant "OOO Ostankinsky Molochny Kombinat" (Moscow, Russia) was used to accumulate *C. vulgaris* biomass. The cultivation was performed in Erlenmeyer flasks (750 mL) containing 100 mL of cultivation medium. The cells were harvested after cultivation using Avanti J 25 centrifuge (Beckman Coulter, USA) at 8,000 rpm for 10 min.

Anaerobic sludge sample formed using the Bogatyr biopreparation (Moscow, Russia). The dry matter (56.4±2.6 g/L), ash content (38.7±1.8 %), and volatile suspended solids (VSS) (34.6±1.6 g/L) were determined according to the published procedure [17-18]. The both acidogenic (1560±30 mg COD/g VSS/d) and methanogenic (80±1 mg COD/g VSS/d) activity of the anaerobic sludge samples were evaluated according to the previously described techniques [17-18]. Nutrition media based on glucose and acetate, respectively, were used in these experiments.

The bacterial strains *Clostridium acetobutylicum* B1787 and *Desulfovibrio vulgaris.* B4053 were obtained from the Russian National Collection of Industrial Microorganisms (www.genetika.ru) for introduction into the methane tank in addition to the anaerobic sludge. The *C. acetobutylicum* strain B1787 was cultivated in the following medium (g/L): glucose – 20; triptone – 10; yeast extract – 5 (pH 6.8). The *D. vulgaris.* B4053 was cultivated in the Postgate medium [19]. Cells were cultivated under anaerobic conditions at 30°C for 1 week. Cultivation of *C. acetobutylicum* and *D. vulgaris* cells was performed under anaerobic conditions in an argon atmosphere at 37°C for 20–24 h and 48 h, respectively.

2.2. *Methods.*

2.2.1. Pretreatment and enzymatic hydrolysis of Chlorella biomass and lipids extraction

Pretreatment and enzymatic hydrolysis of *Chlorella* biomass for its microbial conversion into various products were described in previous work [20]. Biochemical composition of *Chlorella* biomass was done using standard methods described previously [1]. Extraction of lipids from biomass was carried out by the method describe Araujo et al. [21].

IOP Conf. Series: Materials Science and Engineering **525** (2019) 012089 doi:10.1088/1757-899X/525/1/012089

2.2.2. Anaerobic fermentation

The initial inoculum concentration in batch reactors was 10% (v/v) for suspended anaerobic sludge. The quantity of the artificially immobilized anaerobic sludge introduced into the medium was such as to ensure similar concentrations of sludge biomass in the liquid phase. The anaerobic incubation was carried out at 35°C in all the experiments.

To study the bioconversion of all the substrates to methane the 0.1 M phosphate buffer (pH 7.2) (55 mL) loaded into hermetically sealable vials ("anaerobic reactors", 120 mL). The experiments were performed in triplicate.

A control experiment similar to that described above as usual was concurrently conducted to account for the biogas formation due to the possible lysis of the microbial inoculum [17]. The methane content in the biogas in the experimental control batches was subtracted from that obtained in the corresponding test batches to calculate the methanogenesis efficiency.

2.2.3. Accumulation of biogas and determination of its content.

After 2-5 days we measured the total pressure and biogas concentration in the gas phase of each reactor. Gas measurements were repeated until constant methane content was reached in the gas phase of the reactor.

The content of hydrogen, methane, and carbon dioxide in the gas phase was measured with an LKhM 8 MD chromatograph (Russia) Model 3 with a katharometer (the carrier gas was argon with 20 mL/min flow rate). 2 m long columns were filled with Q porapak [17]. Oven temperature was maintained at 50°C, the retention times of hydrogen, methane, and carbon dioxide were 43, 67, and 82 s, respectively.

2.2.4. Determination of adenosine triphosphate (ATP) concentration

The concentration of intracellular ATP in immobilized cells was determined by the bioluminescent luciferin–luciferase method using cell-containing granules immediately after thawing. For this purpose, granules were weighed (0.15 _ 0.05 g), transferred to dimethyl sulfoxide (1 mL) and allowed to stand at 25°C for 2 h to extract intracellular ATP [22].

2.2.5. Calculations

The efficiency of methanogenesis (E) was calculated using the equation:

$$E = (Q/Q\text{max}) \times 100\ (\%); \qquad (1)$$

where Q (mL) is the volume of methane produced in a reactor with a test sample, (equation 2):

$$Q = \{(C/100 \times P_{tot} T_0 V_{gph})/(T_1 P_0)\} \times 1000, \qquad (2)$$

where C is the methane content in the gas phase (%); V_{gph} is the volume of the gas phase in the reactor (L); T_0 is the temperature under normal conditions, 273 K; T_1 is the operating temperature in the reactor (K); P_0 is the pressure under normal conditions, 1 atm; and P_{tot} is the total pressure in the reactor (atm),

and Qmax (mL) is the theoretical maximum volume of CH_4 (equation 3):

$$Q\text{max} = (C_I V_{lph}) \times 0.35 \times 1000\ (\text{mL}), \qquad (3)$$

where C_I is the initial concentration of the organic substances in the used waste sample, (g COD_{tot}/L); V_{lph} is the volume of the liquid phase in the reactor (L); and 0.35 is the volume of methane produced from 1 g COD at 273 K (L).

The specific productivity of methanogenesis was calculated as the volume of methane formed from 1.0 g of COD under the influence of anaerobic microbial inoculum introduced into the reactor (mL CH_4/g COD).

The data were shown as means of at least three independent experiments ± standard deviation (± SD). Statistical analysis was realized using SigmaPlot 12.5 (ver. 12.5, Systat Software Inc., San Jose, CA, USA). The significant ($p \leq 0.05$) differences between obtained results were estimated by one-way analysis of variance (ANOVA).

3. Results and Discussion

The following experiment was carried out to assess the feasibility of methanogenesis under the action of anaerobic consortia in the presence of sulfones. Benzothiophenesulfone (BTO_2) at a concentration of 0.15 mM was used as the sulfur-containing compound. This concentration was chosen based on the analysis of known data on the decomposition of sulfur-containing xenobiotics by anaerobic bacterial cells. Various combinations of anaerobic sludge, sulfate-reducing bacteria *D. vulgaris* [19] and anaerobic producers of hydrogen, ethanol, cellulases *C. acetobutilycum* were used as anaerobic biocatalysts [1, 3, 5]. These cultures were introduced into artificially created anaerobic consortia to intensify the process of methanogenesis.

It is known that the processes of xenobiotics decomposition proceeds much more efficiently in the presence of additional sources carbohydrates. [23]. Glucose is traditionally used as such carbon source. However, the transition to a cheaper source of raw materials could contribute to the implementation of such developments. It was decided to use hydrolysates of *C. vulgaris* biomass, containing polysaccharides and proteins, as an additional carbon source. Such substrate is a waste of lipid extraction from *C. vulgaris* biomass obtained during wastewater treatment. [3, 4, 24]. As a control similar process was performed using glucose as an additional source of carbon (Figure 1).

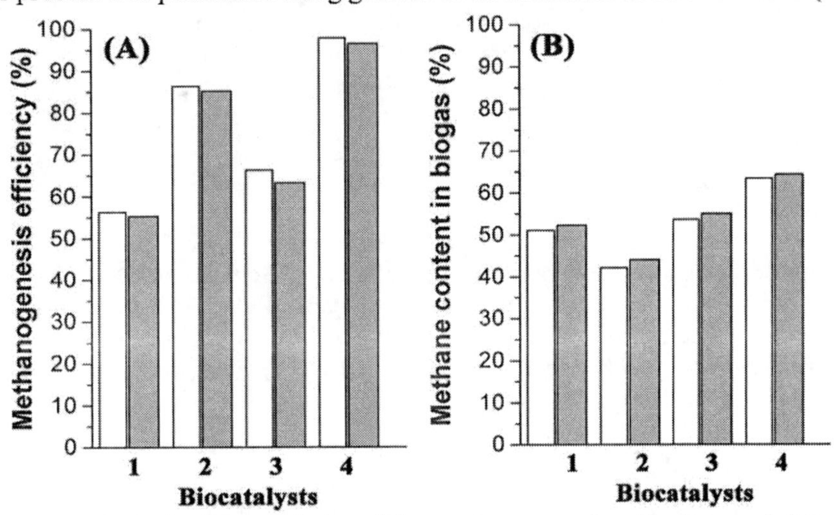

Figure 1. Basic characteristics of the methanogenesis with glucose (white bars) or hydrolysates of *C. vulgaris* biomass (grey bars) and different biocatalysts (1- anaerobic sludge, 2 - anaerobic sludge + *C. acetobutilycum* cells, 3 - anaerobic sludge + *D. vulgaris* cells, *4* - anaerobic sludge + *C. acetobutilycum* cells+ *D. vulgaris* cells*)* in the presence of 0.15 mM BTO_2

The transition from glucose use to hydrolyzed biomass of microalgae *C. vulgaris* as a carbon source does not reduce the main indicators of methanogenesis, carried out under the action of anaerobic consortia in the presence of sulfones. Obviously it is necessary to create anaerobic consortia, which contain additional cells, allowing intensifying the different stages of a complex biotechnological process with a lot of intermediate enzymatic reactions for the most effective methanogenesis in the presence of sulfones.

Evaluation of the energy status (viability) of anaerobic consortia components (Table 1), indicate that, in general, the concentration of intracellular ATP decreases in all cells used in the process in the presence of BTO_2. The degree of BTO_2 influence at various anaerobic cells is varied. The maximum negative effect from sulfones presence was noted for cells of sulfate-reducing bacteria *D. vulgaris*. At the same time, the presence of BTO_2 at a concentration of 0.15 mM practically did not negatively affect at viability of *C. acetobutilycum* cell (Table 1).

The observed decrease of the specific concentration of intracellular ATP is not a limiting factor for considering the process with the participation of anaerobes as a promising and further study and implementation of methanogenesis in the presence of sulfones, possibly simultaneously with their biological destruction. The decrease in the intracellular concentration of ATP even of anaerobes in the experiment conditions, particularly, may be due to including insufficient concentration of nutrients in the media, because in the control (medium without BTO2) also observed a decrease in the intracellular concentration of ATP compared with the initial level of this parameter.

Table 1. The concentration of intracellular ATP ($\times 10^{-12}$, mol/g biomass) in anaerobic bacterial cells used in methanogenesis of biomass *C. vulgaris* hydrolysate in the presence of BTO_2.

Biocatalyst	Initial	At the end of the process		% reduction from initial	
		without BTO_2	with BTO_2	without BTO_2	with BTO_2
Anaerobic sludge	50.2±2.1	15.4±0.6	1.5±0.1	69.3	97.0
C. acetobutilycum	66.4±3.1	4.2±0.1	4.0±0.1	93.7	94.0
D. vulgaris	27.3±0.3	23.6±0.9	12.1±0.5	13.6	55.7

However, if it is necessary to produce a biotechnological process in the presence of sulfur-containing xenobiotics, then may be it is appropriate to use artificially immobilized microbial cells for reduce negative influence of xenobiotics on anaerobic cells. Such approach has long been positively proven itself in different biocatalytic transformation processes of renewable raw materials, in particular, waste [1-5, 22-23, 25].

4. Conclusion

The possibility of methanogenesis process in the presence of sulfur-containing xenobiotic BTO_2 was demonstrated under the action of natural and artificially created anaerobic consortia. It has been shown that, despite a slight decrease in the concentration of intracellular ATP in the used biocatalysts, in general, the methanogenesis process proceeded quite effectively. It has been established that the components of microalgae *Chlorella vulgaris* biomass hydrolyzate, remaining after the extraction of lipids, are an excellent substrate for the production of biogas.

Acknowledgment

The authors are sincerely grateful to the staff of the Department of Petroleum Chemistry and Organic Catalysis Faculty of Chemistry Lomonosov Moscow State University in the person of the head of the department Professor Anisimov A.V. for BTO_2 provided for these studies.

Extraction of lipids from microalgae *Chlorella vulgaris* biomass was carried out with the support of the program of the Presidium of the Russian Academy of Sciences No. 33 “Carbon power engineering: chemical aspects”.

This work was supported by the Russian Foundation for Basic Research (18-29-05064).

References

[1] Efremenko E N, Nikolskaya A B, Lyagin I V, Senko O V, Makhlis T A, Stepanov N A, Maslova O V, Mamedova F and Varfolomeyev S D 2012 Production of biofuels from pretreated microalgae biomass by anaerobic fermentation with immobilized *Clostridium acetobutylicum* cells. *Bioresour. Technol.* **114** 342–348.

[2] Senko O V and Efremenko E N 2017 Highly concentrated populations of *Aureobasidium pullulans* cells in the biocatalytic processes of pullulan production. *Catalysis in Industry*, **9(4)** 344–348.

[3] *Immobilized cells: biocatalysts and processes*. Ed. E. Efremenko (Moscow RIOR) 2018, 499 p.

[4] Maslova O V, Stepanov N A, Senko O V and Efremenko E N 2019 Production of various organic acids from different renewable sources by immobilized cells in the regimes of

separate hydrolysis and fermentation (shf) and simultaneous saccharification and fermentation (sff). *Bioresour. Technol.* **272** 1–9.

[5] Efremenko E N, Stepanov N A, Nikolskaya A B, Senko O V, Spiricheva O V and Varfolomeev S D 2011 Biocatalysts based on immobilized cells of microorganisms in the production of bioethanol and biobutanol. *Catalysis in Industry* **3(1)** 41–46.

[6] Kamal Z M., Nashwa A H., Fetyan A., Ibrahim M A., Sherif E N. 2008 Biodegradation and detoxification of malathion by of *Bacillus thuringiensis* MOS-5. *Aus J Basic Appl Sci* **2** 724-32.

[7] Martínez I, Mohamed M E S, Rozas D, García J L, Díaz E 2016 Engineering synthetic bacterial consortia for enhanced desulfurization and revalorization of oil sulfur compounds. *Metab. Eng.* **35** 46-54.

[8] Nuhu A A 2013 Bio-catalytic desulfurization of fossil fuels: a mini review. *Rev. Environ. Sci. Bio.* **12(1)** 9-23.

[9] Gallagher J R, Olson E S, Stanley D C 1993 Microbial desulfurization of dibenzothiophene: A sulfur-specific pathway. *FEMS Microbiol. Lett.* **107(1)** 31-35.

[10] Kilbane II J J 2006 Microbial biocatalyst developments to upgrade fossil fuels. *Curr. Opin. Biotechnol.* **17(3)** 305-314.

[11] Mohebali G, Ball A S 2008 Biocatalytic desulfurization (BDS) of petrodiesel fuels. *Microbiol.* **154(8)** 2169-2183.

[12] Aggarwal S, Karimi I A, Kilbane II J J, Lee D Y 2012 Roles of sulfite oxidoreductase and sulfite reductase in improving desulfurization by *Rhodococcus erythropolis*. Mol. Biosyst. **8(10)** 2724-2732.

[13] Gray K A, Pogrebinsky O S, Mrachko G T, Xi L, Monticello D J and Squires C H 1996 Molecular mechanisms of biocatalytic desulfurization of fossil fuels. Nature Biotechnol. **14(13)** 1705-1709.

[14] El'kin A A, Grishko V V and Ivshina I B 2010 Oxidative biotransformation of thioanisole by *Rhodococcus rhodochrous* IEGM 66 cells *Appl. Biochem. Microbiol.* **46(6)** 586-592

[15] Kurita S, Endo T, Nakamura H, Yagi T, Tamiya N 1971 Decomposition of some organic sulfur compounds in petroleum by anaerobic bacteria. *J. Gen. Appl. Microbiol.* **17(3)** 185-198.

[16] Lomans B P, Op den Camp H J, Pol A, van der Drift C, Vogels G D 1999 Role of methanogens and other bacteria in degradation of dimethyl sulfide and methanethiol in anoxic freshwater sediments. *Appl. Environ. Microbiol.* **65(5)** 2116-2121.

[17] Gladchenko M A, Kovalev D A, Kovalev A A, Litti Yu V and Nozhevnikova A N 2017 Methane production by anaerobic digestion of organic waste from vegetable processing facilities. *Appl. Biochem. Microbiol.* **53(2)** 242–249.

[18] Kalyuzhnyi S, Gladchenko M, Starostina E, Shcherbakov S and Versprille A 2005 Combined biological and physico-chemical treatment of baker's yeast wastewater. *Wat. Sci. Tech* **52(1-2)** 175–181.

[19] Ismayilov I T, Stepanov N A, Efremenko E N and Abbasov V M 2015 Evaluation of biocidal properties of vegetable oil-based corrosion inhibitors using bioluminescent enzymatic method *Mosc. Univ. Chem. Bull.*. **70(4)** 197–201.

[20] Stepanov N. and Efremenko E 2018 "Deceived" concentrated immobilized cells as biocatalyst for intensive bacterial cellulose production from various sources. *Catalysts* **8** 1–16.

[21] Araujo G S, Matos L J and Fernandes J O 2013 Extraction of lipids from microalgae by ultrasound application: prospection of the optimal extraction method. *Ultrason Sonochem* **20** 95–98.

[22] Stepanov N. and Efremenko E 2017 Immobilised cells of *Pachysolen tannophilus* yeast for ethanol production from crude glycerol. *New Biotech.* **34** 54–58.

[23] Efremenko E N, Lyagin I V, Senko O V, Sirotkina M S and Zavyalova N V 2011 Immobilized heterogeneous biocatalysts for destruction of C-P bond in the hydrolytic products of the organophorous chemical warfare agents *RUDN journal of Ecology and life safely* **1** 61-66

[24] Efremenko E N, Nikolskaya A B, Mamedova F T, Senko O V and Trusov L I 2013 Semicontinuous and continuous process of accumulation of microalgae biomass of *Chlorella vulgaris* cells. *ISJAEE* **2(2)** 44–49.

[25] Senko O, Maslova O, Efremenko E 2017 Optimization of the use of His_6-OPH-based enzymatic biocatalysts for the destruction of chlorpyrifos in soil. *Int.J.Environ. Res. Public Health* **14(12)** 1438.

IOP Conf. Series: Materials Science and Engineering **525** (2019) 012090 doi:10.1088/1757-899X/525/1/012090

Biocatalytic transformation of various mycotoxins: modern problems and existing potential

R Ahundov[1], I Lyagin[1,2], O Senko[1,2], O Maslova[1], T Makhlis[1] and E Efremenko[1,2]

[1]Lomonosov Moscow State University, Moscow, Russia
[2] Emanuel Institute of Biochemical Physics RAS, Moscow, Russia

E-mail: elena_efremenko@list.ru

Abstract. A number of microorganisms and some plants have biochemical mechanisms of protection against the toxic effects of mycotoxins. The transformation of mycotoxins in the cells of these organisms is catalyzed by a number of enzymes, including oxidases, peroxidases, lactone hydrolases, laccases, lipases, amidases, aminotransferases, etc. recombinant variants of enzymes obtained by cloning the genes encoding these enzymes. The enzymes are basis for the development of new effective materials with decontamination activity against different micotoxins. The investigation and knowledge about the properties of the enzymes, sources of their isolation and specificity of their action as well as products of reactions catalyzed by them compose the modern scientific base for creation of new effective protective means (sorbents, fabrics etc.).

1. Introduction

Mycotoxins compose a large group of secondary metabolites of filamentous fungi, affecting agricultural products. (Table 1). As a rule, mycotoxins are highly stable compounds that are not degradable even at high temperatures, so they can be detected in heat-treated foods. These compounds pose a serious danger to humans and animals, have carcinogenic, immunosuppressive and teratogenic effects, causing various kinds of acute and chronic serious diseases (Table 1) [1-4].

The most toxic among the known mycotoxins are aflatoxins (AF), ochratoxin A (OTA), fumonisins (F), deoxynivalenol (DON) and other trichocytes, zearalenone (ZEA), patulin (PAT) and ergot alkaloids [5].

The toxicity of mycotoxins is determined by the combination of chemical groups present in their molecular structures (Table 2), capable of participating in chemical reactions. So, the common toxic link in the molecules of P and AF is the coumarin fragment, and the similarity of the structure of AF, ZEA P is determined by the presence of a lactone ring. Carboxyl derivatives, making a significant contribution to the overall toxicity of mycotoxins, are also often present both in the lactone part and in other fragments of the structure of these compounds.

A number of microorganisms, including pro- and eukaryotes, capable of carrying out biological modification or transformation of mycotoxins, have been isolated and studied [6-8]. Despite the large number of studies on the biological detoxification of mycotoxins, there is no much information about the selected individual enzymes that are capable of catalyzing this process.

Enzymatic catalysis holds a special place among approaches that can be used to detoxify mycotoxins [9]. A distinctive feature of the enzymatic detoxification is its specificity. Enzymes that specifically act on various functional groups of mycotoxins and belong to different classes are presented in Table 2.

Owing to economic point of view, most of the scientific investigations are devoted to enzymatic destruction of most significant mycotoxins: AF, OT, ZEA, DON, F and PAT [10].

It is interesting the use different functional materials that are based on the use of enzymatic nanobiocatalysts that catalyze the detoxification of mycotoxins. They can have functioning under physiological conditions with high efficiency and specificity of action, and, like all catalysts, can be introduced into the reaction medium in much smaller amounts as compared to the substrates transformed by them. The stability of such enzymes can be realized via creation of polyelectrolytic complexes with various polyanions or polycations [11-15]. The activity of enzymatic preparations is usually high enough in such complexes, but noncovalent variants are preferable, since it is not necessary to additionally use different toxic coupling agents.

Table 1. The most commonly detected mycotoxins in various agricultural products [1-4]

Type of toxin	Producer	Pollution objects	Tolerable daily intake	Toxic effect
AF	*Aspergillus flavus, A. niminus, A. parasiticus, A. pseudotamarii*	Grains, tobacco, peanuts, pistachios, cassava, cottonseeds and oilseeds, milk	0.15 g/kg	Liver damage, genotoxicity, oncogenic immunosuppression
OT	*A. alliaceus, A. auricomus, A. carbonarius, A. glaucus, A. melleus, A. niger, A. ochraceus, Penicillium cyclopium, P. verrucosum, P. viridicatum*	Grains, grapes, currants, nuts, dried fruits, coffee, cocoa, spices, red wine	14 ng/kg	Kidney damage, genotoxicity, carcinogenicity, immunosuppression
ZEA	*Fusarium culmorum, F. crookwellense, F. equiseti, F. graminearum, F. porotrichioide*	Grains, walnuts	0.25 μg/kg	Estrogenic, reduced reproductive function
DON	*F. culmorum, F. graminearum*	Cereals, hops	5 μg/kg	Gastrointestinal disorders, reduction of reproductive function, immunosuppressive effect, neurochemical disorders in the brain
FB1	*F. anthophilum, F. dlamini, F. oniliforme, F. napiformeI, F. nygama, F. proliferatum, F. verticillioides*	Cereals, beans, asparagus	2 μg/kg	Central nervous system damage, liver damage, genotoxicity, immunotoxicity
PAT	*P. expansum, P. griseofulvum, Aspergillus sp., Byssochlamys sp.*	Apples, pears, grapes, cherries, blueberries, cereals, greens	0.43 μg/kg	Central nervous system damage

IOP Conf. Series: Materials Science and Engineering **525** (2019) 012090 doi:10.1088/1757-899X/525/1/012090

2. Aflatoxins (AF)

AF are produced by fungi of the genus *Aspergillus*, belong to the class of furanocoumarins (Table 2) [6]. The toxicity of AF is associated with the presence of a lactone ring, the splitting of which produces a series of non-fluorescent products with low biological activity. The groups of enzymes that can detoxify aflatoxins include laccases, peroxidases, oxidases, etc.

Several studies have described the possibilities of modifying the difuran ring *in vitro* under the action of individual enzymes. Manganese-containing peroxidase (Mn-peroxidase) (EC 1.11.1.13) was isolated from the white rot fungus *Phanerochaete sordida* YK-624 [16]. The maximum reached detoxification efficiency of AFB1 was 69.2% in 48 h in the presence of this enzyme. AFB1 is firstly oxidized to AFB1-8,9-epoxide and then is hydrolyzed to AFB1-8,9-dihydrodiol, a compound that is significantly less mutagenic than the original mycotoxin. A similar transformation pathway for AFB1 was observed under the action of enzymes isolated from the fungus *Armillariella tabescens* [17].

Table 2. Mycotoxins and enzymes for their detoxification

Type of toxin	Structure	Enzyme for detoxification	References
AFB1	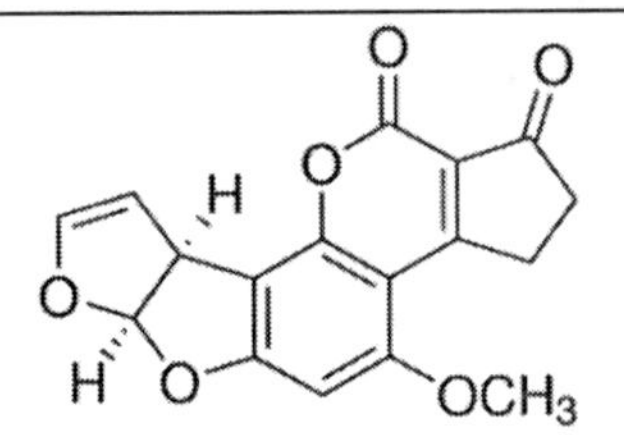	Mn-peroxidase, Aflatoxide oxidase, Laccase, $F_{420}H_2$-dependent reductase	[16-28]
OTA	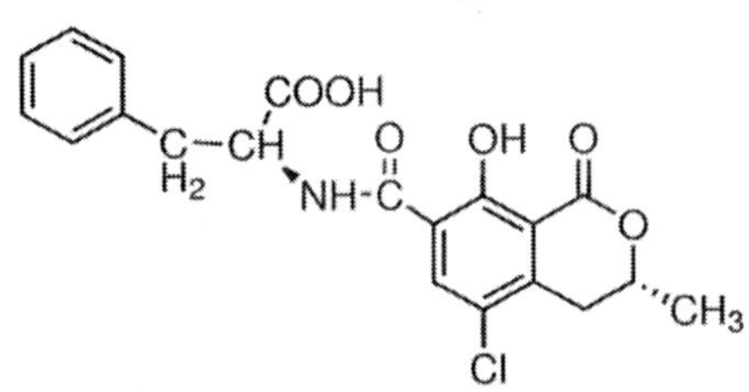	Carboxypeptidase A Lacton hydrolase Lipase	[34-38]
ZEA	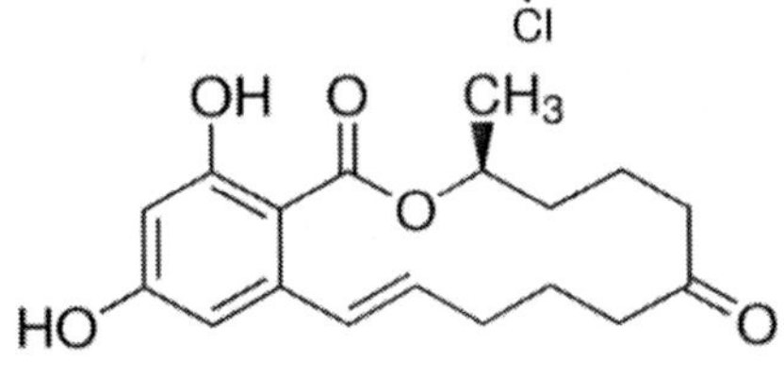	Lacton hydrolase Unspecified a/b-hydrolase Peroxiredoxin Cytochrome Fimbrial protein precursor	[41-47]
DON	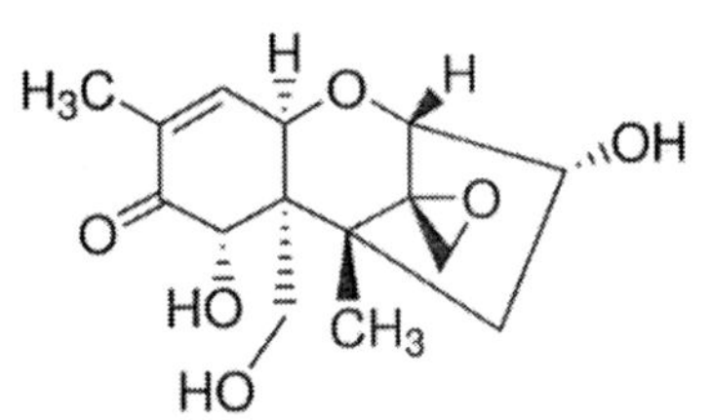	3-O-acetyltransferase UDP glycosyl transferase Cytochrome P-450 FAD-dependent ferredoxin reductase, ferredoxin [2Fe-2S]	[54-57]
F	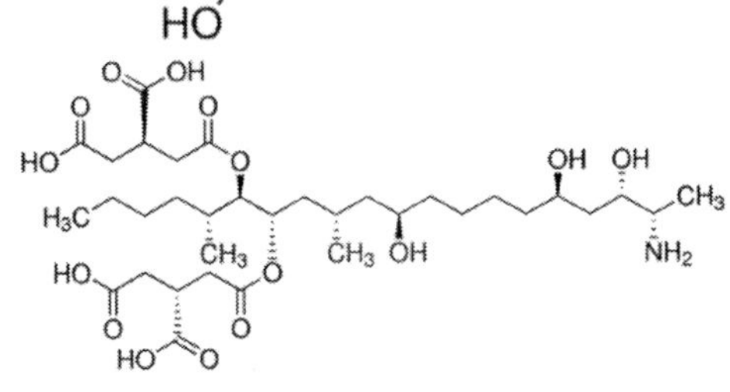	Carboxyl estrasise Aminooxidase Aminotransferase	[63-67]

PAT	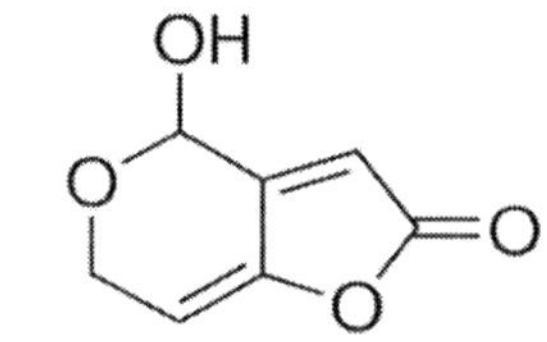	His_6-organophosphorus hydrolase	[76-77]

As a result of incubation in the presence of aflatoxin-detoxifying enzyme (ADTZ), produced by tabescens (E20), toxicity and mutagenicity of AFB1 were completely eliminated [18]. The enzyme "opened" the furan ring AFB1, which led to the subsequent hydrolysis of this mycotoxin [19]. This enzyme is assumed to be a new kind of aflatoxoxidase (AFO), which has more similarity to hydrolases, such as dipeptidyl peptidase III (EC 3.4.14.4), than oxidases that can destroy APV1 such as fungal laccase or horseradish root peroxidase [19]. AFO from the fungus *A. tabescens* can hydrolyze a furan ring of the molecule AFB1 [20]. It is believed that this enzyme acts on the 8,9-unsaturated carbon-carbon bond in the AFB1 molecule. AFO is an oxygen-dependent enzyme, as a result of which hydrogen peroxide is formed, which, apparently, also plays a significant role in detoxifying AFB1. The extremely low Km value (0.33 μM) determined for this enzyme indicates the high specificity of this oxidase in relation to this mycotoxin. Some attempts are known to produce a recombinant AFO enzyme from

A. tabescens [21]. Such recombinant protein with specific activity of 234 units/mg detoxified AFB1 under optimal conditions (pH 6.0 and 30 ° C). Km for this reaction was 39.3 μM. Mn-peroxidase with a specific activity of 78 units/mg isolated from the culture filtrate of the fungus *Pleurotus ostreatus* could effectively detoxifying AFB1. The maximum efficiency of degradation of AFB1 (1 mM) was observed with enzyme activity of 1.5 units/ml and was 90% after 48 h [22].

An enzyme was isolated rom the cultural medium of the fungus *Pleurotus ostreatus*, and it destroied AFV1 in the presence of H_2O_2 [23]. During the incubation of AFB1 with this enzyme, the intensity of fluorescence decreases, which was due to the rupture of the lactone ring.

It is known that laccases (EC 1.10.3.2) are able to detoxify AFB1. Pure laccase preparations: the enzyme isolated from the saprophytic fungus *Trametes versicolor*, as well as the recombinant enzyme produced by *A. niger* D15-Lcc2 # 3, provided 87.34% and 55% degradation of AFB1 during 72 h, respectively [24].

The effectiveness of the degradation of mycotoxin AFB1 under the action of a commercial laccase from the fungus *T. versicolor* was 67% for 2 days, while the prooxidant properties and mutagenicity of the products of degradation were significantly less pronounced than the initial toxin [25].

The efficiency of degradation of AFB1 (1 mg / ml) under the action of the isoenzyme (Lac2) of laccase, produced by the fungus *Pleurotus pulmonarius*, was 23%. In the presence of acetosyringing redox mediator (10 mM), the efficiency of degradation of AFB1 during 72 h of incubation at 250°C reached 90%, and AFM1 (0.05 mg/ml) under same conditions was completely destroyed. The authors of the study suggested that in this case there is oxidative demetoxylation [26].

Aflatoxin-detoxifying enzymes of bacterial origin are also known, for example, F420-dependent oxidoreductase from *Mycobacteria smegmatis* cells, reducing AF and other furanocoumarins, which is followed by the subsequent spontaneous hydrolysis of lactones. The efficiency of degradation of aflatoxins AFG1 and AFM1 after 48 h of incubation in the presence of this enzyme (100 units/ml) was 96.6 and 95.8%, respectively. The mechanism and products of AF degradation have not been established in this case [27].

$F_{420}H_2$-dependent reductases from *Mycobacteria smegmatis* have been identified and characterized. The most active against AFB1 was the recombinant protein FDR-A MSMEG_5998, the catalytic characteristics of which Km and Kcat were 47 mM and 63 min^{-1}, respectively. As compared to AFO [20] this enzyme was less specific to AFB1, but had a higher catalytic activity [28].

Contrary to AFO [16] or Mn–peroxidase [16], catalyzing the destruction of the diffuran site of the AFB1 molecule, $F_{420}H_2$-dependent reductases catalyze α,β-unsaturated double bond between C-2 and C-6, which leads to destabilization of the lactone ring of this structure [28].

IOP Conf. Series: Materials Science and Engineering **525** (2019) 012090 doi:10.1088/1757-899X/525/1/012090

3. Ochratoxins (OT)

OT are a group of mycotoxins, which are coumarin derivatives, produced by some types of microscopic mold fungi of the genus *Aspergillus* and *Penicillium*, which infect grain crops, grapes, coffee, and cocoa, are found in the products of their processing due to the high thermal stability of these compounds [29]. The most common and toxic is OTA, which is a phenylalanyl derivative of isocoumarin. In the OTA molecule, the isocoumarin site - охратоксин α - is linked to an L-β-phenylalanine amide bond (Table 2) [30].

OTA toxicity is mainly associated with the isocoumarin fragment of the molecule. The carboxyl group of the phenylalanine part, as well as the group containing Cl, also contributes to the toxicity of OTA [31]. Biological destruction of OTA can be carried out by hydrolysis of the amide bond under the action of non-specific peptidases, for example, carboxypeptidase A, with the formation of L-β-phenylalanine and OT α, a compound that is much less toxic than OTA [32].

The second, hypothetical, pathway for OTA destruction involves the destruction (rupture) of the lactone ring of this mycotoxin molecule by the action of lacton hydrolase [33]. The first OTA hydrolysis enzyme described in the literature was carboxypeptidase A (EC 3.4.17.1) from the bull pancreas [34]. The kinetic characteristics of this enzyme in the hydrolysis of ochratoxins are Km 5.6 and 266 mM, and Kcat - 36.8 and 2717 min^{-1}, for ochratoxins A and B, respectively.

The enzyme produced by the nonhochogenic strain *A. niger* MUM 03.58 was isolated, hydrolyzing OTA to ochratoxin α, Vmax of which was equal to 0.44 mM /min, and Km - 0.5 mM. The activity of this enzyme, as in the case of carboxypeptidase A, was inhibited by a chelating agent (EDTA), while the serine proteinase inhibitor did not affect the properties of the enzyme, which indicates that this enzyme is a metalloproteinase similar in properties to carboxypeptidase A [35].

The specific activity of recombinant amidase produced by *A. niger* was equal to 900 nmoles of OTA per min/mg protein at pH 7.0 and 40°C with an initial concentration of OTA equal to 1 μg/ml. The efficiency of OTA degradation (160 pg/ml) under the action of this enzyme in 0.5 h of incubation reached 83% [36].

Lipases are known to break down amide bonds [37]. Screening was made among 23 lipases and esterases from various sources for OTA degradation. A commercial lipase preparation from the *A. niger* fungus hydrolyzed OTA with the formation of phenylalanine and OT α. The activity of the purified enzyme with respect to OTA was 2.32 units/mg and the degradation efficiency was 95% for 2 h [38].

4. Zearalenone (ZEA)

ZEA is nonsteroidal estrogenic mycotoxin, secondary metabolite of anamorphic mold fungi of the genus *Fusarium*, developing on plants living in a warm and humid climate, as well as on agricultural products during storage conditions [39]. ZEA toxicity is determined by the presence of a lactone ring and C4-hydroxyl groups in its structure (Table 2) [40]. The ZEA peculiarity is that its transformation products (α-zearalenol and ß-zearalenol), like other members of the ZEA family, have similar or more pronounced estrogenic properties [32].

The destruction of the lactone ring ZEA can be carried out in two ways with the formation of different end products. In the first case, under the action of lacton hydrolase produced by *Clonostachys rosea* IFO 7063 [41, 42] ZEA was transformed into a significantly less estrogenic compound 1-(3,5-dihydroxyphenyl)-10′-hydroxy-1′-undecen-6′-one. This enzyme catalyzed the hydrolysis of the ester bond in the lactone ring, followed by spontaneous decarboxylation.

From the mycelium of the fungus *C. rosea* IFO 7063, a portion of the gene encoding the amino acid sequence of this protein was isolated and cloned into *E. coli* BL21 (DE3). At 30°C, the catalytic constant for the recombinant ZHD101 enzyme was 0.51 s^{-1}. When the mixture consisting of ZEA, α-zearalenol and ß-zearalenol (at a concentration of 2 μg/ml) was incubated in the presence of recombinant *E. coli* cells carrying the zhd101 gene, the concentration of all mycotoxins decreased to zero as a result of 1 h of incubation [43].

The second pathway for destruction of the lactone ring was established for basidial *Trichosporon mycotoxinivorans sp. nov.* cells [44]. The product of degradation of Z cells of *T. mycotoxinivorans*, was identified and designated as ZOM-1. It is presumably in this case, the gap of the lactone ring occurs under the action of by unspecified a/b-hydrolase in the region of the C6-ketone group. At the same time, carboxyl and hydroxyl groups are formed, which is not accompanied by decarboxylation, in contrast to what occurs when ZEA degrades under the action of the fungus *Gliocladium roseum*. The degradation product of ZEA is ZOM-1 and it does not possess estrogenic activity.

From the culture filtrate *A. niger* FS10, capable of effectively detoxifying ZEA, an active fraction was isolated containing 3 proteins with a molecular weight in the range of 30–150 kDa. The effectiveness of ZEA detoxifying in the presence of the indicated fraction was about 90% of the initial concentration of ZEA in the incubation medium. One of the proteins found in the active fraction of the culture filtrate was identified as hydrolase [45]. This fraction was immobilized using rice husk as a carrier [46]. The rates of degradation of ZEA under the action of free and immobilized enzyme preparations were similar, however, the immobilized preparation significantly exceeded the free enzyme in stability at different pH and temperature values, as well as in the level of activity preservation during storage at different temperatures. The effectiveness of detoxifying ZEA with an immobilized enzyme preparation in artificial gastric juice and intestinal fluid was 75.5% and 90.4%, respectively.

The active fraction was isolated from the culture filtrate of the bacterium *Acinetobacter sp.* SM04, able to effectively degrade ZEA with the formation of a number of less estrogenic products [47]. The structure of the two detected products (ZEN-1 and ZEN-2) indicates that under the action of the enzymes of this fraction, the lactone ring of ZEA is split into products containing carboxyl groups. Three enzymes were identified in this fraction, identified as peroxiredoxin and, presumably, cytochrome and fimbrial protein precursor [47].

5. Deoxynivalenol (DON)

DON is one of main mycotoxins produced by some phytopathogenic fungi of the genus *Fusarium* [48], infecting cereals and causing fusariouse [49]. The main structure of the molecule that determines the toxicity of DON, as well as other trichocenic mycotoxins, is epoxy (Table 2) [5,32].

Some microorganisms destructing the DON are known, however, individual enzymes from these bacteria were not yet isolated. However, these strains are definitely of great interest as a source of enzymes and genes encoding them for use in detoxification processes of DON [50-53].

The second toxic determinant in the DON molecule is the hydroxyl group in C3-position. All modifications of C_3-OH in the DON molecule, known to date, reduce the toxicity of this compound. The most studied modification of the site is acetylation, catalyzed by 3-O-acetyltransferase from *Fusarium graminearum* and related species [54], and glycosylation, a reaction catalyzed by a UDP glycosyltransferase from the *Arabidopsis thaliana* plant, which transfers glucose from UDP glucose to the hydroxyl group of the C3- in the DON molecule [55].

The structure and kinetic characteristics of two recombinant acetyltransferases FsTRI101 and FgTRI101, encoded by the TR1101 genes in the cells of the fungi *F. sporotrichioides* and *F. graminearum*, respectively, were investigated. Km, kcat, kcat / Km for FsTRI101 and FgTRI101 with respect to DON were 1,463 μM, 28 s^{-1} and 1.7 x 10^4 M^{-1} s^{-1} and 11.7 μM, 13.5 s^{-1} and 1.2 x 10^6 M^{-1} s^{-1}, respectively. The *in vitro* enzyme specificity of the enzyme (kcat / Km) with respect to DON in TRI101 from *F. graminearum* was 70 times higher than that of this enzyme from *F. sporotrichioides*. Significant differences in catalytic properties are explained by differences in the structure of active centers of these enzymes [56].

Some enzymes (cytochrome P-450, FAD-dependent ferredoxin reductase, and ferredoxin [2Fe-2S].) were isolated from bacterial cells *Sphingomonas* sp. PC. KSM1 and involved in the DON transformation. The catalytic efficiency of such an enzyme system (kcat/Km) was 6.4 mM^{-1} s^{-1},in reaction with DON, and DON detoxification product was identified as 16-hydroxy-DON [57].

IOP Conf. Series: Materials Science and Engineering **525** (2019) 012090 doi:10.1088/1757-899X/525/1/012090

6. Fumonisins (F)

FB1, FB2, FB3 are secondary metabolites of fungi of the genus *Fusarium* that infect cereals, as well as *A. niger* fungi, which are often found on grapes [58]. The toxicity of FB1 is determined by the presence of two side chains in its structure, which are tricarballylic acid, and also by the presence of free amino groups (Table 2) [59].

Biological detoxification of FB1 includes at least two enzymatic stages. The process begins with deesterification, as a result of which the lateral units in the PV1 molecule, which are tricarballylic acid, are cleaved from the main chain, resulting in a compound, hydrolyzed fumonisin (HFB1), also known as aminopentol 1 (AP1), in the second stage deamination occurs received HFB1 [60]. HFB1 is significantly less toxic than FB1 [61], but it is a rather dangerous substance, since the N-acylated metabolites of this compound are even more toxic than the original F [62].

The first step in the hydrolysis of FB1 in pro- and eukaryotic microorganisms is catalyzed by carboxylestrase (EC 3.1.1.1) [63-64]. In the second stage, the resulting GFB1 is deamidated. In the case of the black yeast strain *Exophiala spinifera*, the second stage is catalyzed by aminooxidase, the necessary condition for the functioning of which is the presence of molecular oxygen in the medium, and H_2O_2 is formed as a by-product [63]. Deamination occurs under the action of aminotransferase in bacterial strains [60, 65].

The fumD gene encoding carboxylesterase in the cells of the bacterium *Sphingopyxis* sp. MTA144 was cloned and expressed in *Pichia pastoris* yeast. Km for this enzyme was 0.90 μM, and catalytic constant was 900 s^{-1} [66].

The properties and kinetic characteristics of the recombinant enzyme (aminotransferase) produced by the transgenic *E. coli* strain carrying the fumI gene encoding the synthesis of aminotransferase in *Sphingopyxis sp.* MTA144. The final product of the degradation of HFB1 under the action of the fumI-aminotransferase was 2-keto-HFB1. It was established that the preferential co-substrate and the amino group receptor in the HFB1 deamination reaction for the purified His-tagged FumI enzyme is pyruvate (Km = 490 μM at a HFB1 concentration of 10 μM). Under optimal conditions, in the presence of 5 mM pyruvate as a co-substrate, Km and kcat of this enzyme were 1.1 μM and 104 min^{-1}, respectively [67].

7. Patulin (PAT)

PAT is pyran derivative (4-hydroxyfuropyranone), triketid, water-soluble secondary metabolite of fungi genus *Penicillium, Aspergillus и Byssochlamys spp.* [68-69], affecting fruits, and contaminating the products of their processing. Detoxification of PAT can occur due to the rupture of the pyran ring with the formation of E-ascladiol - the biosynthetic precursor of PAT, and/or its isomer Z-Ascladiol, as well as by breaking the lactone ring with the formation of deoxypatulinic acid (Table 2).

Enzymatic transformation of PAT was detected in a number of bacterial and yeast strains. Among such microorganisms, the yeast *Rhodosporidium kratochvilovae* performs biodegradation of PAT with the formation of desoxypatulinic acid (DPA) a product is not toxic for various microorganisms and the cell culture of human lymphocytes [70]. Bacteria *Gluconobacter oxydans* [71], yeast *S. cerevisiae* [72] and lactobacterium *Lactobacillus plantarum* transforming PAT to E- or Z-ascadiol [73], as well as basidiomycetes, the yeast *Sporobolomyces sp.* strain IAM 13481, belonging to the *Pucciniomycotina subdivision*, destroying PAT to form two different products: DPA and Z-Ascladiol [74]. It is known regarding the isolation and purification of individual enzymes that detoxify PAT [75], which were isolated from the cells of the yeast Kratochvilova RS. LS11, capable of degrading PAT with the formation of DPA [70]. It is known about the recombinant enzyme His_6-organophosphorus hydrolase, which decomposing some micotoxins including patulin [76-77].

8. Conclusion

A distinctive feature of enzymes as detoxifying agents is their high specificity with respect to mycotoxins. Currently, individual enzymes capable of deep and effective degradation of mycotoxins have not been widely used in the development of effective means for detoxification of these compounds.

It should be noted that in the case of using enzymes to create materials capable of detoxifying mycotoxins, it is necessary to create stable forms of these proteins, since the environment in which they should function can be quite aggressive. Because the contaminated product is found, as a rule, not one, but several mycotoxins, means that can help in this kind of pollution needs to be sufficiently universal and active in relation to multiple mycotoxins, similar in structure.

Aknowledgement
This work was financially supported by RFBR (Project 18-29-17069).

References

[1] Gelderblom W C A, Cawood M E, Snyman S D, Vleggaar R and Marasas W F O 1993 Structure-activity relationships of fumonisins in short-term carcinogenesis and cytotoxicity assays *Food Chem. Toxicology* **31(6)** 407-414.
[2] Zain M E 2011 Impact of mycotoxins on humans and animals *J. Saudi Chem. Society* **15** 129-144.
[3] Da Rocha M E B. Freire F D O, Maia F B F, Guedes M I F and Rondina D 2014.Mycotoxins and their effects on human and animal health. *Food Control* **l36** 159–165.
[4] Williams J G, Deschl U and Williams G M 2011 DNA damage in fetal liver cells of turkey and chicken eggs dosed with aflatoxin B1 *Arch. Toxicol.* **85** 1167–1172.
[5] Karlovsky P, Suman M, Berthiller F, De Meester J, Eisenbrand G, Perrin I, Oswald I P, Speijers G, Chiodini A, Recker T and Dussort P 2016 Impact of food processing and detoxification treatments on mycotoxin contamination. *Mycotoxin Res.* **32** 179–205.
[6] Wu Q, Jezkova A, Yuan Z, Pavlikova L, Dohnal V and Kuca K 2009 Biological degradation of aflatoxins. *Drug Metab. Rev.* **41** 1–7.
[7] Hathout A S and Aly S E 2014 Biological detoxification of mycotoxins: a review. Ann Microbiol **64** 905–919.
[8] McCormick S P 2013 Microbial detoxification of mycotoxins. *J. Chem. Ecol.* **39** 907–918.
[9] Efremenko E N, Lyagin I V, Makhlis T A, Maslova O V, Senko O V and Stepanov N A 2017 Enzymatic detoxification of mycotoxins: known and promising solutions. *Acta Naturae* **special issue** 124-125.
[10] Zhu Y, Hassan Y I, Watts C and Zhou T 2016 Innovative technologies for the mitigation of mycotoxins in animal feed and ingredients-A review of recent patents. *Animal Feed Sci. Technol.* **216** 19–29.
[11] Efremenko E N, Senko O V and Maslova O V 2017 Optimization of the use of His_6-OPH-based enzymatic biocatalysts for the destruction of chlorpyrifos in soil. *Int. J. Environ. Res. Public Health.* **14** 1438.
[12] Maslova O V, Stepanov N A, Grigoryeva A I, Bruyako M G and Efremenko E N 2016 New effective His6-OPH-containing mineral carriers pretreated by low-temperature plasma for destruction of organophosphates in different types of soil. *IJPT* **8(4)** 27317-27333.
[13] Maslova O V, Senko O V and Efremenko E N 2018 The influence of enzymatic removal of chlorpyrifos from feed grain mixes on biochemical parameters of rat blood. *Biochem (Mosc) Suppl Ser B* **12(2)** 181–185.
[14] Lyagin I V and Efremenko E N 2018. Biomolecular engineering of biocatalysts hydrolyzing neurotoxic organophosphates. *Biochimie* **144** 115–121.
[15] Sergeeva V S, Efremenko E N, Kazankov G M, Gladilin A K and Varfolomeev S D 1999 Kinetic behavior of phosphotriesterase and its non-covalent complexes with polyelectrolyte in systems with polar and non-polar organic solvents. *Biotechnol. Techniques*, **13 (7)** 479-483.
[16] Wang J Q, Ogata M, Hirai H and Kawagishi H 2011 Detoxification of aflatoxin B-1by

manganese peroxidase from the white-rot fungus *Phanerochaete sordida* YK-624. FEMS *Microbiol. Lett.* **314** 164–169.

[17] Liu D L, Yao D S, Liang R, Ma L, Cheng W Q and Gu L Q 1998 Detoxification of aflatoxinB-1 by enzymes isolated from *Armillariella tabescens*. *Food Chem. Toxicol.* **36** 563–574.

[18] Liu D L., Yao D S, Liang Y Q, Zhou T H, Song Y P, Zhao L and Ma L 2001 Production, purification, and characterization of an intracellular aflatoxin-detoxifizyme from *Armillariella tabescens* (E-20) *Food Chem. Toxicol.* **39** 461-466.

[19] Cao H, Liu D, Mo X, Xie C and Yao D 2011 A fungal enzyme with the ability of aflatoxin B1 conversion: purification and ESI-MS/MS identification *Microbiol. Res.* **166** 475—483.

[20] Wu Y Z, Lu F P, Jiang H L, Tan C P, Yao D S, Xie C F and Liu D L 2015 The furofuran-ring selectivity, hydrogen peroxide production and low Km value are the three elements for highly effective detoxification of aflatoxin oxidase *Food Chem. Toxicol.* **76** 125-131.

[21] Wen S, Guan M, Zhou T, Gao H, Xie C, Liu D and Yao D 2011 Cloning, expression, purification and characterization of an aflatoxin-converting enzyme from *Armillaria tabescens Acta Microbiologica Sinica* **51(9)** 1212-1221.

[22] Yehia R S 2014 Aflatoxin detoxification by manganese peroxidase purified from *Pleurotus ostreatus Braz. J. Microbiol.* **45(1)** 127-130.

[23] Motomura M, Toyomasu T, Mizuno K and Shinozawa T 2003 Purification and characterization of an aflatoxin degradation enzyme from *Pleurotus ostreatus Microbiol. Res.* **158** 237–242.

[24] Alberts J F, Gelderblom W C A, Botha A and van Zyl W H 2009 Degradation of aflatoxin B1 by fungal laccase enzymes *Int. J.Food Microbiol.* **135** 47–52.

[25] Zeinvand-Lorestani H, Sabzevari O, Setayesh N, Amini M, Nili-Ahmadabad A and Faramarzi M A 2015 Comparative study of in vitro prooxidative properties and genotoxicityinduced by aflatoxin B1 and its laccase-mediated detoxification products *Chemosphere* **135** 1–6.

[26] Loi M, Fanelli F, Zucca P, Liuzzi V C, Quintieri L, Cimmarusti L, Monaci M T, Haidukowski M, Logrieco A F, Sanjust E and Mulè G 2016 Aflatoxin B1 and M1 degradation by Lac2 from *Pleurotus pulmonarius* and redox mediators *Toxins* **8(9)** 245

[27] Zhao L H, Guan S, Gao X, Ma Q G, Lei Y P, Bai X M and Ji C 2010 Preparation, purification and characteristics of an aflatoxin degradation enzyme from *Myxococcus fulvus* ANSM068 *J. Appl. Microbiol* **1** 147–155.

[28] Taylor M C., Jackson C J, Tattersall D B, French N, Peat T S, Newman J, Briggs L J, Lapalikar G V, Campbell P M, Scott C, Russell R J and Oakeshott J G 2010 Identification and characterization of two families of $F_{420}H_2$-dependent reductases from *Mycobacteria* that catalyse aflatoxin degradation *Mol. Microbiol* **78** 561–575.

[29] Amézqueta S, González-Peñas E, Murillo-Arbizu M and López de Cerain A 2009 Ochratoxin A decontamination: A review *Food Control* **20** 326–333.

[30] Abrunhosa L, Paterson R R M and Venâncio A 2010 Biodegradation of ochratoxin A for food and feed decontamination *Toxins* **2** 1078-1099.

[31] Xiao H, Madhyastha S, Marquardt R R, Li S, Vodela J K, Frohlich A A and Kemppainen B W 1996 Toxicity of ochratoxin A, its opened lactone form and several of its analogs: structure-activity relationships. *Toxicol. Appl. Pharmacol.* **137** 182–192.

[32] Karlovsky P 1999 Biological detoxification of fungal toxins and its use in plant breeding, feed and food production. *Nat. Toxins* **7** 1–23.

[33] Shimizu S, Kataoka M, Honda K and Sakamoto K 2001 Lactone-ring-cleaving enzymes of microorganisms: their diversity and applications. *J. Biotechnol* **92** 187–194.

[34] Pitout, M.J. The hydrolysis of ochratoxin A by some proteolytic enzymes. Biochem. Pharmacol. 1969, 18, 485–491

[35] Abrunhosa L and Venâncio A Isolation and purification of an enzyme hydrolyzing ochratoxin A from *Aspergillus niger*. *Biotechnol. Lett.* **2007**, *29*, 1909–1914

[36] Yu S, Poulsen C H, Dalsgaard S, Wang H and Nikolaev I 2013 Food additive comprising an amidase for detoxifying ochratoxin, US Patent 20130216654 A1.

[37] Bornscheuer U T and Kazlauskas R J 2006 Hydrolases in organic synthesiss regio and stereoselective biotransformations; Wiley-Blackwell, 368 p.
[38] Stander M A and Bornscheuer U T, Henke E, Steyn P S 2000 Screening of commercial hydrolases for the degradation of ochratoxin A. *J. Agric. Food Chem.* **48** 5736–5739.
[39] Bennett J W and Klich M 2003 Mycotoxins *Clin. Microbiol. Rev.* **16(3)** 497–516
[40] El-sharkawy S H and, Abul-hajj Y J 1988 Microbial transformation of zearalenone. 2. Reduction, hydroxylation, and methylation products. *J. Org.Chem.* **53** 515–519.
[41] Takahashi-Ando N, Kimura M, Kakeya H, Osada H and Yamaguchi I 2002. A novel lactonohydrolase responsible for the detoxification of zearalenone: enzyme purification and gene cloning. *Biochem. J.* **365** 1-6.
[42] Kakeya H, Takahashi-Ando N, Kimura M, Onose R, Yamaguchi I and Osada H 2002 Biotransformation of the mycotoxin, zearalenone, to a non-estrogenic compound by a fungal strain of *Clonostachys sp. Biosci. Biotechnol. Biochem.* **66** 2723-2726.
[43] Takahashi-Ando N, Ohsato Sh, Shibata T and Hamamoto H 2004 Metabolism of zearalenone by genetically modified organisms expressing the detoxification gene from *Clonostachys rosea. App. Environ. Microbiol.* **70(6)** 3239–3245
[44] Vekiru E, Hametner Ch, Mitterbauer R, Rechthaler J, Adam G, Schatzmayr G, Krska R and Schuhmacher R 2010 Cleavage of zearalenone by *Trichosporon mycotoxinivorans* to a. novel nonestrogenic metabolite. *Appl. Environ. Microbiol.* **76(7)** 2353– 2359.
[45] Sun X, He X, Xue K, Li Y, Xu D and Qian H 2014 Biological detoxification of zearalenone by *Aspergillus niger* strain FS10 *Food Chem. Toxicol.* **72** 76–82.
[46] He M, Li Y, Pi F, Ji J, He X, Zhanga Y and Sun X 2016 A novel detoxifying agent: Using rice husk carriers to immobilize zearalenone-degrading enzyme from *Aspergillus niger* FS10 *Food Control* **68** 271-279
[47] Yu Y, Qiu L, Wu H, Tang Y, Lai F and Yu Y 2011 Oxidation of zearalenone by extracellular enzymes from *Acinetobacter sp.* SM04 into smaller estrogenic products *World J. Microbiol. Biotechnol.* **27** 2675–2681.
[48] Audenaert K, Vanheule A, Hofte M and Haesaert G 2014 Deoxynivalenol: a major player in the multifaceted response of *Fusarium* to its environment. *Toxins (Basel)* **6** 1–19.
[49] McMullen M, Bergstrom G, DeWolf E, Dill-Macky R, Hershman D, Shaner G and Sanford D V 2012 A unified effort to fight an enemy of wheat and barley: *Fusarium* head blight. *Plant Dis.* **96** 1712–1728.
[50] Fuchs E, Binder E M, Heidler D and Krska R 2002 Structural haracterization of metabolites after the microbial degradation of typeA trichothecenes by the bacterial strain BBSH797. *Food Addit. Contam.* **19** 379–386.
[51] Binder E M and Binder J 2004 Strain of *Eubacterium* that detoxifies trichothecenes. US Patent 6794175B1.
[52] He J W, Bondy G S, Zhou T, Caldwell D, Boland G J and Scott P M 2015 Toxicology of 3-epi-deoxynivalenol, a deoxynivalenol-transformation.product by *Devosia mutans* 17-2-E-8 *Food Chem. Toxicol.* **84** 250-259.
[53] Islam R, Zhou T, Young J C, Goodwin P H and Pauls K P 2012 Aerobic and anaerobic de-epoxydation of mycotoxin deoxynivalenol by bacteria originating from agricultural soil. *World J Microbiol Biotechnol* **28** 7–13.
[54] Khatibi P A, Newmister S A, Rayment I, Susan P M, Nancy J A and David G S I 2011 Bioprospecting for trichothecene 3-O-acetyltransferases in the fungal genus *Fusarium* yields functional enzymes with different abilities to modify the mycotoxin deoxynivalenol. *Appl Environ Microbiol* **77(4)** 1162-1170.
[55] Poppenberger B, Berthiller F, Lucyshyn D, Sieberer T, Schuhmacher R, Krska R, Kuchler K, Glössl J, Luschnig C and Adam G 2003 Detoxification of the *Fusarium* mycotoxin deoxynivalenol by a UDPglucosyltransferase from *Arabidopsis thaliana. J Biol Chem* **278(48)** 47905-47919.

IOP Conf. Series: Materials Science and Engineering **525** (2019) 012090 doi:10.1088/1757-899X/525/1/012090

[56] Garvey G S, McCormick S P and Rayment I 2008 Structural and functional characterization of the TRI101 trichothecene 3-O-acetyltransferase from *Fusarium sporotrichioides* and *Fusarium graminearum*: kinetic insights to combating *Fusarium* head blight. *J. Biol. Chem.* **283** 1660-1669.

[57] Ito M, Sato I, Ishizaka M, Yoshida S, Koitabashi M, Yoshida S and Tsushima S 2013 Bacterial cytochrome P450 system catabolizing the *Fusarium* toxin deoxynivalenol *Appl. Environ. Microbiol.* **79(5)** 1619–1628.

[58] Tamura M, Mochizuki N, Nagatomi Y, Harayama K, Toriba A and Hayakawa K 2015 Identification and quantification of *Fumonisin* A1, A2, and A3 in corn by high-resolution liquid chromatography- orbitrap mass spectrometry *Toxins* **7(2)** 582-592.

[59] Vanhoutte I, Audenaert K and DeGelder L 2016 Biodegradation of mycotoxins:tales from known and unexplored worlds *Front Microbiol* **7** 561.

[60] Heinl S., Hartinger D, .Thamhesl M, Vekiru E, Krska R, Schatzmayr G, Moll W D and Grabherr R 2010 Degradation of fumonisin B-1 by the consecutive action of two bacterial enzymes. *J.Biotechnol.* **145** 120–129.

[61] Grenierc B, Bracarense A P F L, Schwartz H E, Trumel C, Cossalter A M, Schatzmayr G, Kolf-Clauw M, Moll W D and Oswald I P 2012 The low intestinal and hepatic toxicity of hydrolyzed fumonisin B1 correlates with its inability to alter the metabolism of sphingolipids *Biochem. Pharmacol.* 83 1465–1473.

[62] Humpf H U, Schmelz E M, Meredith F I, Vesper H, Vales T R, Wang E, Menaldino D S, Liotta D C and Merrill A H 1998 Acylation of naturally occurring and synthetic 1-deoxysphinganines by ceramide synthase. Formation of N-palmitoyl-aminopentol produces a toxic metabolite of hydrolyzed fumonisin, AP1, and a new category of ceramide synthase inhibitor. *J Biol Chem* **273** 19060– 19064.

[63] Blackwell B A, Gilliam J T, Savard M E, Miller J D, Duvick J P 1999 Oxidative deamination of hydrolyzed fumonisin B1 (AP1) by cultures of *Exophiala spinifera*. *Nat Toxins* **7** 31–38.

[64] Duvick J P, Maddox J R and Gilliam J T 2003 Compositions and methods for fumonisin detoxification. US Patent 6538177.

[65] Heinl S, Hartinger D, Thamhesl M, Schatzmayr G, Moll W-D and Grabherr R 2011 An aminotransferase from bacterium ATCC 55552 deaminates hydrolyzed fumonisin B1 *Biodegradation* **22** 25–30.

[66] Moll W-D, Hartinger D, Grieβler K, Binder E-M and Schatzmayr G. 2010 Method for the production of an additive for the enzymatic decomposition of mycotoxins, additive, and use thereof US 8703460 B2.

[67] Hartinger D, Schwartz H, Hametner C, Schatzmayr G, Haltrich D and Moll W-D 2011 Enzyme characteristics of aminotransferase FumI of *Sphingopyxis sp.* MTA144 for deamination of hydrolyzed fumonisin B1// Appl Microbiol Biotechnol (2011) **91** 757–768.

[68] Dombrink-Kmizman M A and Blackburn J A 2005. Evaluation of several culture media for production of patulin by *Penicillium* species. *Int. J. Food Microbiol.* **98** 241–248.

[69] Moake M M, Padilla-Zakour O I and Worobo R W 2005 Comprehensive review of patulin control methods in foods *Compr. Rev. Food Sci. Food Saf.* **1** 8-21.

[70] Castoria R, Mannina L, Durán-Patrón R, Maffei F, Sobolev A P, De Felice D V, Pinedo-Rivilla C, Ritieni A, Ferracane R and Wright S A 2011 Conversion of the mycotoxin patulin to the less toxic desoxypatulinic acid by the biocontrol yeast *Rhodosporidium kratochvilovae* strain LS11. *J. Agric. Food Chem.* **59(21)** 11571–11578.

[71] Ricelli A, Baruzzi F, Solfrizzo M, Morea M and Fanizzi F P 2007 Biotransformation of patulin by *Gluconobacter oxydans*. *Appl. Environ. Microbiol.* **73** 785–793.

[72] Moss M O and Long M T 2002 Fate of patulin in the presence of the yeast *Saccharomyces cerevisiae*. *Food Addit. Contam.* 19 387–399.

[73] Hawar S, Vevers W, Karieb S, Ali B K, Billington R and Beal J 2013 Biotransformation of patulin to hydroascladiol by *Lactobacillus plantarum Food Control.* 34 502-508.

IOP Conf. Series: Materials Science and Engineering **525** (2019) 012090 doi:10.1088/1757-899X/525/1/012090

[74] Ianiri G, Idnurm A, Wright S A I, Durán-Patrón R, Mannina L, Ferracane R, Ritieni A and Castoria R 2013 Searching for genes responsible for patulin degradation in a biocontrol yeast provides insight into the basis for resistance to this mycotoxin *Appl. Environ. Microbiol.* **79(9)** 3101–3115.

[75] Zhu R, Feussner K, Wu T, Yan F., Karlovsky P and Zheng X 2015 Detoxification of mycotoxin patulin by the yeast *Rhodosporidium paludigenum Food Chem* **179** 1–5.

[76] Efremenko E N, Lyagin I V, Aslanli A G and Makhlis T A 2017 The method of bioneutralization of mycotoxins. Ru Patent 2634914

[77] Efremenko E N, Lyagin I V, Makhlis T A, Maslova O V, Senko O.V., Stepanov N A 2017 Enzymatic detoxification of mycotoxins: known and promising solutions. *Acta Naturae* **special issue** 124-125.

IOP Conf. Series: Materials Science and Engineering **525** (2019) 012091 doi:10.1088/1757-899X/525/1/012091

Elastic properties of the plasma sprayed Ni coating after the electromechanical treatment

A Yu Ivannikov[1,2], I N Lukina[1], V I Kalita[1], D I Komlev[1], A A Radyuk[1], N V Umnova[1], A S Baikin[1] and A V Alpatov[1]

[1] Baikov Institute of Metallurgy and Material Science, Russian Academy of Sciences, 49, Leninsky ave., 119334 Moscow, Russian Federation
[2] Moscow State University of Technology 'Stankin', 1, Vadkovsky per., 127055 Moscow, Russian Federation

E-mail: ivannikov-a@mail.ru

Abstract. The principal aim of this study was to evaluate the effect of the electromechanical treatment (EMT) on the elastic properties of the air plasma sprayed coating from Ni powder. During the EMT process the boundaries of the splats welded and the quantity of the pores decreased from $8.0 \pm 1.5\%$ to $2.5 \pm 0.5\%$, thus the elastic modulus of the coating increased from 63 ± 5 GPa to 95 ± 5 GPa. The obtained results revealed the high potential of the EMT for post-treatment of the plasma sprayed coatings.

1. Introduction

Re-melting of coatings is a commonly used post-treatment technology of the plasma sprayed coatings [1], but it leads to disintegration of metastable structure with coarsening of second phases. For this reason, for amorphous and nanostructured coatings it is not recommended to use re-melting.

Another type of post-treatment is surface induction heating. The surface induction heating can help improve structure and mechanical properties of the surface [2-4]. However, this treatment has limits on form and size of the samples due to size and forms of the inductor. The surface induction heating is usually applied for treatment of cylindrical samples.

Hot isostatic pressing (HIPing) is another type of the post-treatment of the plasma sprayed coatings [5]. Low porosity level is obtained, and HIP treatment at increasing temperatures and pressures changes the microstructure and increases the microhardness of the coatings. The results show that the elastic modulus of the HIPed coatings is greater than the elastic modulus of the as-sprayed coatings, that can be related to plastic flow, interlamellar diffusion, and creep, that occur at increased temperatures and pressures. Two independent studies, performed by T. Chraska et al. [6], A. A. Almathami and M. Brochu [7], use the spark plasma sintering for post-treatment of amorphous and nanostructured coatings. The post-treated samples have microstructure that is best described as nanocomposite with very small crystallites, by the way near-zero porosity levels were obtained. Meanwhile, there are restrictions on the form and the size of specimens, which are treated by the spark plasma sintering technology, i.e., this method excludes large samples with complex shapes. For this reason, in the industry of the thermal spray coatings one of the actual tasks is determination of the new post-treatment methods of amorphous and nanostructured coatings.

The authors of this investigation used the electromechanically treatment to improve the structure and the properties of the plasma sprayed coatings [8-11]. This approach is based on the simultaneous

IOP Conf. Series: Materials Science and Engineering **525** (2019) 012091 doi:10.1088/1757-899X/525/1/012091

resistive heating of the surface and the deformation of the coating by the electrode. In practice, equipment for the EMT can be mounted on lathes, milling machines, and robotic systems for post-treatment of thermal sprayed coatings. The EMT can replace traditional post-treatment processes, such as heat treatment and mechanical treatment. For this reason, the use of the EMT optimizes the production processes. Also, there is no machine waste, consequently, environmental risks are lower.

In this study feedstock powder of Ni is selected for investigation of the elastic properties of the plasma sprayed coating after the EMT. Nickel has good mechanical properties, which are favorable for the EMT of a plasma sprayed Ni coating, because they allow residual stresses to relax in the cooling cycle. Pure nickel is used as a protective layer for corrosion resistance and as a binder in the cemented tungsten carbide. In addition, nickel is used as a base of Ni-alloys, in which it typically increases tensile strength, toughness, and elastic limit [12]. In particular, T. S. Sidhu et al. [13] show that the Cr_3C_2-NiCr coating has imparted necessary resistance to hot corrosion, which has been attributed to the formation of oxides of nickel and chromium, and spinel of nickel-chromium. Hong-Bin Zhu et al. [14] show that coating deposited from agglomerated and sintered TiB_2–40Ni powder presents better tribological properties, i.e. low friction coefficient and wear rate. Jianbo Lei et al. [15] show that the carbon fibers enhanced the corrosion resistance and the wear resistance of Ni-based coatings. The analysis of the referenced articles show that nickel is an important and demanded material for coatings. Consequently, we can use nickel as a model material for our research. For this reason, it becomes relevant to investigate the effect of the electromechanical treatment on the properties of the plasma sprayed Ni coating.

The purpose of this study is to estimate the influence of the EMT on the elastic properties of the plasma sprayed Ni coating.

2. Materials and methods

Commercial feedstock powder of Ni (chemical composition of ≤ 0.003 wt.% N, ≤ 0.086 wt.% O, Ni constituted the balance) was chosen for plasma spraying. The powder was manufactured by a carbonyl process (Severonickel Combine). The powder particles had irregular shape and size range of 40-56 μm.

Substrates from low-carbon steel were used for plasma spraying of coating and subsequent post-treatment. The sample for determining the effect of the EMT modes was a cylinder 38 mm in diameter and 150 mm in length.

The plasma spraying equipment was the air plasma spray system UPU-3D (JSCo "Electromechanica", Rzhev, Russia) with the plasma spray gun PP-25. The maximum power of the plasma spray gun PP-25 is 25 kW.

Plasma spraying was carried out according to a protocol, which is presented below.

At the first stage, before spraying, the surface of the substrates was cleaned with acetone and roughened by grit blasting with SiC powder (grit size −900 + 700 μm). Further, the samples were placed alternately in a rotator, which was installed under the plasma spray gun, and they moved reciprocatively relative to the sprayed powder with a linear velocity of 1.2 m/min. The samples rotated with the speed of 200 rpm.

For the second stage, Ni powder was plasma sprayed on cylindrical samples. A mixture of Ar and N_2 gases was used as a plasma gas. The plasma gas consumption was 38 l/min. By plasma spraying process the arc current and the voltage were constantly 350 A and 50 V, respectively. The carrier gas was Ar and its feed rate was 3.5 l/min. The powder feed rate was 21 g/min.

The electromechanical treatment was chosen for post-treatment of the plasma sprayed Ni coating. The process of the electromechanical treatment was carried out on a lathe [16].

Three WC–6Co rollers were used for the EMT [16]. The first roller (number 1) had the following dimensions of the working part: 40 mm in diameter and ~1 mm in width. Two diametrically opposite rollers (numbers 2, 3) had the following dimensions of the working parts: 38 mm in diameter and ~4 mm in width. During the EMT process the area of contact between the first roller and the coating was about 1.8 ± 0.1 mm^2, and the area of contact between the second and the third rollers and the coating was about 7.2 ± 0.1 mm^2. Two pneumatic cylinders were used for transmitting efforts to the rollers. A toroid AC transformer with a capacity of 5 kW was used as a power source. The supply voltage to the primary winding of the transformer was limited by a laboratory autotransformer. The measurement of the electric current in the secondary circuit was carried out by a shunt.

The electromechanical treatment was carried out according to the following protocol. The cylindrical sample was installed in the lathe. The speed of the EMT, the density of the electric current, and the voltage were 1.57 m/min, 600 A/mm^2, and 0.5 V, respectively. The longitudinal movement of the roller was determined by the feed rate of the tool holder; in this experiment, the feed rate was constant (0.5 mm/rev). The transmitting efforts to the rollers was 100 kgf. Water was used for cooling of the contact zone during the EMT process, for this reason there was no overheating of the WC–6Co rollers and the samples.

For determining the properties of the coatings, the sample was sectioned to two parts. One part was cut from the areas treated by the EMT, one part had the plasma sprayed Ni coating, not treated by the EMT. Then each part was cut transversely, mounted in plastic, and polished to 1 mm diamond finish. The particles of the feedstock powder were mounted in resin and polished to 1 μm diamond finish, too.

The elastic modulus of the coating was obtained by means of nondestructive method. It was determined in cross-section of the plasma sprayed coating and the treated by means of the EMT coating. For the test a dynamic ultra micro hardness tester (SHIMADZU DUH 211S) with the Vickers indenter at the loads of 1 and 1.5 N were used. Minimum 15 indentations were applied to each sample. The indentation tests were made according to ISO 14577-1:2015 "Metallic materials -- Instrumented indentation test for hardness and materials parameters -- Part 1: Test method".

3. Results and discussion

The analysis of the structure and the mechanical properties was described in the article [16]. The Rz parameter of the plasma sprayed Ni coating decreased from 30 ± 5 μm to 7 ± 3 μm after the EMT. The porosity of coatings decreased from 8.0 ± 1.5% to 2.5 ± 0.5% after the EMT. The tensile strength of the Ni coating was 105 ± 15 MPa, the tensile strength of the coating after the EMT with the load of 100 kgf was 305 ± 20 MPa. The tensile tests showed that the displacement increased from 0.22 ± 0.05 to 0.37 ± 0.05 mm after the EMT.

The indentation tests (Figure 1) show that the indentation depth under applied test force decreases from 5.29 ± 0.05 μm to 5.21 ± 0.05 μm, the elastic modulus of the coating increases from 63 ± 5 GPa to 95 ± 5 GPa, the average relation W_{elast}/W_{total} increases from 24.6 % to 49.3 % after the EMT.

The analysis of the elastic properties of the plasma sprayed Ni coating shows that the elastic modulus of the coating and the average relation W_{elast}/W_{total} are low because of large amount of the residual pores and the low cohesive strength between the splats. During the EMT process of the plasma sprayed Ni coating the combined action of pressure and local Joule heating deforms the splats and welds their boundaries. The amount of the residual pores decreases. For this reason, during the indentation process (Fig.1) the indentation depth (h_{max}) decreases and at the end of the process the permanent indentation depth decreases after removal of the test force (h_p). Thus, the elastic modulus and the average relation W_{elast}/W_{total} of the treated by the EMT coating increase, too.

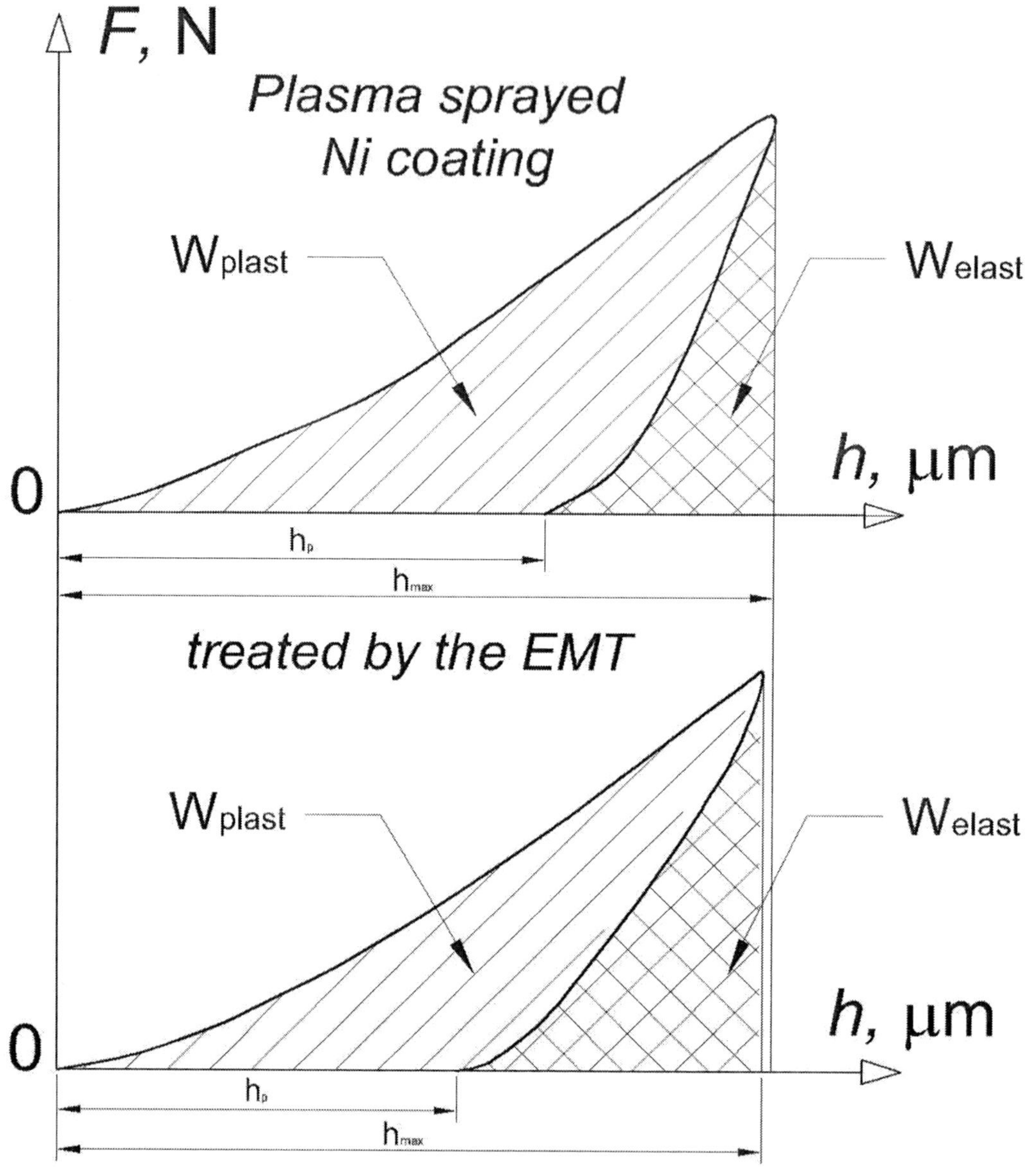

Figure 1. The indentation diagrams of the plasma sprayed Ni coating and of the treated by the EMT plasma sprayed Ni coating

4. Conclusion

1. After the EMT the elastic modulus of the plasma sprayed coating increases from 63 ± 5 GPa to 95 ± 5 GPa because of densification and increase of cohesive strength.
2. The average relation W_{elast}/W_{total} increases from 24.6 % to 49.3 %, this confirms welding processes.
3. During the EMT process the combined action of pressure and local Joule heating deforms the splats and welds their boundaries, for this reason, the EMT has a significant impact on the elastic properties of the treated coating.

Acknowledgments
The reported study was supported by grant RFBR 18-08-00842 a, the mechanical investigation was supported by State target № 075-00746-19-00.

References
[1] L. Pawlowski, 2008, The Science and Engineering of Thermal Spray Coatings, Wiley &Sons, Chichester, p. 627
[2] A.Fomin, et al., 2016 Journal of Physics: Conference Series, 741(1),012170
[3] A.Fomin, et al., 2017 Journal of Physics: Conference Series, 917(8), 082004
[4] V.Koshuro, et al., 2018 Ceram. Inter. 11, 12593-12599
[5] K.Khor, et al., 1994 J. Therm. Spray Technol., 3 (1), 57-62
[6] T.Chraska, et al., 2015 J. Therm. Spray Technol., 24, 637-643
[7] A.Almathami, et al., 2010 J. Mater. Process. Technol.,210 (15), 2119-2124
[8] A.Ivannikov, et al., 2016 Inorganic Materials: Applied Research 7(3), 363-371
[9] D.Komlev, et al., 2013 Inorganic Materials: Applied Research.,4(3), 236-246
[10] V.Bagmutov, et al., 2017 PNRPU Mechanics Bulletin, (1), 233-250
[11] V.Bagmutov, et al., 2007 Vestn. Voronezh. Gos. Tekhn. Univ. 3 (8), 135-140
[12] E.Ezugwu, et al., 1999 J. Mater. Process. Technol.,86(1–3), 1-16
[13] T.Sidhu, et al., 2006 J. Therm. Spray Technol., 15(4), 811-816
[14] H.Zhu, et al., 2013 Surf. Coat. Technol., 235, 620-627
[15] L.Liu, et al., 2017 Surf. Coat. Tech., 325, 548-554
[16] A.Ivannikov, et al., 2019 J. Mater. Process. Technol., 266, 442-449

IOP Conf. Series: Materials Science and Engineering **525** (2019) 012092 doi:10.1088/1757-899X/525/1/012092

Study of the coefficient of heat expansion of TiNbTaZr alloy

K V Sergienko[1], D D Titov[1], S V Konushkin[1], A S Baikin[1], E O Nasakina[1], M I Baskakova[1], A Bespamiatnova[1], E E Baranov[1], L A Shatova[2], A G Kolmakov[1] and M A Sevostyanov[1]

[1]A.A.Baikov Institute of Metallurgy and Materials Science, Russian Academy of Sciences, Moscow, Russia
[2]Voronezh State Technical University, st. 20-letiya Oktyabrya, 84/4, Voronezh, Russia

E-mail: ksergienko@imet.ac.ru

Abstract. This article describes the study of the thermal expansion coefficient of a TiNbTaZr alloy by a dilatometric differential method used to determine phase transformation temperatures. The object of the study was samples of an alloy of TiNbTaZr with a variable composition, which underwent homogenizing annealing and rotary forging. According to the obtained results, all the studied alloys have three temperature regions, which are distinguished in the graphs of uniform elongation with increasing temperature.

1. Introduction

Dilatometry, as a method of studying substances, is based on the effect of changing the linear dimensions of samples with a change in temperature. Initially, dilatometry was used only to determine the coefficient of thermal expansion of mainly solids. However, this method is able to determine the presence of phase transitions, which are characterized by the influence on the sizes of samples of materials, in particular alloys, thereby violating the linear growth of the sizes of samples with increasing temperature. This research method usually determines not the absolute value of the expansion coefficient, but the difference between the expansion of the sample under study and the standard devoid of phase transitions in the temperature range under study. Thus, comparing the mutual change in the size of the sample and the standard, with a change in temperature, various extensions of the sample are revealed, meaning any transformations in the latter. As well as differential scanning calorimetry (DSC), this method cannot give an unambiguous answer because of what process the expansion linearity is disturbed with increasing temperature, however, the temperature at which the phase composition studies, heat treatment, and much more is appropriate.

2. The object of study and used equipment

The object of the study was samples of alloys of the TiNbTaZr system with niobium content from 20 to 30%, tantalum from 10 to 15%, zirconium 5% and titanium by the residual principle. Obtaining samples consisted of melting ingots on a vacuum arc furnace with a non-consumable tungsten electrode, conducting homogenizing annealing, rolling ingots to a cross section of 10x10 mm2, rotary forging to a circular cross section with a diameter of 4 mm, cutting a bar to obtain samples in the form of cylinders 4 mm in diameter and 10 long mm, the ends are polished. Studies were conducted on a high-temperature dilatometer DIL 402 C7G from Netzsch. This equipment allows continuous measurement of linear dimensions, under conditions of programmable exposure to temperature in the range from 20 to 1800 ° C. Studies are conducted in argon or nitrogen. The maximum heating rate can be 25 ° C per minute. The following research conditions were used in the work: temperature from 50 to 1000 ° C, medium — argon, heating rate 10 ° C per minute. The following are graphs of research (Fig. 4-9).

IOP Conf. Series: Materials Science and Engineering **525** (2019) 012092 doi:10.1088/1757-899X/525/1/012092

3. Experimental

When preparing the samples, the structure of Ti20Nb13Ta5Zr ingots after melting (Figure 1), homogenizing annealing at 800°C (Figure 2), and rotary forging (Figure 3) was investigated. The various compositions of the TiNbTaZr system have the same structure at each stage of the study between themselves.

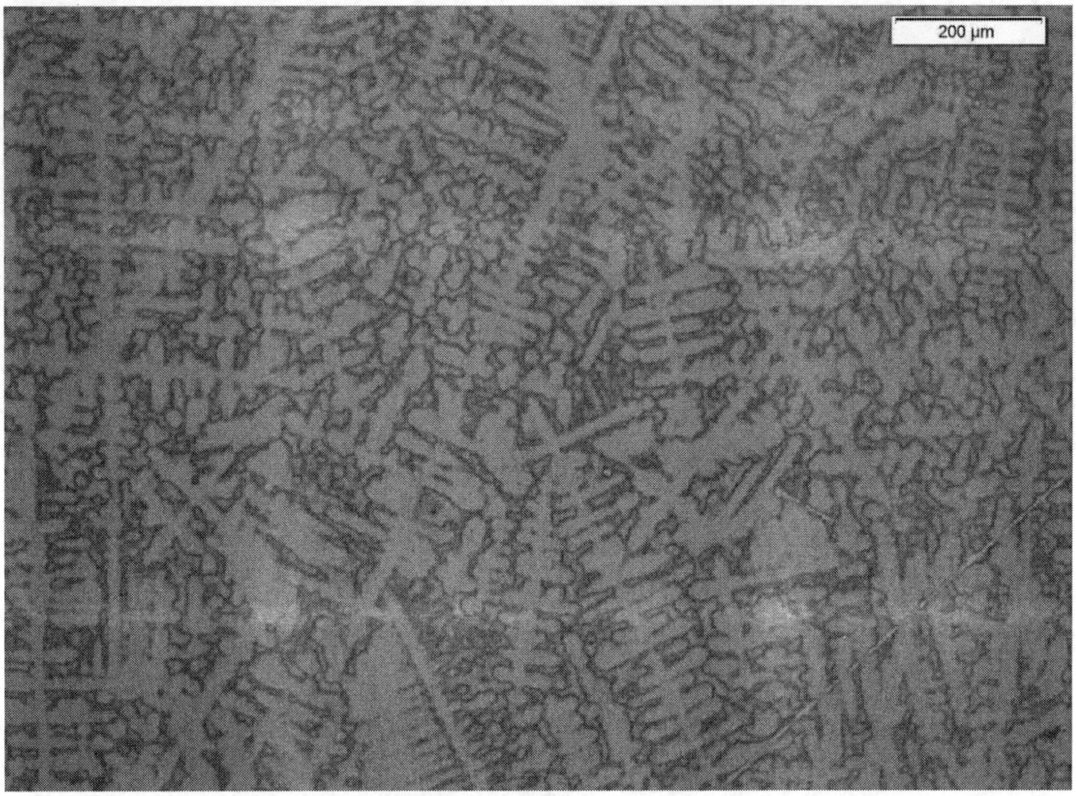

Figure 1. Structure of the Ti20Nb13Ta5Zr ingot after smelting

After smelting, the structure of the ingots (Figure 1.) has a clearly defined dendritic structure, which indicates non-uniformity of the chemical composition and requires homogenizing annealing to level the composition at 800 ° C for 12 hours. After the homogenization annealing (Figure 2) was carried out, the dendritic structure was destroyed with the formation of grains, which were stretched in the direction of the rod during rotational forging (Figure 3).

In the figures (Figure 4-9.) Shows the dilatometric curves of various compositions. Internal transformations in metals and alloys are characterized by changes in volume, linear dimensions and coefficient of expansion. The dilatometric method is based on the study of these changes.

When heated, due to an increase in the interatomic distance in the lattice, there is a uniform reversible increase in the length (volume) of the body - thermal expansion. Phase transformations, on the other hand, lead to an abrupt change in the parameters, which results in a change in the nature of the dilatometric curve.

IOP Conf. Series: Materials Science and Engineering **525** (2019) 012092 doi:10.1088/1757-899X/525/1/012092

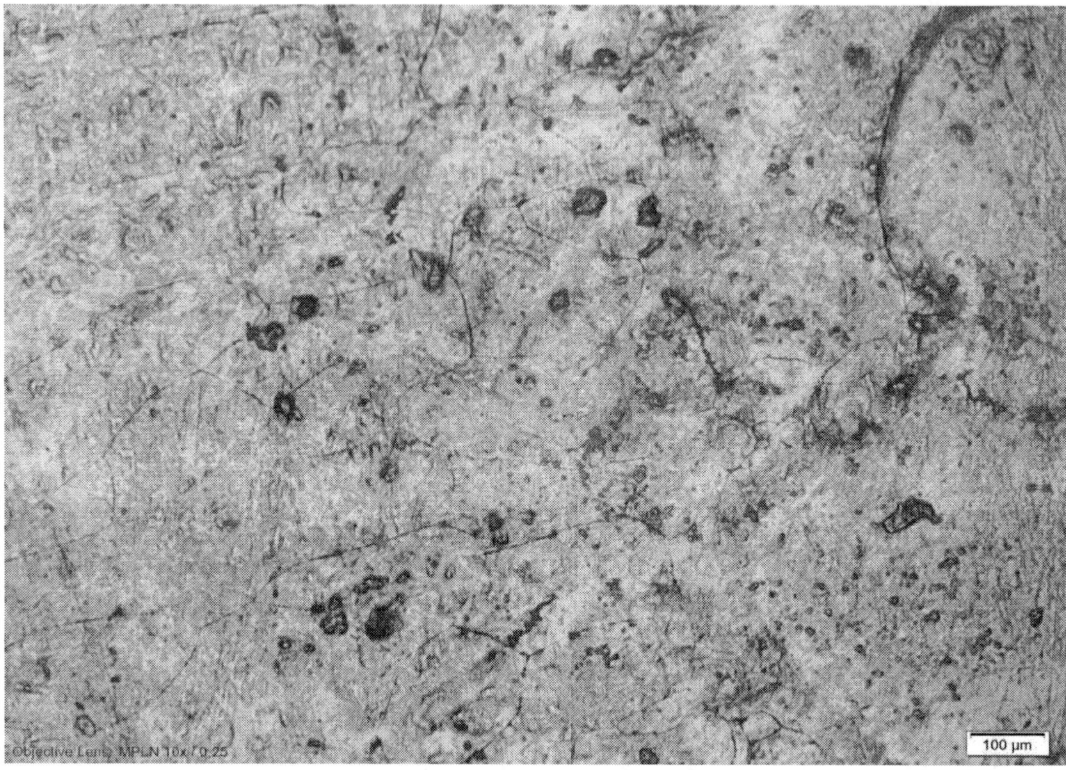

Figure 2. The structure of the ingot Ti20Nb13Ta5Zr after homogenizing annealing

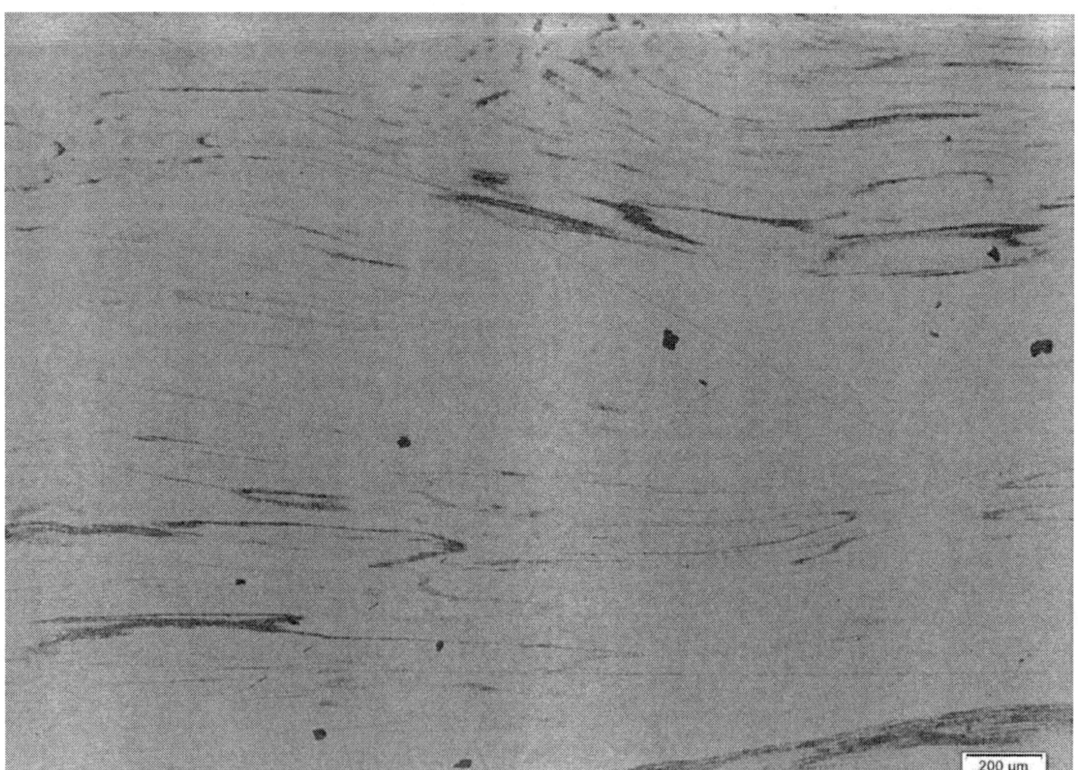

Figure 3. The structure of the ingot Ti20Nb13Ta5Zr after drawing

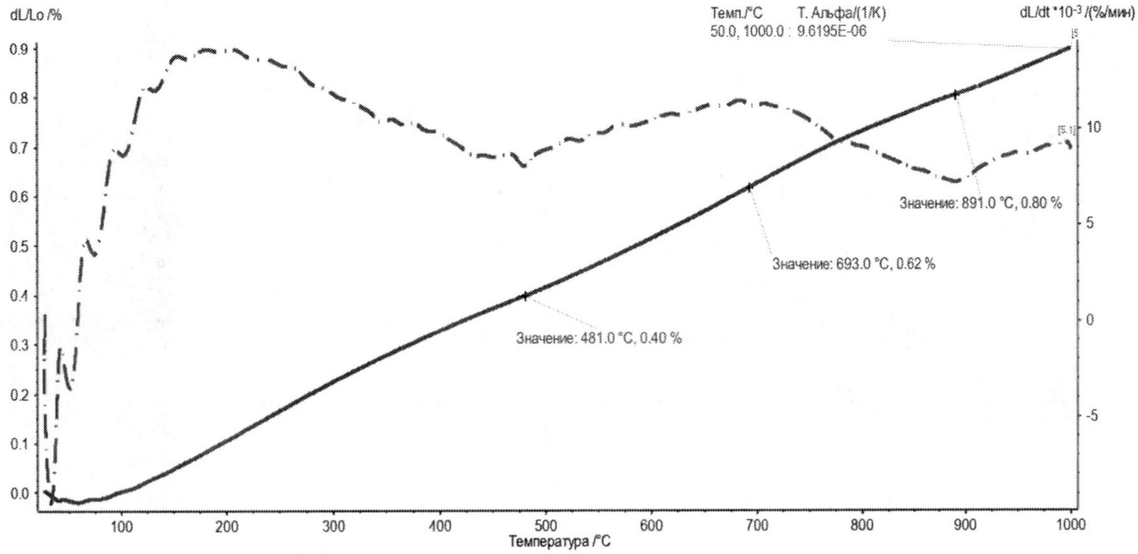

Figure 4. The results of the study of the alloy Ti20Nb15Ta5Zr

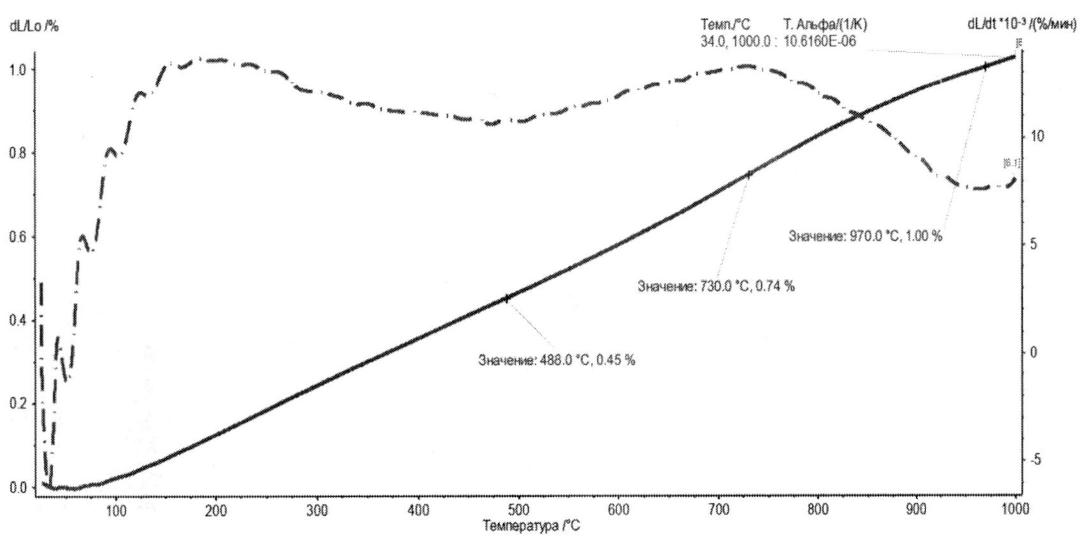

Figure 5. The results of the study of the alloy Ti30Nb13Ta5Zr

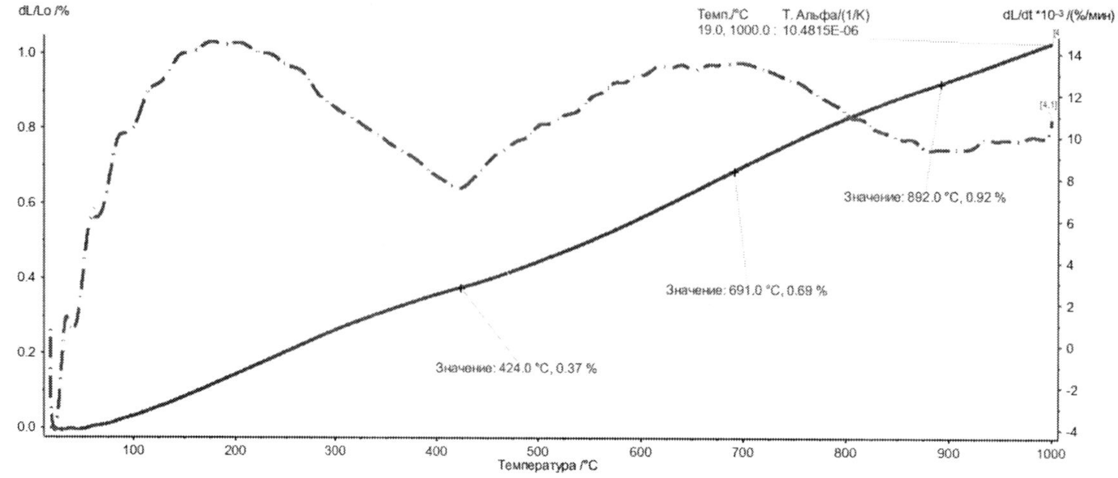

Figure 6. The results of the study of the alloy Ti20Nb10Ta5Zr

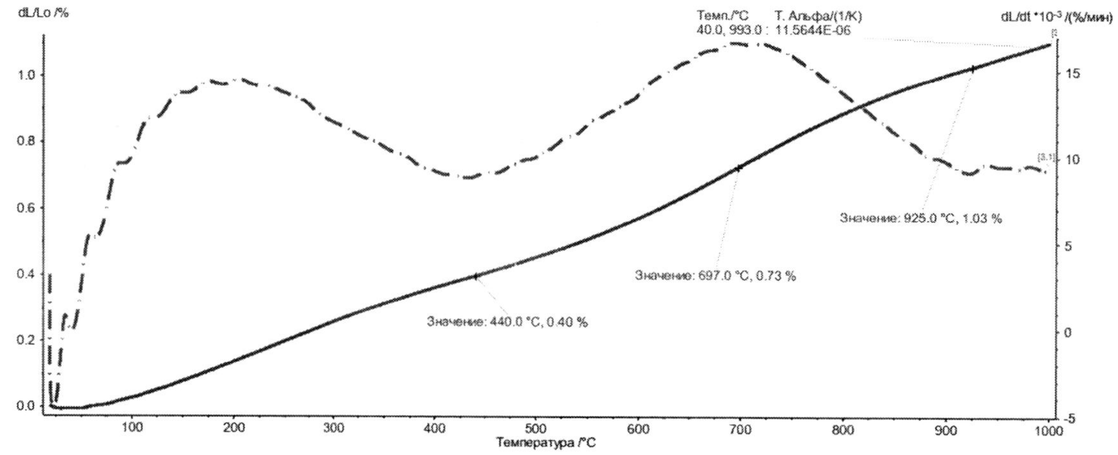

Figure 7. The results of the study of the alloy Ti25Nb10Ta5Zr

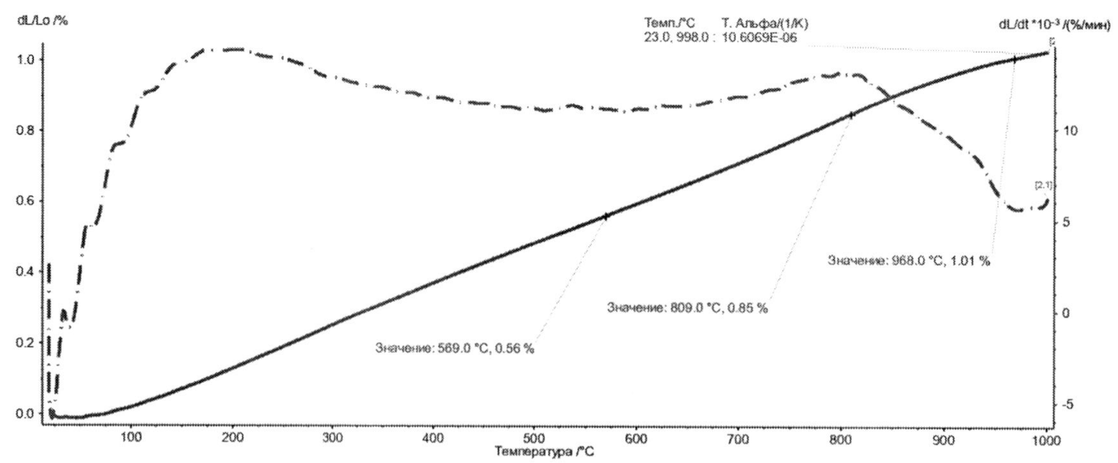

Figure 8. The results of the study of the alloy Ti25Nb13Ta5Zr

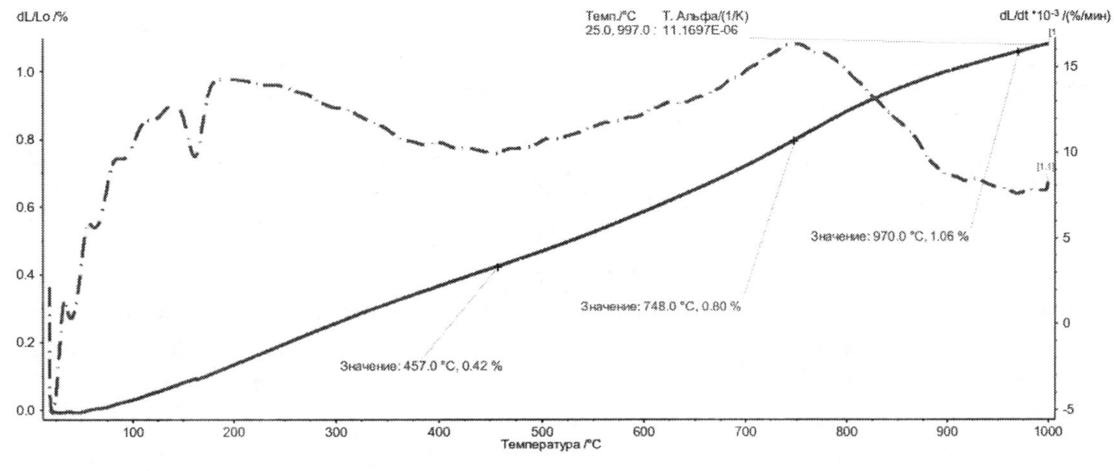

Figure 9. The results of the study alloy Ti30Nb10Ta5Zr

In metals that do not undergo structural transformations, the change in length during heating and cooling occurs monotonously, and the dilatometric curve does not change with a change in the rate of heating and cooling.

When cooled, on the contrary - the reduction of the metal is interrupted by its elongation in the temperature range of structural transformation.

An initial visual analysis of the dilatometric curve is performed. In the general case, there should be no abrupt multiple abrupt anomalies on the curve indicating either that the specimen was not prepared for measurements properly (not the lower and upper planes of the specimen are uneven or not parallel to each other, the specimen surface is not sufficiently clean) or breaks in the process of heating and cooling. If such are present, measurements are usually repeated with newly made samples.

Temperature curves of softening and (or) temperature expansion of samples are separately distinguished on the curve. The temperatures of anomalies corresponding to phase transitions (water evaporation, structural transition, vitrification, melting, and other transitions associated with changes in the mechanical state of the object of study) are noted. In these areas, the magnitude of the change in thickness and the coefficient of thermal expansion of the sample under study are determined.

Phase transitions or other thermal processes associated with changes in the mechanical properties of the sample under investigation on dilatometric curves manifest themselves in the form of kinks or steps. For a clearer determination of the temperatures of the processes, the curve of the first derivative of the TMA signal is also plotted, on which anomalies appear in the form of peaks (Figure 4-9).

4. Results and conclusions

According to the data obtained, it can be concluded that in each composition there are three phase transitions in the temperature range under study: the first starts at a temperature of 440 ° C, the second - 690 - 750 ° C and the third - above 890 - 970 ° C. The phase transition temperatures differ slightly for each alloy. Thus, it is advisable to conduct research by X-ray phase analysis of ingots at temperatures of 400, 600, 800 and 1000 ° C to determine the phase composition.

Acknowledgments

The work was supported by the Ministry of Education and Science of Russia (grant identifier RFMEFI60417X0196).

References

[1] E.O. Nasakina, S.V. Konushkin, M.I. Baskakova, I.M. Fedyuk, K.V. Sergienko, A.S. Baikin, M.A. Kaplan, M.A. Sevostyanov, A.G. Kolmakov. The Production of a Thin Wire of Ti-Nb-

Ta-Zr Shape Memory Alloy for Medical Devices *J Material Sci Eng (Journal of Material Sciences & Engineering) 2018* **7(1)**, 430-436.

[2] Sevost'yanov M.A., Nasakina E.O., Baikin A.S., Sergienko K.V., Konushkin S.V., Kaplan M.A., Seregin A.V., Leonov A.V., Kozlov V.A., Shkirin A.V., Bunkin N.F., Kolmakov A.G., Simakov S.V., Gudkov S.V. Biocompatibility of new materials based on nano-structured nitinol with titanium and tantalum composite surface layers: experimental analysis in vitro and in vivo *Journal of Materials Science: Materials in Medicine*, 2018. **29**. 33

[3] E O Nasakina, A S Baikin, K V Sergiyenko, M A Kaplan, S V Konushkin, A D Yakubov, A V Izvin, M A Sudarchikova, M A Sevost'yanov and A G Kolmakov. Formation of biodegradated polymers as components of future composite materials on the basis of shape memory alloy of medical appointment *IOP Conference Series: Materials Science and Engineering* **347** *(2018) 012016*

[4] M A Kaplan, E O Nasakina, M A Sevostyanov, K V Sergienko, A S Baikin, S V Konushkin, A G Kolmakov, M A Smirnov, A V Izvin and Yu N Ustinova. Influence of annealing on mechanical properties of TiNi (55.8 mass % of Ni) wire made for medical purposes *IOP Conference Series: Materials Science and Engineering* **347** *(2018) 012022*

[5] A S Baikin, M A Sevostyanov, E O Nasakina, K V Sergienko, M A Kaplan, S V Konushkin, A A Kolmakova, A D Yakubov and A G Kolmakov. Investigation of the influence of the composition on mechanical properties of polylactide *IOP Conference Series: Materials Science and Engineering* **347** *(2018) 012026*

[6] A S Baikin, M A Sevostyanov, E O Nasakina, K V Sergienko, M A Kaplan, S V Konushkin, A A Kolmakova, A D Yakubov and A G Kolmakov. Investigation of the influence of the composition on mechanical properties poly(glycolide-DL-lactide) // *IOP Conference Series: Materials Science and Engineering* **347** *(2018)*

[7] K V Sergienko, M A Sevost'yanov, S V Konushkin, E O Nasakina, A S Baikin, L A Shatova, A G Kolmakov. Investigation of static properties of medical alloys Ti-(20-30)Nb-(10-13)Ta-5Zr *IOP Conference Series: Materials Science and Engineering* **347** *(2018) 012049*

IOP Conf. Series: Materials Science and Engineering **525** (2019) 012093 doi:10.1088/1757-899X/525/1/012093

Multifunctional nanomaterials based on ceria solid solution

I V Zagaynov[1], S V Fedorov[1], I V Kulbakin[1] and A A Klimashin[1]

[1] Baikov Institute of Metallurgy and Materials Science

E-mail: igorscience@gmail.com

Abstract. The work is focused on the investigation of the influence of metal oxide promoted ceria based materials with various dopants on the physicochemical, electrical and catalytic properties for CO oxidation in model reaction. The formation of stable solid solutions by doping copper, manganese, zirconium, gadolinium, and bismuth enhances the redox properties, oxygen vacancies, stability, and, of course, catalytic activity and electroconductivity.

1. Introduction

In the context of significant interest in energy efficient technologies and environmental protection, materials based on nanostructured ceria are promising for solving existing problems. The advantage of such materials is due to that CeO_2 has a large oxygen storage capacity (OSC) and high oxygen mobility, which can ensure its high catalytic activity and electrical conductivity (and, as a rule, these data for ceria correlate with explanations of dependencies). Thus, in order to realize an effective process, a crystalline structure with a sufficiently high concentration of oxygen defects is necessary, and an increase in the magnitude of OSC and the number of defects are possible by doping ceria with various oxides.

The present study purpose was the preparation of solid solutions based on doped ceria and subsequent investigation of their physicochemical properties.

2. Materials and Method

$Ce(NO_3)_3 \cdot 6H_2O$, $Bi(NO_3)_3 \cdot 5H_2O$, $Gd(NO_3)_3 \cdot 6H_2O$, $Cu(NO_3)_2 \cdot 3H_2O$, $ZrO(NO_3)_2 \cdot 7H_2O$, $Mn(NO_3)_2 \cdot 4H_2O$ (Acros Organics) were used as metal precursors. Appropriate amounts of salts were dissolved in concentrated nitric acid (68%) with the concentration of salts of 0.667 M. After the dissolution of salts, this mixture was added to distilled water, giving the concentration of 0.1 M (in one experiment after dissolution acetylacetone was added). Then, the co-precipitation was carried out by the addition of 2.5 M KOH solution up to pH 11. Ultrasonic processing (35 kHz, 150 W) was used during all process at 30°C under stirring. The resulting precipitates were filtered, washed with distilled water-ethanol solution (H_2O/C_2H_5OH=9 vol.), dried at 150°C for 12 h, and calcined in static air by heating at a rate of 4°C/min from room temperature to 500°C and kept at 500°C for 1 h in the muffle furnace..

The as-obtained powders were pressed into pellets (with 5 wt.% binder made of 5% aqueous solution of polyvinyl alcohol) with 10 mm in diameter and 1 mm in thickness at 100 MPa. Then they were sintered at 1000 or 1150°C (T_{sint}) for 4 h in air with heating of 4°C/min. To fabricate symmetric cells for the impedance studies (Elins Z-350M impedance meter, the frequency range from 0.1 Hz to 1 MHz at the amplitude of AC signal of 30 mV), platinum paste was brushed onto both sides of the electrolyte pellets, and were dried at 150°C for 1 h and annealed at 900°C for 4 h in air. A platinum wire was used as current conductors.

The catalytic activity of the synthesized samples in the oxidation of CO was determined by the flow method at atmospheric pressure. The process was conducted in a U-shaped quartz reactor at the total flow rate of 60 ml/min within a temperature range of 20-500°C. The temperature was measured with a thermocouple placed in the center of the catalytic bed. The catalyst sample mass was 0.3 g. The model gas mixture has the following composition, vol.%: CO – 1; O_2 – 2; N_2 – balance. The concentrations of gases were determined by a gas chromatograph (Varian 450GC).

All powders and ceramics were characterized by XRD (Rigaku MiniFlex 600, Cu*Kα* radiation), SEM (TESCAN VEGA II SBU with INCA Energy 300 energy dispersive spectrometers), TG-DSC-MS (Netzsch STA449F3, the samples were heated to 1350°C at the rate of 10°C/min in air), TEM (Omega Leo-912AB), nitrogen adsorption-desorption method (TriStar 3000 Micromeritics).

3. Results and discussion

The stability and availability of solid solutions $Gd_xBi_{0.2-x}Ce_{0.8}O_2$ and $Cu_xMn_{0.1-x}(Zr_{0.1})Ce_{0.8}O_2$ were evaluated. According to TG-DSC, the first solid solution is stable up to 1000°C (because of Bi evaporating process after 1050°C) and the second is up to 1400°C. Thus, further obtaining dense ceramics was achieved at temperatures of 1000 and 1150°C, respectively. Typical characteristics for, as will be shown later, the most active samples are shown in Table 1.

Table 1. Some characteristics of initial powders and ceramics.

Sample	d_{XRD} (nm)	S_{BET} (m^2g^{-1})	T(100% conversion of CO, °C)	$\sigma_{600°C}$ (S^1cm^{-1})
$\mathbf{Gd_{0.05}Bi_{0.15}Ce_{0.8}O_2}$	8	23	404	0.005
$\mathbf{Cu_{0.08}Mn_{0.02}Ce_{0.9}O_2}$	9	79	215	0.002
$\mathbf{Cu_{0.08}Mn_{0.02}Zr_{0.1}Ce_{0.8}O_2}$	8	31	310	0.008

According to XRD (Figure 1a), samples have single-phase with fluorite structure (Fm3m, JCPDS-34-0394). Thus, in all cases the solid substitutional solution is formed in which the dopants enter the ceria lattice. In contrast to XRD results, which afford information, related to the cation sublattice, Raman spectroscopy of the catalysts is dominated by oxygen lattice vibrations. The strong band at 460 cm^{-1} is attributed to the F_{2g} Raman vibration mode of the cubic fluorite structure phase, which also suggests that dopant ions have been incorporated into the ceria lattice to form solid solutions, and the broad band at 590 cm^{-1} is attributed to the oxygen vacancies generated by the partial substitution of dopant ions to Ce^{4+} [1, 2].

The shape of the adsorption-desorption curves for all samples (Figure 1b) corresponds to the IV type, which is typical for the formation of the mesoporous structure. For all samples, the shape of the hysteresis loop corresponds to the H2 type: apparently, individual or connected cylindrical and bottle-shaped pores may be present, which is also confirmed by the pore size distribution (Figure 1c).

The average crystallite size calculated using Scherrer formula was about 10 nm for all samples, which corresponds to the individual particles according to TEM (Figure 1d-f).

It is known that pure ceria is a poor oxide ion conductor ($\sigma_{600°C}$~10^{-5} S/cm), and the ionic conductivity can be significantly enhanced by increasing the oxygen vacancies, which can be created by the doping of ceria with about 20% concentration of dopant, because such systems achieve the highest conductivity, but a further increase in the dopant concentration results in decreasing one. The highest conductivity for intermediate temperature application ($\sigma_{600°C}=5\cdot10^{-3}$ S/cm) was found in $Gd_{0.05}Bi_{0.15}Ce_{0.8}O_2$ (in series of $Gd_xBi_{0.2-x}Ce_{0.8}O_2$ systems), and $\sigma_{600°C}=8\cdot10^{-3}$ S/cm was in $Cu_{0.08}Mn_{0.02}Zr_{0.1}Ce_{0.8}O_2$ (in series of $Cu_xMn_{0.1-x}(Zr_{0.1})Ce_{0.8}O_2$ systems) (Table 1). These results demonstrate that co-doping with the optimum ratio can improve the conductivity of ceria based electrolytes. When compared with other electrolytes, sintered even at a higher temperature, the electrical conductivity of the developed systems is comparable or higher than samples doped by REE [3-6] at 600°C.

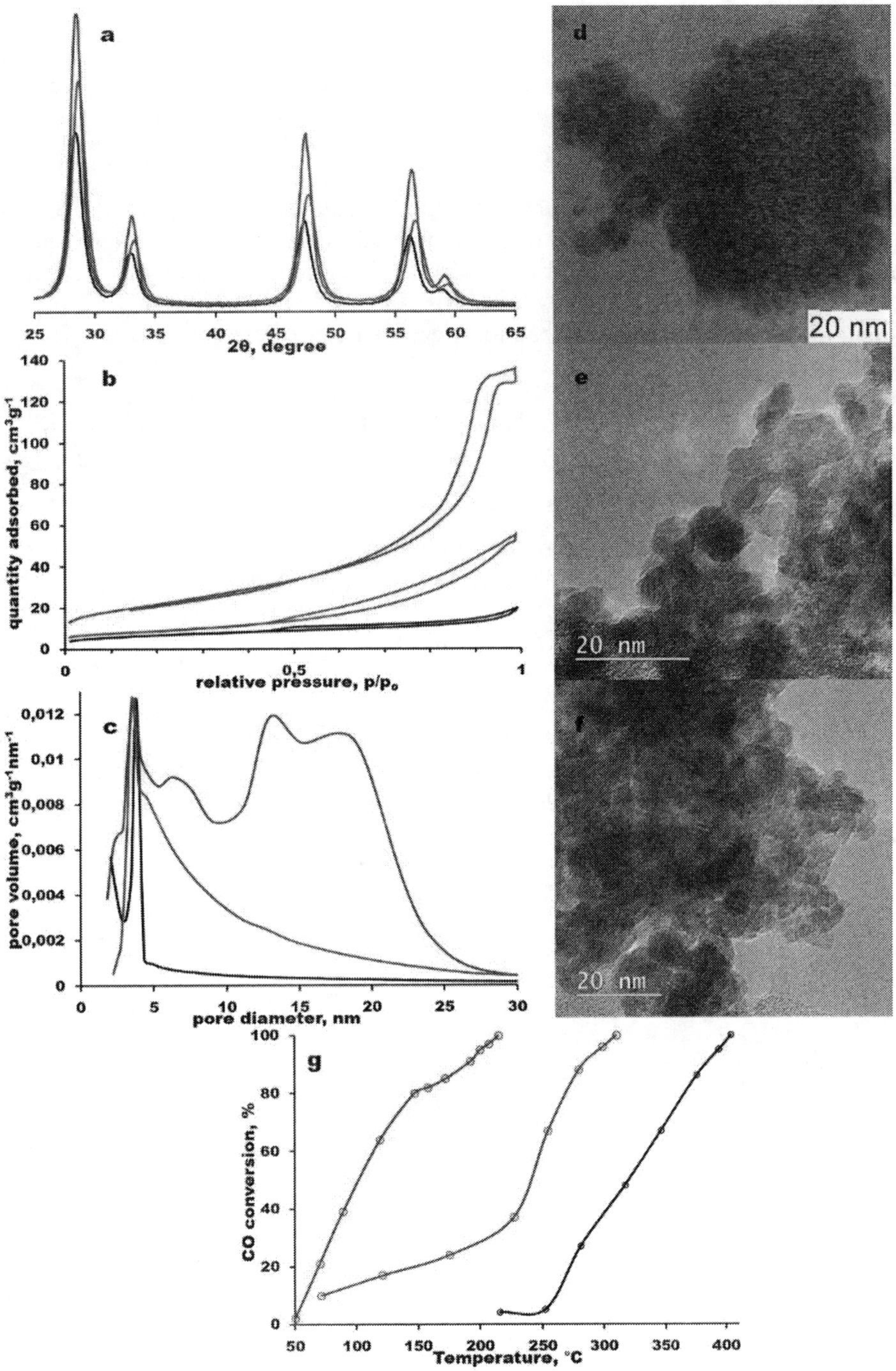

Figure 1. XRD patterns (a), adsorption-desorption isotherms (b), pore size distributions (c), TEM microphotos (d, e, f), and catalytic activity in CO oxidation (g) of calcined powders. $Gd_{0.05}Bi_{0.15}Ce_{0.8}O_2$ – black line, photo d; $Cu_{0.08}Mn_{0.02}Ce_{0.9}O_2$ – red line, photo e; $Cu_{0.08}Mn_{0.02}Zr_{0.1}Ce_{0.8}O_2$ – blue line, photo f.

The catalytic activity of synthesized nanopowders was determined in the model oxidation reaction of CO. Figure 1g shows the increase in temperature the conversion of carbon monoxide increases, while the curve has the S-shaped type. Among the all samples, the same mentioned samples

$Gd_{0.05}Bi_{0.15}Ce_{0.8}O_2$, $Cu_{0.08}Mn_{0.02}Ce_{0.9}O_2$, and $Cu_{0.08}Mn_{0.02}Zr_{0.1}Ce_{0.8}O_2$ had the lowest oxidation temperature. The systems with the addition of zirconium had a conversion temperature higher by 100°C. When comparing the best catalyst with other systems based on ceria ($Zr_{0.12}Y_{0.08}Ce_{0.8}O_2$, $La_{0.12}Y_{0.08}Ce_{0.8}O_2$, etc. [7-9]) under close test conditions, the catalytic activity of the developed system was higher (the temperature of complete oxidation of CO is lower by 50-100°C).

4. Conclusions

The investigation showed the expediency of using such catalysts in the oxidation of CO and these solid solutions could be used as the catalyst or support for the subsequent deposition of an active phase. As noted above, the catalytic activity and electrical conductivity for ceria based materials correlate with each other: samples $Gd_{0.05}Bi_{0.15}Ce_{0.8}O_2$, $Cu_{0.08}Mn_{0.02}Ce_{0.9}O_2$, and $Cu_{0.08}Mn_{0.02}Zr_{0.1}Ce_{0.8}O_2$ had the best electrical conductivity associated with the ionic conductivity of oxygen [10].

Acknowledgments

This research was supported by Russian Science Foundation (No. 17-73-10331). This work was partially conducted within the framework of budget project No. 075-00746-19-00 (XRD).

References

[1] Zagaynov IV, Konovalov AA and Koneva EA 2018 *Letters on Materials* **8** 135.
[2] Zagaynov IV, Naumkin AV and Grigoriev YuV 2018 *Appl. Cat. B* **236** 171
[3] Arabaci A 2017 *Metall. Mater. Trans. A Phys. Metall. Mater. Sci.* **48** 2282
[4] Santos TH, Grilo JPF, Loureiro FJA., Fagg DP, Fonseca FC and Macedo DA 2018 *Ceram. Int.* **44** 2745
[5] Gan Y, Cheng J, Li M, Zhan H, Sun W 2015 *Mater. Chem. Phys.* **163** 279
[6] Zagaynov IV, Fedorov SV, Konovalov AA, Antonova OS 2017 *Mater. Lett.* **203** 9
[7] Zagaynov IV 2017 *Appl. Nanosci.* **7** 871
[8] Mukherjee D, Rao BG and Reddy BM 2016 *Appl. Cat. B* **197** 105
[9] Devaiah D, Tsuzuki T, Aniz CU and Reddy BM 2015 *Catal. Lett.* **145** 1206
[10] Zagaynov IV 2019 *Inorg. Mater.: Appl. Res.* **10** 42

IOP Conf. Series: Materials Science and Engineering **525** (2019) 012094 doi:10.1088/1757-899X/525/1/012094

Water and water-ethanol mixtures (alcoholic beverages) viscosity measurements by dynamic light scattering with use of silica nanoparticles as a seed

V N Kuryakov

Oil and Gas Research Institute of RAS (OGRI RAS), Gubkina Street, 3, 119333, Moscow, Russia

E-mail: vladimir.kuryakov@gmail.com

Abstract. The dynamic viscosity of several alcoholic drinks was studied via determinations of the translational particle diffusion coefficient by dynamic light scattering (DLS). For temperatures from 3 to 80 °C at atmospheric pressure, the dynamic viscosity values of clear water, several brands of vodka, and one brand of cognac were obtained with the uncertainty of less than 10%. Measured viscosity agrees well with literature.

1. Introduction

One of the most intrigue applications of dynamic light scattering (DLS) is the liquids viscosity measurement. Dynamic light scattering is an experimental technique which allows measuring the hydrodynamic radius of Brownian particles through the measurement of their diffusion coefficient by using the Einstein-Stokes equation [1]. In a case when the studied liquid viscosity is known, one can calculate the radius of particles and in opposite case when the radius of particles is known in advance the viscosity can be calculated through DLS measurements. To measure the liquid viscosity by DLS, it is necessary to add the standard sample of nanoparticles with known size into the studied liquid. Here are the main conditions for successful viscosity measurements using DLS: added nanoparticles have to be stable in studied liquid; added dry nanoparticles or added suspension with nanoparticles do not have to influence on viscosity of studied liquid; measured light scattering intensity in studied liquid has to be mostly from added nanoparticles; the particles must be spherical and noninteracting.

One can find several papers with application of DLS for measurements of viscosity of different liquids. Water viscosity measurements were done by M.A. Anisimov and co-workers [2] at different temperatures and by S. Will and co-workers [3] at one fixed temperature. They used latex particles of 234 nm and polystyrene particles of 102 nm respectively. S. Will also applied DLS to measure the viscosity of n-heptane at different temperatures and to measure the viscosity of some alcohols, alkanes, and the refrigerant R123 at fixed temperature using silica particles as a seed [4, 5]. I.K. Yudin and M.A. Anisimov measured temperature dependence of the viscosity of crude oil by DLS using a natural dilute microemulsion of water in oil (stable water droplets of 70 nm) as seed. In this paper measurements of the temperature dependence of the viscosity of several alcoholic beverages and distilled water are presented. The measurements performed by DLS with using SiO_2 nanoparticles of 110 nm diameter as seed.

2. Materials and methods

For this study, the author used the bi-distilled water (medical water for injection sealed in plastic ampoules 5 ml, Solopharm, Russia). Colloidal silica slurry was used as a seed, which has spherical particles and excellent particle uniformity. The content of SiO_2 in initial colloid was 40% weight. Average particle size (hydrodynamic radius) diluted in a water sample measured by DLS was 55 nm. Four-milliliter glass vials with screw caps were used for DLS measurements. Each vial was filled up to the brim with the sample. The one microliter of initially concentrated silica slurry was added into the studied samples of alcoholic drinks in vials. The measurements of temperature dependence of viscosity were done at heating. At each temperature, the time for the equilibrium was about 3 minutes and after that, the correlation function of the intensity of light scattering was measured three times with accumulation time of 100 seconds each. The author used one of them with minimal standard deviation for viscosity determination. The Cumulant Analysis method (DynaLS software, Alango Ltd.) was used for processing the correlation functions of the intensity of scattered light. The value of the viscosity in the Einstein-Stokes equation was variated to obtain the correct size of particles 55 nm as a result of processing. Thus, the viscosity values at each temperature for all studied samples were determined. Dynamic light scattering measurements were performed on particle size and zeta potential analyzer Photocor Compact (Russia). Scattering angle was 90 degree, laser power and wavelength were 25 mW and 654 nm correspondingly.

Three brands of vodka from different manufacturers and one brand of cognac were chosen for the studies. Information on the composition and some properties of the studied alcoholic beverages are presented in Table 1.

Table 1. Properties and composition of the studied alcoholic beverages.

Sample	Alcohol by volume, % (ABV)	Type of the spirit	Calorie in 100 ml, kJ (kcal)	Additives
Vodka Morosha №2	40	Alpha	930 (220)	infusion of alcohol flax, honey
Vodka Derevenka Solodovaya	40	Alpha	930 (220)	sugar syrup, infusion of alcohol cranberries and rye
Vodka Bulbash Honey and pepper	40	Luxe	940 (230)	cognac, sugar, chili pepper, sugar dye №6, alcohol extract of pepper, pollen and honey
Cognac Stone Land №3	40	Cognac spirits of 3 years old	955 (228)	E150a

3. Results and discussion

The intensity of the scattered light in all samples prior to the addition of nanoparticles was 7k-9k counts per second, and no particles were observed by the method of dynamic light scattering. After the addition of nanoparticles, the scattering intensity became about 500k counts per second, and the correlation function of scattered light intensity was measured using the DLS method. Previously, the author carried out the temperature dependence of water viscosity measurements using the DLS method. The obtained results of measuring the water viscosity by the DLS method at different temperatures are in good agreement with the data of the International Association for the Properties of Water and Steam (IAPWS) (see figure 1). Relative viscosity deviation measured by DLS from IAPWS viscosity is 2-6%. The author could not find the published data on the temperature dependence of viscosity for vodka or brandy. In Industrial Solvents Handbook [7], there are data for the temperature

IOP Conf. Series: Materials Science and Engineering **525** (2019) 012094 doi:10.1088/1757-899X/525/1/012094

dependence of water-ethanol mixture viscosity. Figure 1 presents data from [7] for an ethanol-water mixture of 35% by weight, since 40% ABV (alcohol by volume) is about 35% ethanol-water by weight.

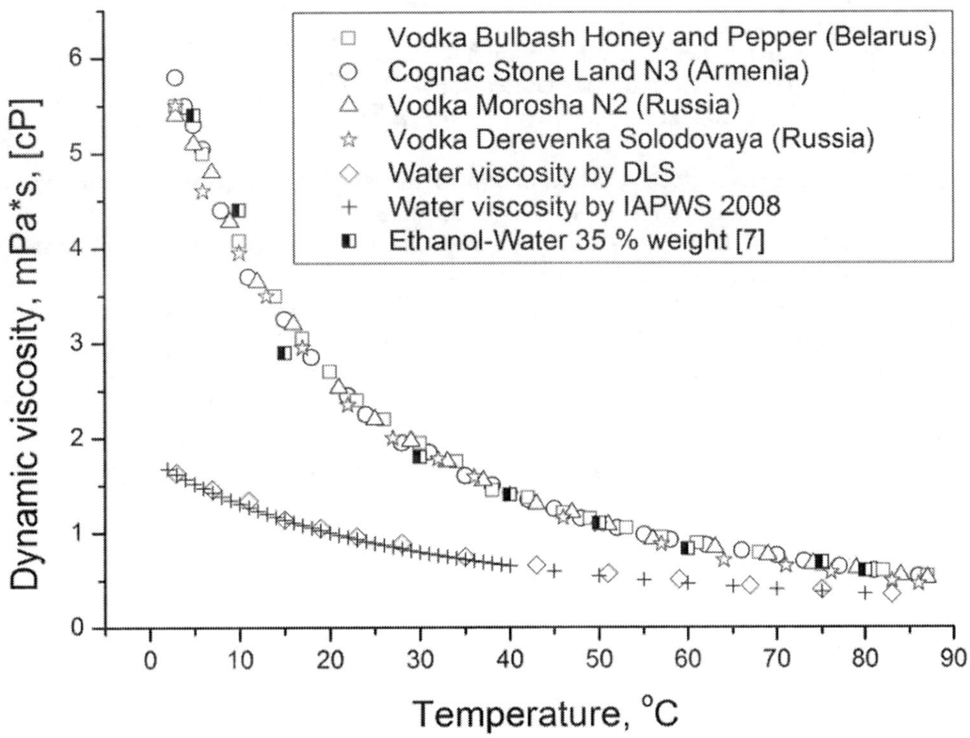

Figure 1. Temperature dependences of water and samples of alcoholic beverages viscosity

For all temperature dependences of viscosity measured, the coefficients in the Vogel equation were determined. The Vogel equation coefficients for the studied alcoholic beverages are presented in Table 2.

Table 2. Vogel equation coefficients for the studied alcoholic beverages.

Sample	A	B	C	Reduced χ^2	Coefficient of Determination (R^2)
Bulbash	-3.80446	629.86574	-162.06264	0.00279	0.99881
Stone Land	-2.66369	342.15898	-198.72735	0.00241	0.99922
Morosha	-3.92005	656.08887	-159.58485	0.00523	0.99821
Derevenka	-4.19805	709.63518	-155.5317	0.00264	0.999

4. Conclusion

Using the method of dynamic light scattering and silica nanoparticles, the temperature dependences of the water and several alcoholic beverages viscosity were measured.

IOP Conf. Series: Materials Science and Engineering **525** (2019) 012094 doi:10.1088/1757-899X/525/1/012094

The volume content of ethanol in all drinks is the same, the type of alcohol and additives are different. It can be concluded that the additives and the type of alcohol in the studied samples do not have a significant effect on the viscosity in the investigated temperature range.

5. References

[1] Bird R B, Stuart V and Lightfoot B 2002 *Transport Phenomena 2nd Edition* (John Wiley & Sons, Inc)

[2] Anisimov M A, Kiyachenko Yu F, Nikolaenko G L and Yudin I K 1980 Measurement of the Viscosity of Liquids and the Dimensions of Suspended Particles by the Method of Correlation Optical-Mixing Spectroscopy *translated from Inzhenerno-Fizicheskii Zhurnal* **38** 651

[3] Will S and Leipertz A 1993 Determination of the dynamic viscosity of transparent fluids by using dynamic light scattering *Appl. Opt.* **32** 3813–21

[4] Will S and Leipertz A 1995 Dynamic viscosity measurements by photon correlation spectroscopy *Int. J. of Thermophysics* **16** 433–43

[5] S. Will S and Leipertz A 1997 Viscosity of liquid n -heptane by dynamic light scattering *Int. J. of Thermophysics* **18** 1339–54

[6] Yudin I K and Anisimov M A 2007 Dynamic Light Scattering Monitoring of Asphaltene Aggregation in Crude Oils and Hydrocarbon Solutions (New York: Springer New York) p 439

[7] Flick E W 1998 *Industrial Solvents Handbook* 5th ed. (USA: William Andrew Publishing)

IOP Conf. Series: Materials Science and Engineering **525** (2019) 012095 doi:10.1088/1757-899X/525/1/012095

Utilization of red mud and boron-containing liquid radioactive wastes of nuclear power plants

M A Fedotov, D V Zinoveev, P I Grudinsky, L V Kovalenko and V G Dyubanov

A.A. Baykov's Institute of Metallurgy and Material Science Russian Academy of Sciences, 49, Leninsky prospect, 119334, Moscow, Russia

E-mail: mfedotov@imet.ac.ru

Abstract. The physicochemical processes of red mud recovery using solid reducing agents were studied. A comparison was made of the effectiveness of obtaining functional materials from red mud with various reducing agents. The extent of recovery of red muds was investigated by X-ray phase analysis. The red mud of the "Alum" plant contains sodium, which has a positive effect when it is used in the process of liquid radioactive waste (LRW) cementation as a result of the partial neutralization of the acidic environment of boron-containing LRW. The use of red mud as a source of raw materials for the recovery process with solid carbon allows not only to solve the problem with the accumulation of technogenic waste, but also to reduce the costs of immobilizing LRW. Replacing the artificially produced nanosized iron hydroxide with red mud will reduce the cost of the feedstock while optimizing the process of obtaining the finished product. The obyaines functional materials can be used for immobilization of LWR of nuclear power plants in order to reduce the hardening time of cement compounds and increase their strength.

1. Introduction

Currently, one of the most important problems of environmental protection is the recycling of ever-increasing technogenic waste from various industries. It can be distinguished two the most hazardous to the environment and significant on the scale of the accumulated volumes groups of waste - radioactive waste and waste resulting from metallurgical production.

Among radioactive waste, a significant proportion is occupied by LRW. Currently, over 550 million m^3 of radioactive wastes have been accumulated in the Russian Federation [1]. 85% of these wastes are LRW. They include various organic and inorganic liquids. LRW are also formed during the recycling of spent nuclear fuel. Possible methods for the immobilization of LRW are: cementing, bituminization, deep evaporation and vitrification. The most common method of disposing of low and medium level LRW is cementing, which makes it possible to obtain a durable product in the form of a cement compound with low capital and operating costs [2]. The resulting compounds are prepared for long-term burial in specially equipped places [3].

Cementation of boron-containing LRW is associated with a number of difficulties. Low pH values of the solutions lead to a significant increase in the hardening time of cement compounds and a decrease in their strength. This leads to an increase in operating costs of the process of LRW recycling. Currently, a physical impact method is used to regulate the acidity of boron-containing

IOP Conf. Series: Materials Science and Engineering **525** (2019) 012095 doi:10.1088/1757-899X/525/1/012095

LRW [4]. A mixture of LRW with cement mortar is subjected to electromagnetic treatment in the vortex layer of ferromagnetic rods (VET). Extraction of ferromagnetic rods is associated with the need for their deactivation, which leads to the formation of secondary liquid radioactive waste and an increase in the total volume of radioactive waste.

Earlier, a technology for producing magnetic activators of the process of LRW cementation by the method of solid-phase carbon reduction of artificially prepared raw materials - nano-sized iron hydroxide, was developed [5]. The use of the obtained mixture in the process of boron-containing LRW cementation has reduced the setting time of the cement slurry with the addition of LRW, as well as improved the strength characteristics of the cement compounds, including due to the accelerated set of the early strength of the cement composition [6].

The obtaining of magnetic activators of the LRW cementation process is also possible from industrial waste with high iron content. The most promising raw material for these purposes is the waste of the aluminum industry - red mud, the iron content of which can reach 50%. Depending on the composition of the initial bauxite and technology, from 0.9 to 2.5 tons of these wastes are produced per 1 ton of alumina [7, 8]. The world has already accumulated from 2.7 [7-9] to 4 [10] billion tons of such waste, and every year the amount increases by 5-8 million tons. The storage volume of only Ural aluminum plants is 100-300 million tons [9]. The annual generation of such waste worldwide exceeds 120 million tons. Despite the large number of developed technologies for the recycling of red mud [11, 12], in most cases, they are unprofitable, with the result that the slimes continue to be dumped, which is dangerous for the environment. Due to the high alkali content for storage of red mud, special storage facilities are used. Each such storage is a reservoir with a height of 30-50 m, occupies 10-50 hectares of land area and holds tens of millions of tons of red mud [13]. Figure 1 shows an example of red mud storage. The cost of building new and maintaining old red mud dumps requires significant financial investments.

Figure 1. Red mud storage

The use of red mud as a precursor for the recovery process by solid carbon allows not only to solve the problem of accumulation of industrial wastes, but also to reduce the cost of obtaining activators of the process of LRW cementing. Replacing the artificially produced nano-sized iron hydroxide with red mud will reduce the cost of raw materials and optimize the process of obtaining the finished product.

2. Materials and Methods

High-alkaline red mud from the "Alum" plant (Romania) and low-alkaline red mud from JSC Ural Aluminum Plant (UAP) (Kamensk-Uralsky), containing up to 45% hematite, and aluminum oxides silicon, calcium and sodium were used. High-alkaline red mud has pH values from 10 to 13, which allows partially neutralizing the acidic environment of boron-containing LRW solutions. The chemical composition of the red mud is given in table 1, and figures 1 and 2 show their radiographs.

Table 1. The composition of the main components of red mud

Component	"Alum" (Romania) (%)	UAP (Kamensk-Uralsky) (%)
Fe_2O_3	43.39	36.9
Al_2O_3	21.51	11.8
SiO_2	9.04	8.71
CaO	4.63	23.8
Na_2O	7.09	0.27
TiO_2	2.92	3.54

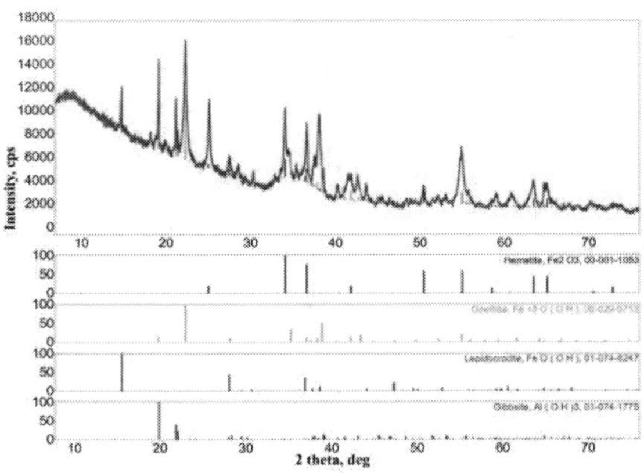

Figure 2. X-ray of "Alum" red mud

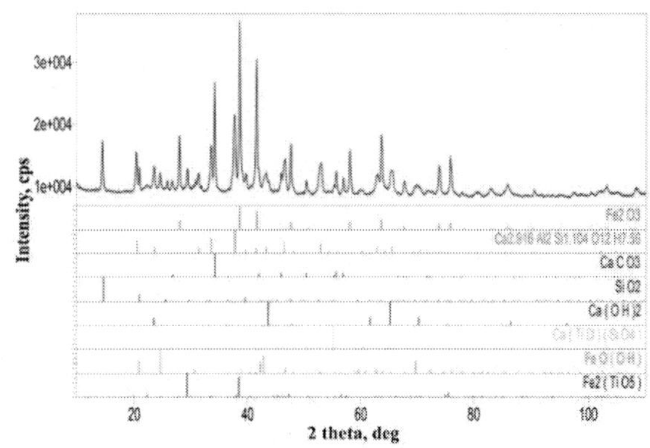

Figure 3. X-ray of UAP red mud

Phase analysis of red muds and recovered samples was carried out using a ULTIMA IV diffractometer (Rigaku, Japan) on Cu-kα X-rays and a Diffray 401 diffractometer (JSC "Scientific Instruments", Russia) on Cr-kα X-rays, a quantitative calculation of the phase content was carried out by Rietveld method. Mössbauer spectroscopy was carried out on a Ms-1104Em spectrometer (RSU, Russia) in the mode of constant accelerations with a Co-57 source in the Rh matrix. Studies have shown that the distribution of iron over iron-containing phases for the red mud "Alum": hematite (α-Fe_2O_3) - 45%, lepidocrocite (γ-FeOOH) - 55%. For the UAP red mud, the iron distribution over the iron-containing phases is as follows: hematite (α-Fe_2O_3) - 80%, iron-alumino-silicate - 15% (ferrous iron) and iron hydroxide of complex composition - 5%.

The original mixture of red mud was mixed with carbon of various grades in different ratios and heated in a stream of argon to a temperature of 700 °C in a tubular furnace, followed by exposure for 30 minutes. The following were chosen as reducing agents: "Medisorb" activated carbon, BAU-A activated carbon, charcoal and sucrose. Table 2 shows the composition of the reducing agents.

Table 2. Composition of reducing agents

Type of reducing agent	Composition, wt.%				
	The main substance, formula	C	H_2O	Starch	Cinder
"Medisorb" activated carbon	carbon	78	5	17	-
BAU-A activated carbon	carbon	84	10	-	6
Charcoal	carbon	85	10	-	5
Sucrose	$C_{12}H_{22}O_{11}$	42	58	-	-

IOP Conf. Series: Materials Science and Engineering **525** (2019) 012095 doi:10.1088/1757-899X/525/1/012095

Thermodynamic calculation of the amount of reducing agent required for the reduction of iron-containing phases to magnetite was carried out by the HSC Chemistry 5.11 program [14].

3. Results

Thermodynamic calculation showed that for complete reduction of the iron-containing phases of the red mud UAP to magnetite at a temperature of 700 °C and atmospheric pressure, 1 wt.% of carbon is sufficient and for red mud "Alum" – 0.5 wt.% of carbon. With such an amount of carbon, the magnetite content in the final mixture is maximal (Figure 4, 5). Along with magnetite, a small amount of iron compounds with titanium (Figure 4) and sodium (Figure 5) and magnetic aluminoferritic phases (Hercynite) may be present in the final product.

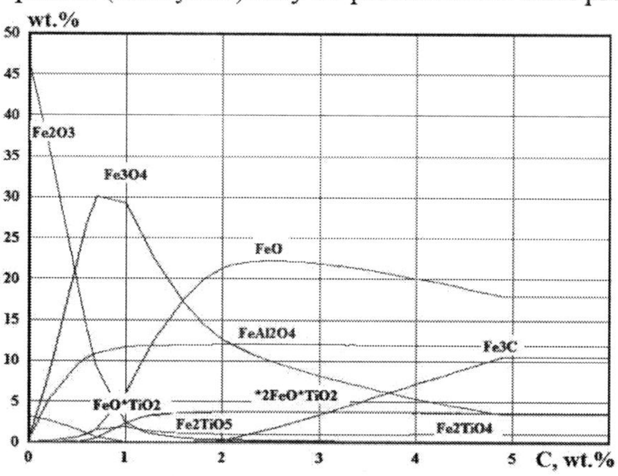

Figure 4. The change of the main iron-containing phases of UAP red mud depending on the change in the amount of carbon at 700 °C.

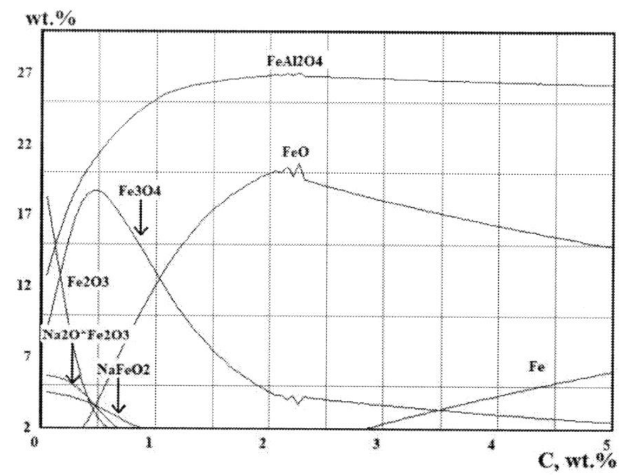

Figure 5. The change of the main iron-containing phases of UAP red mud depending on the change in the amount of carbon at 700 °C.

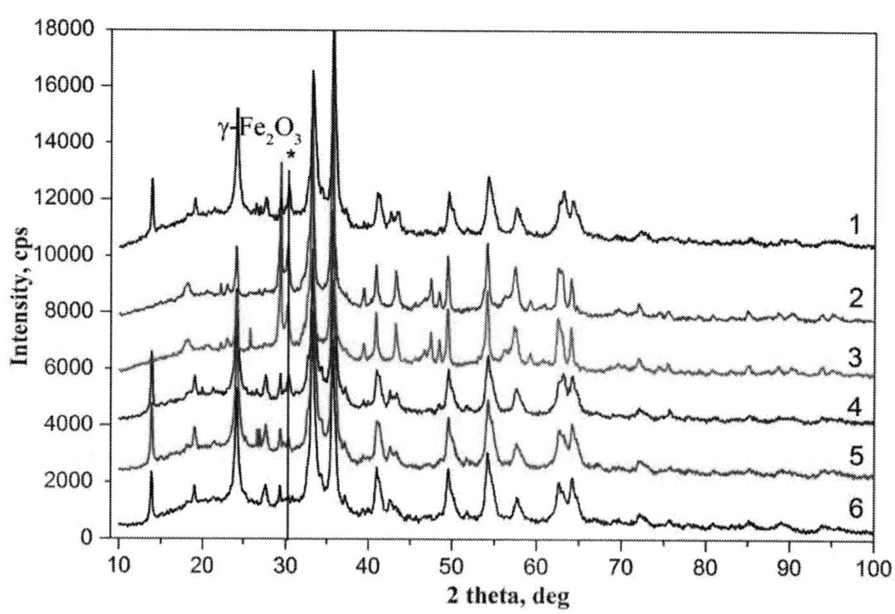

Figure 6. The results of X-ray phase analysis: 1 – "Alum" + 1% C; 2 – "Alum" + 2% C; 3 – "Alum" + 5% C; 4 - UAP + 1% C; 5 - UAP + 2% C; 6 - UAP + 5% C.

For the experiments, samples of mixtures of red mud with “Medisorb” activated carbon were prepared in ratios of 1, 2 and 5 wt.%. Depending on the exposure time and the carbon content in the initial mixture, the samples obtained had a different ratio of magnetic phases and nonmagnetic phases. The results of X-ray phase analysis of the recovered samples are shown in figure 6.

Figure 6 shows the line of the spectrum most characteristic of the maghemite phase (γ-Fe_2O_3).

As a result of the experiments, it was shown that samples recovered using 1% carbon were only partially recovered, and samples with a carbon content in the initial mixture of 5% contain a significant amount of magnetic phase. Samples obtained from the red mud UAP, have in their composition a greater amount of magnetic fraction compared with similar samples of red mud from the factory “Alum”.

An increase in the mass fraction of carbon in the initial mixture to 10% leads to the complete transition of iron oxides to magnetite. Figure 7 shows the X-ray diffraction pattern of the “Alum” red mud sample, and table 3 shows its quantitative phase composition.

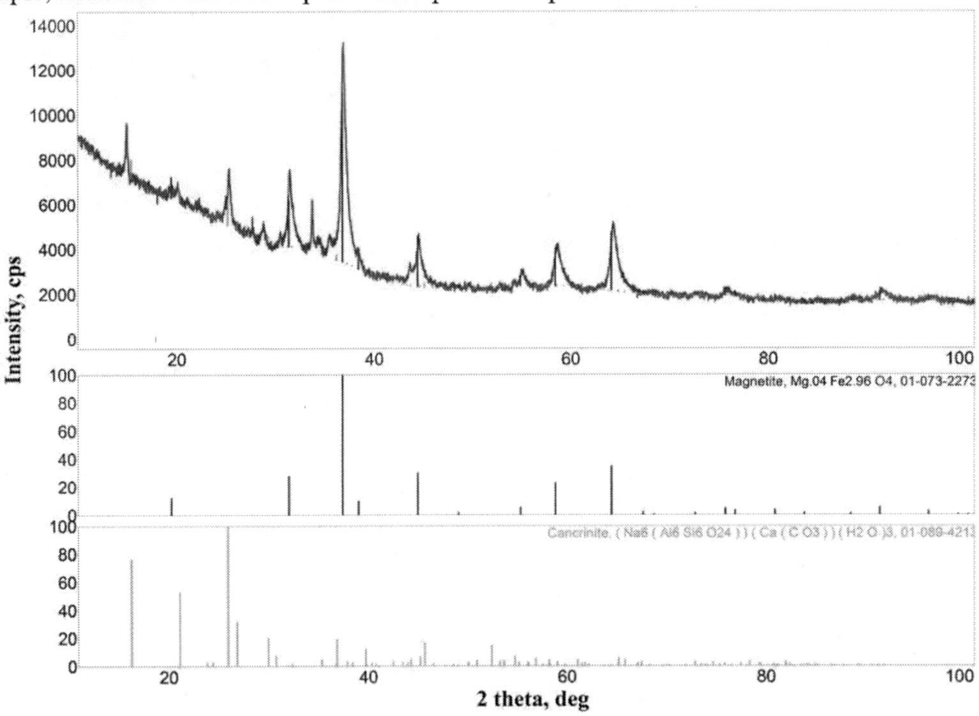

Figure 7. The results of X-ray phase analysis for the sample “Alum” with a carbon content of 10% in the initial mixture.

Table 3. Phase content in the “Alum” red mud sample recovered from 10% carbon.

Structural formula	Quantitative content, %
Fe_3O_4	35.8
$Na_{7.6}(Al_6Si_6O_{24})(CO_3)_{0.93}(H_2O)_{2.93}$	19.2
$Na_{7.262}(CO_3)_{0.932}Al_6Si_6O_{24}$	13.3
$Na_8Ti_{3.5}O_2(OH)_2(SiO_4)_4$	11.8
$Na_{15.6}Ca_{3.84}(Si_{12}O_{36})$	6.9
C	6.6
$CaCO_3$	4.3
$CaTiO_3$	2.1

From table 3 it can be seen that magnetite in the sample is present in an amount of 35.8% of the total mass, the remaining phases are mainly composed of complex sodium aluminosilicates.

The effect of the type of reducing agent on the reduction process of the iron-containing phases of red mud was also investigated. Samples of mixtures of red mud with 10 wt.% of activated carbon BAU-A, 10 wt.% of charcoal and 20 wt.% of sucrose were reduced by the described method.

Figure 8 and 9 shows radiographs of samples obtained with various types of reducing agents.

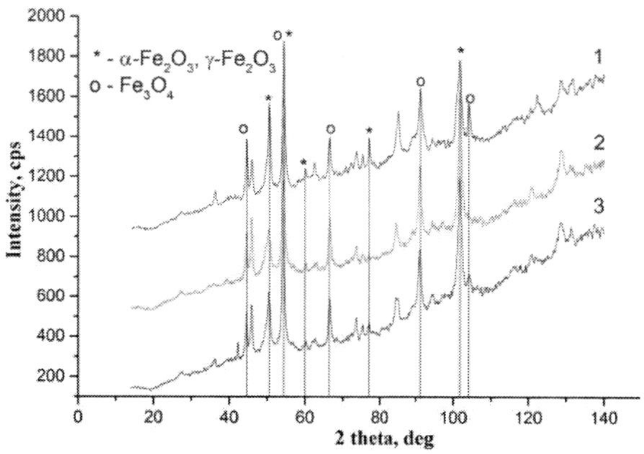

Figure 8. The results of X-ray phase analysis of red mud UAP reduced by BAU-A (1), charcoal (2) and sucrose (3).

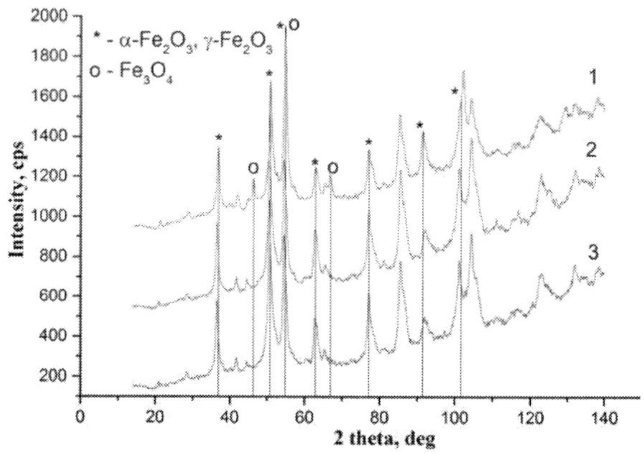

Figure 9. The results of X-ray phase analysis of red mud "Alum" reduced by BAU-A (1), charcoal (2) and sucrose (3).

It can be seen from the figures that the results of the reduction of the iron-containing phases of the "Alum" and UAP red mud are different. In the "Alum" red mud, a noticeable amount of magnetic phases is present only in the sample reconstituted with activated carbon BAU-A, in the remaining samples there are practically no magnetic phases, and the main iron-containing phase is hematite. The UAP red mud recovered better, as evidenced by magnetite peaks on radiographs in all cases.

Taking into account the data of thermodynamic modeling in the samples, along with magnetite, hercinite ($FeAl_2O_4$) may be present. However, due to the similarity of types of crystal lattices of these compounds, it is not possible to separate them by X-ray phase analysis.

4. Discussions

The results of the experiments showed that to obtain magnetic phases, an excess of reducing agent is necessary, several times higher than that required by the results of thermodynamic modeling. This is most likely due to the low content of carbon monoxide in the gas phase, which is the main reducing agent for iron minerals at low temperatures. An increase in the amount of carbon in the mixture leads to an increase in the CO content in the gas phase and an increase in the degree of reduction. Also, a low degree of recovery may be associated with insufficient exposure time.

The influence of the type of reducing agent on the degree of reduction of iron-containing phases is due to the different rate of gasification of the coal under study. The speed of the gasification process is primarily determined by the value of the specific surface area, which is greatest for the activated carbons "Medisorb" (about 400 m^2/g) and BAU-A (about 600 m^2/g).

The low degree of recovery of red mud "Alum" is most likely due to the presence in its composition of a large amount of alkali and calcium, which, when reacting with iron compounds, can form hardly-recoverable iron-containing phases.

The presence of magnetite in this sample in the case of the use of activated carbon "Medisorb", in contrast to samples obtained with other types of carbon, is explained by the composition of the activated carbon itself. Carbon contains up to 20% of organic additives in the form of sucrose and

IOP Conf. Series: Materials Science and Engineering **525** (2019) 012095 doi:10.1088/1757-899X/525/1/012095

starch, which act as binders. When heated, these substances begin to decompose in the temperature range from 170 to 250 °C with the formation of water and black carbon. As a result, the sample is saturated with a dispersed reducing agent in the form of black carbon, which contributes to the formation of the magnetic phase of iron oxide.

As a result, the recovery of the oxide phases of iron in the composition of the red mud provides for obtaining magnetic particles necessary for conducting the VEO. Reduced red mud from the "Alum" plant contains sodium, which has a positive effect when it is used in the process of LRW cementation due to partial neutralization of the acidic environment of boron-containing LRW.

5. Conclusion

It was determined that in the process of reduction with increasing temperature, the structure of iron compounds changes, accompanied by the transition of the nonmagnetic phase to the magnetic one. Thus, the magnetic phase of maghemite (γ-Fe_2O_3) is formed first, followed by its transition to magnetite (Fe_3O_4).

It was found that the most effective reducing agent is activated carbon "Medisorb", containing in its composition binders, in the form of sucrose and starch, the use of which contributes to the most complete recovery process of the iron-containing phases in the red mud with the formation of a magnetic phase in the form of magnetite.

As a result, the recovery of iron-containing phases in the composition of the red mud provides for obtaining magnetic particles necessary for conducting the VEO.

Joint utilization of such industrial wastes as LRW and red mud contributes to a significant reduction in the amount of accumulated waste and the prevention of environmental disasters, as well as lower costs for the disposal of LRW.

Acknowledgement
The work was carried out according to the state assignment No. 075-00746-19-00.

References

[1] Gorbunova O A and Kamaeva T S 2013 Physical and chemical methods of controlling the properties of cement compounds with boron-containing liquid radioactive waste *Physics and chemistry of materials treatment* **1** pp. 69-76

[2] Gorbunova O A 2015 Cementation of liquid radioactive waste with high content of borate salts *Journal of Radioanalytical and Nuclear Chemistry* **304**, 1 pp. 361-370DOI:10.1007/s10967-014-3886-3

[3] Fedotov M A, Gorbunova O A, Fedorova O V, Folmanis G E and Kovalenko L V 2015 Magnetic iron oxides in the cementation technology of the boron-containing radioactive waste *IOP Conference Series: Material Science and Engineering* **81** 012063

[4] Flit V U, Volkov A S, Dmitriev S A and Varlakov A P 1993 The method of solidification of liquid radioactive waste and device for its implementation *Certificate of authorship* No. 1690488 G21F9/16 23.06.1993

[5] Teplov O A, Fedotov M A, Kovalenko L V and Folmanis G E 2015 The processes of nanosized hematite reduction by carbon *Perspektivnye materialy* **10** pp. 73-79

[6] Fedotov M A, Dzidziguri E L, Kovalenko L V and Folmanis G E 2015 Synthesis of nanosized magnetic iron oxides for conditioning of liquid radioactive waste *Nanoengineering* **8** pp. 3-8

[7] Zhaobo L and Hongxu L 2015 Metallurgical process for valuable elements recovery from red mud – A review *Hydrometallurgy* **155** pp. 29-43

[8] Alkan G, Xakalashe B, Yagmurlu B, Kausse F and Friedrich B 2017 Conditioning of Red Mud for Subsequent Titanium and Scandium Recovery – A Conceptual Design Study *World of Metallurgy – ERZMETALL* **70**, 2 pp. 5-12.

[9] Martynov N N, Martynova N A, Chernousov P I and Pyrikov A N 2014 *Actual aspects of environmentally friendly production and recycling of metals* (Moscow: Roliks) p 256

[10] Kaussen F and Friedrich B 2015 Reductive smelting of red mud for iron recovery *Chem. Ing. Tech.* **87**, 11 pp. 1535-1542
[11] Zinoveev D V, Grudinskii P I, Dyubanov V G, Kovalenko L V and Leont'ev L I 2018 Global recycling experience of red mud – A review. Part I: Pyrometallurgical methods *Izvestiya. Ferrous* Metallurgy **61**, 11 pp. 843-858
[12] Power G, Grafe M and Klauber C 2011 Bauxite residue issues: II. Options for residue utilization *Hydrometallurgy* **108**, 1-2 pp. 60-79
[13] Trushko V L, Utkov V A and Bazhin V Yu 2017 Relevance and possibilities of complete processing of red mud of aluminia production *Zapiski Gornogo instituta* **3** pp. 547-553
[14] Roine A 2002 *Outokumpu HSC Chemistry for Windows. Chemical Reaction and Equilibrium Software with Extensive Thermochemical Database* (Pori: Outokumpu Research OY) p 448

IOP Conf. Series: Materials Science and Engineering **525** (2019) 012096 doi:10.1088/1757-899X/525/1/012096

Study of liquid-phase catalytic oxidation of natural renewable raw materials in alkaline media

P A Sakharov[1], A V Khvatov[1], E V Koverzanova[2] and S M Lomakin[1]

[1]Emanuel Institute of Biochemical Physics, Russian Academy of Sciences, Kosygin str., 4, Moscow, 119334 Russia
[2]Semenov Institute of Chemical Physics, Russian Academy of Sciences, Kosygin str., 4, Moscow, 119991 Russia

E-mail: andrei-sakharov@mail.ru

Abstract. The present study is devoted to liquid-phase catalytic oxidation of natural raw materials (polysaccharides, lignin) with the aim of their utilization, obtaining valuable chemical products and development of environmentally friendly intumescent flame retardants for wood. For this purpose, a method of oxidation of raw materials: polysaccharides and lignin, by oxygen in the presence of catalyst is proposed. The main oxidation products of catalytic oxidation (salts of polyoxy- and polyphenoxy acids) are analyzed using the data acquired with the aid of GC-MS experiments.

Oxidation of organic substances is one of the most widespread methods of preparative chemistry, and organic synthesis [1]. Molecular oxygen (O_2), and the other oxidizers, such as H_2O_2, NaOCl, KIO_3, etc. and metal oxides (Cr_2O_3, $Na_2Cr_2O_7$, $KMnO_4$, etc.) are widely used for this purpose. Some compounds of transition metals allows obtaining the high rate of oxidation and products yield. However, such oxidizers do not satisfy the environmental safety of oxidation leading to formation of the toxicants and produce the heavy metals' ions.

The most environmentally friendly and cheap oxidizer is the molecular oxygen. At the same time under soft conditions (up to 150 °C) it is a very weak oxidizer and being in the nature generally in triplet state, is not capable to interact directly with the molecules of the organic substances (in singlet state). In this regard, liquid-phase oxidation by gaseous oxygen demands preliminary activation of reagents (substrate or molecular oxygen). In the most cases for non-catalytic processes, and in the presence of homogeneous and heterogeneous catalysts of liquid-phase oxidation key stage of process is oxidation substrate activation due to its transfer to free radical form. Oxygen reduction in such processes occurs via the series of consecutive one-electron stages [2].

The use of homogeneous catalysts allows to reduce the temperature of oxidation and to change the ratio of reaction products due to possible interaction of transition metals compounds with peroxides, free radicals and other oxidation products. Processes of oxidation of organic compounds by oxygen in the presence of copper and the can be widely used in the processing of vegetable raw materials [2]. In 1995 Skibida and Sakharov [3] published the information of the new process describing the liquid-phase catalytic oxidation of starch-containing raw materials by oxygen (Figure 1).

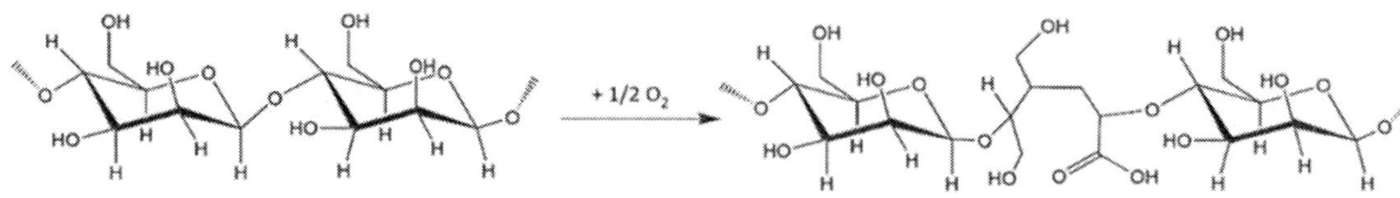

Figure 1. Scheme of oxidation of starch-containing raw materials by oxygen.

The oxidative modification of polysaccharides, leading to polyacids as main products, is widely applied in practice due to the availability of raw materials and high consumer-attractive properties of oxidized polysaccharides. Salts of polyacids are used as water-soluble glues in the production of paper and cardboard, as components of drilling agents, etc. the oxidizing agent, used until recently for the oxidation of polysaccharides, was hydrogen peroxide since gaseous oxygen presents low activity in such procedures [4, 5]. However, when copper complexes and bases are present, not only the simple in structure alcohols and ketones, but also polysaccharides (e.g. starches, dextrans, cellulose) can be oxidized by oxygen, achieving high rates. These high rates of oxidation by oxygen, exceeding the ones achieved by oxidation with hypochlorites, are reached at temperatures 40-90°C [6]. The oxidation of polysaccharides proceeds with the greatest efficiency in the presence of copper and alkali salts. On the contrary, the molecular weight of an initial polysaccharide exhibits a slight decrease during the first minutes [7].

Copper salts are the most effective catalysts of oxidation of polysaccharides in the presence of alkali. In the absence of the catalyst, starch and cellulose practically cannot be oxidized at temperatures below 100°C. However, lignin, which contains many of high reactive phenolic groups (under high rates), can be oxidized by oxygen at measurable rates, even without the catalyst. The role of the catalyst (Cu^{2+} ions) is to activate the deprotonated forms of substrate to react with oxygen. In other words, the role of bivalent copper ions is supposed to be the facilitation of the oxidation of anion forms of substrate (interaction of intermediate radicals or anion radicals with O_2). However, high-molecular polysaccharides with a low content of end aldehyde groups are redox inactive in anaerobic medium and the rate of reduction of Cu^2 to Cu^+- ions (due to electron transfer from substrate anion form to Cu^{2+}) is one hundred times lower than the rate of oxygen adsorption during the process of polysaccharides oxidation in alkaline mediums. Therefore, it seems that direct interaction of oxygen with anions A^- in coordination sphere of copper ion that leads to the formation of hydroxycarboxylates (as the main primary products of oxidation). Absorption of oxygen, on the other hand, is completely stagnated after neutralization of introduced alkali by the polyoxyacids, formed during the oxidation process. Thus, by varying of the amount of introduced alkali the degree of polysaccharides oxidation into polyacids can be altered, giving the possibility to tailor the properties of the final product (such as viscosity, bonding ability, solubility in water, etc.).

Starch can be used as initial raw material for receiving salts of polyoxyacids, as well as any of other starch-containing raw materials (like corns of maize, oats, rice, etc.), including ill-conditioned raw materials (such as grain-crops affected by various fungus diseases, waste of rice slashing, waste of mill houses, etc.). In so doing, the prime cost of the final product can be decreased significantly. Moreover, similar to the polysaccharides behavior, under oxidation in alkali media with or without the catalyst, has been found to be exhibited by lignin. Lignin is amongst the most abundant biopolymers on earth. It is estimated that the planet currently contains 3 x 10^{11} metric tons of lignin with an annual biosynthetic rate of approximately 2 x 10^{10} tons [8]. Lignin constitutes approximately by 30% (of the dry weight) of softwoods and by about 20% of hardwoods [9]. It is absent from primitive plants such as algae, and fungi which lack a vascular system and mechanical reinforcement. The presence of lignin within the cellulosic fibre wall, mixed with hemicelluloses, creates a naturally occurring composite material that imparts strength and rigidity to trees and plants. Lignin, moreover, is one of the most important bio-resources for obtaining raw materials needed for the synthesis of environmentally compatible polymers, since it is non-toxic and extremely versatile in performance. Lignins are derived from renewable resources such as trees, grasses, and agricultural crops. About

30% of wood constituents are lignin. Furthermore, the production of lignin as a by-product of pulping process is worldwide over 30 million tons per year [9]. Lignin is a random copolymer consisting of phenyl propane units having characteristic side chains. Lignin slightly crosslinks and takes an amorphous structure in the solid state. The molecular motion is observed as glass transition by thermal, viscoelastic and spectroscopic measurements. The hydroxyl group of lignin plays a crucial role during its interaction with water. Lignin is usually considered as a polyphenolic material having an amorphous structure, which arises from an enzyme-initiated dehydrogenative polymerization of *p*-coumaryl, coniferyl and sinapyl alcohols [10]. The basic lignin structure consists of two components; one is the aromatic part and the other is the C_3 chain. The only usable reaction site in lignin is the OH group, which is the case for both phenolic and alcoholic hydroxyl groups. In the process of radical-based biosynthesis, the types of reactants employed to produce the lignin polymer are three: 4-hydroxyphenyl (1), guaiacyl (2), and syringyl (3) structures (Figure 2).

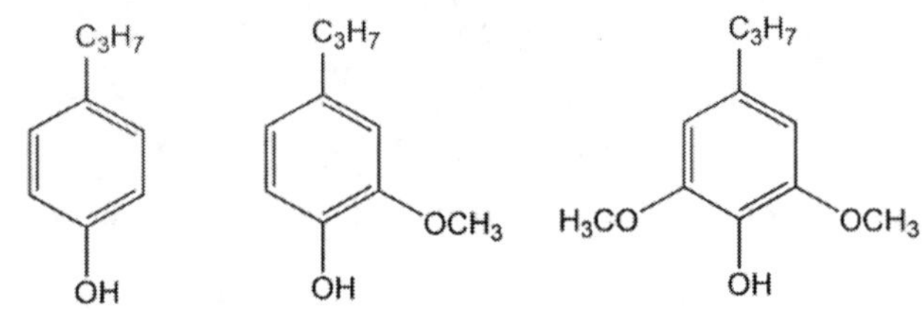

(1) (2) (3)

Figure 2. Chemical formula of 4-hydroxyphenyl (1), guaiacyl (2), and syringyl (3) molecules.

In the present study, we have focused on the study of catalytic oxidative modification of lignin and simple natural saccharides (glucose and sucrose), as model systems, by the chromatography mass spectrometry (GC-MS) method.

GC–MS measurements were performed on a gas chromatograph Trace-1310 and Single Quadrupole MS ISQ Detector. The pyrolysis products were injected into the GC capillary quartz column 15 m × 320 µm i.d. and 0.25 µm film (TG-SQC, Thermo Scientific Phase) under the temperature-programmed conditions in the range of 40 – 290°C with gradient 15°/min. MS-spectra of degradation components were identified and interpreted by matching the results with NIST Mass Spectral.

The GC–MS analysis of hexane solutions of silanated products of initial sucrose (the top chromatogram) and the oxidized sucrose after receiving acids from salts (the lower chromatogram) presented on Figure 3.

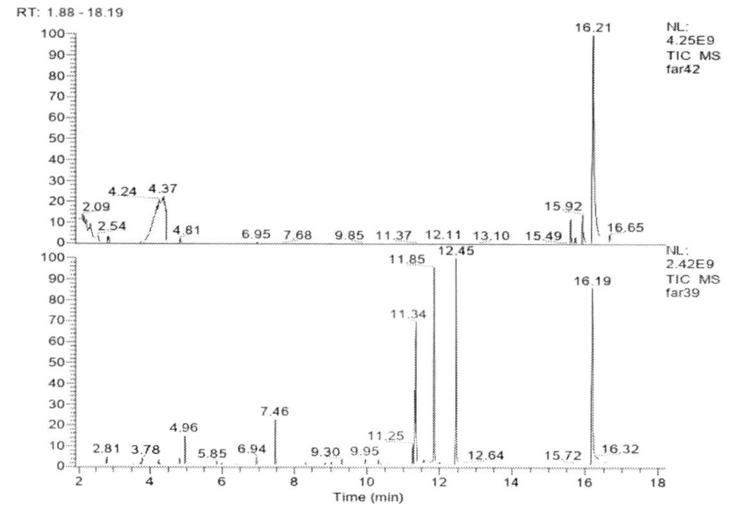

Figure 3. Gas chromatograms of the hexane solutions of silanated products of initial sucrose (the top chromatogram) and the oxidized sucrose after receiving acids from salts (the lower chromatogram).

Gas chromatograms of the hexane solutions of silanated products of oxidized glucose presented on Figure 4.

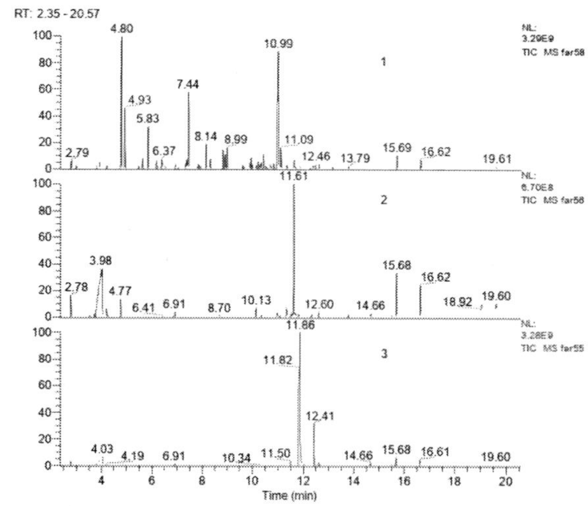

Figure 4. Gas chromatograms of the hexane solutions of silanated products of oxidized glucose: 1 – the oxidized dried-up glucose after receiving acids from salts; 2 - the oxidized dried-up glucose without receiving acids from salts; 3 – initial glucose.

Chromatograms of sucrose are similar except for intensity of peaks. The proposed mechanism of sucrose oxidation leads to formation of pentapyranoses (78.5%), pentafuranoses (11.7%) and acids (8.8%) (Figure 5).

IOP Conf. Series: Materials Science and Engineering **525** (2019) 012096 doi:10.1088/1757-899X/525/1/012096

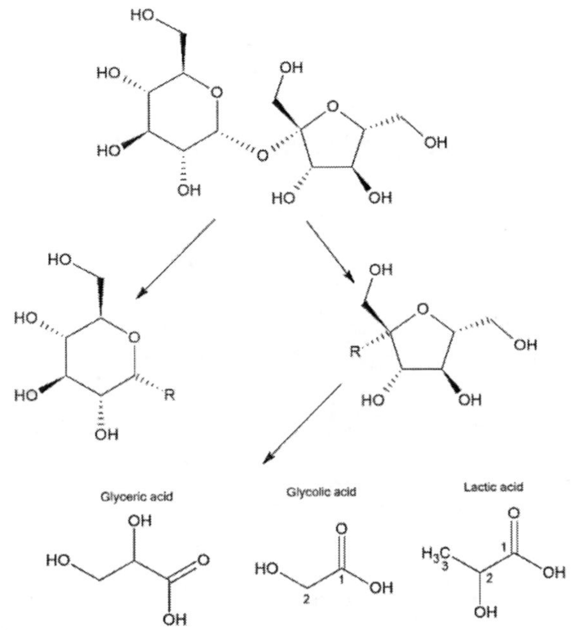

Figure 5. Scheme of proposed mechanism of sucrose oxidation.

The obtained organic acids can be widely used in medicine and food industry.

In this work the way of liquid-phase catalytic oxidation of the lignin oxygen in the presence of salts of transition metals leading to formation of soluble fraction of polyaromatic acids and their derivatives is developed. This way differs from selective pyrolysis in the fact that he answers with the principles of "green" chemistry and represents environmentally friendly single-stage and waste-free process. In addition, main products of processing of lignin in the form of aqueous solutions and emulsions can be received without the big power expenses connected with pyrolysis of lignin containing raw materials. Now homogeneous catalysts allow conducting process in much softer conditions, than classical heterogeneous catalysts, at the same time higher activity and selectivity is peculiar to them.

The oxidized lignin offered for research consisted of viscous, tarry deposit of black color and opaque solution.

Products of lignin oxidation were extracted from obtained solution with use of methylene chloride. The lignin hydrolysates extracted from acidified liquids by methylene chloride are given in Table 1.

The oxidized lignin offered for research consisted of viscous, tarry deposit of black color and opaque solution.

According to provided GC–MS measurements of lignin oxidation products we assume that liquid-phase catalytic oxidation of the lignin by oxygen in the presence of salts of transition metals provides the formation of valuable polyaromatic derivatives (e.g. vanillin).

The process of homogeneous oxidative catalysts in alkaline media allows providing the utilization of lignin, obtaining valuable chemical products and development of environmentally friendly intumescent flame retardants for wood [11].

Table 1. Composition of lignin oxidation products obtained via extraction by methylene chloride.

№	Connection	contents, %
1	Methylbenzaldehyde	3.10
2	2-methyl-5-hydroxycoumarine	10.41
3	2.5-dihydroxy-2.5-dimetoxy furan	7.34
4	Furan carboxylic acid	4.91
5	Propoxybenzene	2.96
6	Methylbenzoic acid	2.43
7	Vanillin	15.03
8	Acetovanilon	3.52
9	Vanillic acid	8.54
10	Isovanillin	4.21
11	Ethers of benzene dicarboxylic acids	28.09
12	Not identified products	9.46

Acknowledgments
The work was carried out within the framework of budget financing on the theme of state assignment No. of state registration 01201253305.

References

[1] Emanuel N M, Zaikov G E and Maizus Z K 1984 Oxidation of organic compounds. Medium effects in radical reactions *Oxford Pergamon Press* 628
[2] Aseeva R M, Sakharov P A and Sakharov A M 2009 The Physicochemical characteristics of oxidized polysaccharides *Russian Journal of Physical Chemistry B* **3** № 5 pp 844–850
[3] Skibida I P and Sakharov A M 1995 Chemical model of oxidases. Cu^I-catalyzed oxidation of secondary alcohols by dioxygen *Russian Chemical Bulletin* V **44** I 10 pp 1872–1878
[4] Santacesria E, Trully F, Brussani G F, Gelosa D and Serio M Di 1994 Oxidized glucosidic oligomers: a new class of sequestering agents — preparation and properties *Carbohydrate polymers* V **23** I 1 pp 35-46
[5] Skibida I P, Sakharov An M and Sakharov Al M 1993 EP 91122164.6
[6] Floor M, Kieboom A P G, van Bekkum H, Greenland W P T and Verhaar L A Th 1989 Preparation and calcium complexation of oxidized polysaccharides Part I: Oxidation of Maltodextrins and Starch with Alkaline Sodium Hypochlorite *Starch* V **41** pp 348–354
[7] Sun R C 2010 *Cereal Straw as a Resource for Sustainable Biomaterials and Biofuels* Elsevier 300
[8] Sakharov A, Sakharov P and Zaikov G 2012 Catalytic oxidation of ethylene glycol by dioxygen in alkaline medium. The new example of one-stage oxidative cleavage of C–C bond *JCCT* V **6** № 1 pp 19-24

[9] Sakharov A M, Silakhtaryan N T and Skibida I P 1996 Catalytic oxidation of polyols by molecular oxygen in alkaline media *Kinetics and Catalysis* **37** pp 368-376
[10] Hatakeyama H and Hatakeyama T 2010 *Lignin structure, properties and applications* Biopolymers ed Abe A, Dušek K and Kobayashi Sh Springer pp 2-64
[11] Lomakin S M, Sakharov A M, Sakharov P A and Zaikov G E 2014 Novel class of ecofriendly flame retardants based on renewable raw materials *Polymer Green Flame Retardants* chapter 7 pp 255-267

IOP Conf. Series: Materials Science and Engineering **525** (2019) 012097 doi:10.1088/1757-899X/525/1/012097

Characteristic of zinc ferrite decomposition by calcium and magnesium oxides

P I Grudinsky[1], A F Semenov[1,2], M A Sevostyanov[1] and V G Dyubanov[1]

[1]A.A. Baikov Institute of Metallurgy and Material Science RAS, 49 Leninsky avenue, Moscow, 119334, Russia

[2]D. I. Mendeleev University of Chemical Technology of Russia, 9 Miusskaya square, Moscow, 125047, Russia

Abstract. Electric arc furnace (EAF) dust generated in steelmaking is a valuable waste material with high content of zinc, lead and iron. Recovery of zinc by current industrial hydrometallurgical methods at zinc plants is difficult due to the presence of zinc in the hardly soluble ferrite form and high contents of chlorine and fluorine that are harmful impurities for further electrolysis. To decompose zinc ferrite and remove halogens from EAF dust there is a promising approach that promote zinc ferrite decomposition to highly soluble zinc oxide and volatilization of halogens in one pyrometallurgical stage, where EAF dust are calcined with lime at 900-1100°C. This paper aims to investigatekinetics of the reaction between pure $ZnFe_2O_4$ and CaO to determine its controlling step and evaluate its characteristics. The reaction between pure $ZnFe_2O_4$ and MgO is also considered. The investigation of the EAF dust sample has confirmed that majority of zinc is in the form of hardly soluble ferrite. The results of experiments have shown that the reaction of zinc ferrite with calcium oxide at 900-1000°C is kinetic controlled, so temperature has a significant effect of on the reaction rate. The degree of conversion of the reaction between zinc ferrite and magnesium oxide is negligible, so the application of MgO during the EAF dust processing is inexpedientat the studied temperatures.

1. Introduction

Electric arc furnace (EAF) dust generated in steel industry is a valuable by-product with high content of zinc, lead and iron. The content of zinc and lead in EAF dust is in large excess over their content in the base metal ores. Currently, about 45% of EAF dust generated in the world is recycled [1], primarily in Western Europe and North America for nonferrous metals extraction.

The EAF dust processing in order to extract zinc by current industrial hydrometallurgical methods at zinc plants is difficult due to the presence of zinc in the hardly soluble ferrite form [2], as well as high content of chlorine and fluorine that creates additional hindrances in the further electrolysis of obtained zinc-containing solutions [3]. Therefore, to decompose zinc ferrite and remove halogens from EAF dust many plants apply the Waelz process [4], which is the pyrometallurgical processing of dust with carbonaceous reducing agent in rotary kiln at 1200-1300°C. During the Waelz process, zinc, lead, chlorine and fluorine volatilize from the charge to Waelz-fume. For the separation of lead and halogens from zinc some producers use additional calcination of obtained Waelz-fume in a rotary kiln [5] or salt-water leaching [6]. Currently, the Waelz process is the most popular method of EAF dust recycling. Today, about 85% of the processed dust are recycled in the Waelz kilns [7]. In the practice of Waelz process, processing of EAF dust consumes up to 350 kg/ton of expensive coke breeze [8]. Due to the high-energy costs of such processing and need of the subsequent calcination of Waelz-fume, it is necessary to look for new effective methods of EAF dust recycling. Other methods with processing of wastes in rotary-hearth furnace [9, 10], shaft furnace [11], multiple-hearth furnace [12], induction furnace [13], flash smelter [14], as well as leaching methods [15-17] have limited application for EAF dust recycling [8]. Thus, it is preferable to develop new methods based on the processing of EAF dust in rotary kiln as are used in the Waelz process. Authors [18-20] proposed a method to transform zinc ferrite to highly soluble zinc oxide and to remove halogens from EAF dust in one processing stage, where the dust iscalcined with lime at 900-1100°C.

In this paper, kinetics of the reaction between pure $ZnFe_2O_4$ and CaO were studied to reveal its controlling step and evaluate its characteristics. Moreover, the reaction between pure $ZnFe_2O_4$ and MgO was analyzed because magnesium oxide also promotes zinc ferrite decomposition [21, 22].

2. Materials and Methods

Elemental composition of EAF dust of Vyksa Steel Works (AO VMZ) was investigated by X-ray fluorescence spectrometer Axios Advanced (Netherlands). The carbon and sulfur contents were analyzed by LECO CS-400 (USA). Phase composition of the dust was studied by diffractometer DRON-4 (Russia) with X-rays of Cu-K_α. The distribution of iron phases was analyzed by Mössbauer spectrometer Ms-1104Em (Russia) in constant acceleration mode with the source of Co-57 in Rh matrix.

Experiments for the investigation of the reactions between pure zinc ferrite and calcium oxide were carried out on the laboratory muffle furnace. Reagent grade CaO and MgO were preliminarily calcined at 900°C for 2 hours to remove the hydration moisture. Pure zinc ferrite was obtained by the following method. Reagent grade Fe_2O_3 and ZnO were mixed in a 1:1.1 ratio (with a 10% molar excess of zinc oxide) in the V-shell blender. The briquettes obtained from mixtures were calcined in the muffle furnace at 1000°C for 6 hours. The obtained sample was milled and washed in aqueous solution of NH_4Cl+NH_4OH to wash the remaining ZnO [23], then dried at 1000°C for 2 hours. All the initial components for the experiments were preliminarily tested by the X-ray phase analysis.

Experiments were carried out as follows. The briquetted mixtures of the $ZnFe_2O_4$–CaO and $ZnFe_2O_4$–MgO systems were placed to the preheated muffle furnace at 900°C or 1000°C in molar ratio 1:1.5, 1:3, 1:4.5 or 1:9. Every 1-15 minutes one sample was taken out from the furnace. Leaching of zinc in the soluble oxide form from the calcined samples was carried out by aqueous solution of NH_4Cl+NH_4OH (22 g of NH_4Cl in a mixture of 80 ml of NH_4OH with density of 0.9 g/cm^3 and 120 ml of water) according to the method [24]. The obtained zinc-containing solution was analyzed by the ICP AES method using Perkin Elmer Optima 3300RL (USA) device.

The controlling step of the reaction between $ZnFe_2O_4$ and CaO was evaluated using the corrected time method by comparison of experimental data with kinetic models based on the Yander (1), the contracting sphere (2), and Erofeev-Avrami (3) equations [25]:

$$[1-(1-\alpha)^{\frac{1}{3}}]^2 = \mathrm{K_Y}\cdot\tau \tag{1}$$

$$[1-(1-\alpha)^{\frac{1}{3}}] = \frac{\mathrm{K_{chem}}}{\mathrm{r_B}}\cdot\tau \tag{2}$$

$$[-\ln(1-\alpha)]^{1/n} = K_{\mathrm{AE}}\cdot\tau \tag{3}$$

In the equations α is the conversion degree, τ is time, K_Y, K_{chem}, K_{AE} are the corresponding constants, r_B is the initial radius of the particle and n is the coefficient that considers shape and type of nucleuses (point nucleuses were taken in account, for which $n = 1$).

3. Results and Discussion

The chemical composition of the EAF dust sample was obtained as follows: 14.6% Zn; 25.4% Fe; 1.38% Pb; 0.62% Al; 1.85% Si; 11.9% Ca; 2.06% Mg; 1.61% Na; 1.7% K; 2% Mn; 0.31% Cr; 0.2% Cu; 3.69% Cl; 0.45% F; 1.63% C; 0.66% S; 0.11% P. Figure 1 shows XRD pattern of the sample.

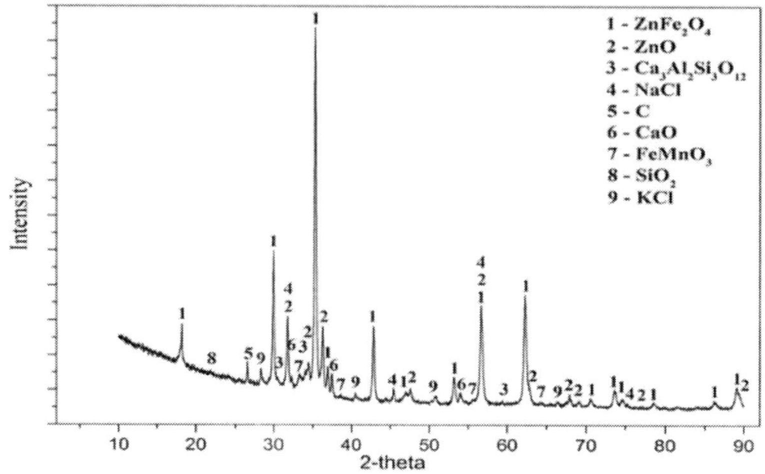

Figure 1. XRD pattern of the EAF dust sample.

The main phases of the dust are franklinite $ZnFe_2O_4$, zincite ZnO, sylvite KCl, halite NaCl. Zinc is in both ferrite and oxide forms. The presence of a few amount of graphite is probably due to the carry-over of carbonaceous materials in the steelmaking process that used for foaming of slag.

Mössbauer analysis indicates that more than 55% of iron is in ferritic form, about 40% is in the form of magnetite, the rest is part of other compounds.

Figure 2 (a, b) shows XRD patterns of the calcined samples of $ZnFe_2O_4$–CaO and $ZnFe_2O_4$–MgO systems at 1000°C during 4 hours with 1:3 and 1:5 molar ratios respectively. The molar ratios are due to thermodynamic probability of at least 95% zinc ferrite decomposition in accordance with the thermodynamic calculation in [26].

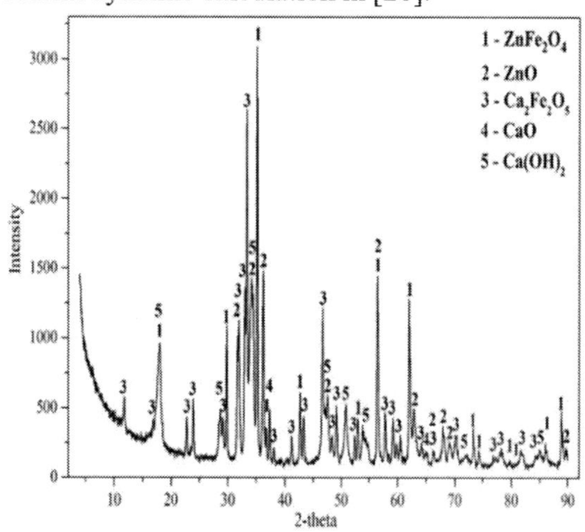

(a)

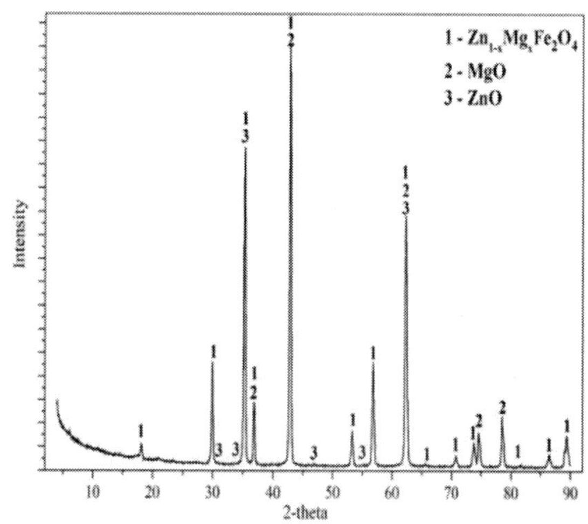

(b)

Figure 2. XRD pattern of the calcined sample $ZnFe_2O_4$–CaO (a) and $ZnFe_2O_4$–MgO (b) systems at 1000°C during 4 hours with 1:3 and 1:5 molar ratios respectively.

XRD patterns indicate that decomposition of zinc ferrite occurs by following reactions:

$$ZnFe_2O_4 + 2CaO = 2CaO{\cdot}Fe_2O_3 + ZnO \quad (4)$$

$$ZnFe_2O_4 + MgO = MgO{\cdot}Fe_2O_3 + ZnO \quad (5)$$

There was a partial decomposition of zinc ferrite by reaction (4) and insignificant by reaction (5). The presence of $Ca(OH)_2$ in one sample is due to the hygroscopicity of unreacted calcium oxide. Leaching of zinc and subsequent chemical analysis has shown that 43% of zinc transforms into the oxide from the calcium-containing sample, and only 1.7% is converted to ZnO from the magnesium-containing sample.

Figure 3 (a, b) shows kinetic curves for the calcium-containing system.

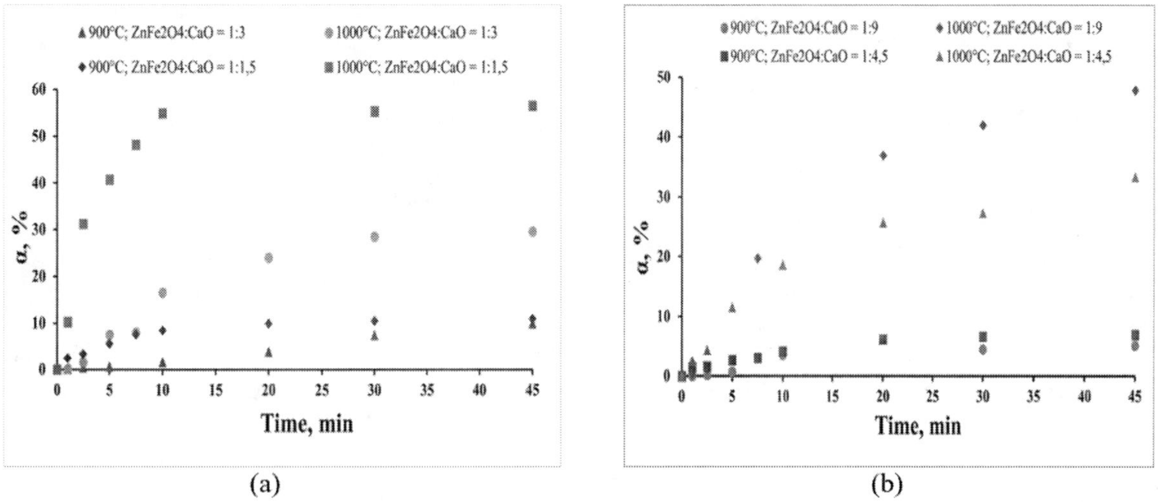

Figure 3. The kinetic curves for 1:1.5 and 1:3 (a), 1:4.5 and 1:9 (b) molar ratios of $ZnFe_2O_4$–CaO system at 900°C and 1000°C

The obtained data indicates that the reactions of the initial components at all studied ratios and temperatures have a similar development trend corresponding to a typical solid-phase reaction. It is possible to distinguish a stage of growing rate associated with the formation of the reaction products and increasing their amount, as well as a stage of reducing rate during which a continuous layer of reaction products is formed and contact of reactants is broken. The kinetic curves point out that increasing of temperature speeds up the reaction by several times.

Figure 4 (a, b) shows experimental and calculated effect of the corrected time on the degree of conversion for a 1:3 ratio at 900°C and 1000°C respectively. There is clear evidence shown in figures 4 (a, b) that the Yander and the Erofeev-Avrami equations corresponding to the diffusion and formation of nucleuses controlling steps respectively are mismatched with the experimental dependences. In the case of using other values of the coefficient n the Erofeev-Avrami equation is even more at variance with the experimental curves. The best match of calculated and experimental data is observed while using the contracting sphere equation. It should be mentioned that similar trends were obtained for all experimental ratios. The calculated values of the activation energy for all selected ratios are in the range of 175-210 kJ/mol.

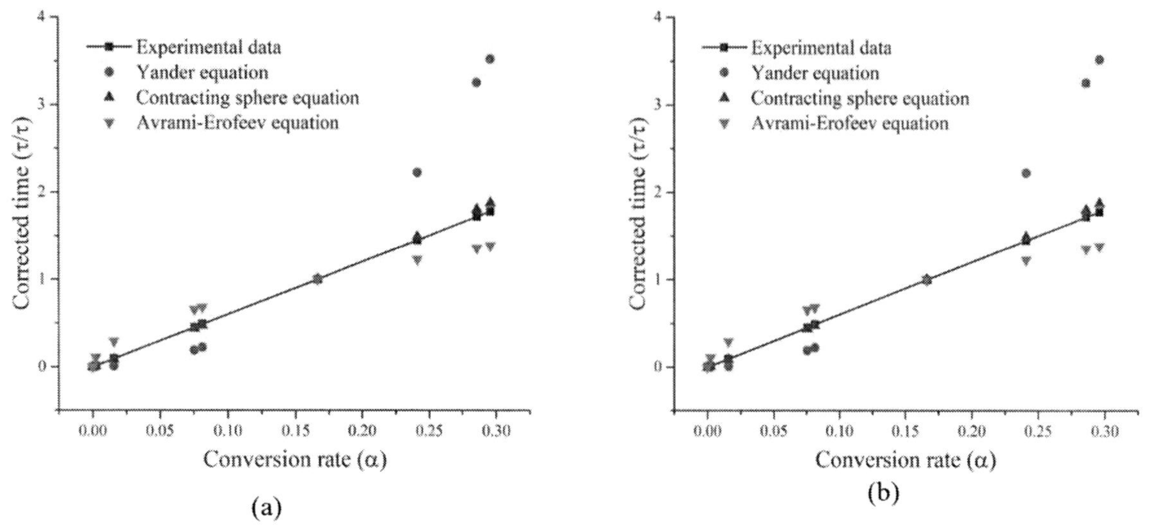

Figure 4. Experimental and calculated effect of the corrected time on the degree of conversion for a 1:3 ratio at 900°C (a) and 1000°C (b).

Table 1 shows the kinetic data for the magnesium-containing system.

Table 1. Experimental data for 1:5 molar ratio of $ZnFe_2O_4$–MgO system at 900°C and 1000°C.

900°C		1000°C	
Time, min	Conversion degree, %	Time, min	Conversion degree, %
10	3.53	10	2.03
20	2.68	20	2.52
30	3.15	30	2.29
45	1.68	45	2.26
60	2.31	60	2.27
75	1.95	75	2.36

The obtained data points out that kinetic dependences for the $ZnFe_2O_4$–MgO system are unreliable. The degree of conversion values are insignificant, which indicates the inefficient application of magnesium oxide during the processing of EAF dust containing zinc ferrite at 900-1000°C. The results obtained in [22] that have shown a quite high efficiency of magnesium oxide are probably due to the method of leaching experiments. For the selective dissolution of zinc oxide from calcined samples the authors have applied 7% sulfuric acid that promotes partial passinginto the solution of zinc ferrite [27].

4. Conclusion

Thus, the analysis of the EAF dust sample has confirmed that a considerable part of zinc is in the form of hardly soluble ferrite. The experiments with pure components have shown that the reaction of zinc ferrite with calcium oxide at 900-1000°C is kinetic controlled, which indicates a significant effect of temperature on the reaction rate and an ability to control the process through its adjustment.

In practice, decreasing of the process rate due to the formation of a continuous layer of reaction products can be avoided by continuous mixing of the charge. The mixing process facilitates the continuous renewal of the contact surface of the reacting particles and therefore increase the conversion degree. Such conditions can be achieved, for example, during the calcination in a rotary kiln. It is necessary to verify the obtained data by experiments using waste material (EAF dust).

The degree of conversion of the reaction between zinc ferriteand magnesium oxide is insignificant that makes the application of MgO at 900-1000°C for the calcination of EAF dustinexpedient.

Acknowledgments
The present study was funded by Program of Basic research of the Presidium of Russian Academy of Science №39, Grant No. AAAA-A18-118031490124-4.

References

[1] FAO data, http://www.globalsteeldust.com/about, 2018 (accessed 15.12.2018)
[2] Stefanova A and Aromaa J 2012 *Alkaline leaching of iron and steelmaking dust* (Helsinki: Aalto University) p 72
[3] Kazanbaev L A, Kozlov P A, Kubasov V L and Kolesnikov A V 2006 *Hydrometallurgy of Zinc.Purification of Solutions and Electrolysis* (Moscow: Ore and metalls) p 176
[4] Kozlov P A 2003 *The Waelz Process* (Moscow: Ore and metals publishing house) p 160
[5] Degtyarev A M, Ivakin D A, Shumilin Yu P and Mayorov S P 2015 Mastering of technology of waelz-oxide calcination in pipe rotary furnace *Base metals* **5** pp 31-35
[6] Iliev P, Stefanova V, Lucheva B and Kolev D 2017 Purification of zinc containing Waelz oxides from chlorine and fluorine *Journal of Chemical Technology and Metallurgy* **52 (2)** pp 252-257
[7] FAO data, http://www.ilzsg.org, 2018 (accessed 23.12.2018)
[8] Madias J 2015 Technologies to transform steelmaking sludge/dust into byproducts *ABM Week*
[9] Tsutsumi H, Yoshida S and Tetsumoto M 2010 Features of Fastmet process *Kobelco Tech.* **12** pp 85-92
[10] Nakayama T and Taniishi H 2011 Dust recycling technology for electric arc furnace using RHF *Nippon. Steel Eng. Tech. Rev.* **2** pp 25-29
[11] Holtzer M, Kmita A and Roczniak A 2015 The recycling of materials containing iron and zinc in the oxycup process *Arch. Foundry Eng.* **15** pp 126-130
[12] Roth J L, Frieden R, Hansmann T, Monai J and Solvi M 2001 Primus, a new process for recycling by-products and producing virgin iron *Rev. Metall.* **98 (11)** pp 987–996
[13] FAO data, http://pizotech.com, 2007 (accessed: 03.04.2017).
[14] Wu L and Themelis N J 1992 The flash reduction of electric arc furnace dusts *JOM* **44 (1)** pp 35-39
[15] Maccagni M G 2016 Integrate process on secondary zinc-bearing materials *J. Sustain. Metall.* **2** pp 133–140
[16] Montenegro V, Agatzini-Leonardou S, Oustadakis P and Tsakiridis P 2016 Hydrometallurgical treatment of EAF dust by direct sulphuric acid leaching at atmospheric pressure *Waste and Biomass Valorization* **7 (6)** pp 1531-48
[17] Dutra A, Paiva P and Tavares L 2006 Alkaline leaching of zinc from electric arc furnace steel dust *Miner. Eng.* **19** pp 478-485
[18] Miki T, Chairaksa-Fujimoto R, Maruyama K, et al. 2016 Hydrometallurgical extraction of zinc from CaO treated EAF dust in ammonium chloride solution *Journal of hazardous materials* **302** pp 90-96
[19] Chairaksa-Fujimoto R, Inoue Y, Umeda N, et al. 2015 New pyrometallurgical process of EAF dust treatment with CaO addition *International Journal of Minerals, Metallurgy, and Materials* **22 (8)** pp 788-797
[20] Yakornov S A, Pan'shin A M, Grudinsky P I, Dyubanov V G, Leont'ev L I, et al. 2017 Thermodynamic analysis of zinc ferrite decomposition in electric arc furnace dust by lime *Russian journal of non-ferrous metals* **58 (6)** pp 586-590

IOP Conf. Series: Materials Science and Engineering **525** (2019) 012097 doi:10.1088/1757-899X/525/1/012097

[21] Dimitrov R and Boyanov B 1984 Investigation of solid state interactions in the systems ZnO–α-Fe_2O_3, $ZnFe_2O_4$–CuO, $ZnFe_2O_4$–CaO *Rudarsko-Metalurski Zbornik* **31 (1)** pp 67-80.
[22] Peltekov A B and Boyanov B S 2013 Study of solid state interactions in the systems $ZnFe_2O_4$ – CaO, $ZnFe_2O_4$ – MgO and zinc cake with CaO and MgO *Journal of Mining and Metallurgy* **49 (3)** pp 339-346
[23] Filippova N A 1975 *Phase analysis of ores and products of their processing* (Moscow: Chemistry) p 280
[24] Osokina G N and Vyatkina L S 1988 The method of phase analysis of zinc compounds in the products of metallurgical production *Pat.1399668*
[25] Guseva A F, Neiman A Ya and Animitsa I E 2005 *Solid phase reactions when receving and operating inorganic materials* (Ekaterinburg) p 42
[26] Grudinsky P I, Shurlova A A, Sevostyanov M A and Dyubanov V G 2018 Characteristics of the process of electric arc furnace dust roasting with calcium and magnesium oxides *Bulletin of science and practice* **4 (12)** pp 309-317
[27] Sinclair R J 2005 *The extractive metallurgy of zinc* (Carlton: The Australasian Institute of Mining and Metallurgy) p 294

IOP Conf. Series: Materials Science and Engineering **525** (2019) 012098 doi:10.1088/1757-899X/525/1/012098

Study of the effect of the introduction of heparin on the mechanical properties of polylactide

M A Sevostyanov[1], A S Baikin[1], E O Nasakina[1], K V Sergienko[1], M A Kaplan[1], A V Leonov[1], A A Kolmakova[1], E A Danilova[1], L A Shatova[2], N P Leonova[1] and A G Kolmakov[1]

[1]A.A.Baikov Institute of Metallurgy and Materials Science, Russian Academy of Sciences, Moscow, Russia

[2] Voronezh State Technical University, st. 20-letiya Oktyabrya, 84/4, Voronezh, Russia

E-mail: baikinas@mail.ru

Abstract. The paper describes a method for the preparation of polymeric composite films for medical purposes based on polylactide with introduced heparin. The effect of the molecular weight of polylactide and the concentration of the drug on the mechanical properties and thickness of the films obtained was studied. The possibility of creating biodegradable films with the introduced drug with given properties is shown.

1. Introduction

The treatment of many diseases in many cases is accompanied by adverse adverse effects on healthy human organs. Modern medicine has long sought to minimize this possible impact, including from the use of various medicines [1-3].

A large number of studies aimed at creating systems of local controlled drug delivery. Such systems provide therapeutic effects directly in the required area, reducing or completely eliminating the effect on the rest of the body [4-6].

As a basis for creating systems of controlled local delivery, biodegradable polymeric materials of both synthetic and natural origin are used. Polyesters such as polylactides, polylactide glycolides, polyethylene glycol, polyanhydrides, polyorthoesters, polysaccharides (starch, dextran, chitosan) have found wide application in this area [7–8].

In this paper, we consider films based on polylactide of various molecular weights. Polylactide is biodegradable, biocompatible and thermoplastic. Polylactide is already used in medicine for the manufacture of surgical sutures and the processing of pins. [9]. The polymer can be obtained with different molecular weights and structure, and can also encapsulate molecules of almost any size. Heparin preparations are chosen as drugs. The created composite material on the basis of polylactide with injected drug can be used to cover medical products of the "kava-filter" type and provide drug therapy during implantation. Thus, using different molecular weights of the polymer, drug concentrations, it is possible to obtain a wide range of coatings with different properties and further selection of the required characteristics. [10]

An important parameter of the resulting composite polymer materials are mechanical properties that must meet the requirements for creating coatings for medical implants.

The dependences of the mechanical properties of the resulting composite materials on the molecular weight of the polymer and the concentration of the drug are studied in this work.

2. Materials and methods

To create films, we prepared weights of polylactide weighing 2 g (± 0.01 g).

Chloroform with a volume of 200 ml was placed in a flask with a volume of 500 ml and heated to 800 ° C on a magnetic stirrer. [eleven]

Next, the obtained weights of polylactide were dissolved to a homogeneous state in chloroform at 800 ° C for 1 hour with constant stirring using an electronic top-wheel mixer.

In the resulting homogeneous solution is kneaded the drug in an amount necessary to obtain 1, 2 or 3% solutions, at a solution temperature of 30 ° C;

The resulting homogeneous solution is poured into a glass mold and set to dry in a thermostat at 37 ° C for 48 hours.

At the end of drying, the films obtained were removed from the glass form.

The resulting samples were assigned ciphers using 2 digits (X.Y), used for independent research in the future.

The first digit "X" indicates the type of polymer from which the film is made: 1 - PLA 45kDa; 2 - 90kD PLA; 3 - PLA 180kDa. The second digit "Y" indicates the concentration of the drug. Values are given in mass percent relative to the weight of the polymer itself: 1 - 1%; 2 - 2%; 3 - 3%.

The main characteristic feature of this method of drug administration is low temperatures, allowing to preserve the integrity of the drug, obtaining a uniform, uniform structure of the material, as well as the possibility of drug administration without using ultraviolet radiation and applying intermediate primer layers, which greatly simplifies the technology for producing films.

Strength studies of polymer films of polylactide under tension were carried out on an INSTRON 3382 universal testing machine with a loading rate of 10 mm / min. Samples of polymer films for testing were made according to GOST 14236-81, in the form of a double blade. The sample was fixed in the grips of the testing machine, which are evenly tightened to ensure that the sample does not slip during the test. Testing of polymer films with the determination of the relative elongation, yield strength and tensile strength was carried out according to GOST 14236-81. Processing of test results in determining the mechanical properties was carried out using the INSTRON Bluehill 2.0 software. The measurement error of the testing machine is less than 1%.

5 samples were tested per experimental point. The values of yield strength , tensile strength and relative elongation were determined.

3. Results and discussion

The results of the study of the mechanical properties of polymer films for each composition based on polylactide (PLA) with different molecular weight and drug concentration are presented in Tables 1-2.

Table 1. Mechanical properties of polymer films without drugs.

Polymer and concentration	Deformation, %	Yield strength, MPa	Ultimate Strength, MPa	Thickness, μm
PLA 45kDa	7,1±0,4	22,9±1,1	35,6±1,8	42±2
PLA 90kDa	58,1±2,9	17,3±0,9	20,1±1,0	38±2
PLA 180kDa	43,9±2,2	16,8±0,8	19,6±0,9	39±2

IOP Conf. Series: Materials Science and Engineering **525** (2019) 012098 doi:10.1088/1757-899X/525/1/012098

Table 2. Mechanical properties of polymer films with heparin.

Code	Polymer and concentration	Deformation, %	Yield strength, MPa	Ultimate Strength, MPa	Thickness, μm
1.1	PLA 45kDa, 1% heparina	2±0,1	17,1±0,9	17,9±0,8	93±5
1.2	PLA 45kDa, 2% heparina	1,8±0,1	15,0±0,8	15,7±0,7	100±5
1.3	PLA 45kDa, 3% heparina	1,7±0,1	16,4±0,8	13,8±1,7	110±6
2.1	PLA 90kDa, 1% heparina	2,4±0,2	23,9±1,2	33,4±1,6	80±4
2.2	PLA 90kDa, 2% heparina	2,1±0,1	34,0±1,7	32,1±0,9	82±4
2.3	PLA 90kDa, 3% heparina	1,8±0,1	16,2±0,8	18,5±1,3	100±5
3.1	PLA 180kDa, 1% heparina	2,5±0,2	19,9±1,0	25,9±1,2	100±5
3.2	PLA 180kDa, 2% heparina	2,2±0,1	22,0±1,1	24,3±1,1	105±5
3.3	PLA 180kDa, 3% heparina	1,9±0,1	20,6±1,0	22,3±	112±6

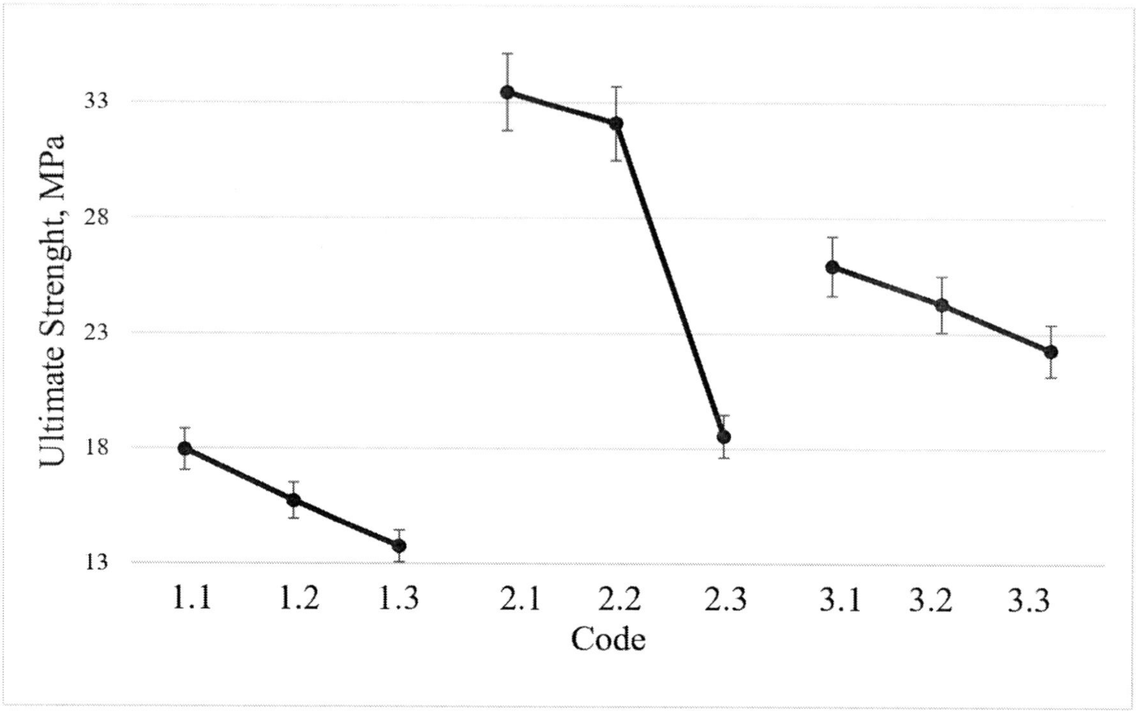

Figure 1. Diagram of deformation depending on the concentration of heparin for PLA

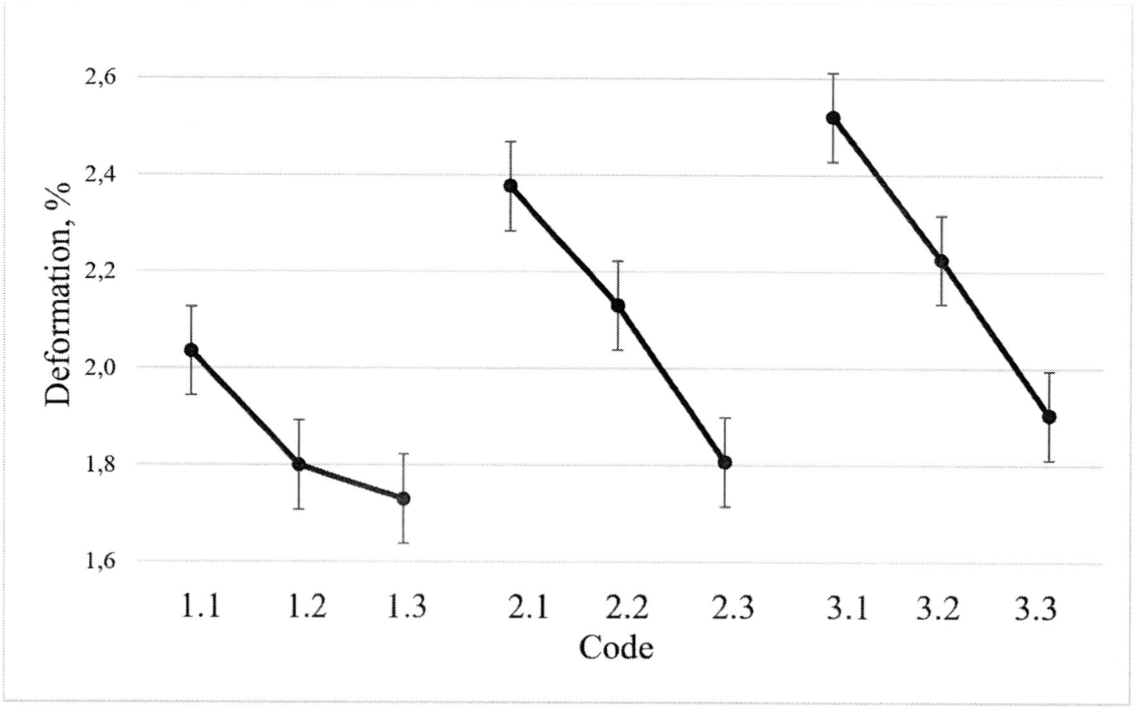

Figure 2. Diagram of the dependence of tensile strength on the concentration of heparin for PLA

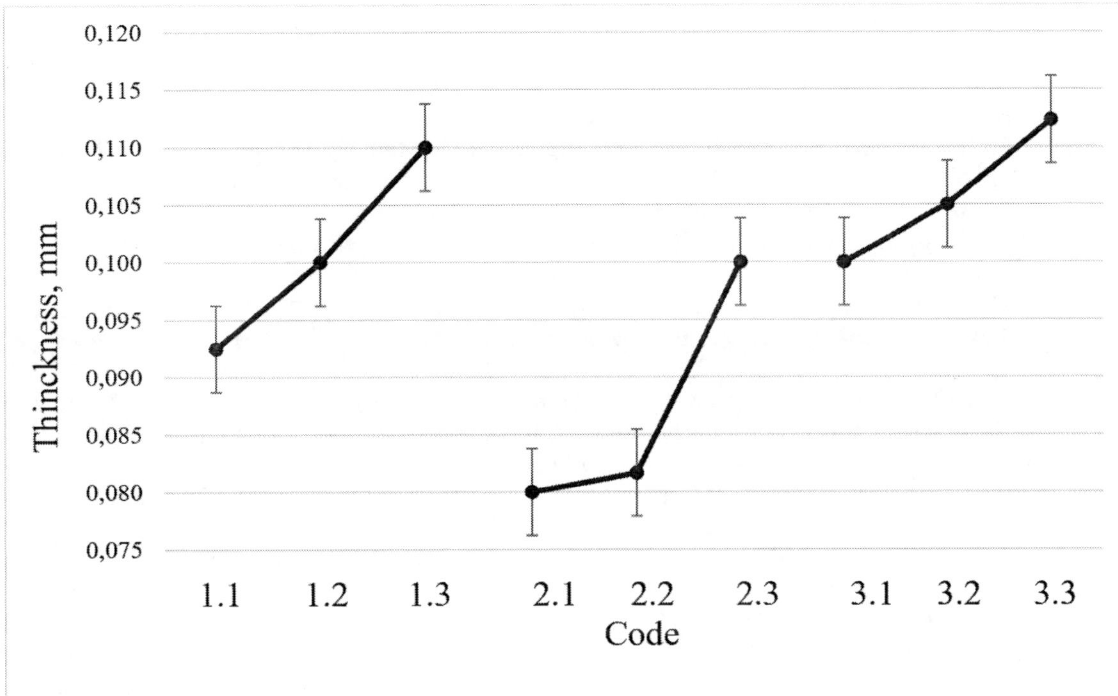

Figure 3. The diagram of the dependence of the thickness of polymer films on the concentration of heparin for PLA

With an increase in the concentration of the drug (heparin) in polylactide polymers (PLA) for all molecular weights, the strength and plasticity decrease and the film thickness increases (Figure 1-3). Comparing the results of mechanical tests with a film without a drug, we can conclude that the plasticity of all samples decreases. The tensile strength, on the contrary, increases in all polymers except PLA 45 kDa, but then there was a decrease in the strength characteristic with increasing drug concentration. The increase in strength is associated with the formation of a composite material in the form of a polymer, as a plastic base, with a strong filler in the form of small particles of a drug with high strength, and the decrease in strength is associated with the accumulation of a large number of these particles and the appearance of fragile zones. In the initial PLA 45kDa (without medication), the tensile strength is 35.63 MPa. With an increase in the concentration of heparin from 1% to 3%, the strength characteristics decrease to 17.9 MPa, 15.03 MPa, 13.75 MPa (49.76%, 57.81%, 61.4%), respectively, which shows a poor correlation between PLA polymer with a molecular weight of 45kDa and heparin.

4. Conclusions

Polymer composite films based on polylactide with introduced heparin were obtained. The dependences of the strength and plastic properties, as well as the thickness of polymeric composite films based on polylactide with heparin injected, on the concentration of the drug and the molecular weight of polylactide were established.

Acknowledgements

The work was supported by the Ministry of Education and Science of Russia (grant identifier RFMEFI60417X0196).

IOP Conf. Series: Materials Science and Engineering **525** (2019) 012098 doi:10.1088/1757-899X/525/1/012098

References

[1] Giessen WV, Lincoff A, Schwartz R, Beusekom HV, Serruys P, Holmes D, et al. Marked inflammatory sequelae to implantation of biodegradable and nonbiodegradable polymers in porcine coronary arteries. Circulation 1996;94(7):1690–7.

[2] Nasakina E. O., Sevostyanov M. A., Mikhaylova A. B., Baikin A. S., Sergienko K. V., Leonov A. V., Kolmakov A. G. Formation of alpha and beta tantalum at the variation of magnetron sputtering conditions// IOP Conference Series-Materials Science and Engineering. 110 (2016) 012042 doi: 10.1088/1757-899X/110/1/012042

[3] E. O. Nasakina, A. S. Baikin, K. V. Sergienko, M. A. Sevost'yanov, A. G. Kolmakov, B. A. Goncharenko, V. T. Zabolotnyi, R. S. Fadeev, I. C. Fadeeva, S. V. Gudkov, A. Solntsev Biocompatibility of Nanostructured Nitinol with Titanium or Tantalum Surface Composite Layers Formed by Magnetron Sputtering // Doklady Chemistry, 2015. V.461. Part.1. P.86-88

[4] Cooper SL, Visser SA, Hergenrother RW, Lamba NMK. Polymers.In: Ratner BD, Hoffman AS, Schoen FJ, Lemons JE, editors. Biomaterials science an introduction to materials in medicine. 2nd ed. San Diego: Elsevier Academic Press; 2004. p. 67–79.

[5] Stack R, Califf R, Phillips H, Pryor D, Quigley P, Bauman R, et al. Interventional cardiac catheterization at Duke medical center. Am J Cardiol 1988;62:3F–24F.

[6] Lincoff AM, Schwartz RS, Giessen WJVD, Beusekom HMMV, Serruys PW, Holmes DR, et al. Biodegradable polymers can evoke a unique inflammatory response when implanted in the coronary artery. Circulation 1992;86(Suppl.4):

[7] Supper S., Anton N., Seidel N., Riemenschnitter M., Curdy C., Vandamme T. Thermosensitive chitosan/glycerophosphate-based hydrogel and its derivatives in pharmaceutical and biomedical applications. Expert Opin. Drug Deliv. 2014;11:249–267. doi: 10.1517/17425247.2014.867326.

[8] A S Baikin, M A Sevostyanov, E O Nasakina, K V Sergienko, M A Kaplan, S V Konushkin, A A Kolmakova, A D Yakubov and A G Kolmakov. Investigation of the influence of the composition on mechanical properties poly(glycolide-DL-lactide) // IOP Conference Series: Materials Science and Engineering 347 (2018) 012042 doi:10.1088/1757-899X/347/1/012042.

[9] Timchenko TV, Shcherbakova LI, Kompantsev VA Poly-D, L-lactide-co-glycolide: methods of preparation, properties and use for the development of drugs with micro- and nanodelivery means. // Modern problems of science and education. 2015. № 4. P.559-570 (in russian)

[10] Kedik SA, Zhavoronok ES, Sedishev IP, Panov AV, \\ Polymers for delayed drug delivery systems (review) (May 3, 2013) (in russian)

[11] Glotova VN, Novikov VT, Izhenbina TN, Titova NG Solubility of lactide and glycolide in organic solvents // Polzunovskii vestnik. 2014. No. 3. P. 145-147. (in russian)

IOP Conf. Series: Materials Science and Engineering **525** (2019) 012099 doi:10.1088/1757-899X/525/1/012099

Study of the effects of the introduction of heparin on the mechanical properties of poly (glycolide-dl-lactide)

A S Baikin[1], K V Sergienko[1], M A Kaplan[1], A V Leonov[1], A A Kolmakova[1], E A Danilova[1], L A Shatova[2], A M Tsareva[1], Yu O Leonova[1], Yu N Ustinova[1], Glinushkin A.P.[3] and M A Sevostyanov[1]

[1]A.A.Baikov Institute of Metallurgy and Materials Science, Russian Academy of Sciences, Moscow, Russia

[2] Voronezh State Technical University, st. 20-letiya Oktyabrya, 84/4, Voronezh, Russia

[3] All-Russian Research Institute of Phytopathology, Bolshiye Vyazyomy, Odintsovsky district, Moscow Region, 143050

E-mail: baikinas@mail.ru

Abstract. A technology has been developed for producing biodegradable films based on poly (glycolide-lactide) containing heparin. The obtained biodegradable films simulate the coatings of cava filters. The effect of the concentration of introduced heparin and the change in the molecular weight of poly (glycolide-lactide) on the strength and plastic properties, as well as on the thickness of the films obtained, is shown. The possibility of creating coatings with introduced heparin for cava filters with the required mechanical properties is shown.

1. Introduction

Deep venous thrombosis of the lower extremities (DVHC) and pulmonary thromboembolism (PE) are common diseases of the human cardiovascular system. Their prevention includes both pharmacotherapy and implantation of kava-filter medical devices. Cava filters are a device for capturing blood clots carried by blood. Currently, in medical practice, operations using these implants are widespread. There is a large number of different cava filters, differing in material of manufacture, construction, and other parameters, possessing both advantages and significant disadvantages. Therefore, there are numerous developments of new products that can minimize possible postoperative complications.

A promising direction is the combination in one device of the immediate function of capturing a thrombus with local pharmaceutical effects, which can be ensured by creating on the implant surface a system of local drug delivery. However, direct application to the surface of the implant leads to rapid washing out of the medicinal agent, which can be solved by using a biodegradable matrix. [1-4] When choosing a material for its creation, the following considerations are guided: the material must be chemically inert, not contain leached impurities, be easily synthesized, and in the case of hydrolysis (or resorption) do not form toxic compounds. These are silicone, polyurethane, polyethylene, polyethylene vinyl acetate. Developed new polymers used in medicine that meet the requirements that impose a decay rate and the physical characteristics of the delivery systems. The need to use biodegradable materials for this purpose led to the active development of biodegradable polymers of synthetic and natural origin, among which polyethers: polylactides, polylactideglycolides, polyethylene glycol, polyanhydrides, polyorthoesters, polysaccharum (starch, dextran, chitosan) were most developed. [5-8]

In this work, poly (glycolide-lactide) of various molecular weights was chosen as the biodegradable matrix. Polyglycolide is the simplest polyether, in which, due to the close location of the ester groups, intermolecular interactions are strongly pronounced, it has a high degree of crystallinity, a high melting point (~ 300 ° C), and extreme hydrolytic instability. Polylactide aliphatic polyether, the monomer of which is lactic acid. It is biodegradable, biocompatible and thermoplastic. Using polylactide / polyglycolide copolymers, the bioresorption rate is varied depending on the type of polymer and its molecular weight [9]. Heparin was used as a drug. Heparin is an anticoagulant of direct action, that is, as a substance that prevents blood from clotting. It is used for the prevention and treatment of thromboembolic diseases, during operations on the heart and blood vessels. [10]

The type, concentration and method of introduction into the polymer matrix of a drug can significantly affect the properties of the resulting composite material, which must meet the requirements for the further development of medical implant coatings.

In this paper, the effect of heparin administration on the mechanical properties of poly (glycolide-lactide) of various molecular weights was investigated.

2. Materials and methods

To create films, weights of polymers weighing 2 g (± 0.01 g) were prepared.

Chloroform with a volume of 200 ml was placed in a flask with a volume of 500 ml and heated to 800 ° C on a magnetic stirrer.

Next, the obtained weights of the polymers were dissolved to a homogeneous state in chloroform at 800 ° C for 1 hour with constant stirring using an electronic overhead stirrer.

In the resulting homogeneous solution, heparin is kneaded in an amount necessary to obtain 1, 2 or 3% solutions, at a solution temperature of 30 ° C;

The resulting homogeneous solution is poured into a glass mold and set to dry in a thermostat at 37 ° C for 48 hours.

At the end of drying, the films obtained were removed from the glass form.

The resulting samples were assigned ciphers using 2 digits (X.Y), used for independent research in the future.

The first digit "X" denotes the type of polymer from which the film is made:

1 - PGLA 45kDa; 2 - PGLA 90kDa; 3 - PSLA 180kDa. The second digit "Y" indicates the concentration of the drug. Values are given in mass percent relative to the weight of the polymer itself: 1 - 1%; 2 - 2%; 3 - 3%.

The main characteristic of this method of drug administration is low temperatures, allowing to preserve the integrity of the drug, obtaining a uniform, uniform structure of the material, as well as the possibility of drug administration without the use of ultraviolet radiation and the application of intermediate layers, which greatly simplifies the technology for producing films.

Strength tests of polymer films of poly (glycolide-lactide) were tested on a universal testing machine INSTRON 3382 with a loading rate of 10 mm / min. Samples of polymer films for testing were made according to GOST 14236-81, in the form of a double blade. The sample was fixed in the grips of the testing machine, which are evenly tightened to ensure that the sample does not slip during the test. Testing of polymer films with the determination of the relative elongation, yield strength and tensile strength was carried out according to GOST 14236-81. Processing of test results in determining the mechanical properties was carried out using the INSTRON Bluehill 2.0 software. The measurement error of the testing machine is less than 1%.

5 samples were tested per experimental point. The values of yield strength , tensile strength and relative elongation were determined.

3. Results and discussion

Table 1-2 presents the average results of mechanical testing of polymer films for each composition based on poly (glycolide-lactide) (PGLA) with different molecular weight and concentration of the drug.

IOP Conf. Series: Materials Science and Engineering **525** (2019) 012099 doi:10.1088/1757-899X/525/1/012099

From the obtained results, dependency graphs were built, which can be used to track the trend of changes in mechanical characteristics (Figure 1-3).

Table 1. Mechanical properties of polymer films without drugs.

Polymer and concentration	Deformation, %	Yield strength, MPa	Ultimate Strength, MPa	Thickness, μm
PGLA 45kDa	140±7	4,0±0,2	5,5±0,3	66±3
PGLA 90kDa	262±13	2,4±0,1	5,2±0,3	63±3
PGLA 180kDa	198±10	5,6±0,3	7,6±0,4	32±2

Table 2. Mechanical properties of polymer films with heparin.

Code	Polymer and concentration	Deformation, %	Yield strength, MPa	Ultimate Strength, MPa	Thickness, μm
1.1	PGLA 30/70 45kDa, 1% гепарина	17,6±0,9	12,8±0,6	17,5±0,9	103±5
1.2	PGLA 30/70 45kDa, 2% гепарина	13,9±0,7	11,0±0,6	14,6±0,7	123±6
1.3	PGLA 30/70 45kDa, 3% гепарина	11,2±0,6	9,7±0,5	11,9±0,6	140±7
2.1	PGLA 30/70 90kDa, 1% гепарина	83,1±4,2	7,8±0,4	10,6±0,5	80±4
2.2	PGLA 30/70 90kDa, 2% гепарина	42,1±2,1	2,1±0,1	4,1±0,2	83±4
2.3	PGLA 30/70 90kDa, 3% гепарина	99,4±5,0	4,2±0,2	4,6±0,2	92±5
3.1	PGLA 30/70 180kDa, 1% гепарина	98,5±4,9	5,6±0,3	9,0±0,4	98±5
3.2	PGLA 30/70 180kDa, 2% гепарина	28,3±1,4	5,8±0,3	8,7±0,4	115±6
3.3	PGLA 30/70 180kDa, 3% гепарина	17,9±0,9	6,6±0,3	8,9±0,4	135±7

IOP Conf. Series: Materials Science and Engineering **525** (2019) 012099 doi:10.1088/1757-899X/525/1/012099

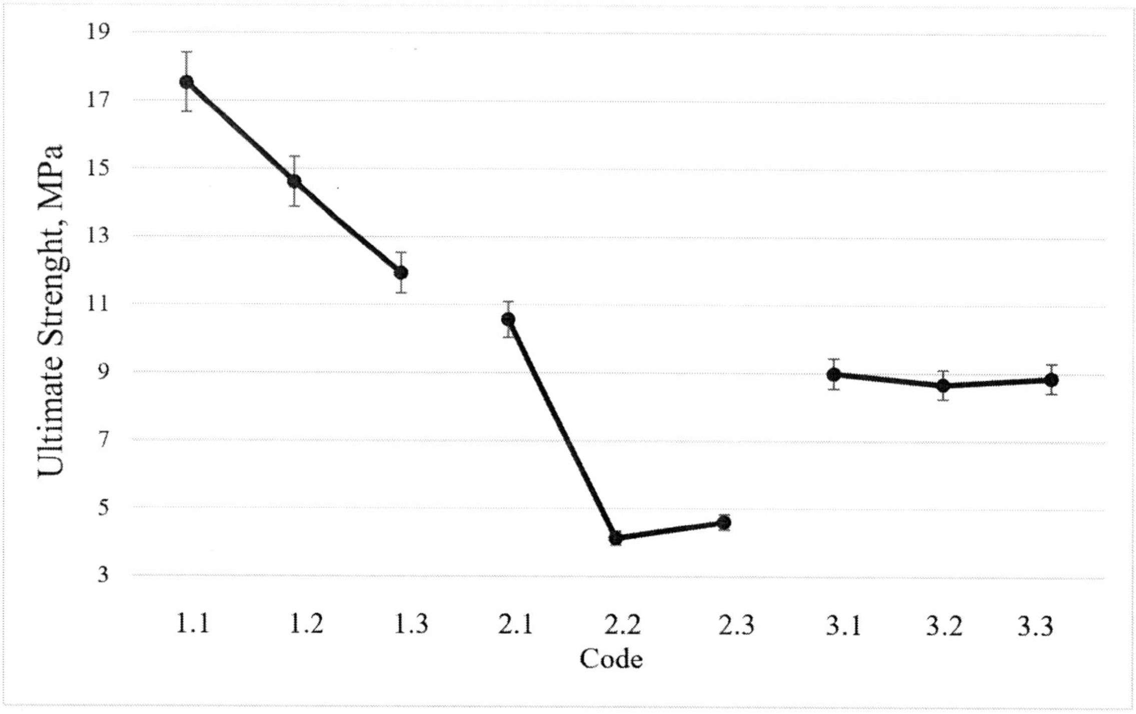

Figure 1. Diagram of deformation depending on the concentration of heparin for PGLA

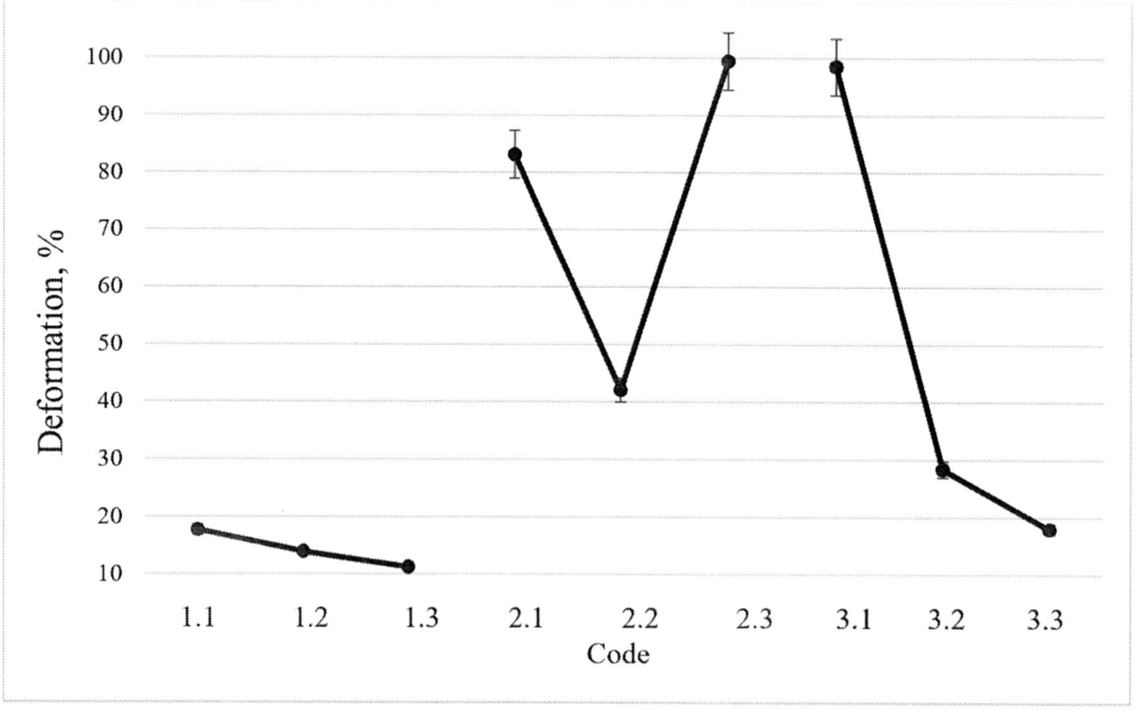

Figure 2. Diagram of the dependence of tensile strength on the concentration of heparin for PGLA

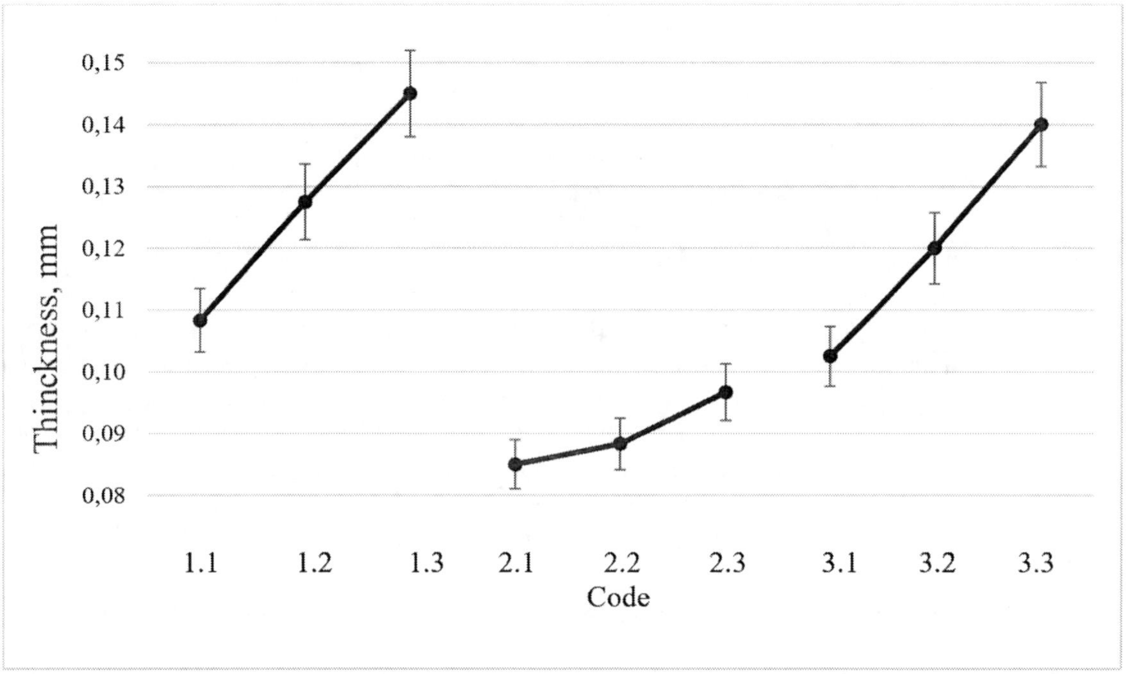

Figure 3. The diagram of the dependence of the thickness of polymer films on the concentration of heparin for PGLA

In Figures 1-3, one can observe the dependence of reducing the strength and plasticity, and increasing the film thickness with increasing drug concentration (heparin) in poly (glycolide-lactide) polymers (PGLA) for all molecular weights. The exception is point 2.3, which is explained by the uneven distribution of the drug in this film. Comparing the results of mechanical tests with a film without a drug, we can conclude that the plasticity of all samples decreases. So the relative elongation of the initial PGLA 30/70 180 kDa (without medication) is 198.32%. The addition of 1%, 2% and 3% heparin leads to a decrease in the relative elongation to 98.46%, 28.31% and 17.94%, i.e. the drop in performance was 50.4%, 85.7% and 90.9%, respectively. In all polymers there is an increase in strength. With increasing drug concentration strength characteristics decrease. The increase in strength is most likely due to the formation of a composite material in the form of a polymer, as a plastic base, with a strong filler in the form of small particles of a drug with high strength, and a decrease in strength is associated with the accumulation of a large number of these particles and the appearance of fragile zones.

4. Conclusions

Studies of the mechanical properties of polymeric composite films based on poly (glycolide-lactide) with introduced heparin were conducted, and the effect of the concentration of the drug and the molecular weight of poly (glycolide-lactide) on the thickness, strength and plastic properties of the resulting films was shown.

Acknowledgements

The work was supported by the Ministry of Education and Science of Russia (grant identifier RFMEFI60417X0196).

IOP Conf. Series: Materials Science and Engineering **525** (2019) 012099 doi:10.1088/1757-899X/525/1/012099

References

[1] Giessen WV, Lincoff A, Schwartz R, Beusekom HV, Serruys P, Holmes D, et al. Marked inflammatory sequelae to implantation of biodegradable and nonbiodegradable polymers in porcine coronary arteries. Circulation 1996;94(7):1690–7.

[2] Nasakina E. O., Sevostyanov M. A., Mikhaylova A. B., Baikin A. S., Sergienko K. V., Leonov A. V., Kolmakov A. G. Formation of alpha and beta tantalum at the variation of magnetron sputtering conditions// IOP Conference Series-Materials Science and Engineering. 110 (2016) 012042 doi: 10.1088/1757-899X/110/1/012042.

[3] Sevost'yanov M.A., Nasakina E.O., Baikin A.S., Sergienko K.V., Konushkin S.V., Kaplan M.A., Seregin A.V., Leonov A.V., Kozlov V.A., Shkirin A.V., Bunkin N.F., Kolmakov A.G., Simakov S.V., Gudkov S.V. Biocompatibility of new materials based on nano-structured nitinol with titanium and tantalum composite surface layers: experimental analysis in vitro and in vivo // Journal of Materials Science: Materials in Medicine, 2018. Vol. 29. P.33 (https://doi.org/10.1007/s10856-018-6039-3)

[4] Mohsin Shaikh, Ganessan Kichenadasse,Namita Roy Choudhury, Ross Butler, Sanjay Garg. Non-vascular drug eluting stents as localized controlled drug delivery platform: Preclinical and clinical experience. Journal of Controlled Release. Volume 172. Issue 1. November. 2013. p. 105–117

[5] Mazid MA, Scott E, Li N-H. New biocompatible polyurethane-type copolymer with low molecular weight heparin. Clin Mater 1991;8(1–2):71–80

[6] Supper S., Anton N., Seidel N., Riemenschnitter M., Curdy C., Vandamme T. Thermosensitive chitosan/glycerophosphate-based hydrogel and its derivatives in pharmaceutical and biomedical applications. Expert Opin. Drug Deliv. 2014;11:249–267. doi: 10.1517/17425247.2014.867326.

[7] Sevost'yanov M.A., Fedotov A.Yu., Nasakina E.O., Teterina A.Yu., Baikin A.S., Sergienko K.V., Kolmakov A.G., Komlev V.S., Ivanov V.E., Karp O.E., Gudkov S.V., Barinov S.M. Kinetics of the release of antibiotics from chitosan-based biodegradable biopolymer membranes // Doklady Chemistry, 2015. V.465. Part.1. P.278-280.

[8] Kimura H, Ogura Y. Biodegradable polymers for ocular drug delivery. Ophthalmologica 2001;215(3):143–55

[9] Tamai H, Gaki K, Kyo E, Kosuga K, Kawashima A, Matsui S,et al. Initial and 6-month results of biodegradable poly-L-lactic acid coronary stents in humans. Circulation 2000;102(4):399–404

[10] Peres C., Matos A.I., Conniot J., Sainz V., Zupančič E., Silva J.M., Graça L., Sá Gaspar R., Préat V., Florindo H.F. Poly(lactic acid)-based particulate systems are promising tools for immune modulation // Acta Biomater. 2017. V.48. P. 41-57.

IOP Conf. Series: Materials Science and Engineering **525** (2019) 012100 doi:10.1088/1757-899X/525/1/012100

Study of polylactide degradation rate in a phosphate buffer solution

A S Baikin[1], E O Nasakina[1], A M Tsareva[1], M A Kaplan[1], A A Kolmakova[1], E A Danilova[1], L A Shatova[2], N P Leonova[1], Yu O Leonova[1], and A G Kolmakov[1]

[1]A.A.Baikov Institute of Metallurgy and Materials Science, Russian Academy of Sciences, Moscow, Russia
[2]Voronezh State Technical University, st. 20-letiya Oktyabrya, 84/4, Voronezh, Russia

E-mail: baikinas@mail.ru

Abstract. The study of the degradation of films based on polylactide in phosphate buffer solution. The effect of the change in the molecular weight of polylactide on the polymer degradation process is shown. With an increase in the molecular weight of polylactide, there is a slight increase in the rate of degradation of the polymer.

1. Introduction

Biodegradable polymers are widely used in medicine. In the body, under the action of the internal environment, they are hydrolyzed, which leads to their gradual degradation [1-5]. It is promising to use biodegradable polymers for controlled drug delivery, and also as implants, which can be gradually replaced in the body with bone or other living tissue. Biodegradable synthetic biomaterials based on polymers of organic acids, for example, lactic (PLA, polylactide) and glycolic (PGA, polyglycolide), were among the first to be used in tissue engineering. In this case, the polymer may include one type of acid residue, and combinations thereof in various proportions. Matrices based on organic acids formed the basis for the creation of such organs and tissues as skin, bone, cartilage, tendon, muscles (striated, smooth and cordial), small intestine, etc. capable of locally delivering drugs and releasing them in a controlled manner, solving possible postoperative complications and providing concomitant pharmacotherapy [6-8].

For the materials used to create systems of controlled drug delivery, the rate of degradation is important, which is influenced by many factors: composition, molecular weight, production method, product shape, environment in which biodegradation occurs, and many others [9-11] . In this work, the rate of degradation of polylactide films of various molecular weights in phosphate buffer solution was studied.

2. Materials and methods

To create films, weights of polymers weighing 10 g (± 0.01 g) were prepared.

Chloroform with a volume of 200 ml was placed in a flask with a volume of 500 ml and heated to 800 ° C on a magnetic stirrer.

Next, the obtained weights of the polymers were dissolved to a homogeneous state in chloroform at 800 ° C for 1 hour with constant stirring using an electronic overhead stirrer.

At the end of the dissolution of the polymer, the volume of the solution was brought to 200 ml with chloroform. The resulting solution was aged for 5 minutes at 800 ° C and poured into glass trays.

Drying was carried out for 2 days in air at 37 ° C in a thermostat.

At the end of drying, the films were removed. (Table 1).

Table 1. Samples of films

№	Polymer	Molecular mass, kDa	Concentration, g per 100 ml of chloroform
1	PLA	45	5
2	PLA	90	5
3	PLA	180	5

Studies of the degradation of the obtained films were carried out according to the method according to GOST R ISO 13781-2011. Preparing samples of square shape measuring 20 by 20 mm. Dried the test sample to achieve constant mass in the desiccator under vacuum at room temperature. The mass of the studied samples was determined using an analytical balance with an accuracy of 1.0% of the total mass.

Next, the samples were placed in an inert plastic container and filled with a solution of 60 ml. Used phosphate buffered saline.

The buffer solution was prepared as follows: a 1/15 mol / l KH2PO4 solution prepared by dissolving 9.078 g of KH2PO4 in 1 l of water and a 1/15 mol / l solution of Na2HPO4 prepared by dissolving 11.876 g of Na2HPO4 · 2H2O in 1 l of water were mixed, in volume fractions of 18.2% and 81.8%, respectively. The pH of the buffer solution was 7.4 ± 0.2. The pH of the buffer solution was measured in at least two different containers in each study period. If in at least one container the pH value was shifted beyond the established limits, the pH level in all tanks was measured and adjusted to pH 7.4 ± 0.2 using a solution with (NaOH) = 0.1 mol / l.

The containers with the solution and the sample were kept in a thermostat at a temperature of 37 ± 1 ° C for 1; 3; 7; 14; 21; 28; 60 and 90 days. For each period 3 samples were prepared.

To study the degraded sample, the filter was dried under vacuum at room temperature until constant weight was reached. Determined the mass of the filter with an accuracy of 1.0% of the total mass. Wash the degraded sample under test in deionized water three times using a filter to hold possible parts of the sample.

After that, the degraded sample and the filter were dried under vacuum at room temperature until constant weight was achieved. Determined the total mass of the sample and filter with an accuracy of 1.0% of the total mass.

3. Results and discussion

Figure 1 (a-c) shows the results of a study of the degradation of polymer films over time.

With an increase in the molecular weight of polylactide (45, 90, and 180 kDa), the degradation rate increases slightly (Fig. 1). So for 90 days for a polylactide with a molecular mass of 45 kDa, the mass loss is 6.5%, for 90 kDa - 7.8%, and for 180 kDa - 8%.

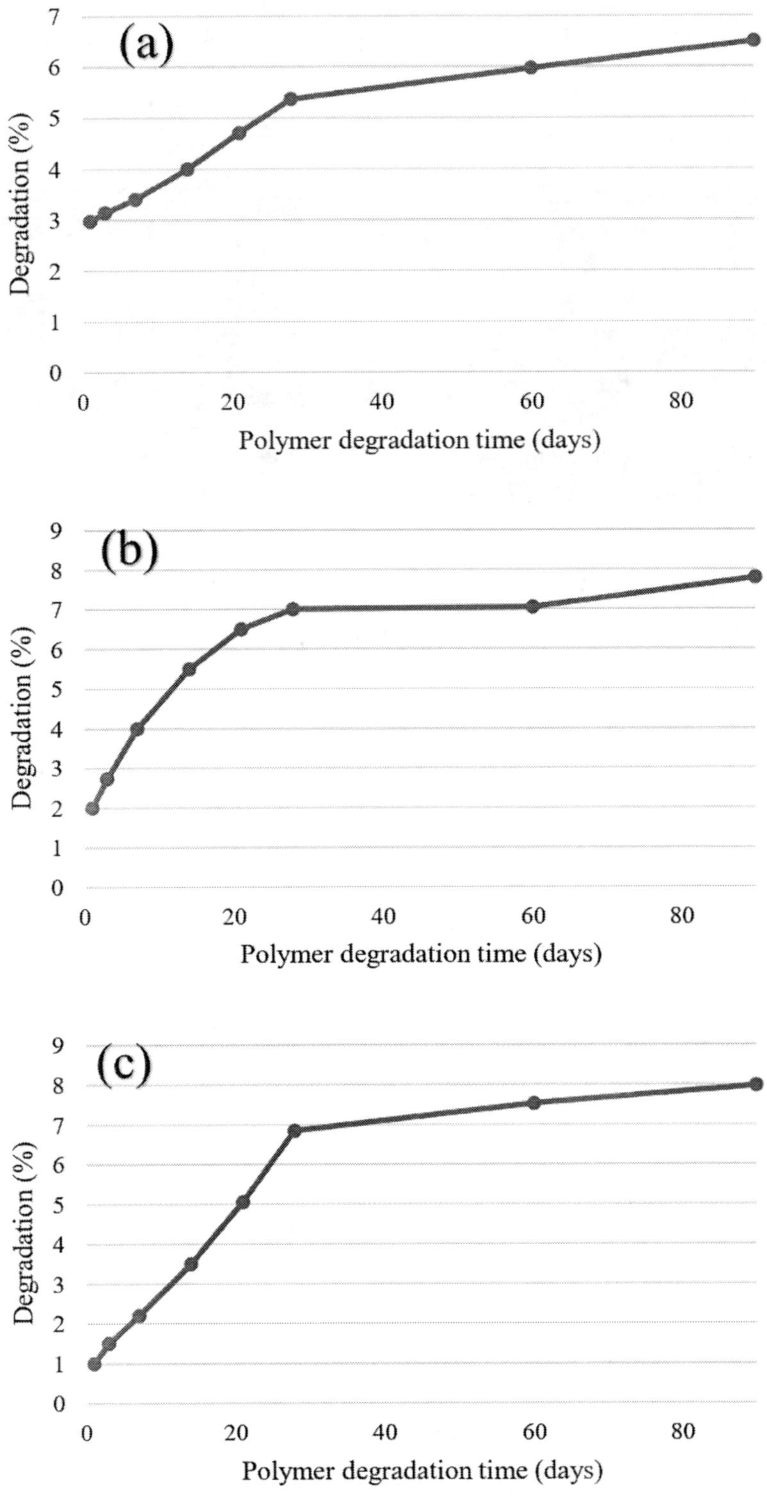

Figure 1. Dependence of the degree of degradation of films on the exposure time in phosphate buffer: a - PLA 45 kDa; b - PLA 90 kDa; c - PLA 180 kDa.

Polylactide films acquired white color and lost transparency on day 60, and remained almost intact for 180 days (Fig. 2). It should be noted that the smallest loss of transparency was observed for samples with the greatest thickness. It can be assumed that the process of swelling, preceding the degradation of the polymer, occurs more slowly with an increase in the volume in which it should occur.

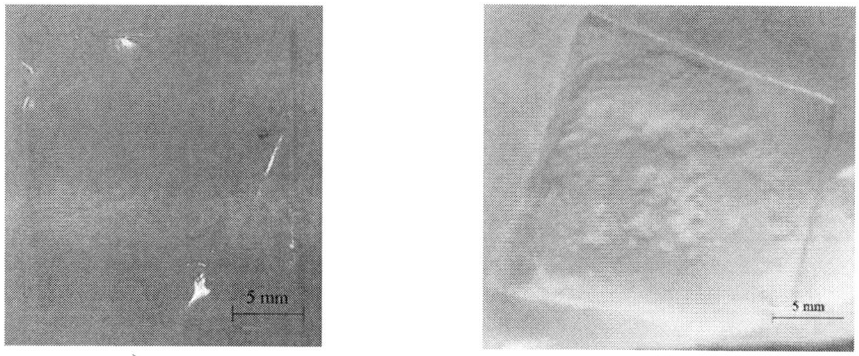

a) b)

Figure 2. The condition of PLA films with a molecular weight of 90 kDa after exposure to phosphate buffer for: a) 1 day; b) 60 days

4. Conclusions

A study of the degradation of polylactidine films of different molecular weights was carried out. An increase in the molecular weight of polylactide does not significantly affect the rate of degradation of the films.

Acknowledgements

The work was supported by the Ministry of Education and Science of Russia (grant identifier RFMEFI60417X0196).

References

[1] Lincoff AM, Schwartz RS, Giessen WJVD, Beusekom HMMV, Serruys PW, Holmes DR, et al. Biodegradable polymers can evoke a unique inflammatory response when implanted in the coronary artery. Circulation 1992;86(Suppl.4):I

[2] Giessen WV, Lincoff A, Schwartz R, Beusekom HV, Serruys P, Holmes D, et al. Marked inflammatory sequelae to implantation of biodegradable and nonbiodegradable polymers in porcine coronary arteries. Circulation 1996;94(7):1690–7.

[3] Sevost'yanov M.A., Nasakina E.O., Baikin A.S., Sergienko K.V., Konushkin S.V., Kaplan M.A., Seregin A.V., Leonov A.V., Kozlov V.A., Shkirin A.V., Bunkin N.F., Kolmakov A.G., Simakov S.V., Gudkov S.V. Biocompatibility of new materials based on nano-structured nitinol with titanium and tantalum composite surface layers: experimental analysis in vitro and in vivo // Journal of Materials Science: Materials in Medicine, 2018. Vol. 29. P.33 (https://doi.org/10.1007/s10856-018-6039-3)

[4] Nasakina E. O., Sevostyanov M. A., Mikhaylova A. B., Baikin A. S., Sergienko K. V., Leonov A. V., Kolmakov A. G. Formation of alpha and beta tantalum at the variation of magnetron sputtering conditions// IOP Conference Series-Materials Science and Engineering. 110 (2016) 012042 doi: 10.1088/1757-899X/110/1/012042.

[5] Shih-Hang Chang, Jia-Jun Huang Biodegradability and anticoagulant properties of chitosan and sulfonated chitosan films coated on TiNi alloys // Surface and Coatings Technology Volume 206, Issue 23, 15 July 2012, Pages 4959–4963.

[6] Basse Asplund, Jenny Sperens, Torbjörn Mathisen and Jöns Hilborn, Effects of hydrolysis on a new biodegradable co-polymer, Journal of Biomaterials Science, Polymer Edition, 10.1163/156856206777346331, 17, 6, (615-630), (2006).

[7] Tamai H, Gaki K, Kyo E, Kosuga K, Kawashima A, Matsui S,et al. Initial and 6-month results of biodegradable poly-L-lactic acid coronary stents in humans. Circulation 2000;102(4):399–404.

[8] Tan LP, Venkatraman SS, Sung PF, Wang XT. Effect of plasticization on heparin release from biodegradable matrices. Int J Pharm 2004;283:89–96.

[9] Sevost'yanov M.A., Fedotov A.Yu., Nasakina E.O., Teterina A.Yu., Baikin A.S., Sergienko K.V., Kolmakov A.G., Komlev V.S., Ivanov V.E., Karp O.E., Gudkov S.V., Barinov S.M. Kinetics of the release of antibiotics from chitosan-based biodegradable biopolymer membranes // Doklady Chemistry, 2015. V.465. Part.1. P.278-280.

[10] Saito N, Murakami N, Takahashi J, Horiuchi H, Ota H, Kato H, et al. Synthetic biodegradable polymers as drug delivery systems for bone morphogenetic proteins. Adv Drug Deliv Rev 2005;57(7): 1037–48.

Calcium phosphate ceramic surface coating via precipitation approach

N V Petrakova[1], E A Kuvshinova[2], A A Ashmarin[1], A A Konovalov[1], Y O Nikitina[1], A A Egorov[1], I K Sviridova[2], S M Barinov[1] and V S Komlev[1]

[1] Institution of Russian Academy of Sciences A.A. Baikov Institute of Metallurgy and Material Science RAS (A.A. Baikov' IMET RAS), Moscow, Leninskiy prospect 49, Russia
[2] P.A. Herzen Moscow Oncology Research Institute – branch of the National Medical Research Radiological Centre of the Ministry of Health of the Russian Federation, Moscow, the 2-nd Botkinskiy proezd 3, Russia

E-mail: petrakova.nv@mail.ru

Abstract. The biomimetic precipitation of calcium phosphate (CaP) coating was performed by soaking of β-tricalcium phosphate (TCP) ceramic scaffolds in supersaturated solution. Surface functionalization was obtained by incorporation of bovine serum albumin (BSA) as a model protein. The structure and phase composition were studied by application of XRD, FTIR, SEM, BET absorption. BSA incorporation content as a result of co-precipitation was evaluated.

1. Introduction

Calcium phosphates (CaPs) hold a significant place in number of biomaterials has been developed for bone tissue engineering. Being similar to mineral component of bone, the CaP based biomaterials are biocompatible, bioactive and can be processed to have osteoconductive properties [1-5]. Porous CaP scaffolds have been designed to permit vascularization and bone tissue ingrowth [2, 6, 7]. Increasing of surface area and porosity elevates cell adhesion and proliferation [8]. Functionalized scaffolds with biologically active molecules and drugs can be used for targeted drug delivery or osteoinductivity enhancement.

Dissolution and precipitation of bone mineral component occur during biomineralization process. It was already proposed that the amorphous calcium phosphate (ACP) forms under the biological conditions [9, 10]. It serves as a precursor to HAp and possess loss in long-range order and lattice imperfections [5]. It subsequently transforms to octacalcium phosphate (OCP), and eventually into HA with high degree of crystallinity [10]. OCP possess greater rate of resorbability and osteoinduction potencies [11, 12].

Based on biomineralization process the biomimetic approach was implicated to obtain the surface structure and composition similar to biological one [13]. Biomimetic precipitation of bioactive CaP coating from supersaturated ion solutions is controlled by interphase boundary processes [14]. The nucleation rate depends on degree of solution supersaturation, surface composition and condition. It was demonstrate, that ACP phase is the first-formed deposit in vitro at neutral pH and moderate supersaturation [15].

The biomimetic approach could be employed for bioactive molecules incorporation such as growth factor and drugs into the crystalline latticework [16, 17]. Salt and organic molecules ions involve in co-precipitation due to electrostatic interaction. It is proposed that drug incorporation into material structure allows its local release of therapeutic concentrations during a long time [18, 19].

This study was aimed to perform scaffolds functionalization by co-precipitation of CaPs and bioactive molecules from supersaturated solution. The tricalcium phosphate (TCP) scaffolds were used to perform biomimetic precipitation in double SBF solution. Scaffolds functionalization was appended by incorporation of bovine serum albumin (BSA) as a model protein. The composition and structure of resulted biomimetic coating at different soaking times were investigated. BSA incorporation effectiveness was evaluated on BSA content in materials.

2. Experimental

2.1. *Fabrication of TCP scaffolds*

TCP porous granules of average size of 500-1000 μm were used as ceramic scaffolds. Detail procedure and technological conditions for preparation of initial β-TCP powder and granules are shown in Figure 1.

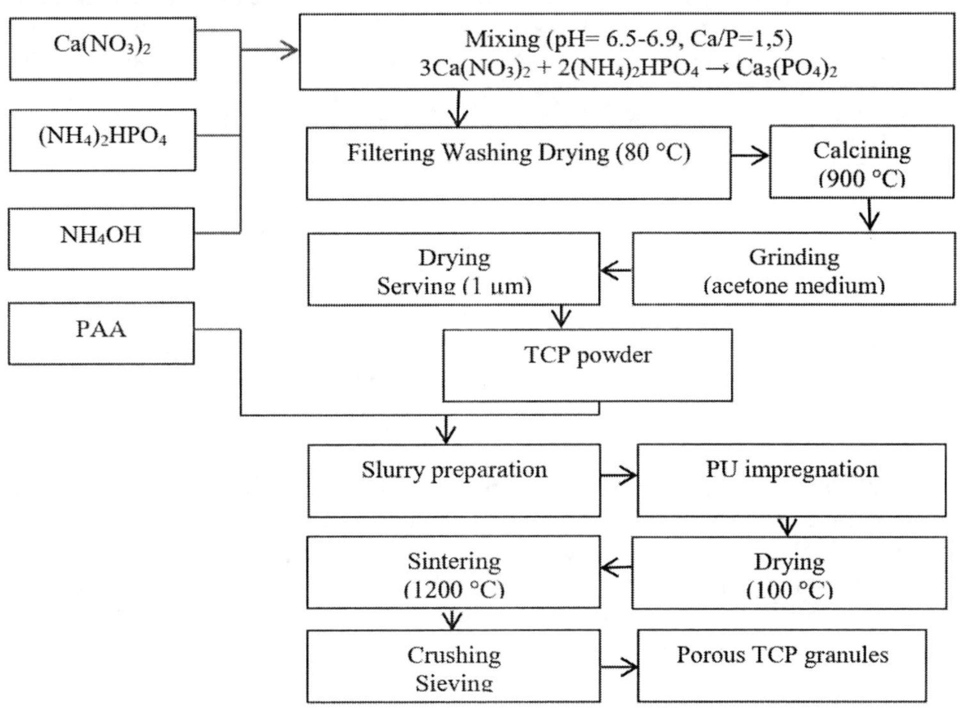

Figure 1. Flowchart of TCP scaffolds fabrication.

2.1.1. *TCP powder synthesis*

TCP powder were prepared by precipitation technique from aqueous solutions according to the reaction [20]:

$$3Ca(NO_3)_2 + 2(NH_4)_2HPO_4 \rightarrow Ca_3(PO_4)_2. \quad (1)$$

Calcium nitrate tetrahydrate ($Ca(NO_3)_2 \times 4H_2O$) and ammonium hydrophosphate (($(NH_4)_2HPO_4$ were used as a source reagents. 0.5 M solution of (($(NH_4)_2HPO_4$ were mixed slowly with a 0.5 M solution of $Ca(NO_3)_2$ in quantities according to the reaction (1) to achieve Ca/P ratio of 1.5. The pH level was kept at 6.5-6.9 by adding drop-wise of ammonia solution. After pH stabilization, the precipitate was filtered and dried at 80 °C during 24 h. Then the compact cake was crushed by using mortar and pestle, calcinated at 900 °C and ground in a planetary mill in acetone medium. The obtained white suspension was then filtered and dried at 60 °C for 24 h. The as-dried powder was crushed and sieved through a sieve with a cell size of 100 μm. Resulted powder samples were calcinated at 1000 °C and then were examined by XRD for the main component and presence of impurities.

2.1.2. Scaffolds preparation

The procedure of scaffolds fabrication based on a method of porous polymeric sponge replica was used [21, 22]. The obtained TCP powder was added to polyacrylamide 10 % wt. water solution to prepare the ceramic slurry. Polyurethane (PU) reticulated foams of average pore size about 0.1-0.5 mm were used as porous matrices. The PU matrices were impregnated with TCP slurry, dried at 100 °C and completely burnt out during sintering at 1200 °C. Then the ceramics was crushed into a pieces and fraction of 500-1000 μm was isolated by sieving on a set of sieves to apply in subsequent experiments.

2.2. Formation of CaP coating

The CaP coating was produced by biomimetic precipitation of calcium phosphates from a double concentrated simulated body fluid (2xSBF, Table 1). The 2xSBF were prepared by successive mixing of preliminary prepared solutions of NaCl, $NaHCO_3$, KCl, $K_2HPO_4 \times 3H_2O$, $MgCl_2 \times 6H_2O$, $CaCl_2 \times 2H_2O$, Na_2SO_4 salts in distilled water. pH value of the mixed solutions was adjusted to 7.2–7.4 using TRIS buffer and HCl. Then the 50.0 mg of the materials were kept in 1.0 ml solution and incubated for 24, 72, 168 hours (1, 3, 7 days) at 37 °C and constant stirring.

Table 1. Inorganic composition of 2xSBF used for experiments (mMol/l)

Ion	Na^+	K^+	Mg^{2+}	Ca^{2+}	Cl^-	HPO_4^{2-}	SO_4^{2-}	HCO_3^-
Content	284,0	10,0	3,0	5,0	295,6	2,0	1,0	8,4

2.3. Characterization of TCP scaffolds

Phase composition of the materials was investigated by X-ray diffraction analysis (XRD, Shimadzu) using CuKα radiation (average wavelength l =1.54183 Å) on powdered samples within the 2θ range of 10–60 deg., step 2θ of 0.02 deg. and counting time 2 s/step. WinFit computer program [23] was used to evaluate the lattice parameters and average crystal size using Sherrer's equation. The surface of calcium phosphate scaffolds was investigated by scanning electron microscopy (SEM, Tescan Vega II SBU) with integrated Energy dispersive X-ray microanalyzer (EDX, INCA Energy). The IR-spectra of powdered samples were collected in the 400 – 4 000 cm^{-1} spectral range with 4 cm^{-1} spectral resolution on standard KBr pallets (FTIR, Avatar 330 Thermo Nicolet). The specific surface area (S_{BET}) of ceramic scaffolds was measured by BET method (TriStar 3000, Micromeritics).

2.4 BSA incorporation procedure

The incorporation of bovine serum albumin (BSA, PAA) on the obtained TCP ceramics surface was performed by biomimetic precipitation from a supersaturated calcium phosphate solution (SCS) (table. 2). For this procedure 50.0 mg of the ceramic material was placed into a 1.0 ml of SCS and BSA solution at the concentration of BSA of 3.0 mg/ml (incorporation solution). The samples in closed plastic pellets were incubated at 37 ° C during continual stirring for 48 h. The effectiveness of BSA incorporation was evaluated by measuring the protein concentration in incorporation solutions before and after materials soaking using BCA Protein Assay Kit (ThermoFisher Scientific, USA).

IOP Conf. Series: Materials Science and Engineering **525** (2019) 012101 doi:10.1088/1757-899X/525/1/012101

Table 2. Inorganic composition of SCS solution (mMol/l)

Ions	Na^{+}	K^{+}	Ca^{2+}	Cl^{-}	HPO_4^{2-}
Content	136,8	3,71	3,1	144,5	1,86

3. Results and discussion

The structure of initial scaffolds is shown in Figure 2. Irregular in shape granules are of size from 500 to 1000 μm with open porosity about 40-50 %. Ceramics is composed of almost rounded shape crystals of 1-2 μm, assembled in a porous net of the solid phase. Porosity is presented at several levels: large extended pores are of about 150-250 μm in cross section, medium pores are up to 10-20 μm, and small pores are up to 0.1-0.5 μm.

Figure 3 demonstrates the surface structure of TCP scaffolds before and after soaking in 2xSBF during 7 days. On the surface of ceramics crystals there is an irregular coating of newly formed phase consisted of amorphous particles of average size from 20 to 100 nm. According to the EDX data, the new-formed layer consisted of calcium phosphate with Ca/P ratio of 1.5-1.7 enriched with sodium and potassium ions. It can be probably supposed the presence of precipitated calcium phosphate, probably ACP.

Figure 2. SEM-photographs of the initial TCP scaffolds (a) x50, (b) x10000.

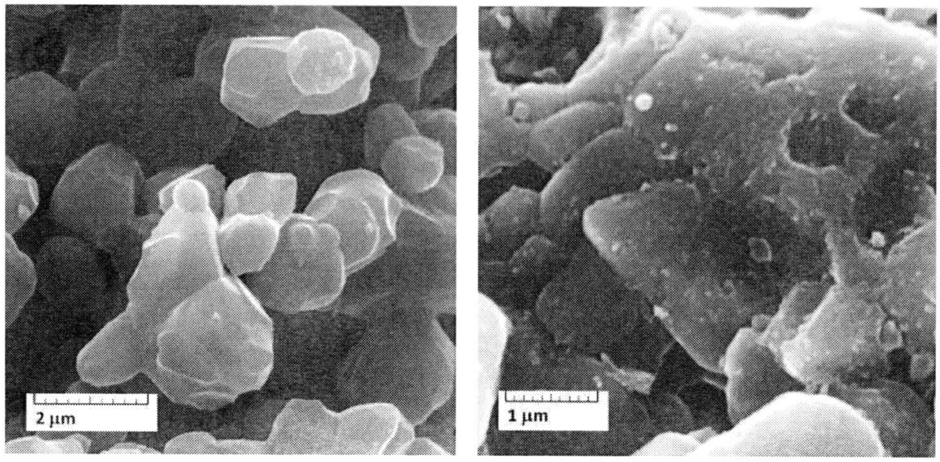

Figure 3. SEM-photographs of TCP surface: (a) initial TCP and (b) TCP soaked in 2xSBF during 7 days.

According to BET measurements, the specific surface area of scaffolds raised with soaking time increasing. Thus, S_{BET} of initial TCP was 0.5 m^2/g. S_{BET} of TCP soaked for 1, 3 and 7 were determined as 0.8, 1.1, and 1.2 m^2/g, respectively. In turn, this confirms the consistent formation of a fine-disperse crystalline layer on the ceramic surface during biomimetic precipitation and thereby enhancement of surface adsorption.

IR spectra of the materials showed a set of absorption bands typical to hexagonal (β) TCP (Figure 4, a). IR lines of all samples displayed the intensive doublet at 550-610 cm^{-1} corresponded to the vibrations of PO_4^{3-} -groups, which are typical for the β-TCP structure. The phosphate group vibrations also showed up in bands at 948, 980, 1048 and 1120 cm^{-1} [24-26]. The bands at 727 and 1214 cm^{-1} corresponded to vibrations of $P_2O_7^{4-}$-groups and indicated the presence of calcium pyrophosphate $Ca_2P_2O_7$ in small amount. However, the increasing of soaking time in 2xSBF up to 3-d day leaded to decreasing of intensity of the PO_4^{3-}group lines, as well as decreasing of distances ratio A_1/A_2 (Figure 4, b). This indicated a reduction of total content of the main TCP phase probably due to precipitation of amorphous calcium phosphate (ACP) on ceramics surface from salt solution. It contributed in material degree of crystallinity reduction [27]. Only the phosphate peak intensity level for the material soaked for 7 days was between the same for 3-d and 1-st day respectively, that probably showed the rising of crystal size up to 7-d day.

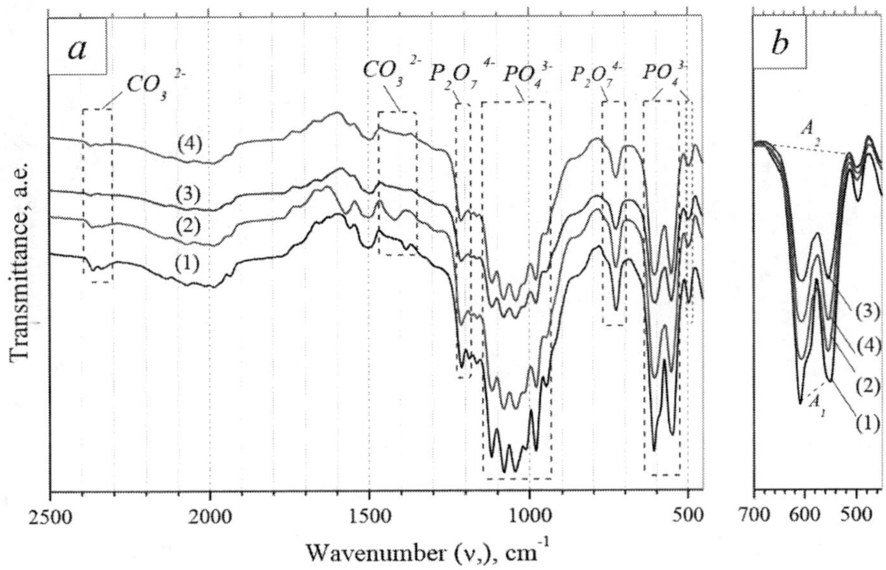

Figure 4. FTIR spectra of the initial TCP granules (1) and soaked for 24 h (2), 72 h (3); (a) – an overall view, (b) – detailing of the phosphate band (ν_4).

XRD analysis data of the initial scaffolds has revealed a single crystalline component β-TCP (Figure 5, a). All soaked in 2xSBF materials displayed an identical diffraction pattern. The coating formed during precipitation procedure was constituted of an amorphous phase that didn't produce diffraction of X-rays. The structure comparison was implemented according the X-ray data by calculating of the lattice constants and average crystallite size (D), determined by Scherrer's equation (Figure 5, b-d). With soaking time increasing the *a*-axis parameter held on value about 10.475±0.002 Å, within the error. The *c*-axis parameter decreased from 10,355 to 10,348 Å within the first day of precipitation and rose strongly to the 7-th day up to 10,366 Å. Difference between lattice constants could be attributed to the prevalent growth direction of the deposited particles that have been along the *c*-axis. The average *D* size has monotonically reduced from 970 to 870 nm by the 3-d day, and then came to the initial level up to the 7-th day of precipitation. It also confirmed by the data of specific surface area of obtained materials, there was more intensive S_{BET} growth up to 3 days, by the 7th day the growth faded.

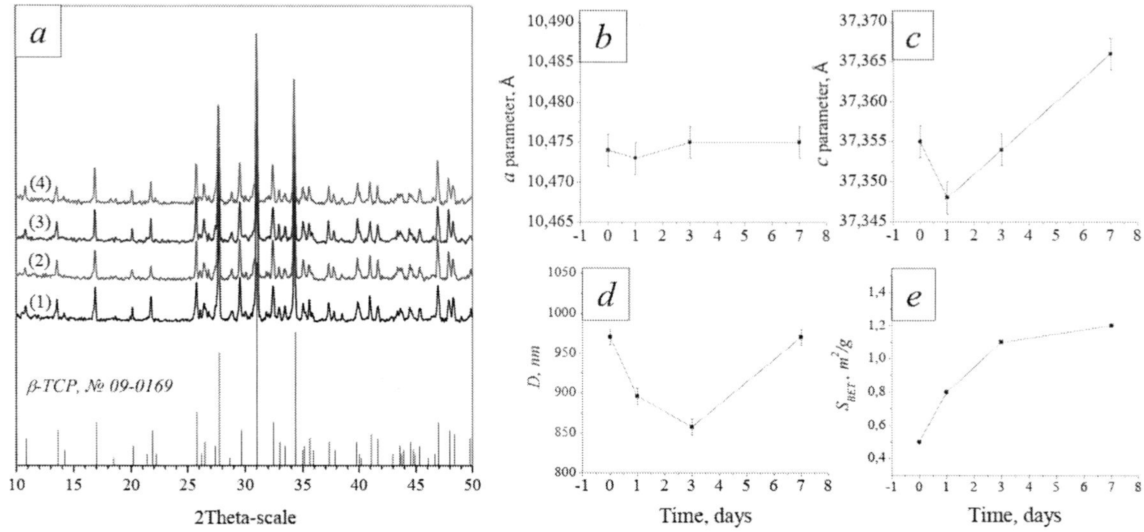

Figure 5. XRD and surface area data of TCP scaffolds at different soaking time.
(a) X-ray diffractogramms of materials: (1) – initial, (2) – 1 day, (3) – 3 days, (4) – 7 days;
(b), (c), (d), (e) – respectively, *a*, *c* parameters, average crystallinity size (*D*) and the surface area data as the dependence of experiment time.

The BSA co-precipitation was performed on initial TCP scaffolds and TCP scaffolds with CaP coating produced by soaking in 2xSBF during 1day. The BSA incorporated content in ceramics was detected. It was found that during biomimetic co-precipitation a larger amount of BSA was incorporated on the surface of TCP scaffolds with CaP coating, 3.5 μg/mg, whereas for the initial it was only 1.1 μg/mg. The difference is over 3 times, that also confirms effectiveness of CaP coating formation for BSA incorporation. We propose that amorphous CaP coating enhance the interaction between ceramics surface and incorporation solution during co-precipitation.

4. Conclusion

In the study the amorphous CaP coating on the surface of TCP ceramic scaffolds was obtained by biomimetic precipitation from 2xSBF. The resulted materials were investigated by SEM, EDX, XRD and BET adsorption. The amorphous CaP coating on the ceramic surface forms from the 1-st day of precipitation. It was revealed that crystallinity level of CaP coating was increased up to the 3-d day and became decrease up to the 7-th. Particle growth took place along the *c*-axis with constancy of *a* parameter during the precipitation process. The CaP coating formation on the ceramic scaffolds surface leaded to 3 times increasing of BSA incorporation effectiveness as a result of biomimetic co-precipitation. It could be conclude that CaP coating via biomimetic precipitation is beneficial approach to obtain the scaffolds functionalization.

Acknowledgements

The reported study was funded by Russian Foundation for Basic Research (RFBR) according to the research project № 18-29-11052.

References

[1] Barinov S M and Komlev V S 2014 *Bioceramics based on calcium phosphates* (Moscow: Science) 201 p.
[2] Dorozhkin S V 2013 *J. Materials* **6** 3840

[3] Navarro M, Michiardi A, Castano O and Pianeil J A 2008 *J. R. Soc. Interface* **5** 1137
[4] Safronova T V and Putlyaev V I 2013 *J. Fiz., Khim., Mat.* **4** 24
[5] Eliaz N and Metoki N 2017 *J. Materials* **10** 334
[6] De Groot K 1993 *J. A review. Ceram. Int.* **19** 363
[7] Barinov S M and Komlev V S 2016 *J. Inorganic Materials* **52** 339
[8] Hing K A, Best S M, Tanner K A, Bonfield W and Revell P A 1999 *J. Mater. Sci.: Mater. Med.* **10** 633
[9] Posner A S, Blumenthal N C, Boskey A L, Betts F 1975 *J. Dent. Res.* **54** 88
[10] Brecevic L J and Furedi-Milhofer H 1972 *J. Calcif. Tissue Res.* **10** 82
[11] Suzuki O, Kamakura S, Katagiri T and et al. 2006 *J. Biomaterials* **17** 2671
[12] Komlev V S, Barinov S M, Bozo I I, Deev R V, Eremin I I, Sergeeva N S and et al. 2014 *J. ACS Appl. Mater. Interfaces* **6** 16610
[13] Ducheyne P and Qiu Q 1999 *J. Biomaterials* **20** 2287
[14] Vasserman I M 1980 *Chemical precipitatiom from solutions* (Leningrad: Chemistry) 208 p.
[15] Rabadjieva D, Gergulova R, Titorenkova R, Tepavitcharova S, Dyulgerova E, Balarew Chr and Petrov O 2009 *J. Mater Sci: Mater Med.*
[16] Lin X, de Groot K, Wang D, Hu Q, Wismeijer D and Liu Y 2015 *J. The Open Biomedical Engineering Journal* **9** 56
[17] Suwanprateeb J and Thammarakcharoen F. 2014 *J. International Journal of Biomedical and Biological Engineering* **8**
[18] Stigter M, Bezemer J, de Groot K and Layrolle P 2004 *J. Journal of Controlled Release* **99** 127
[19] Liu Y, de Groot K and Hunziker E B 2005 *J. Bone* **36** 745
[20] Mirhadi B, Mehdikhani B and Askari N 2011 *J. Processing and Application of Ceramics* **5** 193
[21] Bakunova N V, Komlev V S, Fedotov A Y, Fadeeva I V, Smirnov V V, Shvorneva L I, Gurin A N and Barinov S M 2008 *J. Powder Metallurgy Progress* **8** 336
[22] Kim M, Jang D W, Min Y K, Yang H M, Song H Y, Lee B T *J. Materials Science Forum* **654-656** 2245
[23] Krumm S 1996 *J. Mater Sci Forum* **228**
[24] Berzina-Cimdina L and Borodajenko N 2012 *J. Infrared Spectroscopy - Materials Science, Engineering and Technology* 524
[25] Lin F H, Liao C J, Chen K S and Sun J S 1998 *J. Biomaterials* **19** 1101
[26] Tarasovich B N 2012 *Resource materials* (Lomonosov Moscow state University Moscow) 55
[27] Termine J D and Posner A S 1966 *J. Nature* **211** 268

IOP Conf. Series: Materials Science and Engineering **525** (2019) 012102 doi:10.1088/1757-899X/525/1/012102

Effectiveness of the flame retardant antiseptic properties from plant waste against fungi on alnus glutinosa

A P Glinushkin[1], E M Motasova[1], T P Aysuvakova[1], A V Ovsiankina[1], N S Zhemchuzhina[1], A M Mashenkov[2], A P Nesvat[2], O O Beloshapkina[3] and N N Dubenok[3]

[1]All-Russian Research Institute of Phytopathology, Bolshiye Vyazyomy, Odintsovsky district, Moscow Region, 143050.
[2] Federal state educational institution of higher education Orenburg State Agrarian University, 460014, 18, Chelyuskintsiv street, Orenburg, 460014.
[3] Russian State Agrarian University-Moscow Agricultural Academy named after K.A. Timiryazev, 49, Timiryazev street, Moscow, 127550.

E-mail: glinale@mail.ru

Abstract. The purpose of the research was to study the antiseptic properties of flame retardants. Our studies have shown that Alnus glutinosa wood is highly susceptible to staining. It should be noted that it is possible to distinguish species diversity in the work of antipyrene against the development of fungi. However, this requires deeper scientific research with the change of numerous factors, to reflect the natural conditions of the passage of industrial wood. Studies have shown that the biological effectiveness of the obtained antipyrene solutions reaches up to 88.3% under laboratory conditions. The obtained data allow to judge about the prospects of the direction of innovative use of waste and wood protection.

Wood substance is a construction material for the building of residential, public, industrial, agricultural, livestock, storage and other edifices. This practice is developing quite fast-moving in a number of European countries, Australia, the USA and Canada. The average annual growth rate of demand for wood materials is expected to be in the range of 20–25% by 2020 [1].

Currently, for fire protection of natural wood, such methods as surface and deep impregnation with special compounds containing flame retardants, as well as the application of fire retardant coatings are popular [2 - 13].

In 2000-2015, the development of flame retardants based on biopolymers became widespread [14-17]. Thus, biological substances (fumaric acid [18], phytic acids [19, 20], chitosan [19, 21], sorbitol xylitol [22], cyclodextrin [22, 23], starch [22, 24-28]. Thus, biological substances (fumaric acid [18], phytic acids [19, 20], chitosan [19, 21], sorbitol xylitol [22], cyclodextrin [22, 23], starch [22, 24-28]) are used for the development of intumescent systems.

A new direction in the field of safe export and use of wood substance is the creation and study of fire retardants with antiseptic properties. Environmentally-safe property gives such fire retardants additional benefits. Created from wastes that do not represent food or feed value, fire retardants allow not only to solve the problem of creating new materials with valuable properties, but at the same time to solve the environmental problem of disposing of ballast high-tonnage products [29].

IOP Conf. Series: Materials Science and Engineering **525** (2019) 012102 doi:10.1088/1757-899X/525/1/012102

As such production wastes it is possible to use lignin [30] and paper production wastes and wood hydrolysis. A large number of flame retardant compositions and technologies in which lignin is used as an auxiliary component, for example, as a binder [31 - 33], surfactant [34], lightweight filler [35], dispersing agent and emulsifier [36] and others.

It should be noted that the use lignosulfonate and other unmodified lignin containing waste, does not allow for creation of effective flame retardants, and sometimes leads to the opposite phenomenon, contributing to the development of the flame. To obtain effective lignin-based flame retardants, it is advisable to use lignin-containing wastes subjected to chemical modification [29].

Brominated sulfolignin in the form of an aqueous or alkaline solution can be used for surface treatment or impregnation of combustible materials. Transformation at elevated temperatures forms dioxins, which prevents the use of this product due to environmental insecurity [37].

The basis of the biomass flame retardant properties of antipyrene is the chemical structure of oxidized lignin-containing wastes consisting of phenol-containing structures. Phenol derivatives, being one of the most powerful inhibitors of oxidation processes, interact with free radicals of the branched chain reaction of combustion and block the development of the chain process [40]. The resulting flame retardant can be used in the superficial method of processing by spraying and with the introduction of the dry mixture as a composite additive.

In the case of using woodworking waste as a lignin-containing substrate, for example, sawdust, a complex action flame retardant is obtained, combining the properties of a radical chain inhibitor with the properties of an intumescent (carbon forming) flame retardant, which, according to the invention [41], is formed during oxidation in similar conditions of carbohydrate-containing products, including cellulose. The mechanism of its flame retardant action is associated with the formation on the treated surface during high-temperature heating or with the direct action of a foamed carbonite layer flame. The coked cellular material layer exhibits a heat-shielding and barrier effect during mass transfer of both combustible materials to the flame reaction zone and oxygen of the air to the material surface.

Oxidized starch reagents, as well as oxidized lignin, can be used as effective flame retardant impregnations, flame retardants in the manufacture of various polymeric materials based on water-soluble polymers and latexes, as well as in extinguishing fires. The oxygen index values for various types of paper, sawdust and synthetic materials showed that the oxygen index in all cases exceeds 25% and sometimes 40% [42].

We have not found a scientific description of research aimed at studying the antiseptic properties of flame retardants in the literature.

The purpose of this study was to study the antiseptic properties.

The original flame retardant solution for research was obtained by our team according to the procedure described in [38, 39]. Conducted studies on the flammability of wood samples (150×60×30 mm) showed that the samples treated with a flame retardant solution lose a mass of 5-8% when exposed to an open flame for 2 minutes. It was established that the mass loss depended on the flow rate and the type of impregnation.

To research the antiseptic properties, studies have been carried out on the effectiveness of the flame retardant solution with respect to wood-staining and mold fungi. Wood samples (European black alder (Alnus glutinosa (L.) Gaertn. [X23]) were impregnated with a fire retardant solution. The average flame retardant consumption was 261.9 g/m2.

The main research methodologies consisted of a fifteen-day exposure of wood samples (10×55×75 mm) in wet chambers (a desiccator model with a diameter of 115 - + 5 mm; a depth of 54 - + 3 mm), operating according to the «closed space - moisture reserve» principle with large evaporation surface. Such conditions were created for better growth and development of mold and wood-coloring fungi. After 15 days, an analysis was carried out (definitions: microbial development, area occupied on the sample surface, depth of penetration into the alder; stages of microbial development), compilation of measurement tables and analytics of the materials obtained.

Tests were carried out on fungi, fire retardants were used in three concentrations (groups), all study designs for points, types of fungi and control options are presented in Tables 1 and 2. Part of the work, including reseeding, cultivation, storage of strains of microorganisms, was performed using equipment of State collection of phytopathogenic microorganisms and identifying varieties (differentiators) of pathogenic microorganism strains based in All-Russian Research Institute of Phytopathology http://www.vniif.ru/vniif/page/ckp-gkmf/1373).

For each variant of the study, a separate desiccator was prepared, the placement of which was carried out in the conditions of a common use center at the Artificial-climate laboratory at the All-Russian Research Institute of Phytopathology (http://vniif.ru/vniif/center-cku/). Sterile moistened (80±5%) sawdust from healthy sapwood of alder at 1/4 of the height of the vessel was filled into desiccators. A suspension of fungi was introduced into each desiccator (which was increased in the conditions of a common use center State collection of phytopathogenic microorganisms and identifying varieties (differentiators) of pathogenic microorganism strains based in All-Russian Research Institute of Phytopathology http://vniif.ru/vniif/page/). Sawdust was irrigated with a working suspension of fungi with a volume of 100 ml using a spray bottle. The desiccators were closed with a ground-in lid and placed in a room at a temperature of 21-24°C in the dark. The growth of mycelium was performed for 15 days.

Assessment of the stage of development of fungi on the samples was carried out on a 6-point scale, taking into account the characteristics (consolidated requirements of all-Union State Standard, the generally accepted international system of prevalence and development of plant diseases, trees in points), are given in Table 1.

Table 1. Scale for assessing the stage of fungi development per samples

Rating	Stage of fungi development
0	absolutely clean samples by visual inspection and under a microscope
1	visually clean samples; under a microscope small foci are visible in the form of stains of wood-staining or mold fungi (with a total coverage of> 0 to 25%); the stage of sporulation is absent;
2	superficial development of mycelium of wood-coloring or mold fungi in the form of spots (with a coverage of> 25 to 50%); the stage of sporulation is absent;
3	plentiful growth of mycelium of wood-dyeing or mold fungi (coverage area> 50-75%); the beginning of sporulation stage;
4	clearly visible growth of fungi by visual inspection (coverage area 75-90%); various stages of wood-coloring or mold fungi sporulation;
5	deep damage by wood-painting or mold fungi of the entire sample area (coating> 90-100%); intensive sporulation.

The effectiveness of the concentrations (3; 6 and 9%) of the investigated fire retardant solution was selected in agreement with physicochemical, chemical and biological characteristics according to the methods and conditions of preparation [38, 39]. In particular, to create the studied low-cost environmentally safe effective biomass-based flame retardant (lignin-containing waste from wood processing), it is necessary to use a mixture of oxidation products of the original substance. In the process of liquid-phase oxidation of a lignin-containing substrate in an alkaline medium at a temperature not exceeding 90°C, a mixture of oxidation products is formed. They contain a solid residue insoluble in water and organic solvents of mineral products of destruction of lignin-containing substrates and the liquid fraction used as a flame retardant. The liquid fraction includes: a) salts of polyhydroxyphenolic compounds, which are products of incomplete oxidative depolymerization of lignin, as well as secondary condensation products of low-molecular compounds formed during the oxidation of lignin-containing substrates; b) low molecular weight hydroxyphenolic compounds formed during the oxidation of lignin-containing substrates; c) hydrocarbons. The oxidation can be carried out in the presence of catalysts, or without them under the action of air, oxygen, ozone or hydrogen peroxide. Physicochemical properties are described in more detail [38, 39].

In the Russian forests naturally grows from various sources of 15 of the 50 known species of the genus alder (Alnus, Betulaceae). The most common – European black alder (A. glutinosa (L.) Gaertn.), grey alder (A. incana (L.) Moench), Siberian alder (A. sibirica (Fisch)). The total area of alder forests in the European part of Russia is about 1.6 million hectares, with a timber reserve of more than 170 million m3, including black alder forests - 1.0 million hectares with a reserve of 110 million m3, the rest of the forests are mostly grey alder forest [43]. The dynamic issues of afforestation and the use of industrial wood are diverse in terms of phytosanitary condition such as production, technological preparation (harvesting), «wood ripening», bringing it to the required condition by technological methods [44, 49]. Nevertheless, new pathogens appear in the decade, while the relevance internationally in the commercial quality of wood is also confirmed [50]. In our opinion there is a perspective of preservation of the received wood from defeat by wood-painting or wood-destroying fungi.

However, the choice of remedies in the «List of permitted pesticides and agrochemicals in the territory of the Russian Federation» is very poor. The topical issues are the search for suitable methods, ways to protect the industrial wood, accumulated «wealth» from the loss of quality of industrial wood Alnus glutinosa.

Results.

Studies have established that Alnus glutinosa wood is highly susceptible to staining, and not all genera have successfully manifested themselves.

Table 2. The growth and development of fungi, depending on the concentration of flame retardant (2018)

№	Research Option	Prevalence, %	Development, %
1	Control (wood samples with water, micromycetes)	83,3	70,8
	Concentration 3%		
2	Control (wood samples)	0	0
3	Control (wood samples with micromycetes without water)	83,3	20,8
4	Control (wood samples with water without micromycetes)	33,3	8,3
5	Aspergillus hennebertii Blochwitz.	100	66,7
6	Aspergillus niger Tiegh.	100	50
7	Aspergillus amstelodami Thorn & Church.[=Eurotium amstelodami L. Mangin]	100	50
8	Aspergillus terreus Thorn.	100	45,8
9	Penicillium brevicompactum Dierckx.	100	79,2
10	Penicillium chrysogenum Thorn.	100	79,2
11	Penicillium ochrochloron Biourge.	100	79,2
12	Penicillium purpurogenum Stoll.	100	70,8
	Concentration 6%		
13	Control (wood samples)	0	0
14	Control (wood samples with micromycetes without water)	50	12,5
15	Control (wood samples with water without micromycetes)	33,3	8,3
16	Aspergillus hennebertii Blochwitz.	100	37,5
17	Aspergillus niger Tiegh.	100	45,8
18	Aspergillus amstelodami Thorn & Church.[=Eurotium amstelodami L. Mangin]	100	41,7
19	Aspergillus terreus Thorn.	100	37,5
20	Penicillium brevicompactum Dierckx.	100	29,2
21	Penicillium chrysogenum Thorn.	100	54,2
22	Penicillium ochrochloron Biourge.	100	58,3
23	Penicillium purpurogenum Stoll.	100	45,8

	Concentration 9 %		
24	Control (wood samples)	0	0
25	Control (wood samples with micromycetes without water)	33,3	8,3
26	Control (wood samples with water without micromycetes)	33,3	8,3
27	Aspergillus hennebertii Blochwitz.	100	33,3
28	Aspergillus niger Tiegh.	100	29,2
29	Aspergillus amstelodami Thorn & Church.[=Eurotium amstelodami L. Mangin]	100	33,3
30	Aspergillus terreus Thorn.	100	25
31	Penicillium brevicompacturn Dierckx.	100	41,7
32	Penicillium chrysogenum Thorn.	100	37,5
33	Penicillium ochrochloron Biourge.	100	49,7
34	Penicillium purpurogenum Stoll.	100	33,3

At an antiseptic concentration of 3%, the average affected area of Alnus glutinosa is different for all the tested variants. At the same time, Alnus glutinosa is affected differently in the control variants, creating an essential composition of materials for probit analysis. The prevalence of fungal infection in all application options for the selected fungi on Alnus glutinosa was around 100%. 15 days of growth and development of fungi gave the results. The qualitative indicator of the development of fungi differed significantly. According to studies, our observations found obtaining 4 points for fungal infection. Concentrations significantly altered scoring by fungi.

Efficiency was not considered a control option (wood samples) as the development of fungi did not occur there.

Biological effectiveness from the two controls is given sequentially through a semicolon for each concentration.

Comparison of the development of fungi with application (of all of the studied representatives in the form of a total solution) showed a greater growth of the development of fungi with a decrease in fire retardant concentration. In particular, at 3% development reached 20.8% (biological effectiveness 70.6%); at 6% 12.5% (biological efficiency 82.3%), while at 9% it was 8.3% (biological efficiency 88.3%) is actually a state of biological purity. Since this value was within the accuracy of the experiment in all repetitions, where the fungi were not specifically used. However, conditions were created for development of wood-staining and mold fungi by means of humidity. In our studies, we have achieved the desired concentration in the resulting fire-retardant solution as a working agent against mycosis of wood.

Work concentrations are also approximately characterized by the individual representatives of the two genera of fungi common on wood and reduce its quality.

It should be noted that it is possible to distinguish species diversity in the work of fire retardant against the development of fungi. However, this requires deeper scientific research with the change of numerous factors, to reflect the natural conditions of the passage of industrial wood. In general, the use of the obtained fire-retardant solutions in the laboratory has a biological efficacy of up to 88.3%. This seems to be a promising area for innovative waste management and wood protection.

References

[1] O.V. Artsybasheva, R.M. Aseeva, B. B. Serkov, A.B. Sivenkov, Current trends in the field of fire resistance of wooden buildings and structures. *News SFU. Technical Sciences*, p. 178-196, 2014.

[2] Garcia, D .; Tortosa, E .; Mans, V .; Futterer, T .; Nägerl, H.-D. 2002: A new approach to intumescent clear varnish in wood fire protection. *Third international Woodcoating congress.* **Paper 24**. The Paint Research Association, Teddington, UK.

IOP Conf. Series: Materials Science and Engineering **525** (2019) 012102 doi:10.1088/1757-899X/525/1/012102

[3] Koverninsky I.N. The basic technology of chemical wood processing. *M.: Forest industry*, 1984. 154 p.
[4] Romanenkov I.G., Levites I.G. Fire protection of building structures *M.: Stroyizdat*, 1991 — 320 p.
[5] Strakhov VL, Krutov A.M., Davydkin N.F. Fire protection of building structures. *M .: Information and Publishing Center «TIMR»*, 2000. - 433 p.
[6] Korolchenko A.Ya., Korolchenko O.N. (2006) Fire Protection Products. Reference Guide. *Pozhnauka, Moscow*, 258 p.
[7] Reducing the fire hazard of wood materials, products and structures / A.A. Leonovich, A.V. Sheloumov. - *SPb .: SPbSPU*, 2002. - 59 p.
[8] Ostman BA-L (2010) Fire safety in timber buildings: technical guideline for Europe. *SP Technical Research Institute of Sweden*, Boras. www.jrc.ec.europa.eu
[9] Leonovich A.A. Physico-chemical basis of the formation of wood boards. - *SPb .: Himizdat*, 2003, 192 p.
[10] Bazhenov S.V. et al. (1999) Methods and means of fire protection of wood: *Moscow, VNIIPO*, 50 p.
[11] Deeva O.V. Formulations of flame retardant compositions: Factual information on domestic and foreign patent information for 1992-1998. *Cherkasy: NIITEKHIM*, 1999. - 14 p.
[12] Tychino H.A. Modern fire retardants for wood: research results // *Fire and Explosion Safety, March,* 1999 13-20 p.
[13] Tychino, N.A. Means of fire and bioprotection of construction wood: quality objectives // *Fire and explosion safety*. 2003. - **12**, №6. - pp. 23-25.
[14] Aseeva R., Serkov B., Sivenkov A. Fire behavior. Dordrecht: Springer, 2014, p. 291.
[15] New production processes and applications Nighlighted. Miller N.G. "Plast Sonth. A £ r, 1982, No. **2**, 22, 25-29 p.
[16] Cagliostro D.E. Riccitielio S.K., Clark K.J. Jnlumescent coating modeling. J. Fire and Flammable, 1975, **6, No. 2**, 205-291 p.
[17] http://bent.ru/modules/Articles/article.php?storyid=202&storypage=1
[18] Bourbigot S., Fontaine G., Duquesne S. Intumescence for the flame retardancy of PLA. In Proceedings of the 21st BCC Conference-Recent Advances in Flame Retardation of Polymeric Materials, Stamford, CT, USA, 21–23 May 2012.
[19] Zhang, T.; Yan, H.; Shen, L.; Fang, Z.; Zhang, X.; Wang, J .; Zhang, B. Chitosan / phytic acid polyelectrolyte complex: A green and renewable intumescent flame retardant system for ethylene – vinyl acetate copolymer. *Ind. Eng. Chem. Res*. 2014, **53**, 19199–19207.
[20] Cheng, X.-W.; Guan, J.-P.; Tang, R.-C.; Liu, K.-Q. Phytic acid as a bio-based phosphorus flame retardant for poly (lactic acid) nonwoven fabric. *J. Clean. Prod*. 2016, **124**, 114–119 p.
[21] Laufer, G.; Kirkland, C.; Cain, A.A.; Grunlan, J.C. Clay-chitosan nanobrick walls: Completely renewable gas barrier and flame-retardant nanocoatings. *ACS Appl. Mater. Interfaces* 2012, **4**, 1643–1649 p.
[22] Giraud, S.; Rault, F.; Rochery, M.; Gacquere, L. Use of a flame retardant polyurethane coating for polyester textile fabric. In Proceedings of the 14th European Meeting on Fire Retardant Polymers, FRPM13, Lille, France, 30 June – 4 July 2013 p.
[23] Feng, J.-X.; Su, S.-P.; Zhu, J. An intumescent flame retardant system using β-cyclodextrin as a carbon source in polylactic acid (PLA). *Polym. Adv. Technol*. 2011, **22**, 1115–1122 p.
[24] Reti, C.; Casetta, M.; Duquesne, S.; Bourbigot, S.; Delobel, R. Flammability properties of intumescent PLA including starch and lignin. *Polym. Adv. Technol.* 2008, **19**, 628–635 p.
[25] Reti, C.; Casetta, M .; Duquesne, S.; Delobel, R.; Soulestin, J.; Bourbigot, S. Intumescent biobased-polylactide films to flame retard nonwovens. *J. Eng. Fibers Fabr*. 2009, **4**, 33–39 p.

[26] Wu, K.; Hu, Y.; Song, L.; Lu, H.; Wang, Z. Flame retardancy and thermal degradation of intumescent flame retardant starch-based biodegradable composites. *Ind. Eng. Chem. Res.* 2009, **48**, 3150–3157 p.
[27] Chapple, S.; Anandjiwala, R.; Ray, S.S. Bleach /starch blend / clay composites. *J. Therm. Anal. Calorim.* 2013, **113**, 703–712 p.
[28] Wang, J.; Ren, Q.; Zheng, W.; Zhai, W. Improved flame retardant properties of poly (lactic acid) foams using a natural charring agent. *Ind. Eng. Chem. Res.* 2014, **53**, 1422-1430 p.
[29] RU 2425069 C2, publ. 07/27/2011;
[30] Cayla A., Rault F., Giraud S., Salaün F., Fierro V., Celzard A. PLA with Intumescent System Containing Lignin and Ammonium Polyphosphate for Flame Retardant Textile. *Polymers* 2016, **8 (9)**, 331 p.
[31] WO 2008006779 A2, publ. January 17, 2008.
[32] RU 2113982 C1, publ. 06.27.1998.
[33] US 20070125446 A1, publ. 07.06.2006.
[34] JP 2000001380 A, publ. 01.07.2001.
[35] RU 2003103719/03 A1, publ. July 27, 2004.
[36] US 4,219,605, publ. 08.26.1980.
[37] US 3862854 And publ. 01/28/1975.
[38] RU 2674208 publ. 01.22.2018.
[39] RU 2674210 publ. 01/10/2018.
[40] Emanuel N.M., Denisov E.T. Advances in chemistry, 1958, v. XXVII, issue **4**, p. 365-402.
[41] RU 2204547 C1, publ. May 20, 2003.
[42] Lomakin S.M. and Zaikov G.E., Ecological Aspects of Flame Retardancy, *VSP International Science Publishers*, Utrecht, The Netherlands, 1999, 170 p.
[43] Bacterioses of Black Alder (Alnus glutinosa (l.) Gaertn) in the Western Caucasus / Cherpakov V.V. // *Actual problems of the forest complex*. 2016. No. **44**. P. 143-149 p.
[44] Two new species of anamorphic fungi on the alder from the Novgorod and Pskov regions / Shabunin D.A. // *Mycology and Phytopathology*. 2007. **V. 41. No. 4**. S. 337-342 p.
[45] Ecological features, structure and forecast of changes in the typological structure of black alder tree stands in the forests of Belarus / Baginsky V.F., Katkov N.N. // *Eco potential*. 2013. № **1-2 (1-2).** 84-92 p.
[46] Optimization of the composition of plantations when creating forest crops of Black Alder in the steppe zone of Russia / Turchina TA // *Actual directions of scientific research of the XXI century: theory and practice*. 2015. **Vol. 3. No. 2-1** (13-1). 129-134 p.
[47] Energy-saving regimes of wood drying of Black Alder without artificial humidification in convective drying chambers of periodic action / Chernyshev A.N. // *Actual problems of the forest complex*. 2013. **No. 36**. 112-116 p.
[48] New dangerous disease of Alder in Russia / Gninenko Yu.I., Zhukov AM, Komarova I.A. // *Protection and quarantine of plants*. 2010. **No. 2**. 45 p.
[49] Monitoring of the sanitary condition of Alder Chernoy plantings in arid regions of Russia / Turchina, T.A. // *Actual directions of scientific research of the XXI century: theory and practice*. 2017. **V. 5. No. 1** (27). 285-289 p.
[50] Optimization of the cutting schedule during sanding / Emilia-Adela S. // *Forest Messenger. Forestry Bulletin*. 2017. **V. 21. No. 4**. 70-72 p.
[51] Mycorrhiza on the roots of Black Alder (Alnus glutinosa) in the Orenburg region / Abaimov V.F., Glinushkin A.P., Mashenkov A.M. // *Russian Agricultural Science Review*. 2014. **Vol. 3. No. 3**. 4-12 p.
[52] Method of microclonal propagation of Black Alder in vitro / Sivolapov VA, Sivolapov AI, Tabatskaya TM, Blagodarova TA // patent for invention RUS 2515385 07.12.2012.

Study of antiseptic properties of the flame retardant solution provided by oxidized plant waste with regard to wood staining and mold micromycetes

A V Khvatov[1,4], P A Sakharov[1], S M Lomakin[1], S D Varfolomeev[1], Yu K Lukanina[1], A A Minikh[2], A P Glinushkin[3], L G Seraya[3] and D V Demin[3]

[1]Emanuel Institute of Biochemical Physics, Russian Academy of Sciences, Kosygin str., 4, Moscow, 119334 Russia
[2]Research and Development Center, Ltd. "NTC Tatneft, Ltd.", Nobel str., 5, Skolkovo, Moscow, 143025 Russia
[3]All-Russian Research Institute of Phytopathology, Bolshiye Vyazyomy, Odintsovsky district, Moscow Region, Russia

E-mail: hvatovanatoliy@gmail.com

Abstract. Antiseptic properties of the obtained aqueous solution of pine wood flame retardant with regard to 25 types of micromycetes have been evaluated. In was shown that aqueous flame retardant solution suppressed actively such species as Aspergillus terreus Thorn., Discula pinicola (Naumov) Petr., Leptographium fungi lundbergii Lagerb. & Melin, Penicillium purpurogenum Stoll., Paecilomyces variotii Bainier. The obtained solution of flame retardant in concentration below than 3% showed that the average affected area for all tested groups of fungi was 50.4%. Therefore, this flame retardant solution was attributed to the medium effective antiseptic agent for pine wood.

Wood is widely used in industry as a structural material for the construction of residential, public, industrial, agricultural, livestock, storage and other buildings and structures, as well as in transport infrastructure. In the northern and northeastern climatically cold regions of the country rich in forests, the construction of timbered buildings and structures (mainly in small towns and rural settlements) is still popular. Current trends in the global construction industry indicate that not only one- and two-storeyed building objects can be built from wood (townhouses and individual cottages, agricultural buildings, religious and other objects), but also buildings of apartment and multi-storey houses, hotels and restaurants , shops and offices. This practice is quite dynamic in a number of European countries, Australia, USA and Canada. According to the forecast in the near future low-rise wooden construction could be one of the most promising construction areas in the world. In the near future, the national programs of the countries of the European Union are aimed at ensuring the share of wooden houses at 75–80% in low-rise housing stock. The average annual growth rate of demand for wood is expected to be in the range of 20–25% until 2020 [1].

Wooden buildings and structures have a high fire risk. The process of burning wood is complex, multistage, and takes place in a wide range of temperatures [2].

To prevent fires in buildings and structures with bearing and enclosing structures, as well as finishing and veneering materials made of wood, flame retardants are used. Flame retardants differ in

IOP Conf. Series: Materials Science and Engineering 525 (2019) 012103 doi:10.1088/1757-899X/525/1/012103

the method and quantity of application parameters and the fire protection mechanism. It should be noted that the development of fire protection products and their use are currently reduced, as a rule, to a decrease in the flammability of wood material.

The fire resistance of the wood element is increased with the use of fire-resistant structural panels or fire-retardant intumescent coatings. The effectiveness of wood protection depends on the thickness and durability of coatings. Therefore, fire protection for wood can be developed in two groups: (1) flame retardants, reducing the spread of flame on the surface and (2) refractories, reducing the rate of charring.

Currently, for fire protection of natural wood, such methods as surface and deep impregnation with special compositions containing flame retardants, as well as the application of fire retardant coatings are popular [3].

Intumescence or swelling in the process of burning on the surface occurs under the action of simultaneous foaming and carbonization of the burning material. Intumescent additives are considered to be environmentally friendly fire retardants, due to the practical absence in the environment of highly toxic combustion products [4]. Since the first intumescent coatings (materials) were patented in 1938, the mechanism of fire protection through the formation of coke-forming foam has been explained in terms of the insulating barrier action between the flame and the surface of the material exposed to temperature and combustion.

Particularly attractive is the use of thin layer of intumescent fire retardant coatings. Even when they are relatively thin, they demonstrate high efficiency of fire protection and allow the use of modern mechanized methods of application on the protected item, while maintaining the basic texture of wood and its aesthetic appearance.

Surface impregnation, with its cheapness and manufacturability, is less effective than deep impregnation, but it allows, by wetting the surface with compounds, to protect assembled wooden structures directly on construction sites. To obtain reliable fire protection of wood in a similar way is extremely difficult. This is largely due to the weak penetration of the flame retardant salts into the wood structure. In this regard, to achieve the required performance of flame retardant efficiency requires a high consumption of the composition. Deeper penetration of the fire retardant solution into the surface layers of wood is provided by hot-cold baths, as well as processing in industrial apparatuses — autoclaves in the mode of successive increase in pressure or alternation of vacuum and increased pressure (vacuum impregnation). A number of technological difficulties, the need for a large consumption of flame retardants, the need for special equipment for conducting deep wood impregnation with flame retardants limit the practical use of this method of fire protection on a scale smaller than industrial. At present, a large number of flame retardant impregnating compositions for wood have been developed, which differ from each other in different sets of quantitative combinations of low molecular weight inorganic substances and derivatives of organic compounds exhibiting the properties of flame retardants.

A relatively new direction in the creation and study of flame retardants is the creation of environmentally-friendly flame retardants created from waste that do not represent food or feed value, which can not only solve the problem of creating new materials with valuable properties, but also simultaneously solve the environmental problem of recycling ballast high-tonnage products .

Lignin [5] and lignin-containing wastes that are difficult to biodegrade and accumulate as byproducts of paper production and wood hydrolysis, as well as due to its wide distribution in nature, can be used as such production wastes.

In the case of using woodworking waste as a lignin-containing substrate, for example, sawdust, a complex action flame retardant is obtained, combining the properties of a radical chain inhibitor with the properties of an intumescent (coke-forming) flame retardant that is formed during oxidation of carbohydrate-containing products, for example, starch [6]. The mechanism of its flame retardant action is associated with the formation on the treated surface during high-temperature heating or with the direct action of a foamed coke layer flame. The foamcoat layer exhibits a heat-shielding and barrier effect during mass transfer of both combustible materials to the flame reaction zone and

IOP Conf. Series: Materials Science and Engineering **525** (2019) 012103 doi:10.1088/1757-899X/525/1/012103

oxygen of the air to the material surface. It has been experimentally confirmed that flame retardants based on lignin-containing woodworking waste substrates are effective representatives of a new class of intumescent flame retardants [7,8].

However, there is no information in modern literature on the antiseptic properties of such compounds, which determine the performance characteristics of wood-treated materials with fire retardants.

The purpose of this work was to study the antiseptic properties of the flame retardant solution.

To obtain a flame retardant solution with excellent intumescent properties, both the qualitative and quantitative composition of the initial combined substrate is important. Of the possible starch-containing components, the use of crushed rice grains is optimal, since allows you to get the final product with the best consumer properties. From the point of view of economic feasibility, it is preferable to use unsuitable products - substandard rice grain with expired or food production waste containing rice. Experiments have shown that the best result is achieved by using as a lignocellulosic component of the combined substrate wood flour of coniferous species, and not sawdust.

The viscosity of the final product and the technological conditions of oxidation depend on the total amount of substrate introduced into the reaction, which is chosen empirically depending on the specific characteristics of the feedstock. Experiments on the selection of oxidation conditions showed that the use of the initial mixtures containing 20 - 22% of the mass. oxidizable substrate allows the reaction under optimal conditions, and the final product obtained after filtration is an aqueous composition containing a mixture of products of the joint oxidation of the combined substrate, suitable for use as a flame retardant without further concentration and purification. During bulk processing of samples, it easily penetrates into the pores of the material, and upon surface treatment it exhibits good adhesion to the surface and, after drying in air, forms a durable, almost transparent coating that preserves the natural appearance of the treated wooden surface.

The flame retardant solution is synthesized by the method of liquid-phase catalytic oxidation of the mixture (crushed rice - wood flour) in an alkaline medium. Gaseous oxygen is used as an oxidizer.

The synthesis is in a steel, hermetic reactor. The temperature of the oxidation process is 75 °C. The reaction time was controlled by changing of the solution pH and was 10 hours. The resulting solution was filtered on a nylon filter. The oxidation process is single-stage and waste-free. The flame retardant solution is environmentally friendly, as the products of combustion are no more toxic than the products of combustion of wood and / or cereals.

The joint oxidation of these products leads to a non-additive effect due to the fact that during the joint oxidation of starch-containing and lignocellulosic substrates a complex mixture is formed, including salts of polyoxycarboxylic acids and polyoxyphenolic compounds, low molecular weight hydroxyphenol compounds, etc., as well as their interaction products. Compared with protective coatings containing monosubstrate oxidation products, the coating formed when wood is processed with a synthesized fire retardant solution not only has intumescent flame retardant properties, but is also more resistant to external climatic factors due to the formation of additional chemical crosslinks on the surface of the material. An additional factor contributing to the preservation or improvement of the strength characteristics of fire-protected material is the good penetration of the flame retardant into the wood. Moreover, in addition to forming a surface coating, products formed during the process of joint liquid-phase catalytic oxidation of a mixture of starch-containing and lignocellulosic products diffuse into the wood and form additional hydrogen bonds in the microfibrillar structure of cellulose fibers and the porous structure of lignin. The result of the formation of such weak bonds is the physical hardening of the wood structure.

Studies of the antiseptic properties of the obtained flame retardant solution were carried out on samples of wood with dimensions of 10 × 55 × 75 mm, where the last dimension is indicated along the length of the fibers. Samples were made of straight-layer freshly sawn wood of sapwood with a density of 0.48-0.52 g/cm^3 in the air-dry condition.

To study the antiseptic properties, studies have been carried out on the effectiveness of the flame retardant solution with respect to wood-staining and mold fungi. Samples of wood (pine) were impregnated with flame retardant by immersion in the solution. The average consumption of flame retardant was 261.9 g /m^2. Beyond that, for fifteen days, wood samples (10 × 55 × 75 mm) were kept in wet chambers using the "closed space - moisture reserve" principle with a large evaporation surface, under conditions most favorable for the activity of biological agents (mold fungi), with the subsequent determination of the average area of damage to the surface of the samples and the stage of development of the microorganism.

The tests were carried out on micromycetes, composed in three groups and presented by the following species (table 1). Part of the work, including reseeding, cultivation, storage of strains of microorganisms, was performed using the equipment of State collection of phytopathogenic microorganisms and identifying varieties (differentiators) of pathogenic microorganism strains based in All-Russian Research Institute of Phytopathology (http://www.vniif.ru/vniif/page/ckp-gkmf/1373).

When conducting the experiment used a suspension of spores obtained from pure cultures of fungi grown in bacteriological tubes on beveled agar wort. For the preparation of a spore suspension, cultures of fungi aged from 14 to 28 days were used, counting from the time of subculture, whose colonies filled the entire surface of the agar and had well-developed sporulation.

The spore suspension was prepared separately for each species of fungus by washing off the spores into chemical beakers containing 25 cm^3 of distilled water. The number of fungal spores in the suspension was calculated using a Goryaev counting chamber. The concentration of spores in suspension was (N ± 0.001) ·106 spores/ml. Next, they prepared a working suspension of fungi to infect sawdust in desiccators, taking into account the concentration of spores by mixing suspensions of certain types of fungi belonging to a certain group. To prepare a working suspension, a calculated amount of individual species of fungi was taken from each suspension in order to ensure the content of 106 spores (or 1 million spores) in a calculated volume of 1 ml. Measured volumes of suspensions of each species of fungus were introduced into a beaker and made up to (100 ± 1) ml with distilled water. The shelf life of the suspension did not exceed 1 hour from the moment of preparation.

For each variant of the test (group of fungi 1,2,3 according to table 3) a separate desiccator was prepared (a total of three desiccators - 1E, 2E, 3E). In the desiccators, sterile sawdust (80 ± 5%) sawdust from healthy pine saplings was filled up to 1/4 of the vessel's height. A suspension of fungi of a certain group was introduced into each desiccator (1E-1, 2E-2, 3E-3). Sawdust was irrigated with a working suspension of fungi with a volume of 100 ml using a spray bottle. The desiccators were closed with a ground-in lid and placed in a room at room temperature, in the dark. The growth of mycelium continued for 15 days from the start of the test.

Table 1. List of mold fungi cultures used in the experiment

№	Name of micromycete species
	1 group
1	*Alternaria humicola* Oudem.
2	*Aspergillus niger* Tiegh.
3	*Aspergillus terreus* Thorn.
4	*Fusarium moniliforme* J. Sheld.
5	*Fusarium sporotrichiella* var. *poae* (Perk.) Bilai.
6	*Penicillium brevicompacturn* Dierckx.
7	*Penicillium chrysogenum* Thorn.
8	*Penicillium ochrochloron* Biourge.
9	*Cadophora fastigiata* Lagerb. & Melin. [=*Phialophora fastigiata* (Lagerb. & Melin) Conant.]

IOP Conf. Series: Materials Science and Engineering **525** (2019) 012103 doi:10.1088/1757-899X/525/1/012103

№	Name of micromycete species
	2 group
1	*Aspergillus hennebertii* Blochwitz.
2	*Cladosporium cladosporioides* (Fresen.) G.A. de Vries.
3	*Cladosporium herbarum* (Pers.) Link.
4	*Discula pinicola* (Naumov) Petr.
5	*Exophiala jeanselmei* var. *heteromorpha* (Nannf.) de Hoog.
6	*Fusarium javanicum* Koord.
7	*Fusarium merismoides* Corda.
8	*Paecilomyces marquandii* (Massee) S. Hughes.
9	*Paecilomyces variotii* Bainier.
	3 group
1	*Alternaria tenuis* Nees.
2	*Aspergillus amstelodami* Thorn & Church.[=*Eurotium amstelodami* L. Mangin]
3	*Aureobasidium pullulans* var. *pullulans* (de Bary) G. Arnaud.
4	*Fusarium culmorum* (W.G.Sm) Sacc.
5	*Leptographium lundbergii* Lagerb. & Melin.
6	*Penicillium purpurogenum* Stoll.
7	*Trichoderma harzianum* Rifai.

Then the prepared wood samples were installed in special stands with external dimensions of 60 × 94 × 28 mm and three grooves with dimensions of 22 × 60 × 15 mm. Then, in each of the desiccators in the peripheral zone, at three equal intervals, three supports were placed, on which 6 samples were placed, protected by a protective agent of the same concentration, and two control samples (without impregnation with a protective agent). In total, 18 samples (6 for each of the three groups of fungi) were tested for each variant of the experiment.

The area of damage to the surfaces of the samples by fungi (%) or the state of the samples was assessed visually after 5, 10, 15 days using the "palette" method with a grid step of 5x5 mm.

The assessment of the current state of the samples was recalculated for the average area of fungal damage relative to the area of a rectangular parallelepiped or sample (in percent). The following results are obtained defeat the samples:

- group 1 - on the 5th day: in the experiment - 0%, in the control - 3%; for 10th day: in experience - 3-11%, in control - 27-41%; for 15th day: in the experience - 44-82%, in the control - 40-49%;
- group 2 - for 5th days: in the experiment - 0%, in the control - 0%; for 10th day: in the experience - 7-15%, in the control - 3-15%; for 15th day: in the experience - 31-49%, in the control - 33-77%;
- group 3 - for 5th days: in the experiment - 0%, in the control - 0%; for 10th day: in experience - 1-10%, in control - 1-17%; for 15th day: in the experience - 29-69%, in the control - 65-94%.

In the course of the research, the antipyrene solution actively suppressed such species of fungi as Aspergillus terreus Thorn., Discula pinicola (Naumov) Petr., Leptographium lundbergii Lagerb. & Melin, Penicillium purpurogenum Stoll., Paecilomyces variotii Bainier.

The effectiveness of the test flame retardant solution was equal in the case of a flame retardant solution concentration of less than 3% - the average lesion area for all test groups of fungi was 50.4%, so the product can be attributed to medium-effective antiseptics.

As a result of research, it was found that in order to prevent wood materials treated with synthesized flame retardants from being infected by wood-dyeing and mold fungi the additional treatment with antiseptic agents is not required.

The work was carried out within the framework of budget financing on the theme of state assignment No. of state registration 01201253305.

References

[1] Artsybasheva O V, Aseeva R M, Serkov B B and Sivenkov A B 2013 Modern trends in the field of fire resistance wooden buildings and constructions *Izvestiya SFedU. Engineering Sciences* **№ 8(145)** pp 178-196
[2] Garcia D, Tortosa E, Mans V, Futterer T and Nägerl H D 2002 A new approach to intumescent clear varnish in wood fire protection *Third international Woodcoating congress* (The Paint Research Association, Teddington, UK) paper 24
[3] Ostman B, Mikkola E, Stein R, Frangi A, Konig J, Dhima D, Hakkarainen T and Bregulla J 2010 Fire safety in timber buildings: technical guideline for Europe *SP Technical Research Institute of Sweden* pp 1-140
[4] Camino G and Lomakin S 2001 Intumescent materials *Fire retardant materials* ed A R Horrocks and D Price (CRC Press LLC,Woodhead Publishing) chapter 10 pp 318-335
[5] Cayla A, Rault F, Giraud S, Salaün F, Fierro V and Celzard A 2016 PLA with Intumescent System Containing Lignin and Ammonium Polyphosphate for Flame Retardant Textile *Polymers* **8(9)** p 331
[6] Aseeva R M, Sakharov P A and Sakharov A M 2009 The Physicochemical characteristics of oxidized polysaccharides *Russian Journal of Physical Chemistry B* **3** № 5 pp 844–850
[7] Varfolomeev S D, Lomakin S M, Gorshenev V N, Sakharov P A, Sakharov A M and Demin V L 2011 Pat RU 2425069
[8] Sakharov P A, Lomakin S M, Khvatov A V, Koverzanova E V, Lukanina Yu K, Shilkina N G, Usachev S V, Varfolomeev S D and Minikh A A 2018 Pat RU № 2018102367

IOP Conf. Series: Materials Science and Engineering **525** (2019) 012104 doi:10.1088/1757-899X/525/1/012104

Oxo-degradation of LDPE with pro-oxidant additive

Yu K Lukanina[1], N N Kolesnikova[1], A A Popov[1,2] and A V Khvatov[1]

[1] Emanuel Institute of Biochemical Physics, Russian Academy of Sciences, Moscow, Russia
[2] Plekhanov Russian University of Economics, Moscow, Russia

E-mail: julialkn@gmail.com

Abstract. The article describes polymeric additives accelerating the thermal oxidation of polyolefins at moderate temperatures (oxo-degradation additives), which increase the ability of polyolefins to oxo-decomposition. The oxo-degradation additive includes metal-supported 2-ethylhexanoates or mixtures of metal 2-ethylhexanoates supported on an inert carrier.

1. Introduction

At present, the population of large cities is facing an acute problem of giant landfills for municipal solid waste (MSW) [1]. Most of the MSW is polymeric waste due to its short service life (for example, as a packaging material) and high resistance to environmental factors. So far separate collection of garbage has not found due attention, which would have more effectively solved the problem of recycling and facilitating the disposal of solid waste.

To partially solve this problem, biodegradable polymers are developed [2-4], as well as additives that accelerate the decomposition of polymeric wastes (for example, polyethylene) due to the destruction of long polymer chains under the influence of natural factors such as heat, moisture, light and microorganisms.

To make materials from polyolefins (polyethylene and polypropylene) the ability of biodegradation one of the way is the introduction of natural filler. The resulting material is capable of destructuring in environmental conditions after the expiration of the service. On the other hand the introduction of natural filler reduces the cost of the material as a result of the replacement of the polymer matrix, derived from non-renewable raw materials. It should be noted that in this direction of creation of biodegradable polymer materials, the use of cheap non-nutritive or feed value materials and production wastes is most promising. Thus, to increase the rate of biodegradation, polyolefins are introduced into: cellulose, starch, wood, soy flour rice husks, etc. [2–8].

Another way of giving biodegradability to polymeric materials based on polyolefins is the introduction of pro-oxidant additives into the polymer matrix, which contain oxidation initiators [9, 10]. Due to their presence as a result of the action of light or heat, the oxidation of the polymer matrix is initiated. The ongoing chemical processes contribute to the destruction of polymer macromolecules, the accumulation of oxygen-containing groups, the decrease in molecular weight and the fragmentation of the entire material. The most widely used as oxo-decomposing additives are stearates of transition metals. Despite their wide distribution, their main disadvantage is the low content of metals of variable valence relative to the weight of salt. To solve this problem, we proposed using metal 2-ethylhexanoates as their pro-oxidants (their acid residue is almost two times less).

2. Experimental part

2.1Getting the pro-oxidant additive

The pro-oxidant additive is a variable-valence metal salt deposited on an inert carrier (calcium carbonate) (metal 2-ethylhexanoates). As a 2-ethylhexanoate of metals and their mixtures used sikkativy company LLC "INDLAC", which are solutions of individual salts or their mixtures in Nefras - C4 - 155/200. Calcium nanocarbonate (particle size 70 - 90 nm), obtained by grinding on a ball mill of calcium carbonate brand Omyafilm 750-KA from Omya Madencilik A.S., was thoroughly mixed with a salt solution or a mixture of salts in Nefras - C4 - 155/200 to obtain a homogeneous pasty mass. The resulting mass was dried at a temperature of 90 ° C for 24 hours until complete removal of the solvent and crushed to obtain a homogeneous powder crumbly (80-100 nm).

2.2Getting the masterbatch

A masterbatch was obtained by mixing polyethylene (LDPE of 15803-020 grade (PJSC Kazanorgsintez) with an oxidation initiator in a Brabender-type laboratory mixer for homogenization at 140 ° C, the number of revolutions is 20-30 rpm, the total residence time of the mixture in the mixer chamber is 6 minutes. The resulting mixture was cooled to room temperature, crushed to crumb. The obtained oxo-decomposing thermoplastic additive can be used as an additive to polymer compositions based on polyolefins (polyethylene) or their copolymers. The recipient of the masterbatch: LDPE (50 wt.%), inert carrier (calcium carbonate) 40 wt.%) metal salts of variable valence (2-ethylhexanoate cobalt - 1.9 wt. %, zinc - 2.8 wt.% and zirconium - 5.3 wt.%)

2.3Getting the base polymer with the addition of masterbatch

The content of masterbatch in the base polymer 1 and 3 wt. % The calculated amount of the base polymer was loaded into a Brabender type laboratory mixer (20–30 rpm) at 140–2 ° C, then the required weight of masterbatch prepared as described above was added and mixing continued until the mixture was homogenized (6–8 minutes).). The composition of the base polymers prepared with the addition of masterbatche, are presented in table 1. The material was cooled to room temperature, Production of film samples

Table 1. Recipe base polymers prepared with the addition of masterbatch

	Content of component, wt. %	
№ blend	LDPE	Masterbatch
1	100	-
2	-	100
3	99	1
4	97	3

2.4Preparation of films

Samples of films with a thickness of 80–100 μm were obtained by pressing on a laboratory press at T = 140 ° C, time 60 seconds, pressure 50 kgf / cm2.

2.5Research methods

By the method of IR Fourier spectroscopy, the accumulation of carbonyl groups D1720comp = D1720 / D3640 was estimated. The optical band at 1720 cm-1 corresponds to carbonyl groups, the accumulation of which indicates the oxidation process of the samples under the experimental conditions. The optical band at 3640 cm-1 is an internal standard for LDPE samples; reference to it makes it possible to exclude the effect of the sample thickness on its IR absorption spectrum [11]. Spectrometric studies were carried out using the equipment of the Collective Use Center “New Materials and Technologies” " at the Emanuel Institute of Biochemical Physics of

IOP Conf. Series: Materials Science and Engineering **525** (2019) 012104 doi:10.1088/1757-899X/525/1/012104

Russian Academy of Sciences, Perkin Elmer Spectrum 100, at T = 23 ± 2 ° C in transmitted light in the wavelength range 4600 ≤ ν ≤ 450 cm-1.

In order to identify the strength characteristics of the obtained mixtures were conducted physical and mechanical tests in a laboratory tensile machine PM-10. Testing, training and payment module samples were assayed according to GOST 9550-81.

3. Results and discussion

Structural changes during degradation were investigated by recording FTIR spectra of films as a function of degradation time.

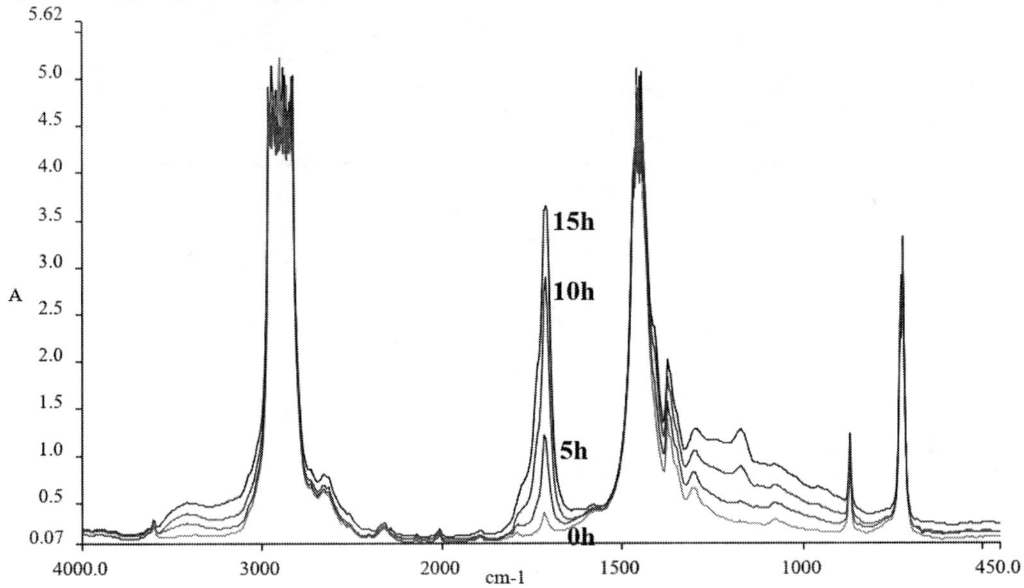

Figure 1. FTIR spectra for LDPE with 3% masterbatch during degradation.

A sharp increase in IR absorption in the carbonyl region (>C=O, between 1800 and 1650 cm^{-1}) was detected with exposure time in all samples containing pto-oxidant additive (as an example, FTIR spectra for LDPE with 3% masterbatch during degradation Figure 1). This region corresponds to different oxidized products, such as carboxylic acids (1708-1698 cm^{-1}), ketones (1723-1713 cm^{-1}), aldehydes (1740-1733 cm^{-1}) and lactones (1786-1780 cm^{-1}) [12-14].

Figure 2 shows kinetic curves of degradation for LDPE at different aging times and for masterbatch and LDPE with 1 and 3% masterbatch. As can be seen from the figure 2, materials containing pro-oxidant additive in their composition are easily oxidizable compared with polyethylene without additives (curve 1). As was to be expected, an increase in the number of ORDs in polyethylene accelerates the process of oxidation. However, it should be noted that the oxidation curve of the oxo-destructive additive itself lies between the oxidation curves of LDPE + 1% MB and LDPE + 3% MB. Apparently, this is due to the high concentration of inert carrier in the superconcentrate, which prevents the oxidation of LDPE in it.

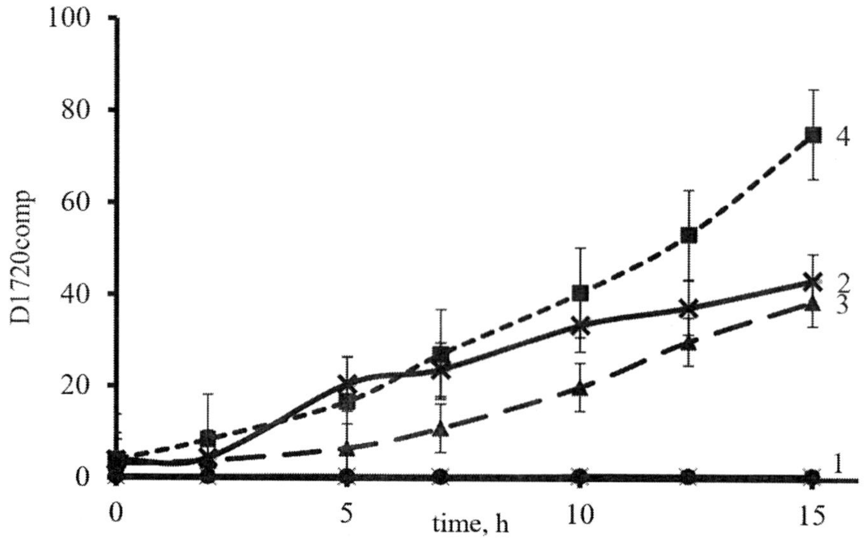

Figure 2. Dynamics of D1720otn changes during thermal oxidation of base polyethylene samples with the introduction of masterbatch (MB), where 1 - LDPE, 2 - MB, 3 - LDPE + 1% MB, 4 - LDPE + 3% MB.

It is well known that thermooxidative destruction of polyethylene proceeds by a free radical mechanism similar to the oxidation of low molecular weight hydrocarbons [15]. Polyethylene films subjected to thermal aging, containing prooxidant additives, demonstrate a higher level of oxidation, which is fixed by an increase in the number of carbonyl groups in the IR spectra. The presence of metals has a catalyzing effect on the oxidation processes due to the occurrence of redox reactions [16, 17]:

$$Me^{n+} + ROOH \rightarrow Me^{(n+1)+} + RO\bullet + OH^-$$
$$Me^{(n+1)+} + ROOH \rightarrow Me^{n+} + ROO\bullet + H^+.$$

The resulting radicals can react with the macromolecules of the polymer to form carbonyl groups, which lead to the breaking of the chain of macromolecules, followed by embrittlement and fragmentation of the polymer material.

Thus, the use of metal 2-ethylhexanoates as initiators of oxidation for polyethylene makes it possible to accelerate polymer degradation processes in the environment, increasing its availability to microorganisms by reducing the molecular weight, increasing the hydrophilicity of the material (accumulation of oxygen-containing functional groups) and fragmentation of the material.

Conventional tensile testing was performed as indicated in the experimental section. Destructive stress under tension and relative extension were used to monitor mechanical changes in the material under with the introduction of masterbatch inti the LDPE.

From the results presented in the table 2 it follows that the physicomechanical characteristics of the film materials of pure LDPE and containing prooxidants in their composition have similar values.

Table 2. Physico-mechanical characteristics of film samples

№ blend	Destructive stress under tension, MPa	Relative extension, %
1	20	320
2	-	-
3	20	322
4	19,5	319

The tensile properties of LDPE containing Oxo additives without any exposure to degradation conditions were identical to the properties of neat LDPE. Therefore, the Oxo additives did not degrade the polymers during the processing stage and did not cause any noticeable structural changes in the LDPE.

4. Conclusion

Thus, as a result of research, it has been shown that polymeric materials containing pro-oxidant additive (2-ethylhexanoates of cobalt, zinc, and zirconium) in low concentrations demonstrate a higher level of oxidation compared to polyethylene. The presence of 2-ethylhexanoates of cobalt, zinc and zirconium leads to an increase in the concentration of carbonyl groups in the polymer material during the thermoaddition process and, as a consequence, to an increase in polarity and adhesion characteristics of the surface, which increases its accessibility to the action of microorganisms. It should be noted that to use this masterbarch as a commercial product, it is necessary to introduce oxidation stabilizers to prevent premature aging of the material during its storage and operation.

Acknowledgments

The work was carried out within the framework of budget financing on the theme of state assignment No. of state registration 01201253305.

References

[1] Green peace of Russia. http://www.greenpeace.org/russia/ru/campaigns/waste/
[2] Lukanina Yu K 2011 The role of the chemical structure of polypropylene in the biodegradation of their compositions with cellulose-containing materials: dis. for the degree of Candidate chemical sciences M.p 06 [in Russian]
[3] Lukanina Yu K, Pantuhov P V, Khvatov A V, Koroleva A V, Kolesnikova N N, Lichahev A N and Popov A A 2014 Biodegradation of materials based on polyethylene and wood flour *All mat. Enc. ref.* **1** pp 2-7
[4] Lukanina Yu K, Khvatov A V, Kolesnikova N N and Popov A A 2015 Biocomposite LDPE - accelerate biodegradation additives *In the collection: Biotechnology: state and prospects of development materials of the VIII Moscow International Congress CJSC Expo-Biochem-Technologies Municipal Chemical Technological University DI Mendeleev* pp 369-70
[5] Balakireva N A 2012 Patent RU 2446191 C1
[6] Kirsh I A, Tikhomirov A A, Frolova Yu V, Pomogova D A, Ananiev V V, Kolpakova V V, Filinskaia Yu A and Chalich T I 2016 Integrated use of food industry waste and packaging to create biodegradable polymeric materials *Food ind* **11** 46-9.
[7] Rogovina S Z, Aleksanyan K V, Kompaniec L V and Pryt E V 2009 Obtaining and studying the properties of biodegradable mixtures based on synthetic and natural polymers *All materials. Enc. ref.*
[8] Mostafa N A, Awatef A Farag and Hala M 2015 Abo-dief Production of biodegradable plastic from agricultural wastes *Arab. J. of chem.* https://doi.org/10.1016/j.arabjc.2015.04.008
[9] Ershova O V and Mishurina O A 2016 Destruction of polymer films with an oxo-biodegradable additive D2W *Succ. of mod. Sc* **11-2** 221-5
[10] Vikhtinskaya A R and Reznichenko O Yu 2015 Biodegradable polymers with the use of oxo-bioadditive "D2W" *In the coll.: International Scientific and Technical Conference of Young Scientists of BSTU. V.G. Shukhov Belgorod State Technological University* pp 445-50
[11] Dehant And, Danz R, Kimmer V and Schmolke R 1976 Infrared spectroscopy of polymers *M.: Chemistry* P 100
[12] Roy P K, Surekha P, Rajagopal C, Chatterjee S N and Choudhary V 2006 Accelerated aging of LDPE films containing cobalt complexes as prooxidants *Polym Degrad Stab* **91(8)** 1791-9.

[13] Roy P K, Surekha P, Rajagopal C, Chatterjee S N and Choudhary V 2006 Effect of cobalt carboxylates on the photo-oxidative degradation of low-density polyethylene Part-I *Polym Degrad Stab* **91(9)** 1980-8.
[14] Chiellini E, Corti A, Antone SD and Baciu R 2006 Oxo-biodegradable carbon backbone polymers - oxidative degradation of polyethylene under accelerated testconditions *Polym Degrad Stab* **91(11)** 2739-47.
[15] Emanuel N M and Buchachenko A L 1982 *Chemical physics of molecular destruction and stabilization of polymers* M .: Nauka p 360
[16] Scott G 1995 Introduction to the abiotic degradation of carbon chain polymers *Degradable Polymers: Principles and Applications* Scott G and Gilead D eds Chapman and Hall London UK chapter 1
[17] Wiles D M 2005 Oxo-biodegradable polyolefins in packaging *Biodegradable Polymers for Industrial* Applications Smith R ed Woodhead Publishing Cambridge UK chapter 3 pp 57–76

IOP Conf. Series: Materials Science and Engineering **525** (2019) 012105 doi:10.1088/1757-899X/525/1/012105

Structural - chemical studies of the destruction of organochlorine compounds

D V Demin[1,2], D Y Aladin[1], N F Deeva[1], S M Sevostyanov[1]

[1]Institute Basic Biological Problems RAS, Pushchino, Moscow Region, Russia
[2]All-Russian Research Institute of Phytopathology, Bolshiye Vyazyomy, Odintsovsky district, Moscow Region, Russia

E-mail: aladin-danila@rambler.ru

Abstract: Polychlorinated biphenyls (PCB) are persistent organic pollutants used worldwide up 1990s. Although their use has been heavily restricted, PCB can be found in contaminated soils and sediments. Conducted structural-chemical studies of the destruction of polychlorinated biphenyl salts of amino acids in the soil. The effect of concentrated solutions of sodium salts of amino acids on soil particles has been investigated using FTIR spectroscopy as a means of identifying PCBs in the soil, controlling the interactions of PCBs with amino acid salts. The reaction products —C— substituted biphenyl polyamino polyacetic acids and their salts are soluble in water and extracted from particles on the surface of which polychlorinated biphenyls are adsorbed.

Polychlorinated biphenyls (PCBs) belong to the group of persistent organic pollutants (POPs), have toxic properties, are persistent and bioaccumulative, and are capable of long-distance transport in various environments. PCBs are biologically one of the most terrible carcinogenic (Pencikova et al., 2018) poisons, and, most importantly, having got into the soil or water bodies, accumulate in fats, for example, in breast milk (Rawn at al., 2017). PCBs were included in the 2002 Stockholm Convention on Persistent Organic Pollutants.

PCBs are used industrially as combustible oils (heat carriers) in capacitors and transformers, in addition to ballast resistances, fluorescent lamps, hydraulic and dielectric fluids, insulation of electrical wires, cables, plasticizers in adhesives, protective compositions of metals, wood, concrete, as a component in varnishes, paints, plastic fillers, paper impregnations, coolant during metalworking, as a flame retardant, insecticides and bacterial preparations (Schlatter C.,1994).

Currently, there are industrial and the surrounding area, where soil polluted with PCBs. They have a significant ecotoxic effect (Needham & Ghosh, 2019), primarily on the microflora. An example is the territory located in the zone of the factory in Serpukhov, which previously used PCBs in its manufacturing (Khakimov et al., 2003). Due to the ban on the use of PCBs and toxicity, the negative impact of the affected areas should be stopped. To do this, the contaminated soil must either be removed and placed in a special repository, or detoxified by transforming the chemical conversion of PCBs into less toxic substances, preferably biodegradable. Chemical detoxification in all respects is preferable to excavation and burial. However, to date there have been no effective reagents for these purposes (Gomes et al., 2013).

It is known that organochlorine compounds interact with derivatives of primary and secondary amines, including simple and complex amino acids. Currently, there is information about the possibilities of chemical detoxification and the disposal of highly toxic organochlorine compounds using NaL (Fridman et al., 2005). An example is the process of transformation of 2,2 / - dichlorodiethylsulfide (mustard gas) and other halogen-containing molecules in building structures. It is very likely that similar processes can be carried out for polychlorinated biphenyls to produce products of the type of complex aromatic amino acids that are soluble in water and are capable to biotransformation.

Concentrated solutions of α-amino acid salts consist of liquid solvatocomplexes (liquid hydrates), which form with the halogen-organic compounds molecules sufficiently strong molecular associates capable of hydration. In such molecular associates intramolecular transformations with the elimination of halide – ion and the formation of salts of complex amino acids are possible. The latter are soluble in water and do not form associates with liquid hydrates of α-amino acid salts. Typically, in an associate, react as many solvates complexes as halogen atoms exist in an organohalogen compound. At the same time, the associates include a larger number of solvates complexes. The process will continue until all of these molecules are converted into salts of the corresponding complex amino acids. The rate of reactions of intramolecular transformations in associates is catalyzed by alkali metal hydroxides, which bind hydrogen ions, which are cleaved from the amino groups of the salts of amino acids simultaneously with the removal of halide ions from the organohalogen molecules. Reactions can be catalyzed by aluminosilicate carriers that form the basis of soil particles.

Salts of amino acids from aqueous solutions are capable of adsorbing onto the surface of silicate particles and aluminosilicates of the soil. In the adsorbed state, they can come into contact with PCB molecules and then react with them in the form of chlorine amines (Demin et al., 2013). The reaction products - carbon - substituted biphenyl polyamine polyacetic acids and their salts may be soluble in water and extracted from particles on the surface of which PCBs have been adsorbed. By structure, they are similar to complex polyaromatic amino acids, which are formed in the metabolic processes of steroids of biosystems, and can be food products of soil microflora and water bodies. In this case, the action of concentrated solutions of sodium salts of amino acids on the soil with adsorbed PCBs can be used as the basis for the methods of detoxification and rehabilitation of the soil in the area of the factory in Serpukhov, previously using PCBs in production.

Soil treatment with reagent can be carried out in different ways - by impregnation with stirring, by injecting liquid reagent into the horizon, by stepwise dredging, suspension, etc. The choice of a particular method of treatment depends on the indicators of the diffusion of the reagent into the soil.

Objective: - To conduct structural and chemical studies of the destruction of organochlorine compounds with salts of amino acids in soils contaminated with polychlorinated biphenyls, including:

- interaction of salts of amino acids with polychlorinated biphenyl molecules sorbed in the soil, structure and solubility of products of transformation
- dynamics of sorption of sodium salts of amino acids in the soil containing polychlorinated biphenyls.

The study of the interaction of amino acid salts sorbed in soil with polychlorinated biphenyl molecules, the structure and solubility of conversion products was performed using FTIR spectroscopy as a means of identifying PCBs in the soil, controlling the interactions of PCBs with amino acid salts to their effects dissolving in water when washing the treated soil.

Studies were carried out on soil samples contaminated with PCBs provided by the Institute of Basic Biological Problems of the Russian Academy of Sciences. For the impregnation of soil contaminated with PCB, the composition of sodium salts of amino acids (composition AK-3E) was used. The concentration of salts of amino acids in the composition is 3.02 mole/l.

Before impregnation, the samples were repeatedly washed with water on the filter in order to remove the components soluble in water.

The water washed soil sample was divided into three parts. One part was used as a sample of soil containing PCBs. Two other parts were placed in a tube and a portion of 3-4 ml of AK-3E composition was added to them. As it is absorbed into the soil, new portions are added until complete saturation. The sample in the bottle was kept for 10 days at room temperature. After that, removed from the bottle and divided into two parts. One piece was used as a sample of the impregnated soil. The other part was washed with water to a pH of less than 8 and used as a sample of washed soil. Before taking the spectra, the samples were dried in air at room temperature to constant mass. The spectra were measured on a

Perkin Elmer – 2000 Fourier spectrometer in the range of 4000-100 cm^{-1}. The diffuse reflection method was modified using a Harrick extension that we modified. The assignment of the bands was performed using reference data (atlas of the IR - spectra of organic and inorganic compounds), and literature data.

Figure 1 shows the IR spectra of soil containing PCB (spectrum 1), soil impregnated with AK-3E (spectrum 2) and washed soil (spectrum 3). Comparing the spectra shows the following. In the spectrum of soil containing PCB, soil impregnated with AK-3E, and washed soil there are a number of identical bands - this is the shoulder at 3697 cm^{-1}, the bands with maxima at 3619, 3406 cm^{-1}; 2229 and 2134 cm^{-1}, shoulder at 2000 - 1950 cm^{-1} strip with maxima at 1158, 816, 702, 570 cm^{-1}. The shoulder at 3697 cm^{-1} and the band at 3619 refer to OH - stretching vibrations of Si-O-H not associated groups, and the band at 3406 cm^{-1} – to OH- stretching vibrations of Si-O-H associated groups. The bands at 2229 and 2134 cm^{-1}, the shoulder at 2000–1950 cm^{-1} are related to the deformation vibrations of different types of hydrogen bonds, including vibrations in water molecules. The bands at 1158 cm^{-1} to $Si-O^-$ stretching vibrations in polysilicic acid anions associated with metal ions (sodium, potassium, calcium) and the bands at 816, 702, 570 cm^{-1} correlate with the oscillation frequencies of the aluminosilicate skeleton.

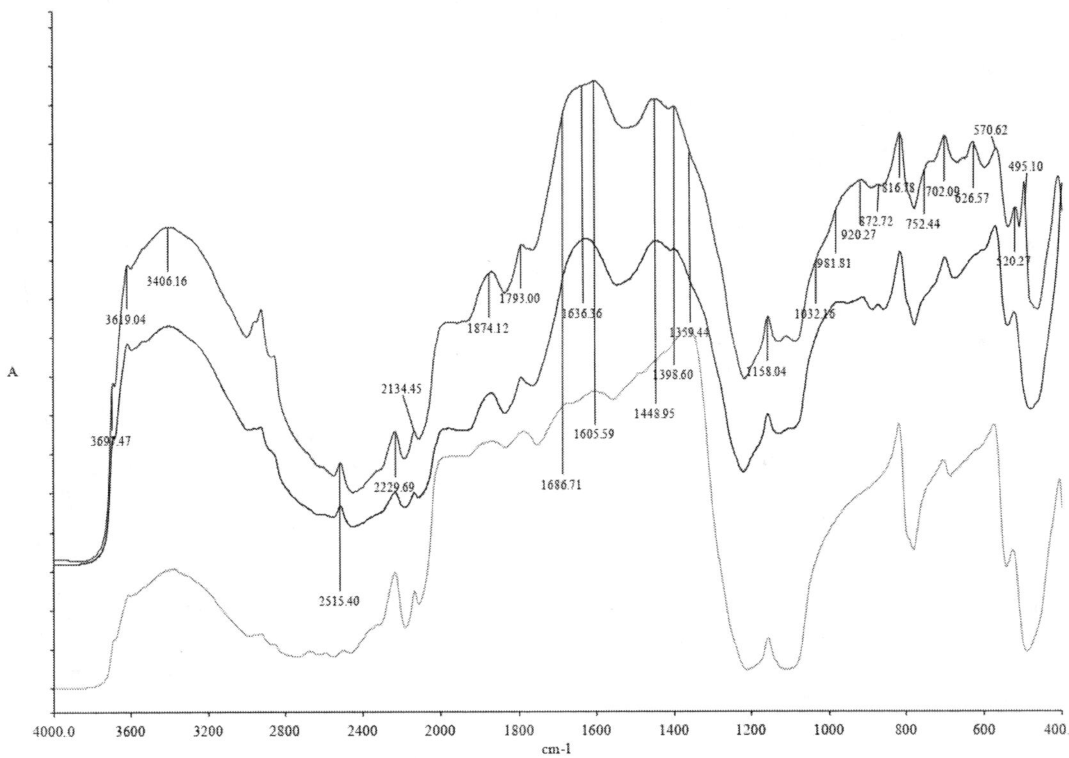

Figure 1. IR - spectra of soil containing PCB (1), soil impregnated with AK-3E composition (2) and washed soil (3).

In the spectrum of soil containing PCB, there is absorption in the region of 2950–2800 cm^{-1}, related to C–H stretching vibrations in biphenyls, there are intense bands at 1625, 1449 and 1359 cm^{-1}, related to vibrations of benzene rings of biphenyls. In addition, it should be noted that absorption in the region of 990–850 cm^{-1}, in which there are Si–O bands, stretching vibrations of Si–O–Si groups of silicate structures (about 985–970 cm^{-1}), deformation and fan vibrations of Si–OH groups (about 870 - 820 cm^{-1}), as well as C-Cl stretching vibrations in PCBs (about 900 - 850 cm^{-1}).

In the spectrum of the soil impregnated with a solution of salts of amino acids, compared with the spectrum of the soil containing PCB, in these areas there are the following differences. The maximum of the band related to the vibrations of the benzene ring at 1625 cm^{-1} for soil containing PCBs shifted to the long-wavelength region to 1605 cm^{-1}. In addition, the ratio of the intensity of the vibration bands of the benzene rings of biphenyls has changed. This may be due to a change in the nature of the substituents in the benzene rings of biphenyls to substituents with greater nucleophilicity than the chlorine atom. In the spectrum of soil impregnated with a solution of salts of amino acids, a shoulder appeared at 1660 - 1650 cm^{-1}, referring to N-H deformation fluctuations in amino acids and a shoulder at 1636 cm^{-1}, referring to antisymmetric vibrations of the carboxyl group of amino acids. In addition, there are certain changes in the region of 2950 - 2800 cm^{-1} C-H oscillations. Instead of one band, two bands appeared, the intensity of which is greater than the intensity of the band in the spectrum of the soil containing PCB. A number of changes in the region of 990 - 600 cm^{-1} should be noted - a decrease in the intensity of absorption at 920 - 870 cm^{-1} where C-Cl stretching vibrations can be found in the spectrum of the soil containing PCB, as well as the appearance of new bands at 752 and 626 cm^{-1} related to skeletal fluctuations of amino acids. The shift of the band related to the vibrations of the benzene ring at 1625 cm^{-1} for soil containing PCBs to the long wavelength region to 1605 cm^{-1} after impregnation with amino acid salts and changes in the intensity ratio of the vibrational bands of the benzene rings of biphenyls may be due to a change in the nature of substituents in the benzene rings of biphenyls on substituents with greater nucleophilicity than a chlorine atom. A decrease in the intensity of absorption at 920–870 cm^{-1} where C–Cl stretching vibrations in the spectrum of a soil containing PCB can be found can be associated with a decrease in the number of C–Cl bonds in PCBs. This suggests that during the impregnation between PCBs and amino acid salts, interactions occurred, as a result of which part or all of the chlorine atoms in the PCBs were replaced by –NH–CHR–COONa radicals. The composition of the products corresponds to the formula $C_{12}H_{12-n}Cl_{n-m}(NH\text{-}CHR\text{-}COONa)_m$, where n is the number of chlorine atoms in the initial PCB molecule, m is the number of –NH-CHR-COONa radicals in the product.

In the spectrum of washed soil, the absorption significantly decreased in the regions 2950–2800 cm^{-1}, 1700– 1400 cm^{-1}, 980– 850 cm^{-1}, and 750–850 cm^{-1}, which have absorption in the spectra of the soil containing PCB and the soil impregnated with salts of amino acids. At 1449 cm^{-1}, where the absorption maximum of the fluctuation of the benzene rings of PCB is located, there is no absorption characteristic of aromatic compounds. The absorption intensity at 1605 cm^{-1} is significantly lower compared with the intensity of the band in this region in the spectrum of the soil containing PCB. Consequently, there are either no PCBs in the washed soil, or their content has not decreased to trace amounts. PCBs are insoluble in water. Therefore, a decrease in their content in the washed soil is a consequence of the transformations of PCB with the formation of products that are soluble in water. For polyaromatic compounds, solubility in water occurs only when there are several hydrophilic radicals in their structure that can bind a significant number of water molecules in the first solvation shell. Products $C_{12}H_{12-n}Cl_{n-m}(NH\text{-}CHR\text{-}COONa)_m$, as N-substituted aromatic amino acids meet the specified requirements for solvation. It should be noted that they are similar in structure to complex polyaromatic amino acids, formed in the metabolism of steroids of biosystems, and may be food products of soil microflora and water bodies.

Conclusions

1. Conducted structural-chemical studies of the destruction of polychlorinated biphenyl salts of amino acids in the soil.
2. The effect of concentrated solutions of sodium salts of amino acids on soil particles has been investigated (basis - silicates and aluminosilicates with sorbed polychlorinated biphenyls; soil samples in the area of operation of the Serpukhov plant, which previously uses polychlorinated biphenyls in capacitor production). It has been established that salts of amino acids enter with molecules of polychlorinated biphenyls in the reaction of the type of chlorination of amines catalysed on aluminosilicate carriers (parts of the developed surfaces of silicates and aluminosilicates). The reaction products —C— substituted biphenyl polyamino polyacetic acids and their salts are soluble in water and extracted from particles on the surface of which polychlorinated biphenyls are adsorbed. By structure,

IOP Conf. Series: Materials Science and Engineering **525** (2019) 012105 doi:10.1088/1757-899X/525/1/012105

they are similar to complex polyaromatic amino acids, formed in the metabolism of steroids of biosystems, and can be food for the microflora of the soil and water bodies.

3. The sorption of concentrated solutions of sodium salts of amino acids on soil particles containing polychlorinated biphenyls was studied. It has been established that salts of amino acids are adsorbed on the surface of silicate particles in the form of solvate complexes from solutions with a deficiency of solvent for ion solvation. The dynamics of the distribution of salts of amino acids in the soil with sorbed polychlorinated biphenyls was studied. It was established that the limiting stage is the wetting of soil particles, including the transfer of solvate complexes from the liquid to the solid phase, and accompanied by a change in the salt concentration in the contact zone. The movement of the fluid front occurs at approximately the same speed regardless of the depth.

References

[1] Katerina Pencikova, Lucie Svrzkova, Simona Strapacova, Jiri Neca, Iveta Bartonkova, Zdenek Dvorak, Martina Hyzdalova, Jakub Pivnicka, Lenka Palkova, Hans-Joachim Lehmler, Xueshu Li, Jan Vondracek, Miroslav Machala In vitro profiling of toxic effects of prominent environmental lowerchlorinated PCB congeners linked with endocrine disruption and tumor promotion. *Environmental Pollution*, 2018, **v.237**, p. 473-486.

[2] Dorothea F.K. Rawn, Amy R. Sadler, Valerie A. Casey, François Breton, Wing-Fung Sun, Tye E. Arbuckle, William D. Fraser Dioxins/furans and PCBs in Canadian human milk: 2008–2011. *Science of the Total Environment* 2017, **v.595**, p.269–278

[3] Schlatter C. Environmental pollution and human health. // *Sci. Total Environment*, 1994, **v.143**, p. 93-101.

[4] Trevor P. Needham, Upal Ghosh Four decades since the ban, old urban wastewater treatment plant remains a dominant source of PCBs to the environment // *Environmental Pollution*, **Volume 246**, March 2019, Pages 390-397

[5] Khakimov F.I., Deeva N.F., Ilina A.A. Pollution with polychlorinated biphenyls of the soils of the city of Serpukhov // *Soil Science*, 2003, **№ 4**, p. 493-498.

[6] Helena I. Gomes, Celia Dias-Ferreira, Alexandra B. Ribeiro Overview of in situ and ex situ remediation technologies for PCB-contaminated soils and sediments and obstacles for full-scale application. *Science of the Total Environment* 2013, **v.445–446**, p.237–260

[7] Fridman A.Ya., Tsivadze A.Yu., Shemyakina E.V., Khakimov F.I., Sevastyanov S.M., Prokuronov V.G., Sokolova N.P., Gorbunov A.M. Interaction of copper ions with sodium salts of amino acids in organic-mineral substrates capable of biotransformation // *Material Science*. **№ 12**, 2005, p.30-34.

[8] Demin D.V., Aladin D.Yu., Tatarkin I.V., Sevostyanov S.M., Deeva NF, Il'ina A.A. Study of the destruction of polychlorobiphenyl by sodium salts of amino acids and humic acids in soils. // *Proceedings of the Samara Scientific Center of the Russian Academy of Sciences*, 2013, **vol.15**, №3-4, p.1282-1286.

Institute of Physics
Dirac House, Temple Back
Bristol BS1 6BE UK

ISSN: 1757-8981
ISBN 978-1-5108-8817-3

9 781510 888173

CURRAN ASSOCIATES INC.
proceedings.com